ADVANCED MACHINE TECHNOLOGY

C. Thomas Olivo

BRETON PUBLISHERS

A Division of Wadsworth, Inc.
North Scituate, Massachusetts

No comprehensive writing is ever done without sacrifice and devotion. Thus, *Fundamentals of Machine Technology* and *Advanced Machine Technology* are affectionately dedicated to my wife, Hilda G. Olivo. Over these many years of inquiry, study, research, and analysis, she has been an outstanding teacher, constant companion, and reviewer. She provided the continuous thread of understanding and support.

C. Thomas Olivo

BRETON PUBLISHERS
A Division of Wadsworth, Inc.

© 1982 by Wadsworth, Inc., Belmont, California 94002. All rights reserved. No part of this book may be reproduced, stored in a retrieval system, or transcribed, in any form or by any means, electronic, mechanical, photocopying, recording, or otherwise, without the prior written permission of the publisher, Breton Publishers, a division of Wadsworth, Inc., North Scituate, Massachusetts 02060.

Library of Congress Cataloging in Publication Data

Olivo, C. Thomas.
  Advanced machine technology.

  Includes index.
  1. Machine-tools.   2. Machine-shop practice.
I. Title.
TJ1185.038   670.42'3   81-10136
ISBN 0-534-01040-7   AACR2

Printed in the United States of America
1 2 3 4 5 6 7 8 9 — 86 85 84 83 82

ADVANCED MACHINE TECHNOLOGY was prepared for publication by the following people: Jean Peck, copy editor; Linda F. Kluz, production editor; Juanita Brown, art coordinator; Joanna Prudden Drummond, interior designer; Juanita Brown, cover designer; Robert Rome, illustrator. Sylvia Dovner supervised production. The sponsoring editor was Jim L. Linville. The book was set in Century by Graphics International; printing by Crest Litho; and binding was by Hamilton Printing Company. Cover photograph courtesy of Lodge and Shipley Company.

ABOUT THE AUTHOR
C. Thomas Olivo is an internationally recognized industrial-vocational educator with extensive experience as a journeyman tool and die maker, machine trades teacher, and technical institute director. Dr. Olivo has also served as Chief of Vocational Curriculum Development and as the New York State Director of Industrial Education.

CONTENTS

PREFACE — xvii
ACKNOWLEDGMENTS — xix
INTRODUCTION — xxii

PART ONE
PRECISION MEASUREMENT AND INSPECTION — 1

SECTION ONE Advanced Measurement: Mechanical — 2

UNIT 1 Linear and Angular Measurement — 2
Steel Gage Blocks 2
Gage Block Accessories 5
Advantages of Black Granite Measurement Products 6
Black Granite Surface Plates 7
Granite Precision Angle Plates and Accessories 8
Parallels and Other Forms of Granite Accessories 8
Precision Angle Measurements 10
Mathematics Applied to Angular Measurement 11
How to Make Direct Calculations for Sine Bar Setups 14
Tables of Constants for Sine Bars and Plates 14
Sine Bar, Sine Block, and Sine Plates 15
How to Set Up a Sine Plate for Layout and Inspection 17
How to Set Up and Use a Permanent Magnetic Sine Plate (Perma-Sine) 18
Safe Practices with Precision Linear and Angular Measuring Accessories and Instruments 19
Terms Used in Precision Linear and Angular Measurement 19
Summary 20
Unit 1 Review and Self-Test 22

UNIT 2 Layout and Measurement Practices in Surface Plate Work — 23
Generating a Flat Surface 23
Accuracy of Surface Plates 24
Calibrating Flat (Plane) Surfaces 24
Techniques for Checking Parallelism, Squareness, Roundness, and Concentricity 26
How to Check Parallel Surfaces by Using a 0.0001″ (0.002mm) Dial Test Indicator 27
How to Check for Flatness and Squareness by Using a Transfer Gage and Dial Indicator 28
How to Check for Concentricity, Axial Runout, and Roundness 30
General Precision Layout and Measurement Practices 31

CONTENTS

Precision Measurement of Angles and Tapers   33
   How to Set Up and Check Angles with Angle Gage Blocks   34
Reading Drawings That Use Datums and Ordinate Dimensions   36
Safe Practices in Precision Layout and Measurement   37
Terms Used in Precision Layout and Measurement Practices   38
Summary   39
Unit 2 Review and Self-Test   40

SECTION TWO   Precision Measurement Instruments   41

UNIT 3   Measurement with Optical Flats and Microscopes   41

Principles of Optical Measurement   42
Shop and Toolmaker's Measuring Microscopes   44
Optical Height Gage   46
Industrial Magnifiers   47
   How to Use a Toolmaker's Measuring Microscope   48
Precision Measurement with Optical Flats   50
Measurement of Surface Flatness   53
Cylindrical Surface Interference Bands   54
Interference Band Conversion Table   55
   How to Measure a Workpiece with an Optical Flat   55
Tabular Dimensioning Applied to Drawings   56
Safe Practices in the Use of Industrial Microscopes and Optical Flats   56
Terms Used in Measurement and Inspection with Optical Instruments   58
Summary   59
Unit 3 Review and Self-Test   60

UNIT 4   Measurement with High Amplification Comparators   62

General Types of High Amplification Comparators   62
High Amplification Mechanical Comparators   62
   How to Measure with a Mechanical Comparator   63
Mechanical-Optical Comparators   64
   How to Measure with a Reed Comparator   66
Optical Comparators   67
   How to Check a Thread Angle by Using an Optical Comparator   68
Pneumatic Comparators   70
Operation of a Flow (Column-Type) Pneumatic Comparator   71
Operation of a Pressure-Type Pneumatic Comparator   73
   How to Measure with a Pressure-Type Pneumatic Comparator   74
Factors Influencing the Design of a Gaging Spindle Plug   74
Electrical Comparators   75
Electronic Comparators   75
Safe Practices in the Use and Care of High Amplification Comparators   78
Terms Used in Measuring with High Amplification Comparators   79
Summary   80
Unit 4 Review and Self-Test   82

SECTION THREE   Quality Control and Surface Texture (Finish)   83

UNIT 5   Quality Control Principles and Measurement Practices   83

Interchangeable Manufacture and Quality Control   83
Fixed-Size Gages   84
Adjustable Thread Ring Gages   87
Allowances for Classes of Fits   88
Interchangeability of American National and Unified Form Screw Threads   89
The Basic Size (Controlling Point) in Thread Measurement   91
Characteristics of the Unified Screw Thread System   92
Basic ISO Metric Thread Designations and Tolerance Symbols   94
Quality Control Methods and Plans   96
Sampling Plans Used in Quality Control   98
Quality Control Charts   100
Safe Practices in Applying Gages for Quality Control Inspection   102
Terms Used in Quality Control and Measurement Practices   103
Summary   104
Unit 5 Review and Self-Test   105

CONTENTS v

UNIT 6 Surface Texture Characteristics and Measurement — 106
Terms and Symbols Used to Specify and Measure Surface Texture 106
Representation of Lay Patterns and Symbols on Drawings 108
Computing Average Surface Roughness Variations 108
Measuring Surface Finishes 111
How to Measure Surface Texture with Comparator Specimens and a Viewer 112
Surface Roughness Related to Common Production Methods 113
Surface Texture Measuring Instruments 114
How to Measure Surface Finish with an Indicating Instrument 116
Safe Practices in Using Surface Finish Analyzers and Instruments 116
Summary 117
Unit 6 Review and Self-Test 118

PART TWO
TURNING MACHINES — 119

SECTION ONE Advanced Engine Lathe Work — 120

UNIT 7 Advanced Thread-Cutting Technology and Processes — 120
Production Methods of Making Threads 121
American Standard Unified Miniature Screw Threads 123
Technology and Cutting Processes for Square Threads 124
How to Cut a Square Thread 126
Technology and Machining Processes for Acme Threads 126
How to Cut a Right-Hand Acme Thread on a Lathe 130
Multiple-Start Threads 131
Safe Practices in Advanced Thread-Cutting Applications 133
Terms Used in Advanced Thread Cutting and Production Methods 134
Summary 135
Unit 7 Review and Self-Test 137

SECTION TWO Turret Lathes: Bar and Chucking Machines — 138

UNIT 8 Classifications, Features, and Applications of Turret Lathes — 138
Evolution of Modern Bar and Chucking Machines 138
Classification of Bar and Chucking Machines (Turret Lathes) 141
Major Components of the Turret Lathe 146
Basic Machine Attachments 147
Care and Maintenance of Turret Lathes 150
Safe Practices in the Care and Maintenance of Turret Lathes 151
Terms Used in Turret Lathe Classifications and Design Features 151
Summary 153
Unit 8 Review and Self-Test 154

UNIT 9 Spindle, Cross Slide, and Hexagon Turret Tooling and Permanent Setups — 155
Kinds of Machine Cuts 155
Headstock Spindle Tooling 155
Universal Bar Equipment for Permanent Setups on Turret Lathes 162
Slide Tools for Turrets 165
Cutting-Tool Holders 169
Cutting-Off Tools 169
How to Mount Cutters on Center 170
How to Start a Bar Turner 172
Safe Practices in Setting Up Spindle, Cross Slide, and Hexagon Turret Tooling 173
Terms Used with Spindle, Cross Slide, and Hexagon Turret Tooling 173
Summary 174
Unit 9 Review and Self-Test 176

UNIT 10 Basic External and Internal Machining Processes 177

Cutting Speeds (sfpm) and Feeds (ipr) for Bar Turners 178
Basic External Cuts 178
   How to Sequence External Cuts 182
Basic Internal Cuts 182
   How to Sequence Internal Cuts 187
Applications of Cutting Speed and Feed Tables to External and Internal Cuts 188
Representing Complex Details of Parts by Sections 189
Safe Practices for Basic External and Internal Machining Processes 193
Terms Used with External and Internal Tooling, Setups, and Cuts 194
Summary 195
Unit 10 Review and Self-Test 197

UNIT 11 Planning Turret Lathe Setups and Tooling 198

Factors to Consider in Tooling Setups 198
Machine Stops 199
   How to Set the Hexagon Turret Stops on a Saddle-Type Turret Lathe 201
The Planned Tool Circle 201
   How to Set Up a Bar Turner 202
   How to Adjust a Bar Turner for Size and Finish 202
Added Machining Capability with the Cross-Sliding Turret 203
Permanent Setups with Chucking Tooling 204
Machine Vibration and Tool Chatter Problems 206
Safe Practices in Turret Lathe Setups and Tooling 206
Terms Used in Turret Lathe Setups and Tooling 207
Summary 208
Unit 11 Review and Self-Test 210

SECTION THREE Single- and Multiple-Spindle Automatic Screw Machines 211

UNIT 12 Screw Machine Technology, Practices, and Troubleshooting 211

Classifications of Screw Machines 211
Major Components of Automatic Screw Machines 212
Tooling for Automatic Screw Machines 215
Guidelines for Automatic Screw Machine Tool Design and Selection 217
Operation of Fully Automatic Screw Machines 217
Functions and Design of Cams 217
Cutting Speeds and Feeds 218
Cutting Fluids 218
Automatic Screw Machine Accessories 220
Automatic Screw Machine Troubleshooting 220
Safe Practices in Setting Up and Operating Screw Machines 224
Terms Used in Screw Machine Technology and Processes 225
Summary 226
Unit 12 Review and Self-Test 227

PART THREE
VERTICAL MILLING MACHINES 229

SECTION ONE Vertical Milling Machines, Attachments, and Accessories 230

UNIT 13 Vertical Milling Machine History and Accessory Design Features 230

Historical Overview of Developments 230
Design Features of Modern Vertical Milling Machines 231
Combination Vertical/Horizontal Milling Machine 233
Major Components of Standard Vertical Milling Machine 235
Vertical Milling Machine Attachments 238
Attachments for Vertical Milling Machine Spindle Heads 239
Vertical Milling Machine Tooling Accessories 242
Vertical Milling Machine Work-Holding and Work-Positioning Devices 245
Safe Practices Using Vertical Milling Machines, Attachments, and Accessories 247
Terms Used with Vertical Milling Machines, Attachments, and Accessories 248
Summary 249
Unit 13 Review and Self-Test 251

SECTION TWO General Vertical Milling Processes 252

UNIT 14 Basic Vertical Milling Cutting Tools, Speeds and Feeds, and Processes 252

Design Features of End Mills 253
Considerations in Selecting End Mills 253
Types of End Teeth 255
End Mill Holders 255
Cutting Fluids for End Milling 256
Cutting Speed and Feed Rates for End Milling Processes 256
Interpreting Cutting Speed and Feed Data from Handbook Tables 257
Basic Vertical Milling Machine Processes 257
  How to Mill a Flat Surface with an End Mill 258
  How to End Mill an External Stepped Surface 260

How to End Mill an Angular Surface 261
How to End Mill a Slot or Keyway 262
How to Cut a T-Slot on a Vertical Miller 263
How to Mill a Dovetail on a Vertical Miller 263
Safe Practices for Basic Vertical Milling Processes 263
Terms Used with Vertical Milling Cutting Tools and Processes 264
Summary 265
Unit 14 Review and Self-Test 266

UNIT 15 Hole Forming, Boring, and Shaping 267

Work- and Tool-Holding Devices for Basic Hole-Forming Processes 267
Factors Affecting Hole-Forming Processes 267
Boring with the Offset Boring Head 268
The Offset Boring Head and Cutting Tools 269
Factors Affecting Boring Processes 271
Shaping on a Vertical Milling Machine 271
Baseline, Ordinate, and Tabular Dimensioning 272
  How to Machine Holes by Using a Conventional Vertical Milling Machine Head 275

How to Bore Holes with an Offset Boring Head 276
How to Use a Shaper Attachment on a Vertical Milling Machine 278
Safe Practices for Using Boring and Shaper Tools and Attachments 279
Terms Used in Hole Forming, Boring, and Shaping 280
Summary 281
Unit 15 Review and Self-Test 283

PART FOUR
HORIZONTAL MILLING MACHINES 285

SECTION ONE Horizontal Milling Machine Technology and Processes 286

UNIT 16 Sawing, Slotting, and Keyseat and Dovetail Milling 286

Metal-Slitting Saws and Slotting Cutters 286
Work-Holding Setups 288
  How to Saw on a Horizontal Milling Machine 289
  How to Mill Narrow, Shallow Slots 291
Keyseat Milling 292
  How to Mill a Keyseat 293
Woodruff Keyseat Cutter Specifications 294
Centering the Workpiece and a Keyseat Cutter 295
  How to Mill a Keyseat with a Woodruff Cutter 296

Characteristics of Dovetails 297
  How to Mill Internal and External Dovetails 299
Safe Practices for Sawing, Slotting, and Milling Keyseats and Dovetails 299
Terms Used for Sawing, Slotting, and Milling Keyseats and Dovetails 300
Summary 300
Unit 16 Review and Self-Test 302

viii CONTENTS

UNIT 17 Helical and Cam Milling 303

- Lead, Angle, and Hand of a Helix 303
- Gearing for a Required Lead 304
- Gearing for Short-Lead Helixes 308
- Methods of Milling a Helix 308
  - *How to Mill a Helix on a Universal Milling Machine* 308
  - *How to Machine the Secondary Clearance Angle* 311
  - *How to Mill a Helix by End Milling* 311
- Functions of Cams in Cam Milling 312
- Descriptions of Cam Motion 312
- Common Types of Cams and Followers 313
- General Cam Terms 315
- Basic Cam Machining Processes 315
- Calculating Cutter and Work Angles for Milling Short-Lead Cams 317
  - *How to Mill a Uniform Rise Cam* 317
- Safe Practices in Setting Up and Milling Helixes and Cams 319
- Terms Used in Helical and Cam Design, Setups, and Processes 320
- Summary 321
- Unit 17 Review and Self-Test 322

SECTION TWO Spur, Helical, Bevel, and Worm Gears and Gear Milling 323

UNIT 18 Gear Developments, Design Features, and Computations 323

- Early Developments and Gear-Cutting Methods 323
- Other Methods of Forming Gears 325
- Generating Machines for Bevel Gears 327
- Gear Finishing Machines and Processes 327
- Gear Tooth Measurement and Inspection 328
- Spur Gearing Drawings (Representation, Dimensioning, and Tooth Data) 331
- Spur Gearing Terminology 331
- Module Systems of Gearing 334
- Characteristics of the Involute Curve 334
- Rack and Pinion Gears 335
- Helical Gearing 337
- Calculating Helical Gear Dimensions 337
- Bevel Gearing 339
  - *How to Calculate Bevel Gear Dimensions* 342
  - *How to Calculate Large- and Small-End Gear Tooth Dimensions* 343
  - *How to Select the Bevel Gear Cutter* 343
  - *How to Compute the Amount of Gear Tooth Taper Offset* 343
- Worm Gearing 344
- Worm Gearing Terminology 345
- Representation of Worms and Worm Gears 346
- Machining Worms and Worm Gears 346
- Gear Materials 347
- Summary 348
- Unit 18 Review and Self-Test 349

UNIT 19 Machine Setups and Measurement Practices for Gear Milling 350

- Involute Gear Cutter Characteristics 350
  - *How to Mill a Spur Gear* 351
- Rack Milling 353
  - *How to Mill a Rack* 353
- Helical Gearing 355
  - *How to Mill Helical Gear Teeth* 355
- Factors to Consider in Bevel Gear Cutting 356
  - *How to Mill Teeth on Miter Bevel Gears* 356
- Worm Thread Milling 358
  - *How to Mill a Worm (Thread)* 358
- Worm Gear Milling 359
  - *How to Mill a Worm Gear* 359
- Safe Practices for Setting Up, Milling, and Measuring Gear Teeth 360
- Terms Used in Machine Setups and Measurement Practices for Gear Milling 361
- Summary 361
- Unit 19 Review and Self-Test 363

PART FIVE
VERTICAL BAND MACHINES 365

SECTION ONE Band Machine Technology and Basic Setups 366

UNIT 20 Band Machine Characteristics, Components, and Preparation 366

- Types of Vertical Band Machines 366
- Advantages of the Band Machine 367
- Types of Band Machine Cuts 367
- Preliminary Considerations for Band Machining 370
- Functions of Major Machine Components 372

Saw Band Welding 376
   How to Prepare the Saw Band 381
   How to Set Up the Machine for Operating 383
   How to Regulate Speeds and Feeds 385
Factors Affecting Feeding Force 385
Safe Practices in Setting Up the Vertical Band Machine 386
Terms Used for Band Machine Features and Processes 386
Summary 387
Unit 20 Review and Self-Test 389

SECTION TWO   Basic Band Machining Technology and Processes   390

UNIT 21   Band Machine Sawing, Filing, Polishing, and Grinding   390

Basic Sawing Operations 390
External Sawing 390
   How to Make External Cuts 392
Internal Sawing 392
   How to Saw Out an Internal Section 394
Common Band Machine Sawing Problems 395
Slotting 395
Splitting 395
Radius and Other Contour Cutting 395
Friction Sawing 398
Friction Sawing Band Machines 399
Friction Saw Band Characteristics 400
Friction Sawing Setup Procedures 402
High-Speed Sawing 402
Spiral-Edge Band Sawing 403
Diamond-Edge Band Sawing 403
Band Filing 404
   How to Set Up and Band File 406
Band Grinding and Polishing 407
   How to Grind and/or Polish with an Abrasive Band 408
Line Grinding 408
   How to Line Grind 409
Electroband Machining 409
Knife-Edge Bands and Processes 410
Safe Practices in Setting Up and Performing Basic Band Machining Operations 411
Terms Used in Basic Band Machine Processes 412
Summary 413
Unit 21 Review and Self-Test 415

PART SIX
SHAPING AND PLANING MACHINES   417

SECTION ONE   Shaping and Planing Machine Technology and Processes   418

UNIT 22   Shaping and Planing Machine Design Features, Tooling, and Setups   418

Basic Types of Shapers 418
Basic Types of Shaper Cuts 420
Conventional High-Speed Steel Cutting Tools 423
Carbide Cutting Tools 423
Early and Modern Planer Designs 425
Major Planer Components 427
Planer Drive and Control Systems 428
Work-Holding Devices for Planer Work 428
Machine and Workpiece Setups in Preparation for Planer Work 430
Planing Flat Surfaces 431
Machining Parallel Surfaces 432
Simultaneous Planing with a Side Head 432
Planer Cutting Tools and Holders 433
Cutting Speeds for Planer and Shaper Work 435
   How to Set Up the Workpiece and Planer and Machine Parallel Surfaces 435
Safe Practices in Tooling Up and Preparing for Shaping or Planing 437
Terms Used for Shaper and Planer Design Features, Tooling, and Basic Processes 438
Summary 439
Unit 22 Review and Self-Test 441

PART SEVEN
GRINDING MACHINES   443

SECTION ONE   Functions and Types of Grinding Machines   444

UNIT 23   Grinding Machine Technology and Processes   444

Characteristics of Abrasives and Grinding   444
Early Types of Abrasives, Grinding Machines, and Accessories   446
Abrasive Machining   447
Basic Precision Grinding   448
Surface Grinding Machines   448
General-Purpose Cylindrical Grinding Machines   451
Centerless Grinding Methods and Machines   453
Tool and Cutter Grinders   455
Gear and Disk Grinding Machines   456
Other Abrasive Grinding Machines   456
Microfinishing Processes and Machines   457
Honing Processes and Machines   457
Lapping Processes and Machines   459
Superfinishing Processes and Machines   460
Polishing and Buffing as Surface Finishing Processes   461
Safe Practices Applied to Grinding Processes and Machines   462
Terms Used with Grinding Machines, Abrasive Machining, and Microfinishing   462
Summary   463
Unit 23 Review and Self-Test   465

SECTION TWO   Grinding Wheels: Standards and Preparation   467

UNIT 24   Grinding Wheel Characteristics, Standards, and Selection   467

Conditions Affecting a Grinding Wheel   467
Other Conditions Affecting the Grade   468
Wheel Types, Uses, and Standards   469
Wheel Marking System   471
ANSI System of Representing and Specifying Grinding Wheels   474
Peripheral Grinding Wheels   474
Side Grinding Wheels   478
Combination Peripheral and Side Grinding Wheels   479
Mounted Wheels (Cones and Plugs)   479
Computing Cutting (Wheel) Speeds, Work Speed, and Table Travel   480
Factors to Consider in Selecting Grinding Wheels   481
How to Select a Grinding Wheel   481
How to Test for Grinding Wheel Performance   484
Recognizing Grinding Problems and Taking Corrective Action   484
Summary of Factors Influencing the Selection of Grinding Wheels   486
Safe Practices for Machine Grinding   486
Terms Used in Grinding Wheel Characteristics, Standards, and Selection   487
Summary   488
Unit 24 Review and Self-Test   489

UNIT 25   Grinding Wheel Preparation and Grinding Fluids   491

Abrasive Wheel Preparation   491
Wheel Dressing Techniques and Accessories   492
Concentric and Parallel Face Dressing (Truing)   492
General Procedures for Using Diamond Dressers   493
How to Prepare Conventional Aluminum Oxide and Silicon Carbide Wheels   495
Effect of Traverse Rate on Wheel Cutting Action   497
Properties and Features of CBN and Diamond Wheels   497
Conditioning Diamond and Borazon (CBN) Grinding Wheels   498
How to Prepare Borazon (CBN) and Diamond Abrasive Wheels   500
Functions and Techniques of Wheel Balancing   501
Grinding Problems Related to Truing, Dressing, and Balancing   503

Grinding Fluids 503
Coolant Systems 504
Systems for Recirculating Clean Grinding Fluid 506
Desirable Properties of Grinding Fluids 507
How to Check and Maintain the Quality of the Grinding Fluid 508
Safe Practices in Grinding Wheel Preparation 508
Terms Used in Grinding Wheel Preparation and Grinding Fluids 509
Summary 510
Unit 25 Review and Self-Test 511

SECTION THREE   Horizontal-Spindle Surface Grinding Technology and Processes   513

UNIT 26   Surface Grinder Design Features and Setups for Flat Grinding   514

Design Features of the Horizontal-Spindle Surface Grinder 514
Operating Features of the Fully Automatic Precision Surface Grinder 518
Wheel and Work-Holding Accessories 519
Problems Encountered in Producing Flat Surfaces 521
Down-Feeds and Cross Feeds for Rough and Finish Grinding Cuts 522
How to Set Up a Surface Grinder and Take a First Cut 523
Wheel Preparation for Finish Cuts 524
Dimensional Grinding Using the Down-Feed and Cross Feed Handwheels 524
Grinding Edges Square and Parallel 525
How to Grind Edges Square and Parallel 525
How to Grind Ends Square with the Axis (Round Work) 526
Holding Thin Workpieces 528
How to Grind Thin Workpieces 528
Safe Practices in Grinding Flat Surfaces and Thin Workpieces 529
Terms Used with Design Features, Setups, and Flat Surface Grinding Processes 529
Summary 530
Unit 26 Review and Self-Test 532

UNIT 27   Grinding Angular and Vertical Surfaces and Shoulders   533

General Types of Vises for Angular Work 533
Angle Plate Accessories 534
Magnetic Work-Holding Accessories 534
How to Grind Angular Surfaces 536
How to Grind an Angle with a Right Angle Plate, Sine Bar, and Gage Block Setup 538
How to Set Up the Magnetic Sine Table to Grind a Compound Angle 538
Shoulder Grinding 539
Types and Functions of Shoulders 539
Undercutting for Shoulder Grinding 540
Truing and Dressing Devices for Ground Shoulders 540
How to Form Dress a Grinding Wheel for Angular Grinding 541
How to Grind Shoulders 541
Grinding Multiple Shoulders 543
Safe Practices in Grinding Angular and Vertical Surfaces and Shoulders 543
Terms Used in Grinding Angular and Vertical Surfaces and Shoulders 544
Summary 544
Unit 27 Review and Self-Test 545

UNIT 28   Form (Profile) Grinding and Cutting-Off Processes   546

Commercial Wheel Forms 546
General Types of Wheel Forming Accessories 546
How to Rough Form a Wheel by Hand Using an Abrasive Stick Dresser 548
How to Form a Convex or Concave Wheel by Using a Radius Dresser 548
Diamond Dressing Tools for Forming Grinding Wheel Shapes 550
How to Form Dress a Wheel with a Radius and Angle Dresser 552
Irregular Form Dressing with a Pantograph Dresser 553
Forming a Wheel by Crush Dressing 554
How to Form Grind with a Crush Dressed Wheel 556
Forming Crush Rolls on a Microform Grinder 557

xii CONTENTS

Effect of Varying the Grinding Conditions on Surface Finish 557
Abrasive Cutoff Wheels and Cutting-Off Processes 558
Safe Practices in Using Form Dressers and in Crush Dressing and Form Grinding 559
Terms Used with Form Dressers and in Crush Dressing and Form Grinding 559
Summary 561
Unit 28 Review and Self-Test 563

SECTION FOUR Cylindrical Grinding Technology and Processes 564

UNIT 29 Cylindrical Grinder Components and External and Internal Grinding 564

General Types of Cylindrical Grinding Machines 565
Applications of Cylindrical Grinding Machines 565
Design Features of the Universal Cylindrical Grinder 565
Major Parts of the Universal Cylindrical Grinder 566
Basic Cylindrical Grinder Accessories 569
Operational Data for Cylindrical Grinding 570
Grinding Wheel Cutting Speed 570
Wheel and Workpiece RPM, Rate of Table Travel, and Depth of Cut 571
High-Speed Cylindrical Grinding 573
Automated Cylindrical Grinding Machine Functions 573
Cylindrical Grinding Problems and Corrective Action 573
Wheel Recommendations for Cylindrical Grinding (Common Materials) 576
Basic External and Internal Cylindrical Grinding Processes 576
How to Set Up and Operate a Universal Cylindrical Grinder 576
How to Grind Straight Cylindrical Work 579
Supporting Long Workpieces 579
How to Set Up and Use Spring Back Rests 579
How to Grind Small-Angle and Steep-Angle Tapers 580
Face Grinding on a Universal Cylindrical Grinder 582
Grinding Internal Surfaces on a Universal Cylindrical Grinder 582
How to Set Up for and Cylindrically Grind Internal Surfaces 583
Cylindrical Grinder Wheel Truing and Dressing Devices 584
How to True and Dress Cylindrical Grinding Wheels 585
Crush Truing on a Cylindrical Grinder 587
Safe Practices in Setting Up and Operating Cylindrical Grinders 587
Terms Used in Cylindrical Grinding Technology and Processes 588
Summary 590
Unit 29 Review and Self-Test 592

SECTION FIVE Cutter and Tool Grinding Technology and Processes 594

UNIT 30 Cutter and Tool Grinder Features, Preparation, and Setups 594

Machine Design Features of the Universal Cutter and Tool Grinder 595
Cutter and Tool Grinder Attachments 597
Centering Gage, Mandrel, and Arbor Accessories 600
Functions and Types of Tooth Rests and Blades 601
Review of Basic Cutter Terms 601
Tooth Rest Applications and Setups 602
Grinding Considerations and Clearance Angle Tables 603
Methods of Checking Accuracy of Cutter Clearance Angles 606
Preparation of the Cutter and Tool Grinder 607
How to Set Up a Cutter and Tool Grinder for Grinding Plain Milling Cutters 608
Safe Practices for Setting Up and Operating a Cutter and Tool Grinder 609
Terms Used with Cutter and Tool Grinder Machines and Setups 610
Summary 611
Unit 30 Review and Self-Test 613

CONTENTS xiii

UNIT 31 Grinding Milling Cutters and Other Cutting Tools   614
Characteristics of Slitting Saws 614
 How to Sharpen Peripheral Teeth on a Slitting Saw 615
Design Features of Staggered-Tooth Milling Cutters 615
 How to Sharpen a Staggered-Tooth Milling Cutter 617
Design Features of Shell Mills 618
 How to Grind a Shell Mill 618
Considerations for Grinding Carbide-Tooth Shell Mills 620
 How to Grind Face Teeth and Chamfer Edges on a Carbide-Insert Shell Mill 620
Design Features of End Mills 620
 How to Sharpen an End Mill 622
Design Features of Form Relieved Cutters 623
 How to Grind the Teeth on a Form Relieved Cutter 624
 How to Use a Gear Cutter Attachment 625
Design Features and Grinding of Single- and Double-Angle Milling Cutters 626
Design Considerations for Grinding Machine Reamers 627
 How to Sharpen the Chamfer of a Machine Reamer 627
Design Features of Adjustable Hand Reamers 628
 How to Sharpen an Adjustable Hand Reamer 628
 How to Grind Cutting, Clearance, and Rake Angles on Flat Cutting Tools 629
Considerations for Sharpening Taps 630
Other Applications of Cutter and Tool Grinding 630
Safe Practices for Machine Setups and Cutter Grinding Processes 630
Terms Used in Machine Setups and Cutter Grinding Processes 631
Summary 632
Unit 31 Review and Self-Test 634

PART EIGHT
NUMERICAL CONTROL SYSTEMS AND MACHINE PROCESSING   637

SECTION ONE   Numerical Control Machine Tools   638

UNIT 32 NC Systems, Principles, and Programming   638
Advantages of Numerical Control Machines 639
Disadvantages of Numerical Control Machines 641
Application of Numerical Control to Standard Machine Tools 641
Foundational Programming Information 642
Special Machine Design Considerations 642
Types of Automated Processes 642
Numerical Control Systems 643
Numerical Control Drawings and Interpretation 644
Binary System Input to Numerical Control 646
Numerical Control Word Language 649
Factors to Consider in NC Programming 650
Positioning the Spindle 650
Numerical Control Tape Programming, Preparation, and Correction 653
Tape Readers 654
Tab Sequential Basic Format Programs 655
 How to Program a Single-Axis, Single-Machining Process 656
 How to Program for Two Axes and Tool Changes 659
 How to Program for Three Axes 661
 How to Program for Milling Processes Involving Speed/Feed Changes 663
Word Address Format Programs 663
Fixed Block Format Programs 663
Mirror Image Programming Feature 664
Safe Practices Relating to Numerical Control Machine Tools 664
Terms Used with Numerical Control Machine Tools 665
Summary 666
Unit 32 Review and Self-Test 668

xiv CONTENTS

SECTION TWO NC Applications and Computer-Assisted Programming for CNC Machine Tools 670

UNIT 33 NC Basic Machine Tools and CNC Programming Applications 670

NC Machines and Controls 670
NC Data Processing 676
Computer-Assisted Numerical Control Programming 677
Functions of the Machine Control Unit (MCU) 679
Functions of a Computer (CPU) in Computer-Assisted Programming 679
Role of the Programmer in Computer-Assisted Part Programming 680
How to Prepare an APT Program Manuscript 682
Basic Components of a CNC System 683
Design Features of a CNC System 685
Expanding Productivity and Flexibility of a CNC System 686
Presetting Numerical Control Cutting Tools 687
Safe Practices for Computer Numerical Control Machine Tools 688
Terms Used with Computer-Assisted Numerical Control Machine Tools 688
Summary 689
Unit 33 Review and Self-Test 692

PART NINE
TOOL DESIGN AND TOOLING PRODUCTION 693

SECTION ONE Applications of Design Principles 694

UNIT 34 Producing Jigs and Fixtures, Dies, Cutting Tools, and Gages 694

Positioning and Clamping (Work-Holding) Methods 694
Types of Locating Devices and Methods 695
Principles of Clamping and Common Design Features 696
Drill Jig Design 698
Drill Jig Design Considerations 699
Additional Jig Construction Features 701
NC Machining Technology in Jig Work 701
Fixture Design 702
Metal Die-Cutting and Forming Processes 704
Power Presses for Stamping and Forming Parts 705
Major Components of the Open-Back Inclinable Press 706
Stroke, Shut Height, and Die Space 707
Die Clearance and Cutting Action 707
Die Sets and Punch and Die Mountings 709
Conventional, Compound, and Combination Dies 709
Sheet Metal Bending, Forming, and Drawing Dies 712
Factors Affecting Metal Flow 714
Work-Holding Devices for NC and CNC Machine Tools 715
Basic Cutting Tools for Numerical Control Machining 717
Conventional and NC Tool Change and Setup Methods 717
Tool Presetting 720
Safe Practices for Applications of Production Tooling 720
Terms Used with Production Tooling 721
Summary 722
Unit 34 Review and Self-Test 725

SECTION TWO Construction and Measurement of Precision Production Tooling 727

UNIT 35 Jig Boring, Jig Grinding, and Universal Measuring Machines and Processes 727

Precision Jig Boring Machine Tools, Processes, and Accessories 727
How to Accurately Position and Set Up a Workpiece on a Jig Borer 731
Design Features of a Locating Microscope 732
How to Use a Locating Microscope 732
How to Set Up and Use a Jig Borer 733
Precision Jig Grinding Machine Tools and Processes 733
Design Features of Jig Grinders 735
Dressing Attachments for Jig Grinding Wheels 737
Factors Influencing Wheel Selection 737
Wheel and Diamond-Charged Mandrel Speeds 737
Material Allowance for Grinding 737
How to Dress a Jig Grinder Wheel 737
How to Grind Holes on a Jig Grinder 738
How to Grind a Tapered Hole 738
How to Grind Shoulder Areas 739
Universal Measuring Machines and Accessories 739
Safe Practices for Operating Jig Borers, Jig Grinders, and Measuring Machines 741

Terms Used with Jig Borers, Jig Grinders, and
  Universal Measuring Machines 742
Summary 743
Unit 35 Review and Self-Test 745

PART TEN
PRODUCTION OF INDUSTRIAL AND CUTTING TOOL MATERIALS 747

SECTION ONE Structure and Classification of Metals 748

UNIT 36 Manufacture, Properties, and Classification of Industrial Materials 749
Manufacture of Pig Iron and Cast Iron 749
Basic Steel Manufacturing Processes 751
Chemical Composition of Metals 752
Composition of Plain Carbon Steel 752
Mechanical Properties of Metals 754
Classification of Steels 755
SAE and AISI Systems of Classifying Steels 755
Unified Numbering System (UNS) 758
Aluminum Association (AA) Designation System 759
Visible Identification of Steels 760
Characteristics of Iron Castings 761
Characteristics of Steel Castings 764
Common Alloying Elements and Their Effects 765
Groups of Nonferrous Alloys 767
Applications of Industrial Materials 768
Safe Practices in the Handling and Use of Industrial Materials and Cutting Tools 770
Terms Used with Industrial and Cutting Tool Materials 771
Summary 772
Unit 36 Review and Self-Test 774

SECTION TWO Basic Metallurgy and Heat Treatment of Metals 776

UNIT 37 Heat Treating and Hardness Testing 777
Heating Effects on Carbon Steels 777
Application of Iron-Carbon Phase Diagram Information 779
Liquid Baths for Heating 782
Quenching Baths for Cooling 782
Quenching Bath Conditions Affecting Hardening 783
Interrupted Quenching Processes 783
Heat-Treating Equipment 784
Hardening and Tempering Carbon Tool Steels 786
  How to Harden and Temper Carbon Tool Steels 788
Heat Treatment of Tungsten High-Speed Steels 789
Annealing, Normalizing, and Spheroidizing Metals 790
  How to Anneal, Normalize, and Spheroidize 791
Casehardening Processes 791
  How to Carburize 792
  How to Carburize in a Liquid Bath 793
Special Surface Hardening Processes 795
Subzero Treatment of Steel 797
  How to Stabilize Gages and Other Precision Parts by Subzero Cooling 797
Hardness Testing 797
Rockwell Hardness Testing 798
  How to Test for Metal Hardness by Using a Rockwell Hardness Tester 800
Brinell Hardness Testing 801
  How to Test for Metal Hardness by Using a Brinell Hardness Tester 802
Vickers Hardness Testing 803
Scleroscope Hardness Testing 803
  How to Test for Metal Hardness by Using a Scleroscope 804
Microhardness Testing 804
Knoop Hardness Testing 804
Safe Practices for Heat Treating and Hardness Testing 804
Terms Used in Heat Treating and Hardness Testing 805
Summary 807
Unit 37 Review and Self-Test 810

APPENDIX Handbook Tables 813

INDEX 844

LIST OF HANDBOOK TABLES

A-1	American Standard Surface Texture Values
A-2	Microinch and Micrometer (μm) Ranges of Surface Roughness for Selected Manufacturing Processes
A-3	Conversion Table for Optical Flat Fringe Bands in Inch and SI Metric Standard Units of Measure
A-4	Constants for Setting a 5" Sine Bar (0°1' to 15°60')
A-5	Suggested Tapping Speeds (fpm) and Cutting Fluids for Commonly Used Taps and Materials
A-6	Suggested Starting Speeds and Feeds for High-Speed Steel End Mill Applications
A-7	ISO Metric Screw Thread Tap Drill Sizes in Millimeter and Inch Equivalents with Probable Hole Sizes and Percent of Thread
A-8	Constants for Calculating Screw Thread Elements (Inches) for ISO Metric Screw Threads
A-9	Major, Pitch, and Minor Diameters (in mm) of Internal SI Metric Threads for Fine, Medium, and Coarse Fits
A-10	Size and Tolerance Limits of Unified Miniature Screw Threads
A-11	Basic Formulas and Dimensions of General-Purpose (National) Acme Screw Threads
A-12	Suggested Starting Points for Cutting Speeds on Turret Lathes for Basic External Cuts
A-13	Suggested Starting Feed Rates on Ram-Type Turret Lathes for External Cuts
A-14	Suggested Starting Points for Cutting Speeds on Turret Lathes for Basic Internal Cuts
A-15	Allowances and Tolerances on Reamed Holes for General Classes of Fits
A-16	ANSI Dimensions (in Inches) for Woodruff Keys and Keyseats
A-17	SI Metric Spur Gear Rules and Formulas for Required Features of 20° Full-Depth Involute Tooth Form
A-18	ANSI Spur Gear Rules and Formulas for Required Features of 20° and 25° Full-Depth Involute Tooth Forms
A-19	Selected Standard Shapes, Design Features, and General Toolroom Applications of Grinding Wheels
A-20	Standard Types of Diamond Grinding Wheels
A-21	Conversion of Surface Speeds (sfpm) to Spindle Speeds (RPM) for Various Diameters of Grinding Wheels
A-22	Feed and Depth of Cut Factors for Calculating Cutting Speeds for Turning Ferrous Metals
A-23	Equivalent Hardness Conversion Numbers for Steel
A-24	Recommended Heat Treatment Temperatures for Selected Grades and Kinds of Steel
A-25	Melting Points of Metals and Alloys in General Shop Use

PREFACE

Machine technology is essential to each individual's existence and to the nation's well-being. Transforming raw materials into goods, products, and services that are necessary to maintain high standards of healthful living, economic stability, and national security depends on machine technology.

Machine technology is used as a generic term. Broadly interpreted, it relates to the study of skills and technical knowledge. These are applied to design, measurement, forming, fabricating, assembling, inspecting, and other processes in machine and metal products manufacturing, machine and tool design, and other mechanical and related industrial occupations.

The clusters of job titles at many employment levels in the occupational constellation comprise a significantly high percent of the nation's work force. Job levels within each cluster range the full spectrum of workers from beginning machine operators to highly skilled machine, tool, and industrial engineers and researchers. The greatest numbers of people are employed as production workers, semiskilled operatives, skilled craftspersons and technicians, and supervisors, as well as middle management and semiprofessional workers.

The worker at each level and in each occupation must master a varying combination of skills and accompanying technical knowledge or theory. In each instance, the skills and technical knowledge are interdependent. They combine the "why-to-do" technology with the "how-to-do" manipulation of tools, machines, instruments, materials, and the like. Skills must also be developed in dimensional measurement. Industrial drawings must be interpreted. Shop and laboratory sketches must be made. Dimensions must be calculated. Materials must be selected, laid out, machined, and fitted.

Equally important is the dependency of these skills on the worker's ability to relate them to mathematics, physical science, blueprint reading, and drafting and design principles.

There are two textbooks in this machine technology series. Each is complemented with a *Study Guide* (workbook) and a *Teacher's Manual*. The textbooks, student's guides, and teacher's manuals are based on extensive labor market studies of job titles, specifications, and levels. Occupational and task analyses were made to inventory common skill and technology elements at each successive stage of development. Curriculum studies were conducted to establish student/learner needs. Functional, effective teaching methods and performance objectives were researched. Curriculum and instructional material resources were studied in actual teaching and learning situations over many years.

Institutional and on-the-job organizational patterns were diagnosed for effective preemployment, retraining, and upgrading programs. Comprehensive, long-term analyses were made of scope, content, organization, difficulty index, and quality of resource materials used by individuals for improvement through self-study. Formalized programs and instruction provided by school systems, industry, the military, labor organizations, and other manpower development agencies were reviewed.

Another important dimension of the machine technology series relates to the application of SI metrics in American and Canadian industry, labor, military, vocational-technical schools, and other postsecondary institutional training programs. The amount and depth to which metric concepts, measurements, and procedures are interwoven into the units is based on the author's firsthand experiences over a number of years in assessing training programs in the United States, Canada, and abroad. Judgments on metrics in machine technology are made from these experiences and the vantage point of a journeyman tool and die maker, former machine trades teacher, technical institute director, and state director of industrial-technical education.

It is obvious that the machine technology series is the end product of extensive study, analysis, and experience. *Fundamentals of Machine Technology* (the first volume) covers basic hand tools, cutting tools, and machine tool setups and processes; measurement and layout instruments and tools; and common bench and assembly practices. Industrial safety is threaded in detail throughout the units. Essential blueprint reading and mathematics principles are applied directly to typical shop jobs. Drilling machines, floor grinders, power hacksaws and band cutoff machines, engine lathes, and horizontal milling machines are the fundamental machine tools that are treated.

Advanced Machine Technology (the second volume) deals in depth with bench, assembly, design, machining, and manufacturing theory and practice on intermediate and advanced levels. The machine tools, setups, and machining practices relate to engine lathes; turret lathes, screw machines, and other production turning machines; vertical and horizontal millers; shaping and planing machines; vertical band machines; and surface, tool and cutter, and centerless grinding machines. Tool design principles and tooling for production are related to punches and dies, jigs and fixtures, and cutting tools. Jig boring, jig grinding, and universal measuring machines are applied to precision machining and measurement. Numerical control technology and processes are demonstrated in computerized, automatic machine tool applications. Ferrous and nonferrous metals and alloys and heat treating and hardness testing are covered. Again, blueprint reading, mathematics, and physical science principles are an integral part of the instructional units.

The two *Study Guides* (workbooks) provide a tremendous range of test items. These simulate actual industrial conditions. Each unit provides for individualized, directed learning experiences. Each study guide interweaves the hands-on development of skills and technology in layout, machining, fabricating, and assembly of a single workpiece or the production of multiple parts.

The two *Teacher's Manuals* provide suggestions for effective teaching, learner-directed activities and assignments, and the solutions to all shop problems in the *Study Guides*. Detailed descriptions of the scope, organization, and applications of the textbooks, student's guides, and teacher's manuals follow in the Introduction.

C. Thomas Olivo

ACKNOWLEDGMENTS

A statement of grateful appreciation is made to the many teachers and directors in postsecondary institutions offering machine and manufacturing technology curriculums, to other occupationally competent teachers in vocational/industrial/technical schools providing instruction in the machine trades, and to the instructors in in-plant training programs who assisted. Recognition is made to teachers and instructors in similar institutional and industrial positions in Europe, the Middle East, Southeast Asia, and South America who made on-site assessments and curriculum and facilities planning possible.

Also recognized are those shop foremen and industrial management persons who permitted in-plant analyses of modern design, process, and control techniques.

The work and contribution of the leaders in the curriculum and instructional materials centers of the Vocational-Technical Education Consortium of States and other regional vocational education research centers is appreciated.

A particular word of commendation is made to the Canadian, British, German, and Austrian machine tool and instrument companies and institutions. They provided technical assistance, advice, relevant data and in-plant experience, particularly in relation to dimensional measurements in SI metrics.

Through an acknowledgment of the participating companies that follows comes a personal "thank you" to each individual who provided photos and other supporting technical documents for that company:

AGF Inc. (formerly American Gas Furnace Company)
Acco Industries Inc., Measurement Systems Division
American Drill Bushing Company
American Iron and Steel Institute
American Precision Museum
American Tool & Grinding Company
Apex Tool & Cutter Company, Inc.
Barber-Colman Company, Machines & Tools Division
Bausch & Lomb, Inc.
Bay State Abrasives, Dresser Industries, Inc.
Bendix Corporation, Automation and Measurement Division
Bendix Corporation, Industrial Controls Division
Blanchette Tool & Gage Manufacturing Corporation
E.W. Bliss Division, Gulf & Western Manufacturing Company
Boston Gear, Incom International Inc.

Bridgeport Machines, Division of Textron Inc.
Brown & Sharpe Manufacturing Company
Cabot Corporation
Carboloy Systems Department, General Electric Company
Cincinnati Inc.
Cincinnati Milacron Inc.
Clausing Machine Tools
Cleveland Twist Drill, an Acme-Cleveland Company
Danly Machine Corporation
DeVlieg Microbore Division, DeVlieg Machine Company
DoALL Company
Dorsey Gage Company, Inc.
Dunham Tool Company, Inc.
Engis Corporation, Diamond Tool Division
Ex-Cell-O Corporation
Federal Products Corporation
Fellows Corporation
Grinding Wheel Institute
Hammond Machinery, Inc.
Hardinge Brothers, Inc.
Harig Products, Bridgeport Machines, Division of Textron, Inc.
Hitachi-Magna-Lock Corporation
Industrial Plastics, Inc.
J & S Tool Company, Inc.
Johnson Gas Appliance Company
Jones & Lamson Metrology Products, Waterbury Farrel Division of Textron Inc.
Landis Tool Company
Linde Division, Union Carbide Corporation
Lodge & Shipley Company
Mahr Gage Company, Inc.
Moore Special Tool Company, Inc.
National Acme, an Acme-Cleveland Company
National Broach & Machine, Division of Lear Siegler, Inc.
National Twist Drill Division, Division of Lear Siegler, Inc.
Norton Company
Pratt & Whitney Machine Tool Division, Colt Industries
Rockford Machine Tool Operations, Ex-Cell-O Corporation
Shore Instrument & Manufacturing Company
Simmons Machine Tool Corporation
South Bend Lathe, Inc.
L.S. Starrett Company
TRW/Geometric Tool Division
TRW/Greenfield Tap & Die Company
TRW/J.H. Williams Division
Taft-Peirce Manufacturing Company
Thermolyne Corporation
Thompson Grinder Company, Waterbury Farrel Division of Textron Inc.
Tinius Olsen Testing Machine Company, Inc.
USM Machinery Division, Emhart Corporation
United States Steel Corporation

Universal Vise & Tool Company, Schwartz Fixture Division
Warner & Swasey Company
Waterbury Farrel, Division of Textron, Inc.
White-Sundstrand Machine Tool Company

Special appreciation is expressed to each of the following persons for "going that extra mile" in providing unusual technical resources: James D. Barber, Supervisor of Customer Training, Industrial Controls Division, Bendix Corporation; Richard J. Koulbanis, Advertising Coordinator, Industrial Products Division, Brown & Sharpe Manufacturing Company; Thomas C. Doud, Technical Writer, Hardinge Brothers, Inc.; and John Meek, Advertising Manager, Landis Tool Company.

Credit is due Lowell Pyles, Senior Products Promotion Specialist, Advertising and Sales Promotion Department, Cincinnati Milicron Inc., for extensive technical assistance; Lincoln Peters, Public Relations Manager, DoALL Company, who provided outstanding services supported by Robert L. Garro, Publicity Manager; William Hoadley, Advertising Manager, Moore Special Tool Company, for many outstanding illustrations of jig borer, jig grinder, and precision measurement machines; Philip L. Starrett, Manager, Advertising and Sales Promotion, the L.S. Starrett Company, for precision tools and gages; John Krisko, Coordinator (Advertising Turning Group), the Warner & Swasey Company, for tremendous assistance in connection with tooling, setups, and procedures for bar and chucking machines and other screw machine production turning processes; and Robert Young, Manager, Marketing Engineering, USM Corporation, for special photomicrographs.

Recognition is made of Carolynn G. Whitehurst, for excellent editorial review work during the developmental and final manuscript preparation stages. Credit is due Genevieve M. Pruscop, for manuscript typing services and to Evelyn Wells, for supportive assistance.

Thanks is expressed to Peter J. Olivo, District Manager, Chevron Hoffman Fuel Company and former master toolmaker, Danbury, Connecticut, for providing industrial consultant services and to Thomas P. Olivo, former associate supervisor, New York State Education Department, for assistance in making occupational analyses and other curriculum planning contributions.

A special commendation is made to Jim Linville, Senior Editor, and to Sylvia Dovner, Production Manager, Breton Publishers, for their persisting, dedicated efforts through successive stages of development and production to ensure technical accuracy and quality.

Collectively, these acknowledgments of organizations and individuals who participated in major activities behind the scenes are the keystones upon which quality and content relevance depend.

C. Thomas Olivo

INTRODUCTION

SCOPE AND CONTENT OF THE MACHINE TECHNOLOGY SERIES

There are two textbooks in the machine technology series: *Fundamentals of Machine Technology* and *Advanced Machine Technology*. Each has a companion *Study Guide* (workbook) and a *Teacher's Manual*. Both textbooks incorporate manipulative skills with complementary technology. These must be mastered to become a productive worker with the potential for advancement within a cluster of job opportunities.

Fundamentals of Machine Technology is a basic textbook. It is organized for persons who plan to enter the field and for others who wish to develop the skills and technological knowledge essential to advance in the occupational cluster. The organization and contents deal with theory ("why-to-do") and practice ("how-to-do"). These aspects are related to fundamental measurements (using line-graduated and micrometer measuring tools) and to layout and other bench cutting, forming, and assembly processes.

The machining processes and technology sections in the basic textbook cover such basic machine tools as power hacksaws and horizontal band machines, drilling machines and accessories, engine lathes, and horizontal milling machines (including the dividing head). Both the SI metric and customary inch-standard systems and units of measure are incorporated in all bench and machine tool layouts, machining processes, and measurements.

Advanced Machine Technology deals with the intermediate and advanced skills and technology. More precise instruments, layouts, and measurements are treated. Fundamental heat treating and metallurgy, quality control, and other fabrication processes are covered in relation to the manufacture of common industrial materials.

Setups of machine tools, accessories, and representative workpieces are related to advanced lathe and production turning machines, vertical and horizontal milling machines, shapers and planers, and band machining equipment. There is also a comprehensive treatment of precision grinding (including surface finishing) with surface, cutter, tool, cylindrical, and other special grinding machines. Fundamentals of numerical control are applied to design, layout, and machining on computerized and automatic machine tools and accessories. Machine and tool design principles are dealt with through advanced shop processes on jigs and fixtures, punches and dies, cutting tools, and other production tools. New electronic forming and machining equipment is introduced.

Related Mathematics, Physical Science, and Blueprint Reading

Both volumes in the machine technology series incorporate fundamental principles of mathematics, physical science, blueprint reading, sketching, and design. These principles are applied in shop, laboratory, and design and development activities. The productive worker must be able at each level of job responsibility to use appropriate formulas and make computations. Design, production, and measurement information must be communicated and interpreted accurately from drawings, sketches, design data, and other technical directions.

Physical and chemical phenomena and principles (identified through occupational analyses) are related to layout, tool, cutter, and machine setups and machining processes. Dimensional accuracy, machinability, surface finish, tool life, safety (to mention a few factors) all depend on a knowledge of basic science and practical applications. Each principle is interwoven into the theory and processes to which it applies.

Special Features of the Machine Technology Series

Industrial Terms. Shop and laboratory terms represent the technical language and method of communicating that is required and used in industry. Thus, each unit contains technical (occupational) terms. These are expressed and interpreted in the special way in which they are used on the job. Important terms are grouped and defined in a *Terms* section at the end of each unit. The purpose is to provide a review and handy reference section.

Step-by-Step Procedures. In each unit considerable attention is devoted to bench work, inspection practices, and basic machine tool setups and processes within *How-to-Do* sections. The step-by-step procedures are carefully developed to show how even complex processes may be simplified and easily carried out. The steps are presented in the same manner as they are performed by craftspersons in small production jobbing shops and in general machine trade work.

Safety. Worker safety and the protection and safe operation of machine tools and accessories are covered throughout each text in relation to a setup and process. *Cautions* within the *How-to-Do* sections highlight the safe use of hand tools and orderly procedures for machine tool operations. General *Safety Practices* are grouped at the end of each unit for careful analysis and easy reference. These sections include precautions that are necessary for personal and group safety and for the protection of sensitive and expensive instruments and equipment.

Use of Examples. *Examples* are used throughout each text to show how formulas are applied in machine technology. The technique of using typical examples simplifies the interpretation of the formulas, technology principles, and other conditions.

Notes. *Notes* are widely used within the *How-to-Do* sections to call attention to additional factors a craftsperson must consider in carrying out the step-by-step procedures. In some instances, a note extends the content covered earlier in the unit. In other instances, notes signify alternate techniques or methods for doing a particular step.

Illustration Program. Photographs are included throughout each book to reinforce the descriptive material. Other art work has been especially prepared as line drawings. These, too, strengthen and extend the written content. The illustration program shows conditions, setups, mechanisms, tools, or operations in a simplified graphic form.

Unit Summary. The technical content of each unit is crystallized by a series of condensed statements brought together in the *Summary*. Each summary brings into focus highlights from within the unit. The summary helps to identify any items that may require further study. Each summary also reinforces learning by presenting essential factors, conditions, values, processes, and so forth, in relation to all other areas in the unit.

Review and Self-Test. A series of *Self-Test* items are included at the end of each unit. These are designed to serve a twofold purpose. The answers provide (1) a review of the technology and processes and (2) a measure of how well the content has been mastered.

Significant Technical Tables. Workers in machine technology and related occupations use handbooks containing industrial reference tables. The tables provide formulas as well as design, engineering, and other important data. These are used for computing dimensions and values that are applied to hand and machine processes. The complete set of the *Handbook Tables* that are used within each text is contained in the *Appendix*. The portions of the tables presented within each text provide examples of how the technical information is applied in actual shop practice.

Study Guide

Each *Study Guide* is designed as a study, assignment, and test item workbook. The units in the *Study Guide* parallel the content and experiences in each textbook. Each *Study Guide* contains test items that are designed to measure performance objectives. Performance relates to the mastery of technology; the ability to perform work processes (skills); the interpretation of shop drawings, prints, and sketches; computational ability and accuracy in applying shop mathematics; and a comprehension of practical physical science applications. The test items complement the end-of-unit review and self-test items in each text and may be used for both pre- and post-testing.

Directed student learning activities are detailed in each unit. These are followed by *Student Test Items*. Both the units and the test items proceed in a logical learning progression.

Teacher's Manual

The units in each *Teacher's Manual* parallel the instructional units in each textbook and *Study Guide*. The organization of the teaching/learning content into units provides flexibility in subdividing the material into a series of lessons and in outlining courses and programs for full- and part-time instruction. Solutions to test items, step-by-step work processes, and other guidelines are provided. Answers to personal, machine, and tool safety practices are analyzed and stressed.

LEARNING PATTERN AND ORGANIZATION OF THE MACHINE TECHNOLOGY SERIES

The same pattern of organization, format, and writing style is used for both volumes in the machine technology series. Each textbook, study guide, and teacher's manual consists of a number of *Parts*. These represent major measurement, bench work, and machine tool categories within the occupation. The parts are subdivided into *Sections*. Each section contains a grouping of significant related activities. The skills and technology that relate to each section are described in detail in what are called *Units*. The relationship to be found among the units, sections, and parts is illustrated by reviewing the contents pages of each machine technology book.

SCOPE OF ADVANCED MACHINE TECHNOLOGY

Advanced Machine Technology contains ten parts. *Part One* has three sections on precision measurement and inspection practices. *Section One* covers fine mechanical measurements and layout work. High amplification comparators utilizing optical, pneumatic, and electronic systems are identified in *Section Two*. Quality control practices and surface texture characteristics and measurements are contained in *Section Three*.

Part Two deals with turning machines. *Section One* is devoted to advanced threading and turning processes; *Section Two*, to tooling, setups, and processes for bar and chucking machines; and *Section Three*, to single- and multiple-spindle automatic screw machines.

Part Three focuses on vertical milling machines. *Section One* describes vertical milling machine features and operating controls; *Section Two*, general vertical milling machine processes.

Part Four centers on horizontal milling machines. Setups and processes for cutting dovetails and keyseats, sawing and slotting, and helical and cam milling are covered in *Section One*. *Section Two* is devoted to gears and gear milling.

Part Five covers vertical band machines. Band machine design features and basic vertical band machine processes are described in *Section One*. *Section Two* covers band machine sawing, filing, polishing, and grinding technology and processes.

Part Six includes information on shaping and planing machines. Standard and vertical shapers (slotters) are treated. Shaping and planing machine features; tooling; setups; and typical horizontal, vertical, angle, and form cuts are illustrated.

Extensive coverage is made of abrasive machining in *Part Seven*. There are five sections. *Section One* provides advance information about the functions and types of grinding machines. *Section Two* covers grinding wheel standards and preparation. *Section Three* deals in depth with horizontal-spindle surface grinders and processes. The theory and practices relate to work-holding and accessory setups; precision truing and dressing devices; and horizontal, vertical, shoulder, and form grinding, including cutting-off processes. *Section Four* emphasizes machine tool features and setups for internal and external grinding machines and accessories. Parallel, shoulder, taper, and plunge grinding are included. A description is provided for centerless grinding techniques. *Section Five* deals extensively with tool and cutter grinding setups and processes. Procedures are established for the grinding of straight and helical-flute cutters, staggered-tooth and form-relieved cutters, end mills, and single-point cutting tools.

Part Eight has two sections covering numerical control technology. This includes basic formats and computer-assisted programming and machining with two-, three-, four-, and five-axes numerical control machine tools.

Part Nine deals with tool design principles and tooling production. *Section One* relates to the production of jigs and fixtures, punches and dies, and cutting tools. Emphasis is placed in *Section Two* on jig borers, jig grinders, and universal measuring machines.

The concluding *Part Ten* covers the production of industrial and cutting tool materials. In *Section One*, consideration is given to the manufacture, properties, and classifications systems of ferrous and nonferrous metals, alloys, carbides, and ceramic materials. The final unit in *Section Two* deals with basic metallurgy and the heat treatment and testing of metals.

In addition to the 37 instructional units in the textbook, there are 25 companion tables in the Appendix. Each table is especially selected to complement the contents within appropriate units in the textbook.

The Machine Technology Series in Skilled Manpower/Womanpower Development

The organization, content, and format of the machine technology series provide flexibility for use at many levels and under differing learning conditions. Effectiveness and efficiency in human resource development are incorporated in the materials. The series is planned for the following, as well as other training applications:

> The machine technology series may be used for full- or part-time instruction in institutional settings, in cooperative work-experience programs, or in-plant or institutional training programs.

> The machine technology textbooks and their study guides are geared for use within technical institutes and junior colleges and within vocational industrial/technical schools that have machine technology, metal products manufacturing, mechanical technology, machine trades, and other related curriculums and programs.

> The machine technology textbooks and their study guides are adapted to apprenticeships and upgrading in-plant industrial training programs, cooperative training programs within union training centers, military occupational specialty programs, and other manpower/womanpower development agency training programs.

> The machine technology textbooks and their study guides provide resource materials for self-study by craftspersons, technicians, supporting personnel in related fields, training directors, management persons, and others who must have an understanding of the theory, processes, practices, and products associated with machine technology.

PART ONE

Precision Measurement and Inspection

SECTION ONE   Advanced Measurement: Mechanical
        UNIT 1   Linear and Angular Measurement
        UNIT 2   Layout and Measurement Practices in Surface Plate Work
SECTION TWO   Precision Measurement Instruments
        UNIT 3   Measurement with Optical Flats and Microscopes
        UNIT 4   Measurement with High Amplification Comparators
SECTION THREE   Quality Control and Surface Texture (Finish)
        UNIT 5   Quality Control Principles and Measurement Practices
        UNIT 6   Surface Texture Characteristics and Measurement

SECTION ONE

Advanced Measurement: Mechanical

High degrees of layout and measurement accuracy are commonly used in advanced machine technology. This section provides advanced technical information, mathematics, and practices related to linear and angular measurements and layouts. Tools and accessories are treated in terms of the high degrees of accuracy used in precision surface plate work.

UNIT 1

Linear and Angular Measurement

Most linear measurements in the machine trades and metal products manufacturing industries are held to tolerances of ±0.001" to ±0.0001" or ±0.02mm and ±0.002mm. Similarly, angular measurements are made to within ± five minutes of arc (±5'). These dimensional accuracies are practical and efficient to produce and are reproducible anywhere in the world.

More precise limits are also common for machining, assembling, and measuring other workpieces. Linear measurement accuracies ranging to the millionth part of an inch and 0.000025mm are used regularly. Also, angular dimensions need to be measured to such degrees of fineness as one minute of arc and fractional parts of one inch and metric equivalents.

Instruments and gages for highly precise measurements are normally made of metal. For example, gage blocks, which are described in this unit, are frequently made of steel. In the case of optical flats, a specially treated glass is used. Design improvements in measurement and layout tools, instruments, and accessories also include the use of granite.

Surface plates, angle plates, universal right angles, parallels, and sine plates—also described in this unit—are all available in granite. The advantages and applications of granite as a base material for linear and angular layout and measuring instruments also are covered in detail in this unit.

STEEL GAGE BLOCKS

Steel gage blocks are precision finished for *size*, *flatness*, and *parallelism*. Size control is obtained from grained master gage blocks. These blocks are calibrated by precision *optical inferometers*. Master gage blocks and calibration instruments are regularly checked against national standards of the Bureau of Standards. The accuracy of the surface shape and finish of gage blocks may be seen through optical flats and recorded in a *micro-interferogram*.

For example, a micro-interferogram will show whether the edges of gage blocks have been burr proofed by the manufacturer. Fine burrs may damage other blocks and affect accuracy of the blocks. Burr proofing smooths microscopic, sharp, ragged edges on the corner break. As shown in Figure 1-1, such an edge is formed where the lapped surface of the block and the blended corner radius meet. Figures 1-2A and B show a micro-interferogram comparison of a burr-proofed gage block and a regularly finished gage block. Note how the fringe lines blend smoothly over the edge in Figure 1-2A.

Hardness and Stability

Gage blocks must also be *dimensionally stable*. Each block must be free of residual internal stresses if it is to maintain dimensional stability throughout its lifetime. While the size of the gage block will vary depending upon temperature, the block measurements must remain stable at any one temperature.

Gage blocks are made of selected alloy steel and heat treated to a high hardness. The heat treatment is controlled. A uniform hardness results in maximum resistance to wear.

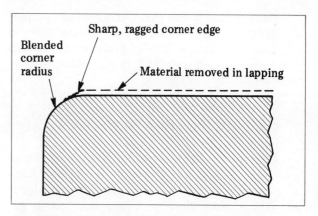

FIGURE 1-1 Effect of Burr Proofing Gage Blocks

Care of Gage Blocks

Gage blocks are furnished in specially designed cases. The size of each block is easily identified. The cover of each case lies flat. It serves as a tray in selecting and holding gage block combinations.

Lint or dust should be brushed away from a gage block with a soft camel's hair brush. The blocks should be wiped clean before and after use with lint-free wiping tissues. An aerosol-propelled cleaning fluid serves as a convenient way to clean blocks as they are selected. After they are used and wiped, blocks should be coated with a fine wax-base film and then returned to their appropriate cases.

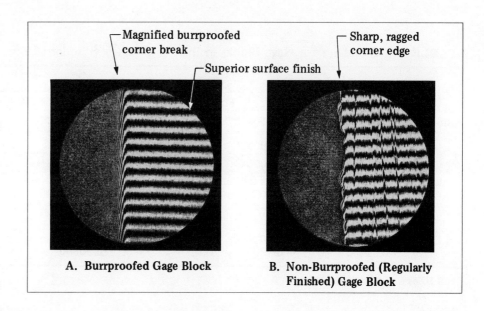

FIGURE 1-2 Micro-Interferogram Comparison of Burrproofed and Regularly Finished Gage Blocks

A. Burrproofed Gage Block

B. Non-Burrproofed (Regularly Finished) Gage Block

Microscopic nicks and burrs prevent *perfect wringing* of gage blocks. A deburring stone is used to remove the fine imperfections. Blocks are selected and handled during rustproofing with plastic-tipped forceps.

Grades of Steel Gage Blocks

Gage blocks of most manufacturers fall into two basic categories. These categories are grade 1 and grade 2. Many older sets of gage blocks are still in use. These sets are classified as AA (grade 1), A+ (grade 2), and B (grade 3).

Grade 1 (AA) blocks are used principally in calibration and for setup operations. The measurements are related to scientific, instrumentation, and the highest level of precision inspection. Grade 2 (A+) blocks are of finer quality than the earlier coarse-finished grade 3 (B) and grade 2 blocks.

Table 1-1 provides the standards for dimensional accuracy (length, flatness, and parallelism) and surface finish of one manufacturer for both inch- and metric-standard grade 1 and grade 2 gage blocks. The metric tolerances applied to length, parallelism, and flatness are sometimes expressed in *microns* (millionths of a meter). For example, a tolerance of ±0.00005mm may be expressed as ±0.05 microns (±0.05µm). Surface finish range is stated in microinches (0-0.4″) for inch-standard blocks and in microns (0-0.10µm) for metric-standard blocks.

Gage Block Sets

Gage blocks are available in sets with either rectangular or square-shaped blocks. Some sets have gage blocks of fractional size that provide for fractional measurements from 1/64″ to 12.000″ in steps of 1/64″. Decimal measurements (in increments of 0.001″) may be made from 0.020″. Setups for measurements in ten-thousandths of an inch are possible in steps of 0.0001″ to 0.120″.

Other sets have blocks that provide for measurements from 200 thousandths of an inch (0.200″) to approximately 25″ in steps of 25 millionths of an inch. Long gage block sets are also available. The long gage block set shown in Figure 1-3 permits the fast, convenient assembly of blocks for measurements from 5″ to 84″.

Metric Gage Blocks

Sets of metric gage blocks are available in series. Increments within a common series range from 0.001mm to 25.0mm. The lengths of the blocks in the series range from 0.50mm through 100.0mm. Table 1-2 gives the increments and the accompanying range of lengths in millimeters for each set of rectangular blocks in the series. The number of blocks in a set may vary up to 112 pieces.

TABLE 1-1 Standards for Gage Blocks in Inch- and Metric-Standard Units

Grade of Gage Blocks*	Dimensional Accuracy					Surface Finish
	Length (Size)	Flatness		Parallelism		
		0.100″ to 2″	Over 2″	0.100″ to 2″	Over 2″	
1 (AA)	±0.000002″	±0.000002″	±0.000004″	±0.000002″	±0.000004″	0-0.4 microinches
2 (A+)	+0.000004″ −0.000002	+0.000004″ −0.000002	+0.000004″ −0.000005	+0.000004″ −0.000002	+0.000004″ −0.000005	
1 (AA)	±0.00005mm (0.05 microns, 0.05µm)	±0.00005mm	±0.0001mm	±0.00005mm	±0.0001mm	0-0.10 microns
2 (A+)	+0.00010mm −0.00005	+0.00010mm −0.00005	+0.00010mm	+0.00010mm −0.00005	±0.00010mm	

*66 Rockwell C scale hardness

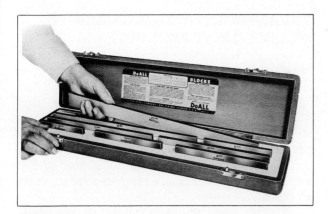

FIGURE 1-3 Rectangular Long Gage Block Set (Steel)

GAGE BLOCK ACCESSORIES

Applications of gage blocks are extended by using accessories. For example, assembled gage blocks and accessories may be used for precision layouts, gaging, and the taking of internal and external measurements. When a number of accessories are included in a set, they are referred to as a *gage block accessory system.*

Figure 1-4 shows a number of components in a rectangular gage block accessory system. Similar parts are available in both inch and metric measurement standards. The inside and outside *caliper bars* shown in the figure each have a flat surface lapped on one side. This surface is used when outside linear measurements are taken. The second side of each caliper bar has a radius. This radius permits the taking of inside linear measurements.

TABLE 1-2 Metric Increments and Length Ranges of Rectangular Gage Blocks

Increment of Blocks in Set	Range of Length per Set
0.001mm	1.001mm to 1.009mm
0.01mm	1.01mm to 1.49mm
0.5mm	1.0mm to 24.5 mm
1.0mm	1.0mm to 9.0mm
10.0mm	10.0mm to 100.0mm
25.0mm	25.0mm to 100.0mm

The gage block accessory system may also serve as a snap gage, as shown in Figure 1-5. The caliper or gage block is assembled in a component called a *channel* (Figure 1-4). The channel is closed at one end. Accessories like the combination caliper and gage blocks drop easily against the back of the channel. The gage blocks are automatically aligned.

A *micro-clamp* (Figure 1-4) is used at the end of the buildup. The clamp exerts the proper force for the stack of gage blocks and accessories. The micro-clamp consists of a clamp screw, a pusher screw, ball, and block. The clamping force applies in-line pressure. This pressure assures rigidity, parallelism, and accuracy. Surface wear on the gage blocks is thus minimized. Wringing the blocks together is eliminated in a micro-clamp setup. An *end standard* (Figure 1-4) may be assembled with gage blocks in a channel.

A *scriber* (Figure 1-4) is an accessory that may be set to permit scribing accurate lines. Figure 1-6A illustrates a scriber set at required precision

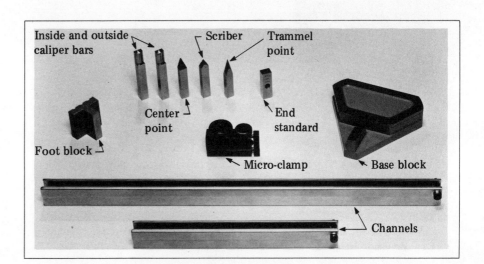

FIGURE 1-4 Rectangular Gage Block Accessory System

6 PRECISION MEASUREMENT AND INSPECTION

FIGURE 1-5 Snap Gage Assembly of Gage Block Accessory System

height for a layout application. A height gage assembly for a measurement application is shown in Figure 1-6B. Gage blocks, the outside (flat) face of a combination caliper, a channel, and a micro-clamp are used in the setup. While accessories and combinations have been described for rectangular gage block applications, a similar system is available for square gage block sets.

ADVANTAGES OF BLACK GRANITE MEASUREMENT PRODUCTS

Measuring instruments made from granite were being used more than five thousand years ago. The Egyptians produced a black granite *cubit stick* as a standard of measurement. Black granite was selected because it possessed important properties such as density, hardness, and durability. Cubit sticks, like the one illustrated in Figure 1-7, were periodically compared for accuracy to the royal master.

Today, there are several basic advantages to using black granite over metal for precision layouts and measurements.

—*Hardness:* Black granite has an 80 Rockwell C scale hardness by test. It is harder than tool steel. Black granite is also tough enough to resist chipping and scratching.
—*Rigidity:* Black granite surface plates have extremely low deflection under load. They are, therefore, almost distortion free and retain their original accuracy.
—*Density:* The uniform density of black granite makes it possible to produce a fine-textured, lapped, and frosted surface. The surface finish prevents gage blocks and precision surfaces from wringing.
—*Low reflectivity:* A black granite surface provides a dark nonglare finish. Light reflected from illumination, the surface plates, tools, instruments, and workpieces

A. Scriber Set at Required Precision Height for Layout

B. Height Gage Assembly Using Outside Face of Caliper Bar

FIGURE 1-6 Applications of Snap Gage Accessory System to Layout and Measurement Processes

FIGURE 1-7 Early Black Granite Cubit Stick Standard of Egyptian Measurement System

is limited. The dark surface is also an excellent background for layout lines that are scribed with a brass rod. Such lines may be removed with an ordinary eraser.
—*Porosity:* The uniform grain structure of black granite gives it a low water (moisture) absorption rate. A black granite surface is not affected by acids, chemical fumes, dust, or dirt. Surface plates and gages are, therefore, rustproof, and rusting and pitting of precision gages that are left on the plates are prevented.
—*Thermal (heat) stability:* Temperature changes have extremely limited effect on the accuracy of black granite gaging products.
—*Easy care and maintenance:* Black granite surface gages and accessories may be cleaned easily. Soap and water, detergent, or a manufacturer's solvent may be used. The surface will not stain or discolor.
—*Low cost:* The initial, maintenance, and relapping costs of a granite plate are lower than for a metal surface plate.

BLACK GRANITE SURFACE PLATES

Black granite surface plates are manufactured in three standard grades: AA, A, and B. *Grade AA* is used in the laboratory. Grade AA surface plates are produced commercially in basic sizes of 8″ × 12″ × 3″ thick or 200mm × 300mm × 75mm. The overall accuracy is 50 millionths of an inch (0.000050″) or 12 ten-thousandths of a millimeter (0.0012mm). One of the largest commercial sizes of black granite surface plates is 72″ × 144″ × 18″ thick (1800mm × 3600mm × 450mm). The laboratory quality of this size of surface plate has an overall accuracy of 0.0011″ or 0.028mm.

Grade A is generally used for inspection work. Grade A surface plate sizes begin at the same basic size as grade AA plates—that is, 8″ × 12″ (200mm × 300mm). However, the surface plate thicknesses are slightly smaller for each size. The overall accuracy of grade A plates is not as precise as the AA laboratory grade. For example, the accuracy of the work surface of the 8″ × 12″ × 2″ thick (200mm × 300mm × 50mm) basic plate is 0.0001″ (0.0025mm). The largest size of plate, 72″ × 144″ × 16″ (1800mm × 3600mm × 400mm) is accurate to within 0.0022″ (0.056mm).

Grade B surface plates (for the same range of sizes) are the least accurate of the three grades. The 8″ × 12″ surface plate is lapped to an accuracy of 0.0002″ (0.005mm); the 72″ × 144″, to within 0.0044″ (0.011mm).

Adaptation of Surface Plates

Granite surface plates are used in many laboratory, inspection, and shop applications. They provide high degrees of flatness and accuracy required in precision layout and machining processes.

Toolmaker's Flats. Handy, lightweight, and portable granite plates, called flats, are used by toolmakers, diemakers, inspectors, and machinists. These plates provide uniform flatness standards. Square or rectangular surface plates are used for layout, assembly, and inspection processes on the bench.

Master Flats. Super-accurate master flats are available for laboratory applications. Master flats are usually cylindrical granite plates. Standard sizes range from 6″ to 48″ (150mm to 1200mm) in diameter. The thickness varies from 2″ to 16″ (50mm to 400mm), respectively. The flat surface is lapped to provide for repeat measurements to 10 millionths of an inch. This degree of accuracy is compatible with laboratory-grade gage blocks.

Modified Surface Plates. Granite surface plates are available with holes, grooves, slots, and metal inserts. Holes may be drilled into a surface plate. The holes permit the layout and measurement of a part that has one or more projecting lugs. A stainless steel threaded insert is useful in clamping a part to the face of the plate. T-slots that conform to dimensions and tolerances approved by the American National Standards Institute (ANSI) may be machined. Metal T-slot inserts may be fitted to a granite plate. Granite surface plates may also be machined with U-slots or to any contour shape. Three of these common design features are shown in Figure 1-8.

GRANITE PRECISION ANGLE PLATES AND ACCESSORIES

Granite *precision angle plates*, such as the plates shown in Figure 1-9, are available for production or for inspection. The difference between production and inspection plates is in the degree of accuracy. Production plates are accurate to within ±0.00005″ in 6″ (0.001mm in 150mm). The accuracy of inspection, or master, plates is ±0.000025″ in 6″ (0.0006mm in 150mm). The accuracies relate to flatness and squareness. Angle plates are lapped on either two surfaces (base and face) or four surfaces (base, face, and two ends). These surfaces are finished square and flat.

Angle plates are available with *metal inserts*. Examples of these plates are shown in Figure 1-10. Some inserts are recessed into one side and the main gaging face. These inserts are used for magnetizing purposes. Threaded inserts are available for clamping applications.

The term *universal* indicates that all faces on a *universal right angle plate* may be used. All six faces of the universal right angle shown in Figure 1-11 are square and parallel.

Since angle plates are a precision accessory, they are usually provided with a protective case. Angle plates, like all other fine tools and instruments, should be cleaned before storing. Any nicks and burrs must be removed with a special deburring stone.

PARALLELS AND OTHER FORMS OF GRANITE ACCESSORIES

Parallels are matched in pairs. Granite parallels serve the same purposes as ground steel parallels. They are machined and finished for production and inspection (master) standards of accuracy. Either the top and bottom or the top and bottom and two sides may be finished with parallel surfaces.

Black granite precision box parallels and a granite straight edge are shown in Figure 1-12. *Box parallels* (Figure 1-12A) are matched pairs of parallels. The base area is smaller in comparison to the height than it is on regular parallels. Workpieces that are difficult to layout and inspect are usually elevated from a surface plate with box parallels. These parallels provide an accurate working surface that is parallel to the surface plate.

The granite *straight edge* (Figure 1-12B) has one precision finished plane face. If it is of

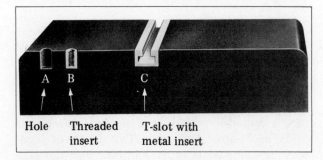

FIGURE 1-8 Granite Block Machined for Layout and Machining Processes

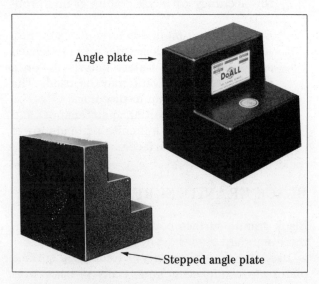

FIGURE 1-9 Black Granite Precision Angle Plates for Production or Inspection

LINEAR AND ANGULAR MEASUREMENT 9

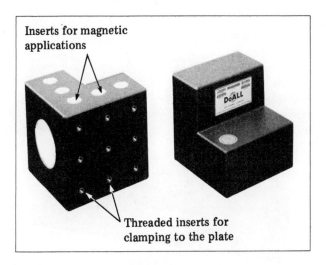

FIGURE 1-10 Angle Plates with Metal Inserts

FIGURE 1-11 Universal Right Angle

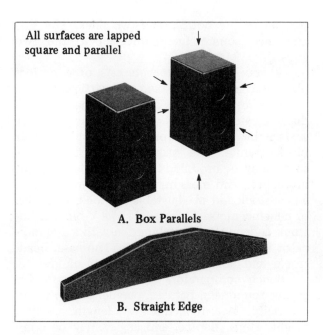

FIGURE 1-12 Black Granite Precision Box Parallels and Straight Edge

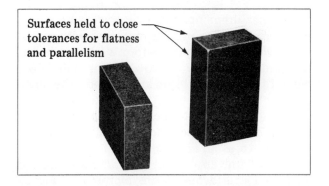

FIGURE 1-13 Black Granite Step Blocks

inspection grade, the plane face is accurate to ±0.00005" (1.0μm) over 12". The more precise laboratory-grade straight edge is finished to ±0.000025" (0.6μm) accuracy over 12". Commercial sizes of black granite straight edges range from 24" × 1-1/2" × 4" (600mm × 38mm × 100mm) to 72" × 2" × 10" (1800mm × 50mm × 250mm). Straight edges are used for such purposes as checking the straightness and parallelism of flat surfaces and the beds of machine tools.

Granite *precision step blocks*, like the blocks shown in Figure 1-13, have flat, parallel, and square surfaces. These surfaces are held to close tolerances for flatness and parallelism. Step blocks are used singly or in pairs for layout, inspection, or measurement processes. A step block is often used as an anvil for a bench comparator.

A *cube* is really a combination of three sets of parallels. The opposite faces are parallel and square. The faces provide perpendicular, flat, and parallel reference surfaces in relation to the plane of a surface plate. Single or matched pair cubes may be finished in production- or master-grade accuracy. The cubes are available with solid faces. Some cubes, like the granite cube shown in Figure 1-14, have imbedded stainless steel threaded inserts. These inserts permit parts to be clamped to a particular face.

V-blocks support cylindrical workpieces during layout, manufacture, or inspection. The 90° V is centered and parallel to the bottom face. Some V-blocks are produced with the 90° V

10 PRECISION MEASUREMENT AND INSPECTION

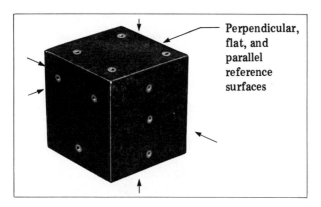

FIGURE 1-14 Granite Cube with Imbedded Stainless Steel Threaded Inserts

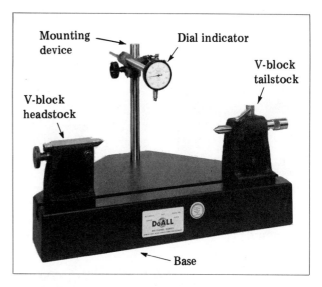

FIGURE 1-15 Bench Comparator Equipped with V-Block Headstock, Tailstock, and Indicator

parallel to the bottom and two sides. The V is machined square (perpendicular) to the two ends.

A *bench comparator* consists of a perfect plane surface (surface plate), a vertical post, a swivel arm, and a fine-feed adjustment with a dial indicator. The surface plate (plane reference surface base) may be produced from metal or black granite. The base may be slotted. Its shape may be modified to accommodate layout, inspection, and measurement accessories such as a headstock and an adjustable tailstock, V-blocks, or bench centers. An example of a V-block headstock and tailstock combination with a dial indicator is shown in Figure 1-15. The base, headstock, and tailstock are made of black granite.

Bench comparators are highly accurate for gaging roundness, flatness, parallelism, and concentricity. Measurements of length, width, height, steps, depths, and angles may be compared against linear standards.

PRECISION ANGLE MEASUREMENTS

Angle measurements to an accuracy within five minutes of arc are usually measured with a universal vernier bevel protractor. While the dial of this protractor may be rotated through 360°, part of the base and blade are covered. The range of use of the protractor is thus limited. An angle attachment is used to increase the range. Since an attachment is needed for certain acute angles, it is known as an *acute angle attachment*. Two models of universal bevel protractors using the acute angle attachment are pictured in Figure 1-16.

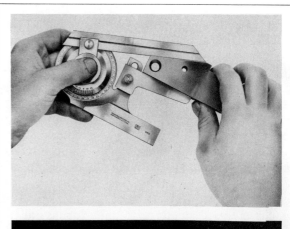
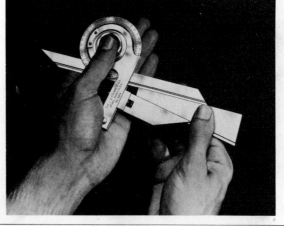

FIGURE 1-16 Measurements of Small Angles with Two Models of Universal Bevel Protractors Using the Acute Angle Attachment

Practices and Factors Influencing Accurate Measurements

An angular dimension is always read directly. Whether a piece part is measured to an accuracy within five minutes of arc depends on the skill of the worker. Two measurement practices must be observed. First, the base and blade of the protractor must be in close contact with the surfaces to be measured (Figure 1-17A). Second, the protractor must be perpendicular to the surfaces to be measured. As Figure 1-17B shows, the blade of the protractor must be at a right angle to the angle plane. The base of the protractor must be at a right angle to the horizontal reference plane.

Accurate, reliable measurements also depend on three groups of considerations: mechanical, positional, and observational. These considerations are summarized in Table 1-3.

MATHEMATICS APPLIED TO ANGULAR MEASUREMENT

Angular surfaces of workpieces or assemblies are often laid out, machined, or inspected to a degree of accuracy within *one minute of arc*. This is a finer measurement than can be taken with a vernier bevel protractor. A sine bar, sine plate, sine block, and compound (universal) sine block are four precision angular layout and measurement

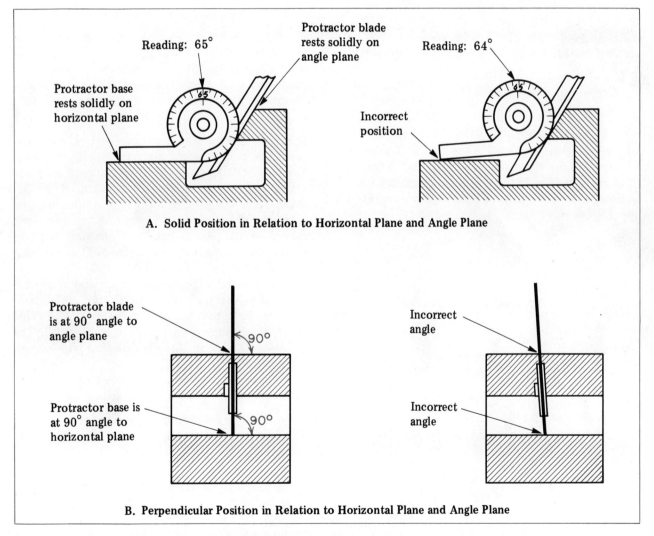

FIGURE 1-17 Positioning Vernier Bevel Protractor Base and Blade Accurately in Relation to Reference Planes of Angle to Be Measured

12 PRECISION MEASUREMENT AND INSPECTION

TABLE 1-3 Mechanical, Positional, and Observational Considerations in Precision Angle Measurement

Mechanical considerations	—The surfaces to be measured are free of burrs, dust, and other foreign particles. —Surface roughness is reduced. The base and blade are brought into close contact with the angle surfaces to be measured. —The movable parts of the instrument have freedom of movement. —The accuracy of the instrument has been checked against a known measurement standard.
Positional considerations (relation between the instrument and the angle to be measured)	—The vertical (X) axis of the vernier bevel protractor falls in the vertical reference plane of the angle. —The horizontal (Y) axis of the instrument base is parallel to the base (surface) of the part to be measured. —The (Z) axis of the instrument is parallel to the (Z) plane of the angle.
Observational considerations	—The vernier reading is taken without parallax error. —Personal measurement bias is not read into the bevel protractor reading. —The angular reading is correct in terms of the required complement or supplement of the angle being measured.

instruments. Each of these instruments may be set to a required angle by applying trigonometry.

Computing the Sine Value (Gage Block Height)

The name *sine bar* is derived from the use of triangles and trigonometric functions. The *sines* of required angles are used to calculate the height of a triangle. The triangle formed consists of the hypotenuse of a known length and the required angle.

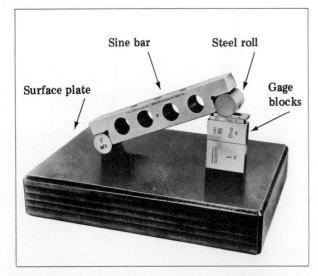

FIGURE 1-18 Typical Angle Setting Using a 5″ Sine Bar, Gage Blocks, and Surface Plate

Figure 1-18 shows a typical angle setting of a sine bar. Gage blocks and a surface plate are used. The center-to-center distance between the hardened steel rolls of equal diameter is a multiple of 5″. The basic center distances are 5″, 10″, or 20″.

Example: Determine the gage block measurement (height) required to set a workpiece at an angle of 20°28′. A 5″ sine bar and surface plate setup are required. The sine value of 20°28′ must first be established. This value may be read directly in a *table of natural trigonometric functions*. A partial listing of natural trigonometric functions is shown in Table 1-4. The sine value for 20°28′ is given as 0.34966.

If a natural trigonometric function table with 10-minute intervals were used, the 28′ would need to be *interpolated* (estimated). The value of 28′ lies between the values of 20′ and 30′. The 28′ value is 8′ more than the natural trigonometric function of 20′.

The steps used to interpolate a minute value (dimension)—28′ in this example—are as follows:

natural trigonometric sine value for

20°30′ = 0.3502
20°20′ = −0.3475

difference between sine values

= 0.0027

8′ (eight-tenths) of the difference

= 0.00216

LINEAR AND ANGULAR MEASUREMENT 13

TABLE 1-4 Partial Table of Natural Trigonometric Function Values for 20° and 159°, by Minutes (M)

20°   159°

M	Sine	Cosine	Tan.	Cotan.	Secant	Cosec.	Vrs. Sin.	Vrs. Cos.	M
0	0.34202	0.93969	0.36397	2.7475	1.0642	2.9238	0.06031	0.65798	60
1	.34229	.93959	.36430	.7450	.0643	.9215	.06041	.65771	59
2	.34257	.93949	.36463	.7425	.0644	.9191	.06051	.65743	58
3	.34284	.93939	.36496	.7400	.0645	.9168	.06061	.65716	57
4	.34311	.93929	.36529	.7376	.0646	.9145	.06071	.65689	56
5	0.34339	0.93919	0.36562	2.7351	1.0647	2.9122	0.06080	0.65661	55
...	...	...	...	...	...	...	...	...	...
18	.34693	.93789	.36991	.7033	.0662	.8824	.06211	.65306	42
19	.34721	.93779	.37024	.7009	.0663	.8801	.06221	.65279	41
20	0.34748	0.93769	0.37057	2.6985	1.0664	2.8778	0.06231	0.65252	40
21	.34775	.93758	.37090	.6961	.0666	.8756	.06241	.65225	39
22	.34803	.93748	.37123	.6937	.0667	.8733	.06251	.65197	38
23	.34830	.93738	.37156	.6913	.0668	.8711	.06262	.65170	37
24	.34857	.93728	.37190	.6889	.0669	.8688	.06272	.65143	36
25	0.34884	0.93718	0.37223	2.6865	1.0670	2.8666	0.06282	0.65115	35
26	.34912	.93708	.37256	.6841	.0671	.8644	.06292	.65088	34
27	.34939	.93698	.37289	.6817	.0673	.8621	.06302	.65061	33
28	.34966	.93687	.37322	.6794	.0674	.8599	.06312	.65034	32
29	.34993	.93677	.37355	.6770	.0675	.8577	.06323	.65006	31
30	0.35021	0.93667	0.37388	2.6746	1.0676	2.8554	0.06333	0.64979	30
31	.35048	.93657	.37422	.6722	.0677	.8532	.06343	.64952	29
...	...	...	...	...	...	...	...	...	...
55	0.35701	0.93410	0.38220	2.6164	1.0705	2.8010	0.06590	0.64299	5
56	.35728	.93400	.38253	.6142	.0707	.7989	.06600	.64272	4
57	.35755	.93389	.38286	.6119	.0708	.7968	.06611	.64245	3
58	.35782	.93379	.38320	.6096	.0709	.7947	.06621	.64217	2
59	.35810	.93368	.38353	.6073	.0710	.7925	.06631	.64190	1
60	0.35837	0.93358	0.38386	2.6051	1.0711	2.7904	0.06642	0.64163	0
M	Cosine	Sine	Cotan.	Tan.	Cosec.	Secant	Vrs. Cos.	Vrs. Sin.	M

110°   69°

Thus, natural trigonometric sine value for

20°28′ = 0.3475 (value of 20°20′)
         +0.00216 (value of 8′)
       = 0.34966

Computing Inch- or Metric-Standard Gage Block Combinations

After the sine value of a required angle has been established, the required gage block combination may be found (Figure 1-19). The sine formula is used. The values from the previous example are used in the formula as follows:

$$\frac{\text{sine of angle}}{(20°28′)} = \frac{\text{height (gage block combination)}}{\text{length of sine bar (5″)}}$$

height = sine of angle (0.34966) × length of sine bar (5″)

= 1.7483″

If metric-standard gage blocks are to be used, the problem may be solved according to the same trigonometric function. The 1.7483″ gage block

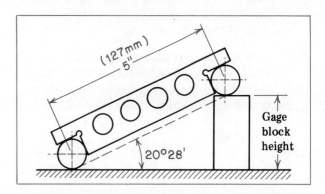

FIGURE 1-19 Typical Sine Bar Angle Setting Application

height may be converted to the metric equivalent of 44.407mm. Metric gage blocks of this height will produce an angle of 20°28' when a 127mm (5") sine bar is used.

Note: If an abridged table of natural trigonometric functions is used, the decimal difference would need to be interpolated. The final angle reading is 20°28'.

HOW TO MAKE DIRECT CALCULATIONS FOR SINE BAR SETUPS

Computing the Gage Block Height (5" Sine Bar)

Note: The mathematical process of determining a gage block combination (height) for setting a sine bar or sine plate may be simplified. The use of a 5" sine bar is assumed in the following procedure.

STEP 1 Look up the natural trigonometric function value of the required angle. In the earlier example, the sine value of 20°28' is 0.34966.

STEP 2 Move the decimal point one place to the right (3.4966).

STEP 3 Divide by 2. The result represents the height of the gage block combination. In the example, 3.4966 ÷ 2 = 1.7483".

Note: The result represents the height to which the rear roll must be set on a sine bar, sine block, or sine table.

Determining the Angle from a Height Setting (5" Sine Bar)

Note: The angle of sine bar setup may be determined by reversing the mathematical steps used in computing gage block height. The 1.7483" gage block combination is again used as an example.

STEP 1 Move the decimal point in the 1.7483" measurement one place to the left (0.17483).

STEP 2 Multiply by 2. The product represents the natural trigonometric function of the required angle. In the example, 0.17483 × 2 = 0.34966.

STEP 3 Refer to a table of natural trigonometric functions. Locate the 0.34966 value in the sine column.

STEP 4 Read the angle corresponding to the 0.34966 value.

Establishing a Height Setting for Different Length Sine Bars

Sine bars with center distances of 5" and 10" are common. For larger layout, inspection, and measurement operations, a 20" sine bar is used. The basic 5", 10", and 20" lengths were especially selected in this unit to simplify the mathematical computations and setups. The height to which a 10" sine instrument is set is established by simply moving the decimal point (in the natural trigonometric function value of the required angle) one place to the left. For example, with the 20°28' angle, by moving the decimal point in the trigonometric sine value of 0.34966, the height becomes 3.4966. This value represents the gage block combination height.

TABLES OF CONSTANTS FOR SINE BARS AND PLATES

The machine tool industry has further simplified layout, inspection, and machining practices with sine bars and sine plates. *Tables of constants* are available in handbooks or are furnished by product manufacturers. Part of a table of constants is reproduced in Table 1-5. Note that values of degrees and minutes are given with the accompanying constant (inch value).

The values in Table 1-5 apply to 5" sine bars or plates. Settings for 10" instruments are found by multiplying by 2. For example, in the 20°28' setting, the constant value shown in the table is 1.7483" for a 5" sine bar. A 10" sine bar is set at 1.7483" × 2, or 3.4966". For a 20" sine bar, the constant value is multiplied by 4. By contrast, the constant value in the table is divided by 2 when a 2-1/2" sine bar is used.

The same table of constants are used to determine the degrees and minutes in an angle. For a 5" setup, the height (gage block measurement) is located in the inch values section of the table. The degrees and minutes corresponding to the inch value represent the angle of the setup.

The gage block measurement for a 10" sine bar setup is divided by 2. The constant inch

LINEAR AND ANGULAR MEASUREMENT 15

value is then checked in the 5″ sine bar table. Similarly, a 20″ sine bar gage block measurement is divided by 4. Again, the angle setting is checked in the table of constants for 5″ sine bars.

SINE BAR, SINE BLOCK, AND SINE PLATES

Sine bars, sine blocks, and simple and compound sine plates and perma-sines are commonly used in layout, inspection, assembly processes, and measurement. They are essential in maintaining accuracies to one minute of arc.

Sine Bar

A *sine bar* is a precise layout and measuring accessory. It consists of two rolls mounted on opposite ends of a notched steel or granite parallel.

All metal parts are hardened, machined to precise limits, and accurately positioned. The center distance between the rolls is fixed at lengths that are a multiple of 5″ (127mm).

Sine Block

A *sine block* is essentially a wide sine bar (Figure 1-20). A sine block (or modification) provides maximum efficiency for layout, inspection, and angle measurements. The exact angular degree that a 5″ or 10″ sine block makes with a plane reference surface is determined by a precise vertical height. As discussed earlier, the height is established by the correct combination of gage blocks. These blocks are placed under the elevated end of the sine block. Like the sine bar, the body of a sine block may be made of hardened, stabilized, and normalized steel or black granite.

TABLE 1-5 Partial Table of Constants for Setting 5″ Sine Bars or Sine Plates (16°0′ to 23°60′)

Minutes	16°	17°	18°	19°	20°	21°	22°	23°
0	1.3782	1.4618	1.5451	1.6278	1.7101	1.7918	1.8730	1.9536
1	.3796	.4632	.5464	.6292	.7114	.7932	.8744	.9550
2	.3810	.4646	.5478	.6306	.7128	.7945	.8757	.9563
3	.3824	.4660	.5492	.6319	.7142	.7959	.8771	.9576
4	.3838	.4674	.5506	.6333	.7155	.7972	.8784	.9590
5	1.3852	1.4688	1.5520	1.6347	1.7169	1.7986	1.8797	1.9603
6	.3865	.4702	.5534	.6361	.7183	.8000	.8811	.9617
7	.3879	.4716	.5547	.6374	.7196	.8013	.8824	.9630
8	.3893	.4730	.5561	.6388	.7210	.8027	.8838	.9643
9	.3907	.4743	.5575	.6402	.7224	.8040	.8851	.9657
...	...	...	...	...	...	...	...	...
20	1.4061	1.4896	1.5727	1.6553	1.7374	1.8189	1.8999	1.9804
21	.4075	.4910	.5741	.6567	.7387	.8203	.9013	.9817
22	.4089	.4924	.5755	.6580	.7401	.8217	.9026	.9830
23	.4103	.4938	.5768	.6594	.7415	.8230	.9040	.9844
24	.4117	.4952	.5782	.6608	.7428	.8244	.9053	.9857
25	1.4131	1.4966	1.5796	1.6622	1.7442	1.8257	1.9067	1.9870
26	.4145	.4980	.5810	.6635	.7456	.8271	.9080	.9884
27	.4159	.4993	.5824	.6649	.7469	.8284	.9094	.9897
28	.4173	.5007	.5837	.6663	.7483	.8298	.9107	.9911
29	.4187	.5021	.5851	.6676	.7496	.8311	.9120	.9924
...	...	...	...	...	...	...	...	...
55	1.4549	1.5381	1.6209	1.7032	1.7850	1.8663	1.9469	2.0270
56	.4563	.5395	.6223	.7046	.7864	.8676	.9483	.0283
57	.4577	.5409	.6237	.7060	.7877	.8690	.9496	.0297
58	.4591	.5423	.6251	.7073	.7891	.8703	.9510	.0310
59	.4604	.5437	.6264	.7087	.7905	.8717	.9523	.0323
60	1.4618	1.5451	1.6278	1.7101	1.7918	1.8730	1.9536	2.0337

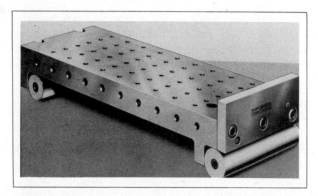

FIGURE 1-20 Sine Block with Tapped Holes and End Plate (Stop)

Simple Sine Plate

Single angles, which lie in one plane, are usually set by using a *simple sine plate* (Figure 1-21A). This sine plate has a precision base, a movable sine plate, and two hardened precision rolls. The base and elevating sine plate are hinged. The roll at the left end is located under the sine plate and is not visible in the illustration.

The simple sine plate is set at the required angle by establishing the overall measurement of a gage block combination. The blocks are wrung together. They are then placed between the precision-finished base reference plane and underneath the left-hand roll. The base has a step ground 0.100" or 0.200" deep. This step permits the use of gage blocks for small angles. Without the step, the height of the gage block would be larger than the required distance from the roll to the base.

Compound Sine Plate

Compound angles are angles formed by edges of triangles that lie in different planes. By comparison, simple angles lie in one plane. Compound angles may be laid out, inspected, or measured by holding a workpiece on a *compound sine plate*.

The compound sine plate illustrated in Figure 1-21B has a series of tapped holes in the top, sides, and end. These holes permit the direct strapping and clamping of workpieces.

A typical compound sine plate has an intermediate hinged plate. Attached to it is the top sine plate. The design of this plate includes the same features as the features on a simple sine plate.

The intermediate plate and the top sine plate are each set at a required angle in one of two planes. The two angles make up the compound angle. The gage block combination for each angle is established. These combinations represent the heights to which the intermediate plate and sine plate are set. The sine plate parts are hardened, stabilized, and normalized to provide for maximum accuracy over a period of time.

Setting a Single Angle When Using a Compound Sine Plate. A single angle operation is normally set up on a simple sine plate. However, under certain conditions, a compound sine plate may also be used. The upper assembly is swung to a closed position. The upper roll rests against the sine plate face.

The first hinged section is elevated the height of the gage block combination. The workpiece is positioned and clamped on the sine plate or held by magnetic force on a chuck face. Machining

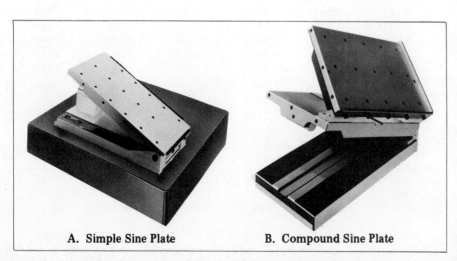

A. Simple Sine Plate   B. Compound Sine Plate

FIGURE 1-21 Simple and Compound Sine Plates

and other measurement or inspection processes may then be performed at the angle of the setup.

Permanent Magnetic Sine Plate (Chuck): Perma-Sine

A *permanent magnetic sine plate* is a simple or compound angle sine plate in which a permanent magnetic chuck forms the working face. The name is shortened so that the combination is often referred to as a *perma-sine* or a *sine chuck*. Simple angle perma-sines (Figure 1-22A) and compound angle perma-sines (Figure 1-22B) are available.

Perma-sines are especially adapted to surface grinder work. The workpiece may be laid out, inspected, or machined. Angles produced are within one minute of arc accuracy. The fine magnetic pole spacing on the compound angle plate permits holding small and large parts securely. The magnetic force is turned on for holding and off for releasing by moving a simple lever.

A compound sine plate (chuck) has a side and an end stop. The stops facilitate setups. They also prevent any movement of the workpiece during machining. When sine plates and sine chucks are used in machining, the processes are usually related to finish grinding.

HOW TO SET UP A SINE PLATE FOR LAYOUT AND INSPECTION

Note: The following procedure requires the setting up of a 10″ sine plate to inspect a workpiece that is finished to an angle of 32° 27′. Wear blocks are to be used.

STEP 1 Move the hand carefully over the working plane of the surface plate. Check for and remove any burrs and nicks.

STEP 2 Wipe the surface clean so that it is free of all dust.

STEP 3 Check and clean the two faces of the base, the top face of the sine plate, and the rear roll.

STEP 4 Position the workpiece on the sine plate. It should be square with the sides and end of the top plate.

STEP 5 Clamp the workpiece securely to the top plate.

STEP 6 Compute the gage block height for the 32° 27′ angle (Figure 1-23):

FIGURE 1-22 Simple and Compound Angle Permanent Magnet Sine Plates (Perma-Sines)

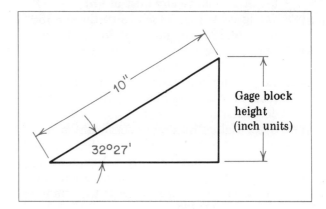

FIGURE 1-23 Known and Required Measurement (Inch-Standard Units)

$$\frac{\text{sine } 32°27'}{(0.53656)} = \frac{\text{height}}{\text{hypotenuse (10")}}$$

$$\text{height} = 0.53656 \times 10"$$

$$= 5.3656"$$

Note: The height may also be established by using a table of sine bar constants. The constant for a 5" sine bar is 2.6828", which is multiplied by 2 for a 10" sine bar. The height, again, is 5.3656".

STEP 7 Select a gage block combination for the 5.3656" height. Use protective wear blocks in the combination.

Note: If metric gage blocks are to be used, multiply the natural trigonometric value by the hypotenuse—that is, by the center distance of the rolls in millimeters (Figure 1-24):

$$\frac{\text{sine } 32°27'}{(0.53656)} = \frac{\text{height}}{\text{hypotenuse (254mm)}}$$

$$\text{height} = 0.53656 \times 254$$

$$= 136.28624\text{mm}$$

The five decimal places are rounded off to the nearest 5 ten-thousandths of a millimeter. In this example, the metric gage block measurement is 136.286mm.

With wear blocks, one possible combination of metric gage blocks for the 136.286mm height is as follows: 2.00 (wear block), 2.006, 2.28, 18.00, 50.00, 60.00, and 2.00 (wear block).

STEP 8 Clean each gage block. Use a camel's hair brush to remove any dust or lint.

STEP 9 Wring each gage block until the 5.3656" (or 136.286mm) combination is formed.

STEP 10 Slide the gage block combination carefully onto the sine plate base.

STEP 11 Lower the top plate gently until the roll rests on the wear block of the gage block combination.

STEP 12 Set a dial indicator, vernier height gage, or other layout instrument to the required dimension. Perform any subsequent layout, gaging, or measurement operations. The sine bar and workpiece in this example are set at 32°27'.

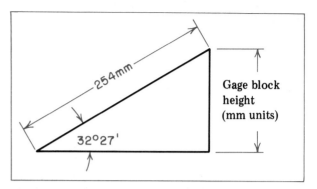

FIGURE 1-24 Known and Required Measurement (Metric-Standard Units)

HOW TO SET UP AND USE A PERMANENT MAGNETIC SINE PLATE (PERMA-SINE)

Setting a Compound Angle

STEP 1 Examine the base of the perma-sine and face of the surface plate or machine table. Remove any burrs or nicks.

STEP 2 Strap the base of the perma-sine to the machine table. Flanges on the base permit clamping.

STEP 3 Wipe and use a fine bristle brush on the base, sine plate, and rolls. They must be clean and free of dust and foreign particles.

STEP 4 Determine the required gage block combination for the first angle and then for the second angle.

STEP 5 Select and wring the gage blocks together for the first setup.

STEP 6 Place the gage block setup between the base and the first hinged plate.

STEP 7 Lower the hinged plate and other parts of the perma-sine. The hardened roll should just set on the wear block of the gage block setup.

STEP 8 Repeat the preceding two steps for the second gage block combination.

STEP 9 Remove any burrs from the workpiece. Clean the workpiece and face of the magnetic chuck (sine plate).

STEP 10 Position the workpiece in relation to the sides of the sine plate. Use parallels between the end plates and the workpiece.

LINEAR AND ANGULAR MEASUREMENT

STEP 11 Turn the magnet lever on. The workpiece is ready for layout, inspection, or machining at the compound angle.

SAFE PRACTICES WITH PRECISION LINEAR AND ANGULAR MEASURING ACCESSORIES AND INSTRUMENTS

—Remove dust and foreign particles from gage blocks and microfinished surfaces with a fine, soft camel's hair brush.
—Check along microfinished surfaces that are to be in contact in order to detect any nicks or burrs. Remove all imperfections properly.
—Check the mechanical movement of parts for freedom.
—Allow time for measuring instruments and workpieces to normalize to the same temperature.
—Avoid working near heat sources that produce varying temperatures of a workpiece and measurement tool.
—Check gage blocks, measurement surfaces, and instruments against master standards to ensure that correct measurements can be established.
—Place granite angle blocks, parallels, master flats, gage blocks, and other precision measurement accessories and instruments in their appropriate cases.
—Apply the amount of force on straps and clamps that is needed to hold parts securely. Any excess force tends to spring or bend an assembly and thus produces inaccurate measurements.
—Clean measuring instruments and dry them thoroughly. Avoid using compressed air, which may drive dust and other foreign particles into an instrument.
—Lubricate moving parts. After an instrument is used and cleaned, apply a fine corrosive-resistant lubricant to it and then store the instrument in an appropriate container.
—Adjust a dial indicator so that the measurement range falls within the maximum movement of the pointer (movable rod).
—Recheck all mathematical computations with the actual overall gage block combination when setting a sine bar or sine plate.
—Use the side and end stops on a magnetic sine plate to position and hold workpieces securely.
—Make sure that the dimensions, shape, and materials of a workpiece permit it to be drawn securely to a perma-sine. The magnetic force must be adequate to prevent any movement of a workpiece during a machining or other working process.

TERMS USED IN PRECISION LINEAR AND ANGULAR MEASUREMENT

Dimensional stability (precision instruments) The ability of an instrument to provide the same precise measurements over a long period of service. The relief of stresses within an instrument to ensure continuing accuracy of measurement.

Black granite A natural black-colored granite. A granite with great strength, hardness, rigidity, thermal stability, and a low (water) absorption rate. An ideal material that can be lapped to extremely precise limits with a fine-textured, nonglare finish.

Granite precision measurement products Instruments and gaging products used in precision measurement. Granite surface plates and other layout and measurement accessories. Angle plates, universal right angles, straight edges, parallels, sine bars, and sine plates made of black granite.

Modified (granite) surface plate A granite precision surface plate with holes or grooves cut into the face that are used to position and hold plane-surfaced and modified-contour workpieces. A granite surface plate with standard stainless steel slide and threaded inserts.

20 PRECISION MEASUREMENT AND INSPECTION

Universal right angle	A precision ribbed steel or granite layout, measurement, and holding block. A right angle block with six faces that are accurately finished at right angles and are parallel.
Box parallel	A metal or granite parallel used to elevate a workpiece so that it may be accurately measured and laid out. A parallel whose area (end on which it rests) is smaller than a regular parallel.
Precision step block	A block that is precisely machined for flatness and parallelism. A precision block used singly or in pairs in layouts, measurement, and inspection.
Bench comparator	An instrument for establishing the accuracy of a measurement against a precise standard. A layout, inspection, and measurement machine. A machine having a precisely finished base, a vertical post, and a universal dial indicator that may be adjusted to a linear dimension.
One minute of arc	A precision angular measurement that is accurate within one minute of an arc of one degree. A higher level of accuracy in measuring an angle than is possible by using a universal bevel protractor.
Sine bar	A combined precision parallel and two hardened and finely finished rolls that are set apart at a fixed center distance. An instrument used to obtain angular accuracies within one minute of arc.
Sine block	A wide sine bar.
Simple (single angle) sine plate	A combination of a sine block and attached base. (The opening may be adjusted to obtain a required angle. The angle is set to an accuracy within one minute of arc.)
Compound sine plate	Two modified sine blocks that are adjustable on a single base. (The angle of each block lies in a different plane. The setting of the two angles produces the required compound angle.)
Table of constants for sine bars and plates	Tables produced by manufacturers and professional organizations for work with sine bars. Tables for directly reading an angle or an inch or metric value. (The value represents the measurement needed to produce a required angle when using a particular size sine bar.)
Perma-sine	A trade designation for a permanent magnetic sine plate. A sine plate with a permanent magnetic table. A sine plate that uses magnetic force.

SUMMARY

The layout, measurement, and machining of workpieces to what were at one time laboratory standards are now accepted as precision shop work.

Steel gage blocks are produced to accuracies within one-millionth of an inch and 0.000025mm.

Metric tolerances in the equivalent of millionths of an inch are often given as microns (μm). For example, a tolerance of ±0.00005mm (±0.000002″) may be written ±0.05μm.

The surface finish of precision gage blocks ranges from 0 to 0.4 microinches (0 to 0.10μm).

Protective wear blocks are usually made of tungsten carbide. Wear blocks prolong the accuracy and life of gage blocks.

LINEAR AND ANGULAR MEASUREMENT

Precision scribers, inside and outside caliper bars, and other accessories extend the range of use of gage blocks. The parts are aligned and secured in a channel with a micro-clamp. The gage blocks and other attachments may be used for layout, inside and outside gaging, parts inspection, and other precision linear measurements.

Black granite with a hardness range to 80 Rockwell C (80 R_C) provides a commercially economical measurement tool and instrument material.

Some of the desirable properties of granite are: rigidity, ensured by a low deflection under load; a low water absorption rate, which prevents the surface from being affected by acids, chemical fumes, dust, and dirt; thermal stability, which helps to maintain continuous accuracy at the same temperature with linear measurements; a fine texture, which permits the production of a precise lapped black surface to which gage blocks and other precision tool surfaces do not wring; and hardness, which minimizes surface scoring.

Granite surface plates are commercially available in sizes ranging from 8" × 12" × 3" thick (200mm × 300mm × 75mm) to 72" × 144" × 16" (1800mm × 3600mm × 400mm).

Granite surface plates are manufactured in three grades: AA, used in laboratories; A, used for inspection; and B, used in shops.

Toolmaker's flats and super-accurate cylindrical master flats provide precision plane surfaces.

Angle plates; universal right angles; adaptations of parallels such as cubes, box parallels, and step blocks; and V-blocks are available in either metal or granite.

The bench comparator is a highly accurate precision instrument. It is adapted to gaging operations. Linear measurements on workpieces may also be compared with fixed standards. A bench comparator combines the design features of a precision base, upright, and a dial indicator that may be swiveled and adjusted. Accurate angular measurements depend on mechanical, positional, and observational factors. These factors relate to working conditions of the instruments and part, the correct placement of the instrument in relation to one or more planes, and possible observational errors in reading.

Angular measurements to accuracies of one minute of arc require the use of more precise instruments and setups than a universal vernier bevel protractor.

The use of sine bars, blocks, and plates requires trigonometry and the computing of gage block heights. Simple formulas and tables of constants provide a simplified method of establishing a sine bar setup.

Sine bars and accessories permit accuracies to one minute of arc. These accuracies are required in precision layout, inspection, machine and workpiece setups, and measurement.

Simple or compound sine plates are used for work setups with angles in one or two planes, respectively.

Perma-sines provide a permanent magnetic chuck mounted on a sine plate. Workpieces are held securely by magnetic force. The magnetic force is released by a simple lever movement.

Safe practices in the care and use of precision linear and angular measuring instruments and accessories and in the mounting and securing of workpieces must be followed. Safe practices are essential to accuracy in layout, gaging, and measurement, as well as in any setting up and machining that may be required.

UNIT 1 REVIEW AND SELF-TEST

1. Describe the meaning of *dimensionally stable* gage blocks.
2. a. Refer to a table on standards for gage blocks.
   b. Examine the specifications of the different grades of SI metric gage blocks.
   c. Make a general statement about the difference in precision (length, flatness, and parallelism) between standard grade 1 (AA) and grade 2 (A+) gage blocks. The gage blocks are smaller than 50mm thick.
3. List three advantages of using granite over similar cast iron and steel layout and measuring instruments and accessories.
4. a. State the difference between a bench comparator and other precision measuring instruments.
   b. Give five typical applications of bench comparators.
5. Cite two factors the worker must consider when measuring an angle with a universal vernier bevel protractor.
6. a. Refer to Table 1-1 or to any other manufacturer's table of standards for gage blocks.
   b. State three measurement tests that are applied to inch-standard or metric-unit gage blocks.
   c. Give the maximum allowable surface finish of inch-standard grade 1 (AA) gage blocks.
   d. State the upper (+) and lower (-) limits of parallelism on 24mm thick grade 2 (A+) metric-standard gage blocks.
7. a. Refer to a table of constants for 5" sine bars (blocks).
   b. Determine the required gage block height for each of the following 5" sine block angle settings: (1) 16°, (2) 18°5′, and (3) 22°57′.
8. a. State four main uses of compound sine plates.
   b. Indicate the design features of a compound sine plate that make compound angle settings possible.
9. State two precautions to take to safely store precision linear and angular measuring instruments and gages.

UNIT 2
Layout and Measurement Practices in Surface Plate Work

The *surface plate* is a precision tool. It provides an accurate reference plane. Workpieces are positioned on the plane surface so that precision measurements and inspection practices may be carried on. The face of the surface plate relates to layout, inspection, and measurement processes that are collectively called *surface plate work*. In surface plate work, workpieces are inspected for parallelism, squareness, roundness, and concentricity. General measurements are taken in relation to length, height, hole locations, angles, and tapers.

Measurement and layout processes using line-graduated and other precision measuring tools and instruments are applied in fundamental machine technology. Precision angle gage blocks are treated in this unit in relation to the extremely precise angle measurement and inspection practices used in advanced machine technology.

Ordinate dimensioning is used on drawings to represent the design or measurement features of a piece part. These features are referenced from a fixed point or surface called a datum. Principles of drafting and practices in interpreting typical shop and laboratory drawings that require ordinate measurements are also treated, in this unit.

GENERATING A FLAT SURFACE

Fine, precise reference planes have been produced commercially for many years. Some early surface plates were made by hand-filing and hand-lapping and then later, by hand-scraping. One technique for producing flat surfaces dates back to Henry Maudslay. In 1797, he used the three-plate method to hand-file and hand-lap the finished face of surface plates. Later, in 1840, Sir Joseph Whitworth replaced lapping with abrasive particles by hand-scraping.

Although precision grinding and machine lapping processes are widely applied in producing highly accurate plane surfaces, the three-plate method is still used. An extremely fine, precise, flat surface may be produced by the three-plate method without a standard reference surface.

Flat surfaces are hand scraped for the following reasons:

—To produce and maintain flat, precise plane surfaces (hand-scraping is important when it is not practical to use a surface plate to manufacture a straight edge, template, angle, or other accurate surface);
—To produce a fine, close-grained surface on a part that has been machined;
—To align two work surfaces so that they may be clamped without warping or bending;
—To generate an excellent sliding, bearing surface;
—To form a perfect seal against the leakage of oil and fluids between two surfaces that are fitted together;
—To frost or flake a machined surface for decorative purposes.

Early Development of Standard Planes

While Sir Joseph Whitworth is credited with the three-plate method, Henry Maudslay had used the method to produce *standard planes* in 1797 in the Maudslay and Fields Company of London. These standard planes were in daily use at each work place. The planes were used for layout, measurement, and testing to the highest degree of accuracy possible.

When Joseph Whitworth worked in the Maudslay company, he served under one of the master craftsmen of the time, John Hampson. At the time Whitworth turned his efforts to "true plane and power of measurement." Whitworth improved on the Maudslay method by substituting hand-scraping for grinding with an abrasive powder. Hand-scraping localized the cutting action to a particular spot where it was needed. Hand-scraping is still used today.

When Whitworth was able to reproduce standard planes to an extremely high degree of accuracy, he devoted the remainder of his life to improving manufacturing methods and systems of measurements that involved line measurements, uniform design standards, and precision systems for manufacturing. By 1850, Whitworth was recognized as the foremost metal-working and machine tool builder in the world.

Three-Plate Method of Producing Flat Surfaces

Whitworth's *three-plate method* of producing flat surfaces requires three accurately machined plates. The highest possible surface finish should be produced. The plates are marked #1, #2, and #3.

There are six basic steps in the three-plate method. These steps are described in Table 2-1. The procedure is essentially the same as the procedure used by Maudslay and Whitworth.

- —Step 1: Scrape plates #1 and #2 alternately to each other. Continue until they conform to each other and an accurate plane surface is produced.
- —Step 2: Use plate #1 as the control plate. Scrape plate #3 to plate #1.
- —Step 3: Scrape plates #2 and #3 alternately to each other. Scrape until the two plates conform to each other.
- —Step 4: Use plate #2 as the control plate. Scrape plate #1 to plate #2 until as accurate a plane surface as possible is produced.
- —Step 5: Scrape plates #1 and #3 alternately to each other. Scrape until these two plates conform to each other.
- —Step 6: Use plate #3 as the control plate. Scrape plate #2 to plate #3.

All three plates are now scraped to the same accuracy of flatness. If a higher degree of accuracy is required, the six steps are repeated. They are continued until the three surfaces conform to the desired measurement standard of flatness.

ACCURACY OF SURFACE PLATES

Many design, development, and production contracts require part specifications that conform to both national and international measurement and design standards. Some manufacturers provide a *certificate of accuracy*. This certificate indicates that the product conforms with government specifications. The calibration is traceable to the National Bureau of Standards. The accuracies to which a surface plate conforms—for example, to Federal Specification CGG-P463c—are certified.

Expressed in technical terms, "all points of the work surface (plane) shall be contained between two parallel planes that are separated by a distance no greater than that specified for its respective grade." There are three general grades of surface plates. *Grade AA* provides for extremely precise measurements, such as measurements required in the laboratory. The tolerance per two square foot area is ±0.000025" (0.0006mm). The flatness tolerance of the *grade A* inspection-practice surface plate is ±0.000050" (0.0012mm). The *grade B* shop accuracy is ±0.0001" (0.0025mm) per two square foot area.

CALIBRATING FLAT (PLANE) SURFACES

The term *calibration*, as used in this unit, relates to the inspection, measurement, and reporting of size, flatness, and parallelism errors of plane surfaces. The degree of accuracy and method of calibration depend on the nature and use of the measuring equipment. For example, federal specifications suggest an annual check of ground master- and laboratory-grade gage blocks. Semi-annual checks of inspection gage blocks and surface plates are common practice.

Under normal conditions, the accuracy of a shop or laboratory surface plate may be checked easily by a simple test for the *repeatability* of a measurement. That is, the same part is measured under the same conditions on a number of different areas of the surface. The plane surface is accurate if the measurement on all areas is the same.

SURFACE PLATE WORK 25

TABLE 2-1 Producing a Plane, Flat Surface (Whitworth Three-Plate Method)

Step	Plate Conditions — Before Scraping	Plate Conditions — After Scraping		Comments
1	#2 on #1	#3 (Exaggerated surface roughness)		—There is no control plate. —Plate #1 agrees with plate #2. —No other plate agrees. —No plate is known to be flat.
2	#3 on #1	#2		—Plate #1 is the control plate. —Plate #1 agrees with plate #2. —Plate #1 agrees with plate #3. —Plates #2 and #3 do not agree. —No plate is known to be flat.
3	#3 on #2	#3 on #2	#1	—There is no control plate. —Plates #1 and #2 do not agree. —Plates #1 and #3 do not agree. —Plate #2 agrees with plate #3. They are known to be flatter than plate #1. —No plate is known to be flat.
4	#1 on #2	#1 on #2	#3	—The control plate is plate #2. —Plate #1 agrees with plate #2. —Plate #2 agrees with plate #3. —Plates #1 and #3 do not agree. —No plate is flat, but all are of about equal flatness.
5	#1 on #3	#1 on #3	#2	—There is no control plate. —Plate #1 agrees with plate #3 but not with plate #2. —Plate #2 does not agree with plate #3. —Plates #1 and #3 are flatter than plate #2. —No plate is known to be flat.
6	#2 on #3	#2 on #3	#1	—The control plate is plate #3. —Plate #1 agrees with plates #2 and #3. —Plate #2 agrees with plate #3. —Plates #1, #2, and #3 are equal to one another. —The entire procedure is repeated to obtain a specified degree of flatness.

Calibration of Surface Plates

Surface plates are commercially calibrated and periodically checked. Their accuracy is checked with a precision instrument called an *autocollimator* (Figure 2-1). This instrument uses lenses, light rays, and micrometer readings. The light source is built into the autocollimator. The lens system causes light rays to leave the instrument in parallel paths. The light rays are directed at a target mirror. The rays are reflected back into the instrument. The rays are viewed in the eyepiece.

A corner target mirror and a reflecting target mirror are required. They are placed so that the corner target mirror is located at one end of the surface plate at a right angle to the instrument. The corner target mirror reflects the light rays from the autocollimator to the reflecting target mirror. The light rays are then reflected from the reflecting target mirror to the corner target mirror and back to the instrument.

A micrometer on the autocollimator is used to read any deviation in the right-angle setting of the mirror target. Errors to within one-fifth second of arc may be read on the micrometer. Readings are taken (1) around the four edges of the surface plate, (2) along the two diagonals, and (3) across the two centerlines. A number of other readings are also taken along each line. Each reading is recorded and plotted on a graph. The readings form a profile along the eight lines of sighting.

The time required for calibrating a surface plate with an autocollimator may be reduced substantially by using a laser-powered surface contour projector. A surface plate may be calibrated using this projector in ten to fifteen minutes (for a 36" × 48" surface). This time is significantly less than the four to six hours required to autocollimate, record, and graph the readings.

TECHNIQUES FOR CHECKING PARALLELISM, SQUARENESS, ROUNDNESS, AND CONCENTRICITY

Extremely precise checking for parallelism, squareness, roundness, and concentricity may be done with standard measuring instruments such as a vernier height gage and/or a dial test indicator. Greater sensitivity is achieved by using electronic and pneumatic gages and instruments, which are considered in Unit 4.

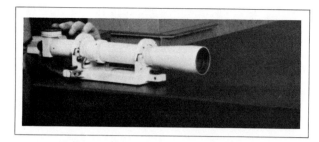

FIGURE 2-1 Calibrating a Flat (Plane) Surface with an Autocollimator

All checking processes start with cleaning. Tools, instruments, workpieces, and the face of the surface plate must be clean. Foreign particles, nicks, burrs, and scratches are removed from contact surfaces. Then, the accuracy of each surface and measuring tool is checked against the required standard of measurement.

Checking for Parallelism

To check for *parallelism*, the workpiece is placed with one of the parallel surfaces resting flat on the face of the surface plate. If the part has a projection on the bottom surface, precision parallels may be used. These parallels accurately elevate the surface to permit accurate measurement.

A test indicator is then selected with either a front- or top-mounted dial. Whether a swivel-point test indicator or a regular dial indicator (with plunger-type stem) is used depends on the location of the surface to be checked. A dial indicator with a magnet base (Figure 2-2) is especially adapted for rigid mounting on any ferrous surface. An indicator with a regular base is adapted to either black granite or metal surface plates.

Two general-purpose dial test indicators are pictured in Figure 2-3. The swivel-point model is available with a short- (1/2") or long-range (1-7/16") contact point. The point may be swiveled on some models through a 210° arc. An automatic reversal feature permits the dial test indicator to measure over, under, outside, or inside a surface (Figure 2-3A).

The smallest division on test indicators such as the fine-measurement model shown in Figure 2-3B is 0.00005". A comparable metric-reading indicator has the smallest division of 0.002mm. The range of movement (measurement) of dial

SURFACE PLATE WORK 27

FIGURE 2-2 Dial Indicator and Magnet Base Set

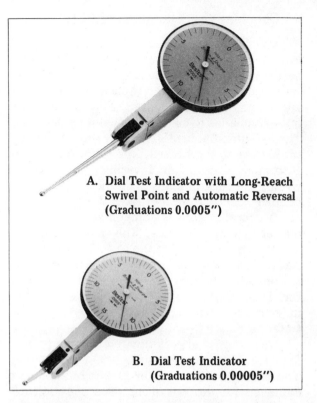

FIGURE 2-3 Two General-Purpose Dial Test Indicators

test indicators is limited. The dials, mechanisms, and contact points on regular sizes of dial test indicators may range from 0.008" (0-4-0) to 0.030" (0-15-0).

Like the dial indicator, the dial test indicator is usually held in an attachment from a universal set. The dial test indicator is mounted on a column, magnetic base, or height gage. The contact point of the indicator is slid into position over one corner of the workpiece. The setup must provide for freedom of movement of the pointer. The dial on the indicator is set at zero. The contact point is then moved carefully over the length and width or around the surface of a round part. The readings on the indicator dial face show any deviation in parallelism. As in all precision measurement work, the readings are rechecked.

HOW TO CHECK PARALLEL SURFACES BY USING A 0.0001" (0.002mm) DIAL TEST INDICATOR

STEP 1 Check the cleanliness of all tools and instruments and of the workpiece. Remove all burrs.

STEP 2 Position the workpiece on a surface plate by using parallels.

STEP 3 Select a 0.0001" dial test indicator with a swiveling point that is long enough to reach the areas of the surface that are to be checked.

STEP 4 Mount the indicator on a permanent magnet or adjustable base. The accessory used should permit the point to be adjusted.

STEP 5 Bring the dial test indicator point down over the work surface. Adjust the point to a zero reading.

Note: If the dial test indicator is brought to the work, the point should be moved carefully by hand and set on the work surface. The range of the dial test indicator must provide the movement needed to take complete measurements.

STEP 6 Move the dial test indicator setup over each corner of the work surface. Check for variations in the readings.

Note: Readings are usually taken near the corners of square and rectangular workpieces. Similar readings are noted near the outer edge (periphery) of round-shaped workpieces.

Checking for Squareness and Flatness

Usually the *squareness* of two surfaces is given in terms of *deviation per six inches of length*, particularly when stating the accuracy of measuring tools. One common practice in determining the amount that a workpiece or other part is out-of-square is to use a master cylindrical square. The cylindrical square is carefully brought against the surface to be checked. The cylindrical square is rotated until no light is visible. Any variation within the range of the cylindrical square may be read directly from the graduations.

A transfer gage is sometimes used with a master cylindrical square. The transfer gage has a button on the front of the base. This button is positioned against the square. The dial indicator is then adjusted to a zero reading while the button is in contact with the square. Checking a universal right angle block for squareness is illustrated in Figure 2-4.

The transfer gage is then moved carefully to the face of the squared surface to be measured. The button on the transfer gage is brought into contact with the workpiece. The dial indicator shows any difference in reading from the original setting.

The transfer gage is moved slowly across the top of the square surface. Direct dial indicator readings show whether the block is square or how much it deviates. Also, the flatness can be determined by any variation in reading. Such a reading indicates that the face is dished inward or bowed outward.

The setting of the transfer gage is repeated at the center height of the right angle. Any movement of the indicator from the original setting of the dial on the transfer gage indicates the amount the block deviates (is off) from a true square. In other words, the block is bowed or dished and is not a flat surface. Any inaccuracy in squareness and flatness of the universal right angle block is transferred to any part that is laid out or held against the right angle.

HOW TO CHECK FOR FLATNESS AND SQUARENESS BY USING A TRANSFER GAGE AND DIAL INDICATOR

STEP 1   Mount a dial indicator on a transfer gage.

STEP 2   Move the transfer gage until the button just touches the base of a master cylindrical square (or a universal square).

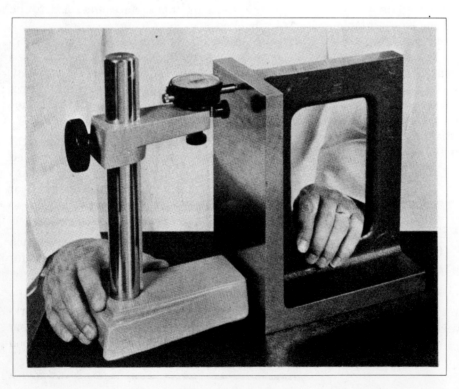

FIGURE 2-4 Checking the Squareness of a Universal Right Angle Block

STEP 3 Set the indicator dial at zero at a height that will be near the top of the workpiece.

STEP 4 Remove from this setting. Place the button against the workpiece that is to be checked for squareness.

STEP 5 Record the reading on the dial indicator. Any variation from the original zero setting incicates the amount the workpiece is out-of-square.

STEP 6 Repeat the process at a few places.

STEP 7 Continue to test for squareness by resetting the dial indicator at midpoint of the workpiece. The indicator point is again set at zero against a cylindrical square.

STEP 8 Check the workpiece with this new setting. The dial indicator readings are compared with the first set of readings obtained in steps 5 and 6. Any difference in readings indicates whether the workpiece is dished or bowed. The amount the workpiece varies from a flat plane surface is found in 0.001″ or 0.0001″ (0.02mm or 0.002mm), depending on the precision of the dial indicator.

STEP 9 Disassemble the setup. Clean and prepare the instruments for storage. Return each instrument to its proper container.

Checking for Concentricity

Concentricity relates to the periphery (outside surface) of a cylindrical part and an axis of rotation. A cylindrical surface is said to be concentric when every point is equidistant from the imaginary axis. The measurement of concentricity is often called *total indicated runout (TIR)*. Testing and measuring for runout are usually done on a metal or black granite surface plate. A bench center with two dead centers and a dial test indicator are generally used. Higher accuracy of measurement is obtained with an electronic height gage and amplifier unit (Figure 2-5).

The center holes must be clean, accurately machined, and burr-free. The workpiece is mounted between the dead centers. The dial indicator is positioned so that the contact point is at the exact center of the part. As stated earlier, the center is the point where the indicator reading is the highest. The pointer is set at zero at the high point. The part is rotated one turn. The total indicated runout, or TIR, is in thousandths or ten-thousandths of an inch. If a metric dial indicator is used, TIR is indicated in two-hundredths or two-thousandths of a millimeter. The procedure is followed in several positions along the periphery of the workpiece. If there are a number of different diameters, the procedure is repeated for each diameter.

Checking for Roundness

Roundness refers to the uniform distance of the periphery of a plug, ring, or hole in relation to the axis. Roundness is a measurement of a true diameter or radius. V-blocks, a surface plate, a dial indicator, and a holding device are generally used for checking roundness.

The workpiece is placed in the V-block. The contact point of a dial indicator (that is mounted securely) is moved onto the outside diameter. The indicator is positioned at the point where the highest reading is obtained—that is, at the centerline of the workpiece.

The dial face is turned until the indicator reading is zero. The part is then rotated carefully so that the position of the V-block, part, and original indicator setting are not changed. Any out-of-roundness from the true diameter shows up on the direct dial readings of the indicator.

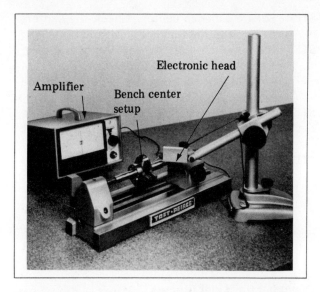

FIGURE 2-5 Measuring Concentricity by Using a Precision Bench Center and Electronic Height Gage and Amplifier

Checking for Axial Runout

Axial runout relates to the amount a shoulder or flange on a cylindrical part is out of alignment with the axis of rotation. A bench center, plane surface, and mounted dial test indicator are needed to test for axial runout.

The centered workpiece with a shoulder or flange is placed between centers on the bench center. The bench center is placed vertically on the surface plate. The contact point of the dial test indicator is positioned at the shoulder face. A dial test indicator with an appropriate length point is used for hard-to-reach surfaces. The contact point is set at zero. The part is turned one revolution. Any runout is indicated by the movement of the contact point. The amount of runout is read directly from the readings on the dial test indicator face. An example of a setup for measuring axial runout is shown in Figure 2-6. The same steps are followed to measure the axial runout of other surfaces that are to be machined square with the axis.

FIGURE 2-6 Setup for Measuring Shoulder (Axial) Runout with Bench Center Positioned Vertically

HOW TO CHECK FOR CONCENTRICITY, AXIAL RUNOUT, AND ROUNDNESS

Checking the Concentricity of a Centered Workpiece

STEP 1 Secure a bench center. Check the centers to be sure that they are clean and burr-free.

STEP 2 Examine the centers of the workpiece. Remove any burrs or foreign particles. Scraping or lapping the centers may be necessary.

STEP 3 Mount the workpiece between the centers. They should be free to turn, but without end play.

STEP 4 Position the dial test indicator. The contact point must be on dead center. Turn the dial to read zero.

STEP 5 Rotate the workpiece one turn and note any movement of the contact point.

Note: The dial reading at the highest point represents the total indicated runout (TIR).

Checking for Axial Runout of a Centered Workpiece

STEP 1 Reset the dial test indicator. The pointer is positioned against the face to be checked. The dial is set at zero.

STEP 2 Rotate the workpiece one revolution. Watch the dial test indicator for any movement of the contact point.

Note: The axial runout is read directly on the dial test indicator.

Checking the Concentricity of a Noncentered Part

STEP 1 Place the workpiece on V-blocks. Whether a single block or pair of blocks is used depends on the length and diameter of the part.

STEP 2 Position a dial indicator on a stand. The contact point must be able to reach the top of the workpiece.

STEP 3 Move the dial indicator over the outside diameter. Set the contact point reading at zero.

STEP 4 Turn the workpiece slowly and carefully one complete revolution. Note any variation (+ or –) from the zero setting. The difference indicates the amount the workpiece is out-of-round.

Note: The steps are repeated to recheck the accuracy.

STEP 5 Clean and replace the instruments in their proper storage containers.

GENERAL PRECISION LAYOUT AND MEASUREMENT PRACTICES

Taking Linear Measurements (Length and Height)

Line-graduated measuring tools such as steel rules, micrometers, and vernier calipers are used with transfer and layout tools. The degree of accuracy of the layouts or measurements ranges from ±1/64" (±0.4mm) to ±0.001" (±0.02mm). More precise linear measurements may be taken mechanically by using precision gage blocks, a height gage, or a dial indicator.

Linear measurements may also be checked, or measuring instruments may be set, with a combination gage block/height gage instrument such as the instrument shown in Figure 2-7. This instrument consists of a base and post. The post supports a carrier with 1.0000" (in the case of a metric instrument, 25mm) thick precision blocks. These blocks are arranged so that the top of one block is on the same plane as the bottom of an adjacent block. On some models, the lower gage block in the column is 0.090" thick. This size permits measurements of 0.100" height to be checked.

The 12" instrument in Figure 2-7 has a super-accurate micrometer head. The thimble graduations may be read directly in 0.0001". A metric micrometer head has readings from the spindle in increments of 0.002mm. The two scales on the sides of the precision blocks show each 0.100" of movement. The direct readings are further simplified by a digital readout. The indicator is available in 12" and 18" lengths (300mm and 450mm). The overall accuracy of the column (12" or 18") is ±0.000050" (±0.001mm). Linear measurements on the 12" instrument may be taken from 0.100" to 12.100" or 2.5mm to 305mm.

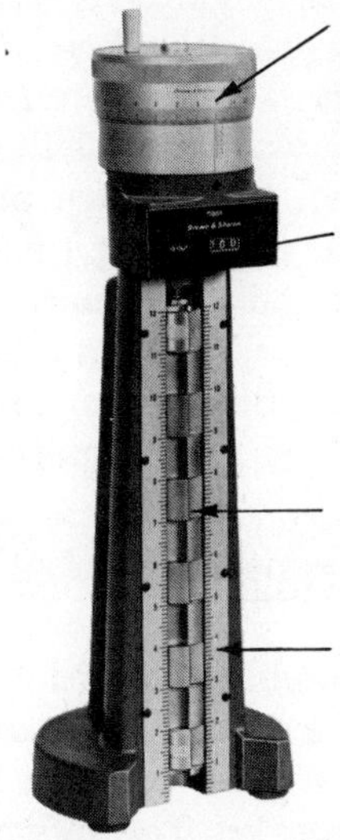

The micrometer height gage is a height setting and height reference gage. The gage block assembly is moved up or down over a 1" (25mm) range by the micrometer head. This large-diameter head is graduated in ten-thousandths of an inch (or 0.002mm on metric micrometer models). The linear setting is read by noting the number of inches on the side of the frame. The thousandths of an inch (0.02mm) value is read on the digital readout. The ten-thousandths of an inch (0.002mm) value is read on the micrometer head. A dial test indicator or an electronic height gage may be set at a precise height directly from the micrometer height gage. In reverse, any setting of a dial test indicator or electronic height gage may be transferred and converted into a micrometer height gage measurement.

Making Standard Length and Height Measurements

Length and height measurements are most commonly made on the surface plate. These measurements may be taken at a number of places around one or more workpieces. A common method of measuring the height of a number of sections is shown in Figure 2-8. A vernier

FIGURE 2-7 Precision Height Gage with Digital Readout and 1.0000" Precision Blocks

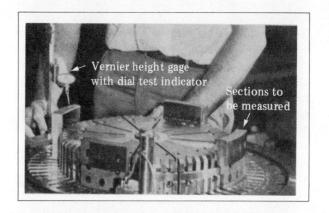

FIGURE 2-8 Measuring a Number of Sections by Using a Vernier Height Gage and Dial Test Indicator Setup

height gage and dial test indicator combination is needed. The distance between the surface and the instrument prevents the vernier height gage from being used directly. In the figure, the dial test indicator is set to height with gage blocks. The contact point is moved onto the surface. The variation in height is noted by the distance the pointer moves and the reading on the dial face. Measurements within ±0.001" and ±0.0001" (0.02mm and 0.002mm) are practical. Accuracy depends on the precision of the dial test indicator.

Another method is to set the dial test indicator and vernier height gage to the required gage block measurement. The contact point of the indicator is set to read zero. The measurement is read on the height gage. The vernier arm is raised so that the contact point clears the surface to be measured. Then, it is lowered onto the surface until the contact point reading is zero. The length measurement is read directly on the height gage.

Laying Out and Measuring Hole Locations

Measuring the exact locations of holes is another common surface plate process. When holes are to be laid out to within ±1/64" (±0.4mm), a steel rule and surface gage are generally used. When the accuracy is increased to ±0.001" (0.02mm), a standard vernier height gage and scriber point are set to the required height. The line or lines are scribed with the base of the workpiece and the height gage resting on the surface plate.

Hole locations within ±0.0001" (±0.002mm) may be laid out with a gage block accessory set. Gage blocks of the required height and the precision scriber point combination are applied.

One older practice for the precise locating of holes requires the use of toolmaker's buttons. The workpiece is laid out with line-graduated tools. A hole is drilled and tapped at the center. A hardened and ground steel button is secured to the workpiece. The height of the button is measured and adjusted until it is at the required height. The button is then secured tightly. The workpiece is transferred to the machine. The button is trued with an indicator. The hole may then be drilled, bored, threaded, reamed, or ground, as the part may require.

Indirect Measurement of a Hole Position

Instruments for laying out and measuring hole locations are also used to measure the location of holes that are already formed. One major difference is that the hole location is not established directly. The measurement is made to the bottom of the hole or to the top of a plug inserted in the hole. Many parts in punch and die, jig and fixture, and job shop work require that the location of one or more holes be measured. Figure 2-9 provides an example of the indirect measuring of a hole position. In this example, the edges of the workpiece are square. The plate is positioned on a surface plate and strapped to a precision angle block with a clamp. A dial test indicator is secured to a vernier height gage.

The contact point of the indicator is set at the required height (the centerline distance less one-half the diameter of the hole). The measurement is made with a gage block combination. The hole position is then checked against the setting of the dial test indicator. Any difference recorded on the dial face represents the variation of the hole measurement from the required dimension. The error is recorded. The same steps are followed with any other hole measurement that is to be checked.

Measuring a Center Distance between Holes

The center distance between holes may be determined by taking the measurement of each hole in relation to a plane surface on the workpiece. The measurement is computed by simple arithmetic.

If the holes are not aligned and perpendicular to the surface plate face, the hole measurements are taken indirectly as just described. The vertical measurement is used as the value for one leg of a right triangle. The workpiece is then turned with a second edge on the surface plate. The holes are measured and the height (distance between centers) is found arithmetically. The measurement represents the second leg of the right triangle. The centerline distance (hypotenuse) is then calculated by trigonometry.

Measuring Small Hole Locations

It is not always possible to position the indicator contact point in a hole. If the diameter of the hole is too small, a press or push fit pin or plug is inserted temporarily. The indicator contact point is set at zero at the required gage block height. This height equals the center distance plus the radius of the pin or plug. The zero reading is

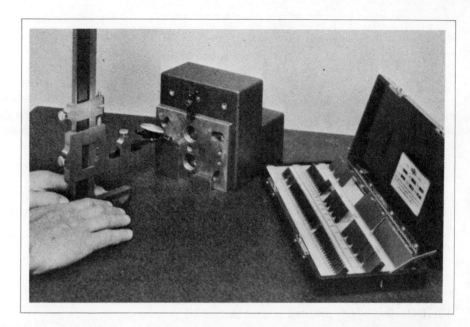

FIGURE 2-9 Measuring a Hole Position by Using a Dial Test Indicator and Height Gage

then compared with the dial reading when the contact point is positioned over the center of the plug. The variation of the dial pointer reading from the zero setting indicates the distance the hole varies from the required dimension. The dial pointer reading is recorded. The actual center distance of the hole is equal to the original setting (plus or minus any variation in the actual measurement) less the radius of the pin.

PRECISION MEASUREMENT OF ANGLES AND TAPERS

Precision Angle Measurement with Angle Gage Blocks

Angles that require an accuracy of ±1/2° may be laid out with a combination of line-graduated measuring tools and transfer layout tools. For example, equally spaced holes around a circle may be scribed with a divider. The divider legs are set with a steel rule to the linear measurement of the arc on the circumference.

Sometimes, the 45° side of a combination square or a protractor head is used. More precise measurements may be taken to within 5′ with a vernier bevel protractor. Rectangular and square precision gage blocks make it possible to make angle measurements with sine bars to accuracies of ±1′.

Angle gage blocks are simpler to use and more precise for measuring angles than rectangular and square gage blocks and the sine bar. Angle gage blocks permit the measurement of angles within limits of a fraction of a second of arc. Angle gage blocks are long rectangular blocks. They are made of extra-hard steel. The blocks are heat treated and dimensionally stabilized. Each angle gage block is tapered. The blocks are checked for accuracy against master gage blocks at a constant temperature of 68°F. Master blocks that are used to calibrate laboratory-grade angle gage blocks have an accuracy of 0.1 second.

Angle gage blocks are available in three accuracy grades. *Toolroom* angle gage blocks are used for angle measurements within 1-second accuracy. *Inspection grade* angle gage blocks are applied to measurements within 1/2-second accuracy. The most precise set is called the *laboratory master*. Laboratory masters are used for measurements within 1/4-second accuracy. The surface finish of the gaging surfaces also varies from 0.6 microinch for the toolroom grade to 0.1-0.3 microinch for the laboratory grade.

A standard 16 angle gage block set has a range of measurement from 0° to 99° in steps of one second. The laboratory master set shown in Figure 2-10 has 16 blocks. The number of blocks in a set and the angles of each block appear in Table 2-2. It is possible, with a laboratory grade set of 16 blocks, to measure 356,400 angles in steps of one second to an accuracy of ±1/4 second. In addition to the angle gage blocks, a 6″ parallel and a 6″ knife edge (form of parallel having one edge that is narrowed to a smaller area) are available.

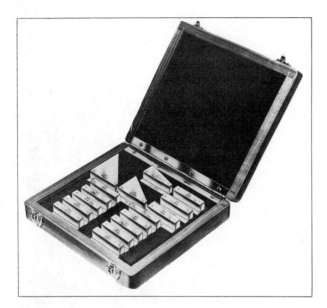

FIGURE 2-10 Laboratory Master Set of 16 Angle Gage Blocks

Like microfinished rectangular and square gage blocks, angle gage blocks are also wrung together in different combinations. The angle gage blocks are designed to be combined in plus and minus positions. One end of a gage block is marked *plus*. The other end is marked *minus*.

Angle combinations are easy to establish. Figure 2-11 shows three angle gage blocks and a parallel block. Toolroom grade blocks are used in the setup with a dial test indicator. The face of a magna-sine is being set to an angle of 38° ±1 second. The angles of the three blocks are (+30°), (+5°), and (+3°). The addition of the three angles of the blocks produces the 38° angle. If an angle of 38°5′ were required, the five-minute block would be used in addition. The plus side would be added to the previous 38° setup.

The blocks may be combined by adding or subtracting. For example, Figure 2-12 shows a workpiece that is being checked with a dial test indicator and angle gage blocks. The three angle gage blocks produce an angle of 13°. The workpiece is resting on a parallel block that is wrung to the angle blocks. The setup is aligned at a right angle by positioning the parts against an angle plate. The 13° angle requires a 15° angle gage block, minus a 3° block, plus a 1° block.

HOW TO SET UP AND CHECK ANGLES WITH ANGLE GAGE BLOCKS

Note: The following procedure requires the setting up of a magnetic sine plate to an angle of 12°20′30″ ±1″ of accuracy. Angle gage blocks are to be used.

STEP 1 Determine the angle gage blocks to use.

Note: One combination includes the 15°, 3°, 20′, and 30″ angle gage blocks. A parallel block may also be used.

STEP 2 Wipe, clean, and check each block.

STEP 3 Slide the 15° block and the 3° block so that they are wrung together. The minus side of the 3° block matches the plus side of the 15° block.

TABLE 2-2 Angles of the Gage Blocks in a 16-Block Set

Number of Blocks	Angles
6	1, 3, 5, 15, 30, 45 degrees
5	1, 3, 5, 20, 30 minutes
5	1, 3, 5, 20, 30 seconds

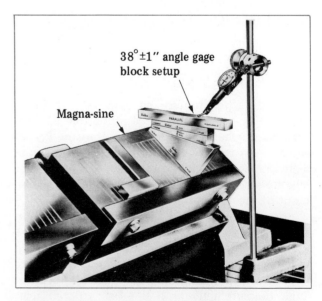

FIGURE 2-11 Setting a Magna-Sine at 38° ±1″ with Angle Gage Blocks and a Dial Test Indicator

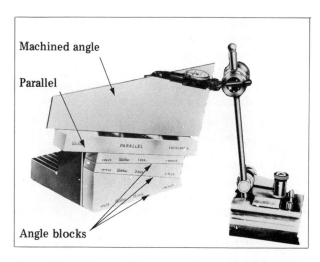

FIGURE 2-12 Angle Gage Blocks and Dial Test Indicator Setup to Measure a Required Machined Angle on a Workpiece

Note: The two blocks, when wrung together, produce the 12° angle.

STEP 4 Wring the 20' block to the combination. The plus side is matched with the 12° of the two wrung blocks.

Note: The combination produces an angle of 12°20'.

STEP 5 Add 30" to the setup by using the plus side of the 30" gage block.

Note: The overall angle is 12°20'30".

STEP 6 Wring the parallel gage block on the total combination.

STEP 7 Place the 15°20'30" angle gage block setup on the cleaned and burr-free surface of the magna-sine.

STEP 8 Place a dial or test indicator on a stand or holder.

STEP 9 Bring the contact point onto the top of the gage blocks at one end. Set the dial at zero.

STEP 10 Move the indicator across the face of the gage block combination. Note any variation in the dial reading.

STEP 11 Adjust the angle of the sine plate until a zero reading is obtained across the entire face of the gage blocks.

STEP 12 Lock the magna-sine at the required angle. Recheck to ensure that the angle position has not changed.

STEP 13 Disassemble the angle gage block combination. Start with the parallel block. Place each gage block on the protective cover.

STEP 14 Clean, wipe, and prepare each block for storage in the case.

Measuring Taper Angles

Tapers that have been machined between centers may be measured with bench centers. These centers are sometimes mounted on a sine bar, as shown in the setup in Figure 2-13. The required gage block height combination is determined for the required angle. The part is carefully inserted between the bench centers. The sine bar is then raised to the height of the gage blocks.

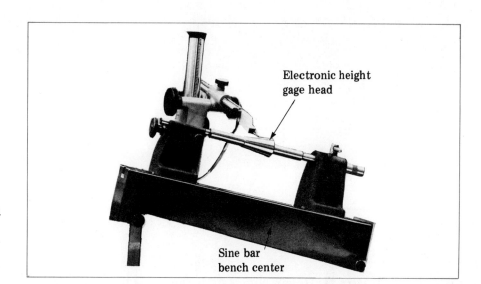

FIGURE 2-13 Measuring a Taper Angle by Using a Granite Sine Bar Bench Center and Electronic Height Gage

36 PRECISION MEASUREMENT AND INSPECTION

While the figure shows an application of an electronic height gage, it is common practice to use a dial test indicator. The dial test indicator is moved across the workpiece at one end of the taper. The highest reading indicates the centerline. The dial face is set at zero at this point. The indicator is then moved across the part at the other end of the taper. Any variation in readings across the small end and the large end of the taper indicates the amount the workpiece varies from a required measurement.

READING DRAWINGS THAT USE DATUMS AND ORDINATE DIMENSIONS

Important dimensioning for layouts and measurement is provided on drawings and prints by using datums and zero coordinates. Linear, angular, and circular surfaces may be located and measurements may be taken in relation to a datum and zero coordinates.

Datums

The term *datum* means an exact point, line, plane, or other reference surface. Two or three datums are usually required to derive all essential information for dimensioning, machining, and inspection. One advantage of using datums is that a series of measurements may be taken from the same reference point. Thus, the accuracy of one dimension does not depend on another. That is, dimensional errors are not cumulative; one error does not add on to another. Dimensions governing each feature are laid out and measured from the datum and not from other features.

Datum Planes and Ordinate Dimensions

Two dimensions are normally required to establish the height, depth, or length of a feature. When datums are used, two or more mutually related *datum planes* are needed. These planes are called *zero coordinates*. The zero coordinates of the two datum planes in Figure 2-14 are easily identified. One zero coordinate is the vertical datum plane; the other is the horizontal datum plane. Note on the drawing that the centerline distances from the zero coordinates are represented by extension lines. There are no dimension lines or arrowheads.

Each center for holes A through G is located at the intersection of two datum dimensions. *Ordinate dimensioning*, rectangular datum dimensioning from zero coordinates, is used in Figure 2-14. An *ordinate table*, the simple table accompanying the drawing, is included.

Many features or dimensions on a conventional drawing are often difficult to read and cause layout, measurement, and machining errors. Ordinate dimensioning and tables simplify a drawing. The machine operator is able to

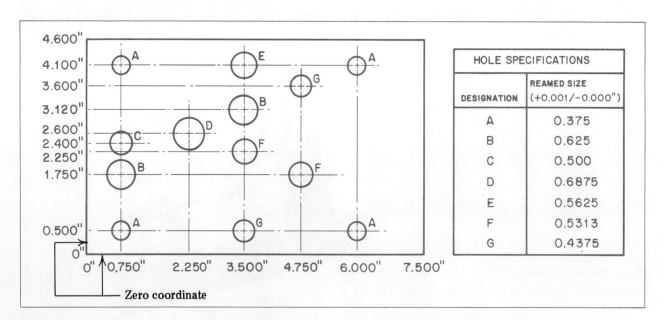

FIGURE 2-14 Locating Hole Positions from Zero Coordinates by Using Ordinate Dimensions

accurately position the workpiece for each center. Horizontal, cross feed, and vertical feed handwheels may be moved a specified distance to position the workpiece for each dimensioned location. The part is moved a specified distance in a required direction. The amount of movement (dimension) is read on the graduated micrometer collar of the feed handwheel. Drawings that use datums and ordinate dimensioning are particularly useful in programming numerically controlled machine tools.

In Figure 2-14, the center of one hole A is located 0.500" up from the zero horizontal coordinate and 0.750" from the zero vertical coordinate. Four holes on the drawing are designated A. As shown in the table, each of these holes is reamed to 0.375" +0.001/-0.000. One B hole falls along the same 0.750" dimension and is 1.750" from the zero horizontal coordinate. The two B holes on the drawing are each reamed to 0.625" +0.001/-0.000. However, the second B hole is located at distances of 3.500" and 3.120" from the zero coordinates. The locations of the remaining holes are measured in a similar manner.

SAFE PRACTICES IN PRECISION LAYOUT AND MEASUREMENT

—Protect the accuracy of a surface plate by not laying unnecessary tools, instruments, and workpieces on it. All items used on a surface plate should be checked for burrs, scratches, and nicks. Remove all imperfections before the items are used.

—Remove the harmful effect of nicks, burrs, and scratches by carefully stoning. Use white Arkansas or granite deburring stones for steel gages. Use sintered aluminum oxide stones for either carbide or steel gages.

—Place items carefully on a precision surface. Any required instrument and setup that must be moved over a plane face is to be slid slowly and carefully.

—Use care in lifting workpieces, instruments, tools, or other accessories off a surface plate. Heavy contact between any object and the edges of a surface plate may cause chipping.

—Check the accuracy of the surface plate and layout and measuring instruments. Measurements taken are to be within tolerances given in specifications.

—Place only the workpiece and the measuring and layout tools and accessories required for a job on a surface plate. Remove any items when they are no longer required for the job.

—Use the full surface to distribute the wear over the entire face of a surface plate.

—Support a large surface plate as it was supported in its manufacture. Level the surface plate and keep it in a level position.

—Elevate parts with projecting lugs, arms, and other surfaces. These surfaces should clear the face of a surface plate. Use protective materials between a rough-surfaced workpiece and a precision-finished surface.

—Bring the contact point of an indicator down on a surface, where possible. If the contact point is to be carefully swung onto a work surface, raise the point gently before making contact. Select a movement range that permits the indicator to cover any variation of the workpiece.

—Recheck all measurements for repeatability of measurement.

—Lap work centers if any burrs or nicks exist on the angular surfaces.

—Follow the practice of noting measurements, particularly where a number of measurements are to be taken.

—Use the protective cover provided with a gage block set. Gage blocks are placed on covers during assembly, disassembly, and use.

—Disassemble wrung combinations as soon as practical. Wipe each gage with a clean chamois dampened with a protective coating. Replace the gage without touching it with bare fingers.

—Use a carbide end test wear block to limit wear on regular gage blocks. The parallel block in an angle gage set may also be used.

—Wring gage blocks and accessories prior to placing them in an outside clamp or before inserting rods and screws. Wringing eliminates errors in size and parallelism.

—Wipe a surface plate frequently. It should be kept clean and free of dust, dirt, and oil. Use the commercial cleaner recommended by the plate manufacturer.

—Spray a dust inhibitor over a cast-iron surface plate when the plate is not in use.

—Check the temperature of instruments and workpieces. The heat absorbed from handling must have a chance to dissipate.

TERMS USED IN PRECISION LAYOUT AND MEASUREMENT PRACTICES

Three-plate method of producing flat surfaces	A technique of scraping flat surfaces. (Each surface forms an accurate plane in relation to the two other plane surfaces.)
Certificate of accuracy (surface plate)	A written statement indicating that the accuracy of the plane surface conforms to federal specifications for flatness tolerances.
Repeatability of measurement	The accuracy of a plane surface. Providing the same measurement under the same conditions at different locations on a plane surface.
Calibration (surface plates)	The inspection and measurement of size and flatness errors. The reported accuracy of a plane surface.
Autocollimator	An instrument designed to use light rays, lenses, and micrometer readings. An instrument designed to check the accuracy (flatness) of a plane surface. A precise instrument used to check flatness errors to within one-fifth second of arc.
Swivel-point test indicator	A standard short (1/2") or long (1-7/16") contact-point test indicator that may be swiveled through a 210° arc. A small end (diameter) point indicator that permits measuring in difficult places and inside surfaces.
Deviations per six inches of length	The trueness of a surface for each six inches of length.
Roundness	A true, closed cylinder where every point on the periphery is equidistant from its center.
Concentricity	The trueness of the periphery of a cylinder in relation to its axis.
Axial runout	The amount a shoulder varies from a true square in relation to its axis.
Angle gage blocks	Heat-treated, dimensionally stabilized, long-tapered rectangular gage blocks. Angle gage blocks of toolroom, inspection, and laboratory grade. Precision angle gage blocks that permit angle settings and measurements to be taken to accuracies of 1 second, 1/2 second, and 1/4 second, respectively.
Bench centers	Precision-machined centers that may be mounted and positioned on a bed or sine bar. Mounted centers used for accurate surface plate work involving centered workpieces.
Datum (datum plane)	An exact line, surface, or reference point.
Ordinate dimensions	A technique used in drafting and design rooms and in shops to relate each dimension to a fixed point or plane. Dimensioning from a datum (zero coordinates). A basic dimensioning system used for numerical control.
Ordinate table	A simple table with specifications, size, and other dimensions relating to each feature or machining process.

SUMMARY

The generating of standard planes by the three-plate method dates back to Henry Maudslay in 1797.

Sir Joseph Whitworth improved the method of producing a "true plane" by hand-scraping, a procedure that "localized" the cutting action. Whitworth also established design and manufacturing systems and standards.

Flat planes may be produced by precision machining that is followed by alternately scraping three plates until they conform to one another.

Many design, development, and production contracts require measurements to specified degrees of accuracy. The tolerances and conditions are contained in job specifications.

Tolerances of surface plates are given as ±0.000025" (0.0006mm) per two square foot area for grade AA laboratory surface plates. The flatness tolerance for grade A inspection-practice plates is ±0.000050" (0.0012mm). Grade B shop plates are accurate to ±0.0001" (0.0025mm) over a two square foot area.

Surface plates and other plane surfaces may be calibrated commercially by using an autocollimator. Applications of light rays, lenses, and a micrometer are involved in calibrating to within one-fifth second of arc.

Light rays are reflected from the autocollimator to corner and reflecting target mirrors and back. The variation in rays is measured for error by noting the micrometer reading. Planned readings are taken and a profile is made along eight lines of sighting.

Precision inspection and measurement of flatness, parallelism, squareness, roundness, and concentricity are typical surface plate processes.

The plunger-type dial indicator and the swivel-point test indicator are common inspection and measurement instruments. These indicators are used with other indicator attachments, a cylindrical square, transfer gage, precision V-blocks, and accessories.

Bench centers are used with indicators to measure total indicated runout (concentricity) and axial runout.

Gage blocks, stacked in a post that is supported on a base, permit taking precise linear measurements. The instruments have a super-accurate micrometer head from which direct readings are taken. Digital readouts further simplify readings. Thimble readings on the micrometer are made directly in 0.0001". Metric-head readings are in increments of 0.002mm.

Hole locations may be laid out or measured with toolmaker's buttons. These buttons are positioned and then measurements are taken with a vernier height gage, gage blocks, and indicator or with other combinations of measuring tools and plug gages.

Angle gage blocks make setting and taking angle measurements to within 1/4 second possible. The three grades of accuracy are toolroom (within one second), inspection (1/2 second), and laboratory master (1/4 second).

A standard set of angle gage blocks includes six blocks that have angles, five that provide variations in minutes, and five that permit measurement settings in seconds. The plus and minus design feature of each gage block permits angle measurement combinations to be built by adding or subtracting the angle of each gage block.

Angle gage blocks are wrung together in a similar manner to other square and rectangular gage blocks.

Taper angles may be measured with angle gage block setups. A bench center and indicator may also be used on the surface plate.

Precision layouts and numerically controlled programming and machine processes involve the interpretation of drawings with datums and ordinate dimensions.

A datum provides a fixed point, line, plane, or reference feature. Ordinate dimensions are identified either on the drawing or in combination with dimensional information found in a table. Ordinate dimensioning and layout simplify reading a print. The possibility of errors that may accumulate from one dimension to another is eliminated.

Precision measurement depends on cleanliness of tools, workpieces, and instruments. They must be free of burrs, nicks, and scratches.

Cases and covers are furnished for precision instruments to provide a safe place for assembling and storing instruments and gage blocks at all times.

Temperature control of workpieces and instruments is necessary. Temperature variations affect dimensional accuracy.

UNIT 2 REVIEW AND SELF-TEST

1. Explain the importance of a perfectly flat plane surface to precision layout and inspection processes.
2. Tell what the difference is between *precision measurement* and *calibration.*
3. List the steps required to check the flat faces of a hardened and ground rectangular fixture plate for parallelism (within ±0.0001″).
4. Explain the difference between the *roundness* and the *concentricity* of a turned part.
5. List the steps to lay out three horizontal lines when a vernier height gage is used. The lines are located 25.4mm, 38.1mm, and 50.8mm from the same ground edge.
6. a. Indicate the degree of accuracy to which angle measurements may be held when (1) laboratory master, (2) inspection, and (3) toolroom grades of angle gage blocks are used.
   b. Cite two advantages of using angle gage blocks for laying out and inspecting angle surfaces in preference to regular gage blocks and a vernier height gage setup.
7. a. Define the term *datum* as used on a shop drawing.
   b. Make two statements about the advantages of using datums for layouts and measurements.
8. State two precautions to take in precision layout and inspection work.

SECTION TWO

Precision Measurement Instruments

Preceding units on measurement have dealt with increasing degrees of accuracy. The range of accuracy has been from ±1/64" (±0.5mm) (using a steel rule) to ±0.0001" (±0.002mm) (using a vernier-scale instrument). Higher degrees of accuracy for linear and angular measurements have been obtained with instruments such as rectangular, square, and angle gage blocks; sine bars; and sine plates. Working accuracies in the hundred-thousandth part of an inch and metric equivalent are within the limits of these instruments. All of the measuring instruments covered thus far are mechanical.

This section deals with even higher degrees of accuracy that are obtained with optical, electrical/electronic, and pneumatic instruments and comparators. General surface plate layout and measuring practices and precision measurements with these instruments are covered in the following units.

UNIT 3

Measurement with Optical Flats and Microscopes

Precision measurements and the inspection of surface finish to higher degrees of accuracy are obtainable with optical flats and microscopes—instruments that depend on physical science principles related to light waves and optics. In this unit these principles are described and applied to optical flats. Measurements for flatness, parallelism, small linear dimension differences, and surface inspection are covered.

Two common designs of microscope are introduced in the unit. Applications are made of the shop microscope and the toolmaker's measuring microscope. A common microscope setup for positioning and measuring on a machine tool is described. Attachments for angular measurements and thread leads are applied to the toolmaker's microscope. The optical height gage is also introduced and provides another application of optics to linear measurement. Simple drafting room techniques of adding tabular dimensioning to ordinate drawings are presented.

PRINCIPLES OF OPTICAL MEASUREMENT

Optical measurement is associated with precision measurement. The instruments used are called *optical instruments*. They are used to magnify an image of an object. Surfaces that are otherwise invisible to the eye can be inspected and measured with these instruments.

Optical instruments control light waves. Light wave and optics phenomena and principles are the foundations upon which measurements with optical flats and microscopes depend.

Light Waves

The action of a light wave on the eye permits it to perform the function of sight. Light travels in a straight line from a source to the eye. The velocity at which the light wave travels is constant unless the wave is intercepted. The medium intercepting the light wave is usually different from the medium in which the wave is traveling. Light is invisible. Light becomes visible when it is reflected to the eye. Ordinary light is reflected to the eye from minute particles in the atmosphere.

Light Rays

Light ray is the term that is used for a light wave from the point at which the wave is emitted or reflected to the observed point. A light ray is produced when the wave is reflected from the surface against which it is directed. The amount of light that is reflected is proportional to the surface finish and the nature of the object. The portion of light that is not reflected either is absorbed by the object or passes through it.

The term *angle of incidence* is used to describe the angle at which a light ray approaches a surface. The *angle of reflection* is the angle at which light is reflected from the surface. Both the angle of incidence and the angle of reflection are equal. The angle is measured perpendicular to the surface. Figure 3-1 illustrates how a light ray is reflected at the same angle as the angle of incidence.

Refraction

A light ray bends when passing from a medium of one density into a medium of another density. The scientific *Law of Refraction* states that light ray is *refracted* when its velocity is increased or decreased by the medium (object) through which it passes. For example, Figure 3-2 shows a light ray passing from air through glass. The ray is refracted when the velocity of the ray is reduced. Since the glass object has parallel surfaces, the light ray is bent upon entering and when leaving the object.

The direction the ray is bent is also illustrated in Figure 3-2. The angle of incidence (A_1) is equal to the angle of reflection at which the ray leaves the glass block (A_2). The refraction through the block is the direct result of the change in velocity that occurs between the top (a and b) and bottom (a_1 and b_1) surfaces of the block. The bending of the light ray is in proportion to the angle of incidence.

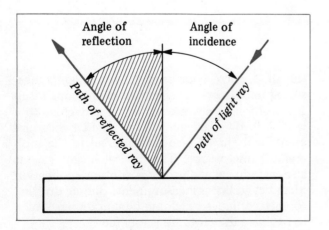

FIGURE 3-1 Angles of Incidence and Reflection

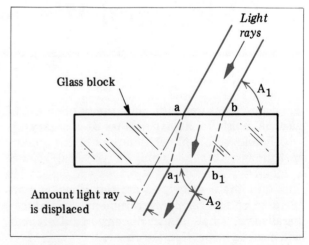

FIGURE 3-2 Refraction of Light Rays through a Glass Block

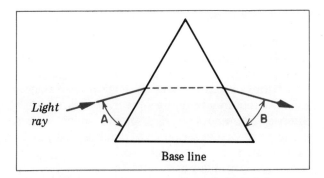

FIGURE 3-3 Path of a Light Ray through a Triangular Glass Prism

Refraction Principles Applied to Lenses

As just discussed, a ray of light that travels through a flat glass block emerges at the same angle as it enters the block. However, the ray is displaced a short distance (Figure 3-2). If the same ray of light is passed through a triangular glass prism, it is bent toward the base line, as shown in Figure 3-3. The light ray path again follows the Law of Refraction. The angles of incidence (A) and reflection (B) are equal. The direction in which the ray is bent and the magnitude of displacement can be predicted.

Glass plates or blocks that are used to do useful work in controlling light rays are known as *lenses*. The four basic forms of lenses are the flat plate, triangular prism, convex (converging) lens, and concave (diverging) lens. These lenses may be designed to any curvature or thickness. Lenses may also include a combination of flat and concave or convex. In addition, one or more lenses may be used together. Figure 3-4 provides examples of convex, concave, and combination lenses. The convergence and divergence of light rays in passing through convex and concave lenses, respectively, is important in the design of all optical instruments.

Convex Lenses. A convex lens may be considered as being made up of two triangular prisms having a common base. These prisms appear in dotted outline in Figure 3-5. Parallel rays of light, as they strike the prisms, converge toward the base. Since rays that pass through the thinnest area of the prism—for example, ray A—pass through quickly, their paths are deflected and thus long. Rays at the center of the prism—for example, ray C—move slowly and are not deflected as much. Note in Figure 3-5 that the rays converge toward a single point. This point is called the *focal point*.

The phenomenon of converging rays has been illustrated here in terms of imaginary straight lines. In physical science, the convergence of light rays through a lens is treated according to the theory of light waves. These waves are pictured in Figure 3-6. The light waves converge as they pass through the lens to the focal point. Beyond this point they diverge in a similar pattern.

Concave Lenses. A concave lens may be pictured as consisting of (approximately) two triangular prisms with their apexes meeting at a common point, as shown in Figure 3-7. The parallel rays of light coming to the lens diverge

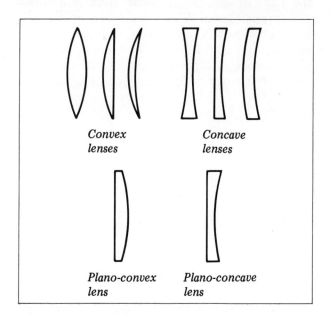

FIGURE 3-4 Sample Convex, Concave, and Combination Lenses

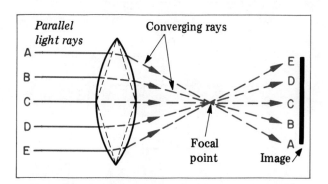

FIGURE 3-5 Convergence of Parallel Rays of Light through a Convex Lens

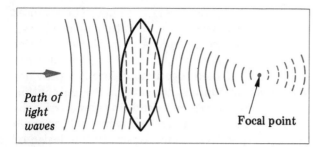

FIGURE 3-6 Concept of Light Rays as Light Waves

as they pass through the changing thickness of the cross-sectional area. At the thick outer edges, the parallel rays are caused to diverge the greatest amount. The rays at the center pass through without diverging.

Magnification

As the eye views an object, a normal impression is gained about size and shape. If a convex lens is placed between the eye and the object, an image is viewed. This image is called the *virtual image* (Figure 3-8). The shape and size of the virtual image is proportional to the shape (curvature) and size (magnification power) of a single lens (or lenses) and the distance between the lens and the object. The combination of lens curvature, size, and distance produces the *magnification power*. This power is marked on an optical instrument by the symbol "X" and a numerical value. A 40X shop microscope has a magnification power 40 times normal vision. The range of magnification of a precision inspection microscope is from 37.5X to 300X.

Illumination

The light from an object is diffused when the object is magnified. The light is absorbed in proportion to the magnification. The light may be natural or self-contained. High-power magnification requires that the object be viewed with intense *illumination*. High-power microscopes incorporate an external light source (lamps) in their design to permit light to be directed from the most desirable position to the object.

SHOP AND TOOLMAKER'S MEASURING MICROSCOPES

Two common optical instruments found in the shop and laboratory are the *shop microscope* and the *toolmaker's measuring microscope*. Practical measurement and inspection procedures may be carried on with these microscopes.

Operation of the Shop Microscope

A shop microscope of simple design is pictured in Figure 3-9. The main features include an adjustable ocular lens and tube, a column and stand, an adjusting ring with a graduated scale, and a light source. The microscope is mounted in a stand. The tube may be raised or lowered by a clamp screw.

The ocular lens is adjusted by turning it to the right or left. Adjustment brings the object into proper focus. The light source in the column is directed to the object through the relieved area in the base. The degree of magnification of the microscope is the product of the magnifications of the ocular lens and the objective lens.

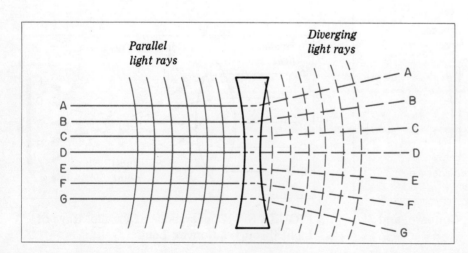

FIGURE 3-7 Divergence of Light Rays Produced by a Concave Lens

OPTICAL FLATS AND MICROSCOPES 45

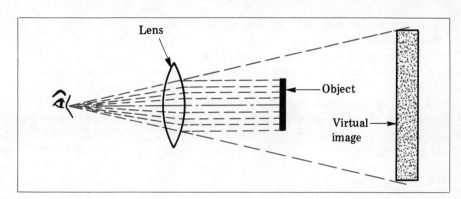

FIGURE 3-8 Magnification of an Object and the Virtual Image

For example, the 40X shop microscope has a power of magnification of 40 (10X ocular lens with a 4X objective lens).

A micrometer scale, like the scale shown in Figure 3-10, is mounted within the ocular. The 150-division scale represents 0.150". Measurements may be made to 0.001". Subdivisions of 0.001" (0.02mm) may be estimated to within 0.00025" (0.005mm). The instrument is thus capable of a high degree of accuracy.

One advantage of the shop microscope is that it is small and convenient to carry to a required location. A disadvantage is that it has a limited field. *Field* refers to the diameter of the area that may be observed in a single position. The model shop microscope, as illustrated in Figure 3-9, has a field of 7/32" (5.5mm) and magnifies 40X.

General Applications of the Shop Microscope

General applications of the shop microscope are as follows:

—Measuring the diameter of small holes having tolerances of ±0.001" (0.02mm) (such holes often require measurement and inspection during machining);
—Measuring small dimensions;
—Checking the thickness of the case on hardened parts;

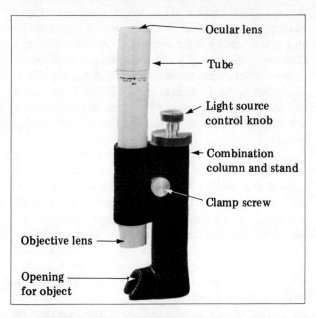

FIGURE 3-9 40X Shop Microscope with 0.150" Scale Graduated in 0.001"

FIGURE 3-10 150-Division (0.150") Scale Graduated in 0.001" (Mounted within a Shop Microscope)

—Measuring the diameter of impression made during hardness testing;
—Checking mechanical parts quickly for wear (welds, for example, may be inspected easily);
—Examining surfaces on-the-spot to detect cracks and flaws (the condition of plated, polished, and painted surfaces may be checked quickly).

Binocular-Head Microscope

Another microscope, which is not identified as a shop or toolmaker's microscope, combines the advantages of a *binocular* (two-eyepiece) head design and a *zoom* feature (Figure 3-11). The head design and zoom feature deliver a bright, three-dimensional stereo image. The image keeps the area being viewed continuously in focus at a constant working distance up to 10.5" (267mm).

The magnification knob is turned to zoom the image to the exact size for optimum viewing. The stage of the model illustrated is 4" X 5" (101.6mm X 127mm). The Stereozoom® microscope uses a *flat field* optical system. The system permits the image to remain clear from one edge to the opposite edge. This type of microscope minimizes operator fatigue.

Toolmaker's Measuring Microscope

The toolmaker's measuring microscope is a ruggedly designed microscope with high-power magnification. It is adaptable to taking linear and angular measurements. The microscope is especially valuable for inspecting small parts and tools. The main features of the microscope are labeled in Figure 3-12.

A number of attachments permit the microscope to be used as a comparator. A thread form or other tool contour may be compared with a charted outline that is superimposed directly upon the virtual image of the object. The contour of the object may thus be checked to within 0.0001" (0.002mm).

Angles may be measured to an accuracy of five minutes of arc on the toolmaker's measuring microscope. The microscope in Figure 3-12 has a *protractor eyepiece* (eyepiece and lens assembly, combined with a vernier protractor) that is used to make angular measurements.

The toolmaker's microscope may also be used to measure pitch diameter, major and minor diameters, pitch, and lead. Lead-measuring and thread-measuring attachments extend the practical applications of this microscope. Some advanced models include a projection attachment that permits making contour and other precision measurements normally performed on comparators.

OPTICAL HEIGHT GAGE

The *optical height gage* is another instrument that applies optics. This instrument combines design and measurement features of gage blocks and a measuring microscope. The optical height gage consists basically of a "stack of gage blocks." These blocks are permanently wrung together. The accuracy of the stack is ±0.00005" per inch (0.0002mm per 25mm) of height. Measurements are read directly in the eyepiece of the microscope.

Some older height gage models are being retrofitted by their manufacturers to replace the optical feature by an easier reading digital readout. One manufacturer's model, the Digi-Chek® II, is shown in Figure 3-13. This particular height gage includes the digital readout component.

The instrument requires the use of a standard reference bar. The reference bar consists of a master stack of alternating 0.300" gage block jaws and 0.700" spacer blocks. The gage blocks are permanently wrung and fastened together to form 1" increments. The master stack has a special bushing arrangement. The bushing allows

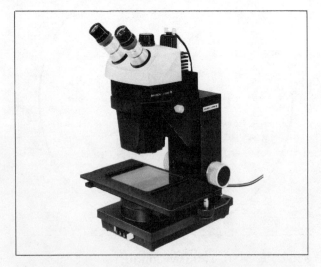

FIGURE 3-11 Binocular-Head Microscope with Stereozoom® Feature and 4" X 5" Stage

OPTICAL FLATS AND MICROSCOPES 47

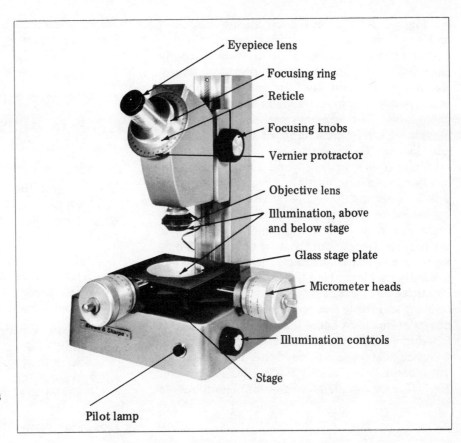

FIGURE 3-12 Main Features of a Toolmaker's Measuring Microscope

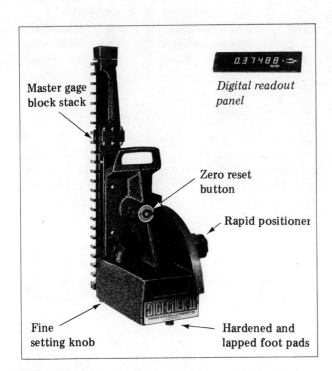

FIGURE 3-13 Optical Height Gage with Master Gage Block Stack (Standard Reference Bars) and Digital Readout

the stack to conform to thermal conditions under actual use.

The linear accuracy of the master stack of gage blocks is +0.000004/-0.00000" per inch of length. The accuracy of a master stack in metric units of measure is +0.00001/-0.0000mm per 25mm of length. The parallelism of the gage surfaces to the base and to one another is 0.000015" (0.0004mm). The measuring system accuracy is 0.00005" (0.00125mm). Most digital systems are equipped with a safety power failure flashing display.

INDUSTRIAL MAGNIFIERS

Industrial magnifiers are used in the shop and laboratory for magnifying lines, surface areas, and design or machining features. Some magnifiers are mounted and provided with an outside source of illumination. They are adjustable. They permit working at a short distance from the workpiece or object. Other magnifiers attach to vernier scales and permit easier and more accurate readings.

Figure 3-14 shows a measuring magnifier and four scales. The magnification of this particular magnifier is 7X. Each transparent body allows adequate illumination to produce a sharp, flat image. Each of the four scales is contained in a separate mount. The scales are readily interchangeable. The general-purpose scale permits measurements in increments of 0.005″, 0.1mm, and 1° and the measurement of 1/16″ to 5/8″ radii and 0.001″, 0.002″, and 0.003″ line thicknesses. A separate inch scale has graduations in 0.005″ intervals up to 3/4″. The metric scale is graduated in 0.1mm intervals over 20mm. The protractor scale is divided in single degrees.

Simple designs of magnifiers with single lenses are identified as hand magnifying readers (magnifying glasses). Magnifiers with one-, two-, and three-lens designs give combinations of single, double, and triple lens combinations. Such magnifiers are called pocket magnifiers. Each lens is fitted in a simple case that serves as a holder.

A regular eyepiece has cross lines that permit measuring angles of 30°, 60°, 90°, 120°, and 150° (Figure 3-15). More precise measurements are made with an angle-measuring protractor eyepiece, a microscope accessory that was described earlier.

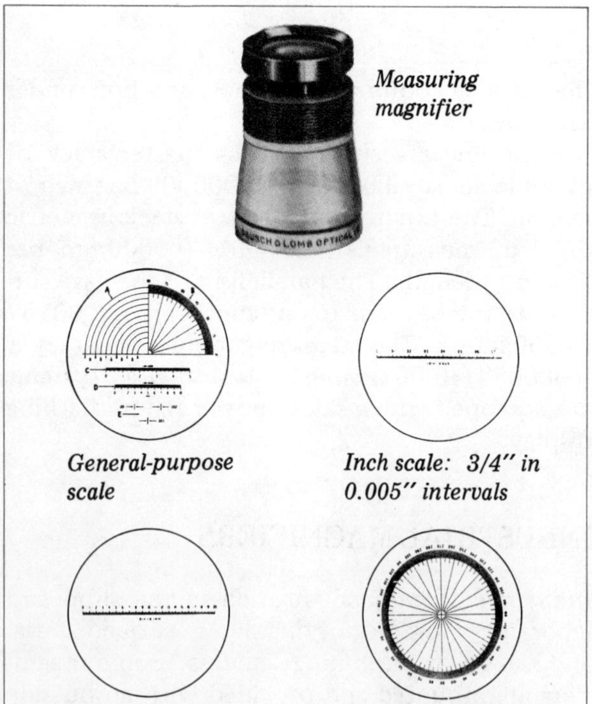

HOW TO USE A TOOLMAKER'S MEASURING MICROSCOPE

STEP 1 Mount the specimen in the center of the stage.

STEP 2 Check to see that the light source is adequate.

Note: A vertical illuminator attachment is sometimes used to direct light through the lenses of the objective to focus on the object from above.

STEP 3 Make a rough adjustment of the optical unit. Loosen the clamp screw. Adjust with the locating pin.

STEP 4 Adjust further by turning the knob until the object is in focus.

Measuring Angles

STEP 1 Align the cross lines in the eyepiece with the angular surface to be measured.

STEP 2 Turn the micrometer heads to adjust the workpiece horizontally. The heads move the stage.

FIGURE 3-14 Measuring Magnifier and Four Scales

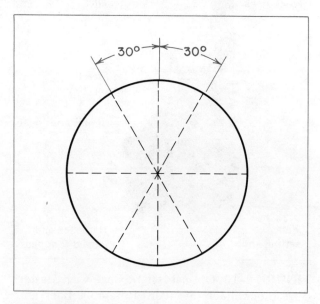

FIGURE 3-15 Cross Lines in an Eyepiece

OPTICAL FLATS AND MICROSCOPES

STEP 3 Read the graduation on each micrometer head at the start of each setting and at each required measurement.

Note: The graduations on the head permit movements and measurements to an accuracy of 0.0001" (0.002mm) in either direction.

STEP 4 Adjust the stage by aligning one side of the part to be measured with a cross line.

STEP 5 Take and record the micrometer-head reading.

STEP 6 Move the stage until the side to be measured is in line with the same cross line.

STEP 7 Take this micrometer reading.

STEP 8 Subtract the two readings. The difference (represented by the distance the stage is moved) represents the linear measurement.

Measuring Angles with a Protractor Eyepiece Attachment

Note: The thread angle of a 60° threaded part is to be checked.

STEP 1 Mount the workpiece on the stage. The axis is set parallel to the movement of the table.

STEP 2 Set the angle between the two cross lines at 60° by turning the knurled adjusting ring on the eyepiece.

STEP 3 Set the eyepiece to one-half the thread angle. The outside scale is used to set the 30° angle.

Note: Since the thread angle is 60°, each slope makes a 30° angle with a line perpendicular to the screw axis (Figure 3-16A).

STEP 4 Turn the micrometer adjustment on the stage to adjust the image.

Note: The thread form must coincide as nearly as possible with the cross lines.

STEP 5 Adjust both cross lines (Figure 3-16A). These lines must coincide with the sides of the thread image (Figure 3-16B).

STEP 6 Read the angle of the thread on the scale in the eyepiece.

Note: When a line bisecting the thread angle is perpendicular to the movement of the stage, the outside scale indicates half the angle between the cross lines.

Note: The outside scale of a 60° thread angle is set to 30°. Any variation in the reading indicates an error in just one-half the thread angle.

Measuring Thread Pitch and Lead

STEP 1 Mount the threaded part so that its axis is parallel to the movement of the stage.

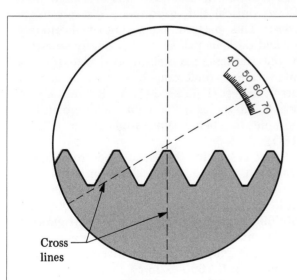

A. Cross Lines on Inside Scale

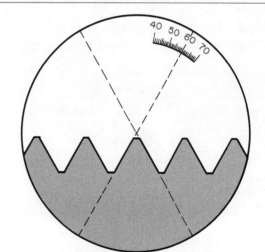

B. Thread Form Coincides with Sides of Thread Image

FIGURE 3-16 Measuring a 60° Thread Angle with a Protractor Attachment

STEP 2 Line up the cross lines with the outline of one thread.

STEP 3 Note the reading on the micrometer head.

STEP 4 Adjust the stage. The cross lines are lined up with the next thread.

STEP 5 Note the micrometer reading. The difference between the two readings is the pitch of a single thread. If the thread is a multiple thread, the difference represents the lead.

PRECISION MEASUREMENT WITH OPTICAL FLATS

Characteristics of Optical Flats

Optical flats are used for precision checking flatness, parallelism, size, and surface characteristics. The optical flats used for measurement and inspection are made of a high-quality optical quartz. This material has a low coefficient of expansion (approximately 0.32 millionths of an inch for each degree Fahrenheit change). Ultra-low expansion fused silica optical flats are also available. One or two faces may be finished as a measuring surface. A band on the side of the optical flat indicates the measuring surface(s). The accuracy of the flat is marked on the band in millionths of an inch. Figure 3-17 shows the common round and square shapes of optical flats and the marking band around each outside edge.

Optical flats are manufactured in three grades. *Working flats* are produced with a degree of flatness of 0.000004" (0.0001mm). No point on a working flat will deviate from any other point by more than 4 microinches (millionths of an inch). *Master flats* are accurate to within 0.000002" (0.00005mm). *Reference flats*, used primarily under close controlled laboratory conditions, are held to an accuracy of 0.000001" (0.000025mm).

Optical flats are also furnished with coated surfaces. A thin film of titanium dioxide on the optical flat sharpens the image for greater accuracy.

Commercial optical flats are available in sizes ranging from 1" to 10" diameter (25mm to 250mm). The thicknesses (depending on diameter) vary from 1/2" (12mm) to 2" (50mm). The common sizes of square optical flats are 2", 3", and 4" on a side by 5/8" and 3/4" thickness. Metric sizes are 50mm, 75mm, and 100mm by 15mm to 18mm thick.

Monochromatic Light Source

White light represents a combination of colored light. Therefore, it is extremely difficult to reproduce uniformly. In order to establish international uniformity in terms of the light used, a monochromatic light source is necessary. *Monochromatic light* refers to light waves that conform to a definite wavelength. Helium serves as standard because it has a definite wavelength. The light is reproducible anywhere in the world.

A monochromatic light source, when used with an optical flat, produces a distinct *fringe pattern*. This pattern, consisting of alternately light and dark bands, shows up clearly to establish the flatness, parallelism, and size (over a limited dimensional range), as well as surface characteristics (Figures 3-18A, B, and C). A portable monochromatic light, or monolight, is adaptable for bench-type applications (Figure 3-18D).

Each color in the solar spectrum has a wavelength that tends to blend with the next color. The wavelengths range from 0.0000157" (0.0003937mm) for ultraviolet rays to 0.0000275" (0.0006985mm) for red rays. The light emitted from a monochromatic (rays of one definite wavelength) light is 23.1 microinches (millionths of an inch) or 0.0005867mm.

A monolight and optical flats are generally used in combination with a plane surface and gage blocks. Precision measurement with optical flats thus depends on the accuracy of the

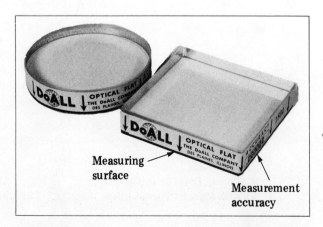

FIGURE 3-17 Round and Square Optical Flats

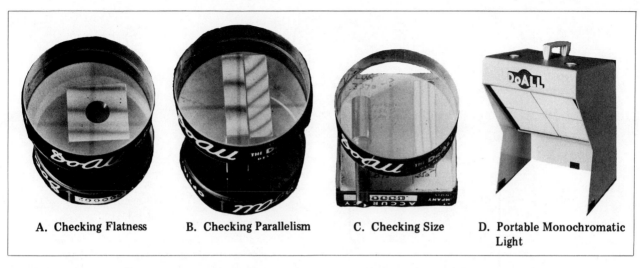

FIGURE 3-18 Application of Optical Flats under Portable Monochromatic Light Source

instruments and accessories that are used with the flats. Accuracy may be maintained in the following ways:
- By brushing the workpiece or part, gage blocks, and optical flats clean;
- By carrying on operations in a dust-free area;
- By normalizing all parts and instruments at the time of measurement at the same temperature;
- By viewing the object as nearly perpendicular to the surface as possible to reduce parallax errors (Figure 3-19 shows that the smaller the angle of viewing, the more accurate the measurement will be);
- By keeping the measurement surface as close to the light source as practical to help in achieving maximum clarity.

Measurement with Light Rays

An optical flat works on the principle that a ray of light of known wavelength penetrates a translucent or transparent object. When the light ray reaches the lower surface of the optical flat, the ray is divided into two components. One component, as stated earlier, is reflected back to the eye. The other component continues down to the surface of the object (work). On hitting the object, the ray is also reflected back to the eye.

Both light ray components have the same wavelength. However, the second component lags behind the first by an amount equal to twice the space between the optical flat and the surface of the work. A set of parallel separation planes formed by the rays are seen as fringe bands. The number of fringe bands and their shape are used in comparison measurement and surface inspection with optical flats.

Example: Make a simple comparison measurement with an optical flat. For simplicity, assume the diameter of a precision ball is required.

All precision parts and instruments used must first be thoroughly cleaned. The equipment required to make the simple measurement includes a toolmaker's flat (or other accurate plane surface), an optical flat, a gage block master, and the part.

The diameter may be established by comparing it against a known measurement of a gage block. A formula is used to establish any variation from a required size:

$$M = n \times \frac{l}{w} \times 11.6$$

where

M = the difference in measurement between the part and the gage block measurement;

n = the number of bands that appear in the optical flat over the gage block;

l = the distance between contact (reference) points of the part and gage block;

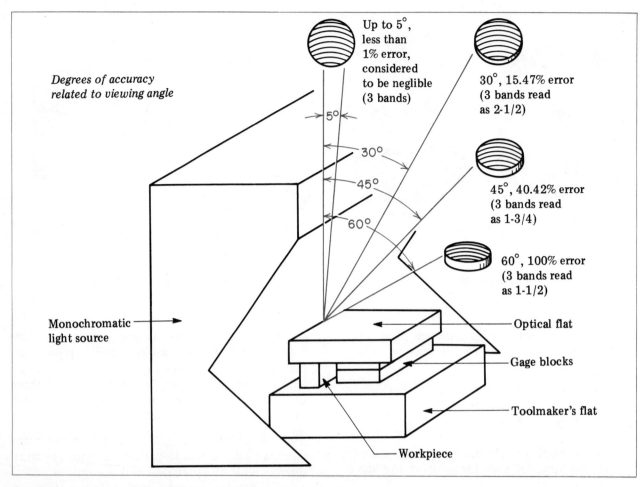

FIGURE 3-19 Perpendicular Viewing to Reduce Parallax Errors

    w = the width of the master (gage) block;
11.6 = a half wavelength of monochromatic light, in microinches.

Figure 3-20 shows the setup, the width of the gage master (1″), and the distance between the contact points (1-1/2″). By using the dimensions in the formula, the measurement (M) is computed to be 104.4 microinches (millionths of an inch) larger than the required size:

$$M = 6 \times \frac{1\text{-}1/2}{1} \times 11.6$$

$$= 104.4 \text{ microinches}$$

Point of Contact. In the preceding example, the determination of point of contact is important in establishing whether the precision ball is over or under the required size. The point of contact must be known before any measurement may be taken. Figure 3-21 shows the interference bands of a precision, flat surface. *Interference bands* indicate there is an air wedge between the optical flat and the workpiece. The thickness (width) of this wedge is 4 (bands) × 11.6 microinches (distance at 1/2 wavelength apart). The air wedge is thus 46.4 microinches (0.000464″) thick. Note that the distance between bands may vary. However, the height difference remains 11.6 microinches. The difference is always counted from the line of contact. It is necessary to establish which edge of the optical flat is making contact with the workpiece.

A simple test to establish the point of contact is shown in exaggerated form in Figure 3-22. A finger is pressed alternately at the opposite ends of the optical flat. In its original position, five bands (one-half wavelength apart) are visible in

OPTICAL FLATS AND MICROSCOPES 53

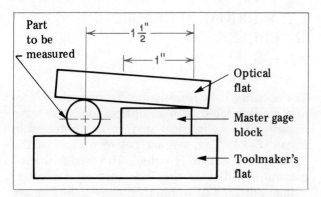

FIGURE 3-20 Setup for Small Dimensional Measurement with an Optical Flat (Exaggerated)

MEASUREMENT OF SURFACE FLATNESS

Thus far, only straight interference bands have been discussed. The straight, parallel, uniform bands relate to a flat surface as shown in Figure 3-23A. Variations in a flat surface produce curved interference bands. Such bands indicate

the optical flat (Figure 3-22A). When the finger is placed at one end, the air wedge becomes wider (Figure 3-22B). The number of interference bands increases and the bands are closer together.

If the optical flat is allowed to return to the original position and the opposite end is pressed, the air wedge becomes narrower (Figure 3-22C). Three bands are now visible. They are spread across the optical flat. The three bands indicate the open end of the air wedge—that is, the point of contact has been found.

FIGURE 3-21 Interference Bands of a Precision, Flat Surface

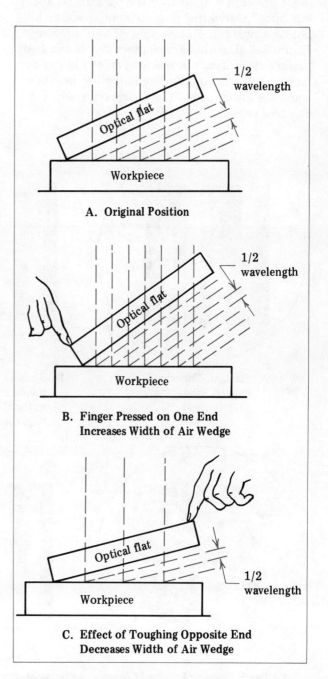

FIGURE 3-22 Establishing the Point of Contact (Conditions Exaggerated)

54 PRECISION MEASUREMENT AND INSPECTION

a concave or convex surface, a slight angle, or a radius.

Figure 3-23B shows interference bands that curve slightly at the ends, showing that the edges of the workpiece are rounded. Figure 3-23C shows interference bands in a series of curved lines that form elliptical circles. This pattern indicates a high spot on the surface of the workpiece. The pattern may be identified when a slight pressure is applied to the optical flat above the spot. Curvature due to a high spot (hill) points toward the open side of the air wedge. Figure 3-23D shows interference bands as a bent pattern indicating a low spot (valley) in the surface of the workpiece. The curvature points toward the closed side of the air wedge when it is due to a valley.

CYLINDRICAL SURFACE INTERFERENCE BANDS

The accuracy of cylindrical and spherical surfaces may be inspected with an optical flat. The wave pattern of a cylindrical surface is illustrated in Figure 3-24. The spacing between each of the interference bands is equal. The wedge-shaped curvature indicates that the surface is convex. Fringe patterns of a spherical convex surface are shown in Figure 3-25A as concentric circles. Note that the center of the pattern follows the point of contact between the optical flat and the work surface as indicated by the off-centered circles that are darker at the center.

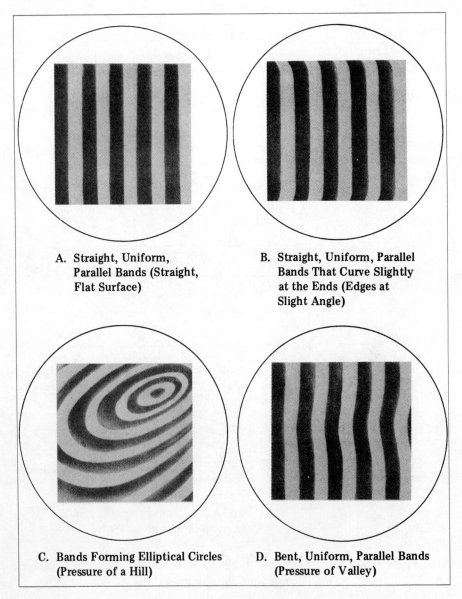

FIGURE 3-23 Interference Band Patterns Indicating Surface Flatness and Variations

OPTICAL FLATS AND MICROSCOPES 55

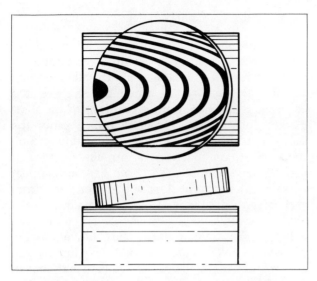

FIGURE 3-24 Wave Pattern of a Cylindrical Surface

If the surface is spherical and concave, as pictured in Figure 3-25B, the amount that the surface varies is established by the number of concentric full-circle bands. The surface may be shown to be concave when any slight pressure near the center causes the bands to increase in width and spacing.

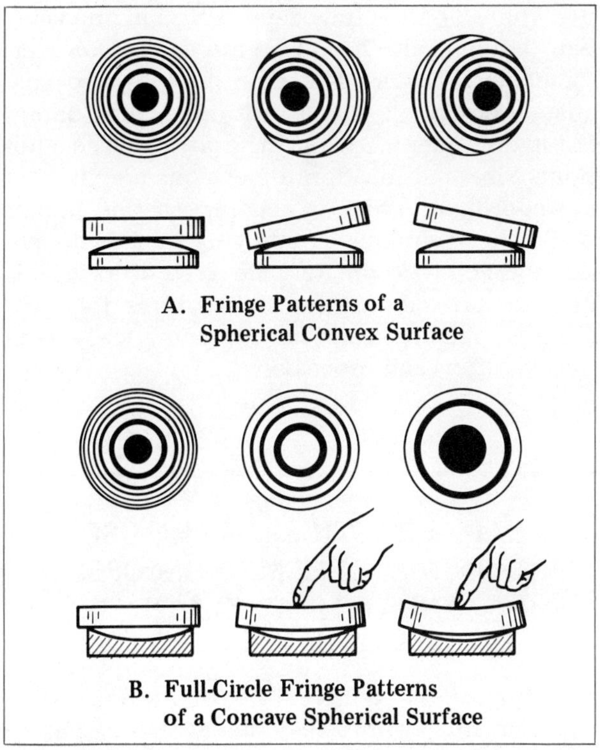

INTERFERENCE BAND CONVERSION TABLE

Part of the full conversion table that appears in the Appendix for converting fringe bands to microinches (or metric equivalents) is reproduced in Table 3-1. Values are given in fractional parts of an inch and of a millimeter. When close approximations are sufficient, one band indicates an accuracy of approximately 10 millionths of an inch (0.00025mm).

Direct measurement with optical flats is restricted to a very narrow range. While optical flats are still used for comparison measurements, some applications are being taken over by high amplification, electronic comparator instruments. These instruments require relatively less skill and are more rapid than optical flats. Optical flats are, however, excellent for surface measurements.

HOW TO MEASURE A WORKPIECE WITH AN OPTICAL FLAT

STEP 1 Select a master gage block (or combination) that is as close to the required size as possible.

TABLE 3-1 Partial Conversion Table for Optical Flat Fringe Bands (Inch and Metric Standard Units of Measure)

Number of Bands	Equivalent Measurement Value		
	Microinches	Inches	Millimeters
0.1	1.2	0.0000012	0.000029
0.2	2.3	0.0000023	0.000059
0.3	3.5	0.0000035	0.000088
...	...	...	...
1.0	11.6	0.0000116	0.000294
2.0	23.1	0.0000231	0.000588
3.0	34.7	0.0000347	0.000881
...	...	...	...
19.0	219.8	0.0002198	0.005582
20.0	231.3	0.0002313	0.005876

A. Fringe Patterns of a Spherical Convex Surface

B. Full-Circle Fringe Patterns of a Concave Spherical Surface

FIGURE 3-25 Inspecting Spherical Surfaces with an Optical Flat

STEP 2 Clean the gage block, workpiece, and optical flat.

STEP 3 Place the part (workpiece) and the master gage on a toolmaker's flat or other plane surface.

STEP 4 Measure the width of the gage and the distance between the reference (contact) points.

STEP 5 Place an optical flat so that it is centered over the master gage and the part.

STEP 6 Establish the point of contact.

STEP 7 Read the number of bands appearing in the width of the master gage.

STEP 8 Use the formula:

$$M = n \times \frac{l}{w} \times 11.6$$

Substitute the known measurements of n, l, and w.

Note: The answer is the amount in microinches (or millimeter equivalent) that the part varies from the required size. Compensation must be made for any variation in the size of the gage block used from the required size.

STEP 9 Disassemble the setup carefully. Clean and place all instruments in proper storage containers.

TABULAR DIMENSIONING APPLIED TO DRAWINGS

One technique of dimensioning, known as ordinate dimensioning, was discussed in Unit 2. Another technique is to leave off location and size dimensions and other specifications from drawings and to include the information in chart form. Each size and location dimension on a chart relates to measurements from datums on a drawing. This technique is called *tabular* (or rectangular) *dimensioning*. It prevents cluttering a drawing with dimensions when there are a large number of features. Tabular dimensioning also simplifies the reading of a blueprint or sketch.

Tabular dimensions start at the origin (datum) of the coordinate—that is, each dimension begins at the point where the coordinates intersect. Inch or metric units of measurement may be used.

Coordinate Charts

A *coordinate chart* contains location and size dimensions. Sometimes, a column is added for specifications that relate to a particular feature. Figure 3-26 shows a line drawing with its related coordinate chart. The three columns of the chart—hole, location, and specifications—show the kind of information needed to machine and inspect the part. Each hole on the drawing is listed in one column. The hole sizes also appear in this column. The location dimensions from each of two datums (X) and (Y) are given in the next column. Specifications appear in the last column. The drawing shows the relative position of each hole. The chart provides the dimensions. The worker works from the drawing and coordinate chart to select the cutting tools; determine the machine setups, processes, and tolerances; and check measurements.

In reading the drawing and the chart, note that the first hole is identified as A_1. The hole size is 1/2" diameter. The centerline of A_1 is 1.000" from datum X and 0.750" up from datum Y. Holes C_1 and C_2 are to be reamed to 0.625" +0.002/0.000, as indicated in the specifications column. The tolerance on holes D_1, D_2, and D_3 is indicated as ±0.001".

Where a part is fully represented in one view, two datums (like X and Y) are used. However, some parts require two or more views to adequately describe a part or feature. A third datum (Z) is then identified on the drawing. The additional size and location dimensions are given in the coordinate chart in relation to this datum. While the technique of applying tabular dimensioning and coordinate charts to drawings is used for general machining operations, the technique is readily adaptable to numerically controlled setups and processes.

SAFE PRACTICES IN THE USE OF INDUSTRIAL MICROSCOPES AND OPTICAL FLATS

—Normalize the temperature of the measuring instruments, workpiece, and accessory before taking a measurement or carrying on surface

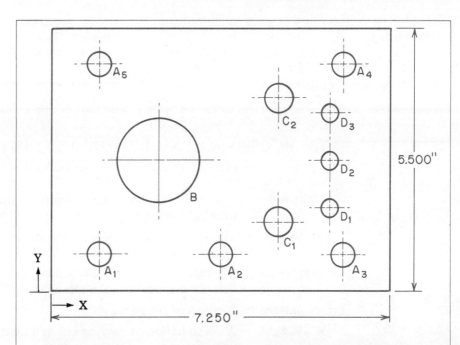

FIGURE 3-26 Line Drawing with Related Coordinate Chart

inspections. Optical flats have a low conductivity of heat; they normalize at a slower rate than metallic parts.
—Use a fine, soft-bristle brush to wipe over the surfaces of an optical flat or microscope and areas of contact.
—Set up the part to be measured for proper viewing. Keep the line of sight as nearly perpendicular to the work surface as possible to reduce parallax in reading.

—Wipe instrument lenses with a clean, soft cloth or lens tissue. Saturate the cloth with an approved cleansing solution. Any dirt rubbed across a lens will scratch the surface.
—Place a part on the stage of a toolmaker's measuring microscope in order to permit movement over the length or width to be measured.
—Take care in adjusting the objective of the instrument so that it does not strike the specimen or part.

58 PRECISION MEASUREMENT AND INSPECTION

—Recheck each linear and angular measurement for accuracy.

—Clean and check each gage block and optical flat at the end of a process. Store each instrument in its protective case.

—Use a precision plane surface, such as a toolmaker's flat, for precision measurements with optical flats and gage blocks.

TERMS USED IN MEASUREMENT AND INSPECTION WITH OPTICAL INSTRUMENTS

Light wave	The transmitting of light energy by pulsations. A wave that transmits light energy.
Interference	The action of one energy pulse on another when two light rays are brought together.
Fringe bands	Alternate light and dark bands produced by interference. Bands that are one-half wavelength apart.
Fringe pattern	A group of fringe bands.
White light	A combination of varying colored light wavelengths.
Monochromatic light	A light source that consists of one wavelength (or color) only. A practical ray of light that penetrates an optical flat. Light emitted from a monochromatic lamp having a wavelength of 23.1 microinches (0.0005867mm).
Flats	Reference planes of almost perfectly flat surfaces.
Toolmaker's flat	A reference surface of steel, granite, or other opaque material. A precise plane surface.
Optical flat	A quartz, glass, sapphire, or other transparent very nearly perfect plane surface.
Master flats	Flats with high-precision surfaces from which other surfaces may be calibrated.
Working flats	Flats used in the shop and laboratory for most routine measurement and inspection.
Coated flat	A coating, such as titanium dioxide, bonded to an optical flat. An optical flat with a coating. A coating that permits faster readings with greater accuracy. A coating to produce sharp, dark fringe patterns.
Air wedge configuration	A fringe pattern produced when an optical flat rests on a part.
Parallel separation planes	A set of planes that are parallel to a working surface. Parallel planes that are one-half wavelength (11.6 microinches) apart. Planes that intersect with the part and are seen as fringe bands.
Microinch	A simplified form of stating a value in millionths of an inch. (For example, 11.3 microinches refers to 0.0000113", "eleven point three millionths of an inch").
Refraction	The bending of a light ray as it travels from one medium to another. A basic law of physics upon which measurement and inspection with optical flats depend.
Convex lens	An optical lens surface that is curved (bulged). The exterior surface of a sphere. A lens through which light rays converge to a focal point.

Concave lens	A curved lens in which there is less divergence of light rays at the center than at the outer edge. The interior (hollow) of a curved lens.
Field	The diameter of the area that may be observed through a magnifier lens (or microscope). A magnified area that is clearly visible when a lens is held at the correct distance from an object.
Focal point	The point at which light rays that pass through a lens intersect to form an image.
Object	A part, particle, surface, or feature of a workpiece that is viewed through a magnifier.
Virtual image	The optical counterpart of an object formed by a lens.
Tabular dimensioning	A simplified system of rectangular dimensioning. Dimensioning in relation to one or more datums. Dimensions that begin at the intersection of the coordinate. Recording dimensional information on a coordinate chart.
Coordinate chart	A chart containing the identification of features, with size and location dimensions. Dimensional information provided in chart form instead of directly on a drawing.

SUMMARY

Optical flats and microscopes permit measurement and inspection processes to be carried out at a high level of precision.

> Optical instruments control light waves to form patterns. These patterns are used to establish a comparison measurement or the accuracy of a surface.

Optical microscopes depend on concave and convex lenses for diverging or converging light rays to magnify an object.

> The shop, wide field, and toolmaker's measuring microscopes are practical measurement and inspection instruments.

The toolmaker's measuring microscope is especially adapted to precise linear and angular measurements and surface finish inspection. Thread measurement applications include measuring pitch; major, minor, and pitch diameters; and lead.

> Angular measurements to within an accuracy of five minutes of arc may be read directly on a toolmaker's measuring microscope.

The optical height gage combines the features of a stack of gage blocks and a measuring microscope.

> Industrial magnifiers are simple, inexpensive, practical, optical devices. Magnifiers are used in the shop and laboratory for magnifying lines, areas, design features, and surface finish.

A light ray is reflected light from a source to an observed point.

> Optical flats depend on the Law of Refraction. A light ray bends through an optical flat. The ray is seen (if there is a variance in measurement) as fringe bands. The number of fringe bands and their shape are used in measurement and surface inspection.

Optical flats are adaptable for measuring flatness, parallelism, and size, as well as for inspecting the quality and accuracy of a finished surface.

Working optical flats are accurate to within 4 microinches (0.0001mm); master flats, 2 microinches (0.00005mm); and reference flats, 1 microinch (0.000025mm).

Monochromatic light has a single wavelength of 23.1 microinches (0.0005867mm). It is used with optical flats.

The formula for computing comparison measurements with optical flats is:

$$M = n \times \frac{1}{W} \times 11.6$$

When the point of contact is established, the answer indicates any variation in microinches from the required size.

Curved interference bands indicate that a surface is convex or concave or has a slight radius or angle.

The range of measurement with optical flats is restricted.

Tabular dimensioning simplifies both a drawing and the reading of dimensional information. Dimensions relating to datums and features on a drawing are put in chart form.

A coordinate chart contains full information on size and location dimensions and other machining specifications.

Precision measurement depends on normalizing the temperature of the instruments, workpiece, and work surface.

Cleanliness is essential in the care, use, and storage of precision instruments.

Safe practices are to be observed to protect optical instruments from rubbing across or hitting the surface of a part.

UNIT 3 REVIEW AND SELF-TEST

1. a. Explain the importance of lenses in the design of toolmaker's measuring microscopes.
   b. Describe a design feature on a toolmaker's microscope that makes it possible to make measurements to accuracies of 0.001" (0.02mm) and finer.
   c. Give two applications of a toolmaker's microscope that is equipped with a vernier angle-measurement scale.
2. List three scale mounts that are interchangeable on an industrial measuring magnifier.
3. Explain light refraction and the principle of interference bands as they relate to optical flats and comparison precision measurements.
4. Interpret what each of the following six patterns of interference bands means in relation to a measurement or the accuracy of a machined surface:
   a. Straight, parallel, uniform
   b. Off-centered circles, darker at the center
   c. Concentric, full circles
   d. Wedge-shaped with a rounded crest
   e. Straight, parallel, uniform bands that curve slightly at the ends
   f. Elliptical circles.

OPTICAL FLATS AND MICROSCOPES 61

5. a. State two advantages of using tabular dimensioning in place of conventional dimensioning on part drawings that have a great number of design features.
   b. Identify the kind of information normally contained in a coordinate chart on a part drawing.

6. Give one reason for normalizing the temperature of optical measuring flats, other instruments, and the workpiece in a measurement, layout, or precision inspection setup.

7. List two precautions to observe when industrial microscopes and optical flats are used.

UNIT 4

Measurement with High Amplification Comparators

Mass production depends on the interchangeability of parts. Machining and inspecting processes for such parts are often carried on independently of the place where the parts are assembled. The parts must function according to specifications and without further fitting. Interchangeability depends on accurate, rapid, and economical methods of measurement. Preceding units dealt with direct measurements taken largely from the scales on precision mechanical and optical instruments.

This unit deals with comparison measurements that require the use of *high amplification comparators*. Comparison measurements play an important function in mass production. The part is compared with a master or standard. The standard represents the basic dimension to be checked.

A comparator is a highly sensitive instrument that checks the accuracy of a measurement for any variation from a standard. It incorporates within its design:

—A device for holding a workpiece,
—A master against which the workpiece is checked,
—An amplification unit that permits small variations from basic dimensions to be observed and measured.

GENERAL TYPES OF HIGH AMPLIFICATION COMPARATORS

Six general types of high amplification comparators are available:

—*Mechanical comparators* include high amplification dial indicators and bench comparators.
—*Mechanical-optical comparators* are combination mechanical-optical instruments. A light beam casts a shadow on a magnified scale, and dimensional variations are indicated on the scale.
—*Optical comparators* are projection and reflection comparators. Features are measured by casting a magnified shadow of a part on a screen and then using a chart for comparison measurement.
—*Electrical-electronic comparators* are measuring instruments that have the added feature of power amplification.
—*Pneumatic comparators* require a flow of air between a part and a measuring gage. Pressure and flow changes indicate any variation in dimensional accuracy.
—*Multiple gaging comparators* are a combination of mechanical, electronic, optical, and pneumatic comparators. Multiple gaging comparators are used when a number of dimensions are to be gaged at one time.

Principles underlying each type of comparator, with general applications, are covered in this unit.

HIGH AMPLIFICATION MECHANICAL COMPARATORS

High Amplification Dial Indicators

The most commonly used dial indicators are graduated for readings in 0.001″ (0.02mm) and 0.0001″ (0.002mm). The mechanical gearing system used limits the range of accuracy. When the rack and pinion is replaced with a precision steel ball that rolls against a sapphire-lapped surface, the normal error within a dial indicator is eliminated. The design permits extremely fine movements of the indicator spindle to be transmitted directly to the remaining gear train, where lost motion is negligible.

Two mechanical comparators, high amplification dial indicators, are shown in Figure 4-1. The dial indicator pictured in Figure 4-1A has a

HIGH AMPLIFICATION COMPARATORS 63

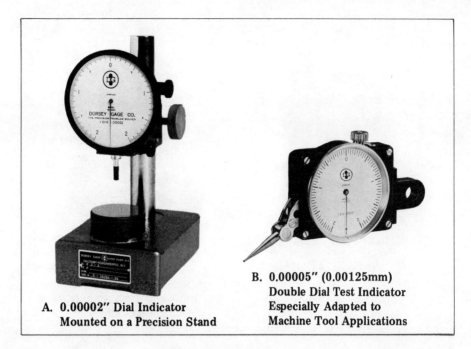

FIGURE 4-1 Mechanical Comparators (High Amplification Dial Indicators)

A. 0.00002" Dial Indicator Mounted on a Precision Stand

B. 0.00005" (0.00125mm) Double Dial Test Indicator Especially Adapted to Machine Tool Applications

discrimination of 20 microinches. Jeweled bearings are used in the instrument to reduce friction and wear and to ensure *repeatability*—that is, the dial indicator reads (after a number of repeated insertions of a workpiece) to a standard of accuracy of ±one-fifth of a division from an original reading. The dial indicator here is graduated so that each division equals 0.00002" (0.0005mm); the repeatability is 0.000004" (0.0001mm). The indicator head is mounted on a precision stand by using a standard mounting attachment. The indicator head may be adjusted vertically to a required height and locked in position.

A precision double dial test indicator, which is particularly suited to machine tool applications, is shown in Figure 4-1B. This indicator has a balanced 0.004" (0.01mm) dial (0.002-0-0.002"). The graduations are in 0.00005" (0.00125mm). The repeatability error of the instrument is ±0.00001" (0.0002mm). The round steel or carbide contact points are available in three diameters: 0.040", 0.080", and 0.120" (1mm, 2mm, and 3mm).

Mechanical Comparators

Two other mechanical comparators are shown in Figure 4-2. The comparator in Figure 4-2A consists of a base, column, adjusting arm, and a gaging head. A mechanical comparator works on the same principles as a dial indicator. It is used for taking accurate linear measurements by comparison with a standard. Any dimensional variation between the part being measured and the standard against which the instrument is set is shown on a magnified scale. The scale is graduated with plus and minus inch or metric units of measurement. The mechanical comparator in Figure 4-2B provides for extremely high amplification.

HOW TO MEASURE WITH A MECHANICAL COMPARATOR

STEP 1 Check the anvil and gage block for nicks or scratches. Clean the instrument and gage block.

STEP 2 Slide the gage block (master) carefully on the anvil.

STEP 3 Lower the gaging head until the contact point just touches this block.

STEP 4 Lock the gaging head in position.

STEP 5 Use the fine adjusting knob to adjust the pointer. Adjust until the scale reading is zero.

STEP 6 Recheck the setting.

STEP 7 Remove the master (standard). Replace it with the part to be measured.

64 PRECISION MEASUREMENT AND INSPECTION

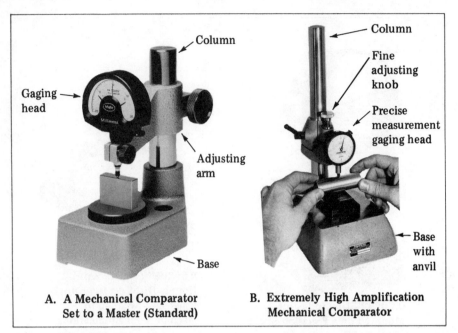

A. A Mechanical Comparator Set to a Master (Standard)

B. Extremely High Amplification Mechanical Comparator

FIGURE 4-2 Mechanical Comparators

Caution: The size of the part must be within the movement range of the instrument.

STEP 8 Note the reading on the scale. A plus reading indicates an oversize measurement. A minus reading shows the amount the part is undersize.

STEP 9 Remove the part carefully. Replace the gage block in a case. Cover the comparator.

MECHANICAL-OPTICAL COMPARATORS

Reed Comparator

A *reed comparator* (Figure 4-3) combines the features of a mechanical instrument and an optical instrument. It consists of a base, column, and a gaging head and bracket. The main design features of a reed comparator are shown in Figure 4-3A. A lamp is required in the magnification system. This lamp is housed in the base. The base supports an anvil, a back stop, and a column. The gaging head is moved by a rack

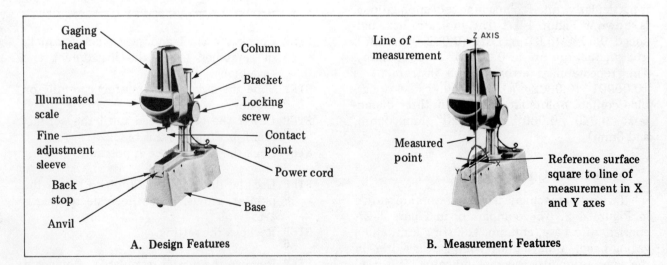

A. Design Features

B. Measurement Features

FIGURE 4-3 Design and Measurement Features of a Reed Comparator

HIGH AMPLIFICATION COMPARATORS 65

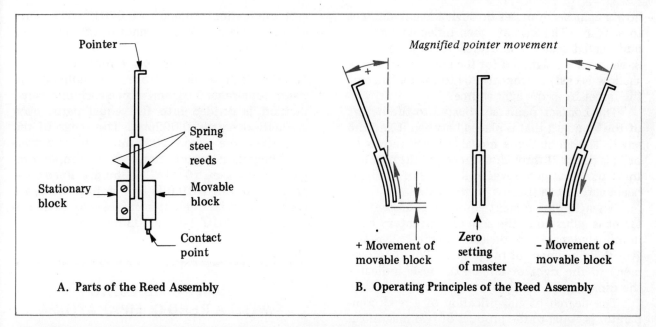

FIGURE 4-4 Schematic of Reed Assembly and Magnified Pointer Movement

and pinion on the column. The locking screw clamps the gaging head in position. The gaging head may also be swung away from the base so that parts that are larger than the working height of the comparator can be measured.

The measurement features are shown in Figure 4-3B. The reference surface for the Z axis is at a right angle to the line of measurement in the X and Y axes.

Reed Mechanism. A reed comparator derives its name from the use of two vertical spring steel strips, called *reeds*, that are attached to two steel block. One steel block is stationary; the other block is movable. The reeds are joined to a pointer. Figure 4-4A shows the parts of the reed assembly.

Figure 4-4B illustrates the principle of operation. When the movable block is moved up, the pointer is deflected to the left. By contrast, when the movable block is in a down position, the pointer is deflected to the right. Since no movement losses due to any gear system occur, it is possible to obtain a higher degree of accuracy with a reed comparator than with a dial indicator.

Another part of the reed comparator is the *light beam (optical) lever*. The optical lever further amplifies the reed motion into a measurement that is read on the scale. Figure 4-5 shows

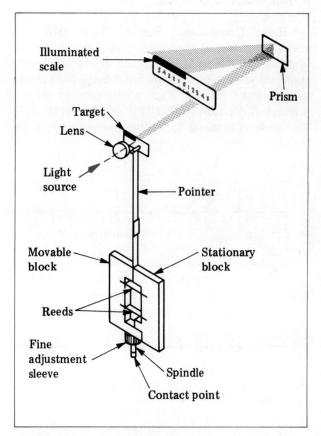

FIGURE 4-5 Amplifying Reed Motion with a Light Beam (Optical) Lever

how a light beam passes through a lens and target to a prism. The rays are then reflected onto an illuminated scale. The degree to which pointer movement is magnified (or the area over which a shadow is cast) is proportional to the distance of the target from the light source.

The contact point is adjusted to the height of the standard that is placed between it and the anvil. The amount the movable block moves (+ or -) is directed through the reeds to the pointer. In turn, the target moves in the field of the light beam and casts a shadow on the scale.

When the instrument is set on a master, the target is adjusted to the zero at the center of the illuminated scale. The divisions on the scale represent the movement of the pointer. Pointer movement to the right or left on the scale indicates the dimensional variation from the required size.

The degree of magnification of a reed comparator is equal to the product of two magnification units. A reed mechanism with a magnification of 100X and a light beam magnification of 40X produces a total magnification of (100 × 40) 4000X. The general range of reed comparators is from 500X to 20,000X.

Reed Comparator Scales. Two different scales for a sample reed comparator are shown in Figures 4-6A and B. The note above the inch-standard scale indicates the instrument has a magnification of 1000X. Each whole plus or minus numerical division (0.001") is divided in tenths. The scale is marked 1/10,000 and indicates that each graduation is equal to 0.0001". The range of the scale is ±2 thousandths (±0.002"), or 0.004" overall.

The metric scale also has a magnification of 1000X. Each whole plus or minus numerical division represents 0.01mm. Since each numbered division is divided into five equal parts, each graduation equals 0.002mm. The range of the scale used here is from ±0.05mm or 0.1mm overall. The plus and minus range of the sample comparator scales are ±0.002" (0.05mm). Reed comparators, having a 10,000 to 1 amplification, are capable of measurement accuracies to 10 microinches (0.000010") or 0.00025mm.

HOW TO MEASURE WITH A REED COMPARATOR

STEP 1   Clean the anvil, contact point, and workpiece.

STEP 2   Unlock the locking screw. Move the gaging head so that the contact point clears the workpiece.

STEP 3   Select a gage block having the required dimension of the workpiece (or gage).

STEP 4   Mount the gage block centrally on the anvil.

STEP 5   Bring the gaging head down carefully until the contact point of the spindle touches the gage block.

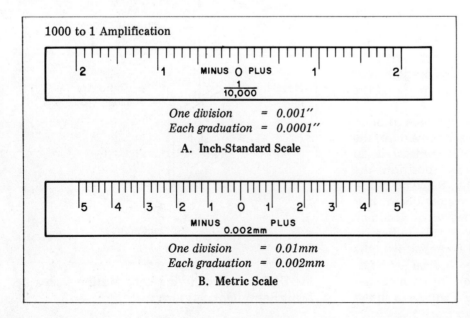

FIGURE 4-6 Sample Reed Comparator Scales in Customary and Metric Units of Measure

STEP 6 Clamp the gaging head to the column. Make further adjustments with the adjusting sleeve. The shadow on the illuminated scale should coincide with the zero on the scale.

STEP 7 Replace the gage block with the part to be measured.

Caution: Slide the part carefully into position under the pointer. The height of the part must fall within the range of the instrument.

STEP 8 Note the reading on the scale. A reading on the plus (+) side indicates the amount the part is oversize. Similarly, a reading on the minus (−) side shows the part is smaller than the standard size.

STEP 9 Remove the workpiece. Turn the electric switch off. Return the gage or standard to a proper storage area. Cover the reed comparator.

OPTICAL COMPARATORS

Optical measuring instruments have great versatility, range, precision, and accuracy. One of the most popular and widely used optical measuring instruments is the *optical comparator*. It is sometimes called a *contour projector*. This projector provides an accurate method of measuring and comparing the contour of irregularly shaped parts. Flat or circular tools, gages, grooves, angles, and radii may be checked for size. Metal, plastic, and soft materials such as rubber may be measured without physical contact. Tolerances within ±0.0005″ (±0.01mm) may be measured accurately.

Principles of Operation

The optical comparator projects an enlarged shadow onto a viewing screen (Figure 4-7). The screen is especially designed for mounting a chart. The chart may be an accurate, enlarged drawing of the contour of the part to be measured. The chart may also include lines, angles, radii, or combinations of drawings and lines. The shadow projected on the screen is compared to the chart. The chart contains information about the limits of the dimensions or contour. The optical comparator is especially valuable in checking extremely small and odd-shaped parts that are difficult to check with other instruments.

Optical comparators are designed in bench or floor models. The principles of operation—*projection* and *reflection*—and the actual use of these models are identical. Each comparator has a light source that is passed through a condensing lens. The light rays are projected against a workpiece. A shadow is produced and transmitted through a projecting lens system. The lens system magnifies the image and projects it onto a mirror. The image is further magnified as it is reflected from the mirror to the viewing screen. Interchangeable lenses provide different magnifications.

Viewing Screen

The viewing screen of an optical comparator has either a comparator chart or master form mounted on the face. The accuracy of the enlarged image of a workpiece is compared with the chart or master form. The viewing screen may be made of clear, ground glass with 90° cross lines and 30° angle lines. The screen may have a rotary protractor ring with graduations. These graduations are used to measure angles to an accuracy within either a five minute or a one minute of arc, using a vernier scale.

FIGURE 4-7 Application of Optical Comparator (with Digital Readouts) for Contour and Size Measurements

Stage and Accessories

The stage of an optical comparator may be moved horizontally or vertically. A work table provides a plane surface for accessories and workpieces. The table is provided with a large-diameter micrometer head that permits horizontal movement of the stage to accuracies within 0.0001" (0.002mm). Vertical travel is obtained with a dial indicator. A work stage that is moved by turning micrometer dials permits the accurate measurement of linear dimensions. Fixture bases with a small rotary vise or adjustable horizontal and center attachments extend the use of the comparator. Fixtures provide an exact location for a part and a master.

Tilting work centers are available. They permit a workpiece to be swung to the required helix angle to check threads. In addition, gage blocks, measuring rods, and dial indicators may be used as accessories to add other measurement capacity. For example, the optical comparator and thread chart are used with some of the accessories to measure thread depth and major and minor diameters and to inspect the crest and root.

The general range of magnification of optical comparators is from 5X to 100X. Some models are available with higher ranges. Note that, with each magnification, it is important to use a chart that has the same scale of magnification as the lens on the comparator.

Comparator Charts

The accuracy of comparator measurements depends on the accuracy of the comparator charts. The lines on a chart must all be the same width to permit taking off accurate measurements on features from corresponding edges of any lines. Sometimes a chart has small breaks in the lines to form an easy-to-follow pattern. Numbers appear in open areas and simplify the reading on the chart. Measurement accuracy is thus ensured.

The charts may be made of plate glass for critical inspection. Plate glass is dimensionally stable and thus provides for the most accurate type of chart. Plastic is recommended for overlay-type charts. These charts are subject to constant handling. The surface of plastic charts is not affected by the oils used in the shop or laboratory. However, because they are thermoplastic (capable of distortion when heated), such charts must be kept away from concentrated heat.

Each chart is identified according to its magnification and linear increment. The magnification—for example, 5X, 10X, 31.25X, and 125X—is followed by the linear measurement (distance between lines on the chart). Sample comparator charts are shown in Figure 4-8. The charts most commonly used are for linear, radius, and angular measurements (Figures 4-8A, B, and C, respectively). The magnification and linear range are given on some of the charts shown in Figure 4-8. Each magnification must correspond with the magnification of the lens system used in the comparator. An American National Form basic thread chart is shown in Figure 4-8D. While almost indistinguishable in reduced size, the chart is furnished in 14" and 30" diameters in magnifications from 10X to 62.5X.

HOW TO CHECK A THREAD ANGLE BY USING AN OPTICAL COMPARATOR

STEP 1 Select a lens of sufficient magnification to permit measuring to the required degree of accuracy.

STEP 2 Position the tilting centers on the table.

STEP 3 Set the centers at the helix angle of the thread.

STEP 4 Mount the workpiece between the centers.

STEP 5 Place a thread form chart, if it is a replacement chart, into the screen recess.

Note: A sealed-image overlay chart is placed on top of the screen.

STEP 6 Align the chart to permit accurate vernier angle readings.

STEP 7 Focus the lens with the light turned on. The part is in focus when a clear image is produced.

STEP 8 Center the thread image on the screen. Move the micrometer cross slide stage to position the thread.

STEP 9 Turn the vernier protractor chart to one-half the thread angle.

STEP 10 Adjust the cross slide until the image of one slope of the thread angle coincides with the protractor angle.

STEP 11 Check the other slope of the thread by repeating steps 9 and 10.

STEP 12 Read any variation of the included thread angle directly from the graduated angles on the chart.

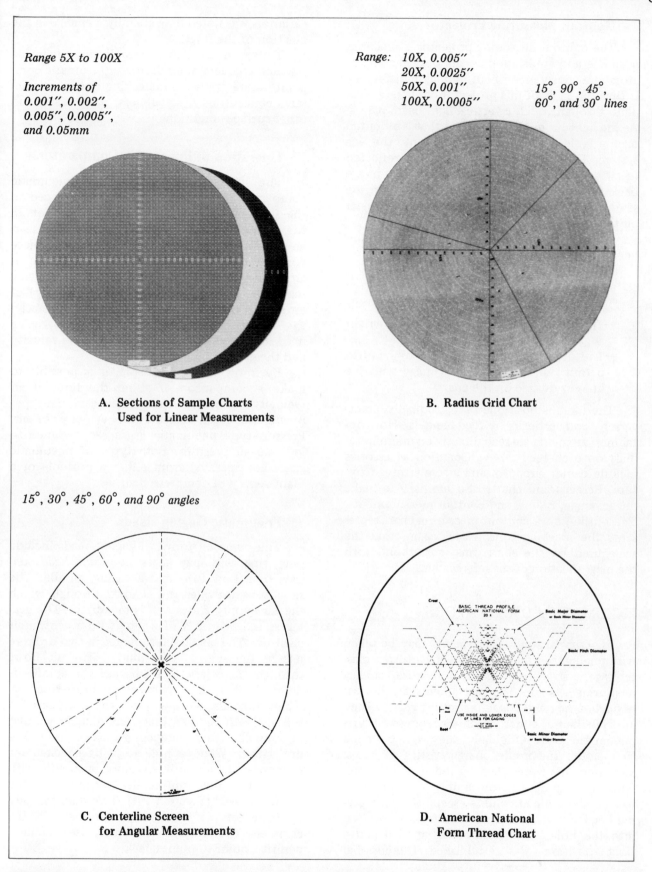

FIGURE 4-8 Samples of Charts Used on Optical Comparators

Contour Measuring Projector

The contour measuring projector is similar in principle and application to the optical comparators previously described. A major difference is that the table of this projector contains a glass plate through which a vertical light beam passes. As the light beam hits the object that is mounted on the table, a shadow is projected into the lens system. The enlarged image produced is reflected to two reflectors that, in turn, reflect the image back to the screen. The image is visible in the same position as the operator views the object on the table.

The contour measuring projector has a number of advantages:

—The need to hold the object in special tools or fixtures is eliminated,
—Setup time is reduced,
—The object is placed automatically at a right angle to the beam of light,
—The object may be positioned by hand to permit rapid setup of the image with the lines or outline on the chart.

The table is positioned longitudinally, transversely, and vertically by feed handles. Horizontal measurements are read directly on micrometer dials on each feed screw. Common accessories include center supports and a screw thread fixture. Screens and charts also are used to make linear, angle, radius, and contour measurements. The outline on a contour measuring chart represents the ideal contour. Note, again, that the scale used on the chart must correspond with the magnification of the lens system.

PNEUMATIC COMPARATORS

Precise comparison measurements may be taken with pneumatic comparators, which utilize a master gage, a stream of air, and a pressure/velocity measuring system. The instruments are known as *pneumatic comparison measuring gages.* They are classified as either air (flow) or pressure-type pneumatic gages. Originally, they were designed to measure the inside characteristics of holes. Measurements were taken by measuring the volume of escaping air from a hole in comparison to a predetermined volume escaping from a gaging head standard. The gaging head was inserted into the hole. Air pressure was applied and a float position was established. Dimensional variation was noted on a scale by a change in the position of the float.

Today, in addition to establishing the inside size and characteristics of holes, pneumatic comparators are applied in checking outside diameters, concentricity, squareness, parallelism, and other surface conditions.

Operation of Pneumatic Comparators

The principle of operation of pneumatic comparators is simple. Dependence is placed on the flow of air between the faces of the jets in the gaging head and the workpiece. The clearance between the gaging head and the workpiece controls both the velocity and the pressure of air. The larger the workpiece is than the gaging head, the greater the flow (higher velocity) of air and the lower the back pressure. Conversely, the smaller the clearance between the gaging head and the workpiece, the slower the velocity and the greater the back pressure.

By knowing this principle, it is possible to make a comparison of either the flow of air (velocity) or its pressure. Flow (column-type) pneumatic gages indicate the velocity of air. Pressure-type pneumatic gages are designed to indicate air pressure. Both types of pneumatic gages are operated from either a portable or a plant supply of compressed air.

Pneumatic Gaging Heads

The common forms of gaging heads include plug, ring, and snap gages. A *gaging head* consists of a flow tube and a gaging spindle. The flow tube provides the channel through which the air supply flows and is measured. The gaging spindle consists of a plug with a central air channel. The channel terminates in two or more jets in the side of the spindle. The air flows through these jets. The form of the spindle is determined by the operation to be performed. When the diameter of a hole is to be measured, a spindle with air jets that terminate in annular (around the circumference) grooves is used (Figure 4–9A). When a hole is to be checked for roundness, bell-mouth, or taper, a spindle with air channels that run parallel to the axis of the spindle is used (Figure 4–9B). Two diametrically opposite jets or more may terminate in the channels. The longitudinal (lengthwise) channels permit a point-to-point check.

Advantages of Pneumatic Comparators

The major advantages of gaging with pneumatic instruments over other measurement instruments are as follows:

— Amplifications range from 1 to 1000 to 1 to 40,000. Pneumatic instruments provide a simple and direct method of high amplification.
— Minute dimensional variations can be controlled within close tolerances.
— Parts may be gaged without contact between the workpiece and the gaging head. Thus, scoring of highly polished surfaces or damage to soft material parts is prevented.
— Less skill is required in using pneumatic gaging instruments than in gaging processes with precision conventional plug, ring, and snap gages.
— The work life of pneumatic gaging heads is longer than for conventional gages.
— Pneumatic gages may be used at a machine or on the bench. The work can be brought to the instrument, or vice versa.
— Multiple diameters may be checked at the same time.
— Pneumatic gages are easier to use than mechanical gages. Holes may be inspected for out-of-roundness, parallelism, concentricity, and other surface irregularities.
— Gaging heads and parts are self-cleaning.

OPERATION OF A FLOW (COLUMN-TYPE) PNEUMATIC COMPARATOR

The flow, or column-type, pneumatic comparator measures air velocity. The major components of such a comparator are shown in Figure 4-10. Air is passed through a filter and a regulator. The air flows through a tapered, transparent tube and reaches the gaging head at about 10 pounds pressure per square inch. The air flow causes a float in the tube to be suspended.

As air flows through a metering gage, it exhausts through the channels of the gaging head and the clearance area between the head and the workpiece. The rate of flow is established by the clearance as indicated by the position of the float in the tube. Figure 4-11 is a phantom drawing of a column-type pneumatic comparator.

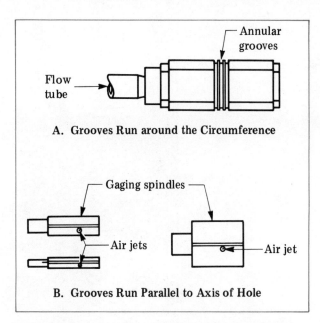

FIGURE 4-9 Features of Pneumatic Gaging Heads

Air flow from the filter through the regulator, column, and plastic tubing to the gaging head is shown clearly. While various manufacturers produce different pneumatic gages, all gages are governed by the same basic principle—that is, the change in flow (or pressure) of air between the workpiece and the gaging head indicates any variation in dimensional accuracy.

Setting Upper and Lower Limits

Master setting heads can be used to set the position of the float. The float is positioned by an adjusting knob. The upper and lower limits of a hole are then set. According to the principle described earlier, any variation from these limits is governed by the amount of air that flows through the gaging head. If the amount of air is great enough to cause the float to rise over the scale marker, an oversize dimension is indicated. Conversely, if less air flows and the float falls below the lower limit marker, an undersize dimension is indicated. The scale on the side of the instrument is graduated in inch or millimeter units of measure. The location of the float in relation to the scale indicates the size of the feature that is being measured in the workpiece.

72 PRECISION MEASUREMENT AND INSPECTION

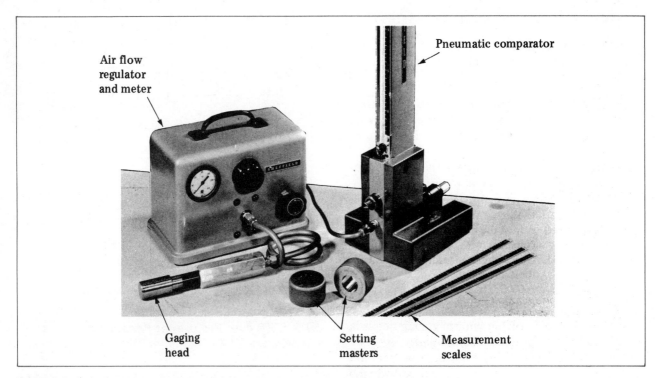

FIGURE 4-10 Major Components of a Flow (Column-Type) Pneumatic Comparator

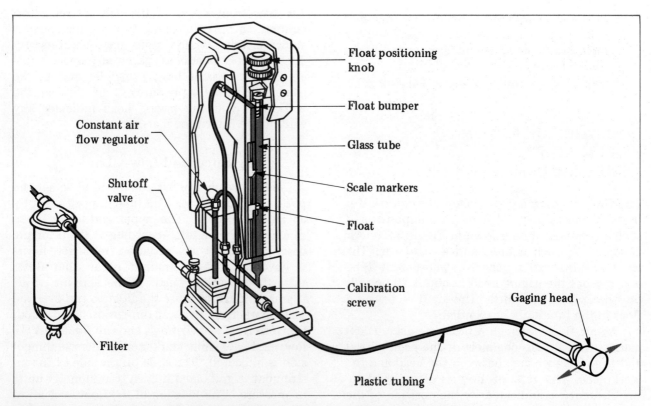

FIGURE 4-11 Phantom Drawing of Air Flow Circuit and Principal Controls of a Column-Type Pneumatic Comparator

OPERATION OF A PRESSURE-TYPE PNEUMATIC COMPARATOR

The major features of a pressure-type pneumatic comparator gage system appear in Figure 4-12. A constant air supply passes through a filter, a pressure regulator, and a master pressure gage. The supply of air is then branched through two channels. The upper branch is identified as the reference channel. The lower branch is identified as the measuring channel. Air escapes from the reference channel to the atmosphere through a zero setting valve. Air from the measuring channel flows through the gage head jets and the part being measured and then into the atmosphere. Note that the two channels are connected through a differential pressure meter.

Setting Up the Gage System for Measurement

The pressure-type pneumatic comparator gage system is set by placing a setting master over the gaging spindle plug (Figure 4-12). The zero setting valve is adjusted until the gage pointer (needle) reads zero. Any variation between the size of the master gage and the workpiece affects the pressure in the measuring channel. The difference is indicated by the movement of the dial gage pointer.

A reading on the right side of the scale indicates how much the diameter of the workpiece is smaller than the master size. If the diameter is oversize, additional air escapes from the gaging plug. Pressure in the measuring channel becomes less and causes the dial gage pointer to move counterclockwise. The scale indicates the amount the workpiece is oversize.

The range of amplification of pressure-type pneumatic comparators is from 1250 to 1 to 20,000 to 1. A common pressure-type pneumatic comparator is pictured in Figure 4-13. The data in Table 4-1 indicates the range of amplification, discrimination, and measurement with this type of instrument in a *balanced system*. With increased amplification, a higher level of discrimination and a lower measuring range result.

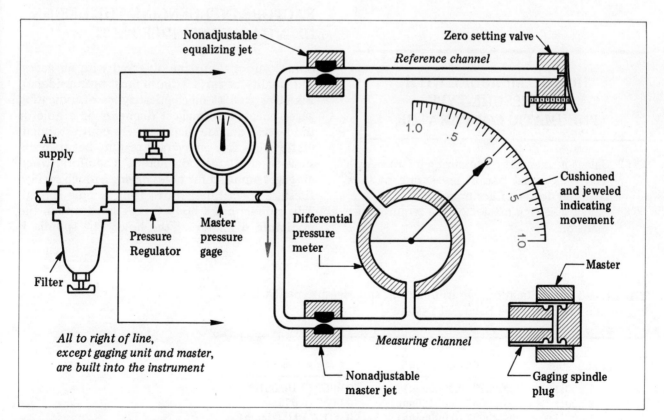

FIGURE 4-12 Main Features of a Pressure-Type Pneumatic Comparator

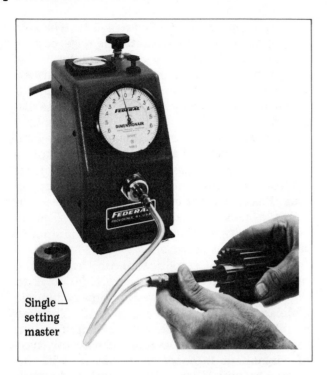

FIGURE 4-13 Measuring a Diameter with a Pressure-Type Pneumatic Comparator and Gaging Spindle

HOW TO MEASURE WITH A PRESSURE-TYPE PNEUMATIC COMPARATOR

STEP 1   Select a pneumatic measuring instrument. The scale must permit measuring to the required degree of accuracy.

STEP 2   Select a setting master of the required dimension.

STEP 3   Mount the gaging head corresponding to the required dimension. The gaging head is attached to the plastic tubing.

STEP 4   Supply a flow of air. Set the pressure gage.

STEP 5   Place the setting master over the gaging spindle plug. Adjust the dial pointer until it indicates zero.

STEP 6   Remove and store the master.

STEP 7   Insert the gaging spindle plug in the hole in the workpiece.

STEP 8   Note any variation from the original zero reading on the dial face.

Note: A clockwise dial pointer movement indicates how much the hole is undersize. A counterclockwise movement indicates an oversize measurement.

STEP 9   Shut off the air supply at the end of the measuring process. Disassemble the gaging spindle plug. Store in an appropriate container.

FACTORS INFLUENCING THE DESIGN OF A GAGING SPINDLE PLUG

Some cautions must be observed with air gages. The quality of surface finish must be considered. Fixed as well as mechanical types of indicating gages check the smallest diameter of a hole or the largest outside diameter. By contrast, pneumatic gages give an average reading between the smallest and largest diameters. Therefore, when air gages are used for holes having a rough surface finish, a special design of gaging spindle is used. This spindle has a body that is smaller than the minimum diameter of the hole. The spindle is

TABLE 4-1 Amplification, Discrimination, and Measurement with a Pneumatic Pressure-Type Balanced System

	Ranges	
Amplification	Discrimination	Measurement
1,250:1	0.0001" (0.0025mm)	0.006" (0.15mm)
2,500:1	0.00005" (0.00125mm)	0.003" (0.075mm)
5,000:1	0.00002" (0.0005mm)	0.0015" (0.0375mm)
10,000:1	0.00001" (0.00025mm)	0.0006" (0.015mm)
20,000:1	0.000005" (0.000125mm)	0.0003" (0.0075mm)

fitted with a hardened steel contact blade. The blade either pivots or floats. The action of the contact blade controls the air flow through the jets. The gage then serves as a regular, solid plug gage.

The features of a hole that are to be measured determine the design of plug to use on a spindle. Five different designs are shown in Figure 4-14:

- A single-jet plug is used to check concentricity, location, flatness, squareness, length, and depth.
- A two-jet plug is applied in checking inside diameters and holes that are out-of-round, bell-mouth, or taper.
- A three-jet plug is used for checking triangular out-of-round conditions.
- A four-jet plug provides average diameter readings.
- A six-jet plug produces average measurements for the conditions indicated for two- and three-jet plugs.

ELECTRICAL COMPARATORS

The three major components in an electrical comparator are a gaging head, a power unit, and a microammeter. The microammeter contains a graduated scale. Any movement of the end of the gage head is magnified greatly by the microammeter.

The spindle (lever) of the gage head bears against an armature. The end of the armature is positioned between two electrical induction coils. Current flows between the two coils. When the armature is midway between the two coils, the circuit is balanced. The scale on the instrument reads zero.

Any minute variation in the size of a workpiece causes the spindle to move and change the circuit balance. The amount of movement on the scale is proportional to the movement of the spindle. The magnification factor of the instrument affects the value of each division of the scale. For example, if the magnification of the instrument is 10,000, each division of the scale equals 0.000010″ (10 millionths of an inch) or 0.00025mm (25 hundred-thousandths of a millimeter). The total measurement range on the scale of such an instrument is usually 0.0005″ (0.01mm).

ELECTRONIC COMPARATORS

The versatility of the electronic comparator makes it an ideal instrument for comparison measurements. Standard components may be used in a great number of measurement problems. The absence of mechanical parts makes electronic instruments comparatively rugged as high precision instruments. Also, their sensitivity and speed reduce the time lag during measurements. Electronic comparators are particularly useful, therefore, when measurements of a moving feature are required. Measurements along a strip passing through a rolling mill are a typical example of *dynamic measurement*.

Other considerations for selecting an electronic comparator are as follows:

- Rapidity of operation even at high amplifications,
- Use of the same instrument for multiple amplification ranges,
- Portability of the instruments, which are also self-contained,
- Easily adjusted and understandable controls,
- High instrument sensitivity in all amplification ranges,
- Limited self-checking,
- Instrument accuracy that compares favorably with other instruments.

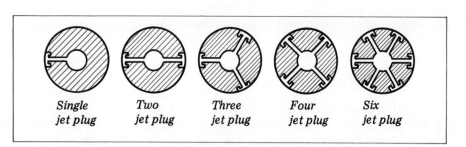

FIGURE 4-14 Cross Sections of Gaging Spindle Plugs with Jets to Meet Specific Measurement and Inspection Requirements

Single jet plug *Two jet plug* *Three jet plug* *Four jet plug* *Six jet plug*

76 PRECISION MEASUREMENT AND INSPECTION

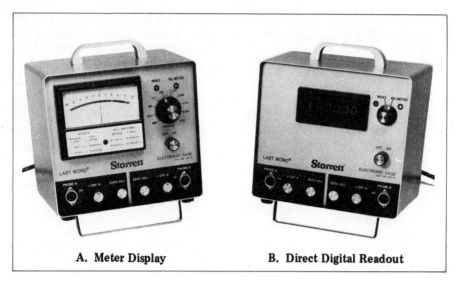

A. Meter Display B. Direct Digital Readout

FIGURE 4-15 Electronic Gaging Amplifiers

Electronic Amplifier and Readout

The two major components in an electronic measurement unit are a gage amplifier and one or more gaging heads. The amplifier and a single gaging head serve as a comparator and height gage. In applications where two heads are used, measurements may be taken of differences in size, length, thickness, diameter, flatness, taper, or concentricity. In such applications, the amplifier measures either the difference or the sum of the gaging head outputs.

Electronic gage amplifiers have magnification ranges from ±0.010″ (±0.200mm) to ±0.0001″ (±0.002mm). They are designed to operate by plugging into a 115/230 volt, 50-60H_z AC outlet or from a self-contained rechargeable battery. The electronic comparator is portable and may be carried to any location in a shop or laboratory.

Amplifier readouts are provided on either a sensitive meter (Figure 4-15A) or a direct digital readout display (Figure 4-15B). The meter version has a precision-calibrated meter dial that is designed for easy readability without parallax error. Both amplifiers contain solid-state circuitry. The result is extreme reliability from variations of temperature, line voltage, and repeatability of measurement. There is instantaneous warm-up.

Electronic Gaging Heads

Input into an electronic amplifier is provided by the movement of one or more gaging heads. The two basic types of gaging heads are identified in Figure 4-16 as a *lever (probe) type* and a *cartridge (cylindrical) type*. The gaging heads are actuated by a contact of 8 to 12 grams of pressure against an object.

Lever (Probe) Type Gaging Head. The lever-type gaging head has a wide range of applications. Two sample applications are shown in Figures 4-17A and B. The head is adjustable to clear obstructions and reach into places that otherwise are inaccessible, as illustrated in Figure 4-18. The four major parts in a gaging head are the outer shell, or housing, the transducer, the probe tip, and the lever assembly. The transducer has matched coils and a core and is designed

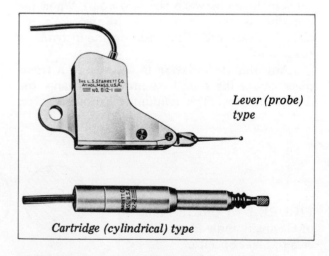

Lever (probe) type

Cartridge (cylindrical) type

FIGURE 4-16 Two Basic Types of Gaging Heads

HIGH AMPLIFICATION COMPARATORS 77

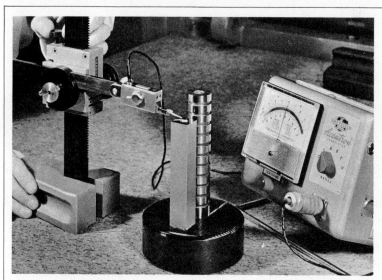
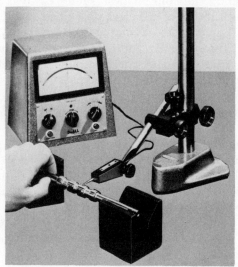

A. Mounting on Height Gage Attachment to Check Gage Block

B. Use of Granite V-Blocks and Height Gage Measurement Unit Setup to Record Concentricity

FIGURE 4-17 Applications of Lever (Probe) Type Gaging Head

as a compact unit. The transducer is sensitive to fine measurements. Movements of the probe tip are converted into electrical energy in the transducer. These movements are amplified in the electronic amplifier. Direct readouts are indicated by the pointer movement on the amplifier scale.

Cartridge (Cylindrical) Gaging Head. The diameter of the cartridge (cylindrical) gaging head has been standardized. Thus, the gaging head may be substituted in holders that accommodate the stem of dial indicators. Thus, the cartridge gaging head may be used interchangeably for many dial indicator applications. The advantage of using the cartridge head is that the contact moves in a frictionless head. The transducer components in the cartridge and lever gaging heads are identical.

Comparator Stands

In many applications, the gaging head of an electronic comparator is held in an adjustable

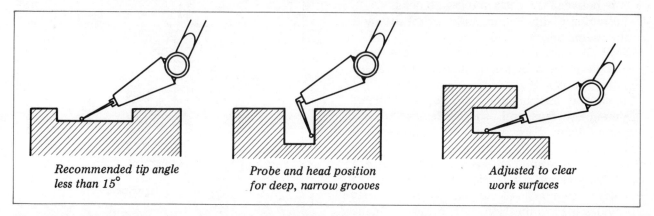

Recommended tip angle less than 15°

Probe and head position for deep, narrow grooves

Adjusted to clear work surfaces

FIGURE 4-18 Positioning of Lever (Probe) Gaging Head

stand. The combination is referred to as a height gage setup or a height gage instrument. Three different types of comparator stands are pictured in Figure 4-19. The bases of the stands shown in Figures 4-19A and B consist of a heavy-duty casting. The columns are hardened, ground, and nickel chrome plated. One stand (Figure 4-19A) has a serrated anvil. The other stand (Figure 4-19B) is equipped with a flat and serrated reversible anvil. The anvils are hardened, ground, and lapped steel blocks. The surfaces of the anvils provide precise reference planes. Both stands provide for maximum rigidity and stability in comparator gaging.

The height gage stand shown in Figure 4-19C is also designed with a solid base and a rugged column. Sometimes the base is made to rest on tungsten carbide lapped pads. An adjustment assembly and gage head holder on the height gage stand permit easy positioning in any angular relationship with the base.

SAFE PRACTICES IN THE USE AND CARE OF HIGH AMPLIFICATION COMPARATORS

—Investigate the potential sources of error before making high precision measurements. Check for the following conditions:
  -Direct heat from sunlight;
  -Electric lights or drafts;
  -Outside vibrations;
  -Heat transfer due to handling the stand, parts, and gages;
  -Differences in materials, finish, and temperature between the parts and gage blocks;
  -Fastening of the locking nuts for the arm and the gaging head.
—Clean all reference surfaces before mounting a workpiece on a comparator anvil or surface plate.
—Check the instrument, workpiece, standard, reference surface, and other parts needed for a measurement. Each must be fully normalized before any precise measurement is taken.
—Repeat a measurement as a check. Compare results. Make adjustments, if required.
—Select a scale on an electronic comparator that provides a magnification of the measurement that is within the tolerance requirements.

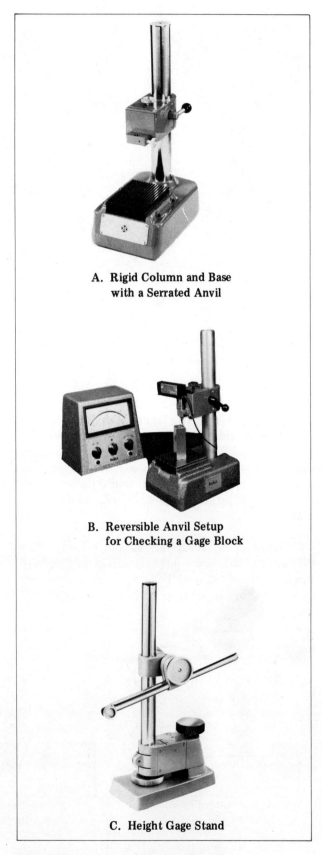

A. Rigid Column and Base with a Serrated Anvil

B. Reversible Anvil Setup for Checking a Gage Block

C. Height Gage Stand

FIGURE 4-19 Comparator Stands

— Use the same scale screen on a contour measuring projector that corresponds with the magnification of the lens system.
— Avoid temperatures above 140°F that may concentrate on plastic charts. Distortion may result.
— Measure holes that have a rough surface finish with a pneumatic spindle fitted with a hardened contact blade.
— Place the measuring instrument and setup on a vibration-free surface. Vibration may cause inaccuracies in measurement.
— Avoid handling the stand, parts, and gages. Handling may produce uneven temperatures.
— Use a rust inhibitor after cleaning each piece of equipment and before storing in an appropriate container.

TERMS USED IN MEASURING WITH HIGH AMPLIFICATION COMPARATORS

Reed-type instrument	A precision measuring instrument. An instrument in which a spring (reed) suspension is substituted for other mechanical systems.
Optical lever	A frictionless, weightless beam of light. A light beam used in a manner similar to a mechanical lever.
Electronic measuring instruments	Instruments in which amplification is obtained by an electronic system.
Amplifier	An intermediate modifying stage. A device for increasing (amplifying) electronic signals. (When used with different heads, it measures length, thickness, diameter, flatness, taper, and concentricity.)
Multiple-scale comparators	Instruments that may be switched from one scale of amplification to another. A range of scales to accommodate different degrees of measurement accuracy.
Lever (probe) gaging head	A small lever arm mounted to an electronic unit for magnifying small movements of the contact point. A lever arm that may probe into small openings for purposes of measurement.
Cartridge (cylindrical) gaging head	A cylindrical gaging head with a standardized body diameter. A moving member that travels within and along the axis of a gaging head.
High amplification comparators	A series of comparative measuring instruments. Instruments in which dimensional variations are multiplied. Instruments for amplifying measurements for purposes of obtaining a high degree of accuracy.
Comparative measuring instruments	Instruments that depend on an initial setting, using an outside master. Highly sensitive instruments for obtaining measurements registering any variation from a standard.
Optical comparator (contour projector)	A comparative measuring instrument. An instrument on which an enlarged object is projected on a screen. A master form against which a projected object is examined for contour conformance and accuracy of measurement.
Tilting work center	An optical comparator accessory. A set of centers that may be mounted on a work stage. Centers that permit a workpiece to be swung to the thread helix angle to measure thread features.
Pneumatic comparison measuring instruments	Instruments that require the flow and recording of the velocity or pressure of air to make comparison measurements. Instruments that translate the flow of air into a measurement in inch and metric standard units.

Gaging head	A combination of a flow tube and a gaging spindle. A head through which air flows out of one or more jets and through annular or longitudinal grooves.
Flow (column-type) pneumatic comparator (gages)	A comparator that measures air velocity. An instrument that establishes a measurement based on air velocity.
Pressure-type pneumatic comparator (gages)	A pneumatic gaging system in which measurement is based on an air pressure differential. A pneumatic gage that translates linear measurements based on a standard according to differences in air pressure.
Electrical comparators	Instruments that make measurements by changing mechanical movement into electrical energy. A system that uses a milliammeter in a circuit with a power unit and a gaging head. Instruments that permit reading a highly magnified measurement on a linear measurement calibrated scale.
Electronic comparator	An electronic measurement unit consisting primarily of a gage amplifier and one or more gaging heads. A device using solid-state electronic components to highly magnify and precisely measure linear dimensions.

SUMMARY

High amplification dial indicators have a discrimination of 20 millionths of an inch (0.0005mm).

The reed comparator works on the principle of two spring steel reeds. One reed is attached to a solid block; the other reed, to a movable block. Fine, precise movements are magnified on a pointer.

Additional magnification on a reed comparator is produced by passing a light beam through an aperture to reflect on a scale. Scale readings within 50 millionths of an inch (0.001mm) are common.

Optical comparators depend on the projection of a large image of the object being measured. The light rays against a workpiece are projected through a lens system into a mirror and onto a viewing screen. Shape and size characteristics are compared to an enlarged measurement chart on the screen.

Optical comparator accessories, such as tilting centers and fixtures, permit threaded and contour-formed workpieces to be positioned in a particular location. These accessories facilitate inspection and ensure measurement accuracy.

Commercially prepared optical comparator charts are available for measuring thread features, angles, radii, and other linear dimension combinations.

Pneumatic comparison measuring instruments depend on the measurement of either the flow or pressure of air.

Flow gages are used to measure the velocity of air. The air escapes between a workpiece and a gaging spindle plug. The larger the diameter of a hole than the gaging plug size, the faster the velocity of air.

Variations from a master size on a column-type pneumatic instrument are indicated by the position of a float in relation to a measuring scale. Master setting heads are used to set the position of the float.

Pressure-type pneumatic comparators are used to identify variations from specific dimensions based on changes in air pressure. A differential pressure meter measures pressure changes. The change is produced between a reference air channel and a measuring channel. The pressure is affected by the volume of air escaping between the workpiece and the gaging plug.

Plug, ring, and snap gages are three common forms of gaging heads.

The nature of the plug (gage) used with a reed comparator depends on the nature of the measurement. A single jet is used to check concentricity, location, flatness, squareness, length, and depth.

Two-jet plugs are needed to check inside diameters and out-of-round, bell-mouth, and taper conditions. Three-, four-, and six-jet plugs are used for other measurements.

The electrical comparator is a high magnification instrument. Measurements within 10 millionths of an inch (0.00025mm) are common.

Electrical comparators require a power unit, a gaging head, and a milliammeter. The position of a gaging head against an armature regulates the current flow between two induction coils. The minute variations in current flow are transmitted to a movable pointer on the milliammeter scale.

Electronic comparators permit the use of the same instrument for measurements at multiple high amplification ranges. The sensitivity and speed of operation permit dynamic measurement of a moving feature.

An electronic amplifier and readout system and one or more gaging heads may be combined in an electronic comparator.

Electronic comparator gaging heads are available in lever (probe) and cartridge (cylindrical) types.

A transducer unit in a gaging head transforms minute movements of the probe tip into electrical energy. The amount of movement finally appears as a readout.

Comparator stands are precisely machined with lapped work surfaces. Precision and rugged construction permits positioning and holding gaging heads securely.

Pressure (pneumatic) gages measure the average area produced by the surface finish. It is necessary in measuring roughly machined holes to use a gaging spindle that has a floating contact blade.

The dependence on an extremely high degree of dimensional accuracy and measurement requires that all instruments be checked thoroughly against master gages.

The work surface, parts, and instruments must be thoroughly cleaned, kept at the same temperature, and properly mounted and secured.

The work area should be vibration free.

The measurement scale selected must provide a reading to the required degree of accuracy.

The chart used on an optical comparator must be of the same magnification as the image projected.

A rust inhibitor (where required) should be applied to all instruments and workpieces before they are stored.

HIGH AMPLIFICATION COMPARATORS

UNIT 4 REVIEW AND SELF-TEST

1. Name four general types of high amplification comparison measuring instruments.
2. Compare the accuracy range of a standard, everyday shop dial indicator with the accuracy range of a high amplification dial indicator.
3. List the steps in measuring a 0.9375" dimension on a flat gage to within +0.000020/-0.000010". Use a mechanical comparator.
4. Cite two design features of a reed comparator that differ from similar features of mechanical comparators.
5. Give two advantages of a contour measuring (optical) projector in comparison to a mechanical comparator.
6. Describe briefly how pneumatic comparison instruments operate.
7. State three advantages of using pneumatic gaging instruments over GO and NO-GO gages and other mechanical measuring methods.
8. Indicate some factors the worker must consider when selecting an appropriate design for a plug to use on a gaging spindle.
9. Give two advantages of a meter readout on an electronic amplifier for precision measurements.
10. Cite three precautions a worker must follow when working with high amplification comparators.

SECTION THREE

Quality Control and Surface Texture (Finish)

The introduction of interchangeable manufacture and the development of accurate measuring instruments in the nineteenth century formed the foundation for mass production. After years of extensive developing and testing, high-quality master gages were produced. These gages permitted the transfer of measurements to working gages. The working gages provided a practical device for controlling dimensional measurements in machining and assembling. As cutting tools, machine tools, instruments, lubricants, coolants, and materials improved, so the accuracy to which parts could be held increased. Consistently, the capability of the worker improved to produce precision surfaces to an ever higher quality of finish. This section deals with the principles and measurement practices related to quality control and with the characteristics and measurement of surface texture (finish).

UNIT 5

Quality Control Principles and Measurement Practices

INTERCHANGEABLE MANUFACTURE AND QUALITY CONTROL

As engineering standards developed, specialization in manufacturing became possible. Specialization permits parts produced by one company to fit into an assembly of parts produced by one or more other companies. For example, spur, spiral, bevel, hypoid, and other gears, in addition to ball bearings and races, are more economical to produce as specialities. Conformance to standards of shape, size, surface finish, and systems of tolerances and fits, guarantee the manufacture of gears, bearings, and other parts to meet rigid specifications.

British, American, and Canadian (inch) standards are accepted worldwide. While there is a movement toward metrication, a considerable number of proposed metric engineering standards still remain universally unaccepted. Accordingly, the skilled craftsperson and technician should be able to convert dimensions from one system to

Importance of Quality Control

The function of quality control is to ensure that the specifications and standards established by design and engineering departments are maintained. Quality control is essential to the maintenance of the quality of a product and to continuous improvements in the output. Quality control requires dimensional control, the testing of materials, inspection of surface roughness, assembly run-in periods, and other checks.

Dimensional Measurement and Control

Precision manufacture depends on dimensional control. Without this control, the rapid assembly of parts is impossible. Years ago, tolerances of ±0.001" (±0.02mm) were acceptable (as in the case of automotive pistons and other engine parts). Currently, tolerances of ±0.0001" (±0.002mm) with higher quality surface finishes are required. These tolerances make it possible to operate new vehicles at high speeds as soon as they pass final assembly inspection.

The tools and instruments used for dimensional measurement are determined by the specified accuracy. Nonprecision linear measurements may be taken indirectly by using squares, dividers, calipers, and surface gages. These instruments may be used with line-graduated tools such as steel rules, bevel protractors, and depth gages. Common precision measurements are taken to tolerances of ±0.001" (±0.02mm) and ±0.0001" (±0.002mm) with inside, outside, depth, and other forms of micrometers and vernier-graduated instruments.

Dimensional Control Using Gages

The line-graduated tools just described are designed for a wide range of measurements and are adjustable. It is more practical and economical in mass production to check parts for size with fixed gages. There is less chance of error when no adjustable parts are involved. Of the many types of gages, six are commonly used. They include the *fixed-size gage*, the *micrometer-feature gage*, a *limit (standard) comparator*, the *indicating comparator*, the *combination gage*, and the *automatic gaging machine*. These gages and instruments are checked against master gages such as precision gage blocks, angle gage blocks, sine bars and plates, and optical flats (all of which were discussed in earlier units). The gages are dimensionally accurate and stable to within a fraction of 0.000002" (0.00005mm).

Dimensional Tolerances

The permissible variation of a part from its nominal dimension (designated size) is identified by the designer, engineer, and craftsperson. The designer establishes the dimensional limits and surface characteristics of a part and mechanism. The more exacting a dimension, the more difficult and expensive a part is to produce. There is a minute range within which a part will function effectively.

On some dimensions (a shrink fit, for example), only a smaller dimension than the nominal size is permissible. Where a running fit is required, the part may be larger than the nominal size. The amount of variation from the nominal size is called *tolerance*. A tolerance may be *unilateral* (only plus or minus) or *bilateral* (both plus and minus). The amount of tolerance depends on factors such as the function of a part, the kind of material used, the required service life, cost, and production methods and systems.

FIXED-SIZE GAGES

Fixed-size gages are widely used in quality control. Although limited in scope, they are used to check either an internal or external dimension. With use, a plug gage becomes smaller; a ring gage, larger. Therefore, the wear life of fixed-size gages is determined by the number of parts that may be inspected before the gage size exceeds the allowable work tolerance. Constant wear requires that the working gages be checked against a master set of gages. A master gage is easily identified and is used as a dimensional control gage.

In the design of fixed-size gages, a *wear allowance* is provided. For example, if it is established that after 2,000 gagings a 1-1/4" plug gage wears 0.0001" (0.002mm), the plug may be finished within the allowable tolerance of the workpiece to 1.2501". The wear life is thus increased to measure 4,000 workpieces. Wear life may be further increased by chrome plating a worn gage and regrinding and lapping to size. Cemented carbide gages are available for long-run production gaging. Such gages resist abrasive action and wear. Wear life of some carbide gages is increased to almost 100 times longer than wear life of steel gages.

GO and NO-GO Gages and XX, X, Y, and Z Classes of (Gage Maker's) Tolerances

Fixed gages are commercially available in a wide range of sizes for GO and NO-GO measurements. Ring, plug, and thread gages are commonly used for such measurements. Some examples of these gages are illustrated in Figure 5-1. These gages are available in inch and metric units of measure and within the same tolerance limits. The accuracy of the gages has been standardized. The tolerances are specified in terms of four classes: XX, X, Y, and Z. Table 5-1 gives the range of nominal sizes from 0.029" to 12" and the four classes of tolerances. These tolerances are known as *gage maker's tolerances* The tolerance classes are generally defined as follows:

- *Class XX* refers to master and setup standards that are precision lapped to laboratory tolerances;
- *Class X* refers to working and inspection gages that are precision lapped;
- *Class Y* refers to working and inspection gages that have a good lapped surface finish with slightly increased tolerances;
- *Class Z* refers to working gages for short production runs where the tolerances are wide or commercially finished gages that are ground and polished but are not fully lapped.

Gage manufacturers recommend that the gage used should have an accuracy of one-tenth

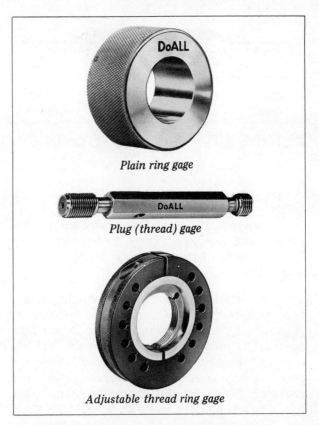

FIGURE 5-1 Examples of Ring, Plug, and Adjustable Gages

the tolerance of the dimension the gage is to control. Thus, the gage used for a part having a tolerance of 0.001" (0.02mm) should have a tolerance of 0.0001" (0.002mm). The surface finish of the gage should also be finer than the workpiece.

TABLE 5-1 Nominal Sizes and Gage Maker's Dimensional Tolerances

Range of Nominal Sizes (inches)	Gage Maker's Tolerance According to Class			
	XX	X	Y	Z
0.029 to 0.825	0.00002	0.00004	0.00007	0.00010
0.826 to 1.510	0.00003	0.00006	0.00009	0.00012
1.511 to 2.510	0.00004	0.00008	0.00012	0.00016
2.511 to 4.510	0.00005	0.00010	0.00015	0.00020
4.511 to 6.510	0.000065	0.00013	0.00019	0.00025
6.511 to 9.010	0.00008	0.00016	0.00024	0.00032
9.011 to 12.010	0.00010	0.00020	0.00030	0.00040

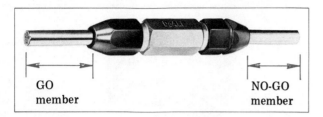

FIGURE 5-2 Pin-Type GO and NO-GO Gage

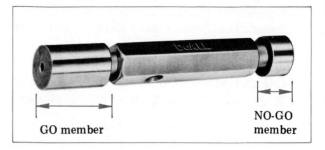

FIGURE 5-4 Taperlock Cylindrical GO and NO-GO Plug Gage

Four Basic Gage Designs

Most types of plug and ring gages are marked for size and tolerance. There are four basic gage designs. Each of the following designs is adaptable within a range of sizes.

A *reversible cylindrical plug gage* has a pin-type gage design with GO and NO-GO members (Figure 5-2). Gage maker's tolerances are plus on the GO member and minus on the NO-GO member. This type of gage is primarily used for checking hole sizes and depths, gaging slots, and checking locations, threads, and distances between holes. Pin-type gages are available in steel, plated steel, and carbide. A common range of sizes is from 0.005″ to 1.000″ (0.125mm to 25mm). Each size is produced in the four tolerance classes. Plug members are available in increments of 0.001″ and are provided with collet bushings.

A *trilock plug gage* has a body design that permits GO and NO-GO members to be secured to the ends by means of a fastener (Figure 5-3).

A *taperlock cylindrical plug gage* has plug ends with a taper shank that fits a taper hole in the body (handle) of the gage (Figure 5-4).

A *progressive plug gage* is a single-end gage (Figure 5-5). The design permits complete inspection of a hole at one time. The entering portion is the GO gage, which is followed by the NO-GO gage.

While only plug gage and holder designs have been described, the same types of holders (body) are used for different types of thread plug gages. Taperlock and reversible types of thread plug gage are shown in Figure 5-6.

Cylindrical and thread gages have been used here as examples; however, many other designs are available. Taper, spline (multiple grooves), flat plug, and special gages are widely used. *Taper plug gages* measure the accuracy of the taper and size. The taper may be one of the standard tapers or a special taper. Figure 5-7 shows an application of a simple solid taper plug gage. The two vertical lines indicate the tolerance and are the GO and NO-GO limits. Figure 5-8 shows two *flat plug gages*. These gages are used to check the width of slots and grooves.

Snap Gages. Another design of a progressive (combination GO and NO-GO) gage is the

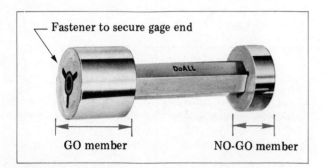

FIGURE 5-3 Trilock Type of GO and NO-GO Plug Gage

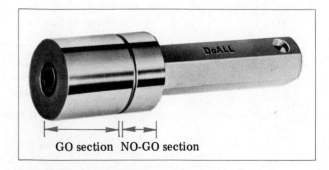

FIGURE 5-5 Progressive GO and NO-GO Plug Gage

QUALITY CONTROL PRINCIPLES 87

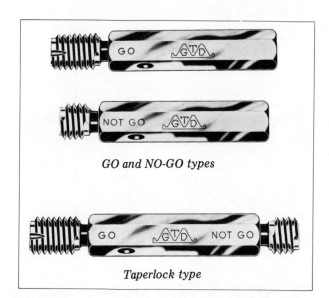

FIGURE 5-6 Taperlock and Reverse Types of Thread Plug Gage

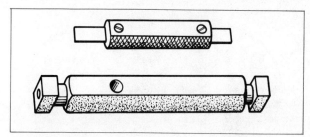

FIGURE 5-8 Flat Plug Gages

The GO and NO-GO buttons of an adjustable snap gage are set to gage blocks or other masters. Precision measurements with snap and other gages depend on taking the measurement at the correct angle with respect to the reference planes. Small parts are measured by holding the workpiece with one hand and the gage with the other hand. The gage is then brought to the workpiece and carefully moved along the line of measurement.

adjustable limit snap gage (Figure 5-9). This gage consists of a structurally strong but lightweight and balanced frame. Adjustable gaging pins, buttons, or anvils are brought to size with adjustment screws. They are then secured to and locked in place on the frame with locking screws. A marking disk is stamped. The disk indicates the dimensions and identifies the gage in relation to a production plan.

ADJUSTABLE THREAD RING GAGES

One common type of *adjustable thread ring gage* is shown in Figure 5-10. This type is available for standard metric thread sizes and pitches and for American National and Unified thread forms. Other gages are available for Acme thread form sizes.

In thread control, *roll-thread gages* are also available. The term indicates that the thread-measuring form is a roll instead of a flat plate. There are even adaptations of roll-thread gages. In some applications, a dial indicator is used to show any variation from the basic size to which the gage is set.

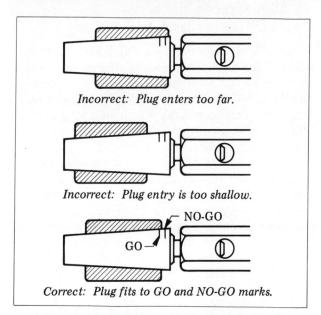

FIGURE 5-7 Application of Taper Plug Gage

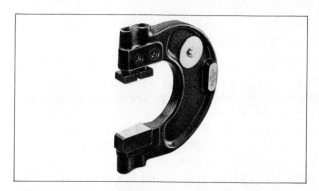

FIGURE 5-9 Adjustable Limit Snap Gage

88 PRECISION MEASUREMENT AND INSPECTION

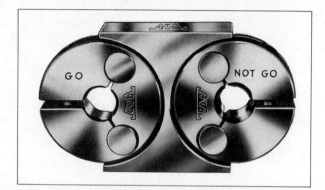

FIGURE 5-10 Adjustable Thread Ring Gages in Holder

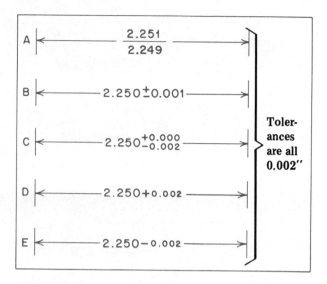

FIGURE 5-11 General Dimensioning Practices

ALLOWANCES FOR CLASSES OF FITS

The term *fit* indicates a range of tightness between two mating parts. This range is expressed on a drawing in terms of the *basic (nominal) size* and the *upper* and *lower (high limit* and *low limit) dimensions*. Tolerance, to repeat, is the total permissible variation from the basic size. Nominal size is the dimension from which the maximum and minimum permissible dimensions are derived.

Figure 5-11 indicates five general practices for dimensioning tolerances. Examples A and B indicate plus and minus tolerances that are the same, and example C indicates plus and minus tolerances that are different. If a tolerance is given in one direction only—as in examples D and E—the craftsperson assumes the other limit is the basic dimension on the drawing.

The dimensions may also be given in metric-standard units (mm) for the basic dimension, allowances, and tolerance. Some drawings may have dimensions given in both inch- and metric-standard units of measure.

Clearance and Interference Fits

Some mating parts must move in an assembly. A *positive clearance* is therefore required. How much clearance to provide depends on the nature of the movement, the size of the parts, material, surface finish, and other design factors. The drawing in Figure 5-12 illustrates limit dimensioning and a positive clearance. There is a 0.002″ (0.05mm) clearance allowance on the punch. The clearance allowance on the block is 0.001″ (0.02mm). The clearance fit has a maximum allowance of 0.0035″ (between 2.0005″ and 1.9970″) and a minimum allowance of 0.0005″ (between 1.9995″ and 1.9990″).

The drawing in Figure 5-13 shows a *negative*, or *interference*, *fit*. The plug is ground to a high limit of 1.253″. The low limit of the mating hole is 1.250″. The size limits of this particular interference fit is 0.001″ minimum and 0.003″ maximum.

Allowances on Holes for Different Classes of Fits

Standard allowances on holes have been developed for different *classes of fits*. Some allowances relate to tolerances for general purposes. Others relate to *forced*, *driving*, *push*, and *running fits*. The term *class* is followed by a particular letter (or combination of letters) that relates to specific design functions.

In the partial Table 5-2, two examples are given of industrial standard allowances and tolerances for reamed holes. A *class A* fit refers to a grade of hole that can be produced by general-purpose reamers. Class A reamed holes are machined to closer dimensional tolerances than *class B* holes. The high and low limits and tolerances are given in this table for a range of diameters from under 1/2″ to 5″. Inch-standard measurements may be converted to equivalent metric-standard units of measure for metric diameters.

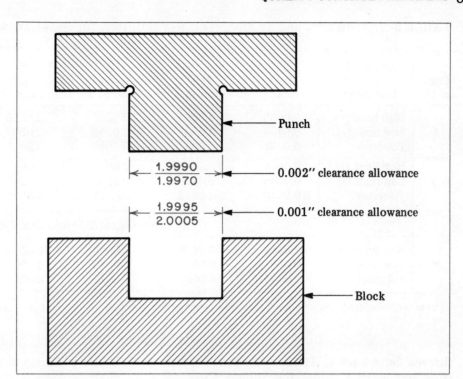

FIGURE 5-12 Allowance on Each Mating Part for a Clearance Fit

The table also shows allowances for *class F* forced fits. A complete table in the Appendix includes allowances for other forced and drive fits, in which the parts must function as a single part. One workpiece that is machined oversize must be forced into another workpiece.

The three grades of running fits are covered in the appendix table. Running fits provide for different motion conditions. *Class X* fits provide the greatest allowance for work parts that need to fit easily. *Class Y* fits are used for high speeds and average machine work. *Class Z* fits require a higher degree of precision and are applied in fine tool work and assemblies.

Tables of allowances provide a general guide. The material in the mating parts, the conditions under which a mechanism operates, the kind of work to be done, and a host of other factors must be considered. Usually, the designer indicates the ideal allowances through the dimensions on a drawing. There are situations, however, where the machinist, toolmaker, and inspector must refer to tables and make decisions.

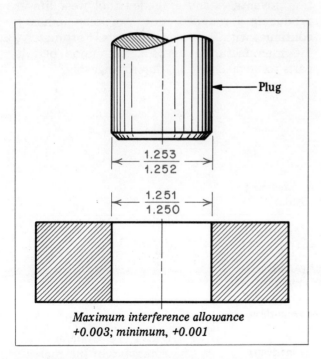

FIGURE 5-13 Negative Allowance for Interference Fit

INTERCHANGEABILITY OF AMERICAN NATIONAL AND UNIFIED FORM SCREW THREADS

Commercial NC and NF American National Form threads are the most widely used form in the United States. The origin of Unified Form screw

90 PRECISION MEASUREMENT AND INSPECTION

TABLE 5-2 Allowances and Tolerances on Reamed Holes for General Classes of Fits (Partial Table)

Class of Fit	*Allowances and Tolerances	*Nominal Diameters					
		Up to 1/2"	9/16" through 1"	1-1/16" through 2"	2-1/16" through 3"	3-1/16" through 4"	4-1/16" through 5"
A	High limit (+)	0.0002	0.0005	0.0007	0.0010	0.0010	0.0010
	Low limit (−)	0.0002	0.0002	0.0002	0.0005	0.0005	0.0005
	Tolerance	0.0004	0.0007	0.0009	0.0015	0.0015	0.0015
B	High limit (+)	0.0005	0.0007	0.0010	0.0012	0.0015	0.0017
	Low limit (−)	0.0005	0.0005	0.0005	0.0007	0.0007	0.0007
	Tolerance	0.0010	0.0012	0.0015	0.0019	0.0022	0.0024
		Allowances and Tolerances for Forced Fits					
F	High limit (+)	0.0010	0.0020	0.0040	0.0060	0.0080	0.0100
	Low limit (−)	0.0005	0.0015	0.0030	0.0045	0.0060	0.0080
	Tolerance	0.0005	0.0005	0.0010	0.0015	0.0020	0.0020

*These inch-standard measurements may be converted to equivalent metric-standard measurements.

threads dates back to 1948, when the standardizing bodies of Canada, the United Kingdom, and the United States signed an agreement. This agreement produced the Unified (UN) Form screw thread as a complete standardized system of threads for fasteners in mechanisms and structures. Such threads have a basic form and limits of size and tolerances.

Interchangeability of Unified and American National Form threads is provided through the standardization of thread form, diameter and pitch combinations, and size limits. The general characteristics of the internal and external Unified Form threads are shown in Figure 5-14.

The form is theoretical. Actually, neither British nor American industry has made a complete changeover. The 60° thread angle, diameter and pitch relationships, and limits are accepted. However, the British continue to machine threads with rounded roots and crests. There are production advantages and extended tool wear life in producing a rounded root. American industry continues with flat roots and crests. Fortunately, the manufacturing limits agreed upon permit parts to be mechanically interchangeable.

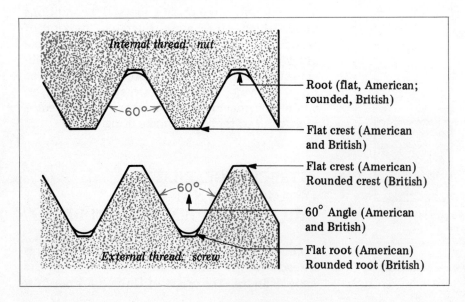

FIGURE 5-14 General Characteristics of Internal and External Unified Standard Form Threads

Importance of Accurate Thread Measurement

With increasingly higher physical demands on machines, it has become necessary to produce stronger and more dependable fastenings. Maximum thread strength and dependability are determined by the following conditions:

— *Proper thread lead* (testing shows that a thread with an error in lead is capable of sustaining only a fraction of the load of a perfectly formed thread);
— *Correct tooth form* (out-of-round and taper threads wear rapidly, come loose in service, and often fail as a fastener).

THE BASIC SIZE (CONTROLLING POINT) IN THREAD MEASUREMENT

Allowances for classes of fit for screw threads are designated differently from diameter sizes. All thread measurements must be derived from a controlling point. This point is the theoretically correct size, or *basic size*. The basic size relates to a full form thread. Thus, for each thread, there is a *basic major diameter*, *basic minor diameter*, and *basic pitch diameter*.

In actual practice, it is impossible to form a screw thread to its precise basic sizes. Limits have been established within which dimensional accuracy and form may be held. The deviations from basic sizes are identified as the *minimum* and *maximum limits*. A product may be accepted when its size limits are within the minimum limit and the maximum limit. Like dimensions for holes, the difference between the upper and lower limits is the tolerance. In general practice, tolerance for an external thread is *below* the basic size. The tolerance for an internal thread is *above* the basic size.

Impact of Tap Drill Sizes and Thread Height on Strength

There are a number of advantages for producing a partially threaded hole. Production efficiency, added tap wear life, reduced tap breakage, and decreased costs are four main advantages. The major requirement of a tap drill is to produce a hole size that permits tapping within specified limits. Too large a tap drill produces a thread that may be too shallow. Too small a tap drill requires excessive cutting force with possible tap breakage. Many production difficulties are caused by the use of tap drills that are too small.

A comparison of thread depth and strength is illustrated by the three drawings in Figure 5-15. After years of testing in practical applications, the 100% depth thread (Figure 5-15A) has proven to be only 5% stronger than a thread cut to

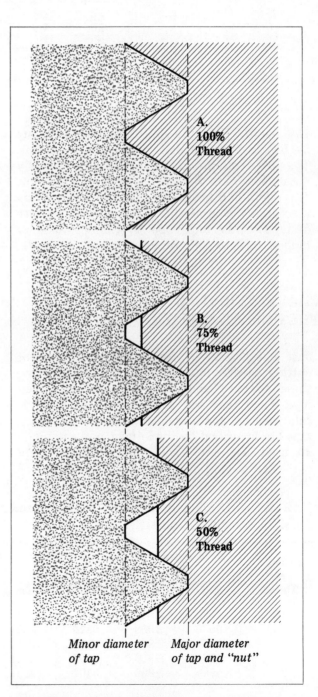

FIGURE 5-15 Comparison of Tap Drill Sizes and Height of Thread

TABLE 5-3 Class of Thread, Tap Drill, and Thread Height Specifications for Tapping 1/4-20 UNC and NC Threads

Tap Suggested		Tap Drills				
		Nearest Available Commercial Drills				
Class of Thread	Type Tap	Nominal Size	Decimal Equivalent	Probable Oversize Will Cut	Probable Hole Size	% Thread
1B	Cut	9	.1960	.0038	.1998	77
2B	H5	8	.1990	.0038	.2028	73
3B	H3	7	.2010	.0038	.2048	70
2	H3	13/64	.2031	.0038	.2069	66
3	H2	6	.2040	.0038	.2078	65

75% depth. Thus, the 75% depth thread (Figure 5-15B) is recommended for average conditions. The 50% depth thread (Figure 5-15C) is generally unsatisfactory in terms of strength requirements. This depth of thread may be stripped easily. Manufacturers recommend that for the vast majority of tapped hole requirements, a minor diameter tapped 55% to 75% of full depth is adequate.

CHARACTERISTICS OF THE UNIFIED SCREW THREAD SYSTEM

Just a few years ago, *serial taps* were used for cutting precise internal screw threads. Today, these taps have been replaced with *ground thread taps*. Such taps are identified by the letter *H* followed by a numeral. For external threads, taps are identified by the letter *L* and a numeral are used. The letters and numbers are a method of designating limits. The letters indicate the range of allowances above (H) or below (L) the *basic pitch diameter*. The numbers correspond to *class of fit*. The series of H1 through H10 is used for taps through 1" inclusive. H4, H5, and H6 ground taps are widely used for sizes over 1" and through 1-1/2" and for 6TPI and finer pitches. The limits start at the basic pitch diameter.

The kind of background information the worker needs to know to supplement drawing specifications is given in Tables 5-3, 5-4, 5-5 and 5-6. A ground thread tap size of 1/4-20 UNC and NC is used throughout as an example. In Table 5-3, the suggested type of tap is given in the first column. As the table shows, an H5 ground thread tap will produce a class 2B fit. The standard (cut) tap will produce a class 1B fit.

Tap drill sizes and the probable hole size for general thread depths (77% to 65%) appear in

TABLE 5-4 Limits and Tolerances for Minor and Pitch Diameters of 1/4-20 UNC and NC Internal Threads

Class of Thread	*Minor Diameter (Basic 0.1850)			Pitch Diameter (Basic 0.2175)		
	Max.	Min.	Tol.	Max.	Min.	Tol.
1B	.2070	.1960	.0110	.2248	.2175	.0073
2B	.2070	.1960	.0110	.2224	.2175	.0049
3B	.2067	.1960	.0107	.2211	.2175	.0036
2	.2060	.1959	.0101	.2211	.2175	.0036
3	.2060	.1959	.0101	.2201	.2175	.0026

*Extreme limits

	Recommended Hole Size Limits for Various Lengths of Engagement				Theoretical Hole Sizes to Give Various Percentages of Threads	
	1B and 2B		3B		% Height of Thread	Required Hole Size
Diameters	Max.	Min.	Max.	Min.		
To 1/3 dia.	.202	.196	.2013	.1960	83-1/3	.1959
1/3 to 2/3	.204	.199	.2040	.1986	75	.2012
2/3 to 1-1/2	.207	.202	.2067	.2013	70	.2046
					65	.2078
1-1/2 to 3	.210	.204	.2094	.2040	60	.2111

the second column. The percent of thread depth is also affected by the length of thread engagement. Thus, maximum and minimum hole size limits are shown for lengths up to 3 times the thread diameter. These limits appear in the third column. The fourth column of Table 5-3 gives recommended hole sizes for thread depths varying from 83-1/3% to 60%.

Additional information for internal threads is provided in Table 5-4. Maximum and minimum limits are given for minor and pitch diameters for class 1B, 2B, 3B, and 2 and 3 fits. Thread tolerances are also provided. Table 5-5 covers similar measurements for external threads.

Table 5-6 provides major and pitch diameter thread limits for ground thread taps H1, H2, H3,

TABLE 5-5 Limits and Tolerances for Minor and Pitch Diameters of 1/4-20 UNC and NC External Threads

Class of Thread	Minor Diameter (Basic 0.2500)			Pitch Diameter (Basic 0.2175)		
	Max.	Min.	Tol.	Max.	Min.	Tol.
1A	.2489	.2367	.0122	.2164	.2108	.0056
2A	.2489	.2408	.0081	.2164	.2127	.0037
3A	.2500	.2419	.0081	.2175	.2147	.0028
2	.2500	.2428	.0072	.2175	.2139	.0036
3	.2500	.2428	.0072	.2175	.2149	.0026

TABLE 5-6 Major and Pitch Diameter Thread Limits for Ground Threads and Cut Thread Taps (1/4-20 UNC and NC)

	Major Diameter (Basic 0.2500)		Pitch Diameter (Basic 0.2175)						
	Ground Thread	Cut Thread	H1	H2	H3	H4	H5	H6	Cut Thread
Max.	.2550	.2557	.2180	.2185	.2190	—	.2200	—	.2200
Min.	.2540	.2532	.2175	.2180	.2185	—	.2195	—	.2180
Tol.	.0010	.0025	.0005	.0005	.0005	—	.0005	—	.0020

and H5 and for a cut thread tap. Note that the thread limit tolerance on the major diameter is 0.0010" for a ground thread and 0.0025" for a cut thread. The pitch diameter thread tolerance limits with the H1, H2, H3, and H5 ground thread taps are all 0.0005". The pitch diameter tolerance of the cut thread tap is 4 times greater (0.0020").

H and L Pitch Diameter Tap Limits

Another specification of ground thread taps refers to the dimensional factors of H and L. H dimensions are *above* the basic pitch diameter. L dimensions are *below* the basic pitch diameter. The two letters are followed by numbers—for example, H1 and H2 or L1, L2, and so on. The H and L dimensions indicate a range of allowances above or below the basic pitch diameter. Individually, each H or L designation indicates a single thread depth tolerance (Figure 5-16).

The use of H and L designating limits for a ground thread tap and the tap limit sizes for a 1/4-20 UNC and NC thread are shown in Figure 5-17. The basic pitch diameter in this thread example is 0.2175". The H1 ground thread tap will produce a thread having a pitch diameter of 0.2185". That is, the tap cuts to a tolerance of 0.001" (or 0.0005" tolerance for the single thread depth).

The right side of Figure 5-17 relates the H range to class of fit. Note the use of the designations 1B, 2B, 3B and 2 and 3. The "B" following a thread notation refers to an inside thread. A drawing marked 1/4-20 UNC-3B provides thread specifications that include the class of fit. The 3B for this particular size thread (and according to the figure) has a tolerance of 0.0036". The allowable range of acceptance of a 1/4-20 UNC thread for a class 3B fit for a tapped hole is from the basic pitch diameter of 0.2175" (minimum) to 0.2211". If within this range the maximum pitch diameter is 0.2190", an H3 tap will produce this class of fit.

BASIC ISO METRIC THREAD DESIGNATIONS AND TOLERANCE SYMBOLS

ISO metric threads are designated by the letter *M*. This letter is followed by the nominal size in millimeters. The size is separated from the pitch size by the sign ×. M16×1.5 is an example of the format. This format is used on a drawing for all ISO metric thread series except the coarse pitch series. Coarse pitch ISO metric threads are designated in a simpler form. Only the letter (M) and the nominal size in millimeters are given, for example, M16. The practice of using this form is common outside the United States. Internally, United States industry includes the pitch in order to avoid confusion.

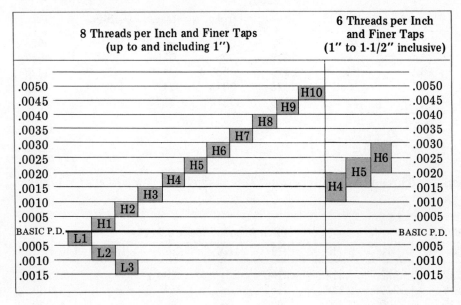

FIGURE 5-16 Method of Designating H and L Ground Tap Limits

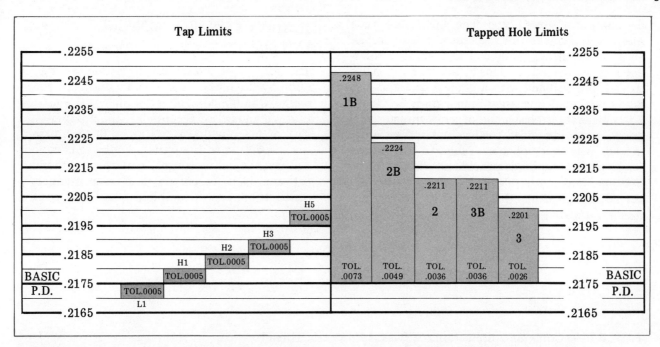

FIGURE 5-17 Application of Tap Limits to Class of Fit for a 1/4-20 UNC and NC Thread

Tolerance Symbols for ISO Metric Threads

Letters and numbers are used with ISO metric threads to identify the *position* and *amount of thread tolerance*. Letters designate the position relative to basic diameters. The lowercase letters *g* and *h* are used for external metric threads. The capital letters *G* and *H* apply to internal metric threads. The position of the tolerance establishes the clearance (allowance) between internal and external threads.

A *tolerance symbol* is a combination of the number indicating the amount of tolerance and the tolerance position letter. The tolerance symbol indicates the actual maximum and minimum limits for internal or external threads. The first number and letter combination usually is the pitch diameter tolerance symbol. The second number and letter combination is the tolerance symbol for the crest diameter.

Figure 5-18 shows how a drawing is dimensioned for an external ISO metric thread. The tolerance symbols 5g and 6g are used together. For the M16-1.5 thread, the 5g indicates that the pitch diameter tolerance is grade 5 and the thread is an external one. The 6g refers to the crest diameter tolerance of grade 6 on the same external thread.

A single combination symbol is used when the crest and pitch diameter tolerances are the same. For example, an ISO metric thread may be marked: M16-1.5 6g. This screw thread has an outside diameter of 16mm and a pitch of 1.5mm. The thread is cut to a pitch diameter and crest diameter tolerance of grade 6.

Identifying the Amount of Tolerance

The amount of tolerance permitted on either internal or external threads is defined by a *grade number*. This number precedes the letter that

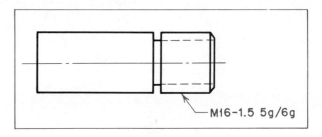

FIGURE 5-18 Application of Pitch and Crest Diameter Tolerance Symbols on an Outside ISO Metric Thread

96 PRECISION MEASUREMENT AND INSPECTION

designates the thread type. The smaller the grade number, the smaller the tolerance. For example, an external thread marked with a grade 4 tolerance is machined to closer tolerances than a thread dimensioned with a grade 6 tolerance. A grade 8 tolerance is larger than a grade 6. Grade 6 is a general-purpose grade of screw thread. This grade is recommended for internal and external threads wherever possible. Tables of ISO metric screw thread standards provide additional specifications and limits. Table 5-7 summarizes ISO thread designations that are most commonly used on drawings.

QUALITY CONTROL METHODS AND PLANS

The control of the quality of a process is agreed upon by the parts designer and industrial management. *Quality control* is the means of ensuring that a standard of acceptance (the product quality) of machined parts is maintained. Further, the inspection of the parts is kept at an economical level. Quality control requires a mathematical approach based on a normal frequency-distribution curve, sampling plans, and control charts.

TABLE 5-7 Common Applications of Basic ISO Metric Thread Designations

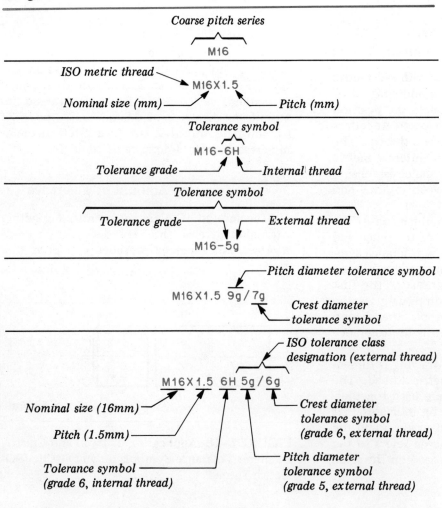

Normal Frequency-Distribution Curve Foundation

Tests carried on over many years indicate that measurement variations occur during manufacturing processes. When they are plotted on a graph, the normal size variations follow a curved pattern. The general shape of the pattern is shown in Figure 5-19A. A limited number of the parts are larger or smaller than the basic dimension. The curve formed is known as the *normal frequency-distribution curve*.

This curve is divided into six zones or divisions (Figure 5-19B and C). Each zone is mathematically equal in width. The average on the curve is represented by the centerline. Note that there are three zones on each side of the centerline. The Greek letter sigma (σ) is used with a number to indicate the range of measurements.

For example, the 50 machined lathe parts in Figure 5-19A were measured by micrometer. The diameters varied from 0.9372" to 0.9378". Note that the greatest number of parts fall along the centerline (the basic diameter of 0.9375"). The next largest numbers are 0.9376" and 0.9374"—either +0.0001" or -0.0001". These parts are said to fall within plus or minus one sigma ($\pm 1\sigma$).

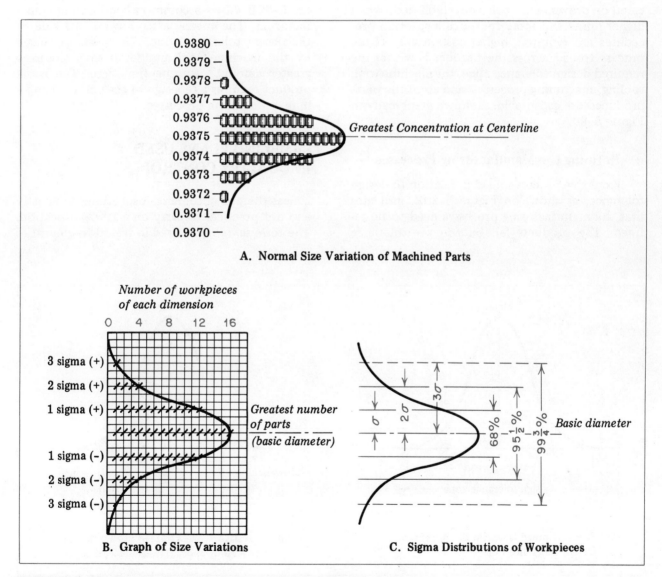

FIGURE 5-19 Graphic Representation of Normal Size Variations of 50 Machined Workpieces

Statistically, 34% of the workpieces fall within this first sigma. Thus, the diameters of 68% of the parts lie between 0.9374″ and 0.9376″. At ±2σ, the diameters of 95-1/2% of the parts fall between 0.9373″ and 0.9377″. At ±3σ, 99-3/4% of the parts measure between 0.9372″ and 0.9378″. The ±3σ limit is the *natural tolerance limit.*

Information on normal size variations is important to design, development, and manufacturing departments. Department persons are responsible for decisions on the limitations of the machining and/or inspection processes. The designer must consider the level at which production may be carried on efficiently and economically. Dimensional tolerances and surface finish indicated on drawings must also be established. When higher limits and more thorough inspection procedures are required, higher costs result. If the process spread across the product is within the required design tolerance zone, the machine tool, tooling, machining processes, and all of the parts produced are acceptable, as shown graphically in Figure 5-20A.

Refining the Manufacturing Processes

Excessive process spread in relation to design tolerance, as shown in Figure 5-20B, indicates that the manufacturing processes need to be refined. The products fall outside the tolerance requirements. More precise tooling is needed. The machined dimensions must be controlled to produce all parts within the specified tolerance. Only accidental deviations need to be inspected. Where the products fall outside the tolerance, the designer sometimes corrects the problem by redesigning the part or mechanism. The design tolerance is then brought within the limitations of the machine and setup.

The conditions shown in Figure 5-19 are based on the fact that the mean (centerline) of the frequency curve of distribution is correct. Many times the mean is on one side of the tolerance. The *mean* represents the point (or line on a graph) where there are an equal number of parts above and below the basic dimension. Figure 5-20B shows a common condition in manufacturing. The mean leans toward one side of the design tolerance zone. The parts produced by the machine tool and setup thus require a greater amount of inspection time. The waste product requires a redesign of the part or changes in manufacturing processes.

SAMPLING PLANS USED IN QUALITY CONTROL

Unless there is automatic inspection, it is costly and not practical to carry on a 100% inspection. The term *sampling* is used in regard to quantities

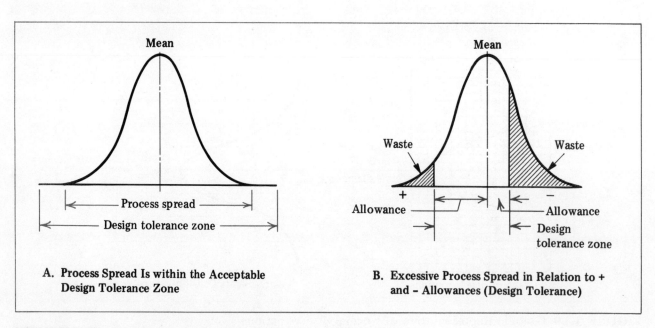

FIGURE 5-20 Acceptable and Unacceptable Process Standards

QUALITY CONTROL PRINCIPLES

(batches) of parts that have been machined, cleaned, and are ready for inspection. The plan for sampling is referred to as a *sampling plan*. Such plans are based on the premise that a definite number of defective parts is allowable. This number is identified as the *acceptable quality level (AQL)*.

The AQL is usually based on previous trial runs on machined parts. For example, the AQL for a particular part that is machined on a bar chucking machine may indicate that 0.8% to 1.4% are defective. This small percent of defective workpieces is allowable.

The same batch may have an *average outgoing quality limit (AOQL)* of 0.9% to 1.6% defective. This AOQL means that the batch of parts leaving the inspection station will average no more than the fixed percent of defects. The batches (lots) meet the sampling requirements. Any lot that fails to meet the sampling requirements is re-inspected. The defective parts are sorted out before they move out of inspection.

There are several different types of sampling plans. The *single-sampling*, *double-sampling*, and *sequential-sampling* plans are most common.

Single-Sampling Plan

Sampling plans for the part and lot sizes are shown in Table 5-8. The first column lists the three most common types of sampling plans. If a single-sampling plan is used, a random number of machined parts is required for inspection. In this example, the single-sampling size is 55, as shown in the third column. As a result of inspection, if there are 3 or less parts defective, the lot is accepted. If 4 parts in the lot are defective, the lot is rejected. The acceptance and rejection numbers appear in the sampling plan in the fourth column.

Double-Sampling Plan

Rejecting the lot means that each part in the sample is inspected. The defective parts are replaced. In a double-sampling plan, Table 5-8 shows that a first double-sampling size of 36 is required. Note that if there is more than 1 defective part within the range between the acceptance number (1) and the rejection number (5), a second screening must take place. In the second screening, 72 additional parts are needed. The acceptance number is now 4; rejection, 5. The lot is accepted if there are 4 or less defective parts. If there are 5, the lot is rejected. The lot is subjected to further screening. In this double-sampling plan, the total number of parts checked is 108 when a second sample is needed.

Sequential-Sampling Plan

The representative sequential-sampling plan in Table 5-8 for the 800 to 1,300 lot parts lists seven samples. The first sample size consists of

TABLE 5-8 Example of Sampling Plans for Inspecting Lot Sizes Ranging from 800 to 1,300 Parts (1.1 to 2.1 AQL)

Type of Plan	Sample and Sequence	Size of Sample	Combined Samples		
			Cumulative Size of Sample	Lot Acceptance Number	Lot Rejection Number
Single-Sampling	First	55	55	3	4
Double-Sampling	First	36	36	1	5
	Second	72	108	4	5
Sequential-Sampling	First	15	15	*	2
	Second	15	30	0	3
	Third	15	45	1	4
	Fourth	15	60	3	5
	Fifth	15	75	3	5
	Sixth	15	90	3	5
	Seventh	15	105	4	5

*Requires inspection of two samples to permit acceptance

15 parts. If 0 or 1 part is defective, a second sample must be inspected. If 2 or more parts in the first sample are defective, the lot is rejected. If the number of defectives in the second sample is between 0 and 3, a third sample of 15 is used. The process continues with additional samples of 15 as long as the combined samples are within the acceptance and rejection numbers given in the table.

Selection of Sampling Plan

Each sampling plan serves a different function. Major factors affecting the choice of plan to use are cost and additional inspection time. Single-sampling plans are used on complicated precision parts that require close dimensional and surface finish accuracy. The example used in Table 5-8 shows that a greater number of parts are required for single sampling than for sequential sampling. Sequential sampling is used with homogeneous, large lots and provides a quick and comparatively inexpensive check. Any lots that fail to meet sampling requirements are reinspected and sorted (screened). The defective parts are replaced with acceptable parts.

QUALITY CONTROL CHARTS

Control charts show graphically the processes undergoing inspection. There are four common types of control charts:

— c charts are used to plot the number of defects in one workpiece;
— p charts show the percent of defective parts in a sample;
— $\overline{X}$ charts provide a graph of the variations in the averages of the samples;
— $\overline{R}$ charts show variations in the range of samples.

Figure 5-21 is a combination $\overline{X}$ and $\overline{R}$ chart. The sample averages are recorded at one-half hour intervals. Usually the size of the sample is 4 or 5 parts. The average dimensions are plotted on the chart. The range (the difference between the largest and smallest dimensions) is also plotted as a bar (Figure 5-21).

The grand average ($\overline{\overline{X}}$) is computed after all the samples are plotted. The $\overline{\overline{X}}$ is marked and distinguished as a colored, dashed line. The range of variations in tolerance ($\overline{R}$) is averaged in a similar manner. The average is drawn as a solid, colored, distinguishing line.

Computing the Upper and Lower Averages Control Limits

The upper and lower averages control limits are computed by using the following formulas:

$$\text{upper control limit (UCL)} = \overline{\overline{X}} + (A_2\overline{R})$$

$$\text{lower control limit (LCL)} = \overline{\overline{X}} - (A_2\overline{R})$$

The symbol A_2 refers to a constant that is based on the sample size. A simple listing of control-chart constants is given in Table 5-9.

Example: Assume a sample size consists of 4 parts. The average range of variation in tolerance is 0.0004". The grand average is 0.9376". Both the upper and lower control limits are required.

$\overline{\overline{X}}$ is given as 0.9376", and $\overline{R}$ is given as 0.0004". The A_2 constant for 4 parts in a sample (from Table 5-9) is 0.7289. By substituting these values in the formula:

$$UCL = \overline{\overline{X}} + (A_2\overline{R})$$

TABLE 5-9 Averages and Ranges Control-Chart Constants (Small Samples)

Number of Inspections in Sample	Factors for Control Limits (Averages)			Factor for Central Line	Factors for Control Limits (Ranges)			
	A	A_1	A_2	d_2	D_1	D_2	D_3	D_4
2	2.1210	3.7590	1.8800	1.1280	0	3.6860	0	3.2680
3	1.7320	2.3940	1.0230	1.6932	0	4.3581	0	2.5740
4	1.5000	1.8800	0.7289	2.0590	0	4.6980	0	2.2820
5	1.3420	1.5959	0.5770	2.3260	0	4.9180	0	2.1140

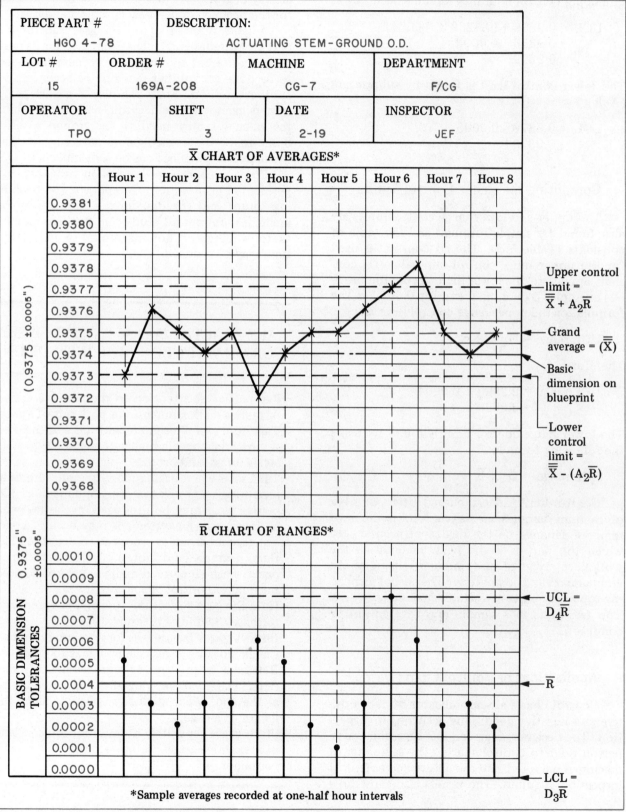

FIGURE 5-21 Average and Range Chart with Inspection Data and Control Limits for a Manufactured Part

The upper control limit may be found as follows:

$$UCL = 0.9376 + (0.7289 \times 0.0004)$$
$$= 0.9376 + 0.0003$$
$$= 0.9379''$$

The lower control limit is found by subtracting $(A_2 \overline{R})$:

$$LCL = 0.9376 - 0.0003$$
$$= 0.9373''$$

Computing the Range Control Limits

The upper and lower range control limits are also found by simple formulas and the table of constants (Table 5-9). The D_4 constant is used for the upper range control limit. The D_3 constant applies to the lower limit. From the previous example, $D_4 = 2.2820$ and $\overline{R} = 0.0004''$. The formula for the upper range control limit is:

$$UCL \text{ range} = D_4 \times \overline{R}$$

Therefore,

$$UCL \text{ range} = 2.2820 \times 0.0004$$
$$= 0.0009''$$

The lower range control limit is found by using the following formula:

$$LCL \text{ range} = D_3 \times \overline{R}$$

Control limits are established after a number of random samples are taken. The number of samples depends on the degree of accuracy to which the part is to be held, material in the workpiece, type of machine and cutting tools, and part size and complexity. The cost of inspecting too many parts in a sample may be excessive. Too few parts in a sample may give an untrue condition.

Applications of Control Charts

Control charts are an indicator of both positive and negative conditions occurring in inspection. The designer, engineer, manager, and craftsperson refer to control charts. The samplings of machined parts indicate the nature and extent of inspection required. The results also show how closely one or more processes must be controlled for the product to meet acceptable standards.

A great number of rejects may call for the part to be redesigned. The tolerance limits may require different tooling. The process may need to be done on another machine or according to another method. For example, if a part being turned cannot be held within a tolerance limit, the diameter may need to be ground to size. Sometimes, the control-chart data suggest that the material specifications be reexamined.

Control charts are helpful in providing experimental information. Comparisons of machining setups and speeding-up techniques may be made. Charts permit study and decision making as to whether or not machining or inspection time may be reduced.

SAFE PRACTICES IN APPLYING GAGES FOR QUALITY CONTROL INSPECTION

—Pay particular attention to the abrasive action of metals on the dimensional accuracy of a gage. Cast iron, cast aluminum, and some of the plastic materials have a higher abrasive action than steels, brass, and bronze.
—Take care when gaging close tolerance holes in soft metals to prevent seizing and loading. Excessive wear may be produced on the gage and an inaccurate measurement may result. A special gage lubricant is produced to coat the gage.
—Burr, degrease, and thoroughly clean machined parts. Wash all parts in a lot that are to be used for quality control.
—Hold the gage to feel the true measurement and to sense the line of measurement.
—Inspect gages frequently for size and scratches. Check working gages against a master gage.
—Return a worn gage for reprocessing or scrapping. The wear on plug and ring gages begins at the end. Wear gradually advances along the length of a gage.
—Store gages separately from other tools. Clean and examine gages before storing in a special container.

QUALITY CONTROL PRINCIPLES 103

TERMS USED IN QUALITY CONTROL AND MEASUREMENT PRACTICES

Quality control	Techniques and measurement systems used in manufacturing to ensure conformity with standards. Dimensional and materials control obtained through sampling plans. Maintenance of quality and the improvement of product resulting from continuous inspection.
Fixed-size gage	A designation of a set plug (inside) or ring (outside) gage used to check a dimension.
Dimensional tolerance	The permissible variation of a workpiece from its nominal dimension. (This variation may be the result of unilateral or bilateral tolerances.)
Classes of tolerances (holes)	Four classes of (gage maker's) tolerances: XX, for laboratory standards; X, for inspection gages; Y, for working gages that are lapped; and Z, for ground working gages.
Fit	A range of tightness between two mating parts. Dimensioned on a drawing by the basic size and the upper and lower limits.
Interference fit	A negative fit. The machining of a part to a dimension that requires the mating parts to be forced together.
Classes of fit (holes)	Standards accepted by manufacturers, designers, and engineers for mating parts. Specifications generally used by the craftsperson for sliding parts, forced fits, average machine fits, fine tool and work assemblies, and others.
Controlling point in thread measurement	The theoretically correct (basic) size from which all thread measurements are derived.
H and L pitch diameter limits	A letter specification for ground thread taps. (The letter is followed by a number corresponding to a class of fit.) A range of allowances above (H) or below (L) the basic pitch diameter.
Tolerance symbols (ISO metric threads)	A number and letter system to identify the position and amount of thread tolerance. The maximum and minimum tolerance limits for internal (G and H) and external (g and h) threads. (The first combination of number and letter is the pitch tolerance symbol. The second combination is the tolerance symbol for the crest diameter.)
Normal frequency-distribution curve	A profile with equal distribution on both sides of the mean (centerline). A symmetrical line graph with six equally spaced zones. A graph rising from zero to a maximum value at the center mean and dropping uniformly back to zero.
Acceptable product	Workpieces that fall within the required tolerance.
Sampling plan	A scheme for inspecting a given number of parts in a batch. Establishing the number of parts to be inspected to meet an acceptable quality level.
Control charts	Types of charts that provide technical information about the parts undergoing inspection. $\overline{X}$, $\overline{R}$, c, and p charts. Charts that show the number of defective parts in a sample or variations in the averages or range of samples.

SUMMARY

Quality control involves dimensional, materials, assembly, and other inspections.

Fixed gages are used in mass production. These gages are set or calibrated against masters or other standards. Fixed gages are commercially available in inch-standard and metric-standard sizes.

The four tolerance classes are: XX, X, Y, and Z. Fixed gages may be used to check parts ranging in precision from lapped surfaces (class XX) to commercially finished gages that are ground but not lapped (class Z).

Fixed, snap, and adjustable plug and ring gages are widely used in the shop.

The upper and lower limits of a basic size indicate the required range of tightness between two mating parts.

Engineering tables provide standards for allowances for different classes of fits. Some relate to standard tolerances for general purposes. Others give high and low limits for forced fits, driving fits, push fits, and running fits.

The thread form, diameter and pitch combinations, and size limits are standardized for Unified and American National Form threads.

Ground thread taps are designed with dimensional factors H and L. H refers to dimensions above the basic pitch diameter; L, below this diameter.

Numbers that follow H and L designations relate to class of fit.

Letters and numbers are used with ISO metric threads to indicate the position and amount of thread tolerance. Lowercase g and h are used for external threads. Capital G and H apply to internal threads.

Precision threads are specified by using a tolerance symbol for actual maximum and minimum limits. Letters and number combinations on ISO metric threads identify the position and amount of thread tolerance.

The distribution of acceptable manufactured parts generally falls within a normal frequency-distribution curve.

Products that fall outside the tolerance requirements indicate a need to refine the manufacturing processes.

The most common sampling plans in quality control include single-sampling, double-sampling, and sequential-sampling plans.

Control charts in quality control present graphically what is happening in inspection. c charts reflect the number of defects in one workpiece. p charts provide the percent of defective parts in the sample. $\overline{X}$ charts show the variations in the averages of the samples. $\overline{R}$ charts indicate variations in the range of the samples.

Upper and lower averages control limits are computed by formula:

$$UCL = \overline{\overline{X}} + (A_2 \overline{R})$$
$$LCL = \overline{\overline{X}} - (A_2 \overline{R})$$

The upper and lower range control limits require the use of constants for D_4 and D_3 limits:

$$UCL\ (range) = D_4 \times \overline{R}$$
$$LCL\ (range) = D_3 \times \overline{R}$$

Machined parts lots must be deburred and degreased before undergoing inspection.

Safe practices for regularly checking gages and instruments against masters for dimensional accuracy must be observed.

Gages and instruments require proper handling during use and protection when stored.

UNIT 5 REVIEW AND SELF-TEST

1. Identify four different kinds of testing that are necessary to quality control.
2. State the importance of *nominal size* to dimensions of mating features.
3. a. Refer to a table of classes of fits and gage maker's dimensional tolerances.
   b. Determine the gage maker's tolerance for parts (1) through (4) according to the class of fit given for each part: (1) 25.4mm, class XX; (2) 1.125", class X; (3) 203.2mm, class Y; and (4) 10.375", class Z.
4. a. Refer to an ANSI table of standard running and sliding fits.
   b. Determine (1) the standard tolerance limit for a 1" nominal diameter shaft and mating hole and (2) the minimum and maximum clearance for the shaft and mating hole. Note that the parts are assembled to an RC_1 fit, which indicates minimum running play.
   c. Give the maximum and minimum clearance between the shaft and the mating hole.
5. Explain why interchangeable Unified and American National Form threads will also interchange with SI metric threads having corresponding pitches.
6. a. Refer to a handbook table that compares metric series thread specifications in the French, German, Swiss, and British metric systems.
   b. Compare (1) the minor diameter of a bolt and (2) the major diameter of a nut in each of the four metric systems. The 16mm bolt has a 2mm thread pitch and a 14.7mm pitch diameter.
7. Give three corrective steps a machine operator takes when a quality control inspection shows a machined part falls outside the tolerance.
8. a. Use a sequential-sampling plan for the inspection of 900 parts at an acceptable quality level of 1.5.
   b. Indicate (1) the size of the next sample and (2) the cumulative size of sample if there are 4 or more rejects in the fourth sample.
9. a. Tell how information from a control chart may affect the part design and/or the manufacturing processes.
   b. Indicate how control charts are valuable in time-motion studies.
10. Tell how machined parts are made ready for quality control inspection.
11. State two safe practices to follow to ensure accurate gaging without damage to gages.

UNIT 6

Surface Texture Characteristics and Measurement

Interchangeable mechanical parts require surfaces to be machined to specified dimensions and quality of surface finish. Every machined surface has a degree of roughness. For example, a mirror-like precision surface is produced by super-finishing; a rough surface results from power cutoff sawing.

The initial approval of surface quality (texture) standards was made by the American Standards Association in 1962. These standards became known as the *ASA B46.1 Surface Texture Standards*. The standards had been developed cooperatively by the United States, Canada, and Great Britain. They were accepted by the three nations. Since 1962, there have been limited modifications in the standards to accommodate increasingly finer surface finishes and more precise instrument capability.

Surface texture standards are given on drawings by designers. The machinist, toolmaker, or machine operator needs to interpret specific terms and symbols. In addition, the worker must know how to measure the surface quality. Surfaces may be compared visually against a standard specimen. More precise measurements may be read directly on dial and surface tracer types of instruments.

This unit covers basic characteristics of surface texture. New terms are described and illustrated early in the unit. Drafting symbols and dimensional specifications, as found on drawings and sketches, are included. Comparator specimens and more precise surface texture measuring instruments are applied.

TERMS AND SYMBOLS USED TO SPECIFY AND MEASURE SURFACE TEXTURE

The terms *surface texture*, *surface finish*, *surface roughness*, and *surface characteristics* are used interchangeably in the shop and throughout this text. However, particular American (ANSI) standard terms, ratings, and symbols are described and illustrated. Use of these terms, ratings, and symbols permits designers, skilled mechanics, and other workers to communicate. Thus, part specifications may be accurately interpreted to produce a given surface quality. Figure 6-1 identifies some common surface texture characteristics and terms.

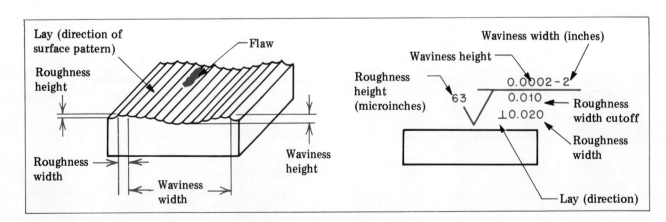

FIGURE 6-1 Common Surface Texture Characteristics and Terms

SURFACE TEXTURE CHARACTERISTICS

Surface Texture. *Surface texture* relates to deviations (from a nominal surface) that form the pattern of the surface. The deviations may be repetitive or random. The deviations may result from roughness, waviness, lay, and flaws.

Surface. According to American (ANSI) standards, an object is bounded by a *surface*. This surface separates the object from another object, substance, or space.

Surface Finish. *Surface finish* is indicated on a drawing by using the modified mathematical symbol √ and a microinch measurement. For example, a roughness height of 63 microinches appears on a drawing as ⁶³√ .

Profile. A *profile* is the contour of a machined surface on a plane that is perpendicular to the surface. The plane may also be at an angle other than 90°, if it is specified. A short section of a surface profile is illustrated in Figure 6-2.

Centerline. The *centerline* is a line that is parallel to the general direction of the profile. Figure 6-2 shows a centerline of an exaggerated profile. Roughness is measured around the centerline. This line lies within the limits of roughness width cutoff. The centerline in the profile serves as the cutoff line between roughness areas on both sides of the line.

Nominal Surface (Mean Line). A *nominal surface* is considered to be a geometrically perfect surface. Such a surface would result if the peaks were leveled to fill the valleys. The theoretical surface is represented by a *mean line*.

Microinch. A *microinch* is the basic unit of surface roughness measurement. A microinch designates one-millionth of an inch (0.000001″). The numerical value is sometimes followed by the symbol Mµ or µ.

Lay. *Lay* is the direction of the predominant surface finish pattern. Lay patterns, symbols, and applications in drawings are described later.

Roughness. The fine irregularities produced by a cutting tool or production process are identified as *roughness (surface roughness)*. Roughness includes traverse feed marks and other irregularities.

Roughness Width. The irregularities produced by production processes extend within the limits of the roughness width cutoff (the distance between successive peaks or ridges). *Roughness width* is measured parallel to the nominal surface. The peaks considered constitute the dominant pattern. The maximum roughness width rating, in inches, is placed to the right of the lay symbol. A rating of 0.020″ appears in Figure 6-3A.

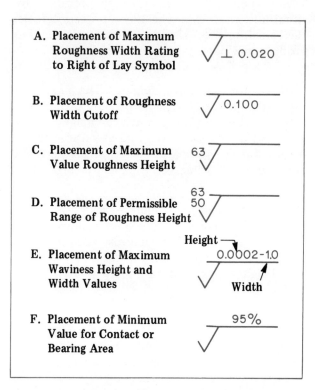

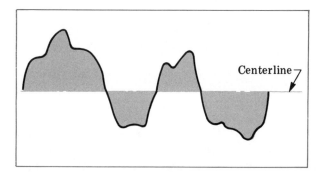

FIGURE 6-2 Exaggerated Short Section of a Surface Profile

FIGURE 6-3 Representation of Surface Characteristics on Drawings and Specifications

Roughness Width Cutoff. A rating in the thousandths of an inch is used for the greatest spacing (width) of surface irregularities that are repetitive. These irregularities are included in the measurement of average roughness height. The *roughness width cutoff* rating appears directly under the horizontal extension of the surface finish symbol. A rating of 0.100" is given in Figure 6-3B. If no value is shown, a rating of 0.030" is assumed.

Roughness Height. *Roughness height* is expressed in microinches. The height is measured normal to the centerline. The height is rated as an arithmetical average deviation. The roughness height value is placed to the left of the surface finish symbol leg. If only one value is given—for example, 63 microinches as shown in Figure 6-3C—it indicates the maximum value. When maximum and minimum values for roughness height are given, they indicate the permissible range. In Figure 6-3D, the range indicates 63 microinches maximum and 50 microinches minimum.

Waviness. *Waviness* is a surface deviation (defect) consisting of waves. These waves are widely spaced. Waviness may be produced by vibrations, chatter, or machine or work defects. Heat-treating processes and warping strains also produce waviness.

Maximum Waviness Height. The *maximum waviness height* is the peak-to-valley distance of a wave. The maximum waviness height rating is placed above the horizontal extension of the surface finish symbol, as shown in Figure 6-3E by the 0.0002" value.

Waviness Width. *Waviness width* is the spacing of successive wave peaks or valleys and is measured in inches. The value also appears above the surface finish symbol but next to the waviness height, as shown in Figure 6-3E by the 1.0" value.

Flaws. *Flaws* are surface defects. They may occur in one area of the unit surface. Flaws may also appear at widely varying intervals on the surface. Casting and welding cracks, blow holes, checks, and scratches are common flaws. Normally, the effect of flaws is not included in measurements of roughness height.

Contact or Bearing Area Requirement. The *contact or bearing area* is expressed as a percentage value. This value is placed above the extension line of the surface finish symbol, as shown in Figure 6-3F by the 95%. The value is the minimum contact or bearing area requirement as related to the mating part or reference surface.

REPRESENTATION OF LAY PATTERNS AND SYMBOLS ON DRAWINGS

Six lay symbols are commonly used in drawings and parts' specimens. Tool or surface finish patterns are indicated by perpendicular (⊥), parallel (∥), angular (X), multidirectional (M), circular (C), and radial (R) symbols. These symbols are further described and illustrated in Table 6-1 in relation to predominant tool patterns and applications on drawings.

Application and Interpretation of Lay Symbols on a Drawing

The manner in which lay symbols and values are applied on drawings is shown in Figure 6-4A. On the one-view drawing, the two symmetrical diameters are indicated by the conventional drafting symbol ⌀.

The surface finish of the smaller ground diameter (1.500 +0.0000/−0.0001) must be held to the limits as specified. The surface texture symbol rests on the object line. The dimensions are interpreted in Figure 6-4B.

COMPUTING AVERAGE SURFACE ROUGHNESS VARIATIONS

Computation of Arithmetical Average (AA)

The *arithmetical average* (AA) of surface roughness is a new unified reading (number) of the average (roughness) variations of a surface above and below a centerline. This average is a measurement of the "hills and valleys" in microinches.

The arithmetical average (AA) is obtained by adding the heights of the surface roughness increments. The increments—for example, a, b, c, and so on in Figure 6-5—represent the distance from the centerline to a point on the profile taken at regular intervals. The values in Table 6-2 give the surface roughness measurements above and below the centerline. These values are added. The sum is then divided by the number of intervals. In the example, the arithmetical average (AA) is 8.86 microinches (Table 6-2).

TABLE 6-1 Predominant Tool Patterns and Applications of Lay Symbols on Drawings

Lay Symbol	Predominant Tool Pattern	Application of Lay Symbol on a Drawing
∥	Parallel tool marks	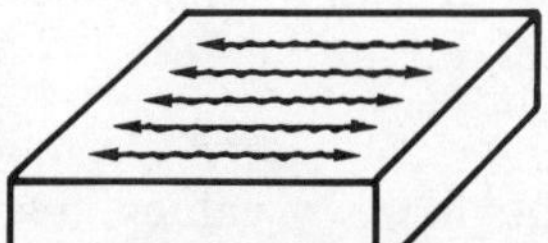 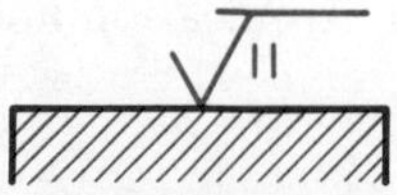 Lay parallel to object line (surface) to which the symbol applies.
⊥	Perpendicular tool marks	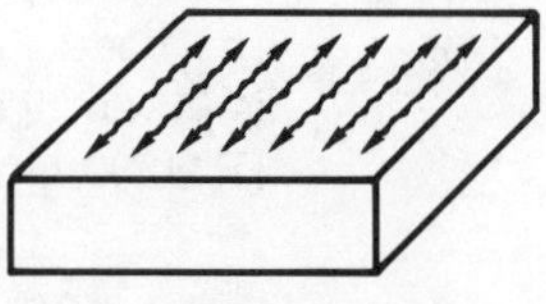 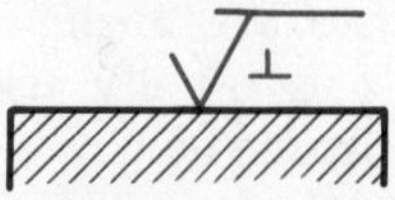 Lay perpendicular to object line to which the symbol applies.
X	Angular direction tool marks	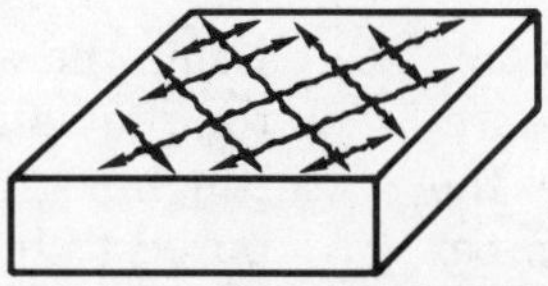 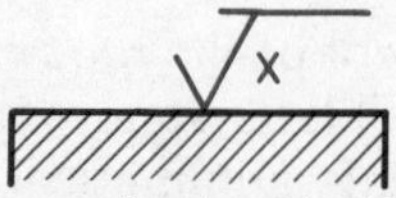 Lay angular to object line.
M	Multidirectional or random direction tool marks	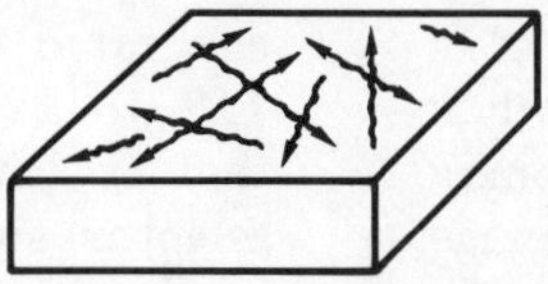 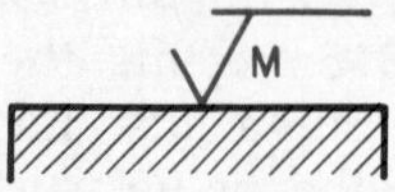 Lay multidirectional to object line.
C	Circular tool marks	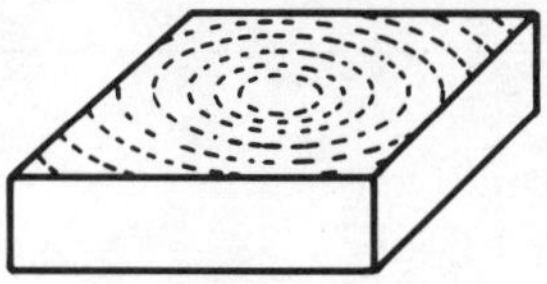 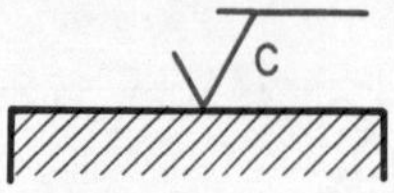 Lay approximately circular to the center of the workpiece represented by the object line.
R	Finish marks radial to the center of the surface to which the symbol is applied	 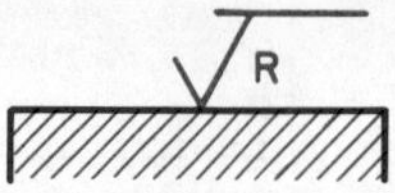 Lay approximately radial in relation to the object line to which the symbol is applied.

110 PRECISION MEASUREMENT AND INSPECTION

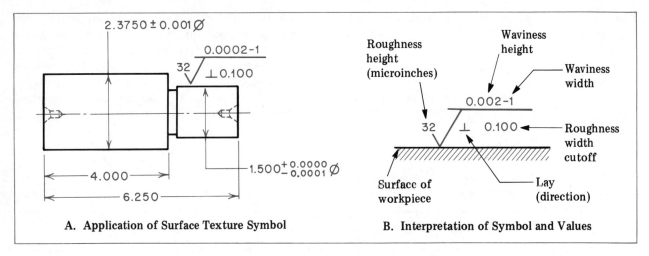

FIGURE 6-4 Application and Interpretation of Lay Symbol and Surface Texture Specifications

The total profile height of the surface roughness has been established from experience. This height for most surfaces is approximately 4 times the measured surface finish in microinches. Thus, the number 4 is usually used to establish the approximate profile height.

Concept and Computation of Root Mean Square Average (RMS)

The concept of the *root mean square average* (RMS) is an older system of roughness measurement than the arithmetical average. Like the AA, the RMS consists of a number. The number is read on the dial of a surface finish analyzer.

The RMS represents the square root of a series of measurements. Each measurement is the distance the actual surface varies from a nominal surface. Like the arithmetical average values, each measurement is taken at equally spaced intervals. Each value is squared and added to the next measurement. The sum of the measurements is divided by the number of intervals (n) used. Finally, the square root of the result is taken. The square root is the root mean square (RMS) value:

$$\text{RMS} = \sqrt{\frac{a^2 + b^2 + c^2 + d^2 \ldots}{n}}$$

Using the microinch intervals (n) and values from the previous example, Table 6-3 gives the squared roughness measurements. The RMS value is computed to be 9.92 microinches.

A roughness measuring instrument that is calibrated for AA values gives an approximately 11% to 12% lower reading than an instrument that is RMS calibrated. Some designs have a selector switch to permit selecting either the

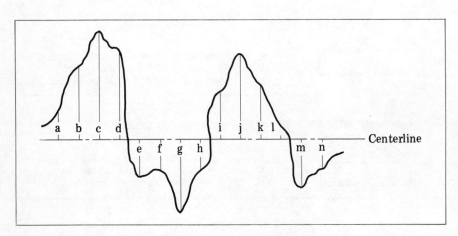

FIGURE 6-5 + and − Microinch Measurements Taken at Regular Intervals (Hypothetical Profile)

SURFACE TEXTURE CHARACTERISTICS 111

TABLE 6-2 Roughness Measurements and AA Value (Intervals a through n)

Interval	Roughness Measurement (Microinches)
a	4
b	12
c	18
d	15
e	6
f	5
g	12
h	5
i	8
j	14
k	9
l	3
m	8
n	5
Totals	
14	124

$$AA = \frac{124}{14} = 8.86 \text{ microinches}$$

TABLE 6-3 Squared Roughness Measurements and RMS Value

Interval	Squared Roughness Measurement (Microinches)
a	a^2 = 16
b	b^2 = 144
c	c^2 = 324
d	d^2 = 225
e	e^2 = 36
f	f^2 = 25
g	g^2 = 144
h	h^2 = 25
i	i^2 = 64
j	j^2 = 196
k	k^2 = 81
l	l^2 = 9
m	m^2 = 64
n	n^2 = 25
Totals	
14	1,378

$$RMS = \sqrt{\frac{1,378}{14}} = 9.92 \text{ microinches}$$

RMS or AA scale. Many manufacturers of parts adjust AA readings without changing RMS values found on old drawings.

MEASURING SURFACE FINISHES

A few American standard measurement values of surface roughness height, roughness width cutoff, and waviness height are given in Table 6-4. The full table of surface texture values is included in the Appendix.

Functions of Surface Comparator Specimens

Surface comparators (specimens) are produced commercially to measure surface roughness. The comparators have scales that conform

TABLE 6-4 American Standard Surface Texture Values

Dimensional Measurement	Measurement Values				
	Microinches				
Roughness height	1	13	50	200	
	2	16*	63*	250*	
	...				
	Inches				
Standard roughness width cutoff	0.003	0.010	0.100	0.300	1.000
	...				
	Inches				
Waviness height	0.00002	0.0001	0.001	0.010	
	0.00003	0.0002	0.002	0.015	
	...				

*Recommended values

precisely with ANSI standards of surface roughness and lay. Surface comparators are generally flat or round. Different scales are designed for gaging common types of machined, ground, and lapped surfaces. The finer microinch-finish surface comparator specimens are often used with a *surface finish viewer* (Figure 6-6).

Figure 6-7 shows a microfinish comparator specimen containing 22 scales of machined surfaces. The roughness height range is from 2 to 500 microinches. The roughness height, lay, and machining process are indicated on each scale. For example, the 2L scale is used to gage the surface finish of lapped parts: The 2 indicates a 2-microinch roughness; the L relates to a lapped finish.

Surface Measurement with Comparator Specimens

In actual practice, a number of different machined surfaces may be measured by comparator specimens. Several surfaces are listed in the first column of Table 6-5. General ranges of surface roughness heights and lay appear in the second and third columns respectively. Roughness widths are given in inches in the fourth column for machining processes that produce roughness heights between 2 and 500 microinches. The actual measured values in microinches are given in the fifth column. The letters in the sixth column tell that the measured value was obtained by either a brush analyzer (B) or a profilometer (P).

Surface roughness specimens are available for electrical discharge machined finishes on specific steels such as tool steel (16 to 250 microinches). Surface finish specimens are also provided for grit-and-shot blast surface finishes (32 to 1,000 microinches). Other specimens are available for die-cast and centrifugal, green-sand, and other casting processes.

HOW TO MEASURE SURFACE TEXTURE WITH COMPARATOR SPECIMENS AND A VIEWER

Direct Inspection

STEP 1 Determine the surface texture specifications from the drawing.

STEP 2 Note the lay pattern, machining process, and the required roughness height and roughness width.

STEP 3 Select a comparator specimen with the same characteristics as the required surface finish.

STEP 4 Compare the surface finishes of the workpiece and the specimen standard.

Note: On the basis of the comparison, the accuracy of the tools is established. If needed, adjustments are then made to produce the desired surface finish.

FIGURE 6-6 Surface Finish Viewer and Comparator Specimen for Lay, Process, and Surface Roughness

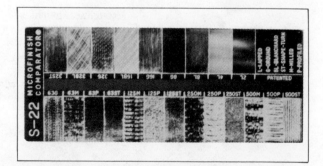

FIGURE 6-7 Microfinish Comparator Specimen Containing 22 Surface Finish Scales

SURFACE TEXTURE CHARACTERISTICS 113

Inspection with Viewer

STEP 1 Repeat steps 1 through 3 for direct visual inspection.
STEP 2 Slide the surface finish comparator specimen into the viewer.
STEP 3 Place the comparator instrument over the surface. The instrument has a magnifier and built-in illumination.
STEP 4 Turn the eyepiece to bring the machined surface into focus.
STEP 5 Match the surface roughness.

Note: A visual check against a standard establishes the quality of surface finish. The skilled mechanic may also use this check to provide information about the material, machine and cutting tools, speeds, and feeds.

SURFACE ROUGHNESS RELATED TO COMMON PRODUCTION METHODS

Different production methods produce variations in surface texture. Figure 6-8 provides ranges of surface roughness for selected

TABLE 6-5 Machined Surfaces That May Be Measured by Comparator Specimens

Machined Surface	Roughness Height (Microinches)	Lay	Roughness Width (Inches)	Actual Microinch Value	Type* of Instrument
Honed, lapped, or polished	2	Parallel to long dimension of specimen	....	2.4	B
	4		....	3.8	B
	8		....	6.5	B
Ground with periphery of wheel	8	Parallel to long dimension of specimen	....	7.8	B
	16		....	14.	B
	32		....	28.	B
	63		....	64.	B
Ground with flat side of wheel (Blanchard)	16	Angular in both directions	....	17.	B
	32		....	28.	B
Shaped or turned	32	Parallel to long dimension of specimen	.002	31.	B
	63		.005	60.	P
	125		010	130.	P
	250		.020	250.	P
	500		.030	530.	P
Side milled, end milled, or profiled	63	Circular	.010	57.	P
	125		.020	125.	P
	250	Angular in both directions	.100	235.	P
	500		.100	470.	P
Milled with periphery of cutter	63	Parallel to short dimension of specimen	.050	55.	P
	125		.075	115.	P
	250		.125	240.	P
	500		.250	520.	P

Tolerances	
Roughness Height (Microinches)	Tolerance (Percent)
2 to 4	+25 −35
8	+20 −30
16	+15 −25
32 and above	+15 −20

*B represents brush analyzer; P, profilometer.

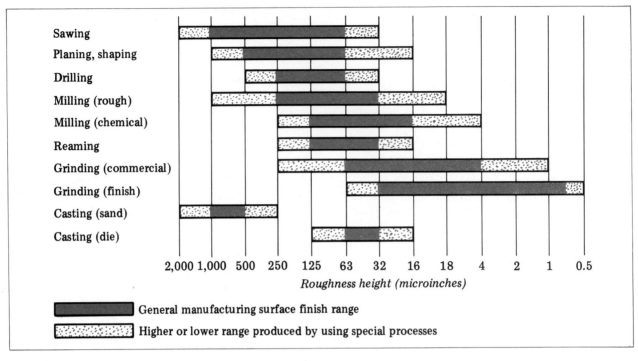

FIGURE 6-8 Microinch Range of Surface Roughness for Selected Manufacturing Processes (Partial Listing)

manufacturing processes. General and nonconventional machining processes and forming methods are considered.

The roughness height range is indicated in microinches. The solid bar for each process covers general applications. The coarser and finer roughness heights (dotted areas) are produced under special conditions.

SURFACE TEXTURE MEASURING INSTRUMENTS

Surface Finish Analyzer

Variations between the high and low points of the work surface may be measured by a *surface finish analyzer*. The three main components of the instrument are a tracer head, amplifier unit, and a recording dial (Figure 6-9).

A diamond stylus is mounted in the tracer head. The tip of the stylus is rounded. The two basic tip radii are 0.0001″ and 0.0005″. The profiles of surface roughness produced by using these two tip radii in four examples of machining processes are presented in Table 6-6. The same

wave forms are shown with readings taken using 0.0001″ and 0.0005″ tip radii.

The tracer head is moved across the work surface either manually or mechanically. More precise surface roughness measuring instruments, such as a *brush analyzer*, for example, are moved mechanically. The stylus moves as it follows the surface profile. The minute movements are

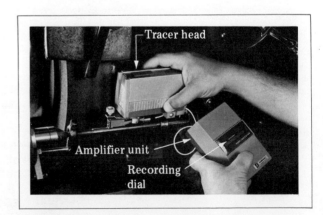

FIGURE 6-9 Surface Finish Analyzer Used to Measure Roughness Variations

TABLE 6-6 Readings of Machined Surface Profiles Obtained by Using Two Radii Stylus Sizes

Machining Process	Surface Profiles and Roughness	Stylus Radius (0.0000")	Average Reading (Microinches)
Shaping	⟵ 0.120" ⟶	0.0005 0.0001	100-105 110-115
Turning		0.0005 0.0001	45-50 40-44
Grinding		0.0005 0.0001	8-9 10-11
Polishing		0.0005 0.0001	1.4-1.7 1.4-1.7

converted in the tracer head into electrical fluctuations. These fluctuations are magnified in the amplifier. The mechanical movement of the stylus is registered by a needle on the surface recording dial.

The reading is given in microinches. It is an arithmetical average (AA) or the root mean square (RMS). This reading indicates the average surface variation from a reference or centerline.

Profilometer

The *profilometer* is another surface roughness measuring instrument. It employs a tracing point or stylus to establish the irregularities. These irregularities are greatly magnified and recorded. The profilometer in Figure 6-10 can be set up for manually checking a ground cylindrical surface. The amplimeter pointer shows the roughness on a microinch scale. Direct readings are given as the instrument tracer moves across the work surface.

The tracer unit of a profilometer may be moved over the work surface by hand or by a motor-driven tracing mechanism. The vertical movement of the tracing point produces a fluctuating voltage. This voltage is amplified so that it actuates the pointer in the amplimeter. As the tracer moves along the work, the pointer moves back and forth. The movement is within a range of roughness values.

Another type of profilometer produces a recorded profile on a tape, as shown in Figure 6-11. The height of the horizontal lines is 25μ (microinches). Waviness, roughness, and waviness/roughness profiles are recorded.

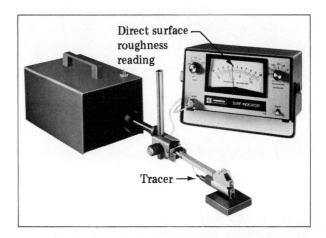

FIGURE 6-10 Profilometer Setup for Manually Checking the Roughness of a Cylindrical Surface

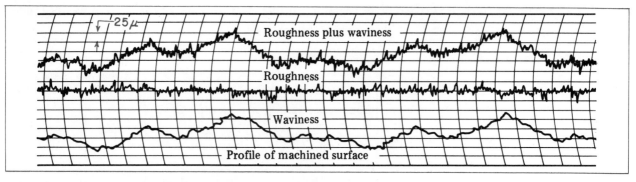

FIGURE 6-11 Recorded Roughness and Waviness Profiles of a Machined Surface

HOW TO MEASURE SURFACE FINISH WITH AN INDICATING INSTRUMENT

STEP 1 Read the blueprint of the workpiece. Determine the required surface finish characteristics.

STEP 2 Turn the switch of the surface finish analyzer on. Allow the instrument to warm up.

STEP 3 Check the machine calibration. Use the 125 microinch test block. Move the stylus on the tracer head over the block. A movement of approximately 1/8" per second is a good speed.

STEP 4 Adjust (calibrate) the instrument. The pointer should register 125 microinches, the same measurement as the test block.

STEP 5 Use the 0.030 cutoff range for surface roughness of 30 microinches or more. Use the cutoff range of 0.010 for surfaces finer than 30 microinches.

Note: Set the range switch to a high setting if the roughness is not known. The switch may then be turned to a finer setting to get the accurate surface finish reading.

STEP 6 Clean the work surfaces to be measured.

STEP 7 Place the tracer over the machined surface. Move the stylus at the rate of 1/8" per second over the surface.

STEP 8 Read the surface roughness by the movement of the pointer. The reading is in microinches.

Note: On instruments like the surface analyzer, the recording device produces an ink graph of surface irregularities. The graph provides an accurate record for study. If necessary, changes may then be made in the production tooling or methods of manufacture.

Surface Finish Microscope

Surface texture may also be measured with a *surface finish microscope*. This microscope depends on a light wave principle of measurement similar to the principle behind optical flats. The surface to be measured is magnified from 125 to 200 times. The surface finish is then compared to a comparator plate of a known standard. The view of one surface is superimposed on the other surface.

The variation indicates the accuracy of the surface finish. Some microscopes have a direct-reading attachment. Variations in surface finish are read on a dial face.

SAFE PRACTICES IN USING SURFACE FINISH ANALYZERS AND INSTRUMENTS

—Set the range switch on a surface finish analyzer to a high setting when the surface roughness is not known. After the initial test, a finer setting may be needed to get an accurate surface reading.

—Allow a surface finish analyzer instrument to warm up before taking a measurement.
—Operate measuring instruments and workpieces in an area that is dust free and away from vibration.
—Protect the precision stylus point from hitting a surface.
—Check surface finish instruments periodically against a master. The instrument should be calibrated at the same AA or RMS as the master.

SUMMARY

Surface texture specifications reflect the ideal surface finish required by the designer.

Repetitive and random deviations in surface finish result from roughness, waviness, lay patterns, and flaws.

Surface roughness refers to the fine irregularities produced by a cutting tool or production process. These irregularities extend within the limits of the roughness width cutoff.

Roughness width cutoff is the rating in inches for the greatest spacing of surface irregularities that are repetitive.

Roughness height is an arithmetical average deviation in microinches. This height is measured normal to the centerline.

The six common symbols used to show the dominant tool or finishing patterns are: ⊥ (perpendicular), ‖ (parallel), X (angular), M (multidirectional), C (circular), and R (radial).

Comparator scales for inspecting surface finish conform precisely with ANSI standards of surface roughness.

Comparator scales are marked to show the roughness height, lay pattern, and machining process.

Comparator specimens are available for conventional machine processes, electrical discharge machined finishes, casting, and other production techniques.

The surface finish analyzer consists of a tracer head, amplifier, and recording dial. Readings are obtained using either a 0.0001" or 0.0005" stylus.

The profilometer also magnifies surface irregularities. The tracer unit may be moved across the work surface either manually or automatically.

Recording units on surface finish instruments produce charts of waviness and roughness and a combination waviness/roughness graph.

The surface finish microscope magnifies a surface finish. The finish is then compared with a comparator specimen or other standard.

Standard precision instrument safety precautions must be followed in the use of all gages in quality control.

Surface finish instruments must be warmed up before measurements are taken. Vibration, dust, and foreign particles affect measurement accuracy.

UNIT 6 REVIEW AND SELF-TEST

1. a. Use a table of surface texture values for general machining processes.
   b. Give the *lay pattern* and the *roughness height in microinches* for the following five machining processes and roughness width values: (1) end milling, 0.005; (2) straight turning, 0.002; (3) shaping, 0.001; (4) cylindrical grinding, 0.0005; and (5) drilling, 0.005.

2. a. Describe the functions served by a surface comparator specimen and viewer.
   b. Identify the kind of information the worker obtains by visual and viewer inspection of surface texture measurement.

3. a. Use a table of surface roughness heights for general manufacturing processes.
   b. Give the lower surface finish range in microinches (where special precision manufacturing processes are used) for the following operations: (1) reaming, (2) finish grinding (cylindrical), (3) drilling, and (4) rough milling.

4. a. Cite two main differences between a surface finish analyzer and a comparator specimen.
   b. State one advantage of using a surface finish analyzer over the comparator specimen method.

5. Tell why work surface profiles are tape recorded on a profilometer.

6. State two advantages of using a surface finish microscope in preference to a profilometer for general surface texture applications.

7. Indicate one check that must regularly be made with surface finish measuring instruments.

PART TWO

Turning Machines

SECTION ONE   Advanced Engine Lathe Work
        UNIT 7   Advanced Thread-Cutting Technology and Processes
SECTION TWO   Turret Lathes: Bar and Chucking Machines
        UNIT 8   Classifications, Features, and Applications of Turret Lathes
        UNIT 9   Spindle and Hexagon Turret Tooling and Permanent Setups
        UNIT 10   Basic External and Internal Machining Processes
        UNIT 11   Planning Turret Lathe Setups and Tooling
SECTION THREE   Single- and Multiple-Spindle Automatic Screw Machines
        UNIT 12   Screw Machine Technology, Practices, and Troubleshooting

SECTION ONE

Advanced Engine Lathe Work

The most widely used thread profile is the 60° American National Form. This form was adapted to be interchangeable and replaceable with British and Canadian screw threads. The modified profile that resulted from design and engineering changes is known as the Unified Form (UN). It has the 60° included thread angle with a rounded root and a flat crest. While ISO metric screw threads conform to the 60° thread angle, the metric pitch series vary from the Unified series. Screw threads in the American, British, Canadian, and metric forms and different series are part of the study of fundamental machine technology. Finer 60° screw threads, as well as square and 29° thread forms, are produced in advanced engine lathe work. Therefore, this section deals with the technology and processes involved in cutting these threads.

UNIT 7

Advanced Thread-Cutting Technology and Processes

The construction of smaller and more compact (microminiaturized) mechanisms requires applications of fine screw threads. Therefore, the American Standard Unified Miniature Screw Thread (UNM) is considered in this unit. Tabular information for this 60° thread is related to the Unified (inch-standard) and ISO (metric-standard) thread form and systems.

The square and the 29° thread form are also important thread forms. Three series in the 29° included angle forms are considered: the general-purpose American Standard Acme and Stub Acme threads and the worm thread (Brown & Sharpe). Characteristics, design features, and the cutting of these threads on an engine lathe are covered in the unit. Production thread-making processes and the technology and cutting of multiple-start threads are also considered.

PRODUCTION METHODS OF MAKING THREADS

The method of producing a thread depends on factors such as use, design features of the part to be threaded, dimensional accuracy and required surface finish, equipment available, and cost. The seven basic methods of producing screw threads are as follows:

—*Casting,*
—*Rolling,*
—*Chasing,*
—*Die and tap cutting,*
—*Milling,*
—*Grinding,*
—*Broaching.*

A brief description of each method follows.

Casting Methods

The most widely used casting methods are: sand casting, die casting, permanent-mold casting, plastic-mold casting, and shell-mold casting. The old sand-casting method produces an extremely rough thread. With refinements in shell molding, sand molding will result in smoother castings in a more accurate form.

Die Casting and Permanent Mold Casting. Threads produced by die casting and permanent mold casting have a high degree of accuracy and a good surface finish. The threads (internal) are cast in parts that are usually fastened together and are not disassembled.

The disadvantage of die cast and permanent-mold cast threads is in the low melting point alloy that is used. The parts are comparatively soft and have limited durability if reused. Many die cast parts are designed with steel and other inserts. These inserts are either cast in place or threaded into a hole. The inserts overcome the problem of rapid wear that results when a soft, die cast metal is used.

Plastic Mold Casting. Plastic mold casting is used on plastic materials only. Metal inserts (aluminum, brass, and steel) may be cast in place. Where a plastic material is strong enough to hold a fastener without stripping, it is economical to tap the threads.

Rolling Method

Rolling threads is the fastest and most economical method of producing external screw threads. This method is not new. Early patents for rolling threads were granted in the 1830s. Cold rolling has many advantages. The rolling and flowing *(plastic deformation)* action reduces mechanical problems such as tension, fatigue, and shear. Also, no chips are produced because no metal is removed. Furthermore, the method produces a *burnished* thread surface. Surface finish accuracies as close as 4 microinches (AA value) are finer than the accuracies produced by any other thread-forming method.

Cold-rolling threads requires the use of thread-roll dies. The dies are made of high-speed steels, high-carbon–high-chrome steels, and silicon carbides. The dies do not require sharpening because the threads are formed by *upsetting*, which is similar to the raising of diamond or straight-line surfaces by knurling.

Rolling threads are formed by displacing metal using flat or round dies. These dies are shaped in the exact form of the finished thread. A sliding motion of the dies burnishes and work hardens the threads. No burrs are left. The accuracy of the thread lead, pitch diameter, and thread angle is maintained over longer runs with cold rolling than with any multiple-point cutting tool process.

Chasing Method

Thread chasing is the process of cutting a screw thread on a lathe. The method is comparatively slow. A series of cuts is required to machine a thread to depth. However, the tool may be adjusted and controlled in thread chasing. Thread chasing provides versatility in cutting a thread to a specific class of fit. Multiple and quick-lead threads, taper threads, and nonstandard or unusual thread sizes may be cut by the chasing method. It is also more economical than keeping taps and dies for large diameters when only an occasional part is to be threaded.

Die and Tap Cutting Method

Taps are used for the internal threading of mass-produced parts. Similarly, dies are used to cut external threads. The quality and accuracy of threads produced with taps and dies meet general commercial requirements.

Milling Method

A milled thread is formed by a revolving milling cutter. The shape of the cutter conforms to the required thread form. Both internal and external threads may be milled. This method produces a more accurate thread than a thread produced by using taps and dies. Several advantages of the milling process are as follows:

—The formation of coarse-pitch or long thread is particularly suited to this process;
—Lead screws may be milled to close tolerances under fast production conditions;
—The full thread depth may be cut in one pass;
—A simple cutter may be used to mill more than one thread size.

Single or multiple cutters are used for thread milling. When a single cutter is used, the workpiece revolves a number of turns. When a multiple cutter is used, a thread is finished to depth during one revolution of the work. A thread is usually rough and finish milled in two steps when it is to be held to close dimensional and form tolerances.

Grinding Method

Hardened parts are threaded by grinding. Grinding is the most precise method of generating a screw thread. Pitch diameters may be held to an accuracy of ±0.0001" per 1" (±0.002mm per 25.4mm). The accuracy of lead may be ground within 0.0003" in 20" (0.01mm in 508mm).

Both internal and external threads may be ground. Single or multiple-form grinding wheels are used. The threads may be completely ground to depth from solid stock or finish ground. Some advantages of grinding are as follows:

—Distortion resulting from heat treating may be eliminated (the threads may be ground from solid stock after hardening);
—Parts that may be distorted by milling can have the threads ground to depth without distortion;
—Hardening and stress cracks from preformed threads are eliminated;
—Grinding threads to close dimensional and form accuracies is practical.

The thread is generated by dressing the wheel edge to the required form. As the work rotates against the direction of the wheel, the complete thread may be formed in one pass. A continuous size control mechanism is included on some types of grinding machines. This mechanism permits the wheel to be trued and adjusted to the work automatically.

The work speed for general thread grinding is from 3 to 10 inches per minute (ipm) or 75mm to 250mm per minute (mm/min). Wheel speeds to 10,000 surface feet per minute (sfpm) or 3,050 meters per minute (m/min) may be used. During each revolution of the work, the work is moved past the grinding wheel a distance equal to the thread pitch.

Thread Grinding Wheels. Resinoid-bond wheels are used when a fine edge must be maintained and where a limited degree of accuracy is required. These wheels operate at 9,000 to 10,000 sfpm (2,750 to 3,050 m/min). Vitrified-bond wheels are more rigid than resinoid-bond wheels. Vitrified-bond wheels are used for extreme accuracy. The recommended speed range is from 7,500 to 9,500 sfpm (2,300 to 2,900 m/min).

Threads are ground in carbide and hard alloys by using diamond wheels. These wheels are of rubber or plastic bond. Diamond chips are set in the bond. Threads may also be produced by centerless grinding. Product examples of such grinding include taps, worms, lead screws, thread gages, and hobs.

Broaching (Scru-Broaching)

Presently, broaching has limited applications to internal threads. The method, often referred to as *scru-broaching*, is used principally in the automotive field. Workpieces such as steering-gear ball nuts, lead screws, and rotating ball-type automobile assemblies may be scru-broached.

An opened positioning fixture and tooling setup for a broaching machine are shown in Figure 7-1. The broach has a pilot end and spiral-formed teeth on the body. The broach is guided by a lead screw. As the broach turns, the workpiece and fixture are drawn up to cut the thread. The threads may be cut in one or two passes. If roughing and finishing cuts are to be taken, rough- and finish-cutting broaches are required. Parts that are threaded by broaching are sometimes heat treated after broaching. The threads are then finish ground on an internal thread grinder. The teeth of a threading broach are

FIGURE 7-1 Opened Fixture, Broach, and Scru-Broach Machine Setup for Broaching an Internal Thread

sharpened by grinding the spiral faces. While the thread size is reduced a small amount by grinding, the thread form remains the same.

AMERICAN STANDARD UNIFIED MINIATURE SCREW THREADS

Sizes in the UNM Series

The *American Standard Unified Miniature Screw Thread Series* (UNM) is an addition to the standard coarse and fine thread series of the American National and the Unified thread series. The UNM thread series provides general-purpose fastening screws for applications on instruments and microminiature (exceedingly small, precise) mechanisms. UNM threads are also known as *Unified Miniature Screw Threads*. The 60° UNM basic thread form and series range from 0.0118″ to 0.0551″ (0.30mm to 1.40mm) in diameter. This UNM screw thread series supplements the Unified and the American National standard thread series, which begin at 0.060″.

Fourteen sizes and pitches have been endorsed as the foundation for the Unified standard. These sizes and pitches coincide with a corresponding range endorsed by the International Organization for Standardization (ISO).

The UNM screw thread form is compatible with the basic profiles of both the Unified (inch-standard) and ISO (metric-standard) systems. Threads in either series are interchangeable.

Characteristics of the UNM Thread Form. The 60° UNM thread profile is identical to the 60° basic thread form in the Unified and American National systems. However, the height and depth of engagement equals 0.52 × pitch, which is used instead of 0.5413 × pitch to permit more precise agreement between metric and inch dimensions.

There is only one thread class in the UNM screw thread standard. This class establishes a zero allowance. The screw threads are identified on drawings by the major (outside) diameter in millimeters and the letters UNM. The designation indicates that the threads are in the Unified Miniature Screw Thread Series. For example, a designation of 1.20 UNM on a drawing gives the major diameter as 1.200mm (0.0472″). The thread is in the UNM thread series. From tables, the metric pitch is established as 0.250mm (or the equivalent 102 threads per inch).

UNM Thread Form Tables. Tables are available that give the *limiting diameters* according to size and tolerances for internal and external threads. The limiting diameters correspond to the major, pitch, and minor screw thread diameters. Table 7-1 gives the size and tolerance limits required in the design, machining, and measurement of UNM screw threads. The complete table for the 14 metric pitches and corresponding inch-standard threads per inch in the series is included in the Appendix.

Note that maximum and minimum dimensions are given in Table 7-1 for the limiting diameters. In practice, the cutting-tool form is often relied on to produce a thread within the specified dimensions. The thread lead angle is a very important item to consider in grinding the cutting-tool point. The lead angle at the basic pitch diameter for each thread in the series appears in the last column in the table.

The minimum flat on a thread-cutting tool equals 0.136 × pitch. The thread height at the minimum flat is 0.64 × pitch. The thread flat and depth may be either computed or read from UNM tables of design data.

TECHNOLOGY AND CUTTING PROCESSES FOR SQUARE THREADS

The *square thread* is still widely used, although it is not as common as it was years ago. Square threads transmit great force. They usually have a coarser pitch than threads that are cut to a 60° form. Unfortunately, it is difficult and often impossible to compensate for wear on square threads. Cutting tools for square threads must, therefore, be ground with clearance for the leading and following sides.

Figure 7-2A and B shows the helix (lead) angle of a thread and the side angles of a tool bit for cutting a right-hand square thread. The helix angle of the thread varies. The angle depends on the thread lead and the thread diameter. The helix angle is needed to calculate the leading and following side angles of the cutting tool. A clearance of 1° is required on the leading and following sides of the cutting tool, as shown in Figure 7-3.

TABLE 7-1 Partial Table of Size and Tolerance Limits of Unified Miniature Screw Threads

Size Designation*	Metric Pitch (mm)	Metric-Standard External Threads (mm)						Lead Angle at Basic Pitch Diameter	
		Major Diameter		Pitch Diameter		Minor Diameter		Deg.	Min.
		Maximum[a]	Minimum	Maximum[a]	Minimum	Maximum[b]	Minimum[c]		
0.30 UNM	0.080	0.300	0.284	0.248	0.234	0.204	0.183	5	52
0.35 UNM	0.090	0.350	0.333	0.292	0.277	0.242	0.220	5	37
0.40 UNM	0.100	0.400	0.382	0.335	0.319	0.280	0.256	5	26

Size Designation*	Threads per Inch	Inch-Standard Internal Threads (0.001")						Lead Angle at Basic Pitch Diameter	
		Minor Diameter		Pitch Diameter		Major Diameter		Deg.	Min.
		Minimum[a]	Maximum	Minimum[a]	Maximum	Minimum[d]	Maximum[c]		
0.30 UNM	318	0.0085	0.0100	0.0098	0.0104	0.0120	0.0129	5	52
0.35 UNM	282	0.0101	0.0117	0.0115	0.0121	0.0140	0.0149	5	37
0.40 UNM	254	0.0117	0.0134	0.0132	0.0138	0.0160	0.0170	5	26

*Boldface denotes preferred series
[a] Basic dimension
[b] Use minimum minor diameter of internal thread for mechanical gaging.
[c] Reference only. The thread tool form is relied on for this limit.
[d] Reference only. Use the maximum major diameter of the external thread for gaging.

THREAD-CUTTING TECHNOLOGY 125

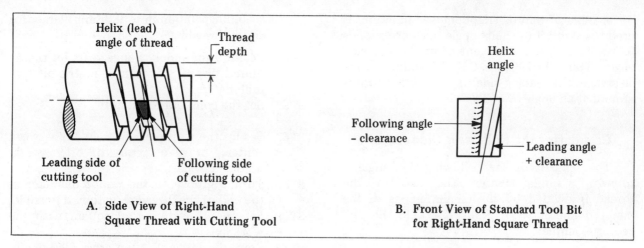

FIGURE 7-2 Leading and Following Side Angles of a Square-Threading Tool

Calculating the Leading Side Angle

Since a right triangle is formed (Figure 7-4), the tangent of the leading side angle equals the lead (l) of the thread divided by the circumference of the minor diameter (D_2). Expressed as a simple formula:

$$\tan \text{ leading side angle} = \frac{\text{lead of thread (l)}}{\text{circumference of minor diameter }(D_2)}$$

Example: A 1-1/4″-4 square thread is to be cut with a single-point cutting tool. Calculate the leading side angle.

Note: The lead of a single depth is equal to 1 divided by the number of threads per inch. In this example, the lead is 0.250″ and the minor diameter is 1.000″:

$$\tan \text{ leading side angle} = \frac{0.250}{1.000 \times \pi}$$
$$= 0.0796$$

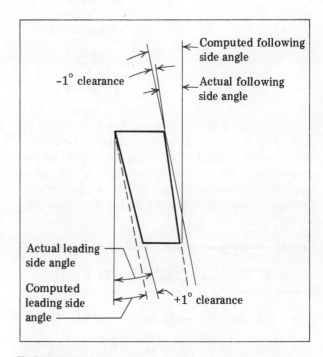

FIGURE 7-3 Clearance Allowance for Leading and Following Side Angles of a Square-Threading Tool Bit

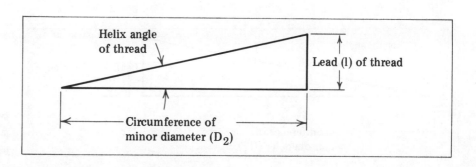

FIGURE 7-4 Triangle Formed to Establish Leading Side Angle

126 TURNING MACHINES

A table of natural trigonometric functions is used to establish the helix angle that corresponds to the value for the tangent of the leading side angle. Thus, 0.0796 = 4°33′. By adding 1° for clearance, the leading side of the cutting tool is ground at an angle of 5°33′.

Calculating the Following Side Angle

The tangent of the following side angle is found in a similar manner. The lead (l) of the thread is divided by the circumference of the major diameter (D_1) (Figure 7-5). In the preceding example:

$$\tan \text{following side angle} = \frac{\text{lead of thread (l)}}{\text{circumference of major diameter } (D_1)}$$

$$= \frac{0.250}{1.250 \times \pi}$$

$$= 0.0637$$

The natural trigonometric function value of 0.0637 as a tangent is 3°39′. The side clearance of 1° is subtracted. Thus, the tool bit must be ground to 2°39′ on the following side. In practice, the craftsperson grinds the tool bit to 2°40′ for ease in reading and checking the ground angle.

HOW TO CUT A SQUARE THREAD

STEP 1 Calculate the width of the tool bit.

Note: For roughing out coarse-pitch threads, grind the tool point width from 0.010″ to 0.015″ (0.2mm to 0.4mm) smaller than the thread groove width. The square cutting tool point is ground 0.002″ to 0.003″ (0.05mm to 0.08mm) larger for finish threading.

STEP 2 Position the quick-change gears for the required threads per inch or metric pitch in millimeters.

STEP 3 Set the compound rest at 0°.

STEP 4 Set up the workpiece in a chuck, between centers, or in a fixture or other work-holding device.

STEP 5 Turn a groove to the minor diameter at the end of the threaded section, if possible.

STEP 6 Set the threading tool square with the work axis.

STEP 7 Start the lathe. Set the cross feed graduated collar at zero.

Note: Some workers feed the cutting tool by using the compound rest.

STEP 8 Proceed to cut the thread to depth. Feed the tool from 0.005″ to 0.010″ (0.1mm to 0.2mm) for each roughing cut.

Note: Use feeds of approximately 0.002″ to 0.003″ (0.05mm to 0.08mm) for each finishing cut.

Note: Use a follower rest to support long workpieces. Apply a cutting fluid, where applicable.

STEP 9 Clean the threads during and at the end of the cutting process. Use a nut, ring gage, or other gaging device to check the threads for the required class of fit.

TECHNOLOGY AND MACHINING PROCESSES FOR ACME THREADS

All Acme screw threads have a 29° included angle. Although not as strong, the Acme thread is replacing the square thread. The American Standards

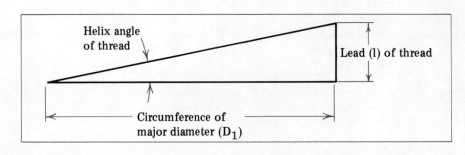

FIGURE 7-5 Triangle Formed to Establish Following Side Angle

for Acme threads provide for two categories of applications: *general-purpose* and *centralizing*.

General-Purpose (National) Acme Thread Form

The *general-purpose Acme screw thread* (National Acme Thread Form) consists of a series of diameters and accompanying pitches. Table 7-2 shows the recommended sizes, pitches, and basic minor diameters for all thread classes (of fits).

Basic Thread Classes. Three classes of general-purpose Acme threads are: 2G, 3G, and 4G. Each class has clearance on all diameters to provide free movement. The class of mating external and internal threads should be the same for general-purpose screw thread applications. Class 2G is used for such assemblies. Classes 3G and 4G provide for finer fits with less backlash or end play.

Basic Dimensions. A number of general-purpose Acme screw thread tables are found in trade handbooks. These tables provide basic dimensions; formulas for calculating other dimensions; limiting diameters (minimum and maximum major, minor, and pitch diameters for 2G, 3G, and 4G thread classes); and pitch diameter allowances and tolerances. The pitch diameter allowance on external threads may be calculated by multiplying the square root of the outside diameter by 0.008 for class 2G, 0.006 for class 3G, and 0.004 for class 4G. For example, the pitch diameter allowance on a 4″ diameter class 3G Acme thread equals $0.006 \times \sqrt{4}$, or 0.012″.

The thread profile drawing in Figure 7-6 shows the basic formulas used by the worker to calculate the basic dimensions of the general-purpose Acme thread form. Table 7-3 provides the formulas and dimensions for a few selected threads. Other formulas for computing diameters for general-purpose Acme screws appear in the partial Table 7-4. Complete Acme thread form tables are included in the Appendix. Additional related design and engineering data are contained in technical journals, manufacturers' literature, and handbooks.

Designations for General-Purpose (National) Acme Screw Threads on a Drawing

Single Threads. Acme screw threads are designated on drawings by the major (outside) diameter, number of threads per inch (or pitch in millimeters for metric threads), Acme, and thread class. Left-hand threads include LH following the thread class. The notation on the drawing in Figure 7-7 tells the worker that the part is to be

TABLE 7-2 Recommended Threads per Inch and Basic Minor Diameters for General-Purpose (National) Acme Thread Series

Size	Number of Threads per Inch	Basic Minor Diameter
1/4	16	0.1875
5/16	14	0.2411
3/8	12	0.2917
7/16	12	0.3542
1/2	10	0.4000
5/8	8	0.5000
3/4	6	0.5833
7/8	6	0.7083
1	5	0.8000
1-1/8	5	0.9250
1-1/4	5	1.0500
1-3/8	4	1.1250
1-1/2	4	1.2500
1-3/4	4	1.5000
2	4	1.7500
2-1/4	3	1.9167
2-1/2	3	2.1667
2-3/4	3	2.4167
3	2	2.5000
3-1/2	2	3.0000
4	2	3.5000
4-1/2	2	4.0000
5	2	4.5000

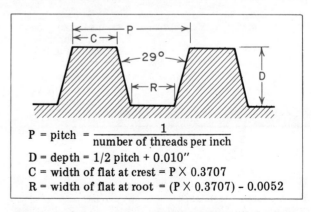

P = pitch = $\dfrac{1}{\text{number of threads per inch}}$
D = depth = 1/2 pitch + 0.010″
C = width of flat at crest = $P \times 0.3707$
R = width of flat at root = $(P \times 0.3707) - 0.0052$

FIGURE 7-6 Profile and Basic Formulas for the General-Purpose Acme Thread Form

128 TURNING MACHINES

TABLE 7-3 Basic Formulas and Dimensions of Selected American Standard General-Purpose Acme Screw Threads

Design Features	Threads per Inch	Pitch	Height of Thread (Basic)	Total Height of Thread	Thread Thickness (Basic)	Width of Flat	
						Crest of Internal Thread (Basic)	Root of Internal Thread
Formula →	N	P = 1/N	P/2	P/2 + 1/2 allowance*	P/2	0.3707P	0.3707P − (0.259 × allowance)*
Basic dimensions (rounded to four decimal places) →	16	.0625	.0313	.0362	.0313	.0232	.0206
	14	.0714	.0357	.0407	.0357	.0265	.0239
	12	.0833	.0417	.0467	.0417	.0309	.0283
	10	.1000	.0500	.0600	.0500	.0371	.0319
	8	.1250	.0625	.0725	.0625	.0463	.0411
	6	.1667	.0833	.0933	.0833	.0618	.0566

*The allowance for 10 or more threads per inch is 0.010"; for less than 10 threads per inch, 0.020".

machined with a general-purpose Acme screw thread. The outside diameter is 1-3/4" (1.750"). The number of threads per inch is 4. The thread is single and turns right-hand (clockwise). The thread class is 2G. The thread is always assumed to be right-hand unless the drawing designates LH (for left-hand).

Multiple Threads. Multiple-start, fast-lead Acme threads are used when a thread is to be advanced faster than is possible with a single lead. Multiple threads are designated on drawings by both the pitch and the lead. Figure 7-8 shows a general-purpose Acme screw thread that has an outside diameter of 2-7/8" (2.875"). The 0.4 (0.400") and 0.8 (0.800") indicates that it is a double thread (right-hand). The thread class is 3G. With this information, the craftsperson is able to establish, by using formulas and tables, all tolerances, allowances, and dimensions required to machine the Acme thread.

TABLE 7-4 Partial Listing of Formulas for Computing Outside and Inside American Standard Acme Threads

External Threads (Screws)		
Major diameter	Maximum =	D
	Minimum =	D minus 0.05P* (but not less than 0.005")
Pitch diameter	Maximum =	E minus allowance for class of thread
	Minimum =	pitch diameter (maximum) minus thread tolerance
Minor diameter	Maximum =	K minus 0.020" for 10 threads per inch and coarser and 0.010" for finer pitches
	Minimum =	minor diameter (maximum) minus 1.5 × pitch diameter tolerance

*When P falls between two recommended pitches (from a handbook table), the coarser of the two values is used instead of the actual pitch value.
D = basic major diameter and nominal size, in inches
P = pitch = 1 ÷ number of threads per inch
E = basic pitch diameter = D − 0.5P
K = basic minor diameter = D − P

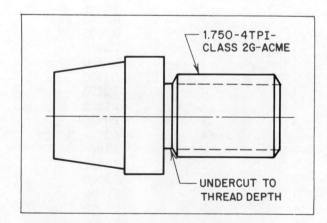

FIGURE 7-7 Notation on a Drawing of a General-Purpose Acme Screw Thread

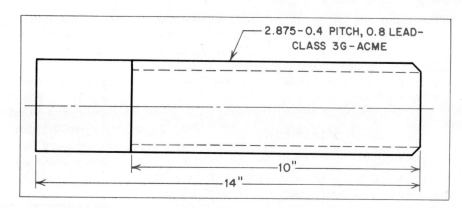

FIGURE 7-8 Designation of a Multiple-Start Acme Thread

General Thread Clearance. A clearance of 0.010" (0.2mm) is generally provided at the crest and root of the mating threads for pitches that are finer than 10 threads per inch (or 2.5mm pitch). The hole diameter on an internal Acme thread, for coarser pitches than 10 threads per inch, is 0.020" (0.5mm) larger than the minor diameter of the screw. The outside diameter of an Acme tap is correspondingly larger than the major (outside) diameter of the screw.

Centralizing Acme Threads

Centralizing Acme threads, another category of Acme threads, have the same design features as do general-purpose Acme threads. However, there are five thread classes: 2C, 3C, 4C, 5C, and 6C. Each class has limited clearance at the major diameters of internal and external threads. The five classes prevent wedging on the thread flanks and ensure closer alignment (centralizing) along the thread axis.

American Standard Stub Acme Screw Threads

The *Stub Acme screw thread* has a 29° included angle with a flat crest and root. The thread is used for unusual applications where a coarse-pitch, shallow-depth thread is needed. While the formula for pitch is the same as for general-purpose Acme threads, the formulas for thread height, tooth thickness, basic width of flat at crest and root, and the limiting diameters are different. Table 7-5 provides formulas for the basic dimensions of two selected Stub Acme threads with sizes of 16 and 14 threads per inch. The full table of basic dimensions of Stub Acme threads and other formulas for computing the limiting diameters are found in trade and technical handbooks.

There are other Stub Acme thread forms. The *Modified Form 1* has a smaller pitch diameter and minor diameter than similar dimensions for the standard Stub Acme form. The same

TABLE 7-5 Basic Formulas and Dimensions of Two Selected American Standard Stub Acme Screw Threads

Design Features	Threads per Inch	Pitch	Height of Thread (Basic)	Total Height of Thread	Thread Thickness (Basic)	Width of Flat	
						Crest of Internal Thread (Basic)	Root of Internal Thread
Formulas	N	P = 1/N	0.3P	0.3P + 1/2 allowance*	P/2	0.4224P	0.4224P − (0.259 × allowance)*
Basic dimensions (rounded to four decimal places)	16	.0625	.0188	.0238	.0313	.0264	.0238
	14	.0714	.0214	.0264	.0357	.0302	.0276

*The allowance for 10 or more threads per inch is 0.010"; for less than 10 threads per inch, 0.020".

130 TURNING MACHINES

Dimensions are larger on the *Modified Form 2* because of the thread height. Another thread profile is the 60° Stub thread. This thread is adapted to design and operating conditions that are better met by a thread form other than the 29° Acme thread.

Designation of Standard Stub Acme Threads

Standard Stub Acme threads are designated on drawings as illustrated in Figure 7-9. The major diameter of the thread is 3/4" (0.750"). There are 6 standard, left-hand, Stub Acme threads per inch.

29° Worm Thread (Brown & Sharpe)

The *worm thread* (Brown & Sharpe) is also a 29° form thread. Worm threads are generally combined with worm gears to provide mechanical movement for transmitting uniform angular motion rather than power. Such applications permit the design of a deeper thread form than on general-purpose Acme threads. The tooth thickness at the crest and root are correspondingly smaller. The 29° worm thread form and simple formulas for the basic dimensions are shown in Figure 7-10.

The *worm thread gage* is available for common pitches. The round gage has a series of slots machined around the periphery. These slots conform to the shape and size of each worm thread pitch. The 29° angle of the cutting-tool sides and the point are checked for accuracy with this gage. The cutting tool is set square with the workpiece by using a *29° setting tool* (template or gage). The back edge of this gage is set against the workpiece. The side face of the cutting tool bit is then checked against the 14-1/2° surface of the gage. When machined on a lathe, the worm thread requires the same steps as other 29° form threads. When a great number of parts are to be threaded, worm threads are usually hobbed (machined with a formed rotary cutter) by using a milling machine.

HOW TO CUT A RIGHT-HAND ACME THREAD ON A LATHE

STEP 1 Grind the tool bit to fit the 29° Acme thread gage. The slot used should correspond with the thread pitch.

Note: If a gage is not available, grind the 29° cutting-tool point to a width equal to 0.3707 divided by the number of threads per inch (0.0052").

STEP 2 Set the quick-change gears to cut the required number of threads per inch or metric thread pitch.

STEP 3 Position the compound rest at 14-1/2° to the right.

STEP 4 Locate the thread-cutting tool at center height and square with the workpiece. (The setup is shown in Figure 7-11.)

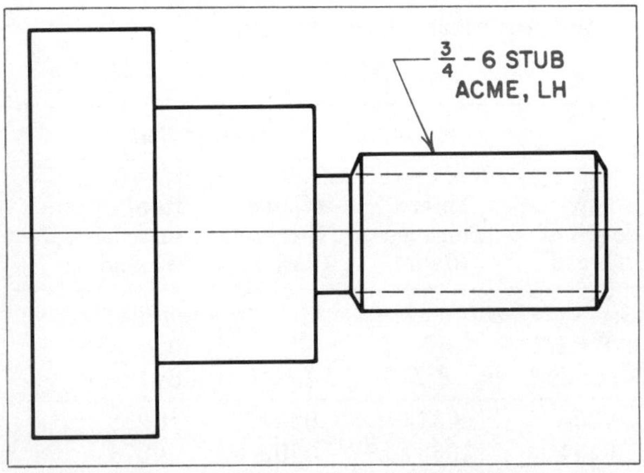

FIGURE 7-9 Representation of Standard Stub Acme Thread

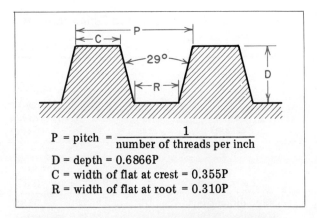

$P = \text{pitch} = \dfrac{1}{\text{number of threads per inch}}$
$D = \text{depth} = 0.6866P$
$C = \text{width of flat at crest} = 0.355P$
$R = \text{width of flat at root} = 0.310P$

FIGURE 7-10 Profile of 29° Worm Thread (Brown & Sharpe) and Formulas for Basic Dimensions

THREAD-CUTTING TECHNOLOGY

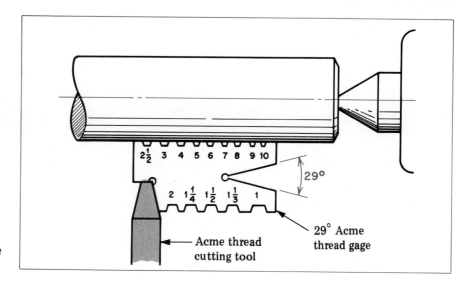

FIGURE 7-11 Positioning Acme Threading Tool Square with Workpiece

STEP 5   Feed the cutting tool to thread depth (minor diameter). Turn the end of the workpiece to this minor diameter for about 1/16″ (1.5mm) long.

Note: The turned diameter is used as a check. It indicates when the thread is cut to depth.

STEP 6   Back the cutting tool out to position it at thread cutting depth for the first roughing cut.

STEP 7   Feed the tool in at the 14-1/2° angle. Take a series of roughing and finishing cuts.

Note: Coarse-pitch Acme threads may be roughed out with a square-point cutting tool. The square point is ground smaller than the thread width at the root. The sides are later finished with the 29° form threading tool.

STEP 8   Clean and burr the workpiece. Use a ring thread gage, Acme nut, or mating part to check the size and fit.

Note: Precision parts often require further checking for the quality of surface finish, thread form, and dimensional accuracy. In such cases, optical, electronic, mechanical, and other comparators or microscopes are used.

MULTIPLE-START THREADS

Multiple-start threads are used to increase the rate and distance of travel along a screw thread per revolution. The lead of a multiple-start thread is increased without increasing the depth to which a thread is cut. The three common multiple-start threads are known as *double, triple,* and *quadruple threads.* The relationship between the pitch and leads of double and triple multiple-start threads and a regular single-pitch screw thread are shown in Figure 7-12.

Multiple-Start Thread Designations

Multiple-start threads are designated on drawings by the diameter, number of threads per inch (or pitch in millimeters for metric threads), the lead, and the number of starts. The working drawing in Figure 7-13 shows how a 4 pitch, square, double thread with a major diameter of 1-1/4″ is dimensioned. The thread depth of multiple-start threads is always calculated in relation to the pitch of a single thread.

Methods of Cutting Multiple-Start Threads

There are several methods of cutting multiple-start screw threads. The positioning of the single-point chasing (cutting) tool, the setting of the quick-change gears, cutting speeds, depths of cuts, and the rough and finish cutting of the thread are the same as for single-pitch screw threads. However, there are two major differences:

—The quick-change gears must be set to cut the thread to the required *lead* (the lead of a double thread equals 2P and the lead of a triple thread equals 3P);

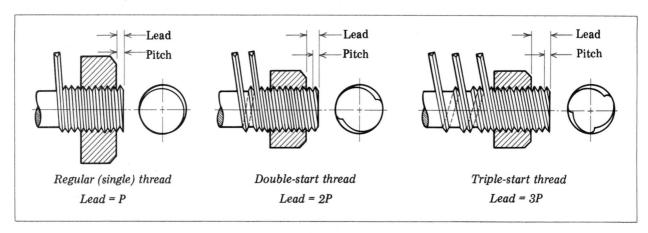

FIGURE 7-12 Relationship between Pitch and Lead for Single-, Double-, and Triple-Start Threads

—The workpiece must be advanced a fractional part of one revolution for a multiple thread (for example, a double thread has two starting places located exactly 180° from each other). At the completion of the first thread, the cutting tool is backed out of the thread. The workpiece is turned 180° for the second thread.

If a thread-chasing dial is used, the cutting tool may be positioned at the next entry position after each cut. Thus, on a double thread, each thread is cut alternately to the same depth.

Another method of positioning for the next start is to advance the cutting tool by using the compound rest. The compound rest is set parallel to the thread axis. The graduated collar on the compound rest is used to advance the tool a distance equal to the pitch. The movement positions the cutting tool at the next entry place in relation to the previously cut thread.

Driver Plate and Gear Train Methods. Workpieces that are mounted between centers may be indexed accurately by using the slots or holes in the driver plate. The workpiece assembly is advanced to the hole or slot on a driver plate corresponding to the number of thread starts. For example, the workpiece with dog attached is turned and engaged in the slot 180° from the driver plate slot used for the first thread.

The gear train method is also used in cutting multiple-start threads. The position of the spindle gear, in relation to the mating intermediate gear, is marked with chalk. The intermediate gear is disengaged. The spindle is advanced a number of teeth to correspond with the fractional turn required to position the second thread. In case of a double thread, the spindle (and spindle gear) is rotated through an arc of 180°. The intermediate and other gears are then reengaged. The guard cover is secured. The second (multiple-

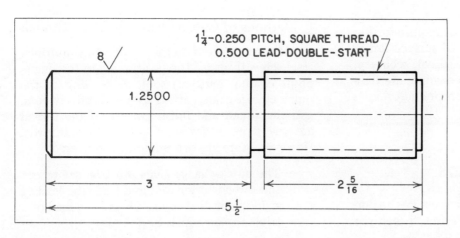

FIGURE 7-13 Designation of a Multiple-Start Thread on a Working Drawing

start) thread is rough and finish cut to single thread depth.

Advantages of the Thread-Chasing Dial Methods. When the thread-chasing dial method is used to cut multiple threads, each thread is cut to the same depth to permit each thread to be rough turned. The roughing-out cutting tool may then be changed to one for finish cutting to shape and size. Multiple threads are cut by setting the tool to depth and alternately engaging the half nut for each thread. The numbered and unnumbered lines on the chasing dial are employed for machining each multiple thread.

SAFE PRACTICES IN ADVANCED THREAD-CUTTING APPLICATIONS

—Limit the number of times that die cast threaded parts are removed. Applications should be made to parts that do not require a great amount of force to screw them together.
—Use metal inserts in plastic-molded parts that are to be held securely with fasteners.
—Operate vitrified-bond thread-grinding wheels within a 7,500 to 9,500 sfpm (2,300 to 2,900 m/min) range. Resinoid-bond wheels may be operated between 9,000 to 10,000 sfpm (2,750 to 3,050 m/min).
—Grind and check the dimensional accuracy and thread form of the cutting tool for Unified Miniature Screw Threads. Due to the fineness of thread pitch and small diameters, the cutting tool is depended upon to produce an accurate thread.
—Check the leading and following side angles on single-point threading tools to permit the cutting edges to cut without interference from rubbing against the sides of the threads.
—Machine the end of a workpiece to the major (outside) or minor (inside) thread diameter for about 1/32" (0.8mm) to provide a checkpoint for cutting to thread depth. Burrs are thus prevented from projecting beyond the face of the workpiece. The small turned area permits mating threaded parts to engage easily.
—Cut a groove at the end of the thread, if possible, to help in backing out the cutting tool quickly at the end of the cut. Sometimes, the threading tool is permitted to end in the groove, while the chasing mechanism and carriage movement are disengaged. The cutting tool is then backed out.
—Use a 29° Acme screw thread instead of a square thread when less force is required and compensation must be made for wear.
—Use a multiple-start screw thread rather than a comparatively coarse and deep thread for applications where the thread depth reduces the strength of the workpieces beyond an allowable limit.
—Use the five C thread classes of centralizing Acme threads for more precise applications than the three G classes of general-purpose Acme threads.
—Use Stub Acme threads on screw thread applications requiring a short-depth, coarse-pitch, physically stronger thread than the 60° thread form.
—Feed the 29° Acme threading tool with the compound rest set at a 14-1/2° angle. Angle feeding for roughing cuts requires simultaneous cutting by the leading side and point of the tool bit. The thread flank (following thread side) is formed by feeding at the 14-1/2° angle.
—Rough cut deep Acme threads with a square cutting tool. Leave enough stock for finishing to the required width of flat at the thread root.
—Remove the thread-cutting tool at the end of each cut. Reset it at the next depth to start each new cut. Check the cutting-tool depth before taking each cut.
—Back the cutting tool out and away from the workpiece before positioning it for the next successive multiple-start thread.
—Remove the fine, wedge-shaped burr at the starting and end threads, particularly if they are square or Acme threads.

TERMS USED IN ADVANCED THREAD CUTTING AND PRODUCTION METHODS

Casting method of making threads	Production of threads by flowing molten metal into a preformed mold. Producing screw metals by using conventional casting processes such as sand casting, die and permanent-mold casting, and plastic molding.
Rolling threads	Displacing metal by forcing a part to be threaded through flat or round thread dies. The fastest, most economical method of producing highly accurate threads. (Threads are burnished to a precise surface finish.)
Thread chasing	Cutting a screw thread on a lathe. Using a single-point threading tool, set to depth, to cut the thread groove as the workpiece revolves. The movement of a thread formed cutting tool in a fixed relationship (pitch) to the revolution of the spindle (and workpiece).
Milled thread	A screw thread that is generated on a milling machine. A thread form produced as a revolving workpiece is advanced past the cutting teeth of a formed milling cutter.
Ground threads	Precision threads formed by a preformed grinding wheel. Threads produced by grinding, usually on a thread-grinding machine.
Broaching (scru-broaching)	Forming a screw thread by using a multiple-tooth cutting tool. (The teeth are spiral-formed on a body that has a pilot end. The broach is turned by a feed screw that draws the workpiece into the cutting tool. Limited applications of scru-broaching are in the automotive field.)
Unified Miniature Screw Thread Series	An American Standard series of 60° form threads that are particularly adapted to instruments and microminiaturized mechanisms. A thread form that is compatible with the Unified inch-standard and the ISO metric-standard screw thread systems. Threads designated on drawings as UNM.
Leading and following side angles	Thread angles resulting from the slope of a thread form around the periphery of the workpiece. (The leading side refers to the thread side of the thread-cutting tool that advances with the thread-cutting process. The following side relates to the opposite thread-cutting tool side.)
National Acme thread form	A 29° thread form with a flat crest and root. A series of standard thread pitches for specific diameters.
General-purpose Acme thread classes	Three classes of fit designated as 2G, 3G, and 4G. Thread classes for general-purpose applications of Acme threads ranging from coarse (2G) to finer, precise fits (4G).
Designation of National Acme screw threads	Screw threads that are designated on drawings by the major diameter, pitch, Acme (thread system), and thread class. (Left-hand threads are so indicated.)
Multiple-start thread	A thread having a pitch that is a multiple of the lead of the thread form. A double-, triple-, or quadruple-entry thread, cut to the depth of the single pitch thread. A thread that engages and advances rapidly during each full or partial revolution of the mating parts.

THREAD-CUTTING TECHNOLOGY

Centralizing Acme thread	A general-purpose Acme thread form having five thread classes (fits): 2C, 3C, 4C, 5C, and 6C. Design specifications given to precise limits to ensure closer alignment (centralizing) along the thread axis.
Stub Acme screw thread	A modified 29° form, shallow-depth thread. Variations in thread height, tooth thickness, basic flat at crest and root, with corresponding changes in the limiting screw thread diameters. A thread form that meets coarse-pitch and shallow thread depth requirements.
Limiting diameters	Major, pitch, and minor diameters applied to screw threads.
29° worm thread (Brown & Sharpe)	A standard series Acme thread form, generally used in a gear train with a worm gear to transmit uniform motion. An Acme thread form having a deeper groove and narrower thread crests and roots.

SUMMARY

Screw threads are manufactured by methods such as casting, rolling, chasing, die and tap cutting, grinding, and broaching.

> Metal inserts are cast into position in materials that are too soft to withstand the forces that need to be applied to a screw thread.

The flow of metal in rolling threads produces a high-quality and dimensionally accurate thread form and size.

> Thread chasing on a lathe is a practical, versatile method of cutting threads of unusual size or limited quantity.

Deep, coarse, long screw threads are often milled with a form cutter on the milling machine.

> Thread grinding is the most precise method of generating a screw thread. It is particularly adapted to finish grinding precut threads in materials that are hardened. Threads may be rough and finish ground from solid stock.

Resinoid-bond wheels are used for general-purpose thread grinding. Vitrified bonds are better suited in producing extremely accurate thread forms.

> Scru-broaching is extending beyond threading applications in the automotive field. A spiral-formed broach may be used for rough and finish threading.

Design trends toward reducing the sizes of mating parts and components in mechanical movements and mechanisms have resulted in the manufacture of miniature products. Further reduction in size is known as microminiaturization.

> With the movement toward metrication, it is necessary to have a compatible, interchangeable miniature screw thread system. The Unified Miniature Screw Thread Series (UNM) provides general-purpose fasteners for applications on instruments and microminiature mechanisms.

The 14 thread sizes in the UNM series range from 0.30mm to 1.40mm. The basic thread form and thread height permit agreement between metric and inch-standard threads in the series.

> Two widely used thread forms, other than the 60° V-thread, include square and 29° Acme threads.

Square threads are used when a coarse (pitch) thread and extreme force are required. The thread form does not permit compensation for wear.

Single-point thread-chasing tools must be ground so that the side cutting faces clear the thread flanks.

The leading and following thread angles are computed by using a tangent formula and locating the angle in a table of trigonometric functions. One degree is added to the lead angle. One degree is subtracted for the following angle. The one degree provides clearance for the cutting tool.

Square threads are usually cut 0.002" to 0.003" (0.05mm to 0.08mm) wider than thread width to provide clearance between the mating threads.

General-purpose Acme screw threads consist of a pitch-diameter series of 29° form threads. The series is standard and has three thread (fit) classes: 2G, 3G, and 4G. Technical tables provide formulas and basic dimensions for the limiting diameters of major, pitch, and minor diameters.

Screw threads are designated on drawings by major diameter, threads per inch or metric pitch, screw thread series, and thread class. The letters LH designate left-hand threads.

Multiple-start threads are designated by following the outside diameter with both the pitch and lead dimensions.

The American Standard Stub Acme screw thread is adapted to parts that require a coarse-pitch, shallow-depth thread of the 29° form. The Stub Acme thread dimensions differ from the general-purpose Acme in terms of smaller thread height, changed limiting diameters, and different sized flats at the crest and root.

The 29° worm thread is used generally as the driver to turn a worm wheel in order to transmit uniform motion. Deeper thread flanks are required, while the crests and roots are thinner than the general-purpose form.

Coarse-pitch and deep Acme threads are chased on a lathe (roughed out) by using a square-point cutting tool. The 29° sides are then formed and the thread is cut to depth.

Double, triple, quadruple, and other multiple-start threads are also referred to as multiple-entry threads. They are applied primarily where a fast rate of travel or a number of starting positions are required per revolution.

Multiple-start threads are cut to the depth of a single thread. When the thread-chasing dial method is used, it is practical to cut each thread by engaging each thread at its fractional position.

Multiple-start threads may be positioned by moving the compound rest, relocating the workpiece in a particular slot on a driver plate, or by advancing the spindle (gear) in relation to the intermediate gear on the gear train.

Where possible, a groove turned at the end of the thread location permits the safe backing out of the cutting-tool point.

The general safety precautions for mounting and holding the workpiece, positioning the cutting tool, backing the tool out to take successive cuts, and using a cutting fluid must be followed.

Any burrs produced during threading are to be removed. The workpiece is cleaned before measuring or checking the threads.

UNIT 7 REVIEW AND SELF-TEST

1. a. Describe two production methods for manufacturing screw threads, other than chasing on an engine lathe.
   b. Cite one advantage of each production method in comparison to the chasing of threads.

2. a. Identify the general purpose for which the American Standard Unified Miniature Screw Thread (UNM) Series was designed.
   b. Describe the basic thread form and size range of the UNM series.
   c. Tell why a UNM screw thread is interchangeable with a same pitch ISO miniature screw thread.

3. Interpret the meaning of the following dimension found on a drawing of a fast-operating clamping screw for a drill jig:
   $$1.500 - 0.25 \text{ PITCH}, 0.75 \text{ LEAD}$$
   $$\text{ACME, CLASS 2G, LH}$$

4. a. Refer to appropriate Acme screw thread tables.
   b. Give the dimensions needed for both machining and inspecting three parts that are threaded to conform to general-purpose (National) screw thread series standards. The outside diameters of the three parts are 0.500″, 1.000″, and 2.000″.

5. Explain the difference in setting up an engine lathe to chase a multiple-start thread as compared to cutting a single-lead regular screw thread.

6. Indicate two features, incorporated into the design of threaded areas, that reduce the possibility of tool breakage during cutting, ensure the thread is not cut too deeply, and compensate for machining burrs.

7. State what precautions must be taken to avoid the following screw-thread cutting problems:
   a. Thread groove rubs against the following side of a square-thread cutting tool.
   b. Burrs project beyond the threaded-end face of the workpiece.
   c. A coarse, deep thread is produced that weakens the part beyond allowable physical property limits.

SECTION TWO

Turret Lathes: Bar and Chucking Machines

The *turret lathe* is a production turning machine tool. It may be used to reproduce any one or combination of internal and external machining processes performed on an engine lathe. One important difference is that the turret lathe uses multiple tooling. When the cutting tools are set up precisely, each tool may be brought into position as needed. Except for shutdown time to resharpen and to set the cutting tools, the manufacture of a part is continuous. Thus, the turret lathe bridges the gap between a conventional engine lathe and an automatic screw machine.

Turret lathes vary in design and size from hand-feed to sophisticated numerically controlled models. This section deals with general-purpose turret lathe classifications, descriptions of major parts and components, and design features. Tooling setups are related to basic processes requiring a full knowledge of work- and cutter-holding devices. Principles of tooling, using single- and multiple-point cutting tools, and the planning of machine setups, cutting speeds, and feeds are considered.

UNIT 8

Classifications, Features, and Applications of Turret Lathes

This unit begins with a brief history of the evolution of the turret lathe. The lathes are then classified under bar and chucking machines of the ram and saddle types. Major design features of the turret lathe and maintenance items with which the machine operator should be familiar are described. Units 9, 10, and 11 deal specifically with accessories, standard cutting tools and holders, work-holding devices, and important setups for internal and external cuts in quantity production.

EVOLUTION OF MODERN BAR AND CHUCKING MACHINES

Early Turret Lathe Developments (1845-1945)

Although the age of interchangeable manufacturing was ushered in around 1800, mass production developed very slowly. It took about 45 years before a new type of metal turning lathe was built that would accelerate the mass production movement. The eight-station turret lathe of Stephen Fitch revolutionized turning processes. The addition of a turret to the iron lathe of the day made it possible to hold and position a number of cutting tools on the same machine. All of the cutting tools mounted in a turret could be used to machine part after part without changing the setting.

One of the early turret lathes was developed by Reed Warner and Ambrose Swasey. Both men were machinists who had worked in the Pratt and Whitney shops in Hartford, Connecticut. They moved to Cleveland, Ohio, to establish a plant for the manufacture of machine tools. The Warner & Swasey Company has served the machine tool industry in making production machines from 1880 until now.

An early development of the original 1880 turret lathe is shown in Figure 8-1. While simple, the features of the 1880 machine are the same as turret lathes produced today. The 1880 model had a step-cone pulley. The spindle was driven from an overhead countershaft by a flat leather belt. The three cones and a belt-reversing shifter permitted three forward and three reverse speeds. The cross slide permitted front and back mounting positions of cutting tools. The round turret was mounted on a ram. Each tool in the turret was positioned and secured by a hand clamp. Each cutting tool was fed by a hand lever. This hand lever controlled the tool movement of the ram. The coolant dripped onto the workpiece from the container on the back of the machine. A chip pan under the bed collected the chips and permitted the coolant to be strained into a reservoir that extended below the chip pan.

The standard turret lathe of the early 1900s is pictured in Figure 8-2. This lathe was built rigidly to withstand increased cutting forces. These forces were due to increased speeds, feeds, and the harder materials that were being machined. Speed changes were made by shifting a clutch lever. Power feed was added to the turret unit. By 1912, a universal cross slide was incorporated. Thus, more tool setups could be mounted and positioned on the cross slide.

During World War I, a 12-speed geared-head turret lathe was in production. This lathe, shown in Figure 8-3, was equipped with a universal cross slide with square turret and power feeds for the hexagon turret and cross slide. A rapid traverse mechanism permitted fast movement of heavy hexagon turrets. A coolant system was added. Coolant flow at a constant rate became automatic. Independent adjustable stops permitted accurate control of linear dimensions. A typical 1-1/2" bar machine of 1914 had a three-horsepower (3HP) drive and a maximum spindle speed of 450 RPM.

Modern Design Changes and Automated Bar and Chucking Machines

As new, harder-to-machine metals were produced, new alloys for cutting tools were developed. In turn, tough alloy steels were used to develop more powerful machines. By the end of World War II, the versatility and productivity of bar and chucking machines (turret lathes) increased significantly.

A modern *fully automatic chucker and bar machine* is shown in Figure 8-4. This machine may be programmed directly. Trip blocks

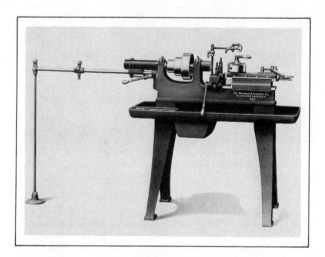

FIGURE 8-1 1890-Model Turret Lathe

FIGURE 8-2 Turret Lathe of the Early 1900s

140 TURNING MACHINES

FIGURE 8-3 Geared-Head and Power-Fed Turret Lathe of World War I

operate microswitches, which control the functions of the carriage, cross slide, vertical slide, and threading head, as well as speed and feed changes. The reverse feed is also independent of the forward feed so that rapid return may be programmed. The machine is capable of being programmed for the following operations:

—Straight turning,
—Taper turning,
—Facing,
—Forming,
—Chamfering (beveling),
—Single-point threading,
—Die-head threading,
—Cutting off,
—Drilling,
—Straight or taper boring,
—Reaming,
—Recessing,
—Tapping.

There are 34 spindle speeds on this machine. The range is from 125 to 3,000 RPM. Four speeds within this range are available throughout the machining cycle. Tolerances within 0.0001″ (0.002mm) are obtainable on the cross slide by using a "tenth" (0.002mm) dial indicator. Similarly, a carriage dial indicator reading in increments of 0.001″ (0.02mm) is used for dimensions of length. This chucker and bar machine is a small model, having one 2 HP spindle drive motor and another 2 HP motor for the hydraulic pump.

Turning Centers (Universal Turning Systems)

Another modern adaptation of the turret lathe is a group of machine tools identified as

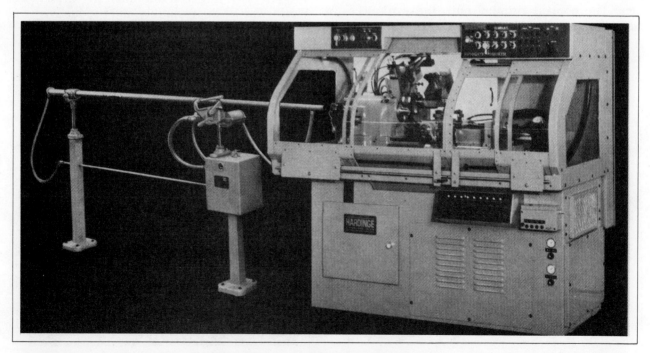

FIGURE 8-4 Modern Fully Automatic Chucker and Bar Machine

turning centers or *universal turning systems*. These machine tools are usually numerically controlled. Part of a universal numerically controlled (NC) turning system is illustrated in Figure 8-5. A multi-tool single turret design and chucking setups are shown. The turret slants 35°. This feature places the tooling for outside diameter operations at a plane perpendicular to the plane of operation. Thus, maximum turning force and interference-free turning are provided.

Note that the inner row of eight tools on the turret are for inside diameter operations, as shown in Figure 8-5A. The boring and inside diameter turning tools are positioned in this inner row of the turret. This single turret accommodates 16 tools. There are three spindle speed ranges: (1) 65 to 2,000 RPM, (2) 80 to 2,500 RPM, and (3) 90 to 3,000 RPM. This model heavy-duty turning machine can accommodate bar stock up to 2-1/2" (64mm) diameter. It has a chucking capacity up to 12" (305mm). Boring bars up to 2" diameter, with a maximum length of 9-1/2" for bar and holder, may be mounted in the turret. Other selected specifications for workpiece and chuck clearance, using a boring bar and standard facing or turning tool, accompany Figure 8-5B. This particular model requires a 40 HP motor to drive the spindle.

CLASSIFICATION OF BAR AND CHUCKING MACHINES (TURRET LATHES)

Multiple tooling is provided on a turret lathe by replacing the usual lathe tailstock spindle and assembly with a hexagonal turret. Also, the regular tool post on a compound rest is replaced by a square turret to hold four separate cutting tools. A square turret may also be positioned at the back of the cross slide. Thus, 14 different cutting tools may be set up at one time.

Horizontal turret lathes may be classified into *bar machines* and *chucking machines*. Bar machines are designed for feeding bar stock through a spindle or for machining castings or forgings that may be held in collets. Chucking machines are used for workpieces that are held and driven by a chuck or work-holding fixture. Chucking machines are used for machining regular and irregular-shaped castings, forgings, and large-size cut bar stock.

A bar machine (ram-type) is illustrated in Figure 8-6A. A chucking machine appears in Figure 8-7A. Although these machines look alike, there are some basic differences in the cutting tools, forces, and processes. The drawing in Figure 8-6B shows a typical tool setup for the bar turner for turning an outside diameter. The *rollers* on the *bar turner* provide a follower rest for turning concentrically and for support against the cutting forces. By contrast, the *multiple turning head* (Figure 8-7B) used on a chucking machine overhangs the workpiece for machining an outside diameter. Plain and adjustable cutter holders are shown. The plain cutter holder is used for roughing cuts. The adjustable holder permits micrometer adjustments for finish cuts.

Ram and Saddle Types of Turret Lathes

Bar and chucking machines may be of the *ram type* or the *saddle type*. On the ram type, the base on which a ram moves is clamped securely on the bed of the turret lathe. The turret is mounted on the slide or ram. The ram movement toward or away from the spindle is controlled by hand or power feed. The ram type is adapted for bar work and chuck work where it is possible to control the overhang of the ram. Since the ram moves outwardly, the overhang increases as the depth (length) of the cut increases. The ram type has a short turret stroke and is an easy machine to operate for short-cycle workpieces.

On the saddle type of turret lathe, the turret is mounted on a saddle. The saddle with its apron and gearbox move back and forth on the machine ways. The saddle type has a longer stroke than the ram type and a more rigid turret mounting. The tool overhang remains constant as the whole carriage and turret setup advances as a unit. The side-hung carriage makes it suitable for heavier chuck work and long turning and boring cuts.

Features of the Ram-Type Turret Lathe. The main features of a ram-type turret lathe are identified in Figure 8-8. The spindle of a ram-type turret lathe may be driven by a motorized head or a geared head. The multiple-speed motor is mounted directly on the spindle inside the headstock housing. There is a wide range of forward and reverse speeds to accommodate different processes, sizes and shapes of workpieces, and varying degrees of machining hardness and toughness.

142 TURNING MACHINES

A. Boring and Inside Diameter Turning Tools on Inner Turret Row

1. Maximum length of inside diameter tool: 9.80"
2. X-axis zero position: 12.25"
3. 12" high-speed chuck: 12.00"
4. Maximum outside diameter of workpiece: 12.00"
5. Maximum diameter of boring bar: 2.00"
6. Width of bar holder: 3.35"
7. Depth of bar holder: 3.35"
8. Maximum out X axis: 12.80"
9. Tool tip travel to center: 6.86"
10. Center of tool shank to next position: 3.38"
11. Workpiece center hole maximum diameter: 3.25"
12. Maximum turret face to spindle nose: 25.19"

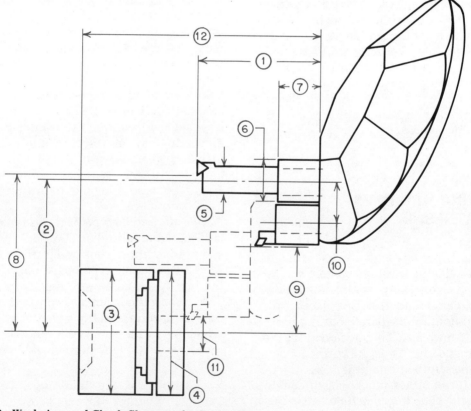

B. Workpiece and Chuck Clearance for Boring Bar and Right Angle Facing Tool on the Same Station

FIGURE 8-5 Multi-Tool 16-Station Slant Turret and Chucking Setups on Universal (NC) Turning System

TURRET LATHES 143

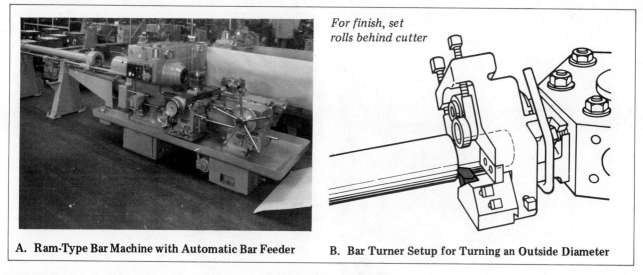

A. Ram-Type Bar Machine with Automatic Bar Feeder

For finish, set rolls behind cutter

B. Bar Turner Setup for Turning an Outside Diameter

FIGURE 8-6 Bar Machine (Ram-Type) with Bar Turner Setup

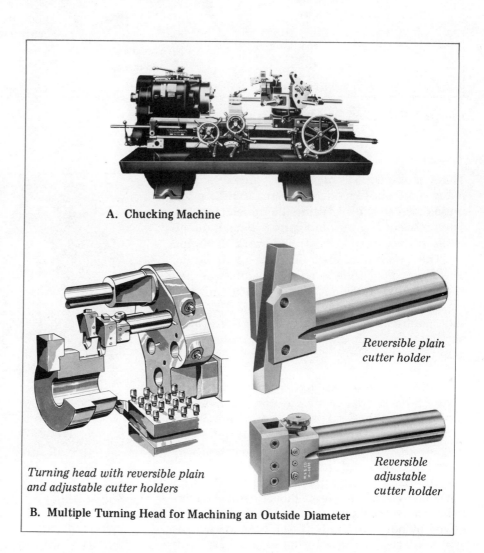

A. Chucking Machine

Turning head with reversible plain and adjustable cutter holders

Reversible plain cutter holder

Reversible adjustable cutter holder

B. Multiple Turning Head for Machining an Outside Diameter

FIGURE 8-7 Chucking Machine and Multiple Turning Head Setup

144 TURNING MACHINES

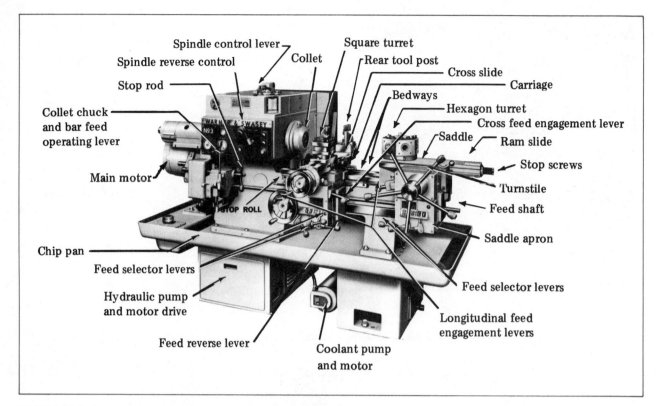

FIGURE 8-8 Main Features of Ram-Type Turret Lathe

There is a direct-reading speed preselector on many geared-head models. It permits shifting to a preselected speed by setting the preselector dial. The preselector shown in Figure 8-9 reads directly in surface speed, RPM, and work diameter.

The cross slide may be *plain* or *universal*. The plain cross slide is adjusted along the bed and locked at a required position. The universal cross slide is mounted on a carriage. This cross slide permits movement in two directions to cut parallel (longitudinally) and at right angles, crosswise to the bed ways.

Classification of Saddle-Type Turret Lathes. Figure 8-10 shows the main features of a saddle-type turret lathe. Feed design features are similar to a ram-type turret lathe. Saddle turret lathes may be of the *fixed-center turret*, *cross-sliding turret*, or *compound cross slide* type.

The fixed-center turret is fixed to align with the spindle axis. The turret moves parallel to this axis. The cross-sliding turret (Figure 8-11) may be fed by power or hand. The movements are parallel and at right angles to the spindle axis. This turret is used for facing deep holes in chucking

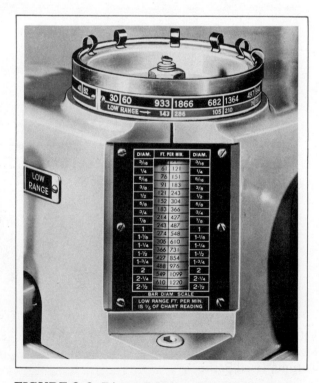

FIGURE 8-9 Direct-Reading Spindle Speed Preselector for Dial Settings

TURRET LATHES 145

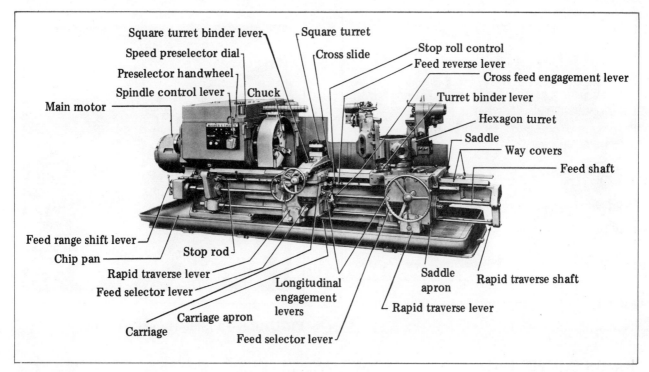

FIGURE 8-10 Main Features of Saddle-Type Turret Lathe

work and on workpieces where most of the cuts are facing and the turning and boring operations are secondary. The cross-sliding turret is also used for turning internal tapers, threads, and contours. It is practical in jobbing shops where limited quantities of a part that may not justify long setup time are to be produced.

The compound cross slide (Figure 8-12) has a compound rest that may be set directly at any angle. This feature permits cutting bevels and steeper tapers than it is possible to machine with a taper attachment.

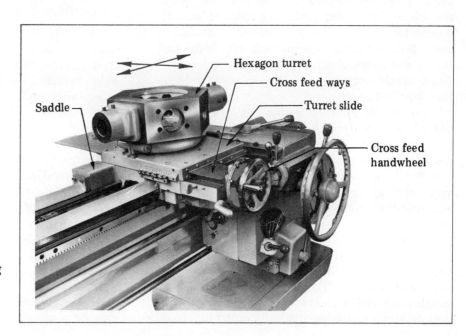

FIGURE 8-11 Cross-Sliding Turret with Movements Parallel and at Right Angles to Spindle Axis

FIGURE 8-12 Compound Cross Slide

FIGURE 8-13 Side-Hung Cross Slide

MAJOR COMPONENTS OF THE TURRET LATHE

Headstock

Modern turret lathe headstocks are designed with heavy-duty spindles that ride in preloaded bearings. These bearings permit freedom of rotation with exceedingly limited end and radial play. The drive may be direct, as in the case of a motorized unit where a variable-speed motor is used. The geared-head is common on some models and provides for speed changes usually within two or three ranges. On other models, a variable-speed drive is combined with high- and low-speed range clutches. This combination delivers maximum torque to the spindle, even at low spindle speeds.

Automatic turret lathes have an infinite number of variable spindle speeds and may be programmed from 150 to 3,000 RPM. The spindle on a turret lathe is designed to run clockwise or counterclockwise. Spindle brakes are available as a machine and personal safety precaution to automatically brake the spindle to a stop in an emergency.

A quickly adjusted *stop rod* is located under the headstock. This stop rod has a series of regularly spaced notches that are changeable for each tool position. The stop rod permits adjustments over several inches in comparison to the limited adjustments permitted by individual feed-stop screws. These screws control each tool position to within 0.001″ (0.02mm).

Speed changes are made on geared-head turret lathes by positioning the speed-change lever for range and a second lever for the desired speed within that range. On machines equipped with a preselector, a single lever or dial engages each spindle speed according to the speed required for each work process.

Carriage or Cross Slide

The two basic types of cross slides are: (1) *side-hung* and (2) *reach-over*. The side-hung cross slide (Figure 8-13) provides maximum swing capacity over the cross slide. This feature is essential on the saddle-type machine where large-diameter work requires maximum swing capacity.

The reach-over cross slide is also called a *bridge cross slide*. This cross slide is supported on both bed ways. It may also be supported by a lower rail. The reach-over cross slide is best suited for the ram-type turret lathe where maximum swing is not a requirement. This cross slide has the advantage over the side-hung type because it is possible to add a tool post or cross slide turret in the rear position.

Cross Slide Position Stops. The cross slide may be positioned for different depths of cut by

TURRET LATHES

adjustable stops. The set of stops on the front end controls the depth of cut of the front tools. The rear stops control the depth of the rear turret tools. The cross slide travel is controlled by dogs, which engage the stops.

Carriage Feeds. Carriage movement is controlled in a manner similar to a lathe. The carriage may be fed longitudinally or transversely, by hand or power. The power feed may be rapid traverse or regular. Trip dogs are used to move the feed lever to the off position. The feed may be reversed to provide tool movement in either direction. Handwheels are provided for cross slide and longitudinal movement of the main turret. Graduated collars enable an operator to continue a precision finish cut by hand, after the power feed is disengaged.

A second set of feed stops, at the left side of the carriage, controls the longitudinal movement of all carriage tools. The feed stops are part of a *stop roll*. This stop roll has a center stop that permits adjusting all of the stops the same amount, without disturbing the individual stop screws.

Main and Secondary Turrets

Main turrets are generally hexagonal or octagonal. Some main turrets are designed with an inside and an outside row. Such turrets accommodate two sets of tools: one set of eight tools for internal operations; another set of eight for external operations. Some main turrets are mounted with the ram or saddle in a horizontal plane. Other turrets are mounted vertically on the ram or are designed with the bed and turrets inclined for easier access and machine tool setups. *Secondary turrets* on the cross slide may be mounted horizontally or vertically.

All turrets are designed to receive adapters and to position each tool along the central axis. Each tool is advanced to a required position by setting an adjustment stop screw. This screw is located in the turret slide. The turret is automatically indexed to the next position by action of the handwheel.

The tool at each turret station is set up to perform a specific work process. After the turret is indexed, a power feed may be engaged. This feed is automatic and is disengaged by stops that are set to a definite length of travel for each particular tool. Turret positioning on bar and chucking machines is done automatically. Thus, turret indexing by hand is eliminated.

Cutting fluids are usually supplied through the center of the turret. The coolant is directed on the individual cutting tool when it is involved in a cutting process. Cutting tools on the cross slide receive a supply of coolant from a source other than the main turret.

The Feed Train and Rapid Traverse

Power to operate the cross slide and turret is transmitted through the *feed train*. The feed train consists of the gearing at the headstock end (lead screw gearbox), the feed shaft, the cross slide carriage apron or gearbox, and the main turret apron or gearbox. Feeds are first selected by setting the possible range in the head-end gearbox. Cross feeds and main turret feeds are set by shifting the levers on the aprons.

The main turret on saddle-type turret lathes usually travels a considerable distance up to and away from the workpiece. A *power rapid traverse* is designed in such lathes to facilitate positioning and removing cutting tools.

BASIC MACHINE ATTACHMENTS

Thread-Chasing Attachments

The two basic kinds of threading attachments are: (1) *leader and follower* and (2) *independent lead screw*. These attachments are used for chasing threads with single-point thread-cutting tools or for accurately leading-on taps and die heads from the main turret. The leader and follower (thread-chasing) attachment shown in Figure 8-14 is attached to the cross slide apron of a saddle-type turret lathe. By contrast, the leading-on attachment in Figure 8-15 is attached to the main turret on a ram-type turret lathe. The attachment feeds the turret at the correct pitch for the thread being cut. An automatic knock-off mechanism, operated by the turret stop screws, provides a depth control for threading cuts to shoulders or in blind holes.

Lead Screw Attachment for Ram-Type Turret Lathe. The lead screw attachment for the ram-type turret lathe is applied to the apron of the cross slide. If taps and dies are to be led on from the main turret, additional parts are used to hook up the cross slide and the ram. The lead

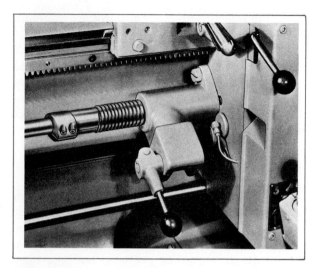

FIGURE 8-14 Thread-Chasing Attachment for Saddle-Type Turret Lathe

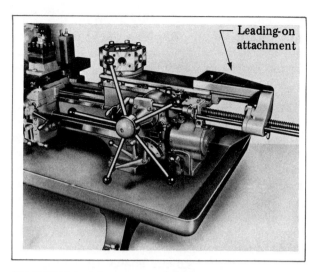

FIGURE 8-15 Leading-on Attachment to Feed Main (Hexagon) Turret on Ram-Type Turret Lathe

screw is part of the feed shaft. The lead screw thread is cut on the left end of the feed shaft only. This end of the feed shaft is never used for feeding—that is, it is used only for threading.

The cross slide is actuated for thread cutting by opening and closing half nuts with a lever located on the side of the cross slide apron. The length of travel is controlled by the automatic knock-off mechanism. This mechanism may be used for right- and left-hand threads. Movement of the apron for thread cutting is controlled by a lever for starting, stopping, and reversing the lead screw and the cutting action. There is an interlock control so that the threading mechanism cannot operate at the same time as the feed.

Lead Screw Attachment for Saddle-Type Turret Lathe. The lead screw attachment for the saddle-type turret lathe may be designed to cut right- and left-hand threads with the cross slide, the main turret, or both. The attachment includes the following main units:

—A lead screw gearbox, which provides a feeding and a threading gear train;
—A selective quick-change gearbox with gears positioned to obtain different thread pitches;
—An independent lead screw with a single thread extending the full length of the turret lathe;
—A motor-driven quick traverse for constant speed and rapid return between threading passes;
—Half-nut boxes, including carriage and saddle control levers, interlock protection, knock-off device, and threading dials;
—Screw or feed selector lever for engaging and disengaging the lead screw or the feed.

Most of these units are identified in Figure 8-16.

Taper Turning Methods and Attachments

Tapers are turned on turret lathes by three general methods. Tapers may be turned by using (1) a formed tool, (2) a roller rest taper turner, or (3) a taper attachment.

Formed cutters may be used to produce short tapers. However, several problems must be overcome. The work must be rigid and supported to withstand the heavy forming cut without chatter. There is the question of the required degree of accuracy and quality of surface finish. If the workpiece cannot support the heavy forming cut without chatter or if it is impractical to produce a taper surface that meets surface finish requirements, another method must be employed.

A *roller rest taper turner* (Figure 8-17) may be used to produce long, accurate tapers on bar jobs. The taper turner can be set quickly for size by adjusting a graduated dial setting. The angle of taper is controlled by the *taper guide bar*. As the cutting tool advances toward the headstock, the cutter and rolls in the roller rest taper turner recede to produce the required taper.

TURRET LATHES 149

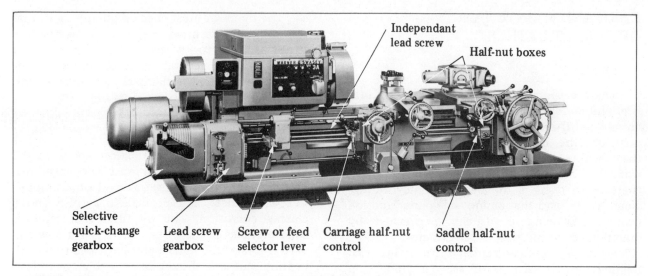

FIGURE 8-16 Identification of Main Units Required in Thread Cutting

A *taper attachment* (Figure 8-18) provides a third general method of turning an internal or external taper. The design and function of taper attachments for turret lathes and engine lathes are similar. When turning a taper, the attachment is clamped to the bed. The cross slide or the cross sliding main turret moves along the bed. The *pivoted guide plate* of the taper attachment is set at the angle of the required taper. The guide plate guides the cross slide and the mounted cutter along the taper angle. This movement produces the required taper. When the attachment is not in use, it is unclamped to permit the attachment to move freely along the bed with the cross slide or the cross-sliding main turret. The extension rod of the taper attachment is usually removed in such cases.

The taper attachment may also be used to cut taper threads. Internal or external threads can be chased by using the taper attachment in combination with a leader and follower attachment or a lead screw thread-chasing attachment.

FIGURE 8-17 Roller Rest Taper Turner

FIGURE 8-18 Taper Attachment for **Ram-Type** Turret Lathe

CARE AND MAINTENANCE OF TURRET LATHES

Leveling

The precision of a machine tool is built into it at the factory where the bed and major units are aligned from a fixed plane. The accuracy of a turret lathe, in particular, is maintained when the machine tool is leveled at its operating position. Periodic leveling is necessary to ensure that there is no twist in the bed, that the bearings wear evenly, and that all movable parts function properly. Leveling is usually performed by maintenance or millwright departments.

Sensitive, graduated-tube, spirit levels with readings to 10 seconds of arc per graduation (0.0006 inch per foot) are used. Readings are taken across the bed at the headstock and at the outer end of the bed ways. Another reading is required parallel to the ways.

When a saddle-type machine is leveled, the saddle is moved as close as possible to the spindle nose at the headstock end. The leg or base leveling screws are adjusted until the bubble reading of the level is in the center. The saddle is then moved by hand to the right end of the bed. The reading is taken again with the level placed across the top of the turret at right angles to the bed ways. The base screws are moved to bring the bubble reading to center. After the machine is secured, the same checks are made to be certain no movement has occurred.

Lock Bolts and Turret Clamps

The accuracy of indexing the turret depends on the proper operation of lock bolts and turret clamps. Springs that operate these tightening devices must be strong enough to push each lock bolt into a particular bushing to position the turret before the clamping ring operates.

The turret mechanism often becomes sluggish. This sluggishness may be due to the oil that has gummed because dirt may have worked into the lock bolt mechanism. Sometimes, a weak spring causes the lock bolt to move slowly. Under these conditions, the turret will not repeat its index accurately. This problem may be corrected by replacing a broken or weakened spring with a new one obtained from the manufacturer. Gummy oil may be washed away with kerosene. When all moving parts are clean, they require coating with fresh lubricating oil, as specified.

The operator must check regularly to see that the automatic or hand-operated clamps hold the turret securely to its base. The turret lock bolt is designed to accurately position the turret. After positioning, the clamp alone must be tightened to maintain the turret position.

Lubrication

The principles of lubrication apply to the turret lathe. Headstock oil must be changed regularly. Harmful condensation and other impurities accumulate and produce a varnish and a glazing condition. The oil filter that removes solid particles needs to be cleaned or a new one used if it is of the replaceable cartridge type. All bearing surfaces should be wiped with a clean cloth.

The turret lathe and attachments require careful oiling and greasing according to the manufacturer's instructions. Before power is turned on and after all parts have been oiled completely, the slides and turrets are moved by hand to see that they work freely.

The operator needs to establish a daily routine of checking carriage and saddle reservoir levels, lubricant systems that have force feeders, and hand oiled parts that require routine care. Lubrication must be maintained for easy operation of sliding units and for the prevention of unnecessary wear.

Cutting Fluid Systems

Most turret lathe operations require the continuous recirculation of cutting fluids. As these fluids cool the cutting tools and workpieces and wash the chips away, the fluid drains through the chips, is strained to remove foreign particles, and is recirculated. Strainers, sediment trays, and other settling devices require regular checks. The circulating system must be kept clean.

The cutting fluid requires a constant test to maintain the original cutting and cooling properties. Different workpieces and processes may require draining off one cutting fluid to replace it with another recommended by cutting fluid specialists. A dirty solution needs to be changed. Additional cutting fluid must be added to a weakened solution.

Checking Adjustments

Saddle and carriage gibs require adjustment to compensate for normal wear and to help maintain the original machine accuracy. Other parts and mechanisms that require checking on a scheduled basis include the square and hexagonal turret clamps, feed and rapid traverse clutches, and feed knock-offs. Operator service manuals provide detailed information on care and maintenance of a particular manufacturer's turret lathe.

SAFE PRACTICES IN THE CARE AND MAINTENANCE OF TURRET LATHES

—Use a sensitive spirit level to check across the bed and parallel to the ways. Adjust the leveling screws on the legs, if required.
—Study the turret lathe operator's manual to establish a routine plan for checking all automatic and manual oiling systems.
—Shut down the turret lathe before checking all automatic oiling systems. Make sure that there is an adequate supply of oil and that the forced feed systems are working.
—Remove chips and foreign particles from all exposed machined surfaces on which mechanical movements take place. Wipe the surfaces with a clean wiping cloth. Manually oil the surfaces that require hand-oiling with the recommended lubricant.
—Check the movement of the turrets. If they are sluggish due to a weakened lock bolt spring, replace the spring with one furnished by the manufacturer. If the turrets are sluggish due to gummed oil, the unit may require disassembly, removal of the thickened and dirty oil, and coating with a new lubricant.
—Go through a complete cycle at reduced speeds to establish that all tools and stops are positioned correctly. Check the machined part for dimensional accuracy and quality of surface finish.
—Examine the overhang of cutting tools and follower rolls. Reduce the lengths to which they extend in order to provide maximum tool support without interfering with the cutting process.
—Check the aligning hole on the turret at each station before inserting or sliding a positioning or cutting tool into the turret. Remove any nicks or burrs from the turret and the cutting-tool shank or sliding holder.
—Check the cutting fluid systems for the main turret, secondary turret, square tool post, and back position toolholder. The nozzles must be capable of delivering a required flow of coolant to the place where it properly serves to cool and aid in the cutting process.
—Test the cutting fluid with a manufacturer's gage to establish the strength of the solution. Run another test to determine the amount of foreign matter that is in the solution. Either replace the solution or add new cutting fluid to bring the solution to a required specification.
—Determine from the material and design features in the piece part what kind of cutting fluid and flow processes are best suited to the job at hand. Then, check the characteristics of the cutting fluid in the coolant system. The system may need to be drained and filled if a different coolant is required.

TERMS USED IN TURRET LATHE CLASSIFICATIONS AND DESIGN FEATURES

Bar machine	A type of turret lathe where bar stock is fed through a spindle or the workpiece is held in a collet or step chuck.
Chucking machine	Usually a turret lathe used for heavier machining operations than a bar machine. A turning machine where forgings, castings, large and irregular workpieces are held in a chuck or a fixture.
Automatic bar and chucking machine	A turret lathe that may be programmed so that trip blocks actuate microswitches. (These switches control the functions of the carriage, cross slide, vertical slide, threading head, speed and feed changes, and forward and reverse directions.)

152 TURNING MACHINES

Turning center (system)	A numerically controlled universal turret lathe with a greater range of speeds and feeds and number of tool stations than a bar or chucking machine or combination.
Ram-type bar and chucking machine	A turret lathe that has a ram on which the main turret is held and positioned. A movable ram that slides on a base to move the cutting and positioning tools toward or away from a workpiece. (The base for the ram is secured at a required location on the ways.)
Saddle-type bar and chucking machine	A movable mechanism consisting of a saddle, apron, and gearbox. A turret lathe with the turret as an integral part of a saddle that may be moved manually or automatically along the bed ways. A complete assembly to position workpieces against a stop, to feed, or to remove cutting tools during a machining process.
Plain cross slide	A cross slide that is locked in a fixed position to provide movement in a crosswise direction.
Universal cross slide	A cross slide that may be moved, by hand or power, lengthwise or crosswise in relation to the bed.
Fixed-center turret	A turret with a single axis that coincides with the spindle axis of the turret lathe.
Cross-sliding turret	A main turret that may be moved crosswise (transversely) from the axis of a turret lathe.
Compound cross slide	A turret lathe cross slide with a compound rest that permits angle settings, as for turning steep tapers and cutting bevels.
Side-hung cross slide	A cross slide, mounted on the front of a carriage and apron, that moves along the front side of the bed. A design feature that provides maximum swing over the cross slide.
Reach-over (bridge-type) cross slide	A cross slide mounted on a saddle that extends across the front and back ways. A cross slide that permits the use of a back toolholder or turret tool post.
Stop roll	A device that contains the individual stops for depth of cut. (When designed with a center stop, all individual stops may be adjusted without resetting the individual stops.)
Leader and follower threading-chasing attachment	An attachment to the cross slide apron. A mechanism in which a half nut (follower) engages a lead screw (leader) to feed the cross slide according to the required thread pitch.
Leading-on attachment	An attachment used for either thread chasing with a single-point threading tool or for leading-on die heads and taps from the main turret.
Roller rest taper turner	An attachment for producing long, accurate tapers on bar work. A taper-producing device that is controlled by a taper guide bar. (As the cutting tool advances, the cutter and the rolls, actuated by the guide bar, recede to produce the required taper.)

SUMMARY

The turret lathe is a multi-tool production turning machine.

Turning processes were revolutionized in the mid-1800s by incorporating additional tooling stations to increase the capability of the lathe to mass-produce interchangeable parts.

A main turret fed by a ram or saddle, a square or hexagon secondary turret on the cross slide, the back position on the cross slide for other cutting tools, and multi-speed geared headstocks or motorized spindles are distinguishing features of the turret lathe.

Fully automatic bar and chucking machines may be programmed for all internal and external turning and thread-forming processes. As precision machines, they are capable of meeting high production requirements while maintaining tolerances within 0.0001" (0.002mm).

Turning centers combine additional tooling stations to increase the number of processes on one machine that otherwise would require a part to be produced on two or more machines.

The multi-tool single turret design includes slanting the turret on an angle. The tooling is placed perpendicular to the plane of operation. Inside diameter tooling is placed in the inside row of the turret; outside diameter tooling, in the outside row. This combination and turret design accommodate 16 different tool stations and allow interference-free part turning.

There are multiple speed ranges from 125 to 3,000 RPM on the small precision toolroom bar and chucking machines. Heavier-duty machines, with bar stock capacities up to 2-1/2" (64mm) diameter and chucking up to 12" (305mm), operate over three speed ranges from a low of 65 RPM to a high of 3,000 RPM.

Bar machines are used for turning processes on bar stock or cutoff pieces that may be held in regular or special-shaped collets.

Chucking machines are designed for the mass production turning of irregular or large-diameter workpieces. Such workpieces must be held in a chuck or fixture.

Ram-type turret lathes have a movable ram. The ram turret is positioned, fed, and returned to a starting position by moving the ram in a base. The base and assembly may be locked in any position along the bed.

Bar and chucking machines are designed with 6, 8, or 16 tooling turret stations. The turret may be placed flat, vertically, or at an angle. The saddle and turret move as a unit along the bed. The saddle type is adapted to chuck work requiring heavy cuts or involving a long overhang.

Cross slides are either plain (one direction) or universal (for movement on a carriage transversely or longitudinally).

Main turrets may be held in a fixed, center position. The cross-sliding turret model permits the additional transverse movement of the turret and permits facing, turning internal tapers, and threading.

The compound cross slide permits cutting bevels and steep tapers. A compound rest may be set at any angle.

The side-hung cross slide permits maximum swing. The cross slide and apron are on the front of the bed.

TURNING MACHINES

The reach-over (bridge) type of cross slide is mounted on a saddle that extends over the front and back ways. The cross slide permits the mounting and holding of additional tools on the back tool post (holder).

Saddle-type turret lathes have the main turret attached to a saddle.

Many turret lathes are designed with a lead screw attachment for thread chasing with a single-point cutting tool or for accurately leading-on die heads or taps.

A leader and follower attachment, mounted on the apron of the cross slide, is a thread-chasing device.

Tapers may be turned on a turret lathe by using a formed tool, a taper attachment (similar in construction and operation to an attachment for an engine lathe), or a roller rest taper turner. The taper guide bar controls the simultaneous movement of the support rolls and the cutter.

All machine tools require leveling to maintain the precision built into them. Leveling readings must be taken periodically across the bed ways and parallel to them.

Machine maintenance requires a routine schedule for checking oil levels within automatic feeder systems. Exposed machined surfaces, over which there is to be movement, must be cleaned and oiled.

Turrets require testing to be sure that each position is indexed accurately with the engagement of the lock bolts. The turret clamp is then used to secure the turret.

Cutting fluid systems are to be checked for freedom of flow. The strength and cleanliness of the cutting fluid must also be tested.

The depth positioning of a cutting tool on the cross slide or main turret is set by adjusting an individual tool stop.

Movement of a cross slide or carriage is controlled by feed engagement levers. Trip dogs are set to automatically control the length or depth over which a cut is taken.

Bar stock may be fed by using the collet chuck and bar feed operating lever.

UNIT 8 REVIEW AND SELF-TEST

1. State how Stephen Fitch's eight-station turret lathe of 1845 revolutionized the production of interchangeable parts and metal-cutting lathe work.
2. Make general comparisons of five features of a turret lathe of the 1914 period and a modern fully automatic chucker and bar machine.
3. a. Describe a bar machine and a chucking machine in terms of holding and feeding workpieces.
   b. Distinguish between a ram-type and a saddle-type bar and chucking machine.
   c. Indicate the difference between a fixed-center turret and a cross-sliding turret on a saddle-type turret lathe.
4. Cite three functions of a secondary turret on a cross slide.
5. Describe briefly the location and operation of a leading-on thread-chasing attachment.
6. Describe how a taper is produced using a roller rest taper turner.
7. Cite three possible problems that may develop with improper care and maintenance of an indexing turret.
8. State three important safety checks the turret lathe operator makes before starting a production run.

UNIT 9

Spindle, Cross Slide, and Hexagon Turret Tooling and Permanent Setups

The basic machine processes performed on a turret lathe are grouped under *internal cuts* and *external cuts*. Internal cuts involve the following processes:

—Center and other drilling,
—Boring (straight and taper),
—Reaming,
—Counterboring,
—Undercutting,
—Beveling (chamfering),
—Threading.

External cuts relate to processes such as:

—Facing,
—Straight turning,
—Beveling and rounding corners,
—Taper turning,
—Grooving,
—Threading,
—Form turning,
—Cutting off.

Combinations of these processes account for the production of most turret lathe work.

These basic machining processes are carried on with tooling positioned in the square turret and the back tool post of the cross slide. Other tooling is mounted on the hexagon ram turret on a ram-type or saddle-type machine. Spindle, cross slide, and hexagon turret tooling and accessories are described and illustrated in this unit.

KINDS OF MACHINE CUTS

The four kinds of cuts that may be taken on a turret lathe are: (1) *single*, (2) *multiple*, (3) *combined*, and (4) *successive*. A single cut involves one cutting tool that performs one operation at one time. Multiple cuts are two or more cuts taken at one time from one turret station. Combined cuts are cuts taken by tools mounted on both the cross slide and hexagon turret at the same time. Successive cuts relate to cuts that are made from successive faces of the hexagon turret in consecutive order.

As a result of being able to use different cutters, work-holding and positioning devices, combinations of movements of the cross slide and hexagon turret, and required speeds and feeds for each process, the turret lathe has the following advantages over the engine lathe:

—Combined and multiple cuts permit several operations to be performed at the same time.
—Tooling setup time is reduced; each tool is set to positive stops; and each cut is identical over the production run.
—Precision machining to close dimensional and surface finish tolerances is possible under production conditions.
—Permanent multi-tooling setups on the machine conserve downtime; combined multiple, and successive cuts are adapted to quantity production requiring common tooling.
—A single turret lathe has the tooling capacity to mass-produce parts that would otherwise require a number of engine lathes as second, third, and so on operation machines.

HEADSTOCK SPINDLE TOOLING

A work-holding device, mounted in the turret lathe spindle, is the connecting link between the turret lathe and the cutting tools. The precision to which the finished job is machined depends upon the accuracy and rigidity in holding a workpiece. Production time depends on the speed and ease with which the work-holding device may be operated.

156 TURNING MACHINES

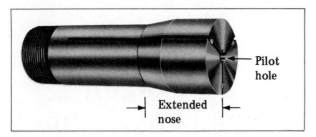

FIGURE 9-1 Extended-Nose Collet

Wherever practical, standard work-holding devices should be used. Bar work is usually held in a collet chuck; chucking work, in a three- or four-jaw chuck. Work of irregular shape and difficult jobs that require heavy cuts or extremely accurate machining may be held in special jaws mounted on standard or pneumatic chucks. Fixtures that are adapted for faceplate mountings are also used. A description of standard collets and chucks, special jaws, and typical examples of fixtures follows.

Collet Chucks

The functions and design features of turret lathe and engine lathe collets are similar. Standard special alloy steel collets have a spring-tempered body. These collets are manufactured in round and other regular shapes in fractional, metric, decimal, letter, and number sizes. Special-shaped collets are available for holding odd-shaped parts and extruded stock.

Extended-nose collets, like the collet shown in Figure 9-1, have a soft face and pilot hole. These collets may be machined to accommodate a special size or odd shape. The collet may be drilled, bored, or stepped out to the exact required size. The extended nose permits deep counterbore and tool clearance for extended work processes.

Positive-stop collets have all the design features of the standard collet plus a precision-threaded section at the back end of the collet bore. The positive-stop collet may be fitted with a *solid positive* stop, an *ejector* stop, or a *long* stop. All stops are threaded into a positive-stop collet. When they are positively shouldered, the stops are locked against the end of the collet. Each stop pin is adjustable to the desired part length.

The three types of stops, shown in Figure 9-2, permit a wide variety of chucking work. The solid positive stop (Figure 9-2A) has a capacity for chucking parts to a depth of 3-1/8" (79mm) from the collet face. The ejector stop (Figure 9-2B) accommodates workpieces to 2-3/4" (69mm) length. The long stop (Figure 9-2C) is used for chucking parts up to 7-1/2" (190mm) from the collet face. This depth accommodates round parts up to 7/8" (22mm) diameter, hexagons up to 3/4" (19mm), and square parts up to 19/32" (15mm).

Dead-length collets and *dead-length step chucks* permit shoulders and faces to be machined to exact length regardless of variations in the outside diameter. An adjustable solid stop is threaded into an inner collet. The inner collet is spring-loaded against the spindle face so that no lateral movement occurs.

Figure 9-3 shows a disassembled dead-length collet and step chuck. The step chuck has soft faces and a pilot hole to permit boring to size in the machine spindle and to ensure concentricity. The location accuracy is maintained by pinning the inner collet to the outer collet.

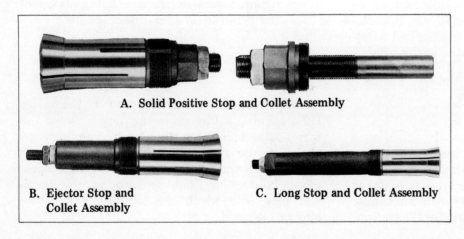

A. Solid Positive Stop and Collet Assembly

B. Ejector Stop and Collet Assembly

C. Long Stop and Collet Assembly

FIGURE 9-2 Positive-Stop Collets

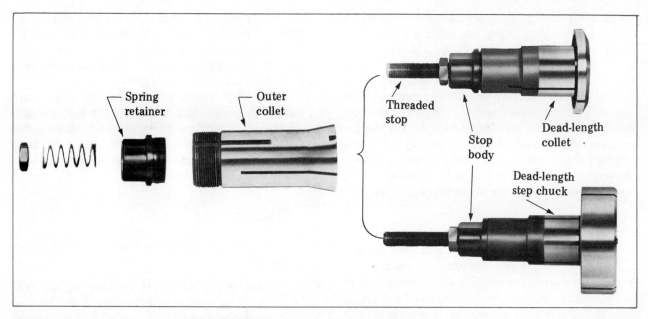

FIGURE 9-3 Parts of Dead-Length Collet and Step Chuck Assemblies

Expanding collets are valuable for close-tolerance machining because there is no lateral movement of the master expanding collet or pads. Figure 9-4 shows the three main parts of a master expanding collet. The master collet (Figure 9-4A) is hardened and ground. The pads (Figure 9-4B) are soft. The shoulder on the machinable pad is faced to locate the work. Repetitive parts are thus faced to the same length. Turning the pad on the machine spindle ensures exact concentricity. The limit ring (Figure 9-4C) is used to hold the pads at nominal size while they are turned to a required outside diameter. The pad segments are marked *A*, *B*, and *C* to permit replacement in the same position. The pads may be remachined to hold subsequent parts of smaller diameter.

Precision expanding collets are used for chucking on internal surfaces. The expanding collet assembly holds the workpiece in a previously machined bore. This assembly permits the part to be machined with concentric and square shoulders, faces, and diameters in relation to the bore.

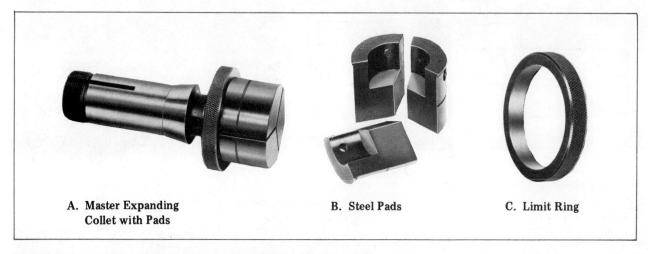

A. Master Expanding Collet with Pads   B. Steel Pads   C. Limit Ring

FIGURE 9-4 Three Main Parts of a Master Expanding Collet

A complete expanding collet assembly and a cutaway section showing the main parts are featured in Figure 9-5. The work-locating stop is faced and bored in position in the assembly to assure an absolutely square locating face. The spindle collar has four adjusting screws. These screws are for concentricity adjustment. The three cap screws are for mounting the work-locating stop. The draw collet controls the movement of the draw plug and the expanding collet. Expanding collets are available in standard round form in fractional, decimal, and metric sizes.

Step Chucks and Closers

Step chucks and *step chuck closers* are designed for accurately holding large-diameter work. The step chuck and closer shown in Figure 9-6 accommodates workpieces up to 6″ (152mm) diameter. Parts such as castings, molding, stampings, and tubing may be held rigidly and accurately, without crushing or distortion.

Step chucks for machining diameters larger than 2″ (51mm) usually have pin holes and pins, which are located beyond the maximum rated capacity of the chuck. Thus, the step chuck may be stepped to its full capacity without cutting into the pins. Once the chuck is stepped to the required diameter, the pins may be removed and reused later for other stepping.

Hardened and ground step chucks are available in regular depth (Figure 9-7A) and for extra-depth capacity (Figure 9-7B). Step chuck closers are designed to seat directly on a taper or threaded-nose spindle.

The step chuck closer and the step chuck have a corresponding taper. As the step chuck is drawn in, the taper surfaces direct the closing force over the stepped area of the chuck. A strong gripping force is thus produced to accurately hold the workpiece.

Chucks

The four prime requirements of chucking are as follows:

—The workpiece must be held rigidly with the least amount of distortion and surface damage, if any;
—Parts must be loaded and unloaded quickly and easily;
—Parts must be positioned accurately;
—The chuck or other work-holding device should be as nearly standard as possible and readily accessible.

The standard chucks used are similar to independent-jaw chucks, universal (scroll) chucks, and combination independent-jaw and universal

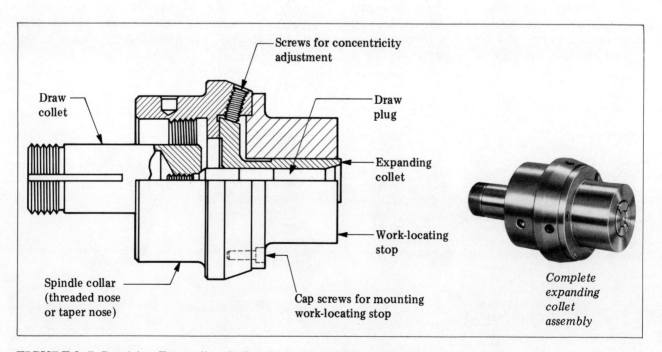

FIGURE 9-5 Precision Expanding Collet Assembly and Cutaway Section

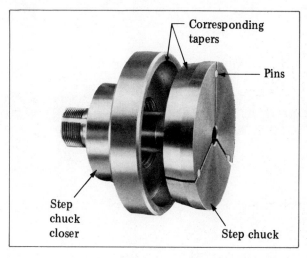

FIGURE 9-6 Step Chuck and Step Chuck Closer

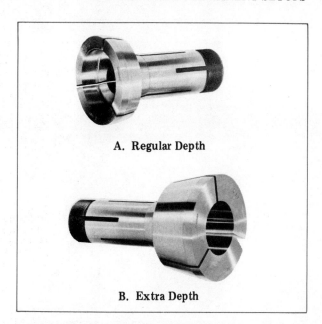

FIGURE 9-7 Hardened and Ground Step Chucks (Regular- and Extra-Depth Capacity)

chucks. The step design of the jaws permits their use over a wide range of diameters.

One main difference between turret lathe and engine lathe chucks is in the use of standard reversible-top soft jaws. These jaws are used for inside and outside chucking operations. There are certain production jobs where a simple set of special jaws permits holding the workpiece more securely than conventional jaws. The special jaws allow heavy cuts to be taken to reduce machining time. Cutter life is lengthened on some jobs due to holding rigidity.

The sectioned line drawing in Figure 9-8A illustrates the limited gripping area of standard chuck jaws. The effect of too large an arc of contact (greater than half the diameter of the workpiece) is shown in Figure 9-8B. The effect of too wide a jaw gripping surface is to pull the workpiece between two jaws and away from the third jaw. The condition is remedied by reducing the contact area, as shown in Figure 9-8C, between the chuck jaws and the work. The corrected condition permits each of the three chuck jaws to grip the work equally in a centralized position.

Chuck Jaw Serrations and Jaw Types. Standard chuck jaws are *serrated*. Grooves are cut at 60° to a medium depth of 0.078" (2mm) and pitch of 1/8" (3mm). For most jobs, the serrations provide a good bite on a rough surface. However, the depth of the serration marks must often be limited to permit the outside diameter to be cleaned up to the required dimension. A limited-depth (fine) serrated jaw, 60° × 3/32" (2.4mm) pitch × 0.050" (1.1mm) depth, must then be used.

Similarly, the exterior depth of the chuck jaw must be considered. The depth is established to permit a minimum overhang with adequate clearance. All surfaces to be machined must be accessible and safely clear of all cutting tools.

Rocking jaws are used on castings, forgings, weldments, and other rough surfaces. Surface roughness and irregularities in the gripping area require the use of rocking jaws. A horizontal rocking jaw, like the one shown in Figure 9-9, provides a secure grip on long, rough castings and irregular shapes. In the setup shown, two chuck jaws are solid. The third jaw is a rocking jaw. The part is aligned by the solid jaws. The rocking jaw adjusts to the uneven surface and grips the part securely. The rocking jaw is undercut so that the clamping forces grip the part at the ends only.

Solid jaws may be relieved to serve the same function as horizontal rocking jaws. In Figure 9-10, the chuck jaws of the two-jaw chuck are relieved to hold the long, rough part. On a three-jaw chuck, two of the jaws are relieved in the center. The gripping areas are on the outside of the chucked part. The gripping areas must not overlap.

160 TURNING MACHINES

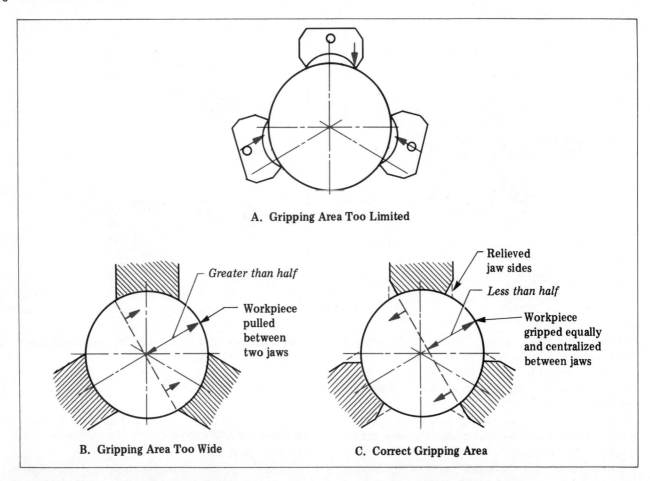

FIGURE 9-8 Effect of Chuck Jaw Contact Area on Centralizing and Gripping Forces

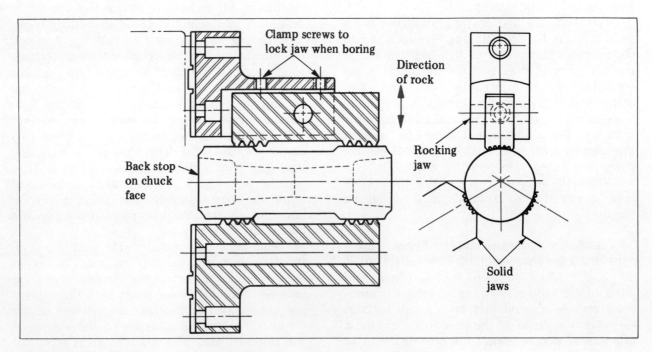

FIGURE 9-9 Rocking Jaw Application on Long, Rough-Surface Part

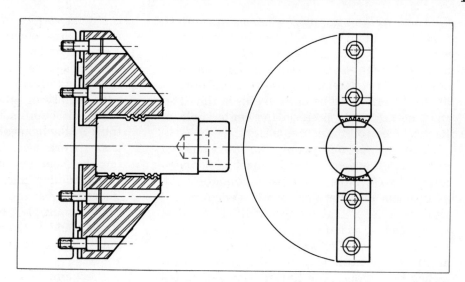

FIGURE 9-10 Relieved Solid Jaws Gripping Rough Part

Wide or wrap-around jaws are often used on second operation work. Thin-wall, fragile parts require wide or wrap-around jaws. These jaws distribute the gripping forces and minimize distortion. Wrap-around jaws also provide adequate friction (force) between the jaws and the workpiece to permit the machining process to be performed accurately.

Sometimes, a driving pin is provided on one jaw. The jaws centralize the part and apply an extremely light force. The pin provides the driving force. The pin usually rests against an opening or protrusion in the part.

It is important that the radius of the chuck jaws be bored to within ±0.001″ (±0.02mm) of the workpiece diameter to ensure that the gripping forces around a fragile part are distributed evenly over the entire chucking area. It is also important to apply a uniform, adequate gripping force, particularly for parts that may become distorted by excessive force. Unevenness may be avoided by substituting a torque wrench for the T-handle wrench of a hand chuck. A power chuck wrench may also be used. This wrench provides a force at a preset torque for internal and external gripping. An air (pneumatic) chuck is used on some turret lathes. The gripping force is adjusted by varying the air pressure.

Care of Chucks. The sides of chuck jaws and the sliding areas on the chuck body require regular checking against the accumulation of foreign particles, burrs, and nicks. The scroll and mating threads on chucks need to be free of dirt and chips. The jaws also require checking and regrinding if they are hardened and sprung. After periodic disassembling and cleaning, the moving parts should be lubricated with a *moly grease* (H–80) after assembling.

Special Fixtures and Faceplates

Special fixtures are often used under the following conditions:

—A difficult-to-machine job must be held very rigidly to withstand heavy or extremely accurate cuts;
—The use of a fixture is more economical on a high-production job;
—The part may be loaded and unloaded faster and easier than by conventional chucking.

Fixture designs vary considerably. Some fixtures are mounted on standard chucks and complement the chuck jaws. Other fixtures are used for second and third operations to nest workpieces on a previously machined surface. Indexing is used on some fixtures for the multiple positioning of a workpiece.

Faceplate Applications. The use of a T-slotted faceplate for holding a fixture is a common shop practice. Angle plates that fasten directly to the T-slots of the faceplate are used. These plates may be supplemented with uniform holes that are drilled and tapped in the turret lathe faceplate. These holes permit strapping fixtures to the face by using standard bolts. Faceplates are available for threaded and flanged spindle noses.

162 TURNING MACHINES

Angle fixtures are used when a surface to be machined is in an exact relation to a previously machined flat surface. Figure 9-11 gives an idea of how a right-angle nesting plate, which is mounted on a faceplate, positions the bore at 90° to the machined base. Stop pins in the right-angle nesting plate position the center hole of the workpiece on the center axis.

Indexing fixtures permit the movement of workpieces that require similar machining operations in two or more locations. A common holding problem in turret lathe work is encountered with workpieces that require identical turning, boring, facing, chamfering, and threading operations on two or more bosses. A sliding or swinging indexing fixture (Figure 9-12) may be used in such operations. The fixture that is mounted on the faceplate may be swiveled between two guide pins. These pins establish the precise distance between two holes.

Fixture plates are available commercially. They are round plates with a flange for direct application to the headstock spindle. Fixture plates may be machined to become a fixture and for mounting special-purpose chucks or other fixtures.

UNIVERSAL BAR EQUIPMENT FOR PERMANENT SETUPS ON TURRET LATHES

Turret lathe manufacturers recommend permanent setups of universal bar equipment (standard tools) for small-lot production jobs. Where there is a great variety of machining processes on different parts, changeover time is reduced significantly. Standard tools are commercially available on ram- and saddle-type turret lathes. The setup in Figure 9-13 provides tooling flexibility over a wide variety of jobs on a ram-type turret lathe. With this combination of holders and cutting tools in the hexagon turret, it is possible to position a workpiece, face, turn, chamfer, undercut (groove), center, drill, bore, ream, and thread. Other tooling may be provided on the square turret and rear tool post to provide for additional single, multiple, or combined cuts. External straight and taper turning, facing, shoulder turning, thread chasing, and cutting off processes are added.

A permanent setup on the hexagon turret permits the rapid changing of shank tools. Job changeover, thus, involves changing collet bushings to accommodate the bar stock, inserting shank-type tools, positioning each cutter, and setting stops.

Bar Turner Functions and Operations

The *bar turner* is the most widely used cutting-tool holder and work-support device for bar work. It is available for both ram- and saddle-type turret lathes. While the designs of bar turners differ, the functions and principles of opera-

FIGURE 9-11 Use of Special Right-Angle Nesting Plate on a Faceplate Fixture

FIGURE 9-12 Application of Swinging Indexing Fixture

TOOLING AND PERMANENT SETUPS 163

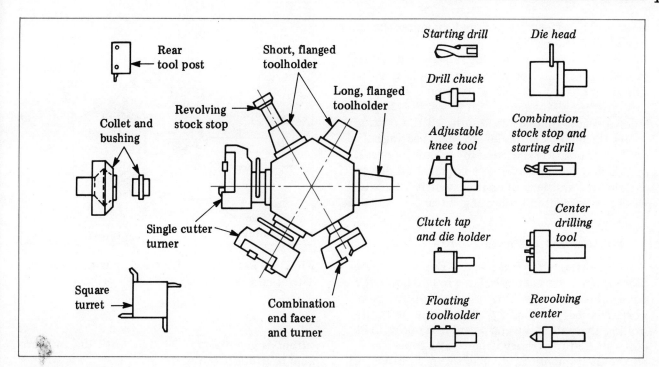

FIGURE 9-13 Permanent Setup of Universal Bar Equipment for a Ram-Type Turret Lathe

tion are similar. Each bar turner consists of a rigid cutter holder and self-contained rolls or flat carbide shoes that support the cutting action and serve as a steady rest. The rolls or shoes also support the workpiece to produce a concentric, turned surface.

The roll-type bar turner shown in Figure 9-14 is equipped with antifriction bearing rolls. The rolls are adjustable to accommodate a wide range of finished diameters. During the cutting action, the cutting forces produce a constant vibrationless relationship between the workpiece and the cutting edge. The rolls are usually set just slightly beyond the turned radius (nose of the cutter). In this position, the cutting forces that keep the turned surface against the rolls burnish the di-

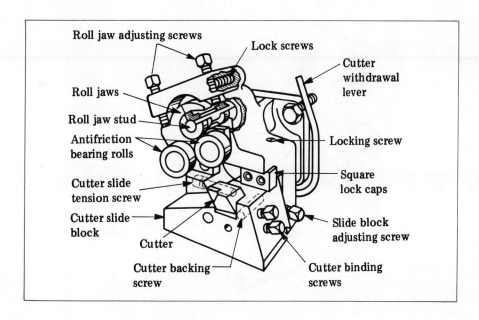

FIGURE 9-14 Principal Parts of Roller-Type Bar Turner

164 TURNING MACHINES

ameter. The effect is a finished diameter that is free of tool marks.

A shank-type cutter of high-speed steel, brazed-tip carbide, or insert-tip carbide is used. The cutter is set on end. It is ground to cut in this position. Maximum cutter rigidity is provided because the cutter absorbs all the cutting forces through its entire length. The bearing rolls and cutter position and assembly permit increased speeds and feeds. Cutter life is increased due to the rigidity and lack of chatter, which otherwise would be present in a vibrating cutter.

Starting Bar Turners Properly

Supporting the Workpiece. Bar turners are used when a bar extends far enough beyond a collet chuck or chuck to require support. Support may be provided by chamfering or center drilling the end and using a center support. The cut is started for a short distance by a tool mounted on the square turret. The setup is shown in Figure 9-15.

After the starting cut is made by the tool on the square turret, the bar turner is positioned to turn the entire outside diameter for the required length. The bar turner must be started properly in order to produce a concentric, accurate, turned surface. Therefore, the end of the workpiece must be prepared. As the bar turner is fed to the revolving workpiece, the cutter of the bar turner is given a gradual start until the rolls are able to support the work.

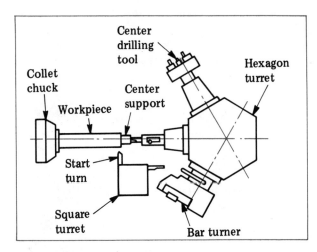

FIGURE 9-15 Setup for Cutting with a Square Turret and Bar Turner

Adjusting the Cutter Slide Block. The bar holder is designed to accommodate variations in cutter thickness. A cutter slot, set at an angle in the block, permits gripping the cutter and automatically establishing front and side clearances.

A pictorial view of a bar turner (Figure 9-16) shows the adjustments of the slide block. The slide block adjusting screw permits adjust-

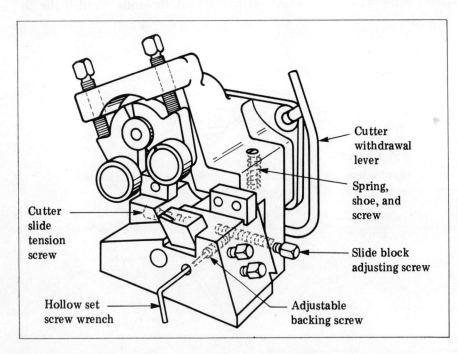

FIGURE 9-16 Adjustments of Cutter Slide Block (Ram-Type Turret Lathe Bar Turner)

ment of the cutter for size. The spring, screw, and shoe provide tension against the adjusting screw. The screw is turned clockwise to feed the cutter toward the work.

Compensation must be made for variations in cutter thickness. An adjustable backing screw repositions the cutter (after it is reground or changed) in relation to the rolls.

At the end of a cut, the cutter rolls must be retracted from the workpiece before withdrawing the bar turner. This procedure eliminates cutter withdrawal marks. The cutter withdrawal lever is turned to retract the cutter.

The cutter slide block is adjusted to increase the tension when turning large diameters. The adjustment is made by tightening the cutter slide tension screw. Similarly, the slide block is adjusted to the extreme forward position for small work where less tension is required.

Selecting Cutters and Materials. The same factors that govern the selection of cutter materials for other machine tools apply to turret lathe cutters. The three recommended bar turner cutter materials are cobalt high-speed steel, brazed-tip carbides, and insert-tip carbides. The insert-tip carbides are recommended for long production jobs where cutters must be changed during the run. Chip breakers are used on insert-tip cutting tools.

Maintaining Chip Control. The large quantity of metal that is machined quickly with a bar turner requires rapid chip removal from the workpiece and rolls. Otherwise, the finished work diameter may be scored by any chips that may work under the rolls. Based on years of experience in chip control, the shape, angles, and

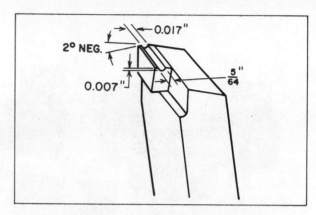

FIGURE 9-17 Specifications of Ground Chip Groove in Brazed-Tip Carbide Bar Turner Cutter

dimensions given in Figure 9-17 are recommended for brazed-tip carbide bar turner cutters. In the case of insert-tip carbide cutters, separate chip breakers are used. The chip breaker is clamped on top of the insert tip.

SLIDE TOOLS FOR TURRETS

A *box tool* (Figure 9-18) is designed for right- or left-hand general turning work on soft metals such as brass and aluminum. It is adaptable for light finishing cuts where a fine-quality surface finish and close tolerances are to be maintained. Carbide back rests are used to extend wear life. These V-rests may be single or double. Box tools are made for right- or left-hand turning with one or two cutters. Some box tools are arranged to hold a centering tool or center drill in the shank.

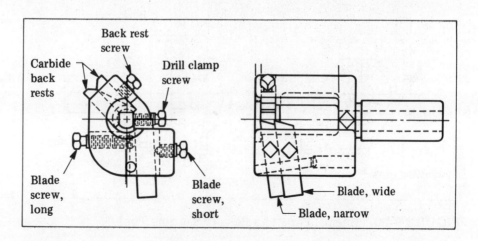

FIGURE 9-18 Right-Hand (Two-Blade) Box Tool with Carbide V-Rest

166 TURNING MACHINES

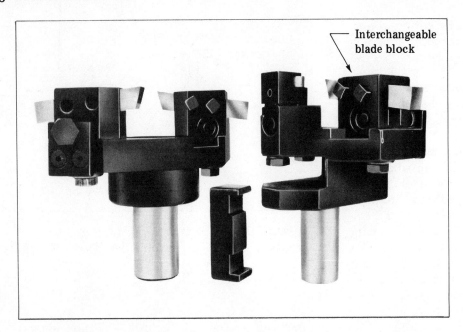

FIGURE 9-19 Roller Back Box Tools

A *roller back box tool* performs similar functions to the bar turner. Some designs require blade block slides. These slides permit straight-line diametral setting of the cutter without changing the blade clearance angle. The blade blocks are interchangeable for use with left-hand or right-hand tools, balance turning tools, and extended roll holders. Figure 9-19 shows two models of roller back box tools with interchangeable blade blocks. Extended roll holders permit turning a diameter concentrically with a previously turned diameter.

A *balance turning tool* (Figure 9-20) consists of blade blocks, a blade block retainer, and fine adjusting screws. These parts are interchangeable with equivalent-sized roller back box tools. One-piece setting gages are available for adjusting each blade. The blades may be set for equal depth or for roughing and finish turning cuts. The equal-depth setting permits machining fine finishes to close tolerances with increased tool life. Setup time is decreased and repeatability of settings is ensured.

A *slide tool* (Figure 9-21) can be used for taper turning, turning behind shoulders, and tracer turning. The names of the principal parts are descriptive of their functions. Slide tools provide a rugged, movable slide that permits the turning of tapers, turning behind shoulders, turning irregular shapes or contours,

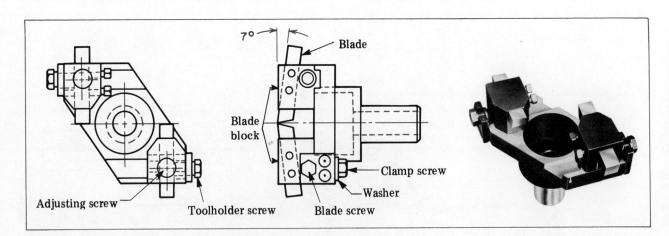

FIGURE 9-20 Principal Parts of a Balance Turning Tool

TOOLING AND PERMANENT SETUPS 167

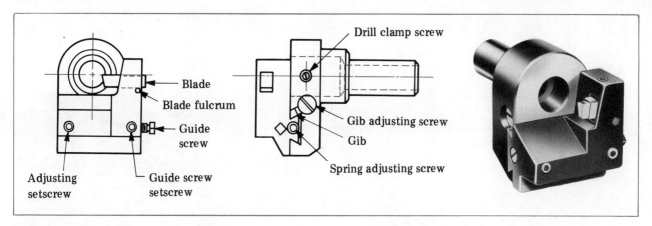

FIGURE 9-21 Slide Tool for Taper Turning, Turning behind Shoulders, and Tracer Turning

necking, forming, or cutting off. The cutting tool remains on center during the full length of slide travel. The tool is actuated by a guide mounted on the front cross slide. The movement of the slide tool controls the diameter and form of the workpiece.

Slide tool bodies accommodate cutting tools for all basic internal and external cuts. A slide tool body with three interchangeable heads is pictured in Figure 9-22. This body holds any one of three heads for turning, knurling, or recessing. The *slide tool turning head* (Figure 9-22A) is used to turn accurate longitudinal tapers, to turn behind shoulders, and to turn irregular shapes. It is also used for necking, forming, or cutting off. The blade may be reversed for right- or left-hand turning. The head can be used with a back rest.

The *knurling tool head* (Figure 9-22B) produces diamond knurls anywhere along a turned part or behind a shoulder. Right- and left-hand diagonal knurl rolls are used. The knurl rolls pivot on a fulcrum for self-equalizing forces on the knurl to ensure knurling to uniform depth. A different style of knurling holder operates from the turret. The knurls are adjustable. The knurl rolls are mounted in swivel holders and may be set at any angle to produce straight or diamond knurling patterns.

The *recessing tool head* (Figure 9-22C) provides constant centerline machining. The straight-line movement of the cutting tool permits accurate grinding for angular or diameter dimensions.

An *adjustable toolholder* (Figure 9-23) accommodates drilling, reaming, counterboring, chamfering, and similar tools. The arm extending from the side of an adjustable holder holds a

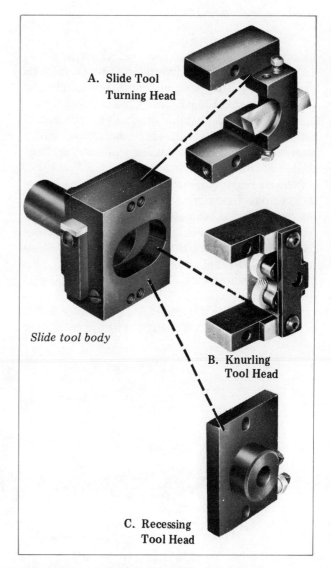

FIGURE 9-22 Slide Tool Body with Three Interchangeable Tool Heads

168 TURNING MACHINES

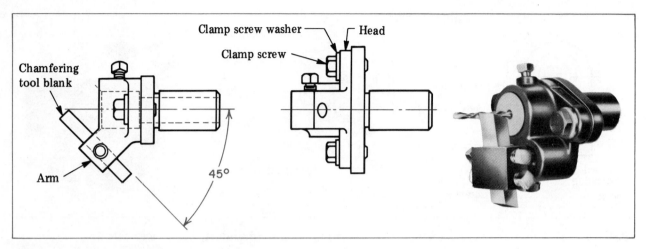

FIGURE 9-23 Adjustable Multiple-Operation Toolholder

square-face chamfering tool bit at an angle of 45°. Round chamfers (rounded corners) may be produced by using a chamfering tool bit ground to the required radius. Drills may be held by a bushing in an adjustable head. The adjustable toolholder permits setting center drills, drills, reamers, and other end-cutting tools at the exact center of the workpiece.

A *self-aligning floating reamer holder* (Figure 9-24) floats freely on antifriction bearings. These bearings provide alignment when the reamer enters a hole. After being adjusted to an approximate alignment with the work, the reamer aligns itself with the hole that is being reamed. This alignment eliminates bell-mouthed and egg-shaped holes.

An *adjustable stub collet holder* (Figure 9-25) permits tool and collet changes without realigning the holder. Stub collets are used. They provide a range of 0.015" (0.4mm): +0.005" (0.12mm) to -0.010" (0.25mm) for variations in tool diameters. Stub collet holders eliminate the need for special bushings for fractional, decimal, letter, number, and metric sizes of tool shanks.

A *standard drill chuck* has wide application in universal bar tooling and other turret lathe setups.

A *releasing "acorn" die holder* (Figure 9-26) is used for cutting standard external threads. This holder is provided with a clutch mechanism that allows the holder to be released at thread length. A *releasing tap holder* is used for tapping internal threads. At the end of the tapping process, the forward progress of the turret and tap holder is stopped. The tap is withdrawn by reversing the spindle direction.

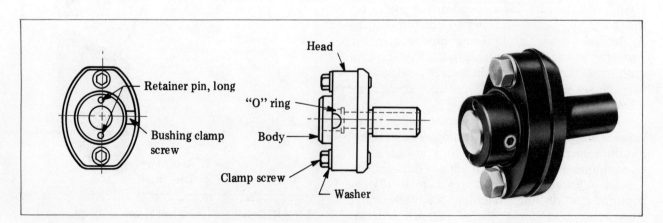

FIGURE 9-24 Self-Aligning Floating Reamer Holder

TOOLING AND PERMANENT SETUPS 169

FIGURE 9-25 Adjustable Stub Collet Holder (with Collet and Drill)

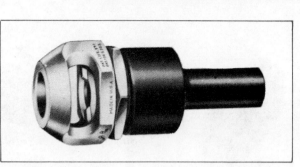

FIGURE 9-26 Releasing "Acorn" Die Holder

A *self-opening stationary die head* (Figure 9-27) is used to cut threads on machines where the die head does not rotate. The chasers (thread form cutters) are adjustable to a required pitch diameter. Die heads are available with an inside or outside trip. The length of thread on an inside trip model is accurately controlled. The die head trips automatically when the workpiece contacts the inside trip gage. An outside trip model (Figure 9-27) is used principally for short-length, fine-pitch, and close-to-shoulder threads. The outside trip relieves the chasers and threads of all stress in tripping. Right- and left-hand threads may be cut by changing the carrier and chaser for the hand of the thread.

CUTTING-TOOL HOLDERS

While the designs of toolholders used on hexagon turrets differ, the functions are similar. The general design features of one manufacturer's cutting-tool holders are shown in Figure 9-28. A *straight extension* toolholder (Figure 9-28A), *boring* and *shank* toolholders (Figure 9-28B), and an *offset turning* toolholder (Figure 9-28C) for turning large-diameter work are displayed. These holders are used for boring tools, square tool bits, and shank-type tooling. One unique feature is the fine adjustment provided by the graduated collar. The graduations are in increments of 0.0002" (0.005mm). Each fifth graduation reads in 0.001" (0.025mm). Each graduation indicates the amount of change on the diameter of the workpiece. A movement between two graduated lines shows the diameter is changed by 0.0002"

(0.005mm). The cutter is locked in position when the required diameter is reached.

A slide tool used for turret turning and boring operations is shown in Figure 9-29. The model shown receives shank tools. The fine pitch adjusting screw permits dial settings of 0.0002" (0.005mm). Each fifth dial graduation is marked to represent 0.001" (0.025mm).

CUTTING-OFF TOOLS

Cutting-off processes are generally performed by mounting the cutoff tool in a holder on the square

FIGURE 9-27 Self-Opening Stationary Die Head (Outside Trip Type)

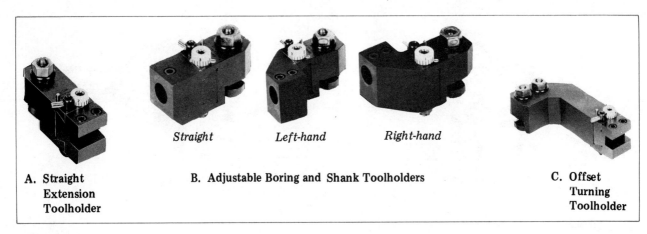

| Straight | Left-hand | Right-hand |

A. Straight Extension Toolholder B. Adjustable Boring and Shank Toolholders C. Offset Turning Toolholder

FIGURE 9-28 Cutting-Tool Holders for Turret Tooling

turret or rear tool post. On some turret lathes, the cutting-off process is done from a vertical cutoff slide. In all instances, the cutter must be on center. One way of checking and adjusting the cutter is to face the end of the bar stock.

Cutoff tools, similar to the tools used for engine lathes, may be ground from standard high-speed steel tool bits. Cutters may also be carbide tipped or of the insert-tip type. The side clearance and back taper designs of rectangular cutoff tools are widely used on cross slide square turrets or back tool post holders.

One grinding technique for increasing tool life on insert carbide cutoff tools that are used for parting solid stock is to grind the face to a 3/4″ (19mm) radius. The edges are ground at a 45° angle chamfer. The radius form stabilizes the cutter. A smoother, flatter surface results.

When tubing is to be parted, a slight lead angle of 3° to 5° is ground on carbide inserts. The lead angle reduces the fine burr normally produced on the workpiece.

HOW TO MOUNT CUTTERS ON CENTER

Turret lathe efficiency depends partly on the speed and accuracy with which tools are set. The distance above or below center affects the front and top cutting angles (Figure 9-30). The correct on-center angles are decreased when the tool is above center and increased when the cutting tool is below center. The steps required to mount cutters on center follow.

Square Turret or Rear Tool Post Cutters

STEP 1 Place the end of a steel rule (scale) on the top face of the cross slide.

STEP 2 Hold the steel rule vertically.

STEP 3 Read the cutting-tool edge (tip) height. Adjust until the scale reading and cutting-tool point coincide with the spindle axis height.

Note: Two alternate methods may be used. (1) The center height may be established by positioning the cutting edge on line with a hexagon turret center (Figure 9-31). (2) The cutter may be positioned by using a right-angle gage. The cutter is adjusted

FIGURE 9-29 Slide Tool (with Collet and Boring Tool)

for height until the cutting edge coincides with the center height as marked on the gage.

Multiple Turning Head Cutters

STEP 1 Grind the cutting tool for the required inside or outside cut to be taken.

STEP 2 Secure the cutter in the overhead cutting-tool holder.

STEP 3 Rotate the shank of the cutter holder until the grooves on the shank and turning head face coincide (Figure 9-32).

Note: A steel rule method of setting the cutter on center may be used. This method is applied when the cutter holder does not have a slotted shank or the cutter is not ground at the same height as the cutter shank. A center or drill may be placed in the center hole of the turning head. The scale is held parallel with the ground cutter edge. The scale is then extended past the cutter. The cutter holder is turned until the edge of the scale aligns with the center.

Bar Turner

STEP 1 Mount a turning cutter in the square turret.

STEP 2 Turn a section to within 0.001" (0.02mm) of the desired diameter for about 1/2" (12mm).

STEP 3 Swing the roll jaws out of position.

STEP 4 Set the bar turner cutter above center.

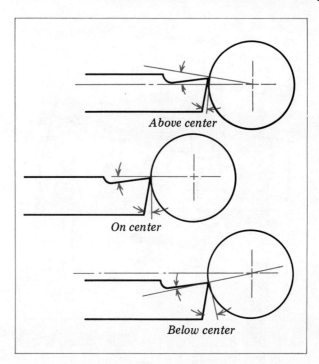

FIGURE 9-30 Effect of Cutting-Tool Height in Relation to Center Height on Front (Face) and Top Cutting Angles

STEP 5 Advance the bar turner with the spindle in a brake position.

STEP 6 Bring the cutter against the turned portion by adjusting the cutter slide.

STEP 7 *Rub* a light shine mark on the turned diameter.

STEP 8 Set the cutting edge of the cutter at the center of the shine mark.

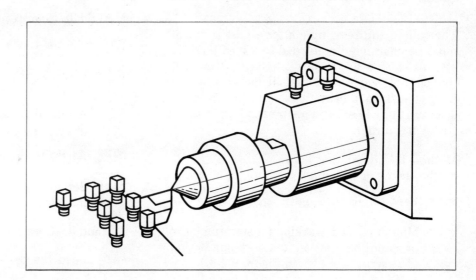

FIGURE 9-31 Aligning Cutter on Center with a Hexagon Turret Center

172 TURNING MACHINES

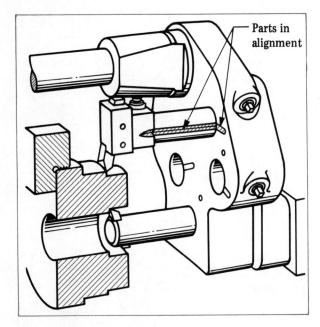

FIGURE 9-32 Aligning Groove in Cutter Holder Shank and Turning Head

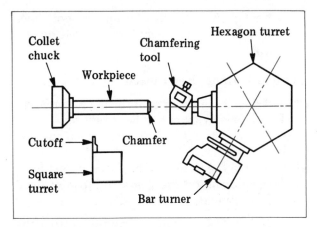

FIGURE 9-33 Use of Chamfering Tool on Hexagon Turret

Slide Tool

STEP 1 Insert a cutter in the slide tool.

STEP 2 Check the cutting edge alignment. Hold a steel rule (scale) to the cutting-tool edge.

STEP 3 Adjust the cutter until the scale cuts the center line of the slide tool and its center.

HOW TO START A BAR TURNER

The square end remaining after a part is cut off must be chamfered or turned to a specific diameter to start a bar turner properly. Otherwise, the end of the roll turned diameter will have *lumps* (surface indentations resulting from uneven starting). Four general methods are used to start a bar turner.

Method 1: Chamfering Tool on the Hexagon Turret

STEP 1 Feed the bar to length.

STEP 2 Mount an end working (chamfering) tool or pointing tool in the hexagon turret. (The setup is shown in Figure 9-33.)

STEP 3 Chamfer the end of the bar.

Note: Chamfer to a slightly smaller diameter than the required size.

STEP 4 Index the turret to the next station on which the bar turner is mounted. Adjust the bar turner to accommodate the workpiece and to take a required cut.

Method 2: Turning the End of the Workpiece

STEP 1 Extend the workpiece a short length from the collet.

STEP 2 Use a turning cutter in the square turret. Turn a small area to the required size.

STEP 3 Feed the stock to full length.

STEP 4 Turn the diameter using a bar turner to support the workpiece as the cut is taken.

Method 3: Chamfering Tool in the Square Turret

STEP 1 Cut off the workpiece.

STEP 2 Chamfer the bar stock by using a chamfering (bevel) cutter.

Note: The bar stock is often cut off and chamfered by using a combination cutter.

STEP 3 Feed the stock to the required length.

STEP 4 Set up the bar turner rolls and cutter to machine to the specified diameter.

Method 4: Center Drilling

STEP 1 Feed the bar stock to the required length.

STEP 2 Center drill the end.

Note: Use the center drilling tool for cold finished bars. On hot rolled stock, use a center drill mounted in a drill chuck.

SAFE PRACTICES IN SETTING UP SPINDLE, CROSS SLIDE, AND HEXAGON TURRET TOOLING

—Clean the spindle bore, collets, and other spindle accessories before mounting. Use a wire hand brush to remove foreign particles from threads. An air hose should not be used. The air pressure forces dirt and grit into bearing surfaces.
—Check all seating areas on holders to be sure each cutting tool rests solidly.
—Clean and lubricate the square and hexagon turrets. Check to see that the turret may be moved easily into the next index position and may be locked securely.
—Keep the chuck or collet and bar feed machanism free of dirt and grit in order to eliminate unnecessary wear and stiff operation.
—Oil chuck slides and mating jaw faces.
—Select chuck jaws so that two jaws grip an area that is less than one-half the diameter of the workpiece. The two jaws thus tend to center the workpiece against the third jaw.
—Change from medium to fine serrated jaws for soft materials or for conditions where the chuck jaws produce indentations that may not be removed by subsequent machining operations.
—Preload the chuck jaws against a chucking ring with the same force that will later be used with workpieces. Then proceed to grind the chuck jaws (or turn them if they are soft) to correct an inaccurate chucking condition.
—Use torque wrenches to tighten chuck jaws uniformly against a workpiece.
—Add counterbalancing weights on off-centered parts, particularly parts held in a fixture.
—Position all cutting tools on center. This position maintains a correct relationship between the cutting angles as ground originally and the machining requirements.
—Turn or chamfer the end of a workpiece. Chamfering provides a bearing surface for the roller bar turner.
—Grind cutting-off tools with a slight convex cutting face. This convex cutting face tends to stabilize the cutter and produce a more accurate turned face.
—Use rocking jaws on rough surfaces (castings, forgings, weldments, and so on). Such jaws compensate for surface inaccuracies and imperfections.
—Turn all setups slowly through one cycle to be sure all parts clear.
—Machine a small area on the end of bar stock to nearly the required diameter or turn a beveled edge to a slightly smaller diameter than the required one. The rolls and cutter of the bar turner will then track and cut concentrically.

TERMS USED WITH SPINDLE, CROSS SLIDE, AND HEXAGON TURRET TOOLING

Headstock spindle tooling	Primarily work-holding devices for bar and chuck work on turret lathes. (Collets, step chucks, regular chucks, and faceplates are mounted in the turret lathe spindle.)
Extended-nose collet	A machinable collet that extends the normal length of a collet to permit deeper counterboring and tool clearance.
Dead-length collet	A collet design that permits inserting an adjustable solid stop. (The collet is spring-loaded so that no lateral movement occurs.)
Step chuck	A soft-face chuck that may be bored to size on the turret lathe. A collet chuck bored to hold large-diameter (short-length) parts.

Rocking jaws	A combination of recessed jaws that grip in alternate locations along the length of irregular or rough-surfaced bars, castings, or welded parts.
Jaw serrations	Indentations cut across the gripping surface of each chuck jaw. Medium or fine pitch and depth cuts that improve the gripping surfaces of chuck jaws.
Faceplate fixture (turret lathe)	A work-holding device mounted on the faceplate of a turret lathe. A device for positioning a workpiece accurately and holding it securely during machining processes.
Universal bar equipment (ram- or saddle-type turret lathe)	A combination of standard cross slide and hexagon turret tools. A setup of tools established by manufacturers after years of extensive experience. A tooling setup that may be left permanently on turret lathes for short-run production jobs requiring basic internal and/or external cuts.
Bar turner	A hexagon turret accessory that is used widely to control concentricity during an external turning process. A cutting tool assembly consisting of adjustable guide rolls that serve as a follower rest and a turning tool.
Diametral setting	Position of a turret lathe cutting tool in a holder so that a theoretical line along the cutting face (edge) passes through the spindle axis.
Roller back box tool	A toolholder design in which a number of different cutter block slides may be interchanged. (The cutter in the setup is always positioned at the diametral setting.)
Adjustable toolholder	A toolholder for drills, reamers, counterbores, and chamfering and similar end-cutting tools. A holder that permits setting each cutting tool on the spindle axis. Also, a toolholder with a graduated collar. (Limited tool movement may be read directly in decimal-inch or metric units of linear measure.)
Self-opening holders	Usually, toolholders for die heads that trip at the end of a required length and release thread chasers.

SUMMARY

Basic internal and external cuts on a turret lathe are the same as cuts performed on an engine lathe.

> Single, multiple, combined, and successive cuts are taken with single-point and multiple-point cutters.

Turret lathe tooling is considered in relation to the headstock, front and rear stations of the cross slide, and the hexagon turret.

> Collets provide an easy, economical, and accurate method of feeding and holding bar stock manually or automatically.

Collets are made in round, square, hexagon, and other regular shapes. They are manufactured commercially in fractional, metric, decimal, letter, and number sizes.

> Some collets are designed with an extended nose. This extended nose permits work processes to be performed over a longer length.

Solid positive stop, ejector stop, and long stop parts collets are used to control the accurate positioning and holding of workpieces.

Step chucks permit shoulders and faces to be turned to exact lengths on parts where the outside diameter may vary. Step chucks are usually used for workpieces of larger diameter than workpieces that may be held in standard collets.

Precision expanding collets are actuated by a draw plug. Concentric and square shoulders, faces, and diameters are possible by holding the part on a previously machined bore.

Work-locating stops in collets permit exact positioning so that workpieces may be machined accurately without any lateral movement.

The diameter and area over which the chuck jaws act must ensure maximum gripping force with the least distortion, if any.

A releasing tap holder turns when the depth to which it is set is reached. The tap and holder are usually removed by reversing the spindle direction.

External threads may be cut with a die head. The self-opening stationary head is tripped at thread depth. The chaser inserts are automatically backed away from the thread. The die head is then moved to the starting position for the next part to be threaded.

Cutting-off tools are generally mounted in the cross slide square turret or back tool post. Some spindles are fitted for a permanently mounted cutting-off head.

Turning head bar turners, slide tools, and other cutters must be mounted on center in order to control the front and top clearance angles.

The ends of workpieces must be chamfered in order to have the rolls and cutter of a bar turner start properly.

The turret lathe may be used to machine a part completely or as a second machine for additional work processes on previously machined parts.

Fixtures mounted on faceplates permit accurate nesting in relation to a fixed axis, with maximum holding force.

Turret lathe manufacturers recommend universal bar equipment for small-lot production runs. Permanent setups permit the rapid changing of tools.

The bar turner is a widely used work-support device and cutting-tool holder. The self-contained rolls, when mounted over a precision-machined surface, ensure concentricity, repeatability of a specific diameter on subsequent parts, and a burnished surface finish without tool marks.

The rigid cutter holder on the bar turner supports the cutting action against the end of the cutting tool. The tool design and supporting rolls produce a vibrationless cut.

Turning tools are generally of cobalt high-speed steel, brazed-tip carbides, and insert-tip carbides.

Manufacturers recommend forming a chip breaker near the cutting edge of each cutter. This chip breaker provides for maximum chip removal.

Slide tools for turrets permit the fast changing and accurate replacement of cutting tools. Slide tool bodies accommodate cutting tools for all basic internal and external cuts.

An adjustable toolholder may have a graduated collar. The graduations on some toolholders are in increments of 0.0002" (0.005mm). Each fifth graduation represents 0.001" (0.025mm). The reading indicates the exact amount by which a diameter is decreased.

Standard machine tool and personal safety precautions must be observed for turret lathe operation. Moving parts of the machine, work-holding accessories, and toolholders must be cleaned, kept free of burrs, and oiled.

Parts must be chucked securely without distortion. Torque or automatic preloaded wrenches should be used where a uniform tightness is required.

All machine parts, cutting tools, and workpieces require checking before power is turned on to ensure that all moving parts clear safely.

UNIT 9 REVIEW AND SELF-TEST

1. Describe the difference between combined cuts and successive cuts as related to turret lathe work.
2. Cite two advantages of an extended-nose collet as compared to a conventional spring collet.
3. a. State the purpose of positive-stop collets.
   b. Name three general types of stops for positive-stop collets.
   c. Explain a common design feature for positive-stop collets.
4. a. Give applications for the use of rocking jaws on a standard independent-jaw chuck on a turret lathe.
   b. Tell what function wrap-around jaws serve.
5. State two conditions under which special fixtures and other faceplate setups are more practical for turret lathe operations than are standard chucks or step chucks and closers.
6. a. Describe how cutter wear life is increased by features that are incorporated in the design of a bar turner.
   b. Name two cutter design features for controlling chips in carbide-tip or brazed-tip insert applications.
7. a. Name two slide tools for turret lathe work.
   b. Describe the function of each slide tool.
8. Tell how to set up the bar turner for straight turning.
9. State why air pressure should be avoided in cleaning the spindle bore, work-holding accessories, and cutting tools on a turret lathe.
10. Cite one safety precaution that is taken in order to prevent damaging a turned workpiece when a bar turner is used.

UNIT 10

Basic External and Internal Machining Processes

Parts are machined on turret lathes by taking basic cuts with single or multiple-tooth cutters. Cuts on the outside diameter of a workpiece are referred to as external cuts or processes. Basic external cuts refer to straight, taper, and bevel turning; facing; grooving; thread cutting; cutting off; and knurling, which, technically, is not a cutting process.

Basic internal cuts relate to center drilling and drilling; straight, angle, and taper boring; counterboring; countersinking; reaming; undercutting (recessing); and threading either chased or tapped threads. Figure 10-1A pictures basic processes that are common to turret lathe work. Typical bar machined parts are shown in Figure 10-1B.

This unit covers turret lathe tooling and set-ups for basic internal and external cuts. *Sequencing* is involved for machining parts that require a combination of these cuts. Therefore, step-by-step

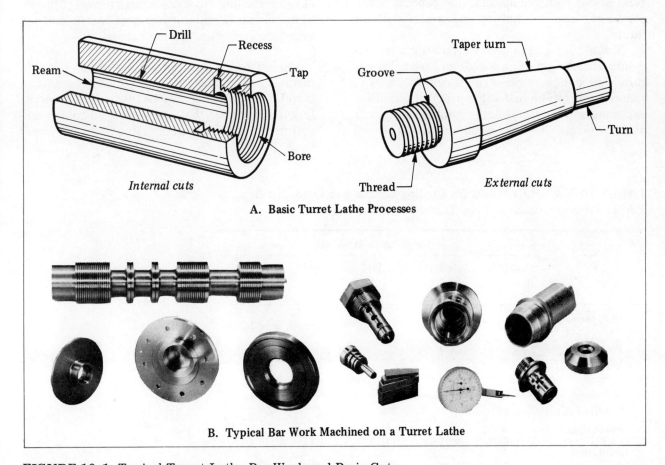

A. Basic Turret Lathe Processes

B. Typical Bar Work Machined on a Turret Lathe

FIGURE 10-1 Typical Turret Lathe Bar Work and Basic Cuts

procedures are given for sequencing both internal and external cuts. Cutting speeds and feeds are considered as they relate to specific cuts.

The unit also covers the representation of cutaway sections. In drafting practice, cutaway sections are used on many drawings of complex parts where the numerous lines on a conventional view are difficult to read.

CUTTING SPEEDS (sfpm) AND FEEDS (ipr) FOR BAR TURNERS

Cutting speeds for internal and external turret lathe cuts are calculated in terms of surface feet per minute (sfpm). Feeds are given in fractional parts of an inch per revolution (ipr). Tables are provided as guides for cutting speeds and feeds, which differ according to the nature of each internal or external cut. Some factors that affect cutting speeds and feeds are: machinability, required surface finish, and tolerances; design features, size, and material of the workpiece; type of cut and tooling; machine capability and condition; and the nature and use of cutting fluids.

Table 10-1 gives one manufacturer's recommended starting cutting speeds and feeds for bar turner processes on ram-type turret lathes. The values were established after years of extensive experience with turret lathe tooling and parts manufacturing. The use of these cutting speeds and feeds provides for maximum machining production rates, quality of surface finish, and dimensional accuracy (concentricity and diametral setting).

All cutting speed and feed tables provide a baseline of values that serve to guide the turret lathe operator. These values may require change according to the experience and judgment of the operator. Compensation may need to be made for factors such as chip control, rate of tool wear, machine horsepower, specifications, and other conditions governing all machine tools and set-ups. Parts of other tables are included in the unit and accompany specific cuts. Each table is representative of the kind of technical information the operator uses from manufacturers' literature and trade handbooks.

BASIC EXTERNAL CUTS

Turning Cuts

Straight turning is generally performed by one or a combination of three different methods: (1) side turning from the square turret; (2) overhead turning from the hexagon turret; and (3) long-bar turning using a bar turner.

Side Turning from the Square Turret. In side turning, the cutting tool is mounted in the square turret. Turning cuts may be taken simultaneously with processes that are carried on from hexagon turret tooling stations. Roughing and finish

TABLE 10-1 Starting Points for Cutting Speeds and Feeds for Bar Turner Processes on Ram-Type Turret Lathes

	Metals to Be Machined				
	C-1070	A-4340	C-1045	C-1020	B-1112
	Cutting Speeds (sfpm)				
High-speed steel cutter	50	70	90	120	150
Premium Carbide cutter	220	245	320	420	490
	Feeds				
Depth of cut	1/8"	1/4"	3/8"	1/2"	5/8"
Feeds (ipr) (0.001" per revolution)	0.030	0.018	0.012	0.0075	0.0045

BASIC MACHINING PROCESSES 179

turning cuts may be taken by positioning the cutting tools at successive square turret stations.

Straight Turning with a Bar Turner. Bar turners are capable of machining a fine surface to tolerances within 0.001″ (0.02mm). Bar turners, equipped with either brazed-tip or insert-type carbide cutters, are used to perform most of the heavy metal turning processes.

Long bars that overhang the collet and require support are turned with a bar turner that is mounted on one turret face. The support rolls on the bar turner permit heavy cuts to be taken at high feed rates. Concentric diameters are produced by mounting the bar turner rolls ahead of the cutter (Figure 10-2). The rolls must ride on a previously finished surface.

A burnished, quality surface finish is produced by positioning the rolls just slightly behind the point of the cutter. At the end of the turning cut, the cutter relieving lever is moved to retract the cutter from the workpiece. The bar turner may then be backed-off the finished, turned surface without leaving cutter marks. The cutter is reset for the next cut by returning the relieving lever to the starting position.

Overhead Turning from the Hexagon Turret. The multiple turning head on the hexagon turret provides rigid tool support that is essential in heavy metal-removing processes. Cutter holders are mounted in the multiple turning head (Figure

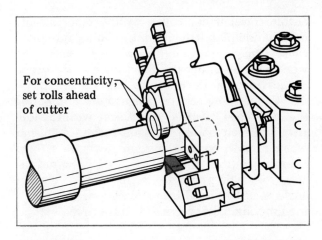

FIGURE 10-2 Bar Turner Rolls Mounted Ahead of Cutter

10-3A). A reversible plain cutter holder (Figure 10-3B) is used for rough turning cuts. A reversible adjustable cutter holder (Figure 10-3C) has a built-in micrometer feature that permits a cutter to be set to size accurately and quickly.

Taper Turning. Internal or external tapers may be machined by using a taper attachment. The cutting tool is mounted in the square turret on ram-type and fixed-center saddle-type turret lathes. The taper attachment is set at the required taper angle and locked to the bed. The

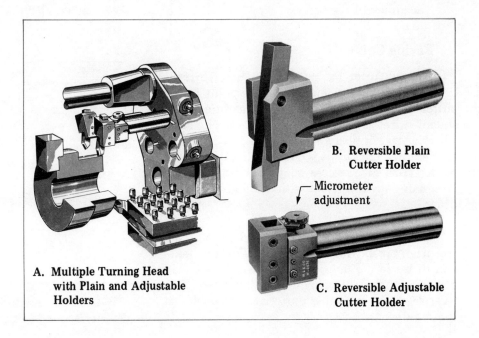

FIGURE 10-3 Overhead Turning from Hexagon Turret with Multiple Turning Head

A. Multiple Turning Head with Plain and Adjustable Holders

B. Reversible Plain Cutter Holder

C. Reversible Adjustable Cutter Holder

cutter travel is controlled by the taper attachment guide plate (slide block). This guide plate causes the mounted cutter to be fed across the workpiece by either the cross slide or the cross-sliding turret, as the case may be.

A taper may also be roughed out combining the cross feed of the square turret with the longitudinal feed of the hexagon turret. A *pusher block* is used on the hexagon turret in a fixed relation to the cross slide feed. Longitudinal and cross slide power feeds are used. The taper angle may be varied by increasing or decreasing the longitudinal or cross slide power feed.

Maintaining Accurate Tapers. Four conditions must be met to maintain and to duplicate accurate tapers when a taper attachment is used:

—The taper attachment must be located in an exact position with respect to the cross slide for each cut.
—The cross slide unit must be in exactly the same position on the bed when the extension rod is clamped and the binder screw is loosened.
—When the binder screw is tightened and the extension rod is loosened, the cross slide must be in the same position in relation to the taper attachment.
—If the taper attachment has a *backlash eliminator*, the nut must be tightened against the feed screw nut to eliminate all play between the feed screw and the nut as the backlash of the cross feed screw affects the accuracy of the taper.

Facing Cuts

Square Turret Facing. The square turret is used for most facing cuts. A single-point facing cutter is generally mounted in one station of the square turret as shown in Figure 10-4. The cross slide carriage is positioned along the bed ways at the required length. It is clamped in position for facing cuts. Workpieces may also be faced by combining the facing cut with drilling, boring, or other processes.

Hexagon Turret Facing. Short facing cuts are often taken with tools mounted on the hexagon turret in a *quick-acting slide tool*. The slide tool is actuated by a hand-operated lever. This lever feeds the mounted cutter at right angles to the spindle axis.

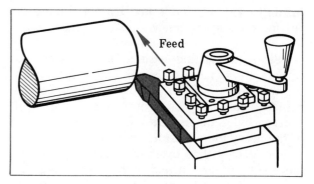

FIGURE 10-4 Facing with a Single-Point Cutter Mounted in the Square Turret

Facing across large-diameter workpieces is performed on cross-sliding turret machines. The cutter is held on the hexagon turret in a through-slot cutter holder, boring bar, or tool-base cutter block.

End Facing. An *end former* or *combination turner and end former* (Figure 10-5) is used to support the end of a workpiece that would otherwise spring under the cutting action. The rolls are set ahead of the cutter to support the workpiece.

Forming and Cutoff Cuts

Forming Cuts. A fast method of producing a finished shape and diameter is by *forming*. The four basic forming cuts shown in Figure 10-6

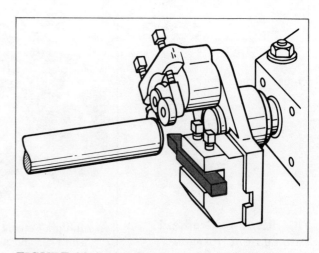

FIGURE 10-5 Application of Combination Turner and End Former to Facing

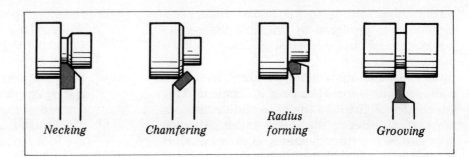

FIGURE 10-6 Four Basic Forming Cuts

are: (1) necking (cutting a recess close to a shoulder), (2) chamfering a beveled surface, (3) radius forming, and (4) grooving. Single forming cuts are usually taken with the cutter mounted in the square turret. A *necking cutter block* mounted on the rear of the cross slide is used when several necking or grooving cuts are required.

The hexagon turret may also be used for forming cuts. A bar and cutter mounted in a quick-acting slide tool (Figure 10-7) may be positioned and fed to cut a groove or chamfer. An internal facing cutter is usually held in a boring bar or tool-base cutter block on cross slide turret machines.

Cutoff Cuts. Cutting-off operations are performed by mounting the cutoff-tool holder in the square turret or in the rear cutter block. High-speed steel and carbide cutoff blades, mounted in a holder, permit the rapid changing of cutters as they wear. Maximum rigidity in the setup ensures fast, smooth cutoff.

Threading

External threads may be accurately cut with a multiple-point threading tool (die head) or chased with a single-point threading tool. The method of threading that is selected depends upon thread size, required accuracy, and quantity.

Threading with Die Heads. A continuous, even force must be applied when tapping so that the hexagon turret moves with the die head. For most threads, the die head may be led on by hand—that is, the turnstile is turned manually in starting and following the thread. Standard-size external threads are cut with self-opening die heads. As described earlier, the die head may be sprung open when the thread is cut to a required length. The trip to spring the die head open and release it automatically may be actuated from within or outside the die head.

Long, accurate threading requires that a positive lead be provided for uniformly feeding the die head across the workpiece. A leading-on attachment, set at the required thread pitch, is used on the hexagon turret of ram-type machines. Turret lathes that are equipped with a thread-chasing attachment on the cross slide may also be set to feed the die head at the thread pitch rate. Saddle-type machines have a leader and follower model of hexagon turret thread-chasing attachment.

Precision Threading with Single-Point Cutters. Single-point thread cutting is used to produce threads that must be concentric with other diameters and meet precise tolerances. The thread-cutting tool is set and held in the square turret. The cutter is advanced each revolution of the

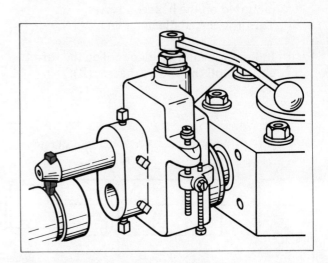

FIGURE 10-7 Taking a Form Cut with a Quick-Acting Slide Tool Setup

182 TURNING MACHINES

workpiece according to the thread pitch. This movement is produced by engaging the revolving lead screw at a particular position for each successive cut.

Some ram- and saddle-type turret lathes are equipped with a thread-chasing attachment. This attachment is fitted to the cross slide and feeds the thread cutter in the square turret. Threads may also be cut from the hexagon turret when it is fitted with a thread-chasing attachment.

HOW TO SEQUENCE EXTERNAL CUTS

It is common practice to use different sequences to take external cuts. However, some processes such as center drilling (to support a long workpiece) and turning the outside diameter of a thread must be performed first. Therefore, a sequence of cuts is established after the part drawing is studied.

Example: Set up a ram-type turret lathe to mass-produce the threaded shaft shown in Figure 10-8.

Note: The thread is to be die cut. The die head is to be led on by hand.

STEP 1 Feed the stock to length against a revolving bar stop.

Note: It is assumed that the bar end is already chamfered.

STEP 2 Position (index) the bar turner that is mounted on the hexagon turret.

STEP 3 Turn the thread diameter (A).

STEP 4 Index the center drilling tool on the turret. Center drill the end of the bar (B).

STEP 5 Index to position the revolving center. Support the workpiece with the center.

STEP 6 Form the V-groove with a cutter mounted in the square turret (C).

STEP 7 Position the shoulder-turning cutter on the square turret. Square the shoulder at the end of the thread (D).

STEP 8 Turn the outside diameter by using a turning tool mounted in the square turret (E).

STEP 9 Remove the center. Position the end former or combination turner and end former on the turret.

Note: The tools have rolls that hold the workpiece securely against the cutter.

STEP 10 Face (F) and chamfer (G) the end of the threaded shaft.

STEP 11 Turn the hexagon turret to the self-opening die head position.

STEP 12 Check the thread length trip stop. It should be adjusted so that the die head opens for the chaser insert to clear the workpiece just before the die head reaches the shoulder. Cut the thread (H).

STEP 13 Feed the cutoff tool mounted on the rear tool post to cut the workpiece to length (I).

STEP 14 Position the chamfering tool in the square turret. Chamfer the end of the next workpiece (J).

The tooling setup and sequencing of cuts required to machine the threaded shaft are illustrated in Figure 10-9.

BASIC INTERNAL CUTS

Internal cutting processes for turret lathe work require cutting tools that are similar in design and principles of operation to the tools used on engine lathes and other machine tools. The

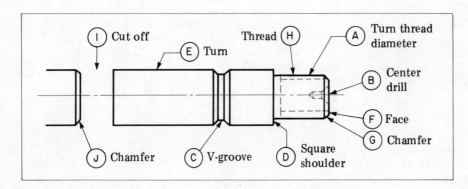

FIGURE 10-8 Threaded Shaft

BASIC MACHINING PROCESSES 183

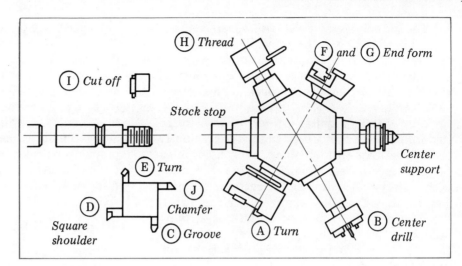

FIGURE 10-9 Tooling Setup and Sequence of Cuts for Machining the Threaded Shaft

differences lie in the way in which the tools are mounted and supported on the turret lathe.

Drilling Cuts

Parts are drilled to produce holes to specifications, to center stock, and to remove material from existing holes. Drilling cuts require different drills. Parts may be center drilled or spotted with a *start drill*. Holes may be produced from the solid by using a *twist drill*. Holes that are larger than 2" (50mm) diameter are drilled with a *spade drill*. Previously drilled, cored, or pierced holes may be enlarged with a *core drill*. Deep-hole drilling sometimes involves a *coolant-feeding drill*.

Drills for turret lathe work are designed with a straight or taper shank. Three common drill holders for straight-shank drills are shown in Figure 10-10A. Three other different holders for taper-shank drills are shown in Figure 10-10B.

Center drilling is performed by using a center drill. The tool is fed into a revolving workpiece from the hexagon turret. The self-centering rollers support the end of the workpiece and thus permit the drilling of a concentric hole.

Start drilling requires the use of a start drill to *spot a cone* on the front face of the part. The shortness and rigidity of a start drill permits it to follow and not deviate from a central path as longer drills do.

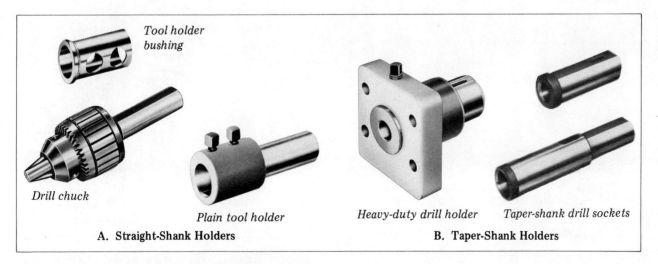

FIGURE 10-10 Common Types of Drill Holders

The principles of *twist drilling* on a turret lathe are the same as the principles that govern efficient drilling on a lathe, drill press, or other machine tool. Also similar precautions must be taken to grind the cutting edges at the correct angle and length.

Core drilling requires a three- or four-fluted cutter. The core drill is used to enlarge previously drilled, cast, or pierced holes. Such holes must first be chamfered. Chamfering may be done with a start drill or a chamfer cutter in a boring bar. In either case, the chamfered surface centers the core drill and thus permits cutting a round hole concentric with the required centerline. Long holes are often started by boring to the core drill diameter for about 1/2" (12mm). Then the core drill is used to drill to the required depth. A core drill may be mounted on the hexagon turret in a short, flanged toolholder as shown in Figure 10-11.

Spade drilling is a practical method of drilling holes that are 2" (50mm) or larger in diameter. The spade drill consists of a flat cutter (blade). This cutter is held firmly in a bar-type holder. The face of the cutter is ground to a cutting angle similar to the angle of a twist drill face. A spade drill has two cutting edges. The reduced length and overhang of a spade drill permits it to be mounted close to the turret. The rigidity thus provided helps to control vibration and permits the drilling of concentric holes.

Deep-hole drilling relates to holes that are four or more times deeper than the drill diameter. As in the case of deep holes drilled on drilling machines and engine lathes, problems of cutting efficiency, drilling accuracy, and tool breakage must be considered in turret lathe work. Most deep-hole drilling is done with drills designed so that a cutting fluid may be pumped to the cutting edges. The coolant floods the cutting area to control the temperature at the cutting edges of the cutter and the workpiece. At the same time, the chips are flowed out of the drill along the flutes and/or ahead of the drill in a previously drilled hole. Chips are cleared from small-diameter holes by withdrawing the drill frequently.

Boring Cuts

Boring cuts are used to turn concentric, straight holes to precise dimensional limits. Boring bars are designed for internal rough and finish turning, facing, counterboring, undercutting, chamfering, bevel and taper turning, and thread chasing. Boring bars hold square high-speed steel tool bits, brazed-carbide or carbide-tip cutters, ceramic inserts, and chip breakers. Many types of solid boring bars exist. They are usually mounted on the square turret.

Straight Boring. *Hexagon turret boring* on a ram-type or fixed-center saddle-type turret lathe requires the cutter to be mounted in a slide tool. The boring bar and cutter are thus permitted to be adjusted to the bore size. The slide tool setup shown in Figure 10-12 permits adjustment of the cutter to the required size. The micrometer

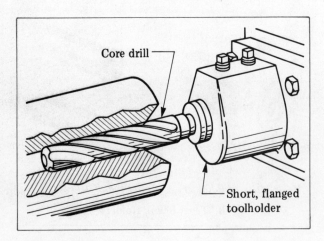

FIGURE 10-11 Core Drilling Setup

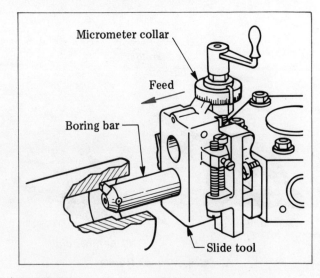

FIGURE 10-12 Slide Tool Setup to Permit Tooling Movement to a Required Size

collar on the slide tool makes it possible to establish the depth of cut and bore diameter. Once the cutter is set to depth, the slide is clamped securely and provides the rigidity necessary to take roughing cuts or to finish bore a hole.

Cross-sliding hexagon turret boring on a saddle-type machine does not require the use of a slide tool. Adjustments for bore diameter are made by moving the cross slide directly. One advantage of this machine over the ram type is that the hexagon turret and cutting tools may be repositioned between cuts so that the boring of several different size holes with the use of the same boring bar is permitted. Fewer tools are required. Setups may be made quickly and easily. Boring bars may be held in flanged toolholders. They may also be mounted directly on the hexagon turret.

Square turret boring is used for limited operations. These include the boring of short holes, taking lighter cuts, and using slower feeds. The solid, forged cutter designed for square turret use is generally less rigid than the heavier, solid boring bars.

Taper Boring. The square turret is also adapted for *taper boring*. The taper attachment is set at the required taper. When the cross slide is positioned and the taper attachment is secured to the bed, the bore size is established by moving the cross slide handwheel.

A precaution must be observed when taper boring. Unlike most lathe operations where the feed is toward the spindle, for accurate taper boring the cutter is fed *away* from the spindle. The cut is in the direction of the largest taper diameter.

Large taper boring operations are usually performed on a cross-sliding hexagon turret machine. This turret permits the use of heavier, rigid boring tools adapted to heavy cuts. The steep-taper boring setup in Figure 10-13 indicates that the cross slide follows the taper on the attachment as the cutting tool is fed longitudinally.

Reaming Cuts

Reaming usually follows boring. Boring corrects out-of-roundness, deviations from straightness, and other inaccuracies of drilled or cored holes. Boring permits machining to a specific size and leaving the proper amount of material to remove in the finish reaming process.

Solid and *expandable reamers* are used in both engine lathe and turret lathe work. However, in turret lathe applications, a *floating reamer holder* is used. This holder is mounted on the hexagon turret and is designed to float. Thus, the reamer is able to align itself with a previously drilled or bored hole. The bored hole and reamer combination produces a precisely machined, dimensionally accurate, finished hole. As in the case with drilling and other turning machines, slower speeds and coarser feeds are used for reaming in comparison to drilling.

Taper reamers are used to finish a taper bore or to produce an internal taper directly. A pair of roughing and finish cutting taper reamers is used, especially when steep tapers are to be cut. Taper reamers may be mounted on the hexagon turret in either a floating reamer holder or an adjustable toolholder. These holders permit the reamer to be aligned accurately.

Adjustable floating-blade reamers are designed for reaming large-diameter holes. There are two adjustable cutting blades in a floating blade reamer. This reamer is mounted in a flanged toolholder. The two reamer blades are aligned in a previously formed hole by the floating action of the cutter blades. As is required with all cutting tools, the reamer blades must be ground to correct cutting angles and kept sharp to produce precision reamed holes.

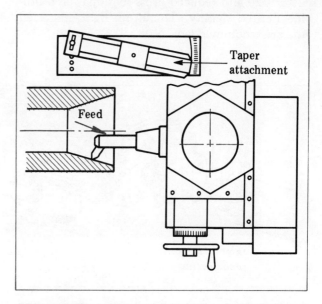

FIGURE 10-13 Taper Attachment Setup for Steep-Taper Boring on Cross-Sliding Hexagon Turret

Internal Recessing (Undercutting), Facing, and Back-Facing Cuts

Relief areas for threading, grinding, shouldering, and taper surfaces must often be produced as internal cuts. A combination of longitudinal and cross slide movements is required. The cutting tools are brought to position inside the workpiece. The tools are then fed by cross feeding to undercut (internally recess) or produce an internally faced surface.

Vertical motion on ram or fixed-center hexagon turret lathes is produced by mounting a boring bar and cutter in a quick-acting slide tool. This setup represents the tooling at one turret station. Cross feed movement is provided by direct feeding.

Hexagon turret setups for undercutting, back facing, and internal facing are shown in Figure 10-14. Each process requires a differently formed cutter. Usually, a square-edge cutting tool is used for end turning a square internal recess (undercut surface). The cutting tool for back facing is ground with a cutting edge that permits taking front-edge cuts. Clearance is provided on the right and front edges of the cutter. For internal facing, side clearance is ground on the cutter in the same manner as for a regular external facing cutter.

Undercutting, back facing, and internal facing cuts may also be taken by mounting a solid cutter in the square turret. Usually the smaller cutter size and reduced cross-sectional area provide limited rigidity. Therefore, light, hand-fed cuts are recommended.

Tapping and Single-Point Thread Cutting

Many of the same tapping practices used on drilling machines and engine lathes apply to threads produced by taps on turret lathes. Sufficient force must be exerted to start the tap in a tap hole. Once started, the tap must be self-feeding. A tap that is forced during cutting, or *dragged* while it is being withdrawn, will produce defective threads. The turret lathe operator must carefully judge and apply the required force to the turret turnstile during the cutting and withdrawing strokes. The turret and holder must be fed at a rate equal to the thread pitch. That is, the hexagon turret, tap, and holder must be fed at the same rate that the tap moves into or out of the workpiece.

Tapping with Solid and Collapsing Taps. *Solid taps* are generally used for thread sizes up to 1-1/2" (38mm). These taps are held on the hexagon turret in a releasing tap holder (Figure 10-15). The forward motion of the turret and tap is stopped by the setting on the turret stop. This setting in approximately 1/8" (3mm) less than the required thread length. The tap holder releases at the end of the cut. The release causes the tap to revolve with the workpiece. The tap is withdrawn by reversing the direction of spindle rotation.

Collapsing taps are used for thread sizes larger than 1-1/2" (38mm). These taps are designed to collapse. The cutting edges pull away from the workpiece at the end of the cutting stroke.

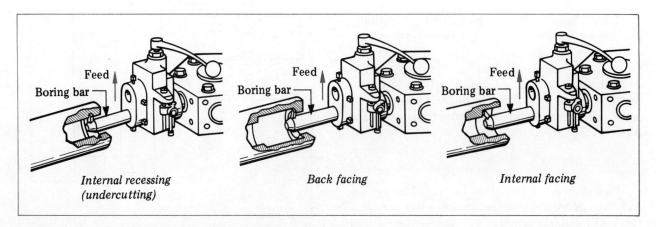

FIGURE 10-14 Slide Tool Setup on Hexagon Turret for Undercutting, Back Facing, and Internal Facing

BASIC MACHINING PROCESSES 187

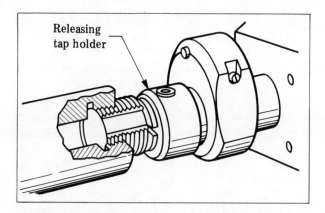

FIGURE 10-15 Releasing Tap Holder Setup on Hexagon Turret

Collapsing taps may be mounted directly on a hexagon turret or in a flanged toolholder. One advantage of a collapsing tap over a solid tap is that the cutter is withdrawn without reversing the spindle direction.

As in thread cutting with a die head, precision threads require accurate feed control. The leading-on attachment provides for a tap to be fed at the thread lead rate. A knock-off mechanism, controlled by the hexagon turret stop screws, permits automatic tapping to thread depth.

Single-Point Thread Cutting. Single-point thread cutting is used to cut precise straight and taper threads that must be concentric with other surfaces. The single-point threading tool is fed at the required thread pitch. Multiple passes are required until the cutter is fed to thread depth. The feed is controlled by using either a thread-chasing attachment or a lead screw attachment.

The single-point threading tool is held in a boring bar that is mounted in a hexagon turret slide tool. A flanged holder is used for mounting the threading tool on cross-sliding turrets. The cutter may also be held in a thread-chasing cutter holder that is mounted on the square turret of the cross slide.

HOW TO SEQUENCE INTERNAL CUTS

Example: Set up the sequence of internal cuts to mass-produce a quantity of the threaded sleeve shown in Figure 10-16.

Note: The chamfered end and outside diameter are turned as external cuts from the square turret.

STEP 1 Chamfer the workpiece (A).

STEP 2 Feed to the required length against a combination stock stop and start drill (B).

STEP 3 Start drill the end (C).

STEP 4 Position the reamer size drill. Drill the hole at least 1/8″ (3mm) deeper than the length of the threaded sleeve (D).

STEP 5 Bore the thread diameter (E).

Note: A slide tool mounting for the boring bar permits size adjustments.

STEP 6 Ream the drilled hole (F).

Note: A fluted reamer is generally mounted in a floating reamer holder to ensure alignment with the drilled hole.

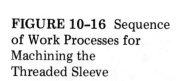

FIGURE 10-16 Sequence of Work Processes for Machining the Threaded Sleeve

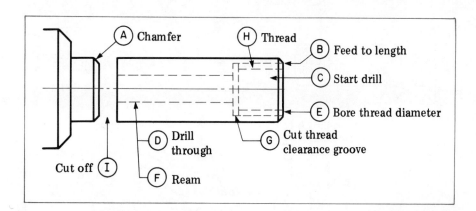

STEP 7 Cut the thread clearance groove by feeding the grooving cutter to thread depth (G). A quick-acting slide tool is used for the cutter and boring bar mounting.

STEP 8 Position the tap and releasing tap holder. Turn the turnstile to feed the tap to the required depth (H).

Note: Nonstandard and special fit threads are generally chased. The single-point thread cutter is fed by engaging the lead screw in connection with either the cross slide square turret or the ram turret.

Note: Remove the tap be reversing the spindle direction.

STEP 9 Cut off the workpiece to length (I).

Note: Remove the outside diameter burr during the cutting-off process. Burr the remaining fine-burred edge from around the reamed hole.

The positioning of the standard tools for the sequence of cuts is diagrammed in Figure 10-17.

APPLICATIONS OF CUTTING SPEED AND FEED TABLES TO EXTERNAL AND INTERNAL CUTS

The turret lathe operator must be able to establish appropriate cutting speeds and feeds. Reference is made to manufacturers' suggested starting points, which provide a baseline of cutting speed and feed values. These values are modified as one of the first considerations in machinability. Metals that are commonly machined on the turret lathe are grouped into *machinability classes* as listed in Table 10-2. Information on machinability class is used with other cutting speed tables. For example, Table 10-3 is a partial table, showing machinability classes A and B, of cutting speeds for a few basic external cuts on turret lathes.

Table 10-4 is a partial table of data for starting feed rates (ipr) for external cuts on ram-type turret lathes. Complete and additional cutting speed and feed tables are included in the Appendix. Four factors must be considered in determining the final feed for roughing cuts:

—The rigidity of the cutter and the tooling strength,
—The rigidity of the work-holding device and the nature and design of the workpiece,
—The condition and capacity of the machine,
—End thrust.

Coarse feed ranges are used for roughing cuts. High cutting speeds and fine feeds are used for high-quality, accurate cuts to produce precision surface finishes.

Table 10-5 provides technical information relating to threading on turret lathes with high-speed taps or chasers. Cutting speeds are listed for a range of threads from 3 to over 25 threads per inch (8.5mm to less than 1mm metric pitches).

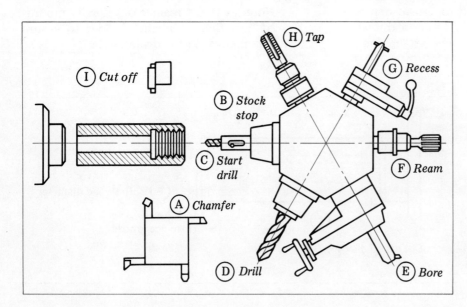

FIGURE 10-17 Tooling Setup for Producing the Threaded Sleeve

BASIC MACHINING PROCESSES

TABLE 10-2 Machinability Class of Commonly Machined Metals

Machinability Class	Commonly Machined Metals
A	B-1112 C-1118 Bronze (phosphor 64)
B	C-1010 and 1015 (tubing) C-1019, C-1020, C-1137, C-1141 Brass (Naval 73) Bronze (Tobin) Cast iron (soft) Stainless (#416)
C	C-1040, C-1045, C-1050 C-4140, A-4150, A-4340, A-4615 Bronze (aluminum 68) Cast iron (hard) Malleable iron Stainless (#302 and #304)
D	E-3310, E-4160 Cast steel Stainless (#440C)
E	C-1070, C-1090 E-52100 Monel
F	Aluminum Brass (free-machining)

REPRESENTING COMPLEX DETAILS OF PARTS BY SECTIONS

As the number of design features increases, the additional lines on a drawing representing these features make the drawing difficult to read. Also, an incorrect reading of dimensions leads to work spoilage. To simplify the part drawings, a view with *cutaway sections* may be drawn. These sections expose the interior of a part and make the design features easy to distinguish.

Cutting Planes

Cutaway sections require the use of an imaginary plane or planes. Such planes cut through part or all of an object. When a section is removed, all the edges appear as object lines on a drawing. Figure 10-18A pictures an imaginary plane cutting through a part. The full-section view of the interior of the part is seen in Figure 10-18B. In a *full-section view*, the cutting plane cuts completely through the part. All of the interior details are exposed.

Cutting Plane Line and Section Lining

Note in Figure 10-18A that a cutting plane line is used. A *cutting plane line* identifies how

TABLE 10-3 Suggested Starting Points for Cutting Speeds on Turret Lathes for Basic External Cuts (Partial Table)

		Starting Cutting Speeds in Surface Feet per Minute (sfpm)							
		Turning							
		Overhead and Cross Slide		Roller Turners		Facing	Forming		Cutting-Off	
Machinability Class	Nature of Cut	Cutter Material								
		Carbide	High-Speed Steel	Carbide		Carbide	High-Speed Steel	Carbide	High-Speed Steel	Carbide
A	Roughing Finishing	490 560	150	490		490 560	150	490	120	500
B	Roughing Finishing	420 490	120	420		420 490	120	420	100	500

TABLE 10-4 Suggested Starting Feed Rates on Ram-Type Turret Lathes for External Cuts (Partial Table)

Depth of Cut	Turning			Facing	Forming		Cutting Off	
	Overhead and Cross Slide		Roller Turners		Steel	Cast Iron	High-Speed Steel	Carbide Insert
	Hexagon Turret	Square Turret						
	Feed per Revolution of Spindle							
1/16"	0.030"	0.019"		0.017"			Stock to 1" (25.4mm) diameter 0.0045" (0.1mm)	Machinability Class: A, B, F 0.003" (0.1mm) C, D 0.004" (0.1mm) E 0.005" (0.1mm)
1.5mm	0.8mm	0.5mm	0.030" (0.8mm)	0.4mm				
1/8"	0.018"	0.019"		0.017"	0.004" (0.1mm)	0.006" (0.2mm)		
3mm	0.5mm	0.5mm		0.4mm				
1/4"	0.012"	0.012"	0.018"	0.010"				
6mm	0.3mm	0.3mm	0.5mm	0.3mm				

Note: Millimeter values are rounded to nearest 0.1mm.

TABLE 10-5 Recommended Starting Cutting Speeds for Threading with High Speed Steel Taps or Chasers on Turret Lathes (Partial Table)

	Threads per Inch (tpi)			
	3 to 7-1/2	8 to 15	16 to 24	25 and up
	Equivalent Metric Thread Pitch (mm)			
	8.5 to 3.5	3.2 to 1.5	1.5 to 1	1 and less
Material	Starting Cutting Speeds in sfpm*			
Aluminum	50	100	150	200
Bakelite	50	100	150	200
Brass				
Bar stock and castings	50	100	150	200
Forgings	25	40	50	80
Iron				
Cast	25	40	50	80
Malleable	20	30	40	50
Wrought	15	20	25	30
Steel				
Carbon 1010-1035	20	30	40	50
Carbon 1040-1095	15	20	25	30
Chrome	8	10	15	20
Stainless	8	10	15	20

*Use 75% of the cutting speed for cutting taper pipe threads.

BASIC MACHINING PROCESSES 191

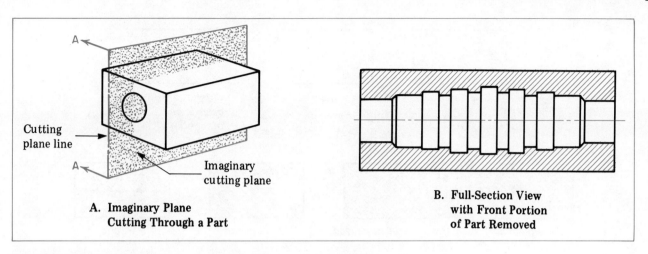

FIGURE 10-18 Application of Cutting Plane and Sectioning

much of the object is cut by one or more planes, where it is cut, and the direction from which the object is viewed. Diagonal (cross-hatch) lines are used on an interior view to identify it as a full-section view (Figure 10-18B).

The cutting plane line is a heavy line (Figure 10-19). It is usually broken at intervals by two short dashes. The cutting plane line shows the plane or intersecting planes that cut through the part. The arrows tell the direction of viewing. The section is indicated on a drawing by the word SECTION followed by an identifying letter. The same capital letter usually appears at the ends of both arrowheads. The section view of a part is identified by the notation SECTION A-A, or SECTION B-B, and so on.

Different combinations of fine, standard lines are drawn to identify particular materials. For example, one set of lines is used for cast iron, another set for steel, still another set for brass, and so on. Samples of the cross-hatch lines that are at a 45° angle are drawn in Figure 10-20 for seven common materials.

Half Sections

Parts with symmetrical details are often represented on a drawing by combining a section view with a regular view. Where one-half of the object is cut away with a plane and the other half is drawn as an exterior view, the drawing is said to be a *half-section view.*

The evolution of a half-section view may be seen in Figures 10-21 and 10-22. Figure 10-21A shows two cutting planes cutting through a part. A pictorial of the part with the half section removed is illustrated in Figure 10-21B. The regular front view of this particular part is shown in the industrial drawing in Figure 10-22A. The half-section right side view appears in Figure 10-22B.

Partial Sections

Many times a drawing may be simplified by the technique of *breaking out a section.* This technique is used to show clearly the design

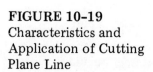

FIGURE 10-19
Characteristics and
Application of Cutting
Plane Line

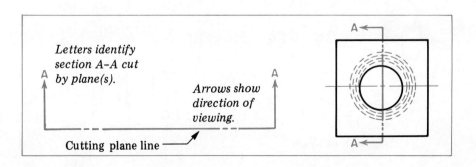

192 TURNING MACHINES

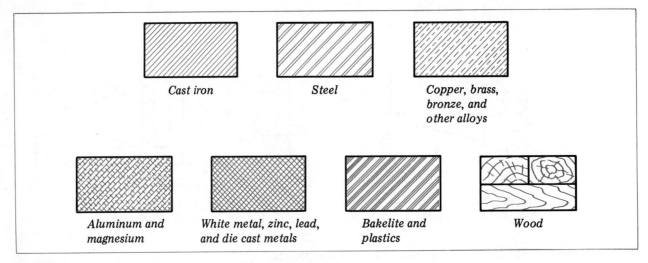

FIGURE 10-20 Section Lining for Common Materials

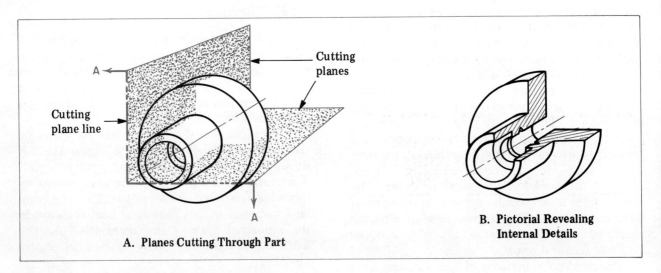

FIGURE 10-21 Evolution of a Half-Section View

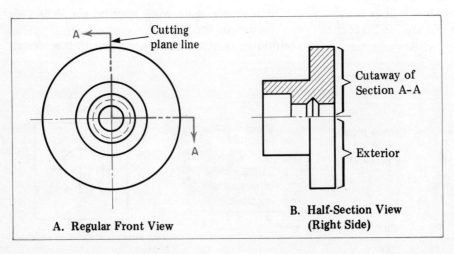

FIGURE 10-22 Typical Drawing Representation of Half-Section View

BASIC MACHINING PROCESSES 193

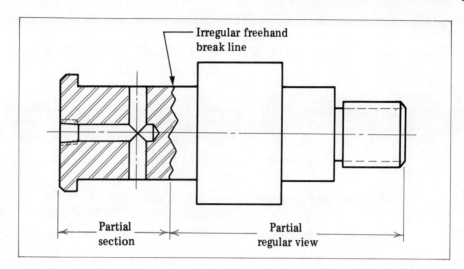

FIGURE 10-23 Example of Partial (Broken-Out) Section

features in a specific area of the part. The partial section is shown with crosshatching. An irregular freehand line is drawn to separate the partial section from the regular view of the outside of the part. An example of a *partial (broken-out) section* is illustrated in Figure 10-23.

SAFE PRACTICES FOR BASIC EXTERNAL AND INTERNAL MACHINING PROCESSES

—Use manufacturers' tables of recommended cutting speeds and feeds. Adjust the recommended values. Personal judgment must be exercised based on the specifications of the piece part, the accuracy and quality of surface texture, rate of tool wear, chip control, and machine capability.

—Check the working clearance between the cross slide and turret tooling.

—Mount bar turner rolls on a finished diameter ahead of the cutter to produce a precise, concentric diameter.

—Mount bar turner rolls behind the cutter to support the cutting force and burnish the turned surface.

—Use a multiple turning head on the hexagon turret to provide rigid tool support for heavy cuts.

—Tighten the nut on the backlash eliminator of the taper attachment to take the play out of the cross slide and the guide block.

—Use a leading-on attachment to cut fine, precise threads. The attachment accurately controls the infeed movement and withdrawal of a tap or die head.

—Use the single-point thread-cutting method to cut extremely accurate and perfectly concentric threads.

—Chamfer or bore the end of a cored or pierced hole for a short distance to center guide the lips of a drill.

—Use a spade drill for drilling holes over 2″ diameter to minimize the overhang of the drill (extension from the hexagon turret of the drill body and shank).

—Plan to remove the drill from the workpiece continuously during deep-hole drilling to quickly remove chips.

—Bore or chamfer a previously drilled, cored, or pierced hole to drill diameter to start a core drill concentrically.

—Use the heaviest possible boring bar with the shortest overhang for heavy, long boring cuts.

—Bore an internal taper on the finish cut with the direction of feed toward the large diameter.

—Set the turret stop about 1/8″ (3mm) less than thread depth to control threading to a shoulder or other fixed point of a workpiece.

—Study the part drawing carefully. Plan the sequence of cuts. Select cutting tools and holders. Make machine setups that provide safe clearance and economical operation.

—Check the composition of the cutting fluid and the rate of coolant flow at each station. Position splash guards to contain the coolant and chips.

—Keep the work area around the turret lathe dry and free of chips.

TERMS USED WITH EXTERNAL AND INTERNAL TOOLING, SETUPS, AND CUTS

Sequencing of cuts	Planning the tooling for each internal and/or external cut and the position of each tool and accessory on the turret or cross slide. Step-by-step scheduling of cuts to produce a desired part.
Side turning	Refers in turret lathe work to turning with the cutting tool mounted in the square turret on the cross slide.
Overhead turning	Supporting one or more cutters in a multiple turning head on the hexagon turret. A method used for heavy stock removal.
Backlash eliminator	A design feature on taper attachments that permits the cross slide and guide block to eliminate end play (backlash).
Quick-acting tool slide	A tool-holding device that makes it possible to adjust a cutter. (Once set, the cutter may be fed quickly to depth by a quick-acting lever on the tool slide.)
Combination turner and end former	A tooling device consisting of work-supporting rolls and a cutter that may be fed to end face a workpiece.
Leading-on attachment	A ram-type turret lathe attachment that may be set to feed a cutter at a rate equal to the pitch of the required screw thread. An attachment that permits the feeding and withdrawal of a threading setup at the pitch rate of the thread to be cut.
Indexing (turret lathe)	Moving a hexagon or square turret to the next position (face). Positioning the tooling for each successive process in relation to the axis of the workpiece.
Start drilling	The process of centering or drilling a hole with minimum deviation from the axis. Using a short, rigid start drill to spot a cone shape in a workpiece.
Core drilling	The use of three- and four-fluted drills to drill a cored, pierced, or previously drilled hole to size.
Spade drilling	Forming holes, usually larger than 2" diameter, with a flat cutter (blade) mounted in a bar holder.
Hexagon turret boring	Boring from a hexagon turret on a ram or fixed-center turret lathe. Boring with the cutter mounted in a slide tool that permits micrometer adjustment to obtain a required dimension.
Floating reamer holder	A holder designed to float a reamer so that it is self-centering and follows a previously drilled or bored hole.
Adjustable floating-blade reamer	A two-blade reamer for reaming large-diameter holes. Floating blades that align with a formed hole.
Internal recessing	The cutting (undercutting) of a groove inside a workpiece. Usually, a clearance area at thread depth or a groove at a shoulder.
Back-facing cut (internal)	Facing an inside surface at a right angle to the axis of the workpiece. Facing the inside right face of a workpiece.
Dragging (thread cutting)	The resistance of a tap or die head to being withdrawn. The cutting or wearing effect of a cutter on a thread by the force against the cutter as it is withdrawn from a revolving workpiece.

Collapsing tap	A tap design generally used for sizes over 1-1/2" diameter. A tap with threaded inserts that move away (collapse) from a thread. A feature that permits a tap to be withdrawn without reversing the spindle direction of rotation.
Knocking off	The process of tripping a hexagon turret or cross slide at a preestablished position. Setting turret or cross slide top screws to trip a preset cutter or process.
Sectioning	A method of representing complex internal details on a drawing. Cutting an object with imaginary planes. Removing a partial area or completely cutting through an object to clearly represent internal features.
Section lining	The use of 45° angle, fine lines to distinguish the material in a workpiece on a section drawing.

SUMMARY

The basic internal cuts in turret lathe work include center drilling and drilling; straight, angle, and taper turning; counterboring and countersinking; reaming; undercutting (recessing); and threading.

The basic external cuts cover straight and taper turning, bevel cutting and chamfering, facing, grooving (for turning), thread cutting, and cutting off.

Cutting speeds are calculated in terms of surface feet per minute (sfpm). Feeds are stated either as a fractional part of an inch per revolution (ipr) or metric equivalent.

Cutting speeds and feeds are established according to work processes, nature of tooling, machine capacity, tool wear life, and specifications and condition of the part. Manufacturers' charts of recommended cutting speeds and feeds provide a guide. Operator judgment based on experience, is needed to determine the actual operating conditions.

General internal and external taper turning processes involve (1) the use of a taper attachment and (2) the combining of the cross slide and hexagon turret feeds.

Accurate tapers are produced when all the backlash is eliminated. The taper cut must be started at the same point for each successive workpiece.

Facing cuts are taken with a cutter mounted and fed from a square turret, a slide tool mounting, or a fixed position on a cross-sliding turret machine.

The four basic forming cuts are: (1) necking, (2) chamfering, (3) radius cutting, and (4) grooving.

Cutting-off cuts are generally made with a cutoff tool positioned in either the square turret or the rear tool post of the cross slide.

Outside threads may be cut with a self-releasing die head.

When precisely machined and concentric threads are required, they are chased. A single-point threading tool is used on machines equipped with a lead screw and thread-chasing attachment.

The sequencing of cuts requires a study of the piece part specifications, an analysis of the work processes, universal permanent tooling on the machine, and work simplification techniques.

Straight turning may be done with the cutter mounted in a square turret on the cross slide or the hexagon turret. Bar turners are used for light turning processes; multiple turning heads, for heavy-duty overhead turning.

Procedures for drilling holes to depths over four times the drill diameter are the same as the procedures for removing chips in deep-hole drilling.

Machine stops are set to trip and stop a work process in time to finish machine a surface to a required dimension.

Boring, as in lathe jig boring and boring on milling machines, is a precision machining process. Holes are bored to produce concentric, straight, accurately machined diameters.

The heaviest boring bar or forged boring tool should be mounted as short as possible. Boring may be done from a cross slide or hexagon turret position. The cross-sliding turret permits a boring tool to be repositioned between cuts.

Taper boring requires the removal of end play and a constant fixed relationship between the starting point of taper cuts and successive workpieces.

Internal taper finish boring cuts are taken feeding toward the large diameter.

Drilling processes usually require center drilling and/or start drilling. Cored holes are core drilled. Large-diameter holes (approximately 2" diameter and over) are spade drilled.

Threads are machine tapped with solid taps for sizes up to 1-1/2". Collapsing taps are used for larger diameter threads. The chasing inserts are retracted when the thread length is reached.

Leading-on attachments provide for a turret and thread-cutting tool feed rate equal to the pitch of the required thread form.

Section drawings are used to expose internal details of a piece part. These drawings simplify otherwise difficult-to-read details and permit accurate interpretation.

Cutting plane lines and section lining are used on a drawing to identify the location of the cutaway section, internal features, and the material in the workpiece.

Safety precautions governing machine tool setup and operation must be observed. While each tooling station should be as close to the work and cutting area as possible, there must be sufficient clearance to work safely.

Chip control at the point of cutting must be planned. Chip removal around the tooling area must be continuous.

Cutting fluids must be contained. The flow rate must be controlled at each tooling station.

Large taper boring operations are performed with greater rigidity when cross-sliding hexagon turrets are used.

Holes are reamed with solid, taper, or adjustable floating-blade reamers.

Internal recessing, grooving, and back-facing cuts require tool positioning inside the workpiece. The cutters are fed by cross slide movement. On a fixed-center turret, the tool is mounted in and fed from a quick-acting slide tool.

After setup, the complete cycle of cuts should be taken carefully and checked for correct cutting action, dimensional accuracy, surface finish, and safe clearance in the work area.

UNIT 10 REVIEW AND SELF-TEST

1. a. Refer to two different tables: (1) machinability class of materials and (2) starting cutting speeds and feeds for various depths of cut. These tables relate to turret lathe operations.
   b. State how the information in both tables interrelates and why it is important to the turret lathe operator.
   c. Set up a shop problem that requires the use of the machinability class and cutting speeds and feeds tables.

2. Explain how a burnished, quality surface finish is produced by using a bar turner.

3. Describe how large-diameter workpieces may be faced on machines equipped with a cross-sliding turret.

4. a. List four basic forming cuts for turret lathe work.
   b. Indicate two general locations for cutting-off tools. State why such locations are functional and desirable.

5. Determine the starting cutting speeds in sfpm for the following three jobs:
   a. Rough facing class A machinability parts with a carbide-tipped cutter.
   b. Cutting off class B machinability parts with a high-speed steel cutter.
   c. Rough cutting class B machinability parts with a high-speed steel cutter by using a bar turner.

6. Determine the starting feed rates for machining on a ram-type turret lathe for each of the following processes:
   a. Turning a workpiece, using a roller turner setup, for a 3mm depth of cut;
   b. Forming steel parts to a 1/4" depth of cut;
   c. Cutting off a 1" diameter workpiece with a high-speed steel cutter.

7. Tell what function is served by each of the following drills: (a) start drill, (b) spade drill, and (c) core drill.

8. Describe two turret lathe setups that permit a boring bar and cutter to be adjusted for each successive cut when straight boring.

9. Explain the difference between the cutting of an internal, square recess and back facing a square shoulder on a turret lathe.

10. Give two advantages of using collapsing taps in preference to solid taps for certain applications on a turret lathe.

11. a. Tell what purpose is served when a part is represented on a drawing by a cutaway section.
    b. Describe a cutting plane line.

12. State three safety precautions the turret lathe operator should take to prevent personal injury or damage to the machine, setup, or workpiece.

UNIT 11

Planning Turret Lathe Setups and Tooling

Efficient machining with reduced worker fatigue is accomplished by planning a practical tool cycle. This unit on bar and chucking machines deals with factors the operator must consider in tooling up to machine a part. The approach is to analyze a shop drawing to establish the specifications, processes, tooling, and machine setups.

The positioning of machine stops and their use for controlling cross slide and hexagon turret tooling travel are covered. Attention is directed to the sequencing of cuts. Setups are considered for representative chucking jobs.

Step-by-step procedures are examined on how to set and how to adjust a bar turner and how to position hexagon turret stops on a saddle-type turret lathe. The unit concludes with a description of machine vibration and corrective steps to overcome tool chatter.

FACTORS TO CONSIDER IN TOOLING SETUPS

The piece part drawing specifications are the base for establishing tooling and machining setups. Figure 11-1 is a typical section shop drawing of a turned part. The worker considers at least five factors in analyzing a drawing, as follows:

—The size and shape of the workpiece,
—The material and stock removal processes,
—The quantity or lot size,
—Dimensional tolerances,
—Surface finish.

Size and Shape of Workpiece

The size and shape of the workpiece dictate how the part is to be chucked, the size of machine to use, and the nature and sequence of the cuts to be taken. Cutting speeds and feeds, as well as the kind of cutter material to use, are influenced by the nature of the workpiece.

Workpiece Material and Stock Removal

The machinability of the workpiece directly affects the material of which the cutting tools are made and the cutting speed and feed rates. Manufacturers' tables of machinability provide recommended starting cutting speeds and feeds for particular materials and processes.

Notes of the amount of material to be left for finishing often appear on drawings. Such notes are usually found on drawings of castings, forgings, and flame-cut and other irregular surface parts. This information helps in planning hogging and finish cuts, as well as determining the capacity of the turret lathe to machine a part.

The workpiece material indicates whether to machine dry or wet and the properties of the cutting fluid (when applied). Cutting, rake, and clearance angles must be related to the workpiece material and processes.

Lot Size

Decisions are made on a production run relating to setup time, the nature of tooling, and processes in terms of cost effectiveness. Long runs of parts usually justify special, more efficient tooling than short runs. While it takes additional time and cost to produce special tools, a lower per unit cost results over the lot run.

Dimensional Tolerance

Acceptable dimensional variations often dictate the number of cuts to take. Scale (steel rule) measurements may require only a single cut. Some dimensional tolerances—for example,

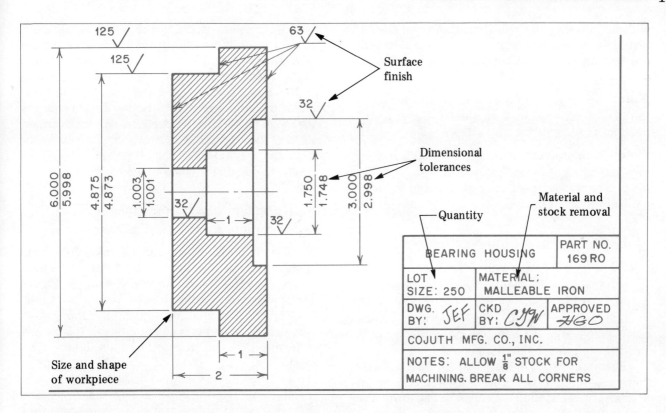

FIGURE 11-1 Shop Drawing Used to Establish Tooling and Machining Setups

0.002″ (0.05mm)—require two cuts. Finer tolerances may make it necessary to take two or more finish cuts.

Surface Finish

A common surface finish range for turret lathe work is from $\sqrt[32]{}$ to $\sqrt[250]{}$ (microinches). Fine turning finishes of $\sqrt[16]{}$ and below are attainable by using finely ground and honed cutters and by accurately controlling speeds and feeds. The microinch symbol and number on a drawing indicate the maximum allowable roughness of the surface. The finer the surface finish, the more precisely the machining is to be controlled. A greater amount of time is required to produce a fine, precise surface finish.

MACHINE STOPS

Machine stops are used on turret lathes when more than three pieces are to be machined. The stops ensure the fast, accurate duplication of parts. Stops are used to control power feed knock-off, to position the hexagon turret along the bed for longitudinal distance settings, and to control cross slide cuts.

Stops on the Cross Slide Carriage

The cross slide carriage has one set of adjustable stop dogs that controls the in and out feed of the cross slide carriage by disengaging the power feed in either direction—toward or away from the spindle axis. On precision dimensions, the common practice is to disengage the power feed and feed the cross slide by hand for the last few thousandths of an inch. The final depth of cut is controlled by a clip on the handwheel micrometer collar.

The longitudinal feed of the cross slide is controlled by a six-position stop roll that is located on the left side of the carriage apron. This stop roll works in combination with a notched adjustable stop roll mounted in the headstock bracket. Each stop roll may be adjusted to an independent longitudinal setting of the cross slide. A master screw in the end of each stop roll permits adjustment of the overall setup without altering the relationship among the individual stop screws. After each screw is adjusted, the setting is secured by tightening the binder screws.

Note: The adjustable stop roll is positioned in the headstock bracket with the least amount of overhang of individual stop screws.

Note: Each stop screw on the stop roll is locked in place by tightening the lock screws. Tightening prevents movement during a run and helps to control linear accuracy.

Hexagon Turret Stop Screws on a Ram-Type Turret Lathe

Stop screws for each turret position on a ram-type turret lathe are mounted in a stop roll extending from the ram. Each stop is set after a cut is taken from the hexagon turret. The turret and stop roll index simultaneously.

Each stop is set by first machining the workpiece according to the process(es) and tooling on a single turret station. A cut is taken from the hexagon turret to establish the required diameter and depth (length). The spindle is stopped. The feed lever is engaged and the turret is clamped. The stop screw for the turret is turned until the feed knocks off. When the feed lever drops out at the knock-off position, it indicates the cutting action is approaching the dead stop. The turning of the stop screw is continued until it touches the dead stop. The binder screw is then turned to lock the stop screw in position. On extremely accurate workpieces, the final movement to length (or depth) is made by hand.

Hexagon Turret Stop Roll on Saddle-Type Turret Lathes

The six-position stop roll assembly for a hexagon turret saddle-type turret lathe is located under the saddle and between the ways (Figure 11-2). The stop roll indexes with the turret. Each stop dog has an adjustable stop screw. When the stop screw is accurately set, it is locked in position on the stop roll by lock screws. The master stop may also be adjusted. If all the stops are to be changed an equal distance without disturbing the relationships among the stops, the master stop adjusting screw is turned to make the change.

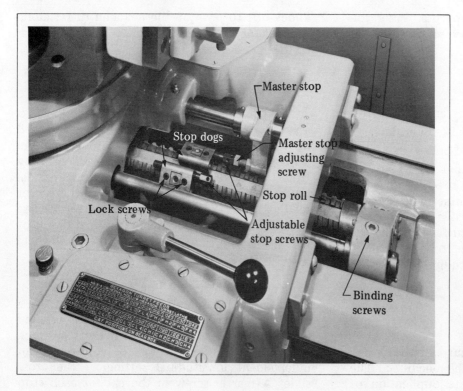

FIGURE 11-2 Major Parts of Six-Position Stop Roll Assembly for Hexagon Turret Saddle-Type Turret Lathe

HOW TO SET THE HEXAGON TURRET STOPS ON A SADDLE-TYPE TURRET LATHE

STEP 1 Move all the stop dogs back to the outer end of the stop roll.

Note: The master stop will engage the loosened stop dog for the particular turret face being used.

STEP 2 Take a trial cut.

STEP 3 Stop the spindle at the end of the cut. Engage the horizontal feed lever. Clamp the saddle.

STEP 4 Adjust the stop dog in the nearest *detent* (groove) *position* of the stop roll. Lock the stop dog in this position by tightening the lock screws.

Note: Line up the first lock screw by bringing the edge of the stop dog flush with the edge of a detent on the stop roll.

STEP 5 Tighten the second lock screw.

STEP 6 Adjust the stop screw on the stop dog until the master stop is moved to the point where it knocks off the feed lever.

STEP 7 Tighten the center lock screw on the stop dog to secure the stop at the correct position.

Note: Steps 1 through 7 are adjusted for rough work where an exact length is later produced by facing.

Note: Machining to a precision length requires adjusting the stop screw until the master stop moves solidly against the dead stop.

STEP 8 Index the turret to the next station. Then repeat steps 1 through 7 to set the tooling on the second and successive faces of the turret.

THE PLANNED TOOL CIRCLE

Production efficiency requires that operator effort and operating time be reduced, where practical. Efficient machining may be accomplished by positioning the cutting tools and setups so that they fall within a *tool circle*. The saddle should also be positioned so that each cutting tool travels the shortest possible distance before machining begins.

In a tool circle, all positioning, support, and cutting tools extend an equal distance from the hexagon turret. Figure 11-3 shows a production setup where all of the tools and holders fall within a tool circle. As each position of the turret is indexed, the longitudinal travel to the end of the workpiece is approximately the same. The operator is able to set up uniform operation timing motions to index and position each tool to the cut. The longitudinal turret movement stroke and the turnstile movement are kept short. Operator fatigue is reduced when the saddle is locked in a position that permits the operator to grip and move the turnstile comfortably.

Note: The distance between the saddle and the end of the workpiece must be great enough for the binder clamp to lock the hexagon turret securely.

Note: Each tool on the turret must be checked for interference. When indexed, each tool must clear the square turret and rear tool post tools.

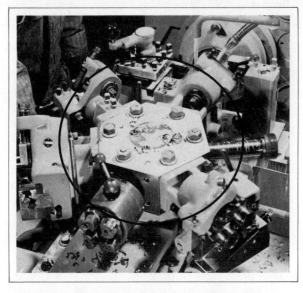

FIGURE 11-3 Planned Production Tooling and Setups That Fall within a Tool Circle

TURNING MACHINES

HOW TO SET UP A BAR TURNER

STEP 1 — Extend the bar stock about 3" from the collet.

STEP 2 — Turn the diameter 0.001" (0.02mm) under the required size for 1/2".

STEP 3 — Position the bar turner cutter slightly above center. Loosen the roll jaw adjusting screws until the rolls clear the turned diameter.

STEP 4 — Move the bar turner so that the cutter is located over the area of the turned surface. Check to see that the withdrawal lever is in the closed position.

STEP 5 — Adjust the cutter slide block until the cutter just grazes the turned diameter.

Note: The spindle must be in the brake position and the cutter on center.

STEP 6 — Adjust the cutter slide block 0.0015" away from the turned diameter.

Note: A thickness (feeler) gage is used to adjust the cutter slide block (Figure 11-4).

STEP 7 — Clear the cutter from the turned surface by retracting the cross slide until the cutter touches the turned shoulder.

STEP 8 — Loosen the roll jaw stud. Adjust the rolls longitudinally. The roll radius is positioned behind the cutter radius.

STEP 9 — Start the spindle. Turn the roll adjusting screws until the rolls lightly contact the workpiece and roll. Clamp the rolls securely.

STEP 10 — Adjust the coolant nozzle and start the flow.

STEP 11 — Advance the cutter to the cutting position. Take a trial cut.

STEP 12 — Return the bar turner to its original position. Stop the spindle.

STEP 13 — Check the accuracy of the turned diameter and the quality of the surface finish.

HOW TO ADJUST A BAR TURNER FOR SIZE AND FINISH

Correcting an Oversized, Finish Turned Diameter

STEP 1 — Check the position of the cutter and binder screws. If these screws were not tightened securely, the cutting force may have moved the cutter below the centerline. Adjust the cutter to the centerline.

STEP 2 — Check the back roll. A loose roll does not provide a firm support; consequently, the workpiece may spring away from the cut. Tighten the roll if it is loose.

STEP 3 — Inspect the cutting edge for wear. Adjust the cutter slide to compensate for wear.

Correcting an Undersized, Finish Turned Diameter

STEP 1 — Check the cutting edge of the cutter for buildup. Remove the buildup by stoning. Readjust the cutter, if needed.

STEP 2 — Check the front roll for adjustment and looseness. These conditions permit the work to vibrate and cause the cutter to chip, flake, or wear rapidly. Correct the problem by tightening the front roll. Both rolls are then checked for proper contact.

STEP 3 — Check the rolls for tightness. Tightness produces an overburnished, undersized diameter. Readjust the tightness of the rolls.

Correcting Variable Diameter Sizes

STEP 1 — Check the looseness of the back roll. Looseness produces a tapered diameter, particularly on long turned parts. The condition is corrected by resetting the back roll with the bar turner positioned near the collet.

FIGURE 11-4 Thickness Gage Used to Adjust Cutter Slide Block and Cutter

TURRET LATHE SETUPS AND TOOLING

STEP 2 Check the spring tension on the cutter slide block. If there is any play due to improper tension, the turned diameter may vary over its length. Tighten the spring adjusting screw at the rear of the slide block to increase the tension.

Correcting Surface Irregularities

STEP 1 Check the rolls for proper setting. A series of undersized grooves along the turned surface indicate that the rolls are improperly set.

STEP 2 Check for alignment, looseness, and wear on the rolls. Any one of these conditions can cause the rolls to dig into a finished surface to produce a rough, irregular surface.

STEP 3 Check the position of the rolls in relation to the nose radius of the cutter. Advance the rolls if they are set too far behind the cutter.

STEP 4 Decrease the feed if the cutter dips in at intervals. Check the cutting speed. Adjust, if needed.

STEP 5 Check the shape and condition of the chip groove. Regrind for the chips to form properly and to be removed without being forced between the rolls and the finished diameter.

ADDED MACHINING CAPABILITY WITH THE CROSS-SLIDING TURRET

The turret of the cross-sliding hexagon turret may be fed longitudinally and transversely (cross feed). Movement may be by hand or power feed. The positioning of the cutting tool for an internal or external cut with cross feed or longitudinal motion is controlled by micrometer dial settings. The cross feeding and longitudinal feeding motions and positioning dials are shown in Figure 11-5.

With a cross-sliding turret, all internal operations are normally performed with the boring bar as the primary tool. The boring bar is mounted on the hexagon turret. The square turret is used for all external cuts like turning, grooving, facing, chamfering, and so on. The cross-sliding turret is ideal for short-run lots. A great range of diameters may be machined by merely setting dial clip positions for each new diameter. Cuts may be taken from both the cross-sliding hexagon and the square turret at the same time.

Added machining capability is provided by a built-in positive stop. This stop permits the cross-sliding turret to operate also as a fixed-center turret. Tooling is then aligned with the centerline of the turret lathe.

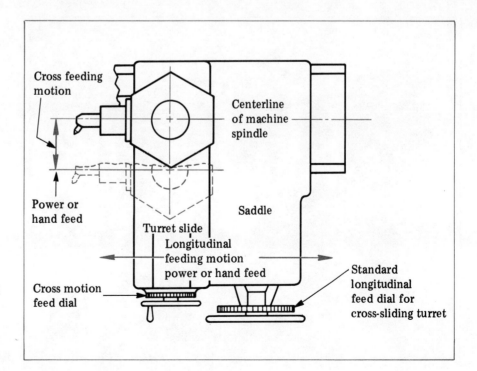

FIGURE 11-5 Longitudinal and Cross Feeding Movements of the Cross-Sliding Hexagon Turret

PERMANENT SETUPS WITH CHUCKING TOOLING

Castings, forgings, large-diameter rolls, collars, gear blanks, and flanged parts may be machined with chucking tooling. Machining usually involves combined and multiple cuts. As many processes as possible are completed in chucking in order to ensure concentricity and squareness between internal and external surfaces. Rough castings and forgings require three distinct cuts:

—A first cut for heavy stock removal,
—A second cut for truing up,
—A third cut for dimensional accuracy.

Concentricity and Squareness in Machining

Three internal cuts are required when diameters and faces are to be concentric and square and held to close tolerances. These cuts are as follows:

—Drilling or hogging out large-diameter holes by boring,
—Semifinish boring to correct inaccuracies such as run-out and out-of-roundness,
—Reaming or finish boring to size and to a specific surface finish.

Under certain conditions, external cuts may be taken at the same time as internal cuts. Roughing external cuts may be combined with such internal cuts as drilling, rough boring, or other hogging cuts. Internal finish boring cuts may be combined with finish facing and turning cuts. Extremely precise finish cuts, however, require that each cut be taken alone. Short stub or shouldered parts, such as gear hubs, are rough turned and faced before being machined to size. Long hubs require additional cuts.

Representative Chucking Tooling (Flanges)

Cast or forged flanges are considered here as a representative sample of a group of chucking jobs. Figure 11-6 pictures an assortment of flanges. Each part may be chucked on a turret lathe and machined on the inside and outside. The multiple turning head is widely used with an overhead pilot bar. This bar provides rigidity. Drilling and rough turning cuts are taken at the same time. Also, cuts are taken simultaneously from the square turret (facing, turning, and grooving) and the hexagon turret. On long runs, turning cuts should be taken from the hexagon turret. Setup time is saved because the turning cutters on the hexagon turret may be preset.

FIGURE 11-6 Samples of Cast and Forged Flanges Requiring Chucking Tooling

TURRET LATHE SETUPS AND TOOLING 205

A permanent setup for machining flanges with chucking tooling is shown in Figure 11-7. The representative flange is shown as a half section. The flange is held in a chuck. The surfaces to be machined are lettered A through G. A description of the square turret and hexagon turret tooling and sequence of cuts follows.

Square Turret Tooling. The rough and finish facing tools for surfaces F and C and the chamfering tool for corners B and E are mounted in three stations of the square turret on the cross slide.

Note the sequence of cuts on both the square turret (2, 4, and 5) and the hexagon turret (1, 2, 3, 4, and 6). The rough facing cuts (#2 on the square turret) are taken at the same time as the core drilling and rough turning cuts (#2 from the tooling sequence on the hexagon turret). Similarly, the finish facing cuts (#4 on the square turret) are taken while outside diameters A and D are finish turned (#4 on the hexagon turret tooling).

Note also that a pilot bar is used for core drilling G, rough turning outside diameters A and D (cuts #2), and finish turning diameters A and D (cuts #4).

Hexagon Tooling and Sequencing Cuts. Usually, the cast or forged blank has a lot of excess metal to be removed. Therefore, the first step in machining the flange is to *true bore* the hole a short distance to the diameter of a core drill. A true-bored hole serves to guide the core drill concentrically. The boring bar on the slide tool holder is used to true bore cast hole G (cut #1 on the hexagon turret).

Hole G is then core drilled (cut #2 on the hexagon turret). The cutter holders, in the same multiple turning head, rough turn diameters A and D. Face F and then face C are machined (cut #2 on the square turret). The longitudinal cross slide stop is set so that face C is completed at the same time as the turning cutter on diameter D. Next, hole G is finish bored with the boring bar and slide tool setup (cut #3 on the hexagon turret).

The finish turning cutters for inside diameters A and D are then set up (cut #4 on the hexagon turret). The cutters are mounted in a multiple turning head. Facing cuts are taken at the same time on surfaces F and C from the facing cutter (cut #4 on the square turret).

Corners B and E are chamfered with the chamfering tool (cut #5 on the square turret). Finally, hole G is reamed to size (cut #6 on the hexagon turret). The reamer is held in a floating reamer holder.

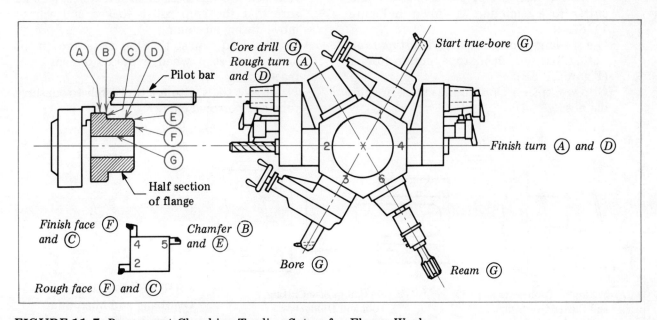

FIGURE 11-7 Permanent Chucking Tooling Setup for Flange Work

MACHINE VIBRATION AND TOOL CHATTER PROBLEMS

Machine vibration and tool chatter affect surface finish and dimensional accuracy. Machine vibration produces a lacework, coarse finish on a diameter or face of a workpiece. Machine vibration may be detected by a low, rumbling sound during the cut. On a new machine, most vibration is traceable to incorrect installation. On an old machine, vibration may be caused by worn spindles or gear and belt defects. Machine vibration problems may be avoided by making the following checks:

—Check to see that the lag screws in each stand are tight (the turret lathe must rest on a solid foundation and must be level);
—Check the alignment of the bar feed to the bed ways (the bar feed unit must be *lagged down* securely);
—Check the balance of the chuck or work-holding fixture (add balancing weights, if required);
—Test the motor armature for balance, particularly on high-precision production work.

Tool chatter produces a high-pitched, shrieking, whistling sound. Most tool chatter problems are caused by excessive tool overhang. The tool setup is not rigid enough to withstand the cutting force and speed. Corrective steps for tool chatter problems are as follows:

—Keep the overhang of the toolholder and cutter to a minimum as shown in Figure 11-8A;
—Use the largest possible size of cutting tool, boring bar, or shank-type cutter holder (Figure 11-8B);
—Grip workpieces close to the chuck or spindle nose;
—Use a revolving center to support heavy cuts;
—Tighten and recheck all clamping screws in cutting-tool holders;
—Position cutters on center;
—Replace a dull cutter with one especially sharpened for a specific job and keep the cutter sharp;
—Reduce the feed rate;
—Eliminate *dwell* so that a cutting tool does not remain in contact at the same machined surface position after the process is completed.

SAFE PRACTICES IN TURRET LATHE SETUPS AND TOOLING

—Study the specifications of the part to be machined. Then select cutters, holders, and square and hexagon turret accessories that provide maximum rigidity for holding, driving, and machining.
—Make sure each stop roll is adjusted to trip the power longitudinal and transverse feeds at the place where the workpiece is machined to the specified dimension.
—Position all adjustable stop rods, cutting tools, and cutter holders with the least possible overhang.
—Recheck the tightness of each lock screw to be sure that the stop roll is locked and cannot move during machining.
—Set the dead stop so that the cutter movement and action stop when a preset dimension is reached.
—Plan to set tools within a tool circle to conserve time and personal energy.

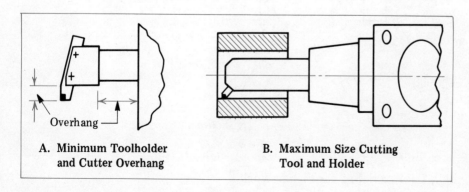

A. Minimum Toolholder and Cutter Overhang

B. Maximum Size Cutting Tool and Holder

FIGURE 11-8 Minimum Overhang and Maximum Cutter Rigidity

- Check for possible interference between the workpiece, tooling setups, and the indexing of each turret face.
- Tighten the cutter and binding screws on a bar turner so that the cutting edge does not move below center.
- Check the back roll to see that it is tight and provides maximum support (rigidity) to the workpiece.
- Stone or hone the cutting edge of a cutting tool to remove buildup.
- Check the adjustment and tightness of the front roll of a bar turner. A loose front roll permits vibration and chatter.
- Loosen the bar turner rolls if they are tight and produce a turned surface that is overburnished and undersized.
- Correct variations in a turned diameter by resetting the bar roll of a bar turner.
- Check the alignment, looseness, wear, position, and condition of the bar turner rolls if the workpiece is being machined with surface irregularities.
- Regrind the chip groove on a bar turner cutter so that chips are formed and removed properly.
- Combine external roughing cuts with internal roughing cuts. Allow sufficient stock for final finish cuts.
- Use a pilot bar with the multiple turning head on the hexagon turret for heavy-duty internal and external roughing cuts.
- True bore to hole size for about 1/2″. A true-bored hole is concentric and guides a core drill in drilling a straight hole that is aligned with the axis.
- Withdraw the cutters and stop the turret lathe if machine vibration or tool chatter occurs. Each condition is detected by a different sound during machining and by examination of the surface texture.

TERMS USED IN TURRET LATHE SETUPS AND TOOLING

Piece part drawing specifications   Complete information contained on a shop drawing from which the operator establishes the required tooling and machine setups, including speeds and feeds.

Lot size   The quantity of a part that is to be mass-produced on a turret lathe. Long runs and short runs of parts that determine whether regular or special tooling or setups are justified.

Stop roll   A device for stopping the cross slide or hexagon turret at the position where a cutting tool reaches a required dimension. A device that permits settings for each hexagon turret or square turret face (station). A multiple setting device for controlling longitudinal and transverse movement of tooling on the cross slide and hexagon turret.

Stop screws   Adjustable screws that control the distance of movement of tooling on each turret face.

Stop dog assembly   A mechanism located under the hexagon turret on a saddle-type turret lathe. Six control dogs that index separately with a turret to control longitudinal movement of the tooling at each position.

Detent   A groove on the stop roll for adjusting a stop dog.

Tool circle   Positioning the tooling on a hexagon turret to approximately the same starting point. Setups on the hexagon turret that provide for uniform starting positions and that permit the operator to develop an easy machining rhythm and to minimize fatigue.

Roll jaw stud   An adjusting screw on a bar turner that permits the rolls to be moved longitudinally.

Buildup (cutter)	The addition of the material being cut on and around the cutting edge of a cutter. Excess material, fused or adhering to the cutter, that results from tremendous force and heat in the cutting area.
Chucking tooling	The positioning of accessories and cutting tools for a hexagon turret and a square turret for large-size chuck work.
Chucking tooling setups	The combination of tooling and positioning cutters and accessories at each turret position to perform required machining processes.
True bore	An initial cut or cuts on a cast or rough surfaced hole for the purpose of providing a concentric straight starting hole for subsequent operations. Boring a concentric hole for a short distance to guide a drill or other cutting tool.
Lagged down	The use of a threaded stud or lag (coarse thread) screw to secure a machine to a firm foundation.
Dwell	Leaving a cutter in contact with a workpiece as it continues to turn and no cutting takes place.

SUMMARY

A shop drawing provides information from which the operator determines processes and setups. Features of the workpiece, amount of material to remove and the work processes, lot size, and dimensional and surface finish accuracy may be established from a drawing.

Machinability tables help to establish the size, shape, and design of the cutter(s) and cutter holder(s); cutting speeds and feeds; the type and size of machine tool; and the composition of the cutting fluid, if required.

Machine stops are used to control the movement of the cross slide and hexagon turret. Distances to which cutters are fed for each process are controlled by screws on a stop roll. Separate stop rolls are provided for each station on the square turret of the cross slide and the hexagon turret.

Micrometer dial collars on the cross slide and cross-sliding hexagon turret permit accurate tool settings.

The stop roll for the hexagon turret indexes automatically for each station of the turret.

Stop dogs for hexagon turret stops on saddle-type machines are adjusted in the detent position and locked with a lock screw. Each stop dog trips the master stop, which, in turn, knocks off the power feed lever. Turret movement stops when the master stop moves solidly against the dead stop.

Cutting tools and holders are mounted on the hexagon turret so that the cutting edges fall within a planned tool circle.

Bar turner rolls are positioned and set on a turned surface. The cutter is set on center. A thickness gage is used to adjust the cutter slide block 0.0015" (0.04mm) from the turned diameter. The roll jaw stud permits longitudinal adjustment of the bar turner rolls. Once set, the stud holds the rolls securely.

A bar turner is adjusted to correct turning oversized, undersized, and variable-sized diameters and surface irregularities.

> A cross-sliding hexagon turret may be fed longitudinally and transversely with hand or power feed. Longitudinal and cross feeding motions may be precisely controlled by micrometer dials and the use of clips to set fixed positions.

Stop screws are set independently for each process. If all cutters are to be moved an equal distance, a single screw or master stop may be turned.

> Stop screws actuate the dead stop at a preset dimension at the place where all cutter feed motion should stop.

The added transverse movement of the cross-sliding turret permits multiple setting and cuts with the same tooling setup, in addition to fixed-center setups of the hexagon turret.

> In common practice, there are a number of common permanent tooling setups for both bar and chucking work. Chucking tooling is used for castings, forgings, large-diameter piece parts, collars, gear blanks, and other flanged parts.

Rough castings and forgings usually require three cuts: heavy stock removal, truing up, and finish cut(s).

> Concentric holes and square faces in chucking work require three basic internal cuts: hogging out by boring (or rough drilling), semifinish boring, and either reaming or finish boring to size.

Cast, forged, or flame-cut holes are true bored to the diameter of a core drill. Rough boring ensures proper drill alignment and concentricity.

> Multiple internal and external cuts may be taken at the same time. However, only rough internal and external cuts or finish cuts are taken together.

Permanent tooling and setups are common shop practices for workpieces that may be grouped according to similar design features and chucking processes.

> A pilot bar is used with the multiple turning head to provide additional tool support (rigidity) for precision machining.

Square turret and hexagon turret setups are required for chuck tooling.

> Machine vibration and tool chatter are conditions that affect surface finish and accurate dimensional measurement control. These conditions may be identified by sound during machining or by checking with instruments. Balance of the workpiece; trueness and condition of the spindle; overhang of the setup, tooling, or workpiece; cutter position and sharpness; and feed rate are some checkpoints.

Standard machine tool safety must be observed. Clearances between the chuck and cross slide and hexagon turret setups must be adequate to permit indexing and all other processes to be carried out without interference.

> Particular attention must be paid to chip formation and removal from the cutting edge area.

Protective goggles or a face shield must be worn. Coolants must be contained by splash guards and other machine enclosures.

UNIT 11 REVIEW AND SELF-TEST

1. List five informational items found on shop drawings that are important in tooling up a turret lathe for a production run.
2. Explain the functions of a stop roll in relation to the cross slide carriage.
3. Tell how a hexagon turret stop roll on a saddle-type machine operates.
4. Explain three characteristics of a planned tool circle in production machining on a turret lathe.
5. Identify three possible causes of turning a finished diameter undersize when a bar turner is used.
6. Tell how to correct two problems of variations in diameter size when a bar turner is used.
7. List two practices to consider when taking external and internal cuts at the same time.
8. State three corrective steps to take to avoid tool chatter problems.
9. List three safety checks the turret lathe operator makes with respect to a tooling circle setup.

SECTION THREE

Single- and Multiple-Spindle Automatic Screw Machines

The automatic screw machine is an adaptation of the engine lathe and the turret lathe. The screw machine is used principally for semi-automatic and fully automatic, high-volume production of identical parts. The three basic types of screw machines are as follows: (1) plain, (2) automatic single-spindle, and (3) automatic multiple-spindle. Each type with the major components and typical tooling setups, including cam features, are described in this section. The descriptions build upon engine and turret lathe technology and machining processes. Cutting speeds, feeds, and cutting fluids are related to general screw machine processes.

Considerable attention is paid to common problems encountered by screw machine operators. Problems relating to feeding, stock stops, basic internal machining processes, turret tooling, and cross slide tooling are grouped within a troubleshooting table. The causes of these problems and corrective actions are suggested.

UNIT 12

Screw Machine Technology, Practices, and Troubleshooting

CLASSIFICATIONS OF SCREW MACHINES

Originally, the screw machine developed from the engine lathe for the economical manufacture of interchangeable screws and bolts. In addition to the feeding of regular bar and wire stock, today's machines permit forgings, castings, and irregular-shaped parts to be fed by hand or automatically by *magazine*.

Like the turret lathe, the tooling for screw machines is mounted and operated from front and rear stations on the cross slide and turret. Attachments provide added capability to cross drill, cross tap, and mill flats and slots. (Cross drilling and cross tapping refer to the drilling and tapping of holes at a right angle to the axis of the turned part.)

Plain Screw Machine

Screw machines are designed to perform a single machining operation or several operations at one time or in a rapid sequence. The main slide, turret, and other slides of the *plain screw machine* are operated manually. By contrast, these components and the feeding of bar stock on the semiautomatic (wire feed) screw machine are controlled by a pair of levers. An operator is required to manipulate the levers.

Automatic Screw Machine

The *automatic screw machine* is fully automatic in operation. Once the cutting tools are set up on the cross slide and turret and the controlling cams are positioned, the stock is automatically advanced and clamped in the chucking device. The turret and tools are automatically indexed in sequence. Following the tooling setup, the main function of the operator is to inspect workpieces regularly for dimensional accuracy and surface finish. Corrective steps are taken when dimensional accuracy or the surface texture is outside the specified limits. One operator usually attends a battery of automatic screw machines.

Single-Spindle Automatic Screw Machine. The *single-spindle* automatic screw machine is a general-purpose machine in which a single piece of stock is fed through the spindle at one time. The workpiece is machined in the sequence in which the tools in the indexing turret and on the cross slide are set up.

Usually, round, hexagon, square, and regular-shaped bar stock is processed on the single-spindle automatic screw machine. The bar stock is securely held in a spring collet chuck. The chuck is actuated by a friction clutch inserted between two pulleys on the spindle. The spindle drive allows the direction of rotation to be changed (reversed).

Multiple-Spindle Automatic Screw Machines. The *multiple-spindle* automatic screw machine has four, five, six, eight, or more work-rotating spindles. Additional production capacity is possible due to the increased number of operations that can be carried on at the same time on one machine.

MAJOR COMPONENTS OF AUTOMATIC SCREW MACHINES

There are many designs for screw machines. A Brown & Sharpe ultramatic screw machine is shown in Figure 12-1. However, all designs include cams, levers, clutches, trip dogs, and collet or chucking devices. Bar stock in 10, 12, and 20 foot (250mm, 300mm, and 500mm) lengths is automatically advanced after each workpiece is cut off.

Spindle Drive Mechanism

The spindle is driven by a constant-speed motor. This motor is mounted on an adjustable bracket inside the base. The motor drives the speed and ratio change gears in a speed case. In turn, the gears drive the spindle through chains and sprockets. A small, friction-brake motor powers the drive shaft, chuck, and feed cam drive at the end of the machine. The speed change gear design is illustrated in Figure 12-2.

Spindle

The spindle is mounted in antifriction bearings. The capacity for the Brown & Sharpe ultramatic screw machine (Figure 12-1) is from 3/4" (19mm) to 1-5/8" (41mm) diameter. Large-capacity #3 machines have either a 2" (50mm) or 2-3/4" (70mm) capacity. Two-speed and four-speed screw machines are available. There are 18 spindle speeds in the two-speed range from 20 RPM to 5,018 RPM for the 3/4" (19mm) capacity machine. The #2 machine, 1-5/8" (41mm) size, has 17 speeds in the two-speed range from 14 RPM to 2,906 RPM. High speeds and low speeds are produced in a clockwise and counterclockwise direction.

Feed Cycle Change Gears

Feed cycle change gears (Figure 12-3) control the rate of production. The gears are installed at the right end of the machine. The safety cover provides for the automatic disengagement of the drive shaft. The drive shaft is stopped whenever the drive shaft handwheel is engaged or the cover is opened.

SCREW MACHINE TECHNOLOGY 213

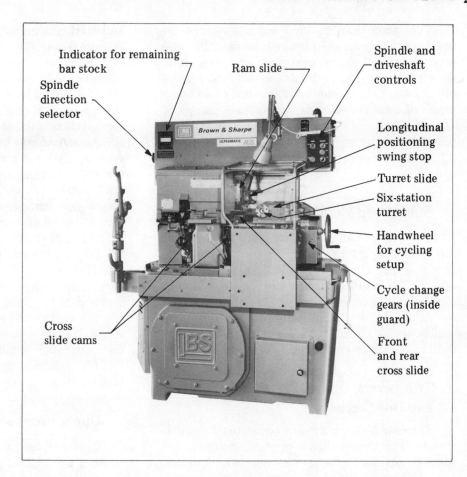

FIGURE 12-1 Ultramatic Screw Machine

FIGURE 12-2 Speed Change Gears That Actuate Drive Shaft, Chuck, and Feed Cam Gears

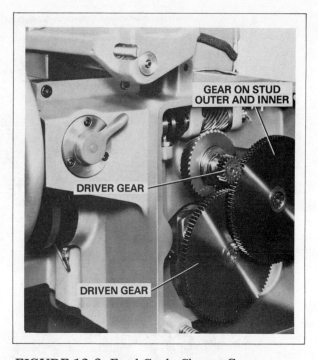

FIGURE 12-3 Feed Cycle Change Gears

214 TURNING MACHINES

The gears that are used are determined by the time (in seconds) required to machine one piece. The feed change gear plate on the machine shows the position of the driver and driven gears on the stud. Gear combinations are given for a range of time (in seconds) to machine one piece from 1.6 to 1.000. The right column on the gear plate shows the gross production per hour from 2,250 (1.6 seconds per piece) to 3.6 parts (1,000 seconds).

Stock Feed Mechanism

The stock feed mechanism controls the feeding and gripping of the bar stock. The bar is firmly gripped in the spindle by a spring collet chuck. The collet permits the cutting action close to the spindle. The bar stock is supported in the feed tube by bushings. These bushings, which are as close to stock size as possible, are used in the outer end of the tube.

Trip Levers, Camshafts, Dogs, and Dog Carriers

The chuck clutch trip lever controls the chuck clutch. The chuck clutch operates mechanisms through gearing to the chuck camshaft. A chuck and feed cam on this shaft operates a chuck fork (Figure 12-4). The fork causes the collet to open and then to close. The stock is automatically advanced and positioned by an adjustable feed slide. Adjustable dogs on dog carriers control the opening of the chuck before the stock is advanced during the stock feeding cycle.

An automatic operating mechanism controls such operational movements as indexing the turret and changing the spindle speeds. One camshaft carries the control drum, upper slide cams, chuck dog carrier, and the cross slide cams. A second camshaft carries the lead cam for the turret slide and the rapid pull-out cam.

Dog shaft carriers control the turret dog carrier (for timing the turret indexes) and the reversing dog carrier (for spindle clutch and spindle speed changes). Adjustable trip dogs on the turret dog carrier are set to move the trip lever for single or double indexing. Single indexing is used to index each position of the hexagon turret. Double indexing is used to index every other tool position.

Upper Front and Rear Tool Slides

Upper slides provide additional capacity for turning and cutoff processes. Figure 12-5 illus-

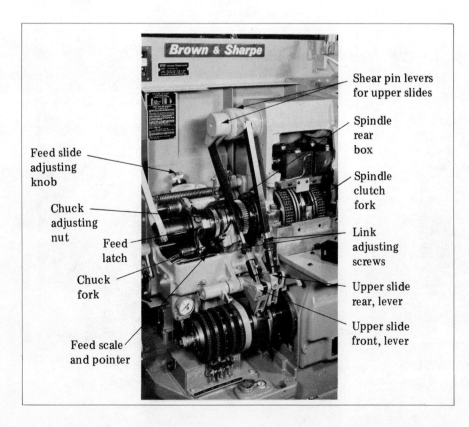

FIGURE 12-4 Spindle and Feed and Upper Slide Adjustments

SCREW MACHINE TECHNOLOGY 215

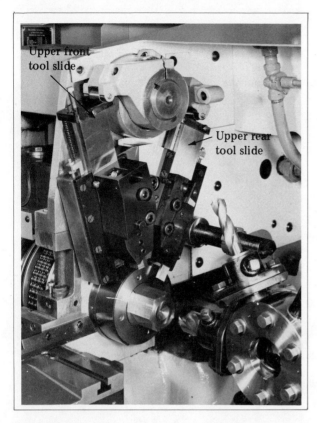

FIGURE 12-5 Upper Front and Rear Tool Slides

trates an upper front slide equipped with a forming tool post. This slide accommodates right- and left-hand turning tools. The upper rear slide has two cutoff toolholders. These are adapted for right- and left-hand "T" type cutting-off blades.

Turret Slide

The turret slide (Figure 12-6) may be moved toward or away from the spindle. Movement is due to the relationship between the lead lever, withdrawal cam, and the rack of the turret slide. The turret slide is provided with a positive stop screw. This screw is used to maintain extreme accuracy of depth to close dimensional tolerances.

TOOLING FOR AUTOMATIC SCREW MACHINES

Many different holders and cutting-tool setups are required for screw machines. Some standard tools include: box tools, hollow mills, and balance turning tools; adjustable and floating holders; and swing tools, slide tools, and tap holders. Figure 12-7 shows these tools and other standard screw machine tools.

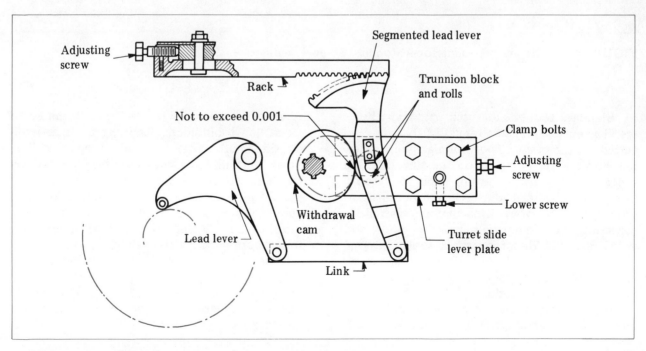

FIGURE 12-6 Relationship of Lead Lever, Withdrawal Cam, and Rack of the Turret Slide Drive

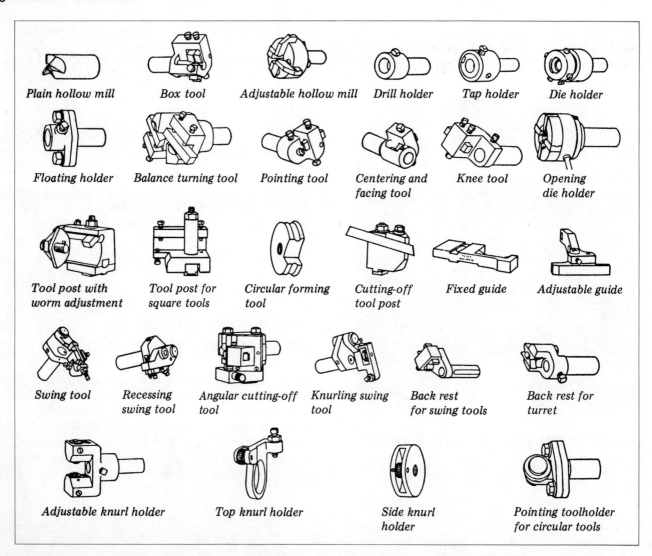

FIGURE 12-7 Standard Tooling for Automatic Screw Machines

Standard tool bits are used with many holders. The cutters may be altered slightly to be converted for direct use. Tools for machining external surfaces are mounted on the turret. These tools include balance turning tools, box tools, knee tools, plain and adjustable hollow mills, and swing tools. End-of-work machining tools include centering and facing tools, pointing tools, and pointing toolholders for circular tools. Internal surfaces are cut with tools such as drills, reamers, counterbores, and recess cutters. These tools may be held in rigid or floating holders.

Internal threads are cut with solid and adjustable taps. External threads are cut with self-opening die heads or thread rolls mounted in cross slide toolholders. Knurling tools and top and side knurl holders are held on the cross slide. Other knurling tools and adjustable knurl holders are mounted on the turret. Circular cutting-off and forming tools are used in the regular cross slide tool post or an adjustable tool post. Thin, straight-blade cutting-off tools and square tools are supported on the cross slide.

The turret-mounted tools include: angular cutting-off tools, support and auxiliary tools, back rest for chuck, fixed and adjustable guides for operating spring tools, and spindle brakes.

GUIDELINES FOR AUTOMATIC SCREW MACHINE TOOL DESIGN AND SELECTION

Maximum efficiency in the operation of an automatic screw machine requires the exercise of judgment in tool selection and design. The operator should consider the following guidelines.

Straight rough-turning operations. The balance turning tool is an all-around cutting tool. Plain hollow mills are used for fixed-size turning. Adjustable hollow mills permit turning within a range of diameters.

Straight finish-turning operations. Turning box tools are adapted to finish turning.

Removing scale on hot-rolled bar stock. The special knee tool is recommended for bar stock that does not have the external finish of screw-machined bar stock.

Straight turning long, slender workpieces. A pointing tool of the box type steadily and rigidly supports the work in a sleeve ahead of the cutting tool. This setup prevents the work from deflecting under the force of the cut.

Taper and form turning a long, shallow, and continuous curve. The swing tool is used for taper turning. It sometimes replaces a circular forming tool for other forming processes.

Regular form turning and straight turning behind shoulders. Circular forming tools are recommended when the length of cut is limited in relation to the diameter of the workpiece. Circular forming tools are mounted on the cross slide for regular form turning and for straight turning behind shoulders.

Centering, drilling, reaming, and counterboring. The particular tool for each process is held in a floating toolholder. This holder permits easy adjustment so that the centerlines of the cutting tool and of the workpiece are aligned.

Rounding off, chamfering, and pointing. These processes are performed with a circular tool. This tool is held in a pointing toolholder. In pointing, the workpiece must be rigid so that it will not spring under the cutting forces.

OPERATION OF FULLY AUTOMATIC SCREW MACHINES

Bars are loaded in each of the hollow spindles on a fully automatic screw machine. Operations at each spindle are carried on simultaneously and in sequence. Figure 12-8 shows five of six spindles on an automatic screw machine. The end-of-work machining tools (except for the die head) are all fed as a unit by movement of the tool slide. The die head is fed independently. Similarly, the tools on the cross slide are fed separately by the drum cams.

At spindle 4, the dovetail form tool is used to rough form the head while the hole is partially drilled. Simultaneously, the previously drilled hole is drilled to depth and size, and the outside diameter is finish turned in spindle 5. Interestingly, at spindle 6, a separate feed is set for the die head. After threading, the cutting-off tool is fed from a vertical cross slide to part the workpiece. All of the end-of-work machining tools are fed as a unit by the movement of the tool slide.

Each spindle rotates as a single-spindle machine. Each spindle has a spring collet chuck. Cutting tools such as drills, reamers, counterbores, and taps are mounted in front of each spindle. Turning, forming, and cutting-off tools are secured on the cross slides. When an operation is finished, each spindle turns to the next position. The spindles are rotated until the cycle is completed.

Since production time depends on the machine time consumed at each position for the longest-time cut, the machine is sometimes set up to do only a portion of this cut. The cut is then completed at the next position. Thus, production time is saved and cutting tool life is extended.

FUNCTIONS AND DESIGN OF CAMS

Cams control the timing and position of all cutting tools, and the action of work-holding devices. Cams impart motion to the turret and the front, rear, and vertical cross slides. This motion controls the cuts. Cams are designed to *rise*, *fall*, or *dwell*—that is, they cause the cutting tools to advance, be withdrawn, or remain stationary.

The cam contour is also designed so that it revolves with the camshaft to enable it to return the tool slide to an inactive position. The operating cam is used to bring the turret to its operating position. Cam designs for cross slides and the turret must provide for starting and stopping each operation at the correct place and time.

218 TURNING MACHINES

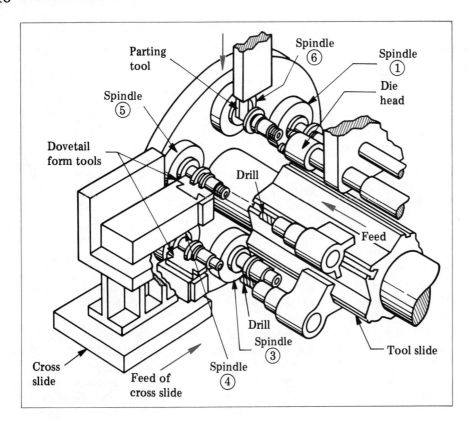

FIGURE 12-8 Schematic of Partial Setup on a Six-Spindle Automatic Screw Machine

A working drawing provides full information on the form of a workpiece, material, sizes, dimensional accuracy, and surface finish. Most drawings specify the type and size of the screw machine to be used. The cam designer takes the piece part specifications and adds the following information:

—All machining operations and required tools,
—Spindle RPM for each operation,
—The RPM needed to produce one workpiece,
—Speeds and feeds,
—Cam and tool throw.

The required cams are then designed based on this known information. The drawings are turned over to the toolroom where the cams are produced. An example of three superimposed cams is shown in Figure 12-9. The cams are used for rough turning, finish turning, and cutting-off operations. This particular cam is used to automatically machine a part on a Brown & Sharpe automatic screw machine. Note that the cam perimeter is divided into 100 parts. The speed of the camshaft determines the number of seconds represented by each division. Many automatic screw machine builders have tables of cam design specifications that are prepared for each different type and size of screw machine.

CUTTING SPEEDS AND FEEDS

Standard engineering handbooks and machine tool manufacturers' literature provide tables of feeds and speeds for automatic screw machines. The values that are included in such tables are compiled from years of extensive testing and analyses of production methods. The speeds and feeds are based on the following conditions:

—Material and shape of the workpiece;
—Type and size of screw machine;
—Material, type, and number of cutting tools;
—Design features of the toolholders;
—Depth of cut;
—Kind of cutting fluid;
—Required quality of surface texture.

CUTTING FLUIDS

As with all other machining processes and machine tools, a cutting fluid serves as a coolant

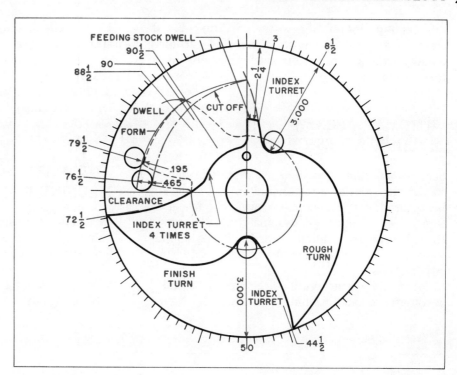

FIGURE 12-9 Cam Features and Design for Machining a Workpiece on an Automatic Screw Machine

and a lubricant. In addition, the cutting fluid must contain properties that permit it to separate rapidly from the chips.

Straight mineral oils are usually used as a blending medium for general-purpose screw machine operations. The mineral oil is mixed with a base cutting oil that possesses the necessary lubricating qualities.

Mineral lard oils are used on parts that require a higher quality surface finish than can be obtained by using straight mineral oils. Mineral lard oils have noncorrosive properties. These properties make mineral lard oils especially suitable for machining copper and copper alloys.

Sulphur cutting oils and *sulphurized oils* are recommended to increase tool life. Sulphur increases the cooling and lubricating characteristics of oils. Sulphurized oils wet surfaces faster than regular cutting oils—that is, sulphurized oils penetrate into remote places. Such oils have excellent film strength and help prevent chips from welding on the tool point.

Mineral oils are adapted for light machining operations on some steels and brass. Mineral oils are used for tapping and threading some nonferrous metals and for other difficult work.

Recently Developed Cutting Oils

The major claims for new, light-colored, transparent, and odorless cutting oils are as follows:

—The transparency of recently developed oils makes the machined part clearly visible to the machine operator during machining;

—The cutting oils have maximum lubricity, which results in reduced heat and tool wear;

—Antiweld qualities minimize buildup at the cutting edge of the tool for a considerable distance from the cutting edge;

—Increased extreme-pressure properties reduce the frictional heat developed by the rubbing chips on the extreme-pressure cutting areas;

—Exceptional cooling characteristics permit heat that is generated by the rubbing of the chips and workpiece to dissipate rapidly;

—The oils are corrosive and are recommended for all kinds of machining operations on all grades of steel (noncorrosive oils are used for nonferrous metals).

Major oil producers have also developed special sulphurized and other cutting oils. These oils are all compounded to be noncorrosive and

TURNING MACHINES

nonstaining. In addition to the previously listed claims, such cutting oils are claimed to produce equally satisfactory results on ferrous and nonferrous metals.

AUTOMATIC SCREW MACHINE ACCESSORIES

Second machine operations are often eliminated by adding special attachments to the machine being used. Attachments are available for slotting screws, burring, tapping nuts, rear-end threading, cross drilling (at right angles to the longitudinal machine axis), slotting, and machining flats.

A machine shaft produced on an eight-spindle automatic screw machine is shown in Figure 12-10. The drilling and tapping of the four holes and the milling of the center slot and the two slots on the end are performed on the same screw machine. Attachments are used for these operations. The production rate for this particular drilled, slotted, and threaded shaft is 200 parts per hour.

AUTOMATIC SCREW MACHINE TROUBLESHOOTING

Efficient operation of an automatic screw machine requires the operator to recognize a problem, diagnose the cause, and take corrective action. Problems may be encountered beginning with the stocking or rod feeding of the machine, positioning the stops, and continuing through all

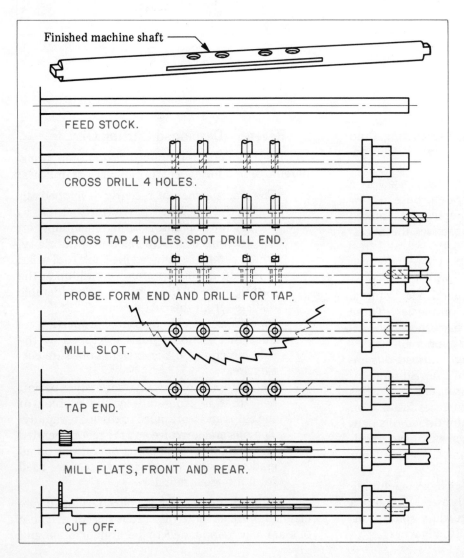

FIGURE 12-10 Applications of Attachments to Eliminate Second Machine Operations on a Shaft (Produced on an Eight-Spindle Automatic Screw Machine)

SCREW MACHINE TECHNOLOGY

successive operations from setups to mechanisms and movements. The recognition, analysis, and correction of a problem is called *troubleshooting*. Table 12-1 provides in capsule form problems commonly encountered by the screw machine operator. The table deals with feeding, stock stop adjustment, hole cutting and knurling, turret tooling, and cross slide tooling. One or more causes of each problem are identified. Corrective action is given for each problem. The problems, causes, and corrective action contained in the table supplement the operator's previous experience in general machining processes performed on basic machines.

TABLE 12-1 Automatic Screw Machine Troubleshooting

Problem	Possible Cause	Corrective Action
Stocking and Rod Feeding		
Stock will not enter	—Chuck in closed position.	—Trip the chucking lever. Turn the handwheel until the chucking cam permits opening the chuck.
	—Bent bar stock.	—Cut off bent section or replace bar.
	—Undersize collet.	—Replace with correct size collet.
	—Misalignment of feed finger and chuck.	—Insert stock in the feed finger and push it through the spindle.
		—Place the chuck with the bar stock in its correct location in the chuck sleeve.
	—Master feed finger and out of position.	—Remove the feed tube. Inspect and repair damaged pads.
Chuck will not open	—Feed trip lever toe in raised position.	—Check the trip lever toe, spring, and the screw and pin in the lever toe.
	—Bent chuck fork, sheared spindle key, cracked chuck sleeve, broken clutch spring.	—Remove and replace defective parts.
Chuck will not close	—Drum cam not in open position.	—Turn the drum to the open position.
	—Feed trip lever toe in raised position.	—Check the trip lever toe, spring, and the screw and pin in the lever toe.
	—Broken chuck lever, chucking roll stud; worn or broken chuck lever fulcrum shoes, bent chuck levers; or worn chuck or sleeve.	—Remove and replace the broken, worn, or bent parts.
Stock Stop Adjustment		
Workpiece length increases	—Stop is moving back in collet.	—Tighten the turret binding bolt.
		—Replace the clamping bushing on the binding bolt if worn.
	—Swing stop arm moves.	—Center the swing stop arm and the cutting-off tool. Tighten securely.
	—Feed slide movement is set longer than required length and there is excessive wear on the feeding finger.	—Reduce the amount of force of the feeding mechanism, which causes the stop to push back.
Variation in length	—Projection left because the cutting-off tool does not remove all of the teat.	—Reset the cutting-off tool on center.
	—Feed-out period on the lead cam is too short.	—Replace with the correct cam.
	—Turret cutting tools push the stock back into the collet.	—Sharpen each cutting tool. Check for correct feed.
	—Worn roll and pin on lead cam lever.	—Remove and replace the worn parts.
	—Chuck or feed finger tension too loose (or worn).	—Adjust the chuck to the required tension or replace a worn feed finger.
Burred or scored end of work	—Stock feed cam lobe is not concentric with the center hole in the cam.	—Refinish the stock feed cam lobe.
	—Center hole in end of stock stop is too large.	—Use flat stock stops without center holes, or face the end of the stock stop.

Basic Hole Forming and Knurling

Drills and Drilling

Problem	Cause	Correction
Hole larger diameter than the drill size	—Drill improperly centered in the holder.	—Adjust the outer body of the holder. Start the drill into the hole by hand feed. Tighten the holder screws.
	—Drill bushing out of line or drill shank is bent.	—Straighten the drill shank or replace both the bushing and the drill.
Holes drilled too rough	—Dull drill; feed too fast; insufficient or wrong coolant.	—Replace drill; reduce the rate of feed; use manufacturer's recommended cutting fluid.
Drill breakage	—Too high or too slow a starting speed; too fast a feed; too small lip clearance.	—Regrind with correct lip clearance; adjust speed and feed.
Drilled hole runs out	—Improper spotting, centering, or lengths of cutting edges.	—Use a spot drill that has a larger angle than the drill point angle; reset the drill on center; regrind the cutting edges evenly.

Reamers and Reaming

Problem	Cause	Correction
Reamed hole larger than reamer size	—Reamer and forming tool cuts start together but the forming tool operation is completed ahead of reaming.	—Withdraw the reamer before the forming tool operation is completed.
Hole reamed too rough	—Excessive float in the holder.	—Reduce the amount the holder floats until the reamer enters without distorting the edge of the hole.
Hole reamed bell-mouthed	—Reamer bushing is misaligned with the workpiece axis.	—Check the bushing for straightness and the fit of the reamer shank. Replace any defective bushing.

Counterbores and Counterboring

Problem	Cause	Correction
Hole counterbored too rough	—Dull counterbore; insufficient chip room; excessive clearance (producing chatter marks); excessive speed.	—Take general corrective actions for tool grinding and controlling the speed.

Threading Tools and Threading

Problem	Cause	Correction
Rough threads	—Improperly ground or dull thread chasers; excessive speed; wrong and/or insufficient coolant.	—Take general corrective steps to correctly sharpen the chasers, reduce the cutting speed, and use the manufacturer's recommended cutting fluid.
	—Insufficient bearing.	—Set the cutting face at least 1/10th of the diameter of the workpiece ahead of center.
Die tears the thread or loads with chips	—Hook or rake angle of the chasers too small.	—Regrind the chasers to increase the rake angle.
Die cutting edges wear rapidly or break off	—Too heavy bearing.	—Grind the chaser face to decrease the bearing. The face should be ahead of center (approximately 1/10th the diameter of the workpiece).
	—Insufficient chamfer.	—Increase the chamfer as much as the job permits (15° to 20°, if possible).
	—Misalignment of die head.	—Adjust the die head to the correct alignment.
	—Die running into shoulder or chuck jaws.	—Use projection chasers; check the setup and chamfer angle.
Thread not long enough	—Slipping shaft or holding clutch fork on friction clutch; chuck tension.	—Adjust to apply a greater force against the movable friction clutch parts.
	—Loose feeding finger and chuck tension too loose.	—Check collet and bar sizes; adjust the chuck tension.
	—Insufficient cam rise.	—Select a cam with a lobe rise that permits thread cutting to length.
Thread too long	—Improper setting or tripping of self-opening die head.	—Check and reset the trip mechanism.
	—Cam rise too high.	—Use only a portion of the cam rise to produce the desired length.
Distorted and out-of-round threads	—Loose lead cam.	—Tighten the lead cam on the camshaft.
Tapered thread	—Tight (drag) or loose (misaligned) turret slide movement.	—Adjust the turret slide.
	—Incorrect threading alignment.	—Make adjustments to correct alignment.

SCREW MACHINE TECHNOLOGY 223

Extensive tap breakage	—Tapping too close to the bottom of the hole; chips packing in the flutes; wrong coolant; tap drill too small.	—Take general corrective action for taps, tap drills, and cutting fluid.
	—Index dog trips before the tap is entirely withdrawn.	—Move the trip dog back to permit the completed withdrawal of the tap.

Knurls and Knurling

Poor knurl	—Knurl binds on the pin.	—Examine the pin. Stone any burrs.
	—Knurl not centered.	—Center each knurl and apply an equal force on multiple knurls.
Rough knurl	—Cam lobe rise too slow, causing chips to crowd and tear the knurl.	—Check and correct the cam design.
Improper tracking	—Knurls not of equal diameter or not mated.	—Check each knurl diameter; select knurl mates.

Turret Tooling

Knee Tools

Cutting with a taper	—Bar too small to use a knee tool; length of cut too long.	—Replace with a box tool, balance turning tool, or hollow mill.
	—Too much force required for the cut.	—Grind the correct clearance on the tool bit.
	—Tool not on center; too much feed; spindle speed too fast.	—Set tool on center or a few thousandths above; correct the feed and/or speed.
Cutting rough	—Incorrect rake (tool digs in); dull tool bit; improper feed and/or speed.	—Grind to correct rake, clearance, and front angles; set to correct speed and/or feed.
	—Incorrect coolant or rate of flow on cutting tool.	—Use manufacturer's recommended cutting fluid; correctly position nozzle and adjust flow rate.

Box Tools

Cutting rough	—Incorrect slide clearance (too little) and rake (too much).	—Regrind to correct clearance and rake angles. Check with an angle gage.
	—Back drag of the blade as it is withdrawn off the work.	—Loosen the back rest. Then, reset the blade to produce a 0.002" (0.05mm) oversize diameter. Bring the back rest against the workpiece with sufficient force to turn the correct diameter.
Cutting taper on an irregular diameter	—Incorrect shank fit in turret hole.	—Check for burrs, nicks, and correct fit.
	—Back rest too tight on workpiece.	—Reset the back rest.
Eccentric diameters	—Support set ahead of the tool bit.	—Reset the support so that it follows the tool bit.

Balanced Turner

Cutting rough	—Front clearance too high; rake too deep.	—Regrind the tool bit to the correct rake and clearance angles.
Prominent shoulder steps	—First cutter advanced too far beyond the center.	—Draw the first cutter back.
	—Second cutter too far in back of the center.	—Push the second cutter out.

Hollow Mills

Cutting taper	—Improper alignment.	—Use floating holder.
Cutting rough	—Stock piled on inside of a lip.	—Reduce spindle RPM or flow on a larger quantity of cutting fluid.
		—Check back clearance of the hollow mill teeth.
	—Chipped, burned, or burred lip.	—Regrind to sharpen the mill.
Cutting too small	—Ring on mill tightened too much.	—Adjust the ring until the mill cuts to the correct size.
Ragged cut at start	—Stock not chamfered.	—Chamfer end of stock to permit the teeth to start cutting concentrically.

Turret and Cross Slide

Cross Slide

Does not go to stop	—Sheared pin in cross slide rack.	—Replace the shear pin. Relieve the tension on the stop screw before readjusting for the correct diameter.
	—Cam lobe cut down too far.	—Replace the cam with a properly machined one.
Cutting with chatter	—Tool too far below center; excessive tool rake; tool and work overhang; cracked cross slide; tool too wide for the cut.	—Reset tool to center height; mount tools as short as possible; replace damaged machine parts.
Irregular size formed diameter	—Worn roller or pin on cam rolls.	—Replace worn parts and reset the cam.
	—Cam lever roll out-of-round.	—Replace worn parts and reset the cam.
	—Insufficient dwell on the forming cam.	—Replace worn parts and reset the cam.
Form diameters fall outside specified tolerances	—Stock diameter too small to withstand cutting forces.	—Use back rest or support.
Cutting off with a burr	—Play in cutoff cross slide.	—Adjust the gib.
	—Cutting tool in too deep during the form turning operation, causing the workpiece to break off.	—Position the cross slide cams to either start forming sooner or cutting off later.

Turret

Clamp bolts do not hold the turret tools	—Turret tool clamp bushing too short; radius worn; or bolt too long.	—Replace the turret tool clamp bushing.
Play	—Turret disc nut loose.	—Tighten the turret disc nut enough to take up the play without causing the turret to bind.
Failure to lock securely after indexing	—Worn locking pin or broken spring.	—Remove and replace worn or broken parts.
Failure to index	—Turret trip lever in raised position.	—Drop the turret trip lever into the working position.
	—Short turret trip dog fails to adequately raise the clutch trip lever.	—Remove and replace the worn turret trip dog.
Locking	—Loose change roll stud.	—Tighten the change roll stud.
	—Incorrect location of the change roll stud.	—Change the two studs on the change roll disc that are used for double indexing to see whether the wrong one was removed.
		—Change the location of the stud.
Failure to advance the same distance each cycle	—Worn cam lever roll or pin; loose cam lead.	—Replace the worn roll or pin; tighten the lead cam.
	—Turret indexes before the lead cam lever reaches the end of the cam lobe.	—Reset the turret index trip dog to index just after the lead cam lever starts down the drop of the lead cam.

SAFE PRACTICES IN SETTING UP AND OPERATING SCREW MACHINES

—Use noncorrosive oils on machine tools where there is a possibility of leakage into the bearing system.

—Place enclosing guards around chucking and machining areas before operating a screw machine to prevent personal injury that can be caused by tool breakage, flying chips, or the workpiece being thrown out of a chuck.

—Check the proper placement of each tool and how securely each is held. In addition to personal injury, work damage may result from a tool working loose or being forced out of a holder.

—Shut down the machine if any tooling changes or setup adjustments are to be made.

—Safeguard bar stock that projects beyond the lathe spindle. Guard rails are placed to cover the extended length in order to shield workers from a revolving workpiece. Note that sections of piping through which bar stock may be fed should be installed to prevent long bars from whipping around.

—Use short sheet metal or plastic screens to intercept and direct the cutting fluid to the chip pan, screens, and oil reservoir.
—Install a fine mesh, nonskid floor mat for worker protection against any oil that may be thrown from the screw machine.
—Change the sawdust or oil-absorbent compound around a screw machine as common practice. (Many departments have adopted nonslip, non-flammable floors for areas surrounding screw machines.)
—Wash hands and face thoroughly at regular intervals. Avoid contact with the cutting fluid as much as possible to prevent skin disorders that are caused by certain cutting fluids.
—Change work clothes frequently. Otherwise, they may become saturated with the cutting fluid.

TERMS USED IN SCREW MACHINE TECHNOLOGY AND PROCESSES

Magazine feeding	A device for holding, positioning, and automatically feeding precut, preformed, and/or premachined parts for chucking in a screw machine.
Single-spindle automatic screw machine	A general-purpose screw machine in which bar stock is automatically fed and machined. A one-spindle machine for multiple turning and cutting processes.
Multiple-spindle automatic screw machine	Four, six, eight, or more work-rotating spindles on one machine. An automatic screw machine that permits operations to be performed simultaneously on the workpiece on each spindle.
Feed cycle change gears	A gear train that permits gear combinations to be set up to control the rate of production. Gear combinations that control the time (in seconds) to machine one workpiece.
Chuck clutch trip lever	A lever that controls the automatic feeding and clamping of stock in a collet-type chuck.
Dog shaft carriers	Carriers that provide control of the turret indexes and spindle clutch and speed changes.
End-of-work machining tools	Cutting tools used for facing, center drilling, reaming, countersinking, and counterboring. Cutting tools that machine or are fed into a workpiece from the end.
Hollow mill	A plain or adjustable, multiple-tooth, end-cutting tool. An end-cutting tool for turning a fixed size diameter or range of diameters.
Cam rise, fall, and dwell	Motion and rest imparted by a cam to operate the turret and front, rear, and vertical cross slides. Cam design features for rough, finish, form turning, and cutting-off operations.
Wet surfaces faster (sulphur cutting fluids)	Properties of sulphur cutting fluids that permit the coolant to penetrate into hard-to-reach places during cutting.
Maximum lubricity	A quality of a cutting fluid to maintain its lubricating properties under severe pressure and temperature machining conditions.
Second machine operations	Additional operations performed on a machine using special attachments. Operations normally performed on a second machine that are possible on a screw machine by adding attachments.
Troubleshooting	Recognizing a machine or tooling setup malfunction, diagnosing the problems and possible causes, and taking corrective steps.

TURNING MACHINES

Stocking and feeding — Mechanism and processes of opening and closing a chuck and automatically advancing stock on an automatic screw machine.

Turret tooling — Positioning and cutting tools mounted directly or held in an adapter in a turret station. (Box, balanced turning, and knee tools are examples of turret tooling.)

SUMMARY

The three basic types of screw machines are: (1) plain, (2) automatic single-spindle, and (3) automatic multiple-spindle.

Screw machine tooling is mounted on a turret, on front and rear stations of the cross slide, and on additional attachments.

The turret, slides, and stock on a plain screw machine are fed manually.

One operation is performed at a time on a single-spindle automatic screw machine. The tooling is indexed for the sequence of operations.

Screw machine spindles are usually mounted in antifriction, preloaded bearings. Spindle speeds range from 20 RPM to over 5,000 RPM.

Screw machine production rates are controlled by feed cycle change gears. The gear combinations permit varying the time (in seconds) to machine each workpiece.

Multiple-spindle automatic screw machines are highly productive. Additional operations may be performed, depending on the number of spindles. Usually, one operation is performed simultaneously for each spindle.

Cams control the advance, withdrawal, and stationary positions of cutting tools on the cross slide and turret.

Chucks and collets are opened and closed after a workpiece is advanced and positioned (by an adjustable feed slide) by a chuck clutch trip lever.

Cam carrier shafts carry the control drum, slide cam, chuck dog carrier, cross slide cams, turret slide lead cam, and the rapid pull-out cam.

The turret is indexed by adjusting trip dogs.

External machining cuts are made with cutting tools held in box tools, balance turning tools, swing tools, slide tools, and knee tools. Fixed and floating holders and adapters are used.

Top and side knurl holders may be held on the turret or cross slide.

The screw machine operator usually troubleshoots problems in five general categories: (1) stocking and rod feeding; (2) stock stop adjustment; (3) basic hole forming processes such as drilling, reaming, counterboring, threading, and knurling; (4) turret tooling problems with knee, box, and balanced turning tools and with the use of hollow mills; and (5) turret and cross slide.

Machine tool safety requires using enclosures, checking tool setups and operations, and controlling the spilling of cutting fluid.

Cam designs are based on the nature of a workpiece, machine processes, spindle RPM for each operation, RPM required to machine one part, necessary tool travel, and feeds.

Speeds and feeds for screw machine processes are governed by the same factors as for similar cutting processes on other machine tools.

Straight mineral oils, mineral lard oils, and sulphurized oils are recommended for general-purpose processes, for machining tough noncorrosive metals, and for deep penetration.

New, light-colored, transparent, and odorless cutting oils provide maximum lubricity, dissipate heat rapidly, maintain properties under severe pressure and temperature conditions, and provide the same cutting and lubricating properties on ferrous and nonferrous metals.

Special attachments permit second machine operations such as slotting, tapping, rear-end threading, cross drilling, and machining flats.

Troubleshooting involves recognizing a screw machine problem and its causes and then, after analysis, establishing the corrective steps that must be taken.

Personal safety relates to the use of guards and protective goggles, protection against revolving overhanging bars, and hygiene in removing cutting fluid from the skin and clothing.

UNIT 12 REVIEW AND SELF-TEST

1. List three attachments that permit additional second machine operations to be performed on a screw machine.
2. Describe briefly the function of (a) feed cycle change gears, (b) the chuck clutch trip lever, and (c) the dog shaft carriers of an automatic screw machine.
3. a. Name five standard tools for automatic screw machines.
   b. Give the major process for which each tool is designed.
4. Indicate five workpiece specifications (in addition to dimensions) that are required for the designer to lay out screw machine cams.
5. State six conditions that control screw machine speeds and feeds.
6. a. Identify four properties of new screw machine cutting oils that make them different from standard mineral oils.
   b. Describe the importance of each property in relation to either tool wear life, quality of surface finish, heat removed, or machining.
7. Record the corrective steps to follow in troubleshooting the following screw machine problems:
   a. Stocking and rod feeding: Chuck not closing; drum cam not in the open position.
   b. Stock stop adjustment: Scored end of workpiece; stock stop center hole too large.
   c. Hole forming: (1) Hole drilled larger than drill size; improper drill centering. (2) Rough reamed hole due to earlier finishing of a form tool operation.
   d. Threading: Torn threads; incorrect rake angle of chasers.
   e. Turret tooling: (1) Knee tool cutting with a taper; tool not on center. (2) Second cutter of balanced turner back of center; shoulder steps produced.
   f. Cross slide: Sheared rack pin preventing cross slide from moving to the stop.
8. Describe three safety precautions an operator should observe before starting a screw machine.

PART THREE

Vertical Milling Machines

SECTION ONE Vertical Milling Machines, Attachments, and Accessories
       UNIT 13 Vertical Milling Machine History and Accessory Design Features
SECTION TWO General Vertical Milling Processes
       UNIT 14 Basic Vertical Milling Cutting Tools, Speeds and Feeds, and Processes
       UNIT 15 Hole Forming, Boring, and Shaping

SECTION ONE

Vertical Milling Machines, Attachments, and Accessories

This section begins with a brief history of major vertical milling machine developments. The major design features of modern machines are then described. These features are related to combination vertical/horizontal machines, the two-axis tracer, the bed-type, and numerically controlled milling machines.

The versatility of the vertical milling machine is extended by attachments such as all-angle, right-angle, multiple-angle, and high-speed milling heads; horizontal milling machine attachment systems; and the cross slide milling head. These attachments as well as several tooling accessories and work-positioning devices such as sine and rotary tables, cutter-holding devices, and a digital readout system, are also described.

UNIT 13

Vertical Milling Machine History and Accessory Design Features

The *vertical milling machine* is used to accurately produce flat, angular, rounded, and multishaped surfaces. These surfaces may be machined in one, two, or three planes (X, Y, and Z axes). The milling processes may be vertical, horizontal, angular (single or compound angle), or spiral.

The longitudinal and cross (transverse) feeds may be by either hand or power or both. A single part may be produced or the machine may be set up for the machining of multiple, duplicate parts. A machine with numerical control may be tooled for mass production.

HISTORICAL OVERVIEW OF DEVELOPMENTS

One of the first milling machines was developed during the 1860s. It combined the vertical spindle of the drill press with the longitudinal and transverse (cross feed) movements of the milling machine. The spindle was designed to permit greater rigidity for the cutting tool. The design features of the early vertical milling machine (Figure 13-1) included a wide face over which

HISTORY AND ACCESSORY DESIGN FEATURES

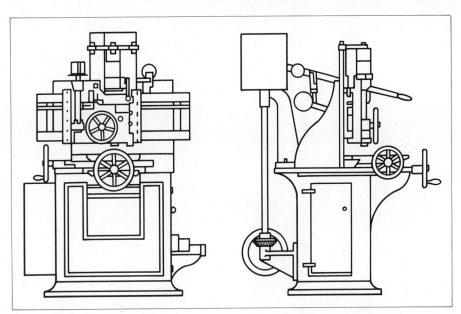

FIGURE 13-1 Design Features of an 1862 Vertical Milling Machine

the belt moved from the back driver pulley to the driven spindle pulley.

The handwheel that actuated the spindle permitted vertical adjustment for positioning cutting tools by hand-feeding. The spindle assembly provided for increased strength to take the side forces resulting from vertical milling. Two other handwheels provided longitudinal and transverse feed movements for the fixed bed and table.

An improvement came in 1883 with the patent of the *knee-and-column* design. This design provided additional space between the table and spindle. The table could be raised or lowered on the column. The spindle head was redesigned to permit setting it at an angle to the horizontal plane of the table.

Power feeds were introduced shortly after 1900 for longitudinal, transverse (lateral, crosswise), and vertical movements. The spindle design, which included a heavy-duty quill, permitted vertical power feed. By 1910, several types of vertical milling machines were available and included design features such as the fixed-position, swivel-head spindle; flat, fixed-bed, or knee-and-column tables; and hand and power feeds.

Electrical, electronic, pneumatic, and optical devices and controls were added later. They extended the applications for milling in three planes by tracing the form and simultaneously reproducing it on a workpiece.

In response to demands for more precise machining and measuring, a *jig borer* was designed for the precision layout and machining of holes. The jig borer is an adaptation of the vertical milling machine. Micrometers, standard length bar and indicator devices, and vernier scales were used to obtain exacting linear dimensions.

DESIGN FEATURES OF MODERN VERTICAL MILLING MACHINES

Modern vertical milling machines are versatile in design. Spindle speeds are usually controlled by variable-speed drives to permit a wide range of cutting speeds and feeds. Rapid traverse power feeds are available to reduce setup time in bringing the work to the cutter. The turret assembly is designed for making precision angular cuts. Handwheels and feed control levers provide for safe and convenient operator control of feeds.

Figure 13-2 shows the design features and manufacturing specifications of a general-purpose vertical milling machine. The specifications for this machine give an idea of the longitudinal feed (37-1/2″ or 950mm), cross feed (15-3/4″ or 400mm), vertical feed (17-5/8″ or 448mm). This particular quill may be fed 5-1/8″ or 130mm.

Spindle Speeds and Feeds

The specifications also indicate that the variable-speed drive provides a low spindle speed range of 60–490 RPM and a high range of 545–

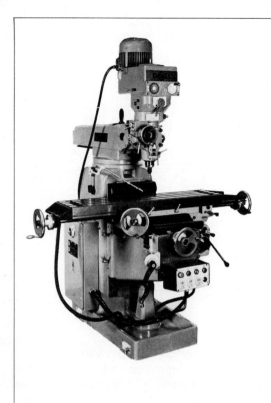

RANGE	DIMENSIONS, in. (mm)
Longitudinal manual feed of table	37-1/2 (950)
Longitudinal power feed of table	31-1/2 (800)
Cross feed	15-3/4 (400)
Vertical feed of knee	17-5/8 (448)
Spindle nose to table (quill up)	
Minimum	0
Maximum	17-5/8 (448)
Quill diameter	3.39 (86)
Quill feed	5-1/8 (130)
Throat depth, spindle center to column	
Minimum	6-5/16 (160)
Maximum	31-1/2 (800)
Rear edge of table to column	
Minimum	1/4 (6)
Maximum	16 (406)
TABLE	
Working surface	12 × 49 (305 × 1,240)
T-slots number	3
and size	11/16 (17.5)
Distance between T-slots	2-1/2 (64)
SPINDLE	
Spindle nose (NS)	#40
Spindle motor (horsepower)	3 HP
Type of drive	variable speed
Spindle speed range	
High range	545–4,350 RPM
Low range	60–490 RPM
Power quill feed,	0.0013 (0.03)
ipr (mm/rev)	0.0025 (0.06)
	0.0040 (0.10)
VERTICAL HEAD (TURRET)	
Longitudinal swing	total 180°
Front-to-back	total 90°
TABLE & SADDLE POWER FEED	optional
Numbers of longitudinal and cross feeds	12
Feed rate, longitudinal and cross, ipm	1/2–20-7/8
(mm/min)	(12–530)
Vertical feed rate	
Rapid advances ipm (mm/min)	optional
Longitudinal and cross	54-5/8 (2,150)
Vertical	42-5/16 (1,075)
Power feed motor	0.75 HP

FIGURE 13-2 Design Features and Manufacturing Specifications of a General-Purpose Vertical Milling Machine

4,350 RPM. There are three power feeds for the quill. These feeds range from 0.0013" to 0.004" (0.03mm to 0.10mm) per revolution. The twelve longitudinal and cross feeds (table and saddle power feeds) range from 1/2" to 20-7/8" per minute (ipm) or 12mm to 530mm per minute (mm/min). The spindle nose has a #40 National Standard taper (NS). The table size is 12" × 49" (305mm × 1,240mm). Rapid longitudinal and cross feed traverses are available at the rate of 84-5/8 ipm or 2,150 mm/min. The rapid vertical advance is half the rate of the other feeds (42-5/16 ipm or 1,075 mm/min).

There are variations among machine tool manufacturers of power feed rates for the quill. Generally, the feeds per revolution are 0.0015"

HISTORY AND ACCESSORY DESIGN FEATURES 233

0.003″, or 0.006″ (0.04mm, 0.08mm, and 0.015mm). Spindle speeds may be *infinitely variable* or back gear and direct drive. Low and high speed ranges extend from 60 RPM to 5,400 RPM with a standard head. Spindles are also designed for a #7 Brown & Sharpe taper, quick-change tapers, and tapers other than National Standard (NS).

Rigid Turret and Swivel Head Assembly

The rigid turret assembly may be swung on its base through 360° (Figure 13-3). The head that is secured to the ram is moved by rack and pinion. The head is counterbalanced so that the operator can easily position it for angular cuts. The head may be positioned to right or left of the 0° vertical axis (Figure 13-4). Some heads are designed for positive locking in position for vertical milling. Angular cuts to the horizontal plane may be made with precision by setting the head at any required angle within a 180° arc. Compound angles or single angles in the front-to-back plane may also be set (Figure 13-5). The head is easily swiveled through a worm gear arrangement.

Handwheels and Feed Control Levers

Most machines are designed for safe, efficient operator control of feeds. Figure 13-6 shows three handwheels to control the longitudinal feed. Two of these handwheels are located on the sides of the table. The third longitudinal handwheel is located on the saddle. A fourth handwheel on the saddle controls the cross feed movement. The vertical movement of the knee is controlled by the crank. For safety, the handwheels are spring loaded—that is, they are *free wheeling* and remain stationary when the rapid traverse is engaged. The direction of table feed is established by the position of the feed lever. This lever operates both the rapid traverse and feed.

COMBINATION VERTICAL/HORIZONTAL MILLING MACHINE

A *combination vertical/horizontal milling machine* may be quickly and easily converted from a vertical to a horizontal mode. The term *mode* indicates an adaptation of a machine tool from one series of major processes to another. For example, by retracting (bringing back) the forearm, loosening the head bolts, and rotating the head, a vertical milling machine may be adapted to a horizontal mode for horizontal milling pro-

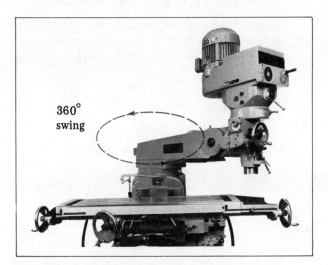

FIGURE 13-3 Rigid Turret Assembly (360° Swing)

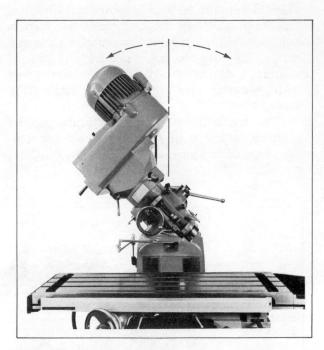

FIGURE 13-4 Positioning of Swivel Head at Angles to the Left or Right of the Vertical Centerline

234 VERTICAL MILLING MACHINES

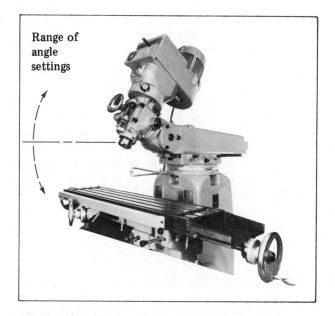

FIGURE 13-5 Front-to-Back Plane Setting of Swivel Head

cesses. Stops at 0° and 90° provide for vertical and horizontal alignment, respectively.

The combination vertical/horizontal milling machine shown in Figure 13-7 has a single swivel head. This head is used for milling from 0° to 180° in any plane. In the horizontal mode, an overarm support may be attached to the knee. This support serves a function similar to the overarm support of a regular horizontal milling machine. The support ensures a high degree of accuracy during heavy-duty horizontal or multiple-cutter cuts by rigidly supporting the arbor.

The combination vertical/horizontal milling machine, as the name suggests, is used for standard vertical and horizontal milling. In addition, drilling, boring, slotting, die sinking, and straddle and other heavy hogging to fine surface finish cuts are performed on this machine. The cuts may be taken by conventional or climb milling. The head has an overload protection to prevent distortion in cutting or damage to the head. Standard accessories used with this machine include the universal dividing head, rotary table with dividing plates, and swivel vise. Cutter-holding devices include arbors and arbor supports, collet holders, and collets.

Tracer Milling Machine

Tracer milling relates to the use of a template or sample part. The template serves as a guide for simultaneously controlling the movement of a cutter to reproduce a desired contour and/or form. A *tracer milling machine* has the versatility of straight-line milling (by locking the cutter at any of the axes) or profiling (by operating the cutter in two or three planes). Feeding may be manually or automatically controlled.

The manufacturer of the tracer milling machine illustrated in Figure 13-8 claims a part-to-template accuracy of ±0.002″ at 5 ipm (0.05mm at 127 mm/min). The accuracy at a higher feed rate of 30 ipm is ±0.0025″ (0.052mm at 762 mm/min).

The part-to-part accuracy is ±0.002″ at both the 5 and 30 ipm (0.05mm at 127 and 762 mm/min) feed rate. The part-to-pattern vertical axis tracing accuracy is within ±0.004″ (0.1mm).

Bed-Type Vertical Milling Machine

A *bed-type vertical milling machine* provides for longitudinal and cross slide (transverse) movement of a workpiece under a heavy-duty head

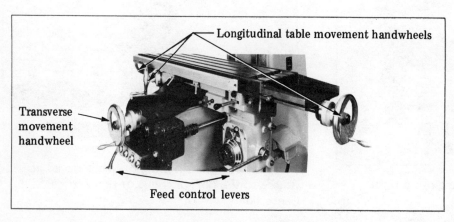

FIGURE 13-6 Handwheels and Feed Control Levers

HISTORY AND ACCESSORY DESIGN FEATURES 235

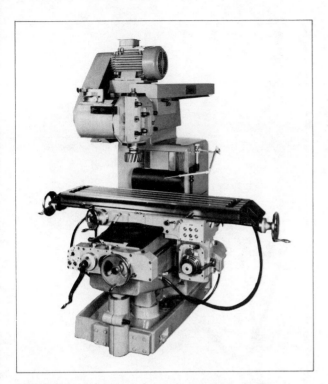

FIGURE 13-7 Combination Vertical/Horizontal Milling Machine

FIGURE 13-8 Tracer Milling Machine

and quill. The machine pictured in Figure 13-9 has a vertical head movement of 22" (560mm) and a quill diameter of 6" (152.4mm). The combination of longitudinal, cross slide, and vertical travel of this machine provides it with a work capacity of 51" × 20-5/8" × 26-3/4" (130mm × 525mm × 680mm).

It is evident that the bed-type vertical milling machine with its #50 NS spindle taper is used on large workpieces. Great power and heavy chip removal capacity are required. The bed-type machine is adapted to either the machining of single parts or the manufacturing of multiple pieces at one setting. The head may be swiveled 45° to the right or left of the vertical axis for angle cuts. The machine is designed with an automatic backlash compensation system. This system extends cutter wear life and assures quality of surface finish and dimensional accuracy.

Numerically Controlled Vertical Milling Machine

Vertical milling machine control movements may be actuated from information that is stored or punched on tape (numerically controlled, NC) or that is recorded in a computer program (computerized numerically controlled, CNC). (Concepts and programming of numerical control for two- or three-axes processes and four- or five-axes centers are examined in detail in Part Eight.)

A *numerically controlled (NC) vertical milling machine* may be used for point-to-point or contour milling with two- or three-axes continuous path controls. Figure 13-10 shows a computerized numerically controlled (CNC) vertical milling machine. This particular machine has a positioning accuracy of ±0.001" (0.02mm) along the full table movement for all axes. The repeatability of positioning is ±0.0005" (0.01mm) for all axes.

Computerized numerically controlled (CNC) machines are available for two- or three-axes processes. The positioning of the quill and feed control may be either manual or CNC.

MAJOR COMPONENTS OF STANDARD VERTICAL MILLING MACHINE

The six major design components of a standard vertical milling machine are as follows:

—Column and base;
—Knee, saddle, and table;
—Turret head and ram;
—Spindle (tool) head;
—Manual, power, and rapid traverse feed drives and controls;
—Cutting fluid and lubricating systems.

236 VERTICAL MILLING MACHINES

FIGURE 13-9 Bed-Type Vertical Milling Machine

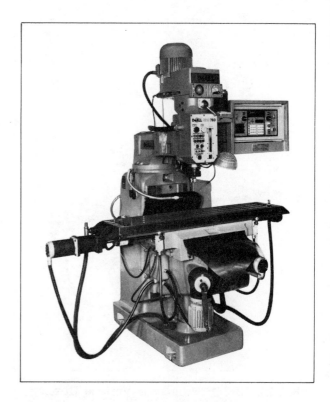

FIGURE 13-10 Computerized Numerically Controlled (CNC) Vertical Milling Machine

The functions of each component and of the control levers for speeds and feeds are similar on standard vertical and horizontal milling machines. The exceptions are the turret head, ram, and spindle (tool) head.

Features of the Turret Head and Ram

The turret swings on a machined base on top of the column. The turret head assembly is graduated through 360°. The spindle (tool) head is secured to the outer end of the ram. The ram is dovetailed to slide in the turret base. The movement of the ram, toward or away from the head, is controlled by a rack-and-pinion device.

Figure 13-11 shows the vertical (0°) position of the turret head assembly. The end of the ram is designed to hold the spindle head so that it may be swiveled through 180° in a vertical plane that is parallel to the column face. The ram is also designed with a graduated holding device that permits the head to be swiveled on a front-to-back plane at an angle to the horizontal axis of the table. The angular positioning of the spin-

End of ram graduated to permit angle settings of spindle head in vertical plane parallel to column face

End of ram graduated for angle (front-to-back) plane positioning

FIGURE 13-11 Vertical (0°) Position of Turret Head Assembly (Spindle Head Mounted on End of Ram)

dle head on the ram makes it possible to set a cutter (the spindle axis) at any single or compound angle.

Features of the Spindle Head

The spindle head serves as a cutter holding unit, the power source for cutting, and houses the cutter feed control mechanism. To serve these functions, the spindle head consists of the power source, speed and feed controls, a quill, and the spindle. The main design elements of the spindle head are shown in Figure 13-12.

Power Source. The three basic types of power sources are: (1) V-belt step pulleys, (2) variable-speed drive, and (3) quick-change gearbox. Each of these power sources serves the function of providing the driving force through the spindle to the cutter. Speed changes between the low- and high-speed ranges for V-belt and gear drives are made when the spindle is stopped. Changing the spindle RPM on a variable drive is done while the spindle is rotating. Electrical switches for reversing the spindle direction are provided on some machines.

Quill and Quill Stop. The spindle is housed in a large quill to provide a rigid support for the spindle and cutter. The quill is moved up or down by the quill feed handwheel or the quill feed hand lever. Once the operator establishes that the cut is set, the power feed is engaged by the feed control lever. The amount of feed is controlled by the power feed change lever.

The depth of a cut may be controlled by adjusting the micrometer depth stop for the quill (Figure 13-13). The stop trips the quill dog on contact. This stop may be set to trip at an upper or lower limit. The micrometer depth stop is calibrated in 0.001" (0.02mm) to permit fine adjustments. Maximum rigidity is obtained by tightening the quill clamp before a process is started.

Spindle Speed and Feed Controls. Plates are mounted on the spindle head to indicate the positions of control levers to obtain different spindle speeds. Usually a dial indicates the spindle RPM on variable-speed machines. The speed is regulated by a variable-speed control handwheel and a high-low speed range change lever. A manual quill feed may be changed to automatic by engaging the power feed change lever.

Spindle. The spindle is tapered at the open end to receive cutter holders and machine accessories, to position them concentrically with the spindle axis, and to transmit power through the cutter in order to carry on milling operations.

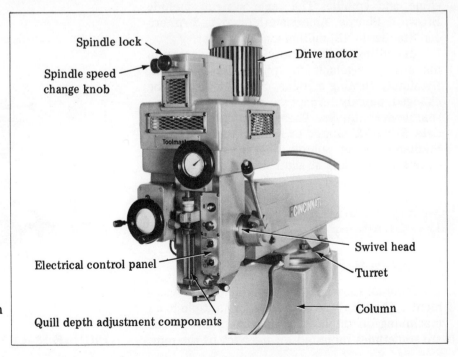

FIGURE 13-12 Identification of Main Design Elements of a Spindle (Swivel) Head

238 VERTICAL MILLING MACHINES

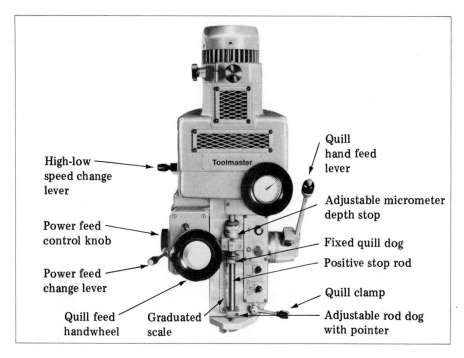

FIGURE 13-13 Stops for Controlling the Quill Depth

The spindle is usually splined to deliver maximum power without slippage between the cutter or holder shank and the spindle. A spindle lock prevents the spindle from turning when holders or tools are being inserted or removed.

Many standard taper systems and taper sizes are used by different manufacturers. The number of the taper varies with the size of the machine and spindle. The taper systems include Brown & Sharpe, American (Morse), and American Standard (NS) milling machine taper.

In addition, a Bridgeport® quick-change spindle nose is available for production purposes. By simply turning a collar, cutting tools may be changed quickly. Adapters are fitted to some spindle-nose tapers. These adapters accommodate Brown & Sharpe or Morse taper end mills. Similarly, some spindle tapers accommodate collets for holding straight-shank cutting tools.

VERTICAL MILLING MACHINE ATTACHMENTS

Spindle Heads

Separate spindle heads are available as attachments. They may also be added for multiple-head machining on duplicating machines. The heads are sometimes mounted on the back of the ram. Extra heads are equipped for two-way power down-feed and up-feed. The spindle and quill are counterbalanced. Figure 13-14 shows a spindle head with a separate lever for rapid manual movement of the quill, handwheel to obtain small movements, and an integral worm and gear to permit angular positioning. A positive quill lock, a switch to reverse the spindle rotation, and a

FIGURE 13-14 Separate Spindle (Swivel) Head

back-gear drive for maximum power are also shown. The micrometer depth stop is graduated in thousandths of an inch on standard machines and 0.02mm on metric machines.

Other spindle heads are available for variable-speed milling, drilling, and boring. These heads have infinitely variable speeds from 60 to 500 RPM in the back-gear range and 500 to 4,200 RPM in the direct-drive range. Heavy models operate at slower speed ranges from 50 to 450 RPM and 450 to 3,500 RPM. Spindle distortion is minimized by a flow-through air cooling system on some heads.

Power Down-Feed

The *power down-feed attachment* (Figure 13-15) controls the quill feed rate. The up-down feeds are infinitely variable from 0.2 to 2.5 ipm or 5 to 64 mm/min. The feed direction is controlled by the up-off-down switch. A pilot light indicates when the control is in operation.

Vertical Shaper Head

Some toolrooms and jobbing shops use a *vertical shaper head* as a milling machine attachment. The vertical tool motion permits shaping processes that normally require special machines or broaches.

The vertical shaper head shown in Figure 13-16 may be used to form shapes in blind holes where broaches cannot be used, cut gear teeth and racks, and produce sharp internal corners.

A shaping tool set, to be described later, is used on this machine head. Shaping may be done at a right angle or at any single or compound angle to the table. This head has a stroke range of 0″ to 4″ (0mm to 102mm). The strokes per minute of a slow speed head are 35, 50, 70, 100, and 200. A standard-speed model produces twice as many strokes per minute.

ATTACHMENTS FOR VERTICAL MILLING MACHINE SPINDLE HEADS

High-Speed Attachment

The *high-speed attachment* to the standard spindle head provides for maximum rigidity and rotational speeds up to 50% higher than the usual speed range. The Quill Master® high-speed head attachment (Figure 13-17) is designed for operations requiring the efficient use of small end mills and drills. The Quill Master is available with a 1/8″ (3mm) collet, a 3/16″ (5mm) spring collet, and a 3/16″ (5mm) solid end mill holder.

One manufacturer's design of the high-speed attachment is capable of operating at 9,000 RPM. Mounted on the quill, the attachment requires one-sixth of the spindle speed. Operating at this lower spindle speed eliminates vibration. At the same time, the high speed permits small size drills

FIGURE 13-15 Power Down-Feed Attachment

FIGURE 13-16 Vertical Shaper Head

240 VERTICAL MILLING MACHINES

and end mills to be run at proper cutting speeds. Thus, feed rates are increased over the rates used for slower speeds.

Right-Angle Attachment (Confined Areas)

Special tooling and fixtures, particularly for milling out pockets and cavities, may be eliminated by using a *right-angle attachment*. A right-angle attachment may be assembled on a high-speed head for internal end milling (Figure 13-18A) or can be mounted directly on the quill of a vertical spindle head (Figure 13-18B). Power is transmitted by a drive shaft that is held solidly by means of a collet and drawbar. The clamping screws on the body of the right-angle attachment hold the unit firmly on the quill.

These right-angle attachments are designed with preloaded ball bearings and precision-lapped, spiral bevel gears that are grease lubricated. An auxiliary arbor support provides added rigidity for horizontal machining operations. The spindle of a right-angle attachment is designed to accommodate a collet.

A supplementary right-angle attachment is used for precision milling operations in confined areas. Figure 13-19 shows a vertical and two horizontal setups for end milling inside a workpiece. The right-angle attachment shown in the figure works equally well on inside or outside cuts on regular parts or irregularly shaped castings. The unit has permanently lubricated bearings and a gear housing. The minimum working space is 2" (51mm) diameter. Milling cuts may be taken to within 1/2" (13mm) of a walled surface.

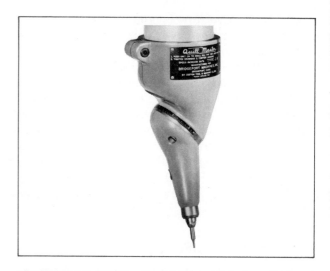

FIGURE 13-17 Vertical Miller High-Speed Head Attachment

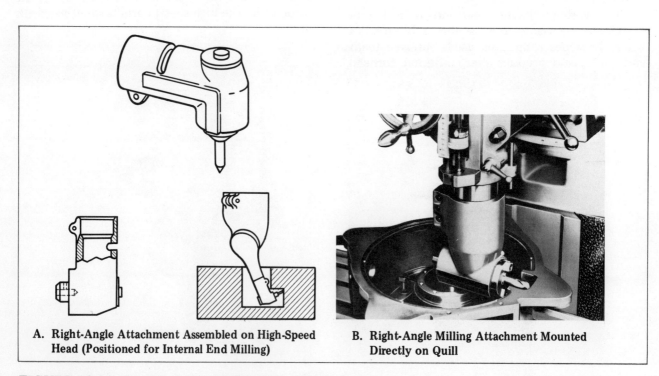

A. Right-Angle Attachment Assembled on High-Speed Head (Positioned for Internal End Milling)

B. Right-Angle Milling Attachment Mounted Directly on Quill

FIGURE 13-18 Two Models of Right-Angle Attachments

HISTORY AND ACCESSORY DESIGN FEATURES 241

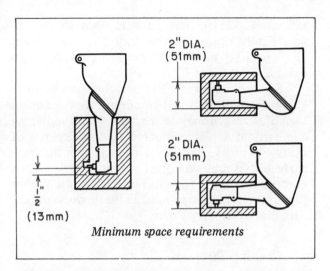

FIGURE 13-19 Internal (Confined Area) Milling, Using High-Speed and Right-Angle Attachments

Rotary Cross Slide Milling Head

The *rotary cross slide milling head* mounts on the vertical head quill. The attachment provides mechanical control of the cutting tool through straight, angular, and radial movements. For example, combined operations such as angles and radii, angles tangent to radii, and one radius blending into another are possible. Applications of the attachment are shown in Figures 13-20A and B. Note the internal and external cuts in the figures. Some are straight; others require accurate corner rounding and precision form milling.

The rotary cross slide milling head may be rotated in a 360° planetary motion. The degrees are subdivided into five-minute graduations. Operation of the head may be by manual or power feed. The head in Figure 13-20 has a 3" (76mm) longitudinal movement range. The lead screw is hardened and ground for close tolerance work. This head is used with workpieces that are mounted on a conventional milling machine table, rotary table, or other work-holding accessory.

Optical Measuring System

Positive, accurate, and fast determination of the table position is possible by a *direct-reading optical measuring system*. Reading accuracies are within 0.0001" with the inch-standard attachment. The operator reads a single line on a scale that is calibrated every 0.010". Settings in 0.001" are obtained through a drum dial that is calibrated in 0.0001" increments. The metric-unit scales are calibrated every 0.10mm; the drum dial, in 0.002mm increments. The optical measuring system is used for longitudinal and cross travel measurements.

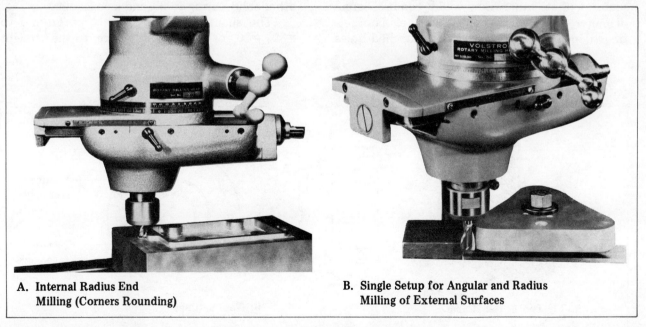

A. Internal Radius End Milling (Corners Rounding)

B. Single Setup for Angular and Radius Milling of External Surfaces

FIGURE 13-20 Applications of Rotary Cross Slide Milling Head

VERTICAL MILLING MACHINE TOOLING ACCESSORIES

Quick-Change Tooling System

The *quick-change tooling system* is designed for fast tooling changes during production. The system is available with #30 or #40 NS taper holders. One manufacturer has the following cutter holders or adapters available for quick-change tooling (the collets are available in both inch and metric sizes):

—Quick-change spindle,
—Chucks for drills and end mills,
—Drill extension chucks,
—Non-pullout end mill collets,
—Floating reamer holders,
—Tenthset boring heads and boring bars,
—Morse taper adapters,
—Jacobs taper adapters,
—Shell end mill adapters,
—End mill adapters,
—Tap holders,
—Preset locking fixture for chucks,
—Spade blade holders,
—Spade blades.

Fly-Tool Cutter Holders

The *fly-tool cutter holder* (Figure 13-21A) accommodates square tool bits that cut singly or in pairs. The fly cutter is used to mill large, flat surfaces that otherwise would require large-diameter or -width cutters to machine. Tool settings depend upon three sets of broached holes (Figure 13-21B). The holes in each set are diametrically opposite. The tool bits may be set at 5° positive rake (holes marked P), at 5° negative rake (N), or at 0° rake (S).

Fly-tool cutter holders take the place of more expensive end mills. The holders have the advantage of saving money in sharpening and replacing standard tool bits compared to the expense of using end mills. Another advantage is in the wide surface that may be milled in a single pass with a comparatively lightweight holder and bits. Other designs of holders are available for single, standard tool bits of different sizes.

Shaping Tool Set

Vertical shaping processes often require the use of a *shaping tool set*. The seven common shapes of cutting tools in such a set are illustrated in Figure 13-22. The shapes and sizes in fractional parts of an inch (with metric sizes in parentheses) appear on the line drawing. Angles are given for cutting tools #6 and #7. The three standard tool bit holders for this particular set are listed as #8, #9, and #10. Their sizes are given in inch and metric units.

Boring Head Sets

Precision boring is done on a vertical milling machine by using a *boring head set*. This set includes a boring head, a number of boring bars, other solid boring tools, and a container.

The shanks of boring heads are ground to #30, #40, or #50 NS tapers or to the straight

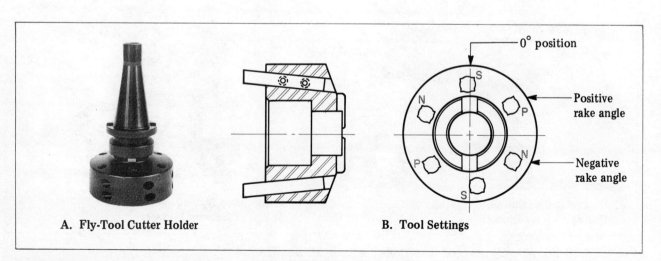

A. Fly-Tool Cutter Holder   B. Tool Settings

FIGURE 13-21 Features of a Fly-Tool Cutter Holder

style (R-8). The boring bit capacity ranges from 1/2" for the R-8 and #30 NS shank sizes to 5/8" for #40 NS and 3/4" for #50 NS.

Boring heads have micrometer dials that permit settings to accuracies of 0.001" and finer (0.0005"). Metric divisions read to 0.01mm. The tenthset boring heads (Figure 13-23) has a direct 0.0001" (0.002mm) vernier-reading adjustment of the bore diameter. The slide and dial are locked simultaneously by a thumb screw. The symmetrical, compact design of the tenthset boring head reduces overhang and provides maximum tool rigidity.

End Mill Holders, Collets, and Adapters

While there are slight variations in construction details among manufacturers of end mill holders, collets, and adapters, three features are standardized: (1) the shank with its taper, (2) the threaded end to receive the drawbar, and (3) the hole to accommodate either straight or taper-shank end mills.

The shank tapers may be #30, #40, or #50; Brown & Sharpe tapers #7, #9, #10, #11, and #12; or the R-8 combination of a straight body and 16°51' included angle taper nose. An *R-8 taper end mill holder* is seen in Figure 13-24. The diameter, taper angle, and spline are shown in the line drawing. This particular end mill holder is available with hole sizes ranging from 3/16" (5mm) to 1" (25mm).

A *Brown & Sharpe taper-shank holder* is heat treated, hardened, and ground. The hole diameters range between 3/16" to 1-1/4" (5mm to 31.75mm). The setscrew in the holder clamps securely against the flat area on the shank of the cutter. There are two setscrews on holders that accommodate 7/8" (22mm), 1" (25mm), and 1-1/4" (31.75mm) diameter mills. Most single-end end mills may be held in this holder.

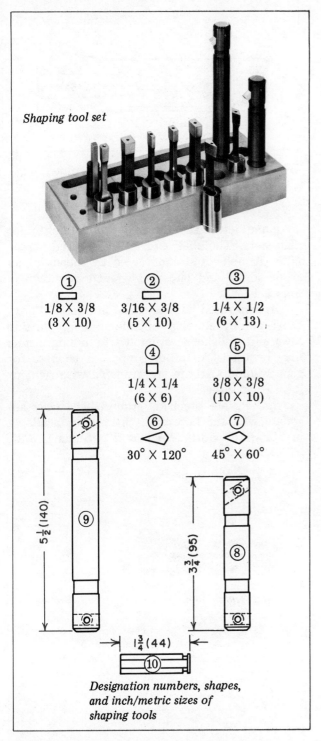

FIGURE 13-22 Common Shapes and Sizes (Inch and Metric) of Shaping Tools

FIGURE 13-23 Tenthset Boring Head with Direct 0.0001" Vernier Reading on Bore Diameter

244 VERTICAL MILLING MACHINES

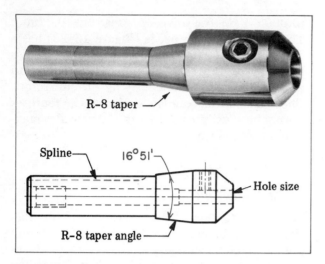

FIGURE 13-24 End Mill Holder (R-8 Taper Type)

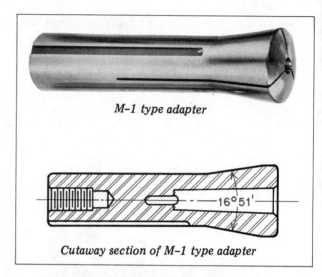

FIGURE 13-26 Vertical Milling Machine Adapter for Taper-Shank Tools (M-1 Type)

Collets for vertical milling machines are available for straight-shank tools in R-8 and N-2 types, Brown & Sharpe tapers of 0.500″ (13mm) taper per foot, and Morse tapers of 0.599″ (15.2mm) taper per foot. One type of collet is shown in Figure 13-25.

The N-2 collet has an included face angle of 20°. The R-8 collet range is from 1/8″ through 3/4″ in increments of 1/64″ (3mm through 19mm in 0.4mm increments). The #7 Brown & Sharpe cutter holds tools from 1/16″ through 1/2″ diameter in 1/64″ (0.4mm) increments.

Adapters are used for taper-shank end mills and drills. These adapters are finished with an R-8 taper. Hole sizes are available for Morse or Brown & Sharpe shanks. An M-1 type adapter and a cutaway section are shown in Figure 13-26. Note that the one end is threaded for a drawbar. Positive drive is provided by the keyseat and key.

Shell end mill holders extend the tooling capability to the use of shell end mills for face and side milling in one operation. These holders are manufactured for a number of different pilot diameters. The shell end mill is secured on the pilot diameter and shouldered by means of an arbor screw and end mill wrench as shown in Figure 13-27.

Arbors (used with the head in a vertical position) permit slotting or side milling. Slotting saws, side mills, or alternate-tooth milling cutters may be used. An arbor support is adapted for gang milling. Arbors are available with right or left thread direction.

Spacers are precision ground. The faces are square with the bore axis. Spacers are available in standard widths such as 2″ (51mm), 3/4″

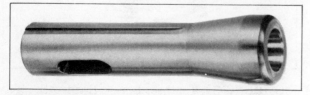

FIGURE 13-25 Collet Design for Vertical Milling Machines

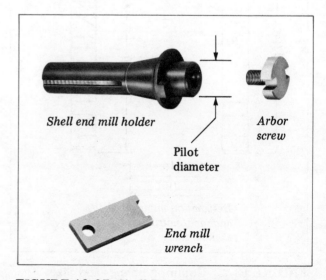

FIGURE 13-27 Shell End Mill Holder

HISTORY AND ACCESSORY DESIGN FEATURES

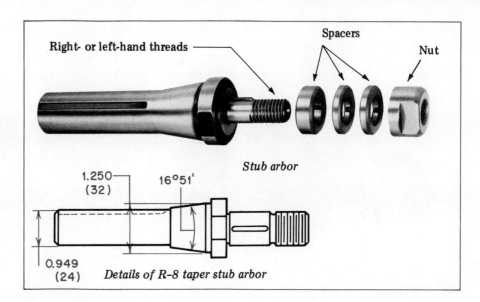

FIGURE 13-28 Features of a Stub Arbor (R-8 Taper Type)

(19mm), 5/8″ (16mm), 3/8″ (10mm), and 5/16″ (8mm).

Stub arbors, as the name implies, are short-length spindle arbors. Figure 13-28 shows the features of an R-8 taper stub arbor. The arbor is provided with narrow-width spacers. Details are illustrated by the line drawing.

VERTICAL MILLING MACHINE WORK-HOLDING AND WORK-POSITIONING DEVICES

Vises

Plain and swivel vises, the same as the vises used on horizontal milling machines, are used on vertical milling machines. The body and base usually are made of gray cast iron. The jaws are heat treated alloy steel. The jaws and the face of the body are precision ground for parallelism and squareness (90° jaw face). The swivel vise is used for angular milling setups through 360°. The base is graduated around the periphery in degrees in ten-minute increments.

The *precision multipurpose vise* (Figure 13-29) is a precision-ground vise with a twin lead screw feature. This feature permits force to be applied uniformly across the entire face of the jaws. The sides of the vise are ground square to within 0.0001″ per inch (0.002mm per 25.4mm). The jaws are square within this tolerance. The vise sides are flat and parallel within 0.0003″ (0.0076mm). The squareness, flatness, and parallelism accuracies simplify the squaring of third and fourth sides of a workpiece. The vise may also be removed to permit workpieces and setups to be inspected and measured. The part and vise may then be remounted if necessary to take any further machining cuts.

Sine Tables

Plain sine tables are used for single angle setups. *Combination sine tables* permit the rapid setting of single and compound angles. Once a table is mounted, the workpiece and table become one rigid unit that may be moved for inspection or additional machining processes on other machines.

The combination sine table in Figure 13-30 becomes a solid angle plate by tightening two screws through the base. The sine bar may be

FIGURE 13-29 Precision Multipurpose Vise

FIGURE 13-30 Combination Sine Table for Single and Compound Angle Setups

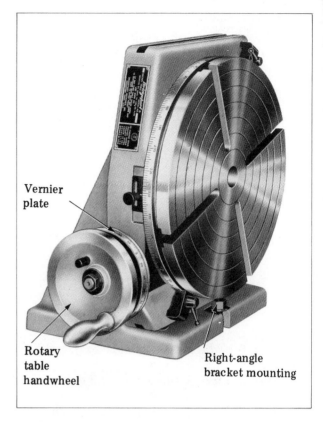

FIGURE 13-31 Rotary Table with Vernier Plate Mounted on Right-Angle Bracket for Working in Vertical Position

detached. It is movable to the horizontal surface of the sine table for single and compound angle setups. The plain sine table has a horizontal table that is hinged to the base. Gage blocks are placed under the sine bar of the table to obtain a precise, correct setting for a required angle.

Sine tables are available in two sizes: 5" (127mm) and 10" (254mm). The horizontal working surfaces are 5" × 7" (127mm × 178mm) and 10" × 12" (254mm × 305mm), respectively. The vertical working surfaces are 3-1/2" (89mm) and 6" (152mm) high. The tables are machined to an accuracy of flatness within 0.001" (0.025mm) and parallelism within 0.002" (0.05mm). T-slots are built into the tables.

Rotary Tables and Attachments

Rotary tables permit a workpiece to be turned in a horizontal plane to precise positions in degrees, minutes, and seconds. With the addition of a right-angle bracket, the rotary table is adapted for work in a vertical position as shown in Figure 13-31. A tailstock is used to accurately support long workpieces between centers.

Some rotary tables are graduated in degrees. An accuracy within 30 seconds of arc is maintained through a complete rotation of the table. There is an adjustable graduated collar on the handwheel. Readings may be taken directly to each minute. Increased accuracy is provided by a vernier plate. Direct readings may be taken with a vernier plate to within 5 seconds of arc.

Other rotary tables have an indexing attachment. The plates for indexing differ from the standard dividing head plate because of the number of holes they contain. The number of holes in the different circles varies from the #1 plate with 159 holes to the #6 plate with 403 holes.

Digital Readout

Table movements along the horizontal (X) axis and the cross slide (Y) axis may be viewed rapidly on a *digital readout* accessory. The dual readout panel has a visual readout (separate display) for the X and Y axes (Figure 13-32). The readouts may be in the inch or metric system. Inch/metric conversion is optional.

Digital readout units are accurate within ±0.0005" (0.01mm). Readings are displayed in color for ease of reading and viewing from any

HISTORY AND ACCESSORY DESIGN FEATURES 247

FIGURE 13-32 Digital Readout Panel with Display for X and Y Axes

working position at the machine. The axis display shows a + or − directional sign, a decimal point, and positions for six-digit readings.

SAFE PRACTICES USING VERTICAL MILLING MACHINES, ATTACHMENTS, AND ACCESSORIES

—Check to see that all machine guards are in place before starting the machine.
—Put coolant splash panels in place. Make sure the channels and end-of-table drains permit the return flow of coolant to the reservoir.
—Wipe any coolant or oil spills in the machine area. Apply an absorbent floor compound to prevent slipping.
—Remove chips when the machine is stopped. Use an appropriate rake and hard-bristle brush.
—Store chips in leakproof containers and dirty wiping cloths in closed containers.
—Secure assistance for mounting heavy machine attachments and/or difficult-to-handle workpieces. Keep the back straight and vertical. Lift with the force exerted by the legs.
—Check the surface condition of the outside diameter of the quill and the inside bore of any attachment that is to be used. Recheck to see that the attachment is properly seated and securely held.
—Exercise care when swiveling the spindle head for angular cuts. Unless the head is counterbalanced or adjusted through gearing, the clamping bolts should be retightened lightly so that the spindle head may be moved when a force is applied to it. Once the head is positioned, the clamping bolts are tightened to secure the head in position.
—Shut down the spindle to check the bore for nicks or burrs. Stone, if necessary.
—Check the mating taper of the spindle and the shank taper of an adapter, collet, or other tool-holding accessory. Remove any burrs.
—Recheck the seating of the workpiece on the positioning or work-holding accessory. Make sure the workpiece is secure before starting the machine.
—Secure cutting tools in the spindle by using an appropriate type and size of collet and/or adapter, when required.
—Select the cutting speed that is appropriate for the operation, size of cutter, shape of the workpiece, and results in the required dimensional accuracy. Set the spindle RPM to the closest cutting speed.
—Determine and set the power down-feed or up-feed that is correct for the setup and operation. The feed must permit machining to the required dimensional and surface finish accuracies.
—Shut down the spindle before any measurements are taken.
—Leave the machine clean. Replace cutters, accessories, and attachments in proper storage areas.
—Check each adapter, collet and arbor, as well as collars and all other accessories, after each setup. Remove any burrs or nicks before the accessories are stored.
—Use a wiping cloth as protection when holding a heated or heavy cutter.
—Dress appropriately. Avoid loose-fitting or long sleeve clothing. Remove personal accessories, such as jewelry and wristwatch, that are dangerous around moving equipment.
—Select and use comfortably fitting safety goggles and/or a protective face shield.
—Use protective footwear.

TERMS USED WITH VERTICAL MILLING MACHINES, ATTACHMENTS, AND ACCESSORIES

Vertical milling machine applications	General milling, drilling, and boring processes for machining flat, angular, circular, and rounded surfaces and contours to full or partial depth. Machine cutting in one, two, or three planes with a cutter that is mounted vertically or at a single or compound angle in a vertical-spindle machine.
Vertical milling machine attachments	Mechanical, pneumatic, optical, or electronic devices, mechanisms, and movements that extend the versatility of a machine. Units that generally are actuated by the turning of the spindle. (Examples of attachments include high-speed, right-angle and multiple-angle, and cross slide heads.)
Vertical milling machine accessories	A group of work-positioning and -holding devices such as plain and combination sine tables, rotary tables, and dividing heads with gear trains. Cutter- and tool-holding devices.
Mode	The positioning of the components of a machine tool to permit another combination of processes to be performed—for example, changing from the vertical mode to the horizontal mode on a vertical milling machine. Conversion of a vertical milling machine (by using attachments) from basic vertical milling processes to serve also as a horizontal milling machine.
Tracer milling	Simultaneous control of cutter movements in one, two, or three planes to reproduce a desired contour or three-dimensional form.
Bed-type vertical milling machine	A heavy-duty, primarily manufacturing type, vertical milling machine. A vertical miller that permits longitudinal and transverse movement of the workpiece. (Vertical movement is controlled by adjusting the vertical head that may also be swiveled.)
Turret assembly (vertical milling machine)	Mechanisms mounted on top of the column. Turret components that may be swung as a unit through 360° in a horizontal plane, 180° in a vertical plane parallel to the column face, and at a single or compound angle toward or away from the column.
Manual feed controls	Handwheels to control the horizontal movement of the table and the cross feed movement of the saddle and table and a crank to control the vertical movement of the knee.
Infinitely variable spindle speeds	Versatility to set the spindle speed at any RPM within the minimum and maximum range of speeds.
Spindle head	The mechanism for driving cutting tools that are mounted in a spindle. A major vertical milling machine component consisting of a power source, speed and feed controls, a quill for positioning cutters and holding attachments, and a spindle.
Quill feed controls	Handwheels or a hand lever for manually feeding a quill. A feed control lever for engaging the quill power feed and the power feed change lever.
Vertical shaper head	An attachment for a vertical-spindle milling machine. A device that changes rotary motion to straight-line (reciprocating) motion. An accessory that permits vertical and single or compound angle shaping and slotting.

HISTORY AND ACCESSORY DESIGN FEATURES 249

Power down-feed attachment	A mechanism for controlling the up- and down-feed rates of the quill.
High-speed head	A spindle head attachment that provides high rotational speeds for effectively and efficiently machining with small-diameter drills and end mills.
Right-angle attachment	A spindle head attachment used in confined spaces for angular, horizontal, slotting, and other milling or drilling processes. A device for machining internal or external surfaces in minimum working areas.
Quick-change tooling system	A wide selection of cutter-holding devices and adapters that permits maximum tool changing flexibility. Holders and adapters to accommodate a wide variety of tools, tool sizes, and shank designs.
Tenthset boring head	A compactly designed boring head having a direct vernier reading. A boring head that permits boring diameter adjustments of 0.0001" (0.002mm).
Multipurpose vise	A precision-ground vise with twin lead screws that uniformly apply force across the faces. A vise with all of its sides precisely ground for squareness, parallelism, and flatness.
Digital readout	An attachment that visually displays table movements along X and Y axes in + or − directions and in inch or metric units of measure.

SUMMARY

Vertical milling machines of the 1860s combined the vertical-spindle feature of the drill press with the longitudinal and transverse (cross slide) movements of the horizontal milling machine table. The table was mounted on a fixed base.

> Early vertical-spindle designs provided for increased rigidity to withstand heavy side cutting forces.

The redesigning in 1883 to include the knee-and-column feature permitted raising and lowering the table in relation to the column.

> The angle-setting head added machine tool capability to machine angular surfaces. A front-to-back swivel head design later made compound angle settings and milling processes possible.

The three general power quill feeds range from 0.0013" to 0.004" (0.03mm to 0.10mm) per revolution.

> Vertical-spindle milling machines are designed for standard and rapid traverse power feeds.

Design changes in the early 1900s included power table and quill feeds. Later, electrical, pneumatic, electronic, and optical measuring and speed and feed controls were added.

> The turret assembly of the standard general-purpose vertical milling machine may be swung through 360°. The vertical head may also be swiveled from 0° to 180° in relation to the 360° horizontal plane. An angle bracket permits swinging the head at an angle toward or away from the column.

Spindle speeds are provided through variable-speed drives and other drives that have a fixed number of speeds in a low-speed and a high-speed range.

Quills are rigidly designed to offset heavy cutting forces. Attachments are also secured to the outside diameter. Depths may be accurately set by micrometer stops.

Speeds are regulated by a variable-speed control handwheel and a change lever for the high- and low-speed ranges.

Handwheels for positioning and feeding (vertically, longitudinally, and transversely) are conveniently located on the front of the machine.

The combination vertical/horizontal milling machine with support arm for the horizontal mode is used for general milling machine processes. In addition, drilling, boring, slotting, and straddle milling cuts may be taken by conventional or climb milling.

The tracer milling machine is primarily a two- and three-plane duplicating machine that follows a preestablished pattern.

The bed-type vertical milling machine is best adapted for heavy machining operations and production manufacturing. The bed is fixed. Vertical movement and feed is provided by the spindle head.

The numerically controlled vertical milling machine is activated from stored machine and tooling setup specifications. The machine may be numerically controlled (NC) or computer numerically controlled (CNC).

The major design components of a vertical milling machine are the column and base; knee, saddle, and table; turret head and ram; spindle (tool) head; feed drives and controls for manual, power, and rapid traverse; and cutting fluid and lubricating systems.

The spindle head is designed to hold cutters, provide the rotary cutting force, house the up- or down-feed control mechanism, and turn the spindle in a quill.

Attachments include the vertical shaping head; power up- and down-feeds; high-speed, right-angle, and rotary cross slide milling heads; direct-reading optical measuring system; and the inch- or metric-standard digital readout system.

Quick-change tool systems permit fast changing of cutting tools. Each system includes a variety of designs and sizes of chucks, collets, holders, adapters, boring heads and bars, and a chuck locking fixture.

Fly cutters are used to machine large, flat surfaces.

Boring heads and tenthset heads permit diametral settings for boring within 0.0001" (0.002mm).

Adapters for shank tapers accommodate NS tapers, B & S tapers, the R-8 straight body and 16°51' taper nose, N-2, and other standard tapers.

Vertical-spindle arbors, stub arbors, and collars are used for standard-hole types of milling cutters.

Plain and combination sine tables position and hold workpieces at single or compound angles, respectively.

Rotary tables position, hold, and index angular work surfaces in a horizontal or vertical plane.

Reporting X and Y axes (+ or -) table movements may be viewed on a digital readout panel attachment.

General and specific safety precautions must be observed in vertical milling machine work. These precautions deal with personal work habits; machine safety checks; the correct mating and fitting of adapters, chucks, collets, and so on; the safe disposal of chips; cleanliness of the cutting fluid; machine cleaning and personal hygiene; and machine operation and shutdown practices.

UNIT 13 REVIEW AND SELF-TEST

1. a. Identify two early (1860 to 1910) machine design features of vertical milling machines.
   b. State the importance of each feature.
2. List four significant design changes in vertical milling machines that have increased its versatility.
3. Give three functions of the quill on a vertical milling machine.
4. State three functions of the spindle on a vertical milling machine.
5. a. Identify two different types of attachments for vertical milling machine spindle heads.
   b. Describe briefly the function of each attachment.
6. Name six different cutter holders or adapters.
7. Tell what the difference is between an NS shank taper holder and an R-8 shank holder.
8. a. Indicate two design features of a multipurpose vise that differ from standard plain or swivel types of milling machine vises.
   b. State the advantage provided by each design feature.
9. a. Explain briefly the purpose served by a digital readout system.
   b. Cite two advantages of a digital readout system.
10. State two vertical milling machine safety precautions to observe before turning on the power for the spindle.

SECTION TWO

General Vertical Milling Processes

Basic vertical milling processes include:

- Milling straight, flat, angular, beveled, and round surfaces in one or two planes;
- Milling continuous flat, angular, beveled, and curved formed surfaces in two or more planes;
- Blind- and through-slotting, grooving, dovetailing, and keyway cutting;
- Hole-forming processes of drilling, reaming, counterboring, countersinking, tapping, and boring;
- Shaping splines, keyways, and other straight surfaces with a shaping head.

This section deals with the cutters used for these processes. Considerations in selecting end mills are covered, as well as the step-by-step procedures and safe practices for the milling of flat, angular, beveled, and round surfaces; stepped surfaces; and slots, dovetails, and keyways. The section also deals with the cutting tools, machines, attachments, and setups for general hole-forming and basic shaping processes on the vertical milling machine when ordinate, tabular, or baseline dimensioning is used.

UNIT 14

Basic Vertical Milling Cutting Tools, Speeds and Feeds, and Processes

The cutters for the basic vertical milling processes are the standard cutters used for horizontal milling and other general machining processes. While high-speed steel cutters are common, cobalt high-speed steel, solid carbide, and carbide-tipped tools are widely used. The carbide tools provide for maximum production, efficient chip removal, and high abrasion resistance. The cutting edges are ground to a high finish for maximum performance and limited machine downtime.

This unit describes design features of end mills and the cutting speed and feed rates for end milling processes. There is a growing need for end mills in improved programming efficiency on tape-controlled milling processes. Changing demands to mill difficult-to-machine materials and complex parts are creating the need for end mills with optimum cutting qualities and a minimum of wear under heavy production conditions.

DESIGN FEATURES OF END MILLS

Like all other cutting tools, the main design features of end mills are designated by technical terms, as shown in Figure 14-1 by the end view and bottom view of a four-flute end mill. Here, the tooth face is ground with a *positive radial rake angle*. End mills are also designed with a *negative radial rake angle*. End mills are formed with *straight* or *spiral teeth* and with *right-* or *left-hand helix angles*. The angles may be *standard* for general-purpose milling or *high* for heavy-duty production milling.

CONSIDERATIONS IN SELECTING END MILLS

Single-end end mills are general-purpose mills. *Double-end* end mills are more economical production milling cutters. Fractional sizes of end mills up to 1″ (25mm) are standard stock items. Larger sizes are also commercially available, but in limited sizes. Figure 14-2 illustrates *stub*, *regular*, and *long-length* double-end end mills.

Sub-Length, Small-Diameter End Mills

Stub-length, small-diameter end mills combine positive rake angles on the flute face, primary relief angles, and short-length teeth. These end mills are designed to machine efficiently at the higher speeds that are required for their small diameters. A medium-high helix angle produces a shearing type of cut that is essential to quality performance when small-diameter end mills are used.

The flutes on stub end mills should be long enough to mill the required surface, but no longer than necessary. Most small-diameter end mills are used until they become broken or dulled. Regrinding of very small diameters is not feasible. However, the larger sizes of small-diameter end mills may be reground because these sizes have a raised margin and primary side relief.

A feed of approximately 0.0005″ (0.01mm) per tooth is recommended for general-purpose

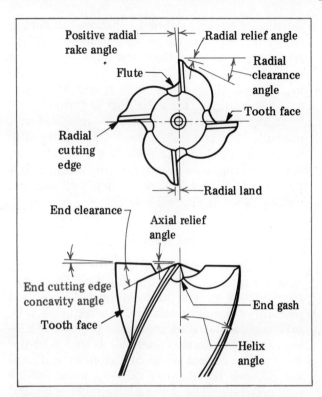

FIGURE 14-1 Four-Flute End Mill Showing Design Features and End Mill Terminology

end milling with small-diameter, high-speed steel end mills. Manufacturers' and handbook tables with other technical data for cutting speeds, RPM, and feeds are described later.

Square- and Ball-End Multiple-Flute End Mills

Square-end, four-flute, single- and double-end end mills are used for general-purpose end milling,

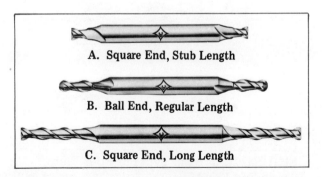

FIGURE 14-2 Double-End (HSS) End Mills

slotting, stepping, slabbing, shallow pocketing, tracer milling, and die sinking processes. When ground to a *ball end*, these end mills are adapted to center cutting for processes such as die sinking, fillet milling, tracer milling, and other processes requiring a radius (ball) to be formed.

Long and *extra-long*, square- and ball-end, multiple-flute end mills permit deep cavity milling due to the long flute (cutting tooth) length. An extra-long, square-end, four-tooth end mill is pictured in Figure 14-3A. A long, ball-end end mill is shown in Figure 14-3B.

High-Helix End Mills

High-helix end mills provide the shear cutting action necessary for producing fine finish surfaces. These general-purpose end mills are used primarily for milling nonferrous workpieces. The rake angle and cutting edge relief permit the milling of soft, nonferrous materials as well as harder alloys. The cutting edges of high-helix end mills are relieved to eliminate buildup when milling soft materials. The cutting edge relief is ground to retain maximum strength and to aid in dissipating the heat behind the cutting edges.

The high-helix angle ensures positive chip ejection in regular and deep pocketing milling applications. When these end mills are used as *slabbing mills* for heavy stock removal, additional support for slabbing cuts is sometimes obtained by using a center in the center hole of the end mill.

Keyway-Cutting End Mills

Keyway-cutting end mills have the same construction details as general-purpose end mills. One difference is that the outside diameter is ground to a tolerance of +0.0000"/-0.0015" (+0.00mm/-0.04mm). The undersize diameter compensates for the tendency of the end mill to cut oversize. The -0.0015" (-0.04mm) permits the end mill to cut a keyway to a precise, nominal size.

Tapered End Mills

Tapered end mills are constructed with cutting edges formed at standard included angles of 10°, 14°, and 15°. The ends are usually ball shaped. The taper end mill is adapted to machining angle surfaces at 5°, 7°, and 7-1/2° and for die sinking processes that require the sides of the relieved surfaces to be cut at an angle.

Cobalt High-Speed Steel End Mills

Cobalt high-speed steel end mills are used for heavy-duty center cutting of high-tensile steels and high-temperature alloys. The extra-heavy-duty end mill in Figure 14-4A has a short flute length, which allows for greater rigidity and less deflection during the cutting of difficult-to-machine materials.

Multiple-flute cutters have the advantage of a smooth, chatter-free cutting action and less cutting edge wear in comparison with two-flute cutters. One of the newer designs for a multiple-flute *roughing end mill* is illustrated in Figure 14-4B. This cobalt high-speed steel end mill is designed for machining low-alloy, high-strength steels and die steels at accelerated feeds and speeds. The small-radius thread form cutting edges permit each flute to cut smaller and thicker chips. The chips are easily carried away from the work.

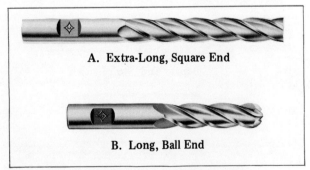

FIGURE 14-3 Basic Types of Four-Flute, Single-End End Mills

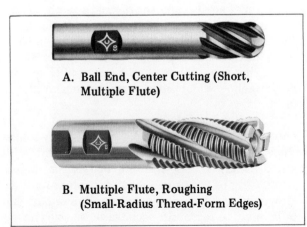

FIGURE 14-4 Heavy-Duty, Cobalt High-Speed Steel, Single-End End Mills

Rapid chip removal allows roughing end mills to be set for deeper depths of cut. The rugged design and chip-breaking characteristics also mean that less power is required to remove large amounts of stock than is required with conventional (HSS) end mills. Since the feed/speed/stock removal relationships are less critical than for regular end mills, roughing end mills are widely used for NC machining operations. In summary, cobalt high-speed steel roughing end mills have the following advantages:

—High material removal rate,
—Minimum horsepower requirement,
—Ability to take heavier cuts at high speeds and with less chatter and vibration,
—Less deflection for the same stock removal rate than conventional end mills,
—Excellent heat dissipation (because of the tooth-form design) under heavy machining operations,
—Fast chip-breaking characteristics,
—Simplified cutter resharpening without affecting the tooth profile.

TYPES OF END TEETH

Two-flute end mills have *center-cutting* end teeth (Figure 14-5A). These end mills are used for plunge and traverse cuts. The end may be ground to a ball (radius) shape or other form.

Multiple-flute end mills (with more than two teeth) are designed with a gashed end or a center-cutting end. *Gashed-end* end teeth (Figure 14-5B) are cut to the center hole or counterbore. The end may be ground square, to a radius, or to another form.

Gashed-end multiple-flute end mills are recommended for the conventional milling of slots and pockets where plunge cutting is not required. Center-cutting, multiple-flute end mills are designed for plunge cutting where a fine surface finish is required. The two center-cutting teeth (Figure 14-5C) cut to the center. Plunge cutting is used as the feeding method in slabbing, pocketing, and die sinking operations.

END MILL HOLDERS

The shank and hole in end mill holders are concentrically ground to ensure maximum machine and cutter efficiency. Possible cutter runout and chatter are reduced to a minimum. End mills are rigidly secured on center because the shank diameters and setscrew flat widths are ground to close tolerances. Holding tools concentrically and rigidly results in less tool breakage, reduced tool wear, and the production of high-quality surface textures.

Some holders are designed for directly mounting and driving end mills. An example of a *solid end mill holder* is shown in Figure 14-6A. This holder has a Brown & Sharpe taper shank. The taper shank is designed with a tang. The holder is also furnished with a tapped shank hole to receive the threaded end of a draw-in bolt.

Similar holders with draw-in and tang shanks are available with American National Standard tapers. These holders accommodate single- and double-end end mills. Other holders have taper shanks to fit #30, #40, and #50 milling machine tapers. These holders are made of tough, heat-treated alloy steel. The setscrew holes are accu-

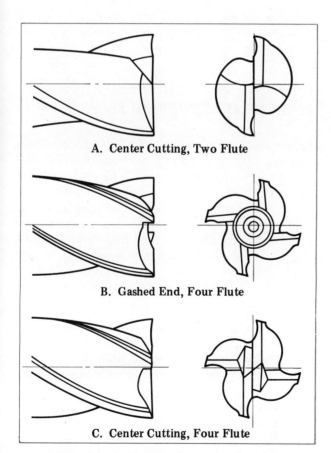

FIGURE 14-5 Center-Cutting and Gashed-End End Teeth on End Mills

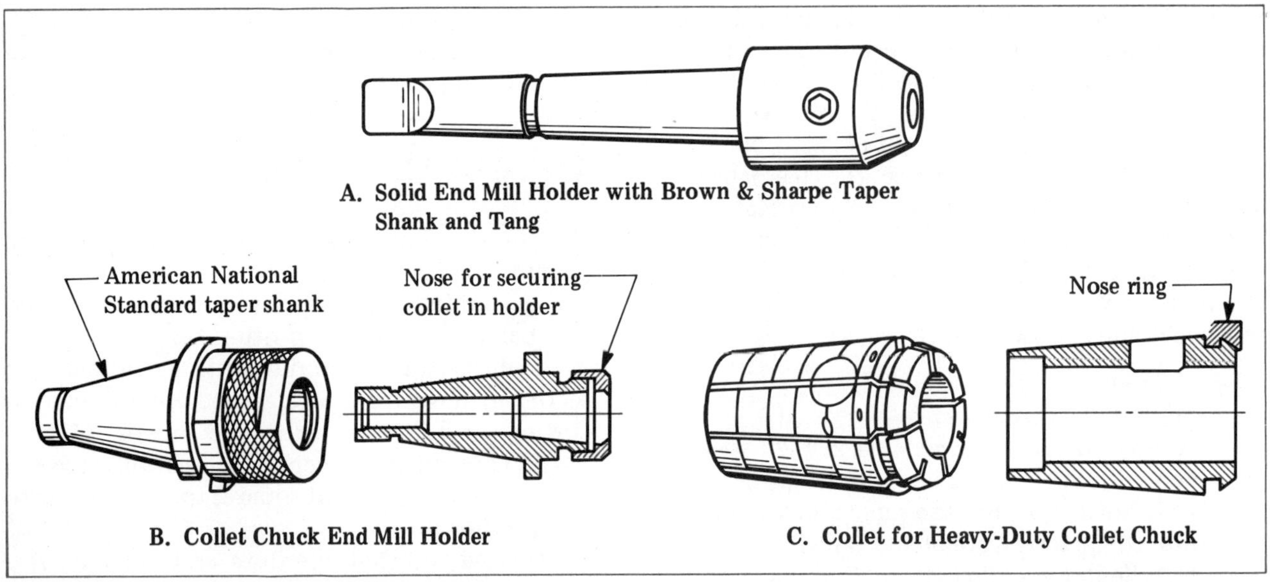

FIGURE 14-6 Types of End Mill Holders

rately located to seat in the close tolerance flats on end mills. The holder design in Figure 14-6B requires a collet and is called a *collet chuck end mill holder.* This holder provides precise mating between the end mill shank and the collet. The collet, shown in Figure 14-6C, has a retaining pin feature that prevents any movement of the end mill during a cutting operation.

CUTTING FLUIDS FOR END MILLING

Cutting fluids are used with end mills to protect the tool and workpiece by dissipating heat that is generated. The fluids also serve the function of lubrication. Mineral, sulphur-base, and soluble oils, as well as water-base synthetics, are common cutting fluids. Water-base cutting fluids have excellent cooling qualities; oil-base cutting fluids produce a high-quality surface finish.

Cutting fluids are especially necessary to clear the large volume of chips produced by carbide cutters.

Note: The teeth on a carbide end mill will chip when milling cast iron or steel unless there is a *continuous supply* of cutting fluid *under pressure.*

In the case of cast iron, compressed air is used. Brass and plastics are also machined dry. The air supply disperses the chips from the cutting area and cools the end mill.

Note: Safety guards or protective chip shields must be placed around the table and cutting area to prevent chips and other foreign particles from flying.

CUTTING SPEED AND FEED RATES FOR END MILLING PROCESSES

Efficient machining with end mills and other hole-forming cutting tools requires considerable judgment on the operator's part. The cutting speed for end milling processes is established from tables of technical data. The data have been gathered over many years by cutting and machine tool manufacturers, designers, and engineers. The cutting speeds found in tables are *suggested starting points.* The speeds may be increased or decreased depending on the following variable conditions:

- —Material being machined,
- —Material of which the cutter is made,
- —Type and condition of the machine itself,
- —Nature of the machining process,
- —Rigidity of the setup and design of the workpiece,
- —Composition and application of the cutting fluid,
- —Required quality of surface texture.

After the cutting speed has been selected for the diameter of cutter to be used, the feet per minute (or equivalent mm/min or M/min) is converted to RPM. The cutting process is then carefully observed. The operator determines whether a cutting speed adjustment is to be made.

Establishing the Feed Rate

Efficient machining also requires appropriate feed rates. Again, the operator uses tables that provide suggested starting feed rates. The feed is expressed as a *chip load per tooth*. The feed is expressed in thousandths of an inch or hundredths of a millimeter. The feed rate for end mills varies for different materials, cutting speeds, and depths of cut.

The feed rate (F) for milling operations is the product of the chip load per tooth (f) multiplied by the number of teeth (n) in the cutter and by the RPM. The feed rate is in inches per minute (ipm) or mm/min. For example, if the chip load per tooth on a 1/4" (6mm) diameter end mill is 0.0005" (0.0125mm) and the spindle speed is 4,000 RPM, the feed rate is 8 ipm (200 mm/min):

$$F = f \times n \times RPM$$
$$= 0.0005 \times 4 \times 4,000 = 8 \text{ ipm}$$

or, in metric units,

$$= 0.0125 \times 4 \times 4,000 = 200 \text{ mm/min}$$

The depth of cut for end milling operations, as a general rule, should be limited to approximately one-half the diameter of the end mill. When this depth is exceeded, the feed rate must be reduced. Otherwise, the cutter teeth and cutter may fracture due to the excess force beyond the breaking point of the end mill. The depth of cut is limited by: (1) the amount of material to be removed; (2) the power available at the machine spindle; and (3) the rigidity of the setup, cutting tool, and workpiece design.

Accuracy in Groove Milling. Inaccurately formed grooves are produced when excessive feed rates are used with end mills. The added cutting force causes the end mill to be deflected from its axis. The sides of the slot that is milled are not perpendicular to the spindle and work axis. The problem is corrected by reducing the feed rate. The distance the end mill projects from the spindle should also be checked and reduced, if possible. An inaccurate groove or slot is also produced when a dull cutter or a worn spindle is used.

INTERPRETING CUTTING SPEED AND FEED DATA FROM HANDBOOK TABLES

The nature of the technical information contained in a table of starting cutting speeds, RPM, and feeds for end mills for selected materials is illustrated by Table 14-1. The full table appears in the Appendix. The left column gives a range of diameters from 1/16" through 3" (1.6mm through 76mm). The separate columns identify different materials, recommended types and styles of end mills, cutting speeds in sfpm, and feed.

In Table 14-1, the low value of the cutting speed range (in sfpm) is primarily for roughing cuts; the high value, for finish cuts. Corresponding ranges are given for chip load per tooth. For example, assume a part made of machine steel in the hardness range of 20-30 on the Rockwell C scale is to be rough and finish machined with a 1/2" (12.5mm) end mill. The information the operator obtains from the table is as follows:

—A high-speed steel end mill with two or more flutes is recommended;
—The cutting speed for end milling ranges from 60 sfpm for roughing cuts to 80 sfpm for finish cuts;
—The feed rate (chip load per tooth) ranges from 0.001" (0.02mm) for finish cuts to 0.003" (0.08mm) for roughing cuts;
—The spindle RPM for the roughing cuts, which equals the surface feet per minute (60 × 12) divided by the circumference of the cutter in inches (1/2 × π), is 458 RPM; for finishing cuts, 611 RPM.

It must be emphasized that each speed, feed, and RPM in the table is a *starting point*. The operator determines when each value should be increased or decreased.

BASIC VERTICAL MILLING MACHINE PROCESSES

The greatest application of the vertical milling machine is in the machining of flat external and internal surfaces. These surfaces may be milled parallel, perpendicular, or at a single or compound angle to the face of the table. Grooving, slotting, and the cutting of T-slots, dovetails, and keyways

TABLE 14-1 Examples of Suggested Starting Speeds and Feeds for High-Speed Steel End Mill Applications

Diameter of End Mills (in Inches)	Materials			
	Machine Steel, Hard Brass and Bronze, Electrolytic Copper, Mild Steel Forgings (20-30C)		Brass, Bronze, Alloyed Aluminum, Abrasive Plastics	
	Types and Styles of End Mills			
	High-Speed Steel End Mills, 2 or More Flutes		High-Speed Steel End Mills of High-Helix Type, 1 or 6 Flutes	
	Speed 60-80 sfpm	Feed	Speed 100-200 sfpm	Feed
	RPM	Chip Load per Tooth	RPM	Chip Load per Tooth
1/16	3667-4888	.0002-.0005	6111-12222	.0002-.0005
3/32	2750-3259	.0002-.0005	4073-8146	.0002-.0005
1/8	1833-2440	.0002-.001	3056-6112	.0002-.001
3/16	1222-1625	.0002-.001	2037-4074	.0002-.001
1/4	917-1222	.0005-.002	1528-3056	.0005-.002
5/16	733-978	.0005-.002	1222-2444	.0005-.002
3/8	611-815	.001-.003	1019-2038	.0005-.003
7/16	524-698	.001-.003	873-1746	.0005-.003
1/2	458-611	.001-.003	764-1528	.0005-.003
9/16	412-543	.001-.004	678-1356	.0005-.004
5/8	367-489	.001-.004	611-1222	.0005-.004
11/16	337-444	.001-.004	555-1110	.0005-.004
...	...	...	...	...
2	115-153	.001-.004	191-382	.0005-.004
2-1/8	108-144	.001-.004	179-358	.0005-.004
2-1/4	102-136	.001-.004	170-340	.0005-.004
2-3/8	97-128	.001-.004	161-322	.0005-.004
2-1/2	92-122	.001-.004	153-306	.0005-.004
2-5/8	88-116	.001-.004	145-290	.0005-.004
2-3/4	83-111	.001-.004	139-278	.0005-.004
2-7/8	80-106	.001-.004	132-264	.0005-.004
3	76-102	.001-.004	127-254	.0005-.004

are common vertical milling processes. The setups and procedures for each of these basic processes follows.

HOW TO MILL A FLAT SURFACE WITH AN END MILL

End Milling a Horizontal Plane Surface

STEP 1 Select an end mill with a diameter and cutter direction appropriate to the job requirements.

Note: It may be necessary to take more than one cut if the width of the surface is greater than the cutter diameter. Large-width surfaces are usually milled with a shell end mill.

STEP 2 Secure the end mill in an end mill holder.

Note: The end mill is held by the fingers of one hand while the draw-in bar is engaged and turned with the other hand (Figure 14-7). A slight force is used to seat the holder and spindle tapers securely.

Note: If a tang end holder with setscrew is used, the holder is inserted and secured in the spindle. Then, the end mill is mounted

CUTTING TOOLS, SPEEDS AND FEEDS, PROCESSES

FIGURE 14-7 Securing an End Mill in a Collet

and the setscrew is turned to seat solidly against the flat area on the end mill.

STEP 3 Set the spindle head at the 0° vertical position.

STEP 4 Locate and secure the workpiece in a workholding device on the table. The surface to be milled should be positioned parallel with the face of the table.

STEP 5 Determine the correct cutting speed for the cutter material, the nature of the operation, and the depth of cut. Set the spindle RPM to produce the required cutting speed.

STEP 6 Set the horizontal table feed travel rate.

STEP 7 Center the workpiece and cutter if the face can be milled with a single pass. Otherwise, position the workpiece so that a minimum number of cuts are taken to mill the flat surface.

Note: A shell end mill, having a larger diameter than the width to be milled, is often used instead of an end mill. The use of a shell end mill overcomes the disadvantage of having to take overlapping cuts.

STEP 8 Lock the spindle for maximum rigidity. Start the spindle and flow of cutting fluid.

STEP 9 Raise the knee until the cutter just grazes the workpiece. Set the micrometer collar on the knee at zero.

STEP 10 Clear the end mill and workpiece. Raise the table to take the first cut.

Note: Roughing and finish cuts are taken when a considerable amount of material is to be removed.

STEP 11 Lock the knee. Hand-feed the table. Mill a small area.

STEP 12 Move the cutter away from the workpiece. Measure the overall height to check the dimension.

STEP 13 Make whatever adjustment is needed to cut to the required depth. Then, engage the power feed.

STEP 14 Return the table to the starting position if the cutter is to be moved to mill a remaining area or if a second roughing or finish cut is required.

STEP 15 Take successive cuts until the surface is machined to the required dimension.

STEP 16 Brush away all chips. Remove the workpiece. File the burrs.

STEP 17 Release the end mill carefully. Return each tool to its proper storage area. Clean the machine.

End Milling a Vertical Plane Surface

STEP 1 Select an appropriate side-cutting regular, long, or extra-long end mill.

Note: The cutter length should permit machining the required width in one pass.

STEP 2 Mount the end mill so that it is held as close to the spindle as possible (Figure 14-8).

Note: The direction of the end mill teeth must be checked. The hand must correspond with the spindle and feed directions.

STEP 3 Set the spindle RPM to produce the recommended cutting speed (sfpm or mm/min). Set the feed rate.

STEP 4 Position the cutter vertically at 0° to mill the required area of the workpiece.

Note: The setup should be checked for minimum tool overhang and adequate clearance between the setup, quill, and workpiece.

STEP 5 Move the revolving end mill into position for the first roughing cut.

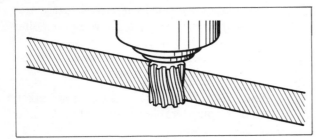

FIGURE 14-8 End Mill Mounted Close to Spindle Head for Side Milling

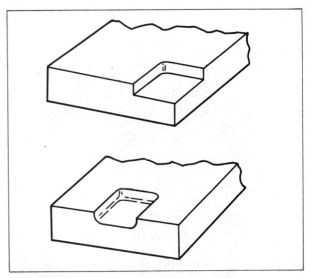

FIGURE 14-9 Two Examples of End Milled Stepped Sections

STEP 6 Hand-feed the end mill to take a cut for about 1/16" (1mm to 2mm) past the center of the cutter. Move the end mill clear of the workpiece.

STEP 7 Stop the spindle. Take a measurement. Make whatever size adjustment is necessary.

STEP 8 Feed the end mill by hand. When all machining conditions have been checked, engage the power feed.

STEP 9 Stop the spindle. Return the cutter to the starting point. Move the cutter into position for other roughing or a finishing cut.

STEP 10 Follow standard procedures for cleaning chips from the machine and workpiece, removing the end mill, and burring.

HOW TO END MILL AN EXTERNAL STEPPED SURFACE

Milling a Right-Angle Step

STEP 1 Lay out the stop or multilevel surface that is to be milled.

STEP 2 Select an appropriate end-cutting end mill.

Note: A stub or regular-length end mill is used to obtain maximum cutting efficiency and cutter strength.

STEP 3 Position the end mill to cut to the required width and position it vertically for depth.

Note: Wide, deep steps may require one or more roughing cuts and a finish cut.

STEP 4 Take a trial cut. Move the end mill horizontally away from the workpiece. Stop the machine. Measure the width and depth.

STEP 5 Make whatever width or height adjustments are needed. Lock the quill and knee.

STEP 6 Engage the power feed and take the cut.

STEP 7 Repeat steps 3 through 6 for subsequent roughing and finish cuts.

Note: When more than one surface is to be milled, repeat the steps for setting the end mill to depth and width.

Milling a Section (Pad)

STEP 1 Lay out the length, width, and depth of the section to be end milled. (Figure 14-9 shows two typical examples of workpiece sections that are commonly end milled.)

STEP 2 Mount the workpiece securely. Check the layout lines to be sure the setup permits the section to be milled to depth.

STEP 3 Select the shortest- and largest-diameter end mill for maximum strength. Mount the end mill. Check to see that the cutter and spindle will clear the workpiece when the final cut is taken.

STEP 4 Position the end mill to depth to take the first roughing cut.

Note: Locate and/or feed the end mill to mill to the required depth, width, or length by using the layout lines or the micrometer collar graduations on the quill.

CUTTING TOOLS, SPEEDS AND FEEDS, PROCESSES

STEP 5 Move the table transversely so that the cutter is positioned to cut to width at the layout line for the length.

Note: Allowance is made if considerable material is to be removed and roughing and finish cuts are to be taken.

STEP 6 Check the cutter depth and the length of the milled area. Then, continue to feed the end mill in until the width is reached.

STEP 7 Proceed to hand-feed the table longitudinally to the next layout line. This line marks the length of the surface to be milled. Lock the table in position at this point.

STEP 8 Feed the end mill out to produce an L-shaped groove when milling a corner pad (stepped surface). A U-shaped groove is milled when the pad location is between the ends of a workpiece.

STEP 9 Remove the excess material from the area between the grooves.

STEP 10 Repeat steps 4 through 9 when the depth of the stepped area requires roughing and finish cuts.

HOW TO END MILL AN ANGULAR SURFACE

Positioning the Workpiece at an Angle

STEP 1 Position the workpiece so that the angular surface may be milled with a square-end end mill held at the 0° vertical axis.

Note: When the angle is laid out, the workpiece is positioned in a vise so that the layout line is horizontal with the table. A surface gage is usually used. Sine plates and tables are used for positioning a workpiece at a precise angle.

STEP 2 Select an end mill that is appropriate for the operation and material.

STEP 3 Mill the angle surface by using the same procedures as for milling a flat surface.

Note: Care must be taken when the first cut is made. If it is started at the small end of the triangle formed by the tapered surface and the horizontal plane, the cut gradually increases. The cutter depth and feed may need to be changed to compensate for this condition.

STEP 4 Take successive cuts to rough and finish mill the angle to size.

Setting the Spindle Head at an Angle

STEP 1 Position and lock the spindle head at the required angle (Figure 14-10).

Note: The cutting action determines whether the head is to be swung parallel with or at a right angle to the column. Small angular surfaces are generally machined by end cutting with a square-end end mill. Large-area angular surfaces are milled by using the side-cutting edges of the end mill.

STEP 2 Position the spindle head so that the centerline of the cutting edges of the end mill falls near the centerline of the angular surface to be milled.

FIGURE 14-10 Spindle Head Setup for End Milling a Required (45°) Angle

STEP 4 Turn the quill handwheel to feed the end mill to depth.

STEP 5 Engage the power feed. Take the first cut. Measure the angle. Make angle adjustments when needed.

STEP 6 Take successive cuts until the angular surface is machined to the required dimension.

HOW TO END MILL A SLOT OR KEYWAY

STEP 1 Select a keyway end mill.

Note: The standard, slightly undersize keyway cutter compensates for the usual oversize milling of the end mill.

STEP 2 Mount the end mill. Check the setup for minimum overhang. The setup must clear the workpiece when the full depth of cut is reached.

STEP 3 Center the keyway end mill (or regular end mill for slotting) with the centerline location on the workpiece.

Note: A closed slot or keyway is machined by positioning the centerline of the cutter at the center of the end radius (Figure 14-11).

STEP 4 Feed the end mill to depth.

Note: Deep keyways and slots may require a number of roughing and finish cuts.

Note: The amount and force of the cutting fluid must be sufficient to flow chips out of the slot before the end mill is repositioned for subsequent cuts.

STEP 5 Take a trial cut just long enough to permit the slot to be gaged for its width measurement. Make adjustments as needed.

STEP 6 Proceed to end mill the keyway or slot to the required dimensions. Figure 14-12 shows an elongated slot milled to size with an end mill.

FIGURE 14-11 Milling an Elongated Slot by Starting with End Mill Centered to Cut End Radius

FIGURE 14-12 Elongated Slot Milled to Size with an End Mill

CUTTING TOOLS, SPEEDS AND FEEDS, PROCESSES 263

HOW TO CUT A T-SLOT ON A VERTICAL MILLER

STEP 1 Select an end mill for first milling a groove to the width of the T-slot opening.

STEP 2 Select a T-slot cutter to form mill the two sides and the top and bottom faces of the T-slot.

STEP 3 Mount the workpiece with the T-slot laid out.

STEP 4 Cut the vertical slot with a standard, straight end mill.

STEP 5 Replace the straight end mill with the T-slot cutter. Center the cutter with the slot. Bring the cutter down so that it just touches the top surface of the workpiece. Then, move the cutter to clear the workpiece. (Figure 14-13 shows the cutter and workpiece setup for machining a T-slot on a vertical milling machine.)

STEP 6 Move the cutter down to the required dimension for the depth of the T-slot.

STEP 7 Feed the cutter into the workpiece by hand to be sure the cutting action is correct. Engage the power feed.

STEP 8 Follow standard procedures for burring the workpiece, cleaning the machine, and storing the tools.

HOW TO MILL A DOVETAIL ON A VERTICAL MILLER

STEP 1 Lay out the dovetail. Mount the work securely in a work-holding device or clamp it to the table.

STEP 2 Select a dovetail cutter that is ground to the required included angle.

STEP 3 Mill out the area between the two side dimensions of the dovetail with an end mill.

STEP 4 Replace the end mill with a dovetail cutter. Position this cutter for depth to within 0.010'' to 0.015'' (0.2mm to 0.4mm) of the base surface of the dovetail.

STEP 5 Position the cutter to take a series of roughing cuts. Rough machine one side of the dovetail.

STEP 6 Raise the workpiece until the dovetail cutter just grazes the base surface. Then,

FIGURE 14-13 Setup for Vertical Milling a T-Slot

move the cutter to the layout line for the angle side. Take the final finish cut.

STEP 7 Machine the second angular side of the dovetail. Repeat steps 3 through 6 for roughing out and finish milling.

STEP 8 Measure the depth and width of the dovetail. Burr the edges.

STEP 9 Clean the machine, cutter, and workpiece. Place all tools and accessories in proper storage areas.

SAFE PRACTICES FOR BASIC VERTICAL MILLING PROCESSES

—Check the direction of rotation of the spindle and the hand of the end mill to prevent damage to the cutter teeth.

—Select the shortest possible length of end mill to reduce tool overhang and to provide adequate clearance for the operation to be performed.

—Start end milling with light cuts. As operator experience increases, the depth and width of cut may be increased for maximum cutting efficiency.

—Check the position of the table trip dogs if power feed is to be used for machining a closed area.

—Take roughing cuts when a great amount of material is to be removed. However, a sufficient amount of stock must be left for finish machining to a required surface texture.

- Start each new end milling process by using the slow speed and feed in the starting range. Based on the variables affecting the rates, gradually increase the spindle RPM and feed rate.
- Use keyway end mills that are undersize to compensate for the tendency to mill slightly larger than the outside diameter of the cutter.
- Flow the chips rapidly and with a sufficient volume of cutting fluid to move them from the cutting area. Otherwise, the keyway cutter will cut to a width that is larger than the nominal size.
- Supply a continuous flow of cutting fluid to carbide end mills when cast iron or steel is milled.
- Feed dovetail cutters slowly into a workpiece. Any sharp contact between the fragile ends of the cutter teeth and workpiece may cause the teeth edges to fracture.
- Place safety guards on the table around the workpiece area if air is used as a coolant or to blow chips away.
- Use a stiff brush and avoid handling the long, needle-like sharp chips that are produced by side milling with an end mill.
- Use safety goggles and/or a transparent shield as protection against flying chips and particles.
- Stop the spindle before cleaning a workpiece, and remove burrs before measuring a workpiece.

TERMS USED WITH VERTICAL MILLING CUTTING TOOLS AND PROCESSES

Center cutting (end milling)	Cutting across the end face with center-cutting end mill teeth. End mill cutting capability from the outside diameter to the center.
Side cutting (end milling)	Milling with the side-cutting edges (teeth) of an end mill.
Square-end end mill	An end mill with cutting teeth that are ground at a right angle to the cutter axis.
Ball-end end mill	An end mill having end-cutting teeth that are ground to a ball (radius) shape.
Keyway cutter	A specially ground end mill having a slightly smaller diameter than the nominal width of a keyway.
Long or extra-long, multiple-flute end mill	An end mill with flutes that are longer than the flutes of a regular-length end mill. An end mill designed to mill to a greater depth that a regular-length end mill.
High-helix end mill	A general-purpose end mill with teeth cut at a steeper helix angle than a standard spiral-tooth end mill.
Stub-length end mill	An end mill designed with short-length teeth that provide greater cutter rigidity and efficient milling.
Gashed-end teeth	End teeth on an end mill that extend from the outside diameter to the center hole.
Collet chuck end mill holder	A vertical milling machine cutter holder and collet combination for holding end mills securely to the spindle.
Stepped surface (end milling)	Machining a flat surface on two or more planes.
Starting speeds and feeds	Manufacturers' recommended speeds and feeds for starting a new process. (Speeds and feeds may be increased or decreased by the machine operator after each variable affecting the cutting action is observed.)

CUTTING TOOLS, SPEEDS AND FEEDS, PROCESSES

SUMMARY

Basic vertical milling machine processes deal with machining flat, angular, beveled, round, and continuous straight and curved surfaces on one or more planes and with slotting, grooving, dovetailing, and keyway cutting.

> End mills are commercially available with straight or spiral teeth, positive or negative radial rake angle, right- or left-hand helixes, and standard or high helix angles.

High-speed steel, cobalt HSS, and carbide cutters are commonly used materials for end mills.

> Single-end and double-end end mills are available in stub, regular, long, and extra-long lengths.

Stub-length, small-diameter end mills provide added strength to operate at the higher speeds required for economical milling.

> Square-end, four-flute end mills are adapted to general-purpose end milling, slotting, machining stepped surfaces, slabbing, shallow pocketing, and die sinking processes.

High-helix end mills produce a shear cutting action and a fine surface finish.

> The outside diameter of keyway-cutting end mills permits the milling of keyways to precise, nominal sizes.

Cobalt high-speed steel end mills premit heavy-duty milling at higher speeds, feeds, and depths of cut than conventional HSS mills. Other advantages include increased strength, long wear life, and fast chip-breaking characteristics.

> The cutting ends on end mills are either center cutting for plunge milling or gashed.

End mills are held concentrically in the spindle by solid or collet-type holders.

> The cutting fluids for vertical milling are the same as the fluids used for horizontal milling of the same materials and similar machining processes.

The starting cutting speeds (converted to spindle RPM) for end mills as given in manufacturers' tables are affected by variables such as required surface texture, condition of the milling machine, nature of the machining process, coolant, type and cutting qualities of the end mill, and the rigidity of the workpiece and setup.

> Work procedures for milling flat (horizontal and vertical) surfaces include the selection and mounting of an end mill, vertical positioning of the spindle head, location of the workpiece in relation to the cutter, feeding of the cutter for roughing and finish cuts, measurement, and machine shutdown steps.

End mills are used for machining right-angle steps and for milling stepped areas.

> The end milling of a beveled or other angular surface may be done by either positioning the workpiece or setting the spindle head at the required angle.

End milling elongated or through slots and keyways requires roughing and finish cuts.

> T-slots are milled by first cutting a slot and then using a T-slot cutter to form mill the rectangular area.

Dovetails are milled on vertical milling machines by finish milling the angular sides with an angle cutter.

Personal safety and safe machine practices in basic vertical milling relate to factors affecting the machine, the work setup, cutter characteristics, and cutting fluid needs of carbide cutters.

UNIT 14 REVIEW AND SELF-TEST

1. Name six design features of standard end mills.
2. a. Identify three design features of stub-length, small-diameter end mills that permit them to cut efficiently at the required higher speeds.
   b. Give the feed rate for general-purpose end milling with stub-length, small-diameter end mills.
3. Give three advantages of high-helix end mills over standard-helix end mills.
4. Explain why a square-end, multiple-flute end mill is a higher production milling cutter than a two-flute end mill.
5. Tell why a keyway-cutting end mill is preferred over a standard end mill for machining a keyway to a precise, nominal dimension.
6. Give two reasons for using a cobalt high-speed steel end mill instead of a conventional high-speed steel end mill for production milling a high-tensile material.
7. Differentiate between a center-cutting and a gashed-end end mill.
8. Describe briefly how a collet chuck holder operates to hold an end mill solidly in the spindle.
9. Use a table to establish (a) the recommended starting spindle speed range for taking roughing and finish cuts and (b) the corresponding feed range for end milling a step in a brass/bronze block. A 1/4" (6mm) diameter high-speed end mill is used.
10. Give two personal safety precautions to take when performing end milling operations.

UNIT 15

Hole Forming, Boring, and Shaping

Design features of the vertical milling machine permit positioning a workpiece precisely by the longitudinal and cross slide movements of the table. These movements fall along the X axis and the Y axis. The vertical movement of the knee makes it possible to set and machine a workpiece according to dimensions on the Z axis. The graduated micrometer collars on the cross slide, table, and knee provide a quick, accurate, precision method of locating cutting tools and a workpiece. The micrometer collars permit direct measurements to accuracies within 0.001″ (0.02mm). Vernier, bar, and indicator attachments extend the accuracies to 0.0001″ (0.002mm).

Because of these and other features, the vertical milling machine is uniquely adapted to *hole-forming processes* such as drilling, reaming, countersinking, counterboring, and tapping with basic cutting tools and to *boring*. Hole and other surface locations and dimensions for milling processes are often specified by ordinate, tabular, and baseline dimensioning. Essential layout and dimensional information is given in relation to the X, Y, and Z axes.

Still other processes may be performed on the vertical milling machine by using standard attachments and accessories. The dividing head, rotary table, sine table, and angle plates are commonly used in toolrooms and custom jobbing shops. The shaper head attachment extends the machine capability to include *shaping processes* that otherwise would have to be performed on a second machine.

WORK- AND TOOL-HOLDING DEVICES FOR BASIC HOLE-FORMING PROCESSES

In basic hole-forming processes, four methods are commonly used to position and hold workpieces on a vertical milling machine. These methods require the use of:

—A standard plain, swivel, or universal vise;
—Straps, parallels, step and V-blocks, T-bolts, and clamps to secure workpieces in an aligned position;
—Fixtures for nesting, positioning, and holding parts;
—Stops that fit snugly into the table T-slots against which a workpiece is aligned and held parallel to the column face.

As in horizontal milling processes, a dial indicator is used to accurately align the solid jaw of a vise. A dial indicator may also be used to align the edge of a workpiece precisely to be parallel or at a right angle to the table axis. In alignment applications, the dial indicator is fastened to the vertical spindle to indicate a machined locating surface.

In drilling, reaming, counterboring, countersinking, and tapping processes, the same construction and sizes of drills, machine reamers, counterbores, countersinks, and machine taps that are applied to drilling machines, horizontal milling machines, lathes, and screw machines are used on a vertical milling machine. These fixed-diameter, multiple-edge cutting tools may be held in a chuck, adapter, or collet. Taps are turned for threading and removing from a workpiece by a tap adapter or a reversing tapping attachment.

FACTORS AFFECTING HOLE-FORMING PROCESSES

In vertical milling, the material of which a hole-forming cutting tool is constructed, its cutting speed for a particular process and workpiece, feed rates, and cutting action are all influenced by the same factors that apply to drill press, horizontal

VERTICAL MILLING MACHINES

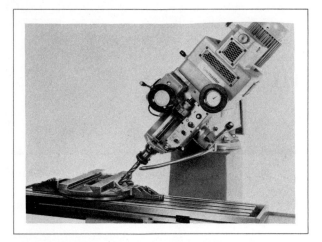

FIGURE 15-1 Vertical Head Positioned to Drill Holes at a Required Angle

miller, and lathe work. However, the following additional factors must be considered:

- The vertical head may be positioned at a single or compound angle, and holes may be formed with the spindle axis (and the axis of the cutting tool) set to cut at a required angle (Figure 15-1).
- The table may be moved longitudinally or transversely to any precise dimension, and hole locations to tolerances of 0.0001" (0.002mm) are possible by using measuring rods and a tenth dial indicator (Figure 15-2).

BORING WITH THE OFFSET BORING HEAD

The *offset boring head* is used to bore a hole to any standard or other precise size. Boring produces straight and concentric holes about a fixed center. The quality of the surface texture may be controlled by the form and sharpness of the boring tool, depth of cut, and feed rate. Boring holes that are larger than 1" in diameter is more economical than maintaining a whole inventory of drills, machine reamers, and other cutting tools to cover a wide range of sizes.

Boring Shouldered Surfaces

A combination boring and facing head is used to form a shoulder as in counterboring. Holes are bored through a workpiece or to a specific depth. Holes are counterbored by boring the side walls to depth. The shoulder area may then be squared by feeding the end-cutting edge of the boring tool toward the center of the hole. A combination boring and facing head is also used to machine a stepped circular area (Figure 15-3). The same cutter is used to bore the hole to size and to machine the flat base.

Machining a Radius

The boring head is also commonly used in applications where an accurate radius is to be machined into a surface. In such applications,

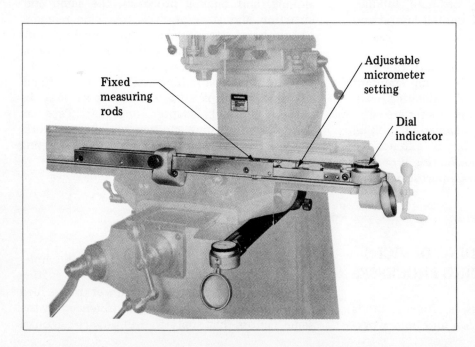

FIGURE 15-2 Precision Coordinate Measuring Attachment System with Measuring Rods and Dial Indicator (0.0001" or 0.002mm)

HOLE FORMING, BORING, SHAPING 269

FIGURE 15-3 Use of Boring and Facing Head to Machine a Stepped Circular Area

the workpiece is strapped to the table or held in a fixture. The centerline for the radius is located at the axis of the boring head and spindle. The table is also positioned. The center of the hole to be bored is centered as close as practical.

A scrap plate is clamped opposite to the workpiece. Its edge is the same distance from the center (boring head axis) as the edge of the workpiece. A trial cut is taken with the offset boring head for about 1/8" (3mm) deep. The diameter is measured and the cutting tool is adjusted for depth of cut. One or more roughing and finish cuts may be needed to produce the required radius.

THE OFFSET BORING HEAD AND CUTTING TOOLS

Offset boring heads are available in a number of different models and sizes. Adjustments for machining diameters in increments of 0.001" (0.02mm) are made rapidly by moving the cutting tool laterally in the holder. The amount of movement is read directly on a micrometer dial on the boring head. Microboring heads permit diametral adjustments within 0.0001" (0.002mm).

Features of the Offset Boring Head

The major construction features of an offset boring head are illustrated in Figure 15-4. These features include the *shank*, *body* with *tool slide* and *locking screw*, *micrometer dial*, and *setscrews* for holding cutting tools, boring bars, and reducer bushings. This boring head is designed for heavy-duty boring. The parts are made of chrome-nickel steel. They are hardened and ground. The micrometer dial is graduated into 50 divisions that permit adjustments to 0.0005" (0.01mm). There is a positive lock when the tool slide is set at a required size.

The boring head set includes three *boring bars*. Two of these bars permit the use of standard-size, round boring tools. The third bar holds the different sizes of solid-shank boring bars in a set. The two *reducer bushings* accommodate two different sizes of boring tools. Like the boring bars, the reducer bushings are held securely in a particular cutting-tool location in the tool slide by a setscrew.

The boring head is a precision tool. The head, boring bars, reducer bushings, and hexagon wrenches are furnished in a case for protection. Each part may be easily located and conveniently stored.

Specifications of Selected Boring Head Features

The boring head in Figure 15-4 is furnished with shanks that fit R-8 and #30, #40, and #50 NS spindles. The diameters that may be bored range from 1/4" to 6" (6mm to 150mm) for the R-8 size. The largest #50 NS shank boring head

270 VERTICAL MILLING MACHINES

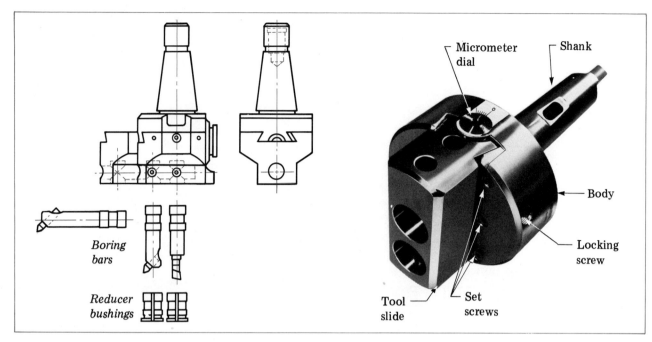

FIGURE 15-4 Construction Features of an Offset Boring Head

accommodates hole sizes from 5/16" to 11-1/2" (8mm to 292mm). The total travel of the tool slide varies for each head. The N-8 size has a travel of 31/32" (24mm). The #50 NS shank boring head has a travel of 1-11/16" (42.8mm).

The R-8 and the #30 NS shank boring heads hold 1/2" (12.5mm) diameter boring bars. The #40 NS and #50 NS heads accommodate 5/8" (15.8mm) and 3/4" (19mm) diameter bars, respectively.

Functions of Boring Head Parts

The shank is an integral part of the boring head body. The shank has a tapered section that mates with the spindle taper. The tapers of the shank and spindle are secured by a draw-in bolt. The body permits the mechanical movements for the micrometer adjusting screw, the tool slide, and the locking screw. The tool slide holds the cutting tool and permits adjustments to be made. A boring tool may be mounted directly in the tool slide or held in a boring bar.

Note that the cutting edge of a boring tool is positioned along the centerline of the boring head (Figure 15-5). Thus, the cutting edge moves in line with the axis of the tool slide. The centerline alignment is the only position where (1) the cutting edge can be moved the same distance as the movement of the tool slide and where (2) the rake and clearance angles of the cutting edge are correct as ground.

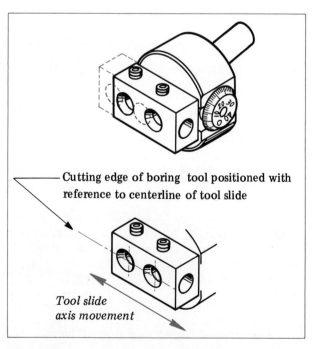

FIGURE 15-5 Position of Cutting-Tool Edge in Relation to Boring Head Tool Slide

HOLE FORMING, BORING, SHAPING

FACTORS AFFECTING BORING PROCESSES

Cutting Speeds and Feeds for Boring Tools

Cutting speeds of high-speed steel, carbide-tipped, and solid carbide boring tools require consideration of the same factors that apply to cutting speeds for other similar machining processes. In addition to the usual rigidity requirements, if an offset boring head produces chatter and excess vibration, the spindle RPM may need to be reduced.

Boring is often used to correct eccentric, roughly formed, bell-mouth holes that may not be parallel with a central axis. The depth of the first cut or cuts, therefore, varies around the periphery of a previously formed hole (Figure 15-6). This variable depth continues until the bored hole is concentric with the spindle axis. The cutting action for the first few cuts may require a reduction in cutting speed.

A boring tool may be hand-fed and/or power-fed by the quill. The coarser quill feeds of 0.003" and 0.006" (0.1mm and 0.15mm) per revolution are used for general roughing cuts. Finish cuts are taken at the finest quill feed of 0.0015" (0.04mm) per revolution. At the end of a finish cut, some experienced operators reverse the feed direction. The boring tool is then fed back up through the workpiece without resetting the cutting-tool position. In general practice, the boring tool is usually withdrawn from the hole at the end of the cut by stopping the spindle and moving the cutting edge clear of the finished surface. Otherwise, a fine, spiral groove may be cut into the work surface as the spindle turns.

Selection of Boring Tools

Operator judgment is required in the selection of boring tools. Selection of the shortest and heaviest boring bar or boring tool should be consistent with the bore diameter and hole depth, the kind of material being bored, the nature and quantity of the cutting fluid, and the process itself.

Consideration of Tool Deflection

The deflection of the cutting tool is an important consideration in establishing the depth of cut, cutting speed, feed, and dimensional and surface finish accuracy. Tool deflection may be caused by any one or combination of the following factors:

—Inaccurately ground or worn and dulled cutting edge;
—Too deep a cut for the body size of the cutter;
—Too heavy a feed;
—Too high a cutting speed for the boring process;
—Improperly supported and secured workpiece;
—Incorrect cutting fluid;
—Improperly positioned or unlocked tool slide;
—Interrupted cutting, as in the case of boring a radius.

SHAPING ON A VERTICAL MILLING MACHINE

The vertical shaper head attachment (Figure 15-7) is found in many toolrooms and custom jobbing shops. It is used for machining straight surfaces that may be cut at a right angle or other angle with the 0° vertical axis of a workpiece. Internal keyways, slots, and splines may be cut. The sides of openings and other intricate forms may be shaped on jigs, fixtures, punches, dies, and other workpieces that are to be precisely machined. Gear teeth and racks may be cut with the shaper head. Some heads are adapted to blind-hole machining processes. The shaper head

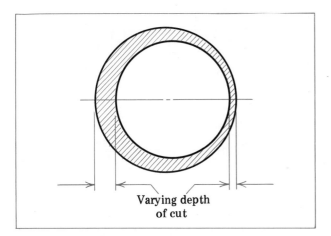

FIGURE 15-6 Varying Depth of Cut Produced by Eccentrically Formed Hole

often replaces a second machine—for example, a broaching machine for broaching.

The cutting tools are formed for side and end cutting, with corresponding rake angles and clearance. The tools are mounted in the tool head. A cutting tool is positioned by the table and cross slide handwheels. Vertical positioning of a workpiece is controlled by elevating or lowering the knee. The length of stroke is adjusted directly on the shaper head. The attachment may be set for shaping at a 0° vertical angle or at any other single or compound angle. The settings are made directly. The angle is read on the graduated shaper head and/or the ram, depending on the job specifications.

The shaper head is designed with an integral power source. The head is fitted on the face of the ram in the same way as a vertical spindle head. Some manufacturers build the vertical milling machine with the shaper head fitted to the rear side of the ram, where the head may be swiveled through 180° over the table and workpiece.

The shaper head is generally used with workpieces that are held stationary on V-blocks, vises, and angle plates or that are strapped to the table. Surfaces that require shaping in a circular direction are positioned, mounted, and secured on a rotary table or by using a dividing head. The accessories permit workpieces to be indexed or turned in relation to other vertical milling processes.

The shaper head has a stroke capacity of 0" to 4" (0mm to 102mm). The length of stroke can be dialed in increments of 1/8" (3mm). The head is available with a slow speed and a standard speed range (strokes per minute. The slow speed range is 35, 50, 75, 100, 145, and 210 strokes per minute. The standard speed range is 70, 100, 145, 205, 295, and 420 strokes per minute.

BASELINE, ORDINATE, AND TABULAR DIMENSIONING

Baseline Dimensioning

Baseline dimensioning is used on part drawings that require precision layout and/or positioning on a machine. All measurements are related to one or more common finished surfaces. These surfaces are called *baselines*. When successive measurements are taken from a single baseline, dimensional errors are not cumulative.

Baseline dimensioned drawings simplify the positioning of a workpiece from corresponding X, Y, and Z machine axes (Figure 15-8). The micrometer collars on the horizontal and cross slide feed screws, on the table and knee, are used to position a workpiece for successive cuts from baseline dimensions.

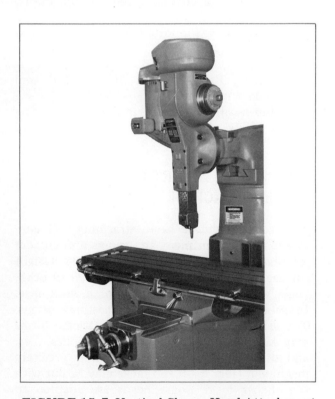

FIGURE 15-7 Vertical Shaper Head Attachment

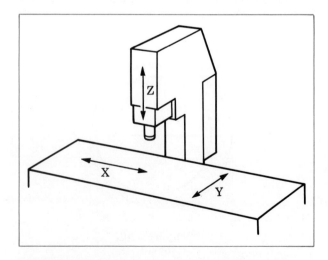

FIGURE 15-8 Three Basic Axes of Vertical Milling and Numerically Controlled Machines and Machining Centers

The drawing in Figure 15-9 shows a simple application of baseline dimensioning. Here, the two baselines are at a right angle to each other. Baseline dimensioning is often applied with the baseline as a centerline (Figure 15-10A). Measurements are then taken from a fixed plane position. Baseline dimensioning may also be applied in laying out an irregular shape (Figure 15-10B).

Datums

A *datum* is a feature from which dimensions are referenced. A datum may be a plane, line, or other exact reference point. Generally, shop and laboratory drawings have two datums. The datums are identified as *zero coordinates*. All measurements are taken from mutually related zero coordinates (datum planes). For example, in baseline dimensioning, measurements are taken from X and Y datums.

Ordinate Dimensioning

Ordinate dimensioning differs from baseline dimensioning in that dimension lines and arrowheads are not used. Figure 15-11 is an example of ordinate (rectangular datum) dimensioning. All dimensions are related to the zero coordinates (datums) along the X and Y axes. Each dimension appears at the end of the extension line of a feature to which it refers. A feature is located by the intersection of two datum dimensions. For example, the center of one A hole in Figure 15-11 is 0.750" from the X axis and 0.625" up from the Y axis.

Ordinate dimensioning, like baseline dimensioning, permits greater reading and machining accuracy than does a conventionally dimensioned drawing. Dimensional errors are controlled and not multiplied. Also, drawings that otherwise would have a large number of confusing dimensions are simplified and may be interpreted quickly and accurately.

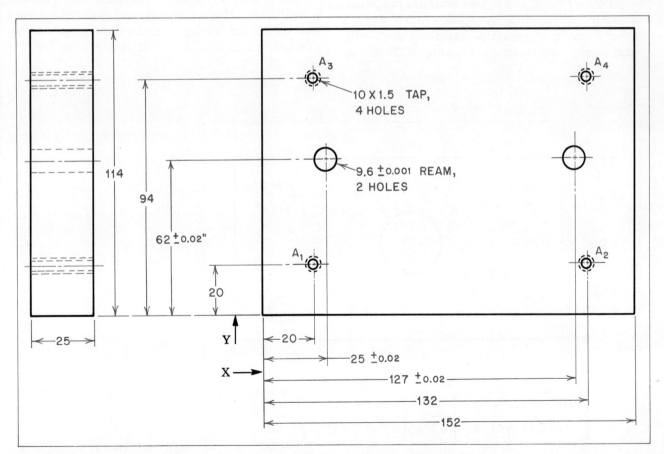

FIGURE 15-9 Drawing Showing Simple Application of Baseline Dimensioning

274 VERTICAL MILLING MACHINES

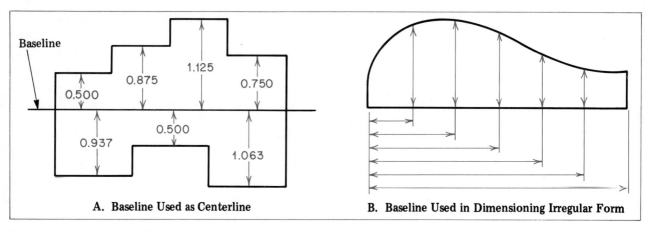

FIGURE 15-10 Additional Applications of Baseline Dimensioning

Tabular Dimensioning

Tabular dimensioning is another form of simplified, accurate dimensioning. All dimensions are related to datums and originate at zero coordinates. Tabular dimensioning is used for milling machine, jig borer, numerically controlled, machining center, and other machining processes. Tabular dimensioning is recommended when a great number of common features make a conventionally dimensioned drawing difficult to read. Figure 15-12 is an example of tabular dimensioning.

The dimensions of each feature on a tabular dimensioned drawing are given in a table, or *coordinate chart* (Figure 15-12). The dimensions of each hole are given as they originate at the X and Y datums. The table also provides sufficient information about a feature to permit accurate machining. Tabular dimensions may be given in inch and/or metric units of measure.

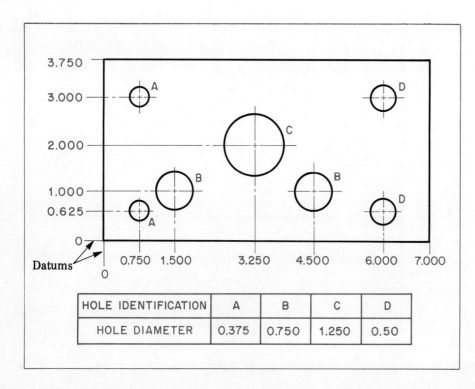

FIGURE 15-11 Example of Ordinate Dimensioning

HOLE FORMING, BORING, SHAPING

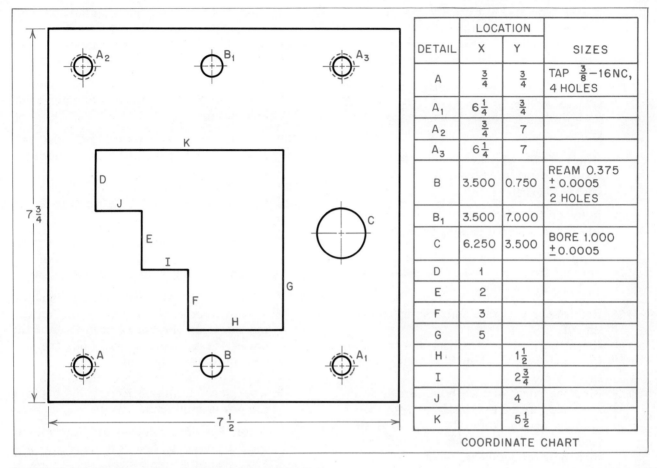

FIGURE 15-12 Example of Tabular Dimensioning

HOW TO MACHINE HOLES BY USING A CONVENTIONAL VERTICAL MILLING MACHINE HEAD

STEP 1 Select the required size and type of hole-forming cutting tool.

Note: The drill, reamer, counterbore, countersink, or machine tap that is selected is based on the same factors that govern its use on other machines.

STEP 2 Establish the cutting speed and feed for the work process. Set the spindle RPM to produce the required cutting speed. Set the quill feed rate.

STEP 3 Select an appropriate cutter holder, adapter, collet, or tapping unit. Secure the tool-holding device in the vertical spindle.

STEP 4 Mount and secure the cutting tool.

Note: A friction clutch or reversing-head tapping holder is used for internal threading.

Note: The torque setting should be adequate for safe tapping. At the same time, the torque should be limited to avoid exerting excessive force, which may cause tap breakage.

STEP 5 Determine the correct cutting fluid, if needed. Set the nozzle to direct the quantity and flow of cutting fluid.

STEP 6 Position the cutting tool at the center of the hole. Lock the knee, cross slide, and table. Set each of the micrometer collars at zero.

Note: Ordinate, tabular, and baseline dimensioned drawings give each dimension from a zero position on the X, Y, and/or

276 VERTICAL MILLING MACHINES

Z axes. The operator usually sets the table, cross slide, and knee micrometer collars at zero. All table movements to position the cutting tool relate to the zero settings.

STEP 7 Proceed to carry on the hole-forming process by following the steps that are used on a drilling machine, lathe, or screw machine.

Note: When it is established that the cutter is cutting properly, the power quill feed is engaged.

STEP 8 Relocate the cutting tool at the center of any other hole that is to be formed.

Note: The dimensional settings are read directly from the zero coordinates of the X, Y, and/or Z axes.

STEP 9 Reduce the feed rate and take a final finish cut.

STEP 10 Clean the workpiece. Check the hole size for dimensional and surface texture accuracy.

STEP 11 Burr the edges of each hole. Brush away the chips from the machine and workholding device. Wipe the machined surfaces clean. Replace each tool and accessory in its proper place.

HOW TO BORE HOLES WITH AN OFFSET BORING HEAD

Boring a Predrilled
or Formed Hole Vertically (0°)

STEP 1 Align the vertical spindle head at 0°.

STEP 2 Select an offset boring head to accommodate the size and type of tool bit needed for the job. Check for and remove any burrs. Mount and secure the boring head in the spindle.

STEP 3 Position and secure the workpiece as near the center of the table as practical.

STEP 4 Mount the boring tool so that the cutting face is aligned with the centerline of the head. Tighten the cutter setscrew.

STEP 5 Align the spindle (and boring head) axis with the zero coordinates on the X and Y axes. Set the cross slide (saddle) and table feed micrometer collars at zero.

STEP 6 Move the table horizontally and transversely to the required centerline dimension.

Note: The dimensional settings may be read directly from the micrometer collar graduations.

STEP 7 Determine the correct cutting speed and feed. Set the spindle RPM to produce the required cutting speed.

STEP 8 Adjust the boring tool for a trial cut. Take the cut for about 1/16" (1mm to 2mm) deep.

Note: The cut should clean up to more than half the diameter to permit taking an accurate diametral measurement.

STEP 9 Set the micrometer dial on the boring head at zero. Turn the micrometer dial adjusting screw on the boring head to feed the cutter for a roughing cut. Lock the tool slide on the boring head.

STEP 10 Bring the cutter into contact with the workpiece. Continue feeding by engaging the power quill feed.

Note: The proper cutting fluid and volume must be used to flow the chips out and away from the area being bored.

STEP 11 Return the boring tool to the starting position. Stop the machine. Remeasure the bored diameter.

Note: Too heavy a cut or a dulled cutting edge may cause the tool to be deflected. Tool deflection produces an inaccurate, smaller-diameter hole that may be bell mouthed.

STEP 12 Set the boring tool for subsequent roughing and/or finish cuts.

Note: The feed rate is decreased to obtain a higher-quality surface finish. Some operators use the down-feed and, at the end of the cut, change to an up-feed without changing the finish cut position of the boring tool.

Note: The cutting edge must be moved clear of the workpiece if it is brought up and out of the bored hole on the final cut.

STEP 13 Brush away all chips. Check the final bored hole for dimensional accuracy, parallelism of the sides, and surface finish.

HOLE FORMING, BORING, SHAPING

STEP 14   Burr the workpiece. Follow standard safe practices for removing chips. Clean the machine. Return all tools and accessories to proper storing places.

Drilling and Boring a Hole

STEP 1   Select a drill that is smaller than the diameter of the hole to be bored.

Note: For holes up to 1" (25mm) that are to be bored, approximately 1/16" (1.5mm) is allowed. The drill size should provide enough stock so that the rough drilled surface cleans up when the required diameter is bored.

STEP 2   Mount the drill in a tool-holding device. Secure it to the spindle.

STEP 3   Set the spindle RPM and feed rate for drilling the first hole. Position the drill at the center of the hole. Drill.

Note: A center drill is usually used, followed by drilling with one or more increasingly larger-size drills (depending on the final drill size).

STEP 4   Remove the drill and adapter. Replace with an offset boring head and boring tool.

STEP 5   Proceed to bore the hole as for conventional boring.

Boring a Hole at a Simple or Compound Angle

STEP 1   Position and lock the vertical milling machine head at the required angle in the plane that is parallel to the column face. Read the angle setting on the graduated quadrant of the head.

Note: The head is positioned for any additional angle by tilting the head in a plane that is at a right angle to the column face. The setting is read directly from the graduated quadrant on the end of the ram.

STEP 2   Set up to center drill or drill a hole. Allow enough stock for boring the hole to size.

STEP 3   Replace the drilling setup with a boring head and appropriately ground cutter.

STEP 4   Proceed to bore the hole to size. Reposition the boring tool for additional cuts.

Note: Follow general steps for boring and checking the diameter.

Boring a Radius or Stepped Circular Area

STEP 1   Strap the workpiece on parallels on the table or hold it securely in a work-holding device.

Note: A scrap piece of stock secured at a position that will permit the cutter to cut an arc on two segments of the circle makes it possible to take an accurate measurement across the measurement line of the diameter.

Note: The workpiece is positioned so that the cutting tool clears at the end of each cut without scoring the parallels or table surface.

STEP 2   Insert a centering rod in a boring head bushing or a chuck. Move the table longitudinally and transversely until the spindle axis is aligned at the laid out centerline at the end of the workpiece.

STEP 3   Move the table longitudinally on the centerline. The distance moved along the X axis should center the boring head at the centerline of the radius to be bored.

Note: The distance from the edges of the workpiece and scrap stock to the center of the boring head should be the same. Otherwise, additional machining time will be needed to take a series of cuts until both edges are bored equidistant from the cutting edge.

STEP 4   Insert a tool bit, with correctly ground rake and clearance angles, in a boring bar. Mount the tool bit so that the cutting edge is aligned with the center axis of the boring head tool slide block.

STEP 5   Set the spindle RPM and feed rate.

STEP 6   Move the boring cutter until it grazes the end of the workpiece. Set the micrometer adjusting screw at zero.

STEP 7   Feed the tool outward for a trial cut. Take the cut for approximately 1/16" (1mm to 2mm) deep.

Note: The cut should form a slight arc on both the workpiece and scrap stock.

STEP 8   Stop the spindle. Depending on the specified accuracy, measure the diameter along the measurement line (centerline) with a steel rule, inside caliper, or vernier inside caliper.

STEP 9 Increase the depth of cut, if possible. Use a coarse feed for a roughing cut.

Note: The depth of cut (amount the boring tool is moved on the tool slide) is established by the micrometer dial reading.

STEP 10 Reset the cutting tool for subsequent roughing and finish cuts.

STEP 11 Clean the workpiece. Check the final bored diameter. Burr the bored hole.

STEP 12 Remove all chips. Clean the machine. Replace each tool and accessory in its proper storage place.

Boring a Counterbored Area or a Flat Shouldered Surface

STEP 1 Mount a boring and facing head on the quill so that the boring tool is driven by the spindle.

STEP 2 Select a boring tool that is ground to produce the specified square, round, or angular corner for the shoulder.

STEP 3 Take a series of cuts to bore the outside diameter to depth and to remove the excess material within the area to form the flat face.

Note: The cutting tool is fed downward to depth by the quill feed handwheel. The flat shouldered surface is cut by turning the boring and facing head handwheel to feed the boring tool inward.

STEP 4 Remove the chips. Clean and burr the workpiece. Measure the diameter and depth. Check the surface finish against a comparator specimen.

STEP 5 Follow standard safe practices for cleaning the machine, tools, and accessories. Return each item to its proper storage place.

HOW TO USE A SHAPER ATTACHMENT ON A VERTICAL MILLING MACHINE

STEP 1 Replace the vertical spindle head on the ram with a self-contained shaper head.

Caution: Place a wooden pad on the table under the vertical spindle. Secure assistance in handling both the vertical head and the shaper head.

STEP 2 Position and lock the shaper head at 0° or whatever simple or compound angle is required.

STEP 3 Select the size and style of cutting tool to accommodate the cuts to be taken. Mount the cutting tool.

Note: Shaper tool sets are commercially available to machine square, angular, and round forms and combinations of these forms as corners and flat or irregular surfaces. Additional tools are forged for special applications.

STEP 4 Determine the cutting speed (number of strokes per minute). Set the shaping head dial at this cutting speed.

STEP 5 Position the cutting tool a short distance up from the top face of the workpiece to provide clearance for taking measurements. (The operator may also easily see the layout lines and cutting action.)

STEP 6 Set the length of the ram stroke.

Note: The distance must permit the cutter to clear the workpiece at the top and bottom end of each stroke. In the case of cuts to a specified depth, the ram is positioned to bottom out at a specific depth.

STEP 7 Locate the cutting edge of the cutter at the depth for the first cut.

STEP 8 Take a trial cut for a short distance. Measure the workpiece. Adjust the cut, if needed.

Note: At this step, the operator must determine whether any adjustments are needed. These adjustments relate to depth of cut, feed rate, and cutting speed.

Note: The heaviest possible cutting tool is used for efficient cutting with the least amount of deflection.

STEP 9 Reset the depth of cut and take successive cuts to rough out the required shape.

Note: Large areas are usually roughed out by ending each successive cut near a corner in a series of steps. When the surface is roughly shaped to depth, the excess corner material is cut away.

STEP 10 Replace the roughing cutter with a cutter ground for finish cuts.

STEP 11 Relocate the cutting edge at the roughed-out edge. Feed to depth.

HOLE FORMING, BORING, SHAPING 279

STEP 12 Take a trial cut. Measure the surface. Adjust the cutter, as needed. Take the finish cut.

STEP 13 Reset the machine for other shaping or additional vertical milling operations that may be required. Then, burr the workpiece.

STEP 14 Follow standard machine and tool cleanup and storage practices.

Shaping and Indexing Radial Surfaces

STEP 1 Mount a rotary table centrally on the vertical milling machine table.

STEP 2 Position the workpiece on the rotary table. The location must permit feeding and revolving the workpiece at a specified radius and angle setting.

Note: The workpiece should be positioned in relation to the zero degree reference graduation setting on the rotary table.

STEP 3 Position the cutter in the shaper-tool holder to permit shaping the required form.

STEP 4 Feed the cutter to depth for a first cut.

Note: A formed cutter is usually used for spline cutting. The cutter is fed to depth by taking a series of cuts in the path of the first cut. The workpiece is then indexed to the next spline, positioned with the rotary table handwheel. The regular table movement and shaping are continued.

Note: The cutter is fed along the required radial path by feeding the rotary table handwheel for each cutting stroke.

STEP 5 Continue to take a series of roughing cuts. Also, shape away any stepped areas that may be formed at the corners.

STEP 6 Replace the roughing cut cutter with a finely ground finish shaping tool.

STEP 7 Reposition the cutter. Feed to the depth of the finish cut. Measure and adjust, if required.

STEP 8 Turn the rotary table handwheel to feed the workpiece along the required radial path.

STEP 9 Remove and burr the workpiece if no additional machining operations are to be performed on the machine.

STEP 10 Leave the machine clean. Return the rotary table and all tools to proper storage places.

SAFE PRACTICES FOR USING BORING AND SHAPER TOOLS AND ATTACHMENTS

—Make sure there is no obstruction in the path of a dial indicator as it is used to align a surface on the machine table. The dial indicator should be moved clear and repositioned beyond any obstruction. Use the micrometer collar on the appropriate table or cross slide handwheel or the knee crank for accurate positioning.

—Start with the recommended starting cutting speed and feed for hole forming with conventional drills, machine reamers, counterbores, countersinks, and taps and for shaping with shaper cutting tools. Adjust the speed and feed rates, as the cutting action may require, after all factors are considered.

—Check the workpiece for effects on surface texture and dimensional accuracy caused by any eccentric forces of the boring head. Reduce the spindle RPM if vibration or chatter occurs.

—Locate the cutting edge of boring tools in line with the axis of the cross slide of the offset boring head.

—Check the boring tool for correctness of the cutting-edge form and of the rake and clearance angles. On small diameters, the heel should be checked to be sure it will not rub against the bored diameter.

—Select the heaviest boring bar or cutter that may be used for the diameter and depth of hole to be bored.

—Use the largest size of drill possible on a hole that is to be bored. The drill size must permit the hole to clean up, be machined straight and concentric with the centerline of the hole, and meet surface texture and dimensional accuracy requirements.

—Set the depth of cut for a boring tool or shaper tool so that the cutting forces will not deflect the cutter and cause an incorrect work surface. The feed may also need to be reduced.

—Rotate the boring tool and head slowly and carefully when the cutting edge is set to take the first cut in forming a radius. Both the workpiece and the scrap stock should be equidistant from the spindle axis.

—Start a cut on a new workpiece by first feeding the cutting tool by hand. When it is established

that the cutting action is correct, engage the power quill feed.
— Move the cutting tool away from the finish bored surface before it is withdrawn from a workpiece.
— Set the torque on a tap adapter or tapping head so that, while it is adequate for tapping, the force is limited to prevent tap breakage.
— Secure the boring tool in the tool slide. Lock the knee, table, and cross slide (saddle) positions to prevent movement during a machining process.
— Place a wooden pad over the table surface to protect it from possible damage when removing or replacing a vertical spindle head with a shaper head. Two persons should be involved in removing any cumbersome and heavy attachment, accessory, or workpiece.
— Set the cutting stroke length to permit the shaping tool to clear the workpiece at the end of the cut. Adequate clearance must be provided in the starting position to take measurements safely.
— Stop the machine spindle before any measurements are taken. The workpiece must be clean and burred.
— Use a stiff brush to remove chips. Clean all machine surfaces. Return all attachments, accessories, and tools to any special box or compartment provided to protect the items.

TERMS USED IN HOLE FORMING, BORING, AND SHAPING

Hole-forming processes	Standard methods of producing holes by machining. The use of drills, machines reamers, countersinks, counterbores, taps, and boring tools to produce holes or segments of holes.
Vertical axis (0°) boring	Positioning the vertical milling machine head in a zero position (at a perfect right angle to the table) and boring.
Offset boring head	A tool-holding device with a shank that fits the machine spindle. (Cutting tools are held in boring bars, reducer bushings, or directly in a tool slide. The cutter is positioned for boring by moving the tool slide with a micrometer adjusting screw. The precise distance is measured on a micrometer dial.)
Boring head set	An offset boring head with three different sizes of boring bars, two reducer bushings, and wrenches. A boring accessory unit that includes the necessary parts for general-purpose boring and comes complete with a storage case.
Boring a shouldered surface	Boring the vertical surface of a hole and a continuing flat surface. (The surface is bored at a right angle.)
Feed increments (boring head)	The amount a tool may be fed for a cut. The distance in 0.001" or 0.0001" (0.02mm and 0.002mm) that represents the movement of the slide between two graduations on the micrometer dial of a boring head.
Diametral measurement	A measurement related to a diameter or the arc of a circle. Measurement of a diameter across the line of measurement.
Cleaning up the surface (boring)	Machining a previously formed surface. Removing tool marks or other surface imperfections while correcting inaccuracies such as eccentricity, bell mouth, and so on.
Stepped circular area	A radial area that is bored below another surface, usually a horizontal plane.

HOLE FORMING, BORING, SHAPING 281

Boring and facing head	An attachment for a vertical milling machine that permits boring and facing with the same boring tool and setting.
Cutting tool deflection	An undesirable machining condition. Springing of a cutting tool away from the surface being machined. A tool movement usually produced by a dull cutting edge, too fast a speed, or too great a feed.
Interrupted cutting (boring)	Cutting action that is not continuous (for example, boring a segment, or arc, or a hole). Starting and ending a cut without cutting around the complete periphery of a hole.
Integral power source (shaper head)	A self-contained shaper head attachment for a vertical milling machine. A shaper head having a ram that is actuated by a power source that is contained within the attachment.
Shaping splines	The process of shaping a series of grooves that are equally spaced around a circumference.
Bottom out (shaper cutting tool)	The point (depth in a blind hole or area at which the shaper ram and cutter reverse direction. The depth of a shaping cut, usually in a blind hole or area.
Baseline dimensioning	Successive measurements that originate from one or more common finished surfaces. A series of continuous dimensions that originate from a fixed reference plane or point.
Datum	A precise point from which all dimensions and measurements are referenced.
Ordinate dimensioning	Dimensions placed at the end of an extension line without using dimension lines or arrowheads. (Each dimension relates to zero coordinates along one or more axes.)
Zero coordinate	A starting point from which all dimensions along a specific axis originate. A fixed reference point for all dimensions on a plane line.
Tabular dimensioning	A drawing on which the dimensions and specifications of each feature are given in table form (coordinate chart).
Coordinate chart	A table accompanying a part drawing. A table that gives dimensions to locate each feature with respect to zero coordinates. (The chart contains other feature specifications.)

―――――――――――――― SUMMARY ――――――――――――――

Conventional work-holding devices are used on vertical millers. Surfaces that provide reference points are aligned with a dial indicator.

Holes may be formed on the vertical milling machine by using standard cutting tools that are held in chucks and adapters in the spindle. Conventional drills, machine reamers, countersinks, counterbores, and taps are used. Each of these cutting tools produces a fixed-size hole. The size is determined by the side and/or end cutting edges.

Machine tap holders and tapping attachments are used for internal threading.

Cuts may be taken at any simple or compound angle from 0° (right angle to the milling machine table) to a horizontal plane angle (180°).

The longitudinal and cross slide table and the vertical knee micrometer collars are used for positioning hole-forming and shaping processes at precise locations.

Measuring bars and dial indicator attachments extend the precision setting potential of the table from 0.001" to 0.0001" (0.02mm to 0.002mm).

The offset boring head is used primarily for machining precise-diameter holes and arcs that are straight and concentric to a fixed center and that are machined to a specific surface texture. Holes of any size may be bored.

The combination boring and facing head extends the range of machining operations. The boring process may be followed by feeding a cutting tool across the face to form a shoulder as in counterboring or producing a stepped surface.

The standard offset boring head set contains boring bars and reducer bushings that accommodate a selected group of boring tools. These tools are adaptable to most general purpose boring operations.

The main features of a boring head include the shank, body with a tool slide that is actuated by a screw, and a direct-reading micrometer dial. Cutting tools, boring bars, or reducer bushings are held securely in the tool slide with a setscrew.

Boring head shanks are commercially available to fit R-8, #30 NS, #40 NS, and #50 NS spindles. The larger the boring head shank, the larger the cutter sizes that may be accommodated to bore larger diameters.

The axis of the cutting edge of the boring tool must be positioned so that it falls along the centerline of the tool slide.

The same factors governing the selection and use of cutting tools for lathe work, drilling machines, and horizontal milling machine processes apply to boring tools. Consideration is given to rake, relief, and clearance angles and the shape of the cutting edge.

High-speed steel, carbide-tipped, and solid carbide cutters are generally used for boring.

Quill feeds of 0.003" and 0.006" (0.08mm and 0.15mm) are used for roughing cuts. The 0.0015" (0.04mm) quill feed is usually used for finish cuts.

The boring tool is moved away from a final finish cut when the cutter is returned to the starting position.

The heaviest boring bar and cutter, greatest depth, and coarsest feeds that are practical should be used for efficient boring. Cutting fluids are recommended consistent with the fluids used for similar machining processes in lathe work.

The following causes of tool deflection must be considered and corrected: Worn or incorrectly ground cutter, too high a cutting speed and/or too coarse a feed and/or too deep a cut, poorly supported or secured workpiece, incorrect quality and quantity of cutting fluid, and interrupted cut and/or movement of the tool slide.

The vertical spindle head and the shaper head are two basic attachments for the vertical miller.

The vertical spindle head is used with cutting tools that require rotary and/or planetary motion, as in the case of a facing process.

HOLE FORMING, BORING, SHAPING **283**

The shaper head actuates cutting tools that depend on a reciprocating motion.

The shaper head attachment permits second machine processes to be performed on the vertical miller. Through- or partial-depth surfaces may be shaped vertically or at any simple or compound angle. The head permits the same angular adjustments as a vertical spindle head.

Shaper tool sets have standard preformed square, round, and diamond-shaped vertically or at any simple or compound angle. The head permits the same angular adjustments as a vertical spindle head.

The shaper head is adapted to perform operations that often require special broaches and to cut gear teeth, racks, and other intricate shapes.

The ram on the shaper head may be adjusted for position in relation to the workpiece. Cutting strokes may be varied from 0" to 4" (0mm to 102mm) on toolroom models.

Shaper heads are designed to operate with any one of six speeds within a slow speed range of 35 to 210 strokes per minute. High speed ranges of from 70 to 420 strokes per minute are also available.

Baseline, ordinate, and tabular dimensioned drawings are used to accurately lay out a part, indicate the amount of table movement, and take measurements. Dimensions are related to datums and X, Y, or Z axes.

Radial surfaces may be shaped or holes may be positioned and formed by using a rotary table or a dividing head.

Standard sine bars and sine plates and other work-positioning accessories and tools, as used for horizontal miller setups, are applied to hole-forming, boring, and shaping processes on the vertical miller.

Safe practices relate to tool, machine, and personal safety. Factors affecting depth of cut, cutting speed and feed, rigidity of cut, and machining accuracy and surface texture must be considered by the operator.

Assistance should be secured in removing or replacing a heavy vertical milling machine or shaper head, rotary table, and/or dividing head.

Measurements should be taken only when the vertical spindle of the shaper ram is stopped.

The table saddle, knee, and offset boring holder, rotary table, dividing head, tool slide, and other positioning devices should be secured as a rigid tooling and machining setup.

UNIT 15 REVIEW AND SELF-TEST

1. a. Identify three basic hole-forming processes that are common to vertical milling machine work.
   b. Indicate the cutting tool that is used for each of the three processes.
   c. Name the tool-holding device for each cutting tool.
2. Explain how the T-slot grooves on a vertical milling machine table may be used to position a rectangular workpiece for work processes that are parallel to the table axis.
3. State three advantages of using a vertical milling machine instead of using a drilling machine for drilling, reaming, counterboring, countersinking, and tapping.

4. a. Name three design features of an offset boring head.
   b. Explain the function served by each feature.

5. a. Secure the specifications of an offset boring head.
   b. Furnish technical information about the following design features: (1) offset boring head manufacturer, (2) shank style and size, (3) draw-in bar thread specifications, (4) boring diameter range, (5) boring bar set lengths, (6) diameter of boring bar, and (7) total tool slide travel.

6. State two practices to observe to ensure maximum rigidity in setting up a boring head and boring tool.

7. Give (a) two possible causes of deflection in a boring tool and (b) the corrective steps that the operator may take.

8. State two unique applications of the shaper head that save second machine setups or special tooling.

9. Tell how the baseline dimensioned, two-view drawing in Figure 15-9 is used by the vertical milling machine operator to position the table for drilling and tapping holes A and drilling and reaming holes B.

10. State two tool or machine safety factors to consider to ensure dimensional accuracy when shaping is performed.

PART FOUR

Horizontal Milling Machines

SECTION ONE Horizontal Milling Machine Technology and Processes
        UNIT 16 Sawing, Slotting, and Keyseat and Dovetail Milling
        UNIT 17 Helical and Cam Milling
SECTION TWO Gears (Spur, Helical, Bevel, and Worm) and Gear Milling
        UNIT 18 Gear Developments, Design Features, and Computations
        UNIT 19 Machine Setups and Measurement Practices for Gear Milling

SECTION ONE

Horizontal Milling Machine Technology and Processes

This section deals with the technology, principles, and mathematics related to horizontal milling processes. General-purpose machine setups and step-by-step procedures are described for sawing, slotting, keyseat milling, and dovetail cutting. The computation and measurement of keyseats and dovetails are included.

The milling of helical forms and cam milling are also described. Typical machine and dividing head setups, cutter selections, calculations of change gears and of angles for milling cams, and methods of machining are covered.

UNIT 16

Sawing, Slotting, and Keyseat and Dovetail Milling

Sawing, slotting, and keyseat and dovetail milling on the horizontal milling machine require different setups and additional safety considerations than general milling processes. Since special job requirements control the form of the cutter, the teeth and cutting areas are not as strong as conventional milling cutters. For example, each tooth on a double-angle milling cutter has limited strength toward the apex of the angle as compared to the teeth of a helical plain or a side milling cutter.

The cutters used for sawing, slotting, and keyseat and dovetail milling are easily damaged and require great care to prevent fracturing. This unit describes the use of these cutters in relation to the work-holding setups, procedures, and safe practices for sawing, slotting, and keyseat and dovetail milling.

METAL-SLITTING SAWS AND SLOTTING CUTTERS

Sawing refers to ordinary slitting, slotting, and cutting-off processes. *Slitting* relates to the use of a saw to cut through a tube or ring or to separate it into a number of parts. *Slotting* involves the machining of a narrow-width groove.

Manufacturers' tables usually provide guidelines for the number of teeth to use for slitting and sawing steel. Saws having two-thirds the number of teeth as saws used for steel are used on brass and deep slots. Copper is one of the more difficult metals to cut on the milling machine. Cutters for copper are specially shaped and have a small number of teeth compared to cutters for steel.

The three general designs of milling cutters used for sawing (slitting) on the horizontal milling machine are shown in Figure 16-1. The cutters are thin and are called *slitting saws*. The *plain metal-slitting saw* (Figure 16-1A) is ground concave on both sides for clearance. The saw is made of high-speed steel with standard keyways. The cutter hub has parallel sides. The hub is the same width as the cutter teeth.

The *metal-slitting saw with side teeth* (Figure 16-1B) provides side chip clearance. The additional side clearance spaces between teeth and the recessed hub permit chips to flow freely away from the cutter teeth. Thus, binding and scoring of the work are prevented and the amount of heat generated is reduced. This saw is suitable for regular and deep slotting and sinking-in cuts.

Metal-slitting saws with side teeth are commercially available in face widths that range from

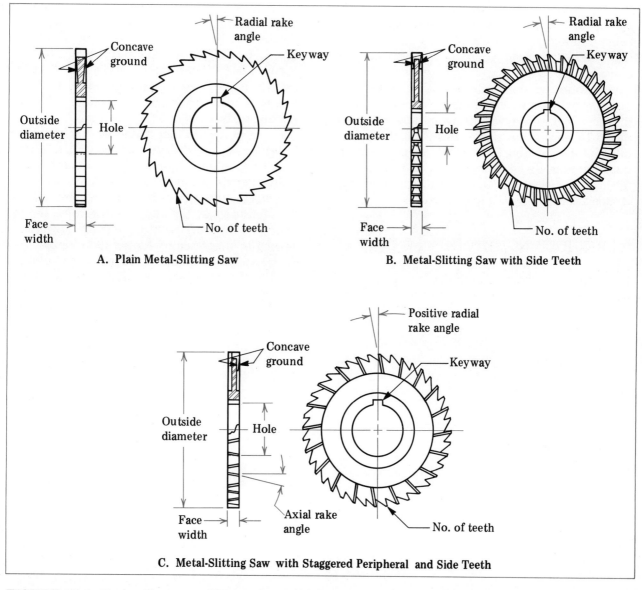

FIGURE 16-1 Design Features of Three General Slitting and Slotting Milling Cutters

1/16" (1.58mm) to 3/16" (4.76mm). The number of teeth on a cutter in this range is from 28 to 48. The outside diameters are from 2-1/2" (64mm) to 8" (200mm).

The *metal-slitting saw with staggered peripheral teeth* (Figure 16-1C) has an axial rake angle. The side teeth on this high-speed steel saw have a positive radial rake angle. The alternate helical (side) teeth help to eliminate chatter. The staggered teeth and side clearance provide the necessary chip space for deep cuts where heavy feeds are required. The side teeth are not designed for cutting.

Screw-Slotting Cutters

The *screw-slotting cutter* (Figure 16-2) is used principally to slot screw heads and to slit tubing, thin-gage materials, piston rings, and similar work. The sides of the cutter are ground concave. Screw-slotting cutters are available in three groups of standard sizes. These sizes and other specifications are listed in Table 16-1. Additional special screw-slotting cutters are produced to meet unusual specifications. Technical information relating to cutter diameter; bore size; number of teeth in cutter; and ranges of face width, wire gage, and screw head diameter is then furnished by the manufacturer.

WORK-HOLDING SETUPS

Rigidity is an important consideration in cutting-off, slitting, and slotting operations. Usually, when screw-slotting cutters are used, the parts to be slotted are nested in a fixture. A common type of positioning and holding fixture to accommodate a number of special screws for slotting is shown in the setup in Figure 16-3.

Sections of a workpiece are often cut off on the milling machine. A solid part, where no deflection would be produced by cutting action, is often mounted directly in a vise. The area to be cut off extends beyond the vise jaw. A plain cutter may be used, or a cutter with straight peripheral and side teeth or staggered peripheral teeth and alternate side teeth may be selected. The depth of the cutting-off operation, the nature and rigidity of the setup, the cutting fluid, and the machinability of the material are factors that guide the craftsperson in cutter selection.

Another common setup is to strap the workpiece directly to the milling machine table. The position of the cut and cutter must be over one of the table T-slots. This setup is illustrated in Figure 16-4. Climb milling is then used. Added rigidity is provided in climb milling as the cutter forces the work against the table or other supporting surface.

Cast iron and other materials that have a hard outer scale may require conventional milling. Conventional milling minimizes the cutting effect that otherwise would require the teeth to be continuously cutting into the scale. A common practice followed in the shop to overcome the tendency of the work to spring during conventional sawing is illustrated by the setup in Figure 16-5. A strap is placed across the face of the workpiece.

Slotting saws are sometimes used to mill narrow and other slots, where a side mill is impractical. If the slot is wider than the cutter face

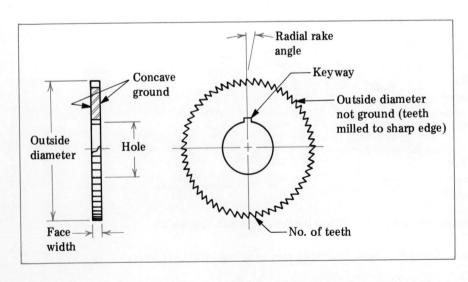

FIGURE 16-2 Design Features of Screw-Slotting Cutter

SLOTTING, SAWING, KEYSEAT AND DOVETAIL MILLING

TABLE 16-1 Specifications of Standard Screw-Slotting Cutters

Cutter Diameter	Bore Size	Teeth in Cutter	Ranges		
			Face Width	Wire Gage	Screw Head Diameter
2-3/4" (69.85mm)	1" (25.4mm)	72	0.020" to 0.144" (0.5mm to 3.7mm)	7-24	1/8" and smaller through 1" (3mm and smaller through 25mm)
2-1/4" (57.15mm)	5/8" (15.9mm)	60	0.020" to 0.064" (0.5mm to 1.5mm)	14-24	1/8" and smaller through 3/8" (3mm and smaller through 9mm)
1-3/4" (44.45mm)	5/8" (15.9mm)	90	0.020" to 0.064" (0.5mm to 1.5mm)	14-24	

width, two or more cuts may be required. The saw is moved to machine the wider slot so that each cut overlaps the preceding one.

Angle pieces requiring a slot to be machined are often mounted on a right-angle plate (for surfaces at 90° to each other). The sine plate is used for cutting precise slots at any angle.

HOW TO SAW ON A HORIZONTAL MILLING MACHINE

Selecting and Mounting the Saw

STEP 1 Select an appropriate, sharp, high-speed steel slitting saw.

Note: The type, diameter, and width must permit the arbor to clear the work-holding straps or device and the cutter and to cut through the workpiece. The direction of the teeth depends on whether the part is to be cut off by climb or conventional milling.

STEP 2 Select the shortest length of type A arbor that will accommodate the cutter and process. Mount the arbor in the spindle.

STEP 3 Position the slitting saw as close as possible to the column. Secure it on the arbor.

Note: A proper fitting key must be used. This key also drives the cutter and the collars next to the cutter. Additional support for thin cutters must be provided. Thin collars with an outside diameter that is larger than the regular collars are used and are placed on both sides of the cutter.

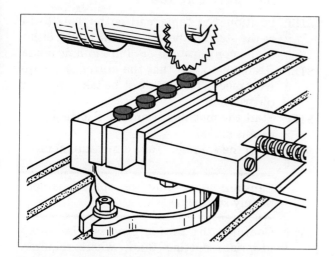

FIGURE 16-3 Simple Fixture for Positioning and Holding Parts to Be Slotted

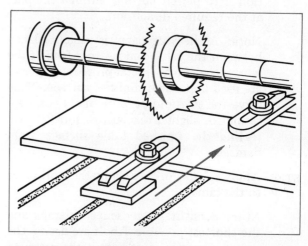

FIGURE 16-4 Workpiece and Cutter Positioned to Cut into Table T-Slot

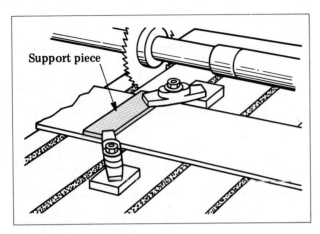

FIGURE 16-5 Support Piece Strapped across Workpiece to Prevent Springing during the Cut

STEP 4 Move the overarm and arbor support over the end of the arbor. Secure the arbor support for maximum rigidity during operation.

STEP 5 Use a table of cutting speeds. Determine the recommended starting sfpm (mm/min) according to the process and the machinability of the workpiece material. Set the spindle RPM according to the sfpm and the cutter diameter.

STEP 6 Set the machine table feed for the cutting-off process.

Mounting the Work on the Table

STEP 1 Clean and remove burrs from the table and workpiece.

STEP 2 Center the workpiece on the table. Position it to permit cutting through the part at the required dimension.

Note: A sheet of paper is usually placed between the table and bottom face of the workpiece. Soft, metal-protecting pieces are used under an unfinished surface. They serve the two-fold purpose of (1) compensating for slight irregularities and (2) protecting the finished table surface from damage.

STEP 3 Apply a slight force to clamp the workpiece to the table.

Note: Straight or gooseneck clamps and the shortest possible T-bolt are used. The cutter diameter must permit the setup to clear the straps and T-bolts.

Note: If sawing by the conventional method is done, a supporting piece is clamped across the top of the workpiece to prevent the workpiece from springing.

STEP 4 Set the workpiece at a right angle to the side of the table.

Note: A combination square is used to make the setup for general-purpose sawing. The workpiece is squared (aligned) by checking the position of one side with the blade of the steel square. The head of the combination square is held firmly against the side of the table (Figure 16-6).

STEP 5 To position the workpiece squarely, tap it with a soft-face hammer. Secure the clamps when the side of the workpiece is parallel with the blade of the square.

Note: A flat, straight piece of material is brought against the aligned edge. This piece is strapped in position as a stop. Other pieces may be positioned against the stop without further checking.

Positioning the Cutter and Workpiece

STEP 1 Move the cutter height so that it clears the workpiece.

STEP 2 Position the saddle so that the cutter is at the line of measurement over a table T-slot. This positioning permits cutting through the workpiece without damage to the table (Figure 16-7). Lock the saddle.

Note: With the spindle stopped, a steel rule is often used to measure the width of the part (Figure 16-8).

STEP 3 Position the table and cutter lengthwise at the start of the cut for machining either in the conventional or climb milling mode.

STEP 4 Recheck to see that the cutter clears the workpiece. The cut must be taken without obstruction.

STEP 5 Start the machine. Hand-feed until a small area is milled.

STEP 6 Stop the spindle. Take a measurement.

Caution: Make sure the spindle is completely stopped.

STEP 7 Continue the sawing process. Use an appropriate cutting fluid, where required. Engage the power feed when the cutting action is correct.

SLOTTING, SAWING, KEYSEAT AND DOVETAIL MILLING 291

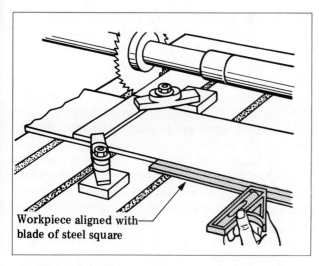

FIGURE 16-6 Setting Workpiece for Cutting-Off Process by Using a Steel Square

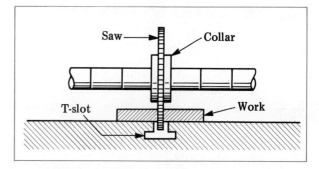

FIGURE 16-7 Cutter and Setup Positioned Clear of Workpiece and Table T-Slot

STEP 8 Stop the spindle. Burr the cutoff part.

Note: Additional pieces may be cut off by using a stop as a guide. The workpiece is simply moved to the next length and strapped.

Note: Thick parts may be held directly on parallels in a chuck.

STEP 9 Disassemble the setup when the required number of pieces are cut. Clean the machine, straps, and bolts. Return each item to a proper storage place. Wipe the milling machine clean.

HOW TO MILL NARROW, SHALLOW SLOTS

Using a Screw-Slotting Cutter

STEP 1 Select a screw-slotting cutter to accommodate the width of slot to be cut.

STEP 2 Select a type A or B milling machine arbor. Mount the arbor securely in the spindle.

STEP 3 Place the cutter in the arbor. Use a key to drive the cutter. Secure the cutter, key, and collars on the arbor.

STEP 4 Determine the correct cutting speed for the process. Set the spindle speed to the correct RPM for the sfpm.

STEP 5 Set the machine feed and the location of the table trip dogs.

Note: The machine feed may be set at a slower rate at first. It may then be increased as conditions warrant.

STEP 6 Secure the workpieces to be slotted.

Note: If a fixture is used, a screw is set in each slot. A slight force is applied on the fixture parts to lock them and the workpieces securely. Then, each screw head is tapped to seat it on the locating surfaces of the fixture. Finally, the vise jaws are tightened to apply adequate force against the screw bodies to hold them securely.

STEP 7 Locate the cutter centrally in relation to the head and body of each part.

STEP 8 Start the spindle. Turn the knee handwheel until the cutter grazes the workpiece. Set the knee graduated collar at zero.

STEP 9 Move the table lengthwise to back the cutter away from the first part.

FIGURE 16-8 Using a Steel Rule to Measure Width of Part to Be Cut Off

STEP 10 Raise the knee to the required depth of cut. Lock the knee.
STEP 11 Start the coolant flow. Take a trial cut. Stop the machine. Check to see that the cut is centered and the slot is machined to the correct depth.
STEP 12 Make adjustments, if needed. Continue the process by hand-feeding. Engage the power feed when all machining conditions are satisfactory.
STEP 13 Stop the machine. Use a hard-bristle brush to remove chips.
STEP 14 Unload the slotted parts. Remove burrs. Recheck the dimensional accuracy of the slots. If necessary, make machine adjustments.
STEP 15 Wipe the fixture surfaces. Reload.
STEP 16 Continue to slot the remaining parts.

Note: The dimensional accuracy of the slotting process should be checked regularly.

STEP 17 Disassemble the setup. Clean the vise and fixture, the table, and the cutter and arbor. Return each item to its proper storage area.

KEYSEAT MILLING

In general practice, *keyseat milling* refers to the machining of a groove in a shaft (keyseat). Three basic forms of keyseats are shown in Figure 16-9. The *plain (open) keyseat* may be milled with a single side milling cutter. When the cutter is the correct width, the keyseat is produced by taking a single cut. Other widths may be milled by taking second cuts. The *Woodruff keyseat* is produced by a fixed-size Woodruff keyseat cutter. The *rounded-end (closed or sunk) keyseat* may be cut at any position along an arbor or drive shaft. This keyseat is produced with an end mill.

In keyseat milling, the manner in which a workpiece is held depends on the method used to produce the keyseat. Four common workholding setups for round parts are illustrated in Figure 16-10. One of the simplest setups is to clamp the workpiece directly in a vise (Figure 16-10A). The workpiece is gripped at two positions that are opposite to each other. When a V-block is used in a vise, the workpiece is gripped at three places (Figure 16-10B).

A workpiece that is too large to be held in a vise is usually mounted directly on the table (Figure 16-10C). The workpiece is placed in the table groove. Straps are used to clamp the workpiece securely. A large workpiece may also be centered on V-blocks and strapped to the table (Figure 16-10D).

Computing the Depth of Cut for a Keyseat

Shop drawings for keyseats are dimensioned from the depth of cut to a theoretically sharp edge. This edge is formed where the vertical side of the keyseat and the outside diameter of the workpiece meet. In keyseat milling, the workpiece must be raised to the keyseat depth dimension plus the vertical height of the arc. The arc represents the additional material to be removed. The height of the arc varies according to the radius of the workpiece and the width of the keyseat. The following formula (illustrated in Figure 16-11) is used to compute the height of the arc:

$$A = r - \sqrt{r^2 - (1/2w)^2}$$

where A = height of arc,
       r = radius of workpiece,
       w = keyseat width.

Example: Determine the total depth of cut to mill a 1/2" (0.500") square keyseat on a 3" diameter shaft. By using the arc height formula and substituting values,

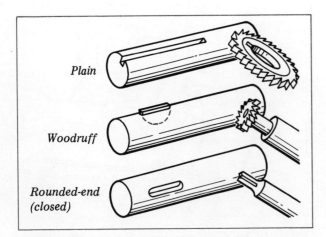

FIGURE 16-9 Three Basic Forms of Keyseats

SLOTTING, SAWING, KEYSEAT AND DOVETAIL MILLING

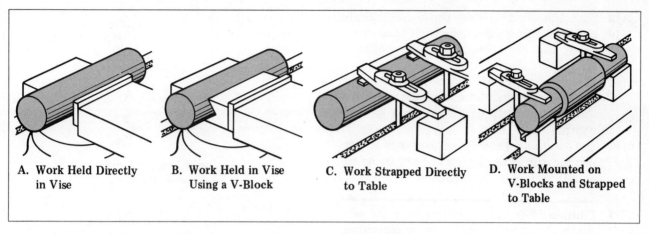

FIGURE 16-10 Four Common Work-Holding Setups for Round Parts

$$A = r - \sqrt{r^2 - (1/2w)^2}$$
$$= 1.500 - \sqrt{1.500^2 - (1/2 \times 0.500)^2}$$
$$= 1.500 - \sqrt{1.479}$$
$$= 0.021'' \text{ (arc height)}$$

The total depth of cut from the top of the shaft to the bottom of the keyseat is equal to one-half the keyseat width plus the arc height. In the example,

$$\text{total depth} = 1/2 \,(0.500) + 0.021$$
$$= 0.250 + 0.021$$
$$= 0.271'' \text{ (distance the table is raised)}$$

HOW TO MILL A KEYSEAT

Using a Standard Milling Cutter

STEP 1 Mount and secure the workpiece near the center of the table.

Note: The location must provide the least overhang and maximum rigidity for the work-holding and cutter setups.

STEP 2 Select a standard high-speed steel milling cutter.

Note: The cutter width must permit machining the keyseat width within the specified tolerance. The cutter diameter must provide adequate clearance between the arbor bushing, arbor support, and the work-holding setup.

STEP 3 Position the cutter on the arbor so that the operation is performed under as rigid a setup as possible. Mount and secure the cutter and collars. Mount and secure the overarm and arbor support to strengthen the cutter setup.

STEP 4 Align the workpiece so that it is centered with the cutter.

STEP 5 Determine the cutting speed. Set the spindle speed at the RPM closest to the recommended cutting speed.

STEP 6 Set the table feed (ipm or mm/min). Use the rate given in a feed table as a starting point.

Note: The feed rate is increased when the cutting action and setup will support a higher feed.

STEP 7 Make a final check for adequate clearance before starting the cut.

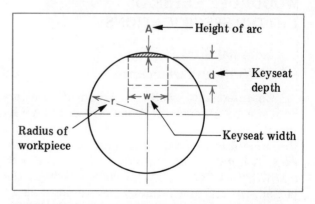

FIGURE 16-11 Depth of Keyseat and Height of Arc

HORIZONTAL MILLING MACHINES

STEP 8 Start the spindle. Raise the knee carefully until the cutter grazes the top of the workpiece. Set the knee micrometer collar at zero.

STEP 9 Move the cutter off the workpiece to the starting position. Raise the knee and bring the workpiece up to the required keyseat depth.

STEP 10 Take a trial cut for a short distance to permit measuring the keyseat. Flow on an adequate supply of cutting fluid.

Caution: Remove the cutter from the keyseat and stop the spindle before measuring the width and depth.

Note: The width may be gaged by a GO-NO-GO gage or a piece of key stock (Figure 16-12A).

Note: A micrometer is used for a precise measurement of keyseat depth. The depth may also be measured with a steel rule from the side of the keyseat (Figure 16-12B).

STEP 11 Make whatever cutter adjustments are necessary for changes in depth or width. Lock the knee.

STEP 12 Cut the keyseat to length by engaging the power feed. Near the end of the cut, disengage the power feed. Feed by hand to the layout line. This line indicates the end of the keyseat.

STEP 13 Brush away all chips. Remove machining burrs from the workpiece. Place each item in its proper storage area.

WOODRUFF KEYSEAT CUTTER SPECIFICATIONS

Nominal Sizes of Width and Diameter

Commercially available Woodruff cutters conform to American National Standard (ANSI) specifications. The number of the cutter and the American Standard key numbers are identified. A notation such as #506 AMERICAN STANDARD WOODRUFF CUTTER indicates both the cutter size and the nominal key dimensions of width and diameter. The last two digits give the cutter diameter in eighths of an inch. The one

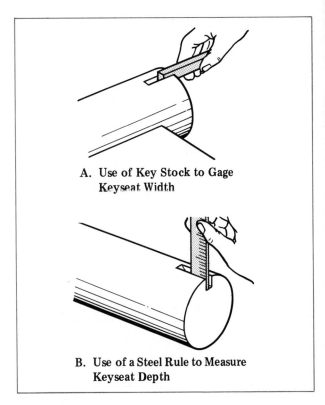

A. Use of Key Stock to Gage Keyseat Width

B. Use of a Steel Rule to Measure Keyseat Depth

FIGURE 16-12 Checking Depth and Width of a Keyseat

or two digits preceding the last two digits give the cutter width in thirty-seconds of an inch. The #506 cutter measures 5/32″ × 6/8″. The cutter is 5/32″ wide × 3/4″ diameter.

Stock cutter sizes range from 1/16″ (1.5mm) to 3/8″ (9.5mm) wide and 1/4″ (6.35mm) to 1-1/2″ (38.1mm) diameter. These sizes are nominal dimensions. The 1/2″ (12.7mm) cutter shanks are of uniform diameter. Shank-type cutters are usually mounted in a collet (Figure 16-13).

FIGURE 16-13 Straight-Shank Woodruff Keyseat Cutter Mounted in a Spindle Collet

Nominal Height above Shaft

Handbook tables provide dimensional data for standard and special Woodruff keys. Key numbers are recommended for different shaft diameters. Woodruff keys project above the shaft diameter a distance equal to one-half the cutter width ±0.005" (±0.12mm).

The depth of the keyseat in a shaft must be held to a tolerance of +0.005"/-0.000" (+0.12mm/-0.00mm). Table 16-2 is a partial table of ANSI dimensions for Woodruff keys and keyseats. The complete table is included in the Appendix, in trade handbooks, and in literature of cutter manufacturers. An important dimension for the machine operator is the depth (B in Table 16-2) of the keyseat from the top of the shaft. This depth represents the distance the workpiece/knee is raised from the initial zero cutter setting.

CENTERING THE WORKPIECE AND A KEYSEAT CUTTER

A Woodruff keyseat cutter must be centered in relation to the centerline of the workpiece (Figure 16-14). The cutter axis is first aligned with the horizontal centerline of the keyseat. The workpiece is then carefully brought to the cutter.

A zero saddle (transverse) position is established by moving the workpiece toward the cutter

TABLE 16-2 Partial Table of ANSI Dimensions (in Inches) for Woodruff Keys and Keyseats (ANSI B 17.2, R1978)

Woodruff		Keyseat Shaft					Key above Shaft	Keyseat Hub	
Key Number	Nominal Size Key	Width A*†		Depth B† +0.005 -0.000	Diameter F		Height C† +0.005 -0.005	Width D† +0.002 -0.000	Depth E +0.005 -0.000
		Min.	Max.		Min.	Max.			
202	1/16 × 1/4	0.0615	0.0630	0.0728	0.250	0.268	0.0312	0.0635	0.0372
202.5	1/16 × 5/16	0.0615	0.0630	0.1038	0.312	0.330	0.0312	0.0635	0.0372
302.5	3/32 × 5/16	0.0928	0.0943	0.0882	0.312	0.330	0.0469	0.0948	0.0529
406	1/8 × 3/4	0.1240	0.1255	0.2455	0.750	0.768	0.0625	0.1260	0.0685
506	5/32 × 3/4	0.1553	0.1568	0.2299	0.750	0.768	0.0781	0.1573	0.0841
606	3/16 × 3/4	0.1863	0.1880	0.2143	0.750	0.768	0.0937	0.1885	0.0997
806	1/4 × 3/4	0.2487	0.2505	0.1830	0.750	0.768	0.1250	0.2510	0.1310
1228	3/8 × 3-1/2	0.3735	0.3755	0.7455	3.500	3.535	0.1875	0.3760	0.1935
1428	7/16 × 3-1/2	0.4360	0.4380	0.7143	3.500	3.535	0.2187	0.4385	0.2247
1628	1/2 × 3-1/2	0.4985	0.5005	0.6830	3.500	3.535	0.2500	0.5010	0.2560
1828	9/16 × 3-1/2	0.5610	0.5630	0.6518	3.500	3.535	0.2812	0.5635	0.2872
2028	5/8 × 3-1/2	0.6235	0.6255	0.6205	3.500	3.535	0.3125	0.6260	0.3185
2228	11/16 × 3-1/2	0.6860	0.6880	0.5893	3.500	3.535	0.3437	0.6885	0.3497
2428	3/4 × 3-1/2	0.7485	0.7505	0.5580	3.500	3.535	0.3750	0.7510	0.3810

*Values for width A represent the *maximum* keyseat (shaft) width that will assure the key will stick in the keyseat (shaft). The *minimum* keyseat width permits the largest shaft distortion acceptable when assembling a maximum key in a minimum keyseat.
†Dimensions A, B, C, and D are taken at the side intersection.

296 HORIZONTAL MILLING MACHINES

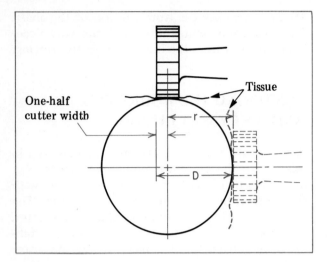

FIGURE 16-14 Centering a Woodruff Keyseat Cutter

HOW TO MILL A KEYSEAT WITH A WOODRUFF CUTTER

STEP 1 Select a Woodruff keyseat cutter with a width and a diameter that meet the dimensional requirements of the job.

STEP 2 Mount the cutter in the machine spindle. Use a spring collet or adapter for a standard Woodruff keyseat cutter. Use an arbor for hole-type cutters.

STEP 3 Align and secure the workpiece on V-blocks in a vise or other suitable work-holding device.

STEP 4 Center the cutter at the required distance from the end of the workpiece and on its vertical axis. Lock the table and saddle.

STEP 5 Start the coolant flow. Set the micrometer collar at zero at the starting position.

STEP 6 Feed the workpiece upward by hand feed to the required depth. Note this depth reading.

Note: The force on a Woodruff keyseat cutter increases from zero to a maximum at the end of the cut. The part is hand-fed to permit the depth of cut and force on the cutter to be decreased as the cutting action increases.

STEP 7 Stop the machine. Lower the knee. Clean the cutter, workpiece, and working area. Remove burrs from the milled edges of the keyseat.

STEP 8 Insert a Woodruff key into the keyseat. Use a soft-face hammer to tap the key slightly to seat it.

face until a piece of fine tissue may be pulled through. The knee is lowered to clear the cutter. The workpiece is then moved a distance (D in Figure 16-14) equal to the radius of the workpiece plus one-half the cutter width plus the thickness of the tissue.

The spindle is started and the workpiece is slowly and carefully moved up until the cutter grazes the outside diameter. The cutter is then fed to depth, and a steady coolant flow is used, where required.

A properly machined keyseat requires the key to be lightly tapped to seat it. The height the key extends and conforms to the specified tolerance may be measured by micrometer (Figure 16-15).

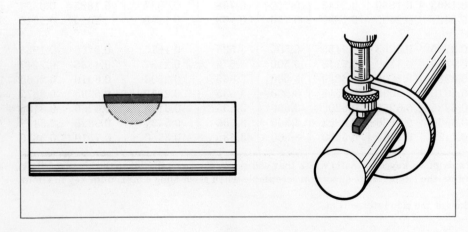

FIGURE 16-15 Measuring Key Height with a Micrometer

Note: The overall height is measured from the shaft diameter to the top of the key. If the keyseat requires further machining, the key must be removed.

STEP 9 Feed the workpiece the additional amount beyond the previous micrometer graduation on the knee elevating screw.

STEP 10 Disassemble the setup. Wipe the cutter and work-holding device clean. Return each item to its correct storage area.

CHARACTERISTICS OF DOVETAILS

Two common methods are used to machine *dovetail slides* in jobbing shops. The corresponding angular sides and the common flat base of mating parts may be formed by feeding a single-point cutting tool at the required angle. This method is the shaper method. The second method is to machine the dovetail on a vertical or horizontal miller. A formed single-angle *dovetail milling cutter* is used. Figure 16-16 shows a vertical milling machine setup for machining a dovetail slide.

On cast parts, dovetail slides are usually preformed. Sufficient material is provided (left on) to permit machining below (under) the scale surface. Machined dovetail slides are produced within precision limits and with low microfinish surface textures.

Other dovetail slides require milling from a solid piece. In such milling, a rectangular area is rough machined. The angle is produced by roughing out with a formed angle cutter. The sides and base are then finish machined with a dovetail cutter. The same cutter is used to machine internal and external dovetails.

Computing and Measuring Internal and External Dovetails

The dimensional accuracy of dovetail slides may be gaged or measured. An optical comparator may be used to establish the accuracy of the angles and other linear dimensions. Dovetail slides may also be measured with a gage or a micrometer.

Figure 16-17 illustrates the dimensional measurements of a dovetail slide. Dimensions X on the external dovetail and Y on the internal dovetail may be measured by micrometer or vernier caliper. Dimension X for the measurement of an external dovetail may be calculated by the following formula:

$$X = d \times (1 + \cot 1/2 \text{ angle } e) + a$$

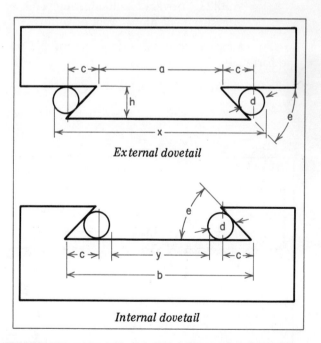

FIGURE 16-16 Milling a Dovetail Slide with a Single-Angle Cutter on a Vertical Milling Machine

FIGURE 16-17 Dimensional Measurements of a Dovetail Slide

Dimension Y for the measurement of an internal dovetail is also found by formula:

$$Y = b - d \times (1 + \cot 1/2 \text{ angle } e)$$ and
$$c = h \times \cot e$$

In these formulas,

X and Y = overall measurements,
- a = the smallest linear measurement of the dovetail,
- b = the largest dimension (overall width) of the dovetail,
- c = the width (base) of the right triangle,
- d = the diameter of the measuring rolls,
- e = the required included angle,
- h = the dovetail height.

Example: An internal dovetail and an external dovetail are to be milled according to the dimensions given on the shop drawing in Figure 16-18. Calculate the overall dimensions X and Y, using 0.500″ (12.7mm) diameter rolls. (Check appropriate tables of trigonometric functions to find the cotangents of angles.)

To calculate external measurement X, use the formula

$$X = d \times (1 + \cot 1/2 \text{ angle } e) + a$$

By substituting values,

$$X = 0.500 \times (1 + \cot 22\text{-}1/2°) + 4.125$$
$$= 0.500 \times (1 + 2.4142) + 4.125$$
$$= 5.8321″ \text{ (overall outside micrometer measurement of external dovetail)}$$

To calculate internal measurement Y, use the formula

$$Y = b - d \times (1 + \cot 1/2 \text{ angle } e)$$

By substituting values,

$$Y = b - 0.500 \times (1 + \cot 22\text{-}1/2°)$$

Determining Y involves the overall width (b) of the dovetail. The overall width is equal to the smallest linear measurement (4.125 in this example) of the dovetail plus the width (c) of the right triangle multiplied by 2. That is, here

$$b = 4.125 + 2c$$

To complete the calculation for b, c must be found. The width of the right triangle (c) is equal to the dovetail height (h) multiplied by the cotangent of angle e. That is,

$$c = h \times \cot e$$
$$= 0.625 \times \cot e$$

From a table of trigonometric functions,

$$\cot e = \cot 45° = 1.000$$

Therefore,

$$c = 0.625 \times 1.000$$
$$= 0.625″$$

Now that c has been found, the substitution of values in the formula for b may be completed:

$$b = 4.125 + 2c$$
$$= 4.125 + 2(0.625)$$
$$= 5.375″$$

Finally, Y may be calculated by substituting 5.375″ for the length of b in the formula for Y:

$$Y = b - d \times (1 + \cot 1/2 \text{ angle } e)$$
$$= 5.375 - 0.500 \times (1 + \cot 22\text{-}1/2e)$$
$$= 5.375 - 0.500 \times (1 + 2.4142)$$
$$= 5.375 - 1.707$$
$$= 3.668″ \text{ (overall inside micrometer measurement of internal dovetail)}$$

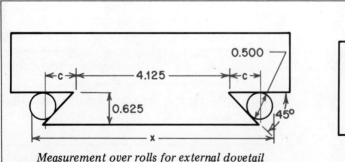
Measurement over rolls for external dovetail

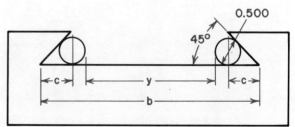
Measurement inside rolls for internal dovetail

FIGURE 16-18 Calculating Dimensions for Measuring Internal and External Dovetails

HOW TO MILL INTERNAL AND EXTERNAL DOVETAILS

Using a Dovetail Milling Cutter

STEP 1 Compute the overall internal and external measurement of the angular surfaces.

Note: A roll diameter is selected that permits easy measurement with a micrometer or vernier caliper.

STEP 2 Mount a dovetail milling cutter of the required cutting angle in the spindle.

STEP 3 Set the spindle RPM and cutting speed (table feed), depending on the machinability rating of the workpiece and other machining factors.

STEP 4 Take a roughing cut on the angular sides and across the face of the dovetail.

Note: The micrometer collar readings are recorded for the cutter setting at each angular side.

STEP 5 Stop the spindle and lock it. Clean the workpiece, working area, and measuring tools.

STEP 6 Measure the dovetail width. The two reference points for a micrometer or vernier reading are the two rolls. These rolls are held against the angular surfaces of the dovetail.

STEP 7 Take finish cuts to machine the dovetail within the laid-out dovetail area.

Note: The finish cuts for a 45° dovetail may be determined by subtracting the micrometer reading from the computed overall measurement. The cutter is set to the correct width by feeding it the difference between the computed measurement and the measured one. Usually an equal amount is taken on each angle side for the finish cuts.

STEP 8 Clean the workpiece, work area, tools, and accessories. Remove burrs.

STEP 9 Remeasure the dovetail slides. Take subsequent cuts if needed.

STEP 10 Check the quality of surface finish. Select a comparator specimen according to the material, process, and required microfinish value for surface texture.

STEP 11 Brush away all chips. Wipe the surfaces clean. Remove burrs.

STEP 12 Take down the setup. Clean and replace all tools and materials and place in proper storage areas.

SAFE PRACTICES FOR SAWING, SLOTTING, AND MILLING KEYSEATS AND DOVETAILS

—Stop the spindle and lock it. The cutter must be at rest before the width and depth to which a slot is being milled are measured.

—Stand to one side and out of line of travel of a revolving saw or slotting cutter. The teeth or cutter occasionally fracture if misused. The flying pieces may cause personal injury.

—Drive the cutter by using a correct size key between the keyway of the arbor, the cutter, and the spacing collars. A sliding-fit key ensures a positive drive.

—Use climb milling, where practical, for sawing, slotting, and Woodruff keyseat cutting operations. The downward force prevents the springing of thin workpieces during machining.

—Check the hand of the cutter against the direction of rotation of the spindle. Fine saw teeth and screw-slotting cutters are easily damaged when fed even for part of a revolution in the wrong direction.

—Make sure there is sufficient operating space between the cutter, arbor and support arm, and the workpiece and work-holding setup.

—Set the position of a saw to run into a table slot. This safe practice applies to workpieces that are to be cut off using a direct table mounting setup.

—Select packing blocks that are of the correct height. The force exerted on a clamp must be directed for maximum holding power.

—Tap the vertical feed crank by hand to control raising the work to starting position. The micrometer collar is set at zero when the revolving cutter just grazes the surface of the workpiece.

—Use a hand feed for milling a Woodruff keyseat. Hand-feeding permits the feed to be decreased as cutting forces increase.

—Begin each machining process with the starting speeds and feeds recommended by the tool manufacturers. These speeds and feeds may be increased when the operator determines that the machining process may safely be carried on at higher speeds and/or feeds.

—Supply a sufficient quantity of cutting fluid, when required. The chips must flow rapidly away from the cutter teeth and workpiece.

—Make sure there is no backlash in the feed screws when a cutter is reset for additional cuts.

TERMS USED FOR SAWING, SLOTTING, AND MILLING KEYSEATS AND DOVETAILS

Slotting	Process of machining a narrow-width groove to divide a piece of stock (or a workpiece) into one or more parts or segments.
Sawing (milling machine)	A cutting-off process. Cutting through a workpiece to remove excess material or to machine-saw parts to specified lengths.
Metal-slitting saw	A milling cutter with straight or staggered peripheral teeth with or without side teeth. A sawing and slotting cutter relieved on the sides for clearance.
Screw-slotting cutter	A milling cutter with concave ground sides and peripheral teeth ground to a sharp edge. A milling cutter used primarily for screw slotting and slitting.
Keyseat milling	Machining a groove in a shaft to receive one-half a square or rectangular key.
Woodruff keyseat	Usually refers to a circular groove milled with a Woodruff keyseat cutter.
Woodruff keyseat numbers	American National Standard (ANSI) specifications for cutter widths and diameters.
Centering the cutter (Woodruff keyseat)	Aligning the vertical axis of a Woodruff cutter with the centerline of the workpiece.
Dovetail slide	An internal or external wide groove having sides machined at the same angle. Mating parts that are fitted with corresponding angular sides and a common flat base surface.
Measuring rolls	Precision-ground rolls of a specified diameter. Rolls placed against the angular surfaces of a dovetail. Rolls that permit the accurate linear measurement of a dovetail.
Dovetail milling cutter	A single-angle milling cutter that simultaneously mills the angular face and the adjoining flat surface of a dovetail.

SUMMARY

Sawing and slitting processes are performed on the milling machine with plain metal-slitting saws, plain saws with side teeth, and slitting saws with staggered peripheral and alternate side teeth.

Standard metal-slitting saws with side teeth range in width from 1/16" (1.58mm) to 3/16" (4.76mm).

High-speed steel saws have an axial rake angle. The side teeth have a positive radial rake angle.

The smallest possible diameter of slotting and slitting saws is used. The size, however, must be large enough to provide safe clearance for the cutting process.

Screw-slotting cutters are used principally to cut screw-head slots and for limited depth sawing and slitting processes.

Narrow slots may be milled with a screw-slotting cutter. A type A or B arbor may be used, depending on the machining requirements and cutter design.

Climb milling is safer and more functional than conventional milling for through slotting and slitting on table-mounted workpieces.

An extra strap is placed across a thin workpiece to provide additional rigidity.

Like all other machining processes, the cutting speed is determined by the process itself, size and features of the workpiece, and the rigidity of the setup. Suggested starting speeds may be increased when the operator establishes that speed increases may be done safely.

Sawing, slitting, flat keyseat, and dovetail cuts are hand fed at the start of the cut. Power feeds are engaged when cutting conditions warrant.

The three basic forms of keyseats are: plain (open), Woodruff, and rounded-end (closed or sunk).

The depth of a keyseat is measured from the outside diameter of the part.

Regular keyseats may be milled with a plain side milling cutter or a keyseat cutter. Keyseats that are wider than the face width of the cutter may be machined by resetting the cutter after the first cut.

Woodruff keyseats are cut with shank-type cutters for American National Standard (ANSI) sizes #202 through #1212.

Arbor-type keyseat cutters are used for face widths from 3/16" through 3/4" (4.7mm through 19mm). The nominal diameters range from 2-1/8" through 3-1/2" (54mm through 88mm).

Woodruff keyseat cutters are hand fed to compensate for increasing cutting forces.

Dovetails may be machined from solid or cast pieces on which sufficient stock is left to clean up the angular surfaces. Single-angle dovetail cutters permit roughing and finish machining each side and the bottom flat surface.

Since mating dovetails may be sliding members, the angular and adjoining areas of the flat base generally require a low microfinish surface texture. Surface finish specimens are used to compare the machined surface against the part specifications.

Dimensions for dovetails are calculated from drawing specifications that relate to angles, depth, and width.

Hardened and ground, special-diameter rolls are used in the measurement of external and internal dovetails.

Sawing, slitting, keyseat milling, and dovetail milling require general machine safety practices to be followed.

The size and shape of the cutter teeth and other design features require that great care is taken to prevent fracturing.

Slow hand starting feeds are important. These feeds permit maximum control, which ensures that the setup is rigid and the cutting conditions are correct.

Adequate flow of cutting fluid must be checked. Chips must be flowed rapidly away from the cutting teeth and the workpiece.

The machine must be completely stopped and the workpiece burred before any measurement is taken.

Each tool, work-holding device, machine accessory, and instrument is to be returned to its proper storage compartment.

UNIT 16 REVIEW AND SELF-TEST

1. State the difference (a) between regular slotting and slitting on a milling machine and (b) between the cutters used for each process.
2. Name one common work-holding setup for (a) sawing off pieces from a rectangular plate on a miller and (b) cutting narrow, shallow grooves that are equally spaced around the periphery of a cylindrical part.
3. a. Secure a milling machine cutter manufacturer's catalog.
   b. Give the specifications of an appropriate slitting saw to use for each of the following jobs: (1) cutting off pieces from 1/8" × 3" (3mm × 76mm) flat, cold-drawn steel plates; (2) sawing sections from cast iron castings where the cuts are 30mm (1-3/16") deep; (3) heavy-duty sawing cuts wider than 3/16" (4.7mm) and 1" to 1-1/2" (25.4mm to 38.1mm) deep.
   c. State the reason for selecting the particular cutter in each instance.
4. Identify the design features of screw-slotting cutters that make them practical for milling narrow, shallow slots.
5. State two special design features of a Woodruff keyseat that make it more functional for certain applications than a square keyseat and key.
6. List three cutter and machine safety precautions to observe in slitting and in keyseat and dovetail cutting on a milling machine.

UNIT 17

Helical and Cam Milling

Many machined parts are designed to include a helical form. A *helical form* is a groove that advances longitudinally at a uniform rate as the part into which it is cut rotates about its axis. The groove (path) is generated on a cylindrical or cone-shaped surface by a cutting tool. This tool is fed lengthwise at a uniform rate. At the same time, the cylinder or cone on which the helical groove is being formed also rotates at a uniform rate.

Helical forms serve many purposes. The grooves, or flutes, on cutting tools such as machine reamers, drills, and taps provide surfaces along which chips flow out of a workpiece. The relieved areas of cutting tools also permit the design of efficient cutting edges. In noncutting applications such as drum cams, the helical groove actuates other machine parts to move at specified rates. Helical forms are widely used in gear design. Here the smooth, sliding action of mating gears that are cut at a helix angle quietly transmits motion and power.

This unit deals with the principles, terminology, and processes related to the milling of helical forms. The cutting of helical forms on universal, horizontal, and vertical milling machines and dividing heads is representative of jobbing shop work. Machine and dividing head setups, cutter positioning, computation of change gears, and techniques of machining are examined.

The unit also deals with common cam and follower motions, designs, terms, machining setups and procedures, calculations for angle settings of vertical-spindle and dividing head axes, and the end milling of a uniform rise cam. A *cam*, classified for design purposes as a machine element, is a device applied to a moving machine part to change rotary motion to straight-line (reciprocating) motion. The cam transmits the motion through a follower to other parts of a machine. There are numerous machines that require some mechanism to transfer motion and power without the usual slippage that occurs with belts. A practical example is a gasoline engine. Cams are needed to control timing for opening and closing the intake and exhaust valves. Cams are also extensively used as quick-locking work-holding mechanisms on jigs and fixtures.

LEAD, ANGLE, AND HAND OF A HELIX

The *lead of a helix* is the distance the helix advances longitudinally on a part as it makes one complete revolution about its axis. The lead of a helix generated by a single-point cutting tool is illustrated in Figure 17-1.

Each helix has an angle. The *helix angle* is the angle formed by the groove (helix) and the axis of the workpiece. The cutting of a helix is influenced by the diameter of the workpiece. Figure 17-2 shows how, if the helix angle remains constant but the work diameter changes, the lead (and the helix angle for a specific lead) changes. The lead of a helix varies with the angle of the helix and the diameter of the part.

Right- and Left-Hand Helix

The direction of a helix is called the *hand of the helix*. A *right-hand helix* (when it is viewed from the end) wraps around a workpiece in a clockwise direction. A *left-hand helix* (when seen with the axis of the workpiece running in a horizontal direction) advances in a counterclockwise direction and slopes down to the left.

Direction for Swiveling the Table

The hand of the helix determines the direction in which the milling machine table is swiv-

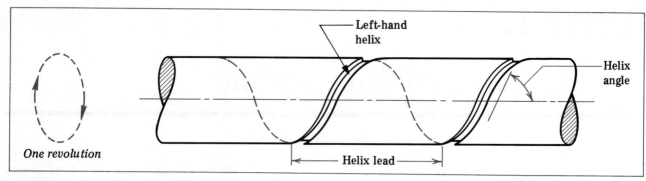

FIGURE 17-1 Lead and Helix Angle (Left-Hand)

eled. Unless the table is swiveled, the helix contour will be milled incorrectly although the lead will be correct, as illustrated in Figure 17-3. The helix contour varies from the incorrect setting and width (Figure 17-3A) to the table setting at the required helix angle (Figure 17-3B). This setting permits milling the correct groove width and form.

A left-hand helix requires swiveling the milling machine table in a clockwise direction when viewed from the front of the machine. Similarly, a right-hand helix requires swiveling the table counterclockwise to the required helix angle.

Calculating the Helix Angle. The helix angle to which the table is set is calculated in relation to the helix lead and workpiece circumference (Figure 17-4).

Example: A 8.16″ lead is to be cut on a 1.500″ diameter machine part. The angle for setting the table is to be established. The following formula is used:

$$\text{tangent of helix angle} = \frac{\pi \times \text{diameter}}{\text{lead}}$$

$$= \frac{3.1416 \times 1.500}{8.16}$$

$$= 0.57736$$

The tangent value of 0.57736 represents a 30° angle. Thus, in this example, the milling machine table is set at a 30° angle.

GEARING FOR A REQUIRED LEAD

A helix is produced by advancing a workpiece past a revolving cutter according to the helix angle and lead. The universal dividing head is generally used to turn the workpiece. The workpiece is fed into the cutter by advancing the table, which has been positioned at the helix angle.

The ratio between the rotation of the dividing head spindle and the longitudinal movement

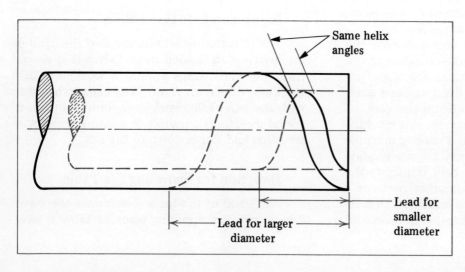

FIGURE 17-2 Effect of Work Diameter on Helix Lead

HELICAL AND CAM MILLING 305

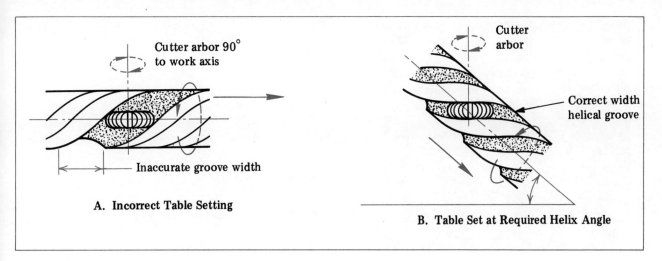

FIGURE 17-3 Effect of Angle Table Setting on Helix Form

of the table requires the use of *change gears*. The change gears provide the required gearing between the worm shaft of the dividing head and the table feed (lead) screw.

Gearing for the dividing head worm shaft and the table feed screw is illustrated in Figure 17-5. An idler gear is inserted in the gearing setup for a right-hand helix (Figure 17-5A). The idler is not a driver or driven gear and is not included in the calculations for change gears. By using an additional idler gear, the direction of rotation of the workpiece is reversed for a left-hand helix (Figure 17-5B).

Calculating Change Gears

The standard dividing head ratio is 40:1. The standard milling machine table feed screw has four threads per inch. The gear on the dividing head worm shaft is driven by the gear on the table feed screw. One revolution of the feed screw moves the dividing head spindle 1/40 of a revolution. It takes 40 revolutions of the feed screw to turn the workpiece (mounted in the dividing head) one complete revolution. In the process, the 40 revolutions of the feed screw advance the table 10" (40 revolutions × 0.250" lead). With equal driver and driven gears—assume, for example, two 48-tooth gears—a lead of 10" is produced.

The ratio of the change gears required to produce a specific lead may be calculated by using a simple formula:

$$\text{gear ratio} = \frac{\text{lead of required helix (driven gears)}}{\text{lead (feed) of machine (driver gears)}}$$

A series of change gears is used in a gear train to accommodate different ratios of gears on the worm shaft and the feed screw. A set of change gears is provided as standard equipment with universal dividing heads and universal milling

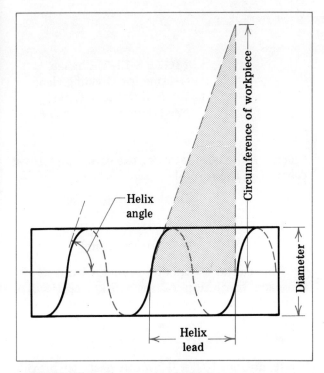

FIGURE 17-4 Calculating a Helix Angle to Establish Correct Table Setting

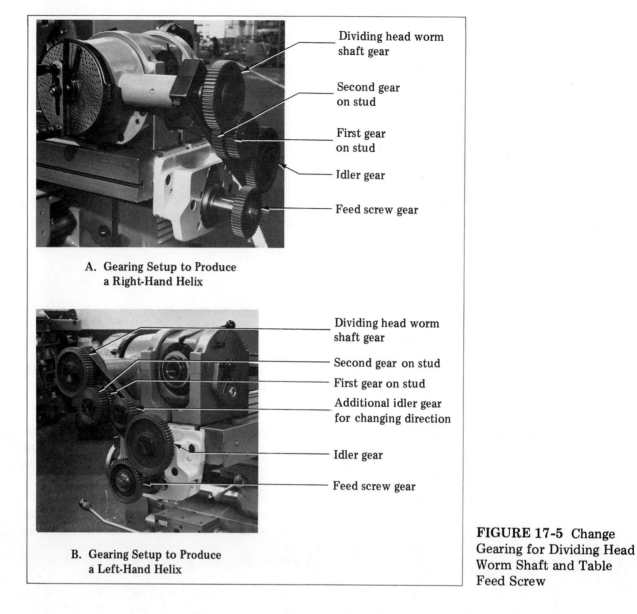

A. Gearing Setup to Produce a Right-Hand Helix

B. Gearing Setup to Produce a Left-Hand Helix

FIGURE 17-5 Change Gearing for Dividing Head Worm Shaft and Table Feed Screw

machines. The gears and the number of teeth on each gear are as follows: 24, 26, 28, 32, 40, 44, 48, 56, 64, 72, 86, and 100. The combination of gears that are available to produce a required ratio may be determined by using the following formula:

$$\frac{\text{lead of helix}}{\text{lead of machine}} = \frac{\text{product of driven gears}}{\text{product of driver gears}}$$

Producing a Helix by Simple Gearing. To produce a helix by simple gearing, the gear ratio is found by dividing the required lead by the machine lead.

Example: A helix lead of 12″ is to be milled on a workpiece. Standard change gears are available. Calculate the gear ratio for the driven and driver gears. Use the formula,

$$\text{gear ratio} = \frac{\text{lead of required helix (driven gears)}}{\text{lead (feed) of machine (driver gears)}}$$

The lead of a standard milling machine is 10″ for each revolution of the workpiece (lead of helix). By inserting the lead values in the formula,

$$\text{gear ratio} = \frac{12}{10}$$

There are no 10- or 12-tooth gears to provide this ratio. However, if each value is multiplied by the same number, appropriate gears may be found. The ratio remains the same. By using 4 as a multiplier,

$$\frac{12 \times 4}{10 \times 4} = \frac{48 \text{ (driven gear)}}{40 \text{ (driver gear)}}$$

Thus, a 40-tooth gear on the feed screw driving a 48-tooth gear on the worm shaft will produce a 12″ helix lead. The gearing setup is called *simple gearing*.

Producing a Helix by Compound Gearing. Many helixes require more complicated setups that involve *compound gearing* in which a number of gears are formed into a train. A compound change gear train is shown in Figure 17-6.

Example: Determine the gearing required to mill a helix lead of 36″. Standard change gears are available.

In the simple gear ratio formula,

$$\text{gear ratio} = \frac{36 \text{ (driven gears)}}{10 \text{ (driver gears)}}$$

Without 36- and 10-tooth gears, it is necessary to either use handbook tables or calculate the compound gears that will produce this ratio. The 36:10 ratio may be factored as follows:

$$\frac{36}{10} = \frac{6 \times 6}{2 \times 5}$$

The combination of gears that are available to produce the required ratio must now be determined. Considering first the 6:2 ratio, the numerator and denominator may be multiplied by 12:

$$\frac{6 \times 12}{2 \times 12} = \frac{72 \text{ (driven gear)}}{24 \text{ (driver gear)}}$$

These two gears are in the set. Continuing with the 6:5 ratio and using 8 as a factor,

$$\frac{6 \times 8}{5 \times 8} = \frac{48 \text{ (driven gear)}}{40 \text{ (driver gear)}}$$

Putting the two sets of gears together, the original 6:2 gear ratio requires

$$\frac{72 \text{ and } 48 \text{ as driven gears}}{24 \text{ and } 40 \text{ as driver gears}}$$

The 24-tooth gear on the feed screw drives the 72-tooth outside gear on the stud. The 40-tooth gear in the inside position on the stud drives the 48-tooth gear on the worm shaft. One or more idler gears may be needed, depending on the hand of the helix.

A word of caution is in order for setting up change gears. The two driver gears and the two driven gears may be interchanged without changing the gear ratio. However, *the gear ratio is changed if one driver gear is interchanged with one driven gear.*

Instead of computing change gears for different leads, the craftsperson refers to handbook tables. Change gear combinations are given for leads ranging from 0.670″ (1.7mm) to 60.00″ (1,524mm). Table 17-1 is a partial table of change gears for different helix leads.

For the 36″ lead used in the preceding example, the table gives another combination of change gears. By referring to the table, note the gears recommended for a 36″ lead. A 40-tooth

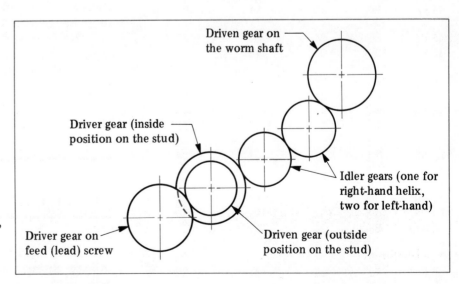

FIGURE 17-6 Identification, Positions, and Functions of Gears in a Compound Change Gear Train

driver gear is secured at the end of the feed screw. This gear drives a 64-tooth gear in the second position on the stud. The 32-tooth gear in the first position on the stud serves as a driver for the 72-tooth driven gear on the worm shaft.

TABLE 17-1 Change Gears for Different Helix Leads (Selected Leads from 32.14″ to 60.00″)

Helix Lead (in ″)	Driven Worm Shaft	Driver Position 1st Gear on Stud	Driven 2nd Gear on Stud	Driver Gear on Feed Screw
32.14	100	56	72	40
32.25	86	48	72	40
32.41	100	24	56	72
32.47	100	28	40	44
...				
35.83	86	32	64	48
36.00	72	32	64	40
36.36	100	44	64	40
36.46	100	48	56	32
36.67	48	24	44	24
36.86	86	28	48	40
...				
38.10	64	24	40	28
38.18	72	24	56	44
38.20	100	24	44	48
38.39	100	40	86	56
38.57	72	28	48	32
38.89	56	24	40	24
...				
43.64	72	24	64	44
43.75	100	32	56	40
43.98	86	32	72	44
44.44	64	24	40	24
44.64	100	28	40	32
44.68	86	28	64	44
...				
45.84	100	24	44	40
45.92	100	28	72	56
...				
46.07	86	28	72	48
46.67	64	24	56	32
46.88	100	32	72	48
47.15	72	24	44	28
...				
55.00	72	24	44	24
55.28	86	28	72	40
55.56	100	24	32	24
55.99	100	24	86	64
56.25	100	32	72	40
56.31	86	24	44	28
...				
59.53	100	24	40	28
60.00	72	24	64	32

GEARING FOR SHORT-LEAD HELIXES

Short-lead helixes, having leads smaller than 0.670″, require a different gearing setup. Short leads are produced by disengaging the dividing head worm and worm wheel. Gearing is direct from the table feed screw to the dividing head spindle. The gears in a direct setup produce a lead that is 1/40 of the lead given in a handbook table. That is, the dividing head spindle turns the workpiece at an increased number of revolutions in relation to the table lead.

The change gear ratio for milling short-lead helixes equals the lead to be milled divided by the lead of the table feed screw. In the case of a standard table feed screw having four thread per inch,

$$\text{gear ratio} = \frac{\text{lead to be milled}}{0.250} = \frac{\text{driven gears}}{\text{driver gears}}$$

METHODS OF MILLING A HELIX

Helical grooves may be machined on a miller by using a double-angle, concave, fluting, or other formed cutter. End mills are also used for forming helical grooves. The universal milling machine is used with standard and special formed grooving cutters. The universal feature permits the table to be swiveled to a required helix angle.

The dividing head is set up to index the required number of grooves, teeth, or cutting edges. The change gear combination produces the required lead. An important consideration is the aligning of the cutter and the workpiece. If the flute being milled is to be further relieved, as in the case of a milling cutter, a secondary angle must be produced.

The flute and secondary clearance angle are milled to depth and form. Stock is left on the outside diameter, ends, and bore to permit internal and external grinding after the part is heat treated.

HOW TO MILL A HELIX ON A UNIVERSAL MILLING MACHINE

The setup and milling of a plain helical-tooth milling cutter are used to demonstrate how design features and part specifications are applied

HELICAL AND CAM MILLING 309

in relation to the setups for the workpiece, dividing head, and cutter and to the processes involved in milling the helix. Table 17-2 identifies design features dealing with angular setups and provides specifications needed for machining the milling cutter. Additional information, usually furnished on a shop drawing, indicates degree of hardness, surface finish, and tolerances.

STEP 1 Face the end. Then, drill, bore, recess (undercut) and ream the hole. Allow 0.005" to 0.010" (0.1mm to 0.25mm) for finish grinding the bore and both faces. Face the opposite end.

STEP 2 Broach the keyway to size.

STEP 3 Color one face and part of the circumference with a layout dye.

STEP 4 Lay out the helix angle of 22-1/2° on the circumference and the 10° positive radial rake angle (Figure 17-7).

STEP 5 Mount the turned and bored cutter on a mandrel between centers of the dividing head and footstock.

STEP 6 Select an index plate that will permit indexing 5-5/7 turns per tooth. Set the sector arms for 5/7 of a revolution.

Note: The locking device for the index plate must be disengaged.

STEP 7 Calculate the helix lead.

$$\text{lead} = \frac{\pi \times \text{diameter}}{\text{tangent of helix angle}}$$

$$= \frac{3.1416 \times 4.000}{\tan 22\text{-}1/2°}$$

$$= \frac{3.1416 \times 4.000}{0.40403}$$

$$= 31.10'' \text{ (helix lead)}$$

STEP 8 Refer to a handbook table of change gears for helical milling different leads. Select the closest lead to 31.10". In this example, the lead is 31.11".

STEP 9 Note the number of teeth and position of each gear in the gear train.

gear on feed (lead) screw: 48 teeth (driver)
second gear on stud: 56 teeth (driven)
first gear on stud: 24 teeth (driver)
gear on worm shaft: 64 teeth (driven)

STEP 10 Position, mount, and secure each gear. Allow minimum clearance between the gear teeth. The gears must turn freely but with limited play.

STEP 11 Swivel the table in a counterclockwise direction to the helix angle of 22-1/2°.

STEP 12 Position the table about an inch from the column face. Check to see that the setup

Table 17-2 Milling Cutter Design Features and Specifications

Design Feature	Specification
Outside diameter	4.000"
Length	3.500"
Bore	1.500"
Material	HSS
Keyway	0.312" × 5/32"
Number of teeth	9
Helix	
Direction	Right-hand
Angle	22-1/2°
Flute	
Angle	55°
Corner radius	1/8"
Depth	1/2"
Rake angle	10° positive radial rake
Secondary clearance angle	30°
Cutting tooth face (land)	3/32"

310 HORIZONTAL MILLING MACHINES

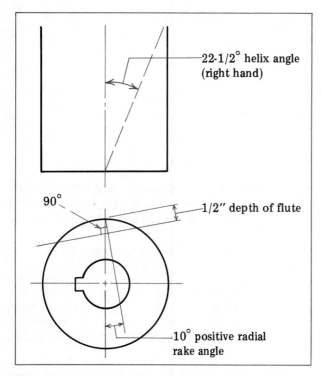

FIGURE 17-7 Layout of Radial and Helix Angles and Flute Depth

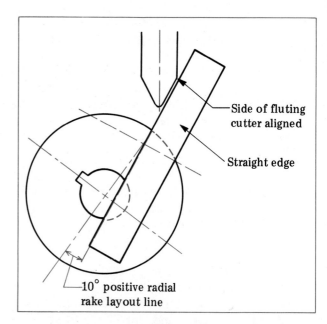

FIGURE 17-8 Positioning of Fluting Cutter in Relation to the Positive Radial Rake Angle Layout Line

clears the column. Then return the table to its zero position.

STEP 13 Mount the flute milling cutter (55° angle with 1/8" radius).

Note: The cutter direction is toward the dividing head.

Note: The cutter is roughly positioned so that one face of the fluting cutter is aligned with the 10° positive radial angle line layout.

STEP 14 Adjust the angle position of the workpiece in relation to the edge of the fluting cutter.

Note: The cutter blank is rotated until the 10° radial layout line of the flute is aligned (same angle) with the angle face of the fluting cutter (Figure 17-8).

STEP 15 Move the table transversely until the centerline of the cutter intersects the center of the layout (Figure 17-9).

STEP 16 Return the table (counterclockwise) to the 22-1/2° helix angle. Secure this angle.

STEP 17 Move the cutter to starting position to mill the first tooth. Raise the table to the 0.500" depth.

Note: A roughing and a finish cut are sometimes taken.

STEP 18 Take a trial cut for a short distance. Stop the machine. Check the helix, tooth angle, and depth of helix. Then, continue to mill the first flute.

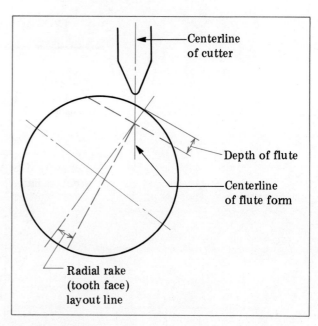

FIGURE 17-9 Cutter Axis Aligned at Center of Tooth Layout

Note: The graduated collar on the cross feed is set at zero to permit making an accurate lateral adjustment, if necessary.

Note: Care must be exercised in starting the cut. The cutter teeth ends cannot withstand severe shock. An appropriate cutting fluid must be applied.

STEP 19 Return the cutter to the starting position.

Note: The table is lowered so that the cutter clears the workpiece at the end of each cut.

STEP 20 Index, reset the cutter, and machine each successive flute.

HOW TO MACHINE THE SECONDARY CLEARANCE ANGLE

The secondary clearance angle (30°) in the example in Table 17-2 extends from the point of intersection with one angular side of the flute to the land. The clearance angle is machined to provide a 1/8″ wide land.

The workpiece must be indexed to rotate it so that the 30° angle is in a correct relation to the side of the flute:

$$\text{indexing} = 90° - 30° + \left(\frac{\text{flute angle of } 55°}{2}\right)$$

$$= 32\text{-}1/2° \text{ or } 32°30'$$

The 32°30′ indexing requires 3 full turns and an additional 11/18 of a revolution. The following steps must be taken to mill the 30° secondary clearance angle for the milling cutter.

STEP 1 Replace the fluting cutter with a plain helical milling cutter.

STEP 2 Index the 3 and 11/18 turns.

STEP 3 Adjust the cutter so that it clears the cutting face of the flute form.

STEP 4 Raise the table to cut the 30° angle face to depth.

Caution: Stop the cutter rotation to check the width of the land.

STEP 5 Continue to machine. Index to mill the secondary clearance angle on each tooth.

STEP 6 Remove the workpiece. Clean and deburr.

STEP 7 Wipe clean all tools, cutters, machine parts, and accessories. Return each item to its proper storage container.

After machining, the cutter is heat treated. It is then ground to dimensional accuracy. The face of each flute may be ground on a tool and cutter grinder. The circumference is ground concentric with the bore. The lands are relieved to produce sharp cutting edges on each tooth.

HOW TO MILL A HELIX BY END MILLING

Cams and other parts—for example, actuators for transmitting motion—are often designed with helical grooves that have vertical sides. End mills are commonly used to mill such grooves on either plain or universal milling machines. Since the width of the groove produced by the cutter does not vary with the helix angle, it is not necessary to swivel the table. Some forms of the helical groove may have straight or angular sides. These sides may have a sharp corner or a radius at the root. Other helical grooves may be concave. The helical groove shape is determined by the form of the end mill. Standard square-face, angular-face, or ball-end end mills may be used. The helix may be milled with the end mill mounted in the spindle of a horizontal or a vertical milling machine. The steps that follow describe the end milling procedures when a vertical-spindle attachment or a vertical milling machine is used.

STEP 1 Determine the change gear setup required to produce the helix lead.

Note: Handbook tables are usually used instead of computing.

STEP 2 Set up the change gears between the table feed screw, studs, and the worm shaft of the dividing head (or the spindle for short leads).

Note: The gear train must produce the correct hand for the helix.

STEP 3 Select the dividing head plate. Set the sector arms if any fractional part of a turn of the crank is required.

312 HORIZONTAL MILLING MACHINES

STEP 4 Determine the appropriate cutting feed. Set the table feed accordingly.

Caution: The longitudinal movement, spindle, and cutter setup must be checked for clearance at the end of the cut. For safety, the table trip dogs are positioned to control the overall movement of the setup.

STEP 5 Position the table. Align the centerline (axis) of the end mill at the axis of the helical groove.

STEP 6 Start the spindle and flow of cutting fluid. Adjust the table setting so that the cutting edge just grazes the work surface. Set the graduated collar at zero.

STEP 7 Set the end mill to cut to depth. Take a trial cut for a short distance and check.

Note: Deep grooves are usually rough milled. The finish cut is then taken at the required depth.

STEP 8 Engage the power longitudinal feed after the cut is started by hand-feeding if the cutting conditions warrant.

STEP 9 Clear the end mill at the end of each cut. Return the setup to the starting position.

STEP 10 Index for the next helical groove.

STEP 11 Position the end mill for the depth of cut. End mill the second helical groove.

Note: If one or more roughing cuts and a finish cut are to be taken, all grooves are rough milled. Then, all of the grooves are finish milled.

STEP 12 Continue to index, position the cutter, and mill the remaining grooves.

STEP 13 Remove the workpiece. Clean and deburr. Return all items after cleaning to proper storage containers.

FUNCTIONS OF CAMS IN CAM MILLING

As a machine element, a cam generates a desired motion. The motion is transmitted to a follower by direct contact. The combination of cam and follower produces a given motion, velocity, or acceleration during a specific portion of a cycle and in a particular direction.

Cam motion depends on timing and type of movement in a part of a cycle or a whole cycle. *Motion* refers to the rate of *movement* or *speed*. The speed of the cam follower is related to the speed of rotation of the cam. *Displacement* refers to *distance*. Displacement is the distance a cam follower moves in relation to the rotation of the cam.

The three most common cam motions deal with:

—Uniform (constant velocity) motion,
—Parabolic motion,
—Harmonic motion.

DESCRIPTIONS OF CAM MOTIONS

Uniform (Constant Velocity) Motion

A mechanism, cutting tool, or device that must rise and fall at a constant rate of speed requires a cam that produces a *uniform motion*. The rate of movement of the follower is the same

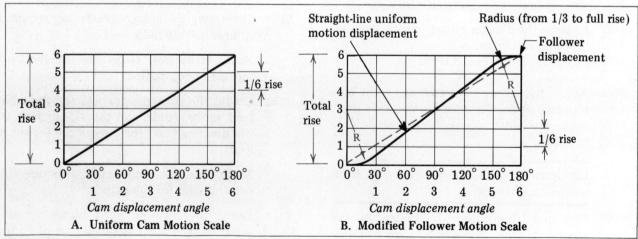

FIGURE 17-10 Layout of Uniform and Modified Cam and Follower Motion

from the beginning to the end of the stroke. For example, if a cutter is to advance 24mm in one-half revolution of a cam and return to zero during the remaining 180°, the follower would feed the cutter at a uniform rate of one-sixth of 24mm (4mm) for each 30° the cam rotates. The motion is uniform. The straight-line graph of Figure 17-10A illustrates uniform cam motion. Figure 17-10B shows the modified motion of the follower.

Parabolic Motion

Parabolic motion is identified as uniformly accelerated or decelerated motion in 180° (Figure 17-11). A curve similar to the curve for accelerated motion follows for the retarded (decelerated) motion in the remaining 180° part of one cycle or revolution.

In Figure 17-11, points are plotted every 30°. The six divisions increase and decrease by a ratio of 1:3:5:5:3:1. Assume a follower is to rise 72mm in 180° and points are plotted every 30° in six proportional divisions of 1:3:5:5:3:1 (18 divisions). In the first 30°, the follower rises 1/18 of 72mm, or 4mm. In the second 30°, the follower rises 3/18 (1/6) of 72mm, or 12mm. The follower rise in the third 30° is 5/18 of 72mm. or 20mm. Similarly, the fourth, fifth, and sixth 30° rises are 20mm, 12mm, and 4mm, respectively.

Uniformly accelerated and decelerated motion starts slowly and then accelerates or decelerates at a uniform rate. The follower is permitted to come to a slow stop before reversal. A cam that produces parabolic motion has smooth operating characteristics and is adaptable for high-speed applications.

Harmonic Motion

Harmonic motion relates to a cam and follower that produce a true eccentric motion (Figure 17-12). The cam moves a follower in a continuous motion at a constant velocity (Figure 17-12A). The cam displacement angle and follower displacement in Figure 17-12B are at 30° intervals. Harmonic motion cams are used in high-speed mechanisms when uniformity of motion is not the essential design requirement.

COMMON TYPES OF CAMS AND FOLLOWERS

Positive and Nonpositive Types of Cams

Plate or bar cams transform linear motion into the reciprocating motion of the follower. Templates that are used to control motion on a tracer miller; profiler; or contour, surface-forming lathe are another form of cam. Cams that impart motion are referred to as *positive* or *nonpositive*.

On a positive design cam, the follower is positively engaged at all times. The cylindrical (drum) cam and grooved recessed plate cam illustrated in Figure 17-13 are positive-type cams. Engagement between the cam and follower is accomplished by a pin or roller riding in the groove.

The plate cam and knife-edge follower and the toe and wiper cam and follower shown in

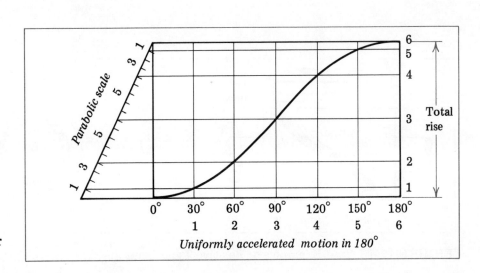

FIGURE 17-11 Layout for Parabolic Motion of a Cam

314 HORIZONTAL MILLING MACHINES

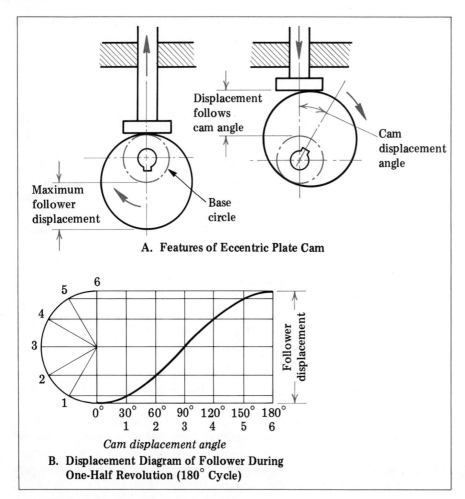

FIGURE 17-12 Features and Layout of Harmonic Motion for an Eccentric Plate Cam

Figure 17-14 are common examples of nonpositive types. The cam produces the direction of motion and acceleration. The follower is pushed against the cam by gravity, a spring, or some other outside force.

Basic Follower Forms

The four basic cam follower designs are illustrated in Figure 17-15. The *roller (flat) type* (Figure 17-15A) has a free, rolling action that reduces frictional drag to a minimum. The *tapered*

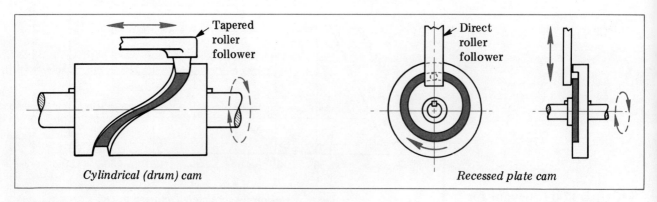

FIGURE 17-13 Two Examples of Positive-Type Cams

HELICAL AND CAM MILLING 315

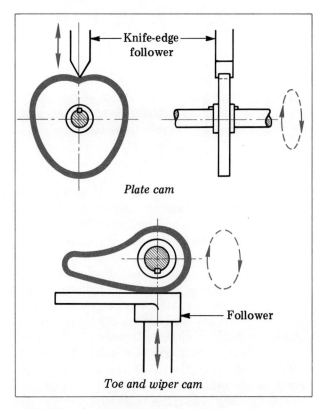

FIGURE 17-14 Nonpositive Types of Cams

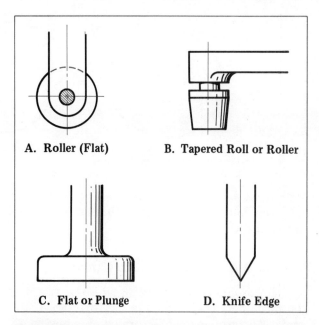

FIGURE 17-15 Four Basic Cam Follower Designs

roller type (Figure 17-15B) is tapered to form a cone shape. The angle sides ride in a corresponding groove form. The *flat or plunge type* (Figure 17-15C) is widely used to transmit heavy forces in actuating a mechanism through a particular motion and time sequence. The *knife-edge and pointed type* (Figure 17-15D) is used for intricate and precise movements that require a limited contact surface. This type permits the follower to accurately transmit motion, particularly from sharp contours on the cam.

GENERAL CAM TERMS

While the field of cam and follower designs and applications is extensive, only six terms used daily in the shop are described here:

- —*Lobe* refers to a projecting area of a cam; the lobe imparts a reciprocal motion to the follower; and single- and double-lobe uniform rise cams (Figure 17-16) are common examples.
- —*Rise* is the distance one lobe on a revolving cam raises or lowers the follower.
- —*Uniform rise* is the distance a follower moves as generated at a uniform rate around the cam; and the uniform rise is produced by a combination of uniform feed and rotation during the machining of a cam.
- —*Lead* is the total distance a uniform rise cam (with only one lobe in 360°) moves a follower in one revolution.
- —*Cam profile* is the actual contour of the working surface of a cam.
- —*Base circle* is the smallest circle within the cam profile.

BASIC CAM MACHINING PROCESSES

Machining Irregular Rise Plate Cams

A number of plate cam applications require an uneven, irregular motion. For these applications, the cams are generally laid out. The cam form is produced by *incremental cuts*. The cam blank is rotated a number of degrees called an *angular increment*. A series of cuts is taken so that each splits the layout line. Each cut is continuously adjusted as necessary to produce the desired form in the angular increment. The forming process is continued in the same manner for each angular increment. The cam outline is re-

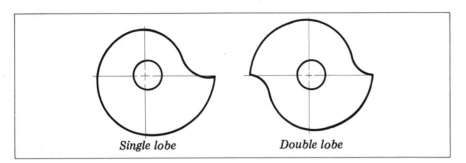

FIGURE 17-16 Single- and Double-Lobe Uniform Rise Cams

produced as closely as possible. The ridge irregularities between the milling cuts are then filed and polished.

Machining Uniform Rise Cams

Drum-shaped cams with a uniform rise may be machined as described in the first part of this unit. The groove may be formed on a horizontal miller using a preformed double-angle milling cutter or a square or special-formed end mill.

Uniform rise cams may also be produced on a vertical milling machine. One common machining method is to use a vertical head, the uniform rotation of the dividing head spindle, and the uniform feed of the machine table. In the vertical position, only cams with the same lead as the change gear setup lead may be cut.

Other uniform rise cams may be machined by positioning the axis of the vertical head or attachment (and the end mill) parallel to the axis of the dividing head spindle (workpiece). With the vertical milling machine setup shown in Figure 17-17, any required cam lead may be cut provided the lead is less than the forward feed of the table in one revolution of the work. That is, the required cam lead is less than the lead for which the milling machine is geared.

In milling a uniform rise helix with an end mill at a right angle to the workpiece (using equal gears on the table feed screw and the worm shaft of the dividing head), the table advances 10″ during one revolution of the workpiece. The 10″ results from dividing 40 (turns of the dividing head crank to revolve the spindle one revolution) by 0.250″ (lead of the table feed screw).

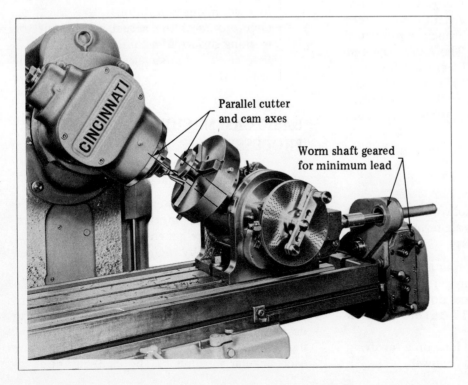

FIGURE 17-17 Vertical Milling Machine Setup for Cam Milling

Assume an imaginary setup in which the end mill and work axes are parallel and in a horizontal (0°) plane (Figure 17-18). It is obvious that if any cut were taken, the cutter would remain at a fixed distance around the axis of the workpiece. As the workpiece is rotated and fed at a particular lead, the end mill theoretically produces a circular form. No lead would be generated.

The 90° and the theoretical 0° settings represent the range of leads that may be milled. The range is from a maximum of 90°, which the milling machine change gears permit, to zero, as produced at the 0° setting. Thus, it is possible to machine a specific lead within the range by positioning the cutter and workpiece axes at a given angle between 0° and 90°. The angle setup for cam milling may be computed from a cam drawing and other specifications. The following formula is used to find the lead of a specified uniform rise cam:

$$\text{lead} = \frac{\text{lobe rise in inches for the full cam circumference}}{\text{space occupied by the lobe in degrees}}$$

$$= \frac{\text{lobe rise ('')} \times 360°}{\text{lobe space (° of circumference)}}$$

Design drawings often represent the circumference as divided into 100 equal parts. Thus,

$$\text{lead} = \frac{\text{lobe rise ('')} \times 100}{\text{lobe space (percent of circumference)}}$$

CALCULATING CUTTER AND WORK ANGLES FOR MILLING SHORT-LEAD CAMS

The shortest possible lead to mill with standard change gears is 0.670″ (17mm). Any shorter leads require the vertical spindle and the divider head (cutter and work) to be positioned at the same specific angle. The relation between cutter and dividing head axes, angle of inclination, and cam and table lead are shown in Figure 17-19. Factors affecting the angular setup for cam milling include (1) table feed, (2) dividing head worm shaft geared for minimum lead, (3) cam rise, and (4) parallel cutter and work axes.

Example: If a lead of 0.400″ is to be cut and the shortest possible lead is 0.670″, the fractional value is the sine of the angle to which the cutter and workpiece are set. Thus,

$$\text{sine of angle} = \frac{\text{required lead}}{\text{shortest machine lead}}$$

$$= \frac{0.400}{0.670}$$

$$= 0.5970$$

From trigonometric tables, a sine value of 0.5970 represents an angle of 36°39′. The vertical head and the dividing head spindles are positioned at this angle.

By setting up the change gears (24, 86, 24, and 100) and positioning the cutter and dividing head at 36°39′, as the table advances 0.670″ in one revolution and the workpiece turns, the cutter mills a cam with a uniform rise of 0.400″.

HOW TO MILL A UNIFORM RISE CAM

Example: Mill a double-lobe uniform rise cam. Each lobe occupies 180°. Each lobe has a rise of 0.250″.

STEP 1 Calculate the cam lead. Use the formula,

$$\text{lead} = \frac{\text{lobe rise} \times 360°}{\text{lobe space}}$$

$$= \frac{0.250 \times 360}{180}$$

$$= 0.500″$$

STEP 2 Refer to a handbook table of change gears for cam milling. Select the gearing for the smallest lead.

Note: A table lead of 0.670″ is produced by positioning an 86-tooth gear on the feed screw (driver), a 24-tooth gear in the first position on the stud (driven), a 100-tooth gear in the second position (driver), and a 24-tooth gear on the worm shaft (driven).

Note: The locking device for the index plate must be disengaged.

STEP 3 Color the cam face with dye. Lay out a centerline to indicate the starting point of each lobe.

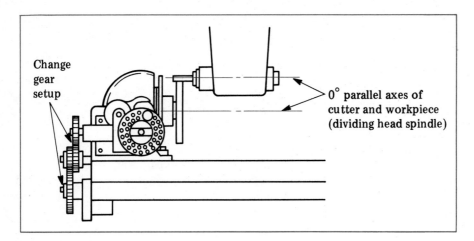

FIGURE 17-18 End Mill and Work Axes at 0°

STEP 4 Set up the workpiece in the dividing head and an end mill in the vertical head spindle.

Note: The end mill must be long enough to provide adequate clearance at the start and finish of each cut.

STEP 5 Calculate the required angle for the dividing head and the vertical head spindle. Use the formula,

$$\text{sine of angle} = \frac{\text{required lead}}{\text{shortest machine lead}}$$

$$= \frac{0.500}{0.670}$$

$$= 0.7463$$

$$= 48°16'$$

STEP 6 Set the dividing head at 48°16'. Set the vertical head spindle at the same angle.

STEP 7 Rotate the mounted workpiece until the scribed centerline is vertical.

STEP 8 Position the centerline of the cutter with the scribed centerline of the workpiece. Adjust the table until the cutter grazes the underside of the cam blank.

Note: A more rigid setup is produced when the cut is taken from this position.

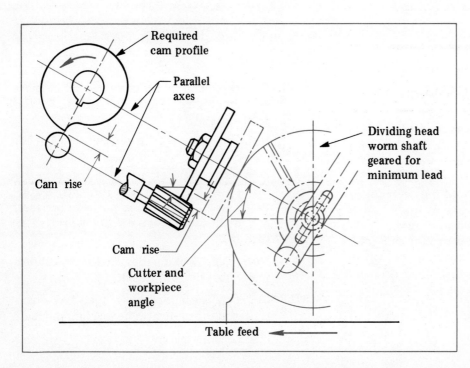

FIGURE 17-19 Factors Affecting the Angular Setup for Cam Milling

Also, chips flow away faster and layout lines remain visible.

STEP 9 Set the vertical feed collar at zero. Start the flow of cutting fluid. Feed the cutter through 180° by rotating the cam blank with the index crank.

Note: The automatic feed may be engaged for leads over 2-1/2″ (63.5mm). For smaller leads, the table may be fed by using either a short-lead attachment or the index head crank.

STEP 10 Lower the table at the end of the cut so that the cutter and workpiece clear. Disengage the gear train or the dividing head worm.

STEP 11 Move the table back to the starting position.

STEP 12 Position the work so that the centerline on the circumference is aligned with the axis of the cutter.

STEP 13 Reengage the gear train or the dividing head worm.

STEP 14 Mill the second lobe.

STEP 15 Shut down the machine. Check the dimensional accuracy of each lobe.

STEP 16 Clean and remove the end mill and workpiece. Deburr the cam.

STEP 17 Replace each tool and accessory in its appropriate storage place.

STEP 18 Disassemble the setup. Clean all parts and accessories. Return each item to its appropriate storage container.

SAFE PRACTICES IN SETTING UP AND MILLING HELIXES AND CAMS

—Check the position of each driver or driven gear in a change gear setup for milling. Changing the position of driver gears or driven gears does not change the lead produced. However, interchanging only one driver and one driven gear does change the ratio.

—Check mathematical computations for change gearing against the combination given in a handbook table.

—Take successive cuts to rough out deeply formed grooves, particularly if the form is being end milled. Use the largest possible size of end mill.

—Reduce the force against the sharp-angle cutting edges of a formed cutter at the start of a helical milling process. Avoid forcing or bringing the cutter teeth into sharp contact with the workpiece.

—Use cutting fluids to flow away chips and to aid in producing a quality surface finish.

—Clear the cutter from the helical groove at the end of the cutting stroke.

—Hand-feed small end mills and formed cutters until all cutting conditions have been checked and the process may be carried on by power feed.

—Stop the spindle and cutter before any measurement is taken.

—Check the clearance between the work-holding, cutter, and machining setups and the helix angle setting of the universal table before a cut is started.

—Set trip dogs to ensure that the table will not be moved accidentally beyond the place where the cutter clears the workpiece.

—Check the length of the end mill to be sure it is long enough to produce the helical curve or cam form for short-lead machining.

—Smooth out the small indentations produced in cam milling and polish the surface for smooth operation and longer wear.

—Disengage the index plate locking device in short-lead milling when the index crank is used to rotate the workpiece.

TERMS USED IN HELICAL AND CAM DESIGN, SETUPS, AND PROCESSES

Cam	A machine element through which a particular pattern of motion is transmitted to a follower.
Follower	A roll, edge, or point on a device that transmits the motion of a cam to the movement of another.
Helix Lead	The distance a curve around a cylinder advances in one complete revolution.
Helix angle	The angle formed by the angular path of the helix and the centerline of the part containing the helix.
Tangent of helix angle	The tangent of the angle formed by dividing the circumference of the workpiece by the required lead.
Hand of helix	The direction a helix advances when turned. (A right-hand helix advances in a clockwise direction. A left-hand helix advances in a counterclockwise direction.)
Compound gearing (change gears)	Gears used to form a gear train between the milling machine table feed screw, intermediate studs, and the worm shaft of a dividing head for purposes of milling spiral grooves.
Change gear tables (lead milling)	Handbook information of change gear combinations for different leads from 0.670" (17mm) to 60" (1,524mm).
Short-lead helixes	Helixes having leads smaller than 0.670". Smaller helix leads than it is possible to generate with standard milling machine change gears.
Secondary clearance machining setup (milling)	Positioning the workpiece by indexing to permit a plain milling cutter to mill clearance between a flute and a land.
Cam motion	Three common forms of motion: uniform, parabolic, and harmonic movement. The motion imparted by a cam. Actuating another part a fixed distance during a specific interval through which a cam moves.
Uniform motion	The rise and fall of a cam and follower at the same rate for the beginning to the end of a cycle.
Parabolic motion	The uniform acceleration and deceleration of movement in one cycle (revolution).
Harmonic motion	A motion produced by an eccentric moving a follower in a continuous motion at a constant velocity.
Positive and nonpositive cam types	Cam designs in which the follower is positively engaged at all times or in which an outside force is needed to keep the follower in contact with the cam.
Lobe	A design feature on a cam over which a follower rides to transmit motion.
Rise	The distance a cam lobe moves the follower.
Cam lead	The total distance a follower is moved in one revolution of a uniform rise cam.
Incremental cuts (milling)	Machining a cam form by taking a series of cuts. Cuts that are stepped off to conform to the required cam shape at different angular positions of the cam. A process of adjusting the cam being milled in relation to the cutter in order to produce the cam form.

Milling short-lead cams   The process of setting up the work and cutter with axes parallel. Machining a cam having a lead smaller than the minimum dimension lead that a standard change gear set may accommodate. A machine setup in which the vertical spindle and the cam are set at the same angle to produce a smaller lead than the minimum change gear capacity.

SUMMARY

A helix is a common form used on cutting tools and machine elements that either move or turn (actuate) other parts.

> A cam imparts motion through a follower or serves to apply force, as in a clamping device.

A helix has a lead and a helix angle. The distance a groove advances in one complete revolution represents the helix lead. The angle formed by the helix and the axis of the part is the helix angle.

> A right-hand helix advances clockwise. A left-hand helix advances counterclockwise.

Helixes with a lead of 0.670" (17mm) and larger may be milled with arbor-type cutters by swiveling the table at the helix angle. Helixes may be end milled without swiveling the table.

> A left-hand helix requires swiveling the table clockwise (when viewed from the front of the milling machine).

The helix angle for setting the table is established by calculating the tangent of the angle (dividing the circumference of the workpiece by the required lead). The tangent value is then converted to the equivalent angle.

> A helix is generated by gearing the dividing head. The workpiece is turned at a particular rate as the table (work) feeds past a cutter the distance of the lead in one revolution.

Standard change gears are provided for cutting leads from 0.670" to 60".

> The gear ratio for the change gears is found by dividing the required lead by the machine lead.

Compound change gearing for helix milling involves a gear train of four gears: feed screw (driver), first (inside) position on stud (driven), second (outside) position on stud (driver), and worm shaft (driven).

> Handbook tables provide gear change combinations with standard gear sets for helical milling.

Short-lead helixes are produced by direct gearing between the table feed screw and the dividing head spindle.

> Helixes are milled by indexing the number of grooves with the index head. The table is fed at a rate that is related to the speed at which the workpiece turns.

The lead of a helix is found by dividing the circumference by the tangent of the helix angle.

> Secondary clearance angles are milled by rotating a workpiece a number of degrees equal to 90° – the secondary clearance angle + 1/2 the included angle of the flute.

Helical grooves may be milled with arbor-type formed cutters or end mills on a horizontal miller or with end mills on a vertical milling machine.

> The machine table setting for end milling helical grooves is 0°.

Cams transmit motion according to a specific pattern that relates to distance and interval of time.

Motion deals with rate of movement. Displacement refers to distance.

Uniform motion is constant velocity motion in which motion and displacement are constant.

Parabolic motion is uniformly accelerated or decelerated during each interval of a cycle.

Harmonic motion produces a continuous motion at a constant velocity.

Cams are of the positive type if they are positively engaged; they are nonpositive if a force is required to keep the cam and follower in contact.

Followers are designed as straight or tapered rollers, flat or plunge type, or knife edge and pointed types for following precise contours.

Irregular rise plate cams are usually laid out, machined by incremental cuts, filed to contour, and polished.

Uniform rise short-lead cams are milled by setting the workpiece and cutter parallel and at the same angle setting. Change gearing is set for the smallest possible lead.

Where design features show a cam divided into 100 parts, the lead equals the lobe rise in inches multiplied by 100 divided by the lobe percent of the circumference occupied by the lobe space.

Machine setup safety precautions must be observed particularly when the table is swiveled at a steep angle. The table, cutter, work-holding, and process setups must be checked to ensure adequate and safe clearance.

The cutter is cleared of the workpiece when being returned to the starting position. The possibility of end play requires the cutter to be moved away from the finished surface.

Standard machine, tool, and personal safety practices must be followed. All equipment, tools, and accessories are to be cleaned and properly stored.

UNIT 17 REVIEW AND SELF-TEST

1. State three basic functions served by helical forms that are milled into a cylindrical surface.
2. a. Describe the lead of a helix.
   b. Tell how the hand of a helix affects the table setting of a milling machine when an arbor (hole) type of milling cutter is used.
   c. Calculate the helix angle setting for a 50.8mm diameter part that has a lead of 254mm.
3. a. Determine the gear ratio for milling a machine part with an 18" lead helical groove.
   b. Select a pair of gears for a direct-drive (simple) gearing setup to cut the 18" lead.
4. a. Refer to a handbook table of change gears for milling different helix leads.
   b. Determine the number of teeth and position of each gear in the change gears that will produce a lead of 190.5mm.
5. Set up a vertical milling machine or prepare a work production plan to mill a three-lobe (120° apart) uniform rise cam. The rise on each lobe is 0.188".
6. State two personal safety precautions to observe when milling helical flutes or cams.

SECTION TWO

Spur, Helical, Bevel, and Worm Gears and Gear Milling

The actual machining of gears involves a knowledge of design features, the interpretation of gear drawings and specifications, and the use of formulas to establish machining dimensions. This section provides information on methods of forming and representing spur, helical, bevel, and worm gears. The craftsperson applies this information to select milling cutters, set up the machine, cut the gear, and check certain gear tooth measurements. The section also describes actual setups for the milling machine, dividing head, and other accessories; cutter selection; and step-by-step gear milling procedures and measurement practices.

UNIT 18

Gear Developments, Design Features, and Computations

This unit begins with brief historical sketches of early gear applications and methods of machining. Other methods of forming, finishing, measuring, and inspecting gears are described next. ANSI techniques of representing spur, helical, bevel, and worm gears are then illustrated and further clarified by gearing specifications. Formulas for these gears are applied to cutting processes on milling machines and gear tooth measurements as they are generally carried out in jobbing and maintenance shops.

EARLY DEVELOPMENTS AND GEAR-CUTTING METHODS

Gears have been used for light-duty purposes, such as clock mechanisms and simple machines, for hundreds of years. Starting with the Industrial Revolution, gears and gear trains for heavy-duty purposes became necessary because machines in such fields as textiles, printing, and simple manufacturing required positive and uniform driving mechanisms.

Form Milling

The first *gear-cutting machines*, dating back to 1800, were used principally to cut gears for simple, low-power drives. Further applications of gears in machine design and the construction of gear-cutting machines accelerated during the 1850s due to increased manufacturing demands for gears capable of transmitting greatly increased power loads at faster speeds. Specialized gear-cutting machines were developed. Most of these machines require a formed milling cutter. The milled form of the tooth and the space between teeth permitted making teeth to fit. The flank of a driver gear tooth form rolled against the flank of a corresponding driven gear tooth produces motion.

Gear Hobbing

By 1900, *hobbing* was widely used as another machining method besides form milling. *Hobbing machines* were produced for highly specialized production gear machining. Hobbing machines are still used today, particularly for extremely large gears of 15 or more feet (4.51M) in diameter. They are constructed with horizontal and vertical spindles.

The hobbing cutter (hob) resembles a worm thread (Figure 18-1B). Cutting edges are formed on the cutter by gashes made parallel to the axis of the bore. The tooth form is the same as a gear rack. A gear tooth is generated when the hobbing (worm) cutter turns one revolution while the gear moves one space (Figure 18-1A). The tooth is formed by feeding the cutter to tooth depth.

The gear hobbing process is the fastest of the generating processes. Several teeth are cut at the same time. The cutter is continuously meshed with the gear blank. This process contrasts with the conventional jobbing shop method of milling one tooth, disengaging the cutter and workpiece, indexing, and then cutting the next tooth.

Spur, helical, and herringbone gears; splines; gear sockets; and other symmetrical shapes (Figure 18-2) may be hobbed. Spur gears are hobbed by setting the cutter with its teeth parallel to the gear blank axis. Helical teeth are cut by setting the hob axis at the angle required to produce the helix. The number of teeth cut on a gear is controlled by the gear ratio between the hob and the gear blank.

Gear Shaping

Gear shapers were first produced around the turn of the twentieth century. They use a cutter formed like a mating gear but relieved to produce

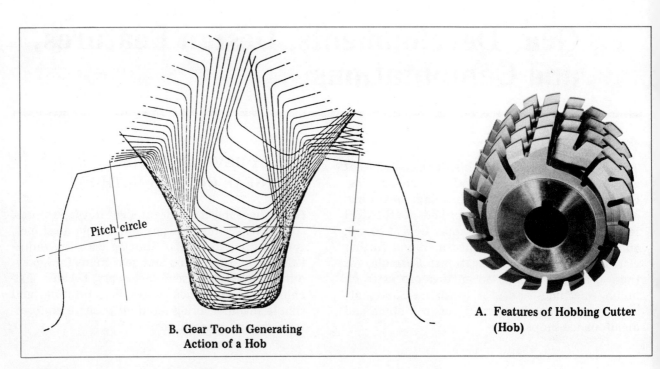

B. Gear Tooth Generating Action of a Hob

A. Features of Hobbing Cutter (Hob)

FIGURE 18-1 Gear Hobbing Cutter Features and Generating Action

GEAR DEVELOPMENT, FEATURES, COMPUTATIONS

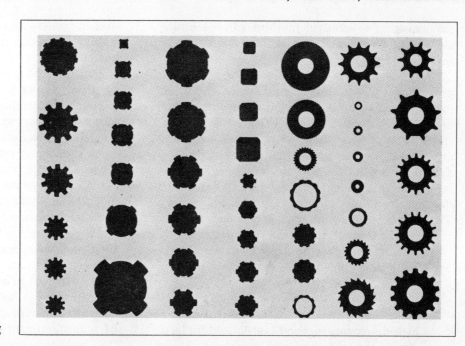

FIGURE 18-2 Examples of Forms Produced by Hobbing

cutting edges. The cutter requires a reciprocating movement. As the cutting progresses and the tool is fed to the required depth, the cutter and gear blank slowly rotate together. A gear is thus produced in one rotation of the blank.

At the end of each cutting stroke, the work rapidly clears the cutter to prevent rubbing. The work is returned to position for the next cutting stroke. Cutters are set to reciprocate from 100 to 2,000 strokes per minute for fine-tooth gears.

Gear shaping is practical for cutting symmetrical spur, helical, and herringbone gears; internal and cluster gears; racks; elliptical gears; and other special shapes. An example of a special external gear form generated on a gear shaper is illustrated in Figure 18-3. Unusual internal forms may also be generated on a gear shaper. An advantage of this machine is that the cutting action permits generating gear teeth up to a shoulder—a common requirement in cluster gears.

The increasing use of *hypoid gears* created a need to produce quiet, accurately meshing gear teeth. A *hypoid gear generator* was developed to meet this need. Hypoid gears are a modification of bevel gears. Spiral teeth are formed on the bevel surface. The axes of the gears are offset for the pinion (small gear) and gear. Hypoid gears are widely used in automotive differentials and other mechanisms where motion is transmitted by a driving shaft at a right angle and offset from the driven shaft.

OTHER METHODS OF FORMING GEARS

The application of a gear determines the required degree of tooth design accuracy and influences the method of production. Some gears are preformed and require second machining to produce an accurate tooth form. Other gears are formed to size as a finished product in a single manufacturing process. Some additional manufacturing methods of producing gears follow.

Broaching and Gear Shaving

Broaching is a fast, economical method of producing individual and nonclustered gears.

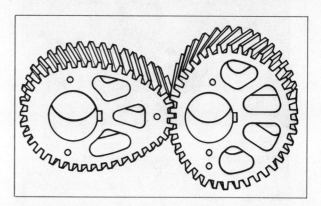

FIGURE 18-3 Special External Gear Form Produced on a Gear Shaper

Broaching is a practical method for machining an accurate straight-tooth or helical-tooth gear form having a pressure angle of less than 21°.

Gear shaving is a technique for producing gear teeth on soft metals, mild steels, and other materials up to a 30 Rockwell C hardness. Surface finishes in the $32R_C$ to $16R_C$ range are possible. Gear shaving requires the use of a rotating cutter that may be positioned at an angle to the axis of a workpiece for diagonal shaving. The teeth are formed as the shaving cutter is fed to depth into a free-rotating blank (Figure 18-4). The cutter revolves at high speed and reciprocates across the entire surface of the gear blank. Gear shaving produces teeth that are accurate in terms of shape, dimensions, and surface texture.

Casting

The *sand casting* method is still used for large gears where the teeth are later machined. Other casting methods include permanent-mold, shell-mold, and plastic-mold casting. Many gears are *die cast* or *lost-wax cast* and require limited, if any, machining.

Hot and Cold Rolling

Hot rolling refers to the production of gear teeth from a solid, heated bar of steel. Gear-shaped rollers are impressed into a revolving heated bar. The force between the rollers and heated material causes the metal to flow and conform to the gear tooth shape and size.

Cold rolling is becoming widely used for the high-volume production of spur and helical gears. Material is saved in producing each gear. Second and third machining operations are eliminated from most gear requirements. Cold rolling produces a smooth, mirror-finish surface texture.

A combination gear shaving and cold rolling machine is illustrated in Figure 18-5. Hardened and precision-ground rolling dies are used. The work parts are loaded into a magazine. They are then picked up and prelocated by a work arbor. The gear teeth are formed by applying a tremendous force (as in hot rolling) for the rolling dies to impress the tooth form in the blank. The cost of cold rolling machinery limits this method of manufacturing gears to high-volume production requirements.

FIGURE 18-4 Gear Shaving a Helical Gear

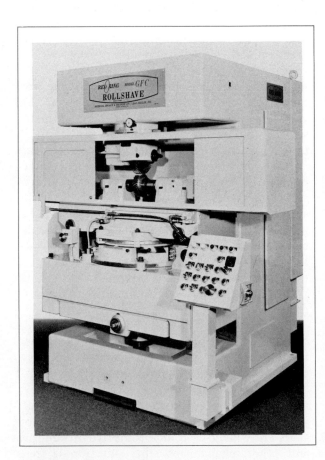

FIGURE 18-5 Combination Gear Shaving and Cold Rolling Machine

Extruding

Extruding is limited to brass, aluminum, and other soft materials. In some cases, bars are formed to gear tooth shape by forcing (extruding) the material through forming dies. Each gear is then cut off to width. Second and other operations are required on extruded parts. For example, the hole must be pierced. The teeth are often finish machined to precise limits.

Stamping

Gears for watches, small-gear mechanisms, and many business and household devices are mass-produced by *blanking (stamping)*. Fine blanking permits the stamping of gear parts up to 1/2″ (12.7mm) thickness within close tolerances and to high-quality surface finish requirements.

Powder Metallurgy

Powder metallurgy is a method that produces gears in three major steps:

— *Mixing*, in which selected metals and alloys in powdered form are mixed in specific proportions;
— *Compressing*, in which the mixture in a particular size of gear-forming die is subjected to extreme force;
— *Sintering*, in which the shaped gear is brought to high temperatures and the metal particles are sintered (fused) together to form a gear of a required form and size.

Large gears produced by this method require the *preform* (gear as it comes from a powder metallurgy press) to be machined to a precise size and surface finish.

GENERATING MACHINES FOR BEVEL GEARS

Generating Straight-Tooth Bevel Gears

Gear tooth generators are built to machine straight-tooth bevel gears that range in size from 0.025″ to 36″ (0.64mm to 914mm) in diameter. Two reciprocating cutting tools are mounted on the generating machine cradle. The cutting tools form the top and bottom sides of a tooth respectively. The cradle and gear blank roll upward together.

At the top of the upward movement, a tooth is completely generated and the cutting tools are automatically withdrawn. The machine indexes while the cradle moves down to the starting position for the next tooth. The tooth cutting cycle continues until all gear teeth are cut.

Generating Spiral-Tooth Bevel Gears

Spiral bevel gear generators produce spiral teeth. Tooth profiles for spiral bevel gears (with intersecting axes) and hypoid gears (with nonintersecting axes) may be generated on the same machine. The tooth profile and spiral shape are produced by combining straight-line and rotary motions between the gear blank and the cutter. Gears requiring large amounts of stock removal are rough machined. Sufficient material is left for the finishing cutters. Rough and finish cuts may be taken on the same machine.

GEAR FINISHING MACHINES AND PROCESSES

More accurate gear tooth shapes and precise surface finishes are produced by such second operations as shearing, shaving, and burnishing. Heat-treated gears are finished by grinding, lapping, and honing.

Shearing

Fini-shear is a new process for accurately producing and finishing gears. It uses a carbide cutter resembling a master gear. The cutter is rotated with and reciprocates against a gear blank. The cutting tool is brought into contact with the work at an angle to the work axis. The shearing, cutting action produces a high-quality surface finish. Fini-shear corrects roughing machine errors. The precision-hardened and -ground gears in the work spindle also contribute to the final work accuracy.

Gear Burnishing

Gear burnishing is a cold-working process. The machine rolls a gear in contact with three hardened burnishing gears. Burnishing produces slightly work-hardened, smooth gear teeth.

Finishing Gear Teeth on Hardened Gears

Two common gear tooth finishing processes for hardened gears include *grinding* and *lapping*. Gear teeth may be form ground or generated. Form grinding is similar to form milling. The grinding wheel is formed by diamond dressing. The diamond holders are guided by templates that conform to the gear tooth form. The formed grinding wheel reciprocates parallel to the work axis to produce accurately formed teeth with a high-quality surface texture.

Gear teeth on hardened gears may also be finish ground on a generating-type grinder. The tooth flanks are ground by the flat face of the wheel. The gear tooth is rolled past the revolving grinding wheel. Precisely formed, dimensionally accurate gear teeth are produced with a high-quality surface finish.

GEAR TOOTH MEASUREMENT AND INSPECTION

Gear teeth are generally inspected by visual comparison on an optical comparator. Gear teeth are usually measured mechanically in the shop by using a gear tooth vernier caliper or an outside micrometer and two wires (pins). Surface texture is usually checked against a comparator specimen.

Gear tooth measurements and tests relate to:

—Concentricity (runout),
—Tooth shape,
—Tooth size,
—Surface texture,
—Lead (on helical gears),
—Tooth angles (on bevel gears).

Tooth Inspection with an Optical Comparator

In gear tooth inspection, as in other uses of an optical comparator, a transparent drawing of the tooth is placed on the screen. The drawing scale conforms to the magnification power of the lens. The enlarged shadow of the gear tooth outline that is reflected on the screen is compared directly with the drawing. The nature and extent of error in tooth shape, size, and pitch are visually established.

Special Testing Fixtures

Special fixtures employing a master gear are used for testing concentricity (runout). The testing relates to out-of-trueness on spur gears (parallelism of the teeth and gear axes) and angular runout on bevel gears. Attachments are used on comparators and other microscopes to measure the lead of helical and worm gears.

Gear Tooth Vernier Caliper Measurement

Gear teeth must be precisely formed and dimensionally accurate if they are to mesh properly. One technique for measuring gear teeth is to use a vernier caliper. The accuracy of the caliper measurement is influenced by the accuracy of the outside diameter of the gear.

The *gear tooth vernier caliper*, illustrated in Figure 18-6, is designed to measure (1) the chordal thickness of a gear tooth at the pitch circle, and (2) the addendum distance from the top of the tooth to the pitch circle. There are two beams on the caliper. The horizontal beam is used for measuring the chordal thickness and has an adjustable sliding jaw. The vertical beam has an adjustable tongue. The jaw and tongue are adjusted independently. Different sizes of gear teeth may be accurately measured when proper allowance is made for any outside diameter variations of the gear blank.

The jaws of the vernier caliper are hardened and ground. The measuring surfaces are lapped. Vernier calipers are also provided with tungsten carbide measuring surfaces. The vernier caliper in Figure 18-6 is graduated in thousandths of an inch (0.001"). It is available in two sizes for measuring gear teeth from 20 to 1 diametral pitch. Metric gear tooth vernier calipers are graduated to fiftieths of a millimeter (0.02mm). The comparable gear tooth range in the English diametral pitch unit system is from 1.25 to 25 millimeter modules.

Two-Wire Method of Measuring Gear Teeth

A micrometer measurement over the outside diameter of two wires is another technique for checking the accuracy of gear teeth. Two cylindrical wires (pins) of a specified diameter are placed in diametrically opposite tooth spaces (Figure 18-7A) for an even number of teeth or

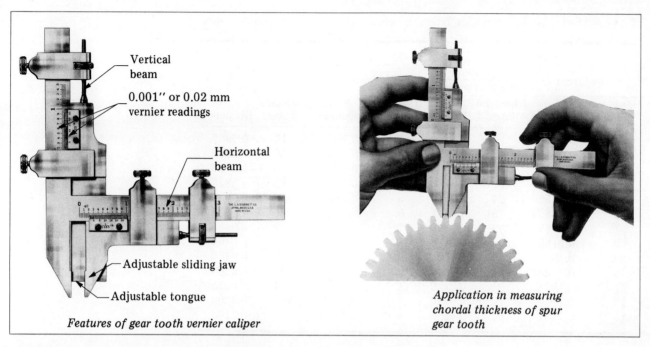

FIGURE 18-6 Gear Tooth Vernier Caliper Features and Application

are centered as closely as possible (Figure 18-7B) for an uneven number of teeth. A standard outside micrometer is generally used to establish the overall measurement (M) on spur gears. Internal measurements may be checked with an inside micrometer or other inside-diameter measuring tool.

Standard wire (pin) diameters are specified in handbook tables according to the diametral pitch of the gear. The diameters of the wire sizes are usually identified. *Diametral pitch* is the ratio of the number of teeth to the number of inches of pitch diameter. The diametral pitch of a gear designed according to ANSI standards equals the number of teeth in each inch of pitch diameter. *Pitch diameter* represents the diameter of the pitch circle at which the tooth thickness (chordal distance) is measured.

Van Keuren Wire Diameters. The Van Keuren wire diameters for external and internal gears are listed in Table 18-1. The Van Keuren standard wire sizes range from 0.8640″ to 0.0216″ diameter for external gears having diametral pitches from 2 through 80, respectively. The wire diameters for internal gears between this same range of diametral pitches is from 0.7200″ to 0.0180″.

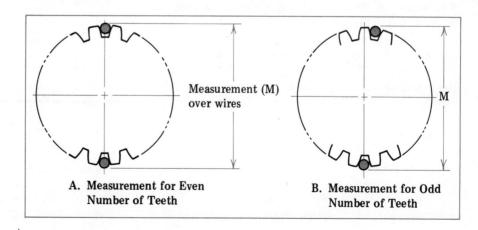

FIGURE 18-7 Two-Wire Method of Measuring Gear Teeth

A. Measurement for Even Number of Teeth

B. Measurement for Odd Number of Teeth

TABLE 18-1 Van Keuren Wire Diameters for External and Internal Gears

External Gears				Internal Gears			
wire diameter = $\frac{1.728}{\text{diametral pitch}}$				wire diameter = $\frac{1.440}{\text{diametral pitch}}$			
Diametral Pitch	Wire Diameter	Diametral Pitch	Wire Diameter	Diametral Pitch	Wire Diameter	Diametral Pitch	Wire Diameter
2	0.8640	16	0.1080	2	0.7200	16	0.0900
2-1/2	0.6912	18	0.0960	2-1/2	0.5760	18	0.0800
3	0.5760	20	0.0864	3	0.4800	20	0.0720
4	0.4320	22	0.0786	4	0.3600	22	0.0655
5	0.3456	24	0.0720	5	0.2880	24	0.0600
6	0.2880	28	0.0617	6	0.2400	28	0.0514
7	0.2469	32	0.0540	7	0.2057	32	0.0450
8	0.2160	36	0.0480	8	0.1800	36	0.0400
9	0.1920	40	0.0432	9	0.1600	40	0.0360
10	0.1728	48	0.0360	10	0.1440	48	0.0300
11	0.1571	64	0.0270	11	0.1309	64	0.0225
12	0.1440	72	0.0240	12	0.1200	72	0.0200
14	0.1234	80	0.0216	14	0.1029	80	0.0180

Tables of Measurements over Wires. Once the standard Van Keuren wire size is determined, information on the measurement over wires is found in other tables. Handbook tables provide measurements for 1 diametral pitch, external and internal teeth, even and odd numbers of teeth in a gear, and a range from 6 through 500 teeth. Table 18-2, for example, lists measurements for a few external spur gears having even or odd numbers of teeth. Values are given for four different pressure angles. The overall micrometer measurement is equal to the value given in the table for the specified number of teeth and pressure angle divided by the diametral pitch of the gear:

$$\frac{\text{measurement over}}{\text{wires (M)}} = \frac{\text{table value}}{\text{diametral pitch}}$$

For example, the overall micrometer measurement for a 16-tooth, 25° pressure angle, 10 diametral pitch requires the use of the measurement formula and wire table. The table value for 16 teeth, 25° pressure angle equals 18.3908. By inserting this value in the measurement formula,

$$M = \frac{18.3908}{10}$$
$$= 1.8391''$$

Allowance for Play between Teeth. The measurement over wires indicates the overall

TABLE 18-2 Sample Dimensions for Checking External Measurement over Wires

	Standard Pressure Angle			
	14-1/2°	20°	25°	30°
Even Number of Teeth	Dimension over Wires*			
6	8.2846	8.3032	8.3340	8.3759
8	10.3160	10.3271	10.3533	10.3919
10	12.3399	12.3445	12.3667	12.4028
12	14.3590	14.3578	14.3768	14.4108
14	16.3746	16.3683	16.3846	16.4169
16	18.3877	18.3768	18.3908	18.4217
18	20.3989	20.3840	20.3959	20.4256
Odd Number of Teeth	Dimension over Wires*			
181	183.5230	183.4469	183.4350	183.4520
191	193.5246	193.4478	193.4357	193.4526
201	203.5260	203.4487	203.4363	203.4532
301	303.5355	303.4538	303.4402	303.4565
401	403.5404	403.4565	403.4422	403.4582
501	503.5433	503.4581	503.4434	503.4592

*The table dimensions are for 1 diametral pitch gear teeth using Van Keuren standard wire sizes:

$$\text{wire diameter} = \frac{1.728}{\text{diametral pitch}}$$

Divide the table dimension by a given diametral pitch when it is more than 1.

dimension when the pitch diameter is correct and no allowance is made for play (backlash) between mating teeth. Compensation must be made for backlash and for cutting the gear teeth slightly deeper in order to reduce the tooth thickness. An equal amount is usually removed from each mating gear, except for gear and pinion combinations. In these combinations, the pinion is machined to a standard size; the total backlash is allowed on the gear.

Handbook tables are available on standard backlash allowances for external and internal spur gears. These tables give the backlash allowances for gears with numbers of teeth ranging from 5 to 200 and for standard pressure angles. The measurement over wires given in such tables is reduced by the amount found in the backlash allowance table for each 0.001" reduction in tooth thickness at the pitch line.

Example: A backlash of 0.0002" is required on a 24-tooth, 10 diametral pitch, external, 25° pressure angle spur gear. The micrometer measurement of standard Van Keuren wires is required. The measurement given in a handbook table for a 1 diametral pitch, 24-tooth, 25° pressure angle gear is 28.4096". The overall micrometer measurement for the 10 diametral pitch is equal to the table value divided by 10:

$$\frac{28.4096}{10} = 2.8410"$$

The backlash allowance as given in a handbook table for each 0.001" reduction in pitch-line thickness for this gear is 0.0019". The overall micrometer measurement is equal to 2.8410" minus 2 times 0.0019", or 2.8372". This micrometer measurement over the two standard Van Keuren wires provides 0.002" allowance for backlash.

SPUR GEARING DRAWINGS (REPRESENTATION, DIMENSIONING, AND TOOTH DATA)

Drawings of gears conform to ANSI standards. Dimensions and notes that accompany the line drawings provide the craftsperson with all the technical information needed for setting up the machine, cutting, and measuring the gear tooth. The drawings contain such information as material and heat treatment (where required), quality of surface texture, gear tooth data, and manufacturer's identifying markings.

The view or views used depend on design features. Gears, pinions, or worms that are machined on a shaft are represented by a view that is parallel with the axis (Figure 18-8). This view is the preferred representation

Gears with holes and hubs are generally drawn as a section view (Figure 18-9). The section cuts through the axis. The section may be partial or full. Usually just one view is used. When a drawing needs to be clarified, a section (front) view and a few teeth may be drawn, as shown in Figure 18-10. The outside circle, root circle, and pitch circle are each represented in the circular (top view), repeated pattern of a long and two short dashes. In this drawing, a position mark is stamped, as noted, in a particular location. Outside diameter, width measurements, and machining notes are included on the drawing.

Gear tooth and other gear measurements appear on drawings in two ways. The drawings in Figures 18-8, 18-9, and 18-10 show only the finished dimensions for a gear blank. Other dimensions relating to the gear teeth are included as *gear tooth data.* Sometimes, *reference dimensions* are included for purposes of computing other needed values. For example, pitch diameter is provided for reference in the tooth data of Figure 18-9.

SPUR GEARING TERMINOLOGY

Standard ANSI definitions are applied to gear terms. The terms identify design features and

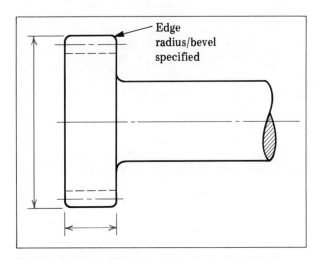

FIGURE 18-8 Full-View Representation of a Spur Gear Machined on a Shaft

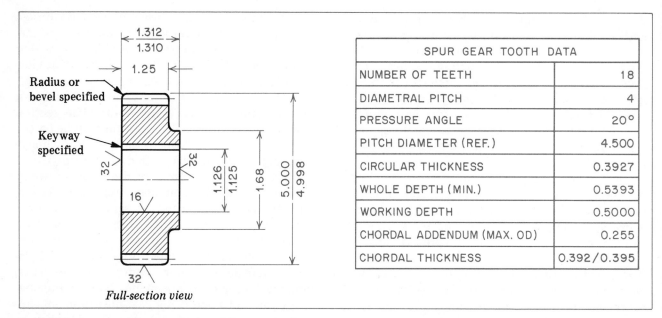

FIGURE 18-9 Representation of Spur Gears with Hubs, Showing Methods of Dimensioning and Required Machining Data (ANSI)

measurements. Common gear tooth terms and measurements are identified in Figure 18-11 for *spur gears*. Additional terms related to helical, bevel, and worm gears are defined later.

The *pitch circle* is an imaginary circle that represents the point at which the teeth on two mating gears are tangent. A gear size is indicated by the diameter of the pitch circle. Most gear

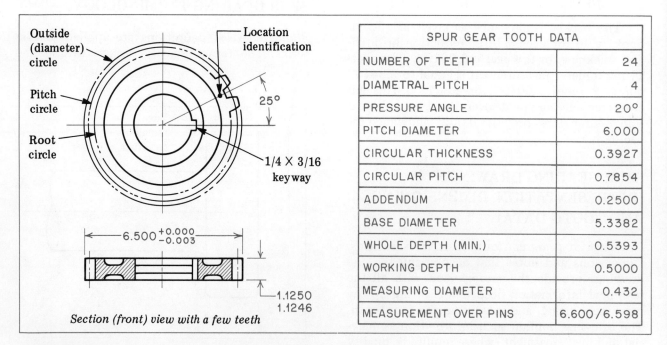

FIGURE 18-10 Representation of Cast Spur Gear, Showing Dimensioning and Required Machining Data (ANSI)

GEAR DEVELOPMENT, FEATURES, COMPUTATIONS 333

definitions are computed in relation to the pitch circle.

The *pitch diameter* is the diameter of the pitch circle.

The *addendum* is a vertical distance that extends from the pitch circle to the addendum circle. The adjustable tongue on the vertical beam of a gear tooth vernier caliper is set to the addendum. At this depth, the gear tooth thickness may be accurately measured at the pitch circle.

The *addendum circle* is the outside (addendum) diameter.

The *dedendum* is the vertical distance that extends from the pitch circle to the root circle. The dedendum accommodates the addendum of the mating tooth. It also provides additional clearance between the outside diameter of the mating gear and the root circle.

The *root circle (diameter)* is located at the root (bottom) of the gear teeth.

The *base circle:* is the circle from which the tooth profile is generated. The diameter of the base circle is found by drawing a line at a 14-1/2°, 20°, or 25° angle through the pitch point. The angle depends on the pressure angle used in the gear design.

The *pressure angle* is the angle that establishes the tooth shape. It represents the angle at which the forces from the tooth of one gear are transmitted to the mating tooth on another gear.

While many 14-1/2° involute and composite system full-depth teeth and 20° involute stub teeth are still used and tables for these teeth are included in handbooks, these tooth forms are no longer recommended. The newer American National Standard (ANSI B 6.1-1968) provides information on 20° and 25° involute spur gear forms. The gear tooth forms are identical except for the different pressure angles and the minimum allowable number of teeth. Under extreme forces, a shorter (stub) tooth involute form gear provides additional strength.

The ANSI 25° standard form provides even greater strength and permits the machining of gears with fewer than 18 teeth. The 25° form produces lower contact compressive strength and greater gear tooth surface durability.

The *circular pitch* is the circular distance that extends along the pitch circle from a point on one tooth to the corresponding point on the next tooth.

The *circular thickness* is the thickness of a gear tooth as measured along the arc of the pitch circle.

The *chordal thickness* is the tooth thickness as measured along a chord of the pitch circle. The measurement is taken by the movable leg on the horizontal beam of the gear tooth vernier caliper.

The *chordal addendum* extends from the outside diameter to the chord at the chordal thickness. This dimension differs from the addendum.

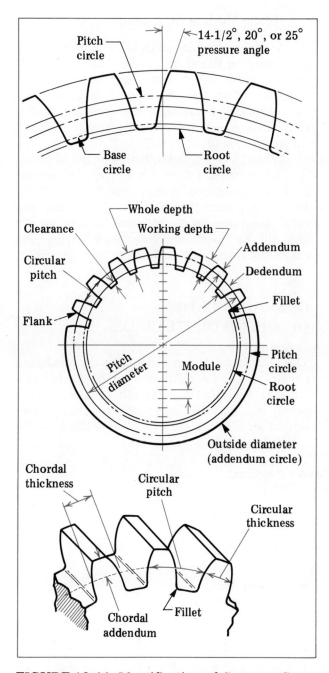

FIGURE 18-11 Identification of Common Spur Gear Terms and Measurements

The *whole depth* indicates the overall height of a gear tooth or the depth to which each tooth is cut.

The *working depth* is the depth to which one tooth extends into the gear tooth space on a mating gear or rack.

The *clearance* is the difference between the working depth and the whole depth of a gear tooth.

The *diametral pitch (system)* is the system that relates number of teeth to pitch diameter. Diametral pitches are standardized. The diametral pitch is equal to the number of teeth per inch of pitch diameter.

The *face width* is the width between the two sides of a gear tooth.

The *number of teeth* identifies the number of teeth to be machined in the gear.

MODULE SYSTEMS OF GEARING

The *English module* represents the *ratio* between pitch diameter in inches and the number of teeth. That is, diametral pitch equals the number of teeth per inch of pitch diameter. A 10 diametral pitch gear, for example, has 10 teeth for each inch of pitch diameter.

The *SI Metric module* is an *actual dimension:*

$$\text{SI Metric module} = \frac{\text{pitch diameter (mm)}}{\text{number of teeth in gear}}$$

For example, a 25-tooth gear with a pitch diameter of 75mm has a module of 3—that is, there are 3mm of pitch diameter for each tooth.

Most of the gearing produced in the United States is designed around the diametral pitch system. This system provides a series of standard gear tooth sizes for each diametral pitch. The system is similar to the standardization of screw threads. For example, in a 10 diametral pitch series of gears, a 40-tooth gear has a pitch diameter of 4"; a 42-tooth gear, 4.2"; a 44-tooth gear, 4.4"; and so on. The variation for successive teeth in a gear equals 1/10" in the 10 diametral pitch series.

Metric Module Equivalent

A metric module equivalent of a diametral pitch may be found by dividing 25.4 by the diametral pitch. Where necessary, the answer is rounded off to the closest module in the standard series. For example, if the metric module equivalent of a 10 diametral pitch is required, the nearest standard module to 2.54 is 2.5.

Table 18-3 is a partial listing of the SI Metric spur gear rules and formulas for required features of a 20° full-depth involute tooth form. Table 18-4 is a partial listing of the ANSI spur gear rules and formulas for required features of 20° and 25° full-depth involute tooth forms. The complete tables are included in the Appendix.

CHARACTERISTICS OF THE INVOLUTE CURVE

The term *involute curve* identifies the almost exclusive profile used for gear teeth. An involute curve is generated as the path of a point. The point, theoretically held taut on a straight line, scribes an involute curve as it unwinds from the

TABLE 18-3 SI Metric Spur Gear Rules and Formulas for Required Features of 20° Full-Depth Involute Tooth Form (Partial Table)

Required Feature	Rule	Formula*
Module (M)	Divide the pitch diameter by the number of teeth	$M = \dfrac{D}{N}$
	Divide the outside diameter by the number of teeth plus 2	$M = \dfrac{O}{N + 2}$
Addendum (A)	Multiply the module by 1.000	$A = M \times 1.000$
Clearance (S)	Multiply the module by 0.250	$S = M \times 0.250$
Whole depth (W^2)	Multiply the module by 2.250	$W^2 = M \times 2.250$
Tooth thickness (T)	Multiply the module by 1.5708	$T = M \times 1.5708$

*Dimensions are in millimeters.

TABLE 18-4 ANSI Spur Gear Rules and Formulas for Required Features of 20° and 25° Full-Depth Involute Tooth Forms (Partial Table)

Required Feature	Rule	Formula*
Diametral pitch (P_d)	Divide the number of teeth by the pitch diameter.	$P_d = \dfrac{N}{D}$
	Add 2 to the number of teeth and divide by the outside diameter.	$P_d = \dfrac{N + 2}{O}$
	Divide 3.1416 by the circular pitch.	$P_d = \dfrac{3.1416}{P_c}$
Whole depth of tooth (W^2)	Divide 2.250 by the diametral pitch.	$W^2 = \dfrac{2.250}{P_d}$
Thickness of tooth (T)	Divide 1.5708 by the diametral pitch.	$T = \dfrac{1.5708}{P_d}$
Center distance (C)	Add the pitch diameters and divide the sum by 2.	$C = \dfrac{D^1 + D^2}{2}$
	Divide one-half the sum of the number of teeth in both gears by the diametral pitch.	$C = \dfrac{1/2\,(N^1 + N^2)}{P_d}$

*Dimensions are in inches.

periphery of the base circle (Figure 18-12). Some of the characteristics of an involute curve are as follows:

- The form of the involute curve depends on the base circle;
- The action of two mating involute gears provides uniform motion;
- The rate of motion is established by the diameters of the base circles of the driver and driven gears;
- A straight line of action is produced by the contact between intermeshing teeth on a driver and a driven gear;
- The pressure angle represents the angle between the line of action and a line perpendicular to the common centerline of the mating gears.

RACK AND PINION GEARS

A *rack* is a straight bar with gear teeth cut on one side. The profile and design features of the teeth on a rack are identical with the mating spur gear *pinion* teeth (Figure 18-13). The addendum, dedendum, tooth thickness, and form are identical. The pitch line of the rack teeth intersects with the pitch circle of a meshing gear. The pitch line and the pitch circle are both one addendum from the top of the gear teeth.

Some different terms are used with rack and pinion gears because all circular dimensions on the rack become linear. The *lineal pitch*, as shown in Figure 18-13, corresponds with circular pitch:

$$\text{lineal pitch} = \frac{3.1416}{\text{diametral pitch}}$$

The *lineal thickness* is equal to the circular thickness of a spur gear.

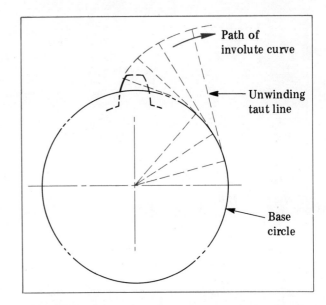

FIGURE 18-12 Generating the Profile of the Involute Tooth

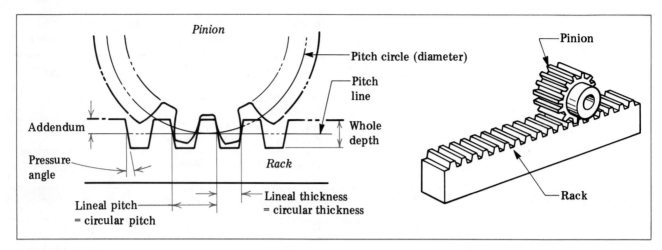

FIGURE 18-13 Rack and Pinion Gear Teeth Design Features

Racks are used with pinion gears to convert rotary motion to straight-line motion. The quill or spindle on drilling and milling machines and the rack under the front way on a lathe are typical examples.

Representing and Machining a Rack

A rack is usually represented and dimensioned on a drawing as shown in Figure 18-14. Tooth data provide additional information for machining and inspecting the gear teeth.

Racks are machined in jobbing, maintenance, and small shops on a milling machine. Short rack lengths may be held on parallels in a vise or strapped to the table. In conventional milling, the workpiece and formed gear tooth cutter axes are parallel. The cutter is accurately moved by the cross slide an amount equal to the circular pitch of the gear. Each tooth is cut by feeding the rack longitudinally past the cutter.

The formed gear tooth cutter may be mounted in a rack-cutting attachment. The table is moved a distance equal to the lineal (circular)

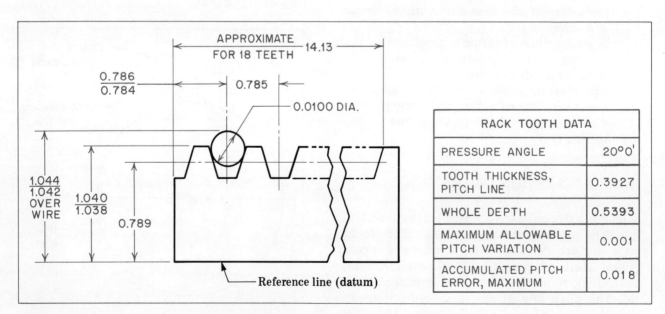

FIGURE 18-14 Representation of a Rack, Showing Methods of Dimensioning and Required Machining Data

pitch to position each gear tooth. An indexing attachment is usually used to position (index) the workpiece to the required lineal pitch. The teeth may also be indexed by using the graduated dial on the table feed screw.

A straight-tooth rack is cut by mounting the gear cutter axis parallel with the gear axis. The teeth on a helical rack require the workpiece to be set at the gear helix angle in relation to the cutter axis.

HELICAL GEARING

Helical gears transmit rotary motion between two shafts. The shafts may have parallel axes that lie in the same plane (Figure 18-15A) or the axes may be at any angle up to 90° to each other (Figure 18-15B).

Helical gear teeth permit continuous, smooth, quiet operation as compared to spur gear teeth. More than one helical gear tooth is engaged at one time. Thus, the gear strength is increased for teeth of equal size and pitch.

One disadvantage of helical gears in relation to spur gears is that end thrust is produced. The use of thrust bearings is therefore required. The end thrust problem is often overcome in gearing in much the same way as for helical milling cutters. A left- and a right-hand helical gear are used in combination. The *herringbone (double-helical) gear* used on parallel shaft movements (Figure 18-15C), combines the features of a right- and left-hand helical gear in one gear. Since the end thrusts are equalized, no end thrust bearings are required.

CALCULATING HELICAL GEAR DIMENSIONS

The basic features, symbols, rules, and formulas for helical gear calculations and measurements are given in Table 18-5.

Pitch Diameters for Mating Gears of Different Sizes

The pitch diameters of mating helical gears may be calculated when the gear ratio and center distance are given. The combined pitch diameter of the two gears equals twice the given center distance. The pitch diameter of each gear is calculated by dividing the combined pitch diameters by the ratio value of each gear.

For example, if two mating helical gears have a 6.000" center distance and a ratio of 3:2, the combined pitch diameter equals 6.000" × 2 or 12.000". Each gear occupies its ratio value of the combined pitch diameters. The pitch diameter of the larger gear is 3/5 of 12.000", or 7.2000". The pitch diameter of the smaller (pinion) gear is 2/5 of 12.000", or 4.800".

Normal Circular and Diametral Pitches

The terms for helical gears are similar to the terms used for spur gears. Two terms, *normal circular pitch* and *normal diametral pitch*, require further discussion. The normal diametral pitch represents the diametral pitch of the helical gear tooth cutter.

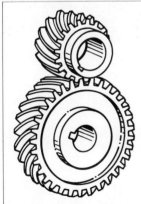
A. Parallel Gear Axes

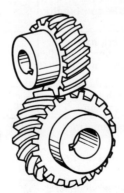
B. Gear Axes at Right Angle

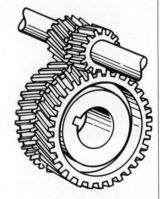
C. Herringbone Gear

FIGURE 18-15 Common Applications of Helical Gearing

TABLE 18-5 Helical Gearing Features, Symbols, Rules, and Formulas

Feature	Symbol	Feature	Symbol
Addendum	A	Circular pitch	P_c
Center distance	C	Normal circular pitch	P_{nc}
Cutter number	C#	Diametral pitch	P_d
Helix angle of teeth	H	Normal diametral pitch	P_n
Lead of tooth helix	L	Pitch diameter (D)	
Number of teeth	N	Gear 1	D_1
Outside diameter	O	Gear 2	D_2
		Tooth helix angle	
		Cosine	cos a
		Cotangent	cot a
		Tangent	tan a
		Normal tooth thickness at pitch line	T_n
		Whole depth of tooth	W

Required Feature	Rule	Formula
Pitch diameter (D)	Divide the number of teeth by the product of the normal diametral pitch and the cosine of the tooth helix angle.	$D = \dfrac{N}{P_n \times \cos a}$
Center distance (C)	Add the pitch diameters of the two gears and divide by 2.	$C = \dfrac{D_1 + D_2}{2}$
Number of teeth for selecting the formed cutter (C#)	Divide the number of teeth by the cube of the cosine of the tooth helix angle.	$C\# = \dfrac{N}{(\cos a)^3}$
	Divide the normal circular pitch by the circular pitch.	$\cos a = \dfrac{P_{nc}}{P_c}$
	Divide the diametral pitch by the normal diametral pitch.	$\cos a = \dfrac{P_d}{P_n}$
Helix angle of the gear teeth*	Divide the number of teeth by the product of the normal diametral pitch and the pitch diameter.	$\cos a = \dfrac{N}{P_n \times D}$
	*Each quotient equals a natural trigonometric cosine function, which is converted into the helix angle of the gear teeth (H).	
Lead of tooth helix (L)	Divide the product of the number of teeth and the circular pitch by the tangent of the tooth helix angle.	$L = \dfrac{N \times P_c}{\tan a}$
	Multiply the pitch diameter by 3.1416 times the cotangent of the tooth helix angle.	$L = D \times (3.1416 \times \cot a)$
Normal diametral pitch (P_n)	Divide the number of teeth by the cosine of the tooth helix angle, add 2, and divide the sum by the outside diameter.	$P_n = \dfrac{\dfrac{N}{(\cos a)} + 2}{O}$
Normal circular pitch (P_{nc})	Multiply the circular pitch by the cosine of the tooth helix angle.	$P_{nc} = P_c \times \cos a$
Addendum (A)	Divide 1 by the normal diametral pitch.	$A = \dfrac{1}{P_n}$
Whole depth of tooth (W)	Divide 2.157 by the normal diametral pitch.	$W = \dfrac{2.157}{P_n}$
Normal tooth thickness at pitch line (T_n)	Divide 1.157 by the normal diametral pitch.	$T_n = \dfrac{1.157}{P_n}$
Outside diameter (O)	Add twice the addendum to the pitch diameter.	$O = D + 2A$

The normal circular pitch of a helical gear is the distance between a reference point on one gear tooth and the corresponding reference point on the next tooth. The measurement is taken at the pitch circle on a reference plane that is at a right angle to the tooth helix angle (Figure 18-16).

Selecting a Helical Gear Cutter

The formed gear cutter used on a helical gear is the same type as is used on a spur gear. However, the cutter size differs since the diametral pitch of the spur gear changes to be the normal diametral pitch of the helical gear. This change in cutter size involves the number of teeth to be cut and the gear tooth helix angle.

The number of teeth for which a cutter is to be selected ($C_\#$) is determined by dividing the number of gear teeth (N) by the cube of the cosine of the helix angle. Expressed as a formula,

$$C_\# = \frac{N}{(\cos a)^3}$$

Example: Determine the cutter number to use to mill a 32-tooth helical gear having a helix angle of 30° by substituting the values in the formula,

$$C_\# = \frac{32}{(0.86603)^3} = 49.26$$

A reference is made to a handbook table of gear milling cutters for different numbers of gear teeth. From such a table, a #3 gear milling cutter is selected because the 49 teeth fall within the 35 to 54 tooth range of this cutter.

A simplified technique of selecting a cutter for helical gear milling is to use a table or a chart such as the one shown in Figure 18-17. The cutter number to use is determined by drawing a straight line from the number of teeth in the helical gear to be cut (through the wide band of cutter numbers) to the helix angle. For the previous example, the straight line would extend from 32 teeth to the 30° helix angle. The line, drawn on Figure 18-17, cuts through the cutter #3.

The chart shows the number of teeth in the range for each cutter size from 12 teeth through 135. Where greater accuracy of tooth shape is required, an intermediate half-number series of cutters is available. These cutters are listed in manufacturers' catalogs and handbooks.

BEVEL GEARING

Design Features

Bevel gears are used to transmit motion and power between two shafts whose axes intersect. Most bevel gear axes intersect at 90°. The axes may also be at any angle. The design features of a bevel gear and the accompanying terms appear in Figure 18-18.

The teeth on bevel gears conform to standard 14-1/2°, 20°, and 25° pressure angles. The tooth form follows the same involute form as a spur gear. However, the teeth taper toward the apex of the cone. Two right-angle (90° axes) bevel gears of the same module, pressure angle, and number of teeth are called *miter gears*. The small gear in a bevel gear combination is identified as a *pinion*.

Working Drawings of Bevel Gears

A one-view, cross-sectional drawing is usually used to represent a bevel gear. A second view is included when the gear has spokes. The relationship of a bevel gear and pinion combination is shown in a single-view drawing of the two gears. Gear teeth are generally not drawn, except on an assembly drawing. The dimensions for machining a bevel gear blank are included on the gear drawing. Cutting dimensions and data appear in a table to supplement the information contained on the drawing. Figure 18-19 is the representation of a bevel gear showing methods of dimensioning with required machining data.

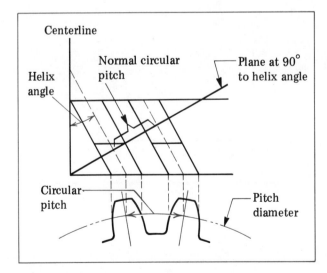

FIGURE 18-16 Relationship of Normal Circular Pitch of a Helical Gear to Circular Pitch

340 HORIZONTAL MILLING MACHINES

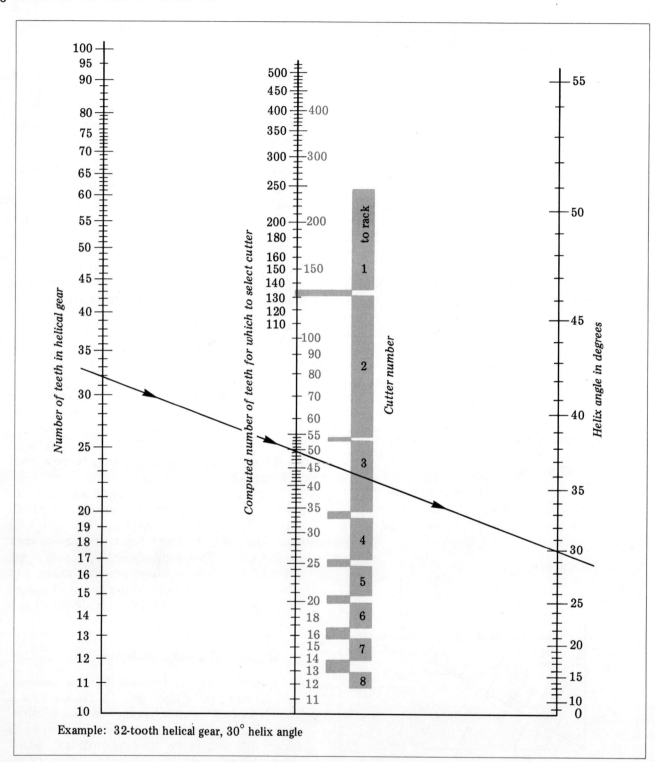

FIGURE 18-17 Cutter Selection Chart for Helical Gear Cutting

GEAR DEVELOPMENT, FEATURES, COMPUTATIONS 341

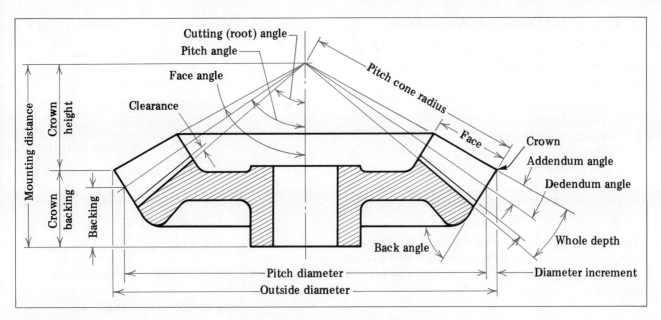

FIGURE 18-18 Design Features and Terms Applied to Bevel Gears

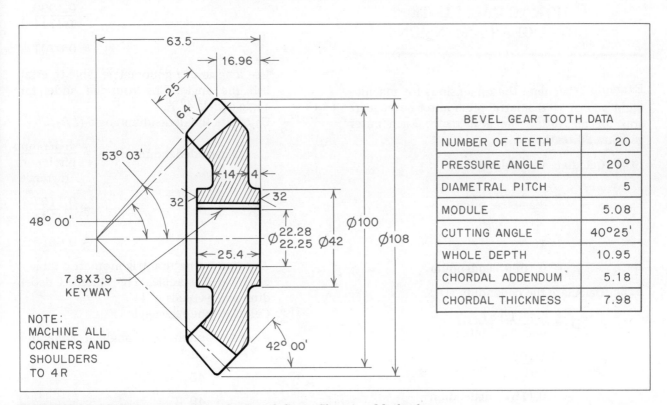

FIGURE 18-19 Representation of a Bevel Gear, Showing Methods of Dimensioning and Required Machining Data

Computing Bevel Gear Dimensions

The formulas for computing spur gear dimensions are also used for the following bevel gear features:

—Addendum,
—Dedendum,
—Whole tooth depth,
—Module,
—Diametral pitch,
—Circular pitch,
—Choral thickness,
—Circular thickness.

The remaining dimensions required for the design and manufacture of bevel gears may be computed by using the appropriate rules and formulas.

HOW TO CALCULATE BEVEL GEAR DIMENSIONS

Example: Calculate the dimensions for machining a pair of 10 diametral pitch miter bevel gears and for measuring the gear teeth. Each miter gear has 24 teeth.

STEP 1 Calculate the addendum (A):

$$A = \frac{1}{\text{diametral pitch}}$$

$$= \frac{1}{10}$$

$$= 0.1000'' \text{ addendum}$$

STEP 2 Calculate the dedendum (D):

$$D = \frac{1.157}{\text{diametral pitch}}$$

$$= \frac{1.157}{10}$$

$$= 0.1157'' \text{ dedendum}$$

STEP 3 Calculate the pitch diameter (D_p):

$$D_p = \frac{\text{number of teeth}}{\text{diametral pitch}}$$

$$= \frac{24}{10}$$

$$= 2.0000'' \text{ pitch diameter}$$

STEP 4 Calculate the pitch cone distance (P_{cd}):

$$P_{cd} = \frac{\text{diametral pitch}}{2 \times \text{ sine pitch cone angle of } 45°}$$

$$= \frac{2.0000}{2 \times (0.707)}$$

$$= 1.4144'' \text{ pitch cone distance}$$

STEP 5 Calculate the diameter increment (D_i):

$$D_i = 2A \times \text{ cosine pitch cone angle}$$

$$= (2 \times 0.1000) \times 0.707$$

$$= 0.1414'' \text{ diameter increment}$$

STEP 6 Calculate the addendum angle ($\angle A$):

$$\text{tangent addendum angle} = \frac{\text{addendum}}{\text{pitch cone distance}}$$

$$= \frac{0.1000}{1.4144}$$

$$= 0.0707$$

Use a natural trigonometric table to establish the angle. The rounded addendum angle equals $4°3'$.

STEP 7 Calculate the dedendum angle ($\angle D$):

$$\text{tangent dedendum angle} = \frac{\text{dedendum}}{\text{pitch cone distance}}$$

$$= \frac{0.1157}{1.4144}$$

$$= 0.0818$$

Again, use a natural trigonometric table to establish the angle. The rounded dedendum angle equals $4°41'$.

STEP 8 Calculate the face angle ($F\angle$):

$$F\angle = \text{pitch cone angle} + \text{addendum angle}$$

$$= 45° + 4°3'$$

$$= 49°3'$$

STEP 9 Calculate the cutting (root) angle ($C\angle$):

$$C\angle = \text{pitch cone angle} - \text{dedendum angle}$$

$$= 45° - 4°41'$$

$$= 40°19'$$

GEAR DEVELOPMENT, FEATURES, COMPUTATIONS 343

STEP 10 Calculate the face width (F_w):

$$F_w = \frac{\text{pitch cone radius}}{3}$$

$$= \frac{1.4144}{3}$$

$$= 0.4713''$$

HOW TO CALCULATE LARGE- AND SMALL-END GEAR TOOTH DIMENSIONS

Large End

STEP 1 Calculate the whole depth:

$$\text{whole depth} = \frac{2.157}{\text{diametral pitch}}$$

$$= \frac{2.157}{10}$$

$$= 0.2157''$$

STEP 2 Calculate the tooth thickness at the pitch line:

$$\text{tooth thickness} = \frac{1.571}{\text{diametral pitch}}$$

$$= \frac{1.571}{10}$$

$$= 0.1571''$$

Small End

Note: The size of the small end is approximately two-thirds of the large-end measurement.

STEP 3 Calculate the whole depth:

$$\text{whole depth} = \text{whole depth at large end} \times 2/3$$

$$= 0.2157 \times 2/3$$

$$= 0.1438''$$

STEP 4 Calculate the tooth thickness at the pitch line:

$$\text{tooth thickness} = \text{tooth thickness at large end} \times 2/3$$

$$= 0.1571 \times 2/3$$

$$= 0.1047''$$

HOW TO SELECT THE BEVEL GEAR CUTTER

STEP 1 Calculate the number of teeth (N):

$$\text{number of teeth for cutter selection} = \frac{\text{number of teeth in gear}}{\text{cosine of pitch cone angle}}$$

$$= \frac{32}{0.707}$$

$$= 45.26$$

Note: The nearest whole number of teeth is used: N = 45 teeth.

STEP 2 Refer to a handbook table of gear milling cutter sizes.

STEP 3 Select the number of the gear cutter that includes 45 teeth in its cutting range.

Note: A #3 formed involute gear milling cutter is selected. This cutter has a tooth cutting range of 35 to 54 teeth.

Note: While the formed cutter number for milling bevel gear teeth is the same as for a spur gear, there is a design difference. The bevel gear cutter is thinner to permit the cutter to pass through the narrow tooth space at the small end of the bevel gear. The bevel gear cutter number is selected from a spur gear milling cutter table.

HOW TO COMPUTE THE AMOUNT OF GEAR TOOTH TAPER OFFSET

The milling of bevel gear teeth with a formed rotary cutter requires two cuts to be taken for each tooth. The gear blank is slightly offset from the centerline. After the first cut is completed on all teeth, the table is moved in the opposite direction an equal amount beyond the centerline.

Precision Method of Computing

STEP 1 Compute the ratio to be used in determining the gear tooth cutter factor:

HORIZONTAL MILLING MACHINES

$$\text{ratio} = \frac{\text{pitch cone radius}}{\text{face width}}$$

$$= \frac{1.4144}{0.4713}$$

$$= \frac{3}{1} \text{ or } 3:1$$

Note: The 3:1 ratio of the pitch cone radius to width of face is used to determine the gear tooth cutter factor for the #3 cutter. This factor is found in handbook tables for obtaining offset for milling bevel gears.

STEP 2 Calculate the amount of table offset:

$$\text{table offset} = \frac{\text{thickness of gear cutter at pitch line at small end of tooth}}{2}$$

$$- \frac{\text{gear tooth cutter factor}}{\text{diametral pitch}}$$

$$= \frac{0.1047}{2} - \frac{0.266}{10}$$

$$= 0.05235 - 0.0266$$

$$= 0.0258'' \text{ table offset}$$

Approximate Method

The approximate table offset is calculated as maximum offset plus minimum offset divided by 2. The maximum offset is equal to one-sixth of the thickness of the tooth at the large end. The minimum offset is one-seventh the thickness of the tooth at the large end.

STEP 1 Calculate the maximum offset. First determine the thickness at the large end. For the 10 diametral pitch tooth,

$$\text{tooth thickness} = \frac{1.571}{\text{diametral pitch}}$$

$$= 0.1571''$$

Thus,

$$\text{maximum offset} = \frac{0.1571}{6}$$

$$= 0.0262''$$

STEP 2 Calculate the minimum offset:

$$\text{minimum offset} = \frac{0.1571}{7}$$

$$= 0.0224''$$

STEP 3 Determine the approximate offset:

$$\text{approximate offset} = \frac{0.0262 + 0.0224}{2}$$

$$= 0.0243''$$

WORM GEARING

Worm gearing is used to transmit rotary motion and power between two nonintersecting shafts. The shafts generally are at right angles to each other. There are two gears in a worm gearing setup. The *worm* serves as the driver gear; the *worm gear*, as the driven gear. The teeth in the worm are similar to a coarse-pitch Acme screw thread with a single-, double-, or triple-lead thread. The teeth in the worm gear conform to the worm thread form. The concave shape in the worm gear produces maximum contact area between the thread of the worm and the gear teeth.

Classes of Worm Gearing

The two general classes of worm gearing are the *fine-pitch* and the *coarse-pitch* class. Fine-pitch worm gears are largely used to transmit uniform, accurate angular motion rather than power. Gear terms and design features conform to American National Standard (ANSI) specifications. There are eight standard axial pitches (0.030'' to 0.160''). Fifteen standard lead angles (30' to 30°) have been established. A standard 20° pressure angle cutter or grinding wheel angle is used.

Industrial (Coarse-Pitch) Worm Gearing

Industrial worm gearing is generally of coarse pitch. Such gearing meets one or more of the following requirements:

—Transmits power efficiently,
—Transmits power at a large reduction in velocity,
—Provides a high mechanical advantage between driving and driven (resistance) forces.

The usual materials for power transmission applications are hardened and ground steel worms and phosphor bronze wheels. Low-carbon steels, 3-1/2% chromium steels, high nickel-chromium case hardening steels, and 0.40% to 0.55% carbon steels are generally used as worm materials.

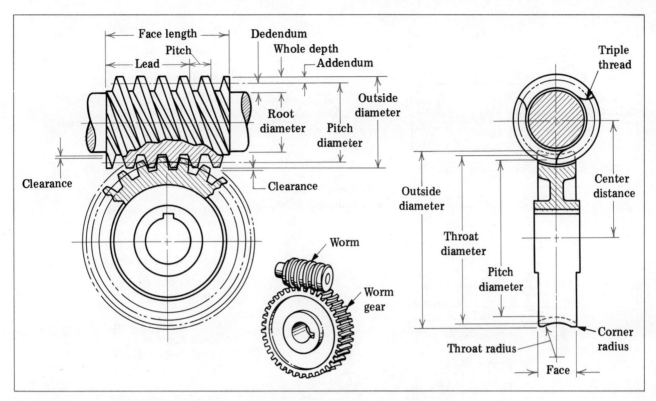

FIGURE 18-20 Example of Worm and Worm Gear and Common Terms

Single- and Multiple-Thread Worms and Velocity Ratio

A single-thread worm advances the worm wheel one tooth space (lead) during one revolution. The worm gear is advanced two teeth for each revolution of a double-thread worm; three teeth, for a triple-thread worm.

The primary use of a single-thread worm is to obtain the highest possible speed reduction in a gear combination. The single-thread worm is comparatively inefficient. A multiple-thread worm has a higher lead angle than a single-thread worm. This angle permits the multiple-thread worm to transmit power more efficiently.

A worm and worm gear combination is capable of carrying greater loads than helical gears and at a *high velocity ratio*. This ratio depends on the thread lead on the worm and the number of teeth on the gear. For example, the velocity ratio of a single thread (single pitch lead) to a 36-tooth gear is 36:1; of a double thread, 18:1; a triple thread, 12:1. The worm gear and worm permit high speed reductions in confined areas.

WORM GEARING TERMINOLOGY

Most of the gear terms and formulas previously described apply to worm gearing. The thread addendum, dedendum, whole tooth depth, outside diameter, and pitch diameter of the worm and worm gear are the same as for spur gears. A worm and worm gear and common terms are illustrated in Figure 18-20. The following terms and formulas are specific to worm gearing:

The *axial pitch* is the distance that extends between corresponding reference points on adjacent threads in a worm. The axial pitch of a worm is equal to the circular pitch of the mating worm gear.

The *axial advance (lead)* is the distance the worm thread advances in one revolution. A single thread advances a distance equal to the thread pitch; a double thread, twice the pitch; and so on.

The *lead angle* is formed by the tangent to the helix of the thread at the pitch diameter and a plane perpendicular to the worm axis. By formula,

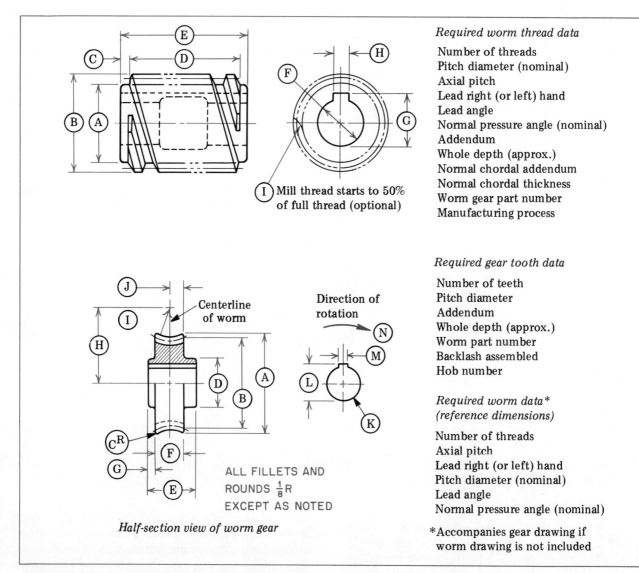

FIGURE 18-21 ANSI Working Drawings of Worms and Worm Gears with Thread and Tooth Data Required for Machining

$$\text{lead angle (tangent to thread helix angle)} = \frac{\text{lead}}{\pi \times \text{pitch diameter of worm}}$$

The *throat diameter of a worm gear* is the overall measurement of the worm at the bottom of the tooth arc. The throat diameter equals the pitch diameter plus twice the addendum.

REPRESENTATION OF WORMS AND WORM GEARS

A one-view partial or half-section view of a worm gear and a single view of the worm are used to represent a worm gearing combination. ANSI standards for gear and worm dimensions are indicated in Figure 18-21. The thread and tooth data needed for machining accompanies the drawings. Complete rules and formulas for calculating dimensions, ratios, and other design features are contained in trade handbooks.

MACHINING WORMS AND WORM GEARS

Gear hobbing machines are used in the production milling of worms, particularly worms with multi-

FIGURE 18-22 Worm Thread Milling Setup Using a Universal Milling and Short-Lead Attachments and a Wide-Range Dividing Head

ple threads. During hobbing, all (multiple) threads are machined simultaneously instead of by taking separate cuts and indexing for each thread.

Worm thread generating machines are also used. A helical, spur gear type of cutter generates a single or multiple thread as it turns in mesh with the workpiece.

Worms may also be milled or cut on a lathe. A setup similar to helical milling is used when milling the worm thread (Figure 18-22). A thread milling cutter is mounted on a universal milling attachment. The axes of the cutter and attachment are set at the required helix angle of the worm and for the thread direction (right- or left-hand). The gear train ratio between the table feed screw and the worm of the dividing head rotates the worm blank as it advances to produce the required lead. Double- and triple-lead threads require the worm to be positioned at the next thread starting location by indexing after each pitch thread is milled to depth.

Hardened worm threads are first roughed out by milling with a formed cutter. They are then precision ground. The disk-shaped cutter has straight edges; the grinding wheel, straight sides. The straight-sided grinding wheel produces a thread having convex sides.

Worm gears are generally hobbed on a gear hobbing machine. The teeth may also be gashed or hobbed on a milling machine.

GEAR MATERIALS

Gears are made of many different materials. Large and small gears used for light- to medium-duty applications are generally made of cast iron. Where one gear is subject to great force or revolves a greater number of times in relation to the mating gear (as in the case of a pinion and bevel gear combination), two different materials are used. Steel and cast iron, steel and bronze, and nonmetallic material combinations are common.

The three categories of materials used on gears are *ferrous*, *nonferrous*, and *nonmetallic*. Cast iron and steel are widely used. Low-carbon steels are used if gears are left soft or are to be case hardened. Higher-carbon steels are used for gears that are to be hardened, tempered, and ground.

The nonferrous materials are generally applied to light-duty operations or where corrosion resistance is required. Bronze; die cast metals; and aluminum, copper, and magnesium base metals are the most common.

Nonmetallic gears that are corrosion resistant to air and chemicals are used for quiet, high-speed mechanisms. Such gears are often meshed with metallic gears. Gears are made of plastics (where temperature is controlled), thermoplastics (for example, nylon), and resin-impregnated canvas (for example, formica).

SUMMARY

The Industrial Revolution accelerated the development of gear-cutting machines.

Gear manufacturing processes include form milling in jobbing shops to production gear tooth generators to preformed casting processes.

Gear teeth may be machine milled, hobbed, shaped, broached, and shaved.

Gears may be hot rolled, cold rolled, extruded, stamped, and formed by powder metallurgy processes.

Straight- and spiral-tooth bevel gears may be produced on a gear generator. The helical tooth profile is generated by a combination of straight-line and rotary motion between the cutter and gear blank.

Gear tooth shapes are precisely finished by shearing, shaving, and burnishing processes. Hardened gear teeth are ground to form and lapped or honed.

Carbide cutters are used in fini-shearing gears to precise tooth size and shape.

Most terms for spur gears are applicable to helical, bevel, and worm gears. However, there are additional terms (helix angle, pitch cone angle, lead, and so on) that are specific to gears other than spur gears.

Gear tooth measurements are usually taken mechanically by vernier caliper or micrometer. Tooth profile accuracy is checked on an optical comparator.

The two-wire method of measuring gear tooth accuracy requires a micrometer measurement over (for external) or inside (for internal) two wires.

Tables are used to establish Van Keuren wire sizes and measurements for specified numbers of gear teeth and pressure angles.

Gear teeth must be cut with adequate allowance for backlash. Standard allowances are given in handbooks for each 0.001" reduction in pitch-line thickness.

Gears are represented in drawings according to ANSI standard terms, representation techniques for views and sections (where used), dimensioning practice, and tables of specifications.

The pitch circle is the imaginary point at which the teeth in mating gears are tangent.

The three common pressure angles used in gear design are 14-1/2°, 20°, and 25°. The latest 20° and 25° angle designs provide greater load-carrying qualities and smoother, quieter, operation.

The English module gear tooth system provides a ratio between the pitch diameter in inches and the number of teeth.

The SI Metric module is an actual dimension obtained by dividing the pitch diameter (mm) by the number of teeth in the gear.

The gear tooth profile manufactured in the United States is almost exclusively an involute curve.

A rack is a lineal layout of a circular gear.

Helical gear teeth are cut at an angle to the gear axis to permit mating gear teeth to mesh with their axes at any angle up to 90°.

Standard ANSI terms, formulas, and rules for spur, helical, bevel, and worm gears are contained in handbook tables.

Other manufacturers' tables are used in the shop for establishing dimensions, selecting cutters, and making angular settings for the table and workpiece.

Bevel gear cutting on a milling machine requires two cuts. Each cut is slightly offset beyond the centerline.

The teeth in a worm gear resemble a coarse-pitch thread with a single, double, or triple lead.

The two general classes of worm gearing are fine and coarse pitch.

Gear materials depend on conditions of use and design features such as number of teeth and gear size. The three basic materials groups are: ferrous (high- and low-carbon steels and cast iron); nonferrous (die cast metals, bronze, brass, aluminum); and nonmetallic (nylon, formica, and other plastics).

UNIT 18 REVIEW AND SELF-TEST

1. Tell what effect the Industrial Revolution had on gearing and gear production machines.
2. a. Describe gear hobbing and gear shaping.
   b. List four different gear forms that may be cut on a gear shaper.
3. List five basic categories of gear manufacturing methods.
4. Give two distinguishing design features of a bevel gear generating machine as contrasted to a conventional milling machine.
5. List three important measurements or tests a craftsperson makes in milling a spur gear.
6. Describe briefly the shop practice of measuring the thickness of standard spur gear teeth with a gear tooth vernier caliper.
7. Calculate the pitch diameter, outside diameter, and whole depth of tooth for the following 20° full-depth spur gear tooth forms: (a) 4 diametral pitch, 48 teeth; (b) 16 diametral pitch, 89 teeth; (c) 6 module SI Metric, 24 teeth; and (d) 1 module SI Metric, 127 teeth.
8. State the difference between normal circular pitch and normal diametral pitch as applied to helical gearing.
9. a. Use a table for cutter selection for helical gear cutting.
   b. Determine the cutter number to use for cutting the following helical gears: (1) 20 teeth, 22-1/2° helix angle; (2) 25 teeth, 30° helix angle; (3) 36 teeth, 45° helix angle; and (4) 50 teeth, 30° helix angle.
10. Determine the following dimensions for a pair of 36-teeth, 10 diametral pitch, miter bevel gears: (a) pitch diameter, (b) pitch cone angle, (c) cutting (root) angle, (d) tooth thickness at large end, and (e) tooth thickness at small end.
11. State two design features of worm gearing that differ from bevel gearing features.

UNIT 19

Machine Setups and Measurement Practices for Gear Milling

The four basic types of gears that are machined in a custom shop include spur (and rack), helical, bevel, and worm gear forms. They are usually cut on a horizontal milling machine. An involute form cutter is used to mill the tooth profile. Tooth spaces are indexed with a dividing head. Typical machine setups and gear-cutting and measurement practices are described in this unit. Consideration is given to cutting gear teeth in the English module (diametral pitch) and SI Metric module systems and for pressure angles of 14-1/2° and of 20° and 25° (newer standard form angles).

INVOLUTE GEAR CUTTER CHARACTERISTICS

An involute gear cutter produces the gear tooth profile. This formed cutter is sharpened on the tooth face. Thus, the exact tooth form may be duplicated over the life of the cutter.

Standard involute gear cutter sizes range from 1 to 48 diametral pitch. Figure 19-1 shows examples of gear tooth sizes from 4 to 16 diametral pitch (P_d). Special cutters are available for teeth that are smaller in size than 48 diametral pitch.

A *set* of gear cutters is available for each diametral pitch. Each set consists of eight cutters (#1 through #8). Sometimes, these eight *full-size* cutters are combined with seven *half-size* cutters (#1-1/2 through #7-1/2). The gear tooth range of full- and half-size standard involute form gear cutters is given in Table 19-1.

A gear cutter is selected according to (1) the diametral pitch and (2) the number of teeth in the gear. The gear cutters in a set are made so that each numbered cutter has a slightly different shape. Shape variations ensure that the gear teeth in a pair of gears having the same diametral pitch will mesh properly. For example, the tooth shape on a pinion must be machined with a greater curvature to permit accurate, smooth matching with the teeth on a larger mating gear.

The gradual change in the cutter tooth form between each cutter in the set is illustrated in Figure 19-2. Note the change in the involute profile from the #1 cutter, which has slightly curved sides, to the #8 cutter, which has a greater tooth curvature.

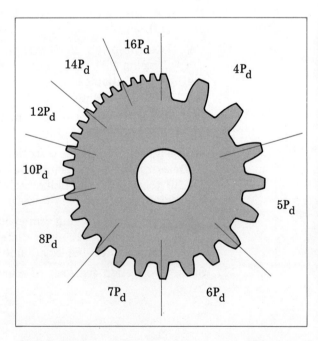

FIGURE 19-1 Examples of Comparative Gear Tooth Sizes from 4 to 16 Diametral Pitch (P_d)

TABLE 19-1 Range of Teeth Cut by Full- and Half-Size Standard Involute Form Gear Cutters

Cutter Number		Range of Teeth
Full Size	Half Size	
1		135 to rack
	1-1/2	80 to 134
2		55 to 134
	2-1/2	42 to 54
3		35 to 54
	3-1/2	30 to 34
4		26 to 34
	4-1/2	23 to 25
5		21 to 25
	5-1/2	19 and 20
6		17 to 20
	6-1/2	15 and 16
7		14 to 16
	7-1/2	13
8		12 to 13

HOW TO MILL A SPUR GEAR

The necessary gear dimensions must be established for machining and measuring the gear teeth. The formulas and handbook table information that are used depend on the gearing system (English or SI Metric) and the pressure angle (14-1/2°, 20°, or 25°). The milling machine setup and the machining and measurement steps that follow are general practices.

STEP 1 Select the involute form cutter.

Note: The cutter size and number depends on the required module and the number of teeth to be cut.

STEP 2 Mount the dividing head so that the cutting action takes place with the table positioned centrally.

Note: A footstock is mounted if the gear blank is secured on a mandrel. Then, the alignment of the headstock and footstock centers is checked.

STEP 3 Position the gear cutter on the arbor as close to the column as the setup will permit.

Note: The cutter must rotate in the correct direction. Also, the cutting forces during milling should be in the direction of the dividing head.

STEP 4 Mount the gear blank and mandrel between centers. Tighten the footstock center and lock.

Note: The large end of the mandrel should be mounted on the dividing head center.

Note: The tail of the dog must clear in the driving fork. The dog is then secured with the setscrews so that there is no play.

STEP 5 Align the centerline of the cutter teeth with the vertical centerline of the gear blank.

Note: One general method of alignment is to move the blade of a steel square until the edge just touches the circumference of the gear blank. The table is then moved transversely until the side of the cutter also touches the edge of the blade. The distance between the centerline of the cutter and the centerline of the gear blank is then determined. The distance is read on the graduated cross slide collar. The table is moved to align the two centerlines.

352 HORIZONTAL MILLING MACHINES

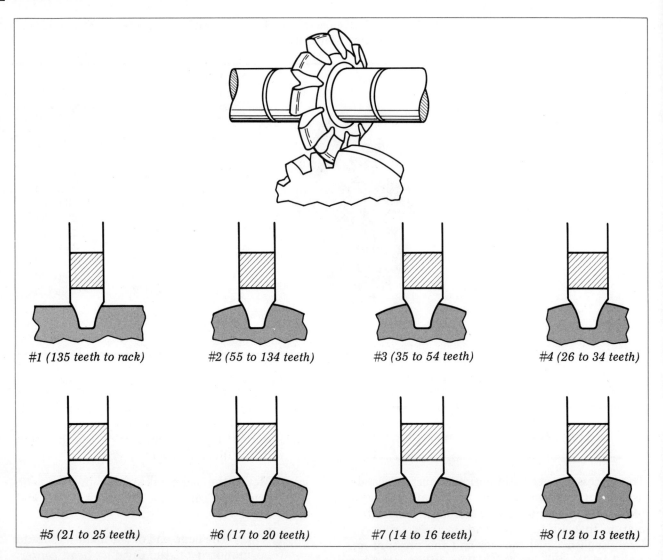

FIGURE 19-2 Gradual Change in Cutter Tooth Form from #1 to #8

STEP 6   Lock the cross slide.

STEP 7   Check to see that the cutting teeth clear the circumference and face of the gear blank.

STEP 8   Select and mount the required index plate. Set the index pin to slide freely into one of the holes on the correct hole circle. Count the number of holes (partial turn) to be indexed and set the sector arms.

STEP 9   Position the spindle speed and cutting feed dials and the coolant nozzle.

STEP 10   Set the table trip dogs.

Note: The cutter and setup must clear the workpiece and the dog during successive cuts.

STEP 11   Start the coolant flow and the spindle. Raise the table until the cutter teeth just graze the gear circumference of the gear blank.

STEP 12   Set the graduated knee feed collar at zero.

STEP 13   Move the cutter clear of the gear blank. Then, raise the table to three-fourths of the tooth depth.

Caution: Lock the knee clamp.

Note: Where a greater amount of material is to be removed, a roughing cutter is used to hog out the teeth.

GEAR MILLING SETUPS AND MEASUREMENTS 353

STEP 14 Rough out the first tooth.

> Caution: Hand-feed the cutter until it is determined that the machine is safely set up, the gear blank is securely held, and the cutting action is correct. Then, engage the power feed.

STEP 15 Return the cutter to the starting position. Index the next tooth. Rough cut this tooth.

STEP 16 Repeat the cutting, clearing, and indexing steps for the remaining teeth.

STEP 17 Loosen the knee clamp. Raise the table to the working gear tooth depth. Lock the knee clamp.

STEP 18 Reduce the cutting feed and/or increase the cutting speed to produce the desired surface finish.

STEP 19 Finish cut the first tooth. Stop the spindle. Return the cutter to starting position.

STEP 20 Continue to index and finish cut the remaining teeth.

> Note: Many craftspersons use a gear tooth vernier caliper to measure the chordal thickness after the first complete tooth is produced.

> Note: Sometimes, the table is set to machine within 0.002″ (0.05mm) of finish size. Adjustments may then be made, after measurement, to machine the gear teeth within specified limits of accuracy.

STEP 21 Stop the machine spindle. Carefully remove the workpiece assembly. Clean the workpiece and remove burrs. Wipe the machine, tools, and accessories. Return each item to its proper storage area.

RACK MILLING

A *gear rack* is a form of spur gear having all teeth in one plane. As for standard gear tooth measurements, measurements on a gear rack are taken from the *straight pitch line*. This pitch line for English module gears is located one addendum (1/diametral pitch) below the top face of the tooth.

An important consideration in rack tooth milling is the pitch. The pitch is the linear distance the workpiece is moved to mill each tooth. This *linear pitch* must be equal to the *circular pitch* with which the rack teeth are to mesh.

Teeth on short racks may be indexed by moving the cross slide the distance equal to the circular pitch of the gear. The teeth on racks longer than the movement of the cross slide will permit cutting are positioned by moving the table longitudinally. The circular pitch distance is read directly on the graduated collar of the table lead screw.

Rack Milling and Indexing Attachments

Milling cutters are often held, positioned, and driven by a rack milling and indexing attachment. Figure 19-3 provides an example of a horizontal milling machine setup with rack milling and indexing attachments for cutting teeth on a rack.

The *rack milling attachment* is driven by the spindle. The attachment permits setups for milling straight teeth or teeth at an angle. The cutter is held at 90° to the position used in conventional milling. Helical teeth are being milled with the setup as illustrated.

The *rack indexing attachment* ensures accuracy in positioning the workpiece for successive teeth to be milled. The indexing attachment consists of an indexing plate and a locking pin. The indexing plate has two diametrically opposite slots. When the indexing requires a complete turn, one of the slots is closed off.

The table movement (increment) for positioning each tooth is obtained by two change gears. These gears are selected from a set. The linear pitch of the rack determines the gear ratio and whether one-half or one complete turn of the index plate is required. A standard rack indexing attachment may be used for indexing as follows:

—From 4 to 32 diametral pitches;
—From 1/8″ to 3/4″ in increments of 1/16″;
—For table movements of 0.1429″, 0.1667″, 0.2000″, 0.2857″, 0.1333″, and 0.4000″.

HOW TO MILL A RACK

STEP 1 Use handbook tables or formulas to calculate the whole depth of the tooth and the linear pitch.

354 HORIZONTAL MILLING MACHINES

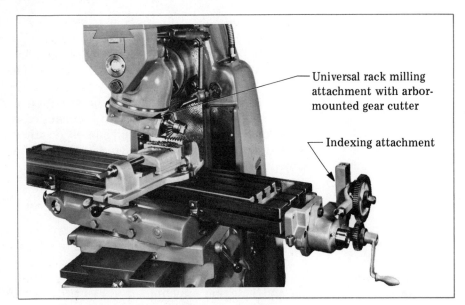

FIGURE 19-3 Horizontal Milling Machine Setup with Rack Milling and Indexing Attachments for Cutting Teeth on a Rack

STEP 2 Determine the change gears and the indexing setup if the rack length requires the use of a rack milling attachment and/or a rack indexing attachment.

STEP 3 Mount and secure the change gears and the indexing setup.

STEP 4 Determine the spindle speed and cutting feed. Adjust the machine speed and feed accordingly.

STEP 5 Position the rack milling attachment at 0° or any other required angle.

STEP 6 Mount and secure the selected form milling cutter.

STEP 7 Align and secure the workpiece.

Note: If the workpiece is held on parallels in a vise, it must extend high enough above the jaws to permit safe cutting of the teeth to depth.

Note: Where a workpiece is strapped to the table, the distance between the straps, nuts, and the rack attachment must be checked. There must be adequate clearance to safely machine the teeth.

STEP 8 Position the work so that the centerline of the first tooth to be milled is aligned with the centerline of the cutter.

STEP 9 Start the coolant flow and spindle. Raise the table slowly until the cutter just grazes the work surface. Move the cutter clear of the workpiece.

STEP 10 Set the graduated collar on the knee at zero. Raise the workpiece the distance of the whole depth of the tooth.

Note: When the size and depth of the tooth require considerable material to be milled, the teeth are usually rough cut. Sufficient stock is left for a finish cut and a finer feed. The finish cut and feed depend on the required surface texture and accuracy of the gear tooth profile.

STEP 11 Feed by hand. Engage the power feed when it is established that the setup and cutter action are correct.

Caution: Lock the table or cross slide (depending on how a cutter is fed) before a tooth is milled. Release for indexing.

STEP 12 Stop the spindle. Return the cutter to starting position.

STEP 13 Index (or use the reading on the graduated collar) for the next tooth.

STEP 14 Start the spindle and machine the tooth.

Note: Some craftspersons take a trial cut for a short distance to form one tooth. The spindle is then stopped. The partial tooth is burred and cleaned. The chordal thickness is checked with a gear tooth vernier caliper. Adjustments are made in the working tooth depth to compensate for any error.

STEP 15 Continue steps 13 and 14 until all teeth are machined. The table or cross slide is locked before machining a tooth.

STEP 16 Stop the machine and coolant flow. Clean all accessories, the machined rack, and the machine.

STEP 17 Remove the burrs formed in milling the rack teeth.

STEP 18 Disassemble the setup if necessary. Place the attachments in appropriate storage compartments. Return the cutter and arbor, other accessories, and tools to appropriate storage areas.

HELICAL GEARING

Helical grooves, teeth in helical gears, flutes, and worm threads are often milled on a horizontal milling machine. Two conditions must be met in order to mill a helix:

—The machine setup must permit a rotating workpiece to move past a revolving cutter (the lead is produced through change gears positioned between the worm shaft on the dividing head and the table feed screw);

—The workpiece and table must be set at the angle of the helix.

HOW TO MILL HELICAL GEAR TEETH

STEP 1 Compute the helix angle of the workpiece:

$$\text{tangent of helix angle} = \frac{\text{circumference of workpiece}}{\text{lead of the helix}}$$

STEP 2 Determine the direction in which to swing the table.

Note: A right-hand helix is cut by moving the table in a counterclockwise direction. A clockwise positioning produces a left-hand helix.

STEP 3 Swivel the table to the helix angle and for the direction of the helix.

STEP 4 Calculate the change gears that will produce the required lead. Use the formula:

$$\frac{\text{lead of helix to be cut}}{\text{lead of table feed screw}} = \frac{\text{product of driven gears}}{\text{product of driver gears}}$$

Note: Handbook tables may also be used for the change gears.

STEP 5 Determine the required direction of rotation of the worm gear shaft on the dividing head in relation to the table feed screw.

Note: Both gears revolve in the same direction to cut a right-hand helix. A left-hand helix is produced when these gears turn in opposite directions.

STEP 6 Mount the required change gears between the table feed screw and the worm gear shaft on the dividing head. Use an idler to change direction, if needed.

STEP 7 Lay out the first tooth, groove, or flute on the face of the workpiece. Scribe a line on the periphery to indicate the direction of the helix.

STEP 9 Mount the workpiece between centers or hold it securely in a chuck or other workholding attachment on the dividing head.

Note: The workpiece should be positioned so that the large-diameter end of the mandrel is centered on the dividing head center.

STEP 9 Position the selected milling cutter on the arbor as close to the column as possible.

Note: Clearance must be provided to permit the cutter, workpiece, and machine setup to safely clear at the start and finish of each cut.

STEP 10 Center the cutter approximately over the groove, flute, or tooth layout.

STEP 11 Set the spindle speed, cutting feed, and table trip dogs according to job requirements. Adjust the cutting fluid nozzle.

STEP 12 Start the milling machine. Raise the table until the cutter just grazes the workpiece. Clear the cutter and workpiece.

STEP 13 Raise the table to the work depth of the tooth. Lock the knee.

Note: Deep-cut teeth or grooves may require rough and finish cuts to be taken.

Note: Many gear cutter operators allow a few thousandths by taking a trial cut (for a distance far enough to permit measuring the working depth). Further depth adjustments are then made.

STEP 14 Hand-feed until the complete setup is checked for secureness and correct cutting action. Then, engage the power feed.

STEP 15 Return the cutter to starting position.

Note: The table is lowered after each cut to clear the cutter and workpiece and to ensure that the backlash will not produce a cutting action on the quick return stroke. Otherwise, the teeth may be machined oversize or surface finish may be damaged.

STEP 16 Index for the next tooth. Reposition the cutter at the work depth. Machine the tooth.

STEP 17 Stop the spindle. Clean the workpiece. Remove burrs from the machined surface.

STEP 18 Measure the normal circular pitch with a gear tooth vernier caliper. Determine whether any working depth adjustment is needed to meet dimensional requirements.

STEP 19 Repeat step 16, (and then step 15) until each tooth is milled.

STEP 20 Stop the machine. Clean the machined gear and mandrel, accessories, tools, and the milling machine. Disassemble and place all items in appropriate storage areas.

STEP 21 Remove burrs produced by milling from the sides and circumference of the gear teeth.

FIGURE 19-4 Setup for Machining Bevel Gear Teeth with a Gear Blank Mounted on a Dividing Head Spindle

FACTORS TO CONSIDER IN BEVEL GEAR CUTTING

Generating-type bevel gear cutting machines are widely used to produce correctly formed bevel gear teeth. Each tooth has the same cross-sectional shape throughout its length, except for the uniform size reduction from the large to the small end of each tooth.

There are many cases where straight bevel teeth must be produced by milling. Applications of milled gears, however, are limited in comparison to gears generated by other methods. For example, milled gears, having teeth that are not as precisely machined as generated gears are not suited to high-speed mechanisms or where angular motion is to be transmitted within a high degree of accuracy.

The bevel gear cutter has the same profile as the shape of the teeth at the small end. Bevel gear milling cutters are the same as the cutters used for milling spur gears. However, they are thinner to conform to the tooth size at the small end.

One common setup technique for machining bevel gear teeth is to mount the bevel gear blank on an arbor. A straight-shank arbor is usually held in the dividing head spindle (Figure 19-4). An arbor with corresponding taper shank may be secured directly in the spindle.

HOW TO MILL TEETH ON MITER BEVEL GEARS

STEP 1 Set the dividing head at the cutting angle.

STEP 2 Determine the required indexing. Mount the correct index plate. Set the sector arms for any fractional part of a turn.

STEP 3 Secure the bevel gear blank and arbor in the dividing head chuck or directly in the spindle.

STEP 4 Mount the selected gear cutter on the arbor.

GEAR MILLING SETUPS AND MEASUREMENTS 357

Note: The cutter is positioned as close to the spindle nose as practical. The arbor bearing is moved near the cutter. The overarm is then secured.

STEP 5 Center the gear blank so that it is aligned axially with the centerline of the gear cutter. Lock the cross slide.

STEP 6 Set the cutting speed and feed dials. Position the coolant nozzle. Adjust and lock the table trip dogs. Start the machine and coolant flow.

Note: The direction of rotation of the cutter and spindle must be checked. The cutting forces must be away from the dividing head spindle.

STEP 7 Position the cutter centerline over the crown of the gear blank. Set the knee graduated collar at zero. Clear the cutter and gear blank.

STEP 8 Raise the table to the whole tooth depth at the large end of the teeth. Lock the knee.

Note: Teeth from 1 to 10 diametral pitch are usually rough cut and then finish cut.

STEP 9 Mill the first tooth. Return the cutter to starting position.

Caution: Lock the dividing head spindle before each cut is taken.

STEP 10 Index and cut the next tooth. Stop the spindle.

Note: Many milling machine operators rough machine all teeth. The sides are then trimmed to size.

STEP 11 Measure the tooth thickness at the large and small ends.

Note: Each measurement is subtracted from the actual gear tooth dimension. The difference represents the additional amount of material that may need to be trimmed from both sides of the tooth form (Figure 19-5).

STEP 12 Move the gear blank off center the distance calculated for the table offset (Figure 19-6).

Note: The cross slide micrometer collar is set at zero. All backlash must be removed before moving the table.

STEP 13 Pull out the index crank pin. Rotate the gear blank layout one-half the distance required to trim one side of the tooth.

STEP 14 Start the spindle. Trim the one side of the first gear tooth. Stop the machine.

Note: The tooth thickness is measured at the small and large ends. This thickness must equal the tooth dimension plus one-

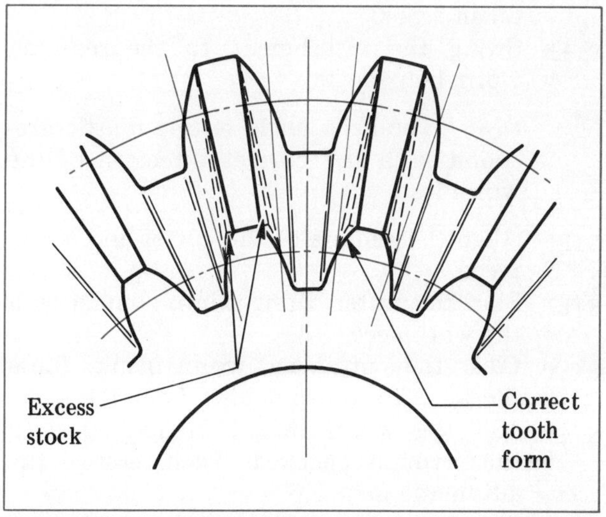

FIGURE 19-5 Excess Stock to Be Trimmed from Both Sides of the Bevel Gear Tooth Form

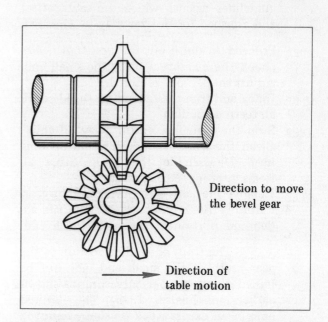

FIGURE 19-6 Dividing Head and Cutter Setup for Table Offset

HORIZONTAL MILLING MACHINES

half the thickness of the stock left to clean up the second side of the tooth.

STEP 15 Index and cut each successive tooth until all teeth have been milled on the one side. Stop the spindle.

STEP 16 Unlock the cross slide. Return the table and indexing plate to the original setting.

STEP 17 Move the table the calculated offset distance in the opposite direction.

Caution: Remove all backlash before moving the table.

STEP 18 Turn the index crank the same amount as was required for trimming during the first cut.

Note: The crank must be turned in the opposite direction.

STEP 19 Take a trial cut just far enough to permit measuring the large end of the tooth. Make whatever adjustments are needed.

Caution: Stop the spindle completely and move the cutter free of the workpiece before taking any measurements.

Note: The first tooth is often coated with a layout dye. The tooth thickness is then laid out on the face. The layout provides guidelines against which the table offset and finished tooth sizes may be checked.

STEP 20 Proceed to finish cut the first tooth. Recheck the measurement of the small end. Adjust the setup if required.

STEP 21 Index and trim the second tooth side until all teeth are milled.

STEP 22 Stop the machine. Brush away all chips. Clean the workpiece, cutter, and dividing head. Disassemble the setup. Store all items in proper storage areas.

STEP 23 Remove the machining burrs from the gear. Carefully file any excess stock in the addendum portion at the small tooth end.

Note: The (top) addendum section of the gear teeth at the small end may not be milled to the proper curvature (as shown earlier in Figure 19-5). The limited amount of excess stock is usually removed by carefully filing to the correct form.

WORM THREAD MILLING

Worm threads may be milled on a plain or universal milling machine. A dividing head is used for indexing and to serve as the mechanism for turning the workpiece past a revolving thread milling cutter. The required helix movement is produced by the gear train mounted between the table feed screw and the worm shaft on the dividing head.

When the helix angle to be milled exceeds the swivel angle of the machine table, a universal *spiral milling attachment* is used. Figure 19-7 shows such an attachment.

HOW TO MILL A WORM (THREAD)

STEP 1 Calculate the lead, pitch, whole depth, and thread angle. Determine the required change gears and the indexing. Select the thread milling cutter.

STEP 2 Position and secure the dividing head and footstock.

Caution: Disengage the index plate lock.

STEP 3 Assemble the gear train.

STEP 4 Secure the spiral milling attachment.

STEP 5 Mount the selected thread milling cutter on an arbor.

STEP 6 Swing the attachment to the required worm helix angle.

Note: The helix angle setting must correspond with the correct direction of the worm lead.

STEP 7 Align the centers of the cutter and workpiece.

STEP 8 Start the cutter. Bring it into contact with the workpiece.

STEP 9 Clear the cutter and worm blank. Raise the table to thread depth. Lock the knee.

STEP 10 Mill the worm thread. Hand-feed until the setup is checked. Then, engage the automatic feed.

STEP 11 Stop the machine. Clean the workpiece. Measure the worm thread.

GEAR MILLING SETUPS AND MEASUREMENTS 359

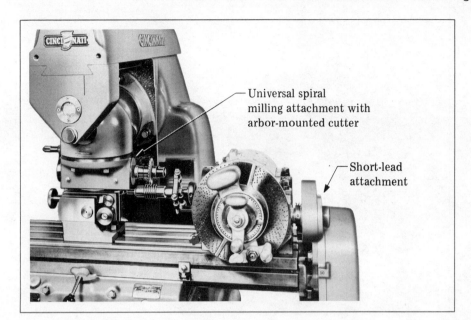

FIGURE 19-7 Application of Universal Spiral Milling Attachment for Cutting a Worm Thread

STEP 12 Disassemble the setup. Remove the burrs produced during the milling process.

WORM GEAR MILLING

Worm gears are mass-produced on hobbing machines. Replacements of worm gears are often machined on a miller. The teeth are first indexed and cut by *gashing*. A formed cutter and a dividing head are used for gashing. The teeth of the worm gear (wheel) are then finish milled to size and shape by *hobbing*. A hob is used with cutting teeth conforming to the shape of the worm teeth.

HOW TO MILL A WORM GEAR

STEP 1 Check the root diameter of the worm gear blank and its radius.

STEP 2 Establish the machining dimensions from the drawing specifications and/or the use of handbook tables.

STEP 3 Mount the selected milling cutter and support it on the arbor as close as practical to the spindle nose.

Note: The cutter must have the same pitch and outside diameter as the mating worm.

STEP 4 Set the gear blank assembly between the dividing head and footstock centers.

STEP 5 Swivel the table to the helix angle and thread direction of the worm.

STEP 6 Align the centerline of the cutter with the vertical axis of the mounted gear blank. Also, center the cutter in relation to the concave radius of the gear.

STEP 7 Start the spindle and flow of cutting fluid.

STEP 8 Raise the table until the cutter grazes the concave radius. Set the knee graduated collar at zero. Lock the table in position.

STEP 9 Feed the cutter to within 0.015" (0.4mm) of the whole tooth depth.

STEP 10 Lower the table. Index for the next tooth. Gash the tooth.

STEP 11 Continue to index and gash until all teeth are cut. Stop the spindle.

STEP 12 Replace the gear cutter with the required hob. Remove the dog from the mandrel. Adjust the centers so that the mandrel turns easily on the centers, but without end play (Figure 19-8).

STEP 13 Center the hob in relation to the outside diameter of the gashed gear blank. Align the axis of the hob so that it cuts through the center of the throat radius.

STEP 14 Start the hob revolving while carefully engaging the hob teeth in the tooth spaces.

STEP 15 Feed the table up until the hob cuts the teeth to the required depth.

FIGURE 19-8 Milling Machine Setup for Hobbing the Teeth on a Worm Gear

STEP 16 Lower the table. Stop the machine. Disassemble the setup. Clean and store all items.

STEP 17 Remove any burrs produced during the gashing and hobbing processes.

SAFE PRACTICES FOR SETTING UP, MILLING, AND MEASURING GEAR TEETH

—Check the direction of the spindle and the gear cutter rotation before gear teeth are milled. Incorrect rotation of the gear cutter may damage the teeth and the workpiece.

—Select the gear cutter according to the number of teeth in the gear to be cut and the diametral pitch. The curvature of the gear tooth cutter changes gradually from a #1 to a #8 cutter.

—Mount the gear cutter as close to the column as the machine setup permits. Use the overarm to provide arbor bearing support.

—Use a dividing head dog with the proper fit in the driving fork. Secure the mandrel assembly by tightening the setscrews in the driving fork.

—Align the centerline of a formed spur gear cutter with the centerline of the gear blank. The teeth sides will then be milled axially for correct mesh.

—Set the trip dogs to control the table movement and to prevent possible damage to the dividing head and/or other gear cutting attachments.

—Follow standard machining practices of hand-feeding at the start of a new process. Engage the power feed when the cutter, setup, and other machining conditions are judged to be safe.

—Clamp the knee and cross slide before any longitudinal machining cut is taken. The setup must be rigid to permit accurate machining.

—Take up all lost motion before the direction of a dividing head or the machine table is reversed.

—Check the safe operating clearance between a milling machine attachment, the dividing head setup, the workpiece, and the work-holding device or clamp screws.

—Recheck the angle setting of each milling attachment and/or the universal table. Each must conform to the direction (hand) of the thread or teeth to be milled.

—Mill bevel gear teeth so that the cutting action and forces are toward the dividing head.

—Clear the formed gear tooth cutter or stop the spindle rotation when the cutter is returned to the starting position for the next tooth.

—Stop the spindle, clean the workpiece, and remove any machining burrs before a precise gear tooth measurement is taken.

—Standard machine safety precautions must be observed. Trip dogs are required to control table movement. Backlash must be compensated for in helical gear cutting and when reversing the direction for settings. Setup and adequate machining clearance are essential.

TERMS USED IN MACHINE SETUPS AND MEASUREMENT PRACTICES FOR GEAR MILLING

Gear cutter set (involute form) A series of formed cutters for milling tooth sizes from 1 to 48 diametral pitch. A set of eight full-size (#1 to #8) and seven half-size (#1-1/2 to #7-1/2) cutters. Commercially available cutters for machining from 12 to 135 teeth in a gear.

Driving fork A slotted plate for mounting on a dividing head. A device for holding a dog securely and driving a workpiece setup.

Straight pitch line An imaginary line on a rack corresponding to the pitch circle on a mating gear. A line located one addendum (English standard) or module (metric) below the top face of the tooth.

Linear pitch The distance on a rack equal to the circular pitch of a gear.

Rack milling attachment A milling machine attachment mounted for driving by means of the spindle on a horizontal milling machine. A mechanism for changing the position of and driving a gear cutter. An attachment for right-angle and other angle positioning of a head to machine a straight rack or to take cuts at other helix angles.

Rack indexing attachment A mechanical device for quickly and accurately indexing a workpiece. A mechanism for advancing a workpiece and positioning it at a specified distance corresponding to the linear pitch of the gear teeth.

Table movement increments (rack gear cutting) Standard distances a milling machine table may be moved during each indexing. A uniform distance a milling machine table may be moved using a rack indexing attachment. Uniform table movement controlled by the ratio of the change gears.

Table offsetting (bevel gear cutting) The amount a milling machine table is moved (offset) transversely to permit machining bevel gear teeth accurately.

Rolling a bevel gear Rotating a bevel gear blank after it is offset. Positioning a gear blank so that the formed cutter removes additional material from the sides of gear teeth.

Trimming bevel gear teeth The process of removing excess material from the sides of milled teeth. Finishing gear teeth to more closely conform to the required tooth profile.

Gashing worm gear teeth Knee-feeding a worm gear blank into a revolving gear cutter. Up-feeding to cut a gear tooth to the whole tooth depth.

Hobbing (worm gear teeth) Use of a formed multiple-width tool cutter to track in gashed teeth. Milling worm gear teeth to shape and size by using a hobbing cutter. Machining as a hob automatically engages each tooth on a free-to-rotate gashed gear.

SUMMARY

The four basic gear designs—spur (and rack), helical, bevel, and worm gears—are produced in jobbing shops.

The horizontal milling machine is commonly used for milling the four basic gear designs.

362 HORIZONTAL MILLING MACHINES

In the English module system, standard involute gear cutter sizes range from 1 to 48 diametral pitch. A similar range of sizes is available in the SI Metric module system.

The tooth form on gear cutters changes slightly from the #1 to #8 cutters. The shape variations permit gear and pinion combinations with the same diametral pitch (module), but with different numbers of teeth, to mesh accurately.

The gear cutter size to select depends on the diametral pitch (module) and the number of gear teeth.

All cutters should be positioned on an arbor for maximum stability.

One simple technique of aligning the gear cutter axis centrally with the gear blank is to use a steel square. The table is moved transversely the distance equal to one-half the diameter of the gear blank less one-half the width of the cutter.

Roughing cuts are taken when large amounts of material are to be removed. All teeth are roughed out before the finish cut is set at the working tooth depth.

Increased spindle speed and/or decreased feed rate permit milling teeth to a higher quality of surface finish.

Table trip dogs are positioned to reduce excess table travel and machining time and to safeguard against cutting into any parts in the setup.

Gear tooth measurements on a gear rack are taken along the straight pitch line. The linear pitch corresponds to the circular pitch.

A rack milling attachment provides greater flexibility for positioning the cutter to mill grooves or straight gear teeth or gear teeth at a helix angle.

A rack indexing attachment simplifies and speeds up the positioning of the table for successive teeth. At the same time, the use of the attachment prevents indexing errors. The table movement is controlled by the ratio of the change gears that are positioned between the table feed screw and an indexing plate.

Short lengths of rack teeth may be milled conventionally with either transverse or longitudinal table movement. Longer lengths require the use of a rack milling attachment.

Precision-milled gears usually require a roughing cut and a finish cut at the working tooth depth.

Helical gear teeth are milled by setting a gear cutter at the helix angle. The workpiece is rotated under the gear cutter so that it advances the required lead distance. The lead and its direction are controlled by the change gear combination between the table feed screw and the worm shaft on the dividing head.

A helical gear tooth thickness (normal circular pitch) is generally measured with a gear tooth vernier caliper.

When bevel gear teeth are cut on a milling machine, the gear blank must be offset slightly.

Bevel gear teeth are trimmed by removing excess material from the sides of each gear tooth. After the gear blank is offset, the gear teeth are rotated a distance that permits cleaning up the excess stock.

Bevel gear teeth require a final touching up at the small end of the teeth. The experienced craftsperson carefully files away the limited amount of excess material to ensure accurate meshing.

GEAR MILLING SETUPS AND MEASUREMENTS 363

The operator must continuously check against any backlash in the longitudinal or transverse movements of the table or from indexing. Checking is especially important when the direction of movement is reversed.

Short helix angle worm threads may be milled by swiveling the table to the required angle. Steep helix angles are milled by using a special milling attachment. The thread lead is produced by the change gearing rotation of the dividing head and the workpiece.

Worm gears are milled in a jobbing shop by gashing, followed by hobbing.

A gear-cutting hob of the same shape and size as the worm engages in the rough cut teeth. Cutting continues as the teeth on the free-to-rotate gear blank are hobbed to working tooth depth.

Machine, tool, and personal safety precautions must be observed.

UNIT 19 REVIEW AND SELF-TEST

1. Give one reason why a set of full- and/or half-size gear cutters is required for each diametral pitch or metric module.
2. Set up a work processing plan to mill a spur gear.
3. State two milling problems that may be overcome by using a rack milling attachment instead of a conventional horizontal spindle setup.
4. Cite the principal advantage for using a rack indexing attachment instead of indexing teeth with a dividing head.
5. Identify two milling machine setups for milling helical gears that differ from the milling of spur gears.
6. Assume that all teeth on a miter bevel gear are rough cut to depth. List the work processes for finish trimming the gear teeth.
7. Give one advantage to the milling of a coarse-pitch, double-lead worm thread in comparison to cutting the same thread on a lathe.
8. Explain the difference between gashing and hobbing the teeth on a worm gear.
9. State three special safety precautions to follow when setting up and milling gears.

PART FIVE

Vertical Band Machines

SECTION ONE   Band Machine Technology and Basic Setups
        UNIT 20   Band Machine Characteristics, Components, and Preparation
SECTION TWO   Basic Band Machining Technology and Processes
        UNIT 21   Band Machine Sawing, Filing, Polishing, and Grinding

SECTION ONE

Band Machine Technology and Basic Setups

Vertical band machines are used for intricate and precise sawing and filing processes and in other processes requiring fast, economical removal of material. This section deals with:

- Principles of band machining and types of machines;
- Functions of major band machine components;
- Factors that affect cutting speeds and feeds;
- Machine and saw band setups for preparing the saw band, setting up the band machine, and regulating the speeds and feeds.

The technology and preparation of the band machine are applied in the next section to basic band machining processes.

UNIT 20

Band Machine Characteristics, Components, and Preparation

TYPES OF VERTICAL BAND MACHINES

In the study of machine technology, the band machine is identified as the last basic machine tool to be developed. While band saws were used prior to 1930 for cutting wood and soft materials, band machines with new features were introduced around that time. The new machine tools featured improved metal cutting bands, heavier construction features, greater range of cutting speeds, coolant and air supply systems, and differently designed band guides. The machines also featured a device for clamping and welding and then grinding the saw band to thickness. Accessories included additional guides for band filing, grinding, and polishing. Automatic work-feeding mechanisms and hydraulic tracing accessories were developed.

ADVANTAGES OF THE BAND MACHINE

Band machining has a number of advantages over other methods of removing material to produce finished parts. These advantages are:

- —Limited material is cut away because a comparatively narrow saw band is used (the amount of material normally wasted as chips is reduced);
- —Less time is required to remove excess material;
- —Angle cutting permits removal of a blank with enough material left to finish machine it as a mating part (the punch and die block of a blanking or forming die set are a typical example);
- —Intricate shapes may be cut on a single machine with no limit to the angle or direction and length of a cut;
- —The wear on the cutting teeth is distributed uniformly over the great number of teeth on the file band;
- —The uniform chip load per tooth and narrow tooth kerf require a relatively light cutting force (less horsepower is required);
- —Cutting time is reduced;
- —Band saw teeth dissipate the cutting heat faster (each tooth is in contact with the workpiece for only a short time and can cool before the next cutting pass).

These advantages do not relate to machining processes requiring the use of shapers, planers, slotters, vertical and horizontal millers, and other machines. These machines use cutting tools that are held in rigid toolholders. The cutting tools follow the fixed path of the toolholder. A holder limits the depth of cut and direction that a cutter can be fed into the workpiece. By contrast, the saw band (as the cutting tool and holder) has complete freedom to cut materials in intricate contours. The workpiece may be manually guided into the saw band or the table may be hydraulically fed.

TYPES OF BAND MACHINE CUTS

Band machines are used primarily for sawing functions such as cutting off, shaping, and slotting. Figure 20-1 illustrates how, by combining internal and external straight and corner radius cuts in two planes, it is possible to produce a part that otherwise would be complicated to machine. In this illustration, the term *three-dimensional cutting* is used. This term means that sawing cuts are made in two or more planes to alter the shape of a part.

Slicing refers to a cutoff operation that is usually performed with a knife-edge blade. *Slabbing* is another cutoff operation. It refers to cutting a thin section or several thin pieces from heavy workpieces.

Shaping, as applied to band machining, differs from the process performed on a shaper. *Band machine shaping* refers to cuts that alter the shape of a part by removing excess portions.

FIGURE 20-1 Three-Dimensional Band Sawing to Produce a Complicated Machined Part

Fixed-Table and Power-Table Band Machines

The two fundamental design groups of vertical band machines are the *fixed-table* type and the *power-table* type. The machine design and operating features of a fixed-table band machine typify the basic machine design features of all vertical band machines (Figure 20-2). This machine may be used to perform all operations required of a *band machinist*.

Structural Features. The three principal structural features of a vertical band machine are identified as the column, head, and base. The machine base houses the lower saw band carrier wheel, drive mechanism, and air pump. A hydraulic tank and pump for operating the hydraulically powered feed table of a power-table machine are also housed in the base. A hydraulic pump and a coolant tank for circulating cutting and cooling fluid are included for all power-table feeds. The base also accommodates the work table and machining accessories.

The column supports the head and provides a safe return route for the saw band. The saw band cutter, welder, and flash grinder; start/stop, blade, and table feed controls; other electrical controls; and band tension mechanisms are conveniently located on the column.

The machine head houses the upper saw band carrier wheel and the adjustable saw guide post. The head encloses the air jet and wet coolant piping and the wiring for the work lamp. The job selector is located on the guard.

Machining Specifications

Band machine specifications include dimensions for *throat* and *thickness* and *transmission speeds*. The dimensions indicate the work space bounded by the column, head, and table. The throat indicates the maximum width or radius of work that may be accommodated. The throat is the distance from the column to the saw band. The thickness is the maximum work thickness that may be handled.

Some manufacturers use model numbers. For example, in the model number 2014-3, the first two digits indicate a throat of 20" (508mm). The next two digits give the maximum work thickness of 14" (3.56mm). The last digit refers to a three-speed transmission. Other important information besides machine size is provided on the manufacturer's name plate. For example, the length of saw band or file band to use and other facts such as lubrication are given.

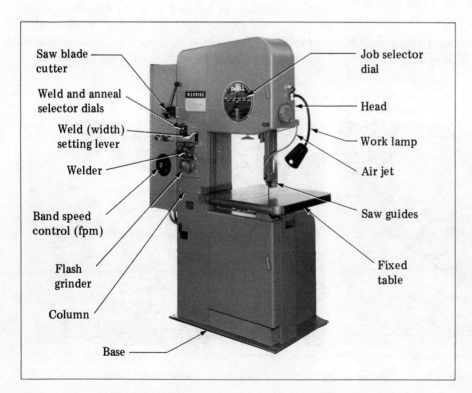

FIGURE 20-2 Basic Design Features of the Fixed-Table Band Machine

CHARACTERISTICS, COMPONENTS, PREPARATION 369

is limited to 5°. The table is designed for settings up to 45° clockwise. The workpiece may be fed by hand or power. An example of a mechanical work-feeding mechanism is shown in Figure 20-3. A *hydraulic tracing accessory* is available. The cutting path is guided by a stylus as it follows a pattern or template (Figure 20-4).

Heavy-duty band machines have power-feed work tables (Figure 20-5). The table is hydraulically fed. The path of the cut is produced according to a steering mechanism setup. The operator guides and controls the movement by turning a handwheel. *Heavy-duty production band machines* are developed to accommodate special needs.

High-tool velocity band machines (also called high-speed band machines) are adapted to the cutting and trimming of nonmetallic products such as nonferrous metals, plastic laminates, paper, wood, and other fibrous materials. Band speeds range as high as 15,000 feet per minute (fpm) to cut these comparatively soft materials. High-speed band machines are also adapted to continuous friction sawing applications. These machines are designed for vibration-free operation at high tool speeds.

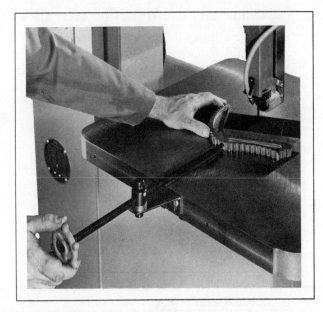

FIGURE 20-3 Mechanical Work-Feeding Mechanism

Sizes and Types of Band Machines

General-purpose band machines usually have nonpower-feed work tables. The table can be tilted for angle cutting. Angle cuts up to 10° may be taken when the table is tilted 10° counterclockwise. On large models, the cutting angle

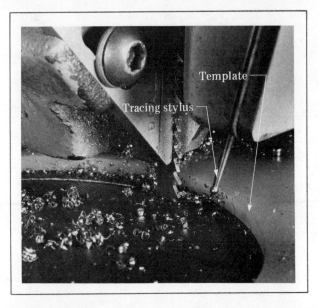

FIGURE 20-4 Application of Hydraulic Tracing Accessory

FIGURE 20-5 Heavy-Duty Band Machine with Power-Feed Work Table

PRELIMINARY CONSIDERATIONS FOR BAND MACHINING

Decisions need to be made by the machine operator before band machining. Consideration must be given to:

—Material machinability and characteristics,
—Job specifications,
—Equipment capacity,
—Work processes to be performed.

Machinability and Work Characteristics

The band cutting tool must be capable of efficiently and economically cutting the material with minimum tool wear. That is, the machinability of the parts to be cut must be within the recommended capacity of both the cutting tool and the band machine.

The surface condition of the work must be considered. Excessive scale or rust may need to be removed to protect the saw band from dulling rapidly.

TABLE 20-1 Partial Table of Manufacturers' Recommendations for Band Machining High-Speed Steel and Carbon Tool Steel Parts

Material	Work Thickness (in)	Work Thickness (mm)	Conventional Sawing — High-Carbon Saw Bands				Conventional Sawing — High-Speed Steel Saw Bands				Coolant/ Lubricant#
			Tooth Form	Pitch	Band Speed (fpm)	Band Speed (m/min)	Tooth Form	Pitch	Band Speed (fpm)	Band Speed (m/min)	
High-speed steel											
Tungsten base types T_1, T_2	0-1/4	0-6.4	P	18	140	43					240
	1/4-1/2	6.4-12.7	P	14	110	34	P	10	160	49	240
	1/2-1	12.7-25.4	P	10	90	27	P	8	135	41	240
	1-3	25.4-76.2	P	8	70	21	P	6	110	34	240
	3-6	76.2-152.4	C	3	55	17	P	4	85	26	240
	6-12	152.4-304.8	C	3	50	15	C	3	60	18	240
Molybdenum base types M_2, M_3	0-1/4	0-6.4	P	18	135	41					240
	1/4-1/2	6.4-12.7	P	14	100	30	P	10	140	43	240
	1/2-1	12.7-25.4	P	10	80	24	P	8	115	35	240
	1-3	25.4-76.2	P	6	60	18	P	6	90	27	240
	3-6	76.2-152.4	C	3	50	15	P	4	70	21	240
	6-12	152.4-304.8	C	3	50	15	C	3	50	15	240
Carbon tool steel											
Water hardening	0-1/4	0-6.4	P	18	175	53					240
	1/4-1/2	6.4-12.7	P	14	150	46	P	10	270	82	240 HD-600
W-1 special	1/2-1	12.7-25.4	P	10	125	38	P	8	225	69	240 HD-600
W-1 extra	1-3	25.4-76.2	P	8	100	30	P	6	180	55	240 HD-600
W-1 standard	3-6	76.2-152.4	C	3	75	23	P	4	140	43	240
	6-12	152.4-304.8	C	3	50	15	C	3	100	30	240

Key to Complete Table

Tooth Form
P = Precision
B = Buttress
C = Claw tooth

Band Tool
SA = Short angle
B = Bastard
L = Line grind
S = Scallop edge
K = Knife edge
W = Wavy edge
F = Friction
D = Diamond edge
A = Abrasive

Coolant
KK = Particular manufacturer's product

Grit
AC = Aluminum oxide
SC = Silicon carbide

CHARACTERISTICS, COMPONENTS, PREPARATION

The finished product and the excess amount of material must be analyzed. Before any sawing is started, there must be adequate stock to permit cutting with the least waste. When multiple pieces are to be sawed, consideration must be given to stacking to cut several at one time.

Job Specifications and Equipment Capacity

The shop print must be examined to determine both the required dimensional accuracy and the quality of surface finish. These two requirements influence the nature of the cut in terms of saw band characteristics, cutting speed, and feed.

The worker needs to determine how the workpiece is to be fed and guided to machine to a particular layout. Accessories, work-holding clamps, and fixtures may need to be selected.

Equipment considerations relate to such machine sizes as throat width and work thickness, if large parts are to be band machined. The available band tools (high-carbon tool steel, high-speed steel, and tungsten carbide) need to be compared with manufacturers recommendations.

Manufacturers' Recommendations. Table 20-1 is a partial table of band machine tool and cutting tool manufacturers' recommendations

Friction Sawing				Filing			Grinding				Coolant		Polishing			
Band Tool	Speed (fpm)	(m/min)	Pitch	Band Tool	Speed (fpm)	(m/min)	Band Tool	Speed (fpm)	(m/min)	Grit	Kind	Data*	Band Tool	Speed (fpm)	(m/min)	Grit
F	8000	2438	14	B	70	21	L	4000	1219	AO	KK	50:1	A	3500	1067	AO
F	1200	3658	10	B	60	18	L	3500	1067	AO	KK	50:1	A	3000	914	AO
F	14000	4267	10	B	50	15	L	3000	914	AO	KK	50:1	A	2500	762	AO
				B	50	15	L	2500	762	AO	KK	50:1	A	2500	762	AO
				B	40	12	L	2000	610	AO	KK	50:1	A	2000	610	AO
F	8000	2438	14	B	70	21	L	4000	1219	AO	KK	50:1	A	3500	1067	AO
F	12000	3658	10	B	60	18	L	3500	1067	AO	KK	50:1	A	3000	914	AO
F	14000	4267	10	B	50	15	L	3000	914	AO	KK	50:1	A	2500	762	AO
				B	50	15	L	2500	762	AO	KK	50:1	A	2500	762	AO
				B	40	12	L	2000	610	AO	KK	50:1	A	2000	610	AO
F	8000	2438	14	B	80	24	L	4000	1219	AO	KK	50:1	A	3500	1067	AO
F	12000	3658	10	B	70	21	L	3500	1067	AO	KK	50:1	A	3000	914	AO
F	14000	4267	10	B	60	18	L	3000	914	AO	KK	50:1	A	2500	762	AO
				B	50	15	L	2500	762	AO	KK	50:1	A	2500	762	AO
				B	50	15	L	2000	610	AO	KK	50:1	A	2000	610	AO

Note: mm values are rounded to one decimal place.
    m/min values are rounded to the nearest whole number.
*Mixture

372 VERTICAL BAND MACHINES

for conventional sawing; friction sawing; and filling, grinding, and polishing. Data is provided for selected band machining high-speed steel and water-hardening carbon tool steel parts. The working data relates to information the machine operator must know for each of the five basic processes. This information is provided for parts ranging in thickness from 0″ to 12″ (0mm to 304mm). Note the tremendous difference in speed (fpm) particularly between the saw band speeds and the grinding and friction sawing speeds.

Work Process to be Performed

Caution must be taken when large pieces are to be handled. A protective guard must be used when a part projects an unsafe distance from the table.

Unfinished and finished multiple pieces are put on auxiliary movable work tables. These tables are located so that the operator can conveniently handle the work. Workpieces that extend a considerable distance beyond the machine table require support. Support may be provided by using a table that moves along at the same rate as the workpiece that is being cut. Provision must also be made for handling scrap pieces. These pieces should be cleared off the table as quickly as possible in order to provide the maximum working area.

FUNCTIONS OF MAJOR MACHINE COMPONENTS

Basic Drive Systems

Fixed-speed machines have a belted or a direct drive. Intermediate speeds may be set in steps of 1,000 feet per minute (fpm) between low and high range. Variable-speed machines may be set for intermediate band speeds within either a low-speed or a high-speed range.

In the drive system of a vertical band machine, power is transmitted from a drive motor to a *variable-speed pulley* (Figure 20-6). Power is then transmitted to the *input pulley* of a transmission. The transmission drives the lower band wheel. This wheel, in turn, drives the saw band. The upper band wheel serves as an idler, aligning, and tension control wheel.

The controls for setting the band speed are shown in Figure 20-7. The band speed is set by

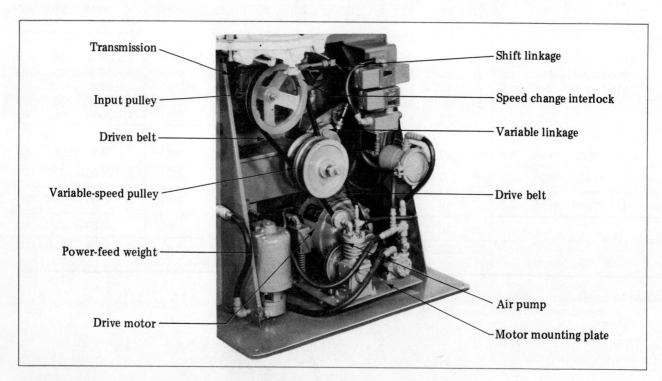

FIGURE 20-6 Drive System of a Vertical Band Machine (Safety Disconnect Covering Removed Only for Parts Identification)

CHARACTERISTICS, COMPONENTS, PREPARATION

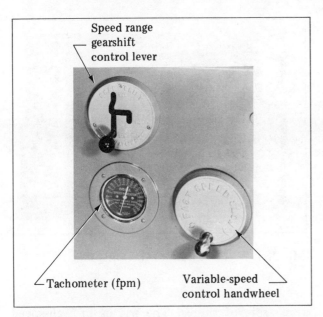

FIGURE 20-7 Range, Speed Control (fpm), and Speed Indicator

FIGURE 20-8 Removable Filler Plate and Center Plate Setup for Sawing

first positioning the *gearshift control lever* to the range of speeds appropriate for the job. The *variable-speed control handwheel* is then turned to increase or decrease the speed within the range. The band fpm (m/min) to which the controls are set is read directly on the *tachometer*. The variable speed handwheel is turned clockwise to reduce the speed; counterclockwise, to increase the speed. The band speed is established by reading it directly on the job selector.

Work Tables

On fixed-table band machines, the work slides on the table and is fed into the saw band. The table has a removable *filler plate* and *center plate* (Figure 20-8). The filler plate is removed from the table slot to mount a saw band. The same slot is used for other accessories to ride in. When secured in position, the filler plate forms an unbroken plane table surface.

The filler plate has a narrow slot for sawing. A plate for filing operations has a larger cutaway section, which accommodates the width of the file band. Filler plates are also provided for use with abrasive, diamond, and line-grinding bands.

A special power accessory is available for feeding and guiding work on fixed-table machines. Work is pulled with a steady mechanical force into the saw band. The force (rate of feed) is regulated by turning the power-feed handwheel.

The feed pressure is released by stepping on the foot pedal of the power feed.

The work table is mounted on and secured to the base by a *trunnion*. A protractor scale and a pointer are attached to the trunnion cradle for angle settings of the table (Figure 20-9).

Power-table band machines are designed with forward and reverse table workstops. The work table rests on hardened and ground steel *slide rods*. These rods move freely in ball bearings. The bearings are housed in a rigidly constructed trunnion support. This design assures uniform, smooth forward and reverse table movement. The table is moved by hydraulic force. Power-table band machines have a coolant return trough. The trough reaches around the edge of the table.

Table feed controls (Figure 20-10) are located in the column. The controls include a *vernier control knob* (feed-pressure) and a *positioning lever*. The control knob regulates feeds from 0 to 12 fpm (0 to 3.6 m/min) on lighter machines and 0 to 8 fpm (0 to 2.4 m/min) on heavy-duty models. Table movement is controlled by turning a directional handle to the power-feed or stop position. The forward position is used for rapid forward traverse; reverse is for rapid reverse traverse away from the work.

Carrier Wheels and Band Assembly

The saw band, as the cutting tool, forms an endless band that rides on the *upper* and *lower*

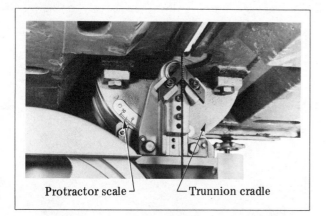

FIGURE 20-9 Trunnion Cradle and Protractor Scale Used for Angle Settings of the Table

FIGURE 20-10 Table Feed Controls

carrier wheels. The band is driven by the lower carrier wheel. The upper wheel is free wheeling.

The wheels are flanged and tapered and have a replaceable flanged back. The tapered face permits *tracking.* When correctly adjusted, the saw band rides up the taper, onto the rim, and against the flange at the back of the wheels.

Filing, polishing, and special bands work only on rubber-tired wheels. These wheels are standard equipment on fixed-table and small sizes of band machines. They are available for power-table machines. The rubber tire has a convex shape and is carried toward the center. As the carrier wheels revolve, the band moves toward and correctly rides on the top of the crown.

Heavy-duty production machines require steel-rimmed, flat-steel, flanged wheels. These wheels permit the saw band to be tightened for more positive drive and nonslip gripping.

Tracking and Saw Band Adjustments. The upper carrier band wheel tilts in or out to position and track a saw band (Figure 20-11). The tilt angle is adjusted by turning the tilt adjusting screw until the belt tracks. The upper carrier band wheel may also be moved up or down to change the band tension. A removable hand crank is turned clockwise to move the carrier band wheel upward to tighten the tension of the band. The tension is reduced in the band by turning the hand crank counterclockwise.

Accurate band machining partly depends on band tension. The amount of and changes in band tension are given on the *band tension indicator* (Figure 20-12). The indicator face shows three sets of values. The widths of high-speed steel bands are given on the outside circle. The widths of carbon steel bands are shown around the middle circle. Units of tension for different band machining processes appear on the center circle. Tension requirements for band tools are given in manufacturers' handbooks.

Saw Guide Post and Saw Band Guides

The *saw guide post* (Figure 20-13) serves two prime functions. It supports the upper saw band guide. The post is adjustable up or down to permit machining workpieces of different sizes.

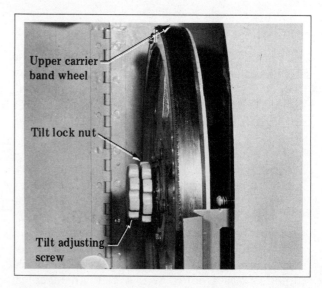

FIGURE 20-11 Positioning and Tracking a Saw Band

CHARACTERISTICS, COMPONENTS, PREPARATION 375

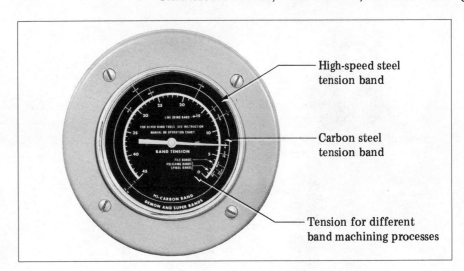

FIGURE 20-12 Band Tension Indicator

Some posts have a calibrated scale to indicate the distance between the table and the section of the saw guide. The post is positioned close to the width of the workpiece in order to provide maximum support for the saw band.

As its name implies, the *saw band guide* functions to provide a bearing for a saw band. During sawing, the saw band is subjected to the force applied by feeding the work into the band. A second force is produced by the resistance of the workpiece to the cutting action. This force tends to bend (laterally deflect) the saw band sideways. The saw band guides help to overcome these two forces.

The guide must be able to support the back edge of the saw band without causing damage. Therefore, the guide is designed with a *backup bearing*. This bearing takes up the back feed thrust and permits the band to move at high speed. The lateral deflection is controlled by *saw band guide inserts*. They are brought in close to the saw band, but with enough space left to permit the saw band to travel freely during a cut. The two common types of saw band guides are *roller* and *insert*.

Roller-Type (High-Speed) Saw Band Guides. Roller-type saw band guides (Figure 20-14) are

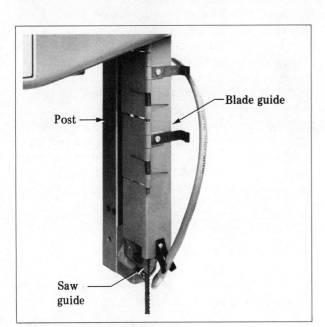

FIGURE 20-13 Saw Guide Post

FIGURE 20-14 Adjusting Roller-Type Saw Band Guide to Accommodate Saw Band Thickness

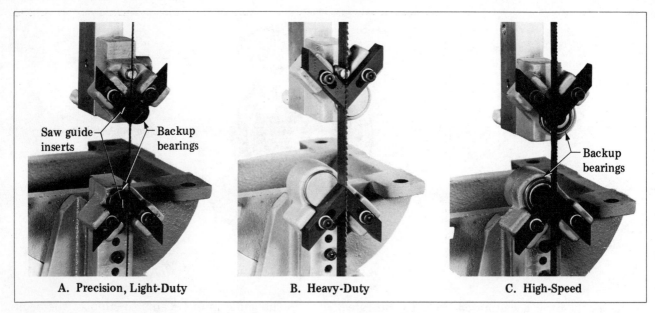

FIGURE 20-15 Three Models of Insert-Type Saw Band Guides

preferred for continuous high-speed sawing. Rollers are available for use with different widths of saw bands from 1/4" to 1" (6.35mm to 25.4mm) and for speed ranges from 6,000 to 10,000 fpm (1,820 to 3,048 m/min). The ball bearing construction reduces friction and wear on both the band guides and saw bands. The back of the saw band is supported by the flanged back on one roller.

Insert-Type Saw Band Guides. Three models of insert-type saw band guides are illustrated in Figure 20-15. The *light-duty* saw guide insert (Figure 20-15A) is a precision sawing guide for bands from 1/16" to 1/2" (1.5mm to 12.7mm) wide. It operates at comparatively low band speeds not exceeding 2,000 fpm (610 m/min).

The *heavy-duty* saw guide insert (Figure 20-15B) is used with bands from 5/8" to 1" (15.8mm to 25.4mm) wide for heavy-duty continuous sawing. The design of the guides is similar to the high-speed type with the exception of the backups, which are faced with carbide to reduce wear.

The *high-speed* saw guide insert (Figure 20-15C) is used for band sizes from 1/16" to 1/2" (1.5mm to 12.7mm) wide and for band speeds in the 2,000 to 6,000 fpm (610 to 1,830 m/min) range. Guides are available for all different band sizes. The guides on the high-speed type are constructed with a large antifriction bearing and a hardened, wear-resistant steel, thrust roller cap.

Coolant Systems

Heat generated in band machining may be dissipated by either reducing the band speed and cutting feed or using a cutting fluid. In some applications where cutting oils are not required, the chips may be removed by a jet air stream. The stream blows the chips clear of layout lines. Fixed-table machines generally are not designed with built-in coolant circulating systems. An air pump and a coolant (cutting fluid) pump provide both air pressure and a coolant on power-table, heavy-duty, and high-speed machines.

Cutting fluids flow from the table drain troughs into a drip pan, through a chip-collecting strainer, and back into the coolant tank. The coolant may be flowed on through a manifold. Where a mist is preferred, the coolant and air are mixed in the same manifold. The controlled mist is forced by air pressure onto the saw band teeth and workpiece. The nature and amount of coolant flow is regulated by turning the knob on the spray manifold. This knob also may be turned to cut off the coolant flow completely.

SAW BAND WELDING

Versatility of band machines for sawing processes is due to design features that permit:

—Welding saw blade stock into a saw band,
—Repairing a saw band,
—Welding a saw band that has been moved

CHARACTERISTICS, COMPONENTS, PREPARATION 377

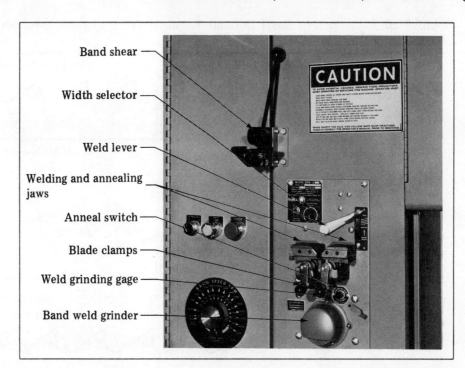

FIGURE 20-16 Major Controls on a Band Blade Butt Welder

through a starting hole in order to remove an internal section.

Four separate components are needed to weld a saw blade. They include a *cutoff shear*, *butt welder*, *annealing unit*, and *grinder*. Certain checkpoints must be observed by the machine operator during saw band welding:

—The weld in the saw band must be as strong as the saw blade material itself;
—The butted ends must be welded squarely (there should be no overlap in width, gage, set, or pitch of teeth);
—The flash must be ground to correct gage (excessive kerf from incorrect grinding may cause unnecessary band wear and produce a poor surface finish).

Cutoff Shearing

The manufacturer's name plate on the machine provides information about the required length of band. This length is laid out and accurately marked. The saw blade is placed (teeth facing outward) in the throat of a cutoff shear. A continuous, firm force is applied to the handle to make a sharp, clean cut.

With the blade cut to length, it is necessary to grind off one or more teeth on each side of the cut. This grinding is required on blades of 4 to 10 pitch because about 1/4" (6.35mm) of material is consumed in making the butt weld. Table 20-2 shows the position of the cut relative to the gullet and the number of teeth to grind off for blades 4, 6, 8, and 10 pitch.

Note: Grinding is performed only to the depth of the teeth. Additional grinding weakens the blade.

Butt Welding

The major operating controls on the band blade butt welder are shown in Figure 20-16. The edges of the upper and lower welding jaw inserts are beveled. The narrow beveled edge is used for blades up to 1/2" (12.7mm) wide and fine-pitch blades 3/4" (19mm) and wider. The wide, beveled edge accommodates coarse-pitch blades of 3/4" (19mm) or wider. The upper and lower jaw inserts, with matching bevels, are positioned to hold the saw blade segments with the teeth facing inward toward the column.

During the butt welding process, particles of excess metal are blown in the area of the welding jaws and inserts. Therefore, the jaws, clamping assembly, and inserts must be cleaned after every weld.

Using a Control Setting Chart. A *control setting chart* (Table 20-3) is provided by band machine manufacturers. The chart gives the jaw

378 VERTICAL BAND MACHINES

TABLE 20-2 Position of Shear Cut and Required Teeth to Be Ground for Proper Spacing after Butt Welding

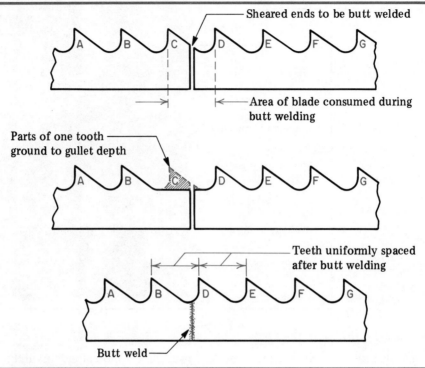

Saw Blade Pitch	Equivalent Teeth to Be Ground to Gullet Depth*	Position of Cut Relative to the Gullet of the Teeth	
4	1	1/2 tooth — 1/2 tooth	1 tooth ground off
6	1-1/2	3/4 tooth — 3/4 tooth	1-1/2 teeth ground off
8	2	1 tooth — 1 tooth	2 teeth ground off
10	2-1/2	1-1/4 teeth — 1-1/4 teeth	2-1/2 teeth ground off

*Tooth form is ground equally 1/8" (3mm approximately) on both sheared blade ends.

TABLE 20-3 Control Setting Chart for Jaw Gap, Jaw Pressure, and Annealing Settings

Carbon Alloy Steel Bands						
Width		Gage		Jaw Pressure	Jaw Gap	Annealing Setting
(mm)*	(inch)	(mm)*	(inch)			
1.6	1/16	.6	.025	2	1	1
6.4	1/4	.6	.025	3	4	1
9.5	3/8	.6	.025	3	4	1
12.7	1/2	.6	.025	3	4	1
15.9	5/8	.8	.032	3	4	2
19.1	3/4	.8	.032	3	5	2
25.4	1	.9	.035	4	5	3
25.4	1	1.0	.050	5	6	3
High-Speed Steel Saw Bands						
6.4	1/4	.6	.025	2 turns less than maximum	6	1
12.7	1/2	.6	.025	maximum	6	1
Friction Saw Bands						
12.7	1/2	.8	.032	3	5	2
19.1	3/4	.9	.035	4	5	2
25.4	1	.9	.035	4	5	2
Spiral Band/Rod Welding						
Diameter				Jaw Pressure	Jaw Gap	Annealing Setting
(mm)*		(inch)				
1.6		1/16		2	1	1
2.4-3.2		3/32-1/8		3	3	1
4.8		3/16		4	4	2
6.4		1/4		6	6	3
7.9		5/16		6	6	3

*mm values rounded to one decimal place

FIGURE 20-17 Positioning the Jaw Pressure Selector at a Required Control Chart Number

gap and jaw pressure for the width, gage, and material on the saw band. The setting for annealing the welded area is also given.

A number on the *jaw pressure selector* corresponds to a number on the chart for the blade to be welded. Figure 20-17 shows the jaw pressure selector being turned to a particular setting.

The control knob on the opposite side of the butt welder is called the *jaw gap control*. The control knob is turned for the proper jaw gap according to the width, gage, and material in the saw blade.

Making the Weld. If an internal cut is to be made, the blade is moved through the workpiece. The blade is aligned so that the ends are clamped as they touch and are centered between the two jaws.

Caution: Use safety goggles or a shield. Step to one side to avoid the welding flash.

The weld switch is pressed in to its stop. The weld is made automatically. The movable jaw of the butt welder advances the jaw and blade. The feed rate is equal to the rate at which the blade material is consumed by the welding process. The welded joint possesses the same metallurgical properties and strength as the band.

At the time welding takes place, there is a consistent flash. It is sharp and bright in color. A dull color or *sputtering* indicates an error in setting the welder controls.

The *reset lever* is raised when the weld is completed. This lever releases the spring tension. The saw band clamp handles are turned down to release the clamping pressure. The saw band is then removed and inspected at the weld.

Inspecting and Treating the Weld. The *flash* is the buildup material that is compressed during heating and flows on both sides of the band at the weld position. The characteristics of a good and bad butt weld are shown in Figure 20-18.

The color of the upset material should be blue-gray and of equal intensity across the flash.

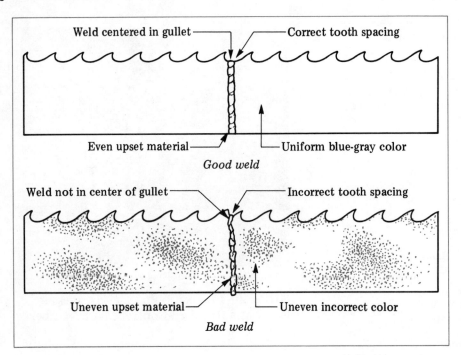

FIGURE 20-18
Characteristics of a Good and Bad Butt Weld

The spacing of the teeth must be uniform. The weld should be located in the center of the gullet. After grinding, the straightness of the band is checked with a straight edge. If the blade sections are misaligned, the band must be broken at the weld. The edges are ground square and the ends are rewelded. Common saw band welding problems, causes, and corrective action are given in Table 20-4.

Annealing the Weld

During saw band welding, the metal around the joined area hardens and becomes brittle. The band must be annealed in the area. Annealing restores the joint to its original heat-treated hardness.

The butt-welded area is clamped in the butt welder jaws. The teeth are positioned to the rear. The welded section is in the gap midway between the two jaws. The reset lever is moved to the anneal position and then back toward the weld position. This movement positions the movable jaw about 1/16" (1.5mm) toward the stationary jaw and allows for expansion of the band upon heating.

The *anneal selector switch* is set to the annealing heat number found in the control setting chart. The anneal switch button (Figure 20-19) is jogged intermittently until the saw band reaches a dull cherry red for both carbon and high-speed steel bands.

Note: Overheating, which causes the band to harden and become brittle, must be avoided.

The switch button is then released. As the welded area starts to cool, the switch button is again jogged and released. The cooling process slows down and the annealed area is thus produced.

Grinding the Weld Flash

A grinder unit is included on the band machine. The unit has a grinder wheel and a thickness gage (Figure 20-20). After annealing, the excess flash is removed. The welded area is ground to the blade thickness. The band is held with the teeth facing outward. Care must be taken to prevent grinding into the teeth. The band is moved continuously to prevent overheating or burning the ground area. If the saw band is burned, it will need to be rehardened.

Grinding starts at the tooth gullet. The band is inclined at a slight angle with the teeth higher than the back edge. The band is brought into contact with the bottom of the grinding wheel on its front edge. The band is moved forward with a light, uniform force. The positioning and grinding are repeated until the flash is removed.

The band is then flipped over and turned around. The flash is ground off the other side. The blade weld is checked for thickness as the ground area approaches the band thickness. The blade is checked with the thickness gage.

CHARACTERISTICS, COMPONENTS, PREPARATION

TABLE 20-4 Common Saw Band Welding Problems, Probable Causes, and Corrective Action

Problem	Probable Cause	Corrective Action
Overlapped or crooked welds	—Misalignment —Dirty or worn jaw inserts —Wrong jaw pressure setting	—Check and adjust jaw alignment —Clean or replace inserts —Recheck control setting chart
Brittle welds; hard spots in weld; spots of lighter color	—Incorrect jaw gap setting —Low voltage	—Increase jaw gap setting by 1/2 notch —Check incoming voltage
Incomplete or partially joined weld	—Low voltage —Burned or pitted weld switch contacts —Wrong electrical cutoff timing —Wrong jaw pressure setting	—Check incoming voltage —Replace contacts —Check timing control —Reset according to control setting chart
Blue color around weld area	—Dirty or worn jaw inserts —Wrong jaw gap setting	—Clean or replace inserts —Reset according to control setting chart
Too much flash or upset formed when welding 1/16" to 1/8" (1.6mm to 3.2mm) bands	—Jaw travel too great for small bands	—Clamp blade ends in jaws and leave a gap of 1/64" (0.4mm) between them to reduce jaw travel
Contacts of weld switch remain closed after welding	—Retracting spring of weld switch broken or weak —Wrong electrical cutoff timing	—Replace spring —Have weld switch checked electrically
Jaws stay together and fail to return to the welding position	—Low line voltage; material too heavy for setting of jaw gap	—Move reset lever halfway between weld and anneal positions. (When lever is in full anneal position, the weld switch is mechanically blocked). Jog weld switch momentarily. Immediately move reset lever to weld position. Run welder through simulated weld. If the jaws still come almost together, check the brake adjustment.

HOW TO PREPARE THE SAW BAND

Preparing the Blade

STEP 1   Determine the saw blade requirements. Check the machinability of the material, machine capability, and job specifications.

STEP 2   Select the saw blade that closely meets the manufacturer's data.

STEP 3   Check the required blade length on the machine name plate.

STEP 4   Pull out slightly more than the required length of saw blade from its container.

Caution: Use gloves to protect the hands from the sharp blade teeth.

STEP 5   Place the saw blade in the throat of the cutoff shear.

Note: The teeth face outward and the blade is held against the squaring bar.

STEP 6   Position the blade at the required length. Apply a firm, continuous force on the shear handle. Cut the blade to length.

STEP 7   Grind off the number of teeth on each end so that the teeth will be uniformly spaced after welding.

FIGURE 20-19 Jogging the Anneal Switch Button for Gradual Temperature Buildup and Slow Cooling Rate

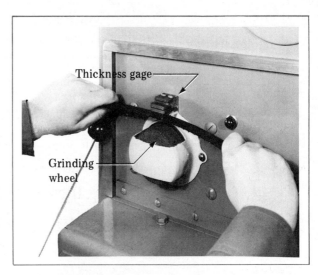

FIGURE 20-20 Band Grinder Unit

Making the Weld

STEP 8 Read the control setting chart for the jaw gap, jaw pressure, and annealing settings.

STEP 9 Set the jaw gap.

STEP 10 Turn the jaw pressure selector knob to the number given on the control chart.

STEP 11 Set the correct beveled edges of the jaw inserts in the inner position.

Note: The welding jaws must be clean. Any protective coating on the saw blade must be removed. An oil film prevents good electrical contacts.

STEP 12 Position the jaw inserts for the kind of blade, blade width, and pitch.

STEP 13 Place one end in the right jaw insert. Point the teeth to the rear and against the back aligning surface of the jaw.

Note: The end of the blade must be centered between the two jaws. Then the blade may be clamped in the jaw.

STEP 14 Place the second end in the left jaw insert. Butt both ends. Clamp.

Note: The blade ends must be checked to be sure they are held securely, touch, and are located midway between the two jaws. The ends should not be offset or overlap.

STEP 15 Press the weld switch button in to a full mechanical stop. Then, immediately release the switch button.

Caution: Stand away from the machine to avoid the flash and sparks.

STEP 16 Remove the saw band from the vise jaws. Inspect the weld for a uniform flash across the band, a blue-gray color, uniform teeth spacing at the weld, and straightness.

Note: Any spattered metal particles must be wiped or scraped from the jaws and inserts after the weld.

Annealing the Weld

STEP 17 Move the reset lever to the anneal position. Then, move it back about 1/16" (1.5mm) toward the weld position.

STEP 18 Position the band so that it is centered between the jaws. The saw teeth point inward. Clamp the band.

STEP 19 Set the anneal selector switch at the required annealing heat. Push and jog the anneal switch button in at short intervals until the band reaches the correct color (dull cherry red) and temperature.

STEP 20 Press the anneal switch button occasionally to slow down the cooling process. This produces the right temper. Remove the saw blade.

Grinding and Inspecting

STEP 21 Grind the excess flash off the top of the band. Use the bottom side of the machine grinding wheel.

STEP 22 Repeat grinding the flash on the second side. Check the blade thickness at the weld. Regrind and recheck if needed.

HOW TO SET UP THE MACHINE FOR OPERATING

Installing Saw Guides and Inserts

STEP 1 Remove the upper and lower saw guides (Figure 20-21). Clean and inspect the backup bearing.

Note: Any chips around the bearing must be removed. If chips enter, the bearing may freeze or stall, causing damage to the bearing and saw band.

STEP 2 Select a set of inserts to match the band size. The insert size appears on the side.

Note: Inserts that are too wide will damage the saw teeth set. Inserts that are too narrow will not provide adequate side support.

STEP 3 Mount and set the left-hand insert (Figure 20-22). Turn the holding screw just far enough to prevent sliding.

STEP 4 Place a saw gage corresponding in thickness to the saw band to be used in the right-hand slot. Unloosen the holding screw. Adjust the insert until it touches the edge of the saw gage.

STEP 5 Replace the upper and lower saw guides on the machine. Add the right-hand insert. Adjust to within 0.001" to 0.002" (0.02mm to 0.05mm) of the band. Secure.

Note: Roller-type guides are mounted with a flanged backup roller and a flangeless side roller in each block. The rollers

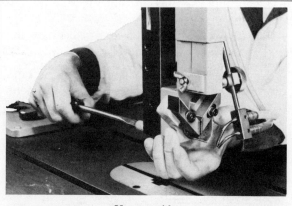

Upper guide

Lower guide (below saw table)

FIGURE 20-21 Removing Upper and Lower Saw Guides

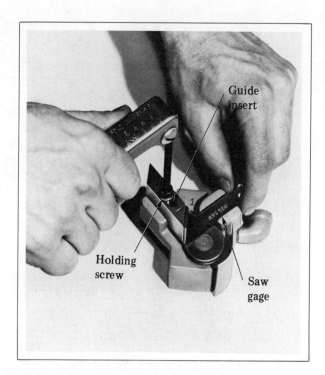

FIGURE 20-22 Mounting and Setting the Left-Hand Insert

are brought up to a correcting tracking band by turning the eccentric bearing shaft. Adjustment is made until the rollers are free to turn without touching the band.

Tracking and Applying Correct Tension

STEP 6 Open the upper wheel door, post saw guard, and the lower wheel door; remove the filler plate.

Note: The table connecting bar on power-table models must be removed.

Caution: Check the position of the gear-shift lever before opening the doors.

STEP 7 Lower the upper wheel with the band tension control.

STEP 8 Grasp the saw band with both hands. The teeth face out toward the operator. The teeth on the right-hand side face downward.

Caution: Wear protective gloves to handle saw blades and bands.

STEP 9 Place the saw band under the lower and over the upper carrier wheels.

Note: The band is centered to ride on the center of the wheel crown. The band on power-table machines with flanged wheels is pushed back against the wheel flanges.

STEP 10 Position the band between the upper and lower guide inserts. The back of the band must touch the backup bearing.

STEP 11 Turn the tension hand crank to take up the slack in the band.

STEP 12 Check the band for correct tracking. Adjust the upper carrier band wheel by turning the tilt handwheel screw clockwise to move the band toward the backup bearings (Figure 20-23). Turn counterclockwise to position the band away from the upper and lower bearings.

Note: The tilt lock nut is loosened to permit tilt adjustments. The wheel is locked in place when the band tracks correctly.

Note: The saw band tracks correctly when it rides the center crown of the upper wheel. In this position, the back edge of the band just grazes the thrust bearings on the saw guides.

STEP 13 Close the post saw guard and the upper and lower wheel guard doors. Replace the filler plate on the connecting bar (depending on the machine model).

STEP 14 Read the required tension on the band tension indicator or chart. Turn the band tension crank until this tension is reached. Lock the upper wheel at this position.

Note: With experience, the operator may adjust the tension recommended by the manufacturer. Generally, the tension is reduced for lesser pitch and lighter gage blades. Heavier bands may require increased tension.

Caution: Check the tension indicator regularly, especially with new saw bands that may stretch. Adjust the tension when it falls off.

STEP 15 Lower the saw guide post as close as practical to the workpiece.

Selecting and Using Cutting Fluids

STEP 16 Select the cutting fluid, if required, from the manufacturer's tables.

STEP 17 Check and make necessary changes in the cutting fluid and level in the coolant tank.

Note: The tank is emptied if the cutting fluid is to be changed. Any sediment in the chip basket, is removed.

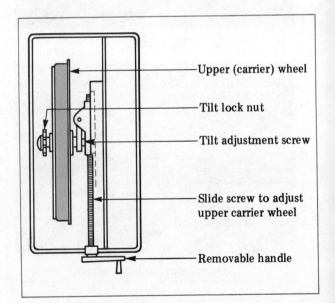

FIGURE 20-23 Features for Adjusting the Upper Carrier Wheel

CHARACTERISTICS, COMPONENTS, PREPARATION 385

Note: The screened intake, tank, nozzles, and screen in the filler caps are cleaned. The entire system is flushed when changing from one type of coolant to another.

STEP 18 Position the air and/or coolant nozzle(s) to direct the flow to the saw band.

Note: Grease and solid-type lubricants are used on machines that do not have coolant systems.

STEP 19 Turn the air valve to blow chips away from layout lines. Leave the coolant motor switch off if a cutting fluid is not recommended.

Note: Both air and coolant are used at the same time to produce a mist. The coolant and air valves on the control panel are regulated to produce the required type of mist.

STEP 20 Turn on the coolant motor and the coolant valves. Adjust the quantity of flow.

HOW TO REGULATE SPEEDS AND FEEDS

Setting Band Speed

STEP 1 Check the band tension and saw band tracking. Move the gearshift lever to the range recommended on the job selector.
STEP 2 Press the start button for the drive motor.
STEP 3 Turn the variable-speed control handwheel until the band reaches the required speed. Read the speed on the indicator.

Using Table Feed Controls

Pressure Control Method
Note: This method is used for general light-duty sawing of irregularly shaped parts.

STEP 4 Move the table feed control handle to the stop position.
STEP 5 Turn the table pressure knob counterclockwise to zero.
STEP 6 Rotate the table feed vernier knob clockwise to the end point.
STEP 7 Move the table feed control handle to the required feed position.
STEP 8 Turn the table pressure knob clockwise. Adjust the knob to produce a normal chip formation.

Vernier Feed Control
STEP 9 Move the table feed control handle to the stop position.
STEP 10 Turn the table pressure knob to the maximum setting.
STEP 11 Turn the table feed vernier knob to the zero setting.
STEP 12 Move the table feed control handle to the feed position.
STEP 13 Rotate and adjust the table feed vernier knob until the required feed rate is reached and chip is formed.

Combined Pressure and Vernier Feed Control Methods
The skilled operator may employ pressure and vernier feed controls to feed the work into the saw band at optimum cutting rates. Judgment is required in determining when force applied to a workpiece may be increased or decreased by changing the rate of feed. For example, sawing through a part that gradually changes in cross section requires maintaining a uniform cutting rate. Therefore, the force is decreased. The rate of feed is increased.

FACTORS AFFECTING FEEDING FORCE

Feeding the work with too great a force produces three basic problems:

—The band may be deflected (unless the cut is through the workpiece, a curved cut results);
—The backup bearings may be damaged;
—The saw band may be bent back.

Saw pitch affects feeding force. The total force required to penetrate the work depends on the number of teeth cutting at the same time. Too fine a pitch requires a greater feeding force and creates the possibility of inducing a welding action of chips in the gullets. Too coarse a pitch produces an added load on each tooth. This condition may cause the teeth to break or dull faster than is necessary.

SAFE PRACTICES IN SETTING UP THE VERTICAL BAND MACHINE

—Close the guard on the saw guide post and upper and lower wheel door guards before applying tension or starting the machine band.
—Position the upper and lower saw guides close to the work to provide maximum support against cambering or deflection during machining.
—Reduce the band speed before shifting gears, even on band machines that have a gearshifting safety feature.
—Push the drive motor start and stop buttons intermittently to mesh the gears that will produce the recommended band speeds. Change gears must mesh without applying force.
—Adjust the cutting feed to prevent work hardening, vibration, and bending of the saw band.
—Maintain a feed rate that produces a chip with a tight curl. At too light a feed, the saw band teeth will not penetrate the work. The abrasive effect between the workpiece and teeth cuases them to dull rapidly.
—Use auxiliary work tables to store parts when a quantity is to be band machined.
—Set the band speed within the manufacturer's recommendations for the job at hand.
—Make sure the table and workpiece are free of burrs and chips.
—Set the forward and reverse table workstops on power-table machines. These workstops prevent table movement beyond fixed limits.
—Stay within the recommended cutting feed rate. With experience, it may be possible to safely increase the feed.
—Point the teeth of a saw band down toward the machine table in the direction for cutting. The saw teeth dull quickly if reversed.
—Track the saw band to ride centrally on center-crowned wheels or against the rim of the upper carrier wheel. Check to see that the back edge of the blade just grazes the roller bearing on the saw guide.
—Keep the band tension constant as recommended. Read the tension indicator during machining. Adjust as needed.
—Use roller-type saw band guides for high-speed ranges of 6,000 to 10,000 fpm (1,820 to 3,048 m/min).
—Use a jet air stream positioned close to the workpiece to clear the chips away from layout lines for easy viewing.
—Check each flash area after welding and grinding. The butt weld must be solid across the width of the blade. The blade color must remain unchanged. A crack or incomplete weld or softening of the blade is dangerous and requires corrective action.
—Remove particles of excess metal deposited on the jaw inserts during butt welding.
—Inspect each butt weld at the flash joint for cracks and uneven welding of the ends.
—Use protective gloves for handling saw stock or bands. Safety goggles or eye shields are required during machining.

TERMS USED FOR BAND MACHINE FEATURES AND PROCESSES

Band machinist	Job classification of a worker who has knowledge and all-around experience as a specialist on a band machine.
Three-dimensional cutting	Altering the shape of a part by band machining in two or more planes.
Shaping (band machines)	Band machining cuts that alter the shape of a part by removing excess portions.
Fixed table	Band machine table that may not be moved. (Work is moved on and fed into the cutter from a stationary table.)
Center and filler plates	Removable table plates that permit changing the cutting tool band. Plates to provide a continuous plane working table surface.

CHARACTERISTICS, COMPONENTS, PREPARATION 387

Band tension indicator	Gage that indicates the tension on the cutting tool band during setup and machining.
Saw guide post	Vertically adjustable post to which saw band guides are attached.
Saw band guide	Consists of a base to which a back bearing roller and two flat faced guides or rollers are attached. A device mounted on the guide post for purposes of guarding the saw band sides and supporting the back edge.
Cutoff shear	Device mounted on the column frame for cutting saw stock square and to length.
Control setting chart (butt welder)	Manufacturer's chart for jaw gap and jaw pressure to use in butt welding different widths, thicknesses (gage), and saw band materials. A chart guide that gives the settings for a jaw pressure selector and jaw gap control knob.
Butt welder	Built-in welding accessory on vertical band machines. Welder designed to weld saw band stock ends together to form a band.
Flash	Uniformly ridged excess metal formed on both sides of a saw band by butt welding the ends of saw stock.
Annealing (butt weld)	Heating the flash area of the saw band to a dull cherry red color and cooling it slowly. The process of reducing blade hardness caused by fairly rapid air cooling of the butt weld.
Tracking	Positioning a saw band on the upper and lower carrier wheels so that it rides centrally on the upper crowned wheel. Correct positioning of a saw band. (The back edge just grazes the thrust bearings on the upper and lower saw guides.)
Tensioning	Applying force to a correctly tracking saw band. Adjustment of the upper carrier wheel to permit it to transmit motion and prevent damage to the saw band for cutting.
Cambering	Arcing or bending of the cutting edge, or back edge, of a saw band. (Positive camber is backward arcing of the cutting edges of a saw band. Negative camber is forward arcing of the cutting edges.)
Chip welding	Fusing of a chip to a tooth face. Extreme heating between the teeth and the material in the part.
Feeding force	Force (pressure) exerted as a workpiece is fed into the cutting teeth of a saw band.

SUMMARY

The vertical band machine provides for the fast, economical removal of material. Straight-line or contour sawing, filing, grinding, polishing, friction sawing, and electroband machining are prime processes.

> The two fundamental design groups of band machines are the fixed-table and the power-table types. Controls on power-table feed machines provide a wide range of cutting speeds.

Band machines include features for mounting, guiding, and driving different cutting tool bands. A welding, annealing, and grinding attachment is provided to form a saw band.

> Four factors to be considered in band machining are (1) characteristics or machinability of the part, (2) job requirements, (3) equipment capacity, and (4) work processes to be performed.

Variable band speeds may be obtained by positioning the gearshift control lever.

The work table contains a filler plate. This plate is removable for threading a saw or file band through a workpiece. Other plates are used for different machining operations.

Tables are adjustable to permit machining at an angle.

The two basic forms of carrier wheels are (1) the center-crowned, rubber-tired wheel and (2) the flat-steel wheel with a flanged inner end.

The tension on a cutting tool band is given on the band tension indicator.

The saw guide post provides a track for the band and an adjustable support for the saw band guides.

Saw band guides provide a bearing (against which the back of a saw band rides) and flat inserts or rollers to guide and to support the sides of the cutting tool.

Coolant and air supply systems permit the use of air to blow chips away at the cutting area, to produce a mist, or to use the air jet together with a cutting fluid.

Saw bands from a roll may be cut to length with a cutoff shear.

Part or whole teeth must be ground off at the weld to compensate for the amount of material consumed in butt welding and to assure equally spaced teeth around the band.

Jaw inserts on the welder are beveled to accommodate the tooth set. The jaw gap and jaw pressure for butt welding depend on the width, gage, and blade material.

The flash on a saw band is inspected for uniformity of metal flow and blue-gray color and the straightness of the band at the weld area.

The flash is annealed and ground to blade thickness. Grinding begins in back of the tooth and at the gullet.

The preparation of a saw band involves selecting the correct blade according to the job requirements, cutting the blade to length, butt welding, and grinding.

Band machine setup processes include: selecting and installing the correct size and type of saw guides and inserts; inserting the saw band, tracking, and applying the correct tension; replacing the band guide and upper and lower wheel guards; positioning the guide and inserts and adjusting them; and selecting the appropriate cutting fluid, adjusting the air and coolant pressure and flow, and positioning the nozzles.

A mist coolant is used for high-speed sawing of free-machining, nonferrous metals and tough, hard-to-machine alloys.

Table feed controls include controls for pressure and vernier feed. The skilled operator may employ a combination of these controls.

The operator must be aware of the effect of too great or too small a force and saw tooth pitch on the rate of cutting feed.

Hand-feeding is best for cutting small dies, templates, radii, and most light work operations.

Some band machines are provided with a safety shutdown in case a blade breaks. However, the operator needs to observe safe machining practices to be protected against personal injury and machine damage. Protective gloves are required when the blade is being handled. Safety goggles and eye protection precautions must be observed.

UNIT 20 REVIEW AND SELF-TEST

1. Inspect a vertical band machine. Identify three safety features that are incorporated in the design of the machine.
2. a. State the purpose served by grinding off teeth at both ends of a blade where the blade is to be butt welded.
   b. Tell how many teeth are to be ground on a (1) 6-pitch blade, (2) 8-pitch blade, and (3) 10-pitch blade in preparation for butt welding.
3. a. State one function of a roller-type and another function of an insert-type saw band guide.
   b. Identify two construction features that distinguish the roller-type from the insert-type saw band guide.
4. Tell what purpose is served by annealing the weld joint area of a saw band.
5. The four parts listed below are to be band sawed and filed. The material and thickness of each part is also listed.

	Material in Part	Thickness of Part
Part A	High-speed steel T-2	20mm
Part B	High-speed steel M-2	1-1/2"
Part C	Carbon tool steel W-1	8mm
Part D	Carbon tool steel W-2	1-1/2"

   a. Prepare a chart with five vertical columns in addition to the three above. Three of the new columns relate to conventional sawing with HSS saw bands. Mark these three columns: Tooth Form, Pitch, and Band Speed. The two remaining new columns are for recording the Band Tool and Band Speed for filing the parts.
   b. Refer to a manufacturer's table of recommendations for band machining high-speed and carbon tool steel parts to complete the chart.
6. Describe briefly how a mist spray is produced and regulated.
7. State four precautions to take for grinding the flash on a butt-welded saw band.

SECTION TWO

Basic Band Machining Technology and Processes

This section applies the technology, basic principles, and setup procedures of vertical band machining to straight and contour sawing processes, filing, polishing, and grinding. Spiral ban cutting, the use of diamond-edge saw bands, friction sawing, line grinding, and electroband machining are also described. Tables are included on common band machining problems; setups for cutting internal and external sections in one operation; and band velocity, pitch, and cutting rates according to work thickness.

UNIT 21

Band Machine Sawing, Filing, Polishing, and Grinding

BASIC SAWING OPERATIONS

The vertical band machine is designed for three basic sawing operations: cutting off, shaping, and slotting (Figure 21-1). One or more of these operations are part of every sawing job. To review, the term *cutoff sawing,* as applied to horizontal and vertical band machines, refers to the dividing of stock (or a part) into sections of specific size. *Shaping cuts* are taken to remove unwanted portions. *Contour sawing* describes band machining operations that include more than a straight-line cut—that is, a part is contour sawed when an irregular shape is required.

EXTERNAL SAWING

External sawing refers to the removal of an external section of a workpiece. The cut begins and ends on the outside of the workpiece.

Techniques of External Sawing Corners

Before sawing to a layout line, the machine operator must read the part print to determine the degree of accuracy of the sawed surface. If the specified quality can be met by sawing alone, the cut is taken to the layout line. Allowance is made for the saw kerf from the waste side (Figure

SAWING, FILING, POLISHING, GRINDING 391

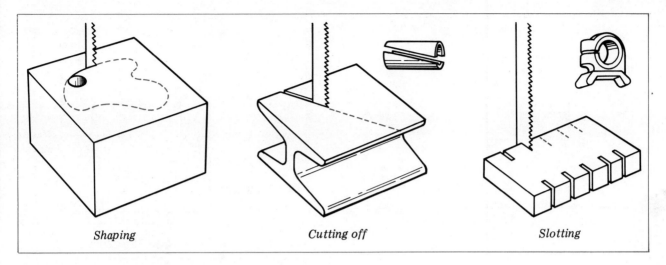

FIGURE 21-1 Basic Band Machine Sawing Cuts

21-2). If the surface is to be further machined, an allowance of about 1/64" (1.5mm) is made for finish filing; 0.010" to 0.015" (0.25mm to 0.4mm), for grinding. Allowance for the width of the saw kerf and an additional width for machine finishing are taken on the waste side.

For cutting corners that have a radius, a hole is drilled to match the radius (Figure 21-3). If the hole size is large enough, the workpiece may be fed to cut to the first layout line. The workpiece is turned at the corner formed by the radius of the drilled hole.

A square corner may be cut to shape by either drilling a hole or making straight cuts. For the square corner in Figure 21-4, a corner hole is first drilled. Then, the waste stock may be removed by cutting parallel to one straight layout line and continuing to cut parallel to the second line. The excess stock at the corner radius is removed by *notching*. The same square corner may also be produced by cutting along one layout line almost to the second line. The workpiece is carefully removed and fed so that a cut is taken along the second layout line.

Figure 21-5 illustrates the sawing technique for cutting a square section without first drilling a corner hole. The waste material is removed. Then, the excess metal left by the curvature of the first cut is removed by notching.

FIGURE 21-2 Cutting Allowance for Saw Kerf

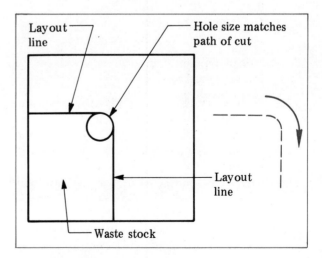

FIGURE 21-3 Hole Drilled to Match Corner Radius and Tangent to Horizontal and Vertical Layout Lines

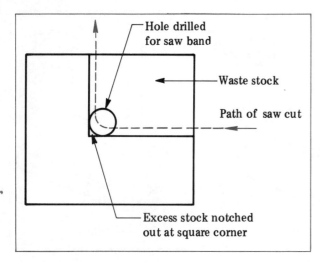

FIGURE 21-4 Removing Excess Corner Stock by Notching Technique

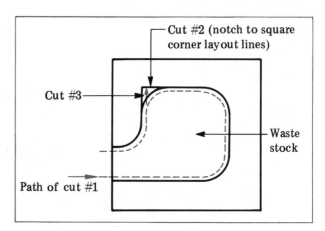

FIGURE 21-5 Technique of Sawing Combination Straight and Curved Surfaces, Including a Square Corner

The width of blade to use when a hole is first drilled is determined by the size of the radius to be cut and the thickness of the workpiece. The widest possible blade should be used. The other specifications for determining the correct saw blade are given on the job selector.

HOW TO MAKE EXTERNAL CUTS

STEP 1 Check the part layout against the shop drawing.

STEP 2 Determine the correct type of saw band teeth, pitch, blade material, and saw width.

Note: This information is given on the job selector.

STEP 3 Mount, track, and apply the necessary tension to the saw band.

STEP 4 Lower the upper saw guide until it clears the workpiece by approximately 1/4" (6mm). Secure the guide at this position.

STEP 5 Check the job selector for the saw band speed and feed. Set the speed at the required fpm (m/min).

STEP 6 Determine the correct cutting fluid, if required. Position the air jet and the coolant nozzle.

Note: The air stream blows the chips away from the layout lines. The cutting fluid keeps the blade and workpiece cool and adds to cutting efficiency. One or both systems are started as needed. The flow is regulated.

STEP 7 Move the workpiece up to the moving saw band.

STEP 8 Start the cut on the waste material side. The distance of the cut from the layout lines is determined by the allowance needed to produce the specified surface finish.

STEP 9 Feed the workpiece into the saw band.

Note: If a power table or feed device is used, the feed rate is determined from a chart. The feed is set accordingly. The power feed is engaged when the saw is cutting properly.

Caution: Use a pusher block on small parts. Slow down the feed rate toward the end of the cut.

STEP 10 Stop the machine. Use a brush to remove chips. Burr the workpiece.

INTERNAL SAWING

Internal sawing refers to the removal of an internal section of a workpiece. The cut begins and ends on the inside. The steps in contour sawing inside a workpiece are shown in Figure

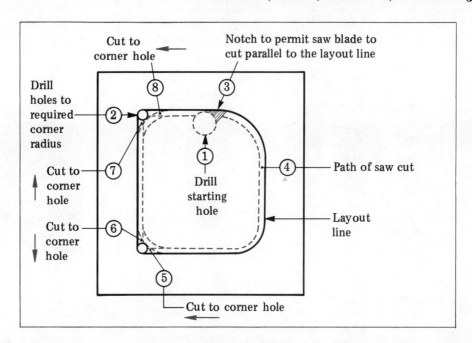

FIGURE 21-6 Steps in Contour Sawing Inside a Workpiece

21-6. One or more starting holes are drilled tangent to the layout lines. The drill diameter must be large enough to permit the correct width blade to be used.

The saw blade is cut and threaded through a starting hole. The blade ends are butt welded and the flash area is ground to saw band thickness. The saw band, which is free to move through the workpiece, is tracked and then tensioned.

The saw cut is started parallel with the layout line. In cases of internal contour sawing (Figure 21-6), the stock is first notched out. Notching permits feeding the workpiece so that the cut follows the curved layout line.

The corners are then notched to remove the excess material. A stock allowance is made for machine finishing the inside surface. When the machine is stopped, the tension is released and the saw band is removed. The saw band is cut at the point of the weld to completely remove the weld.

Contour Sawing Mating Parts (Internal and External Sawed Sections)

The internal cutting process makes it possible to use a contour-sawed inner section. By cutting at an angle, enough material is left to finish mating parts to the same size. For example, internal band machining is used extensively for sawing a punch and die block from the same workpiece in one operation (Figure 21-7). The process requires that a starting hole be drilled at a specified angle and that the table be set at a smaller angle (Figure 21-7A). Note that the starting hole is eliminated in the straight section of the punch and die. After finishing to size, the straight sides of the punch and die are both 9/16″ (14.3mm) (Figure 21-7B).

A starting hole is first drilled at a specified angle. Table 21-1 provides the specifications for laying out, drilling, machine setting, and band sawing internal and external sections from one workpiece. Full information about the drill angle and saw cut angle is given for die thicknesses from 1/2″ (12.7mm) to 6″ (152.4mm). The starting hole begins inside the die layout line. Theoretically, it intersects the layout line at the center of the die block. The hole emerges from the die block on the opposite side of the layout line.

For a 1-1/2″ (38.4mm) thick die block, Table 21-1 gives the distance from the die layout line to the center of the starting hole as 9/64″ (3.6mm). A 9/64″ (3.6mm) starting drill is used. The workpiece is set at an angle of 11°. After the hole is drilled, a 1/8″ wide starting saw blade is inserted through the hole, welded, and ground to band thickness. The band machine table is set to an angle of 8°. A sawing path is followed inside the layout line of the punch. When the inner section is completely sawed at the 8° angle,

394 VERTICAL BAND MACHINES

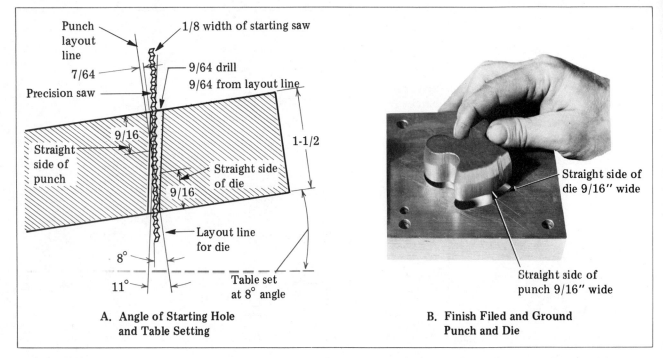

FIGURE 21-7 Internal Band Sawing to Produce a Punch and Die Block from the Same Piece of Stock

there is excess material inside the die and on the outside of the punch to machine the sides straight for 9/16" (14.3mm). The starting hole is almost eliminated from the straight sides of both the punch and die.

HOW TO SAW OUT AN INTERNAL SECTION

STEP 1 Secure a band machine table of angle settings for sawing internal and external sections. Determine the layout distance from the layout line to the center of the starting hole.

STEP 2 Drill the starting hole with the diameter drill and to the angle given in the table.

STEP 3 Set up the table of the band machine at the required angle for the saw cut.

STEP 4 Thread the precision saw blade of specified width (as given in the table) through the workpiece.

STEP 5 Weld and grind the flash to blade thickness. Anneal and inspect.

STEP 6 Mount, track, and tension the saw band.

STEP 7 Replace the table filler plate.

STEP 8 Set the band speed according to the job selector.

STEP 9 Direct the air and coolant nozzles and start the flow.

Caution: Use protective goggles or a safety shield.

STEP 10 Feed the workpiece carefully so that the blade cuts at the specified distance from the layout line.

Note: A cutting line is usually laid out to assist in cutting at the correct distance from the layout line.

Note: Unless enough material is left to machine straight sides on the two parts that are to be accurately fitted together, one or both parts may be spoiled.

STEP 11 Band file and band grind the two sections to fit within the specified tolerance and surface finish.

STEP 12 Stop the machine at the end of the cut. Remove the blade.

STEP 13 Cut the blade at the weld. Use the cutoff shear.

TABLE 21-1 Specifications for Laying Out, Drilling, Machine Setting, and Band Sawing Internal and External Sections from One Workpiece

Die Thickness		Distance from Die Layout Line to Starting Hole		Diameter of Drill		Angle of Starting Hole	Distance from Die Layout Line to Center of Saw Kerf		Angle for Saw Cut	Width of Starting Saw		Amount of Straight Sides of Punch and Die		Minimum Outside Radius		Minimum Inside Radius	
(in.)	(mm)	(in.)	(mm)	(in.)	(mm)	(deg.)	(in.)	(mm)	(deg.)	(in.)	(mm)	(in.)	(mm)	(in.)	(mm)	(in.)	(mm)
1/2	12.7	3/32	2.4	1/8	3.2	21	5/64	2.0	18	3/32	2.4	3/16	4.8	9/32	7.1	1/8	3.2
3/4	19.1	1/8	3.2	1/8	3.2	18	3/32	2.4	15	3/32	2.4	9/32	7.1	5/16	7.9	1/8	3.2
1	25.4	1/8	3.2	9/64	3.6	14	3/32	2.4	11	1/8	3.2	3/8	9.5	7/16	11.1	1/4	6.4
1-1/4	31.8	1/8	3.2	9/64	3.6	12	3/32	2.4	9	1/8	3.2	15/32	11.9	7/16	11.1	1/4	6.4
1-1/2	38.4	9/64	3.6	9/64	3.6	11	7/64	2.8	8	1/8	3.2	9/16	14.3	15/32	11.9	1/4	6.4
2	50.8	3/16	4.8	13/64	5.2	10	1/8	3.2	7	3/16	4.8	13/16	20.6	11/16	17.5	7/16	11.1
3	76.2	1/4	6.4	17/64	6.7	9	5/32	4.0	6	1/4	6.4	1-1/8	28.6	7/8	22.2	9/16	14.3
4	101.6	9/32	7.1	17/64	6.7	8	7/32	5.6	6	1/4	6.4	1-5/8	41.3	1-1/32	26.2	9/16	14.3
5	127.0	5/16	7.9	17/64	6.7	7	1/4	6.4	6	1/4	6.4	2-1/4	57.2	1-1/16	27	9/16	14.3
6	152.4	5/16	7.9	17/64	6.7	6	1/4	6.4	5	1/4	6.4	2-1/2	63.5	1-1/8	28.6	9/16	14.3

Note: All mm dimensions are rounded off to one decimal place.

STEP 14 Clean the machine table with a brush. Carefully wipe with a cleaning cloth.

COMMON BAND MACHINE SAWING PROBLEMS

The band machine operator must be aware of problems caused by incorrect pitch, velocity of the saw band, and feed rate. Saw band tracking and tension, the rigidity with which the workpiece is held, and the type and features of the saw band all affect quality and machining accuracy. The condition of the weld, the positioning of saw band guides, and the selection of a cutting fluid are important.

These and other factors may cause premature dulling of the teeth, vibration in the cut, teeth chipping or loading, welding of chips, and other damage to the saw band or the workpiece. Table 21-2 summarizes common band machine sawing problems. Probable causes and corrective action the machine operator may take are included.

SLOTTING

Slotting is an adaptation of the band sawing process. It requires the use of a saw band having a kerf equal to the required width of the slot. Slots may be sawed to a tolerance of +0.002″ (+0.05mm). The advantage in slotting on a band machine over a milling machine is the shorter distance the table travels to complete the cut.

The cutting of a corner relief between two mating surfaces is a form of slotting. The relief simplifies the finishing of a corner on straight parts to permit accurate nesting. Machining time is lost when the relief is produced on a shaper, planer, or miller.

SPLITTING

Splitting is another typical band sawing process. It is used widely to separate parts of castings or forgings. The separated parts are then further machined and held together with fasteners. Other parts such as bearings and bushings are first machined as a single piece and then split. Fixtures are used in such applications to seat and locate the workpiece and hold it securely.

RADIUS AND OTHER CONTOUR CUTTING

Contours may be sawed by using any one of three common methods: (1) hand-feeding, (2) sawing

VERTICAL BAND MACHINES

TABLE 21-2 Common Band Machine Sawing Problems, Probable Cause, and Corrective Action

Problem	Probable Cause	Corrective Action
Saw band vibrating in cut	—Velocity of saw band is too slow or too fast	—Adjust fpm (m/min) control according to job requirements
	—Feed rate is too light or too heavy	—Adjust rate of speed
	—Saw band tension is too loose	—Turn tension adjusting handwheel to correct tension
	—Saw band pitch is too coarse	—Use finer-pitch saw blade
	—Workpiece is insecurely held	—Tighten clamp or work-holding device
Saw band breaking prematurely	—Velocity of saw band is too slow	—Adjust fpm (m/min) control according to job requirements
	—Feed rate is too heavy	—Adjust rate of speed
	—Saw band tension is too great	—Turn tension adjusting handwheel to correct tension
	—Saw band pitch is too coarse	—Use finer-pitch saw blade
	—Saw band weld is too brittle	—Use longer annealing time; decrease heat gradually
	—Saw band gage is too heavy	—Select lighter gage of saw band
	—Incorrect guiding of band by saw guide inserts and backup bearings	—Inspect; make necessary adjustments
	—Wrong type of cutting fluid and/or improper flow	—Change cutting fluid to meet job requirements
Teeth dulling prematurely	—Velocity of band is too fast	—Adjust fpm (m/min) control according to job requirements
	—Feed rate is too light	—Adjust rate of speed
	—Saw band pitch is too coarse	—Use finer-pitch saw blade
	—Saw guide inserts are too large (teeth strike insert)	—Use inserts of smaller thickness
	—Wrong type of cutting fluid and/or improper flow	—Change cutting fluid to meet job requirements
Saw band teeth chipping or breaking	—Velocity of band saw is too slow	—Adjust fpm (m/min) control according to job requirements
	—Feed rate is too heavy	—Adjust rate of speed
	—Saw band pitch is too coarse	—Use finer-pitch saw blade
	—Wrong type of saw blade	—Refer to job selector for correct type of saw blade
	—Workpiece insecurely held	—Tighten clamp or work-holding device
	—Wrong type of cutting fluid and/or improper flow	—Change cutting fluid to meet job requirements
Loading of gullets of saw band teeth	—Velocity of saw band is too fast	—Adjust fpm (m/min) control according to job requirements
	—Saw band pitch is too fine	—Use coarser-pitch saw blade
	—Wrong type of saw blade	—Refer to job selector for correct type of saw blade
	—Wrong type of cutting fluid and/or improper flow	—Change cutting fluid to meet job requirements
Welding of chips to saw band teeth	—Feed rate is too heavy	—Adjust rate of speed
	—Wrong type of cutting fluid and/or improper flow	—Change cutting fluid to meet job requirements
Wandering of saw band	—Feed rate is too heavy	—Adjust rate of speed
	—Saw band improperly tracking	—Adjust carrier wheel tracking nut
	—Saw band tension is too loose or too great	—Turn tension adjusting handwheel to correct tension

	—Dull saw band on one side	—Replace with sharp saw band
	—Incorrect guiding of band by saw guide inserts and backup bearings	—Inspect; make necessary adjustments
Teeth not cutting	—Velocity of saw band is too slow or too fast	—Adjust fpm (m/min) control according to job requirements
	—Feed rate is too light	—Adjust rate of speed
	—Saw band tension is too loose or too great	—Turn tension adjusting handwheel to correct tension
	—Saw band pitch is too fine	—Use coarser-pitch saw blade
	—Wrong type of saw blade	—Refer to job selector for correct type of saw blade
	—Dull saw band on two sides	—Replace with sharp saw band
	—Incorrect guiding of band by saw guide inserts and backup bearings	—Inspect; make necessary adjustments
Poor surface finish	—Velocity of saw band is too slow or too fast	—Adjust fpm (m/min) control according to job requirements
	—Feed rate is too heavy	—Adjust rate of speed
	—Saw band tension is too loose or too great	—Turn tension adjusting handwheel to correct tension
	—Saw band pitch is too coarse	—Use finer-pitch saw blade
	—Wrong type of saw blade	—Refer to job selector for correct type of saw blade
	—Dull saw band on one or two sides	—Replace with sharp saw band
	—Incorrect guiding of band by saw guide inserts and backup bearings	—Inspect; make necessary adjustments
Cutting rate too slow	—Velocity of saw band is too slow	—Adjust fpm (m/min) control according to job requirements
	—Feed rate is too light	—Adjust rate of speed
	—Saw band pitch is too fine	—Use coarser-pitch saw blade
	—Wrong type of cutting fluid and/or improper flow	—Change cutting fluid to meet job requirements
Positive camber developing in saw band	—Feed rate is too heavy	—Adjust rate of speed
	—Saw band pitch is too fine	—Use coarser-pitch saw blade
	—Incorrect guiding of band by saw guide inserts and backup bearings	—Inspect; make necessary adjustments

with a disk-cutting attachment, and (3) cutting with a contour sawing accessory. *Hand-feeding* requires the workpiece to be guided by hand and cut to layout lines. On power-feed tables, the forward feed is engaged and the workpiece is guided past the saw band. On fixed-table models, the work is both guided and fed by hand.

The *disk-cutting attachment* (Figure 21-8) simplifies single-radius contour sawing. This attachment has an adjustable center point. The center is positioned at the radius distance from the saw band. The workpiece center provides a bearing surface for the attachment center. The workpiece is fed at the radius distance past the saw band.

The *contour sawing accessory* (Figure 21-9) combines the forward hydraulic table motion and the rotary motion produced by the turning of the control handwheel. The feed force is regulated by one hand by turning the table pressure control knob. Rotary motion is produced by manipulating the control handwheel with the other hand. The combination of the power and rotary hand feed controls the movement of the workpiece so that the saw band cuts to the required contour layout.

398 VERTICAL BAND MACHINES

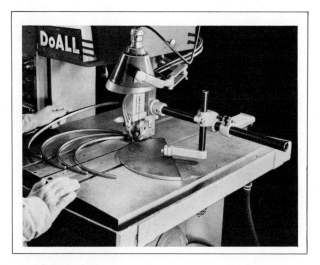

FIGURE 21-8 Disk-Cutting Attachment for Band Machining Radii

FIGURE 21-9 Contour Sawing Accessory

FRICTION SAWING

Principles of Friction Sawing

Friction sawing requires the instant generation of heat immediately ahead of the saw teeth. The heat produces a temperature that causes a breakdown of the crystal structure of the metal. The cutting edge of the saw band then removes the material affected by the friction heat. The saw teeth, traveling at extreme velocities between 6,000 to 15,000 fpm (1,828 to 4,572 m/min), generate the heat. During friction sawing, sparks and extremely hot metal chips are produced.

Figure 21-10 illustrates the heating and cutting action that occur in friction sawing. Figure 21-10A shows the confined heating area. The maximum heat penetration into the sides of the

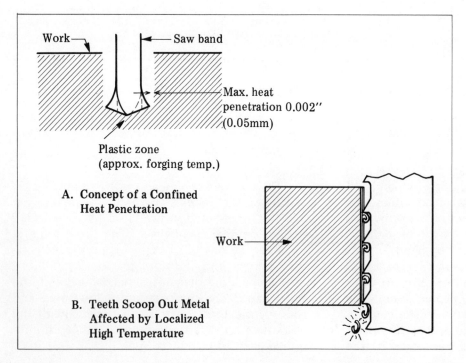

FIGURE 21-10 Heating and Cutting Action in Friction Sawing

cut is 0.002″ (0.05mm). The heat is further confined to the small area ahead of the kerf. The metal immediately ahead of the saw teeth is referred to as the *plastic zone*. In this zone, the approximate forging temperature of the workpiece is reached. The action of the saw blade in forming the chips is shown in Figure 21-10B.

The saw band remains relatively cool during friction sawing. This coolness is due to the fact that a small number of teeth are only momentarily in contact with the workpiece. As the band continues through the cycle, each tooth cools before it again contacts the metal.

Applications of Friction Sawing

Friction sawing is used to cut hard-to-machine steel alloys without annealing them. Other alloys that work-harden during conventional sawing may be cut efficiently by friction sawing. Friction sawing also replaces slower grinding and chemical machining processes on ferrous metals having a high hardness rating.

Friction sawing often replaces cutting methods that produce distortion in the workpiece. Friction sawing may be used on thin-walled tubes, thin parts, and thin-sectioned workpieces. This sawing process may be applied to straight, radius, and intricate contour shapes.

Limitations of Friction Sawing

Friction sawing does produce a 1/32″ to 1/16″ (0.8mm to 1.6mm) thick burr on the underside of the work. As usual, the burr is sharp. The workpiece must be handled carefully. The burr may be removed by filing or grinding.

Some metals, such as aluminum, copper, and brass, are not suitable for friction sawing. The high *heat (thermal) conductivity* quickly dissipates the heat ahead of the teeth and throughout the workpiece. It is impossible to concentrate the heat to the area immediately ahead of the saw band and teeth. This causes the metal to gum the teeth and weld to the blade. Most thermoplastic materials produce these same effects and are not adapted to friction sawing. Steels containing tungsten and most cast irons will not friction saw. The grains in the structure of these metals break off before softening. Friction sawing is generally limited, therefore, to ferrous alloys.

Work thickness is also a limiting factor. Friction sawing is practical for workpieces 1″ (25.4mm) and smaller in thickness. Within this range, it is easy to control the high unit forces between the saw band and the work.

Larger wrok sizes may be accommodated if the workpiece may be rocked. A rocking vertical movement of the workpiece helps to limit the contact area to approximately 1″ (25.4mm) or less. This 1″ (25.4mm) size limitation needs to be cleared in relation to cutting tubing. In such cutting, the saw band reaches the maximum thickness as it friction saws through the inner wall. This amount of cutting is only momentary. It is thus possible to cut greater thicknesses and large-size tubing.

Friction sawing is not recommended for stack cutting. The intense heat generated tends to fuse the many layers.

FRICTION SAWING BAND MACHINES

Friction sawing requires (in comparison with standard band machines):

—Greater machine rigidity,
—Heavier motors,
—Higher band velocities,
—Larger-diameter carrier wheels.

The larger-diameter carrier wheels are important. Their increased diameter reduces rapid, sharp flexing; minimizes band fatigue; and increases saw band life. While special friction saw bands are used, there are extraordinary *flex strains*.

Generally, just the heavy-duty power-table models have adequate capacity for continuous friction sawing. Toolroom and general machine shop standard models of band machines may be used for an occasional friction sawing job.

Machines especially designed for friction sawing have large roller-bearing saw guides; preloaded wheel bearings for continuous, heavy-duty friction sawing; higher band velocities; and more powerful motors. The process requires rigidity, increased power, wheel balance, and heavier guides and moving components. The machines must be designed to deliver vibration-free band machining capability.

Design of Safety Features

The high band speeds required for friction sawing require the saw band to be completely guarded to the point of the cut and at all times. Heavy-constructed telescoping saw band guards are designed and provided to cover the space between the upper carrier wheel and the work table.

Hydraulic brakes are included on both the upper and lower carrier wheels. The band tension is adjusted by a handwheel (Figure 21-11). Adjustments are made according to band width from 1/4″ to 1-1/4″ (6.35mm to 31.75mm). If the band breaks, the tension release automatically triggers the hydraulic system. Hydraulic brakes are applies to immediately stop the carrier wheels.

The saw band guides, which provide adequate backing and side support for the band tool, are mounted on a heavy, precision-ground post. A slide block provides the bearing on which the post is mounted. A set of high-speed, roller-bearing saw guides is shown in Figure 21-12. This particular set turns the saw band 45° to permit sawing workpieces larger than the throat capacity of the machine. The guides are adjustable for the width of the band.

FRICTION SAW BAND CHARACTERISTICS

Tooth sharpness is not a critical factor in friction sawing. Dull teeth are superior to sharp teeth because the generate heat faster. The blunt cutting edges produce greater friction and are able to withstand severe feeding force and fast band speeds as compared to the sharp teeth required for conventional sawing. The saw blade must also be able to withstand maximum flexing. It must have high resistance to abrasion at ultra-high velocities.

The friction saw blade is heavier in gage than the standard carbon steel alloy blade. The blade has extra strength to compensate for the heavy forces required for machining. The saw teeth are securely anchored and locked in the back of the band. The cutting edges of the teeth are specially formed. The features of the friction saw blade (Figure 21-13) enable it to overcome the extreme flexing action, stresses at high speeds, and heavy feed forces.

Blade Selection

The standard widths for friction saw blades are 1/2″, 3/4″, and 1″ (12.7mm, 19mm, and 25.4mm). A standard carbon alloy, precision-type blade is used for applications requiring a narrower saw band width than 1/2″ (12.7mm).

Pitch

Most friction sawing is done with the 10-pitch blade. A 14-pitch blade is used for thin work. The teeth are *raker set.* The pitch is affected by the work thickness and the amount of frictional

FIGURE 21-11 Band Tension Adjustment on Friction Sawing Band Machine

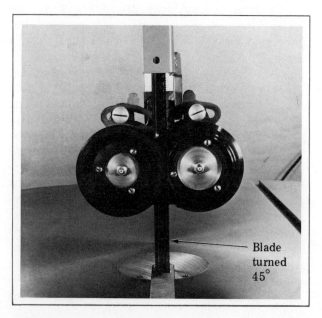

FIGURE 21-12 High-Speed, Roller-Bearing Angle Saw Band Guides

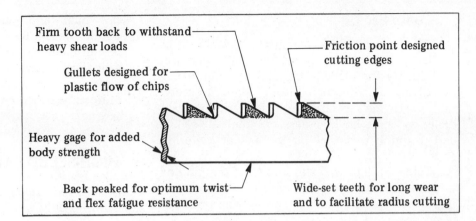

FIGURE 21-13 Principal Features of a Friction Saw Blade

heat to be generated in relation to the number of teeth that are removing the chips.

Table 21-3 compares the effects of too fine, too coarse, and correct saw blade pitch. With too fine a pitch, the heat penetration is excessive. An unnecessarily high feed force (pressure) is required. The chips clog the gullets and cause excessive tool wear. If the pitch is too coarse, the material is only partly softened. The teeth cut into cold metal and reduce cutting efficiency as well as the band life. A proper pitch blade is fast cutting and economical and has the best wear life. A minimum of feed force is required. Also, minimum heat is generated and little power is needed. A minimum burred edge is produced.

TABLE 21-3 Effects of Saw Blade Pitch

Saw Blade Pitch	Effects
Teeth too fine	—Excessive heat penetration —Unduly high feed pressure —Slow cutting —Excessive saw band wear
Teeth too coarse	—Teeth cutting inefficiently into cold metal —Excessive saw band wear —Short saw band life
Correct pitch	—Efficient, fast, economical cutting —Minimum heat penetration and feed pressure —High saw band life

Tooth Set

Tooth set provides clearance so that the body of the saw band may move easily through the kerf cut by the blade. The set is also important in cutting a radius or contour. Manufacturers' tables provide specifications for friction saw blade widths, gage thickness, nominal set for 10- and 14-pitch raker set blades, and an ordering code.

Table 21-4 provides examples of the kind of information the operator needs for friction sawing. Band speeds, pitches, and cutting rates are given for work thicknesses from 1/16" to 1" (1.6mm to 25.4mm). The approximate linear cutting rates are given in inches and meters per minute (m/min).

FRICTION SAWING SETUP PROCEDURES

Procedures for setting up the friction sawing band machine are similar to the procedures used in conventional sawing. There are two changes, however. First, no coolant is used. Second, roller guides are required unless only an occasional single part is to be sawed. Then, the clearance between the band and backup bearing should be increased due to the heavy feeding pressure. The gear shift lever is moved to the high speed range. The band speed handwheel is turned until the required band velocity is reached.

Extra safety precautions must be observed. The plastic guard must be positioned around the saw band and guides. The machine operator should wear a helmet with transparent visor. A pusher block should be used to feed small pieces. The friction-sawed chip is stubby, wrinkled, and has numerous cracks. The chips differ from the uniformly curled chips produced by conventional sawing.

HIGH-SPEED SAWING

High-Speed sawing refers to the speed of the saw band, usually within the range of 2,000 to 6,000 fpm (610 m/min to 1,830 m/min). This cutting range is between the low band speed used for normal band machining ferrous metals to the high band speeds used in friction sawing.

Applications of High-Speed Sawing

High-speed sawing is practical for such soft materials as plastics, wood, paper, and other fibrous products. It is also widely used on non-ferrous metals, such as aluminum, brass, bronze, and magnesium. Common forms that are cut include bar stock and plate and sheet stock, as well as extruded, cast, and tube forms.

As in all other band machining, the machine must operate in the required high speed range. The correct saw blade must be selected and the band speed, feed, coolant, and feeding setup must be determined. The tooth form must deliver the maximum cutting rate. The gullet spacing must be designed for heavy chip formation and rapid chip flow. Standard carbon alloy steel and

TABLE 21-4 Examples of Band Velocity, Pitch, and Cutting Rate for Friction Sawing According to Work Thickness for Selected Metals

Material (AISI-SAE Designation)	Work Thickness in Inches (mm)									
	1/16-1/4 (1.6-6.4)	1/4-1/2 (6.4-12.7)	1/2-1 (12.7-25.4)	1/16-1/4 (1.6-6.4)	1/4-1/2 (6.4-12.7)	1/2-1 (12.7-25.4)	1/16 (1.6)	1/4 (6.4)	1/2 (12.7)	3/4 (19.1)	1 (25.4)
	Saw Band Velocity			Saw Band Pitch			Approximate Linear Cutting Rates in/min (m/min)*				
Carbon steel (plain) 1010-1095	6,000	9,000	12,500	14	10	10	1,400 (35.6)	60 (1.5)	30 (0.8)	8 (0.2)	6 (0.2)
Free-machining steel 1112-1340	6,000	9,000	12,500	14	10	10	1,400 (35.6)	60 (1.5)	25 (0.6)	7 (0.2)	3 (0.1)

*Rounded to nearest tenth of a meter

high-carbon alloy steel blades with spring-tempered backs are best adapted for high-speed sawing.

Insert-type guides are used for short runs of high-speed sawing. These guides permit the greatest accuracy. However, production high-speed sawing requires the use of roller-type guides. Since high-speed sawing is used largely in production, special holding, positioning, and feeding fixtures are used.

SPIRAL-EDGE BAND SAWING

Spiral-edge band sawing requires the use of a round, continuous spiral-tooth band of small diameter. The spiral-edge saw band is unique. A continuous tooth is formed by a spiral cutting edge. The saw band is round as opposed to conventional flat-back blades. The spiral tooth design provides 360° of cutting. Intricate-pattern contours may be formed in thin-gage metals and other soft materials. Spiral-edge saw bands are practical for cutting small-radius contours that cannot be sawed by a conventional saw band because of the saw band width.

Sizes and Design Features of Spiral Saw Bands

There are four basic sizes for spiral saw bands. The outside diameters are 0.020″, 0.040″, 0.050″, and 0.074″ (0.5mm, 1.0mm, 1.27mm, and 1.88mm). A close-up of the tooth form is illustrated in Figure 21-14. Spring-tempered spiral-edge saw bands are used for plastics, wood, and soft materials. Hard spiral-edge saw bands are adapted to cutting metals. The sharp, spiral cutting edges permit the cutting of true, precise contours to the minimum radius of the spiral band.

Spiral-edge bands are used only with center-crowned, rubber-tired carrier wheels. A special saw guide design is required. The guides are mounted on ball-bearing shafts. The wheels are rubber tired and grooved to protect the teeth and provide a groove to guide the spiral-edge band.

Spiral-edge bands also require V-grooved plates for butt welding. The plates are made of a soft metal that serves as a good conductor. The plates protect the sharp, spiral teeth while the ends are held securely for welding and, later, annealing.

DIAMOND-EDGE BAND SAWING

Diamond-edge band sawing requires the use of an industrial diamond band. A diamond-edge saw band has a uniformly distributed concentration of industrial diamonds fused to form the cutting edge. The fusing process prevents the diamonds from chipping or peeling as the band flexes over the upper and lower band carrier wheels.

Diamond-edge saw bands, like other diamond-impregnated cutting tools, are designed to cut superhard materials. Straight, angular, and contour cutting is done on abrasive plastics, ceramic products, hard carbons, quartz, glass, granite, and other difficult-to-cut materials. The diamond-edge saw band cuts rapidly and produces an excellent surface finish and parallel faces.

Preparation for Diamond-Edge Sawing

Diamond-edge saw bands operate at maximum velocities between 2,000 and 3,000 fpm (600 to 900 m/min). The saw bands are available in 1/4″, 1/2″, 3/4″, and 1″ (6.4mm, 12.7mm, 19.1mm, and 25.4mm) widths. Tapered, rubber-tired, flanged carrier wheels and roller-type saw band guides are required.

The diamond band is welded in a manner similar to a conventional saw blade. However, the band must be turned to face outward so that it fits in and is protected by the jaw insert bevel. The back edge of each band end is aligned by eye. The edges are then butt welded. The flash is ground away and the butted area is annealed.

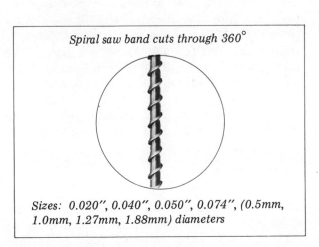

Spiral saw band cuts through 360°

Sizes: 0.020″, 0.040″, 0.050″, 0.074″, (0.5mm, 1.0mm, 1.27mm, 1.88mm) diameters

FIGURE 21-14 Four Basic Sizes of Spiral Saw Bands

404 VERTICAL BAND MACHINES

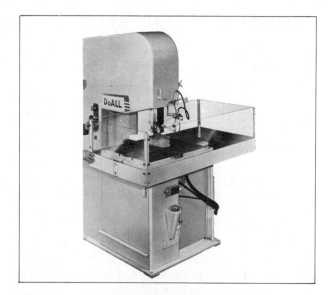

FIGURE 21-15 Band Machine Specially Designed for Diamond-Edge Band Sawing

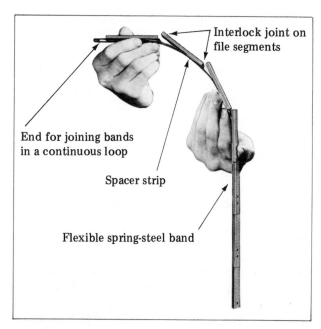

FIGURE 21-16 Design Features of a File Band

Diamond band sawing generates heat. The work area is generally flooded with a coolant to flush the cutting edges of the saw band and kerf and to keep the edges cool. The coolant must have excellent cooling, lubricating, and wetting properties. Power-table band machines, designed for use with a flood coolant, are best adapted to diamond band sawing.

Special Band Machine for Diamond-Edge Sawing

The increasing need to machine superhard and abrasive materials has brought about a special band machine for diamond-edge band sawing (Figure 21-15). The saw carrier wheel size provides maximum flex life of the saw band. The carrier wheels and drive system are dynamically balanced to permit vibration-free operation. The insert saw band guides are carbide faced to maintain sawing accuracy. The coolant spray is contained by the splash guard around the work table.

BAND FILING

The vertical band machine is used extensively for band filing. *Band filing* refers to the filing of internal or external surfaces to a uniformly smooth, accurate surface. Band filing may be done on almost any kind of metal.

The cutting tool consists of a series of file segments. These segments are riveted on one end to a flexible spring-steel band (Figure 21-16). As the flexible band travels around the carrier wheels, the loose ends of the file segments lift away. When the flexible band returns to the vertical plane, the loose end of one file segment fits into the next segment. A special interlocking joint between file segments permits the formation of a continuous, flat filing band (Figure 21-17).

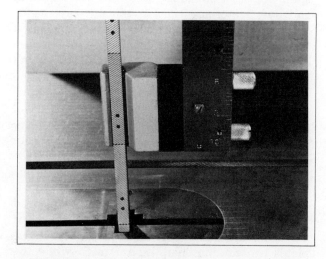

FIGURE 21-17 Continuous File Band Formed by Interlocking File Segments

Kinds of File Bands

Three factors must be considered in selecting a file band:

—Material in the workpiece,
—File shape and width,
—The cut of the file.

Cut, as related to file bands, refers to the short angle or bastard pattern; grade (such as coarse, medium-coarse, or medium); and the number of teeth. The job selector is used to identify the type of file. For example, a short-angle file is used for aluminum, brass, cast iron, copper, and zinc products.

Table 21-5 shows the three common file band segment shapes: oval, flat, and half-round. The widths of these files are 1/4", 3/8", and 1/2" (6.3mm, 9.5mm, and 12.7mm).

The file cuts illustrated in Table 21-5 are for 10, 12, 14, 16, 20, and 24 teeth. The short-angle coarse cuts with 10 teeth are for band filing soft materials; the medium-coarse cuts, for general use on mild steels; and the medium cuts for filing tool steels.

TABLE 21-5 Common File Band Segment Shapes, Cuts and Applications

Sample File Segment	Cut of File Segments			Width	Application
	Tooth Form	Grade	Number of Teeth		
Flat — Short angle, coarse cut, 10 teeth (1/2" / 12.7mm); Bastard, medium-coarse cut, 14 teeth (1/2" / 12.7mm)	Short angle	Coarse	10	3/8", 1/2" (9.5mm, 12.7mm)	Aluminum, brass, copper, cast iron, zinc
	Bastard	Medium-Coarse	14	1/2" (12.7mm)	General use on steel
	Bastard	Coarse	12	3/8" (9.5mm)	General use on cast iron and nonferrous metals
	Bastard	Medium-Coarse	16	3/8" (9.5mm)	General use on tool steel
	Bastard	Medium	20	3/8" (9.5mm)	Medium finish on tool steel
	Bastard	Medium	20	1/4" (6.3mm)	General use on tool steel
Oval — Short angle, coarse cut, 10 teeth (3/8" / 9.5mm); Bastard, medium cut, 24 teeth (1/4" / 6.3mm)	Short angle	Coarse	10	3/8", 1/2" (9.5mm, 12.7mm)	Aluminum, brass, copper, cast iron, zinc
	Bastard	Medium-Coarse	14	1/2" (12.7mm)	General use on steel
	Bastard	Medium-Coarse	14	3/8" (9.5mm)	General use on mild steel
	Bastard	Medium	24	1/4" (6.3mm)	General use on tool steel
Half-Round — Short angle, coarse cut, 10 teeth (3/8" / 9.5mm); Bastard, medium-coarse cut, 16 teeth (3/8" / 9.5mm)	Short angle	Coarse	10	3/8" (9.5mm)	Aluminum, brass, copper, cast iron, zinc
	Bastard	Medium-Coarse	16	3/8" (9.5mm)	General use on mild steel

Band Filing Speeds and Feeding Force

Band speeds between 50 to 100 fpm (15.2 to 30.5 m/min) are used for all-purpose band filing applications. However, recommendations on the job selector and in manufacturers' tables on band filing should be followed.

A continuous, steady, light force (pressure) is applied. The workpiece is moved with an even movement across the width to be filed. When the pattern of file lines on the workpiece shows perpendicular lines, the cutting action is correct. Diagonal and criss-cross patterns indicate the workpiece is being moved too fast from side to side.

Heavy and excessive feed pressure may load the tooth gullets, break or stall the file band, or groove the lower carrier wheels. Light pressure produces a better surface finish in the same time.

Band filing operations are performed from the right side of the table. The file segment design requires them to flex from a spring-steel band. The cutting teeth face the right side. All cutting is done on the face of the file.

HOW TO SET UP AND BAND FILE

Replacing the Saw Band and Guides

STEP 1 Remove the center plate for sawing.

STEP 2 Release the tension in the saw band. Remove and store the band in its proper storage area.

STEP 3 Remove the saw band guides.

STEP 4 Mount the adapter and file guide support on the keeper (lower) block (Figure 21-18).

Note: The slot in the backup support must accommodate the width of the file band.

STEP 5 Lower the upper guide post to within working distance of the workpiece.

Note: If the work thickness is more than 2″ (50.8mm) for a 1/4″ (6.2mm) file band or more than 4″ (101.6mm) for a 3/8″ or 1/2″ (9.5mm or 12.7mm) file band, a special set of larger guides must be inserted.

STEP 6 Install the upper file guide. Secure it to the saw band guide post.

Installing the File Band

STEP 7 Select the correct file band to meet the job requirements. Grasp the file band in both hands. One hand holds the painted end.

STEP 8 Hold the two ends at an angle (Figure 21-19).

STEP 9 Set the rivet head into the slotted hole. Slide the rivet head toward the small end of the elongated slot.

Note: For internal filing, the band section is threaded up through the workpiece. The two file band ends are then joined at the machine.

STEP 10 Straighten the file band to allow the spring-steel end to snap over the dowel.

Note: The ends of the file segments must be checked to be sure they interlock properly.

STEP 11 Place the file band on the carrier wheels so that the band is aligned with the file guides.

STEP 12 Install the center plate used for band filing.

STEP 13 Track the file band in the center of the carrier wheels.

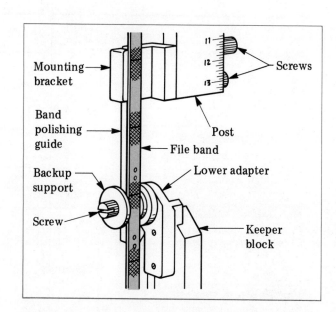

FIGURE 21-18 File Band Guides and Supports Mounted on Keeper Block

SAWING, FILING, POLISHING, GRINDING

Note: Center-crowned, rubber-tired wheels are recommended for band filing.

STEP 14 Close the carrier wheel and the guide post guards.

STEP 15 Apply the band tension specified by the blade manufacturer.

Note: A light band tension produces more accurate filing results.

Caution: Too much tension may cause the file segment rivets to shear when a heavy filing force is applied.

STEP 16 Shift the carrier wheel (file band) speed to the low gear range. Start the drive motor. Check the band tracking.

Note: A correctly tracked file band runs freely in the file guide channels.

Band Filing

STEP 17 Set the file band speed according to the job requirements.

STEP 18 Position the air jet, coolant nozzle, or spray mist combination.

Note: The workpiece and file band must be kept cool. The teeth and grooves in the file segments must be clean.

STEP 19 Start the machine. Stand at the right side. Apply a steady, light force and move the workpiece across the area being filed.

Note: If the teeth load, stop the machine. Use a file card to remove the file particles from each segment.

Note: Loaded teeth cause irregular filing and produce a scratched surface.

STEP 20 Stop the machine at the end of the job. Shift the transmission to neutral. Inspect and clean the file band with a file card.

Note: Every file segment must be cleaned.

STEP 21 Clean the machine table. Remove and return the file band to storage.

BAND GRINDING AND POLISHING

Band grinding and/or polishing produces an excellent grained or polished surface in contrast to the coarser surface produced by band sawing or filing. Abrasive polishing bands are used. The bands are 1″ (25.4mm) wide. They are coated with an aluminum oxide abrasive in grain sizes of 50, 80, or 150. This abrasive is used for grinding and polishing ferrous metals. A silicon carbide, cloth-backed polishing band in the same 50, 80, and 150 grain sizes is used to grind nonferrous metals.

Application of Grit Sizes

The 50 grit size of aluminum oxide abrasive bands is adapted to heavy stock removal operations. The operating band speed should be checked on the job selector. The 80 grit size is for general surface finishing or coarse polishing operations. Band speeds for these operations run to 1,000 fpm (305 m/min). The 150 grit size is for light stock removal and high polish applications. Band speeds range between 800 and 1,500 fpm (244 and 457 m/min) for these abrasive bands.

Abrasive Band Guide Components

Abrasive bands require a rigid backup support and an adapter. The abrasive band adapter replaces the grooved face adapter used for the file band support. Figure 21-20 shows how the band polishing guide (which is secured in the upper mounting bracket) rides against the polishing guide support. The guide usually has a graphite-impregnated facing, which serves as a dry lubricant. A graphite powder is usually rubbed in the polishing band fabric to lubricate it and to increase wear life.

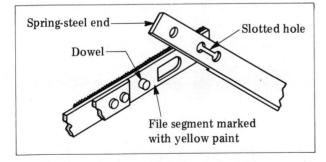

FIGURE 21-19 Position of File Band Segments for Joining the End Sections

The center plate on fixed-table band machines is replaced with a special band polishing center plate. Cutting is done on the coated face of the abrasive band. The operator works from the right side of the table, the same side as for band filing.

Silicon Carbide Abrasive Grinding Band

Silicon carbide abrasive bands are used for grinding tough, hard metals such as carbide-tipped cutting tools. Grit sizes of 50, 80, and 150 are common. The band width is 1″ (25.4mm). A coolant is generally used.

Since a steady, heavier force is required, a *tool finisher attachment* is available. The tool is gripped and held securely in the attachment. The movement of the cutting tool is controlled against the abrasive band. The combination of cutting with a silicon carbide abrasive band, holding the part securely, and applying a constant force across the area being polished produces a superior finish.

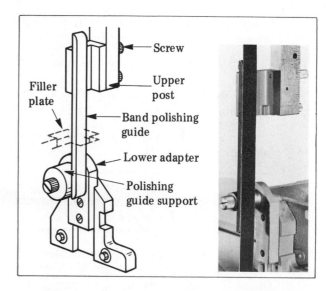

FIGURE 21-20 Guide Components for Band Polishing

HOW TO GRIND AND/OR POLISH WITH AN ABRASIVE BAND

STEP 1 Select the abrasive grain and grit size appropriate for the grinding or polishing process and quality of surface finish.

STEP 2 Set up the rigid backup band polishing guide and the lower adapter support.

STEP 3 Replace the center plate with the polishing band center plate.

STEP 4 Lower the saw band guide post to within 4″ (100mm) of the work table.

STEP 5 Mount and track the abrasive band.

STEP 6 Check a band tension chart for the correct abrasive band tension. Adjust the tension control handwheel.

STEP 7 Set the band speed according to the job selector or manufacturers' recommendations.

STEP 8 Bring the part to be ground or polished to the abrasive band. Apply a steady force as the workpiece is moved slowly across the surface area.

Note: The tool or workpiece is held and fed with a tool finisher attachment when a silicon carbide band is used and a superior surface finish is required.

LINE GRINDING

Line grinding requires a line-grind band. This band tool has a continuous abrasive cutting edge bonded to a steel band. The abrasive coating may be silicon carbide or aluminum oxide. Carbide-coated bands are recommended for grinding quartz, granite, ceramic, glass, and other hard nonferrous materials. Bands coated with aluminum oxide are best for grinding metals, heat-treated alloy steels, heat-resistant steels, wear-resistant nonferrous alloys, and other difficult-to-machine materials. Figure 21-21 shows an application of line grinding to machining the interior surfaces of a hardened die.

Qualities of Line-Grind Bands

Line-grind bands produce a clean, smooth finish surface. The grinding capacity (width) is limited only by the working depth of the machine.

Tapered, rubber-tired carrier wheels and roller-type saw guides are used with line-grind bands. The bands are welded the same as diamond bands. The coated edge is protected by the beveled jaw inserts that face the operator. Before welding, 1/8″ (3.2mm) of abrasive is

SAWING, FILING, POLISHING, GRINDING

FIGURE 21-21 Line Grinding the Interior Walls of a Hardened Die

Note: The band is threaded through the section when an internal area is to be ground.

STEP 3   Install the band around the carrier wheels. See that it tracks correctly around the tapered, rubber-tired carrier wheels and the roller guides.

STEP 4   Set the band speed according to the material to be ground and the nature of the operation.

STEP 5   Position the coolant nozzle and control the flow to deliver a heavy stream.

STEP 6   Use a light hand-feeding force. Grind to the required layout line or dimension.

Note: A steady force is applied as the workpiece is moved across the width of the section being ground.

STEP 7   Dress the abrasive band when the grains dull and for final finish cuts.

removed from each end of the band. When this amount of the metal band is consumed in welding, the abrasive edge forms a continuous band.

Heat Generation and Band Speeds

Considerable heat is generated by the cutting action of the abrasive grains. Therefore, a flood coolant application is needed. Special qualities are required in the coolant. It may be necessary to flush the system to change to the coolant recommended for line grinding a particular metal.

Band speeds for line grinding between 3,000 to 5,000 fpm (900 to 1,500 m/min) are best. Bond abrasives are sharpened by dressing the band with a diamond dressing stick. The stick is moved lightly across the abrasive cutting edge. Light hand-feeding is recommended to permit the operator to accurately control the process.

HOW TO LINE GRIND

STEP 1   Select the type of abrasive and grit size according to the job requirements.

STEP 2   Cut the abrasive band to length. Grind, butt weld, remove flashing, and anneal.

ELECTROBAND MACHINING

Electroband machining operates on the scientific principle of disintegrating the material in front of a cutting edge without having a knife-type band ever touch the material. A low-voltage, high-amperage current is fed into a knife-type band. The electrical input causes the knife edge to discharge a sustained arc into the material being cut. The arc disintegrates the material at the same instant a flood of coolant quenches the arc. The coolant prevents burning and damaging the material.

Figure 21-22 provides an example of electroband machining thin-walled tube sections. The electric arc disintegrates a 1/32" (0.8mm) wide path of material to cut all tubes to length. The flood of coolant prevents burning or damaging the tubes. The ends are burr free and have a surface finish equal to the finish produced by grinding.

Electroband Tool Characteristics, Operation, and Feed Rates

The band tool used in electroband machining must have properties to withstand arcing without cracking and to maintain long flex life. Electrobands are welded in the same manner as knife-edge blades.

410 VERTICAL BAND MACHINES

FIGURE 21-22 Application of Electroband Machining Thin-Walled Tube Sections

FIGURE 21-23 Electroband Machine

The electroband machine (Figure 21-23) is an adaptation of the vertical band machine. Since the stresses produced are limited, machine and holding devices are designed for lighter operating conditions. However, with relatively high cutting rates, hand-feeding is impractical. The tables are hydraulically powered. The band speed is fixed at 6,000 fpm (1,828 m/min). Splash guards contain the coolant without obstructing the operation.

Surfaces requiring a fine finish may be electroband machined at a rate ranging from 5 to 50 square inches (32.2 to 322 sq. cm) per minute. This rate can be increased (when the quality of the work edge is not important) to 150 to 200 square inches (968 to 1,290 sq. cm) per minute.

KNIFE-EDGE BANDS AND PROCESSES

Thus far, blades with cutting teeth have been described and applied primarily to metal-working jobs and processes. *Knife-edge blades*, as the name implies, have a knife edge. The edge may be straight (knife), wavy, or scalloped. When formed into a band, the knife-edge blade is especially adapted for cutting soft, fibrous materials. Other types of cutting edges would tear or fray such materials.

The knife-edge saw blade is produced with a single- or double-bevel cutting edge. The wavy-edge and scallop-edge blades have a double-bevel cutting edge.

The three types of saw blades come in different widths and gages. The sizes of the knife-edge blades range from 1/4" to 1-1/2" (6.4mm to 38.2mm) wide and from 0.018" to 0.032" (0.46mm to 0.8mm) gage thickness. The sizes of the scallop- and wavy-edge saw blades are from 0.015" to 0.032" (0.4mm to 0.8mm) thick and from 1/4" to 2" (6.4mm to 50.8mm) wide.

Knife-edge bands cut efficiently at relatively high band speeds. The bands must withstand maximum *flexation* (bending) over the carrier wheels. In addition, the blades must be ground to razor sharpness and be able to maintain it over long usage. Most knife-edge cutting is done dry and at high speeds. The knife-edge bands slice through to separate material without producing a kerf. As a result, no chips are produced.

Wavy- and scallop-edge forms are used on materials that offer increased resistance to cutting. Materials are cut with a minimum of compacting. High-speed band machine models used extensively for cutting soft, nonferrous materials are designed with automatic knife-type blade sharpness.

SAFE PRACTICES IN SETTING UP AND PERFORMING BASIC BAND MACHINING OPERATIONS

—Position the upper saw band guide as close to the workpiece as practical to provide added support for the cutting tool in the area where the greatest force is exerted.
—Use the machine roller guide shield on high-speed saw cuts. In addition, wear protective safety glasses when working at or near a band machine.
—Replace the table filler plate with the plate designed for the operation (sawing, filing, grinding, and so on) to be performed.
—Wear gloves for handling saw bands during mounting and dismounting. Then, remove the gloves unless the condition of the workpieces or operation justifies their use.
—Check each cutting band before mounting for possible weld defects, the condition of the teeth or cutting edge, and signs of band fatigue.
—Use the correct type of carrier wheels to ensure correct tracking and as a check against band slippage and unnecessary wear on the cutting edges.
—Adjust the band speed and cutting feed rate according to the job requirements. Too high or too low a speed and/or feed reduces tool wear life.
—Apply less force when hand-feeding as the blade starts to cut through. Use a block or pusher on short workpieces.
—Check the feed rate at which the machine table or feed accessory is set against the job selector or a manufacturer's technical data. Checking is done before starting the machine.
—Recheck the distance, angle setting, and size of a starting drill. Sufficient material must be left on the top and bottom of the workpiece to permit straight sides to be filed for a specified height on the internal and external pieces.
—Be aware of the common band machine sawing problems and their probable causes. Be prepared to take the necessary corrective actions if a problem develops.
—Recheck band tracking and the application of the recommended band tension on all band machine cutting tools. Close all doors and secure all guards before starting the machine.
—Adjust the automatic hydraulic carrier wheel brakes for the band tool being used.
—Use a specially prepared saw band for friction sawing. A saw band with sharp cutting teeth will be dulled rapidly and the teeth stripped at a high band velocity and under a heavy feed force.
—Use only center-crowned, rubber-tired carrier wheels for driving spiral-edge saw bands.
—Face diamond-edge saw bands outward and in the beveled jaw inserts of the butt welder to protect the diamond-edge from damage.
—Remove chips from the grooves of each segment of a file band. Otherwise, scratches and a poor-quality filed surface will be produced.
—Apply a continuous, light force when band filing or grinding. Light force is preferred to excessive force, which may load the file band teeth or abrasive band or cause the file band units to shear.
—Use a flood coolant to remove the excessive heat produced by grinding with line-grind bands.
—Set the splash guards to collect the flood of coolant that is required for electroband machining.

TERMS USED IN BASIC BAND MACHINE PROCESSES

Internal sawing	The removal of a section by starting and ending the saw cut inside the workpiece.
Starting hole	A hole drilled primarily to permit a saw band to pass through a workpiece. In the case where an internal and external section are later finish formed to size, a hole drilled at a specified angle.
Slotting	A band machining process of sawing slots into a workpiece. Sawing slots to the width of a saw band kerf.
Splitting	A band machining process of separating parts by band sawing.
Disk-cutting attachment	A band machine attachment used to position a workpiece at a specified distance from the saw band. A table attachment designed primarily for sawing radii.
Friction sawing	A sawing process where heat is instantly generated ahead of the saw teeth. The scooping out of metal at forging point temperature.
Plastic zone	The metal in a limited area immediately ahead of the saw teeth. Metal heated by friction to a temperature where the crystals break down and the plastic metal is removed by the saw teeth.
Thermal conductivity	The ability of a material to conduct heat. (The low thermal conductivity of steels permits limited heat penetration to 0.002" (0.05mm) on the kerf sides of a cut).
High-speed sawing	A method of band sawing at cutting speeds within a range of 2,000 to 6,000 fpm (610 to 1,830 m/min).
Spiral-edge saw band	A circular saw band formed with a continuous spiral cutting edge. A round, spiral-cutting saw band for sawing a precise contour with a minimum radius.
Diamond-edge saw band	A uniformly distributed concentration of industrial diamonds fused to a steel band. An industrial diamond band, dressed to cut hardened metals and other superhard materials.
Band filing	The process of filing on a band machine. Using a file band consisting of overlapping segments that are flexibly attached to a spring-steel band.
Cuts of band files	Specifications for band file segments related to the tooth form, grade, and number of teeth.
Abrasive band	A continuous fabric band coated with grains of aluminum oxide or silicon carbide. A band used for polishing a previously file finished surface.
Band grinding/polishing	A band machine process for producing a fine-textured surface. Use of an abrasive band for purposes of grinding and polishing a surface to a required degree of accuracy.
Rigid backup support and adapter	The support in back of the abrasive band. An abrasive band guide that permits grinding a flat surface. An adapter below the work table against which the backup support rests.
Tool finisher attachment	A band machine attachment for the work table. A device for holding, positioning, and uniformly feeding a cutting tool against a silicon carbide abrasive grinding band.

Line grinding	A band machine grinding process using a steel band that has one edge coated with either silicon carbide or aluminum oxide abrasive grains.
Electroband machining	A band machine cutting process requiring a low-voltage, high-amperage current to be fed into a saw band. The discharge from a saw band producing an arc that disintegrates the material to be cut at the same instant the arc is cooled by a flow of coolant.
Knife-edge bands	Saw bands formed by a single- or double-bevel edge cutting tool. Cutting bands of straight, wavy, or scallop forms, without conventional cutting teeth.
Flexation	The ability of a metal band to flow at high speeds from a straight-line form to a curved form around the carrier wheels and back again.

SUMMARY

The three basic sawing operations performed on the band machine are: cutting off, shaping, and slotting. They are a part of every sawing operation.

An allowance of 1/64" (1.5mm) is made on a sawed surface for finish filing; 0.010" to 0.015" (0.25mm to 0.4mm), for band grinding.

Square corner surfaces that are joined by a radius require a hole to be drilled before the straight band cuts are made.

Internal contour sawing requires a starting hole to be drilled. The saw band is then inserted through the hole.

Square internal corners are sawed by notching the excess material left from a drilled hole or the first saw cut.

The band machine operator must determine the following saw band requirements from a part print and the job selector: the type of saw band teeth, pitch, kind of blade material to use, and the blade width and thickness.

Knowing the kind of material, thickness, and surface finish requirements, the operator needs additional information relating to the machine setup as follows: the cutting speed (fpm or m/min) of the saw band, the band tension, the recommended feed rate, the type of liquid and/or air coolant, and quantity (flow) of coolant (when used).

The technique of contour sawing mating parts from the same workpiece requires a starting hole to be drilled at a specified angle. The table is set at a smaller angle. The saw band is started a short distance from the layout line. The technique of angle sawing provides sufficient stock to finish both sections to the same size and contour.

Common band machine problems are caused by incorrect saw band velocity or feed rate; improper tracking or tensioning; wrong type of blade or pitch; poor weld or annealing; use of incorrect saw guide inserts, rollers, or incorrect spacing; and application of the wrong cutting fluid or flow.

Slotting requires the use of a saw band having a kerf the same size as the required width of the slot.

Radii and other contours may be cut by hand-feeding or by using a disk-cutting attachment or a contour sawing accessory.

Friction sawing is done by using a special saw band traveling at velocities of 6,000 to 15,000 fpm (1,828 to 4,572 m/min). The heat generated by friction produces a breakdown of the crystal structure immediately ahead of the saw band teeth. The teeth remove the affected metal in the form of short, stubby, fused metal particles.

Friction sawing is ideal for hardened and tough-to-machine alloys. Cutting is done without distortion. However, a sizable burr is produced. Friction sawing is most practical for workpieces 1″ (25.4mm) and smaller.

Friction sawing band machines are built heavier than regular jobbing shop machines. Increased rigidity, power requirements for higher speeds and heavier feeds, larger roller-bearing saw guides, and different carrier wheels are some of the main design changes.

The friction saw blade has raker set teeth. The teeth are specially formed and locked to an extra-strong back. Manufacturers' charts provide specifications for type of blade, width, pitch, tooth set, and machining speeds and feeds.

High-speed sawing is used primarily on soft and fibrous materials. Saw band speeds range from 2,000 to 6,000 fpm (610 to 1,830 m/min). The process produces a heavy chip formation and requires a rapid chip flow from the teeth.

Spiral-edge band machining requires the use of a round, continuous spiral-tooth band of small diameter (0.020″, 0.040″, 0.050″, and 0.074″ (0.5mm, 1.0mm, 1.27mm, and 1.88mm). Spiral-edge bands are used in center-crowned, rubber-tired roller guides and specially grooved jaws on the butt welder. Sprial-edge bands permit sawing intricate contours in thin-gage metals or limited thickness parts.

Diamond-edge saw bands have a concentration of industrial diamonds distributed uniformly and fused on a band. Straight, angular, and contour cuts are possible on superhard and abrasive-action ferrous and nonferrous materials. Cutting is done at maximum velocities of 2,000 to 3,000 fpm (600 to 900 m/min). A flood coolant is needed to dissipate the heat generated by the grinding action.

The band machine is widely used for band filing. File segments are riveted on one end to a flexible steel band. This band permits movement over the carrier wheels. As the band and segments return to the vertical position, they interlock and form a continuous file band. The three common shapes of file segments are square, half-round, and oval. The file widths are 1/4″, 3/8″, and 1/2″ (6.3mm, 9.5mm, and 12.7mm). Different angles, tooth forms, and pitches are available for filing nonferrous metals, mild steels, and tool steels. Slow speeds and limited feeds are used.

Band grinding and polishing produce finely grained surfaces as a second operation. Abrasive bands, coated with either aluminum oxide or silicon carbide grains, are used. Three basic grain sizes (50, 80, and 150) produce coarse, medium, and fine surface finishes. A backup support is provided behind the abrasive band. Light feeds are required on aluminum oxide bands. Silicon carbides are used for heavier-duty applications. A tool finisher attachment is used for accurately positioning and feeding precision tools.

SAWING, FILING, POLISHING, GRINDING 415

Line grinding is done with a band cutting tool. The tool has a continuous abrasive cutting edge bonded to a steel band. The aluminum oxide and silicon carbide grains permit grinding internal and external surfaces of hard, tough-to-machine ferrous metals and nonferrous materials. Flood coolant is needed because of the heat generated during grinding.

Electroband machining permits the high-speed cutting of thin materials and sections without touching the part with a knife-type saw band. A flood coolant prevents the arc that disintegrates the material in front of the blade from burning the part. A fine surface finish is produced.

Straight, wavy, or scallop knife-edge blades are used for cutting extremely soft, fibrous materials. The cutting is done dry and at high speeds. Such cutting is carried on outside the machine industry.

Standard machine and tool handling safety precautions must be observed. Particular attention is directed to the use of gloves for handling sharp saw bands. Guards must be adjusted as close as practical to the workpiece. Goggles and safety shields are required for all band machine work.

UNIT 21 REVIEW AND SELF-TEST

1. Referring to a job selector on a band machine, list the kind of information an operator needs to know about conventional sawing a 25.4mm thick block of free-machining steel.

2. a. Study the following working drawing of a form guide. This part is to be band machined.
   b. List the work processes required (1) to saw the square corner, relief, and the contour; (2) to file; and (3) to grind the 0.750 ±0.001 area. Assume the workpiece has the one edge ground to $\sqrt[32]{}$, the plug end is square, and the part is laid out.

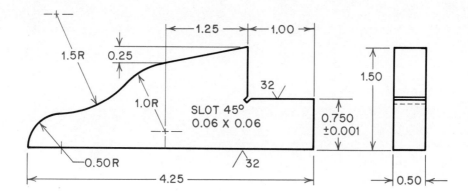

3. List three differences between high-speed sawing and friction sawing.

4. a. Identify two distinguishing design features of a spiral-edge saw band.
   b. Cite two types of jobs for which spiral-edge band sawing is uniquely adapted.

5. Compare electroband machining with conventional band sawing in terms of (a) principles of machining and (b) a typical application.

6. a. State one similarity of a line-grind band and an abrasive band.
   b. Give two major differences between the two bands.

7. Give three personal safety precautions to take prior to operating a band machine.

PART SIX

Shaping and Planing Machines

SECTION ONE   Shaping and Planing Machine Technology and Processes
   UNIT 22   Shaping and Planing Machine Design Features, Tooling, and Setups

SECTION ONE

Shaping and Planing Machine Technology and Processes

Shapers and planers are machine tools that are used for machining internal and external plane and contour surfaces. Shapers and planers generate a required linear surface or form by a reciprocating straight-line cutting motion. Material may be removed to produce a horizontal, vertical, or angular plane surface or a varying-contour form. This section provides an overview of shaping and planing machine design features, fundamental tool geometry, cutting speeds and feeds tables and computations, and basic setups and machining processes.

UNIT 22

Shaping and Planing Machine Design Features, Tooling, and Setups

The cutting action of shapers and planers varies. In shaper work, the part is securely held and the cutting takes place as the machine ram forces a cutting tool across the workpiece. The cutting action for planer work is just the opposite. Cutting takes place as the workpiece moves past a rigidly held cutter. While most shaping and planing processes require external cuts, a number of applications require the removal of material from inside surfaces or sections.

Other machine tools are now sometimes used to perform processes that formerly were done on shapers and planers. However, shaping and planing machines are still economical to operate; are versatile in being easily tooled for jobbing and specialty shop work; and are extremely practical for such heavy machine processes as the machining of beds, ways, and long or massive parts.

BASIC TYPES OF SHAPERS

There are two basic types of shapers: *vertical* and *horizontal*. As the names indicate, the movement of the cutting tool is in either a vertical plane or a horizontal plane.

The Vertical Shaper (Slotter)

The vertical shaper is sometimes called a *vertical slotter*. It is designed primarily to shape internal and external flat and contour surfaces, slots, keyways, splines, special gears, and other irregular surfaces. The design features of a vertical shaper are shown in Figure 22-1. The two principal components include a *base* with a bed and a *column*. The base houses the drive and feed transmission system. The bed ways are part of the base and support the carriage. The carriage may be fed transversely toward or away from the column. A rotary table is mounted on the carriage. The table is also movable longitudinally. The combination of three directions of movement permits the table and workpiece to be positioned and fed for straight-line and rotary cuts.

Motion for cutting is produced by the reciprocating movement of the *vertical ram*. The ram and mechanism for producing motion are housed in the column. Cutting tools are secured in a toolholder on the forward section of the ram. The ram position is adjustable. The cutting tool is set in relation to the starting and ending of a cut. The distance the ram travels is called the *length of stroke*. The ram on some machines may be adjusted at an angle to take angular shaping cuts on vertical surfaces. Power feed is provided for longitudinal and transverse feeding. Generally, the rotary table is turned manually. Cutting speeds are controlled through drives in different speed ranges.

The vertical shaper is of heavier design and construction than most machine tools used in toolrooms and jobbing shops. Workpieces are usually strapped to the work table. Many of the same types of straps, clamps, and T-bolts used for milling machine setups are used with the vertical shaper. Parallels, step blocks, angle plates, and adjustable jacks are required for positioning and supporting overhanging and other sections. Special end-cutting tools and other tools similar to shaper cutting tools are used for vertical cuts.

The Horizontal Shaper

In contrast to the reciprocating cutting movement of the vertical shaper, cutting action takes place on the *horizontal shaper* by feeding the workpiece for the next cut on the noncutting return stroke. The cutting tool removes metal on the forward longitudinal stroke of the *horizontal*

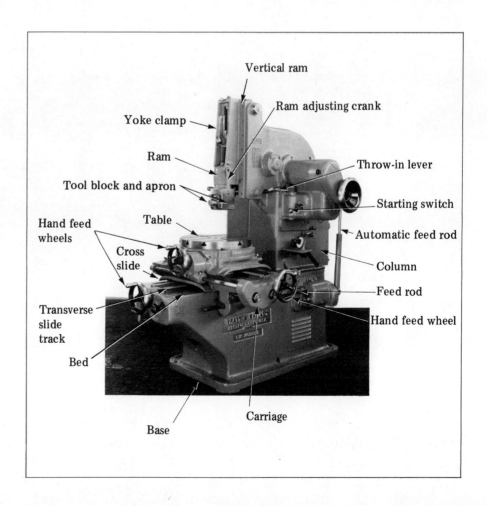

FIGURE 22-1 Design Features of a Vertical Shaper (Slotter)

ram. A surface is shaped by taking a series of successive cuts. For each cycle of the ram, cutting is done on the forward stroke; feeding, on the return stroke.

Shaping of small workpieces is usually done on *bench shapers*. While many of the same processes performed on larger, heavier floor models are also performed on bench shapers, bench shaper work is, nevertheless, more limited. However, the principles of operation, cutting tool geometry, setups, and basic processes are similar.

Positioning, Feeding, and Cutting Movements. The different positioning, feeding, and cutting movements on a universal horizontal shaper are shown in Figure 22-2. Note on this model that the *universal table* may be rotated through a 90° arc. The vertical face (plane surface) has T-slots that permit strapping work-holding devices directly on the table face. The smaller table face to which a vise is bolted may be tilted at an angle. Thus, parts may be held and machined at compound angles.

The *tool head* may be swiveled to feed the cutting tool at a required angle. The angle settings are read directly from the angle graduations on the end of the ram. The clapper box is adjustable to permit the cutting tool to clear the workpiece on the return stroke.

The table is movable vertically and transversely. The cutting tool may be moved up or down by turning the handwheel on the vertical tool-head slide. Cutting takes place during the forward stroke of the ram. On the reverse stroke, the cutting tool is fed for the next successive cut.

BASIC TYPES OF SHAPER CUTS

Shaping Vertical Surfaces

The tool head and clapper box are positioned vertically for machining flat and regular curved surfaces. However, the clapper box and tool head must be swiveled for angular and vertical cuts. Otherwise, the toolholder and/or the cutting tool may not clear the workpiece. Swiveling

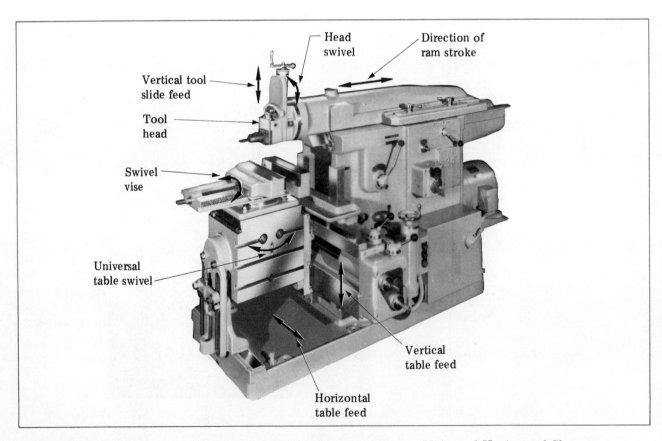

FIGURE 22-2 Positioning, Feeding, and Cutting Movements on a Universal Horizontal Shaper

the head prevents the cutting tool from dragging along and scoring the machined surface during the return stroke.

A right-side vertical surface is shaped, using a down-feed, by swiveling the clapper box assembly as shown in Figure 22-3. The toolholder is first secured in a vertical position on the clapper block. The clapper box is then moved clockwise. The angle to which the clapper box is positioned depends on the vertical height to be machined and the toolholder clearance. The clapper box and toolholder combination are positioned counter-clockwise to shape a left-side vertical surface.

Shaping Simple and Compound Angles

Angular surfaces are shaped either by positioning the workpiece or the work-holding device or by setting the tool head at the required angle. Common setups for machining angular surfaces are shown in Figure 22-4. Figure 22-4A illustrates a common setup for machining an angular surface by *horizontal shaping*. Sometimes, the workpiece is set on tapered parallel bars. The layout line is then in a horizontal plane. Some parts require the machining of surfaces at a compound angle. The vise is set at the first angle; the standard table, at the second angle; and the universal table, at the third angle. The compound angle is then produced by horizontal shaping.

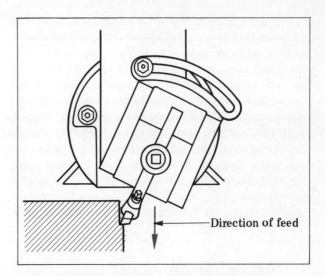

FIGURE 22-3 Position of Toolholder and Clapper Box to Shape a Right-Side Vertical Surface

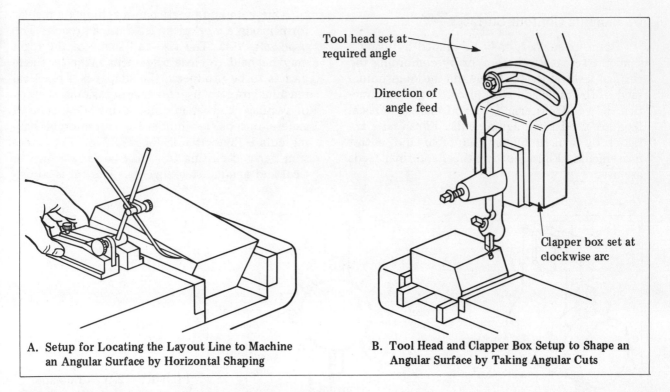

A. Setup for Locating the Layout Line to Machine an Angular Surface by Horizontal Shaping

B. Tool Head and Clapper Box Setup to Shape an Angular Surface by Taking Angular Cuts

FIGURE 22-4 Setups for Machining Angular Surfaces

Figure 22-4B illustrates the tool head and clapper box setup to shape an angular surface by taking *angular cuts*. These cuts are taken with the tool head set to a required angle to the right or left of the 0° vertical angle position of the tool slide unit. The clapper box and toolholder are positioned at an opposite angle or arc to permit the cutting tool to swing away from the workpiece on the return stroke. In addition to standard straight and offset shaper toolholders, an adjustable-type universal cutting-tool holder is available (Figure 22-5). The design permits setting a tool bit for horizontal, vertical, angular, and contour cuts.

Shaping Dovetails

When dovetails are cut on shapers, the tool head is swiveled at the angle of the dovetail. The cutting tool is ground at a smaller included angle than the angle of the dovetail. The tool is ground for cutting on the front face and either the left- or right-hand edges, depending on whether a right- or left-hand dovetail is to be cut. The front cutting edge is formed to the required corner size and shape. Usually, the corners of an internal dovetail are relieved.

Shaping Contour Surfaces

Contour surfaces may be shaped by using a shaper tracing attachment or by combining the vertical feed on the tool slide and the longitudinal feed of the table and workpiece. In general practice, power transverse feed and manual vertical feed are used for roughing cuts. Finish cuts are taken by synchronizing the table and cutter movements. Finish cutting tools and finer feeds are used.

Shaping Multiple Grooves or a Spline

A *spline* is a series of equally spaced grooves or tongues. The grooves are cut into a shaft. The tongues (mating grooves) are machined around the inside diameter of a gear. Splines provide a positive line drive between two mating members. At the same time, lateral movement of either member may occur to engage or disengage the parts. For example, a gear in a gearbox may be moved along a splined shaft to change its position.

While external splines are generally milled, special ones are often shaped. The sides of a spline may be straight or produced at an angle. A form tool is used for shaping angular sides. Each spline is indexed on a shaper by using an indexing head. Workpieces are mounted between the indexing head center and a footstock. The cutting tool is centered in relation to the workpiece. The cutter is fed by manually turning the tool slide handwheel. Splines are usually rough shaped. Stock is allowed to clean up the sides to dimensional limits and to produce the required surface finish during the finish cut.

Shaping Internal Forms

One common practice in machining a regular form inside a workpiece is to use a broach. The process is fast. The size and shape of the form may be held to close tolerances. When a single part is to be produced, the shaper is a practical machine tool to use for the occasional shaping of regular or other internal forms. One type of general-purpose toolholder for four internal shaping cuts is illustrated in Figure 22-6. The extension bar and cutting tool may be turned for vertical and angular shaping cuts. The bar is adjust-

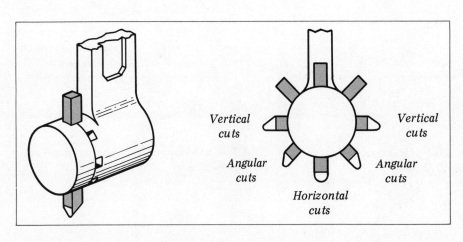

FIGURE 22-5 Adjustable-Type Universal Cutting-Tool Holder

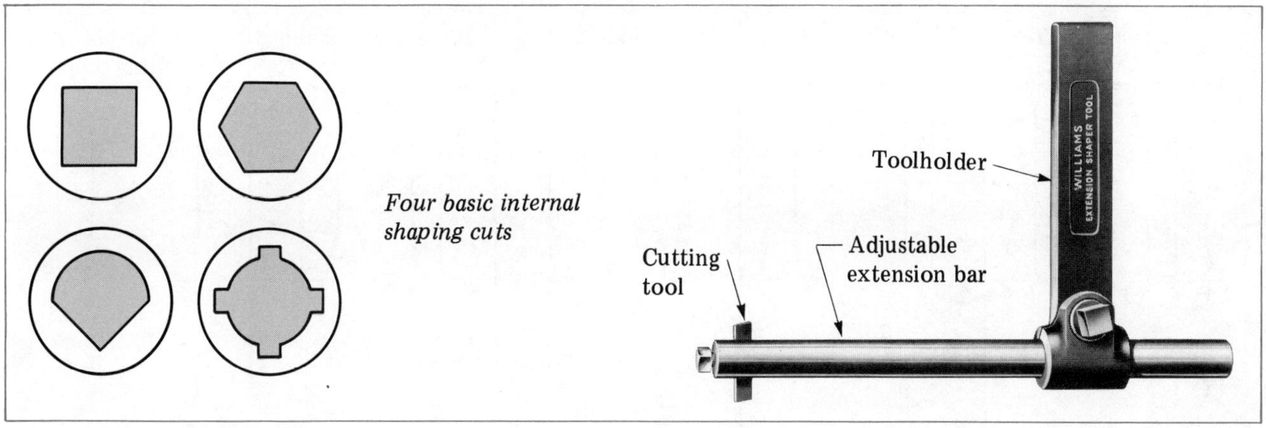

DESIGN FEATURES, TOOLING, SETUPS 423

FIGURE 22-6 Toolholder and Extension Bar for Four Basic Internal Shaping Cuts

able, depending on the length of the cut. Four basic internal forms may be shaped using the extension bar.

Like all other cutter holders, the extension bar should be set as short as possible. Maximum rigidity is especially important in internal shaping. The tool setup must be checked carefully to be sure the cutting stroke permits adequate, safe clearance at the start and end of each cut. A slower ram speed is used than for external shaping. The internal toolholder is not as rigid as a solid tool or toolholder.

CONVENTIONAL HIGH-SPEED STEEL CUTTING TOOLS

Square, high-speed steel tool bits are widely used for roughing and finish shaper cuts. Some tool bits are used for horizontal, flat, or contour shaping. Other tool bits are formed for down-feeding for shaping vertical surfaces or for parting or slotting. Still others are used for shaping internal angles.

All cutting tools have one common angle. The clearance angle back of the cutting edge is 4°. This minimal clearance provides maximum support for the cutting edges. The shape and cutting angles of high-speed steel tool bits vary for cutting machine steels, cast irons, and other ferrous and nonferrous metals and materials.

Tool Geometry for Standard Tool Bits

The term *tool geometry* refers to design features such as shape and angles. Figure 22-7 shows the tool geometry for some regular shaping cut. Figure 22-7A gives the shapes and rake and clearance angles for machining low-carbon steels. While left-hand side roughing and left-hand roughing tools are illustrated, the same tool geometry applies to right-hand cutting tools. Figure 22-7B shows the shapes and angles for machining dovetails in cast iron. The side-rake angle is reduced from 12° to 3°. The top face is ground flat with a 2° front-rake angle to replace the 3/8" (9.4mm) radius.

In general practice, roughing cuts are taken to within 0.010" to 0.015" (0.2mm to 0.4mm). Enough stock is left to machine the surface to a required dimension and surface finish accuracy. Roughing cuts are taken at the maximum depth and highest cutting rate possible. Consideration must be given to the malleability of the material, rigidity of the setup, nature of the operation, and conditions governing the cutting tool and machine. If the cutting forces are too great, the feed rate should be reduced before the depth of cut. Finishing cuts up to 0.015" (0.4mm) are usually taken in one cut.

CARBIDE CUTTING TOOLS

Carbide cutting tools permit machining at faster cutting speeds and about the same feed rates as the speeds and feeds used for shaping with most high-speed steel tool bits. However, the greater cutting efficiency depends on three conditions:

—The cutting speed of the shaper must exceed 100 fpm (30 m/min);

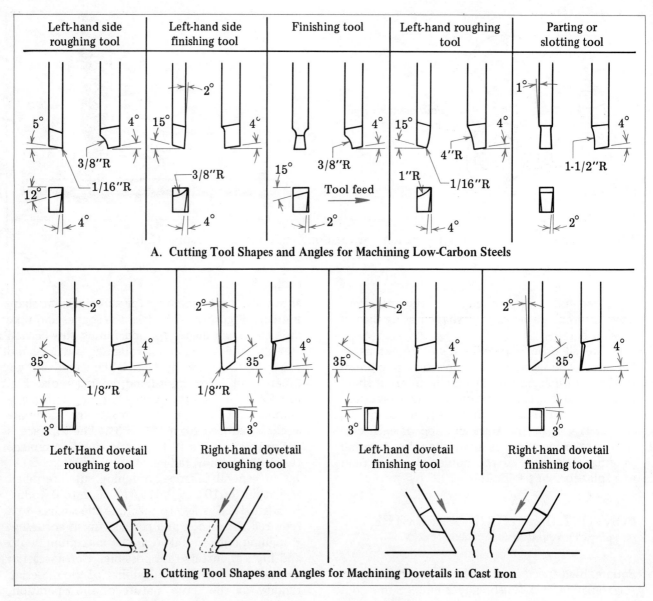

FIGURE 22-7 Tool Geometry for Selected Regular Shaping Cuts

—A constant speed and feed rate must be maintained;
—The tool head should be fitted with a tool lifter to permit the tool to clear the work on the return stroke.

Cutting speeds of carbide cutting tools are usually two or three times faster than for high-speed steels. In using carbides, it is better practice to take heavy, deep cuts and to use a lighter feed. This practice distributes the chip force over a longer cutting area.

Common Designs of Carbide Shaper Tools

Two designs of carbide shaper cutting tools are in general use. Brazed-tip tools are applied to light-duty cutting and fine cuts. The carbide tip is brazed onto a solid shank. Tools with replaceable insert tips are used for deep cuts, large-area cuts, and interrupted cuts. The tips are nested and secured to a solid shank. The advantages of using inserts include ease of changing and replacing, rigid support in the holder, and availability in preformed shape.

DESIGN FEATURES, TOOLING, SETUPS 425

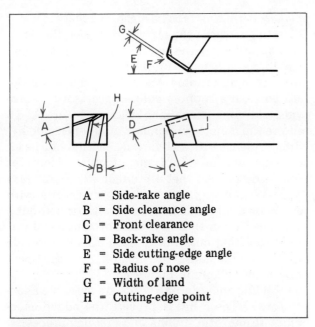

A = Side-rake angle
B = Side clearance angle
C = Front clearance
D = Back-rake angle
E = Side cutting-edge angle
F = Radius of nose
G = Width of land
H = Cutting-edge point

FIGURE 22-8 Tool Geometry of a Carbide Cutting Tool

FIGURE 22-9 1817 Roberts's Planer

Tool Geometry for Carbide Tools

The tool geometry relating to the shape and angles of a carbide shaper cutting tool is shown in Figure 22-8. Tool manufacturers furnish tools having specified angles and tool forms. These tools produce efficiently, machine to particular surface finishes, and have maximum tool life.

Like the high-speed steel tool bits, the front clearance on carbide cutters is 4°. A 0° to 20° negative back-rake angle is used for roughing cuts on steel and cast iron. The back-rake angle permits the cut to start above the cutting edge. This action protects the cutting edge (as the weakest part) from the shock of starting or intermittent cuts. The back-rake angle is increased to a positive angle on finish cutting tools. The positive angle prevents tool chattter when light finish cuts are taken. A side cutting-edge angle of 30° to 40° with a large nose radius should be used on roughing tools.

EARLY AND MODERN PLANER DESIGNS

A *planer* is designed to generate accurate, flat surfaces by the linear travel of a workpiece past a cutting tool. Flat, vertical, angular, and contour surfaces may be machined. A planer is generally used to machine long surfaces on single or multiple units of castings, fabricated parts, heavy and thick plates, and standard metal stock. The planer is widely used to machine flat surfaces, ways, and sides on castings and weldments of machine beds and tables.

The 1817 Roberts's Planer

Richard Roberts of Manchester, England, is credited with introducing the first planer. The 1817 Roberts's planer (Figure 22-9) contributed significantly to the machine tool industry. The Roberts's planer made it possible to plane flat surfaces that were needed for ways and other machine elements. With subsequent refinements, the ways and newly designed elements (when combined) were incorporated into the manufacture of other machine tools.

The early Roberts's planer consisted of a manually operated rack and gear that actuated a table (platen), which was guided along the bed of the machine. Workpieces were secured to the platen. The cutting tool was adjusted and mounted on a vertical (rail) head. This rail head was part of a saddle on a cross rail. Horizontal movement of the head was thus possible. The horizontal movement permitted feeding the cutting tool across the workpiece for successive cuts.

The cross rail was positioned at different heights by adjustment of the vertical support braces.

Modern Planer Design

Although certain basic tooling setups and processes have remained constant over the years, the design of the small, manually operated, 1817 planer has changed radically. The single rail head has been replaced by a more rigidly constructed rail head that may be fed manually or by power and either vertically or at an angle. Machines with one or two rail heads are common. Additional machining capability is provided by a vertically powered side head. The design of the modern planer thus permits simultaneous machining in two planes along a workpiece.

Planers are ruggedly constructed to take deep roughing or finish cuts on large-size workpieces. For example, a 6′ × 6′ × 34′ (1.8m × 1.8m × 10.4m) planer is capable of machining single or multiple workpieces with similar overall dimensions.

Some planers are designed with their upright columns joined together at the top to increase machine rigidity. Such planers are called *double-housing planers*. Other planers are designed with a single column on the right side. These *open-side planers* permit parts to be machined that are wider than the width of the platen. However, the size is still specified by the width, height (depth), and length measurements on planers with two permanent columns. Some open-side planers are designed with a support column accessory to provide additional machining rigidity. Figure 22-10 shows the major components of an open-side hydraulic planer with a single rail head and side head.

One machine tool manufacturer identifies open-side planers that are under 42″ × 42″ (1m × 1m) as *shaper-planers*. They are available with single or double rail heads and with or without a side head mounting. Typical horizontal and vertical roughing cuts taken by both rail and side heads simultaneously are illustrated by the open-side shaper-planer in Figure 22-11.

Milling and grinding heads are available as accessories. These heads provide the added capability to perform milling and grinding processes, respectively. Template and contour following, indexing, and footstock accessories are also available. Cutting tools are sometimes set up on multiple heads for gang planing. Some cutters are actuated by a stylus to permit two- and three-axis duplicator processes to be carried on. The stylus controls the depth of cut as the table and workpiece advance past a cutting tool.

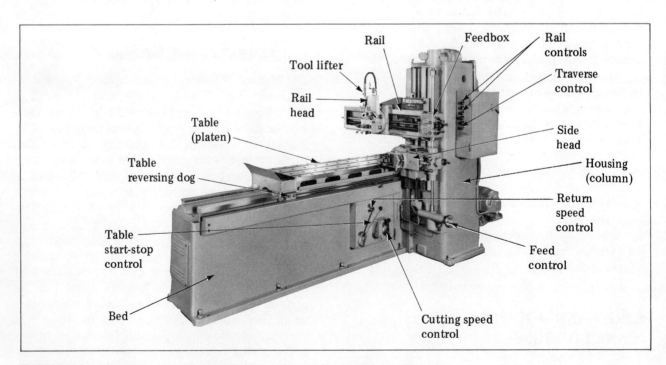

FIGURE 22-10 Major Components of an Open-Side Hydraulic Planer with a Single Rail Head and Side Head

One of the greatest advantages of a planer is its versatility and lower costs for certain machining operations. Single-edge cutting tools that are primarily used for planer operations may be easily modified. These tools often replace more highly expensive multiple-point cutting tools when one part or a small number of parts are to be machined.

Universal planers permit cutting during both the forward stroke and the return stroke of the platen. A double-edge cutting tool is mounted in a fixed holder that indexes between the forward and return strokes. The apron and clapper block are securely fastened to prevent lifting.

On the forward stroke, one cutting tool edge is firmly held against the tool block during the cut. The second tool edge is then automatically indexed at one end of the stroke. With the second cutting edge in position, a cut is taken during the return stroke. Machining time is reduced as the lost return stroke time is eliminated.

MAJOR PLANER COMPONENTS

The open-side shaper-planer is used here to describe the five basic construction features of planers and the operating controls. The *bed*, *platen* (table), *column* (housing), *cross rail*, and *heads* (rail and side) are included in the basic machine design.

Bed Features

The bed serves as a support for the platen. It is usually a cast form that is heavily ribbed for strength, for rigidity, and for dimensional stability. The bed houses the mechanical or hydraulic components that drive the platen. Precision ways are machined in the top surface of the bed. The ways guide and provide support for the platen.

Platen (Table)

The platen has four principal design features. First, the underside is machined with precision ways that match the ways on the bed. Planers are designed so that the platen moves on a thin film of pressurized oil that coats the bed ways. Many high-speed planers are fitted with laminated plastic bearing surfaces. These surfaces permit high sliding velocity between the platen and the bed. Second, an underside component houses the platen devices for engagement of the table with the drive mechanism. Third, the top of the table has a series of slots and holes that permit the use of T-bolts, other positioning and holding clamps and straps, stops, and additional work-holding devices. Fourth, the end has a chip pocket for convenience in depositing chips.

FIGURE 22-11 Simultaneous Planing Using Rail Head and Side Head on an Open-Side Shaper-Planer

Column (Housing)

The column is also rigidly constructed. It is solidly secured to the bed. The positioning and feed drive mechanisms for the side head, cross rail, and rail head(s) are housed in the column. The column is designed to withstand the weights of all components in extreme positions, together with the possible cutting forces. Vertically machined guide surfaces permit adjustment of the side head and the horizontal rail.

Cross Rail

The cross rail is fitted to the vertical column guide surfaces. The right end has an apron used for elevating and lowering the cross rail. The feedbox for the cross rail is fitted to the apron. The cross rail is accurately machined so that the horizontal travel of the rail head(s) and cutting tool is parallel with the top table surface.

Rail and Side Heads

The functions and operating features of the rail and side heads are similar. Each head is mounted on a saddle. The horizontal or vertical movement (according to position) is produced by a feed screw and worm thread engagement. Each saddle is fed with a tool head similar to the tool (swivel) head on a shaper. The bases of the tool heads are graduated in degrees. The heads may be swiveled for angle settings on both sides of the centerline. Each tool head consists of a tool slide, apron, clapper box, clapper block, and tool holding straps.

The rail and side heads are designed for independent operation. Large machines provide for independent as well as synchronized (together) movements. Since the range of cutting speeds permits machining at high ranges, carbide cutting tools are widely used. Open-side shaper-planers are provided with automatic tool lifters. The clapper block and cutting tool are lifted on the return stroke.

PLANER DRIVE AND CONTROL SYSTEMS

Forward and return table movement is produced by one of two basic systems: *mechanical* or *hydraulic*. In both systems, the rate of table return is faster than the cutting stroke.

Mechanical systems use a rack and pinion, herringbone, or spiral bevel transmission drive. On the mechanical system, the motion designs permit quick reversal for the rapid return drive.

The hydraulic system for an open-side shaper-planer is diagrammed in Figure 22-12. The table is driven by the double-acting piston and piston rod. The rod is attached to a handle through an anchor bolt. The system control pilot valve is actuated by the reversing dogs. The hydraulic oil is contained in a reservoir. The reversing hydraulic oil is contained in a reservoir. The reversing hydraulic table drive pump is driven by a constant-speed main drive motor. The lubricating pump for the ways works off the table drive pump.

The system is also designed with a feed cylinder. The feed control handwheel on the column regulates the feed cylinder and determines the speed of the shaft for feeding the tool heads.

The speed control relief valve governs the forward and return speeds of the planer table. The tool lifter, as part of the hydraulic system, is activated by the tripping of the pilot valve.

Larger-size planers have a three-speed range drive system. The system provides for very slow to very high speed table movement.

WORK-HOLDING DEVICES FOR PLANER WORK

Parts to be machined on a planer are either strapped to the table or held in a special fixture or positioning device. In practically all applications, the workpiece is secured against any possible movement by a *stop*. Figure 22-13 shows one pair of stops turned in order to be used as *stop pins*. The screws on the two other right-side stops are adjustable. The screws apply force against the workpiece and secure it to the table. Poppets and toe dogs are widely used in planer setup work.

Plain, gooseneck, U-shaped, and single-end and double-end finger-pattern strap clamps are widely used work-holding devices. "Planer furniture" also includes riser blocks, step blocks, parallels, and planer jacks. Descriptions and applications of these items were covered under milling machine work. In the case of planer operations, larger sizes are required. Plain and swivel-base vises with single or multiple screws are also used as in shaper work.

DESIGN FEATURES, TOOLING, SETUPS 429

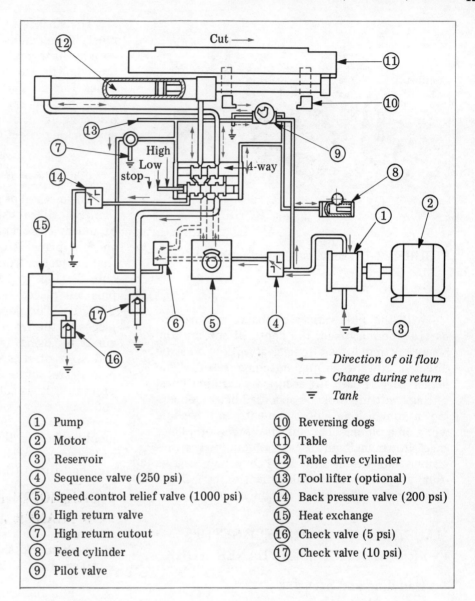

FIGURE 22-12 Hydraulic Drive System on an Open-Side Shaper-Planer

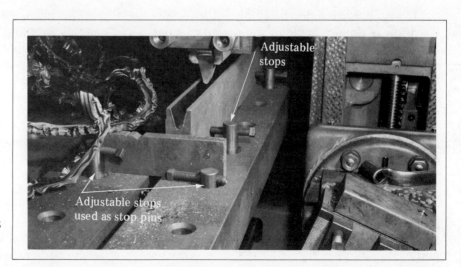

FIGURE 22-13 Applications of Stops and Stop Pins against the Sides and Front End of a Workpiece

430 SHAPING AND PLANING MACHINES

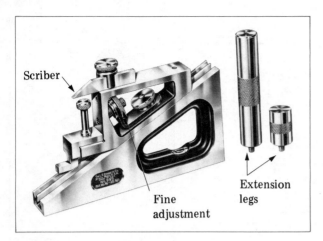

FIGURE 22-14 Universal Precision Planer Gage

With the wide variety of shapes and sizes of parts to be machined, the depth of cut, and the heavy cutting forces, special attention must be directed toward proper machine setups. The *planer gage* is used for accurately scribing lines, setting a cutting tool at a specified height, checking a dimensional measurement, and leveling work on a platen. The universal precision planer gage shown in Figure 22-14 is equipped with a scriber and extension legs. The planer gage is usually set to a specified dimension by using a height gage or gage block combination.

MACHINE AND WORKPIECE SETUPS IN PREPARATION FOR PLANER WORK

Handling and Securing Heavy Workpieces

Heavy parts require handling with a crane. Provision is usually made in the design of a part for handling, placing, and disassembling. A hole, lug, or surface around which a sling may be formed is generally provided. Sometimes, a heavy-duty C-clamp is secured against a surface that can support the heavy weight of the part.

Heavyweight and industrial-quality ropes and steel cables are used to form the sling. The size of the rope or cable must be adequate to safely support the full weight of a part and its suspension and movement. The sling must be positioned so that a center ring or the location into which a crane hook passes (or the crane hook alone) is located at the center of gravity of the part.

The surface to be placed on the planer table must be examined before the workpiece is secured. Burrs and surface irregularities are removed. The surface is wiped free of grit and other foreign particles. Once the workpiece is positioned on the table, jacks are placed under overhanging sections. Shims are placed to remove any rocking effect due to an irregular surface.

Placing the Workpiece

The method of positioning workpieces against planer stops is illustrated in Figure 22-15. Single or multiple workpieces are placed on the table so that planer stops may be inserted in holes that are just in front of the workpieces to be machined (Figure 22-15A). Poppets and planer pins are placed behind the workpieces (Figure 22-15B). The poppets and pins force the workpieces to the front stops. The parts are then secured by using T-bolts and appropriate shapes and sizes of straps. Where two or more workpieces are set up for simultaneous machining processes, it usually is possible to use U-straps and T-bolts across inside sections.

The number of straps to use and their placement depend on the shape, construction, and size of the part, as well as the nature and depth of the cuts and the cutting forces. Once all straps have been placed and a slight amount of force is applied, each strap is tightened with uniform force to avoid uneven stresses that may distort the workpiece. Each strap, stop, poppet and planer pin, jack, block, and other fasteners in the whole setup are given a final, careful rechecking.

Adjusting the Rail

The rail is set at a height that permits the greatest support for the cutting tool with minimum overhang of the tool head. There must, however, be adequate, safe clearance between the workpiece and the cutter. When a rail height needs to be changed, the rail locking clamps and bolts need to be tightened. Open-side shaper-planers have back rail clamps that are loosened for rail adjustment and tightened to rigidly support the cross rail apron to the front of the column. Other planing machine designs require the tightening of bolts to lock the cross slide apron at a fixed height. Consideration is given to taking up backlash on the elevating screw. The rail is lowered to slightly lower than the required height and then raised to position and locked.

A. Inserting Stops in Platen Holes in Front of Workpieces

B. Using Poppets to Force Workpieces Against Front Stops

FIGURE 22-15 Method of Positioning Workpieces against Planer Stops

Preparing the Tool Head

Original cuts require the tool head to be set vertically. The same procedures and cautions are followed as for shaper work. The extension of a tool bit from a standard shaper-planer holder is limited, as is the extension of a solid cutting-tool shank or toolholder below the clapper block: The setting is as short as practical. The clamping nuts on the clapper block are uniformly tightened to prevent tool movement.

Adjusting the Tool Head

The tool head is moved downward almost to the surface to be planed. Once the table stops and cutting speeds are set, the tool slide is moved to locate and to feed the cutting tool to the depth of the first cut. The same procedure is followed to position the cutting tool on a second tool head.

Somewhat similar steps are taken to mount, secure, position, and locate the cutting tool on a side head. The difference is that the cutting action will be vertical and with down-feed. This vertical cut is usually combined with a right (hand) horizontal cut.

Setting the Cutting Speeds and Table Stops

Table stops are adjusted for the approximate length of the cutting stroke and end clearances. The table speed is then set to move at a minimum rate. Tool and workpiece setups are given a final check for clearance. The planer is started. The table lever is engaged, and the table is stopped at the locations where the table stops are positioned and secured. Consideration is given to the stops at the beginning and end of the cut. Adequate end clearance must be allowed to permit feeding the tool for subsequent cuts.

With the table stops positioned for the starting and ending of each cut, the cutting and return speeds are next set. Some hydraulic planers have adjusting valves on two sides of a hydraulic pump. The valve on one side controls the rate of the cutting stroke. The independent valve on the other side regulates the return stroke. Other hydraulic planers have a single cutting-speed control lever.

Adjusting the Feed Rate

Feed rate on a hydraulic planer is controlled by a valve. The feed control valve on the side of the machine column is adjusted to the required feed rate.

PLANING FLAT SURFACES

The table movement is stopped at the start of the cut. The tool head is manually fed along the rail and positioned at depth to take a roughing cut. The tool slide is locked in position at this depth. The table is started again and the feed is engaged. Where a cutting fluid is required, a stream should be directed from a safe distance. Cutting fluid must not be applied with a brush.

Roughing cuts are taken to within 0.005" (0.12mm) of the finished dimension. As the roughing cuts are completed, the tool slides are raised. The cutting tools are replaced by finish cutting tools that have a broad contact surface.

Finish cuts at a depth of about 0.005" (0.12mm) are taken. The feed rate is increased to about two-thirds of the cutting face width. Usually, the same table speed is used for roughing and finish cuts. If a cutting oil is to be used, the entire surface is coated before there is any table movement. The combination of a flat tool form, fine depth of cut, and wide feed rate results in a smooth, finely finished plane surface.

FIGURE 22-16 Setting a Cutting Tool at a Preset Planer Gage Height

MACHINING PARALLEL SURFACES

Once the surface is finish planed to size and burrs are filed from the edges, the whole setup is disassembled. Clamps are fastened to the part so that it may be removed by crane from the planer table. Generally, a shallow wooden skid is brought next to the planer to provide a platform on which the part may be placed and turned after the sling and clamps are removed. If the workpiece design does not permit lifting the part without the aid of clamps, they are once more secured in position.

The platen covering is removed and the top surface is wiped clean. The part is repositioned against the front stops and thus located in the desired position on the platen. Once planer jacks, riser blocks, step blocks, or other supports are placed in position, the sling and clamps are removed.

After turning, the part is again positioned against the front stops. The poppet screws apply force against the back of the part. Straps and other clamping devices are again positioned to hold the workpiece flat against the table. Uniform force is applied to all bolts, nuts, and clamping devices.

A check is usually made to see whether the surface to be planed is set parallel to the table. If the surface has been machined, a dial indicator and height gage are used for checking. Readings are taken near the corners. The jacks may require adjustment if there is a variation in the dial indicator reading.

As the straps and work-holding devices are tightened, any changes in parallelism may be easily noted by variations in the dial indicator readings. The amount of stock to be removed may also be determined by noting the beam reading on the height gage. The height gage slide and indicator are then checked against a known planer gage setting (Figure 22-16).

The reading on the beam at the planer gage height is subtracted from the reading on the workpiece. The result indicates the approximate height. Once a roughing cut is taken for a short distance, the measurement and checking processes are used to establish the precise height of the workpiece. Roughing and finish cuts are taken in the same sequence as for machining the first side. At the end of the finish cut to the required dimension, the burred edges produced by planing are removed.

SIMULTANEOUS PLANING WITH A SIDE HEAD

The preceding discussion focused on the planing of parallel top and bottom surfaces. Attention was directed toward locating the front edge against stops. When a side head is used, steps must be taken to position the workpiece against side stops. The surface to be machined must be parallel to the edge (side face) of the planer table. The front stops are used to prevent any forward movement under heavy cutting forces. The side stops position the workpiece in relation to the cutting action of the side head cutting tool. Once the workpiece is positioned, the side head cutting tool feeds and speeds are similar to the feeds and speeds used with the rail heads.

PLANER CUTTING TOOLS AND HOLDERS

Planers are used for machining large, heavy workpieces and gang planing multiple parts. The operations require cutting tools and holders that are designed to withstand the machining conditions produced by considerable stock removal at greater depths and at higher cutting speeds than most shaping processes.

Tool geometry for planer cutting tools is similar to shaper tools. Clearance angles are reduced to a minimum of 4° to 5°. This angle provides maximum support behind the cutting edge. Carbide tools are ground from 0° to negative top rake angles away from the working area of the cutting tool edge. The negative rake distributes the initial shock at the start of a cut so that it is absorbed a short distance from the cutting edge.

Planer toolholders are generally drop forged of nickel chromium steel. The holders are heat treated for maximum toughness. Some holder designs have a separate *taper seat*. The seat provides a reference surface for holding, adjusting, and replacing a tool bit as the cutting tool. The seat is adjusted to pull the tool bit into position until it is securely locked on three sides: back, top, and bottom. The seat is *serrated* with a series of evenly spaced V-grooves. The serrations prevent the tool bit from moving sideways. The seat is drop forged and tempered and may be adjusted for wear or replaced if damaged.

Four general-purpose planer toolholder designs are illustrated in Figure 22-17. *Straight* planer toolholders (Figure 22-17A) assure rigidity and permit the heaviest cuts to be taken. The straight design makes it possible to position and secure the holder as short as possible (with the least amount of overhang).

Gooseneck planer toolholders (Figure 22-17B) provide an "underhung" tool height. This design of toolholder permits the cutting tool to lift under severe cutting forces and to minimize chatter.

Spring-type planer toolholders (Figure 22-17C) are used for finishing operations. A chatter-free finish is produced because the holder design absorbs vibration.

Universal clapper box planer toolholders (Figure 22-17D) are adjustable to any angle. The head parts may be loosened from the shank to position the cutting tool from 0° (vertical center-

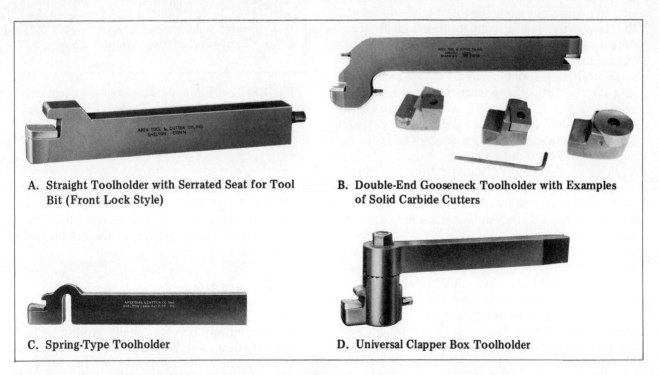

A. Straight Toolholder with Serrated Seat for Tool Bit (Front Lock Style)

B. Double-End Gooseneck Toolholder with Examples of Solid Carbide Cutters

C. Spring-Type Toolholder

D. Universal Clapper Box Toolholder

FIGURE 22-17 Basic Designs of Planer Toolholders

line) for horizontal and simple contour forming cuts to 90° for angle cutting and vertical cuts. The tool bit is secured on the model illustrated by turning a rear lock screw.

Other planer toolholders, classified as special-purpose holders, are designed for specific processes. *Adjustable* toolholders are used for planing T-slots. *Multiple* toolholders permit gang planing or simultaneously planing machine bed V-ways. *Double* toolholders provide for cutting at two depths at the same time. *Reversing* toolholders are used for planing on the forward and return strokes on planers designed for such cutting action.

Adjustable, Replaceable, Serrated Planer Tool Bit Geometry

The tool bits for planer toolholders with serrated seats have a corresponding series of grooves. The grooves permit a tool bit to be adjusted sideways in steps of 0.060″ (1.5mm) and to compensate for wear. Also, the tool bit can be changed without disturbing the original setup.

Tool bits are available in high-speed steel, cobalt high-speed steel, cast alloys, and all grades of carbides. Standard steel serrated tool bits are hardened, tempered, and ground. The end of each tool bit is preformed to a particular shape. Tool bits are supplied in sets in sizes such as #4, #6, #7, #8, and so on. The higher the number, the larger the features are proportionally. The design of each numbered size provides for interchangeability with all styles of holders of the same size. Each tool bit may be inserted and held in a standard straight, gooseneck, or spring-type holder. The cutting tools in a set include right- and left-hand forms, a straight parting tool; full-width, flat-nose tool; diamond-point tool; and straight round-nose tool.

The manufacturer's specifications for three selected size #8, replaceable, serrated-base planer tool bits are shown in Figure 22-18. Clearance and rake angles, form, and sizes are indicated to show the tool geometry of a right-hand, hog-nose tool bit; an offset, flat-nose tool bit; and a right-hand, dovetail tool bit. The cutting tool materials are high-speed steel and cobalt steel.

Full-radius and other special-form tool bits are available in high-speed steels or cobalt steels or as carbide-tipped tool bits. The term *plug* is used with carbide inserts (cutting tool bits). The plugs are generally square or round and are designed with either positive or negative clearance.

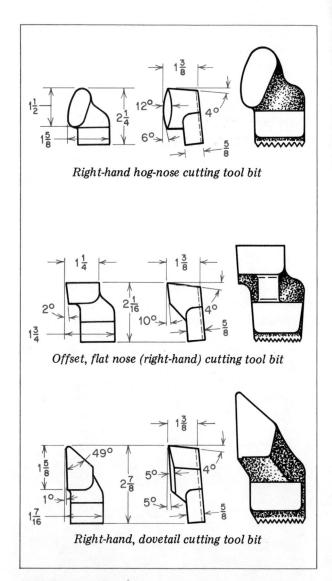

FIGURE 22-18 Manufacturer's Specifications for Selected Replaceable, Serrated Planer Tool Bits

Still other designs of tool bits and holders include *lock-type serrations*. These serrations run at 90° to each other. The cutter serrations match similar serrations in the top of the holder and in the locking seat.

Standard Holders for Tool Grinding

Serrated tool bits are usually ground by locking them in a hand grinding holder for offhand grinding. Tools that are to be ground on a surface grinder or a tool and cutter grinding machine are held in a grinding fixture. Fixture grinding is preferred as it is possible to maintain correct, constant cutting angles with the least amount of stock removal from the cutting tool.

CUTTING SPEEDS FOR PLANER AND SHAPER WORK

Cutting speeds for planer and shaper work are computed in the same manner. The cutting speed in fpm (or m/min) is equal to the product of the strokes per minute and the length of the table stroke in feet (or meters). For example, given the length (L) of a planer table stroke of 20′ (6.1m) and 6 strokes (S) per minute, the cutting speed (CS) is calculated as follows:

$$\begin{aligned} CS &= L \times S \\ &= 20 \times 6 \\ &= 120 \text{ fpm } (36.6 \text{ m/min}) \end{aligned}$$

Occasionally, there is a shop problem where the strokes per minute and the length of the cutting stroke are not known. In the absence of planer and shaper tables, cutting speeds must be computed by using handbook tables for turning or milling ferrous and nonferrous metals. Factors for depth of cut and feed rate are applied to the cutting speed table values:

$$\begin{pmatrix}\text{required}\\\text{cutting}\\\text{speed}\end{pmatrix} = \begin{pmatrix}\text{table value}\\\text{in fpm (or}\\\text{m/min) for}\\\text{turning or}\\\text{milling}\end{pmatrix} \times \begin{pmatrix}\text{table}\\\text{factor}\\\text{for}\\\text{depth}\\\text{of cut}\end{pmatrix} \times \begin{pmatrix}\text{table}\\\text{factor}\\\text{for}\\\text{feed}\\\text{rate}\end{pmatrix}$$

The cutting speeds as given in the tables already include the structure, machinability, hardness of the workpiece itself, and the material in the cutting tool. The planer operator must be concerned with other factors as well. Personal judgments must be made about cutting speeds based on the following factors:

—Rigidity and power of the planer;
—Size and nature of the workpiece to withstand great cutting forces;
—External and internal surface condition and the structure of the casting, forging, weldment, or other heavy-mass part;
—Rigidity of the setup;
—Shape and size of the cutter and holder;
—Tool contact.

Planer operators usually start a new operation at a lower cutting speed. When it is established that the setup and cutting action permit operating at a higher cutting speed, the table strokes per minute may be increased.

When a decision is to be made between reducing the depth of cut or the feed rate, it is good practice to cut down the feed rate. The wider cut distributes the load over more of the tool cutting edge. Thinner chips are formed and are controlled more easily than chips produced by a coarser feed rate.

HOW TO SET UP THE WORKPIECE AND PLANER AND MACHINE PARALLEL SURFACES

Preparing to Machine the First Surface

STEP 1 Determine the safest method of handling and placing the workpiece.

STEP 2 Clean and remove any burrs or grind away any surface imperfections from the surface opposite the one to be machined.

Note: If required, a protective metal or hard plastic material is placed over the platen area in which a rough-surfaced part is to be located.

STEP 3 Attach clamps to a heavy workpiece when there is no other way of lifting it. Use a cable or heavy rope sling. Handle and place the workpiece on the platen by using a crane.

Caution: Attach the cable so that the eyes swivel freely to avoid twisting the clamps.

STEP 4 Place one or more stops in front of the position at which the workpiece is to be located.

STEP 5 Position the workpiece on the platen as close as possible to the stops.

STEP 6 Disassemble the sling and clamps when used. Locate poppets and planer pin screws in back of the part. Tighten the pins to force the part against the front stops.

Note: If vertical cuts are to be taken, the part must be aligned parallel to the side of the platen before the poppet screws are tightened.

Note: Steps 2 through 6 are repeated if multiple parts are to be machined at one time.

STEP 7 Select suitable straps, T-nuts and bolts, and packing and step blocks. Place them between workpieces and at positions where the greatest amount of force may be applied.

Note: The setups must be tightened uniformly. Shims are used to level off any uneven surface area. The tightness of each strap and the rigidity of the setup should be rechecked.

Setting the Tool Head

STEP 8 Set the tool head vertically for general horizontal cuts. If angle cutting is required, set the tool head at the specified angle.

STEP 9 Select the appropriate cutting tool and holder, if necessary. Place and secure the cutting tool to the clapper box.

Note: A tool bit and holder or solid-shank cutting tool is held with the least extension below the bottom edge of the clapper box. The tool head is raised on the saddle so that it extends a minimal distance.

STEP 10 Move the cutter away from the workpiece. Release the rail locking clamps.

STEP 11 Lower the rail so that the cutting tool clears the top surface of the workpiece. Then, raise the rail slightly to take up backlash on the adjustment screw.

STEP 12 Tighten the back rail clamps on the cross rail locking bolts to secure the cross rail apron to the front of the column.

Setting the Overall Travel Stops and Platen Speeds

STEP 13 Set the cutting and return stroke speeds at a minimum. Start the planer.

STEP 14 Set the stop at the beginning of the stroke with adequate clearance to permit feeding the tool before each cut begins.

STEP 15 Set the return stop so that the cutter clears the end of the cut.

STEP 16 Start the platen in motion.

STEP 17 Adjust one hydraulic cylinder control valve for the cutting rate. Turn the other control valve for the rate of return.

Tooling Setups and Feed Rate

STEP 18 Continue with the platen in motion to set the rate of feed. On hydraulic planers, adjust the feed rate valve until the dial registers the required feed.

STEP 19 Stop the planer. Feed the rail head manually across the workpiece to clear the starting position. Turn the tool slide handwheel to lower the cutting tool to a roughing cut depth.

STEP 20 Lock the tool slide in place. Start the planer. Take the first cut.

Caution: If a cutting fluid is to be used, direct a stream from a squeeze container. Under no circumstances is cutting oil to be brushed on.

STEP 21 Continue the cut across the second part if a number of workpieces are mounted side by side.

STEP 22 Stop the machine after the roughing cuts. Replace the roughing cutter with a broad (slightly curved cutting edge) finish cutting tool. Reset the cutter and holder.

STEP 23 Change the feed rate to from one-half to two-thirds the width of the cutter. Leave the platen speed at the same rate as for the roughing cuts.

STEP 24 Spray or coat the cutter surface with a cutting oil, if required. Then, start the planer. Take the finish cut.

STEP 25 Stop the planer. Clean the workpiece and machine area. File any burrs produced by machining.

Planing a Parallel Surface

STEP 26 Remove all clamps, poppets and planer pins, and all other work-holding accessories.

STEP 27 Rig the workpiece for removal to a wooden platform. Turn the workpieces over. Remove the platen covering. Return the workpiece to the platen.

STEP 28 Use stops, straps, and work supports and secure the parts to the platen.

Note: The top surface is checked to be sure the straps are not disturbing the part.

STEP 29 Proceed to take roughing cuts and a finish cut. Follow steps 19 through 25.

Note: The overall height to which the part is to be machined may be determined with a height gage and a planer gage.

STEP 30 Remove all work-holding tools and accessories. Return each to its proper storage compartment. Set the table travel and feed rates to minimum as a safety precaution for the next operator.

SAFE PRACTICES IN TOOLING UP AND PREPARING FOR SHAPING OR PLANING

—Stop the machine. Under no circumstances are workpiece adjustments or checking to be made while the shaper ram or planer table is moving.
—Check the position of a rope or cable sling and the secureness of the clamps before hoisting a heavy, bulky part on the planer table.
—Hook the sling so that it is close to the center of gravity of the workpiece.
—Stand clear of a part that is being moved to or set on the platen. Keep hands away from the underside of the workpiece. Be ready to instantly move safely away from the setup in the event of danger.
—Use stops against the front end of a workpiece to take up the cutting thrust. This practice is preferred to relying on clamps to absorb such forces.
—Protect the platen, table, or other work-holding device against scratching, burring, or denting. These conditions are caused when rough castings, weldments, and rusted or rough surfaces are secured to the machined face. Use a hard, fibrous material between a rough surface of the workpiece and a finish machined surface.
—Use protective strips between jacks and any finished surface of a workpiece.
—Position the safety dogs on a planer to keep the table from moving past the table length.
—Move the shaper ram or planer table manually (or at an extremely slow speed) through at least one forward and reverse stroke. Check the clearance of the cutting tool at the start and end of the cutting cycle. Make a special check for the clearance between the tool head and/or ram head, the workpiece, and/or parts of the machine when making angular machining setups.
—Determine the fpm or m/min cutting speed limitations. Use carbides at the cutting speeds required for effective operation.
—Select a carbide cutter with negative rake angle to permit the impact of the cut to be absorbed over a wide area of the cutting edge. The point of the carbide tool is its weakest area.
—Set the depth for the first roughing cut on a cast iron casting below the outer scale.
—Check for excessive cutting speeds, particularly on long strokes, where tremendous momentum and forces act to shift the setup or move the cutting tool, resulting in possible damage to the workpiece and/or machine.
—Place a screen or chip guard in front of a shaper to contain the hot chips.
—Use a protective screen for shaper operations when the ram at its extreme limit of reverse travel projects into an aisle.
—Remove all excess tools, fastenings, and other parts from a table, vise, or platen prior to manually moving through the first cycle or the startup of an operation.
—Keep areas around a shaper or planer clear of chips and free of oil. Use a liquid absorbent compound on the floor around the machine to keep the area dry and free from slippage.
—Wear safety glasses or a protective shield during setup and machining processes.

TERMS USED FOR SHAPER AND PLANER DESIGN FEATURES, TOOLING, AND BASIC PROCESSES

Vertical shaper — A machine tool for shaping parts by taking a series of cuts with a reciprocating vertical ram. A machine tool designed to machine internal and external flat, angular, circular, and contoured surfaces.

Universal horizontal shaper — A shaper model having a secondary table that may be rotated to position a workpiece at a compound angle to the horizontal plane.

Tool head (shaper) — A complete mechanical component mounted on the head end of a ram that consists of a slide block actuated by a lead screw and handwheel, a clapper box with a clapper block, and a tool post.

Positioning the clapper box (shaper) — Moving the clapper box at an angle to permit a cutting tool to clear the surface being machined on the noncutting stroke.

Offsetting the apron (shaper) — Positioning the apron and clapper box at a small arc on the tool slide. An angle position of the apron that permits the cutting tool edge to clear the workpiece on the return stroke.

Shaping a dovetail — Parallel angular machining of the sides of a groove. Angular shaping of the sides of a groove (internal) and a mating (external) part. Precise angular shaping of internal and external angular features.

Contour shaping — Producing a dimensionally accurate formed surface by shaping. Controlling the down-feed and/or up-feed of the cutting tool in relation to cross feeding a workpiece.

Double-housing planer — A planer with the two vertical columns connected at the top.

Open-side shaper-planer — A planer with one vertical column and a horizontal cross rail. A planer adapted to machining wider workpieces than can be accommodated by a double-housing planer. A manufacturer's designation of a popular size (42″ width × 42″ height, or approximately 1m × 1m) of open-side planer.

Platen — The work surface of a planer on which parts are mounted and held securely for machining. A flat, precision-machined table. A planer table fitted with T-slots to receive T-nuts and bolts and with holes in which stops and poppets are placed.

Cross rail — A horizontal machine element upon which one or more rail heads are mounted. The horizontal ways on which rail heads may be moved transversely.

Rail and side heads — Machine components each consisting of an apron, tool head, tool slide, and clapper box. A vertical-mounted cutter-holding head and travel unit and a side-mounted head.

Speed control relief valve — A valve that controls the cutting and return speeds of the platen.

Planer gage — An adjustable, precisely ground, rectangular gage used to set planer tools and to lay out and gage parts for dimensional accuracy.

Feed control valve — A hand adjustment valve for controlling the cutting tool feed rate on a shaper or planer.

Gooseneck planer toolholder — A toolholder with an offset end. An underhung holder that permits the cutting edge to be moved slightly backward under extreme cutting forces.

Serrated tool bit — A replaceable, specially formed cutting tool with a serrated bottom face. A tool bit with serrations to match the serrations of the seat.

DESIGN FEATURES, TOOLING, SETUPS 439

Universal clapper box toolholder (planer) — A toolholder that provides for the adjustment of a replaceable cutting tool from 0° to 90° to the plane of the platen. An adjustable toolholder used on planers for taking horizontal, vertical, angular, and contour form cuts.

Toolholder seat (planer) — A separate part of a toolholder for seating, adjusting, and securely holding a replaceable tool bit.

SUMMARY

The vertical shaper is adapted to shaping flat, angular, round, and contour shapes by cutting in vertical planes.

> Table feed movements on a vertical shaper are toward or away from the column, transverse, and rotary.

Horizontal shapers may be plain or universal, depending on the number of planes in which surfaces are to be shaped.

> The four basic adjustments of a shaper include vertical movements of the saddle and table, transverse movements of the table, horizontal movement of the ram, and vertical and angular movements of the tool slide.

Standard clamps, step blocks, parallels, V-blocks, jacks, and T-bolts are used to position and secure parts to the table. Hold-down wedges force the workpiece onto parallels or a vise or table surface. The wedges permit holding a workpiece securely.

> The clapper box is positioned at a slight arc to permit a cutting tool to clear a machined surface on the return, feed stroke.

Standard straight and right- and left-hand offset toolholders are used to secure tool bits on the shaper. The tool slot is at 0°. In addition, a solid-shank multiposition shaper-planer toolholder is available. This toolholder permits multiple tool settings for straight, vertical, and angular cuts.

> Vertical surfaces are shaped on workpieces that generally are held in a vise. The cutting tool is fed vertically for roughing and finish cuts. The toolholder and apron are set to permit a cut to be taken and the cutting tool to clear the machined surface on the return stroke.

Contour shaping requires the positioning of the cutting tool in relation to the transverse movement of the table and workpiece. The down-feed or up-feed of the cutting tool is coordinated with the transverse movement of the workpiece.

> Tracer attachments permit the precise forming of interchangeable contour-shaped parts.

Compound angle shaping requires a combination of angle settings of a swivel table and/or the universal table rocker block.

> Angle cuts are produced by swiveling the tool head on a shaper to the required angle, offsetting the apron at a small arc, and down-feeding.

Cutting of multiple grooves and splines requires the use of a direct-positioning index head.

> The capability of the 1817 Roberts's planer contributed significantly to the growth of the machine tool industry.

The open-side shaper-planer permits the machining of parts wider than the platen width.

Planers are designed with single or double rail heads, a side head, and milling and grinding heads for milling and grinding in addition to the regular planer processes.

Contour and spiral work surfaces may be generated by using templates, indexing and footstock accessories, and special attachments.

The major planer components include a bed, platen, column(s), cross rail, and rail and side heads.

Hydraulic systems provide smooth driving power for table (platen) movement, cross feeds, vertical feeds, and side head feeds. Simultaneous cuts may be taken from one or more rail heads and a side head.

Tool lifters may be hydraulically operated to clear the cutting tool on the return stroke.

Stop pins and poppets are placed in holes in the platen. Workpieces are held against the pins with toe dogs to prevent movement during the cutting stroke.

The planer gage is used with other measuring and layout tools to position the cutting tool and measure work heights.

The handling of heavy workpieces requires the safe use of work clamps, slings, and crane equipment near the center of gravity of a workpiece.

Caution must be observed when mounting tools. After positioning, the rails, rail head, and side head are locked before a cut is taken.

Cutting speeds can be computed by using standard turning or milling handbook tables and applying depth of cut and feed rate factors.

One or a set of hydraulic valves on a shaper or planer (depending on the model) are turned to regulate the cutting and return platen speeds.

Flat surfaces are finish planed with a broad, slightly crowned cutting edge. A coarse feed that is about two-thirds the width of the cutting edge, is used. The cutting speed is the same as for a roughing cut.

Possible distortion due to clamping is checked as force is applied to the fastening devices. A height gage and dial indicator are generally used on smooth surfaces to check against any movement.

Planer toolholders are generally designed with seats. The seats permit adjustments and replacement at an original setting without changing the holder position.

Straight, gooseneck, spring, and universal clapper box planer toolholders are used for general-purpose planing. The combination provides the versatility for taking horizontal, vertical, angular, and form cuts.

Adjustable, replaceable, serrated planer tool bits are furnished in numbered sets. Tool bits of the same number are interchangeable in any type of holder within the set.

Planer tool bits are preformed and ground and are available in standard grades of high-speed steel and cobalt high-speed steel and as carbide-tipped cutters.

The main processes to be performed in setting up for planer work include the safe positioning and securing of the parts on the platen, setting the tool head, setting the overall travel stops and adjusting the platen speed, tooling up and setting the feed rate, and disassembling and setting up for planing a second side.

DESIGN FEATURES, TOOLING, SETUPS 441

Machine safety relates to securely holding work against cutting forces, controlling machine speed, protecting cutting tools against shock or scraping, and following other general safe machine operation practices.

UNIT 22 REVIEW AND SELF-TEST
---

1. List three operations that are best performed on each of the following three machine tools: (a) shaper, (b) planer, and (c) slotter.
2. a. Name three major parts of a tool (swivel) head on a shaper.
   b. List the steps to follow in swiveling a tool head to cut a 30° angle surface.
3. Explain why most carbide tools used for roughing shaper and planer cuts have a negative back-rake angle.
4. The corner pad shown below is to be rough machined by shaping all surfaces and allowing 0.4mm for each finish cut. Carbide tools are used.
   a. Compute the cutting speed in m/min and ram strokes per minute for rough machining the 148.50mm length.
   b. Shape all sides square and parallel within the tolerances specified.

   Note: The two ends are to be vertically shaped. The tool head is swiveled to cut the 45° angle.

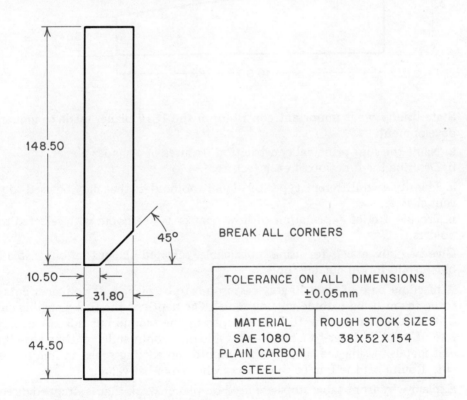

5. a. Prepare a work production plan for shaping the cast iron angle form block shown on page 442. A standard shaper is used. The workpiece is to be positioned and held in a swivel vise.
   b. Determine the angle for setting the vise.
   c. Calculate (1) the cutting speed in fpm and (2) the ram strokes per minute.

The casting size allows for one approximately 3/16" deep roughing cut and a 0.010" finishing cut on each surface. The feed for the roughing cut is 0.060". The feed factor of 1.00 may be used for computing the cutting speed of the final cut. High-speed roughing and finish cut cutting tools are used.

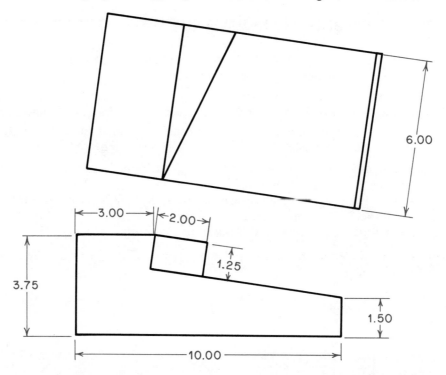

6. State briefly what important contribution the 1817 planer made to industrial development.
7. a. Name the four principal construction features of a shaper-planer.
   b. Describe the function of each feature.
8. a. Identify two different types of planer toolholders that may be used to prevent chatter.
   b. Provide a brief explanation of how chatter is overcome with selected toolholders.
9. Give two advantages for using replaceable, serrated planer tool bits instead of older, solid-forged cutting tools.
10. A machine base of class 40 gray cast iron is to be planed. Two planer cuts are to be taken using carbide cutting tools. The depth of the first roughing cut is 3/8" at a feed rate of 1/16". The depth of the second cut is 3/16" at a feed rate of 1/32". Refer to a handbook table of recommended cutting speeds for cast metals. Using the table fpm, compute the cutting speed in fpm for each cut. Round off the fpm to the nearest whole number value.
11. Explain why front table stops are used on planer work that is strapped directly on the platen.
12. Cite one machine or tool safety precaution to take for each of the following conditions: (a) setting the ram or platen stroke, (b) setting a solid-shank tool or toolholder, (c) swiveling a tool slide at a steep angle on a shaper, and (d) machining at an angle on a shaper.
13. State three personal safety precautions the operator follows after completing a shaper or planer job.

PART SEVEN

Grinding Machines

SECTION ONE   Functions and Types of Grinding Machines
         UNIT 23   Grinding Machine Technology and Processes
SECTION TWO   Grinding Wheels: Standards and Preparation
         UNIT 24   Grinding Wheel Characteristics, Standards, and Selection
         UNIT 25   Grinding Wheel Preparation and Grinding Fluids
SECTION THREE   Horizontal-Spindle Surface Grinding Technology and Processes
         UNIT 26   Surface Grinder Design Features and Setups for Flat Grinding
         UNIT 27   Grinding Angular and Vertical Surfaces and Shoulders
         UNIT 28   Form (Profile) Grinding and Cutting-Off Processes
SECTION FOUR   Cylindrical Grinding Technology and Processes
         UNIT 29   Cylindrical Grinder Components and External and Internal Grinding
SECTION FIVE   Cutter and Tool Grinding Technology and Processes
         UNIT 30   Cutter and Tool Grinder Features, Preparation, and Setups
         UNIT 31   Grinding Milling Cutters and Other Cutting Tools

SECTION ONE

Functions and Types of Grinding Machines

Grinding machines are machine tools that use a natural or artificial abrasive to cut material. Grinding is a practical, efficient, and essential process for both rough and finish machining.

This section deals with the types and construction features of grinders used in abrasive machining and precision grinding processes. The characteristics of abrasives and feeding techniques to accommodate a wide variety of work parts are described. Principles of cutting action relating to surface, centerless, cylindrical, form, plunge, and tool and cutter grinding are examined. Microfinishing, honing, lapping, and superfinishing processes are covered.

UNIT 23

Grinding Machine Technology and Processes

Grinding is one of seven basic machining processes. The other six processes (and related machine tools) include turning (lathe), drilling, (drill press), boring (boring mill), shaping (shaper or planer), milling (milling machine), and sawing and filing (band sawing machine). These seven machining processes are illustrated in Figure 23-1A through G. Each machine tool is unique in terms of the following:

—Construction features,
—Types of tools employed to do the cutting,
—Methods of bringing the workpiece into contact with the cutting tool,
—Geometry (principles) of cutting action.

This unit describes the design features, tools, feeding methods, and cutting principles applicable to grinding machines.

CHARACTERISTICS OF ABRASIVES AND GRINDING

Grinding, as an essential machining process, is used on both hard and soft metals and nonmetallic parts. A part may be rough ground taking hogging cuts to depths of 1/2" (12mm) in a single pass. By contrast, final finish grinding processes are performed to accuracies in the millionths of an inch and metric equivalents.

FUNCTIONS AND TYPES 445

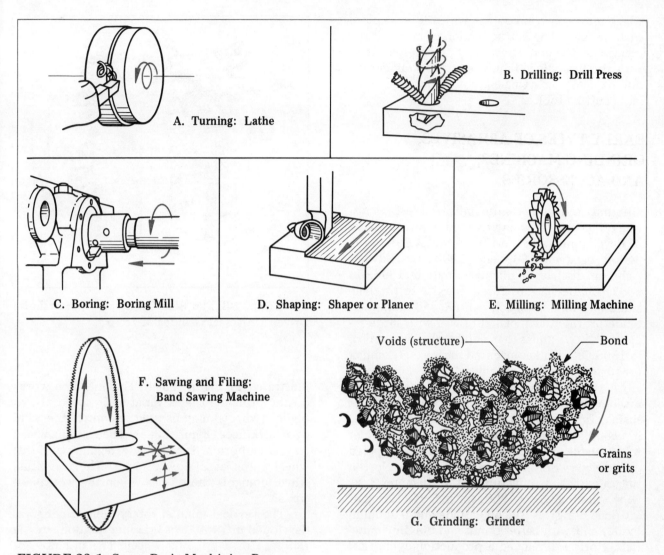

FIGURE 23-1 Seven Basic Machining Processes

Abrasive cutting tools have thousands of cutting edges (grains). Each sharp grain that comes in contact with the workpiece cuts away a chip. Chips magnified 350X are shown in Figure 23-2. The many cutting edges provide continuous cutting points. Abrasive cutting tools are capable of generating a smoother surface than any other machine tool. Also, the surface generated is dimensionally precise; therefore, grinding processes are widely used in finishing workpieces.

Grinding wheels operate at speeds of 6,000 to 18,000 sfpm (1,830 to 5,490 m/min). These speeds are roughly equivalent to 1.1 to 3.4 miles (1.8 to 5.4km) per minute. Usually, the distance between abrasive grains is less than 1/8" (3mm). On a 1" (25mm) wide grinding wheel, over 4,500,000 (at 6,000 sfpm) tiny, sharp cutting

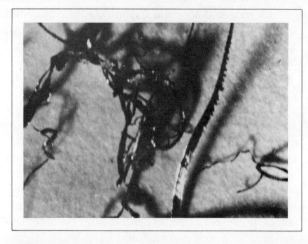

FIGURE 23-2 Chips Produced during Grinding (Magnified 350X)

edges may come in contact with the workpiece each minute. The cutting action of many of the grains and the comparatively fine chips contrast with the heavy feed (chips per tooth) taken by a milling cutter or a single-point shaper, planer, or lathe cutting tool.

EARLY TYPES OF ABRASIVES, GRINDING MACHINES, AND ACCESSORIES

Grinding, today, meets the industrial tests of cutting efficiency, productivity, and quality because of developments in the United States during the period between 1860 and 1900. Dr. Edward Acheson discovered silicon carbide in 1896. In 1897, Charles Jacobs fused bauxite to produce aluminum oxide. Silicon carbide and aluminum oxide provided artificial abrasives with more reliable abrasive grains than natural abrasives. Also, in 1896 O.S. Walker invented the magnetic chuck for surface grinding machines.

A grinding machine that produced accurately machined cylindrical unit pieces with excellent surface finish was first made in the United States in 1860. Charles Norton's cylindrical grinder was capable of grinding cylinders and tapered cylindrical workpieces. In the 1880s, Norton suggested the manufacture of a 1′ diameter × 1″ wide grinding wheel. With the manufacturing of artificial abrasives (silicon carbide and aluminum oxide) and with better bonds and bonding materials, Norton developed a production cylindrical grinder in the 1890s. However, on June 9, 1862, the Brown & Sharpe Manufacturing Company had already patented a universal grinding machine. The universal features extended the applications of external and internal cylindrical grinding and tool and cutter grinding to one machine. An 1876, overhead flat-belt driven universal grinder is pictured in Figure 23-3.

Early in the 1900s, grinding machines became more ruggedly built and more versatile. At the same time, industrial production was moving toward extremely fine surface finishes held to close dimensional tolerances. Rapid improvements of grinding machines, abrasives, grains and wheels, and machine accessories made it possible to meet precision high-production needs of automotive and other industries. The outside diameter of parts were production ground on a centerless grinder developed around 1915 by L.R. Heim. A modern, high-production centerless grinder is

FIGURE 23-3 Flat-Belt (Overhead Countershaft) Driven Universal Grinder of 1876

illustrated in Figure 23-4. This machine works on the principle that a variation in speed of two bodies (wheels) may be used to control the RPM of a workpiece (Figure 23-5). The grinding wheel travels at normal speeds. The regulating (feed) wheel turns at a slower speed. The regulating wheel controls the rate at which the workpiece turns.

The vertical-spindle, rotary-table grinder was developed in 1900. The grinding was done by the flat face of a grinding wheel. The principle of grinding on a vertical-spindle, rotary grinder is shown in Figure 23-6. Large sizes of grinding wheels are designed with segments of abrasive blocks. They are strapped to a wheel (chuck) that is attached to the vertical machine spindle. This type of grinder is a widely used tool for the fast grinding (rapid removal) of large volumes of metal and other materials. The rotary table is a work-holding device. It provides an ideal machine feature for many shapes of workpieces.

Developments of the horizontal-spindle surface grinder also occurred in the last quarter of the 1800s. This machine had many similar design features to the milling machine. The Brown & Sharpe surface grinder of 1887 (Figure 23-7) had a knee elevating screw and a handwheel for adjusting the table height. The table crank permitted comparatively rapid longitudinal movement of the workpiece past the grinding wheel. The handwheel at the top of the column controlled

FUNCTIONS AND TYPES 447

FIGURE 23-4 Modern, High-Production Centerless Grinder

the transverse feed of the wheel. This basic form of grinder evolved into the modern type I surface grinder.

ABRASIVE MACHINING

Abrasive machining refers to heavy metal-removing operations that previously were performed with conventional cutting tools and machines. Abrasive machining may be done using abrasive grinding wheels or coated abrasive belts.

Three general criteria are used to classify abrasive machining:

—Abrasives remove material more economically to a required finish and tolerance than regular cutting tools;

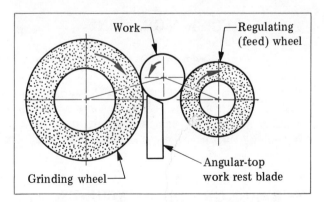

FIGURE 23-5 Principle of Controlling RPM of Workpiece on a Centerless Grinder (Work Set above Center)

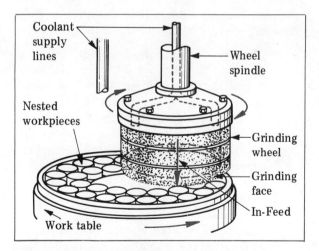

FIGURE 23-6 Principle of Grinding on a Vertical-Spindle, Rotary Grinder

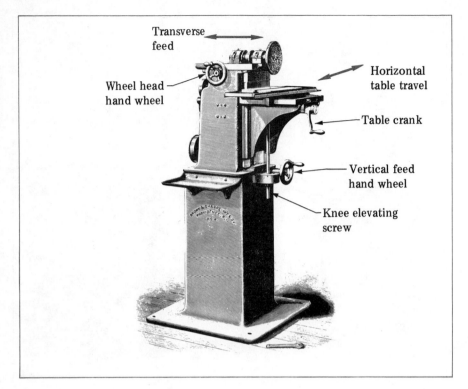

FIGURE 23-7 Design Features of 1887 Brown & Sharpe Surface Grinder

—Metal removal from 1/16" to 1/2" is required;
—Abrasives are used on a rough casting, forging stock, or a welded area to produce the finished surface.

Advantages of Abrasive Machining

Many cost comparison tests have been made between abrasive machining and cutting operations on different types of milling machines, planers, shapers, and lathes. Six advantages are generally cited for abrasive machining:

—Machine tools are designed with *automatic sizing* of the abrasive wheel to compensate for cutting tool wear (downtime or machine shutdown, normally required for sharpening cutting tools is thus eliminated);
—Parts handling, loading, and unloading costs are reduced (magnetic chucks practically eliminate the need for clamping forces, and the downward force of the abrasive wheel on a surface grinder serves as an additional clamping force);
—The costs of fixtures may be reduced or eliminated (for example, the magnetic table of a surface grinder serves as a practical clamping surface for many jobs);
—Abrasive wheels cut through the outer scale of castings and forgings, hard spots, and burned and rough edges due to welding processes (single abrasive machining operations may replace multiple roughing cuts on such rough and hard surfaces);
—Thin-walled areas of workpieces may be abrasive machined with less springing or breakage;
—Casting and forged part design may be simplified (normal allowances of extra stock in order to cut under the scale may be reduced).

BASIC PRECISION GRINDING

Precision grinding processes are generally grouped into six categories: surface, cylindrical (external and internal), form, plunge, centerless, and tool and cutter grinding. (Nonprecision grinding is usually associated with manual, hand control of a tool or cutter.) Bench and floor grinders are used for precision grinding processes.

SURFACE GRINDING MACHINES

Essentially, *surface grinding* deals with the generating of an accurate, finely finished, flat (plane)

FUNCTIONS AND TYPES 449

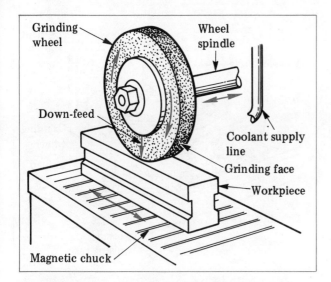

FIGURE 23-8 Typical Setup and Principle of Grinding on a Type I Surface Grinder

surface. It is also possible to form the wheel to conform to a particular shape. The work may also be positioned at different angles. These setups produce formed, angular, and other "straight" surfaces.

The four basic types of *surface grinders* are:

—*Type I* (horizontal spindle, reciprocating table);
—*Type II* (horizontal spindle, rotary table);
—*Type III* (vertical spindle, reciprocating table);
—*Type IV* (vertical spindle, rotary table).

Type I (Horizontal-Spindle, Reciprocating-Table) Surface Grinder

The type I surface grinder is the most commonly used surface grinding machine. Flat, shoulder, angular, and formed surfaces are ground on the machine. Workpieces are generally held by a magnetic chuck on the table. The table is mounted on a saddle. Cross feed for each stroke is provided by the transverse movement of the table. The workpiece is moved longitudinally, under the abrasive wheel on the reciprocating table. The typical setup and principle of grinding on a type I surface grinder are shown in Figure 23-8. The grinding wheel down-feed is controlled by lowering the wheel head. The amount of feed is read directly on the graduated micrometer collar.

FIGURE 23-9 General-Purpose 6" × 12" Type I Surface Grinder

Depending on the material used and nature of the operation, grinding may be done dry or wet. All general-purpose machines are designed either with or for the attachment of a coolant supply system. Dry grinding requires an exhaust attachment to carry away dust, abrasives, and other particles.

On some models, the grinding head and wheel are moved across the work surface. This movement is similar to the movement of the 1887 Brown & Sharpe grinder (pictured earlier in Figure 23-7). Most type I surface grinders (Figure 23-9) are 6" × 12" (150mm × 300mm). This designation indicates that workpieces up to 6" (150mm) wide × 12" (300mm) long may be surface ground. Standard heavy-duty type I machines are produced to grind surfaces 16' and longer.

By dressing the wheel, the face and edges may be cut away in reverse form to grind to a required shape. Radius- and other contour-forming devices and diamond-impregnated formed blocks are used. The formed wheel then reproduces the desired contour.

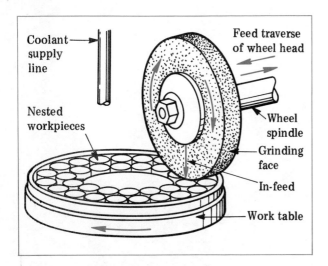

FIGURE 23-10 Principle of Grinding on a Type II Surface Grinder

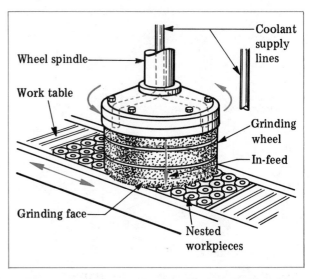

FIGURE 23-11 Positioning of Workpieces and Cutting Action of a Type III Surface Grinder

Type II (Horizontal-Spindle, Rotary-Table) Surface Grinder

The type II surface grinder produces a circular scratch-pattern finish. This finish is desirable for metal-to-metal seals of mating parts. Workpieces are held on the magnetic chuck of a rotary table. The parts are nested so that they rotate in contact with one another. The grinding wheel in-feed is measured by direct reading on the wheel feed handwheel micrometer collar. The traverse movement of the wheel head feeds the grinding wheel across the workpieces. The principle of grinding on a type II surface grinder is shown in Figure 23-10. The table may be tilted to permit grinding a part thinner either in the middle or at the rim.

Type III (Vertical-Spindle, Reciprocating-Table) Surface Grinder

The type III surface grinder, designed with a vertical spindle and a reciprocating table, grinds on the face of the wheel. The work is moved back and forth under the wheel. The positioning of workpieces on a magnetic table and the cutting action of a type III surface grinder are pictured in Figure 23-11.

Type III grinders are capable of taking comparatively heavy cuts. The wheel head is designed so that it may be tilted a few degrees from a vertical position. This design permits faster cutting and metal removal as the rim of the grinding wheel contacts the workpiece. In a vertical position, the surface pattern produced is a uniform series of intersecting arcs. A semicircular pattern is generated when the wheel head is tilted.

Adaptations of Type III Surface Grinders. Many production type III grinders are fitted with auxiliary spindles. These spindles permit a great number of different surface grinding operations to be performed without changing the setup of the workpiece. Figure 23-12 shows examples of surface grinding processes using auxiliary spindles positioned vertically, at an angle, or with the wheel dressed. Straight, angular, and dovetailed surfaces may be ground. Auxiliary spindles are used largely with heavy-duty or special multiple-operation surface grinders.

Type IV (Vertical-Spindle, Rotary-Table) Surface Grinder

The type IV surface grinder, designed with a vertical spindle and a rotary table, also grinds on the face of the wheel to generate a flat (plane) surface. Figure 23-13 shows a typical nesting of workpieces on a magnetic chuck and the movements of the wheel head on a type IV surface grinder. The wheel feed handwheel is used to adjust the depth of cut. The grinding wheel is moved across the workpieces by the transverse movement of the wheel head.

FUNCTIONS AND TYPES

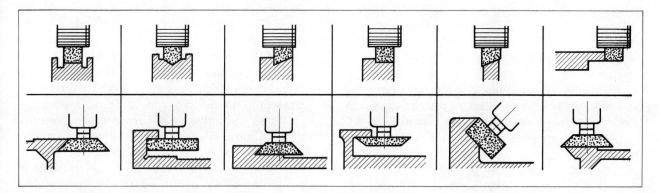

FIGURE 23-12 Examples of Surface Grinding Processes Using Auxiliary Spindles Positioned Vertically, at an Angle, or with the Wheel Dressed

GENERAL-PURPOSE CYLINDRICAL GRINDING MACHINES

Cylindrical grinding basically involves the grinding of straight and tapered cylindrical surfaces concentric or eccentric in relation to a work axis. Such grinding is generally performed on either *plain* or *universal cylindrical grinding machines*.

Form grinding refers to the shape. The grinding of fillets, rounds, threads, and other curved shapes are examples. Straight form grinding processes are performed on surface grinders.

Cylindrical grinders are also used for producing straight, tapered, and formed surfaces by moving the grinding wheel into the workpiece without table traverse. The process is referred to as *plunge grinding*. However, there are occasions when the grinding wheel is moved a short distance horizontally across the workpiece.

Plain Cylindrical Grinding Machines

Cylindrical grinding requires the mounting and driving of a workpiece. Mounting a workpiece between centers is the most widely used method (Figure 23-14). The workpiece is mounted so that it rotates on two dead centers. The dog attached to the workpiece at the headstock end is driven by a plate. This plate rotates with the headstock center.

The plain cylindrical grinder is used primarily for grinding straight and tapered cylindrical surfaces and shoulders. The functions and operation

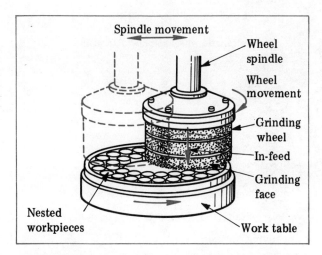

FIGURE 23-13 Nesting of Workpieces and Movements of Wheel Head on a Type IV Surface Grinder

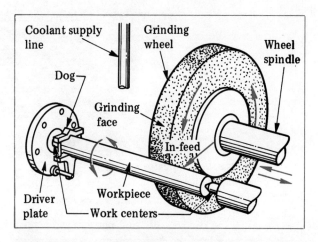

FIGURE 23-14 Principle of Cylindrical Grinding with Workpiece Mounted between Centers

of plain and universal grinding machines are described in detail in later units. One important design feature of the plain grinder is that the wheel head is permanently set at a right angle to the table travel. The headstock may also be stationary. On some models, the headstock may be swiveled up to 45° to permit grinding steep, cylindrical, tapered, and beveled surfaces.

Table travel speed may be controlled manually or by using table speed-selector levers. Plain grinding machines are equipped with table reverse dogs for automatic table reversal. Other design features include an automatic cross feed and controls, wheel slide rapid travel, and six or more rates of power travel.

Universal Cylindrical Grinding Machines

The universal cylindrical grinder has the capability to perform all the operations of the plain cylindrical grinder. In addition, both external and internal cylindrical grinding operations, face grinding, and steep taper and other fluted cutting tool grindings may be performed.

The principal parts of a standard universal cylindrical grinder are shown in Figure 23-15. A strong, heavy base adds to the stability of the machine. The table may be swiveled for taper grinding. The headstock and wheel spindle head may also be swiveled for grinding steep angles and faces. The footstock is provided with a center for holding work between centers. On some machines the internal grinding unit is mounted directly above the wheel spindle head. The versatility of the machine is extended by a number of work-holding devices and machine accessories.

Plain and universal toolroom and commercial shop cylindrical grinders are capable of grinding diameters to within 25 millionths of an inch (0.0006mm) of roundness and to less than 50 millionths of an inch (0.001mm) accuracy. Surface finishes to 2 microinches and finer are produced in the cylindrical grinding of regular hardened steel as well as tungsten carbide parts.

Internal Cylindrical Grinding Machine

The setup for and principle of internal cylindrical grinding are shown in Figure 23-16. A close-up view of the grinding wheel head and revolving workpiece and wheel spindle is shown in Figure 23-16A. The machine features and principle of internal cylindrical grinding are illustrated in Figure 23-16B.

Concentric workpieces may be held in collets and chucks. Irregularly shaped parts, such as an automotive connecting rod, may be positioned and held in a special fixture. Internal grinding

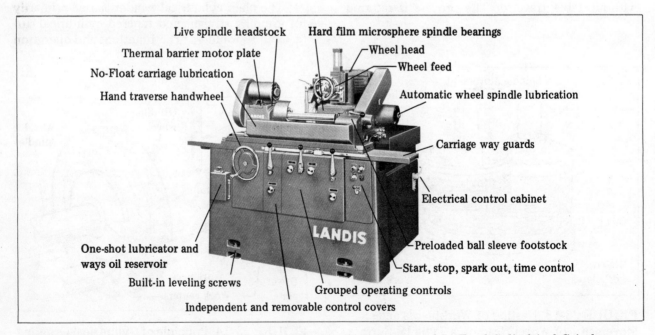

FIGURE 23-15 Principle Parts of a Universal Hydraulic Traverse and In-Feed Cylindrical Grinder

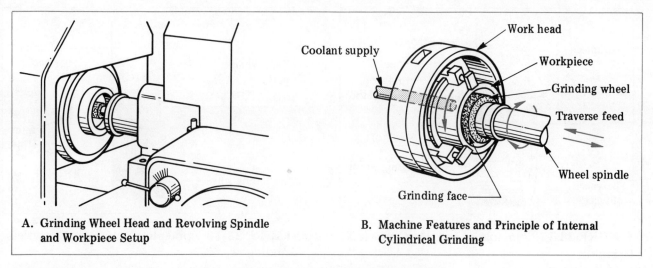

FIGURE 23-16 Setup for and Principle of Internal Cylindrical Grinding

may be done on small-diameter holes with mounted grinding wheels. The diameter should be at least three-quarters of the finished hole diameter. Frequent dressing is required unless a *cubic boron nitride* abrasive is used. This abrasive is a recently developed product of the Specialty Materials Department, General Electric Company. The abrasive bears the trademark of BOROZON™ CBN. While cubic boron nitride abrasives are almost as hard as diamonds, they are an exceptionally good abrasive on hardened steels. Cubic boron nitride wheels require infrequent dressings.

In custom shops where there is only an occasional internal grinding job, the old practice of using a tool post grinder is followed. The direct motor-driven grinder is secured in the tool post of the lathe. The workpiece is mounted on a faceplate, chuck, fixture, or collet. The headstock spindle is set at the required speed to accommodate the hole size and nature of the operation. The grinding wheel is fed into and out of the revolving workpiece.

CENTERLESS GRINDING METHODS AND MACHINES

Centerless grinding machines are primarily used for production machining of cylindrical, tapered, and multiple-diameter workpieces. The work is supported on a *work rest blade*. The blade is equipped with guides suited to the design of the workpiece.

As described earlier (Figure 23-5), the principle of centerless grinding requires the grinding wheel to rotate and force the workpiece onto the work rest blade and against the regulating wheel. This wheel controls the speed of the workpiece. When the wheel is turned at a slight angle, it also regulates the longitudinal feed movement of the workpiece. The rate of feed is changed by changing the angle of the regulating wheel and its speed.

The center heights of the grinding wheel and the regulating wheel are fixed. The work diameter is, thus, controlled by the distance between the two wheels and the height of the work rest blade. There are three common methods of grinding on centerless grinders:

—In-feed centerless grinding,
—End-feed centerless grinding,
—Through-feed centerless grinding.

In-Feed Centerless Grinding

In-feed centerless grinding is used to finish several diameters simultaneously or to grind tapered and other irregular profiles. The principle of in-feed centerless grinding is illustrated in Figure 23-17. The work rest blade and the regulating wheel are secured in a fixed position in relation to each other.

The workpiece is placed on the work rest and against the regulating wheel and the end stop. The work is plunge ground. It is fed into the grinding wheel by moving the in-feed lever through a 90° arc. At the end of the feed lever

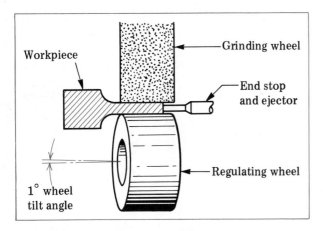

FIGURE 23-17 Principle of In-Feed Centerless Grinding

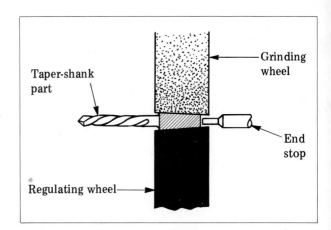

FIGURE 23-18 Principle of End-Feed Centerless Grinding

travel, the predetermined size is reached. The part is thus ground to the specific dimension. The work rest and regulating wheel are then moved back to the starting position by reversing the feed lever movement. The end stop serves to manually or automatically eject the ground part. A part that is longer than the grinding and regulating wheels is supported on one end on rollers that are mounted on the machine. The other end of the part is supported on the work rest.

End-Feed Centerless Grinding

End-feed centerless grinding is used to grind tapered parts. The principle of end-feed centerless grinding is shown in Figure 23-18. Both the grinding wheel and the regulating wheel are formed to the reverse shape of the tapered part. The regulating wheel is turned at a small angle to keep the workpiece against the end stop. The grinding wheel, the regulating wheel, and the work rest are each set in a fixed position. The work is fed either mechanically or manually up to the end stop.

Through-Feed Centerless Grinding

Through-feed centerless grinding involves feeding the work between the grinding and regulating wheels, as illustrated in Figure 23-19. The cutting action takes place as the cylindrical workpiece is fed by the regulating wheel to the grinding wheel. Here, again, the work feed rate is controlled by the speed and angle of the regulating wheel.

Advantages of Centerless Grinding

The four prime advantages of using centerless grinding processes and machines over other methods are as follows:

—Long workpieces that would otherwise be distorted by other grinding methods may be ground accurately and economically by centerless grinding (axial thrust is eliminated);
—Generally, less material is required for truing up a workpiece (the work *floats* in a centerless grinder, in contrast to the eccentricity of many workpieces that are mounted between centers);
—Greater productivity results as wheel wear and grind time are reduced (with less stock removal);
—Workpieces of unlimited lengths may be ground on a centerless grinder.

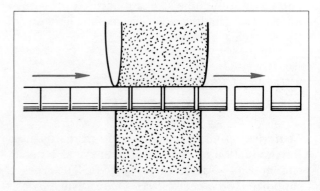

FIGURE 23-19 Principle of Through-Feed Centerless Grinding on a Cylindrical Grinder

TOOL AND CUTTER GRINDERS

Tool and cutter grinders are especially designed for grinding all types of milling cutters, reamers, taps, drills, and other special forms of single- and multiple-point cutting tools. Tool and cutter accessories, machine setups, and grinding processes are covered in detail in later units.

Universal and Tool and Cutter Grinder

The *universal and tool and cutter grinder* (or *universal tool and cutter grinder*) is a general-purpose grinding machine. It is particularly adaptable to custom work in small jobbing shops and toolrooms and in training programs. The universal features and surface grinding attachments permit internal and external grinding operations that otherwise would be done on separate grinders or on a universal cylindrical grinding machine.

Many other attachments are available for the machine. Magnetic chucks, collet chucks, universal scroll chucks, index centers, and straight and ball-ended mill sharpening attachments are produced for tool and cutter grinding. While many cutter sharpening operations are performed dry, wet grinding attachments are available.

The principles of operation of the universal tool and cutter grinder are a composite of the separate principles of grinding on surface grinding machines, universal cylindrical grinding machines, and tool and cutter grinding machines.

The major features of a universal tool and cutter grinder include the *base, saddle, column, wheel head, table,* and *coolant reservoir and system*. These features and other control parts are pictured in Figure 23-20.

The box-like construction of the base provides rigidity. Ways are machined on top of the base. The saddle is positioned on and moves along these ways. The saddle permits transverse movement toward or away from the column. Cross feed handwheels on the front of the machine are turned to move the saddle. Hardened ways are on the upper part of the saddle. They are machined at right angles to the ways that are on the top of the base.

Usually a round column is mounted in the back of the base. The column supports the wheel head and permits it to be swiveled. The wheel head contains the spindle. The wheel head spin-

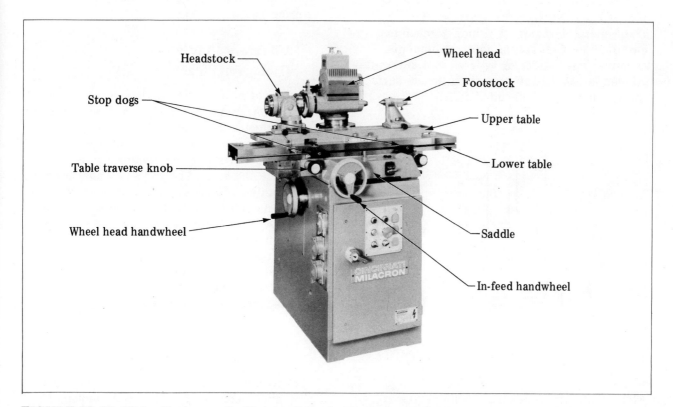

FIGURE 23-20 Major Features of a Universal Tool and Cutter Grinder

dle is designed with tapered and threaded ends to accommodate grinding wheel collets. The spindle is supported in antifriction bearings. The spindle speed is usually controlled by varying the step-cone pulley drive combinations, which provide a range of speeds to meet the requirements of particular sizes and types of abrasive wheels. The complete wheel head mechanism may be raised or lowered by handwheels on both sides of the base. The wheel head may also be swiveled through 360°.

The lower table is mounted and rests on the upper saddle ways. It moves on antifriction bearings. The upper table is fastened to the lower table. The upper table may be swiveled to grind tapers.

The upper and lower tables as a unit may be moved longitudinally or locked in place. The rate of movement is adjusted by regular traverse knobs as well as by a slow-speed traverse crank. The length of table traverse is controlled by a positive stop pin and a spring-loaded plunger. The pin and plunger provide a positive and a cushioned stop of the table on each stroke.

GEAR AND DISK GRINDING MACHINES

The general features of grinding machines are modified to meet special job requirements. For example, one particular type of tool and cutter grinder is capable of generating helical surfaces and of sharpening ball-ended mills.

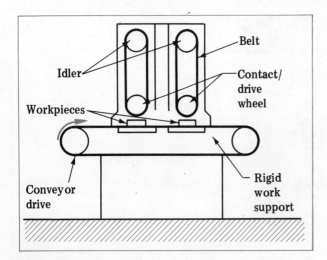

FIGURE 23-21 Schematic of Simple, Double-Belt, Coated-Abrasive Finishing Process

Gear grinding machines are usually of two types. The first type, a *form grinder*, has the grinding wheel dressed to a slope opposite to the slope of the gear tooth form. The second type of gear grinding machine generates the tooth form. The form is produced by the combination movement of the workpiece and the cutting action of the abrasive wheel. Gear grinding machines are designed for the range of general straight-tooth spur gears to hypoid gears.

The Beasly *disk grinder* of 1904, with subsequent improvements, still continues as a high-production machine. Parts may be surface ground for thickness (parallel) and to a finished dimension. Parts are fed by different feed devices between the faces of two grinding wheels. The devices consist of fixtures, moving arms, and guide bar rolls.

OTHER ABRASIVE GRINDING MACHINES

Many surfaces are finished by using abrasive belts. Workpieces, as pictured in Figure 23-21, are moved under the abrasive belts by a conveyor drive unit. The abrasive finishing process is efficient on soft metal, plastic, and other parts requiring a smooth surface.

Vibratory Finishing (Deburring) Machine

In addition to the bonding of abrasives into sheet, roll, grinding wheel, and other forms, the grains themselves are used in many production finishing and deburring processes. For example, machined parts may be deburred in a *vibratory finishing machine*. The parts and abrasive grains are vibrated. At the end of a fixed period of time, the parts and the grains are moved from the vibrator. The grains and parts are separated through a conveyor belt mechanism. The abrasive is returned to the vibrator. The parts move to a staging area for removal. During this process, the tumbling abrasive reaches and deburrs inaccessible areas of workpieces.

Free-Abrasive Grinding Machine

Flat surfaces may be generated by using free abrasive grains in a free-abrasive grinding machine. The grains are made into a slow moving *slurry* and fed into a hardened steel plate. The plate is water cooled. The grains circulate along the plate. As the grains roll around the workpiece, they cut into the material.

Electrochemical Grinding Machine

Single-point, carbide cutting tools are generally sharpened in an *electrochemical grinding machine*. Figure 23-22 shows the major features of an electrochemical grinder.

MICROFINISHING PROCESSES AND MACHINES

The term *microfinishing* is used here to identify four selected high-quality surface finishing processes. Each process requires the removal of limited amounts of material. The ultimate finely finished and highly accurate surface is specified in microinches.

Honing, lapping, and superfinishing processes and machines are discussed here. The microinch range of surface roughness for these surface finishing processes is reproduced in Figure 23-23. Polishing is included although the process is not intended to control the size or shape of a finished part.

The black areas indicate the general range. The dotted areas show finer and coarser ranges that are produced under special conditions. For example, surfaces may be superfinished to a roughness height of 2 to 8 microinches. Under more precise superfinishing conditions and control, the roughness height may be decreased to a 0.5 to 2 microinch range. Conversely, a coarser finish (8 to 32 microinches) may be produced.

The quality of the surface finish of a product is the responsibility of the design engineer or technician. Not all parts require smooth surfaces in low microinch values. The surface required is determined according to the functions, shape, and conditions under which a part is to operate.

HONING PROCESSES AND MACHINES

Honing is a grinding process in which two or three different low-pressure cutting motions of a honing stone produce a microfinished surface.

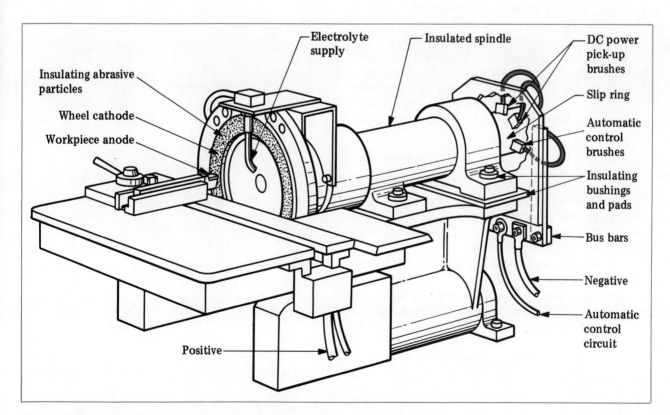

FIGURE 23-22 Major Features of an Electrochemical Grinder

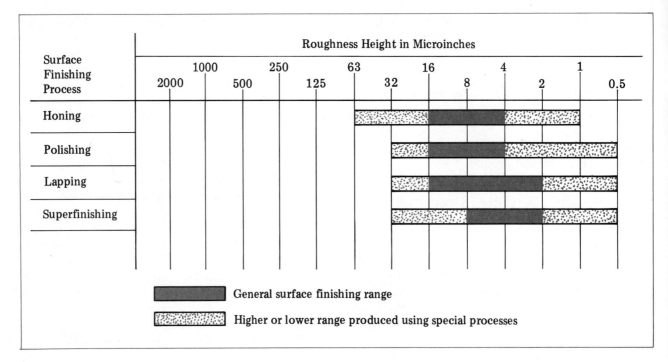

FIGURE 23-23 Microinch Range of Surface Roughness for Selected Surface Finishing Processes

The stock removal rate of honing, up to 0.030" (0.8mm), is the highest of the microfinishing process rates. Honing can correct out-of-roundness and taper and axial distortion. Usually, from 0.004" (0.1mm) to 0.010" (0.25mm) is left for honing. The general range of honing finishes is from 8 to 10 microinches. However, a fine surface finish to 1 microinch is obtainable. Honing may be employed on soft or hardened materials.

Honing Stone Holders and Abrasives

Honing requires the use of bonded abrasive *honing stones*. Some honing stones are held loosely in holders. They may be regularly spaced or interlocked. The interlocked design provides a continuous honing surface. An example of a typical honing tool (holder and stones) is shown in Figure 23-24.

The abrasive stones are expanded against a bored hole. This action is produced mechanically or by hydraulically operated mechanisms. In production, automatic controls to within 0.0001" (0.0025mm) to 0.0003" (0.0075mm) of a specified size are used with hydraulic types of honing tools.

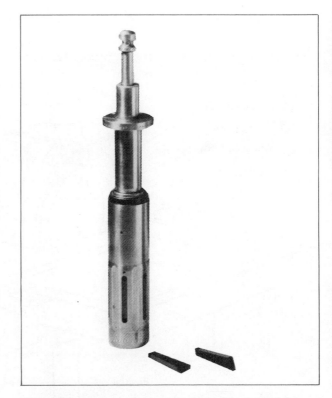

FIGURE 23-24 Typical Honing Tool (Holder and Stones)

Aluminum oxide, silicon carbide, and diamond grain abrasives are used. The usual bonds are resinoid, shellac, vitrified, or metallic. The cutting action and wear life of the stones are increased and controlled by using a "fill" of sulfur, resin, or wax.

Honing Cutting Action, Speeds, and Coolants

Honing, like grinding, requires multiple-point cutting tools. The abrasive grains do the cutting. They fracture upon dulling. There is a large area of contact between the stone and the workpiece. The cutting action is produced by two or three basic motions and low pressure. In a two-motion process, the honing tool (holder and stones) rotates and reciprocates (up-and-down motion) in the bore. A crosshatch pattern surface finish is produced by this cutting action. Machines are designed so that either the honing tool spindle or the workpiece reciprocates. A third reciprocating motion is added to break up the regular pattern. This is necessary to decrease friction and heat between two sliding parts and to increase the lubrication between the surfaces.

Two different speeds (sfpm) are used: *rotation speed* and *reciprocating speed*. The rotation speed for cast iron ranges from 110 to 200 sfpm; the reciprocating speed, from 50 to 110 sfpm. The rotation speed for steel ranges from 50 to 110 sfpm; the reciprocating speed, from 20 to 90 sfpm. The effect of too high a speed is to cause the abrasive grains to dull faster. They then produce a burnishing (rubbing) action rather than a cutting action.

A coolant is required in honing. The coolant serves the usual function of flowing away chips and used abrasives particles. It also prevents the abrasive stones from loading and thus reduces the temperature of the stones and workpiece and improves the cutting qualities. The general coolants include kerosene, turpentine, lard oil, and manufacturers' trade brand names. Water-soluble cutting oils are not recommended.

Horizontal and Vertical Honing Machines

Many honing processes are done by mounting the honing tool on a drill press spindle. However, special *horizontal and vertical honing machines* are available. The horizontal type is most adaptable to honing large-diameter, heavy parts requiring long strokes.

The more common type is the vertical honing machine. Honing machines may have a single spindle or multiple spindles. In mass production, a number of hones operate simultaneously. It is common practice to hone many holes at one time to size variations within 0.0005" (0.01mm). Honing machines are decelerated rapidly at the end of the bore. This speed reduction prevents honing bell-mouthed holes.

Completely round holes or interrupted holes such as keyways may be honed, provided the stones overlap the cutaway area. Honing may be used on ferrous and nonferrous parts, carbides, bronzes and brass, aluminum, ceramics, and some plastics.

LAPPING PROCESSES AND MACHINES

Lapping is a process of *abrading* (removing) material by using a soft-metal, abrasively charged lap against the surface to be finished. Loose-grain *abrasive flows* are used. They are mixed with oil or alcohol to form a grease that is smeared on the lap. Lapping serves three prime purposes:

—To correct minor surface imperfections of two mating surfaces,
—To generate close-fitting mating surfaces,
—To produce smooth, fine-grained finished surfaces.

Characteristics of Laps and Lapping

Laps are made of a material that is softer than the parts to be lapped. Soft cast iron, brass, copper, lead, and soft steels are commonly used materials. They easily receive and retain the abrasive grains. Soft, close-grained gray cast iron laps are common.

An open-grained lap will cut faster than a close-grained lap. The harder the lap, the slower the cutting action. Also, the harder the lap, the duller the finish and the faster the wear. The faces of laps are serrated (grooved) to permit the working surface to remain clean and free of abrasive and other particles.

Lapping is used as a finishing process, not as a metal-removing process alone. The amount of material removed in rough lapping is only 0.003"

(0.08mm). In finish lapping, the material removed can be as little as 0.0001" (0.002mm). The common practice in commercial lapping processes is to grind to within 0.001" (0.02mm) of the basic size. Parts, such as gage blocks, are then finished by lapping. Blocks under 1" (25mm) are lapped to the three standard degrees of dimensional accuracy. Lapping is also used to correct minute errors in flatness caused by magnetic chucks.

Lapping machines use a retainer instead of a chuck or other holding device. The retainer serves two purposes:

—To ensure the correct travel of the workpiece during each lapping cycle,
—To move each piece so that the largest part or all of the abrading surface of the lap comes in contact with the workpiece.

Abrasives and Workholders (Retainers or Spiders) for Lapping

The three most commonly used abrasives for lapping include aluminum oxide, silicon carbide, and diamond flours. Emery, rouge, levigated alumina, and chromium oxide are also used. Aluminum oxide is employed for soft metals. Silicon carbide is a practical abrasive for hardened parts. Diamond flours are used for small precision work and extremely hard workpieces.

The abrasives are used with a *vehicle*, or lubricant. The vehicle partly controls the cutting actions. It serves also to prevent scoring the work and the abrasive from caking. Some of the common lubricants include a mixture of kerosene and a small quantity of machine oil, olive oil, lard oil, and a soap and water combination. Laps are cleaned with naptha and other commercial solvents.

The lapping workholder is known as a *spider* or *retainer*. The workholder is usually a flat plate designed with holes to accommodate work pieces of a particular shape. In addition to nesting workpieces, the retainer confines them. Workpieces are moved by the retainer within a limited area and are brought into contact with every area of the lap during the cycle. The retainer revolves at a fraction (one-half) of the difference between the wheel (lap) speeds. The upper wheel (lap) on the lapping machine and the bottom wheel rotate in opposite directions. The wheel speeds are approximately one-tenth the surface feet per minute of regular grinding wheels. The combination of movements produces an ever-changing cutting pattern and an even wear on the laps. On some machines, the retainers provide a planetary motion to the workpiece.

Types of Lapping Machines

The most common type of lapping machine is the *vertical-spindle (-axis) lapping machine*. This machine employs two rotating laps. Lapping is done with an abrasive and vehicle appropriate to the material and process. Workpieces are lapped between the horizontal rotating laps.

Another type of vertical-axis machine uses fine, bonded abrasive wheels. While manufacturers market such a machine as a lapping machine, the process it performs is closer to grinding. No loose abrasives are used with this type of machine.

Lapping has been discussed here primarily in terms of finishing work surfaces of hardened and soft steels and cast iron. Equally important, the lapping process is used for the final finishing of optical lenses and optical flats.

SUPERFINISHING PROCESSES AND MACHINES

Superfinish™ is the trade name of the process that was invented and developed by the Chrysler Corporation. *Superfinishing* is a refined honing process in which bonded abrasive stones at low cutting speeds and constantly changing speeds produce a scrubbing action. Forces applied on the abrasive are low. From three to ten different motions are involved. As a result, a random, continuously changing surface pattern is produced.

Superfinishing is primarily an external surface refinement process. Only a limited amount of stock is removed. The quality of a superfinished surface depends on the sharpness of the abrasive stone and its cutting action. The low cutting force (pressure) and low cutting speeds (compared to honing) make superfinishing a practical process for refining the surfaces of delicate parts. Superfinishing to 2 and 3 microinches is common practice.

Superfinishing Machines

On superfinishing machines, the workpiece is rotated about a horizontal or vertical axis. The three major features of a superfinishing machine include a headstock, tailstock, and a vertical spindle (Figure 23-25). The workpiece is rotated

FUNCTIONS AND TYPES

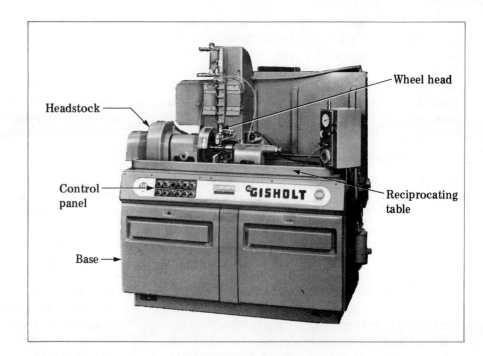

FIGURE 23-25 Major Features of Horizontal Cylindrical-Spindle Superfinishing Machine

around a horizontal or vertical axis, as required. The abrasive stone is mounted above the work. It is oscillated and moved by short strokes across the workpiece. The abrasive stone may be shaped with a diamond-type boring bar. The vertical spindle superfinishing machine is usually used for flat surfaces.

Characteristics of Superfinishing Processes

As a refined honing process, superfinishing possesses the following characteristics:

—The range of speeds is from 3 to 50 sfpm;
—Force applied on the abrasive stone is from 1 to 30 psi for internal surfaces and from 3 to 50 psi for external surfaces;
—The scrubbing action is produced by short strokes (1/16" to 1/4") at from 300 to 3,000 reversals per minute;
—Limited heat is generated due to the low speed and low forces (a light viscosity oil serves as a lubricant and stops abrasive action);
—Only a small amount of material is removed (thus, superfinishing is not a dimensioning process);
—Superfinishing is a finishing process that produces a fine, smooth crystalline finish.

POLISHING AND BUFFING AS SURFACE FINISHING PROCESSES

Polishing

Polishing is intended to remove tool marks and other fine surface scratches. The process is not used primarily to control the shape of the workpiece. Parts are machined to almost the precise dimension with only a thousandth of an inch or less left for polishing. (The hand process was described earlier in relation to lathe work.)

Polishing wheels are used in production work. These wheels are made of leather, canvas, felt, and wool. The abrasive grains are glued to the wheel face. Workpieces are held against the revolving polishing wheels and rotated until the desired polished surface is obtained. Some machines are equipped with endless abrasive belts. These belts conform to the shape of the part as it is brought into contact with the belt.

Buffing

Buffing produces a high luster and removes a minimum amount of material. Buffing follows polishing. The luster is produced by feeding and moving the surface areas of a workpiece to a buffing wheel revolving at high speed.

Buffing wheels consist of a number of layers of soft fabric. They may be linen, cotton, wool, or felt. The wheels are charged with extremely fine (flour size) grains of rouge or other abrasives. The abrasive and a wax are formed into a stick or cake. In this form, the abrasive is applied by force against the revolving buffing wheel. Buffing is not a widely used jobbing shop process.

SAFE PRACTICES APPLIED TO GRINDING PROCESSES AND MACHINES

—Clean and oil all finished surfaces of a grinder or microfinishing machine regularly.
—Refer to the machine tool manufacturer's chart on lubrication and maintenance. Check all machine lubricating systems for correct levels. Make sure all parts that require manual lubrication are lubricated.
—Check the grinding wheel (or spider), workpiece, and machine rotating parts before starting the machine. All should clear without interference.
—Check all work-holding accessories, particularly magnetic chucks. They must be held securely in place to receive a workpiece. In turn, test the workpiece to be certain the chuck is turned on.
—Stand to one side of a grinding wheel or any rotating disk to be out of line in the event that a wheel fractures and to avoid fine particles that may not be contained by the guards.
—Use protective goggles or a shield at all times when working on all moving machines.
—Warm up the machine to normal grinding conditions before dressing and truing.
—Set the wheel head as near as possible to the position at which it will operate.
—Dress and true a wheel for dry grinding dry. Use a dust collector system when dressing, truing, or grinding dry.
—Use plenty of coolant on the wheel when truing, dressing, or wet grinding.
—Test the coolant characteristics. Be sure the coolant is clear of dirt and chip particles and clean.
—True and dress wheels at the required operating speed or slower.
—Round off wheel edges of fine finishing wheels, provided they are not required for shoulder grinding. Rounding off these edges helps to keep them from chipping. A hand stone or precision dressing tool may be used to round off the edges.
—Check the setting of the diamond if a diamond dresser is used. Occasionally, diamonds loosen in the holder. Also, check diamonds for wear and for cracks.
—Point a diamond at a 10° to 15° angle in the direction of rotation of the grinding wheel to prevent the diamond from gouging the abrasive wheel.
—Use a protective corrosion-resistant film whenever a grinding machine is on downtime.
—Clean the faces of laps periodically to remove any grimy sediment of abrasives or dirt.

TERMS USED WITH GRINDING MACHINES, ABRASIVE MACHINING, AND MICROFINISHING

Grinding	The abrading (cutting away) of material by feeding a workpiece against a moving abrasive wheel. A machine process using multiple cutting edges (abrasive grains).
Abrasive machining	The faster removal of material to a specified finish and dimensional sizes, shapes, and tolerances than by using conventional cutting tools.
Automatic sizing (abrasive wheel)	Design feature incorporated into grinding machines to compensate for abrasive wheel wear.
Basic precision grinding	Six categories of grinding to precise dimensional and surface finish requirements. Surface, cylindrical, form, plunge, centerless, and tool and cutter grinding.

Types I, II, III, and IV surface grinders	Four basic types of grinders used to generate plane, angular, and contoured surfaces on flat workpieces. Classifications of horizontal or vertical grinding spindles with reciprocating or rotary tables.
Centerless grinder principle	A variation in the speed of two turning bodies (grinding wheel and regulating wheel) that produces and controls the direction and speed (RPM and sfpm) a workpiece rotates.
Cylindrical grinding machine	A grinding machine with a headstock, footstock, table, and wheel head. A machine tool that is designed to grind parallel and tapered surfaces.
Universal cylindrical grinding machine	A versatile cylindrical grinder for helical, internal, thread forming, and other grinding processes, in addition to processes done on a plain cylindrical grinder.
Plunge grinding	Direct feeding of a formed grinding wheel into a revolving workpiece.
Work rest blade	A positioning device used on a centerless grinder that controls the work diameter by its height in relation to the distance between the grinding wheel and the regulating wheel.
In-feed, through-feed, and end-feed	Three basic techniques of feeding parts for centerless grinding. Feeding techniques to accommodate a wide variety of cylindrical bars, shouldered forms, and other shapes requiring cylindrical grinding.
Tool and cutter grinder	A grinding machine consisting of a headstock, footstock, wheel head, adjustable table, and a variety of accessories.
Vibratory finishing machine	A machine designed to vibrate abrasive grains against the surface of a workpiece for purposes of deburring or producing a particular surface finish.
Microfinishing	The removal of limited amounts of material to produce a finely finished and dimensionally precise surface. As treated in the text, honing, lapping, and superfinishing processes and machines.
Honing	A grinding process using two or three different motions of a honing stone in relation to a workpiece under low pressure to cut away material and produce a dimensionally accurate and high-quality surface.
Lapping	Abrading (removing) material by moving an abrasively charged lap in a random pattern in relation to the surface of a workpiece.
Retainer (spider)	A workholder used on a lapping machine. A nest for holding and moving workpieces between laps.
Superfinishing	A refined honing process used to remove minute quantities of material to produce a smooth, fine crystalline finish.

SUMMARY

Grinding is a practical, efficient machining process for both rough and finish machining. Hogging cuts to depths of 1/2" (12mm) may be taken on heavier grinders in one pass.

Microinch finishes that are dimensionally accurate are produced by grinding. Commercially manufactured grinding machines have such built-in precision machining capability.

Abrasive machining designates heavy stock removal grinding processes. It also refers to single grinding processes on castings, forgings, and welded parts to produce the required finished surface.

During the period between 1860 and 1900, the cylindrical and universal grinder, horizontal-spindle surface grinder, and magnetic chuck were developed within the United States.

The discovery and production of artificial aluminum oxide and silicon carbide abrasives accelerated the applications of grinding processes. Mass production demands required greater versatility; higher productivity; and finer, more precisely machined surfaces.

Many grinding machines are provided with automatic sizing devices that compensate for wheel wear.

Some of the advantages of grinding include: elimination of many workholding fixtures and reduced handling costs, ability to cut through the outer scale of castings and hardened parts and maintain efficient machining, and simplification of parts design because thin-walled sections are machined with less spring than other machining processes.

Surface grinders, a machine tool since the 1860s, incorporate four basic combinations of motions. Types I, II, III, and IV include a vertical or horizontal, rotary spindle motion. The workpiece is fed either on a reciprocating or rotary (turning) table.

Precision parallel and circular surface finish patterns are produced by surface grinding. Flat, angular, and straight formed surfaces are commonly ground.

Plain and universal cylindrical grinders are used for straight, taper, shoulder, and form grinding. The universal cylindrical grinder is also adapted to internal grinding operations. Cylindrical grinding may be done by horizontal or transverse feeding. Certain processes require direct in-feeding (plunge grinding) with a preformed grinding wheel.

The development and use of cubic boron nitride abrasives, particularly on mounted grinding wheels, produce high wear resistance and cutting qualities, which are both important in internal grinding.

Centerless grinders are a production machine for grinding cylindrical, tapered, and multiple-diameter workpieces. The process may be continuous, as in through-feed grinding. Processes requiring a workpiece to be moved up to a stop include in-feed and end-feed centerless grinding.

Centerless grinding is adaptable to the machining of long, cylindrical workpieces. It requires less removal of stock with longer wheel wear life.

The universal and tool and cutter grinder combines the features of the cylindrical grinder, surface grinder, and the tool and cutter grinder.

Straight, end, helical, tapered, form, and other multiple-tooth cutters may be ground. Additional attachments permit the grinding of flat surfaces and single-point cutting tools.

Abrasive machining includes processes and machines that use abrasive grains. The vibratory finishing machine is used for deburring and producing a particular quality of surface finish.

Free-abrasive grinding machines are used to generate flat surfaces.

Microfinishing relates to finish grinding processes whereby limited quantities of material are removed. Dimensionally precise and high-quality surface finishes are produced and are measured in microinches.

The honing process can be used to correct out-of-roundness and taper and axial distortion in a bored hole. Rotating, bonded abrasive stones and two or three reciprocating motions produce the cutting actions. Microfinishes in microinches are possible.

Lapping is used to correct minor surface imperfections and to produce close-fitting mating surfaces with a fine-grained surface finish. Lapping is a finishing, not a metal-removing process.

Lapping requires the use of a retainer to move the workpiece between abrasive-impregnated surface of a soft-metal lap.

Fine abrasive grains of aluminum oxide, silicon carbide, and diamond flour, as well as some of the natural abrasives, are used in lapping.

Superfinishing requires the use of a formed abrasive stone, low cutting forces, low cutting speeds, and an oscillating movement with short strokes. Superfinishing produces a fine, smooth crystalline finish. Superfinishing is not a dimensioning process.

Polishing is intended to remove tool marks and surface scratches.

Buffing produces a high luster and removes extremely limited amounts of material.

Dressing, truing, and grinding processes must be carried on within the manufacturer's recommended sfpm.

General machine safety and maintenance precautions must be observed. The controlled temperature of the machine, workpiece, and grinding wheel or abrasive particles is important to the production of microinch surface finishes.

Chip, dust removal, and coolant systems must be operational and hygienic.

UNIT 23 REVIEW AND SELF-TEST

1. State two differences in the cutting action of abrasive grains as compared to the cutting action of multiple-tooth milling cutters.
2. Give two reasons why grinding processes are replacing many former shaping, milling, and turning processes.
3. Describe briefly the surface texture (scratch) patterns produced on each of the following grinding machines: (a) offhand grinder; (b) horizontal-spindle surface grinder; (c) vertical-spindle, reciprocating-table surface grinder; (d) vertical-spindle, rotary-table surface grinder; (e) cylindrical grinder (regular grinding); and (f) cylindrical grinder (plunge grinding).
4. a. Differentiate between in-feed and end-feed centerless grinding.
   b. Give one application of in-feeding and another of end-feed centerless grinding.
5. a. Cite three differences between tool and cutter grinding and cylindrical grinding.
   b. List three general processes performed on a tool and cutter grinder.
6. Describe briefly what is meant by (a) a soft, abrasively charged lap and (b) the lapping process.

466 GRINDING MACHINES

7. State three differences between superfinishing and honing.
8. Give one difference between buffing and superfinishing as surface finishing processes.
9. Refer to the Appendix Table on the microinch range of surface roughness for selected manufacturing processes.
   a. Indicate the general manufacturing surface finish roughness height for the following processes: (1) finish grinding, (2) honing, (3) lapping, and (4) superfinishing.
   b. Compare the surface textures produced by finish turning and by commercial grinding in terms of (1) measurement and (2) quality of the surface texture.
10. State two machine and/or cutting tool precautions the operator must take when starting up the spindle of a grinding machine.

SECTION TWO

Grinding Wheels: Standards and Preparation

Knowledge of the properties of aluminum oxide and silicon carbide, the general characteristics of abrasive wheels, and the use of straight grinding wheels in offhand grinding on bench and floor (pedestal) grinders is essential to basic machine grinding work. More advanced grinding processes require familiarity with other grinding wheel characteristics; wheel standards; and wheel selection, preparation, and uses.

This section deals with the technology of grinding wheels in relation to wheel types and wheel marking systems (ANSI representation and dimensioning). Simple mathematical applications are made to wheel speeds and feeds. Factors used in selecting wheels and step-by-step procedures for determining abrasive type, grain size, grade, grain structure, and bond type are described. General grinding problems, causes, and corrective steps and factors influencing the selection of grinding wheels are summarized.

This section also deals with industrial grinding practices, cutting fluids, and coolant systems. A general discussion of the selecting, mounting, dressing, truing, and maintaining of grinding wheels is provided.

UNIT 24

Grinding Wheel Characteristics, Standards, and Selection

CONDITIONS AFFECTING A GRINDING WHEEL

Conditions of use effect the nominal grade of a grinding wheel. Conditions of use cause a grinding wheel to act harder or softer than its nominal grade. If the cutting forces on the individual abrasive grains increase, the wheel acts softer the grains tear away more rapidly. If the cutting force is decreased, the wheel retains the grains longer and thus acts harder than the nominal grade. The cutting force is partly controlled by the grain depth of cut.

The rate of feed also causes changes in the action of the wheel. As the feed increases, a greater force is applied by the grains as they cut. Again, this force produces a softer-acting wheel.

The diameter of the grinding wheel also affects the cutting action. If constant spindle RPM is maintained, as the wheel wears the sfpm decreases. The cutting force on each grain then increases. The grinding wheel acts softer.

Conditions of use, as they affect cutting forces and cutting action, are illustrated in Figure 24-1. The depth of cut, chips, and feed are greatly enlarged. An abrasive grain, represented at A, moves from A to B during its cutting action. The distance is called the *arc of contact*. The *grain depth of cut* is from C to the arc of contact.

The feed of the workpiece during the arc of contact is from C to B. The shaded area represents the chips taken by the force of the grain against the workpiece. The *wheel depth of cut* is from C to D.

In all machining processes, as the depth of cut is increased, the force on each cutting edge increases. Thus, if the depth of cut of a grinding wheel is increased, the force on the individual abrasive grains causes more of them to tear away. The grinding wheel then acts softer. When the conditions of use are reversed to a lighter cut, fewer grains will tear from the wheel. The wheel acts harder.

When the sfpm is increased, or the feed is decreased, the grinding wheel leaves the workpiece before point B. The grain depth of cut is decreased. As a result, the wheel acts harder. The same harder action takes place if the feed is decreased.

OTHER CONDITIONS AFFECTING THE GRADE

The increasing or decreasing of the wheel depth of cut, feed rate, wheel sfpm, and wheel diameter are four conditions of use that affect the grade of a grinding wheel. However, three other conditions must be considered by the worker as follows:

—Area of contact (area represented by the arc of contact and the width of the cut);
—Kind and application of a coolant;
—Hardness, softness, and composition of the workpiece.

As always, the nature of the operations, the hardness of the material, the conditions of the machine, the amount of stock to be removed, the required surface finish, and the machine operator's skill are involved in grinding quality parts economically and efficiently.

Area of Contact

Operations involving a large area of contact with a workpiece require a soft abrasive wheel. Over a large area the force on each individual grain is reduced. When the area of contact is small, a large unit force is needed. A hard wheel

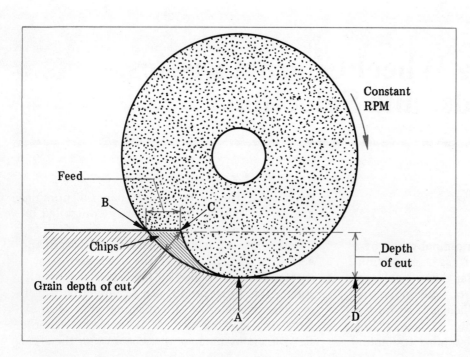

FIGURE 24-1 Conditions of Use Affecting Cutting Forces and Cutting Action

CHARACTERISTICS, STANDARDS, AND SELECTION

is then used. The hard wheel prolongs the cutting (wear) life of individual abrasive grains.

Effect of a Coolant and the Material

The frictional forces on abrasive grains are decreased when a coolant is applied for wet grinding. This coolant makes it possible to use a harder wheel and to extend its wear life.

As far as materials are concerned, a soft-grade wheel is used for hard materials. The soft grade permits the grains that dull to fracture off. New, sharp cutting edges are thus exposed continuously. Conversely, a hard-grade wheel holds the grains longer when grinding soft materials. The hard wheel prolongs the cutting life of each grain.

Grade and Abrasive Type

No one factor may be isolated when determining the wheel characteristics for a particular job. Grade must be related to the kind of abrasive. Grain size, bond, and structure must also be considered.

Iron, steel, steel alloys, and other ferrous metals usually require wheels with aluminum oxide. Copper, brass, aluminum, and other soft nonferrous metals, and ceramic parts are ground with silicon carbide wheels. Cemented carbides and nonmetallic materials such as granite, glass, quartz, and ceramics are ground with diamond wheels. Cubic boron nitride wheels are efficient for grinding hard steel and other high-speed steels. Cubic boron nitride wheels are usually more expensive than aluminum oxide wheels, therefore, they are used only when there is a demonstrated advantage, as for example, the grinding of very hard steel alloys. Many cubic boron nitride wheels are made of a CBN layer around a core of other material.

WHEEL TYPES, USES, AND STANDARDS

Wheel types, uses, and standards depend upon the hardness of the material to be ground, the required degree of surface finish, the area of wheel contact, and the wheel speed. The judgment of the craftsperson is based on a working knowledge of wheel types and standards.

Hardness of Material

Coarse and hard-grade abrasive grains are used on ductile materials. The grain size for general-purpose grinding is usually in the 36 to 60 range. Finer grain sizes in the 80 to 120 range and softer grains are used for hard materials. The hardness range of soft wheels for ordinary grinding is from F to I. The medium-hard wheel range includes J, K, L, and often M and N. Grade P, Q, and R are hard wheels.

Grinding dry or wet influences the grain hardness. Generally, when a coolant is applied, a one-grade-harder grinding wheel may be used. However, if a wheel is too hard, the workpiece may be *burned* due to overheating. Burning produces surface discoloration and may cause internal stresses, distortion, or checking.

Required Surface Finish

Grinding wheel selection is based on the amount of stock to be removed and the required degree of surface finish. For heavy stock removal, a very coarse, resinoid-bonded wheel is recommended. Where fine production finishes are specified, and a coolant is used, a fine-grain, rubber- or shellac-bonded wheel is suggested. It is practical, with proper dressing, to produce an excellent surface finish by using a coarser grain.

Area of Wheel Contact

Grinding wheels are used in both *side grinding* (grinding with the *side* of the wheel) and *peripheral grinding* (grinding with the *face* of the wheel). The area of wheel contact, however, varies. Figure 24-2 compares the areas of wheel contact in side and peripheral grinding. In side grinding, as in the case of a vertical-spindle grinder, the area of contact between the side of the wheel and the workpiece is large (Figure 24-2A). Soft-grade wheels that are coarse grained are recommended. Cup wheels and segmented wheels are side grinding wheels. Coarser and softer grain sizes—for example, 301 and 401—are used.

In peripheral grinding, the face of the wheel makes (compared to side grinding) a small line contact with the workpiece. Cylindrical, tool and cutter, and surface grinders provide examples of peripheral grinding. The area of contact, however, varies in each instance. A small area of contact is made in cylindrical (peripheral external) grinding (Figure 24-2B). The area is generated by the arc of contact on the periphery of the workpiece and the wheel. Grain sizes from 54 to 80 and hardness grades of K, L, and M are used in general cylindrical grinding processes.

There is a larger area of contact in surface grinding than in external cylindrical grinding (Figure 24-2C). The area is produced by the arc of the wheel in relation to a flat surface. A coarser grain size, such as 46, and a softer grade,

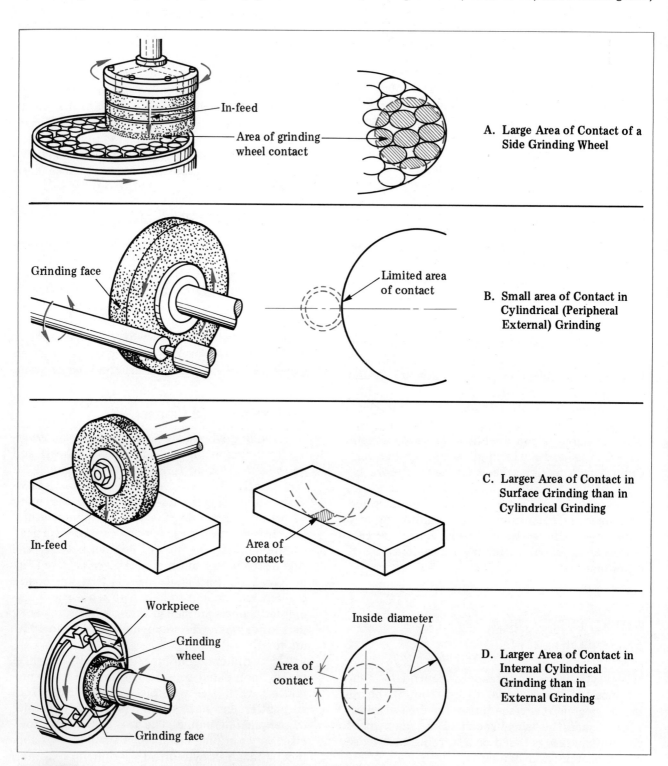

FIGURE 24-2 Comparison of Areas of Wheel Contact in Surface and Cylindrical Grinding

such as I or J, may be used.

On internal cylindrical grinding, the area of contact is formed by the inside diameter of the workpiece and the grinding face of the wheel. The area of contact is larger than in external grinding (Figure 24-2D). A softer grinding wheel is usually used.

Wheel Speed

In all instances, wheel speed is a factor. Every grinding wheel is marked with the safe limits of sfpm or RPM. The safety range of vitrified wheels is around 6,500 sfpm. Resinoid-, shellac-, and rubber-bonded wheels may be operated at speeds up to 16,000 sfpm. Designers of abrasives and grinding machines have set operating speed standards based on years of tested laboratory and shop experiences.

WHEEL MARKING SYSTEM

The grinding wheel marking system incorporates a series of letters and numbers. The standard marking system for abrasive (wheel) types A, B, and C is summarized in Table 24-1. The kind of information that a manufacturer generally includes on the blotter of a grinding wheel is given in Figure 24-3. This information relates to operator safety checkpoints, the safe operable RPM, wheel composition (according to ANSI codes), wheel dimensions, and a repeated safety precaution. The operator must be able to check the meaning of each number and letter against job specifications and the grinding process prior to taking a cut. After considerable use and wear on the outside diameter of the wheel, the RPM may be increased *provided the original sfpm* (at the specified RPM) *is not exceeded.*

Although basic markings between two manufacturers may be identical, the properties of the grinding wheels may vary. Each manufacturer has a range within which each wheel characteristic may be maintained. There are also some variations in the markings. Wheel markings for diamond and cubic boron nitride are slightly different than the markings used with aluminum oxide and silicon carbide.

There are five major symbols in the ANSI standards for grinding wheel markings relating to wheel composition. Each symbol designates a

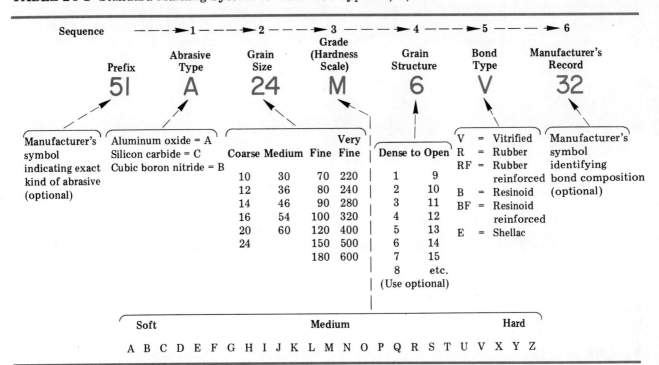

TABLE 24-1 Standard Marking System for Abrasive Types A, B, and C

design feature. The design features (in order of sequence in the ANSI code) are as follows:

—Abrasive type,
—Grain size,
—Grade (hardness),
—Grain structure,
—Bond type.

Symbol 1, Abrasive Type

A letter indicates the type of abrasive material. The letter A, for example, indicates aluminum oxide; B indicates cubic boron nitride; and C indicates silicon carbide. At the grinding wheel manufacturer's option, the letter may be preceded by other numbers or letters. For example, in Figure 24-3, the 9 in the 9A designation for abrasive type pinpoints the wheel as white aluminum oxide.

Symbol 2, Grain Size

A number symbol in the standards refers to the grain size. This number represents the openings per linear inch in screens through which the grains pass. The term *grit size* is often used. Grit sizes (as indicated in Table 24-1) are generally grouped in four categories. Coarse grains run from 10 through 24. Medium grains range between 30 and 60. The range of fine grain sizes is from 70 through 180. The very fine grain sizes, often called *flour sizes*, range from 220 through 600. The fine grits are sized by passing them

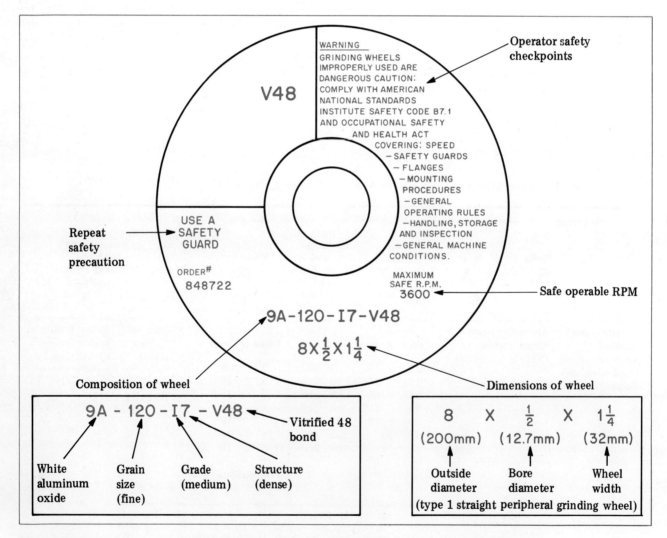

FIGURE 24-3 Information Supplied on a Grinding Wheel Blotter by the Manufacturer

over successively finer vibrating screens. The very fine grit sizes are separated by flotation through centrifugal separation processes.

Symbol 3, Grade (Hardness)

The grade (hardness) of aluminum oxide and silicon carbide wheels is designated by the letters A through Z. The letters from A to I indicate soft-grade wheels. The grains are torn easily during grinding. The letters from J to P provide for medium-grade wheels. These wheels are the most commonly used. The bond holds the grains tightly so that they do not tear away easily.

Thus, the grade really indicates the hold that the bond has on the abrasive grain. The greater the proportion of bond to grain, the harder the wheel. In principle, the ideal grade for a particular grinding job is one where the grains cut until they lose their sharpness. As the grains dull, the bond should release the grains so that the wheel is self-sharpening.

Figure 24-4 shows, by enlarged graphics, the grades of wheels and the holding power between the bond and the abrasive grains. A fine bond surrounding and holding abrasive grains provides only weak holding power. This bond (grade) is representative of a soft wheel—for example, a D or F grade wheel. A medium bond on the abrasive grains produces a stronger holding power. A wheel may be of medium grade such as K, M, or N. A heavy coating of bond on abrasive grains provides the strongest holding power and is representative of the hardest wheel. In short, the grade of the wheel depends on the composition of the bond.

Symbol 4, Grain Structure

Grain structure relates to *porosity*, or spacing, between grains. The spacing is artificially produced in an open, porous wheel by putting a spacing material into the mix. When the wheel is fired, the spacing material burns out and leaves pores, or voids, between the grains. This structure functions in providing chip clearance. The chips of ground material fit into the open areas between the grains. As each grain clears the workpiece, centrifugal force causes the chips to be thrown out of the wheel. The coolant serves a similar function, helping to wash out the chips.

Figure 24-5 compares open, medium, and dense grinding wheel structures. Without structure, the grinding wheel would become loaded and glazed and cutting action would stop.

A number is attached to the grade letter in a wheel marking to indicate the structure. The numbers run from 1 (the densest) to 15 (the least dense). This last number (15) represents an extremely porous structure with wide spacing between the grains. Usually, dense structures and hard grades go together. Soft grades are usually associated with open structures. The use of the grain structure number is optional; the number is left off by some manufacturers.

Symbol 5, Bond Type

A letter is used to designate bond type. Vitrified (V), resinoid (B), rubber (R), and shellac (E) are the most common bonds. Some wheel manufacturers use a number following the letter

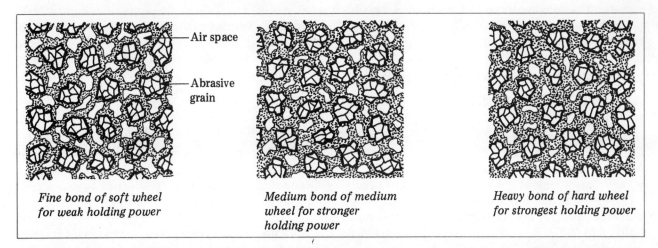

Fine bond of soft wheel for weak holding power

Medium bond of medium wheel for stronger holding power

Heavy bond of hard wheel for strongest holding power

FIGURE 24-4 Three Grades of Wheels and the Holding Power between Bond and Abrasive Grains

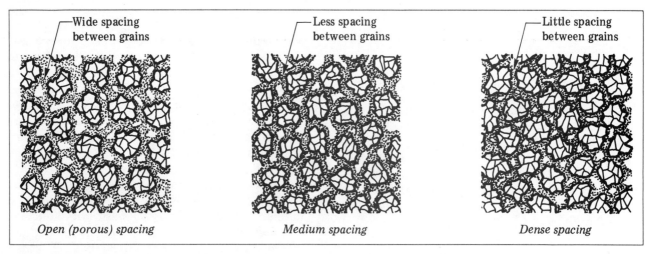

FIGURE 24-5 Comparison of Three Grinding Wheel Structures

of the bond. This number identifies the composition of the bond of a particular manufacturer. Sometimes, a second letter is used if the base material is reinforced. Resinoid-reinforced bonds are identified as BF; rubber-reinforced bonds, as RF.

Vitrified- and resinoid-bonded wheels are the most widely used in industry. The vitrified wheels are applied to precision grinding. The resinoid wheels are used for heavy stock removal in rough grinding processes carried on at high wheel speeds. Thin, rubber-bonded wheels are produced by passing rubber impregnated with abrasive grains between rolls. Thin, rubber-bonded and wider, resinoid-bonded wheels are used in very fine cut-off operations.

The type of bond determines safe operating speeds. The maximum sfpm for vitrified wheels is approximately 6,500. Resinoid wheels may be operated at higher sfpm.

ANSI SYSTEM OF REPRESENTING AND SPECIFYING GRINDING WHEELS

Grinding wheel manufacturers use the American National Standards Institute system for grinding wheel specifications. Engineering and other working data that are of importance to a grinding machine operator are contained in the ANSI bulletin B74.2-1975, entitled "Specifications for Shapes and Sizes of Grinding Wheels."

In all, there are 28 different types of ANSI-coded grinding wheels. The term *type* refers to (1) the wheel shape and (2) the wheel use—whether the wheel is intended for peripheral (outside) or side (flat) grinding. The wheels are numbered according to type beginning with the number 1. The different types of grinding wheels with principal and secondary applications are summarized in Table 24-2.

PERIPHERAL GRINDING WHEELS

Types 1, 2, 5, and 7 are straight wheels used primarily for peripheral grinding. A number of standard straight *wheel faces* are obtainable from grinding wheel manufacturers. Twelve standard shapes of wheel faces are pictured in Figure 24-6. General descriptions are given for each particular shaped face. It is also possible to form and dress a wheel to a particular shape. However, preformed wheels are preferred because no forming time is needed.

Surface Grinding Wheels

The design features of three types of peripheral grinding wheels used for surface grinding are shown in Figure 24-7. The wheel types include types 1, 5, and 7.

Type 1 Wheel (Straight). The type 1 straight wheel grinds on its periphery. The size specifica-

CHARACTERISTICS, STANDARDS, AND SELECTION 475

TABLE 24-2 Applications of Wheel Types to Various Grinding Machines and Processes

Wheel		Precision Grinding							Rough Grinding						
		Cylindrical			Surface		Tool Cutter						Snagging		
		Centertype	Centerless	Internal	Horizontal Spindle	Vertical Spindle	Cutter Grinder	Floor Stand and Bench	High-Speed Hand Grinding	Saw Gumming	Cut Off	Weld Smoothing	Portable	Floor Stand	Swing Frame
Type 1		P	P	P	P	P	P	P	S	P	P	S	P	P	P
Type 2					P	P	S	S							
Type 3													P	P	
Type 4							S						P	P	
Type 5		P	P	P	P		S						P	S	
Type 6				S	S	P	P						P	P	
Type 7		P	P		P		S								
Type 11					S	S	P						P	P	
Type 12							P								
Type 13							S			P					
Type 16													S	P	
Type 17 and 17R													S	P	
Type 18 and 18R				S									S	P	
Type 19 and 19R													S	P	
Type 20		P	S												
Type 21		P			S										
Type 22		P			S										
Type 24		P			S										
Type 25		P			S										
Type 27												S	P	P	
Plate mounted, nut inserted						S		P							
Mounted points				P					P				S	P	
Segments					P	P									

P = Principal use   S = Secondary use   Cincinnati Milacron

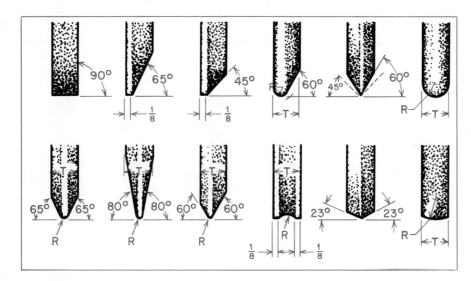

FIGURE 24-6 Twelve Standard Shapes of Wheel Faces

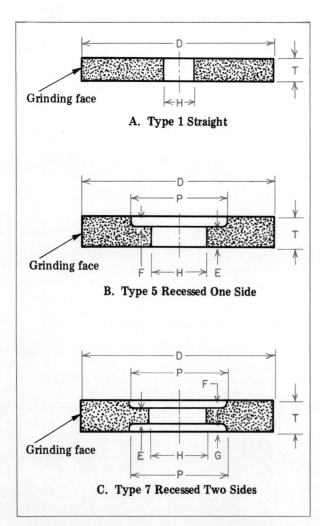

FIGURE 24-7 Types and Design Features of Peripheral Grinding Wheels Used for Surface Grinding

tion of this wheel is represented in Figure 24-7A by three dimensions. The outside diameter is indicated by the letter D; the bore, by H; and the thickness, by T. The letter codes for the dimensions used on all types of wheels are given in Table 24-3.

Type 5 Wheel (Recessed One Side). The type 5 wheel, a variation of the type 1 straight

TABLE 24-3 Standard Code of Letters for Dimensions on All Types of Grinding Wheels

Letter Code	Dimension
A	Radial width of flat at periphery
B	Depth of threaded bushing in blind hole
D	Outside or overall diameter
E	Thickness of hole
F	Depth of recess on one side
G	Depth of recess on second side
H	Diameter of hole (bore)
J	Diameter of outside flat
K	Diameter of inside flat
N	Depth of relief on one side
O	Depth of relief on second side
P	Diameter of recess
R	Radius
S	Length of cylindrical section (plug type)
T	Overall thickness
U	Width of edge
V	Face angle
W	Wall (rim) thickness of grinding face

wheel, is pictured and dimensioned in Figure 24-7B. One side of the wheel is *recessed*—that is, a portion of the wheel from the bore outward is reduced. Wheels are recessed to permit the use of a wheel that has a wider face than the adapter. Here, the thickness (T) is wider than the capacity of the adapter (E).

Type 7 Wheel (Recessed Two Sides). The characteristics and dimensions of a Type 7 wheel are given in Figure 24-7C. As shown, the two sides of the wheel are recessed.

Cylindrical Grinding Wheels

Many grinding wheels have one or two sides *relieved*—that is, the side of the wheel is cut away. The side tapers from the diameter of the inside flat (K) to the radial flat at the periphery (A). The design features of seven types of peripheral

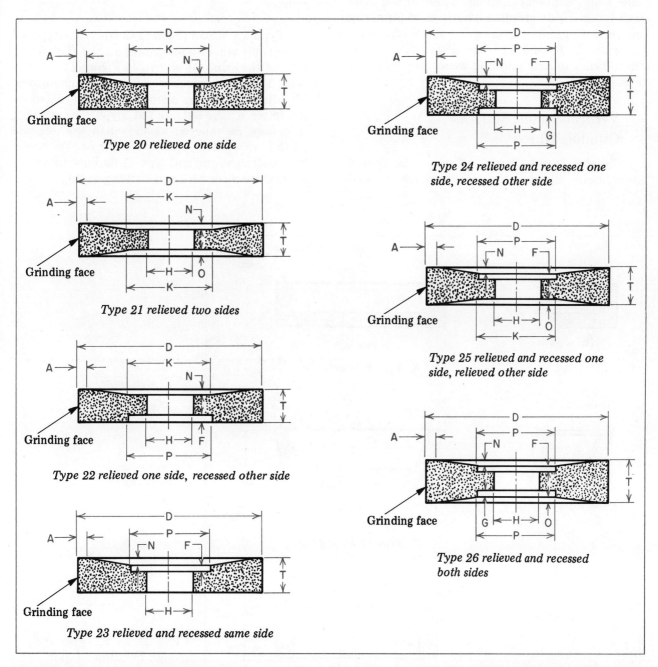

FIGURE 24-8 Types and Design Features of Peripheral Grinding Wheels Used for Cylindrical Grinding

grinding wheels used for cylindrical grinding are shown in Figure 24-8. The wheel types include types 20 through 26. Note that some wheels are relieved on two sides. Other wheels are relieved and recessed.

SIDE GRINDING WHEELS

Side grinding, to repeat, means grinding with the side of the wheel in contrast to grinding with the face as in peripheral grinding. Three of the most widely used side grinding wheels used primarily on vertical-spindle surface grinders are identified as types 2, 6, and 11. The design features of these types of side grinding wheels are shown in Figure 24-9.

Vertical-Spindle Surface Grinding Wheels

Type 6 Wheel (Straight Cup). The type 6 straight-cup wheel is the most extensively used on vertical-spindle grinders. It is specified by four to seven dimensions. As shown in Figure 24-9A, the type 6 wheel has an outside diameter (D), wall thickness (W) at the grinding face, thickness of the hole (E), and bore diameter (H).

The side width of the wheel is sometimes relieved to permit setting the vertical spindle at a slight angle for faster stock removal. The partial section on the left shows the wall thickness (W) relieved to the radial width (A) at the periphery.

Type 11 Wheel (Flaring Cup). The type 11 flaring-cup wheel also grinds on the flat rim or wall of the cup. As shown in Figure 24-9B, the type 11 wheel has two additional dimensions. The diameter of the inside flat is represented by the letter K; the outside flat, by J. The grinding face may be relieved as shown in the left-side partial view. The side walls taper. Both the type 6 straight-cup and type 11 flaring-cup wheels are also used for tool and cutter grinding.

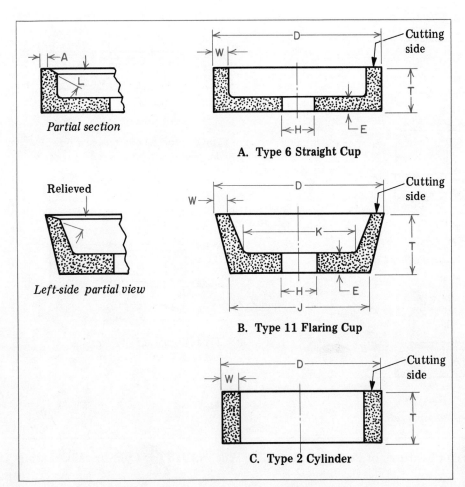

FIGURE 24-9 Types and Design Features of Side Grinding Wheels (Vertical-Spindle Surface Grinding)

Type 2 Wheel (Cylinder). The type 2 cylinder wheel is used extensively for vertical-spindle grinding. As shown in Figure 24-9C, the diameter (D), wall thickness (W), and overall thickness (T) indicate the dimensions for specifying the type 2 wheel size.

COMBINATION PERIPHERAL AND SIDE GRINDING WHEELS

The type 12 dish wheel, illustrated in Figure 24-10, is designed for straight peripheral (face) grinding in addition to side (wall) grinding. This wheel is similar to a flaring-cup wheel. However, there are two essential differences. The dish wheel is shallower. The periphery is dressed as a secondary cutting face. The dish wheel is especially designed to permit grinding on the periphery and side at the same time. The dish wheel is primarily used for tool and cutter grinding processes. Additional consideration must be given in selecting a dish wheel. Since the side grinding face provides a larger arc of contact, a softer wheel may be used.

MOUNTED WHEELS (CONES AND PLUGS)

Types 16 through 19 are referred to as *cones* and *plugs*. They are manufactured to be threaded on a bushing or they are made with a standard-diameter shank. The cones and plugs may be formed to any shape. Figure 24-11A shows the shapes and dimensional features of cone and plug types 16 through 19. An assortment of mounted wheels used for burring and general grinding are illustrated in Figure 24-11B. Cutting is done on any face. Some cones and plugs

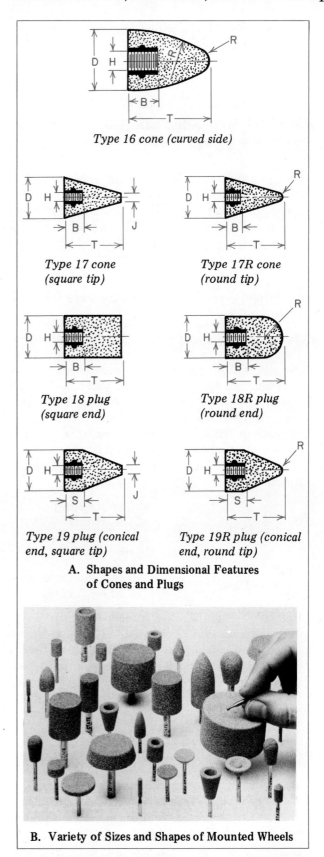

A. Shapes and Dimensional Features of Cones and Plugs

B. Variety of Sizes and Shapes of Mounted Wheels

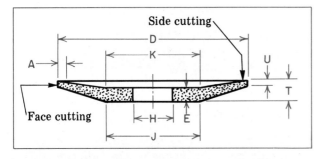

FIGURE 24-10 Face (Peripheral) and Side Cutting Type 12 Dish Wheel

FIGURE 24-11 Cones, Plugs, and Mounted Abrasive Wheels

are used for internal grinding. Most often, they are used in portable grinders for deburring, breaking the edges, or rough grinding.

COMPUTING CUTTING (WHEEL) SPEEDS, WORK SPEED, AND TABLE TRAVEL

The grinding machine operator must establish that the wheel speed (or spindle RPM), work speed, and table travel are correct before starting to grind. Cutting efficiency depends upon appropriate cutting and surface speeds of the wheel and the correct rate of table travel.

Wheel Speed (sfpm) and Spindle RPM

Specifications on a grinding wheel usually indicate the maximum RPM for safe operation. Tables often provide supplementary information about the sfpm at which a particular wheel may be used. As the periphery of the wheel wears down, the wheel acts softer. Cutting efficiency may then be below what it should be. With an increase in RPM, the sfpm may again be brought to maximum operating conditions.

Two basic formulas are used to calculate the spindle RPM or wheel speed (sfpm):

$$\text{spindle RPM} = \frac{\text{sfpm (surface speed of the wheel)}}{\text{circumference of wheel (in feet)}}$$

$$\text{wheel speed (sfpm)} = \text{RPM} \times \text{circumference of wheel (in feet)}$$

Example: Calculate the spindle RPM for a 10″ outside diameter straight vitrified-bonded wheel. It has been established from the manufacturer's data that 6,200 sfpm is a safe operating wheel speed. The wheel circumference in feet equals $10 \times \pi$ divided by 12, or 2.62 feet. By formula,

$$\text{RPM} = \frac{6,200}{2.62}$$
$$= 2,366 \text{ RPM}$$

The grinding wheel spindle should be set to turn at the closest (but slower) speed to 2,366 RPM. This setting will keep the wheel speed within the recommended safety range.

Example: Calculate the wheel (sfpm) of a 12″ diameter silicon carbide grinding wheel. The spindle RPM is 2,000. The circumference of the wheel in feet equals $12 \times \pi$ divided by 12, or 3.14 feet. By formula,

$$\text{wheel speed (sfpm)} = 2,000 \times 3.14$$
$$= 6,280 \text{ sfpm}$$

The sfpm must be checked against the recommended safe operating (wheel speed) range of the grinding wheel.

Work (Surface) Speed

Work (surface) speed relates to the feet (or meters) per minute that a workpiece moves past the area where cutting takes place. Work speeds vary according to the machine process, required surface finish and dimensional accuracy, and the material in the workpiece. In cylindrical grinding, the headstock is set to turn a workpiece at the RPM that will produce the required work surface speed. By contrast, for surface grinding, the longitudinal table travel speed represents the work speed.

The RPM of the headstock for cylindrical grinding processes is equal to the work speed (fpm) divided by the circumference of the work in feet:

$$\text{RPM} = \frac{\text{work speed (fpm)}}{\text{work circumference (in feet)}}$$

If the work speed is given in meters per minute,

$$\text{RPM} = \frac{\text{work speed (m/min)}}{\text{work circumference (in meters)}}$$

Example: A 25.4mm diameter steel pin is to be machined on a cylindrical grinder. Calculate the headstock speed (RPM) to produce a work speed of 30 meters per minute (30 m/min). By formula,

$$\text{work circumference} = \frac{25.4 \times \pi}{1,000}$$
$$= 0.08 \text{ meters}$$

Apply the known values to the RPM formula:

$$\text{RPM} = \frac{30 \text{ (m/min)}}{0.08 \text{ (meters)}}$$
$$= 375 \text{ RPM}$$

Rate of Table Travel

Table travel rates of grinding machines are generally given in inches per minute (ipm). The rate of table travel varies from two-thirds of the width of the grinding wheel down to one-eighth. The smaller rate of travel is used to produce a very smooth surface finish. Longitudinal table travel rate changes are made by adjusting the table speed-selector knob.

The rate of table travel in inches is determined (for cylindrical grinding processes) by multiplying the desired work speed in RPM by the distance (in inches) the workpiece should travel each revolution:

$$\text{table travel} = \frac{\text{work speed (RPM)} \times \text{table travel per revolution}}$$

Example: A 2″ wide wheel is operating at a work speed of 400 RPM. The table travel is one-fourth the width of the wheel. Compute the table travel rate in inches per minute. By formula,

$$\text{table travel} = 400 \times 1/2$$
$$= 200 \text{ ipm}$$

If the computed table travel rate exceeds the machine rate, use the highest rate possible. Then, decrease the RPM of the work accordingly.

FACTORS TO CONSIDER IN SELECTING GRINDING WHEELS

Two major production factors govern the selection of a grinding wheel for manufacturing purposes. These factors relate to quality of finish and the cost involved. The surface finish should be no better than is necessary for the proper functioning of the part. The cost of producing the surface should be as low as possible.

However, there are a number of other production factors to consider in the selection of a grinding wheel. These factors are as follows:

—Material to be ground;
—Amount of stock to be removed;
—Area of contact between the wheel and the work;
—Type and condition of the grinding machine;
—Wheel speed, work speed, and required finish;
—Nature and application of a coolant during the grinding process.

The ability and experience of the operator are as important as these production factors. The worker must interpret production factors precisely to determine the size and the abrasive of which the wheel is made and the grain, grade, structure, and bond.

Manufacturers' recommendations provide specifications based on successful use in production work. However, any change in wheel specifications must be based on the user's knowledge of the factors governing wheel performance. Most difficulties in grinding are due to a lack of knowledge of the principles of grinding wheels and cutting action.

In selecting a wheel, a final choice should not be made until the grinding wheel has been tested under actual working conditions. The product (ground workpiece) that conforms to the specified quality of surface finish and dimensional accuracy must be considered first. Then, maintenance, wheel cost, and labor costs are considered.

When the correct wheel has been selected, a complete performance record should be kept. Similar workpieces may later be ground with a wheel that has been found to be satisfactory. In the following procedures for selecting grinding wheels, it is assumed that the wheels will be operated at the recommended sfpm.

HOW TO SELECT A GRINDING WHEEL

Determining the Shape

STEP 1   Use a type 1 or straight wheel for grinding flat horizontal surfaces or slots or for cutting-off processes.

STEP 2   Use a straight wheel with a formed face for radius work, profile grinding, or angular surface grinding.

STEP 3   Use a straight wheel with a relieved side for taking light cuts on a vertical surface. This same wheel may be used for grinding adjoining vertical and horizontal surfaces.

STEP 4   Use either a straight-cup or flaring-cup wheel for grinding a vertical surface.

STEP 5   Use a type 12 dish wheel for grinding square shoulders. This wheel may also be used to grind adjoining vertical and horizontal surfaces.

Determining the Size

The wheel size for a particular grinder is limited by several basic factors. These factors include the diameter of the spindle, the width or space between wheel flanges, the RPM of the spindle, and the maximum outside wheel diameter that will fit inside the protective guard. Each grinding wheel has its three nominal dimensions printed on the blotter.

STEP 1 Check the wheel size with the specifications on the blotter. The dimensions are given in the following order: outside diameter, thickness, and bore (hole) diameter.

STEP 2 Measure the diameter of the wheel. A worn wheel will not run at the correct circumferential speed unless the spindle RPM is increased. The RPM is adjusted to bring the wheel (peripheral) sfpm up to the specified speed.

STEP 3 Measure the maximum and minimum distance between flanges. The thickness of the wheel must be within these limits.

STEP 4 Check the bore diameter of the wheel with the size of the wheel collet or the wheel spindle. There must be a sliding fit between the wheel and the spindle.

Selecting the Type of Abrasive

STEP 1 Examine the blueprint or sketch of the workpiece. Determine the material to be ground and the general specifications of the ground surface(s).

STEP 2 Check with a wheel manufacturer's chart or specifications. Determine whether a regular aluminum oxide abrasive or a modified form of this abrasive is recommended.

Note: An aluminum oxide abrasive is used for materials of high tensile strength. Metals usually ground with this type of abrasive are as follows: alloy steels, stellite, monel metal, annealed steels, case-hardened steels, hardened steels, high-carbon steels, high-speed steels, and manganese steels; forgings and steel castings; malleable iron, wrought iron, and hard bronze.

Regular aluminum oxide wheels are used for general-purpose grinding operations on surface, cylindrical, and tool and cutter grinding machines. *Friable aluminum oxide wheels* are used for grinding operations on extra-hard, heat-sensitive tool steels. The heat generated at the area of contact is reduced because the grains of these wheels break down readily. *Semi-friable aluminum oxide wheels* are used for finish grinding hard steels. These wheels are adaptable for grinding operations where there is a large area of contact. *Heavy-duty aluminum oxide wheels* are applied in rough foundry snagging processes or when heavy cutting forces are involved.

STEP 3 Select a silicon carbide abrasive wheel for grinding materials of low tensile strength. The sharp cutting edges of silicon carbide grains cut brittle materials rapidly without breaking down prematurely.

Note: A wheel manufacturer's chart is used to see which type of silicon carbide is recommended. Silicon carbide abrasive wheels are usually used to grind cast iron, gray iron, and chilled iron; tungsten, tantalum, and carbide alloys; glass and ceramic and other nonmetallic materials; and nonferrous metals such as brass, soft bronze, copper, and aluminum.

Black silicon carbide wheels are used for general grinding purposes and the grinding of nonferrous and nonmetallic materials such as marble, stone, and rubber. *Green silicon carbide wheels* are widely used for grinding single- and multiple-point cemented carbide cutting tools.

STEP 4 Select a cubic boron nitride wheel if there are a number of very hard tool steel alloy or high-speed parts to be ground. CBN has the advantage over aluminum oxide and silicon carbide because of its hardness, sharp and cool cutting action, and long wheel wear life.

Selecting the Grain (Grit)

STEP 1 Use a coarse-grained wheel for coarse finishes and rough grinding.

Note: A coarse-grained wheel is used, as a general rule, to take deep cuts to remove large amounts of stock rapidly and for soft ductile materials. The softer and more ductile the material to be ground, the coarser the grain size should be.

STEP 2 Use the coarse-to-medium grain (grit) sizes for grinding the common metals indicated in each instance in Table 24-4. This table gives the coarse-to-medium grain sizes for grinding a few selected metals.

STEP 3 Select a fine-grained wheel on tungsten and tantalum carbides. Use this grain size for

TABLE 24-4 Grain Sizes for Grinding Selected Metals

Metal	Recommended Coarse-to-Medium Grain Sizes
Soft steel	20 to 46
Hardened steel	16 (roughing cuts)
	36 to 46 (finishing cuts)
High-speed steel	36 to 46
Cast iron	16 to 36
Aluminum	20 to 46
Brass	20 to 24
Copper	36 to 54

both finish grinding and the rapid removal of stock.

Note: Rapid removal of stock is accomplished by deep penetration of the grains into the material being ground. Hard, brittle, close-grained material prevents deep penetration of the coarser abrasive grains. Consequently, there is no advantage gained by using these grains for rapid removal of stock.

STEP 4 Use a fine-grained wheel for fine finishes, for form and angular grinding, or whenever it is important to retain the shape and corners of the wheel.

Note: A fine-grained wheel may be used with an oil-base lubricant.

Selecting the Grade (Hardness)

The general rule for grade selection is to use a wheel in the soft range for hard materials; in the medium range, for soft materials. The correct grade is usually determined by carefully observing the cutting action and wheel wear during the grinding process.

STEP 1 Use a wheel from E to G grade for surface grinding cemented carbides; from H to N grade for high-speed steel, hardened steel, and cast iron; and from J to N grade for soft steel.

STEP 2 Select a harder-grade wheel for cutting-off operations. For example, use a wheel from M to S grade for cutting off aluminum parts; from P to R grade for cast iron; from O to Q for hardened steel; and from P to W grade for soft steel.

STEP 3 Determine whether the amount of stock to be removed is a factor in selecting the grade of the wheel. A slightly harder grade is sometimes used. The same effect can also be obtained if the speed of the work is reduced.

Note: Deeper-than-average cuts used for roughing out increase the area of contact. As a result, the wheel acts hard. Decreasing the wheel speed or decreasing the work speed tends to compensate for this condition. Frequently, a one-grade-softer wheel is used for roughing operations.

STEP 4 Determine the sfpm of the wheel in relation to the table speed. Remember that the wheel acts softer as it wears. The table speed may be varied and decreased to reduce wheel wear. This reduction in work speed may increase cost unless the feed can be increased. If a harder wheel is selected, the work speed can be increased. Then, as the wheel wears, it can be dressed. This machining technique tends to equalize costs.

STEP 5 Select a wheel of slightly harder grade when it is used on a machine of light construction or when worn parts cause vibration.

STEP 6 Use a wheel of softer grade on heavy, rigid machines that are in excellent condition.

STEP 7 Select a slightly harder-grade wheel when a coolant is used.

Determining the Structure

STEP 1 Use wide grain spacings for soft, ductile materials. Closer grain spacings are used for hard, brittle materials. Cemented carbides are the exception to this practice. Cemented carbides require a wide grain spacing.

STEP 2 Check the structure numbers recommended by the manufacturer of the grinding wheel. Individual manufacturers use different structure numbers.

STEP 3 Select a wheel with an open structure (such as number 12) for the rapid removal of stock. The open structure provides adequate chip clearance and cool cutting action. A denser structure is used for heavy, plunge cuts.

STEP 4 Determine the condition of the machine. The best practice is to use a denser wheel on light machines or where the parts are worn. A wheel with a more open structure is used on a machine that is in good condition or that is rigid in construction.

STEP 5 Use a close (dense) grinding wheel for fine surface finishes.

Determining the Bond or Bonding Process

STEP 1 Select a wheel made by the vitrified process for general precision grinding on all metals.

Note: This bond is not suitable for wheels over 30" in diameter, for thin wheels, or for wheels where lateral stresses may cause wheel breakage. Large wheels are made either with silicate or resinoid bonds. The resinoid bond is also used in many applications in place of the vitrified bond.

STEP 2 Use wheels bonded with resinoid, rubber, or shellac for grinding narrow grooves and for work involving lateral stresses.

Note: Wheels bonded with shellac are not suitable for heavy grinding operations where considerable heat is generated. The heat softens the bond and impairs the cutting quality of the grinding wheel.

STEP 3 Select wheels with resinoid, rubber, or shellac bonds for cutting-off operations.

STEP 4 Specify wheels bonded with resinoid, rubber, or shellac for high finishes.

STEP 5 Use vitrified wheels for speeds up to 6,500 sfpm. Use resinoid, rubber, and shellac bonds for speeds above 6,500 sfpm.

STEP 6 Use silicate-bonded wheels for milder or cooler cutting action, such as tool and cutter grinding. Silicate wheels are practical for wet grinding operations when water or an emulsion is used as a coolant.

STEP 7 Select a vitrified wheel if an oil-base cutting fluid is used. Resinoid wheels are substituted in some cases.

HOW TO TEST FOR GRINDING WHEEL PERFORMANCE

Grinding wheels should be selected based on the manufacturer's recommendations and on an analysis of the work processes and operating factors. The kind of abrasive and the bond can be determined from this information. Usually, the grit, grade, and structure are determined by testing.

STEP 1 Test the selected wheel under the same operating conditions and on the machine on which the work is to be ground.

STEP 2 Record the performance of the wheel for future reference.

STEP 3 Select another wheel. Change the grain, grade, structure, or any combination of these characteristics, based on the performance of the first wheel.

STEP 4 Test the second wheel and record its performance.

STEP 5 Test a number of wheels until one wheel proves to be the most satisfactory. When only one or two wheels are available, and if only a few pieces are to be ground, vary the operating factors until the best results are obtained.

Note: Records of the performance and grinding costs should be kept for production processes. Such detailed records are usually not required on custom toolroom grinding processes, short-run repair jobs, and other work where grinding costs are less important. Records of performance are advisable because a slight change in grinding wheel characteristics often improves the quality or results in better wheel performance.

RECOGNIZING GRINDING PROBLEMS AND TAKING CORRECTIVE ACTION

Three groups of conditions must be recognized as contributing to grinding problems:

—Machine systems (the surface texture quality and dimensional accuracy of a workpiece are partially controlled by the correct functioning and precision of the pneumatic, electrical, and mechanical systems);

—Machine processes (a number of problems are grouped around the condition of the grinding wheel, coolant system, removal of chips, and basic processes);

—Operator responsibilities (these duties range from machine preparation such as lubrication and checking to proper work setups, grinding wheel selection, dressing, and so on).

CHARACTERISTICS, STANDARDS, AND SELECTION 485

TABLE 24-5 Common Grinding Problems, Probable Causes, and Corrective Action

Probable Cause and/or Corrective Action	Chatter Marks	Scratches	Spiral Marks	Burning/Checking	Burnishing	Not Ground Flat	Out-of-Parallel or round	Not Sizing Uniformly	Loading	Glazing	Spindle Running Too Hot
Grinding Wheel											
Grain size											
Too fine				X	X				X	X	
Too coarse		X									
Structure											
Too dense				X					X	X	
Too hard	X			X	X	X			X	X	
Grade too soft	X	X									
Dressing too fine				X		X			X	X	
Out-of-balance	X										
Dull, glazed, loaded	X			X							
Not trued properly	X	X									
Speed too high				X							
Workpiece											
Work speed too slow				X							
Out-of-balance	X										
Not adequately supported	X										
Centers worn or require lubrication	X						X	X			
Insufficient number or improper adjustment of back rests	X						X	X			
Work speed or table travel too high	X	X									
Machine Adjustment or Operation											
Excessive cross feed or cut too heavy	X			X			X	X			X
Diamond truing point too high above work centers			X								
Work sliding on chuck		X									
Defective belts	X										
Machine vibrates on floor	X										
Dirty coolant		X							X		
Insufficient coolant at point of contact or wrong coolant				X		X	X		X	X	
Inaccurate setting of swivel table							X				
Improper seating of headstock or footstock on table or chuck out of alignment							X				
Out-of-round center points or not properly seated							X				
Radial play in footstock spindle							X				
Improper adjustment of spindle clamp or back rests							X	X			
Improper adjustment of cross feed thrust bearing								X			
Insufficient amount or incorrect lubrication fluid for spindle											X
Overtightness of spindle belts											X
Loose particles inside guards		X									
Improper chucking	X	X					X	X	X		

486 GRINDING MACHINES

Many of the common problems encountered in grinding, their probable causes, and corrective action are summarized in Table 24-5. Note that in most cases a faulty system, machine process, or error in judgment creates a number of problems.

SUMMARY OF FACTORS INFLUENCING THE SELECTION OF GRINDING WHEELS

Grinding wheels are designated according to type of abrasive, grain (grit) size, grade, structure, and bond. The selection of a grinding wheel is affected by operator experience and by machine and work characteristics and properties. Table 24-6 indicates a number of major factors influencing the selection of grinding wheels and the grinding wheel characteristics that are affected.

SAFE PRACTICES FOR MACHINE GRINDING

Note: Many provisions of the American National Standards Institute's "Safety Code for the Use, Care, and Protection of Abrasive Wheels" are covered in detail in the next unit. However, general and basic safety practices are emphasized here.

— Inspect each grinding wheel visually for cracks and chips. Then, make a *ring test*. Usually, a small pin is placed through the bore on wheels of 10" (250mm) or smaller in diameter. The wheel is tapped lightly. Sound vitrified- or silicate-bonded wheels produce a clear metallic ring. A cracked wheel has a dull sound.
— Store grinding wheels in a storage rack. Use protective corrugated shims between wheels. Place each wheel in its proper compartment.
— Handle wheels carefully. Avoid hitting a wheel, particularly on its edges.
— Check the outside diameter of the wheel (if worn) and the recommended maximum sfpm on the wheel blotter. Adjust the spindle to keep within the manufacturer's range.
— Use a sliding fit between the bore and the machine spindle or adapter flange.
— Check the mounting flanges for flatness (burr free). Wipe the faces clean.
— Use a guard that covers at least half the diameter of the wheel. Adjust the spark guard and work rest on a high-speed grinder.

TABLE 24-6 Selection Factors and Wheel Characteristics Affected

Factors Influencing Grinding Wheel Selection	Wheel Characteristics Affected				
	Abrasive Type	Grain Size	Grade	Structure	Bond Type
Type of grinding machine and operation (surface, cylindrical, tool and cutter; form, shoulder, and so on)	X	X	X	X	X
Characteristics of material to be ground (tensile strength and hardness)	X	X	X	X	
Machine features and condition (heavy-duty, light; worn bearings, loose fitting parts)			X		
Operator work practices (tendency toward light or heavy cuts)			X		
Area of contact		X	X	X	
Stock removal (severity of heavy or light cuts)		X	X	X	X
Wheel speed (sfpm)			X	X	X
Work speed (ipm)			X		
Feed rate		X	X	X	
Quality of surface finish required		X	X	X	X
Wet or dry grinding process			X	X	X

CHARACTERISTICS, STANDARDS, AND SELECTION 487

- Inspect the manufacturer's machine lubrication requirements. Lubricate the machine. Check the level of all reservoirs.
- Check the coolant system. It should be hygienically clean. The type and quality of the cutting fluid should be tested against the job requirements. Bring the solution up to the specified strength (if needed).
- Stand to one side of the grinding wheel, particularly during start-up.
- Bring the wheel speed up to operating speed. Run at this speed for about a minute before taking a cut.
- Use only the recommended face(s) of a grinding wheel as designed and specified for each process.
- Shut off the coolant at the end of the operation before stopping the spindle to prevent the coolant from collecting at the bottom of the wheel. Such a collection produces an out-of-balance condition.
- Feed a grinding wheel carefully to cutting depth. Any jamming or forcing causes more abrasive grains to fracture than are needed and thus sometimes affects the trueness of the wheel.
- Take light cuts when grinding thin, flat surfaces or small-diameter workpieces to prevent distortion produced by grinding stresses.
- Adjust the back rests and use an adequate number to accommodate the length and diameter of the workpiece to be ground.
- Select the proper type of coolant nozzle. Direct an adequate flow as close as possible to the point of contact.
- Use safety goggles or a protective shield. Position safety glass shields (when they are provided) on the grinding machine.

TERMS USED IN GRINDING WHEEL CHARACTERISTICS, STANDARDS, AND SELECTION

Nominal grade	The actual grade of a grinding wheel. (The cutting condition of the nominal grade is affected by depth of cut, rate of feed, sfpm as the wheel wears, and other factors.)
Area of contact (grinding)	A surface represented by the arc and width of contact of the grinding wheel.
Manufacturer's specifications	Specific information contained on each grinding wheel blotter. A guide to a grinding machine operator on safety limits. Four general sets of suggestions: (1) operator safety precautions, (2) maximum operable RPM (sfpm), (3) composition of the grinding wheel, and (4) wheel dimension.
Grinding wheel markings (composition)	The use of five basic symbols. (Each symbol designates a design feature such as type of abrasive grain, grain size, grade, structure, and bond.)
"Specifications for Shapes and Sizes of Grinding Wheels"	A set of engineering standards accepted by grinding wheel manufacturers. A standard publication of the American National Standards Institute. Twenty-eight basic types of grinding wheels described according to shape and function. Dimensioning practices for uniformly specifying grinding wheel shapes and sizes.
Peripheral grinding	Grinding with the grains at the periphery of the grinding wheel.
Side (wall) grinding	Grinding on the side or wall of the grinding wheel.
Production factors in grinding wheel selection	Judgments usually made by a grinding machine operator and craftsperson. Consideration of factors such as nature and material of the workpiece, area of contact, wheel speed, work speed, required surface texture, and dimensional accuracy.

Grinding wheel characteristics — Working knowledge the worker must have in order to select a grinding wheel. (This knowledge includes determining the shape of the wheel, size, type, grain or grit, grade or hardness, structure, bond, and testing the grinding wheel performance.)

SUMMARY

A grinding wheel acts softer than its nominal grade when the cutting forces on each grain are increased. Depth of cut, rate of feed, diameter of the wheel, and sfpm are factors that produce a softer or harder cutting action.

- Cutting force and wheel action also depend on the grinding arc and width of cut.

The kind of coolant, point of application, and the rate of flow affect the cutting action and the quality of surface finish. A harder abrasive wheel with extended wheel life is possible when a coolant is used.

- Aluminum oxide grains are usually used for grinding ferrous metals. Silicon carbide wheels are practical for grinding soft, nonferrous metals and ceramics parts.

Very hard alloy steels require the tougher cubic boron nitride grains. These grains are often included as a layer around a core of other abrasive material.

- The grain size for general-purpose grinding ranges from 36 to 60.

Hard materials are ground with softer grains within the 80 to 120 grain sizes.

- The soft wheel hardness range for ordinary grinding is from F to I. The hard wheel range extends from J through N.

Cylindrical, tool and cutter, and surface grinding are examples of peripheral grinding. The arc of contact is comparatively small.

- In general, as the area of contact increases, a softer-grade wheel may be used.

Safe operating speeds for vitrified wheels is 6,500 sfpm. Speeds up to 16,000 sfpm are possible with resinoid-, shellac-, and rubber-bonded wheels.

- The information supplied by grinding wheel manufacturers covers safe operating speeds, composition according to ANSI codes, and wheel size.

Grinding wheel symbols relate to (1) abrasive type, (2) grain size, (3) grade or hardness, (4) grain structure, and (5) bond type.

- Vitrified wheels are generally used for precision grinding; resinoid wheels, for rough grinding.

Within the 28 types of ANSI-coded grinding wheels are included peripheral and side grinding wheels. These wheels may be straight or recessed on one or two sides. Some wheels are of a cup type; others, of a dish type.

- Grinding wheels (such as the type 12 dish wheel) are also designed for both peripheral and side grinding.

Mounted wheels are identified in types 16 through 19. Cutting is done on any face. These wheels (cones and plugs) are adaptable for portable and rough grinding operations.

Cutting efficiency depends on maintaining the correct work speed, table travel, and wheel speed (or spindle RPM).

The following formulas are used to compute wheel speed, work speed, and rate of table travel:

$$\text{RPM} = \frac{\text{sfpm of wheel}}{\text{circumference of wheel (in feet)}}$$

sfpm = RPM × circumference of wheel (in feet)

table travel = work speed (RPM) × table travel per revolution

Grinding wheel selection depends on both the manufacturer's recommendations and the judgment of the craftsperson. Decisions have to be made relating to the following factors: most appropriate type of abrasive, shape and size of the wheel, size of the grains or grit, hardness grade, wheel structure, and bond or bonding process.

Performance records permit comparisons of quality, quantity, cost, and other deciding factors.

Tables are provided on selection factors and wheel characteristics that are affected.

The machine operator must be able to spot grinding problems. Tables are available that identify probable causes and provide corrective steps that may be taken.

The American National Standards Institute's Safety Code provides guidelines for safe procedures on selecting, testing, mounting, and storing grinding wheels; machine preparation and grinding operations; and work setups.

UNIT 24 REVIEW AND SELF-TEST

1. State three conditions of use that affect the selection of a grinding wheel. (As a result, the wheel for a particular job may differ from a manufacturer's standard recommendations.)

2. Give one advantage of using a coolant (where practical).

3. a. Explain the term *area of contact* as applied to grinding processes.
   b. Tell what effect the area of contact has in selecting a grinding wheel for cylindrical grinding as compared to surface grinding.

4. a. List the kind (category) of information that is covered by each of three major symbols in ANSI standards for grinding wheel selection.
   b. Describe the wheel construction features for each symbol.

5. a. Indicate the design features of a type 7 surface grinding wheel that differ from a type 1 wheel.
   b. Describe a surface grinding application where a type 7 wheel should be used instead of a type 1 wheel.
   c. Identify a grinding process requiring the use of (1) a type 6 flaring-cup wheel and (2) a type 2 straight-cup wheel.

6. a. Cite two reasons why a grinding machine operator must be able to compute cutting speeds and spindle speeds.
   b. Compute the wheel speed (in meters per minute) of a 450mm diameter grinding wheel that is traveling at 1,500 RPM. Use $\pi = 3.14$.
   c. Determine the spindle speed for a 12″ diameter grinding wheel that is traveling at 6,200 fpm. Use $\pi = 3.14$.

7. a. Define the term *work (surface) speed* as related to cylindrical grinding.
   b. Compute the RPM at which a 6" diameter workpiece is to be turned on a cylindrical grinder to produce a work speed of 84 fpm. Use $\pi = 3.14$.
   c. Determine the table travel rate for a 1-1/2" wide wheel that is operating at 400 RPM. The table advances one-sixth of the wheel face width each revolution.
8. Secure a grinding wheel having its specifications recorded on the blotter.
   a. State the dimensions of the wheel.
   b. List each set of symbols.
   c. Identify the material, grain size, grade, structure, and bond as determined from the information provided by each symbol.
9. List three types of wheels that are used for precision grinding on (a) a vertical-spindle surface grinder and (b) a tool and cutter grinding machine.
10. State three machine and/or personal safety precautions to follow while operating a grinding machine.

UNIT 25

Grinding Wheel Preparation and Grinding Fluids

A grinding wheel is a multiple-grain cutting tool. Precision grinding requires every grain on the periphery to be the same distance to the center. The grains must remain sharp. Theoretically, each abrasive grain, as it becomes dull, is pulled out of the wheel. The wheel must be kept in balance.

Dimensionally accurate parts may be produced with high-quality surface finish only when efficient cutting conditions are maintained. Grinding wheel preparation, therefore, is extremely important. The wheel must be mounted, trued, dressed, and balanced. This unit examines the principles and practices related to this preparation. Several wet coolant systems and procedures for maintaining the quality of the grinding fluid are described. Many key items in the ANSI "Safety Code for the Use, Care, and Protection of Abrasive Wheels" are summarized.

ABRASIVE WHEEL PREPARATION

Aluminum Oxide and Silicon Carbide Abrasive Wheels

During a grinding process, the forces between the workpiece and the grains in an abrasive wheel produce a cutting action. Material is cut away and most of the dulled abrasive grains are torn out of the wheel. However, after considerable use, a wheel may become dull. Such wheels (grains) are then sharpened by *dressing*. Dressing is particularly necessary with silicon carbide and aluminum oxide wheels. Cubic boron nitride and diamond wheels do not require as frequent or as severe a dressing. The grains on these abrasive wheels are harder. They do not wear away as fast as aluminum oxide and silicon carbide.

In the design of aluminum oxide and silicon carbide wheels, the few thousandths of an inch tolerance between the bore (hole) size and the spindle diameter is enough to produce an out-of-true wheel. In the case of diamond or cubic boron nitride wheels, the hole in the wheel core is fitted to the spindle to a closer tolerance, which is another reason why these wheels do not require severe dressing or truing. An out-of-true, or out-of-balance, condition produces surface irregularities.

Precision dressing is usually done with either a *single diamond* or a *diamond cluster*. A cluster-type dresser contains a number of small-size diamonds that are imbedded in a metal matrix (Figure 25-1). This is the process for dressing a flat face. A single diamond is usually traversed across the face of a grinding wheel. The cluster-type dresser is often wider than the grinding wheel. In this case, the wheel is dressed without traversing the dresser.

Good grinding room practice involves dressing a small radius between the face and sides of the wheel. This practice avoids surface lines produced by a sharp-cornered wheel. Generally, a *dressing stick* is swung in a freehand arc lightly against the corner of a revolving wheel. This process is called *breaking the corner*.

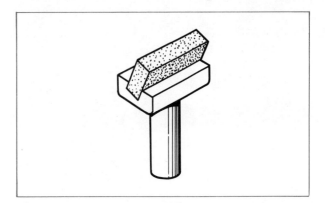

FIGURE 25-1 Cluster-Type Dresser with Small-Size Diamonds

Rough dressing, such as for offhand bench and floor grinding with coarse wheels, is usually done with a *star-* or *Huntington-type dresser.* It is also common practice to use a mounted grinding wheel to dress another wheel. Dressing a metal-bonded diamond wheel with another abrasive wheel is shown in Figure 25-2. This rotary diamond wheel dresser is friction operated. It may be bolted directly to a table or mounted on a magnetic chuck. This model is used on surface, cylindrical, and tool and cutter grinders.

WHEEL DRESSING TECHNIQUES AND ACCESSORIES

Form Dressing

More and more form work is being precision ground to shape and size. Slots, grooves, and contours may be ground by *form dressing* the grinding wheel. The form produced in the wheel is the reverse of the required contour. A deep or large-area contour is often premachined by another machining process. Parts may be rough milled and then finish ground to a precise shape. However, many contour machining processes are performed completely by grinding.

Wheels may be form dressed by a crush roll or with diamond rolls or dressing blocks. Commercially produced form dressing devices are also available. These devices are adjustable and may be set to generate any combination of forms in a grinding wheel.

Crush Forming

A *crush roll* is a duplicate form of the required shape to be ground. The crush roll is brought against a slowly revolving grinding wheel (100 to 300 sfpm) with considerable force. The roll *crushes the form* into the grinding face of the wheel. Figure 25-3 shows two applications of crush rolls in crush forming wheels. Figure 25-3A illustrates the setup for forming a surface grinder wheel with a crush roll. A heavy-duty application to form grinding a cylindrical grinder wheel is shown in Figure 25-3B. Note the rigid support for the crush roll. While crush forming produces very sharp abrasive grains, the resulting surface finish may not meet precise microfinish requirements.

Diamond Roll Form and Plated-Block Dresser

The action of a *diamond roll* form dresser is not as severe as a crush roll. As a result of less force, the abrasive grains are cut instead of crushed. The form generated on the wheel by using a diamond roll form dresser produces a finer-quality surface finish.

A *diamond-plated dressing block* (Figure 25-4) is often used for form dressing a wheel. As shown in the figure, the dressing block is preformed to the desired shape. It is then positioned on the table of the surface grinder. The revolving grinding wheel is brought down to depth while being traversed forward and backward over the block.

CONCENTRIC AND PARALLEL FACE DRESSING (TRUING)

Single-Point and Diamond Cluster Dressers

The *single-point diamond dresser* is widely used to produce a true wheel face. The face is usually concentric and parallel with the spindle centerline. Applications of a *single-point diamond dresser* are shown in Figure 25-5. The *single-point diamond dresser* may be mounted in

FIGURE 25-2 Dressing a Metal-Bonded Diamond Wheel with Another Abrasive Wheel

WHEEL PREPARATION AND GRINDING FLUIDS 493

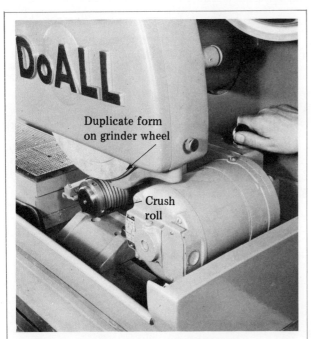

A. Crush Forming a Surface Grinder Wheel

B. Crush Forming a Cylindrical Grinder Wheel

FIGURE 25-3 Applications of Crush Rolls in Crush Forming Wheels

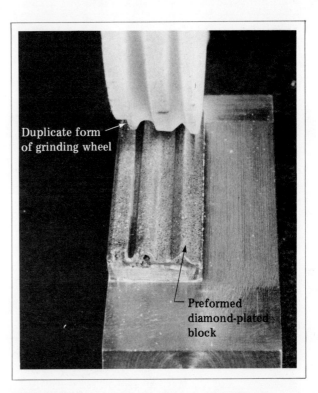

FIGURE 25-4 Preformed Diamond-Plated Dressing Block Used to Form a Surface Grinder Wheel

a simple holder (Figure 25-5A). Some surface grinders have the diamond mounted in an *overhead dresser* (Figure 25-5B). The diamond is fed to depth by means of a micrometer dial control. A dresser traverse handle is moved to the right and left to produce a true and parallel face. Figure 25-5C indicates the positioning of a diamond dresser for dressing a cylindrical or tool and cutter grinding wheel.

There are many types of diamond dressers besides the single-point dresser just described. Three common designs of diamond tools for straight face dressing are shown in Figure 25-6. The multi-point dresser is used primarily in production machines. It has a faster dressing rate than the single-point diamond dresser generally used in the toolroom or jobbing shop.

GENERAL PROCEDURES FOR USING DIAMOND DRESSERS

The following characteristics are illustrated in Figure 25-7:

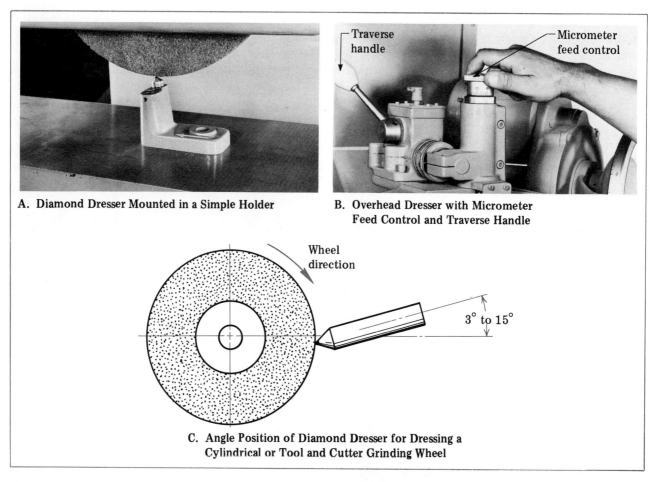

FIGURE 25-5 Applications of a Single-Point Diamond Dresser

—The diamond is set at a negative angle, often called the *drag angle;* a 10° to 15° angle is general practice. (*Note:* The purpose of setting the cutting diamond at an angle is to prevent digging in, chatter, and damage to the diamond.)

—The diamond is positioned from 1/8" (3mm) to 1/4" (6mm) past or just below the center line of the grinding wheel.
—The position and distance depend on the gringing process—for example, cylindrical grinding is below, while surface grinding is

For dressing extremely hard, coarse grinding wheels

For general purpose dressing

Cluster type for high finish and wide traverse dressing

FIGURE 25-6 Basic Designs of Multi-Point Diamond Tools for Straight Face Dressing

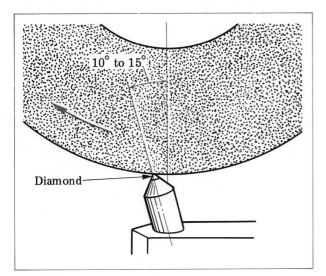

FIGURE 25-7 Relationship of Diamond Dresser and Wheel Center Lines (Dressing a Surface Grinder Wheel)

HOW TO PREPARE CONVENTIONAL ALUMINUM OXIDE AND SILICON CARBIDE WHEELS

Mounting a Wheel and Sleeve (Adapter) Unit

Many grinding wheels are mounted permanently on a sleeve (adapter) in order to permit the wheel to be removed quickly as a unit. As wear and/or a flaw develops, the wheel is taken off and replaced on the adapter. The adapter also reduces the number of truing operations and prolongs wheel life.

STEP 1 Select the grinding wheel to fit the job requirements. The hole size must fit the sleeve diameter. The width of the sleeve must be adequate to accommodate the width of the wheel.

STEP 2 Run the usual checks for wheel soundness and for condition of the washers, flanges, spindle, sleeve, and so on.

STEP 3 Assemble the sleeve (adapter) in the grinding wheel.

Note: The bore of the wheel must be inspected and measured against the diameter of the sleeve. A close fit is important.

STEP 4 Place the washer on the wheel sleeve.

STEP 5 Screw the wheel sleeve nut (outer flange) by hand until it tightens against the washer.

STEP 6 Hold the rim of the wheel with one hand. Tighten the wheel sleeve nut with a pin or spanner wrench. Use enough force to prevent the wheel from turning during the grinding process.

Caution: Overtightening may cause the wheel to crack.

STEP 7 Place the assembled wheel and sleeve unit on the tapered end of the spindle. Align the sleeve unit keyway with the sleeve key.

STEP 8 Draw the wheel and sleeve unit tight on the spindle by turning the spindle nut. Apply the right amount of force to hold the unit securely.

STEP 9 True and dress the wheel.

past the vertical center line. (Note: Diamond dressers that are designed as a permanent part of a machine are positioned on center.)

—The dresser is brought carefully into contact with the high point of the revolving wheel; the first cutting is done across this high point; and more and more of the wheel periphery is dressed at each pass (traverse) across the wheel face. (Note: The diamond is generally infed about 0.001" (0.02mm) per pass until the complete face is true and parallel.)

—It is recommended that diamond dressing to be done wet—that is, there must be a continuous stream of coolant.

—If the wheel is dressed dry because the grinding processes are to be performed dry, it is important to pause after every two or three passes across the wheel face to permit the diamond to cool down. (Note: If coolant reaches a hot diamond, there is a possibility of shattering it.)

—The diamond must be turned frequently to prevent grinding a flat surface on the diamond, thereby reducing its efficiency. (Note: When a diamond has numerous sharp edges, it is possible to produce a free-cutting wheel face.)

—The diamond must always be pointed in the direction of rotation of the grinding wheel.

Remove a Grinding Wheel and Sleeve (Adapter) Unit

STEP 1   Use a wheel sleeve puller appropriate to the size and design of the sleeve (adapter).

STEP 2   Loosen and remove the spindle nut.

    Spindle nut threads are normally left-hand.

STEP 3   Thread the outer body of the wheel puller into the threaded wheel sleeve. Turn to full depth to engage the maximum number of threads.

STEP 4   Tighten the cap screw in the wheel puller. The force applied against the end of the spindle causes the wheel sleeve and spindle tapers to disengage.

Caution: Heavy assembled wheel and adapter units should be lifted by mechanical devices. These units must be stored in suitable compartments to prevent damage. The spindle nut should be replaced on the spindle.

Truing and Dressing a Surface Grinder Wheel

STEP 1   Select a sharp single-point diamond.

STEP 2   Examine the diamond to see that it is not flat, cracked, or loose in its setting.

STEP 3   Mount the diamond in a block or wheel truing fixture. The diamond mounting should bring the centerlines of the diamond and wheel at a 10° to 15° intersecting angle.

STEP 4   Mount the grinding wheel (or combination unit) on the spindle.

STEP 5   Position the fixture or diamond holder. A fixture is usually strapped to the table. Place a holder block on a magnetic chuck across as many magnetic laminations as possible.

Note: A thin piece of paper slightly larger than the area of the holder may be placed under the holder to permit it to be removed easily without scratching the chuck face.

STEP 6   Energize the magnetic chuck. Test to see that the holder is properly seated and secure.

STEP 7   Replace and secure the wheel guard.

Caution: Be sure to use safety goggles or a protective shield.

STEP 8   Start the grinder. Run it free for a minute or so. Position the grinding wheel so that the centerline is 1/8″ (3mm) to 1/4″ (6mm) to the right of the diamond.

STEP 9   Bring the grinding wheel down slowly until contact is made at the high point.

Note: In wet grinding, use a coolant. In dry grinding, be sure the diamond is air cooled after each two or three passes.

STEP 10   Feed the diamond across the wheel face by using the cross feed handle.

Note: A rapid traverse rate produces a fast cutting wheel for rapid stock removal. A high-quality surface finish is sacrificed. A slow traverse rate produces a slower cutting wheel with a high-quality surface texture. A medium traverse rate is customary for general-purpose grinding.

STEP 11   Continue to feed the diamond 0.0005″ (0.01mm) for each pass across the wheel face. The final feed for precision grinding may be reduced to 0.00015″ (0.004mm).

STEP 12   Dress the wheel corners with a hand dresser.

STEP 13   Stop the coolant flow a minute or more before the machine. Clean the machine and remove the diamond fixture or holder.

Truing and Dressing a Cylindrical Grinder Wheel

STEP 1   Secure the diamond holder on the table or footstock.

STEP 2   Set the center of the holder on the horizontal centerline of the wheel. The vertical angle of the holder should be between 3° to 15°. (The position of a diamond holder in a footstock is shown in Figure 25-8).

STEP 3   Replace all guards. Start the machine.

Caution: Use an eye protection device. Stand to one side. Allow the bearings and machine to reach operating temperature.

STEP 4   Adjust the wheel until the highest point is in contact with the diamond.

WHEEL PREPARATION AND GRINDING FLUIDS 497

FIGURE 25-8 Diamond Dresser Setting in Footstock for Dressing a Cylindrical Grinder Wheel

coarser-grained wheel. Unless the machine is equipped with a mechanical traverse mechanism, the operator must control the traverse rate. It must be uniform across the wheel face.

In general practice, traverse rates from 12 to 24 ipm produce sharp cutting grains. The feed rate depends on the nature of the operation, the grinding requirements for stock removal, dimensional tolerances, surface finish specifications, the grinder, and the wheel. A medium cutting grain results from using traverse rates from 6 to 12 ipm. A fine cutting grain is produced at a traverse rate of 2 to 6 ipm.

PROPERTIES AND FEATURES OF CBN AND DIAMOND WHEELS

Cubic Boron Nitride (Borazon CBN) Wheels

The hardness range of CBN is between silicon carbide and the diamond. CBN has a Knoop hardness reading of 4,700. This reading compares to 7,000 for the diamond, which is the only material harder than CBN. The extremely hard CBN crystals have sharp corners, remain sharper longer, and normally do not load. Therefore, in many machine grinding operations where CBN wheels are used, the following note appears: USE AS IS. DO NOT DRESS THE WHEEL. The cooler cutting action of CBN produces less thermal damage to the workpiece.

Tests show that CBN wheels are capable of removing a constant quantity of material before compensation must be made for wheel wear. Cutting tool life is also increased for cutters that are sharpened with CBN wheels. This is the result of sharper and cooler cutting action which reduces surface damage to cutters. While the initial cost of CBN and diamond wheels is many times greater than conventional wheels, overall unit costs are considerably less on production runs.

Like diamond grinding wheels, CBN wheels consist of a core. This core is surrounded by a bond in which abrasive crystals are impregnated. The core is precision machined to fit the spindle within a closer tolerance than is allowed for conventional wheels. The close fit helps the wheel to stay in balance.

Diamond Grinding Wheels

Diamond grinding wheels are generally produced in resinoid, metal, and vitrified bonds.

Note: The coolant supply must be flowing before there is any contact. Otherwise, it is possible to crack and damage the diamond.

STEP 5 Feed the diamond 0.0005" (0.01mm) each pass.

STEP 6 Continue to feed and traverse the wheel each pass.

Note: As a rule, a medium traverse rate is used to produce grain sharpness for general-purpose grinding.

EFFECT OF TRAVERSE RATE ON WHEEL CUTTING ACTION

Skill as a grinding machine operator or craftsperson depends, in part, on judgment in wheel dressing. The same wheel may be dressed for rough grinding and finish grinding. The faster the dresser traverses the face of a revolving grinding wheel, the sharper and more open the grains that are produced. Sharp grains are better suited for fast cutting action and heavy cuts. The slower the traverse, the finer the grains. With a slow traverse, the diamond dresser cuts the abrasive and dulls the grains slightly. Such grains cut less and produce a finer surface finish.

Thus, with experience, the worker may change the cutting qualities of a wheel. A coarse-grained wheel may be dressed to cut as a finer-grained wheel by controlling the speed of traverse while dressing. However, it is not possible to use a fine-grained wheel and dress it to cut as a

TABLE 25-1 Diamond and Cubic Boron Nitride Grinding Wheel Markings

$$\underset{\text{Abrasive}}{\text{ASDC}} \quad \underset{\text{Grit Size}}{100} \quad \underset{\text{Grade}}{\text{N}}$$

Abrasive		Grit Size			Grade		
					Res.	Metal	Vit.
D = Natural Diamond	24	120	500	H R	L R	J R
SD = Manufactured	36	150	500S	J Q*	N	L T
ASD = Armored (nickel)	46	180	600S	L T*	P	N
ASDC = Armored (copper)	60	220	800S	N W*	Q	P
CB = Borazon	80	240	1200S			
	100	320	1500S	(*For Borazon only)		
	100S	400	2000S			
		400S				

The core is molded into one of the wheel type forms. The core material may be steel, an aluminum-filled resin composition (aluminoid); or bronze, copper, or plastic. Three common bonds are generally used in the manufacture of diamond grinding wheels:

- —*Resinoid bond*, which produces fast and cool cutting action for grinding carbide cutters and for other precision grinding processes;
- —*Metal bond*, which is a harder bond that resists the wearing effect produced, for example, by offhand grinding single-point parts and tools;
- —*Vitrified bond*, which combines grinding characteristics of both the resinoid bond and the metal bond to produce an intermediate fast and cool cutting action and a high resistance to wear.

Diamond Wheel Markings

The depth of the diamond section varies from 1/16" (1.6mm) to 1/4" (6.4mm). The greater depths are used for heavy stock removal, large areas of wheel contact, and heavy production demands. Diamond wheel markings differ from the markings used on conventional wheels. Diamond and CBN grinding wheel markings are given in Table 25-1. The only explanation needed deals with concentration. The *concentration number* represents the ratio between the diamond grains and the bonding material. A high concentration (100) is desirable for heavy stock removal.

A low concentration (25) is adequate for light cuts and where low heat generation is a requirement. The 25, 50, and 75 concentrations give the relationship of abrasive grains to bond by comparison to the 100 concentration.

CONDITIONING DIAMOND AND BORAZON (CBN) GRINDING WHEELS

Great care must be taken in conditioning diamond grinding wheels. If the wheel is not concentric, the edges tend to chip, greater wheel wear occurs, and surface finish and dimensional accuracy are impaired. Therefore, diamond wheels—and borazon(CBN) wheels—must run true. As a general rule, the tolerance for outside diameter wheel runout in cylindrical grinding must be within 0.0005" (0.01mm). The tolerance for extremely precise work is 0.00025" (0.006mm). The grinding face of a cup wheel should be held to 0.0001" (0.002mm) minimum. Testing the trueness requires the use of a *tenths dial indicator*.

Diamond and cubic boron nitride wheels are never trued or dressed with a diamond dresser. A coolant should always be used when truing a diamond wheel. A small amount of water as a spray is applied to CBN. Three common devices for truing diamond wheels are shown in Figure 25-9. A standard dressing stick (Figure 25-9A) or brake-controlled dresser (Figure 25-9B) is used on surface grinder applications. On cylindrical and tool and cutter grinding, a toolpost grinder (Figure 25-9C) is occasionally used for truing the wheel.

WHEEL PREPARATION AND GRINDING FLUIDS 499

75	B	69	1/8
Concentration	Bond Type	Bond Modification	Depth of Diamond Section
Low = 25 50 75 High = 100 (Not shown for Borazon)	B = Resinold M = Metal MC = Metal carbide V = Vitrified BA = Borazon, wet BB = Borazon, dry	Numeral to designate special bond modification. Example Resinoid—76 and 69. (Symbol optional)	1/16 1/8 1/4

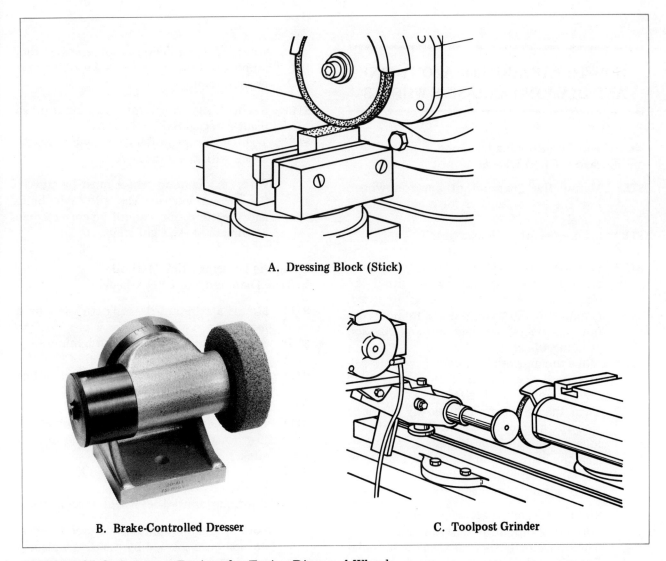

A. Dressing Block (Stick)

B. Brake-Controlled Dresser

C. Toolpost Grinder

FIGURE 25-9 Common Devices for Truing Diamond Wheels

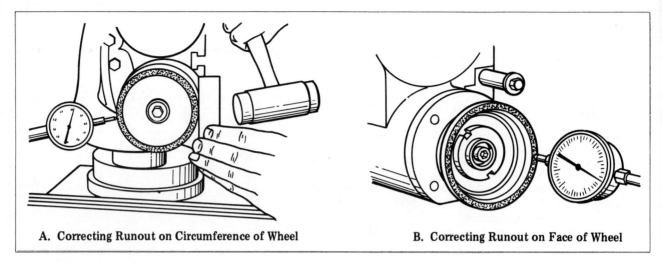

FIGURE 25-10 Correcting Runout on Diamond and Borazon Grinding Wheels

HOW TO PREPARE BORAZON (CBN) AND DIAMOND ABRASIVE WHEELS

Truing and Conditioning Diamond and Borazon (CBN) Wheels

STEP 1 Mount the diamond or borazon wheel. Use the same steps as for a conventional wheel.

STEP 2 Secure the wheel flange nut.

STEP 3 Position a 0.0001″ (0.002mm) dial indicator to read the trueness of either the circumference or the face, as required. (Figures 25-10A and B illustrate how runout is corrected on diamond and borazon grinding wheels.)

STEP 4 Move the indicator toward the wheel. Adjust until the pointer is ready to record any variation. Set the reading at zero.

STEP 5 Turn the grinding wheel slowly by hand through a complete revolution. Note the location and amount of runout, if any.

STEP 6 Tap the high spot on the circumference of the wheel carefully and lightly. Use a soft block against the wheel.

Note: Steel or brass shims are used for correcting runout on face grinding wheels. The shims are placed between the shoulder flange and the back of the wheel core.

Note: If the indicator is not removed, then tapping is done away from the indicator to prevent damage.

STEP 7 Recheck the wheel until no runout movement is registered.

STEP 8 Tighten the wheel flange nut securely. Recheck with the indicator.

Note: A diamond wheel must be trued if the runout exceeds the allowable limits and the condition cannot be corrected according to these eight steps.

Using the Dressing Stick Method to True Diamond and CBN Wheels

STEP 1 Mount a standard dressing stick in a holding device.

STEP 2 Start the spindle and run for a minute.

STEP 3 Position the coolant nozzle. Direct a flow of coolant between the wheel and dressing stick.

STEP 4 Bring the wheel so that the high spot just touches the surface of the dressing stick.

STEP 5 Take cuts of 0.0001″ (0.002mm) each pass.

STEP 6 Continue until the wheel face is trued.

Note: Resinoid-bonded diamond wheels are sometimes trued by grinding a piece of low-carbon steel.

Using a Brake-Controlled Device to True Diamond and CBN Wheels

STEP 1 Select the appropriate grade and grain size to accommodate the diamond or CBN wheel to be trued and dressed.

Note: A metal- or vitrified-bonded diamond wheel is trued with a 100 grain, M grade, silicon carbide vitrified wheel. A resinoid-bonded diamond wheel is trued with a 100 grain, M grade, aluminum oxide vitrified wheel. A resinoid-bonded borazon (CBN) wheel is trued with an aluminum oxide or silicon carbide stick or wheel. A fine grit (400 or finer) and G hardness bond are recommended.

STEP 2 Position the brake-type truing device. The holder must have about a 30° angle with the vertical centerline of the wheel.

STEP 3 Run the spindle for a minute. Move the diamond or CBN wheel into contact with the brake-controlled dresser wheel.

STEP 4 Retard the dressing wheel speed by applying the spindle wheel brake. This action results in scrubbing and truing of the diamond or CBN wheel face.

Note: This truing is usually followed by light stick dressing. This dressing improves the cutting action of the wheel.

Using the Toolpost Grinder Method to True Diamond and CBN Wheels

STEP 1 Select an appropriate dressing wheel for the diamond or CBN wheel to be trued and dressed.

STEP 2 Secure the wheel in the toolpost grinder spindle.

STEP 3 Position and mount the grinder in the toolpost.

STEP 4 Reduce the operating speed of the diamond or cubic boron nitride wheel spindle to about a quarter of its normal speed.

STEP 5 Feed the toolpost grinder wheel until contact is just made with the diamond or cubic boron nitride wheel.

STEP 6 Start the flow of cutting fluid before contact is made.

STEP 7 Take light passes across the wheel face. Continue until the diamond or CBN wheel is trued.

STEP 8 Dress the diamond wheel after truing. Use either a dressing stick or lump pumice. Although a CBN wheel seldom requires dressing, there are special dressing sticks used for CBN and diamond wheels.

FUNCTIONS AND TECHNIQUES OF WHEEL BALANCING

In *wheel balancing*, the weight of a grinding wheel is distributed evenly so that no centrifugal forces are set up at high speed. Grinding wheels are balanced to eliminate vibration and the resulting bad effects. Vibration of a grinder that has been operating smoothly usually indicates the wheel is out-of-balance. The formation of wide, evenly spaced chatter marks in a checkerboard pattern results from the vibrating action of the grinding wheel. Under these conditions, fine surface finishes and dimensionally accurate ground surfaces are impossible to produce.

From a safety point of view, the forces set up in a high speed out-of-balance grinding wheel may cause it to fracture. Damage to the operator, machine, and workpiece are possible. Unbalanced wheels require more frequent dressing, shorten wheel life, and increase tooling and machining costs. Additional wear is produced on the spindle and other moving parts of the grinder.

All large sizes of grinding wheels are balanced to produce fine surface finishes and parts that are machined to extremely close tolerances. General-purpose wheels that are 10" (250mm) and smaller in diameter are customarily not balanced after they are trued and dressed. The manufacturer's balancing is usually adequate.

Parallel and Overlapping Disk Balancing Ways

Parallel and overlapping disk balancing ways are two common devices used in balancing grinding wheels. The *parallel balancing ways* (Figure 25-11) consist of two parallel knife-edge surfaces. These ways are part of a solid frame. The device is leveled so that the two parallel ways are on a horizontal plane.

A grinding wheel is balanced by placing the wheel and its adapter on a *balancing arbor*. The assembly is placed carefully on the ways. The wheel is allowed to turn slowly until it comes to rest. The heavy spot is on the underside of the arbor. This spot is marked with chalk. Two other lines are then drawn at a right angle to the heavy spot.

502 GRINDING MACHINES

FIGURE 25-11 Parallel Balancing Ways

Balancing weights in the flanges of the adapter (Figure 25-12A) are positioned for a *trial balance.* The wheel is retested. Further adjustments of the balancing weights are made until the wheel remains at rest in any position on the balancing ways. Wheel mounts are available with two, three, or more balancing weights (Figure 25-12B).

The overlapping disk balancing ways consist of four perfectly balanced and free-turning overlapping disks. The grinding wheel assembly and balancing arbor are placed across the overlapping edges of the disks. The assembly is allowed to turn until the heavy spot comes to rest at the lowest point. The balancing weights are then adjusted the same as when parallel balancing ways are used. The overlapping disk device has the advantage in that it does not require precise leveling.

Production grinders are now designed so that the grinding wheel may be balanced without removing it from the machine. At any sign of imbalance, the operator raises a balancing lever on the automatic wheel balancer to unclamp the balancing mechanism and free up the moving parts. These parts respond to the forces set up by the imbalance. When the steel balls in the balancing device settle in position (which bring the wheel in balance), the lever is lowered and locks the balancing balls or rolls in the balanced position. When the pointer of the vibration gage on the machine remains almost stationary, the wheel is balanced.

FIGURE 25-12 Positioning of Weights to Balance a Grinding Wheel

TABLE 25-2 Truing, Dressing, and Balancing Problems with Probable Causes and Corrective Action

Problem	Probable Cause	Corrective Action
Chatter marks	—Out-of-balance	—Rebalance after truing operation
		—Balance carefully on own mounting
	—Out-of-round	—True before and after balancing
	—Acting too hard	—Use faster dressing feed and traverse
Scratching of the work	—Foreign matter in wheel	—Dress the wheel
	—Coarse grading	
Checking or burning of the work	—Wheel too hard	—Use a faster dress rate to open the wheel face
Diamond lines in work	—Dressing too fast	—Slow dress the wheel face
Inaccuracies in work	—Improper dressing	—Check alignment of dressing process
Rate of cut too slow	—Dressing too slow	—Increase the dressing rate to open the wheel face
Wheel acting too soft (not holding size)	—Improper dressing	—Slow down the traverse rate
		—Use lighter dressing feed
Wheel loading	—Infrequent dressing	—Dress the wheel more often

GRINDING PROBLEMS RELATED TO TRUING, DRESSING, AND BALANCING

Any one or combination of grinding wheel, machine, and operator variables affects grinding. Eight common truing, dressing, and balancing problems with probable causes and corrective actions appear in Table 25-2. If the problem is not corrected, other variables must be checked. These variables deal with the characteristics of the grinding wheel, the workpiece, cutting fluid, machine condition, and work processes.

GRINDING FLUIDS

Functions of Grinding Fluids

The terms *grinding fluids*, *grinding coolants*, and *cutting coolants* are used interchangeably in the shop. These fluids serve three prime functions:

—To remove the heat generated during the grinding process, particularly on the plane (and points) of contact where the heat is most intense (Figure 25-13A);
—To reduce the amount of power required for grinding, to produce finer surface textures, to limit wheel wear, and to prevent cracks from forming in hardened steels (Figure 25-13B);
—To lubricate the work and wheel particles and chips so that they are not drawn between the wheel and work surface (Figure 25-13C).

Kinds of Grinding Fluids

Three general kinds of grinding fluids are used. *Water-soluble chemical grinding fluids* are the most common. They are particularly useful in medium to heavy stock removal processes. These fluids are transparent (a valuable aid to the operator) and have excellent cooling properties.

Rust inhibitors and detergents are added to the solutions to provide good adhesion and cleaning qualities. The lubricating properties are improved by adding water-soluble polymers. Fluid life is increased by adding bactericides and disinfectants in the solutions. They control bacteria growth and retard rancid conditions in the fluids.

Water-soluble oil grinding fluids combine the cooling qualities of water and the lubricating advantages of oil. These oil/water fluids form a milky solution. An emulsifying agent is added to the water and oil (either natural or synthetic).

Caution: The additives must be controlled. Too strong a disinfectant solution may produce skin irritation or possible infection.

Water-soluble oil grinding fluids are used for light to moderate stock removal processes. Parts may be held to reasonably accurate dimensional and surface finish tolerances.

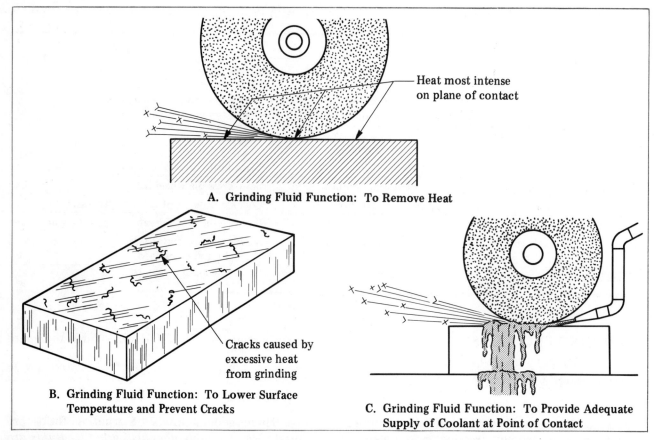

FIGURE 25-13 Grinding Conditions and Applications of Grinding Fluid

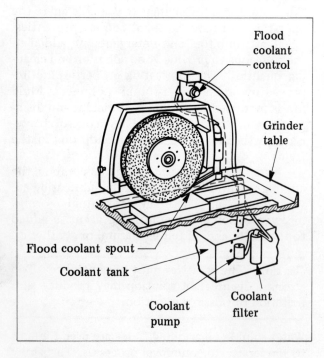

FIGURE 25-14 Features of Flood Coolant System

Straight oil grinding fluids are the most costly of the three fluids. These fluids provide excellent lubrication. However, they are not as effective as the water-soluble fluids in dissipating heat. Straight oil grinding fluids are used primarily on very hard materials requiring high dimensional accuracy and surface finish and long wheel wear life. Oil grinding fluids are especially practical in thread and other heavy form grinding processes.

COOLANT SYSTEMS

Wet Grinding Methods

There are three principal methods of supplying grinding fluids at the point of cutting action between the workpiece and the grinding wheel. These methods of wet grinding include the *flood*, *through-the-wheel*, and *mist coolant systems*.

Flood Coolant System. The basic features of the flood coolant system are identified in Figure 25-14. The grinding fluid is forced from

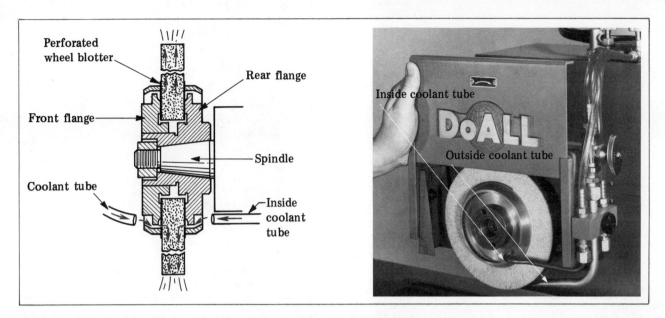

FIGURE 25-15 Features of Through-the-Wheel Coolant System

a reservoir through a nozzle. The fluid floods the work area and is then recirculated through the system. Both the volume and the force of the coolant are controlled to overcome air currents around the wheel. Otherwise, these currents tend to force the fluid away from the required point of contact. The fluid system provides for straining out foreign particles and producing a recycled, clean coolant.

Through-the-Wheel Coolant System. The features of the through-the-wheel coolant system are illustrated in Figure 25-15. The grinding fluid is pumped through two coolant tubes. These tubes are on both sides of the grinding wheel. The wheels have flanges. As the fluid is discharged into a recess on each flange, the centrifugal force of the revolving wheel carries the fluid through a series of holes in the flanges, washers, and blotters to the center of the wheel. The flow continues with force to the periphery of the wheel. The porous feature of vitrified wheels is applied in flowing the coolant through the wheel.

This system keeps the wheel face flushed free of chips and abrasive particles so that there is a cool cutting action. Also, the operator is able to see the grinding operation better than with the flood system. The flood and through-the-wheel coolant systems are sometimes combined to produce a highly efficient coolant system.

Mist Coolant System. An application of the mist coolant sytem to a surface grinder process is shown in Figure 25-16. This system works on an atomizer principle. Air at high velocity flows through a T-connection. One branch includes the coolant supply. By passing the air through the connection, a small volume of liquid is atomized into a large mass with air. The coolant stream emerges from the nozzle as a mist. Cooling action is produced by the air and the

FIGURE 25-16 Application of Mist Coolant System to Surface Grinder Process

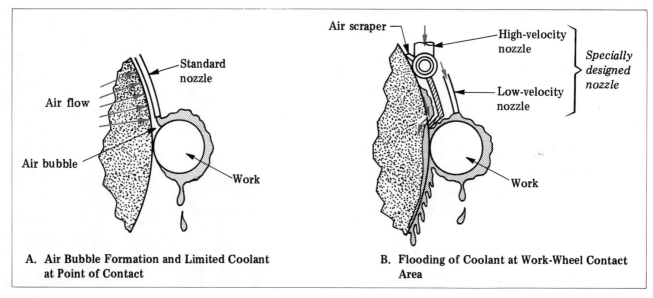

FIGURE 25-17 Effect of Coolant Nozzle Design on Coolant Application at Point of Contact

evaporation of vapor. The material and abrasive particles are blown from the grinding area. One advantage of this system is that the operation is unobstructed.

Importance of Coolant Nozzle Design

Nozzle designs are important in concentrating the cutting fluid at the *work-wheel interface*, the immediate area of cutting action between the wheel and the workpiece. Figure 25-17 illustrates the effect of coolant nozzle design in the application of a coolant at the point of contact. With a standard nozzle (Figure 25-17A), the air flow produced by a highly revolving grinding wheel causes an air pocket or bubble to form. This bubble causes the fluid to be blown away from the area where it is needed. With a specially designed nozzle consisting of a high-velocity nozzle, a low-velocity nozzle, and an air scraper (Figure 25-17B), flooding of the work-wheel interface and the immediate adjacent area occurs. This condition is needed for effective cooling, lubricating, and chip removal.

Exhaust Systems for Dry Grinding

Dry grinding refers to operations that are carried on without a liquid fluid. An exhaust system is needed to remove grinding dust, abrasive material, and chips from the work, grinding wheel, and surrounding area. This system is important for the protection of all machines in the work area. Removal of the dust from the air also constitutes an industrial safety requirement to protect the operator.

The exhaust system usually consists of a nozzle attached to the wheel guard. A flexible tubing leads from another part of the guard to the exhaust attachment. The foreign particles that are removed (drawn) from the grinding area and around the wheel and guard are sucked through a spiral separator. In this separator, the heavier particles are removed by centrifugal force (the swirling action in the separator). The air mixture is then passed through filters. The filters remove most of the finer particles that are in the air mixture. The heavier particles accumulate in a container on the machine and are removed. The fine particles cling to the filters. As they become loaded, the filters are replaced.

The exhaust nozzle is placed as close to the wheel as possible to permit the most effective exhausting. An air supply is sometimes blown at the contact area for air cooling purposes. Care must be taken in adjusting the nozzle to pervent interference between the guard, workpiece, wheel, and table.

SYSTEMS FOR RECIRCULATING CLEAN GRINDING FLUID

Quality grinding depends on the condition and cleanliness of the grinding fluid. The *swarf* (particles of abrasive, chips, and other foreign matter) must be removed before the fluid returns to the point of grinding action. Some grinding machines have a *settling tank*. The heavy particles settle on the bottom of the tank. The clean fluid is pumped through screens back into the system. On heavy-duty grinding operations, the tank is fitted with a conveyor that moves the chips out of the tank.

Some other cleaning systems require a *removable filter*. When filled with swarf, the filter is replaced. Filters retard the volume of coolant flow as the openings must be restricted to remove minimum sizes of particles.

Production grinding machines are also designed with centrifugal, cyclonic, or magnetic separators. The *centrifugal separator* spins at high speed. The grinding fluid is spun in an inner bowl. The centrifugal action causes the swarf to spin outside. The clean coolant stays in the center section of the bowl. It is then recirculated through the coolant system.

The *cyclonic separator* (Figure 25-18) separates the swarf from clean coolant by centrifugal force. The dirty coolant is fed into the filter. The coolant is swirled around in a cone. The clean coolant is fed out through the top of the cone. The swarf is drawn off at the bottom. This type of fluid cleaner is available for light- or heavy-duty cleaning. The larger units have automatic sludge removal, large settling tanks, and large dirt discharge openings.

The *magnetic separator* is restricted to grinding iron and steel. A magnetized drum rotates through the coolant flow. Chips of metal are picked up around the drum. They are then scraped off by a blade and deposited in a sludge container.

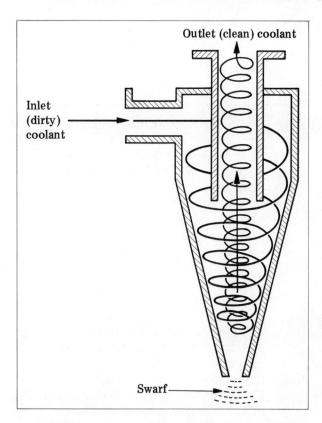

FIGURE 25-18 Schematic Section of a Cyclonic Separator

DESIRABLE PROPERTIES OF GRINDING FLUIDS

A summary of the desirable properties of a good grinding fluid follows. The fluid:

—Has good wetting properties and lubricates the work-wheel interface and surrounding areas;
—Absorbs and dissipates heat at a rapid rate to maintain a comparatively cool temperature under working conditions;
—Permits the settling out of chips, abrasive grains, and other particles fast enough to prevent their recirculation;
—Provides no health hazard to the machine operator or possible damage to the machine parts or grinding wheel;
—Provides a protective, noncorrosive coating for the surface of the ground part and other machined surfaces;
—Breaks down foam rapidly for easy flow-off and recycling through the coolant system;
—Emulsifies readily and without the application of heat;
—Is noninflammable so that there is no possibility of any burning action;
—Has high resistance to becoming rancid or producing objectionable cutting or storing odors.

These properties may be extended to meet special requirements for dimensional and surface accuracy.

HOW TO CHECK AND MAINTAIN THE QUALITY OF THE GRINDING FLUID

STEP 1 Check the odor of the grinding fluid. If it smells sour or rancid, it is necessary to drain the supply from the system.

STEP 2 Determine whether the coolant is clean, dirty, or old. A simple test is to rub a drop of the solution between the fingers. The presence of foreign particles may be felt easily. A gummy feeling indicates an old solution.

STEP 3 Check the amount of chips and other foreign particles in the settling compartment. All screens and filters require checking for cleanliness. If needed, remove the sediment. Clean the filters, nozzle, tank, and trays and other parts of the entire system.

STEP 4 Cycle the coolant through the system. Check for the proper filtering and settling of abrasive grains, chips, and other foreign matter. Recheck the coolant to see that it is clean.

STEP 5 Check the coolant level. Losses result during machine operation and from evaporation. If the level is not maintained, the fluid circulates faster and becomes warmer. There is also the possibility of recirculating swarf.

STEP 6 Check the strength of the solution with a cutting fluid gage, particularly when water-based solutions are used. Water and/or concentrates should be added to bring the solution up to strength and coolant level.

Note: Too rich a solution may affect cutting efficiency and surface quality. Too lean a solution may produce higher scrap losses.

STEP 7 Keep a continuous check to be sure machine lubricating oils are not being fed into the coolant system, especially when water-based coolants are used. Small quantities of oil severely affect coolant efficiency and the surface texture.

STEP 8 Aim and adjust the coolant nozzle to reach the point and plane of contact between the workpiece and wheel. The flow must be checked under actual conditions to be sure that air currents are not flowing the coolant away from the required area.

SAFE PRACTICES IN GRINDING WHEEL PREPARATION

The following safe practices are summarized from the ANSI "Safety Code for the Use, Care, and Protection of Abrasive Wheels."

—Check each grinding wheel for cracks and other surface imperfections that can produce an out-of-balance condition and possible wheel fracturing.

—Fit the wheel to the adapter mounting flange or spindle. A close sliding fit is needed for close wheel balance. All contacting surfaces of the wheel and mounting must be burr free and perfectly clean.

—Place a compressible washer wheel blotter of proper thickness and diameter between each flange and each side of the wheel.

—Use wheel flanges that are at least one-third the diameter of the grinding wheel. These flanges must be recessed from the center. Force may then be applied away from the center of the wheel (which is structurally the weakest point). The amount of force must be adequate to hold the wheel from turning during grinding.

—Tighten any retaining screws on an adapter so that they hold the wheel securely. However, they should not be so tight as to crush the wheel or spring the flanges.

—Use a collet or wheel adapter wherever practical to save setup and truing time and to reduce wheel wear.

—Secure the wheel guard. It must cover more than half the wheel.

—Ring test each grinding wheel. Check the recommended sfpm. Be sure the spindle RPM does not exceed the speed limitations.

—Stand to one side and out of line with the wheel when starting a grinding wheel.

—Bring the machine up to operating speed. Run it in for a minute or more. Also turn on the coolant supply before starting a grinding operation.

—Wear only approved safety glasses or other face protective devices.

—Grind only on the face(s) that is specifically designed for a particular grinding operation.

—Remove jewelry or any item of loose clothing that may be caught by moving parts.

—Stop the coolant flow a short time before the grinding wheel is stopped to prevent a heavy side of the wheel from developing and throwing the wheel out of balance.

- —Keep at the machine only the wheels that are to be used. These wheels should be mounted on a pegboard or stored in a compartment especially designed for each type and size of wheel.
- —Use a corrugated, soft material to separate wheels. Protect them from touching each other and from coming into contact with metal tools or parts.
- —True and dress a wheel with a diamond or other dresser pointed away from the centerline of the wheel. Mount diamond dressing point holders at an angle to the grinding wheel face.
- —Start to true and dress a wheel by carefully bringing the wheel face so that its high point makes contact. Successive cuts are then taken to produce a concentric wheel face.
- —Balance wheels for highly accurate machining, production of a high-quality surface finish, and protection against undue wheel wear.
- —Place the wheel assembly carefully on the balancing ways. Avoid dropping the unit. Dropping damages the ground ways and impairs balancing accuracy.
- —Set the arbor axis at 90° to the ways to ensure that the wheel will not strike against the sides of the balancing device.
- —True and dress diamond wheels with aluminum oxide or silicon carbide. A diamond is never used to dress another diamond wheel.

TERMS USED IN GRINDING WHEEL PREPARATION AND GRINDING FLUIDS

Form dressing	Process of forming a wheel in a reverse contour to the form to be ground. Preparation of a grinding wheel to grind grooves, slots of varying shapes, and other profiles.
Crush roll	A duplicate roll that crush forms a desired shape in a grinding wheel.
Diamond-plated dressing block	A block preformed to conform to a required contour. A formed block with diamond grains. A block used to produce a reverse form in a grinding wheel for purposes of form grinding.
Diamond cluster dressers	Multiple-point diamond dressers used primarily on production grinding machines.
Drag (negative) angle	The angle formed by the centerlines of a diamond holder and the grinding wheel. The angle at which a dresser is held in relation to the centerline of the wheel.
Wheel sleeve (adapter) mounting	The practice of mounting a grinding wheel permanently (during its wear life) on a wheel sleeve (adapter). A technique that saves setup and truing time and extends wheel productivity and work quality.
Dressing traverse rate	The speed with which a dressing tool is moved across the face of a grinding wheel. A factor in controlling the sharpness of abrasive grains and the quality of surface finish.
Diamond and CBN wheel mark	A system of standard wheel symbols and values. Manufacturer's specifications of the characteristics and structure of diamond and CBN wheels.
Concentration	A ratio between the diamond grains and bonding material. Four common ratios (25, 50, 75, and 100) for designating the relationship of grains to bond in a diamond wheel.
Conditioning diamond and CBN wheels	Steps dealing with mounting, testing, and correcting any runout of diamond or cubic boron nitride wheels.

Brake-controlled dressing device	A grinding wheel truing device that removes abrasive grains by a scrubbing action and that retards the spindle speed of a grinding wheel that is in contact with such a device.
Work-wheel interference	Generally refers to the immediate area of cutting action between a grinding wheel and a workpiece.
Wheel balancing	Distributing the weight (mass) of a grinding wheel so that no centrifugal forces set up at high speed. Neutralizing a heavy spot in a grinding wheel. An internal condition of a grinding wheel where it revolves in balance. (A grinding wheel that is balanced cuts effectively and produces a dimensionally accurate surface.)
Overlapping disk balancing ways	A wheel balancing device. A series of four overlapping metal disks that revolve freely as a grinding wheel and arbor assembly are tested for balance.
Flood, through-the wheel, and mist cooling systems	Three basic grinding machine coolant systems. (The coolant floods over a specific work-wheel area, or is fed through the wheel, or a mist of air and coolant is supplied.)
Swarf separators	Machine attachments for wet grinding. Devices for separating out particles of abrasive, chips, and other foreign matter in a dirty coolant. Separating devices that produce a clean coolant that may be circulated.

SUMMARY

Dressing is the process of sharpening a wheel. A glazed or loaded surface is cut away. Sharp, new abrasive grains are exposed.

The periphery of a trued wheel is concentric with the center of the machine spindle. Truing removes high spots and centers a mounted wheel.

A grinding wheel should be trued on its adapter (mounting) before being balanced.

A trued and dressed wheel is essential to grinding accuracy, good surface finish, and high production.

A ring test should be performed before a new or used wheel is mounted.

A grinding wheel should be run at operating speed at least a minute before any grinding is done.

Grinding wheels are generally form dressed by crush forming. Preshaped rolls, a diamond roll, or plated block dressers are used.

Single-point and diamond cluster dressers are widely used for concentric and parallel face dressing.

Diamond dressing tools are set at a 3° to 15° negative angle and either at the center (cylindrical grinding) or beyond the center (surface grinding) for truing.

Wheels are dressed under operating conditions. Coolants are applied for wet grinding. In dressing with a diamond for a dry grinding operation, the diamond must be air cooled to prevent fracturing and breaking down.

Dressing begins at the high point of a wheel. A feed of from 0.001" to 0.0005" (0.02mm to 0.01mm) per pass is used. A final speed of 0.0002" (0.005mm) is taken to produce a wheel capable of grinding to high dimensional and surface finish accuracy.

Mounting a grinding wheel permanently on a sleeve (adapter) unit reduces setup and truing time and production costs.

A wheel sleeve puller is designed to remove a wheel assembly from the spindle without damage to any parts.

Silicon carbide and aluminum oxide wheels are generally trued with a diamond or an abrasive holding device. Diamonds are not used to true other diamond or CBN wheels.

Diamond and CBN wheels are trued with a dressing stick or a brake-controlled device and conventional abrasives and by grinding against a low-carbon steel.

Diamond and CBN wheels have metallic cores. The diamonds are bonded around the core by a resinoid, metal, or vitrified bond.

Diamond wheel markings include the type of diamond abrasive, grain (grit) size, grade, concentration, bond type (and sometimes bond modification numeral), and depth of diamond section.

The trueness of diamond and borazon (CBN) wheels is tested with a dial indicator. Wheel runout is held to 0.0005" (0.01mm) on general surface and cylindrical grinding. Tolerances of 0.0002" (0.005mm) and finer are required for highly precise work. The runout accuracy of cup wheels is held from 0.0001" to 0.0005" (0.002mm to 0.01mm).

General-purpose grinding wheels under 10" (250mm) diameter are balanced by the manufacturer. After truing and dressing, they are used without further balancing unless damaged. Larger-size wheels (after truing and dressing) are balanced for precision grinding and cutting efficiency.

Parallel ways and overlapping disk balancing ways are two common balancing devices and methods.

Three general kinds of grinding fluids are used: Water-soluble chemical, water-soluble oil, and straight oil. Rust inhibitors, detergents, bactericides, and other additives are mixed in different solutions to improve the properties of grinding fluids and for hygienic reasons.

Wet grinding fluid systems include the flood, wheel penetration (through-the-wheel), and mist coolant systems.

Proper design and selection of the coolant nozzle and system are important in concentrating the cutting fluid at the work-wheel interface and immediate area.

Three basic systems attached to the machine for recirculating clean grinding fluid include centrifugal, cyclonic, and magnetic separators for removing swarf.

The application of the ANSI "Safety Code for the Use, Care, and Protection of Abrasive Wheels" is a requirement. These guidelines are designed to protect workers against injury; to prevent machine, grinding wheel, instrument, and work damage; and to provide hygienic working conditions.

UNIT 25 REVIEW AND SELF-TEST

1. Explain briefly why cubic boron nitride and diamond wheels require less frequent dressing than aluminum oxide and silicon carbide abrasive wheels.
2. State the difference between crush forming and preformed diamond dressing block contour forming on a grinding wheel face.

3. List three general conditions the grinding machine operator must meet in mounting and positioning diamond dressing tools.
4. List the steps for truing and dressing an aluminum oxide abrasive wheel for grinding flat surfaces with a horizontal surface grinder.
5. Describe briefly how the grinding machine operator may change the cutting action of an abrasive wheel during dressing.
6. a. Name two common bonds for diamond grinding wheels.
   b. Indicate the cutting characteristics of each bond.
7. a. State a probable cause of each of the three following grinding problems: (1) scratches in the surface finish, (2) dimensional variations of the finished surface (the dimension is not held uniformly), and (3) chatter marks.
   b. Give the corrective action to take for each problem.
8. a. Name three additives to cutting fluids used in grinding.
   b. Make a general statement about the functions performed by the cutting fluid additives.
9. Describe the operation of one of the principal wet grinding systems.
10. State three checks the surface grinder operator makes to prevent damage to a workpiece due to grinding wheel faults.

SECTION THREE

Horizontal-Spindle Surface Grinding Technology and Processes

The horizontal-spindle, reciprocating-table surface grinder is a basic machine tool. It is widely used to grind flat and formed surfaces to close tolerances. Such surfaces are machined by moving the workpiece past a grinding wheel. The longitudinal movement is 90° to the wheel head spindle axis.

The versatility of the surface grinder is extended beyond flat grinding. Special machine accessories used on surface grinders permit cylindrical, centerless, and internal grinding. However, these processes are not common to general surface grinding work.

The greatest percentage of surface grinder work requires the grinding of flat surfaces. These surfaces may be in a horizontal, vertical, or angular plane. The work may be mounted directly on a magnetic chuck, nested in a fixture, held with standard work-holding devices, or mounted on other accessories. Dimensions of machined surfaces may be measured indirectly or controlled by the down-feed handwheel.

This section deals with the major construction and control features of standard and automated precision surface grinders. The technology and shop practices are related to work-holding accessories, machine setups, and flat grinding processes. Steps are given on how to rough and finish grind flat and parallel surfaces, flat surfaces on round parts, and thin workpieces. Common grinding problems of discontinuous surfaces, distortion caused by stresses, and warped workpieces are treated. This section also deals with accessories and procedures for grinding angular and vertical surfaces and shoulders, as well as with truing and dressing devices and methods of form (profile) grinding and cutting off.

UNIT 26

Surface Grinder Design Features and Setups for Flat Grinding

DESIGN FEATURES OF THE HORIZONTAL-SPINDLE SURFACE GRINDER

The horizontal surface grinder, regardless of whether it is hand operated or partially or totally automatic, is designed with five basic construction features: *base*, *column*, *saddle* (or other cross feed movement mechanism), *wheel head*, and *table*. These construction features are shown on the saddle-type horizontal surface grinder in Figure 26-1. A one-shot lubrication system automatically delivers oil to all ways, screws, gears, and bearings. The lubricant is used once and is then discarded. The system prevents the recirculation of contaminated oil.

Machine Base

The base provides a rigid foundation for the machine. Some bases are made of cast iron. Others use a rigid preformed steel base. An expanding concrete is cast into these structural steel members. Tests show that this newer construction has about four times the vibration-absorbing capacity of cast iron. This construction combines the strength and versatility of steel

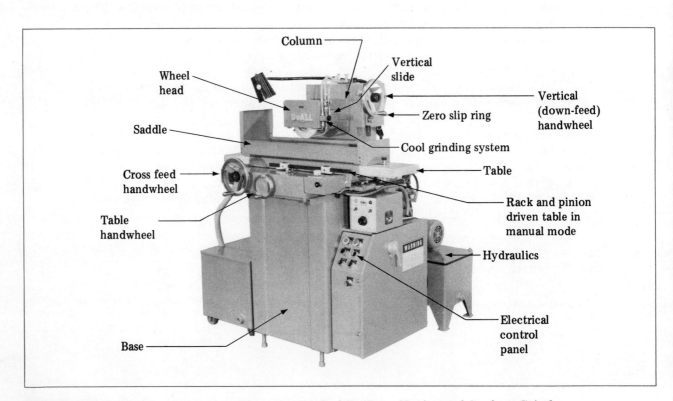

FIGURE 26-1 Basic Construction Features of a Saddle-Type Horizontal Surface Grinder

with the mass and stability of concrete. The vibration absorption, strength, and stability characteristics add to the smoothness of the grinding process.

The base ways for regular machines and models of newer construction are of nickel-alloy iron castings. These ways provide uniform structure surfaces that are precisely machined and hand scraped. Some ways are chrome plated or hardened. Plating and hardening improve wear life, ensure continuing precision bearing surfaces, reduce friction, and permit smooth table travel.

Column

The column contains the vertical ways. These ways are machined at a 90° angle to the base ways. Figure 26-2 shows a cast base with the column and bed as integral parts of the casting.

On machines of steel/concrete-compound construction, the square machined surfaces are replaced by precisely machined cylindrical openings. Centerless ground, heavy-walled cylinders that are hard-chrome plated slide in bearing sleeves. The sleeves fit into the cylindrical openings. The heavy-walled cylinders are filled with concrete matrix for ultimate rigidity and performance.

The column serves two main purposes. It supports the spindle housing and wheel head. It permits vertical movement for positioning and controlling the depth of cut and performing all grinding operations.

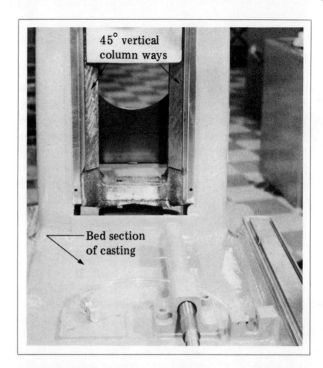

FIGURE 26-2 Column and Bed as Integral Parts of a Cast Base

Saddle

The saddle is a heavily ribbed, H-shaped casting (Figure 26-3). The saddle is fitted to the bed ways. Cross feed movement is included in the design features. The top of the saddle has another

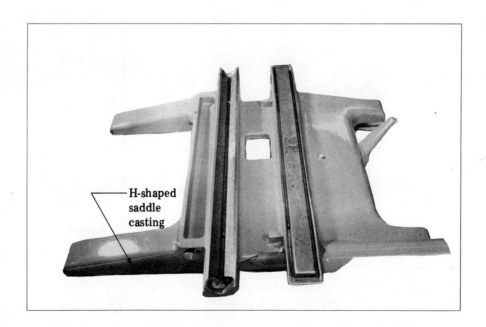

FIGURE 26-3 Ribbed H-Shaped Saddle Casting with V and Flat Way for Table

set of precision ways. These ways are machined at a 90° angle and provide for the longitudinal movement (traverse) of the table.

The cross feed movement of the saddle toward or away from the column may be controlled manually or automatically. A cross feed handwheel is used to move the saddle manually. Saddle movement on machines equipped for automatic operation is produced by the power cross feed control. Larger and production surface grinders permit the quick positioning of the table and workpiece by a rapid traverse attachment to the cross feed.

Table

The underside of the ribbed, cast iron table is fitted with V and flat ways. These ways correspond to the saddle ways on which the table moves. The reciprocating direction is parallel to the face of a type 1 wheel. The table provides the back and forward movement of the workpiece under a grinding wheel.

Transverse table movement is provided, as just discussed, through the saddle. A table handwheel provides for manual control. Table movement may be automatically controlled by a power table control. The length of table travel is also controlled by table reverse (trip) dogs. There are two table reverse dogs for general surface grinding operations. One dog is used to trip the table reverse lever at the end of the table stroke (travel). This causes the table to reverse its direction.

The second trip dog trips the table at the end of the reverse stroke. Where fixtures are used on production grinding, more than one set of dogs may be required to control various table travel positions. Some machines have four dogs. Two dogs are adjustable. The other two are fixed safety dogs that prevent over-travel of the table.

Wheel Head

The wheel head unit contains the wheel spindle and drive mechanism. The wheel head shown in Figure 26-4 has a motor rotor. This is an integral part of the spindle. The spindle housing and support are housed in the same casting. The wheel head and spindle are aligned in relation to the table by the column ways.

The down-feed handwheel on toolroom grinders is graduated in 0.0001" or 0.002mm (Figure 26-5). A handwheel may be fitted with a *zeroing slip ring* or mounted pointer, which permits the operator to set the handwheel to a zero reference point. This setting simplifies the dimensional information the operator must remember. The zeroing ring also reduces setting errors.

Rack-and-Pinion Table Drive

A rack-and-pinion table drive provides a smooth, accurate, positive drive. Surface grinders

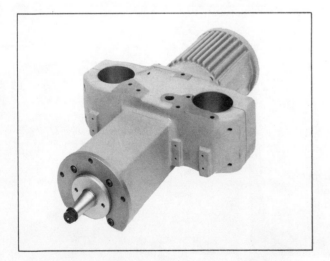

FIGURE 26-4 Wheel Head with Motor Rotor Drive for Mounting on Ground, Heavy-Walled, Vertical Tubes

FIGURE 26-5 Down-Feed Handwheel and Zeroing Slip Ring (0.0001" Graduations)

are available for full manual operation or with combinations of drive features. These features include hydraulic table drive, hydraulic cross feed, and automatic down-feed.

The main components and operating handwheels on a horizontal-spindle, reciprocating-table surface grinder are identified in Figure 26-6. The newer 612 and 618 models of surface grinders employ both inch and SI metric units of measure. The model numbers 612 and 618 designate the maximum magnetic chuck sizes that can be accommodated on the table. The numbers indicate that magnetic chucks of 6″ × 12″ and 6″ × 18″ (150mm × 300mm and 150mm × 450mm) may be accommodated.

The usual table travel of a 612 model is 14″ (356mm). Table feeds with hydraulic controls for this size of machine are infinitely variable from 0 to 78 sfpm (0 to 21.3 m/min). The range of automatic cross feed increments is from 1/64″ to 1/4″ (0.4mm to 6mm). Cross feed reference graduations are usually in 0.001″ (0.025mm). However, a fine-feed cross feed attachment is available. The attachment is graduated in 0.0001″ (0.002mm) increments.

The hydraulic system is operated by a hydraulic power unit. The reservoir capacity for the 612 model is approximately 10 gallons (39 liters). The hydraulic pump volume is 6 gallons per minute (gpm) or 23 liters per minute. The maximum table height from the largest size of grinding wheel to the table is 12″ (300mm).

Additional Grinder Equipment

Additional grinder equipment includes such mechanisms as the hydraulic table control, hydraulic cross feed and automatic down-feed; the over-the-wheel dresser; and dry or wet grinding attachments.

Hydraulic table control equipment includes the valve, hydraulic power unit, piping and cylinder, and the table reversal trip dogs. The hydraulic cross feed permits variable cross feed movements from 1/16″ to 1/4″ (1.6mm to 6mm). The increment is actuated at each table reversal. The direction of the cross feed is controlled by an on-off selector switch. Continuous cross feed for dressing a wheel is provided through a push button.

The automatic down-feed is applied to plunge and automatic surface grinding. A selector switch is employed for automatic down-feed at each table or cross feed reversal. Maximum down-feeds up to 0.125″ (3mm) are possible. The increments of down-feed are variable from 0.0002″ to 0.002″ (0.005mm to 0.05mm).

Belt-driven spindles are available. Belt drives are used for nonstandard spindle speeds and wheel sizes.

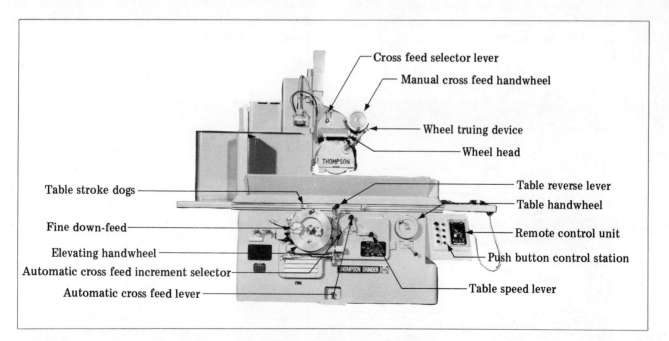

FIGURE 26-6 Main Components and Operating Handwheels on a Horizontal-Spindle, Reciprocating-Table Surface Grinder

One type of over-the-wheel dresser has an increment micrometer dial and knob and a wheel traverse lever. The dial increments may be either 0.001" (0.02mm) or 0.0001" (0.002mm).

OPERATING FEATURES OF THE FULLY AUTOMATIC PRECISION SURFACE GRINDER

Plunge and straight surface grinding may be done automatically on production runs. Dimensional accuracies may be held to within ±0.0001" (±0.002mm) with finishes of 8 to 10 microinch a.a. The five basic features included in an automatic precision surface grinder (Figure 26-7) are as follows:

- A complete digital control system,
- Automatic down-feed (plunge grinding),
- Automatic surface grinding,
- Automatic wheel dressing,
- Automatic digital position readout.

Digital Control System

The digital control system panel on a fully automatic precision surface grinder (Figure 26-8) provides the following controls:

- Vertical feed, a four-position switch for manual, automatic, down, and up movements;
- Single step, a control that provides up or down wheel head movement at 0.000050" (0.0001mm) increments;
- Full feed down, a control that permits the operator to conserve on setups and on productive machine time;
- Down-feed table, the down-feed selector control for plunge grinding;
- Down limit set, a control used for referencing the down-feed counter to the zero plane;
- Auto dresser, an off-on selector switch;
- Single step offset, a control that moves the wheel head in single 0.000050" (0.0001mm) increments to correct any accumulated error;
- Reset up, a control that raises the wheel head to the up limit;
- Cycle start and cycle stop controls.

Other switches on the control panel are used with the coolant system, lubricant control, and for an emergency stop.

Automatic Down-Feed

Down-feed may be controlled for each table reversal or every other table pass. The maximum depth of feed or the full work height is usually 15" (380mm). The feed increments may range from 0.0001" (0.025mm) to 0.0099" (0.25mm).

The grinding wheel is first referenced to the zero plane. The up limit thumb switches on the console panel are set to the desired dimension. The coarse-feed increment (0.0001" to 0.0099" or 0.02mm to 0.25mm) is set. Then the fine-feed increment (0.0001" to 0.0009" or 0.02mm to 0.025mm) and the point at which the fine feed is to begin are set.

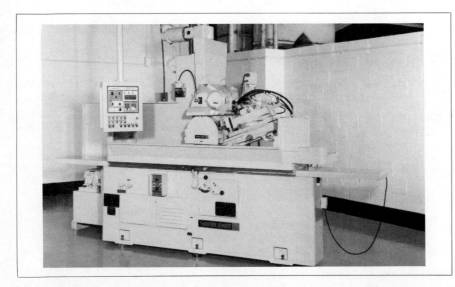

FIGURE 26-7 Precision Surface Grinder with Automatic Down-Feed, Cross Feed, Wheel Dressing, Digital Control System, and Wheel Height Position Readout

DESIGN FEATURES AND SETUPS FOR FLAT GRINDING 519

Additional thumb switches are set for the number of spark-out passes, automatic wheel dressing and compensation (if required), and when the down-feed is to occur. The vertical feed selector is set to auto. Finally, the cycle start button is pushed.

Automatic Surface Grinding

Selector switches are used for machine cycling. The down-feed increments and maximum depth are established as for plunge grinding. Both the cross feed reversal and wheel down-feeds are controlled automatically. They may be set for each or every other reversal.

Automatic Wheel Dressing

The operator selects when the grinding wheel is to be automatically dressed. The digital control is set to the number of feeds to wheel dress. The dressing may take place at any given number of grinding passes from 1 to 999.

The diamond dressing tool in the over-the-wheel dresser is fed automatically down to a preselected depth. Following this dressing pass, the wheel height is set to lower automatically. This lowering is to compensate for the reduction of wheel diameter resulting from dressing. Normally, increments of from 0.0002″ (0.005mm) to 0.001″ (0.02mm) are used.

Digital Position Readout

The position of the freshly dressed wheel above the table top or any other reference surface is shown on the digital position readout. The accuracy of a readout is within 0.0001″ (0.002mm). The digital readout tells the operator at a glance the wheel height above the reference plane. As the wheel down-feeds automatically, position changes are readily observed. The changes of position represent the amount of material remaining to be ground. A single step movement of 0.0001″ (0.002mm) will change the position display.

WHEEL AND WORK-HOLDING ACCESSORIES

General grinding wheel accessories include wheel adapters, a wheel balancing arbor, and parallel ways or an overlapping disk balancing stand. Wheel adapters, as shown in Figure 26-9, may be used for wet or dry grinding wheels. The adapter consists of a rear flange that fits the

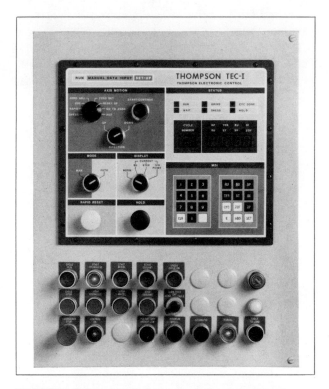

FIGURE 26-8 Digital Control System Panel on Fully Automatic Precision Surface Grinder

FIGURE 26-9 Wheel Adapters (Front and Rear Flanges) for Dry or Wet Grinding

520 GRINDING MACHINES

spindle nose. The front flange holds the wheel in place. Adapters are available with or without balancing lugs.

Magnetic chucks are the most versatile workholding devices for surface grinding ferrous metals. Magnetic chucks may be of the *permanent magnet* or *electromagnetic* type. The common chuck shape for horizontal-spindle surface grinders is rectangular. However, a rotary magnetic chuck is adapted to operations that require a circular, scratch pattern. This chuck has an independent power and magnetizing source. Applications of magnetic chucking principles are to be found in magnetic V-blocks, vises, and parallel blocks, and sine chucks.

Magna-vise clamps, the toolmaker's vise, and the three-way vise are commonly used in combination with magnetic accessories. The three-way vise may be set quickly at any compound angle. Adjustments up to 90° may be positioned around the 360° movement on the base. The angles are adjustable individually. The workpiece setting may be locked at any compound angle.

Electromagnetic Chucks

Electromagnetic chucks stand higher on the table than permanent magnet chucks. A standard electromagnetic chuck is shown in Figure 26-10. The top face is laminated and usually has holding power across its entire area. Ferrous metal workpieces may be held directly on the face or by using other accessories such as magnetic V-blocks and parallels and vises.

Electromagnetic forces are controlled by an on-off position switch. Direct current is used to produce the electromagnetic effect. Residual magnetism remaining in the workpiece is removed on some machines by using a selective chuck control (Figure 26-11). This chuck control serves three functions:

—To rectify AC to DC current;
—To provide variable holding power that is determined by the nature of the part to be ground (thus, certain workpieces may be held with less force to reduce the possibility of distortion);
—To demagnetize both the chuck and workpiece after a grinding operation.

Magna-Vise (Perma-) Clamps

Nonmagnetic workpieces that do not have a large bearing surface may be held for grinding with magna-vise clamps, as illustrated in Figure 26-12. Each clamp is made of a flexible, comb-like, thin, flat bar. The bar is attached to a solid bar by a spring-steel hinge.

The clamps, also called *perma-clamps*, are used in pairs (Figure 26-12A). The solid edge of one clamp is positioned against the backing plate of the magnetic chuck. The work is brought against the toothed edge of the clamp. The toothed edge of the second clamp is fitted snugly against the side of the workpiece (Figure 26-12B). As the magnetic chuck is energized, the toothed edges of the perma-clamps pull down. The force exerted by the perma-clamps holds the workpiece against the chuck face (Figure 26-12C).

FIGURE 26-10 Standard Electromagnetic Chuck

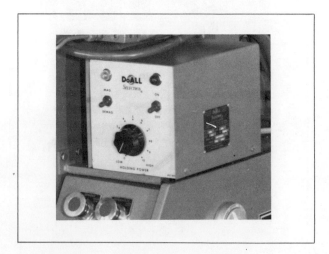

FIGURE 26-11 Selective Chuck Control for Electromagnetic Chucks

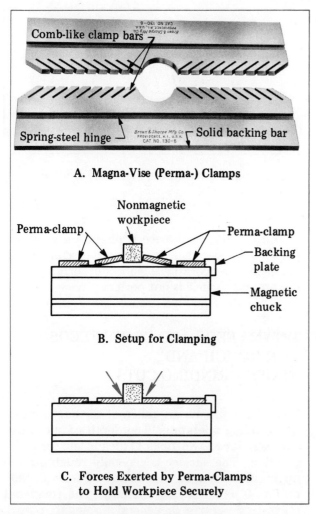

FIGURE 26-12 Application of Perma-Clamps to Hold Nonmagnetic Workpieces

Laminated Plates, Parallels, and Blocks

Parts with bosses or other projections must often be positioned at a distance from a magnetic chuck face. The flux path must extend to hold such parts securely. As the distance between the workpiece and magnetic chuck increases, the strength of the flux lines decreases. Laminated blocks do not have the holding capacity at any distance from the chuck face. Therefore, they are used principally for precision, light grinding operations.

Chuck parallels are referred to as *laminated blocks*. Usually, narrow strips of low-carbon steel, alternated with thin separators (spacing strips), are welded into a positive nonshift unit.

The chuck parallels are then precisely ground for parallelism.

Round, square, and irregular-shaped workpieces are sometimes held on precision laminated V-blocks, also called *chuck V-blocks*. An adapter (auxiliary top plate) is used to hold small and thin workpieces securely.

As in other machine setups, parallels and V-blocks must be aligned in relation to the spindle axis and the nature of the machining operation. Backing plates on the sides and ends of magnetic chucks provide ideal locating surfaces for many grinding setups.

Vacuum Chucks

Vacuum chucks hold work against a plane surface by exhausting the air from between the part and the chuck face. This technique of holding applies to magnetic and nonmagnetic materials. The vacuum chuck has been found to be a practical device for holding paper-thin (0.002" to 0.003") workpieces.

PROBLEMS ENCOUNTERED IN PRODUCING FLAT SURFACES

Discontinuous Surfaces

Flat surfaces are *discontinuous* if they include holes, ridges, grooves, or slots. When many workpieces having a space between each one are ground at the same setting, the flat surface is interrupted. A discontinuous surface grinding setup is shown in Figure 26-13. As the grinding wheel approaches an interrupted section, the reduced area of wheel contact permits the wheel to cut slightly deeper. This dipping of the wheel is due to the fact that a grinding wheel normally exerts force during the cutting action. Dimensional and surface flatness inaccuracy may be produced. It follows that the more rigid the surface grinder, the less the tendency to cut deeper. The wheel cuts to a given plane under ideal grinding conditions.

The problem of discontinuous surfaces is overcome by (1) keeping the wheel sharp and cool cutting at all times and (2) taking light finish cuts. A dull wheel causes excessive force to build up. As the area of contact decreases near the interrupted space, the increased force causes the grinding wheel to tend to plunge downward and cut a little deeper.

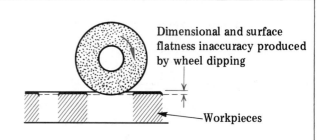

FIGURE 26-13 Discontinuous Surface Grinding Setup Requiring Operator Judgment and Control

Work Stresses and Distortion

Castings; forgings; and heat-treated, cold-formed, extruded, or other manufactured parts have internal stresses and strains produced by chilling or mechanically working the material. Such stresses and strains are relieved, or unlocked, when the top surface or skin is machined away. A relieved surface changes shape. The grinder operator needs to recognize that the workpiece is *distorted* (warped or bent) away from its nominal shape. Compensation must be made to overcome this distortion (warp or bend). This may be done at the machine or by subjecting the part to heat treatment. Severely distorted parts require straightening by force on arbor and other presses.

Warped Workpieces

Thin shims are often used to facilitate holding and grinding a warped piece. The top surface is carefully ground to flatness. The part is then turned over and held without the use of shims. The second surface, when ground, should be parallel to the flat top surface. If a warped workpiece is placed on a magnetic chuck with the bowed ends up, the part will rock. This condition is corrected by placing shims under each raised area. Shims permit the chuck to exert maximum magnetic force without further distortion of the workpiece.

One of the easiest ways of checking the flatness of a surface is to place a precision straight edge on the ground surface. A white paper or light source is held against the back side of the straight edge. Light shines through places where there is any unevenness or variation from a plane surface.

Often, the workpiece is placed on a surface plate. If a feeler gage can be inserted at any point between the workpiece and the surface plate, the ground surface is not perfectly flat.

DOWN-FEEDS AND CROSS FEEDS FOR ROUGH AND FINISH GRINDING CUTS

Coarse, rough cuts are taken at the beginning of the grinding operation. The depth of successive cuts is reduced. Sufficient stock is left for finish grinding. The surface is generally rough ground by using cross feed increments of from 0.030" to 0.050" (0.8mm to 1.2mm). The rough-cut down-feed per pass on general jobbing shop work is 0.003" to 0.008" (0.1mm to 0.2mm) for average rough grinding. A heavier down-feed may be used on cast iron and soft steel. The down-feed may be increased still further if the work area being ground is small.

Fine down-feeds are used for the finishing passes. An excessive number of fine down-feed passes tends to load the wheel face. The cutting action is then inefficient and overheating may result. It is important to redress the wheel for light finish cuts of 0.0005" (0.01mm) and finer. Where the down-feed is very light, fast table speeds and large cross feed increments are used. The workpiece must be passed under the grinding wheel fast enough to eliminate the possibility of overheating. A grinding wheel is *touched up* with an abrasive stick to produce a small radius on the edges of the wheel. The radius helps to eliminate fine grinding lines that may be produced on the finished workpiece by sharp wheel edges.

HOW TO SET UP A SURFACE GRINDER AND TAKE A FIRST CUT

STEP 1   Select the best available grinding wheel to suit the job and machine requirements.

STEP 2   Clean the wheel spindle and wheel bore. Remove any burrs, if necessary.

STEP 3   Ring test the wheel. Mount on the wheel spindle.

STEP 4   Replace the wheel guard. True and dress the wheel.

Note: The diamond or other wheel dresser must be positioned ahead of the centerline of the wheel at a 10° to 15° drag angle. A coolant is used if the part is to be ground wet.

STEP 5   Wipe the chuck face. Remove any burrs. Rub the palm of the hand across the chuck to remove all traces of dust.

STEP 6   Clean, burr, and place the workpiece as near to the center of the chuck as practical.

Note: In shoulder grinding and other flat grinding, the part may be aligned against the side or end plates on the chuck. In other cases, the part is accurately aligned with the longitudinal travel of the table by indicating.

STEP 7   Turn the chuck control lever or switch on.

Caution: Check the workpiece to be sure it is held securely. Make certain all feed control levers are disengaged and are in the neutral position.

STEP 8   Set the automatic trip dogs to accommodate the longitudinal travel of the workpiece and table.

STEP 9   Start the motors for the spindle, coolant system, and hydraulic fluid pump for the longitudinal feed. If dry grinding, turn the exhaust motor on.

Note: The appropriate coolant nozzle is positioned so that maximum flow is provided at the area of contact.

Note: A minute or more must be allowed for running-in time if fine, finish cuts are to be taken.

Caution: Stand out of line with the plane of rotation of the grinding wheel.

STEP 10   Down-feed the grinding wheel carefully until it just touches (sparks) the high spot. Then, move the wheel toward one edge of the workpiece. Set the depth of cut for the first rough grinding pass.

STEP 11   Turn the table traverse feed control knob to the selected feed. In a similar manner, set the horizontal table travel speed to correspond with the sfpm recommended for the material being ground and the required surface finish.

STEP 12   Engage the power cross feed and longitudinal table feed.

Note: During the first pass, the operator must check the clearance between each end of the work and the centerline of the wheel. This checking is usually done manually, without the power feed.

STEP 13   Stop the table after a few strokes. Check the quality of the surface finish and the cutting action.

Note: The table movement should be stopped when the wheel is out of contact with the workpiece.

Note: Any undue rise in the temperature of the workpiece requires a recheck of the grinding wheel, the coolant, and general operating conditions.

STEP 14   Take the first cut (pass). Feed the down-feed handwheel from 0.008" (0.2mm) to 0.003" (0.1mm) for successive roughing cuts. Leave from 0.002" to 0.005" (0.05mm to 0.12mm) for finish grinding.

STEP 15   Shut down the coolant flow. Then, stop the wheel at the end of the cutting stroke.

Note: The wheel must be out of contact with the workpiece.

WHEEL PREPARATION FOR FINISH CUTS

The wheel becomes partially loaded and dulled during rough grinding. To produce a high-quality surface finish, the grinding wheel must either be dressed or replaced with another wheel suitable for finish grinding. The preferred method is to dress the wheel. However, it must be emphasized that the condition of the wheel is only one factor in producing a desired finish. The speed of the work, the amount of cross feed, the depth of cut, and the coolant are other factors to consider.

Dressing is usually done with a diamond dresser mounted on the magnetic chuck. If the operation permits, the diamond dresser is sometimes positioned on the magnetic chuck so that it does not interfere with the grinding process. When it becomes necessary to dress the wheel for finish grinding, the wheel is cleared of the workpiece.

Once contact is made between the diamond nib and the wheel, the diamond is traversed across the wheel face. The wheel is down-fed about 0.0002" (0.005mm) each pass until the operator feels the wheel is dressed to cut sharply. In the last pass or two across the wheel face, there is no feed—that is, the wheel is allowed to *spark out* as it is traversed over the diamond. The wheel must be dry for dry grinding; wet, for wet grinding. The edges of the wheel are slightly rounded about 1/16" (1.6mm). An abrasive stick is used to form the radius freehand.

The dressed wheel is then positioned over the workpiece. The wheel is lowered to the down-feed setting at the end of the last rough grinding pass. Compensation is made for the amount the wheel is dressed. With the longitudinal table travel engaged (and the coolant flowing), the wheel is brought to the ground surface. The zeroing slip ring on the down-feed handwheel is reset at zero. The wheel is fed for successive finish cuts from 0.001" to 0.0005" (0.02mm to 0.01mm) and finer. The depth depends on the surface finish requirement. The cross feed traverse and longitudinal (table) speed rates are increased to produce surface finishes in the low microinch range.

DIMENSIONAL GRINDING USING THE DOWN-FEED AND CROSS FEED HANDWHEELS

The down-feed handwheel is graduated in 0.001" and 0.0001" (0.02mm and 0.002mm). One complete revolution of the handwheel (with 250 graduations of 0.0001") moves the wheel spindle 0.025" (0.6mm). A zeroing slip ring permits the handwheel graduation to be set at zero.

The down-feed handwheel may thus be moved to any required depth to correspond to a dimension specified on a drawing. For example, if a step is to be ground at a depth of 0.015" (0.4mm), the handwheel is turned until the 15 graduation is aligned with the zero index line. It is possible to grind to size or dimension by using the calibrated handwheel directly. The handwheel, in this case, eliminates the intermediate step of gaging or measuring the part after each grinding pass.

Similarly, the cross feed handwheel on some surface grinders is graduated into 100 divisions. If the cross feed screw has a pitch of 0.100" (2.5mm), each graduation represents 0.001" (0.02mm). Many grinders have an auxiliary (vernier) scale on the zero index plate or a fine-feed cross feed attachment as shown in Figure 26-14. The graduations permit the handwheel readings that are in 0.001" (0.02mm) to be further divided into 0.0005" (0.01mm). With the fine-feed cross feed attachment (Figure 26-14), the knob is graduated in increments of 0.0001" (0.002mm). If the auxiliary scale model is used, the movement from one handwheel graduation to the next line on the index scale is 0.0005" (0.01mm).

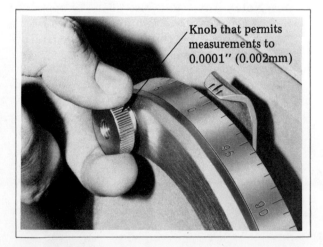

FIGURE 26-14 Fine-Feed Cross Feed Attachment Graduated in 0.0001" (0.002mm) Increments

The zeroing slip ring is an optional feature on both the down-feed and the cross feed handwheels. In either case, the use of a slip ring eliminates the need to continuously add or subtract the amount of movement. Thus, the possibility of mathematical errors and inaccurate wheel settings is reduced.

GRINDING EDGES SQUARE AND PARALLEL

In addition to having flat surfaces ground, most rectangular workpieces require that the edges be ground parallel and square. Such workpieces are usually rough machined slightly oversize. Stock is left to correct any inaccuracy in squareness, parallelism, and machining surface irregularity and to permit the sides to be ground square and to the required dimensional accuracy for surface finish and size. The general shop practice is to allow 0.010" (0.2mm) on each side for rough and finish grinding.

A common method of grinding a workpiece square involves the use of a precision vise. The vise base surface is first checked for parallelism in relation to the spindle. The dimensional accuracy requirements (length, width, and height) for grinding the edges of a workpiece square and parallel are illustrated in Figure 26-15. The workpiece is placed on a parallel on the vise base. The first edge (1) is ground. The adjacent edge (2) is ground next. The squareness of its position in the vise is checked vertically (at 90°) with a dial indicator.

After these two edges are ground square, each edge is used as a reference plane. Edge 1 is placed on a parallel. The workpiece is secured and edge 3 is ground to the required dimension. Then edge 2 is positioned on a parallel and clamped securely. Edge 4 is then ground parallel, square, and to the specified dimension. The squareness of edges 1 and 2 may be checked with a solid-steel square. A more precise measurement may be taken by using a cylindrical square.

A second method of grinding square edges is to use an angle plate. The large, flat surfaces are first ground parallel. The faces then act as plane reference surfaces. The workpiece, machine setup, and machining sequence follow.

HOW TO GRIND EDGES SQUARE AND PARALLEL

Using an Angle Plate

STEP 1 Check the size of the workpiece. There must be sufficient stock left to clean up the sides so that they may be squared and brought to size and required finish.

STEP 2 Clean the workpiece, magnetic chuck, and angle plate. Remove any burrs.

STEP 3 Place a piece of paper under the angle plate and between it and the magnetic chuck.

STEP 4 Center one of the ground faces against the angle plate. Position the workpiece by resting one edge on a parallel or directly on the magnetic chuck.

STEP 5 Place a protecting shim between the outside finished face of the workpiece and either a C-clamp or a pair of parallel clamps. Secure the workpiece against the angle plate.

STEP 6 Magnetize the setup. Check to be sure the angle plate and workpiece are secure.

STEP 7 Start the spindle, hydraulic fluid system, and the coolant system. Position the coolant nozzle for maximum flow at the area of contact.

STEP 8 Set the trip dogs. Start the table. Bring the wheel down until it contacts the workpiece.

STEP 9 Set the down-feed handwheel at a zero reading. Feed the wheel for the first roughing cut.

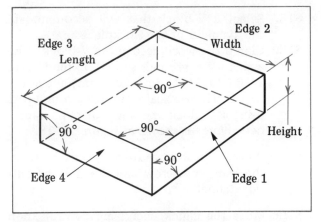

FIGURE 26-15 Dimensional Accuracy Requirements for Grinding Edges Square and Parallel

Note: Roughing cuts from 0.003" to 0.008" (0.08mm to 0.2mm) and finishing cuts from 0.001" to 0.0005" (0.02mm to 0.01mm) are used.

STEP 10 Stop the table at the end of the first few strokes. Check the dimension, cutting action, and the quality of surface finish.

STEP 11 Continue to take roughing cuts. Allow from 0.002" to 0.005" (0.05mm to 0.12mm) for the finish cut.

STEP 12 Dress the wheel before taking any finishing cuts (if necessary).

STEP 13 Increase the table speed and traverse feed. Set the down-feed for each finish cut. Machine to the required dimension.

STEP 14 Stop the machine. Wipe all surfaces dry and clean. De-energize the magnetic chuck.

STEP 15 Remove the clamps from edge 2. Relocate them on edge 1 (Figure 26-16). Remove the clamps on edge 4. Turn the angle plate 90° so that the adjacent edge rests firmly on the magnetic chuck.

Note: It is necessary to start with two sets of clamps to ensure adequate holding force against the angle plate. It is good practice to use a parallel and shims under edge 4 as a precaution against the workpiece moving.

STEP 16 Energize the setup. Rough and finish grind the second edge.

STEP 17 Stop the machine. Wipe the surfaces dry and clean. Disassemble the setup.

STEP 18 Turn the workpiece 90° to position the third edge (Figure 26-17A). Follow similar positioning and holding steps as for the first edge. Then, rough and finish grind the third edge.

Note: Workpieces having a sufficient bearing surface are set directly on the magnetic chuck (Figure 26-17B). Such workpieces do not require an angle plate setup for grinding the third and fourth edges.

STEP 19 Stop the machine when an area of the workpiece is ground and a measurement is possible. Measure the part. Set the graduated dial at zero.

STEP 20 Recheck the workpiece for size. The part should be dimensionally accurate. Use standard linear measuring instruments as a further check.

Note: The surface finish may be checked against a comparator specimen.

STEP 21 Shut down the machine. Break the edges of the workpiece, if required. Clean the machine. Return all tools and accessories to proper storage places.

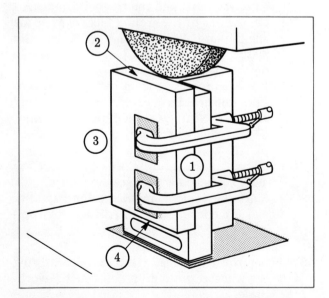

FIGURE 26-16 C-Clamps Relocated on Top Area of Angle Plate (Both Plate and Part Turned 90° to Permit Grinding Second Edge)

HOW TO GRIND ENDS SQUARE WITH THE AXIS (ROUND WORK)

STEP 1 Select a V-block that will accommodate the diameter of the workpiece.

STEP 2 Clamp the workpiece in the V-block. The end to be ground should extend from 1/16" to 1/8" (2mm to 3mm) above the block. The vertical setup for a round part using a V-block and angle plate on a magnetic chuck is shown in Figure 26-18.

Note: A protective strip is placed between the clamping screw and the finished outside diameter.

STEP 3 Move the chuck energizing control knob to the on position. Check on how securely the assembly is held.

DESIGN FEATURES AND SETUPS FOR FLAT GRINDING 527

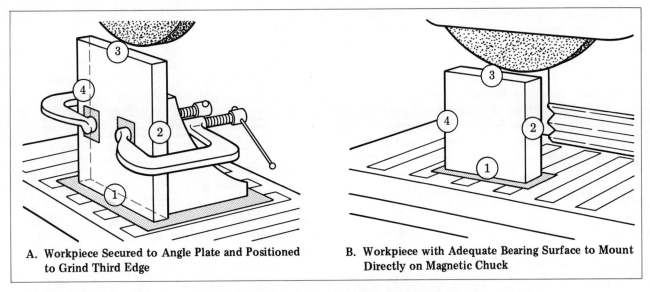

A. Workpiece Secured to Angle Plate and Positioned to Grind Third Edge

B. Workpiece with Adequate Bearing Surface to Mount Directly on Magnetic Chuck

FIGURE 26-17 Position and Holding of Workpiece to Grind Third and Fourth Edges

Note: When additional support is required for the V-block setup, either a square block or an angle block is placed ahead of the V-block and tightly against it.

STEP 4 Adjust the table reversing (trip) dogs. Position the coolant nozzle. Start the machine, hydraulic fluid system, and coolant system.

STEP 5 Lower the grinding wheel until it grazes the workpiece. Position the wheel to start the cut. Move the wheel down and take the roughing cut.

Note: The wheel is usually fed across the workpiece by hand when the diameter is small.

STEP 6 Measure the length of the workpiece. Set the down-feed handwheel graduated collar at zero.

STEP 7 Take roughing and finishing cuts to clean the end to a particular size.

STEP 8 Shut down the operation. Remove the workpiece. Wipe the chuck, V-block, clamps, and workpiece dry and clean. Remove (break) any fine wire edge around the ground end.

STEP 9 Reverse the workpiece in the V-block. Position the second end above the top of the V-block. Use an accessory block against the V-block, if required.

STEP 10 Energize the magnetic chuck. Check the setup to be sure it is secure.

Note: The grinding wheel must clear the workpiece, and the trip dogs must be positioned correctly.

STEP 11 Start the machine and coolant flow.

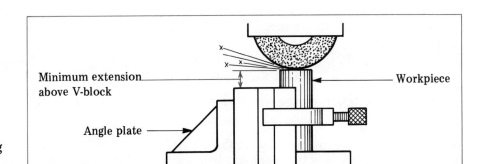

FIGURE 26-18 Vertical Setup for Round Part Using V-Block and Angle Plate on a Magnetic Chuck

STEP 12 Position the wheel for depth. Set for the first roughing cut. Turn the handwheel setting to zero. Take the cut. Stop the machine.

STEP 13 Measure the workpiece. Take successive cuts to grind the piece to size. Use the graduations on the down-feed handwheel to feed the wheel for each successive cut until the required dimension is reached.

HOLDING THIN WORKPIECES

Use of Magna-Vise Clamps

Serrated magna-vise clamps provide a convenient work-holding device for small, thin parts (that are slightly thicker than the height of the clamps). A thin part should be shimmed, if necessary. With this setup, the workpiece is mounted parallel to the side faces of the magnetic chuck. Nonferrous metals are also held securely in a magnetic chuck with magna-vise clamps.

Use of Adapter Plates and Fine Pole Magnetic Chucks

Thin workpieces are also held on standard or finely spaced laminated magnetic chucks. A part is placed across as many laminations as possible. Again, shims are used to prevent the magnetic chuck forces from drawing any bent portion of the part down against the face of the chuck. One disadvantage to this method is the distortion that the full magnetic force may produce.

An adapter plate is specially designed to hold small, thin workpieces. The adapter plate is a precision ground rectangular plate. It is set directly on a magnetic chuck as illustrated in Figure 26-19. The laminations run lengthwise and are finely spaced to allow more, but weaker, lines of force to act on the workpiece. The adapter plate is energized by turning the magnetic chuck on. Small, thin workpieces are positioned at an angle of 15° to 30° to minimize the heat generated during each pass.

HOW TO GRIND THIN WORKPIECES

STEP 1 Select a cool-cutting, open-grained wheel that will produce the required surface finish.

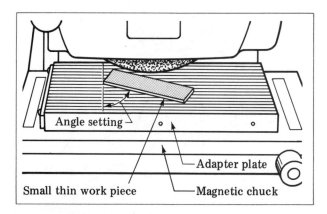

FIGURE 26-19 Adapter Plate Set Directly on Magnetic Chuck for Holding Small, Thin Workpieces

STEP 2 Mount, true, and dress the grinding wheel. If the nature of the work process requires, relieve part of the wheel.

STEP 3 Prepare the machine for grinding, including lubrication.

STEP 4 Test the workpiece for straightness. Determine whether there is adequate material left for grinding (if a distorted part is shimmed) or the part requires straightening before grinding.

STEP 5 Straighten the work, if necessary. Mount, support, and secure the work on a magnetic chuck (with or without magna-vise clamps) or an adapter plate.

STEP 6 Set the necessary machine controls, speeds, feeds, and coolant flow.

STEP 7 Feed the wheel slowly down to the workpiece. When the wheel sparks at the high point, feed it from 0.0005" (0.01mm) to 0.001" (0.02mm). Use a finer cross feed (traverse) than for rough grinding thick workpieces.

STEP 8 Continue with fine cuts until the surface is cleaned up.

STEP 9 Demagnetize the chuck and workpiece. Test for flatness.

Note: It may be necessary to plan to grind the second side almost to size and go back over the first side if the workpiece is still distorted.

STEP 10 Stop the coolant flow, all controls, and the motors. Wipe the parts dry and clean. Stone or use a fine file to remove any grinding burrs.

SAFE PRACTICES IN GRINDING FLAT SURFACES AND THIN WORKPIECES

—Straddle the workpiece over as many laminations as possible on a regular or fine pole magnetic chuck or adapter plate to provide maximum force in drawing the workpiece to the chuck or plate.
—Add a square block or angle plate in front of a V-block to give added support against the cutting forces of a grinding wheel. Nest small workpieces against end and side blocks.
—Check the clearance between the workpiece, mounting device, wheel, and other machine parts before starting the machine.
—Bring the grinding wheel into contact at the high spot of a surface to be ground by slowly feeding the grinding wheel.
—Place thin workpieces that may be warped as a result of heat generated during grinding, diagonally across the magnetic chuck or adapter plate. It may be necessary to relieve the face of the wheel to reduce the area of contact.
—Increase the traverse rate per pass and the table speed when using a dressed wheel. This combination is essential to producing a high-quality surface finish.
—Dress the wheel after a series of finish cuts are taken to provide an open face and grains that cut cool without burnishing.
—Bent or warped work must be straightened before grinding.
—Use shims, particularly under thin-sectioned parts, to prevent distortion from grinding.
—Recheck workpieces that are ground for straightness. While parts may be ground parallel and are dimensionally accurate, the surfaces may be distorted. This distortion may be caused by heat generated during machining or stress relieving the part.
—Allow warm-up time for the machine, accessories, and workpiece to reach a uniform temperature, particularly for precision processes where dimensional control depends on down-feeding with the graduated handwheel.
—Recheck how securely the diamond holder is held on the magnetic chuck for dressing and truing operations. Maintain a sharp diamond dresser by rotating the point from 10° to 20° after each dressing. Use a coolant during dressing with a diamond.
—Replace the machine guard and wear protective goggles or a shield. Observe general machine safety precautions. Stand out of line of the wheel travel.
—Check the position of splash guards. The coolant must be contained on the machine and flow back to the reservoir. Any spilled grinding fluid or other oils are to be wiped from the machine area.

TERMS USED WITH DESIGN FEATURES, SETUPS, AND FLAT SURFACE GRINDING PROCESSES

Basic machine components	The base, column, saddle, wheel head, and table, which comprise the machine tool. In addition, the control handwheels and mechanisms that produce circular, vertical, horizontal, and traverse movements.
Wheel head	The wheel spindle and drive mechanism, which control the grinding wheel. Mechanisms that are housed as a single unit. The spindle and drive component that is adjustable to the work height and for down-feeding.
Fine-feed cross feed attachment	A further refinement of the micrometer dial (collar) on the cross feed. (Cross feed movements are dial graduated in vernier 0.0001" or 0.002mm increments rather than standard increments.)

Digital position readout	Usually, a numerical value that tells the operator the relationship between a cutting tool and a reference plane. In the case of a surface grinder, the amount of material that remains to be ground.
Discontinuous surface	A plane surface that is interrupted by a slot, groove, ridge, hole, or other cutaway section.
Work stresses	Stresses within a material or part. Stresses produced by a manufacturing process such as casting, forging, or cold forming. Stresses within a workpiece that are relieved as a result of a machining process or heat treatment.
Distortion or buckling (warped part)	The twist or bend of a part away from its nominal shape and size.
Increment (down-feed or cross feed)	The amount (distance) the grinding wheel and/or the workpiece is fed uniformly at the completion of each pass over the work face.
Dimensional grinding	Controlling the dimension to which a part is to be ground by feeding to depth (size) with the handwheel graduations. Grinding to within 0.0001″ (0.002mm) tolerance directly from calibrated handwheel settings.
Sparking out	Uniformly traversing the ground surface of a workpiece after the last cut without further down-feed.
Cleaned-up surface	A ground surface with continuous grinding marks. The absence of surface finish marks produced by other manufacturing or machining processes than are produced by grinding.

SUMMARY

The five basic construction features of a surface grinder are the base, column, saddle, wheel head, and table.

Surface grinder movements are controlled by three sets of handwheels and systems that relate to down-feed, cross feed (transverse), and longitudinal (horizontal) travel.

Handwheels may be equipped for vernier ("tenth": 0.0001″ or 0.002mm) settings. A zeroing slip ring permits the operator to set a handwheel to a zero reference point.

Hydraulic table systems provide infinitely variable table feeds such as 0 to 70 sfpm (0 to 21.3 m/min).

Hydraulic cross feed systems permit feeds within a 1/64″ to 1/4″ (0.4mm to 6.0mm) range.

A fully automatic precision surface grinder includes five supportive systems: complete digital control, automatic down-feed increments from coarse to fine feed, automatic cross feed reversal and wheel feed controls, automatic wheel dressing, and automatic digital position readout.

Common work-holding accessories for surface grinder work may be magnetic or nonmagnetic. Magnetic chucks, adapters, parallels, V-blocks, magna-vise clamps, and sine blocks are widely used. Toolmaker's and conventional universal vises, straps and clamps, and vacuum chucks are also common.

The dimensional accuracy of a ground flat surface is influenced by such factors as surface continuity, distortion, release of internal stresses produced by certain manufacturing methods, and general grinding practices.

Grinding discontinuous surfaces requires machine and work rigidity and a sharp cutting wheel.

Unless there is adequate material to remove from a warped workpiece that is shimmed to rest solidly on a chuck, it is necessary to first straighten the part.

As coarse a cut as possible is taken. Enough stock is left for final finish cuts to produce a dimensionally accurate surface. The finish must be within the specified requirements.

The wheel is dressed between roughing and finish cuts, especially where a low microfinish surface is required.

The down-feed is decreased to 0.0005" (0.01mm) and finer for final finish cuts. For extremely light down-feeds, the table speed and cross feed on the final cut are increased.

The wheel sides are dressed with a stick dresser to produce a small radius. This radius helps to eliminate the fine lines produced during each pass by a sharp-edged wheel.

The trip dogs are set so that the workpiece clears the wheel from 1/2" (12mm) to 1" (25mm) at each end.

The spindle RPM is checked before the operation. It is brought as near as possible to the RPM required to develop the recommended grinding wheel sfpm.

The cut is set as deep as practical. The highest table speed and cross feed increment are used (consistent with the job requirements and the machine capability).

Flat surfaces and edges may be ground square and parallel either by holding the part in a vise directly on the chuck or by mounting on an angle plate.

Sufficient stock must be left on each edge to permit it to be cleaned to size and squareness.

When edges are ground square, the first two edges are ground at 90°. These edges may be positioned directly on the magnetic chuck. The third and fourth edges are then ground parallel and square, respectively.

The wheel is often dressed during the grinding of a workpiece. In some cases, a diamond holder is positioned on the chuck away from the grinding process. After a number of cuts, the wheel is positioned over the dresser and dressed.

In dimensional grinding, the down-feed handwheel graduated dial (zeroing slip ring) is set at zero at the start of the cut. The wheel is fed over a series of cuts the distance required to machine to a specified dimension.

Dimensional grinding simplifies the work process. Intermediate measurements may be eliminated.

Flat surfaces may be ground on round parts by holding the workpiece in a vise or fixture or on V-blocks.

A V-block is usually supported against an angle plate or square block when a workpiece is held vertically and the end is to be ground square.

V-blocks are positioned parallel with the spindle to grind a flat area on a round part (at a right angle to its axis).

Lighter roughing cuts are usually taken with workpieces that are held in V-block setups.

Distortion, buckling, and warpage are conditions that must be corrected or compensated for in grinding thin workpieces.

Diagonal positioning and relieving part of the grinding wheel face reduce the heat generated during the grinding of small, thin workpieces.

Grinding machine and personal safety precautions must be followed. The grinding wheel must clear all parts and be brought carefully into contact with the workpiece. Machine controls, speeds, feeds, and coolant flow must be considered before the process is started. Hygienic machine operation conditions must be observed.

UNIT 26 REVIEW AND SELF-TEST

1. a. List three design features of a fully automatic surface grinder, exclusive of the digital control system.
   b. Describe briefly the function of each of the design features listed.
2. Cite three advantages of using magnetic work-holding accessories (where practical) over conventional fasteners and clamps.
3. State three mechanical factors that affect grinding efficiency and machining accuracy.
4. a. Describe briefly the relieving of stresses from castings, forgings, and cold formed and other manufactured parts that results from grinding.
   b. Tell what effect stress relieving during grinding has on the accuracy of a machined plane surface.
   c. State what provision must be made when grinding plane surfaces on parts that are stress relieved during grinding.
5. a. Tell what common practices are followed with table speeds and traverse feeds when fine finish cuts of 0.0005″ (0.01mm) are taken.
   b. Identify two problems encountered by the machine operator (1) when an excessive number of fine cuts (passes) are taken and (2) what corrective steps may be followed.
6. List the step-by-step procedures for preparing a surface grinder to machine a flat surface.
7. Identify the steps for setting up and grinding the two ends square and to accurate length for a short-length die block. An angle plate and magnetic chuck setup are to be used.
8. a. Tell what effect relieving part of a wide-face abrasive wheel has on the generation of heat during grinding.
   b. Explain how the opposite faces of thin parts that are bent or warped may be ground parallel.
9. State two machining techniques that may be used to prevent the distortion that may be produced when thin workpieces are ground.
10. List six safety precautions the surface grinder operator must check for the setup, prior to taking the first cut.

UNIT 27

Grinding Angular and Vertical Surfaces and Shoulders

Angular surfaces are usually ground by holding the work at an angle. Flat surfaces are then ground at a required angle to a line or other work-locating surface. The angular surface is produced by passing the work under the grinding wheel as in flat or form grinding.

There are a number of accessories for positioning and holding a workpiece. The most practical work-holding device to use depends on the number of parts to be ground, size, shape, dimensional tolerances, properties of the workpiece, and the available accessory. The manner in which the work is aligned with surfaces that have already been ground or finish machined must also be considered.

Accessories provide practical, efficient work-holding devices for grinding cutting tools, machined parts, gages, tools, and other parts. Either a single part or a quantity may be produced.

Common accessories that are used to hold workpieces at an angle and to generate a flat (angular) surface are grouped as follows:

- —Vises (plain, adjustable, universal, and swivel);
- —Angle plates (adjustable and sine angle plates);
- —Magnetic work-holding accessories (V-blocks, toolmaker's knee, swivel chucks, and sine table);
- —Grinding fixtures for production work.

This unit describes these accessories with applications to angular surface grinding. Step-by-step procedures are given for machine setups and the actual grinding of angular surfaces. The latter part of the unit deals with the types, functions, and grinding of straight, filleted, and angular shoulders and vertical surfaces.

GENERAL TYPES OF VISES FOR ANGULAR WORK

Plain Vise

The plain general-purpose machine vise is provided with a tongue (key) in the base. The tongue fits the T-slot of the table. The vise may be positioned with the jaws parallel or at a right angle to the spindle. Angular surface layout lines on the workpiece are located parallel to the machine table. A height gage may be used to locate the line for positioning the part. The part is then held securely by the hardened and ground vise jaws.

Adjustable Vise

The adjustable vise has tongues in the base. These tongues align with and fit into the table T-slots. The vise may also be held directly on a magnetic chuck without using the tongues. The vise part is mounted on a hinged base (Figure 27-1). The workpiece may be positioned and clamped at any angle from 0° to 90° in a vertical plane. The angle setting is read directly on a quadrant plate. This plate is graduated in degrees. The setting is not a precise one but it is a reference point.

Universal Precision Vise

The universal precision vise permits the workpiece to be set at a compound angle quickly and easily. A job may also be positioned at any angle in a vertical or horizontal plane for angular grinding. The universal precision vise has two bases. The lower base is designed for direct mounting

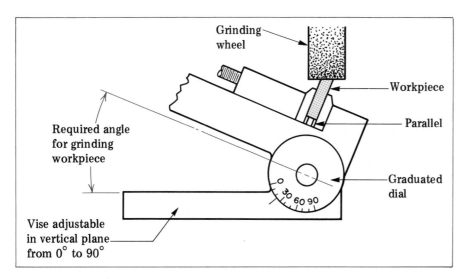

FIGURE 27-1 Adjustable Vise for Angular Grinding Applications

on a table or a magnetic chuck. The degree graduations permit the vise to be positioned through 360° in a horizontal plane.

The upper part combines a swivel base with the features of an adjustable vise. The workpiece may be set at any angle from 0° to 90° in a vertical plane. A reference angle reading is taken directly from the graduated index plate. The vise may also be turned on its base to produce a third angle setting. When the universal vise and the workpiece are positioned at the required compound angle (in two or three planes), the knee clamps are tightened to provide a rigid support.

Adjustable Swivel Vise

The adjustable swivel vise (Figure 27-2) may be clamped directly on the table or held on the magnetic chuck. The jaws may be set at any horizontal angle in relation to the reciprocating table movement. The work, held between the vise jaws, may be positioned at any angle up to 45° on each side of a horizontal plane. The setting is read directly on a graduated collar. When the base and adjustable vise are both set at respective angles, the setting permits the grinding of a compound angle.

ANGLE PLATE ACCESSORIES

Adjustable Angle Plate

The adjustable angle plate provides greater versatility in clamping special-shaped workpieces for grinding compound angles than a right angle plate. The adjustable angle plate consists of a tilting table mounted on a rotary base. The base is graduated through 360° to permit the table to be set at any angle in a horizontal plane. The table may be positioned vertically at any angle from 0° to 90°. Work is strapped directly to the table by using T-bolts and straps.

Sine Angle Plate

The sine angle plate is used for precision applications. The plate is made so that the workpiece may be positioned at a single or compound angle, depending upon whether a single or compound sine angle plate is used. Sine angle plates are designed with or without a magnetic chuck.

MAGNETIC WORK-HOLDING ACCESSORIES

Magnetic V-Blocks

As discussed in Unit 26, the use of a magnetic V-block for angular grinding processes eliminates the use of clamping devices that may interfere with the actual grinding. The 90° V-blocks are most common. Round and flat workpieces may be nested and held securely against the sides of the angular surfaces.

Magnetic Toolmaker's Knee

The magnetic toolmaker's knee (Figure 27-3) is especially designed to hold odd-shaped parts

ANGULAR AND VERTICAL SURFACES AND SHOULDERS 535

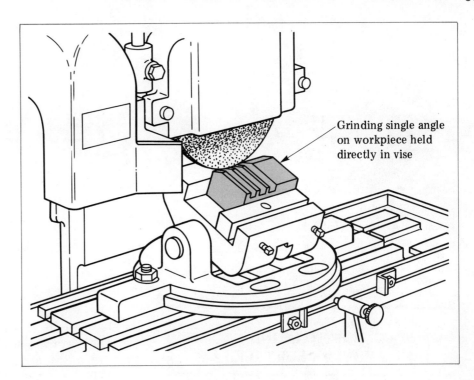

FIGURE 27-2 Adjustable Swivel Vise for Angular Grinding Applications

for grinding. The model illustrated is a permanent magnetic tool. The magnetic surface is 4" × 6-1/2" (100mm × 165mm) and is parallel within 0.002" (0.05mm) overall. The tolerance for squareness is 0.00005" per inch.

Swivel-Type Magnetic Chuck

This horizontal magnetic chuck has a swivel feature. This feature makes it possible to position the chuck at an angle from the usual horizontal plane. The top face on which the workpiece is positioned and held by magnetic force is set at the required angle. The accuracy of the angle may be checked with a vernier bevel protractor. Finer angle setups are checked against an angle sine block combination.

Magnetic Sine Tables

Plain and compound electromagnetic sine chucks are shown in Figure 27-4. A plain electromagnetic sine chuck (Figure 27-4A) has two hinged sections. The first section includes a base and angle plate. Angles from 0° to 90° may be set. Gage blocks are used to establish the vertical height for a required angle.

The second section is hinged to the accessory table. Again, gage blocks are used to set the second table (a magnetic table) at a specified angle from 0° to 90°. Workpieces are positioned directly on the magnetic chuck face.

A compound electromagnetic sine chuck (Figure 27-4B) includes a third hinged section. This design feature permits setting the electromagnetic chuck face at a compound angle. Each section is set precisely at one of the required angles by using gage blocks.

FIGURE 27-3 Permanent Magnetic Toolmaker's Knee for Inspection, Layout, and Grinding Setups

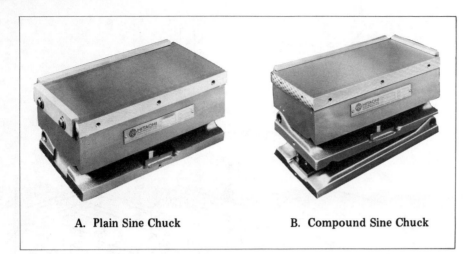

FIGURE 27-4 Plain and Compound Electromagnetic Sine Chucks

HOW TO GRIND ANGULAR SURFACES

Using a Plain Vise

STEP 1 Select an appropriate size and type of vise. Clean and burr all locating surfaces.

STEP 2 Place the vise on the magnetic chuck. Use a parallel to position the vise jaws at a right angle to the wheel spindle. Position the workpiece. Tighten the vise jaws. (The vise setup for holding a workpiece to be ground at an angle is illustrated in Figure 27-5.)

Note: The method of locating the work at a specific angle depends on the nature of the job and the required accuracy. A Surface gage or a vernier height gage may be used to locate layout lines.

STEP 3 Proceed to select the best wheel available. Mount, true, and dress the wheel.

STEP 4 Set the speeds, feeds, length of table movement, and so on.

STEP 5 Start the machine and coolant system. Bring the wheel head down until the grinding wheel just grazes the area to be ground.

STEP 6 Rough and finish grind the angular surface. Check for dimensional accuracy and surface finish.

STEP 7 Proceed to remove any grinding burrs. Disassemble the work-holding setup. Clean and replace each part in its proper storage place.

Using an Adjustable Vise (Single Angle)

STEP 1 Remove the keys from the vise base. Position and hold the vise on the magnetic chuck.

STEP 2 Set the workpiece on a parallel in the vise. Secure the workpiece between the vise jaws.

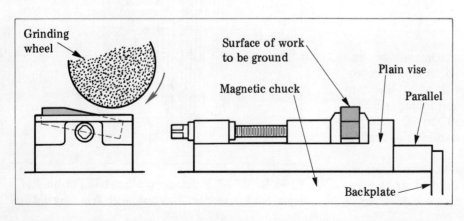

FIGURE 27-5 Vise Setup for Holding a Workpiece to be Ground at an Angle

ANGULAR AND VERTICAL SURFACES AND SHOULDERS 537

STEP 3 Loosen the hinged-base clamping nut. Swing the vise to the desired angle. Use the graduated quadrant to obtain a direct reading.

STEP 4 Secure the clamping nut. Recheck the angle setting.

Note: The angle setting is often checked by using a dial indicator in combination with angle gage blocks.

STEP 5 Proceed with the standard machine setup. Take roughing and finish grinding cuts.

STEP 6 Measure the angle and surface finish. Shut down the operation after the part is ground to size, deburred, and measured.

Using a Right Angle Plate

STEP 1 Select an angle plate of a suitable size. Set the angle plate on the chuck. Align the vertical face.

Note: The surface to be ground should be set parallel with the table travel.

STEP 2 Place the surface of the work against the one face of the right angle. Position the work at the required angle. Use a vernier bevel protractor, gage blocks, or other measuring device. (The angle plate setup for grinding an angle is shown in Figure 27-6.)

STEP 3 Clamp securely. Check to be sure the wheel clears any clamps or other accessory.

STEP 4 Set up the machine. Take roughing and finishing cuts. Check the angular measurements. Grind to size and surface finish.

Using a Magnetic V-Block

STEP 1 Select the magnetic V-block that accommodates the part to be ground. Clean the base and the work contacting surfaces.

STEP 2 Locate the magnetic V-block centrally on the chuck. Position it parallel with the side of the chuck. (The setting and use of a magnetic V-block to position and hold a workpiece are illustrated in Figure 27-7.)

STEP 3 Nest the workpiece in the magnetic V-block. Turn the magnetic chuck control lever to the on position.

Caution: If heavy roughing cuts are to be taken, add an extra block against the end of the workpiece and the V-block.

STEP 4 Follow general machine starting-up and operating practices. Start with the proper lubrication of the machine; condition of the cutting fluid; selection, mounting, and truing of the wheel; and positioning of the table movement.

STEP 5 Set the zero indexing device on the handwheel at zero. Take roughing and finishing cuts.

The wheel is dressed before the final finish cut(s) are taken. Some operators rough grind both angular faces. This

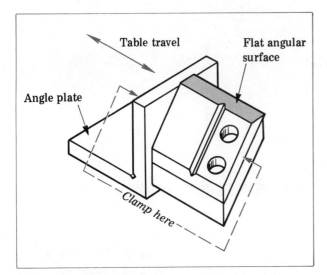

FIGURE 27-6 Angle Plate Setup for Grinding an Angle

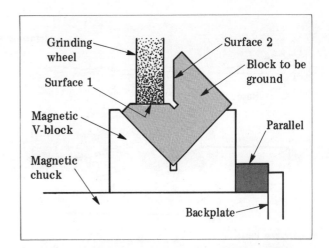

FIGURE 27-7 Setting and Use of Magnetic V-Block to Position and Hold Workpiece

process is then followed by finish grinding each face.

STEP 6 Disassemble the setup. Remove any grinding burrs. Clean and replace all tools and instruments in proper storage areas.

Note: When the adjacent faces of a V-block are to be ground at a 90° angle, a 90° magnetic chuck is used.

Caution: The side of the wheel must clear the vertical surface. During the last pass or two, disengage the automatic traverse feed to provide better machine operator control in feeding the wheel as it contacts the side of the workpiece at the shoulder.

HOW TO GRIND AN ANGLE WITH A RIGHT ANGLE PLATE, SINE BAR, AND GAGE BLOCK SETUP

STEP 1 Place the angle plate on a surface plate. Set up the angle sine bar combination for the required angle. (Figure 27-8 shows the sine bar and angle plate setup commonly used for laying out, holding, and grinding a precision angular surface. As shown, the workpiece is set at a 15° angle.)

STEP 2 Mount the workpiece on the angle sine bar. Clamp in position.

STEP 3 Lubricate and prepare the machine. Take roughing and finish cuts.

STEP 4 Check the dimensional accuracy of the machined surface.

STEP 5 Deburr the workpiece, if necessary. Wipe the machine, accessories, and workpiece clean and dry. Place each item in its proper storage container.

HOW TO SET UP THE MAGNETIC SINE TABLE TO GRIND A COMPOUND ANGLE

STEP 1 Select and bring together the gage block combination for the vertical height of the first angle setup.

Note: The sine formula or a sine table may be used to establish the gage block height.

Caution: Use wear blocks as the end blocks.

STEP 2 Place the gage blocks between the sine table base and the hardened end roll. Bring the sine table down so that it rests on the gage blocks. Tighten the hinge swivel clamping nut.

STEP 3 Align the sine table with the side plate of the magnetic chuck. The sine table face

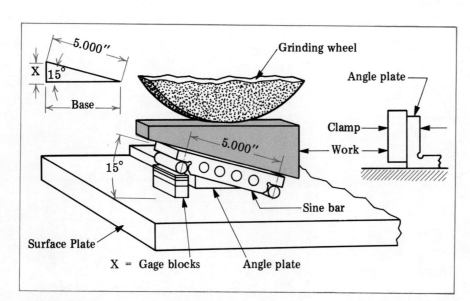

FIGURE 27-8 Sine Bar and Angle Plate Setup for Grinding a Precision Angular Surface

ANGULAR AND VERTICAL SURFACES AND SHOULDERS 539

will then be parallel to the horizontal movement of the table. Strap the base to the table.

STEP 4 Mount the burred and cleaned workpiece on the sine chuck face.

STEP 5 Energize the magnetic chuck and the sine chuck. Check to be sure that the workpiece is held securely to the sine chuck. Similarly, the sine chuck must be secure on the magnetic chuck.

STEP 6 Lubricate the grinder. Select the appropriate grinding wheel. Determine feed rates, table speed, and coolant composition and flow rate.

STEP 7 Set up the surface grinder. Take roughing and finishing cuts.

STEP 8 Check for dimensional accuracy and surface finish.

STEP 9 Disassemble the setup after the machine is shut down. Return each tool and instrument to its proper storage place.

Fixtures for Production Grinding Work

Grinding fixtures serve the same functions as fixtures used in milling machine, drilling machine, and other machine setups. They are designed as work-positioning and work-holding devices. Fixtures are simple to operate and are fast, accurate, and dependable. They may be operated manually, pneumatically, or electronically or by combining hand and automatic controls. Fixtures are essential in production work, particularly where a large number of parts require precision grinding to the same size, shape, and surface finish.

Some fixtures for ferrous parts contain special magnetic holding devices. The workpiece may be located accurately, held securely by magnetic force, and released quickly and easily. Fixtures are widely used to nest irregularly shaped workpieces where, because of the shape, it is difficult to position and hold the part in a standard work-holding accessory.

SHOULDER GRINDING

A shoulder is the step or area that joins two sections of different sizes. A shoulder may be formed on round or flat parts. Shoulder grinding requires an understanding of the functions of shoulders, the setup and operation of the horizontal-spindle surface grinder to produce each shoulder, the workpiece, and abrasive wheels.

In addition, shoulder grinding technology relates to the translation of size, shape, finish and dimensional accuracy information found on a print or drawing; a knowledge of science (for example, heat and magnetism); and the ability to use mathematics to compute required dimensions.

Shoulder grinding includes, also, the step-by-step processes used industrially to produce a finished shoulder. A discussion of the technology, processes, and practices of grinding square, filleted, and angular shoulders follows.

TYPES AND FUNCTIONS OF SHOULDERS

Square Shoulder

The adjoining faces (planes) of a *square shoulder* are at right angles (90°) to each other. The square shoulder is the simplest form of shoulder. It is used on parts that are not subjected to excessive strain or where the mating part has a square shoulder.

Square shoulders may be ground on the horizontal-spindle surface grinder without using any special fixtures. The side of a type 1 straight grinding wheel may be used when it is properly trued and dressed. The side of the wheel, except for a narrow band on its periphery, is dressed toward the hub for clearance. A dish wheel may be used on work that does not require considerable rough grinding.

Filleted Shoulder

A *filleted shoulder*, as the term implies, has a fillet or radius between the two adjoining planes of the shoulder (Figure 27-9). This type of shoulder is best suited for parts that require maximum strength or that must be hardened. The fillet helps to prevent cracks from developing during heat treating. The filleted shoulder adds to the appearance of the job.

The filleted shoulder may be produced by dressing a flat-face grinding wheel to the desired radius and relieving the side of the wheel for clearance. When the filleted shoulder is produced by this method, radial wheel stresses are greatly increased. The in-feed against the shoulder must be done carefully. A minimum amount of in-feeding for the vertical face is 0.0002″ to 0.0003″ (0.005mm to 0.008mm) per wheel pass.

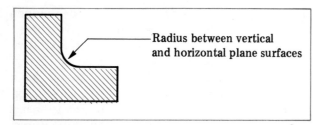

FIGURE 27-9 Distinguishing Features of a Filleted Shoulder

Angular Shoulder

An *angular shoulder* has two adjacent planes that come together at an angle to each other. Angular shoulders may be ground by positioning the workpiece at the correct angle in a fixture. Angular shoulders may also be positioned by setting a tilting magnetic chuck at the required angle to the spindle axis. The surface is ground with the face of the wheel. The angle may also be ground by dressing the side face to the required angle. This method involves excessive dressing of the grinding wheel.

UNDERCUTTING FOR SHOULDER GRINDING

Wherever practical, the juncture of the two flat planes of a square shoulder should be undercut by prior machining processes to eliminate a sharp corner (Figure 27-10). Undercutting is important for the following reasons:

— To preserve the shape of the wheel at the edge.
— To permit the wheel to produce a true surface over the entire area without leaving a slight ridge in the corner.
— To help to eliminate wheel stresses at the corner contact area.
— To increase the life of the abrasive wheel and thus decrease the time consumed in truing and dressing operations.

TRUING AND DRESSING DEVICES FOR GROUND SHOULDERS

The straight-face and the dish-type wheels illustrated in Figures 27-11 and 27-12 are commonly used for grinding square and vertical face shoulders. The redressing or relieving of the side of the wheel may be done by using an abrasive stick offhand. A diamond dresser or other truing and dressing tool, mounted in a suitable holding device, is used for more accurate truing and dressing.

Figure 27-13 illustrates the grinding of square, angular, and filleted shoulders by using straight wheels dressed to shape. Filleted shoulders are ground by using the face of the wheel to grind the flat horizontal face. The formed radius on one edge of the wheel produces the fillet and vertical surface of the angular shoulder. The side is relieved as shown in the figure. The wheel radius may be formed with an abrasive stick or similar dressing tool. The form is checked with a radius gage. Where a precise radius is required, the wheel should be trued and dressed accurately with a radius truing device.

Similarly, one or both sides of a straight-face wheel may be dressed to the required angle for an angular shoulder. The correctness of the angle

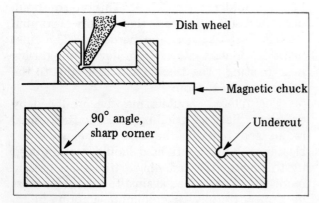

FIGURE 27-10 Undercut Feature on a Square Shoulder

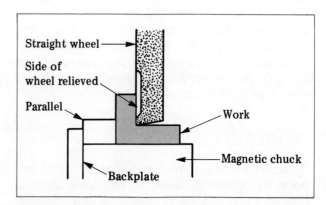

FIGURE 27-11 Setup of Workpiece for Square Shoulder Grinding

ANGULAR AND VERTICAL SURFACES AND SHOULDERS

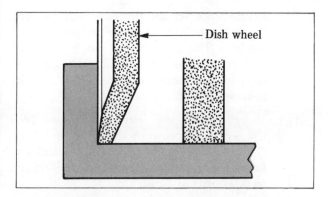

FIGURE 27-12 Use of Dish Wheel to Grind Vertical Surface

is usually checked with a bevel protractor or angle gage. When a wide horizontal surface and angle are being ground simultaneously, the face (periphery) is sometimes relieved as shown in Figure 27-13.

HOW TO FORM DRESS A GRINDING WHEEL FOR ANGULAR GRINDING

STEP 1 Select the grinding wheel having the characteristics needed to produce the shoulder on the specified material. Mount, true, and dress the wheel.

STEP 2 Select the wheel dressing tool. Mount it in a fixture or other holding device on the magnetic chuck.

Note: The sides of the wheel may be formed at the correct angle for angular shoulder grinding. The dressing tool or stick is set in a fixture at the designated angle. The angular side of the wheel is formed by feeding the wheel truing diamond or other abrasive cutter across the side face.

Note: Start at the periphery to form the wheel to the angle shape. A series of cuts will be required to rough and finish true the wheel.

STEP 3 Check the setup to be sure the wheel turns freely. Then, start the machine and coolant flow. Bring the dressing tool into contact with the wheel.

STEP 4 Increase the dressing tool contact a few thousandths of an inch at a time. Continue to dress the side of the wheel until the correct angle or radius is formed and the cutting faces of the wheel are trued.

STEP 5 Stop the wheel spindle. Check the radius with a radius gage or template. Check the angular side face with a bevel protractor or other gage.

HOW TO GRIND SHOULDERS

Setting Up the Workpiece for Shoulder Grinding

STEP 1 Remove burrs from the workpiece and the magnetic chuck. Wipe the chuck face and workpiece to remove abrasive particles and dust.

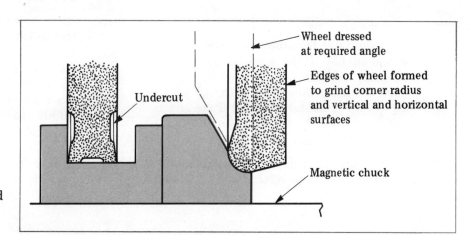

FIGURE 27-13 Grinding Square, Angular, and Filleted Shoulders by Using Straight Wheels Dressed to Shape

Note: In some shops, a thin sheet of paper is placed between the work and the chuck.

STEP 2 Position the work near the center of the magnetic chuck. If possible, use parallels between the workpiece and the side plate of the chuck.

Note: The face of the workpiece may require checking for parallel alignment. This checking is done by attaching a dial indicator to the grinding wheel. The indicator is then moved longitudinally across the length of the part. Either method will assure work alignment in relation to the longitudinal travel of the table.

STEP 3 Switch the magnetic chuck control lever to the on position.

Setting Up the Grinder for Shoulder Grinding

STEP 1 Lower the wheel head.

Note: The grinding wheel must clear the vertical section of the shoulder and permit grinding of the flat horizontal face.

STEP 2 Traverse the table so that the wheel clears the work from 1/2" to 1" at each end. Lock the table reversing dogs.

STEP 3 Start the wheel spindle. Put the automatic table traverse in motion.

STEP 4 Feed the grinding wheel by hand until it grazes the workpiece and sparks at the high points.

STEP 5 Feed the work (with the cross feed handwheel) toward the dressed side of the wheel. Sparking indicates that contact has been made at the high point of the vertical shoulder, fillet, or angular surface.

Rough Grinding Shoulders

STEP 1 Start the table reciprocating. The coolant is turned on, if the work is to be wet ground.

STEP 2 Feed the wheel down to start rough grinding the horizontal surface. Take from 0.001" to 0.002" (0.02mm to 0.05mm) per table stroke as a cut.

Note: This operation requires the manipulation of both the vertical and crossfeed handwheels to set the depth of cut and to traverse the area to be ground, respectively. The graduations on both handwheels are used to continue to feed and adjust the amount of stock that is to be removed.

STEP 3 Rough grind the vertical surface. Feed the work face toward the grinding wheel. Continue to rough grind the vertical surface to within 0.001" to 0.0015" (0.02mm to 0.04mm) of finish size.

Note: A dish-type wheel is preferred over a straight-face wheel when a square shoulder with a narrow adjacent horizontal face is ground.

STEP 4 Stop the coolant flow and then the wheel spindle.

STEP 5 Measure the workpiece. Determine how much more material is to be rough ground from both the horizontal, vertical, angular, or radial surfaces. Continue to rough grind, allowing enough material for finish grinding.

Finish Grinding Shoulders

STEP 1 Traverse the table to the extreme end. True and dress both the face and side of the wheel to true the wheel, produce the necessary relief, and restore sharp abrasive cutting edges. Dress the wheel with a hand abrasive stick, if possible.

Note: Care must be taken when switching the magnetic chuck lever to the on and off positions to avoid disturbing the work location. Also, the chuck face must be clean to properly hold the diamond nib (point) holder.

STEP 2 Repeat the steps used for rough grinding to position the wheel for finish grinding.

STEP 3 Feed the grinding wheel down about 0.0005" (0.01mm) after the first sparks and grind the horizontal surface.

STEP 4 Feed the vertical face toward the wheel at 0.0002" to 0.0003" (0.005mm to 0.008mm) per stroke.

STEP 5 Stop the table at the extreme end of the stroke. Turn off the coolant flow and wheel spindle.

STEP 6 Check for size and surface finish. Use a micrometer, depth gage, radius gage, or bevel protractor, depending on the type of shoulder. The accuracy of the surface finish may be compared against a set of surface finish specimens or work samples.

STEP 7 Continue to machine to size and shape according to the required degree of dimensional and surface finish accuracy.

Note: A coolant is used constantly for wet grinding. The down-feed should be 0.0005" (0.01mm) for each cut.

Note: The in-feed is reduced to 0.0002" to 0.0003" (0.005mm to 0.008mm) for each traverse across the vertical face of the shoulder. The micrometer graduations on the vertical and cross feed handwheels are used for dimensional control.

STEP 8 Allow the wheel to spark out in the final finish grinding step.

STEP 9 Stop the coolant flow and the machine. Remove the work. File or stone the edges to remove the fine fin or burr caused by grinding.

STEP 10 Clean the surface, chuck or work-holding accessory, measuring and hand tools, and the machine itself.

GRINDING MULTIPLE SHOULDERS

The workpiece illustrated earlier in Figure 27-13 combines the grinding of all three types of shoulders. The angular surface is produced by down-feeding a straight-face wheel that has been dressed to the correct angle.

Similarly, the filleted shoulder is ground by using another straight-face wheel that has one side formed to the required radius. Both the face and side of the wheel have been relieved.

The groove has a right and a left square shoulder. It is ground with the same straight-face wheel. In this instance, the wheel is relieved on both sides and the two faces.

SAFE PRACTICES IN GRINDING ANGULAR AND VERTICAL SURFACES AND SHOULDERS

—Check the parallel alignment of the solid jaw of a precision vise. A dial indicator is generally used to test for parallelism.

—Tighten the clamping nuts on the knees of the compound vise to secure the tables from possible movement during grinding.

—Check each precision angle setting with an instrument or gage. Each setting should measure to a finer degree of accuracy than the finest tolerance of the workpiece.

—Measure the distance from the grinding wheel face to the outside diameter of the spindle. This space must permit the grinding of shoulders without interference.

—Disengage the automatic transverse feed upon approaching a vertical, filleted, or angular shoulder. Disengagement permits hand control in order to take fine, precise cuts in forming a corner and the adjacent faces.

—Use wear blocks for gage block combinations for height settings of sine plates.

—Eliminate all end play in the spindle when the side of the wheel is used to accurately grind a shoulder.

—Use a softer-grade wheel to rough grind a side surface. Side grinding tends to produce a harder grinding action because of the increased contact area.

—Judgment is needed to avoid excessive side force on the grinding wheel. Such a force may cause distortion, uneven wheel rotation, and even fracturing.

—Avoid a sudden tap or hitting the side of the grinding wheel. A fractured or damaged wheel may result from such action.

—Work from a side position and never directly in line with the revolving wheel. When a grinding wheel fractures, the spindle speed and the centrifugal force cause the broken pieces to fly at a terrific speed. These pieces can cause serious injury to the operator and damage to the machine.

—Check each setup to ensure that the work is held securely, the grinding wheel is free to rotate and travel without interference, and the trip dogs are tightly fastened.

TERMS USED IN GRINDING ANGULAR AND VERTICAL SURFACES AND SHOULDERS

Universal precision vise	A precision vise that permits a workpiece to be swiveled to any horizontal angle. (The upper plate and chuck may be positioned at any vertical angle from 0° to 90°.) A precise vise that permits compound angle settings in three planes.
Swivel-type magnetic chuck	An adaptation of the magnetic chuck. A magnetic chuck that may be adjusted from a horizontal plane.
Magnetic sine plate	A work-holding accessory having two hinged sections at right angles to each other. A sine plate with a magnetic top plate. An accessory that is set at precise angles by using gage block combinations.
Undercutting	The removing of material from the corner of a shoulder. Cutting a groove deeper than the intersecting surface.
Filleted shoulder	Two adjacent machined surfaces that are connected with a fillet or radius.
Spark out	Minimal contact between a grinding wheel and a workpiece.
Sparks	The visible sign of grinding action between a revolving abrasive wheel and a workpiece.
Relieving	Reducing the area of grinding wheel contact. Recessing the wheel from near its periphery to the hub. Cutting away part of the wheel face to reduce the area of contact.
Rough grinding	The process of rapid removal of material to conserve time and cut costs while at the same time leaving a sufficient amount for finish grinding.
Finish grinding	Precision grinding of a workpiece to a required shape, dimensional degree of accuracy, and surface finish.

SUMMARY

Four groups of accessories are used as common work-holding devices: (1) vises; (2) angle plates; (3) magnetic blocks, chucks, and tables; and (4) fixtures for production work.

Single-angle workpieces are first laid out. The layout line is then positioned in a horizontal plane by using a surface gage or a height gage.

The adjustable vise is used to mechanically hold a workpiece. Angle settings from 0° to 90° are set easily.

The universal precision vise is adjustable to any degree in the horizontal plane. The upper section permits settings to be made at any vertical angle from 0° to 90°. Rigid support is provided by tightening the knee straps.

The adjustable angle plate consists of a swivel base and a hinged table. Workpieces may be set at a compound horizontal and vertical angle.

Single-angle workpieces may be ground at the required angle by direct setting of the swivel magnetic chuck table.

Workpieces held on a right angle block are often positioned at a required angle by using a sine bar and gage blocks or a combination of angle gage blocks.

ANGULAR AND VERTICAL SURFACES AND SHOULDERS 545

Round, square, or rectangular blocks are often positioned in V-blocks for grinding flat surfaces at standard angles.

The magnetic sine table consists of two hinged plates. Each is set for a different angle. Gage blocks are used in each angle setting.

Production parts require the use of grinding fixtures. Multiple parts are set in a fixed position to the grinding process.

The three basic types of shoulders are square, filleted, and angular.

The type of shoulder, dimensional accuracy, and surface finish depend on the function of the part and the required appearance.

Shoulders may be ground on the horizontal-spindle surface grinder. The grinding wheel is trued and dressed to a desired shape and with necessary clearance.

Expertise in shoulder grinding requires the craftsperson to combine technical knowledge with complementary manipulative skills that relate to setting up the workpiece; selecting, truing, and dressing the grinding wheel; setting up the machine tool; and rough and finish grinding processes.

Machine safety precautions, such as clearance between the workpiece and work-holding accessories, must be observed.

Fine burrs produced by grinding must be removed.

Care must be exercised when feeding a grinding wheel against a shoulder.

UNIT 27 REVIEW AND SELF-TEST

1. Name three different setups and/or accessories that may be used to grind single-angle surfaces on workpieces.
2. State three different work-positioning requirements for setting a workpiece to grind a compound angle. A toolmaker's universal vise is to be used.
3. Identify two different setup conditions where the use of a magnetic toolmaker's knee is more functional than an angle plate.
4. Prepare a work production plan for grinding an angular surface on a workpiece. A standard angle plate setup is used. The workpiece is positioned by using a sine bar and gage blocks.
5. State two reasons why the use of a dish wheel is often preferred to the use of a type 1 wide-face abrasive wheel when grinding two surfaces of a shoulder simultaneously.
6. Cite three reasons for undercutting a square shoulder at the intersecting corners.
7. List the steps for form dressing a grinding wheel to a required cutting angle.
8. State three personal safety precautions the surface grinder operator must observe, particularly when grinding vertical shoulders and angular surfaces (with the side face of a grinding wheel).

UNIT 28

Form (Profile) Grinding and Cutting-Off Processes

Form grinding is becoming a widely accepted, effective machining process. Form grinding is especially important for producing regular angles and radii, as well as irregular profiles on hardened metals. Form grinding describes the process of changing a flat surface. The formed surface may be beveled or slotted or may have a concave or convex radius or any combination of angles, radii, and curves. The contoured surface that is generated runs parallel to the reference line of the work-holding device or the table travel (and spindle). The form is produced by grinding the workpiece to the reverse profile of a formed grinding wheel.

Form grinding on the surface grinder usually refers to the grinding of regular angles, radii, and other complicated shapes on flat surfaces. Filleted and angular shoulders generated with a formed wheel were discussed in the preceding unit. Angle grinding was considered as grinding a flat surface with the workpiece positioned at the required angle. Grinding was done with a flat-face wheel.

Form grinding, as described in this unit, deals with dresser accessories and dressing tools, machine setups, and form dressing processes. Grinding wheels may be formed by hand. Simple abrasive dressing sticks of aluminum oxide, silicon carbide, and boron carbide (Figure 28-1A) are used for simple hand forming and dressing. These hand-guided dressing tools are practical for roughly shaping a wheel to a limited degree of accuracy. Parts that are to be ground within close dimensional tolerances for shape and size require that the wheel be precision dressed. Diamond dressing tools (Figure 28-1B) are mounted in toolholders or fixtures for accurate dressing. Radius, angle, and pantograph (contour forming) dressers and the precision microform grinder are described in this unit. Setups with these devices and step-by-step procedures are given on how to form grinding wheels and crush rolls and how to grind simple and complex contours.

COMMERCIAL WHEEL FORMS

A number of wheels are preformed and are available from grinding wheel manufacturers. Some wheels are dressed to specific angles, radii, and combination radius and angle forms. Specifications for some of these preformed wheels appear in Figure 28-2. Each wheel produces the reverse form in the workpiece. The forms may be modified by hand for rough form grinding or with a precision truing device.

GENERAL TYPES OF WHEEL FORMING ACCESSORIES

The three general types of wheel forming accessories are as follows:

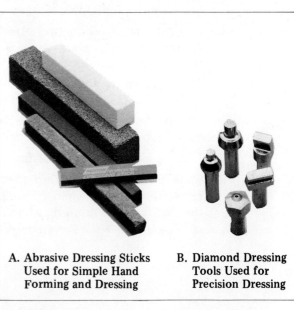

A. Abrasive Dressing Sticks Used for Simple Hand Forming and Dressing

B. Diamond Dressing Tools Used for Precision Dressing

FIGURE 28-1 Hand-Guided and Precision Dressing Tools

FORM GRINDING AND CUTTING-OFF PROCESSES 547

—Radius truing and dressing devices (they form a concave or convex radius);
—Angle truing and dressing devices (they form angles, shoulders, and bevels);
—Combination radius and angle truing devices (they form both radii and angles as used singly or in combination).

Radius Dresser

While differing in design and construction, all radius truing and dressing devices of the *cradle type* follow the same principles. Each device supports a diamond dressing tool that is used to form a required radius on the grinding wheel. The radius is produced by rotating a curved arm that extends from an upright member on the dresser body. The curved arm may be rotated on its axis to swing the diamond in an arc.

The concave or convex radius produced is determined by the relationship of the diamond nib (point) to the axis of the arm. A concave form is generated in the wheel face when the diamond is set above the axis and is swung in an arc while in this position. Conversely, a convex form is produced when the diamond is set below the axis.

The position of the diamond with relation to the axis also governs the radius dimension. The greater the distance from the diamond to the axis, the larger the radius. The exact radius dimension may be determined by measuring the distance the diamond extends beyond a locating surface on the arm. An accurate setting of the diamond may be obtained by using a micrometer. The diamond is set on some models by using a height setting standard and gage block of a particular size. The diamond is held vertically on some dressers; horizontally, on others. Figure 28-3 shows the relationship of the diamond position and the axis of a radius dresser.

The form is produced by feeding the diamond and rotating it as contact is made with the grind-

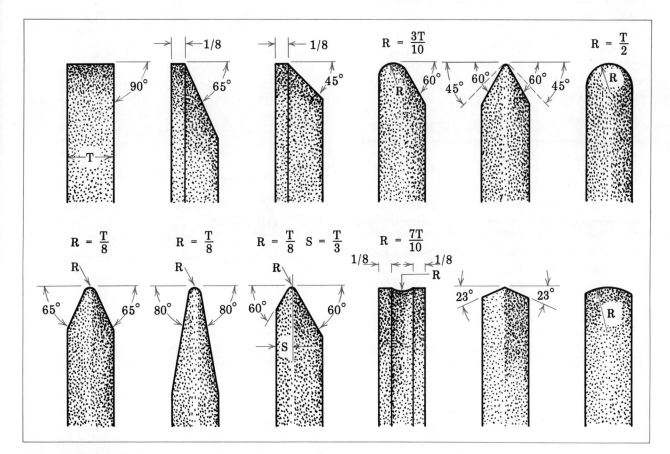

FIGURE 28-2 Specifications of Straight, Angular, Radius, and Combined Forms of Commercially Available Grinding Wheels

548 GRINDING MACHINES

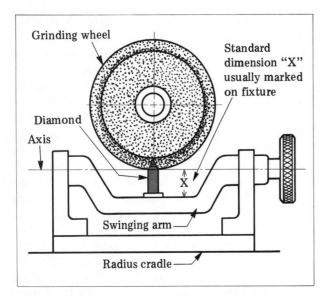

FIGURE 28-3 Relationship of the Diamond Position and the Axis of a Radius Dresser

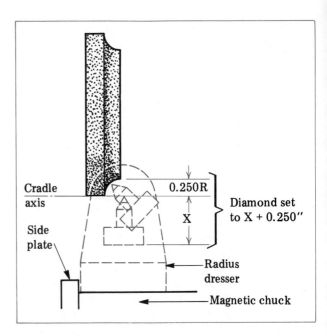

FIGURE 28-4 Radius Dresser Setup to Form a Corner Concave Radius

ing wheel face. A coolant should be used if the form grinding process is to be performed wet. Figure 28-4 shows the radius dresser setup to form a corner concave radius. The radius is formed on the wheel by first setting the diamond and positioning the setup at the centerline of the wheel. The diamond is then rotated by means of the knurled knob.

Note: The wheel is stopped to compare the shape and size with a radius gage, if required.

Note: The wheel may also be dressed to an angle form. The abrasive stick is fed into the face by using an angular movement. The angle is then checked with a bevel protractor or other gage.

HOW TO ROUGH FORM A WHEEL BY HAND USING AN ABRASIVE STICK DRESSER

STEP 1 Select an appropriate abrasive stick. Selection depends on the abrasive wheel characteristics, form to be produced, and nature of the work process.

STEP 2 Secure the abrasive stick in a holder, if necessary. Start the machine, wheel, and flow of coolant (for wet grinding processes).

STEP 3 Hold the dresser firmly. Support one hand on the table or work-holding device.

STEP 4 Advance the abrasive stick toward the face of the wheel. Contact the grinding wheel while swinging the dresser in an arc from the face to the side of the wheel. Take a series of cuts to form the desired radius.

HOW TO FORM A CONVEX OR CONCAVE WHEEL BY USING A RADIUS DRESSER

STEP 1 Select an appropriate grinding wheel for the job. Mount the wheel. Prepare the grinder for operation.

STEP 2 Select a suitable cone-point diamond dressing tool to fit the radius dresser.

STEP 3 Align the radius dresser squarely on the magnetic chuck.

STEP 4 Set the stops on the dresser to control the arc over which the radius is to be swung. (Figure 28-4 illustrates a wheel that is to be formed with a 0.250" radius over 90° of the wheel face and edge. The stops in this case are set 90° apart.)

FORM GRINDING AND CUTTING-OFF PROCESSES 549

STEP 5 Position the diamond point at the required radius. (In the setup of Figure 28-4, the diamond is set at the 0.250" radius. This setting is measured with a micrometer. The measurement equals the distance of the diamond to the axis of the dresser (X) plus the 0.250" radius.)

Note: When a concave radius is to be formed (as in Figure 28-4), the diamond is set above the dresser axis.

Note: To form a convex radius, the diamond point must extend below the axis.

STEP 6 Locate the positioned diamond under the center of the wheel. Lock the grinder table in this position.

STEP 7 Start the machine and flow of coolant. Rotate the dresser arm so that the diamond is in a horizontal position.

STEP 8 Move the diamond carefully toward the wheel until it just touches the side. Then, lock the saddle.

STEP 9 Raise the wheel head until the diamond clears the wheel. Start rotating the radius dresser to make contact with the face.

STEP 10 Feed the wheel from 0.002" to 0.003" (0.05mm to 0.08mm) for each swing (pass) of the diamond. Continue to form the radius. Finish dress the wheel by reducing the depth of cut for each pass. The radius is formed completely when the diamond touches the face and side of the wheel within the 90° arc.

STEP 11 Raise the wheel head to clear the wheel from the dresser. Stop the machine. Clean and remove the radius dresser and diamond tool. Return each tool to its proper place.

Angle Dresser

In the preceding unit, workpieces were positioned for angle grinding by mounting on a magnetic sine chuck, an adjustable angle vise, a compound sine chuck, or other work-holding accessory. After positioning, the angular surface was formed by using a flat-face wheel. In this unit, it was mentioned that commercially formed angle grinding wheels are available. Another method of producing an angular surface is to form the wheel at a required angle with an *angle dresser*. The *precision sine dresser* shown in

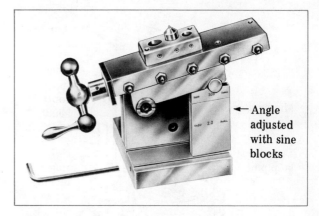

FIGURE 28-5 Precision Sine Dresser

Figure 28-5 is adjustable to 45° either side of a horizontal (0°) position. This device is set by using gage blocks to accurately obtain the required angle.

When an angle dresser or a sine dresser is not available, the diamond dresser may be guided on a parallel that is clamped to an angle plate. The angle of the parallel is set with the aid of sine blocks. The diamond is located under the centerline of the wheel. The table and saddle are locked in position. The wheel head is lowered from 0.002" to 0.003" for each pass across the wheel to rough form the angle. A 0.001" to 0.0005" (0.02mm to 0.01mm) feed is used for the last few passes. The wheel is dressed until the angular surface is the required width.

Combination Radius and Angle Dresser

A *radius and angle dresser* combines the processes performed by the separate dressers. A radius and angle dresser is designed to form convex and concave outlines and angles on grinding wheels. Moreover, this device makes it possible to generate combinations of radial and angular shapes that are continuous (tangential to each other). On this type of dresser, the diamond dressing tool must be located at the horizontal centerline of the wheel. The wheel is formed with the diamond and holder in a right-side position instead of under the wheel.

The principal parts of a precision radius and angle dresser are identified in Figure 28-6A. The base carries a sliding platen that is moved horizontally by a micrometer collar. This movement permits the dressing of angles. Concave

550 GRINDING MACHINES

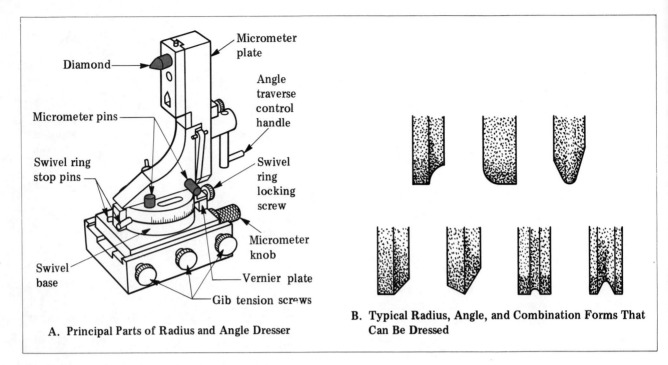

A. Principal Parts of Radius and Angle Dresser

B. Typical Radius, Angle, and Combination Forms That Can Be Dressed

FIGURE 28-6 Radius and Angle Dresser Features and Examples of Form Dressing

and convex radii are formed by using a swivel base movement. The stop pins are located to permit dressing each radius and angle according to specified dimensions.

The radius and angle dresser is adaptable to the precision grinding of punches, sectional dies, cutting tools, special forming tools, and other work involving hardened steels. Typical radius, angle, and combination forms that can be dressed are illustrated in Figure 28-6B. In addition to the functions of the dressing unit for forming angles, radii, and other profiles, the diamond dressing tool itself must be considered.

DIAMOND DRESSING TOOLS FOR FORMING GRINDING WHEEL SHAPES

No one diamond is equally appropriate or efficient for all form dressing. Usually there is no problem in finding a suitable diamond to dress a convex radius or a bevel. The holder is not likely to strike the wheel during the dressing process. However, the diamond point that forms a small concave radius is usable within a limited range of radius sizes. Small, 180° concave radii are usually formed by plunge dressing. The diamond is accurately lapped to shape. It is mounted in correct relation to the wheel. The wheel is formed by feeding the diamond into the face to the required depth.

The size and shape of both the diamond and holder depend on the contour to be generated in the wheel, the wheel diameter and characteristics, and the nature of the application (light, medium, heavy, or extra-heavy). A *cone-point* diamond cutting tool, as pictured in Figure 28-7, is used for the precision dressing of intricate radii and complex combination forms. Cone-point dressing tools are ordered by the type of cone point, the included angle on the diamond, the radius of the diamond, and the application. This information is required by the manufacturer.

The *full-ball* radius dressing tool (diamond dresser) shown in Figure 28-8 is designed to plunge dress a full 180° concave radius in a grinding wheel. This dressing tool is available in standard radii between 0.010″ to 0.050″ (0.25mm to 1.25mm). Smaller and larger radii dressers are available to meet special requirements. Frequent turning of the tool is necessary to provide longer tool life. When a diamond becomes worn with use, it may be relapped.

FORM GRINDING AND CUTTING-OFF PROCESSES 551

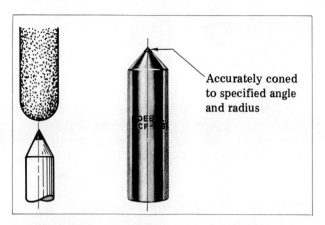

FIGURE 28-7 Cone Point for Precision Dressing Intricate Radii and Combination Forms

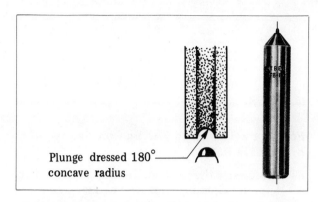

FIGURE 28-8 Full-Ball Radius Diamond Dressing Tool

Figure 28-9 shows three different forms of *radius diamond dressing tools*. The first form (Figure 28-9A) is available in six standard sizes to form concave radii from 0.010" to 0.250" (0.25mm to 6.25mm). The second form (Figure 28-9B) generates convex radii on small-diameter grinding wheels. There are three general sizes of this dressing tool. The sizes accommodate radii from 0.020" to 0.500" (0.5mm to 12.5mm). The third form (Figure 28-9C) is applied when dressing half-circle concave radii. Radii of 0.032" (0.8mm), 0.062" (0.16mm), and 0.125" (3mm) are standard sizes. While only three shank designs are illustrated, other designs are available for use with radius, angle and combination dressers, and different pantographs.

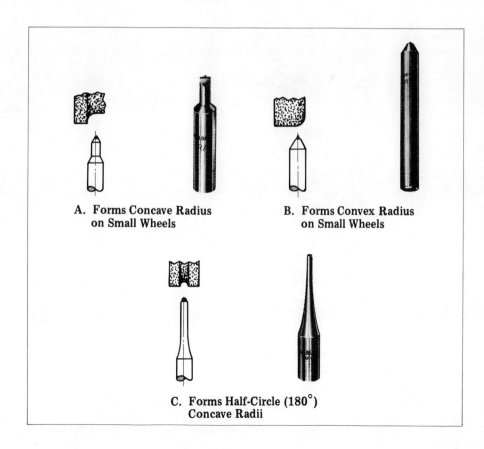

FIGURE 28-9 Radius Diamond Dressing Tools for Forming Concave and Convex Radii

HOW TO FORM DRESS A WHEEL WITH A RADIUS AND ANGLE DRESSER

Setting the Dresser to Form a Convex or Concave Radius

STEP 1 Measure the distance from the selected diamond dressing tool point to the micrometer plate on the dressing device (Figure 28-10A).

STEP 2 Add the required radius to the micrometer reading.

Note: A convex form is generated by subtracting the required radius.

STEP 3 Turn the micrometer base lead screw until the measurement over the micrometer pins equals the combined radius plus micrometer reading (Figure 28-10B).

Note: The knurled gib screws are first loosened to position the diamond and then tightened.

Setting the Dresser to Form One or More Angles

STEP 1 Loosen the swivel ring locking screw.

STEP 2 Bring the two swivel ring stop pins together. Move them until the desired angle for the lower pins is read on the vernier plate.

STEP 3 Move the dresser in the opposite direction to the second required angle. This movement sets the position of the upper swivel ring stop pin.

STEP 4 Lock the pins in the desired position by tightening the swivel ring locking screw.

Note: The positions of the stop pins are the points at which the angles are tangent to the radius.

Dressing the Wheel (to a Radius and Two Angles)

STEP 1 Align the dresser on the magnetic chuck. Secure it.

STEP 2 Position the grinding wheel spindle so that its axis coincides with the horizontal axis of the dresser diamond.

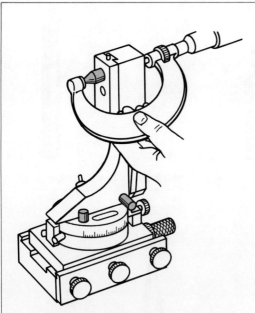

A. Measurement over Diamond Point and Micrometer Plate

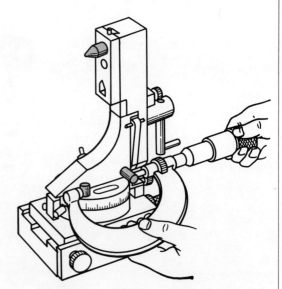

B. Measurement over Micrometer Pins

FIGURE 28-10 Micrometer Settings of a Radius and Angle Dresser

STEP 3 Adjust the machine table so that the diamond almost touches the periphery of the wheel. Lock the table in position. Position the diamond in relation to the width of the wheel.

STEP 4 Lock the saddle. Start the machine and coolant flow if the wheel is to be dressed wet.

Caution: Use the exhaust system if the wheel is to be dressed dry.

STEP 5 Move the diamond until it grazes the wheel. The micrometer base lead screw knob is turned to produce this movement.

STEP 6 Feed the diamond for 0.002" to 0.003" (0.05mm to 0.08mm) and swivel the dresser against one stop.

STEP 7 Dress one angle by turning the angle dressing handle in the direction the diamond is to move. Then, return the handle to the center position.

STEP 8 Dress the radius by swiveling the dresser upright to the other stop.

STEP 9 Produce the other angle by turning the angle dressing handle.

STEP 10 Take successive cuts, repeating steps 6 through 9, until the full form is produced on the wheel.

Note: The feed and speed (movement of the diamond) are reduced for the last two or three passes, if a finely ground surface finish is required.

IRREGULAR FORM DRESSING WITH A PANTOGRAPH DRESSER

The forming of a grinding wheel to produce complex forms or contours in a workpiece requires the use of a *pantograph dresser* (Figure 28-11). This accessory is mounted on a surface grinder. It consists of a tracer that follows a template of the required contour and a linkage system that guides the diamond dresser to reproduce the form accurately. This linkage serves two functions: (1) The system of arms transfers the form on the template to the diamond and (2) the arrangement of the arms permits the outline (form) on the template to be reduced.

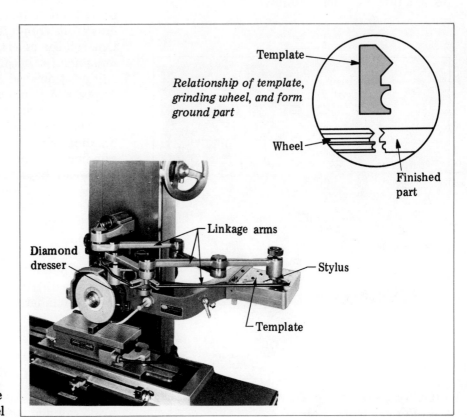

FIGURE 28-11 Pantograph Dresser Attached to Surface Grinder for Dressing Intricate Contour on a Grinding Wheel

Usually, the ratio between the outline and the form reproduced on a grinding wheel is 10:1. This ratio makes it possible to form a wheel to a high degree of accuracy. For example, any error in the template form is reduced to one-tenth of the inaccuracy on the wheel. The stylus which traces the template, must be the same shape as the diamond and, in this case, ten times larger.

FORMING A WHEEL BY CRUSH DRESSING

Crush dressing is another method of forming the periphery of a grinding wheel. The crush method employs a hardened high-speed steel roll or carbide roll, similar to a circular form tool. The shape of the roll is identical to the contour to be ground into a workpiece. Therefore, the crush roll must be machined accurately.

The material to use for the crush roll is determined by the number of parts to be ground. Carbide rolls are excellent for long production runs. These rolls are harder and more wear resistant than high-speed steel hardened rolls. Some crush rolls have gashes cut across the face at an angle to the axis of rotation. This design produces a better cutting action.

Crush dressing is accomplished by forcing a revolving crush roll against a slowly turning grinding wheel. This action produces a reverse profile on the wheel face. There are two major devices on which crush rolls are mounted and driven for crush dressing. Each device is mounted on the surface grinder table with the crush roll positioned under the grinding wheel. The crushing devices are either of the idler or the power-driven type. The *idler crushing unit* is designed so that the crush roll is driven by the grinding wheel. This device must be used with a surface grinder that has a variable or two-speed spindle. The grinding wheel must travel between a 100 to 300 sfpm range during the crushing process.

The *power-driven crushing unit* (Figure 28-12) drives the crush roll. The slowly revolving crush roll is brought into contact with the stationary grinding wheel and causes the wheel to revolve. The crush roll is uniformly fed into the wheel to reproduce the reverse form of the roll. The crush-formed wheel is then ready to grind the required form in the workpiece.

Maintaining the Required Wheel Form with Crush Rolls

One crush roll is used when a limited number of parts are to be ground to a specific form. Two crush rolls are preferred for quantity production. One roll serves as the *work roll;* the other, as the *reference* (or *master*) *roll*. The work roll is usually positioned at one end of the table. The reference roll is mounted at the other end. Both rolls are aligned.

The work roll is used for crush forming the grinding wheel periodically, following the form grinding of a number of parts. When the work roll wears beyond the allowable tolerance, it is reground by form dressing the grinding wheel with the reference roll. The reformed wheel is then brought back into contact (while revolving at grinding wheel speed) with the work roll. The work roll is rotated in an opposite direction under its own power. It is, thus, reground to its original form. In this condition, the work roll may be used a number of times to dress and reform the wheel, as required.

Form Grinding by Crush Dressing with Work and Reference Rolls

Figures 28-13A through F summarize how a work roll and reference roll are used to control

FIGURE 28-12 Power-Driven Crushing Unit with Formed Crush Roll

FORM GRINDING AND CUTTING-OFF PROCESSES 555

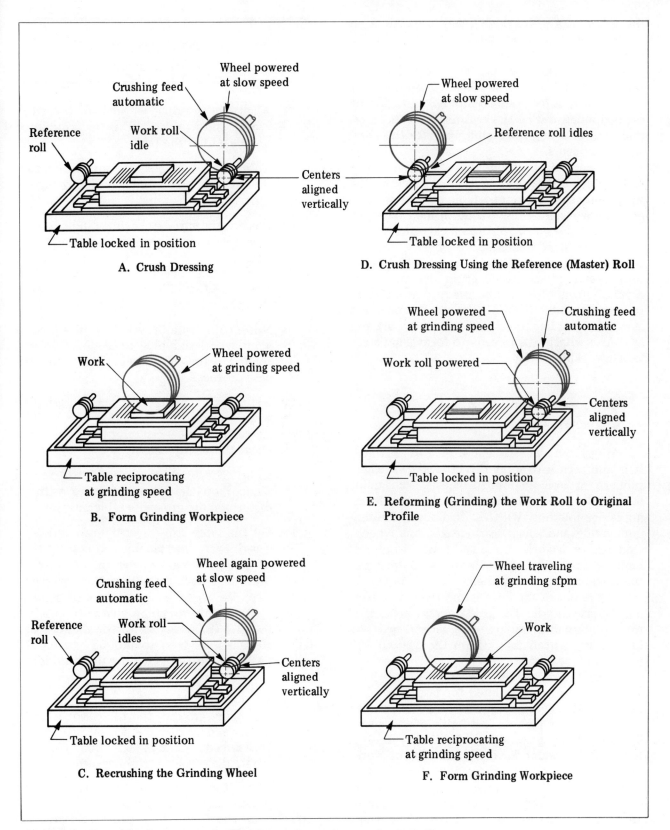

FIGURE 28-13 Crush Dressing a Grinding Wheel Using a Work Roll and Reference Roll Setup to Control Form Accuracy for Quantity Production

the wheel form for high-quantity production grinding. The grinding wheel is crush dressed as it is fed automatically into the work roll (A). The feed rate ranges from 0.0001″ to 0.001″ per revolution of the grinding wheel. It is important that a constant feed rate is employed. A grinding oil is applied during the crushing process to reduce the crushing force and minimize the abrasion that occurs.

In operating position, the wheel revolves at normal grinding speed and moves across the workpiece at the required table speed (B). The grinding wheel is redressed to form and to sharpen it (C). As the work roll becomes worn after a number of form dressings, the wheel is brought over the reference roll. The wheel, powered at slow speed, is accurately crush dressed to its original form (D). This accuracy is transferred to the work roll. This time the wheel is powered at regular grinding speed to regrind the work roll (E). The wheel is then ready to form grind additional workpieces (F).

Considering Grinding Wheel Characteristics

All grinding wheels cannot be crush formed. It is impractical to apply this process to shellac-, rubber-, or resinoid-bonded wheels. The grains tend to break out in clusters. Thus, crush forming is possible only with vitrified-bonded aluminum oxide and silicon carbide grinding wheels. Organic-bonded wheels are too elastic to permit building up enough force between the roll and the wheel to crush form them.

The radius of the form is important in selecting the size of grit. The grit size must be smaller than the smallest radius. The general grit size of grinding wheels falls within the 120 to 400 medium-grade range.

Crush dressing produces a sharp cutting wheel. The grains are sharp pointed for fast removal of material. However, the quality of the surface finish is not as high as the finish produced by diamond dressing a wheel. The grinding wheel is fed a small amount for each pass over the workpiece. Feeding by automatic down-feed is desirable, especially when a large surface area is being form ground. However, many small workpieces are ground to finish size in one pass. The wheel is set to the full depth of cut. An especially slow table feed is used for such grinding.

HOW TO FORM GRIND WITH A CRUSH DRESSED WHEEL

STEP 1 Align the grinding wheel with the crush dresser and work roll. This roll is mounted on one end of the table.

Note: On a high-quantity production run, the reference roll unit is aligned on the other end of the table.

STEP 2 Prepare the surface grinder spindle to turn at a relatively slow speed of about 200 sfpm.

STEP 3 Position the wheel on center over the work roll. Lock the table and saddle. Crush form the grinding wheel.

Note: A nonsoluble grinding oil is used to keep wheel chipping to a minimum and for forms that have deep, sharp, and intricate profiles.

STEP 4 Return the spindle to normal grinding wheel speed.

STEP 5 Lower the wheel head until the wheel just clears the workpiece.

Note: The workpiece is mounted on and secured on the magnetic chuck without disturbing the crosswise table position.

STEP 6 Set the table trip (reverse dogs) and table speed. Start the machine and coolant flow (if wet grinding). Lower the wheel head until the wheel just touches the workpiece.

STEP 7 Feed the wheel with the automatic down-feed set for roughing cuts of 0.002″ to 0.003″ (0.05mm to 0.08mm) per pass. Reduce this feed to 0.0005″ to 0.001″ (0.01mm to 0.02mm) for finish grinding.

Note: The exact feeds to use depend on the size and shape of the form, the material being ground, the dimensional tolerance, and the required surface finish. Down-feeds of 0.004″ to 0.005″ (0.1mm) are used for coarse crushing feeds.

STEP 8 Check the required depth, accuracy of the profile, and surface finish.

STEP 9 Stop the flow of coolant. Disengage the table feed at the end of a pass. Shut down the machine.

STEP Clean the machine area and workpiece.
10 Return all accessories to their proper storage places. Remove any fine grinding burrs from the workpiece.

FORMING CRUSH ROLLS ON A MICROFORM GRINDER

The *microform grinder* is a precision grinding machine. While it is used extensively to grind complex profiles on crush rolls, other circular and flat forms may be ground to shape. The microform grinder is used to grind form tools, templates, cams, gages, punches, dies, and other intricate forms. Such grinding is done on hardened materials, as well as on carbides. The principal design features of a microform grinder are the viewing screen, scope, pantograph, template, stylus, and drawing table (Figure 28-14A).

The pantograph feature provides for the stylus to follow the outline of a profile drawing. The pantograph is mounted on the drawing table. The arms permit the transfer, through a control mechanism, to form dress the crush roll to the required shape. The combination of viewing screen and scope permits the operator to follow the profile lines on the layout drawing and to control the form grinding process. The viewing screen has cross hairs that move in an exact relationship to the profile. The cross hairs indicate that a point on the grinding wheel has been fed to form a cutting edge that corresponds with the profile at the same position on the drawing or template. The grinding wheel is fed manually until the entire profile is ground on the crush roll. The ratio between the enlarged drawing size and the reproduced profile on the crush roll is 50:1 on the model illustrated. This ratio helps to contol the dimensional accuracy and the reproduced profile on a ground crush roll. An example of a complex profile that may be ground is shown in Figure 28-14B.

EFFECT OF VARYING THE GRINDING CONDITIONS ON SURFACE FINISH

The machine operator is accountable for recognizing the effect on flatness and surface texture produced by varying the grinding conditions. Probable causes of an imperfection must be recognized and corrective steps for remedying the defect must be taken. Imperfections relate to marks due to vibration, burnishing and burning,

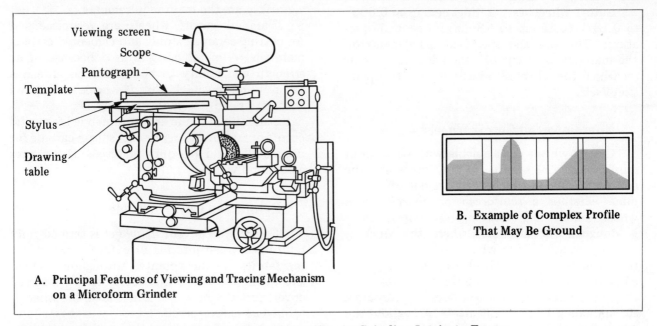

A. Principal Features of Viewing and Tracing Mechanism on a Microform Grinder

B. Example of Complex Profile That May Be Ground

FIGURE 28-14 Microform Grinder and Application in Grinding Intricate Forms

surface scratches, and patterns of lines produced by imperfect grinding.

There is no single scale of comparison that indicates an exact degree of finish that will be obtained by varying a condition in grinding. For example, any variation of grit size, bond, grade, coolant, wheel speed and feed, and the machine itself affects surface finish. The effect may be limited or it may be extreme. The three variables that have the greatest effect on surface finish relate to diamond truing, grit (grain) size, and chatter.

ABRASIVE CUTOFF WHEELS AND CUTTING-OFF PROCESSES

Cutting off (while not a widely used surface grinding process) means severing a material at a specific dimension. The cutting is done with a cutoff abrasive grinding wheel or abrasive saw. The process is particularly suited to cutting off hardened steel, ceramic, stainless steel, new high-temperature metals, and other difficult-to-cut materials. Unusual-shaped parts and extruded forms are held and positioned in fixtures for cutting-off processes.

The process is fast cutting. The temperature may be controlled so that the parts are not work hardened. Cutting-off accuracy is better controlled than heat (flame) or arc cutting processes. The quality of surface finish is also far superior.

Cutoff wheels vary in thickness from 0.006" to 0.187" (0.15mm to 4.8mm) for general operations. The size and specifications depend on the material to be cut off, the grinding machine on which the wheel is to be used, and the operation itself.

Nonreinforced Cutoff Wheels

Cutoff wheels are available with or without reinforcing. The severity of the forces and the conditions under which parts are cut off determine whether a *reinforced* or *nonreinforced* wheel is used. The *nonreinforced cutoff wheel* is designed for machines where the work is clamped securely, the wheel operates in a controlled plane, and the wheel is adequately guarded. The surface grinder meets these requirements.

The nonreinforced wheel does not contain a strengthening fabric or filaments throughout the wheel diameter or reinforcement in special areas of the wheel. General cutoff wheels are made of aluminum oxide or silicon carbide for cutting ferrous metals, except titanium. Silicon carbide cutoff wheels are used for titanium and nonmetallics such as ceramics, plastics, and glass.

Specific Bond Recommendations

All cutoff wheels are type 1. They are furnished for both wet and dry grinding operations. Resinoid, rubber, and shellac cutoff wheel bonds are available for a wide variety of operations. For example, resinoid wheel bonds permit applications that range from the high production cutting of bar stock to the cool cutting action on nonproduction, heat-sensitive materials. Resinoid wheels are adapted to dry grinding.

Rubber wheel bonds range from free-cutting ferrous and nonferrous metals (either wet or dry) to the soft cutting action required for cutting glass and ceramics. Rubber-bonded wheels produce a good surface finish with a minimum of burrs and discoloration. Rubber wheels produce smooth side (vertical) surfaces and square corners. Thus, they are good to use for grooving and slotting operations.

Shellac wheel bonds are adaptable to highly critical applications where no burrs, checking from heat, or discoloration is permitted. Some shellac-bonded wheels are adaptable to extreme applications requiring fine grit sizes and soft grades. Shellac wheels are generally used in toolroom applications where the quality of cut and versatility are the main requirements.

Diamond cutoff wheels are recommended for cutting ceramic, stone, and cemented carbide materials. A *metal bond* is used because of its strength and ability to hold diamond abrasives securely in the wheel. Cutoff wheels are marked with letters to designate the abrasive: *A* designates aluminum oxide as recommended for metal cutoff operations; *C* represents silicon carbide for nonmetallic and masonry materials; *D* indicates a diamond wheel.

Reinforced Cutoff Wheels

The *reinforced cutoff wheel* is termed *fully reinforced* when internal or external strengthening fabrics are incorporated throughout the entire diameter. When the strengthening material covers just the hole and flange area, the wheel is termed *zone reinforced*.

Reinforced wheels are designed for use where the grinding machine is hand guided or the workpiece is not clamped securely. Reinforced wheels resist breakage that normally results from severe cross bending. Reinforced cutoff wheels are manufactured in aluminum oxide or silicon carbide. These wheels may be used for surface speeds up to 16,000 sfpm for wheel diameters up to 16" (406mm) and 14,200 sfpm for wheels over 16" (406mm) diameter.

SAFE PRACTICES IN USING FORM DRESSERS AND IN CRUSH DRESSING AND FORM GRINDING

—Select a commercial wheel form, where practical, that requires a minimum of dressing to form the wheel to a specified profile.
—Select a diamond dressing tool that meets the size, shank type, and cutting requirements for the wheel and form grinding operation.
—Check the dressing tool to see that the type permits dressing the wheel to the required form without grinding into the tool shank.
—Remove all dust and abrasive particles during the form grinding of the wheel. The air exhaust system on the wet grinding system on the machine must be adequate to remove the volume of dust and particles produced when form grinding the wheel.
—Recheck the position of the diamond point in relation to the axis of the dresser unit. Set the diamond above the centerline to form a concave area in the grinding wheel; below, to produce a convex surface.
—Position the diamond point at the centerline of the wheel to reproduce the true form at which the dresser is set.
—Run a contour form check periodically. The diamond face (cutting edge) must be sharp and not worn beyond the allowable tolerance.
—Crush dress only on a surface grinder that is designed with a spindle that compensates for the severe forces required during the process.
—Select a grinding wheel with appropriate characteristics for crush grinding without damage to the wheel. Also, the wheel must have properties that permit it to reproduce the form accurately with maximum operator safety and wheel wear life.
—Select a cutoff wheel with adequate reinforcement to withstand the cutting and bending forces required for a specific set of conditions.
—Use nonreinforced cutoff wheels for securely clamped parts on a rigid surface grinder, where the wheel is properly guarded.
—Analyze the cutting action, burring, and surface finish produced with a cutoff wheel. Too soft a wheel adds to cutting costs, reduces wheel life, and produces larger burrs. Too hard a wheel creates excessive heat, discoloration, burning the bond, and causes the wheel to chip and break.
—Extend the workpiece (with a minimum overhang) so that the section to be cut off drops away from the grinding wheel onto a protective tray.

TERMS USED WITH FORM DRESSERS AND IN CRUSH DRESSING AND FORM GRINDING

Form grinding (surface grinder)	The process of grinding an angle, radius or curved surface, or any combination of these forms. (Form grinding requires the use of a formed grinding wheel and a reciprocating horizontal table movement.)
Radius dressing (truing) device	A mechanism for holding a diamond dressing tool to generate an accurate radius form in a grinding wheel. A fixture for holding a diamond dressing tool in a fixed radius relationship to permit an accurate concave or convex radius form to be produced.
Angle dressing (truing) device	A mechanism serving a similar function to the radius dresser, except that the settings permit forming a grinding wheel at a precise angle.

Forming a concave or convex wheel	A specific regular curve form grinding process. Setting up a radius dresser and positioning a diamond dressing tool above center to produce a concave arc; below center, a convex arc.
Sine dresser	An adjustable angle device that is set with the aid of gage blocks to produce a precise angle setting.
Combination angle and radius dresser	A form dressing accessory for a surface grinder. The dresser that may be set to form a continuous profile of straight-line and curved surfaces on a grinding wheel.
Swivel ring (combination dresser)	A circular movable part on a form dresser that permits setting the device to reproduce a form over a specific distance.
Cone-point, full-ball radius, and radius diamond dressing tools	Standard diamond-point holder designations. Terms used to identify the shape and size of the shank in the area surrounding the cutting edge of the diamond.
Pantograph dresser	A surface grinder accessory incorporating a tracer linkage system for reproducing a template profile to form dress a grinding wheel.
Crush dressing	Rolling the form of a slowly turning work roll (with great force) into a grinding wheel. Using a hardened high-speed steel or carbide preformed work or reference roll to impress the required form (in reverse) in a grinding wheel. Forming a grinding wheel that has properties appropriate to withstand great force and to efficiently maintain its shape in form grinding.
Work roll and reference roll (crush dressing)	A set of hardened high-speed steel or carbide rolls that are precision machined to reproduce a form on a grinding wheel by crush dressing.
Reference roll	A duplicate formed roll of a work roll. A crush roll used to control the accuracy of the form on a worn work roll. A crush roll used in production work to precision dress the grinding wheel. (In turn, the grinding wheel is used to regrind the original form on the worn work roll.)
Microform grinder	A precision grinding machine for form grinding complex forms on precision tools and parts, including work and reference rolls. A form-generating machine that uses a stylus, tracer, and pantograph arrangement. (Design features on a profile drawing are reduced and the movement of a diamond dresser is controlled to produce a required profile on a crush or reference dressing roll or other part.)
Cutting off (grinding)	The severing of materials at a particular point or dimension.
Reinforced grinding wheel	A cutoff wheel that is laminated with tough, flexible fibres. A grinding wheel capable of withstanding extreme forces in offhand processes.
A, C, and D wheel markings	Designations by grinding wheel manufacturers. (A denotes aluminum oxide; C, silicon carbide; and D, diamond abrasives.)
Variable factors	A material, machine component, workpiece, or process that, if changed, results in a change in the quality of a surface finish (product).

SUMMARY

Formed grinding wheels are commercially available for regular radii, angles, and combinations of these forms.

> Grinding wheels are generally dressed to a required precision form by one of three wheel forming accessories: a radius dresser, an angular dresser, or a combination radius and angle dresser.

Simple radii and angles may be formed by using an appropriate abrasive stick. This offhand method is limited to rough dressing.

> The diamond point on a radius dressing accessory is set above the axis of the cradle to form a concave arc in a wheel. A below-the-axis position is required to form a convex arc.

The diamond-point setting in a dresser is established by measurement with a micrometer.

> Stops on a radius dresser control the arc over which the diamond is swung to produce the required radius.

The diamond point is positioned at the centerline of the grinding wheel to produce the precise form to which the dresser is set.

> A feed from 0.002" to 0.003" (0.05mm to 0.08mm) per revolution is used in rough dressing to form a grinding wheel. This feed is reduced to 0.001" to 0.0005" (0.02mm to 0.001mm) for the final dressing cuts.

An angle dresser or a sine dresser is used to dress an angular surface on a wheel to a close tolerance.

> The combination angle and radius dresser may be used to form an individual angle or radius, although it is designed to generate a continuous profile of curved and angular surfaces.

Angle and radius dressers are designed to form wheels for grinding precision cutting, stamping, forming, shearing, and other tools and work parts. Such parts are usually made of hardened steels.

> The type, size, and shape of a diamond dressing tool depend on the specifications of the form to which a wheel is to be ground.

Three common shapes of diamond dressing tools include the cone-point, full-ball radius, and special point for concave and convex radii on small-wheel diameters.

> A small, 180° concave radius is usually formed by plunge cutting to the full depth.

The pantograph dresser is a versatile form dressing accessory. The form on an enlarged template is duplicated at a reduced actual size on the face of the wheel.

> Grinding wheels that are formed with a diamond dresser usually cut harder. However, the form produced is within closer dimensional and surface finish tolerances than form wheels that are crush dressed.

The size of the abrasive grains in a form grinding wheel must be smaller than the finest radius that is to be form ground. The wheel grade should be harder than conventional grinding. Such a wheel maintains the required form to grind a number of workpieces.

> Crush dressing requires a preformed work roll, a low-speed spindle (between 100 and 300 sfpm), a device that idles a work roll, and a surface grinder spindle that can withstand great force.

Crush dressing feeds must be uniform and constant for each revolution of the grinding wheel.

> A work roll and a reference roll are used to maintain quality form control on production runs. The reference roll serves as the master to correct the form of the work roll when it wears beyond the allowable tolerance.

Care must be taken to align the workpiece, grinding wheel, work roll, and reference roll (when used).

> The automatic feed rate for general crush grinding ranges from 0.0001" to 0.001" (0.002mm to 0.02mm) per revolution. Fine feeds are used for the final forming revolutions.

All grinding wheels may not be crush formed. The process may be carried on with vitrified-bonded aluminum oxide and silicon carbide wheels. Resinoid-, shellac-, and rubber-bonded wheels are too elastic to build up the force required to crush form them.

> Crush dressing rolls are made of high-speed steels or harder and tougher carbides, depending on production demands.

Crush dressed wheels cut faster than wheels dressed with diamonds. The sharper cutting grains produce a rougher surface finish and a profile that is ground to a coarser dimensional tolerance.

> Crush rolls, and other parts requiring the grinding of a complex profile, are formed on a microform grinder. An accurate enlarged drawing is used as a guide for a stylus to position a grinding wheel to generate a precise form.

Difficult-to-cut ferrous and nonferrous metals and hard nonmetallic materials may be cut off economically and accurately on a surface grinder.

> A superior surface finish quality is produced by abrasive cutting off than with other heat severing processes.

Shellac wheels are suited to toolroom applications requiring a high-quality surface finish and accurate cut. Resinoid wheels are adapted to dry cutting-off processes in production. Rubber wheels are used extensively for production wet grinding.

> Shellac bonds are used for toolroom cutoff and grooving operations where precision dimensions and surface finish are important considerations.

A metal-bonded diamond cutoff wheel is recommended for cemented carbide, ceramic, stone, and other hard and/or abrasive materials.

> The limited thickness of cutoff wheels requires that they be dressed squarely, the feed rate be controlled, and (where required) a clean and adequate coolant flow is directed over the cutting area.

The nonreinforced cutoff wheel is generally used on a surface grinder. The type 1 nonreinforced wheel is practical for cutting-off and grooving processes. It is adapted to processes where the cutting action is controlled along a fixed path and the workpiece is held securely.

> Surface imperfections such as glazed spots, scratches, grinding marks, burning, surface discoloration, chatter patterns, and burnishing require a methodical check of the machine, work setup, grinding wheel, and the machining process.

While safe general machine setups and operation practices must be followed, four precautions are essential: (1) The possibility that small sections of a grinding wheel may chip or break out is increased in form grinding; (2) crush grinding must be performed at a reduced grinding wheel speed and a slow crush roll speed; (3) the grinding wheel spindle design must be capable of withstanding crush dressing forces without distortion or damage to bearings or mating surfaces; and (4) the dust and abrasive removal system must be adequate to draw away the increased amounts of particles produced during form dressing.

UNIT 28 REVIEW AND SELF-TEST

1. a. Make a simple sketch of four contour formed grinding wheels that are available commercially.
   b. Add two identifying design features or specifications for each grinding wheel.
2. a. Explain the relationship between the position of a diamond dressing tool on the swinging arm of a cradle-type radius dresser and a concave or convex radius produced on a grinding wheel.
   b. Tell how a diamond nib (point) is set to generate a specified radius.
3. List the steps that must be taken to form grind hardened workpieces that have an internal corner radius of 0.75" (18.8mm). The grinding wheel is to be formed with a radius dresser. Stock is left on the part for a series of roughing cuts and a finish cut.
4. a. Cite two advantages to using a sine dresser for dressing a grinding wheel at angles up to 45° from either side of a 0° horizontal plane.
   b. Calculate the gage block height for setting the sine dresser at a 7°15" angle above the 0° horizontal plane.
   c. Select a set of gage blocks that will produce the required height.
5. Form dress the side and front faces of a type 1 grinding wheel to an angle of 45° × 6mm. Use any available angle dresser.
6. a. Describe the special functions of a pantograph dresser in relation to a radius form dresser.
   b. Provide a brief explanation of how a pantograph dresser works.
7. a. Explain why it is impractical to crush form rubber- or resinoid-bonded grinding wheels.
   b. Compare (1) the cutting action and (2) the surface finish of a crush formed wheel with a contour formed on an abrasive wheel when a diamond nib and wheel dresser are used.
8. Cite two advantages of a microform grinder over other contour form grinding machines.
9. a. Give two probable causes of (1) a limited cutting rate and (2) a poor quality of cut when a cutoff wheel is used.
   b. Cite two corrective actions the machine operator may take in each case.
10. Secure an abrasive wheel manufacturer's technical manual on cutoff wheels. Also, check the specifications of a surface grinder in the shop with respect to spindle features, size, and speed range.
    a. Select the wheel size and determine the specifications for a wheel that is adapted to cutting off 20 pieces 4mm thick from a 25mm square bar of SAE 1020 mild steel.
    b. Give the wheel size and specifications.
    c. Compute the spindle speed (RPM) and indicate the rate of table feed.
11. State two personal safety practices the grinder machine operator must follow (a) before crush dressing a wheel and (b) when mounting a cutoff wheel.

SECTION FOUR

Cylindrical Grinding Technology and Processes

Cylindrical grinding involves a series of abrasive machining processes. This section begins with a general description of basic types of cylindrical grinding machines, design features and their functions, work-holding devices, and accessories. Cutting speeds and feeds are applied to basic cylindrical grinding processes. Prior knowledge about grinding wheel selection, mounting, truing, and dressing and safe practices according to the ANSI Code is extended to applications for cylindrical grinding. Certain information on peripheral speeds, problems and corrective action, and basic cylindrical grinding process data are applied.

Step-by-step procedures are given for setting up and operating the machine, grinding external straight and taper cylindrical surfaces, angle grinding, face grinding, and the internal grinding of straight and slight and steep taper cylindrical surfaces. Procedures are also given for using truing and dressing devices for producing angle and radius forms in grinding wheels.

UNIT 29

Cylindrical Grinder Components and External and Internal Grinding

Cylindrical grinding covers a cluster of grinding methods in which material is removed from a revolving workpiece by abrasive machining. The surface produced is controlled in relation to the axis of rotation. The cylindrically ground surface may be straight or tapered or of curvilinear (contour-shaped) profile. The surface formed, while generally parallel to the axis of rotation, may also be at an angle to or perpendicular to the axis. Several surface sections may be ground simultaneously of equal or different diameters and as continuous or interrupted sections. The ground surface may be located in a parallel plane or mutually inclined planes. The common characteristic is the axis of rotation.

CYLINDRICAL GRINDING

GENERAL TYPES OF CYLINDRICAL GRINDING MACHINES

The *plain cylindrical grinder* continues to be the basic type. The plain cylindrical grinder is primarily used for straight and taper cylindrical grinding. The machine may be adapted to plunge grinding and to the producing of contour-formed cylindrical surfaces.

The *limited-purpose cylindrical grinder* is designed for special grinding requirements. Crankshaft, camshaft, and bearing race grinders are examples. The machine is for high-volume production but has limited flexibility for general-purpose cylindrical grinding.

The need for versatility in grinding accelerated the development of the *universal cylindrical grinder*. The term *universal* indicates that multiple processes may be performed in addition to plain cylindrical grinder processes. Slight and steep tapers (angles), faces, and combined internal and external grinding processes are performed on the universal cylindrical grinder. The versatility required in the toolroom for plane and contour form grinding is met by using a number of accessories with a universal cylindrical grinder.

APPLICATIONS OF CYLINDRICAL GRINDING MACHINES

General-purpose plain, special, and universal cylindrical grinders are designed for regular traverse grinding with limited applications to plunge grinding. In *traverse grinding*, the revolving workpiece traverses the face of the grinding wheel in a fixed relationship with the work axis. At the end of each stroke, the wheel slide is advanced to position the grinding wheel for the depth of each successive cut. The length of the cylinder that may be ground is limited by the length of the table movement (stroke).

Plunge grinding requires the in-feeding of the abrasive wheel into a revolving workpiece. The wheel slide on which the abrasive wheel is an integral unit advances at a fixed rate. The feed is stopped at the point where the required diameter is reached. The length of the section to be plunge ground is limited to the width of the grinding wheel face. Plunge grinding is required for profiling cylindrical surfaces where there is no lateral movement of the abrasive wheel. Multiple surfaces may also be simultaneously ground to different diameters by plunge grinding.

A finer surface finish is generally produced by traverse grinding than by plunge grinding. However, productivity is usually higher for plunge grinding.

DESIGN FEATURES OF THE UNIVERSAL CYLINDRICAL GRINDER

The versatility of the universal cylindrical grinder makes this machine ideal for identifying its design features, operating controls, and basic grinding processes. The six principal units of the universal cylindrical grinder are as follows:

—Base, which houses motors for the coolant system, lubrication system, and drive components for the table (the machined ways accommodate the saddle);
—Saddle, swivel table, and drive mechanism for longitudinal travel;
—Headstock;
—Footstock;
—Wheel head, carriage, and cross feed mechanism;
—Operator control panel.

Toolroom universal cylindrical grinders are capable of grinding perfectly round cylinders to less than 25 millionths of an inch (0.0006mm) of roundness. Diametral sizes may be held to less than 50 millionths (0.0013mm). Surface finishes may be ground to less than 2 microinches and metric equivalent.

The universal cylindrical grinder is usually identified by size and whether the table traverse and/or the wheel feed is hand or hydraulically operated. Two popular toolroom sizes are 6″ × 12″ and 10″ × 20″ (152mm × 305mm and 254mm × 508mm). The grinder may be designed for:

—Hand traverse and hand wheel feed,
—Hydraulic traverse and hand wheel feed,
—Hand traverse and hydraulic in-feed,
—Hydraulic traverse and hydraulic in-feed.

The model size—for example, 10″ × 20″ (254mm × 508mm)—indicates the nominal swing over the table—10″ (254mm) diameter—and the nominal distance—20″ (508mm)—between centers.

MAJOR PARTS OF THE UNIVERSAL CYLINDRICAL GRINDER

The major parts of the universal cylindrical grinder (exclusive of operator controls which are described later) are given in Figure 29-1. The model shown is designed for hydraulic traverse and infeed.

Saddle and Table

The saddle is machined with ways on the underside. These ways match with the bed ways. The ways provide for the workpiece and/or accessory mounted on the table to be moved longitudinally. The top face of the saddle receives a precision table. The table may be swiveled to position the work axis at a short angle to the grinding wheel face. The swivel table is adjustable from 0° to 20° included angle and up to 4″ per foot or metric equivalent.

The table may be moved by hand for positioning the workpiece in relation to the grinding wheel or for traversing the workpiece past the wheel face. Trip dogs are provided to control the length of table movement and to actuate the table reversing lever. Angle settings are made by turning an adjusting screw to position the table at the required angle indicated on the swivel table scale and tightening the table clamping screws (Figure 29-2). It should be noted that the taper angle and the table setting depends on the location of the taper on the workpiece, the work diameter, the length to be ground, and the grinding wheel diameter.

Longitudinal Table Travel

Table travel is controlled manually or automatically. The table travel rate is selected with a table speed selector knob. On some 10″ × 20″ (254mm × 508mm) models, the traverse speed is from 2″ to 240″ (50mm to 6,096mm) per minute. Models are designed so that the table may *dwell* from 0 to 2-1/2 seconds at the end of one or both table reversals.

Headstock and Footstock

The headstock and footstock units on cylindrical grinders serve similar functions to the units on an engine lathe. The headstock spindle is mounted on precision antifriction bearings. These bearings are grease packed and sealed for life. The spindle is usually belt driven for smooth rotation. The belt drive provides adequate torque

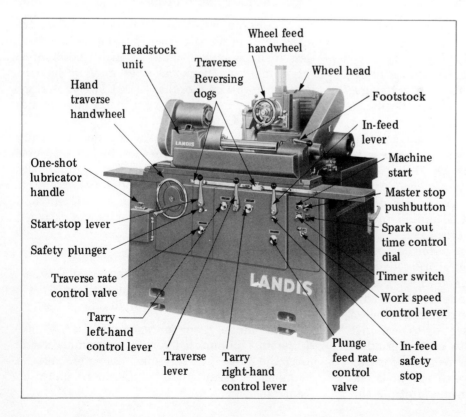

FIGURE 29-1 Major Components of a Universal Cylindrical Grinder (Hydraulic Traverse and In-Feed Model)

CYLINDRICAL GRINDING 567

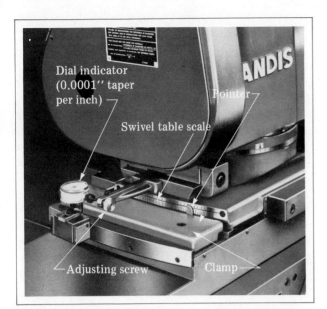

FIGURE 29-2 Adjustments for Setting Swivel Table for Grinding a Taper

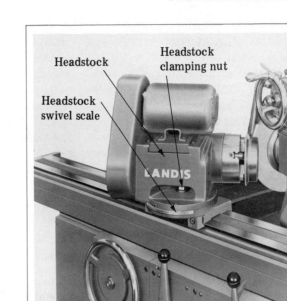

FIGURE 29-3 General Design Features of a Headstock

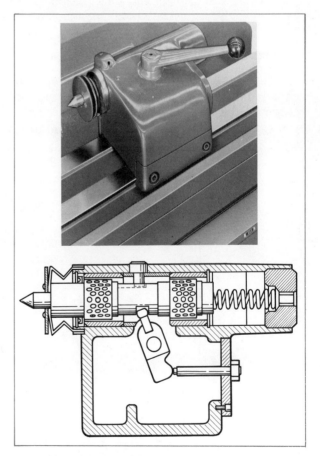

FIGURE 29-4 General Design Features of a Footstock

at low speeds. Figure 29-3 illustrates the general design features of a headstock. On the model illustrated, there are six work speeds of 60, 90, 150, 225, 400, and 600 RPM. The work speeds are controlled by the headstock speed control knob.

The headstock is powered by its own motor. The headstock also includes a work-driving plate and arm that are mounted on the spindle. The base is graduated so that the headstock may be swiveled through a 180° arc. The swivel feature permits steep angle grinding and face grinding on workpieces that are held in a mechanical or magnetic chuck or collet or that are mounted in a fixture on the headstock spindle.

The rotation of the headstock spindle (and workpiece) is controlled by a lever or a handwheel control. The start, stop, jog, and brake of the headstock spindle (workpiece) is controlled by a spindle control lever. The headstock spindle for the model illustrated has a 1″ (25.4mm) diameter hole through the spindle. The headstock work center has a #10 Jarno taper; the footstock, a #6 Jarno taper.

Footstock

The general design features of a footstock are shown in Figure 29-4. The hardened steel spindle is mounted in precision balls. The containers are hardened steel bushings. The design permits axial movement of the spindle with no radial clearance. The preloaded sleeve design eliminates clearance that is normally required for sliding fits. Also, the preloaded balls are packed and sealed for life to protect the balls against coolant and grit.

The work center is carbide tipped for wear life and to provide an efficient bearing surface for the workpiece and grinding processes. The force on the work center is adjustable. The center movement is controlled by the quick-acting control lever.

Wheel Head and Cross Feed Mechanism

The wheel head is a self-contained unit. It includes the spindle and a swivel base on which are mounted the drive motor, spindle, wheel feed, lubricating pump, oil reservoir, and pressure switch. The wheel head has flat and V-shaped ways. These ways are designed for maximum alignment stability. The ways are lubricated with filtered oil. Hand wheel head machines are lubricated at each feed cycle.

Handwheel Feed of the Wheel Head

The feed mechanisms are also equipped with antifriction bearings for easy operation and precision feeding. The handwheel is graduated for a work diameter reduction of 0.100″ (2.5mm) per revolution. The work diameter reduction per graduation of the regular feed knob is 0.001″ (0.025mm). The fine wheel feed graduations are in increments of 50 millionths (0.000050″) of an inch (0.00125mm) on the work diameter.

Graduated Base. The wheel feed unit on the universal cylindrical grinder may be swiveled when the wheel head is swiveled. The wheel head may be swiveled 90° to position it at a required angle either side of center. Figure 29-5 shows universal wheel head settings at 45° each side of center. The large, circular graduated swivel base is graduated in degrees. Graduated machine scales are used only for general reference. Other precision instruments and gages are employed to measure within highly precise limits of accuracy.

Internal Grinding Attachment

An internal grinding attachment permits grinding outside diameters and inside diameters without removing the part, grinding wheel, or spindle. The attachment design permits it to be mounted on the wheel head. It may be swung to a down position and locked for internal grinding. When not in use, the attachment is swung up and out of the way and locked securely. The spindle speed range for internal grinding is up to 16,500 RPM. Hole sizes to 4-3/8″ (111mm) diameter × 2-1/2″ (63mm) deep may be ground without disturbing a regular 12″ (304mm) external grinding wheel.

Cross Feed Mechanism

The cross feed mechanism provides for feeding the grinding wheel into the workpiece and automatically stopping the in-feed movement when the required diameter is produced. In-feeding is controlled manually or by power feed. The power feed is engaged by moving the cross feed control lever to a start position. At the end of the grinding cycle, the cross feed is stopped automatically by a fixed pin that actuates a switch. The switch throws the cross feed control lever to a stop position. The wheel is then withdrawn by the handwheel.

The cross feed mechanism is started by the cross feed control lever. Positive stop mechanisms differ among machine-tool builders. However, the function in each case is to stop cross feeding when the workpiece is ground to a required diameter. The positive stop is set for the desired

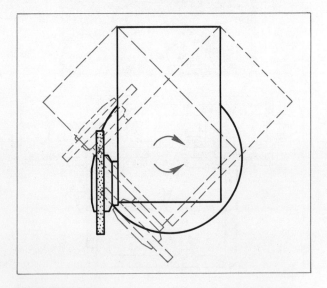

FIGURE 29-5 Universal Wheel Head Maximum Angular Settings at 45° Each Side of Center

CYLINDER GRINDING 569

diameter by in-feeding the grinding wheel until it grazes the work surface. The handwheel disengaging knob is pulled out to free the handwheel. The handwheel is then turned until the fixed stop pin comes in contact with the right side of the switch operating slide. The cross feed is engaged so that the work is ground to the size to which the stop is set.

When the grinding has sparked out at the footstock end, the cross feed is disengaged. The work diameter is then measured. The index dial is set for any material that still is to be removed. The cross feed is reengaged to take successive cuts until the cross feed is stopped by the positive stop. The cylindrical grinder operator usually sets the stop to grind from 0.001" to 0.002" (0.025mm to 0.05mm) oversize. Further adjustments are made in a second setting to bring the workpiece to the required size and within surface texture limits.

Operator Control Panel

An operator control panel is shown in Figure 29-6. While each builder of cylindrical grinders may locate the operator controls in different positions, the functions remain the same. The headstock speed must be selected and controlled. Table movement is controlled for adjustments by manually turning a handwheel. The traverse rate is adjusted by regulating a valve. Motion is started by engaging a traverse reverse lever. The operation may be *tarried* to dwell for a short interval at the end of the right or left position of a cut. On hydraulic in-feed models, in-feed is controlled by a lever. This control lever complements the positioning and feed controls on the wheel head. The panel also has a sparkout timer dial for automatically sparking out at a required diametral dimension.

BASIC CYLINDRICAL GRINDER ACCESSORIES

Work-supporting accessories for cylindrical grinders serve the same functions as the accessories used for lathe work. Long and small-diameter workpieces need support against the cutting forces. Added support must also be provided for plunge grinding where there are increased cutting forces. One or more *universal back rests* (Figure 29-7) may be used to support a long workpiece against deflection. The back rests are positioned along the workpiece, secured to the table, and adjusted to the diameter of the part.

Standard jaw and magnetic chucks and collets are accessories that are mounted on and secured to the headstock spindle. Although chucks are primarily used to hold parts for internal grinding, such chucking permits external grinding. Face grinding operations require holding the workpiece in a chuck or on a magnetic faceplate or fixture.

Other accessories include a series of wheel truing, dressing, and forming devices with func-

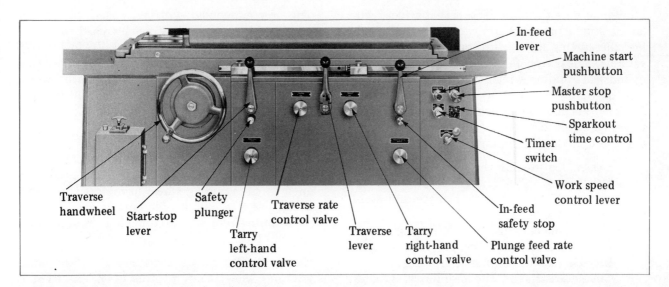

FIGURE 29-6 Operator Control Panel

570 GRINDING MACHINES

FIGURE 29-7 Supporting a Long Workpiece with One or More Universal Back Rests

tions similar to surface grinding devices. Main differences relate to design modifications of the device for mounting on the table and wheel forming from a different position. The diamond or dressing abrasive is held, as discussed in an earlier unit, at a slightly different angle and below the axis of the grinding wheel. Applications of the angle and radius dressing devices are considered later.

OPERATIONAL DATA FOR CYLINDRICAL GRINDING

While many of the same factors that affect regular machining processes apply to cylindrical grinding, the cutting speed of the cutting tool is not a variable. The peripheral speed for general-purpose cylindrical grinding is maintained at the optimum rate of 6,500 feet (1,980 meters) per minute. As for other machining processes, the following factors are considered to establish the operational data for cylindrical grinding:

—Design features of the workpiece and material;
—Machinability of the workpiece and state of hardness or softness;
—Type of machining operation (roughing or finishing cuts), hand-feeding or automatic traverse feeding, or plunge grinding;
—Characteristics of the cylindrical grinder (stability and type);
—Rigidity and balance of the workpiece;
—Wet or dry conditions of grinding;
—Specifications of the grinding wheel.

With so many combinations of factors, the machine operator must interpret grinding data and recommendations of manufacturers. This information is referred to as *operational data*. The following operational data is commonly found in handbook tables and other technical manuals. For cylindrical grinding purposes, basic process data must be known for:

—The work surface speed in feet or meters per minute,
—The in-feed in fractional parts of an inch or millimeter per pass for roughing and finish grinding,
—The fractional part of the wheel width that traverses the workpiece for each revolution.

Table 29-1 provides operational data for traverse and plunge grinding of common workpiece materials on a cylindrical grinder. Using the same work surface speed for a specific material, the information needed for plunge grinding deals with the amount of in-feed—that is, the in-feed per revolution of the workpiece—for roughing and finishing cuts. The speeds and feeds given in a table may need to be adjusted, especially during machine setup when a new part is to be produced to different specifications. Manufacturers' tables provide the starting point.

GRINDING WHEEL CUTTING SPEED

The recommended cutting speed range for vitrified-bonded grinding wheels used for cylindrical grinding is from 5,500 to 6,500 feet (1,676 to 1,980 meters) per minute. Cutting speeds are computed in the same way as for lathe work. Simple formulas are used. If the revolutions per minute are known,

cutting speed = RPM × wheel circumference

The metric cutting speed in meters per minute (m/min) is calculated by inserting the circumference of the grinding wheel in meters (or a fractional part). Similarly, the cutting speed in feet per minute (fpm) requires using the circumference value in feet.

Example: The cutting speed of a 12" (304.8mm) grinding wheel that revolves at 2,069 RPM is calculated by formula as follows:

CYLINDRICAL GRINDING 571

$$\text{cutting speed} = 2{,}069 \times \frac{12 \times \pi}{12}$$

$$= 6{,}497 \text{ fpm (customary unit)}$$

$$\text{cutting speed} = 2{,}069 \times \frac{304.8 \times \pi}{1{,}000}$$

$$= 1{,}980 \text{ m/min (metric unit)}$$

WHEEL AND WORKPIECE RPM, RATE OF TABLE TRAVEL, AND DEPTH OF CUT

Grinding Wheel RPM

The RPM of a grinding wheel is equal to the cutting speed divided by the circumference of the wheel:

TABLE 29-1 Operational Data for Traverse and Plunge Grinding of Common Workpiece Materials

		\multicolumn{6}{c}{Traverse Grinding}		
\multicolumn{2}{c}{Workpiece Material}	\multicolumn{2}{c}{Surface Speed of Work}	\multicolumn{4}{c}{In-Feed per Pass}	\multicolumn{2}{c}{Fractional Wheel Width Traverse per Revolution of Workpiece}
				\multicolumn{2}{c}{Roughing}	\multicolumn{2}{c}{Finishing}		
Kind	Machinability (Hardness) Characteristics	(fpm)	(m/min)*	(0.001")	(0.01mm)	(0.0001")	(0.001mm)	Roughing	Finishing
Carbon steel (plain)	Annealed	100	30	0.002	0.05	0.0005	0.013	1/2	1/6
	Hardened	70	21	0.002	0.05	0.0003 to 0.0005	0.008 to 0.013	1/4	1/8
Alloy steel	Annealed	100	30	0.002	0.05	0.0005	0.013	1/2	1/6
	Hardened	70	21	0.002	0.05	0.0002 to 0.0005	0.005 to 0.013	1/4	1/8
Tool steel	Annealed	60	18	0.002	0.05	0.0005 max.	0.013 max.	1/2	1/6
	Hardened	50	15	0.002	0.05	0.0001 to 0.0005	0.003 to 0.013	1/4	1/8
Aluminum alloys	Cold drawn or solution treated	150	46	0.002	0.05	0.0005 max.	0.013 max.	1/3	1/6
Copper alloys	Annealed or cold drawn	100	30	0.002	0.05	0.0005 max.	0.013 max.	1/3	1/6

		\multicolumn{4}{c}{Plunge Grinding}
\multicolumn{2}{c}{Workpiece Material}	\multicolumn{4}{c}{In-Feed of Grinding Wheel per Revolution of Workpiece}
		\multicolumn{2}{c}{Roughing Cuts}	\multicolumn{2}{c}{Finishing Cuts}
Kind	Machinability (Hardness) Characteristics	(0.0001")	(0.001mm)**	(0.0001")	(0.001mm)
Steel	Soft	0.0005	0.013	0.0002	0.005
Carbon steel (plain)	Hardened	0.0002	0.005	0.000050	0.001
Alloy and tool steel	Hardened	0.0001	0.003	0.000025	0.0006

*Values rounded to closest meter.
**Values rounded to closest 0.001mm.

572 GRINDING MACHINES

$$\text{wheel RPM} = \frac{\text{cutting speed}}{\text{wheel circumference}}$$

For metric setups, the cutting speed in meters per minute and wheel circumference in meters are used with the formula. When the cutting wheel speed is given in feet per minute, the wheel circumference in feet is used. Once the RPM is computed, the grinding wheel spindle speed range selector is set to the closest lower spindle speed.

Workpiece RPM

The workpiece must revolve at a definite surface speed. A review of Table 29-1 shows a range of from 50 to 150 fpm (15 to 46 m/min) for the traverse grinding of common workpiece materials. The headstock spindle speed (RPM) for rate of table traverse is calculated by using the following simple formula:

$$\text{workpiece RMP} = \frac{\text{work surface speed}}{\text{work circumference}}$$

The circumference of the work in metric (meters) units is equal to

$$\frac{\text{work diameter in millimeters} \times \pi}{1{,}000}$$

The circumference of the work in customary (feet) units is equal to

$$\frac{\text{work diameter in inches} \times \pi}{12}$$

The RPM of the workpiece is set at the closest headstock spindle RPM.

Rate of Table Travel

Table 29-1 also gives the recommended fractional width of wheel traverse per revolution of the workpiece. Note that the width is reduced from 1/2 to 1/3, with the narrower width traverse used for finishing cuts. It should be noted that the values given in Table 29-1 and other manufacturers' technical manuals are for start-up. Operator judgment is required to modify the RPM and speeds and feeds depending on factors that apply to the job at hand.

The rate of table travel is calculated by multiplying the speed at which the workpiece is revolving (RPM) by the distance the table advances (width of the grinding wheel traverse) each revolution. The rate of table travel is set by adjusting the traverse rate valve.

Example: Calculate the rate of table travel for rough grinding a hardened tool steel part, using a 2″ (50.8mm) wide wheel face. Use the operational data for traverse grinding as contained in Table 29-1.

The table shows that the work surface speed for hardened tool steel is 50 fpm. The wheel traverse per revolution of workpiece is 1/4 wheel width. The following formulas are used:

$$\text{rate of table travel} = \text{RPM} \times \text{wheel width advance per revolution}$$

$$\text{workpiece RPM} = \frac{\text{work surface speed}}{\text{work circumference in feet}}$$

The rate of table travel is found by substituting values in the formulas:

$$\text{RPM} = \frac{50 \text{ fpm}}{\frac{2}{12} \times \pi}$$

$$= 96$$

$$\text{rate of table travel} = 96 \text{ RPM} \times 0.050″ \text{ wheel width}$$

$$= 48 \text{ ipm (inches per minute)}$$

The rate of table travel may also be calculated in terms of meters per minute (m/min) by dividing the circumference in millimeters by 1,000 and multiplying by the RPM for the workpiece. In the preceding example, the equivalent m/min rate of table travel is calculated as follows:

$$\text{rate of table travel} = 96 \text{ RPM} \times \frac{1/4 \text{ of 50.8mm wheel width}}{1{,}000}$$

$$= 96 \times 0.0127$$

$$= 1.22 \text{ m/min (meters per minute)}$$

Depth of Cut

The in-feed per pass for roughing cuts on the five common workpiece materials given in Table 29-1 is 0.002″ (0.05mm). Finishing cuts are taken at depths of from 0.0001″ to 0.0005″ (0.003mm to 0.013mm) maximum. The depth of cut may be increased, depending on the operation, material, design features of the part, and other machining conditions. In general practice, the heaviest and least number of cuts possible are taken to cylindrically grind a part to specifications.

HIGH-SPEED CYLINDRICAL GRINDING

Standard universal cylindrical grinding machines are designed for commonly used grinding wheels that operate in the range of 6,500 feet (1,980 meters) per minute. Productivity increases of grinding methods are partly due to increasing the peripheral speed of a different group of abrasive wheels. These wheels are capable of operating at speeds of 12,000 feet (3,658 meters) per minute and higher. Grinding at these peripheral speeds is known as *high-speed grinding*.

High-speed cylindrical grinding requires specially designed machines. The headstock, wheel head and spindle, table, and other machine features must have a high degree of ruggedness and stability to withstand higher machining forces. Spindles and bearings must be extrastrong. Wheel guards must be reinforced. The abrasive wheels must be designed and tested for operating at high peripheral speeds for faster stock removal. Any attempt to use a standard machine for high-speed grinding at speeds greater than the safe limits of the machine constitutes a serious hazard for personal and machine tool safety.

AUTOMATED CYLINDRICAL GRINDING MACHINE FUNCTIONS

Cylindrical grinding has thus far been considered in terms of all-around processes and limited manufacturing as performed by a skilled craftsperson in the toolroom or custom shop. Some of the procedures—for example, in-feeding and traverse feeding and sparking out—are mechanized on machines with automatic controls. Production grinding, however, requires more of the processes to be automated. Automation is possible in at least seven different categories, briefly described as follows:

- —*Work loading and positioning*, which involves work-feeding devices, loading and unloading, and work positioning;
- —*In-feed*, which involves automatic presetting of advance feed rates, cutoff points, and duration of time-process functions (in-feed functions relate to rapid approach rates, feeds for roughing and finishing cuts, and the sparkout period that follows);
- —*Table traverse tarry (dwells)*, which involves continuation of wheel and workpiece contact (exposure) at the ends of a cut for a preset time to ensure uniform cutting conditions;
- —*Size control*, which provides for automatically adjusting the cutoff points of the in-feed to control the advance movement of the grinding wheel (provision is also made to automatically compensate for wheel size variations through signals originating with a master gage);
- —*Automatic cycling*, which relates to the actuating and shutting down of all mechanisms dealing with work rotation, in-feed, table traverse, and coolant supply as required through a complete machining cycle (at the completion of the operation, cycling refers to the retracting of the wheel slide and grinding wheel and the stopping of the table movement, work rotation, and coolant supply);
- —*Wheel dressing*, which specifies wheel dressing at a preset frequency (compensation is made for wheel size reduction through automatic in-feed movement);
- —*Computer numerical control operations*, which involve advanced levels of automating cylindrical processes and are designed to improve the speed, accuracy, and efficiency of the setup and to check out the program and troubleshoot problems.

CYLINDRICAL GRINDING PROBLEMS AND CORRECTIVE ACTION

The four fundamental cylindrical grinding problems that confront the machine operator are as follows:

- —Chatter as related to the wheel, workpiece, and machine units;
- —Work defects;
- —Improperly operating machine;
- —Grinding wheel conditions.

The probable causes of these major problems and recommended corrective action are summarized in Table 29-2.

TABLE 29-2 Common Cylindrical Grinding Problems, Probable Causes, and Corrective Action

Cylindrical Grinding Problem	Probable Cause	Corrective Action
Chatter		
Wheel	—Out-of-balance	—Run the wheel without coolant to remove the excess or unevenly distributed fluid in the wheel
		—Rebalance the wheel before and after truing and dressing
	—Out-of-round	—True the wheel face and sides before and after balancing
	—Too hard	—Change to coarser grit, finer grade, and a more open bond
	—Improperly dressed	—Check the mounting position and rigidity of the diamond dressing tool and its sharpness
Work and machine units	—Faulty work support or rotation	—Add work rests and position them for maximum support
		—Check the quality of the headstock and footstock work centers
	—Vibration	—Reduce the work speed and check the workpiece for balance
	—Faulty coolant	—Replace dirty and contaminated coolant solution
		—Clean the coolant tank and lines
	—External vibration transmitted to cylindrical grinder	—Check machine leveling, secureness with which the machine is bolted to the floor, and use of vibration absorption material between the base and floor
	—Interference (operating clearance)	—Check the operating clearance between guards, work-holding devices, the grinding wheel, and the wheel head
	—Wheel head (unevenness of motion)	—Check the belt tension, pulleys, and motor drive for any conditions producing unevenness
		—Replace belts, where required, with belts that are of equal length and required shape
	—Headstock and center	—Adjust and correct the work speed, if needed
		—Check the fit of the work center
		—Replace worn belts with belts of equal length and required shape
Work Defects		
Check marks	—Improper cutting action	—Dress the grinding wheel to act softer
		—Increase the coolant flow and position the nozzle for maximum cooling and cutting efficiency
	—Incorrect wheel	—Use a softer-grade wheel and check the correctness of grain size, abrasive, and bond
	—Incorrect dressing	—Use a sharp, uncracked, securely held, and correctly positioned diamond dresser
Burning and work discoloration	—Improper grinding	—Decrease the in-feed cutting rate
		—Retract the wheel before stopping the rotation of the workpiece
	—Improper wheel	—Use a softer wheel or dress to produce a softer cutting action
		—Increase the coolant flow
Surface imperfections	—Deep marks	—Change to a finer-grit wheel and check the abrasive wheel specifications for the workpiece material
	—Fine spiral marks	—Add work rests, if needed
		—Redress the wheel face parallel to the work axis
		—Reduce the traverse rate
	—Wide, varying depth, irregular marks	—Use a harder-grade wheel
	—Spots	—Remove glazed areas or oil spots on the wheel by turning and dressing
	—"Fish tails"	—Change to a cleaner coolant
		—Flush the wheel guards after dressing
	—Irregular marks	—Clean the machine to remove and flush away any loose chips or foreign particles
		—Check the condition of the blotters and the secureness of the grinding wheel
Improperly Operating Machine		
Dimensional inaccuracies	—Out-of-roundness	—Relap the work centers
		—Check the condition of the machine centers, accuracy of mounting, and the footstock force against the workpiece

Cylindrical Grinding Problem	Probable Cause	Corrective Action
	—Out-of-parallelism (taper)	—Recheck the alignment and secureness of the headstock, footstock, and table setting
	—Inconsistent work sizing	—Check the quality of the centers
		—Add work supports, if required, and space them for maximum support
		—Check the grinding wheel for loading and balance
		—Take up backlash in the wheel head and carriage when readjusting stops for rapid and slow feeds
		—Check the traverse hydraulic system
		—Use a greater volume of clean coolant
Uneven traverse (in-feed) of the wheel head	—Scored ways on the carriage and wheel head	—Check the lubrication of the ways and the recommended oil for the lubricating and hydraulic systems
	—Leakage and gumminess of the hydraulic system fluids	—Check the valves, pistons, and system for oil leakage and gummy lubricant; flush the system and replace the lubricant
	—Unbalanced drive parts	—Replace an unbalanced drive motor, loose pulley, or uneven driving belts
Grinding Wheel Conditions		
Wheel defects	—Faulty wheel dressing	—Incline the dressing tool at least three degrees from the horizontal plane
		—Reduce the depth of the dressing cuts
		—Round off the wheel edges
	—Wheel acting too hard (problems of loading, glazing, chatter, burning, and so on)	—Increase the rate of in-feed, the RPM of the work, and the traverse rate
		—Decrease the wheel speed or width
		—Select a softer wheel grade and coarser grain size or dress the wheel to cut sharper
	—Wheel acting too soft (problems of wheel marks, reduced wheel life, and inconsistent work size)	—Decrease the in-feed and work and traverse speeds
		—Increase the wheel fpm (m/min) or cutting width of the abrasive wheel
		—Dress the wheel to act harder
		—Select a harder wheel, less fragile grain, or both
	—Wheel loading and glazing	—Use a coarser grain size with a more open bond or a softer wheel grade
		—Dress the wheel with a faster traversing feed and a deeper dressing cut
		—Soften the action of the wheel by controlling the cutting speed
		—Use less in-feed to prevent loading and more in-feed to correct glazing
		—Use a cleaner coolant and change the coolant composition to prevent gumming
		—Mount the wheel on the spindle with a sliding (not a force) fit
Wheel breakage	—Radial breaks (two pieces)	—Bring the wheel up to operating speed and carefully move it to grinding position without jarring or striking the wheel against the workpiece
		—Discard a wheel that has been damaged in handling
	—Radial breaks (three or more pieces)	—Check the wheel for proper use of blotters, the presence of chips and foreign particles, good mounting fit, and the application of equal flange force
		—Reduce the in-feed; avoid jarring the wheel and work
		—Reduce the wheel speed to below the maximum rated fpm or m/min
		—Increase the coolant flow to prevent overheating

WHEEL RECOMMENDATIONS FOR CYLINDRICAL GRINDING (COMMON MATERIALS)

The sequence of markings on a typical cylindrical grinding wheel (according to ANSI designation) is the same as for markings described in earlier units. Letters and numbers appear in sequence to designate six categories of information: (1) *abrasive type* (letter), (2) *grain size* (number), (3) *grade* (letter), (4) *structure* (number), (5) *bond type* (letter), and (6) the optimal manufacturer's record (number).

In addition, the shape of the grinding wheel is identified according to another set of standards (ANSI Standard B74.2-1974). Within this standard are thirteen different shapes of grinding wheel faces. The range of shapes and sizes are specified in handbooks and manufacturers' technical literature. Examples of other standard types of regular and diamond grinding wheels are generally used for cylindrical grinding:

—Type 1 (straight wheel),
—Type 5 (straight wheel with one side recessed),
—Type 7 (straight wheel recessed on two sides),
—Types 20 through 26 (straight wheels relieved and/or recessed on one or both sides).

Variables of machines, abrasives, materials, processes, and other conditions make it impractical to recommend a precise grinding wheel. Therefore, manufacturers' and handbook tables are based on typical applications with conditions assumed for common practices. Each table provides for a starting selection. Under actual conditions, the selection is subsequently refined. Table 29-3 provides an example of wheel recommendations of one manufacturer for the cylindrical grinding of common metals. In some instances, grinding processes are identified.

BASIC EXTERNAL AND INTERNAL CYLINDRICAL GRINDING PROCESSES

Once the cylindrical grinder is set up, five basic processes are performed in toolroom work or for small production runs. These processes include:

—Grinding parallel surfaces,
—Grinding tapered surfaces,
—Grinding contour-shaped (curvilinear) surfaces,

TABLE 29-3 Wheel Recommendations for Cylindrical Grinding of Materials Generally Used in the Custom Shop

Material	Wheel Marking*
Aluminum parts	SFA 46-18 V
Brass	C 36-K V
Bronze	
Soft	C 36-K V
Hard	A 46-M 5 V
Cast iron (general, bushings, pistons)	C 36-J V
Cast alloy (cam lobes)	
Soft	
Roughing	BFA 54-N 5 V
Finishing	A 70-P 6 B
Steel	
Forgings	A 46-M 5 V
Soft	
1" (25.4mm) diameter and under	SFA 60-M 5 V
Over 1" (25.4mm) diameter	SFA 46-L 5 V
Hardened parts	
1" (25.4mm) diameter and under	SFA 80-L 8 V
Over 1" (25.4mm) diameter	SFA 60-K 5 V
Stainless (300 series)	SFA 46-K 8 V
General-purpose grinding	SFA 54-L 5 V

The prefix BF with the aluminum oxide designation (A) indicates *blended friable* (a blend of regular and friable). SFA indicates *semifriable*.
*The grain size, grade, and structure are the recommendation of one leading manufacturer.

—Grinding at a steep angle to form beveled surfaces and shoulders up to 90° to the work axis,
—Face grinding.

While most surface forming on a cylindrical grinder is on external surfaces, the same applications are made to internal grinding. The balance of this unit deals with step-by-step procedures for setting up and performing the basic external and internal processes.

HOW TO SET UP AND OPERATE A UNIVERSAL CYLINDRICAL GRINDER

Setting Up the Wheel and Wheel Head

STEP 1 Determine the wheel specifications for the material to be ground and the work processes.

STEP 2 Select the grinding wheel. Make safety checks for soundness, fit on the spindle, condition of the blotters, and flanges.

Note: The wheel and balance are also checked, if necessary.

STEP 3 Mount and secure the wheel on the spindle.

STEP 4 Start and stop the spindle several times if the machine has been idle. This action lubricates the spindle bearings before the spindle is run at operating speed.

STEP 5 Check the graduation on the base of the swivel head. If necessary, adjust the position to a zero reading. The wheel head spindle is then set parallel to the zero table and work center axes.

STEP 6 Mount and position a diamond dressing tool on the table. True and dress the wheel according to job specifications and the kind of cut to be taken. Check the spindle driving belts and adjust for proper tension, if needed.

STEP 7

Setting the Swivel Table

STEP 1 Determine the nature of the table setting.

Note: Parallel cylindrical surfaces require a zero setting. Small-angle surfaces may be produced by setting the table at the required angle or taper. Steep-angle grinding may require both the table and wheel head to be set at complementary angles.

STEP 2 Loosen the clamping bolts at each end to set the table at the required angle. If the table is not graduated directly to read in taper per foot or metric equivalent, the given taper may need to be converted to the customary or metric graduated units.

STEP 3 Turn the table adjustment nut until the table is set at the required angle. Tighten the clamping bolts.

STEP 4 Check the accuracy of the table setting.

Note: The test bar and dial indicator method may be used for checking parallelism. Angle table settings are usually checked by taking a trial cut on the workpiece and measuring or gaging the included angle.

Positioning the Headstock and Footstock

STEP 1 Clean the table face and the bases of the headstock and footstock.

STEP 2 Position the headstock and footstock so that they are centered on the table. The distance between centers must permit mounting and removing the workpiece. Secure in position.

Note: The design of some units permits the headstock to be aligned by means of a base flange.

STEP 3 Check the spindle bore and the condition of the centers. If worn or damaged, the centers should be reground.

Note: Carbide-tipped centers are commonly used to provide greater wear life and an accurate bearing surface for center work.

Checking the Coolant System

STEP 1 Determine what cutting fluid (coolant) is required for the job.

STEP 2 Check the cutting fluid in the system. Test for viscosity and to see that the solution is clean. Make whatever changes are needed.

STEP 3 Position the nozzle to deliver the fluid for maximum cutting and cooling efficiency.

STEP 4 Attach the cutting fluid guards to the table.

Mounting the Workpiece

STEP 1 Check the angle and bearing condition of the center holes. Redrill, grind, or lap, if required.

STEP 2 Select a driving dog. Secure it on one end of the workpiece.

STEP 3 Move the wheel head so that the workpiece may be placed between centers without interference from the grinding wheel.

STEP 4 Make sure the centers are free of particles and clean.

STEP 5 Move the footstock lever to permit mounting the workpiece between centers.

Note: The footstock is moved clear of the work area for chuck, faceplate, fixture, and collet work that may not require a center.

STEP 6 Place the driving dog and the drive end of the workpiece on the headstock center.

STEP 7 Move the spindle lever of the footstock to permit inserting the center in the center hole of the workpiece. The tension (force) exerted by the footstock center may require adjusting. The center must maintain adequate force against the workpiece during the grinding process.

Setting the Table Position and Travel

STEP 1 Move the table to the left until the grinding wheel extends beyond the footstock end of the workpiece. The distance should equal the width of the traverse cut of the grinding wheel (from 1/6 to 1/2 wheel width, depending on the nature of the cut).

STEP 2 Move the table reversing lever as far to the left as it will go.

STEP 3 Move the right-side table trip dog until it touches the table reversing lever. Secure the trip dog in position.

STEP 4 Use the table handwheel to move the table to the right-end location where the cut is to end.

Note: There must be adequate clearance between the driving dog and plate, grinding wheel, and workpiece.

STEP 5 Move the table reversing lever to its extreme right position.

STEP 6 Bring the left-side table trip dog against the reversing lever. Secure the trip dog in position.

STEP 7 Recheck for operational clearance between the driving dog, headstock, workpiece, and grinding wheel and wheel head and tailstock. Then, engage the power traverse to recheck the table movement between the beginning and end of the cut. Make further adjustments of the table reversing dogs, if needed.

STEP 8 Set the dwell (sparkout timer) control to permit adequate time at each reversal for grinding to depth.

Cross Feeding Manually and Automatically

STEP 1 Position the switch-operating cross slide throwout lever to disengage the automatic cross feed.

STEP 2 Start the grinding wheel, coolant, system, and headstock spindle. Engage the table traverse.

STEP 3 Feed the grinding wheel until it grazes the workpiece. Set the cross slide handwheel at zero.

STEP 4 Move the cross slide handwheel to advance the wheel to the depth of the first roughing cut. Take the cut until the wheel sparks out at the footstock end of the cut.

STEP 5 Stop the work rotation and table traverse at the end of the cut. Measure the workpiece. Determine what the final depth reading should be on the graduated handwheel.

STEP 6 Set the cross feed positive-stop mechanism to permit feeding to within 0.001" (0.025mm) of the diameter to which the workpiece is to be ground.

STEP 7 Continue to in-feed for the roughing cuts until the positive-stop mechanism stops further in-feeding.

STEP 8 Spark out the final cut at the footstock end of the workpiece. Then, stop the table travel and headstock spindle (work rotation).

STEP 9 Measure the work diameter. Determine precisely how much additional stock is to be removed.

STEP 10 Start the work rotating. Feed the wheel for the first of several light finish cuts.

Note: The in-feed is set by using the fine-feed knob, which is graduated on toolroom models in increments of 0.000050" (0.001mm).

STEP 11 Engage the table travel. Take successive cuts and permit the final cut to spark out with the grinding wheel clear of the workpiece.

STEP 12 Make a final check for dimensional accuracy.

Note: The quality of surface texture may be established by checking with a portable surface analyzer. An optical check may be made using a comparator scale and optical viewer.

Note: The wheel itself or the speed and sharpness to which the wheel is dressed may need to be changed to produce the required surface texture.

STEP 13 Back the wheel away from the workpiece when the grinding is completed. Stop the table travel and spindle (workpiece) rotation.

STEP 14 Move the footstock lever to release the workpiece.

CYLINDRICAL GRINDING

Feeding Automatically

The procedure for in-feeding automatically is the same as for hand-feeding. The difference is in the engagement of the in-feed control lever. When more than one piece is to be ground, the in-feeding is automatically stopped by the positive-stop mechanism. Compensation is made for wheel wear by hand-feeding. The wheel is fed to final depth (to grind to the required dimension) by using the fine-feed dial.

HOW TO GRIND STRAIGHT CYLINDRICAL WORK

STEP 1 Select the appropriate cylindrical grinding wheel. Check it for balance. Mount the wheel on the spindle and true and dress it.

STEP 2 Set the wheel head at zero.

STEP 3 Attach a driving dog on the workpiece so that there is clearance for grinding.

STEP 4 Check the center holes (on center work) for quality of surface and accuracy of the included angle.

Note: Some workpieces may be held in a chuck when there is limited overhang. Longer chucked pieces may be supported by the footstock center. Collets and fixtures may also be used in the headstock.

STEP 5 Position the grinding wheel. Set the table trip dogs fo the length to be ground. Set the dwell control to spark out each cut. Adjust the traverse rate. Set the spindle speed at the RPM required for rotation of the workpiece.

STEP 6 Feed the grinding wheel into the revolving workpiece until it just grazes the surface. Set the cross feed handwheel at zero.

STEP 7 Set the cross feed in-feed trip to permit grinding to the required finished dimension.

STEP 8 Take successive roughing cuts. Stop the grinding wheel and workpiece rotation. Measure the diameter.

STEP 9 Continue to finish grind, using the fine-feed knob.

STEP 10 After shutdown, remeasure the diameter.

STEP 11 Disassemble the setup. Leave the machine clean and dry.

SUPPORTING LONG WORKPIECES

Long, slender workpieces need to be supported by spring back rests or steady rests. The number and placement of the back rests depend on the work diameter and the length and nature of the grinding process. Universal spring back rests (Figure 29-8) are designed to be positioned along and secured to the table. The *shoes* (bearing surfaces) are usually made of bronze to protect the work surfaces. Shoes are available in increments of 1/16" (1.5mm) for work sizes beginning at 1/8" (3mm) through 1" (25.4mm). Each shoe remains in contact with the workpiece automatically, giving constant support.

In general practice, long workpieces that are 1/2" (12.7mm) in diameter are supported at 4" to 5" intervals along the length. Fewer spring back rests are needed to support stronger, larger-diameter workpieces. Steady rests are used primarily as an intermediate work support at places along a workpiece where the steady rest will not interfere with a grinding process.

FIGURE 29-8 Universal Spring Back Rest for Work Support

HOW TO SET UP AND USE SPRING BACK RESTS

STEP 1 Select one or more spring back rests with a size of shoe that is appropriate to the work diameter.

STEP 2 Mount each back rest and secure it to the machine table with the clamping bolts. Turn the adjusting screws to each shoe so that it clears the workpiece.

STEP 3 Mount the workpiece between centers.

STEP 4 Fit the shoe against the workpiece and hold it in position by hand. Turn the adjusting screw until each bears against a shoe to hold it in position.

STEP 5 Start the workpiece and the wheel spindle rotating. In-feed for a light cut. At the same time, turn the adjusting screws for the shoes to maintain contact with the part.

Note: Each spring back rest is adjusted independently.

STEP 6 Stop the work and spindle rotation after the first light cut. Measure the diameter at several places. Compensate for variations in diameter by turning the lower adjusting screw at the large-diameter end of the work. The action advances the shoe and causes the work diameter to be ground slightly smaller.

STEP 7 Take another light cut. Measure and adjust the shoe(s) until the same micrometer reading is obtained along the ground length of the workpiece.

Note: The adjusting screws are turned just enough to keep the shoes in contact with the workpiece. The jaws must exert only a slight force against the ground part to prevent chatter and springing during grinding.

STEP 8 Take successive roughing and finishing cuts. The adjusting screws are advanced for each cut. The shoe in each instance is thus kept in contact with the workpiece.

Note: The force (pressure) of each shoe against the workpiece is regulated by a spring tension knob on the work support.

When a solid or compensating type of back rest is used, sections along the workpiece are spot ground. The shoe on a solid back rest is kept in continuous contact with the work as it is reduced in diameter. The shoe is advanced by turning the upper adjusting screw.

HOW TO GRIND SMALL-ANGLE AND STEEP-ANGLE TAPERS

Grinding Small-Angle Tapers up to 8°

STEP 1 Release the swivel table clamping bolts.

STEP 2 Swivel the table to the required angle or taper per foot (or metric unit). Read the setting on the graduated scale on the end of the table (Figure 29-9). Convert from metric units, if the dimensions are in metric and the scale is graduated in cutomary inch units.

STEP 3 Tighten the table clamping bolts to secure the table at the angle setting.

STEP 4 Select an appropriate grinding wheel. Balance, mount, true, and dress the wheel.

STEP 5 Check the condition of the work and the machine centers. Then, mount the work between centers.

STEP 6 Set the headstock spindle RPM, the rate of table traverse, in-feed rate, and dwell time. Adjust the flow rate of the cutting fluid and the position of the nozzle. Secure the wheel and splash guards.

STEP 7 Run the machine for a minute or so to ensure that all parts are lubricated and the wheel is at operating speed.

FIGURE 29-9 Table Angle Setting Read Directly from Graduated Scale

Caution: Use safety goggles or a protective shield.

STEP 8 Bring the grinding wheel into position to take a first cut. Set the wheel head positive-stop mechanism to automatically trip the in-feed at the required position.

STEP 9 Take several cuts until the ground taper length is sufficient to be measured.

STEP 10 Adjust the table setting to compensate for any variation between the taper being ground and the required one.

STEP 11 Take successive roughing cuts by using power table traverse. Leave the workpiece from 0.001" to 0.002" (0.025mm to 0.050mm) oversize.

STEP 12 Set the feed rate on the fine-feed dial to remove the remaining material. Take several light finish cuts.

Note: The craftsperson generally measures or gages the taper and makes angle adjustments during roughing (if required) before proceeding with the finishing cuts.

STEP 13 Spark out at the footstock end so that the grinding wheel clears the workpiece. Then, stop the table travel and rotation of the part.

STEP 14 Remove the workpiece. Clean all tools, the taper gage, and the machine.

Grinding Steep-Angle Tapers

Steep tapers may be ground by three fundamental methods:

—Swiveling the universal wheel head on its swivel (angle-graduated) base to the required angle;
—Swiveling the headstock for workpieces that are held directly in a chuck, fixture, collet, or the tapered spindle (the angle setting of the headstock is read directly from graduations on the swivel base);
—Swiveling the universal wheel head and headstock spindle axes (as shown in Figure 29-10).

The following steps describe the method of grinding a steep taper by swiveling the wheel head.

STEP 1 Loosen the clamping bolts on the wheel head slide base.

STEP 2 Swivel the wheel head to the required degree. The angle setting is read directly on the degree-graduated swivel base.

Note: The angle setting permits the wheel to grind the work surface to the required angle. For example, if a 22-1/2° included angle is required, the wheel head is set at 67-1/2° (90° - 22-1/2°).

STEP 3 Tighten the clamping bolts to secure the wheel head at the angle setting.

STEP 4 Set the speed, feed, and cutting fluid flow controls according to the job requirements.

STEP 5 Check to see that the wheel and splash guards are secure.

Caution: Use a protective shield or safety goggles.

STEP 6 Bring the machine through a warm-up period to operating speed.

STEP 7 Move the wheel head to position the grinding wheel in relation to the workpiece.

STEP 8 Hand-feed the table to bring the work into contact with the grinding wheel.

STEP 9 Continue to feed the workpiece to the grinding wheel, if the width of the angle surface to be ground is less than the wheel width.

STEP 10 Advance the table at the end of each cut for each successive cut.

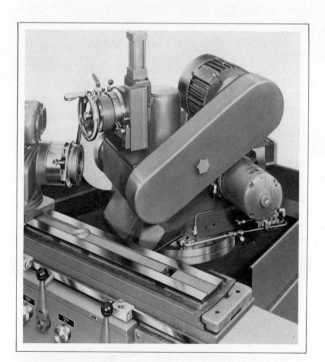

FIGURE 29-10 Wheel Head and Headstock Swiveled to Grind a Steep Angle

Note: Where the angle surface being ground is wider than the width of the wheel face, the cross feed handwheel is used to traverse the grinding wheel across the work surface.

STEP 11 Continue taking roughing and finishing cuts until the angle surface is ground to size.

FACE GRINDING ON A UNIVERSAL CYLINDRICAL GRINDER

Face grinding refers to the grinding of a plane surface at a right angle to the axis of the part. Face grinding on a universal cylindrical grinder requires the workpiece to be set in a plane that is parallel to the axis of the grinding wheel spindle. The headstock is swiveled 90° for face grinding. The grinding wheel is hand fed to the depth of cut against the revolving part. The part is fed across the face of the grinding wheel either by hand or automatically by using the table traverse feed. The wheel is fed into the part for successive cuts.

GRINDING INTERNAL SURFACES ON A UNIVERSAL CYLINDRICAL GRINDER

Straight (Parallel) Grinding

The steps that follow relate to the setting up and grinding of straight, angle, and shoulder surfaces using an internal grinding attachment that is permanently mounted on the grinding wheel head (Figure 29-11). The internal grinding wheel is advanced and retracted by the same controls as are normally used during external grinding. Chucked workpieces are ground using normal traversing controls.

Machine design features that include a wheel head unit for external grinding and a built-in internal grinding attachment permit grinding both externally and internally with the same chucking setup. This setup is particularly important where precision concentricity limits of external and internal ground surfaces are required.

The setup in Figure 29-11 relates to the grinding of an outside diameter. Figure 29-12 shows the internal grinding attachment swung into operable position to grind a straight hole. Since smaller-diameter wheels are used, considerably

FIGURE 29-11 Internal Grinding Attachment Locked in Inoperable Position (Setup Permits External Grinding)

higher spindle speeds are required. On the model illustrated, the belted motor drive produces spindle speeds up to 16,500 RPM. Smaller machines are designed to deliver infinitely variable spindle speeds to 45,000 RPM through a dial control on an electrical panel.

The spindle setup in Figure 29-12 permits grinding a hole to 4-3/8" (110mm) diameter × 2-1/2" (63mm) deep without removing a standard 12" (304mm) diameter external grinding wheel. Spindle quills extend the depth to which an internal surface may be ground.

Internal Grinding Problems

Internal grinding spindles, by the nature of their size and other design limitations, are not as strong or rigid as spindle assemblies for external grinding. As a result, three common grinding problems may occur. Unless the wheel is dressed correctly and run at the required cutting speed with proper in-feed and traverse feed rate, the internally ground hole may have one or more of the following dimensional defects:

—The hole may be bell-mouthed (This defect is caused by the wheel overlapping the hole

CYLINDRICAL GRINDING

FIGURE 29-12 Internal Grinding Attachment Locked in Operable Position (Setup for Internal Grinding)

too much and may be corrected by reducing the overlap at each end of the hole. The bore length may also be too long and outside the capacity of the spindle to accurately machine within the required tolerance limits.);
—The hole may be out-of-round (This defect may be caused by overheating during the grinding process or through distortion in chucking or holding the workpiece. The problem may be corrected by flowing on a greater volume of cutting fluid, reducing the depth of cut and feed rate, and dressing more frequently to cut sharper. Chucking distortion is corrected by more careful chucking and applying only enough force to ensure that the workpiece will not move during grinding.);
—The hole may be tapered (This defect may be caused by using a wheel that is too soft to hold its size, too fast a feed, or the setting of the workpiece axis at a slight angle.).

Other general grinding problems, probable causes, and recommended corrective action were summarized earlier in Table 29-2.

HOW TO SET UP FOR AND CYLINDRICALLY GRIND INTERNAL SURFACES

Grinding Straight Internal Surfaces

STEP 1 Move the internal grinding attachment into operating position. Lock the attachment.

STEP 2 Check the belts for wear and uniform drive.

STEP 3 Determine the spindle RPM required to grind at the specified fpm (m/min). Set the wheel speed controls accordingly.

STEP 4 Select an appropriate diamond dressing tool. Mount and position to dress the wheel.

STEP 5 Bring the internal grinding wheel spindle to operating speed. Dress the wheel and use a cutting fluid (coolant).

STEP 6 Turn the cross feed selector knob to the internal grinding position. The cross feed handwheel may then be turned counterclockwise to advance the wheel into the workpiece.

Note: The setting of the selector knob for internal grinding permits using the wheel head positive-stop mechanism, power traverse feeds, and graduations on the handwheel in the same manner as for external grinding.

STEP 7 Mount the workpiece in a suitable chucking device on the headstock spindle.

STEP 8 Calculate the required speed (RPM) for the workpiece and for the grinding wheel and determine the rate of table travel.

STEP 9 Adjust the headstock, internal grinding spindle controls, and the table traverse feed rate control.

STEP 10 Disengage the power cross feed control lever.

Caution: Make sure the in-feed safety plunger is pushed in at all times on grinders equipped for hydraulic in-feed to prevent accidental in-feed of the wheel head.

STEP 11 Adjust the table trip dogs.

Note: The length of table travel is adjusted to permit only part of the wheel to move past the workpiece at the beginning of a stroke.

Note: For grinding to an internal shoulder, the trip is set to reverse the table travel at the point where the wheel reaches the shoulder.

STEP 12 Position the wheel partly into the revolving workpiece. Bring the wheel into contact with the hole surface. At the same time, feed the workpiece so that the grinding wheel grinds toward the back of the hole.

STEP 13 Withdraw the workpiece to clear the wheel at the end of the cut.

STEP 14 Measure the inside diameter. Set the graduated dial at zero.

STEP 15 Take repeated cuts until the hole is ground to the correct diameter.

STEP 16 Clear the work and the wheel at the end of the grinding process by moving the right-hand trip dog.

STEP 17 Recheck the finished diameter. Clean the machine, tools, and workpiece. Leave the machine in a safe, operating condition.

Grinding Small-Angle Internal Tapers

STEP 1 Check the wheel head to be sure it is set at 0°.

STEP 2 Check the 0° setting of the headstock.

STEP 3 Set the swivel table at the required taper angle, taper per foot, or metric equivalent.

STEP 4 Proceed to grind internally following the same steps as for grinding a straight internal cylinder.

Note: The correctness of taper is checked with a taper plug gage or the mating part.

Grinding Steep-Angle Internal Tapers

STEP 1 Check the wheel head to be sure it is set at 0°.

STEP 2 Check that the swivel table is set at the zero position.

STEP 3 Loosen the clamping bolts on the headstock base.

STEP 4 Swivel the headstock to the required angle. Secure the headstock in position.

STEP 5 Proceed to grind the steep taper. Follow the same steps as were used for small-angle internal taper grinding and measuring.

CYLINDRICAL GRINDER WHEEL TRUING AND DRESSING DEVICES

There are three basic types of wheel truing and dressing devices for cylindrical grinders. The

A. Truing and Dressing Fixture Mounted on Footstock

B. Truing and Dressing Fixture Mounted on Table

FIGURE 29-13 Wheel Truing and Dressing Fixtures for Generating Flat Wheel Face

most common, shown in Figure 29-13, are the wheel truing and dressing fixtures for generating a flat wheel face. These fixtures are designed for mounting either on the footstock (Figure 29-13A) or on the table (Figure 29-13B).

Grinding wheels may be dressed at an angle by using a table-mounted *angle truing attachment*. Concave and convex radii are formed by using a *radius wheel truing attachment* (Figure 29-14). The *combination angle and radius wheel truing attachment* is used to accurately form wheels to different radius and angle combinations.

The principles for setting these attachments are the same as described for similar surface grinding attachments. There are, however, some modified design features, primarily for mounting. The following steps describe mounting and dressing cylindrical grinding wheels using the straight, angle, and radius attachments.

HOW TO TRUE AND DRESS CYLINDRICAL GRINDING WHEELS

Producing a Straight Grinding Wheel Face

STEP 1 Select and mount a wheel truing fixture.

Note: Depending on the model, the fixture may be mounted on the footstock or secured to the table. The diamond-tool holder is designed as part of the truing fixture.

STEP 2 Position the footstock and grinding wheel with adequate, safe working clearance.

STEP 3 Position the diamond dresser. A new face (sharp edge) of the diamond must be exposed for efficient truing and dressing. The diamond point is set just below the axis of the wheel and at a slight downward angle to prevent the diamond from digging into the wheel.

STEP 4 Position the reversing trip lever to set the right-side and left-side dogs. The setting must permit the diamond to clear both sides of the grinding wheel at each pass.

STEP 5 Set the rate of table travel to a fairly fast one.

STEP 6 Set the grind-true switch in the *true* position. This setting controls the constant flow of cutting fluid during the dressing process.

STEP 7 Bring the wheel spindle to operating speed.

STEP 8 Move the diamond dresser until it is in contact with one edge. Round this edge of the grinding wheel slightly by manipulating the longitudinal (traverse) and cross feed movement handwheels.

Note: Slightly rounding the corners prevents grains from breaking out during dressing and prevents grinding imperfections produced by a sharp-edged wheel.

STEP 9 Repeat the process of rounding the other edge of the grinding wheel.

Note: The edges are often slightly rounded by using an abrasive stick. More precise rounding is done with a radius truing attachment.

STEP 10 Start the table movement with the start-stop lever.

STEP 11 Advance the wheel to start the first cut.

Note: The diamond dresser is generally brought into contact with the grinding wheel at about the center (width), which is usually the high-spot area.

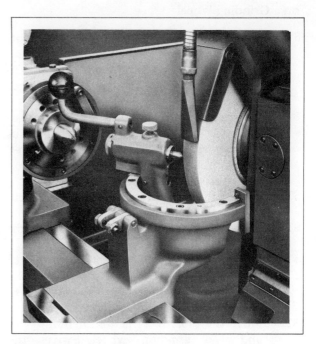

FIGURE 29-14 Radius Wheel Truing Attachment

STEP 12 Move the diamond past the edge of the wheel. Turn the cross feed handwheel to feed the wheel from 0.0005" to 0.001" (0.0125mm to 0.025mm).

STEP 13 Take the first dressing cut.

STEP 14 Continue to in-feed the wheel and take successive cuts until the wheel is trued.

STEP 15 Finish dressing the wheel. The rate of table travel and the depth of the finishing cuts are reduced.

Note: Finish cuts are taken starting at 0.0005" (0.0125mm). The final one or two cuts are at depths of from 0.0001" to 0.0002" (0.0025mm to 0.005mm).

Note: The rate at which the diamond moves across the face and the sharpness of the cutting action influence the surface texture that the wheel will produce.

STEP 16 Remove the truing and dressing device.

STEP 17 Clean the machine thoroughly. All abrasive and other foreign particles from truing and dressing must be removed from the machine. After use, the table and machine parts should be clean and dry. All tools are replaced in proper storage areas.

Forming a Corner Radius on a Grinding Wheel

STEP 1 Select a radius wheel truing attachment. Position it to shape the right- or left-hand corner of the wheel. Clamp the attachment to the table. The knob should face the operator.

STEP 2 Determine the radius setting of the dressing tool. The setting for an external radius is established by subtracting the required radius from the measurement shown on the dressing-tool holder.

STEP 3 Set the diamond dressing tool at the calculated setting.

STEP 4 Position the diamond-tool holder at a right angle to the wheel face.

STEP 5 Move the truing attachment forward until the diamond tool touches the face of the wheel. Tighten the clamping bolt nut in order to secure the slide in position.

STEP 6 Move the table to clear the diamond tool and the grinding wheel. Swivel the diamond-tool holder again. This time the diamond tool is at a right angle to the side of the wheel.

STEP 7 Move the table until the diamond tool just grazes the wheel. This second setting correctly positions the diamond tool in relation to the face and side of the wheel.

STEP 8 Move the table away from the wheel to permit the diamond tool to swing without hitting the corner (edge) of the wheel.

STEP 9 Feed the table carefully by hand and swing the holder back and forth in a 90° arc.

Note: The table is advanced a small amount by hand each pass to prevent the cutting diamond from digging in and fracturing and thus producing a ragged, irregular wheel edge.

STEP 10 Keep advancing the diamond tool by moving the table until the complete radius is formed.

Note: The depth of cut is reduced and the diamond is swung at a slower rate during the final radius finishing passes.

STEP 11 Clean the attachment and machine thoroughly. Check the condition of the diamond tool. Return the attachment to its proper storage compartment.

Forming an Angular Face on a Grinding Wheel

STEP 1 Select an appropriate angle wheel truing attachment. Position and secure the attachment to the table.

STEP 2 Swivel the attachment to the required angle. Read this angle directly on the graduated base of the swivel section.

STEP 3 Position the wheel by turning the longitudinal table handwheel and the cross feed handwheel.

STEP 4 Move the table so that the diamond tool is fed into position for the first light cut of from 0.0005" to 0.001" (0.0125mm to 0.025mm).

Note: The table must remain stationary while the attachment slide and diamond are moved carefully by the hand crank across the face of the wheel.

STEP 5 Continue to feed the wheel and to traverse the diamond across the wheel face for each pass. Move the angle slide and diamond with the fixture hand crank.

Note: The same steps are continued until the wheel is formed at the specified angle and depth.

STEP 6 Clean the machine and attachment. Again check the condition of the diamond. Return all items to proper storage compartments.

Forming a Concave or Convex Face on a Grinding Wheel

STEP 1 Select an appropriate radius wheel truing attachment. Position the attachment so that the axis of the diamond cutting tool is aligned with the centerline of the grinding wheel. Secure the attachment to the table.

STEP 2 Postion the diamond tool in the holder so that contact is made when the diamond is positioned slightly below the axis of the wheel (and spindle) and at a small downward angle.

STEP 3 Position the diamond point. Use the setting gage on the front of the toolholder. Secure the tool in this starting position.

STEP 4 Set the slide on which the diamond tool is mounted to the required radius.

Note: The diamond is positioned to form a convex radius in the grinding wheel by moving the diamond back past the zero setting. The size of the radius is read directly on the graduated scale of the index slide.

The truing attachment is adjusted to form a concave wheel face by moving the diamond forward the radius distance beyond the zero setting. Again, the radius measurement is read on the index slide. The slide clamping screw is tightened at this setting.

STEP 5 Check the setup for the alignment of the radius truing attachment with the wheel centerline and the position of the diamond and check the position of the slide for forming the convex or concave radius.

STEP 6 Bring the spindle up to operating speed. Advance the wheel to the diamond tool for the first cut either at the center of the wheel (for a concave form) or across the ends of the wheel (for forming a convex face).

STEP 7 Continue to advance the wheel from 0.0005" to 0.001" (0.0125mm to 0.025mm) each time the diamond tool sweeps across the face for a roughing cut.

STEP 8 Take a series of light cuts for finish grinding the radius to the required depth. The wheel is advanced from 0.0003" to 0.0005" (0.007mm to 0.0125mm) for the finish cuts.

STEP 9 Clean and remove the attachment at the completion of the operation. Properly store the attachment and tools.

CRUSH TRUING ON A CYLINDRICAL GRINDER

Vitrified bond wheels are often formed on heavy-duty cylindrical grinders by *crush truing*. Hardened steel or cast iron formed rolls are forced under extreme pressure against the revolving grinding wheel to reproduce the reverse form in the grinding wheel. Crush truing is faster than diamond truing, but requires a more rigidly designed machine, particularly the wheel head and spindle.

SAFE PRACTICES IN SETTING UP AND OPERATING CYLINDRICAL GRINDERS

—Consider the fact that vitrified wheels, which comprise the greatest part of the wheels used in industry, are brittle and breakable unless used safely. A damaged wheel can disintegrate due to shock, unusual resistance (grinding forces) of the work material, and improper use.
—Observe the ANSI "Safety Code for the Use, Storage, and Inspection of Abrasive Wheels." Visual inspection is accompanied by ring testing for possible cracks.
—Examine the construction and clearance between the guard and the grinding wheel. The clearance should be adequate to avoid interference with the grinding operation.
—Check the balance of the grinding wheel. Mount each cylindrical grinding wheel (even new ones)

- with wheel and balancing flanges on a short balancing arbor. Support the ends in a balancing stand. Rotate the wheel and position the weights until the wheel is balanced.
- Establish from the manufacturer's specifications what the safe operating (fpm or m/min) speed is for the grinding wheel. Convert the surface speed into the RPM at which the grinding wheel is to revolve.
- Remove abrasive grains and foreign particles from all machine areas after truing and dressing and at the completion of the cylindrical grinding process.
- Test the viscosity of the cutting fluid and the degree to which it is free of chips, abrasive grains, and other particles. Clean the system, change the fluid or bring it up to standard, and check the position and rate of flow, if necessary.
- Examine the diamond-point dressing tool for sharpness and to see that it is not cracked.
- Avoid hitting into a grinding wheel during setup or when positioning and operating a dressing attachment. Position the wheel dressing tool attachment so that the diamond does not take an increasingly deeper cut as it is swept across a corner or face of the wheel at the beginning of the truing process.
- Position splash guards to contain the cutting fluid. Immediately wipe up spilled cutting fluid and use a floor surface compound to keep the machine area clean.
- Lock the in-feed lever when the wheel head is in the retracted position and the internal grinding head is being used. Locking prevents the wheel head from accidently moving during internal grinding.
- Control the in-feed of the grinding wheel (wheel head) by carefully adjusting with the in-feed handwheel. Power in-feed is not recommended until the first cut is set by hand, clearances are checked, and everything is operating correctly.
- Recheck the settings of the trip dogs by moving the table longitudinally by hand. Finer adjustments are made before the power feed is engaged.
- Keep the grinding wheel face sharp and fast cutting. Apply a generous flow of cutting fluid to quickly dissipate the heat at the point of contact and to permit efficient grinding.
- Spark out the finish cut so that the grinding wheel is free of the workpiece.
- Refer to the manufacturer's manual and establish whether there are any hand lubrication or pump systems to operate before starting the cylindrical grinder.
- Make periodic checks on the wear and operating surfaces and evenness of lengths of the formed drive belts.
- Lock the internal grinding attachment when it is swung out of operating position.
- Set the wheel head travel movement so that it is automatically stopped when the required ground diameter is reached.
- Review the table of common cylindrical grinding problems, probable causes, and corrective action (Table 29-2). Study the potential machining problems and surface and dimensional defects that may result from unsafe practices.
- Support small-diameter and long, thin workpieces against grinding forces by using back rests or a steady rest.
- Use eye protection devices against flying chips and abrasive particles.
- Leave the machine tool in a safe, operating condition.

TERMS USED IN CYLINDRICAL GRINDING TECHNOLOGY AND PROCESSES

Cylindrical grinding	A series of abrasive machining processes. Grinding a cylindrical or plane surface by traversing a face grinding wheel parallel with or at an angle (0° to 90°) to the axis of the workpiece.
Universal cylindrical grinder	A cylindrical grinder designed so that the wheel head (grinding wheel) and table may each be swiveled. A machine with built-in features that permit a greater variety of cylindrical grinding processes to be performed using hand or power feed.

CYLINDRICAL GRINDING 589

Traverse grinding	A form of cylindrical grinding where the revolving workpiece is moved across the grinding wheel face. Generally related to the grinding of a straight or taper cylindrical surface by reversing the direction of table travel feed for each cut across a workpiece.
Plunge grinding (cylindrical grinding)	Forming a flat or other contour cylindrical surface by in-feeding the wheel head and grinding wheel into a workpiece without moving the table longitudinally.
Major components (cylindrical grinder)	Principal design features on a cylindrical grinder. The base and table ways; saddle, table, and drive mechanism; headstock and footstock; wheel head, carriage, and cross feed mechanism; cutting fluid system; and control panel.
Dwell (tarry)	The time interval at the end of a grinding stroke that permits the grinding action to complete the pass and adjustment to the next depth of cut before power traverse feed is again applied.
Sparking out	Traversing the wheel across the workpiece without further depth setting until there are no sparks, signifying the cutting action is complete.
Cross feed control	A positive stop (fixed pin) that actuates a switch to stop the cross feeding when the required diameter to which the cross feed is set is reached.
Headstock	A unit designed for mounting on the swivel table. The cylindrical grinder unit for holding, mounting, and positioning a workpiece and rotating it in a fixed relationship (fpm or m/min) to the revolving wheel.
Footstock	A machine device adjustable along the table to support workpieces. A base, movable spindle, and center for supporting center work. (On some models, the footstock is used for mounting and supporting a wheel dressing accessory.)
Wheel head	The entire mechanism for housing, driving, and controlling the grinding wheels for external and internal grinding. The grinding spindle head that is mounted on a carriage to permit swiveling the wheel at an angle to the axis of the workpiece.
Short- (slight-) angle taper	A taper usually within the range of table settings (graduations).
Spring back rests or steady rests	Work-supporting devices. Supports that prevent small-diameter or proportionally long, thin workpieces from becoming distorted during grinding. (Soft metal shoes are used to support the revolving workpiece against the cutting forces of a grinding wheel.)
Operational data	Technical information that the machine operator must know about machining conditions. Data on which the worker makes judgments as a starting point for one or more cylindrical grinding machine operations.
Automated functions	Machine setups, processes, and systems that when once programmed are cycled automatically. Seven basic categories of automatic cylindrical grinding activities related to work, feed, size, cycling, wheel dressing, checkout, and troubleshooting.
In-feeding	Advancing the grinding wheel into the workpiece for purposes of positioning the wheel at a cutting depth.

SUMMARY

Plain, special, and universal are the basic types of general-purpose cylindrical grinders.

Traverse grinding provides for a revolving workpiece to be moved at cutting depth across the face of a grinding wheel. Traverse grinding cuts may be taken parallel to or at an angle from 0° to 90° to the work axis.

Plunge grinding relates to the reducing of one or more diameters (conforming to the shape of the wheel) by continuous in-feeding.

The principal design components of a universal cylindrical grinder include the base; saddle, swivel table, and drive mechanism; headstock and footstock; wheel head, carriage, and cross feed mechanism; cutting fluid system; and operator control panel.

Cylindrical grinders are designed for hand and/or hydraulic traverse and in-feed.

Longitudinal table travel controls permit the table to dwell up to a few seconds at each reversal of the table.

Controls are provided for revolving the workpiece at the required RPM. The controls permit starting, stopping, jogging, and braking.

Carbide-tipped centers are widely used to provide a smooth, precision bearing surface for center work.

Handwheel graduations permit work diameter reductions of 0.100" (2.5mm) per revolution. The fine wheel feed graduations are in increments on the work diameter of 50 millionths (0.000050") of an inch (0.00125mm).

The wheel head is adjustable on its swivel base. Angle settings are made directly according to the angle graduations.

Power in-feed is controlled by a positive-stop mechanism. When set to grind a workpiece to a required diameter, movement of the wheel head (grinding wheel) is automatically tripped (stopped).

Cylindrical processes are sparked out at the footstock end of the workpiece.

The normal allowance for finish grinding is from 0.001" to 0.002" (0.025mm to 0.05mm).

The basic accessories for cylindrical grinders include work-supporting devices (back rests); standard chucking and work-holding devices; and wheel truing, dressing, and forming devices.

Operational data for controlling machine speeds, feeds, and the grinding process are based on factors such as design features and characteristics of the workpiece; machine stability; conditions of grinding; and the grinding wheel.

The grinder operator must establish the surface speed and RPM of the workpiece, in-feed, and rate of traverse.

Cutting speeds may be given in feet or meters per minute or may be calculated by standard formula when the spindle RPM and wheel diameter are known. Similarly, the RPM may be computed.

High-speed cylindrical grinding is performed on machines that are designed to operate at spindle speeds in excess of 6,500 fpm (1,980 m/min). High-speed machines are more ruggedly constructed to withstand the increased cutting forces.

In addition to the automatic controls for feed, speed, and traverse on
standard machines, other cylindrical grinder functions may be automated.
These functions may be grouped to include the loading and positioning
of the workpiece; presetting, cutoff, and duration of in-feed; table traverse for dwell and sparking out; size control with compensation for variations in wheel size; cycling for setup, machining, and shutdown; frequency
dressing of the grinding wheel; and computerized numerical control.

> Cylindrical grinding problems fall into four groups: (1) chatter, (2) work
> defects, (3) machine operation, and (4) grinding wheel condition.

ANSI wheel markings are used for abrasive type, grain size, grade, structure, and bond type. Following the ANSI grinding wheel and machine
process code of safety is an industry requirement.

> Manufacturers' tables are used for the suggested fractional width of grinding wheel to use for roughing and finishing cuts and traverse or plunge
> cutting rates.

Tables provide feeding rate information for roughing and finishing cuts.
In general nonproduction grinding, the in-feed for roughing cuts is 0.002"
(0.05mm). Finish cuts are reduced to 0.0001" to 0.0005" (0.0025mm
to 0.0125mm) per pass.

> The five basic cylindrical grinding processes are parallel, taper, form,
> steep-angle, and face grinding.

The universal cylindrical grinder is prepared for grinding by setting up
the grinding wheel and wheel head, setting the swivel table at 0° or any
required angle, positioning the headstock and footstock (if used), checking the coolant system, mounting the workpiece, setting the table position and stroke, and cross feeding.

> Workpieces may be held between centers or in a chuck or collet or may
> be mounted on a faceplate or fixture. Work supports are used for long,
> small-diameter workpieces or where heavy cuts may cause distortion.

Small-angle tapers are ground by direct setting of the swivel table.

> Steep angles may require a combination of angle settings of the table,
> swivel head for the grinding wheel, or swivel of the workpiece in the
> headstock.

Face grinding requires the angle relationship between the grinding wheel
and workpiece axis to be 90°.

> Universal cylindrical grinders are designed with an internal grinding head.
> This head is swung into position and locked. The internal grinding head
> is an integral unit on the wheel head. The use of smaller-diameter wheels
> requires the internal grinding spindle to rotate at higher speeds than for
> external grinding.

The internal grinding spindle is moved into position and locked. The
spindle RPM is set to the recommended spindle speed requirement. The
distance of travel (depth) inside the workpiece is set by using the table
trip dogs.

> Improper wheel dressing, the rate, and depth of feed may result in a bellmouthed, out-of-round, or tapered hole. These inaccuracies may be partially corrected by a sharp-cutting wheel and by changing the feed and
> speed rates.

Wheel truing devices are similar in design and function to the devices
used for surface grinding. Essentially, the truing devices are for straight,
angle, radius, and combination truing and dressing.

592 GRINDING MACHINES

Concave and convex radii are formed with the same truing device. The distance of the diamond point in relation to the axis of the swivel determines the amount of an external or internal radius that will be generated.

The depth of cut and the speed with which the dressing tool is moved across the face of the wheel must be considered in relation to the required cutting action and surface finish.

Abrasive and foreign particles produced by truing must be cleaned away from the work area to protect the machine and the surface texture of the workpiece.

Standard personal, machine, and workpiece precautions must be observed in setting up and operating cylindrical grinders.

Particular attention is directed to the careful positioning of the table trip dogs for lateral movement. This positioning is especially important in grinding up to a shoulder.

UNIT 29 REVIEW AND SELF-TEST

1. State two differences between traverse grinding and plunge grinding on a cylindrical grinder in terms of (a) feeding, (b) surface area limitations, and (c) design features of the grinder.
2. a. List four principal units of a cylindrical grinder.
   b. Cite the function of each unit.
3. a. State when (1) regular feed knob graduations and (2) fine wheel feed graduations are used in cylindrical grinding.
   b. Indicate the customary inch and metric graduation increments for reducing the work diameter on (1) a regular feed knob and (2) the fine-feed graduated handwheel.
4. Identify the function of each of the following: (a) sparkout timer dials, (b) positive-stop cross feed mechanism, and (c) universal spring back rests.
5. List five important factors the cylindrical grinder operator must consider before using tables to obtain basic process data such as surface speed, feed, and traverse rates.
6. Cast steel parts are to be rough and finish ground using the table traverse method.
   a. Refer to a table of operational data for traverse and plunge grinding.
   b. Establish (1) work surface speed in m/min, (2) roughing in-feed per pass in mm, (3) finishing cut in-feed per pass, and (4) fractional wheel width traverse per revolution of the work for roughing cuts, and (5) fractional wheel width traverse for finishing cuts.
7. Calculate the headstock spindle RPM at which a 3.625" diameter hardened tool steel punch may be ground. Round off the work speed at which the headstock speed control is set.
8. List the major components and accessories that must be set up in preparation for cylindrical grinding.

9. Set up a cylindrical grinder and grind the mandrel shown below to the required taper and surface finish.

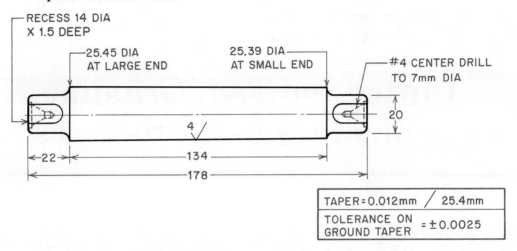

| TAPER = 0.012mm / 25.4mm |
| TOLERANCE ON GROUND TAPER = ±0.0025 |

10. Dress a type 1 wheel for parallel (straight) grinding on a cylindrical grinder.
11. State three personal safety precautions the cylindrical grinder operator must observe with respect to the grinding wheel.

SECTION FIVE

Cutter and Tool Grinding Technology and Processes

The maximum efficiency of cutting tools depends primarily on the quality of the cutting edge and the accuracy to which each cutter tooth is ground. Manufacturers provide technical information about design characteristics of cutter teeth and machining conditions. The machine on which cutting tools are usually ground is the cutter and tool grinder (or the older tool and cutter grinder).

This section covers the functions of the major components of the cutter and tool grinder, its attachments and accessories, and basic machine setups in preparation for grinding. Procedures for grinding straight and helical teeth on a plain milling cutter are described. Simple calculations illustrate how to establish the offset required to produce specific angles.

The section also describes the grinding of common groups of milling cutters and other cutting tools. Step-by-step procedures are included for grinding the teeth on plain slitting saws; staggered-tooth side milling cutters; shell and small end mills; inserted-tooth face milling cutters; angle cutters; and form relieved cutters. Calculations are required to establish the offset for grinding relief, clearance, and rake angles. Considerations for grinding solid and adjustable machine reamers and starting teeth on taps are also included.

UNIT 30

Cutter and Tool Grinder Features, Preparation, and Setups

One of the most versatile grinders is the *universal cutter and tool grinder*. It accommodates a wide variety of accessories. These accessories permit the grinding of flat cutting tools, external and internal cylindrical grinding, surface grinding, and additional special tool grinding.

This unit covers the universal cutter and tool grinder and common accessories, tooth rest supports, and work-holding devices. A review is included of primary relief and secondary clearance angles and tooth form terms, computations for offset, and the measurement of ground cutter angles. Applications are made to reference tables and factors to consider for variations from grinding wheel recommendations. Safety checks are identified for preparing the machine, carrying on grinding processes, and providing machine maintenance. Steps are included for setting up the grinder in preparation for specific grinding processes that are covered in the next unit.

MACHINE DESIGN FEATURES OF THE UNIVERSAL CUTTER AND TOOL GRINDER

One popular size of toolroom universal cutter and tool grinder has a 10" (254mm) swing capacity over the table and a 16" (406mm) longitudinal table movement. The major components and controls of a cutter and tool grinder are illustrated in Figure 30-1. As for most machine tools, the base provides the support for all machine elements. The base also houses electrical, mechanical, and pneumatic controls and a cutting fluid system (when included). The major design components are the table, work head and tailstock, and wheel head.

Table

The cross (traverse) movement of the table is 10" (254mm). The table swivels 180°. A taper setting device permits taper settings toward or away from the wheel head up to 5" (127mm) per foot on the work diameter. The swivel table is graduated in degrees from either side of center.

The table is supported on the base by a series of precision hardened steel balls that roll in hardened steel V-ways (Figure 30-2). The table slide movement produced is accurate, smooth, and straight-line traverse.

The table is fitted with spring-cushioned table dogs (Figure 30-3). The dogs govern the length of table traverse, absorb the shock of table reversal, and provide positive table stop, where required.

Power table traverse is optional. Power table drive may be quickly disengaged for hand-operated control. The addition of power table traverse permits light toolroom production work requiring internal or external cylindrical grinding, surface grinding, and special applications where a steady uniform feed is needed. Uniform feed assures machining to finer finish and closer tolerance requirements.

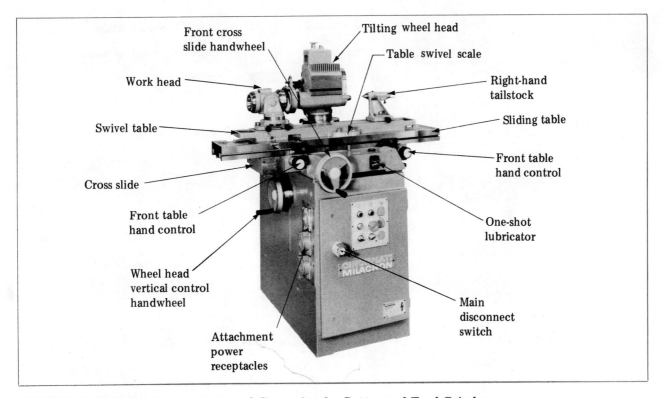

FIGURE 30-1 Major Components and Controls of a Cutter and Tool Grinder

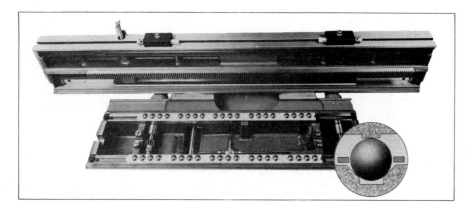

FIGURE 30-2 Hardened Steel Ball and V-Way Feature Provides Accurate, Smooth, Straight-Line Table Traverse

Fine Adjustment Setting Device. The swivel table can be adjusted inward or outward for tapers up to 2″ (50mm) per foot. Taper settings are made directly by turning a swivel adjusting screw. The taper is read on the swivel table scale.

"Tange Bar" Taper Setting Device. The "tange bar" taper setting device (Figure 30-4) uses a standard setting block and a gage block. The center distance from the pivot point of the table to the center of the block setting combination is fixed at 12″ (304.8mm). The gage block size is determined by the height measurement of 1/2 the included angle (Figure 30-4A). This method is used to set the swivel table to the required angle for precision taper grinding (Figure 30-4B).

Work Head

The standard universal work head (Figure 30-5) is designed with two tapered holes running through the spindle. One end accommodates a

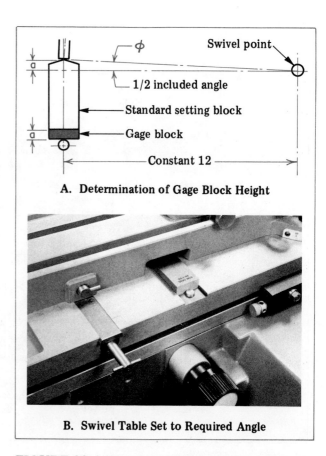

A. Determination of Gage Block Height

B. Swivel Table Set to Required Angle

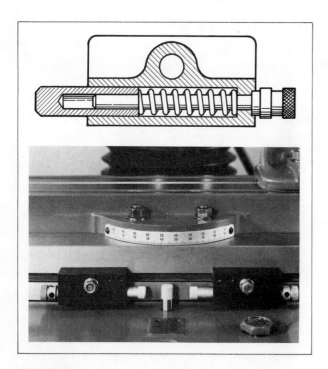

FIGURE 30-3 Spring-Cushioned Table Dogs

FIGURE 30-4 "Tange Bar" Taper Setting Device

CUTTER AND TOOL GRINDING

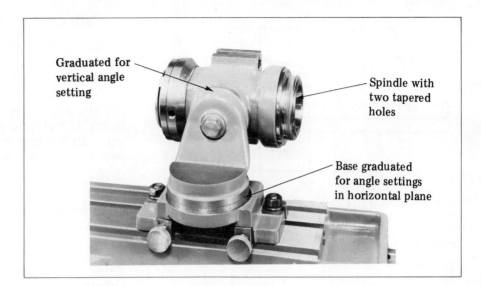

FIGURE 30-5 Universal Work Head

Morse or Brown & Sharpe taper. The other end is designed for a standard—for example, a #50—National Series taper. The base is graduated in 1° increments. The base permits swiveling in a horizontal plane through 360° of arc. The work head swivel is also graduated in 1° increments for angle settings in a vertical plane. The work head is designed to be positioned along a table and held securely with T-bolts.

Tailstock

The tailstock serves work-centering and work-supporting functions similar to the tailstock of a lathe or the footstock on a universal grinder. The tailstock is designed for positive alignment when used in pairs. A retractable center may be used with a tailstock positioned at the right or left end of the table.

Wheel Head

Wheel heads are designed with a graduated base that permits settings through 360°. Wheel heads may also be swiveled to an angle in the vertical plane. An added feature on some machines is called an *eccentric column*. The eccentric column is also graduated through 360° of arc. This standard wheel head and eccentric column combination provides for an infinite number of angular adjustments. Also, with the eccentric column, the grinding wheel may be positioned over the work for operations such as surface grinding.

Since the table saddle may also be swung through 180°, the traverse range of the swivel table on a 10" × 16" (254mm × 406mm) universal model is extended by 3-1/2" (89mm). Thus, the offset table, wheel head, and eccentric column provide for machining over an extended cross (traverse) range. The normal 10" (254mm) cross range is extended to 13-1/2" (343mm) by swiveling the table 180°. The cross range is increased to 17" (431mm) by swiveling the eccentric column to the rear position.

Standard toolroom model wheel heads are operable at spindle speeds of 3,834 RPM for 6" (152mm) diameter wheels having a maximum wheel surface speed of 6,022 fpm (1,835 m/min). The spindle speed of 6,425 RPM is used for 3-1/2" (89mm) diameter grinding wheels at a maximum surface speed of 5,887 fpm (1,794 m/min).

Forced Lubrication System

The cutter and tool grinder is equipped with a one-shot forced lubrication system (Figure 30-6). A single *actuating plunger* is located on the front of the machine on the cross slide. Lubrication is provided to the moving parts on the cross feed screw and the cross slide bearings.

CUTTER AND TOOL GRINDER ATTACHMENTS

A number of standard attachments are available for cutter and tool grinders. The most

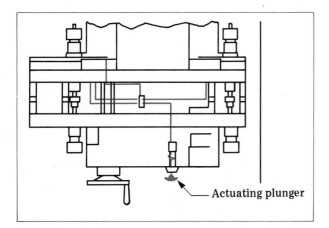

FIGURE 30-6 One-Shot Forced Lubrication System

widely used attachments are illustrated and described.

Surface Grinding Attachment

The *surface grinding attachment* (Figure 30-7) is applied to the grinding of flat forming tools for turning and other machine tool processes. This attachment consists of a swivel base, a vise, and an intermediate support between these two units. The base and intermediate support are graduated and are designed to be swiveled 360° in both the horizontal and vertical planes. The attachment permits the work head to be removed and used between the vise support and the base. This arrangement makes it possible to use the vise in three planes.

Cylindrical Grinding Attachment

The *cylindrical grinding attachment* provides for the grinding of outside diameters on work that is held in a chuck or fixture or between live or dead centers. The attachment is adapted to straight and taper cylindrical grinding and for facing operations. The attachment is provided with an electrical control panel. A reversing switch controls the right- or left-hand rotation of the workpiece.

Internal Cylindrical Grinding Attachment

Internal diameters may be ground by the addition of an *internal grinding spindle*. The spindle is used with the work head drive unit of the cylindrical grinding attachment. The application of a work head drive spindle and an internal grinding spindle to internal cylindrical grinding is shown in Figure 30-8. The grinding spindle is driven by a positive-drive belt from a gear tooth pulley mounted on the wheel head spindle. Spindle RPM up to 23,000 is within the range of this attachment. This speed provides for the use of small-diameter grinding wheels to grind smaller-diameter holes than are normally produced.

Gear Cutter Sharpening Attachment

The *gear cutter sharpening attachment* (Figure 30-9) is designed for sharpening form relieved cutters by grinding the face of the teeth. The cutter is supported on a bracket. The bracket may be swiveled to accommodate the angle of

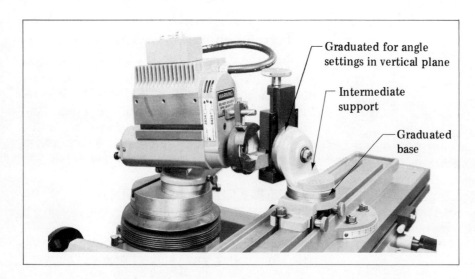

FIGURE 30-7 Surface Grinding Attachment Positioned for Grinding a Clearance Angle on a Flat Cutting Tool

CUTTER AND TOOL GRINDING 599

FIGURE 30-8 Application of Work Head Drive Spindle and Internal Grinding Spindle to Internal Cylindrical Grinding

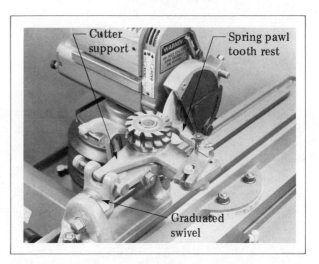

FIGURE 30-9 Application of Gear Cutter Sharpening Attachment for Grinding the Tooth Face on Form Relieved Cutters

the tooth face. The spring pawl tooth rest lies in a horizontal position so that the edge of the spring locates against the back of the tooth. Gear and other formed cutters up to 8-3/4" (222mm) outside diameter with hole diameters to 2" (50mm) may be accommodated on this attachment.

Small End Mill Grinding Attachment

The body of a small end mill is cylindrical. For cylindrical grinding purposes, small end mills are held in collets. The *small end mill grinding attachment* (Figure 30-10) includes an intermediate support, a 24-division master index plate, and a plunger-type indexing mechanism. The spindle is designed to take straight cylindrical and taper collets.

Radius Grinding Attachment

Ball-shaped end mills with straight or helical flutes and cutters that are to be ground to an accurate 90° radius require the use of a *radius grinding attachment* (Figure 30-11). The design features include a base plate with two mounted table slides and a swivel plate. Movement of each slide is controlled by a micrometer adjustment knob (or graduated collar). A 24-notch index

plate may be mounted at the back of the work head to permit direct indexing of straight-fluted cutters.

The starting position is established by a micrometer gage. Radius grinding attachments are available with capacity to position cutters for

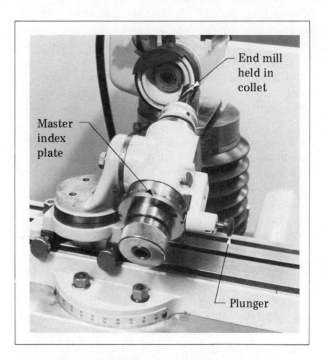

FIGURE 30-10 Small End Mill Grinding Attachment

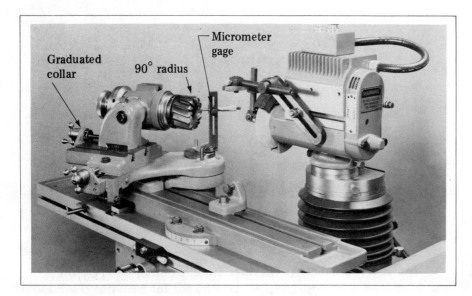

FIGURE 30-11 Radius Grinding Attachment for a Universal Cutter and Tool Grinder

grinding radii from 0" to 2" (0mm to 50mm), cutter diameters up to 12" (305mm), and to a maximum width of cutter face of 3" (76mm).

Additional Attachments

A number of other attachments are optional, depending on the variety of toolroom or small manufacturing cutter and tool grinding operations. For example, the standard work head may be equipped with an *indexing attachment.* This attachment eliminates the tooth rest as the teeth may be indexed according to the notches on an index plate.

A *micrometer table positioning attachment* adds the feature of a precision lead screw for the machine table. A *sine bar attachment* permits the work head to revolve at a predetermined lead without a tooth rest. A *heavy-duty tailstock* increases the swing capacity over the table. A special *extended grinding wheel spindle* is interchangeable with the conventional spindle. The extended spindle provides added range. A *draw-in collet attachment*, having similar design features to draw-in attachments for turning machines, may be used with the work head. Collets are available in sizes from 0.125" to 1.125" in increments of 1/64". Metric sizes range from 3mm to 28mm in increments of 1mm. Small, straight-shank cutters are conveniently held in draw-in collets.

CENTERING GAGE, MANDREL, AND ARBOR ACCESSORIES

Centering Gage

The *centering gage* consists of a base, arm, and center. As the name indicates, it is used to position the wheel head (and spindle axis) at the center height of the work head and tailstock centers. It is also used to locate the cutter teeth to coincide with the center axis (when this setup is required). This accessory is ground on the base for direct placement on the table.

Grinding Mandrel

The *grinding mandrel* serves to hold a cutter so that it may be accurately mounted between centers. The main difference in design from a lathe mandrel is that the body is ground straight (the lathe mandrel has an 0.0005" per inch taper). The diameter of the straight-length portion provides a sliding fit for the cutter. The cutter grinding mandrel has a slight taper for about one-third of its length. When pressed against this slight taper, the cutter is held securely for grinding.

Cutter Grinding Arbor

A *cutter grinding arbor* is used when there is considerable grinding to be done. The general design of the arbor includes a centered, ground shaft to accommodate the bore diameter of the cutter and a series of spacing collars, washer, and nut. Where practical, cutters—for example, shell end mills—should be sharpened on the arbor that will be used for milling.

FUNCTIONS AND TYPES OF TOOTH RESTS AND BLADES

Functions of Tooth Rests

A tooth rest serves three functions:

—To position the tooth or surface to be ground in a fixed relationship to the grinding wheel,
—To provide a support (the top face of the tooth rest blade) for the tooth during the grinding process,
—To permit the tool or cutter to be indexed to the next tooth (the tooth rest springs back into working position after each index).

Types of Tooth Rests

Tooth rests are of two general types: (1) stationary and (2) adjustable. The adjustable type is designed with a micrometer adjustment. The blade (and tooth rest) may be adjusted to more accurately position the cutter for grinding the relief angle. The two types of cutter tooth rests may be fastened to the wheel head or the table.

Forms of Tooth Rest Blades

Different forms of tooth rest blades are available. The blade form depends on the shape of the cutter teeth that rest against the blade. Five common shapes of tooth rest blades are shown in Figure 30-12.

Straight-tooth milling cutters are supported by a *plain tooth rest blade*. Shell end mills, small end mills, taps, and reamers are ground with the support of a *rounded tooth rest blade*. An *offset tooth rest blade* is applied in the grinding of coarse-pitch helical milling cutters and large face mills with inserted cutter blades. A *hooked blade* is used with straight-tooth plain milling cutters having closely spaced teeth, end mills, and slitting saws. An *inverted V-tooth blade* is used for grinding staggered-tooth cutters.

REVIEW OF BASIC CUTTER TERMS

Continuous reference is made throughout cutter and tool grinding to four basic terms: *primary relief (clearance)*, *secondary clearance*, *cutting edge*, and *land*. The basic design features described by these terms are illustrated in Figure 30-13 for review and application to the grinding (sharpening) of milling cutters.

Primary Relief (Clearance) Angle

The primary relief (clearance) angle is formed by a plane tangent to the outside diameter of the cutter starting at the tooth face and the plane of the surface that slopes away at the periphery of each tooth for the land distance. This angle provides for a correctly designed efficient cutting edge and long wear life. Primary relief also prevents the land behind the cutting edge from rubbing aginst the work. Primary relief angles are designed according to cutting requirements.

Secondary Clearance Angle

The secondary clearance angle extends from the heel of the land to the gullet or flute of a tooth. Secondary clearance provides additional chip clearance area and controls the width of the land. The primary relief (clearance) is ground first.

Cutting Edge

The cutting edge is the area formed by the face of the tooth and the edge of the land. The cutting edge may be straight as in straight-fluted

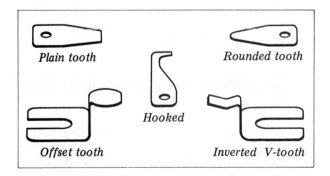

FIGURE 30-12 Five Common Shapes of Tooth Rest Blades

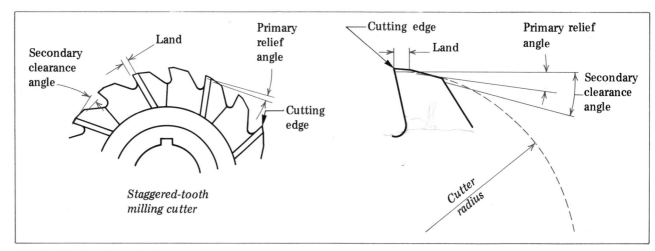

FIGURE 30-13 Basic Design Features Applied to the Grinding (Sharpening) of Milling Cutters

reamers, taps, and milling cutters. Other cutting edges, like the teeth on helical milling cutters, are at a helix angle. Cutting edges may also be on one or both sides of a cutter as well as on the periphery.

Land

The land area is of special importance in cutter and tool grinding. Outside of form cutters that are sharpened by grinding the front face, other cutter teeth are sharpened by grinding on the land. The land area is bounded by the face of each tooth and the plane where the secondary clearance begins. The land on face mills refers to a face edge.

TOOTH REST APPLICATIONS AND SETUPS

Applications of Tooth Rests

Cutters are divided for sharpening purposes into two general groups. The first group includes cutters that are sharpened by grinding primary relief and/or secondary clearance angles behind the cutting edge of each tooth. Examples of cutters that are ground on the periphery are plain and helical milling cutters, cutoff saws, and reamers. Cutters that are ground on the sides or ends are also included.

The second group of form-relieved cutters are sharpened by grinding the cutting faces of the teeth. Such cutters have a definite profile for form machining. Gear tooth cutters, combination radius and angle form cutters, and taps are examples.

Cutter Grinding with Tooth Rest Mounted on Table

The height of the tooth rest depends on the type of grinding wheel and its direction of rotation in relation to the cutting edge of the cutter and the location of the cutter and machine centerline. One typical setup of a tooth rest for hollow grinding a plain milling cutter with a straight-type wheel is illustrated in Figure 30-14. The direction of the grinding force is clockwise. The grinding process produces a counterclockwise force on the cutter to hold it against the tooth rest.

While there is no tendency for the tooth to dig into the wheel during grinding, a slightly burred cutting edge is produced. In addition, the heat generated in grinding moves toward the smallest area, which is the cutting edge. If too much heat is generated, the temper of the teeth at the cutting edge may be drawn.

Cutter Grinding with Tooth Rest Mounted on Wheel Head

The setup for cutter grinding with the tooth rest blade positioned above the cutter is shown in Figure 30-15. The centers of the wheel and cutter are offset to produce the required clearance angle. Here, the wheel is offset below the center-

CUTTER AND TOOL GRINDING 603

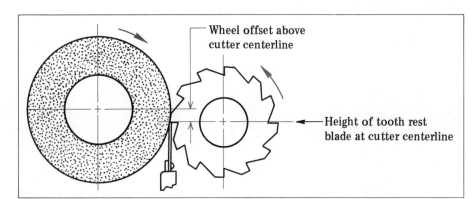

FIGURE 30-14 Typical Setup of Tooth Rest for Hollow Grinding Plain Milling Cutter with Straight Wheel

line of the cutter. A caution is in order when using this setup. The cutter must be held securely against the tooth rest. The cutting action tends to move the cutter away from the tooth rest. Thus, there is the possibility of personal, wheel, and/or cutter damage.

With the wheel and cutter properly set up, there are two advantages of using this method. First, burr-free cutting edges are produced. Second, the possibility of overheating the cutting edges is reduced.

Positioning of Flaring-Cup Wheels

Flaring-cup wheels are widely used for cutter and tool grinding. The two general methods of positioning the tooth rest are similar to the setups used with standard straight grinding wheels. The main difference is that the axes of the cutter and wheel fall on the same centerline. The clearance angle for grinding the cutter teeth is set by adjusting the tooth rest.

The setups for grinding the primary relief and/or secondary clearance angles with a flaring-cup wheel are shown in Figure 30-16. The positioning of the tooth rest below (Figure 30-16A) and above (Figure 30-16B) center at the required angle depends on whether the tooth rest is mounted on the table or the wheel head.

There usually is a greater area of contact when grinding with cup wheels. Therefore, the cuts are lighter than cuts taken with a straight type 1 wheel.

GRINDING CONSIDERATIONS AND CLEARANCE ANGLE TABLES

The clearance angle produced by a straight grinding wheel depends on the diameter of the *wheel*.

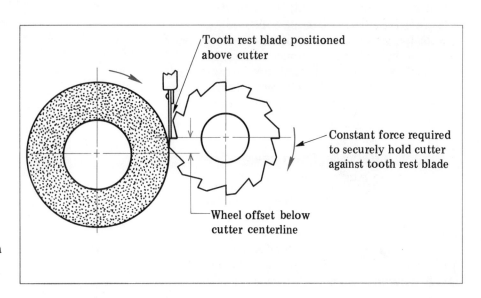

FIGURE 30-15 Setup for Cutter Grinding with Tooth Rest Blade Positioned Above Cutter

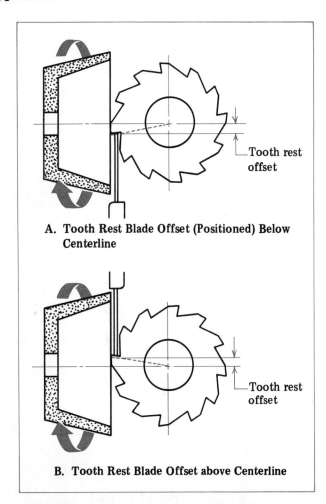

A. Tooth Rest Blade Offset (Positioned) Below Centerline

B. Tooth Rest Blade Offset above Centerline

FIGURE 30-16 Setups for Grinding Primary Relief and/or Secondary Clearance Angles with Flaring-Cup Wheel

When a cup wheel is used, the diameter of the *cutter* determines the cutting angle to be ground. The distance the tooth rest is set above or below center determines the clearance angle.

Clearance is not a problem for the sharpening of formed cutters. The clearance remains the same because the teeth are ground radially on the face.

Primary Relief and Secondary Clearance Angle Considerations

The area immediately behind the cutting edge of a cutter is relieved for the *land distance*. If the land width is too wide, cutter teeth tend to drag over the work. Considerable heat is generated and the cutting action is not efficient. Too much clearance behind the cutting edge may produce chatter and increase wear on the cutter. Land widths vary from 1/64" (0.4mm) or smaller for small, hand cutting tools to 1/16" (1.5mm) and wider for large-sized milling cutters. The land on face mills is called the *face edge*.

Summary tables provide a guide for grinding clearance angles. Table 30-1 gives recommended clearance angles for high-speed steel and cemented carbide cutters. Note that different cutting angles depend on the type of cutter and whether the cutting edge is on the periphery, corner, face, or a combination.

Establishing the Offset

Grinding wheel sizes generally range from 3-1/2" (88mm) to 6" (152mm). The offset for grinding the primary relief with straight-type wheels depends on the wheel diameter. The amount of offset is established from tables. Table 30-2 lists the offset for sharpening milling cutters with straight (type 1) wheels. The amount of offset shown is for standard clearance angles from 3° to 7° for wheel sizes ranging from 3" (76.2mm) to 7" (177.8mm).

Table 30-3 gives the offset for sharpening milling cutters with flaring-cup wheels. Here, the cutter diameter influences the offset. While the table covers cutter diameters to 3-1/2" (88.9mm), the values given in Table 30-2 for larger straight-wheel diameters may be substituted.

Amount to Raise or Lower the Wheel Head

When the wheel head is raised or lowered (depending on the grinding method) to produce the required primary relief angle, the amount may be calculated by formula. The cutter tooth level must be on center. The wheel head offset measurements originate from the same spindle and work center axes. The formula is as follows:

$$\text{wheel head offset} = \text{sine of required primary relief angle} \times \text{grinding wheel radius}$$

Example: Calculate the wheel head offset required to grind a 6° primary relief angle on a cutter. A 6" (152mm) diameter grinding wheel is to be used.

$$\begin{aligned}\text{wheel head offset} &= \text{sine of } 6° \times 3'' \\ &= 0.10453 \times 3 \\ &= 0.314'' \ (7.98\text{mm})\end{aligned}$$

CUTTER AND TOOL GRINDING

TABLE 30-1 Recommended Clearance Angles for High-Speed Steel and Carbide Milling Cutters

Recommended Relief and Clearance Angles for High-Speed Steel Milling Cutters

Material to Be Machined	Primary Relief Angle	Secondary Clearance Angle
Gray cast iron Malleable cast iron	4° to 7°	9° to 12°
Plain carbon steel Cast steel Tool steel High-speed steel Alloy steel	3° to 5°	8° to 10°
Stainless steel	6° to 10°	11° to 15°
Medium and hard bronze	4° to 7°	9° to 12°
Brass and other copper alloys	5° to 8°	10° to 13°
Nickel alloys	6° to 10°	11° to 15°
Magnesium alloys	10° to 12°	15° to 17°
Aluminum alloys	10° to 12°	15° to 17°

Recommended Primary Clearance Angles for Carbide Milling Cutters

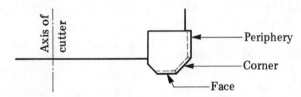

Type of Milling Cutter	Periphery Angle			Corner Angle			Face Angle		
	Steel	Cast Iron	Aluminum	Steel	Cast Iron	Aluminum	Steel	Cast Iron	Aluminum
				Primary Clearance Angle					
Face or side	4° to 5°	7°	10°	4° to 5°	7°	10°	3° to 4°	5°	10°
Saw and slotting	5° to 6°	7°	10°	5° to 6°	7°	10°	3°	5°	10°

TABLE 30-2 Offset for Sharpening Milling Cutters with Straight (Type 1) Wheels

Grinding Wheel Diameter		Cutter Clearance Angle								
		3°		4°		5°		6°		7°	
		Offset Required to Grind Cutter Teeth to Clearance Angle									
(Inches)	(mm)	(0.001")	(mm)	(0.001")	(mm)	(0.001")	(mm)	(0.001")	(mm)	(0.001")	(mm)
3	76.2	0.079	2.01	0.105	2.67	0.131	3.33	0.158	4.01	0.184	4.67
3-1/2	88.9	0.092	2.34	0.122	3.10	0.153	3.89	0.184	4.67	0.215	5.46
4	101.6	0.105	2.67	0.140	3.56	0.175	4.45	0.210	5.33	0.245	6.22
4-1/2	114.3	0.118	3.00	0.157	3.99	0.197	5.00	0.236	5.99	0.276	7.01
5	127.0	0.131	3.33	0.175	4.45	0.219	5.56	0.263	6.68	0.307	7.80
5-1/2	139.7	0.144	3.66	0.192	4.88	0.241	6.12	0.289	7.34	0.338	8.59
6	152.4	0.157	3.99	0.210	5.33	0.262	6.65	0.315	8.00	0.368	9.35
6-1/2	165.1	0.170	4.32	0.227	5.77	0.284	7.21	0.342	8.69	0.399	10.13
7	177.8	0.183	4.65	0.245	6.22	0.306	7.77	0.368	9.35	0.430	10.92

GRINDING MACHINES

TABLE 30-3 Offset for Sharpening Milling Cutters with Flaring-Cup Wheels

Cutter Diameter		Cutter Clearance Angle								
		3°		4°		5°		6°		7°
		Offset Required to Grind Cutter Teeth to Clearance Angle								
(Inches)	(mm)	(0.001″)	(mm)	(0.001″)	(mm)	(0.001″)	(mm)	(0.001″)	(mm)	
1/2	12.7	0.013	0.33	0.017	0.43	0.022	0.56	0.026	0.66	0.031 0.79
3/4	19.1	0.019	0.48	0.026	0.66	0.033	0.84	0.040	1.02	0.046 1.17
1	25.4	0.026	0.66	0.035	0.89	0.044	1.12	0.053	1.35	0.061 1.55
1-1/2	38.1	0.039	0.99	0.053	1.35	0.066	1.68	0.079	2.01	0.092 2.34
2	50.8	0.052	1.32	0.070	1.78	0.087	2.21	0.105	2.67	0.123 3.12
2-1/2	63.5	0.065	1.65	0.087	2.21	0.109	2.77	0.134	3.40	0.153 3.89
3	76.2	0.079	2.01	0.105	2.67	0.131	3.33	0.158	4.01	0.184 4.67
3-1/2	88.9	0.092	2.34	0.122	3.10	0.153	3.89	0.184	4.67	0.215 5.46

Amount to Raise or Lower the Tooth Rest

The distance the tooth rest is lowered or raised to position the edge of the cutter in relation to the centerline of a flaring-cup wheel used to grind the required primary relief angle may be calculated by formula. The amount the tooth rest is lowered or raised is equal to the sine of the required angle multiplied by the cutter radius:

$$\text{tooth rest offset} = \begin{matrix}\text{sine of required}\\ \text{primary relief}\\ \text{angle}\end{matrix} \times \begin{matrix}\text{cutter}\\ \text{radius}\end{matrix}$$

Example: Calculate the (tooth rest) offset required to grind a primary relief angle of 6° on a 6″ (152mm) diameter cutter.

$$\begin{aligned}\text{tooth rest offset} &= \text{sine of } 6° \times 3″\\ &= 0.10453\\ &= 0.314″ \, (7.98\text{mm})\end{aligned}$$

METHODS OF CHECKING ACCURACY OF CUTTER CLEARANCE ANGLES

Dial Indicator "Drop" Method

Cutter clearance angles may be checked by using the *dial indicator "drop" method*. The term *drop* is derived from the fact that when a dial indicator is used to measure the ground clearance angle, the pointer end "drops" (moves downward) as the cutter is revolved. The pointer measures the movement from the front to the back of the land. Tables are available that give the dial indicator movement for different cutter diameters, land widths, and relief angles.

Figure 30-17 illustrates how the dial indicator "drop" method is used to check a primary relief angle. As a general rule of thumb, for each degree of relief on a 1/16″ (1.5mm) land, there is a 0.001″ (0.025mm) movement (drop) on a dial indicator. For example, if the land width of a cutter is 1/16″ and a 6° primary relief angle is required, the "drop" as registered on the dial indicator is 0.006″ (0.15mm). When this method is used, the cutter diameter does not affect the measurement.

Cutter Clearance Gage Method

Cutter clearance angles may also be checked by using the *cutter clearance gage method*. There are two common designs of cutter clearance angle gages. One design consists of two hardened steel

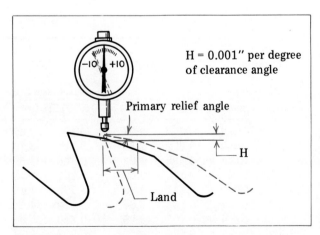

FIGURE 30-17 Checking Primary Relief Angle by Dial Indicator "Drop" Method

CUTTER AND TOOL GRINDING 607

arms that are at right angles to each other. The arms are placed on top of two teeth on the cutter. A hardened sliding center blade, ground on the end with the required angle, is brought into contact with the face of a ground tooth. The tooth is ground to the required angle when the ground angle and the blade angle coincide. A number of blades are provided with this type of cutter clearance angle gage. The gage itself is adaptable to measuring clearance angles on cutters with eight or more teeth within a diameter range of 1/2" to 8" (12.7mm to 203mm).

Figure 30-18 illustrates how the second design of cutter clearance angle gage is applied. The general features of this hardened tool steel gage consists of a graduated frame with a fixed foot and beam (Figure 30-18A). The graduated scale ranges from 0° to 30°. An adjustable foot slides on the beam. A movable blade is adjustable vertically and angularly to contact the ground clearance angle of a cutter tooth.

The gage is set by positioning the feet (fixed and adjustable feet) on two alternate teeth of the cutter and at a right angle to the tooth face (Figure 30-18B). The blade is set to the clearance angle of the cutter. The reading is taken directly from the graduated scale on the frame. The gage may be used to read angles from 0° to 30° on straight, helical, side, or inserted-tooth milling cutters; straight or helical end mills; saws and slitting cutters; and T-slot cutters. The gage is practical for measuring or checking cutter diameters from 2" (50mm) to 30" (762mm). Angles may be measured on end mills (if the teeth are evenly spaced) in diameters that range from 1/2" (12.7mm) to 2" (50mm).

PREPARATION OF THE CUTTER AND TOOL GRINDER

One of the most common cutter grinding operations requires the sharpening of plain milling cutters. Procedures are examined in this unit for setting up to grind such cutters. The flaring-cup wheel is used for the flat grinding method; the straight wheel, for the hollow grinding method. (The next unit covers a wide range of typical custom shop machine setups and grinding procedures for sharpening other circular and single-point cutting tools.)

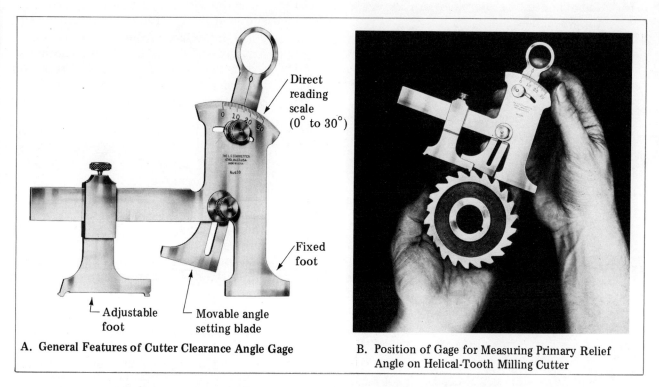

A. General Features of Cutter Clearance Angle Gage

B. Position of Gage for Measuring Primary Relief Angle on Helical-Tooth Milling Cutter

FIGURE 30-18 Checking Primary Relief Angle by Cutter Clearance Gage Method

HOW TO SET UP A CUTTER AND TOOL GRINDER FOR GRINDING PLAIN MILLING CUTTERS

Using a Straight Wheel
(Hollow Grinding Method)

STEP 1 Select the largest possible straight wheel.

Note: The larger the diameter, the smaller the radius (hollow) ground behind the cutting edge.

STEP 2 Run the standard grinding wheel checks. Mount the wheel. Change the spindle drive belts (if required) to bring the wheel within the speed range. True and dress the wheel.

Note: The wheel diameter must be checked with the spindle RPM. The fpm or m/min must be within the maximum recommended wheel speed.

STEP 3 Center the tailstock (or a work head and tailstock) on the table. Secure them at positions that will accommodate the mandrel or arbor assembly.

STEP 4 Use a test bar between centers to check the alignment of the table with respect to the wheel head.

STEP 5 Center the wheel head at the height of the tailstock (work head) center. Use a centering gage.

STEP 6 Set the wheel head at the zero graduation.

STEP 7 Position the height of the tooth rest blade edge on center by using the same centering gage.

Note: The tooth rest attachment is secured to the table for straight-tooth cutters. The tooth rest is attached to the wheel head post for grinding spiral-tooth (helical) plain milling cutters.

Note: The angle of the tooth rest blade must be the same as the helix angle of the spiral-tooth cutter. The angular face of the cutter rests on the angle slope of the tooth rest blade.

STEP 8 Set the wheel guard in position.

STEP 9 Set the vertical wheel head control handwheel at zero.

STEP 10 Calculate the wheel head offset to produce the required primary relief angle.

STEP 11 Offset the wheel head.

Note: The wheel head is lowered to the offset measurement if the tooth rest is attached to the wheel head. The wheel head is raised if the tooth rest is mounted on the table.

STEP 12 Set the arbor and cutter between centers. Position the tooth rest blade on the cutter face just behind the cutting edge.

Note: The cross slide handwheel must be moved carefully to advance the cutter toward the grinding wheel. There must be adequate clearance for the work to traverse past the grinding wheel.

STEP 13 Start the spindle. Allow for a short start-up time at spindle speed.

STEP 14 Move the workpiece for the start of the first cut on the first tooth.

STEP 15 Traverse the cutter past the straight grinding wheel.

Note: A continuous force must be applied to keep the cutter face against the tooth rest blade as the cutter is guided past the grinding wheel.

STEP 16 Clear the wheel from the workpiece at the end of the cut. Rotate the cutter 180°. Position the face of the tooth opposite the first ground tooth against the tooth rest.

STEP 17 Grind this tooth at the same cross feed handwheel setting as the first tooth.

STEP 18 Stop the spindle. Clean the workpiece and wheel. Measure the outside diameter at both ends of the cutter. Any variation in the micrometer readings indicates the amount of taper. The table is adjusted (aligned) for any taper.

STEP 19 Continue to grind each tooth in succession to the same depth.

STEP 20 Make a final sparking-out pass without infeeding.

STEP 21 Measure the primary relief angle by using either the dial indicator drop method or a cutter clearance angle gage.

Note: The angle is usually measured after the first cut is taken across three teeth. Any variation between the ground angle and the required clearance angle is corrected by changing the grinding wheel and workpiece axes.

CUTTER AND TOOL GRINDING

Note: The cutting teeth are stoned to remove the burrs produced by grinding.

STEP 22 Determine the required offset to grind the secondary clearance angle.

Note: If the width of the land is greater than is recommended, it is necessary to grind the secondary clearance until the required width is reached. The same steps are followed to reposition the grinding wheel in relation to the axis of the workpiece.

STEP 23 Reset the grinding wheel offset for the secondary clearance angle.

STEP 24 Proceed to grind the secondary clearance on all teeth.

STEP 25 Stop the machine. Remove the cutter and arbor or mandrel assembly. Carefully stone the fine, sharp, feathered ground edges with a hand abrasive stick.

Caution: Protect the hands and fingers from rubbing against the razor-sharp cutter edges.

STEP 26 Disassemble the machine setup. Clean the machine and leave it in a safe operating condition.

Using a Flaring-Cup Wheel (Flat Grinding Method)

STEP 1 Select an appropriate flaring-cup wheel.

STEP 2 Check the wheel for proper fitting. Mount on the wheel spindle. Lock the guard in place. True and dress.

Note: Many operators dress the wheel grinding face to a narrow land by using a dressing stick.

STEP 3 Set the spindle axis at center height by using a center gage.

STEP 4 Select the tooth rest blade form that is suitable for the cutter (straight or helical teeth). Set the tooth rest attachment so that the blade is at the right height.

STEP 5 Set the graduated ring on the vertical control handwheel at zero.

STEP 6 Calculate or check tables for the offset required to produce the primary relief clearance angle.

Note: It may be necessary to use two tables. The first table provides recommended cutting angles based on cutter, workpiece, and machining processes. The second table furnishes operational data on the amount of offset depending on the diameter of cutter and different primary relief clearance angles.

STEP 7 Position and set the tooth rest blade at the offset distance.

Note: If the tooth rest is attached to the table or is on the wheel head just below the wheel, the blade is positioned below the centerline. The blade is positioned the offset distance above the centerline when the tooth rest is mounted so that the cutting takes place starting at the face and continuing back to the secondary clearance angle.

STEP 8 Proceed to take a first cut on two opposite teeth. Measure the outside diameter for parallelism.

STEP 9 Proceed to grind successive teeth by taking one or more cuts.

Note: The primary relief angle is measured after a few teeth are ground. Then, offset adjustments are made to grind the correct angle.

STEP 10 Finish grind by sparking out the last cut.

Note: If, as a result of grinding, the land width becomes too wide, the offset for the secondary clearance is reset. The secondary teeth are ground until the required land width is reached.

STEP 11 Stop the machine. Remove the cutter and arbor assembly. Stone the grinding burrs off the cutting edges.

STEP 12 Return all tools and items to proper storage areas. Clean the machine. Leave it in a safe operating condition.

SAFE PRACTICES FOR SETTING UP AND OPERATING A CUTTER AND TOOL GRINDER

—Make standard wheel checks for soundness, balance, truing, and dressing.
—Test the wheel fit on the spindle, correct mounting, and condition of the wheel for grinding.

—Check the grinder stops and tripping devices.
—See that the blade and tooth rests are securely attached to the table or wheel head column.
—Check the centerline or offset position (depending on the type of wheel used and the grinding process) of the tooth rest blade in relation to the angle to be ground.
—Examine the work-holding device and the setup. Adequate, safe working space must be provided. The operator must be sure the cutter is held securely against the work rest during grinding.
—Check the spindle guard to ensure that there is adequate wheel protection and that the grinding operation may be performed without interference.
—Test the condition of the tailstock center and the tension (force) of the adjustable center against the arbor. The force exerted must be sufficient to hold the work securely.
—Pull the actuating plunger once a day to flow the lubricant throughout the one-shot system.
—Check the diameter of the grinding wheel and the spindle speed before starting the spindle. The fpm or m/min must be within the maximum recommended wheel speed. In the case of small-diameter wheels, it is equally important to bring the fpm up to maximum wheel speed in order to grind efficiently.
—Regulate the amount of in-feed per cut. Avoid undue force on the grinding wheel, particularly when flaring-cup, saucer, and relieved types of wheels are used.
—Check the quality of surface finish. A fine surface texture on a sharp cutting edge produces a high-quality, efficiently machined surface. Use a coarse-grained wheel if there is a considerable amount of material to be reground.
—Feed the cutter carefully so that the wheel is not brought into sharp contact with the workpiece.
—Protect machine surfaces when dressing the wheel.
—Leave the cutter and tool grinder clean and in safe operable condition for the next operator.
—Use safety goggles and/or a protective shield at all times.

TERMS USED WITH CUTTER AND TOOL GRINDER MACHINES AND SETUPS

Cutter and tool grinder	An abrasive machine primarily designed to generate primary relief and secondary clearance angles on hand and machine single- and multiple-point cutting tools. A grinder that may be adapted to internal and external cylindrical grinding and special tool grinding by adding special accessories.
Tange bar	A taper setting device for positioning the swivel table of a cutter and tool grinder to a precise angle. A standard setting block and a gage block used to obtain a precise measurement for setting the swivel table.
Work head (cutter and tool grinder)	A work head with graduated base and upright arm for housing the spindle. A work-holding device that may be swiveled through 360° in horizontal and vertical planes. (One end of the spindle accommodates tools and collet sleeves having Morse or Brown & Sharpe tapers. The other end is designed for chucks and plates having a standard National Series taper.)
Eccentric column	A design feature that adds to the cross feed capacity of a wheel head. An eccentric column designed as part of the wheel head in order to provide for a greater range of applications.
Cutter grinder attachments	Mechanisms that when added to the basic machine, increase the capacity to perform additional processes. (Examples include micrometer table positioning, sine bar, and heavy-duty tailstock.)

CUTTER AND TOOL GRINDING 611

Centering gage	A device having a base and upright mounting for a pointer (center). A gage used for positioning the wheel head and tool rest blade at center height.
Tooth rest	An adjustable holder for a tooth rest blade against which cutter teeth are positioned and held securely during grinding.
Tooth rest blade forms	Plain, rounded, offset, hooked, and inverted V-tooth formed blades. A variation of blade shapes to accommodate different cutter teeth forms.
Offset (wheel head or tooth rest)	The distance a wheel head or tooth rest blade is moved in relation to the axis of the cutter or the grinding wheel. The distance between axes needed to grind a primary relief or secondary clearance angle.
Cutter clearance angle measurement	Measuring the angle at which tool or cutter relief or clearance angles are ground. The use of a dial indicator or a cutter clearance angle gage.
Hollow ground angle	The radial contour generated by grinding the primary relief or secondary clearance angle on a cutter tooth with a straight-type grinding wheel.
Flat grinding method	Use of a flaring-cup wheel and offsetting the cutter to grind a flat primary or secondary angle surface.

SUMMARY

The capacity of the cutter and tool grinder is increased by using attachments to perform internal and external cylindrical grinding, surface grinding, and other special grinding processes.

> The swivel table with fine adjustment settings provides for accurately grinding straight and taper surfaces.

The work head spindle design permits direct mounting on one end of parts with Morse or Brown & Sharpe tapers. Accessories and other parts with large standard tapers in the National Series are mounted on the opposite end.

> Motorized attachments for revolving workpieces (used in cylindrical grinding) are designed to fit the regular work head.

The retractable tailstock center may be used with a right or left tailstock.

> Toolroom model grinders are designed so that the table saddle may be swung through 180°. An eccentric column increases the traverse (cross slide) range of the table.

The range of high and low spindle speeds permits the largest and smallest diameters of wheels to operate within safe wheel speed limits.

> The gear cutter sharpening attachment permits grinding tooth faces on form cutters.

A radius grinding attachment is used to accurately blend a 90° radius edge with the periphery and face of a cutter.

> The use of an indexing attachment eliminates the need for a tooth rest.

A lead screw feature for table movement may be added by a micrometer table positioning attachment.

The tapered sleeve, draw-in bar, and collet combination are used in the work head. This setup is useful for holding straight-shank cutters. Customary inch and metric sizes of collets are standard.

The wheel head (spindle) and tooth rest blade heights are set on center with the aid of a centering gage.

Cutter arbors and mandrels are designed for a slide fit for accurately positioning the cutter concentric with its axis.

Tooth rests may be stationary or adjustable. They are fitted for table, workhead, or wheel head mounting.

Five basic forms of blades accommodate the standard forms of cutters and tools used in the jobbing shop. The blade forms are: plain, rounded, offset, hooked, and inverted V-tooth.

Four principal design features of multiple-tooth cutters to which the grinder operator relates include primary relief (clearance) angle, secondary clearance angle, cutting edge, and land.

The direction of the cutting action determines the placement of the tooth rest. Sharpening a cutter tooth starting at the heel of the land and continuing to the cutting edge produces a burred cutting edge. The heat generated flows toward the limited area around the cutting edge. Excessive heat may cause the temper of the cutting edge to be drawn.

Grinding from the cutting edge back toward the heel of the land produces a burr-free cutting edge. Grinding heat is distributed away from the cutting edge.

Straight grinding wheels produce an angle surface ground to a slight radius (hollow ground). The largest-diameter wheel is used to reduce the curvature.

Flaring-cup and dish wheels are used to produce a flat surface. The grinding area on a flaring-cup wheel is reduced with an abrasive stick.

Clearance angle tables provide manufacturers' recommendations of primary relief and secondary clearance angles according to cutter material, size, and work processes.

The amount of offset of the wheel head (spindle) or cutter may be computed as the product of the sine of the primary relief angle and the radius of the grinding wheel or the cutter.

Cutter angles may be measured by the dial indicator drop method. A more functional shop practice is to use a cutter clearance angle gage.

One of the most widely used cutter grinding processes is the sharpening of straight or helical-tooth plain milling cutters.

The amount of wheel offset determines the angle at which the primary or secondary clearance angle is ground in a cutter.

Flaring-cup wheels are used for flat grinding relief and clearance angles. The distance the tooth rest blade (and position of the tooth) is offset from the center determines the angle to which cutter teeth are ground.

Light first cuts are taken on opposite teeth. The cutter is then measured for diametral accuracy. Any taper is compensated for by adjusting the swivel table.

The final finish cut is sparked out.

Standard safety checks are made for grinding wheels and cutting speed limits, machine parts and controls, accessories, and work-holding setups.

Particular attention is paid to the need to apply continuous force to hold the cutter securely against the tooth rest blade during cutting.

Personal safety requires the use of safety glasses or other eye protection shields. The operator is to stand clear of the revolving grinding wheel at all times.

UNIT 30 REVIEW AND SELF-TEST

1. a. State one function of each of the following cutter grinder parts: (1) work head spindle, (2) fine adjustment taper setting device, and (3) angle graduations on the vertical wheel head support.
   b. Examine a cutter and tool grinder manufacturer's specifications.
   c. Give the sizes or other data about the capacity of the three cutter grinder parts mentioned in part (a).

2. a. Identify the major parts of a small end mill grinding attachment.
   b. Describe briefly how the attachment works.

3. State three functions of tooth rests.

4. Differentiate between the functions of a primary relief (clearance) angle and a secondary clearance angle.

5. Make a simple freehand sketch of a setup to flat grind a primary relief angle on a cutter. A flaring-cup wheel is to be used. The cutting action is from the face of the cutting edge back across the land.

6. Calculate the tooth rest offset required to grind (a) a 6° primary relief angle and (b) an 11° secondary clearance angle on a 100mm diameter helical-tooth milling cutter.

7. List the steps for measuring a ground secondary clearance angle on a helical-tooth milling cutter by using a cutter clearance angle gage.

8. a. Select a straight face milling cutter from the toolroom.
   b. Use a cutter clearance angle table. Establish the smallest primary relief angle required for the machining of annealed gray cast iron workpieces.
   c. Calculate the offset between the centers of the cutter and a straight (type 1) grinding wheel.
   d. Set up the cutter grinder to use a type 1 wheel to hollow grind the primary angle. The grinding wheel is to rotate in a direction toward the cutting edge of the cutter teeth.

9. Give two reasons why it is important to maintain a constant force to hold the cutter securely against a tooth rest blade when grinding.

UNIT 31

Grinding Milling Cutters and Other Cutting Tools

Cutter grinding and tool sharpening as carried on within jobbing shops and toolrooms require setups and dimensional requirements that vary from one cutter to another. These variations are produced by factors such as: type of cutter, shape of tooth face, cutting angles for particular machining processes, cutter size, and design of the part. There are, however, fundamentals that apply in general.

The types of cutters may be grouped to include the grinding of peripheral, side, face, and end cutting teeth. Therefore, the underlying technology and processes described in this unit relate to sharpening side milling cutters, face mills, end mills of shell or solid-shank form, angle cutters, and form relieved milling cutters—cutters the machine operator usually handles. In addition, there are conventional machine reamers, taps, and other fluted cutters. Tool grinding also includes the sharpening of flat forms of cutters used for lathe, planer, and shaper work.

Cutter grinding operations deal with tooth cutting angles; primary relief, secondary clearance, and radial rake angles; axial movements; and straight or helical-angle teeth. The setups in this unit deal with hollow and/or flat ground teeth.

CHARACTERISTICS OF SLITTING SAWS

Slitting saws are used for medium-width slotting and cutoff operations. The saws may be plain or designed with side chip clearance for deeper slotting and sawing applications. An example of a metal-slitting saw with staggered teeth and side chip clearance is shown in Figure 31-1. This cutter has alternate right- and left-hand, side-cutting teeth. These teeth provide chatter-free cutting action and minimize chip removal problems. The narrow lands of the alternate side teeth serve to maintain the cutter width when slotting and to reduce rubbing between the workpiece and the cutter.

As with all other cutters, the tooth form, numbers of teeth, cutting edge angle, and primary

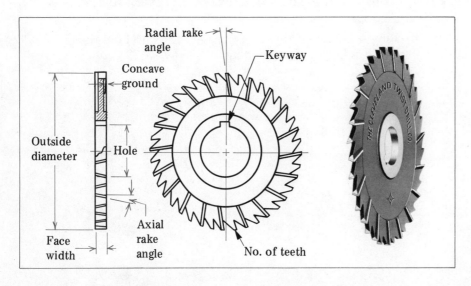

FIGURE 31-1 Metal-Slitting Saw with Staggered Teeth and Side Chip Clearance

relief and secondary clearance angles vary according to the job requirements. Slitting saws are made of high-speed steel or have teeth that are carbide tipped. The basic procedures for setting up and using a straight (type 1) grinding wheel to grind a plain milling cutter are used for sharpening slitting saws.

HOW TO SHARPEN PERIPHERAL TEETH ON A SLITTING SAW

STEP 1 Set the edge of the tooth rest blade and the wheelhead at center height with the centering gage.

STEP 2 Position the micrometer collar on the wheel head at zero.

STEP 3 Calculate the wheel head offset required to produce the primary relief angle.

Note: The following formula is used:

$$\text{wheel head offset} = \text{sine of required relief or clearance angle} \times \text{grinding wheel radius}$$

STEP 4 Raise the wheel head the amount of offset.

Note: The offset may be read directly from the graduated wheel head handwheel.

STEP 5 Mount the arbor and cutter between centers.

STEP 6 Position the tooth rest on the table so that the blade supports the face of the tooth to be ground. Secure the tooth rest to the table.

Note: Hook or L-type rest blades are adapted to the grinding of slitting saws. A flicker-type tooth rest support permits each tooth to be ratcheted into position for grinding.

STEP 7 Start the spindle and run it at operating speed for a minute.

Caution: Stand clear and work from the side of the revolving grinding wheel.

STEP 8 True and dress the wheel face.

STEP 9 Bring the grinding wheel into position for the first cut.

Note: The cutter must be held securely against the tooth rest blade throughout the grinding of each tooth.

STEP 10 Grind each tooth by traversing the table and cutter past the grinding wheel. Use the back machine controls for table traverse.

STEP 11 Grind all teeth. Take successive cuts alternately on opposite teeth, if required.

STEP 12 Take the final cut without additional infeed to spark out the cut. All teeth must be the same distance from the center of the cutter.

Grinding the Secondary Clearance Angle

STEP 1 Calculate the height needed to produce the secondary clearance angle.

STEP 2 Raise the wheel head the additional amount.

Note: A smaller-diameter wheel may be needed to grind the secondary clearance angle clear of the primary relief angle.

Note: Many times, just the primary relief angle is ground.

DESIGN FEATURES OF STAGGERED-TOOTH MILLING CUTTERS

The functions of the alternate right- and left-hand helical teeth on staggered-tooth milling cutters and on staggered-tooth slitting saws are similar. The design features of peripheral and side teeth on a staggered-tooth side milling cutter (Figure 31-2) provide for smooth cutting action, minimize chip removal problems, help to maintain cutter width, and reduce rubbing. Each successive tooth is sharpened on a different helix. Staggered-tooth milling cutters are sharpened by using a type 1 straight wheel. The wheel face is trued and formed to a narrow width.

An inverted V-tooth rest blade, with a slightly smaller included angle, is used to support the cutter teeth. The blade requires accurate positioning at the center of the narrow, flat area on the grinding wheel. Unless the V is centered,

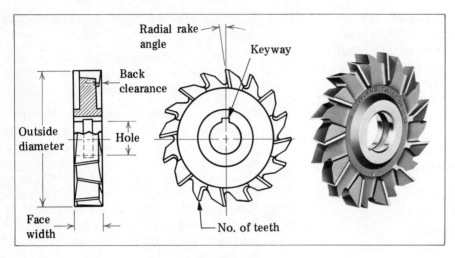

FIGURE 31-2 Design Features of Peripheral and Side Teeth on a Staggered-Tooth Side Milling Cutter

every other tooth will be ground slightly higher or lower than the adjacent tooth. The tooth rest support is mounted on the wheel head assembly. The relief angle and the secondary clearance angle are produced by lowering the wheel head.

Checking the Cutter Teeth for Concentricity

The teeth must be checked for concentricity on both helixes of the cutter. The position of the apex of the V-shaped tooth rest blade determines concentricity. Concentricity is checked with a tenth dial indicator (0.0001" or 0.0025mm). A tolerance of 0.0003" (0.0075mm) is accepted for general milling purposes. When the difference between two teeth exceeds this amount, it is necessary to adjust the location of the blade apex. The apex (blade) is moved in the direction of the helix with the higher tooth. The blade adjustment moves the tooth slightly higher. The result is that a slight amount of additional material is ground away.

Grinding Secondary Clearance

The secondary clearance may be ground by following steps similar to the steps used for grinding the primary relief angle. The wheel head is lowered to permit grinding at the required secondary clearance angle.

Another method is to use a tooth rest with a micrometer adjustment. The blade is lowered for the required offset. The table is swiveled for grinding the peripheral teeth at the right- and then the left-hand helix. Grinding the secondary clearance at the helix angle also produces a uniform land width for the primary relief angle (Figure 31-3).

Considering Side Tooth Land Width

Generally, the smaller the side tooth land width, the less heat is generated through contact. A slight (about 1/2°) back clearance (Figure 31-2) further reduces the contact area of the land. The design of land width, small back clearance angle, and minimal relief angle produces a high-quality surface finish on the sides of deeply milled slots. A flaring-cup wheel is used for grinding the side teeth. The wheel face is narrowed to a small, flat width.

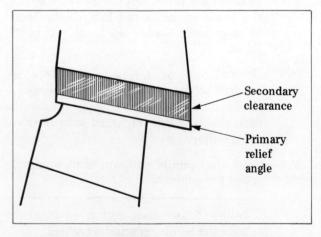

FIGURE 31-3 Appearance of Correctly Ground Secondary Clearance and Uniform Land Width for Primary Relief Angle

HOW TO SHARPEN A STAGGERED-TOOTH MILLING CUTTER

Grinding the Teeth Peripherally

STEP 1 Mount, true, and dress a straight (type 1) wheel to a narrow-width grinding face.

STEP 2 Use a tooth rest with an inverted V-tooth rest blade. Secure the unit to the wheel head.

STEP 3 Set the wheel head and the tooth rest blade edge at center height.

STEP 4 Position the handwheel graduated collar at zero.

STEP 5 Calculate the offset required to produce the primary relief angle.

Note: The following formula is used:

$$\text{wheel head offset} = \text{sine of primary relief angle} \times \text{cutter radius}$$

STEP 6 Lower the wheel head the required offset.

Note: The offset is read directly on the graduated collar.

STEP 7 Mount the cutter carefully so that each tooth is supported in correct relation to the narrow-width grinding area of the wheel.

STEP 8 Start the spindle and run it at operating speed for part of a minute.

STEP 9 Take a fine cut across two successive teeth.

STEP 10 Stop the spindle. Set up a dial indicator and check for grinding concentricity on two teeth of the right- and left-hand helix (Figure 31-4).

Note: If adjustment is needed, the V-blade apex is moved slightly in the direction of the helix with the high (ground surface) tooth.

STEP 11 Take one or more cuts on all teeth.

STEP 12 Finish grind with no additional in-feed of the cross slide on the last pass.

Sharpening the Side Teeth

STEP 1 Mount the milling cutter on a stub arbor. Secure the assembly in a universal work head.

STEP 2 Set one of the side teeth so that it is positioned in a horizontal plane. Lock the work head spindle.

STEP 3 Use a hook or L-shaped tooth rest blade. Mount the tooth rest with micrometer adjustment on the work head.

STEP 4 Set the work head at the required relief angle for the side teeth.

Note: The relief on the land of the side teeth of staggered-tooth milling cutters is usually minimal. On other cutters, a narrow uncleared land on the side teeth maintains the original cutter width and reduces rubbing.

STEP 5 Swivel the wheel head until the spindle axis is parallel with the work head axis. Center the flaring-cup wheel with the centering gage.

STEP 6 Move the cutter carefully up to the revolving grinding wheel for the first cut.

STEP 7 Hold the cutter securely against the tooth rest blade (Figure 31-5). Traverse the cutter across the wheel face. Continue to index and grind the primary relief angle on all teeth.

STEP 8 Make a final sparking-out pass without additional in-feed.

STEP 9 Clear the cutter and grinding wheel.

FIGURE 31-4 Checking Concentricity of Teeth with Right- and Left-Hand Helix on a Staggered-Tooth Milling Cutter

FIGURE 31-5 Counterclockwise Force Applied to Hold Cutter Tooth Securely against Tooth Rest Blade during Grinding

STEP 10 Swing the tool head to the setting for the secondary clearance angle.

STEP 11 Take one or more cuts on all side teeth until the required width of land is reached.

STEP 12 Remove the cutter from the stub arbor. Stone any burrs on the side teeth.

STEP 13 Measure the width of the cutter to determine how much the teeth on the second side are to be ground in order to keep both sides even.

STEP 14 Remount the cutter and secure the arbor.

STEP 15 Swivel the work head downward for the primary relief angle. Reposition the tooth rest so that the tooth face of the cutter is held against the blade.

STEP 16 Proceed to take a series of cuts until the required amount is ground from the face of the side teeth. Spark out the final cut.

STEP 17 Reposition the work head to the secondary clearance angle. Continue grinding each tooth. Take a series of cuts until the land is ground to the required width.

STEP 18 Remove the cutter. Stone burrs on the teeth edges.

STEP 19 Clean the machine and leave it in a safe operable condition.

DESIGN FEATURES OF SHELL MILLS

Shell mills (Figure 31-6) are designed for both face and end milling operations and slabbing or surfacing cuts. These cutters are available with square, chamfered, or round corners. Right-hand cutters with right-hand helix are standard. Left-hand cutters with left-hand helix are available on special order. Shell mills are held and driven by arbors and adapters. The cutter material is usually high-speed steel. Shell mills are also furnished with carbide teeth.

HOW TO GRIND A SHELL MILL

Grinding the Peripheral Teeth

STEP 1 Swivel the wheel head to 90°. Mount a flaring-cup wheel. True and dress the wheel.

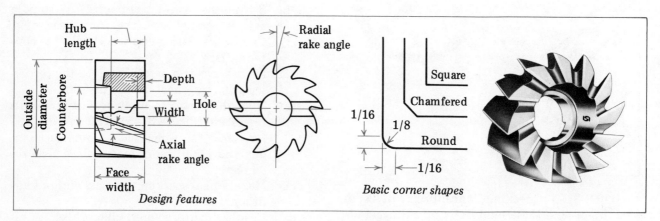

FIGURE 31-6 Shell Mill Designed for Face or End Milling Operations

Note: The sharpness and coarseness to which the grinding wheel is dressed for grinding the primary relief angle will, in part, determine the surface finish of the ground cutter teeth. The quality of the teeth, in turn, determines the surface finish produced on a milled workpiece.

STEP 2 Set the table at 0°. Mount the cutter on a mandrel between the tailstock centers.

STEP 3 Attach the tooth rest with an offset blade to the wheel head. Center the high point of the tooth rest blade with the center of the narrow width (grinding area) on the wheel.

STEP 4 Center the wheel head and the height of the tooth rest blade with the centering gage.

STEP 5 Lower the wheel head and tooth rest to the offset required to produce the primary relief angle.

STEP 6 Hold the cutter firmly against the tooth rest blade.

STEP 7 Bring the cutter and grinding wheel carefully into contact for the first cut.

Note: The depth of cut is checked at several places along the cutter tooth to ensure that the first cut is taken at a depth that will not increase due to any high spots or irregularities on the cutter teeth.

STEP 8 Traverse the table in the direction of the tooth angle.

Note: The first tooth is usually marked for identification with a marking pen.

STEP 9 Bring each successive tooth into position.

Note: Each tooth must be kept in constant contact with the tooth rest blade.

STEP 10 Spark out the final cut.

STEP 11 Reposition the wheel head and tooth rest. Lower them to the required offset to produce the secondary clearance angle.

STEP 12 Start the first cut. Then, continue to take successive cuts until the primary land width is ground to size.

STEP 13 Remove any grinding burrs from the cutting edge.

Caution: Grinding burrs are razor sharp. Use a cloth or other protective covering when handling the cutter.

Grinding the End Teeth

STEP 1 Mount the shell mill on a stub arbor. Draw and secure this assembly in the work head spindle.

STEP 2 Position a micrometer adjustable tooth rest and secure it to the machine table. The blade is located near the periphery of one of the side teeth.

The blade is positioned so that the first face tooth (and all other teeth) is held in a horizontal plane.

STEP 3 Lower the wheel head until the grinding wheel clears the tooth above the one being ground.

STEP 4 Swivel the work head to the required angle setting for the primary clearance angle.

The wheel head is lowered so that the grinding wheel cuts near the top area of the wheel. The primary relief angle provides clearance for the wheel to traverse across the tooth face.

STEP 5 Take a series of cuts until all teeth are ground to a sharp cutting edge and the land is ground to the correct width or wider.

Where specified, the quality of the ground cutting edge is checked against a comparator specimen or other surface texture check. For fine-quality (low microinch) cutter teeth textures, the grinding wheel is precision dressed for a fine finish. Finer finish cuts are taken across the cutting teeth on the periphery and face. After grinding, the cutting edges are stoned with a hand abrasive stick. The cutting edges are finally honed.

STEP 6 Reset the work head to the secondary clearance angle.

STEP 7 Proceed to take one or more cuts at the secondary clearance angle until the land is ground to width.

Grinding a 45° Beveled Corner

STEP 1 Swivel the work head to the 45° graduation on the base.

Note: The angle to which the work head is swiveled on its base depends on the bevel (chamfer) angle between the side and face of the cutter teeth.

GRINDING MACHINES

STEP 2 Set the work head at the 0° (horizontal plane) graduation.

STEP 3 Use the centering gage to bring the face tooth level horizontally and on center.

STEP 4 Rotate the cutter the number of degrees required for the clearance angle. Lock the work head spindle in position.

STEP 5 Swivel (tilt) the work head to the same angle.

Note: The two angle settings (work head and shell mill) are necessary. The clearance angle produced by grinding must be greater than the single angle setting in order to prevent heel drag.

STEP 6 Hold each tooth firmly against the tooth rest blade. Bring the cutter carefully into position for the first sharpening cut.

Note: The wheel head and the grinding wheel must be lowered to clear the tooth above the one being ground.

STEP 7 Continue to grind the beveled edge on each tooth.

STEP 8 Spark out the final cut. Shut down the machine. Remove the cutter from the arbor. Stone the edges to remove grinding burrs.

STEP 9 Clean the machine and accessories. Leave the machine in a safe operable condition.

CONSIDERATIONS FOR GRINDING CARBIDE-TOOTH SHELL MILLS

Similar setups and procedures are used for grinding primary relief and clearance angles on the side, bevel, and face on a carbide-tipped (inserts) shell mill. A diamond-grit cup wheel is generally used for sharpening purposes. Carbide cutters may be ground to positive or negative radial rake. A coolant is used to prevent overheating and checking at the cutting edge. The depth of each cut is usually restricted to 0.001" (0.025mm) or less.

It is important that the runout of the diamond wheel must be checked and corrected. The diamond wheel must cut as it rotates in the cutting edge. Indexing methods are used, if required, to control the equal spacing of the cutter teeth. Cutter teeth are usually honed with a 400-500 grit hand hone after they have been stoned with an abrasive stick.

HOW TO GRIND FACE TEETH AND CHAMFER EDGES ON A CARBIDE-INSERT SHELL MILL

STEP 1 Mount a diamond-grit cup wheel. Set the wheel head at 0°. Use a dial indicator and true up the face of the wheel.

STEP 2 Mount the cutter on an arbor. Secure the assembly in the work head spindle.

STEP 3 Proceed to position the tooth rest and blade and the tooth face in the same manner as for grinding a high-speed steel shell mill.

Considerations for Grinding the Teeth Peripherally

When grinding the teeth on the periphery of the cutter, a tooth rest blade finger that is at least 1-1/2 times longer than the width of the carbide tip is required. The tooth rest is located on the wheel head for grinding the primary angle on the periphery of the teeth. The blade (finger) extension permits the cutter to be held securely against the blade from the start to the finish of each cut.

The front face of each tooth is ground by mounting the tooth rest on the table. The radial angle of the tooth is set parallel with the table. The work head is repositioned to grind the front face of the teeth. The work head is tilted to the required clearance angle.

Carbide inserts, which require grinding at a specific bevel (chamfer) angle, are produced by using a setup similar to the one for high-speed steel cutters.

DESIGN FEATURES OF END MILLS

The grinding of peripheral and end teeth on an end mill is one of the most common cutter grinding processes. On small-size end mills, no outside diameter regrinds are considered. Design features provide for common two-, three-, four-, and six-flute end mills; single-end general-purpose and double-end end mills; regular, long, and extra-long end mills; standard and high-helix end mills; and straight and ball-end end mills. Heavy-duty

TABLE 31-1 Recommended Radial Relief Angles for High-Speed Steel End Mills

Material to Be End Milled	End Mill Diameter							
	1/8" (3.2mm)	1/4" (6.4mm)	3/8" (9.5mm)	1/2" (12.7mm)	3/4" (19.0mm)	1" (25.4mm)	1-1/2" (38.1mm)	2" (50.8mm)
	Radial Relief Angle* (in Degrees)							
Carbon steel	16	12	11	10	9	8	7	6
Tool steels	13	10	9	8	7	6	6	5
Nonferrous metals	19	15	13	13	12	10	8	7

*For secondary clearance angles, multiply the radial relief angle by 1.33.

premium cobalt high-speed steel end mills are used for machining typical exotic metals such as high-temperature alloys and high-tensile steels.

Radial Relief Angles

The general range of outside diameters of end mills is from 1/8" to 2" (3.2mm to 50.8mm). The primary relief angle range for these diameters is from 16° for a 1/8" (3.2mm) diameter high-speed steel end mill for machining carbon steels to 6° for a 2" (50.8mm) diameter mill. Table 31-1 provides recommended radial relief angles for high-speed steel end mills (mild carbon steels, tool steels, and nonferrous metals). The angles are for the conventional grinding of radial relief and accompanying secondary clearance angles.

Eccentric Relief and Method of Sharpening

Side cutting edges of end mill flutes are also designed with eccentric relief. Figure 31-7 shows the characteristics of eccentric relief on a two-flute end mill. The cutter relief starts at the cutting edge and continues in an arc until it joins the body of the end mill. Generally, there is no secondary clearance. End mills that are ground with eccentric relief are especially adapted to milling low tensile strength materials requiring high surface finishes and to numerically controlled machining where the end mill is not reused for a particular process once it wears.

The eccentric relief is generated by swiveling the wheel head at a specified number of degrees. The wheel head angles for eccentric relief grinding of end mill cutter diameters from 1/8" (3.18mm) to 1-1/4" (31.7mm) are given in Table 31-2. An *air-spindle grinding device* with a support finger is attached to the table. The air spindle provides both rotary and axial movement of the end mill. The face of the grinding wheel is trued parallel to the table. The face is then dressed to narrow it to the width across the end mill land (Figure 31-8).

The wheel head is rotated counterclockwise according to the diameter of the end mill and the cutter helix angle (20°, 30°, or 40°). The cutter tooth and wheel head are set at center height. The tooth rest is set to make contact at the right edge of the grinding wheel. Unlike conventional cutter grinding, the cut is started from the end of the cutter. The air spindle turns the end mill while advancing it axially along the helical flute. The cutting edge is generated by the smaller right side of the wheel (due to the

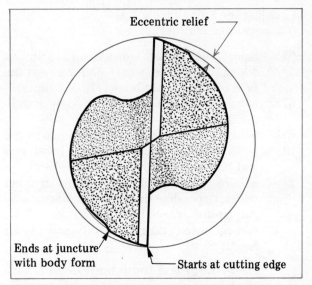

FIGURE 31-7 Characteristics of Eccentric Relief on a Two-Flute End Mill

TABLE 31-2 Radial Relief and Wheel Head Angles for Eccentric Relief Grinding of End Mills

Diameter of Cutter		Recommended Radial Relief Angle	Cutter Helix Angle		
			20°	30°	40°
(in.)	(mm)		Required Wheel Head Angle		
1/8	3.18	13°	4°30′	7°	11°
1/4	6.35	10°30′	3°45′	6°	10°
1/2	12.70	10°	3°15′	5°15′	9°15′
3/4	19.05	9°	3°	4°40′	8°30′
1-1/4	31.75	8°	2°30′	4°	7°30′

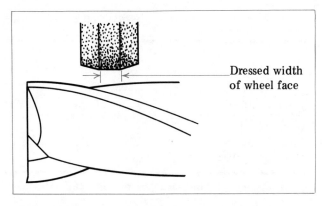

FIGURE 31-8 Grinding Wheel Face Dressed to Width Across End Mill Land

wheel head angle position). A greater amount of clearance is generated from the cutting edge to the heel of the land. The relatively larger left side of the grinding wheel produces the greater clearance as the cutter (supported on the tooth rest) axially traverses the grinding wheel during the rotation of the end mill. The quality of the eccentric relief grind depends on the correct dressing and angle setting of the grinding wheel, narrowing the cutting area of the wheel, the speed of radial and axial movement of the air spindle and mounted cutter, and the shape and correct positioning of the tooth rest.

HOW TO SHARPEN AN END MILL

Grinding the Side Cutting Edges (Conventional Method)

STEP 1 Select, mount, and true a straight grinding wheel. Dress the wheel and narrow the face to a width of about 1/16″ (1.5mm).

STEP 2 Mount the small end mill grinding attachment.

STEP 3 Select a narrow-width blade to match the helix of the end mill. Install the tooth rest and blade assembly on the wheel head.

STEP 4 Secure the end mill in a straight or taper collet, depending on the body shape.

STEP 5 Adjust the wheel head, tooth rest blade, and mounted end mill to the same center height.

Note: Each end tooth must be in a horizontal plane.

STEP 6 Set the micrometer collar on the wheel head column at zero.

STEP 7 Calculate the required offset for lowering the wheel head:

$$\text{wheel head drop} = \text{sine of primary relief angle} \times \text{cutter radius}$$

STEP 8 Lower the wheel head and tooth rest to grind the required primary relief angle.

STEP 9 Move the fixture away from the grinding wheel to clear the end mill.

STEP 10 Advance the cutter forward along the tooth rest blade until the cutting edge at the shank end is in position for the start of the cut.

Note: While an experienced operator may start the cut from the end of the shank area, it is safer (in terms of possible damage to the face end of the end mill if there is accidental contact) to start at the shank end.

STEP 11 Take a light first cut on each flute. Increase the feed increments until the primary relief area of the land is ground to correct any cutter wear or damage.

STEP 12 Spark out the final cut.

STEP 13 Determine the amount of wheel head offset required to grind the secondary clearance angle.

STEP 14 Lower the wheel head the additional amount.

STEP 15 Grind the secondary clearance until the land width is ground to the required size.

STEP 16 Remove the end mill. Stone and then hone the cutting lips carefully.

STEP 17 Clean the machine of abrasive and other foreign particles. Leave the machine in a safe operable condition.

Grinding the End Teeth
(Universal Work Head Method)

STEP 1 Select a small-diameter flaring-cup wheel. True the wheel and dress it to produce a narrow 1/8″ to 3/16″ (3mm to 5mm) grinding area (Figure 31-9). Stop the spindle.

STEP 2 Swivel the workhead counterclockwise past the 90° graduation on the base to 88°.

Note: An additional 2° to 3° is recommended so that the teeth are ground slightly lower at the center than at the outside edge.

STEP 3 Tilt the work head spindle to the number of degrees specified for the relief angle of the axial end teeth.

STEP 4 Attach a flicker-type tooth rest support with micrometer adjustment to the work head to provide ratchet indexing (Figure 31-10).

STEP 5 Insert the end mill in a collet and level the tooth. Adjust the tooth rest support in relation to the master index plate.

STEP 6 Traverse the first end tooth so that the grinding wheel completes the cut at the center of the end mill.

Note: The table stop is set to prevent feeding with possible damage to the opposite tooth face.

STEP 7 Return the grinding wheel to clear the cutter. In-feed about 0.003″ (0.08mm). Traverse the cutter across the wheel face.

STEP 8 Index (ratchet) the spindle to position the next tooth. Continue the grinding and indexing until all teeth have been ground to correct cutter wear or damage.

STEP 9 Spark out the final cut without additional in-feed.

STEP 10 Calculate the additional offset to which the work head is to be set to grind the secondary clearance angle.

STEP 11 Adjust the work head angle and wheel to permit grinding the secondary clearance.

STEP 12 Take successive cuts until the required land width is reached for the primary relief angle.

STEP 13 Stop the machine. Remove the end mill. Stone and hone the cutting edges of the end teeth.

STEP 14 Clean the machine. Return all items to proper storage.

DESIGN FEATURES OF FORM RELIEVED CUTTERS

A distinguishing design feature of form relieved cutters is that the profile of the teeth, starting at the radial face, remains the same. This means the form cutter is sharpened by grinding on the face. The reverse profile to the one that is milled in a workpiece is a permanent design feature of the

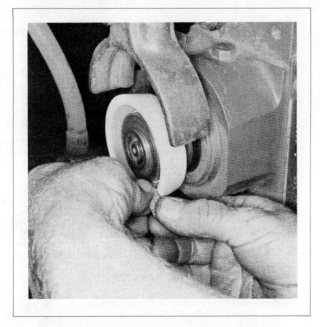

FIGURE 31-9 Dressing the Face of a Flaring-Cup Wheel to a Narrow Grinding Area

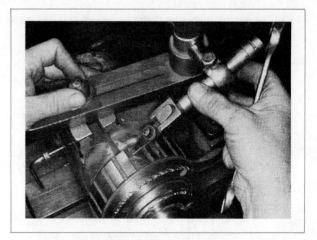

FIGURE 31-10 Use of Flicker-Type Tooth Rest with Micrometer Adjustment for Ratchet Indexing

624 GRINDING MACHINES

form cutter. The continuous profile is produced because the relief is constant and permanent.

Form cutters are usually marked with the radial rake. Some of the most frequently used form cutters that require grinding on the tooth face include gear tooth milling cutters, convex and concave cutters, and multiple-form gang milling cutters. Before the faces of the cutter teeth are ground, it is necessary to check the uniformity of each tooth from the cutting face to the back of the tooth form. If there is a variation in micrometer measurement, it is necessary to accurately grind a reference area on the back of each tooth for indexing. Dish (type 12) wheels are used to grind the radial face of a cutter tooth down to and including part of the tooth gullet.

The two general methods of grinding the face of the teeth include mounting the cutter between centers and positioning the cutter horizontally in a gear cutter attachment or form relieved cutter holder as illustrated in Figure 31-11. The method described in this unit is the method generally used in jobbing shops where the cutter is mounted between centers. A type 12 dish wheel is selected to permit grinding the radial face and part of the gullet and still clear the back side of the preceding tooth.

FIGURE 31-11 Form Relieved Cutter Holder and Setup for Grinding the Face of the Teeth

HOW TO GRIND THE TEETH ON A FORM RELIEVED CUTTER

STEP 1 Select, test, and mount a type 12 dish grinding wheel on the spindle. Replace the wheel guard.

STEP 2 Use an abrasive dressing stick to shape the curved side of the wheel to clear the back side of the cutter teeth.

STEP 3 Shape the peripheral face of the wheel to conform to the shape of the tooth gullet. Dress the side cutting face of the wheel.

STEP 4 Mount the form relieved cutter on an arbor between centers.

Note: The cutter is positioned to grind the reference area on the back of the teeth if there are variations of more than 0.002" (0.05mm) in the width of the teeth. A similar grinding wheel setup and indexing are used as for face grinding the teeth.

STEP 5 Position the radial face of one tooth face so that it is in a vertical plane (Figure 31-12).

Note: The tooth rest blade is adjusted to support the cutter in this position.

STEP 6 Position the dish wheel so that the face just clears the radial face of the cutter at the gullet depth.

FIGURE 31-12 Face of Form Tooth Positioned in Vertical (Radial) Plane

MILLING CUTTERS AND CUTTING TOOL GRINDING 625

STEP 7 Set the graduated micrometer collar on the cross slide handwheel at zero.

STEP 8 Clear the setup. In-feed the cutter by turning the micrometer screw on the tooth rest support.

STEP 9 Set the wheel head (and wheel) for the depth of the down-feed roughing cut (Figure 31-13).

STEP 10 Traverse the cutter for each increment of down-feed.

Note: It may be necessary to dress the grinding wheel a number of times if a considerable amount of the tooth face is to be ground. In such cases, several final cuts are taken at reduced in-feed to the full depth in order to uniformly grind the radial face of each tooth.

Note: The side face of the wheel is usually relieved to reduce the contact area.

STEP 11 Dress the wheel and relieve the side face in preparation for finish cuts.

STEP 12 Take additional finish cuts by making in-feed increments with the micrometer tooth rest adjustment.

Note: The initial tooth is marked with a color marker.

STEP 13 Spark out the final finish cut.

STEP 14 Stop the spindle. Use an indicator to check the concentricity of each tooth. A second check is also made to measure the quality of texture of the tooth face against a comparator specimen.

STEP 15 Disassemble the setup. Clean the cutter grinder, arbor, and other accessories. Return all items to proper storage areas.

HOW TO USE A GEAR CUTTER ATTACHMENT

While the gear cutter attachment setup is described here with respect to grinding gear cutters, similar procedures are followed for the sharpening of other form relieved cutters.

STEP 1 Set up and secure the attachment near the center of the machine table.

STEP 2 Select a type 12 dish wheel. Mount it so that the grinding wheel extends on the spindle far enough to permit grinding each tooth. Dress the wheel to relieve the side face and to form part of the gullet.

Note: On some models, it is necessary to use a spindle extension adapter.

STEP 3 Set the cutter on the cutter fixture (attachment) stud. Align the face of the cutter with the centering gage of the attachment.

STEP 4 Adjust the pawl of the attachment so that the finger rests against the back side of the tooth to be ground.

Note: A reference area may need to be ground on the back side of each tooth to obtain uniform setting if the teeth vary in depth.

STEP 5 Adjust the face of the wheel to almost touch the tooth face to be ground. Clear the cutter and wheel.

STEP 6 In-feed the cutter by moving it to the depth by the micrometer adjustment of the pawl.

STEP 7 Grind the first tooth. Clear the cutter and wheel. Turn the cutter on the stud and back the next tooth firmly against the pawl.

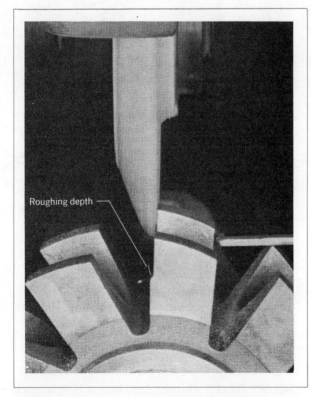

FIGURE 31-13 Wheel Head (and Wheel) Set for Depth of Down-Feed Roughing Cut

STEP 8 Continue to index and grind until all teeth are ground at the same setting. Take successive roughing cuts.

STEP 9 Dress the wheel and again relieve the side face in preparation for finish cuts. Spark out the final cut.

STEP 10 Check the quality of surface finish. If necessary, correct any grinding conditions that prevent meeting surface texture requirements.

STEP 11 Disassemble the setup. Return all items to proper storage areas. Clean the machine and leave it in a safe operable condition.

DESIGN FEATURES AND GRINDING OF SINGLE- AND DOUBLE-ANGLE MILLING CUTTERS

The two general forms of angle milling cutters are single angle and double angle. The design features of single- and double-angle milling cutters are shown in Figure 31-14. Single-angle milling cutters are commercially available with 45° and 60° included angles and are either right hand or left hand. The teeth of a single-angle cutter are on the angular face and on the vertical face (Figure 31-14A). The teeth on the vertical face are ground with a slight back clearance. Single-angle milling cutters are used primarily for milling dovetails and angular surfaces.

Single-angle cutters are ground with the work head set at 0° while the table is swiveled to the included angle. Angle cutters are ground with a flaring-cup wheel. The teeth are set parallel to the table. Usually, the cutter is mounted on an arbor and held in the work head. The wheel head and grinding wheel spindle are set at the back clearance angle for grinding the vertical face of a single-angle milling cutter. The work head is tilted to the primary relief or secondary clearance angle for the vertical face teeth.

Double-angle milling cutters (Figure 31-14B) are designed with included angles of 45°, 60°, and 90°. These cutters are used for milling bevel edges, angular notches, serrations, and so on.

The teeth on the angular surfaces of single- and double-angle cutters are ground by setting the work head at 0° and swiveling the table to each required angle setting. The blade of the tooth rest is aligned at center with the angular cutter. The work head is tilted to the required primary relief or secondary clearance angle when the cutter is mounted on the work head. On cutters that are mounted between centers, either the spindle is tilted or the cutter is offset to produce the relief and clearance angles.

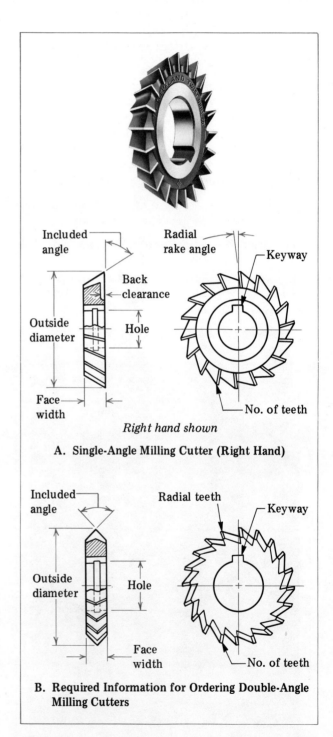

A. Single-Angle Milling Cutter (Right Hand)

B. Required Information for Ordering Double-Angle Milling Cutters

FIGURE 31-14 Design Features of Single- and Double-Angle Milling Cutters

DESIGN CONSIDERATIONS FOR GRINDING MACHINE REAMERS

The cutting on most machine reamers is done on the chamfer edge of each tooth. The small land width (margin) serves to guide the cutting edges to produce a concentric, parallel hole to a precise dimension. When the lands on solid reamers become worn, they are cylindrically ground to a smaller size. The lands are ground to a slight taper of 0.0002" (0.005mm) per inch to provide clearance (longitudinal relief), especially on deep-hole reaming. One caution must be observed in cylindrically grinding of the lands. The direction the reamer turns must be opposite to the direction of normal rotation. The back of the land margin contacts the grinding wheel first. A second caution requires that the design of the cutter grinder must permit the direction of spindle rotation to be reversed and operated safely without loosening the wheel setup.

Secondary clearance is primarily provided to reduce the margin area along the flutes of a reamer. The procedure is basically the same as for grinding secondary clearance on a plain milling cutter or end mill. Either a straight wheel or a flaring-cup wheel may be used. The amount of offset (using the type 1 straight wheel method) is equal to the sine of the secondary clearance angle multipled by the radius of the grinding wheel. In the case of solid-type machine reamers, there is no primary clearance angle to the ground narrow band margin.

HOW TO SHARPEN THE CHAMFER OF A MACHINE REAMER

Sharpening the Chamfer Edge of a Cutter Mounted between Centers

Tilting the Spindle

STEP 1 Swivel the wheel head eccentric to the back side of the cutter grinder (Figure 31-15) to produce maximum clearance.

STEP 2 Loosen the table clamping nuts. Swivel the table to a 45° (or other) reamer chamfer angle. Secure the table in position.

STEP 3 Locate the tailstock centers and mount the reamer between centers.

STEP 4 Mount a micrometer adjustable tooth rest on the table. Adjust the blade so that the cutter tooth rests on center (at center height).

STEP 5 Tilt the spindle assembly down to the primary relief angle for the chamfer on the reamer teeth.

STEP 6 Mount a type 12 dish wheel. Secure the wheel guard. True and dress the wheel to shape with an abrasive dressing stick.

STEP 7 Start up the wheel. Feed the cross slide (saddle) toward the wheel head by turning the back cross slide handwheel.

Note: The table stop is set as a safety precaution to prevent damage to the wheel or center.

STEP 8 Take the first cut. Coordinate the movements of the chamfer edge of each tooth traversing the cutting area of the wheel with the ratcheting (indexing) for each chamfer edge.

STEP 9 In-feed for each successive cut except for the final sparking-out passes.

STEP 10 Shut down the machine. Remove the reamer. Stone and hone the face of each chamfer cutting edge and longitudinally along the cutting face of each tooth.

STEP 11 Disassemble the setup. Replace all items. Leave the machine in a safe operable condition.

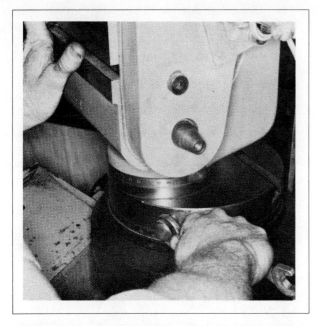

FIGURE 31-15 Wheel Head Eccentric Swiveled to Back Side of Cutter Grinder

628 GRINDING MACHINES

Sharpening the Chamfer Edge by Offsetting the Wheel Head

STEP 1 Set the wheel spindle parallel with the table ways. Position the wheel head at center height.

STEP 2 Select and mount a type 12 dish wheel. Secure the guard.

STEP 3 True and dress the wheel to a narrow margin width and slightly relieve the vertical face.

STEP 4 Swivel the table to the 45° (or other) chamfer angle. Secure at this angle position.

STEP 5 Locate the tailstock centers to accommodate the reamer length.

STEP 6 Mount a micrometer-type tooth rest support on the machine table.

STEP 7 Adjust the chamfer cutting edge with the tooth rest blade to center height.

STEP 8 Feed the chamfer of the reamer almost to the outer edge of the grinding wheel.

Note: The table stop is set to avoid possible damage to the cutter. The stop prevents additional movement toward the tailstock center.

STEP 9 Raise the wheel head according to the offset as calculated by the formula:

$$\text{wheel head rise} = \text{sine of clearance angle} \times \text{grinding wheel radius}$$

STEP 10 Start the wheel spindle. Feed the reamer to the grinding wheel by using the back cross slide handwheel.

STEP 11 Take roughing cuts until all teeth are ground sharp clear to the outside diameter of the machine reamer.

Note: The movement of the table is coordinated with the indexing of each individual chamfer.

STEP 12 Dress the wheel and take the first finish cut.

STEP 13 Spark out the final finish cut. Stop the machine. Remove the reamer.

STEP 14 Stone to remove grinding burrs and to produce finer cutting edges. Hone the chamfer edges and the faces of each land.

STEP 15 Check the quality of the surface finish against a comparator specimen, if required.

STEP 16 Clean the machine and accessories. Return all items to appropriate storage areas.

DESIGN FEATURES OF ADJUSTABLE HAND REAMERS

Adjustable hand reamers permit regrinding to accommodate reaming standard and special diameters of holes. With use, adjustable hand reamers require accurate grinding to produce concentric, accurate holes. To grind a machine reamer to ream accurately, the reamer blades are adjusted oversize. Enough stock is allowed to grind the blades to the correct diameter, starting taper, and correct margin (width) of the land. Table 31-3 gives the primary relief and secondary clearance angles for reamer diameters ranging from 1/2" (12.7mm) to 2" (50.8mm). It shows a range of margins from 0.005" (0.12mm) for reaming steel to a maximim width of 0.030" (0.75mm) for bronze and cast iron. It is important to grind the land width to the recommended margin. Otherwise, the primary relief may be insufficient to prevent land interference when reaming.

HOW TO SHARPEN AN ADJUSTABLE HAND REAMER

STEP 1 Mount a type 1 straight or type 12 dish wheel. Secure the guard.

STEP 2 Dress the outside diameter of either wheel to produce a narrow-width (about 1/16" or 1.5mm) cutting area.

TABLE 31-3 Recommended Primary Relief and Secondary Clearance Angles for Hand Reamers

Diameter of Reamer		Material to Be Reamed			
		Steel		Cast Iron and Bronze	
		Width of Margin			
		0.005" to 0.007" (0.12mm to 0.17mm)		0.025" to 0.030" (0.61mm to 0.75mm)	
		Grinding Angle			
(in.)	(mm)	Primary Relief	Secondary Clearance	Primary Relief	Secondary Clearance
1/2	12.7	3°	12°	7°30'	17°
1	25.4	1°30'	10°30'	4°30'	14°
1-1/2	38.1	1°	10°	3°30'	13°
2	50.8	45'	10°	3°15'	12°30'

MILLING CUTTERS AND CUTTING TOOL GRINDING 629

STEP 3 Attach a tooth rest with a narrow blade to the wheel head.

Note: The tooth rest is mounted to the wheel head instead of to the table because the reamer blades may not be perfectly straight. Any irregularities are compensated for as the blade follows a uniform path past the support blade.

Note: The tooth rest blade is positioned directly in line with the narrow grinding area of the wheel.

STEP 4 Set the wheel head and tooth rest blade at center height.

STEP 5 Prepare the adjustable blades on the reamer. The blades are checked for proper seating. The reamer is adjusted oversize from 0.008" to 0.010" (0.2mm to 0.25mm).

STEP 6 Refer to Table 31-3 for recommended primary relief and secondary clearance angles. Compute the distance the wheel head is to be lowered to generate each required angle.

STEP 7 Mount the reamer on centers. Hold the reamer blade securely against the tooth rest blade.

STEP 8 Take a first cut by grinding a narrow margin across the length of one blade.

STEP 9 Take the same cut acorss the opposite blade.

STEP 10 Stop the wheel. Check the diameter for size and taper.

Note: The table is adjusted to correct for any taper.

STEP 11 Continue to finish grind the primary relief angle.

Note: Spark out the final cut without additional in-feed.

STEP 12 Lower the wheel head the additional distance to produce the secondary clearance angle.

STEP 13 Continue to take successive cuts across each reamer blade until the correct margin is reached.

STEP 14 Reposition the reamer at center height in order to grind the taper portion of the reamer in correct relation to the margin and primary relief angle on each blade.

STEP 15 Loosen the table swivel locking nuts. Swivel the table to a taper of 1/4" (6.25mm) per foot.

Note: This taper is the customary taper at the starting end of hand reamers.

STEP 16 Offset the wheel head to again produce the primary relief angle.

STEP 17 In-feed for the first cut. Start at the end and traverse the blade until the primary starting taper extends along approximately one-fourth the length of the blade (flute) length.

STEP 18 Spark out the final finish cut to ensure concentricity.

STEP 19 Change the wheel head offset (if necessary) to grind the secondary clearance angle for the starting taper portion.

Note: It is not customary practice to grind a secondary clearance over the starting taper section, except for reaming steel and in other cases where the primary relief angle is small (Table 31-3).

STEP 20 Stop the machine. Remove the reamer. Stone and then hone the face of the blades to produce a fine cutting edge.

STEP 21 Clean the machine. Return all items to proper storage areas.

HOW TO GRIND CUTTING, CLEARANCE, AND RAKE ANGLES ON FLAT CUTTING TOOLS

Flat cutting tools such as thread chasers often require the grinding of a chamfer or face, top rake, lip rake, and clearance angles. While these processes are often performed on surface grinders, the use of a surface grinding setup extends the versatility of the cutter and tool grinder to cover the grinding of flat cutting tool forms. The step-by-step procedures that follow use a milled thread chaser as an example for flat cutter grinding.

STEP 1 Use a standard type 1 straight wheel with a spindle extension, if needed.

STEP 2 Swivel the wheel head 90° and up so that it clears the work height.

STEP 3 Secure a thread chaser grinder attachment to the machine table.

STEP 4 Set the thread chaser in the fixture. Then, swing the fixture to the required sharpening angle.

630 GRINDING MACHINES

> Note: The angle may be read directly on the fixture graduations.

STEP 5 Lower the grinding wheel for the first cut. Traverse the chaser until the chamfer angle is ground across the required number of teeth. Note the handwheel graduation reading for the final depth of cut.

STEP 6 Remove the first chaser. Replace with the second chaser.

> Note: Each chaser must be clean, free of burrs, and seated accurately in the fixture.

STEP 7 Grind to the same depth as the first chamfer.

STEP 8 Continue to set and grind each of the remaining chasers.

STEP 9 Change the fixture setup to permit each chaser to be positioned horizontally so that the face (where required) may be ground to a particular rake angle.

STEP 10 Dress the wheel to produce a corner radius.

STEP 11 Feed the chaser for the first cut while traversing the cutter face past the grinding wheel.

STEP 12 Take additional cuts until the teeth along the face are sharp.

> Note: The micrometer collar reading is taken at the final depth.

STEP 13 Remove the first chaser. Proceed to follow the same steps to sharpen the other chasers in the set.

STEP 14 Follow standard procedures for machine cleanup.

CONSIDERATIONS FOR SHARPENING TAPS

Small taps that are used for rough threading operations are sometimes ground by the skilled craftsperson by hand. The grinding is done along the starting position of the flutes where practically all of the cutting action takes place. Where precision tapping is required, using either hand or machine taps, the taper cutting portion is sharpened by machine grinding. Usually a grinding fixture is used. The tap is mounted between centers. The finger rest is set behind each flute. Small taps are held in a chuck and are positioned with a ratchet stop. The taper at the front end of the tap and the clearance are produced by adjusting the attachment and turning the tap by a handle on the tap grinding attachment.

A type 1 grinding wheel is used. The wheel head is set at 1° more than the angle of the attachment setting to reduce the width of the grinding face. The taper section of each flute is indexed and secured in position by the finger rest.

OTHER APPLICATIONS OF CUTTER AND TOOL GRINDING

The applications in this unit relate to basic setups and cutter grinding (sharpening) processes. The cutters and tools described are among the most widely used in custom jobbing shops. In addition, there are many special fixtures and setups for form relief and clearance grinding. For example, combination drills and countersinks may be ground to required cutting angles. Step drills require special positioning and sharpening techniques.

Other tool cutting-off processes are performed to remove sections of broken or damaged cutters. Cutoff wheels are used in such applications. The face end on end mills is often gashed to sharpen the end teeth.

SAFE PRACTICES FOR MACHINE SETUPS AND CUTTER GRINDING PROCESSES

— Select the shape of the tooth rest blade that provides maximum support at the cutting action area.
— Apply a continuous force against the cutter during grinding to hold it securely against the tooth rest blade.
— Use a wheel head mounting for the tooth rest if there is any curvature in a replaceable cutter blade.
— Allow the wheel spindle to reach its operating speed when starting up for a grinding operation. Stand clear of the revolving grinding wheel at all times.
— Check the operating clearance carefully during setups for secondary clearance angle grinding. The wheel must clear the next tooth.

—Make a test cut on wide cutters to correct for unusual wear. It may be necessary to reduce the depth of the first cut.

—Reduce the land width of a cutter to the recommended size in order to prevent the heel from rubbing, particularly when cutters are ground with minimal primary relief angles.

—Dress wheels for cutter grinding to a small area of contact on the peripheral and side faces (as applicable) to reduce the cutting forces and the amount of heat generated.

—Spark out a final cut without additional in-feed to ensure concentricity and accuracy.

—Check all grinding wheels, especially form cutter grinding wheels, for soundness and possible chipped edges.

—Stone and hone the razor-sharp grinding burrs produced by grinding.

—Use a cloth as protection against burrs when removing a cutter from the setup or during burring or other handling.

—Grind toward the cutting edge of carbide-tipped cutting tools. Use a coolant to avoid overheating and checking.

—Sharpen the face of the formed teeth on a form relieved cutter at the radial rake angle indicated by the manufacturer. Any change in this angle affects the accuracy of the tooth profile.

—Keep cutting tools sharp. Continuing to use a cutter after it becomes dull may require excessive grinding to restore the teeth to maximum cutting efficiency.

—Stop the wheel spindle when measuring the ground angle areas or making adjustments to the tooth rest setup.

—Check the safety features of the spindle head when it is necessary to change the direction of wheel rotation. The direction of rotation of a reamer (when grinding the margin) must be opposite to its normal rotation.

—Set the table stop when grinding close to a center (as in the case of grinding the chamfer relief or clearance). The setting is to prevent extra movement that may cause damage to the wheel or setting.

—Replace guards after a wheel is mounted and before the spindle is started.

—Use safety goggles and/or a protective shield at all times.

—Clean the work station so that all machine surfaces are clear of abrasive and foreign particles. Leave the machine in a safe operable condition.

TERMS USED IN MACHINE SETUPS AND CUTTER GRINDING PROCESSES

Alternate side teeth	A design feature of staggered-tooth side mills, slitting saws, and other milling cutters. An alternate tooth pattern where the narrow land area reduces friction and permits machining to a controlled width.
Inverted V-tooth rest blade	A special-form blade for use with staggered-tooth, angle, and other cutters that require precise location for grinding. A blade that is positioned to support each tooth in relation to the center of the wheel grinding area.
Stub arbor	A work-holding arbor that centers and holds the cutter to be ground. (The taper shank is secured in the work head spindle.)
Sparking-out pass	A final pass across the cutter teeth without additional in-feed. Continuation of a grinding process until the wheel traverses the cutter almost without cutting (sparking).
Chamfer cutting edge	The angular surface formed between the face and sides of a cutter. An angular-formed cutting edge on a milling cutter.
Eccentric relief	Relief ground as a continuous arc from the cutting edge of an end mill to the flute form.

Air-spindle grinding device	A machine attachment with a cutter support arm and finger. A work-holding device that feeds and advances the cutter axially along the helical flute past the grinding wheel. A holding, indexing, and feeding device usually used when grinding a quantity of end mills.
Permanent form relief	A design characteristic for relieving form teeth so that the tooth profile remains unchanged after numerous sharpenings. Grinding the face of a form cutter at a constant radial rake and maintaining the same tooth form.
Type 11 flaring-cup wheel	A side grinding (hollow-cup) wheel with shell tapered from the back outward. (Shapes and sizes are based on ANSI standards.)
Type 12 dish wheel	A dish-shaped wheel designed for grinding on the side or face. (Features conform to ANSI standards.)
Longitudinal (back taper) relief (machine reamers)	Grinding the margins of a machine reamer to a slight taper of 0.0002″ (0.005mm) per inch. A slight clearance from the cutting end to the shank end of the reamer margin (land).
Back cross slide handwheel	The handwheel position for actuating the cross slide from the back of the cutter grinder.
Land interference	A condition where the heel of the land rubs against the wall of a workpiece during grinding. Insufficient relief at the heel of the land.
Milled thread chaser	A flat thread formed cutting tool used in sets for machine cutting a thread. A formed cutter requiring sharpening on the starting chamfer, along the edge, and across the face.

SUMMARY

The cutting tools ground in jobbing shops and toolrooms generally require the grinding of side, face, and end teeth and chamfer edges. The cutters most commonly ground are made of high-speed steel or have carbide inserts.

The cutter shapes are plain, helical or staggered, or form relieved.

Primary relief, eccentric relief, secondary clearance, and rake angles are ground on milling cutters, formed chasers, and other flat cutting tools.

The cutters usually ground (sharpened) include slitting saws, staggered-tooth milling cutters, shell mills, carbide-insert end mills, form relieved cutters, single- and double-angle cutters, machine reamers, adjustable hand reamers, and flat cutting tools.

Characteristics of each type of cutting tool and design features provide information needed for correctly grinding the cutting teeth and lands.

The alternate narrow lands on staggered-tooth cutters, particularly slitting saws, serve to maintain the cutter width and reduce rubbing on deep slotting and sawing operations.

The cutting edge of each slitting saw tooth and the wheel head are set at center height. Primary relief and secondary clearance are produced by offsetting (raising) the wheel head.

Each tooth on a staggered-tooth milling cutter is alternately ground to a right- or left-hand helix. Accurate tooth grinding requires the inverted V-tooth rest blade to be positioned at the exact center of the narrow, flat grinding area on the wheel face.

Relief and clearance angles for staggered-tooth milling cutters are generated by lowering the wheel head to the required offset.

> The wheel head offset equals the product of the sine of the relief angle and the radius of the cutter.

Two successive teeth are ground. The spindle is stopped and the concentricity of the teeth is checked with a dial indicator. If necessary, the blade is adjusted with respect to the flat area of the grinding wheel.

> Side teeth are positioned at a horizontal plane. The work head is set at the required relief angle and with a slight back clearance. The side teeth lands do limited, if any, cutting. A flaring-cup wheel is used for grinding the side teeth.

Shell mills require sharpening on the periphery. The wheel head and cutter tooth height are set with a centering gage. The tooth rest is lowered to the required offset.

> The final cut in cutter grinding is sparked out to ensure concentricity and more accurately ground cutting edges.

The secondary clearance angle is designed to limit the width of the land for maximum cutting efficiency and wear life.

> The end teeth of a shell mill are ground by mounting the cutter on a stub arbor secured in the work head spindle.

Beveled corners (cutter teeth chamfer) are ground by swiveling the work head to the required chamfer angle.

> Clearance angles for chamfers on shell and face mills require two angle settings: tilting the work head, and rotating the cutter to the clearance angle. The double angle permits grinding to prevent heel drag.

Diamond-grit cup wheels are used for grinding carbide-tooth shell mills with positive or negative rake. A coolant is used to prevent overheating and checking at the cutting edge. The diamond wheel must rotate into the cutting edge.

> Radial relief angles for end mills range from 16° for 1/8" (3.1mm) diameter to 6° for 2" (50.8mm) diameter. The side cutting edges are ground in the conventional manner. The end teeth are ground by swiveling the work head to 88° to provide a small end clearance. The work head is also tilted to the required axial relief angle. The cutter is traversed across the wheel face. The in-feed for roughing cuts is about 0.003" (0.08mm).

The tooth width is checked and, if needed, a reference area is ground on the back of a form tooth cutter. The ground reference permits accurate positioning for grinding the faces.

> Form cutter faces are ground at the radial rake angle indicated on the cutter. Dish (type 12) wheels are formed to the gullet radius. The grinding wheel face is dressed to provide a small cutting area on the wheel.

Form cutters may be held between centers or on a stud in a gear or form cutter attachment or fixture.

> After positioning the cutter and wheel at the radial rake angle, the cutter is in-fed by adjusting the tooth rest support with the micrometer screw.

The faces of cutter teeth are stoned and honed to remove grinding burrs and to produce a high-quality surface finish. The finished surface is measured against a comparator specimen.

A form cutter is positioned against and indexed by using a pawl on the gear attachment. The cutter is in-fed by micrometer adjustment of the pawl.

Angle cutters are ground with flaring-cup wheels. On single-angle cutters, the face teeth are set in a horizontal plane. The work head is adjusted to the primary or secondary clearance angle. The angle teeth require setting the table to the specified cutter angle. The teeth for the second angle of a double-angle cutter are set by swiveling the table to the second angle setting.

The chamfer edges of machine reamers require accurate grinding for fast, dimensionally accurate reaming. The thin margins are relieved longitudinally.

The side cutting edges of the adjustable blades on hand reamers do the cutting. A starting taper section is ground for about one-fourth the length of the blade.

Thread chasers provide an example of flat cutting tools that require grinding of a chamfer or face, top rake, lip rake, and clearance angles. Flat cutting tools are sharpened or reshaped by positioning the cutter and holding it in a fixture and by traversing the table (and cutter) under the grinding wheel.

Hand and machine taps are generally ground by sharpening the starting (cutting) threads. Sharpening by grinding longitudinally along the flutes reduces the dimensional accuracy of the teeth.

Gashing; cutting off damaged ends of cutters; reforming; and grinding drills, countersinks, and other flat and circular cutters are performed on cutter and tool grinders.

Grinding wheels are selected, tested, mounted, trued, and dressed according to the ANSI safety code.

Standard machine maintenance and safe operating procedures must be observed. The spindle guard is secured before operating; table dogs are used where required; and the cutter is held securely against the tooth rest blade for grinding operations.

Safety goggles and/or a protective shield are worn for all setting up, disassembly, and machine operations.

UNIT 31 REVIEW AND SELF-TEST

1. Identify five design features of flat and/or circular cutting tools that require sharpening or grinding on a cutter and tool grinder.
2. Give the formulas for calculating the offset of the wheel head to grind the primary relief (or secondary clearance angle) on the peripheral teeth of (a) slotting saws and (b) staggered-tooth milling cutters.
3. a. Describe briefly what effect the positioning of the tooth rest blade apex has on the concentric grinding of the peripheral cutting teeth edges on a staggered-tooth milling cutter.
   b. Tell how nonconcentric grinding of the cutter teeth may be corrected.

MILLING CUTTERS AND CUTTING TOOL GRINDING   635

4. Refer to a manufacturer's table of wheel recommendations for grinding selected materials.
   a. Look up and record the recommendations for sharpening high-speed steel milling cutters with the following wheels: (1) type 1 straight wheel (dry), (2) type 11 flaring-cup wheel (wet), and (3) type 12 dish wheel (dry).
   b. Give the recommendations for sharpening cemented carbide cutting tools with (1) a type 11 cup wheel for offhand grinding (roughing cuts) single-point cutting tools and (2) a cup wheel for backing-off (finishing cuts) cutters.

5. a. Select a staggered-tooth side milling cutter that requires sharpening of the peripheral and side teeth.
   b. Determine the most suitable types, sizes, and specifications of the required grinding wheels.
   c. Establish the primary relief and secondary clearance angles for the cutter and the amount of offset required.
   d. Set up the workpiece, machine, and attachment to sharpen the peripheral and side teeth. Grind the lands to the specified size.
   e. Sharpen the cutter. Test for concentricity. Stone and hone the cutting edges. Check the angles.

6. a. Select a shell mill with dull side and end teeth.
   b. Sharpen the side and end teeth.
   c. Check the accuracy of the angles, parallelism of the teeth, and surface texture of the cutting edges.

7. a. Identify two design characteristics of form relieved cutters in comparison to double-angle cutters.
   b. Name two particular cutter grinder setups for sharpening form relieved cutters.

8. a. Use either a gear or form tooth grinding attachment to sharpen a worn gear cutter or other form relieved cutter.
   b. Set up the cutter grinder, attachment, and so on.
   c. Grind the cutting faces of the form cutter to the specified radial rake angle.

9. a. Select a dull adjustable hand reamer or one with unevenly worn or out-of-parallel blades.
   b. Grind the outside diameter.
   c. Regrind the primary relief and secondary clearance on each blade. The reamer is to be used for reaming holes in cast iron plates.
   d. Regrind the starting taper.
   e. Check the concentricity and parallelism of the blades, margin width, and starting taper.

10. a. Give two examples of cutting tools for turning machines that may be ground on a cutter grinder.
    b. State two differences in machine setups between the grinding of circular milling cutters and the grinding of flat, solid or carbide-insert cutting tools.

11. Identify two personal safety factors the cutter grinder operator observes in relation to the use of tooth rests and blades.

PART EIGHT

Numerical Control Systems and Machine Processing

SECTION ONE   Numerical Control Machine Tools
        UNIT 32   NC Systems, Principles, and Programming
SECTION TWO   NC Applications and Computer-Assisted Programming for CNC Machine Tools
        UNIT 33   NC Basic Machine Tools and CNC Programming Applications

SECTION ONE

Numerical Control Machine Tools

Numerical control incorporates all of the technology and processes related to layout, measurement, and hand tools: drawing interpretation, computations and the use of technical tables, and setups of machine tools and accessories for basic machining operations. Electrical/electronic commands required to sequence each setup and process are programmed according to basic machine programming languages. The functions of the skilled craftsperson or machine technician are then performed in the context of ever-increasing automated processes. Therefore, numerical control is interlocked in this section and the next with automation.

This section begins with a brief overview of advantages and disadvantages of numerical control (NC) machine tools. Major components of basic numerical control systems and applications on fundamental machine tools are described. The kind of information the machine or tool designer provides for the manufacturer of a part or mechanism is covered. Principles, codes, systems signals, control word language, and the producing of other feed-in information for machining and/or inspecting a part are examined. Three basic format NC programs are considered: tab sequential, word address, and fixed block.

UNIT 32

NC Systems, Principles, and Programming

Numerical control provides industry with the capability to increase productivity and still maintain high-precision requirements. Numerical control applies to designing, setting up, machining, single-parts or large-volume mass production, and inspection. Numerical control is a complete system of taped or computerized instructions. The basic functions of the system include the following:

—Controlling movements to position cutting and forming tools in relation to a fixed reference point,

SYSTEMS, PRINCIPLES, PROGRAMMING

—Controlling movements of cutting tools for setups and machining,
—Establishing sequences of operations and time intervals,
—Setting feeds and speeds,
—Monitoring accuracy and cutting tool performance or other machining functions,
—Providing readouts of machining accuracy,
—Changing the nature and sequencing of processes,
—Actuating shutdown or recycling.

The term *numerical control* designates a numbering system of letters and numbers that regulate an entire machining process and the sequence of operations to machine a workpiece. Movements of the workpiece and tools are programmed on tapes or the storage banks of computers. *Programming* means that, *on command* (when required), impulses of energy move by electrical means from an electronic controller to interface at the machine where electronic/electrical energy is converted into mechanical energy.

The interfacing of energy from electrical/electronic to mechanical or pneumatic energy actuates motors, switches, clutches, and brakes. The actuating devices control, guide, and time every movement of the processing tools and the machine elements. All machine tools are capable of being controlled by tape or computer. Drilling, turning, shaping, abrasive machining, band machining, and production equipment are examples of NC applications.

ADVANTAGES OF NUMERICAL CONTROL MACHINES

Numerical control equipment is designed as an integral part of many modern machine tools. For example, the feeds and speeds required for the control of table, workpiece, and spindle movements are actuated by input from components of a numerical control system. Control functions formerly performed on machine tools by the machine operator are now translated into functions of the numerical control system. The longitudinal distance and direction movement of a machine table (X axis), the traverse cross feed movement (Y axis), and the vertical or angular movement of a spindle (Z axis) may be numerically controlled. The two basic axes (X, cross slide; Z, longitudinal) of a NC lathe are shown in Figure 32-1. Figure 32-2 illustrates the three basic axes (X, table; Y, saddle; Z, spindle) of a NC vertical milling machine.

Speed and feed rates for particular cutting tools, cycling for each process, and controls of cutting fluids may also be numerically controlled. In each of these examples, the machine control functions are performed by synchronized motors that respond to pulse commands.

Numerical control machines have many advantages. Five significant advantages include (1) productivity, (2) repeatability, (3) flexibility, (4) reduced tooling, and (5) increased machining capability.

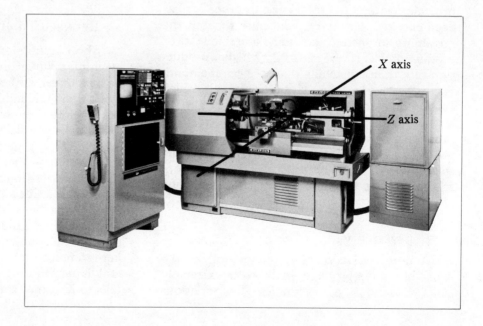

FIGURE 32-1 Two Basic Axes of an NC Lathe Equipped with a CNC System

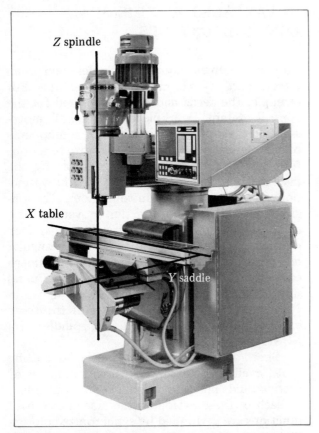

FIGURE 32-2 Three Basic Axes of a NC Vertical Milling Machine

Productivity

Once a numerical control machine has been programmed and the total cycle of processes has been checked, machining continues without the possibility of operator error and at a fixed rate of production. This rate is usually a higher production rate than a machine operator can maintain.

A factor of productivity is the ratio between cutting and noncutting time. Some noncutting steps that can be handled more efficiently by programming than by manual operation include: positioning for different operations; changing spindle speeds and tool feeds; machine start-stop; releasing, retracting, and returning the cutting tool to the starting point; and indexing.

Repeatability

A numerical control machine tool has the capability to reproduce parts with extremely small variations of accuracy. The variations within close dimensional measurement and surface texture requirements result from tool wear. Since cutting tools are constantly monitored and replaced when worn by accurately sharpened tools, the accuracy is repeated throughout a production run. The tape instructions are unvarying in controlling all processes. Therefore, with precision repeatability and the reliability of the system, inspection costs may be reduced.

Once programmed, point-to-point positioning accuracies are obtainable for each linear axis to within ±0.0005″ (±0.013mm). The repeatability accuracy is within ±0.0003″ (±0.008mm). Moreover, point-to-point errors are not cumulative.

Flexibility (Versatility)

Numerical control machines may be programmed to produce a single workpiece, a short run, or a production run. Machining functions are changed by using a new tape. The tapes provide permanent programming and are stored between runs. With computer numerical control (CNC), the programs are stored in computer memory. CNC machines can be operated directly without the use of a tape. CNC machines are discussed in the next unit.

With tape or computerized programming, it is possible to interrupt a production run, set up to produce an altogether different workpiece, and then return to the machining of the original part. The changeover is accomplished in a minimum of setup time. In many cases, jigs and fixtures are eliminated.

Reduced Tooling

Shop layouts are required for conventional machining when one part or a few parts are to be machined without jigs or fixtures. On NC machines, jigs and fixtures are often replaced because the accuracy programmed into a tape or computer permits precise machining. Programming and tape preparation and duplication require less time than the construction of jigs, fixtures, and positioning gages. Further, a tape has a longer wear life.

NC machines have a built-in check system. At the end of a block of commands, the control returns the spindle (cutting tool) to a zero starting reference point. On most systems, the operator is alerted by a warning system if the control fails to return to zero.

Increased Machining Capability

Tool changers are widely used on NC machine tools. Tool changers serve the function of holding the tooling required for machining a particular part. A *tool storage drum* may be turned until the required tool is in position so that it may be pivoted 90° downward. A *tool changing arm* grasps the pivoted tool and places it in the spindle. Simultaneously, the tool in the spindle is released to the tool changing arm, is returned to the pivot arm, and replaced in the drum.

The tool drum may be located vertically or it may be top mounted. Figure 32-3 shows a tool changing application with a top-mounted tool drum and horizontal spindle. A drill is being inserted in the spindle and the previously used shell mill is in position to be returned to the tool drum.

The cutting tools are usually held in specially designed taper-shank toolholders. The taper shank fits the spindle taper (Figure 32-3). The locking mechanism serves to draw the toolholder and tool concentrically into the spindle taper and hold it securely. On command, after the machining process, the mechanism releases the holder and tool to the tool changing arm.

DISADVANTAGES OF NUMERICAL CONTROL MACHINES

Numerical control machines do have some disadvantages. Unless the nature of the machining processes fully utilizes the capability of NC tooling and machines, the initial cost is higher than for conventional machines. Consideration must be given to necessary complementary equipment such as the tape writer, electronic control console, and other accessories. NC and CNC equipment requires considerably more floor space. Also, NC machines must be built more rigidly than conventional machines in order to withstand the accelerations dictated by servocontrols. Small jobbing shops, not able to justify a programmer, may need to farm out the programming of the system.

Maintenance for the NC system requires knowledge and skills that cut across mechanical, electronic/electrical, and pneumatic systems. Machine downtime on NC equipment can be more expensive than the maintenance of regular machine tools and production machines.

FIGURE 32-3 Tool Changing with Top-Mounted Tool Drum and Horizontal Spindle

Considerable analysis must be made to establish that a train of parts may be moved through an automated series of processes economically for profit. It must also be established that the market is ready to receive the products of NC or CNC machine tools.

APPLICATION OF NUMERICAL CONTROL TO STANDARD MACHINE TOOLS

Applications of numerical control vary from a single-spindle drill press requiring positioning along X and Y axes to complex machining on a multipurpose machining center. The drill press applications may include machine capability to follow tape instructions for cycling the quill and spindle. Speeds and depth on the drill press may be hand controlled.

Some numerical control systems are designed to be added to existing lathes, drilling machines, abrasive machining machines, milling machines, and other machine tools. This adaptation is called *retrofit*.

In order to combine a wide variety of machining processes that otherwise would require movement of workpieces among several machines, the whole movement toward numerical control has brought on the concept of the *machining center*. Total machining is performed on a single composite machine tool—that is, a machining center. With the machining center, the several spindle speeds and feed rates; turret indexing and workpiece indexing; positioning of machine components along three, four, and five axes; and continuous-path machine cuts for profile cutting in three dimensions are functions that may be controlled from tape instructions.

FOUNDATIONAL PROGRAMMING INFORMATION

The starting point for NC programming is the part drawing. The part drawing supplies information about design features and dimensional and other machining specifications. The numerical control programmer is concerned with control functions and machine positioning information. This information is coupled with a knowledge of machine technology. The programmer needs to know the capability of the NC machine tool to be used, work-holding devices, tooling, speeds, cutting feeds, coolant flow, and the processes and setups needed to manufacture the part.

The NC program requires information to position a spindle and work table. The tape instructions then specify the desired machining processes. For example, the specific tape letter code used in one system identifies machining tasks as *M functions*. This letter code is followed by a numeric code such as 01, 02, or 06. The following are a few common M functions and their meaning:

M01	index
M02	end of program (calls for tape rewind)
M06	tool change
M07	coolant #2 on
M08	coolant #1 on
M09	coolant off
M10	clamp
M11	unclamp
M50	spindle up
M51	spindle down
M57	override milling rate

SPECIAL MACHINE DESIGN CONSIDERATIONS

Under conventional conditions of machine tool operation, the craftsperson is responsible for compensating for tool wear. On numerical control machines, since final movements are not subject to human controls, the manufacturer must include many features to maintain and prolong the accuracy of the machine.

One of the common problems in table, saddle, and knee movements on a milling machine or in saddle, cross slide, and compound rest movements on lathes and other machine tools that depend on a screw and nut movement is backlash. When numerical control is added to standard machine tools, electromechanical backlash compensation is built into the machine control unit. Where numerical control is an integral part of the design, backlash is overcome by a *recirculating ball screw*.

The limitations of using thread mechanisms for movement of machine components is taken care of on other machine tools by using hydraulic positioning systems instead of thread mechanisms. The introduction of hydraulic, pneumatic, and electronic controls to automated machining was accelerated during the early 1940s. However, some of the concepts of automatically controlling movements of machine components date back to the first quarter of the eighteenth century. At that time, punched cards were used in England to control weaving and knitting machines. Today's added machine design features relate to computerizing numerical control systems and to the establishment of multipurpose machining centers.

TYPES OF AUTOMATED PROCESSES

All machining requires either *constant* or *intermittent movement (travel)*. Intermittent means the workpiece travels at different lengths of time for a number of processes. When the entire processing is operated by mechanisms, the machine is automated and the workpiece is automatically processed.

Constant Cycling

Grinding, milling, and turning are operations that are adapted to constant cycling travel. These operations may require straight-line or circular movement. Normally, the workpiece is held in a fixture while one or more operations are performed. In constant cycling processing, the workpieces are removed from moving fixtures. The workpieces move in a designated sequence for machining.

Intermittent Cycling (Travel and Work Station Indexing)

Drilling, boring, reaming, and counterboring operations in a single workpiece provide an example of intermittent cycling. Uneven time periods are required for each operation, and, within an operation, holes of different sizes may be machined.

Some parts are straight-line indexed. The workpiece moves from one station to the next according to function, sequence, time, and machining requirements. Production machines are arranged according to function and sequence of processing. A number of machines are self-contained, making it possible for a machined part to be processed to completion. In other production, a series of machines may be grouped and the workpieces fed by automatic feeding devices.

NUMERICAL CONTROL SYSTEMS

Automation requires every type of movement of the cutting tools and the workpiece to be plotted and programmed into a controlling unit. Automation also requires the mechanization of every movement according to time factors in response to impulses of energy. The predetermined manner in which machining processes, machine components, and tooling respond to and are controlled by impulses of energy is established through the use of tapes and computers. In general, there are two numerical control systems: *closed loop* and *open loop*.

Closed Loop NC System

The major components of a closed loop NC system are illustrated in Figure 32-4. A signal from the numerical control unit is fed through the machine control unit (MCU) to provide a specific instruction (motion command) to the servo drive unit. Note in the illustration that the lead screw is actuated by a servomotor. If the signal to the MCU directs the servo drive to feed a machine table 6″, the table is moved this distance. A *sensor* on the servomotor (drive motor) feeds back a

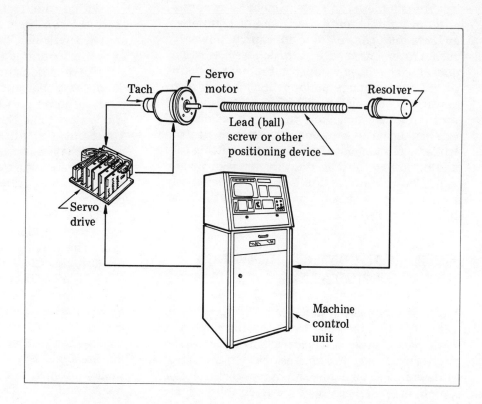

FIGURE 32-4 Major Components of a Closed Loop NC System

signal (through the encoder) to the MCU to indicate the table has moved the instructed distance of 6". In the closed loop system, the machine control unit is provided with a check on the accuracy of the machine movement.

Open Loop NC System

The open loop NC system is used on many installations of numerical control to existing machine tools. In the open loop system, *stepping motors* are used to control the movements of the machine components. There is no feedback system.

The MCU supplies the electric current impulse to the stepping motor. The number of pulses of the MCU is determined by the number of fractions of a revolution required to turn a coupled lead or feed screw to advance or return a table a specific distance within a time limit. Each current pulse causes the motor rotor to turn a fraction of a revolution. Some stepping motors advance a machine table 0.001" (0.025mm) each pulse. For example, if the table is to advance 1.000" (25.4mm), the MCU directs 1,000 pulses to the stepping motor.

Additional Circuits and Loops

The simplest NC system includes a *control unit*, an *information feeder* (tape or computer), an *actuating/control unit* to supply power to each movement, and a *feedback device (transducer)* that tells how much movement is taking place. More circuits or loops may be added to control more motors. The motors drive the tools or move tables on any axis the machine can handle. The movements may also be controlled simultaneously to permit the cutting tool and/or table to be moved in any direction on any axis. Control information is used instantaneously to make any cut in any contour within the capability of the machine.

NUMERICAL CONTROL DRAWINGS AND INTERPRETATION

Numerical control systems rely on drawings to furnish full engineering/design information. The craftsperson produces a part according to a drawing to meet exacting dimensional requirements, surface finish standards, and other specifications. Common guidelines have been established for preparing NC drawings for a programmer to produce the required part program. The NC machine tool technician must also interpret the drawing to establish that all operations are being performed according to requirements. Dimensional and other machining processes on a drawing may also be coded in numerals and letters that are foundational to programming and numerical control.

Rectangular Coordinate System

Numerical control part drawings are based on what is called the *rectangular (cartesian) coordinate system*. Coordinate dimensions are used on drawings to represent the part. The coordinates are also readily adapted to define the position of machine slides that are designed to move in mutually perpendicular directions. The system of rectangular (cartesian) coordinates makes it possible to describe any position of a point in terms of distance from an original reference point. The distance may be along two or three mutually perpendicular axes.

X, Y, and Z Axes and Zero Reference Point

Within the rectangular coordinate system, there are two basic axes: X and Y. These axes are perpendicular, lie in the same plane, and are known as *coordinate axes*. A third *spindle axis*, Z, is perpendicular to the X-Y plane. A three-dimensional part (mass) can be described accurately according to relationships with the X, Y, and Z axes. The three axes intersect at a point. The point is identified as the *origin* or *reference point*. The numerical value assigned to this origin point is *zero*. Figure 32-5 illustrates the basic X and Y axes, spindle axis Z, and zero reference point.

The X, Y, and Z notations are applied to machine tools as identification of the basic machine axes. The Z axis is reserved for the machine spindle axis regardless of whether the spindle is positioned horizontally or vertically.

The two basic axes of a lathe are identified as X axis for cross slide movements and Y axis for longitudinal carriage movements. The spindle axis is the Z axis. On a vertical-spindle milling machine, the X axis relates to longitudinal table movements; Y, to saddle (transverse) movements; and Z, to spindle movements.

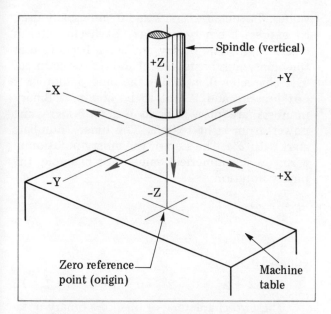

FIGURE 32-5 Basic X and Y Axes, Spindle Axis Z, and Zero Reference Point

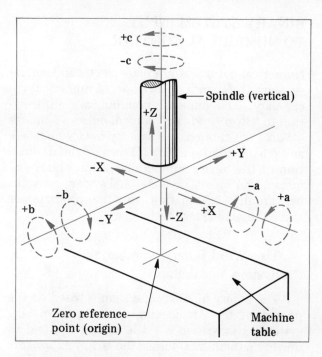

FIGURE 32-6 Rotational Axes a, b, and c

Rotational Axes

Numerically controlled motion around the basic X, Y, and Z axes may be related to *rotational axes*. A rotary table or indexing mechanism may be operated from tape or computer instructions around a basic axis. Information must be provided for (1) direction of rotation and (2) the basic axis around which rotation takes place. Lowercase letters a, b, and c identify the rotational axes (Figure 32-6).

Two- and three-axis numerical control machine tools are used in this unit to explain NC principles, systems, and applications. Common applications of two-axis numerical control machine tools include drilling machines (standard spindle and turret heads), engine lathes, jig borers, and horizontal boring mills. Examples of three-axis numerical control machine tools include turret lathes, vertical turret lathes, planer mills, and horizontal-spindle milling machines. Machining centers are designed and operated as three-, four-, and five-axis numerical control production machines.

Quadrants and NC System Point Values

The X and Y coordinate axes, which are at 90° to each other and are in the same plane, form four quadrants. The quadrants are numbered QI, QII, QIII, and QIV. The origin point is zero.

Any point (position or dimension) within a quadrant has a plus or minus value depending on the direction a measurement or distance is taken from the point of origin. Figure 32-7 shows the four quadrants with plus and minus X and Y point values as follows:

Quadrant I	+X and +Y
Quadrant II	−X and +Y
Quadrant III	−X and −Y
Quadrant IV	+X and −Y

Since the X and Y axes are in the same plane, there are no measurement points above or below the plane.

Quadrant point values are important because they are used as directions or specify locations for particular operations or setups. These directions are programmed on tape and fed into the MCU. The commands control the movements of the machine tool spindle or work table. The movement in one direction is positive; in another quadrant it may be negative. Similarly, spindle movement in the Z axis has positive and negative directions. When the spindle (cutting tool) moves toward the machine table, the movement is defined in terms of a negative (−Z) direction. Conversely, any movement of the spindle away from the machine table is in a positive (+Z) direction.

BINARY SYSTEM INPUT TO NUMERICAL CONTROL

Numerical control is a precise electronic control system. It depends on a two-word numerical vocabulary. The numerals communicate information to "stop" and "go" control pulses. Numerical data are entered on a tape by *binary numbers* and *binary notation*. The following brief description of the binary numbering system, binary notation, and binary code decimal system provides background information for the programmer and NC machine tool technician.

Two-Word Binary Numbering System Vocabulary

The binary numbering system is based on the powers of the Arabic number 2. In the binary system, 2 is written 2^1, 4 is written 2^2, and so on. In mathematical terms, the *law of exponents* requires that any number raised to the zero power equals 1. Thus, $2^0 = 1$.

Any conventional number may be written as the sum of particular binary numbers. The numbers 1 through 10 are expressed as binary numbers as follows:

Arabic Numbers	Binary Numbers
1	2^0
2	2^1
3	$2^1 + 2^0$
4	2^2
5	$2^2 + 2^0$
6	$2^2 + 2^1$
7	$2^2 + 2^1 + 2^0$
8	2^3
9	$2^3 + 2^0$
10	$2^3 + 2^2$

The ten basic numbers in the standard Arabic numbering system use a binary code. Numbers 1 through 10 fit into two-word binary number combinations. The binary numbers are punched according to a particular pattern into one of the channels of a tape. The punched binary numbers transmit control information to the tape reader. The pulses that are produced control the machine tool motors or devices for positioning and machining processes.

Binary Notation

The binary numbers are arranged in an organized pattern to express the original numerical values. Two characters, 1 and 0, are usually used to express a binary number. Each character is called a *bit*. The binary character 1 shows that the binary number *is present* and is to be counted. The character 0 indicates the binary number *is not present* and is not to be counted. Binary numbers are placed in an order of increasing power from right to left. The binary numbers start with 2^0. The sum of the binary notation in a row is its numerical value. For example, the binary notation

$$2^4\ 2^3\ 2^2\ 2^1\ 2^0$$
$$0\ \ 1\ \ 1\ \ 0\ \ 1$$

equals 13:

$$0 + 2^3 + 2^2 + 0 + 2^0 = 0 + 8 + 4 + 0 + 1 = 13$$

The two characters, or bits, of a binary notation may be represented electronically by a switch. The switch may be open or closed. The bits may have a plus (+) or minus (-) charge of a ferrite core or magnetic film in a computer.

Binary notation is also used on punched tape. A portion of a tape for positioning a spindle by using a straight binary number is illustrated in Figure 32-8. In this figure, channel 5 is used to program the input information needed to advance a workpiece along the X table axis for 1" (25.4mm). In this case, each unit of X motion is 0.0002". Thus, 5,000 units of X motion are required to produce the 1" of slide motion. A hole in the tape in channel 5 equals a binary 1. The lack of a hole in the channel equals a binary 0. The binary notation for the 5,000 units of movement is expressed as:

1 0 0 1 1 1 0 0 0 1 0 0 0

Thirteen lines are used in channel 5 on the tape. Binary numerals are identified by a punched hole or no hole. The binary notation for the 5,000 units of motion is punched between the beginning and end-of-block holes in channel 6 of the tape.

Binary Code Decimal (BCD) System

Tape formats for numerical control provide standardized information. The EIA (Electronics Industrial Association) and the ASCII (American Society for Computer Information Interchange) tape formats use the *binary code decimal (BCD) system* of digit coding. Each channel on the tape is assigned a value, as follows:

SYSTEMS, PRINCIPLES, PROGRAMMING 647

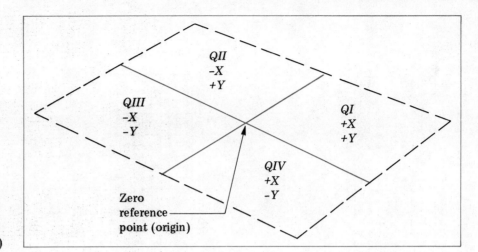

FIGURE 32-7 Quadrants with X and Y Values (Two-Dimensional System of Rectangular Coordinates)

Channel	Assigned Numerical Value
1	1
2	2
3	4
4	8
6	0

These five channels may be used to designate any number between 0 and 9.

Numerical quantities are expressed in binary notation running along the length of the tape. Each number is expressed as a number of digits, usually six. While a decimal point is not shown, it is understood to be between the second and third digit.

EIA tapes are encoded with numbers (0 through 9), letters (A through Z), signs, and other symbols. The codes are arranged in horizontal rows. Two examples of EIA tape formats are illustrated in Figure 32-9.

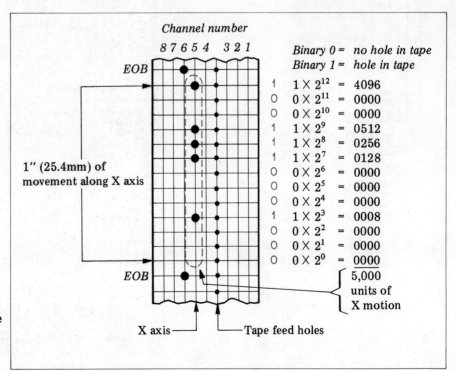

FIGURE 32-8 Portion of a Tape for Positioning a Spindle One Inch (25.4mm) along X Axis by Using a Straight Binary Number

648 NC SYSTEMS AND MACHINE PROCESSING

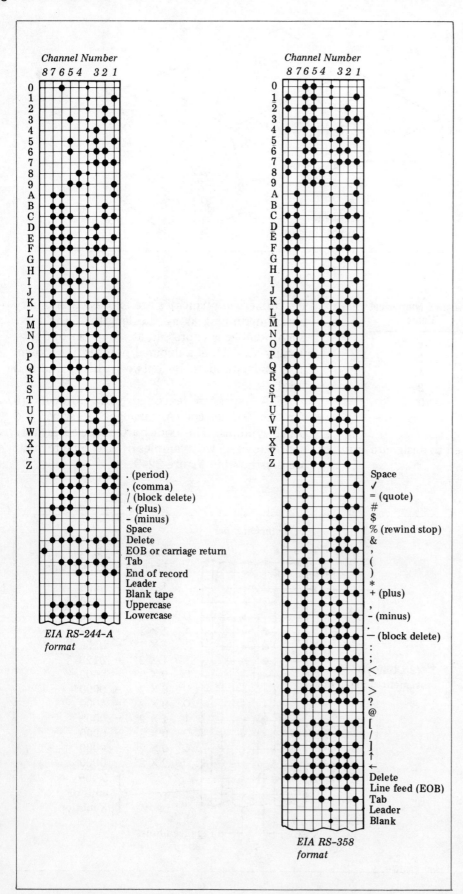

FIGURE 32-9 Sample EIA Tape Formats

NUMERICAL CONTROL WORD LANGUAGE

A *NC word* is a set of letter and numeric characters arranged in a prescribed way. The two basic types of words in a NC word language are (1) *dimension* and (2) *nondimension*.

Dimension Words

Each dimension word begins with the axis address code (letter). Each dimension word has three parts: the axis address code (letter), the appropriate (+ or -) directional sign, and the distance of movement. If there is no + directional sign in the word, the control interprets the word as positive.

Nondimension Words and Functions

Nondimension (direction) words fall into categories such as: (1) sequence, (2) preparatory function, (3) feed rate, (4) spindle speed, (5) tool (turret) selection, and (6) miscellaneous function. In each case, an *address character* (designated letter) is followed by a given number of digits.

The *sequence number* is designated by the address character N and three numeric digits. The word indicates the start of a specific sequence of operations. It is the first word for the programming sequence within the block.

The *preparatory function* is designated by the character G and two numeric digits. This word immediately follows the sequence number word. The G word prepares the numerical control unit for a specific mode of operation. Examples of G words include:

G00	point-to-point positioning
G01	linear interpolation
G02	circular interpolation
G03	circular interpolation, Arc CW (arc clockwise)
G03	circular interpolation, Arc CCW (arc counterclockwise)
G04	dwell
G13-G16	axis selection
G33	thread cutting, constant lead
G40	cutter compensation cancel
G80-G89	fixed cycles 1 through 9

Dimension words follow the preparatory function word. For multiple-axis systems, the dimension words follow in order: X, Y, Z; U, V, W; P, Q, R; I, J, K; A, B, C, D, E.

The *feed function* is designated by the letter F and a maximum of eight numeric digits. This word follows the last dimension word. The F is programmed on tape as a coded feed rate number. The EIA feed rate system consists of a series of two-digit code numbers, each representing the linear motion feed rate in inches per minute (ipm).

The programmer follows the same guidelines for establishing feed rates as are applied to conventional machining. Material to be cut, the machining process, and characteristics of the machine tool are considerations. The control units of NC machines are designed with automatic, smooth acceleration or deceleration to a new higher or lower programmed feed rate.

The *spindle speed* is designated by the letter S and three numeric digits. This word follows the last dimension word or feed rate word. The spindle speed is expressed in a coded three-digit number.

The *tool (turret) function* is designated by the letter T and a maximum of five numeric digits. This word immediately follows the spindle speed word. The digits selected must be compatible with the particular numerical control system being used.

The *miscellaneous function* is designated by the letter M and two numeric digits. This word follows the tool function word and immediately precedes the end-of-block (EOB) character.

Other Selected NC Functions

Arc clockwise (Arc CW) specifies the path of curvature generated by coordinating the movement of a cutting tool in a clockwise direction along two axes.

Arc counterclockwise (Arc CCW) specifies the path of curvature generated by a cutting tool whose movements along two axes are coordinated in a counterclockwise direction in the plane of motion.

Automatic acceleration (G08) means accelerating the feed rate from the starting feed rate within a block. A starting feed rate, in any block in which the G08 code is used, of 10% of the programmed feed rate may be accelerated, according to a time constant, to 100%.

Plane selection (G17, G18, or G19) is generally used for X-Y, X-Z, or Y-Z plane selection for circular interpolation and cutter compensation functions.

Program stop is an M00 word that stops the spindle, coolant flow, and feed after completion of all commands in the block. The remainder of the program may continue after the machine operator pushes a button.

Spindle clockwise is an M03 command that starts the machine spindle rotation to advance a right-hand screw into the workpiece. An M04 command starts the spindle to retract a right-hand screw from the workpiece.

Spindle off is an M05 command that stops the spindle as efficiently as possible and turns off the coolant.

FACTORS TO CONSIDER IN NC PROGRAMMING

Linear Interpolation

In programming a contour, the type of interpolation directly affects the calculations involved in traversing an arc or circle. When an arc is generated in a series of straight lines, the greater the number of lines computed to traverse a given arc, the finer the arc.

Linear interpolation relates to the control of a travel rate in two directions. The travel rate is proportional to the distance traveled. The axis drive motors must be capable of operating at different rates of speed. Thus, in linear interpolation, the stepping motors on the axes drives permit a cutting tool to move along an angular path.

Circular Interpolation

Circular interpolation is defined as the ability of a control unit to generate a circular arc of maximum 99.99° span in one block. A circle or an arc is generated as a continuous curve rather than as a series of straight lines. In circular interpolation, the start and the end of an arc is programmed in only one block of tape. One type of circular interpolation is the EIA standardized method. Some of the newer CNC machines are designed to generate a 360° arc in one block.

Dimension Command Pulse Weight

All dimension words must be divisible by the command pulse weight of the numerical control system. The *pulse weight of the system* is the smallest increment of a machine slide movement caused by one single command pulse. For example, if the control unit has a command pulse weight of 0.0002" (0.005mm), each electronically generated command pulse causes a movement of a machine slide of the same magnitude.

Acceleration and Deceleration for Cuts

The NC programmer must recognize that upon approaching the end of a cut, such as an inside corner, the cutting tool may have to be slowed down (decelerated) to prevent *overshoot*. Overshoot causes the cutter to cut deeper than required and to leave an undesirable indentation in the workpiece. The deceleration is block programmed with a reduced feed in order to machine precision inside corners.

POSITIONING THE SPINDLE

Incremental Measurement

Incremental measurement means that the spindle measures the distance to its next location from its last position. Incremental measurement (positioning) utilizes positive and negative directions. Spindle positioning using incremental measurements is illustrated in Figure 32-10. For example, the dimensions show the first drilling location (centerline) to be 1" (25.4mm) from the point of origin along the X axis. The hole is 2" (50.8mm) along the Y axis. The second hole is centered an additional +2" (50.8mm) along the X axis (same direction of movement). This hole is also 1" (25.4mm) farther along the Y axis in the same direction of movement.

In this example, then, to position a cutting tool by the incremental method, the spindle moves +X 1" (25.4mm) and +Y 2" (50.8mm) from the zero point of origin for the first operation. To reach the second operation, the spindle moves +X for another 2" (50.8mm) and +Y for an additional 1" (25.4mm). That is, to position the spindle for the second operation, location 1 (the first operation) is used as the new point of origin from which movements are measured.

After completing the second operation in this example, the spindle is returned to the original zero point by a -Y movement of 3" (76.2mm) and a -X movement of 3" (76.2mm). Note that the second location is the new origin from which the measurements are made back to point zero.

It should be reemphasized that in numerical control language the spindle position is

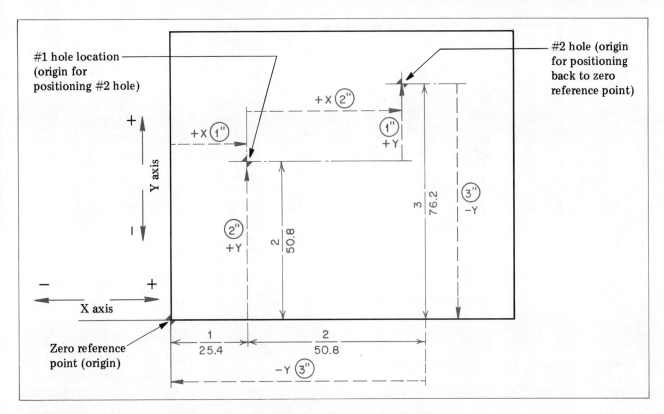

FIGURE 32-10 Spindle Positioning Using Incremental Measurements

considered to be moved. Actually, X and Y movements and positions result from moving the table.

Absolute or Coordinate Measurement

The spindle of a NC machine tool may also be positioned by *absolute or coordinate measurement*. One system is identified in relation to a fixed zero—that is, all measurements are taken from the same reference point. The advantage of using a fixed zero is that the spindle operates only in quadrant I. All movements for positioning locations have positive values. All coordinate location points are specified in relation to distance from the coordinate axes.

Spindle positioning using absolute or coodinate measurements is illustrated in Figure 32-11. The zero reference point is located at the bottom left corner of the workpiece. Location 1 is identified at point 1X, 2Y from point zero. Location 2 is at point 3X, 3Y from the same point zero. The spindle is instructed in NC to position at coordinate location 1X, 2Y. At the completion of this operation, the spindle is instructed to position at coordinate location 3X, 3Y. The third instruction is for the spindle to return to the point of origin, which is coordinate location 0X, 0Y.

Floating Zero Point

NC machine tools may also be programmed to permit a *floating zero* to be used as absolute zero. The floating zero may be established as any point that will make programming easier. Spindle positioning using a floating zero point is shown in Figure 32-12. Figure 32-12A is an ordinate drawing of a workpiece that requires the drilling of four holes symmetrically located. Figure 32-12B shows that the absolute zero may be floated to the intersection of X and Y axis centerlines. The four holes are positioned by coordinate locations in each of the four quadrants. Thus, the X and Y values change.

Absolute positioning has the same advantages over incremental positioning with respect to machining accuracy. Errors that accumulate from incremental positioning are not a problem in absolute positioning.

652 NC SYSTEMS AND MACHINE PROCESSING

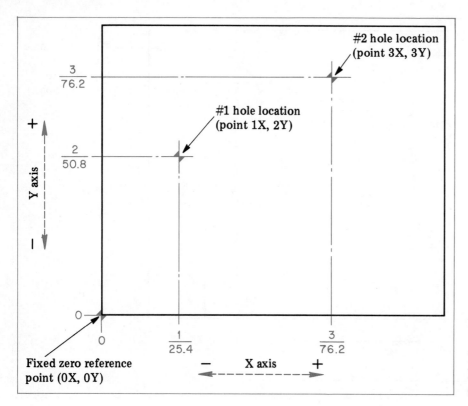

FIGURE 32-11 Spindle Positioning Using Absolute or Coordinate Measurements

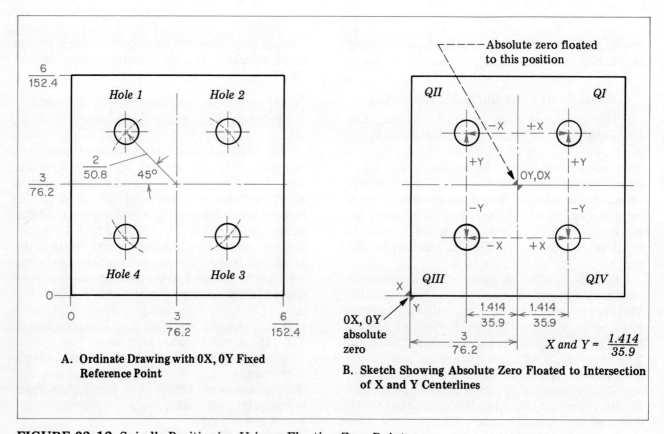

A. Ordinate Drawing with 0X, 0Y Fixed Reference Point

B. Sketch Showing Absolute Zero Floated to Intersection of X and Y Centerlines

FIGURE 32-12 Spindle Positioning Using a Floating Zero Point

NUMERICAL CONTROL TAPE PROGRAMMING, PREPARATION, AND CORRECTION

Manual Programming

The first event that must take place to machine a part by using a numerical control machine tool is the preparation of the program. The programmer reads the part print. With expert knowledge and skills, the programmer translates every machine and cutting movement and setup to machine the part according to the specified dimensions and requirements. All cutting tool and machine tool functions are listed in a logical sequence to produce a numerical control program *manuscript*. The program is written either (1) manually or (2) with computer assistance.

Manual programming requires that all cutter, machine function, and numerical coordinate data be given. The cutter positions must be calculated and specified on the manuscript.

The manuscript is then processed by a *tape preparation unit*. The program for the part is encoded in English letters and Arabic numerals on a tape. The preparation unit also prints a copy or *printout*, of the program. The printout is corrected or changes are made in the tape. Once the tape is accurate, it is fed to the machine control unit (MCU). The MCU feeds control information to the actuating and movement systems, devices, and mechanisms incorporated in the machine tool. The events in manual programming and machining a part are shown in Figure 32-13. The events for computer-assisted programming are described in the next unit.

Tape Material and Features

A NC tape is a common method of giving control instructions to a NC machine. NC tape materials include durable paper, paper-plastic, aluminum, and plastic laminates. The nonpaper tapes are able to withstand greater usage with less wear and are not subject to soil from materials used in the shop. Regardless of material, tape sizes are standardized and are manufactured to close tolerances.

The dimensions and design features of a standard NC tape appear in Figure 32-14. The 0.046" (1.17mm) diameter feed holes are punched off center to permit easy alignment of the tape in the tape reader. Note that eight positions are available for code holes. The positions, or *channels*, are numbered 8, 7, 6, 5, 4, 3, 2, 1. The feed holes between channels 4 and 3 assure the proper positioning of the tape for both punching and reading in the tape reader. The feed holes are in line with the code holes.

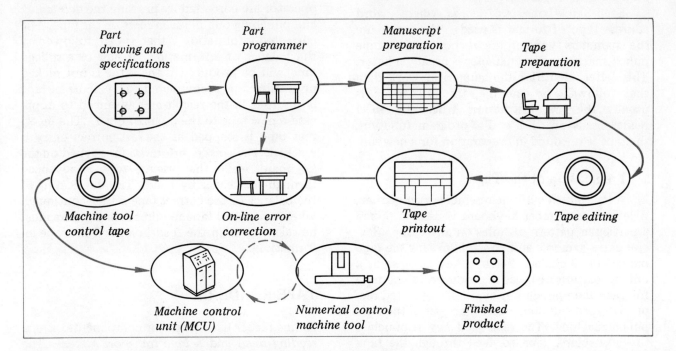

FIGURE 32-13 Events in Manual Programming and Machining a Part

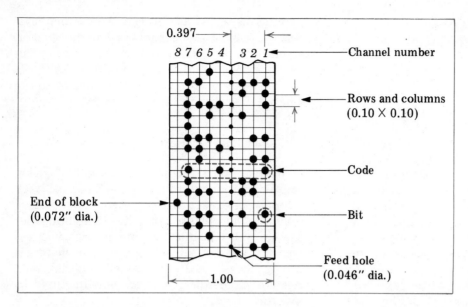

FIGURE 32-14 Dimensions and Design Features of a Standard NC Tape

Tape Blocks and Rewind Stop Code

The information coded on the tape provides input data to the MCU. The MCU directs the machine tool through its various functions. The input coded information on the tape is sectioned in units referred to as *blocks*. Each block represents a complete entity: a machining operation, machine function, or a combination. Each block is separated from a succeeding block by an *end-of-block (EOB) code*. The EOB code is punched on channel 8 (Figure 32-14).

Each block consists of *words* (where a word address type of format is used). The words are the characters typed on the keyboard of the tape punch machine. The character is usually a letter. This letter identifies the numerical work data that follows. For example, the M02 code is usually the first instruction. When the MCU reader reads an M02 end-of-program function, the tape is rewound in preparation for a new run.

Preparation of the Tape

A *tape punch machine* resembles a typewriter. A letter and number keyboard is used to punch a particular pattern of holes on a NC tape. Several extra symbols and control keys for the tape punch are included. When a key is pressed, a unique pattern of holes is produced in the tape for each tape punch key symbol. A printed record is typed simultaneously on paper in the tape punch machine. The *tape feed key* is punched to cause blank tape to feed through the tape punch machine as feed holes are punched.

A *tape reading head (reader)* is an integral part of the tape punch machine. The function of the reading head is to operate the tape punch machine typewriter from the punched tape. This operation provides a printed record of the tape or produces duplicate tapes. The punched tape actuates the typewriter to prepare a record of information.

Correction of Errors and Changes

Any typing errors detected by the tape punch operator are corrected by pressing the *delete key* and punching out all seven rows on the tape. The corrected information is then typed, followed by the balance of information. A new or a spliced tape will be produced if the error is not picked up until a tape is completed. The incorrect tape is inserted in the reader and advanced to duplicate a new tape to the point of error. The incorrect tape is stopped as the last correct entry is reached. The correct information is typed on the keyboard while the original tape is advanced through the reader by hand. The duplicating of the balance of the correct tape is then continued. When a punched tape requires splicing, care must be taken to align the feed holes and the tape information.

TAPE READERS

A *tape reader* has two major components: a *tape reading head* and a *tape transport system*. The transport system includes mechanisms for moving the tape and tension arms to maintain uniform tautness of the tape as it passes across the reading head. The tape reader is generally designed as part of the machine control unit. The MCU may be designed as an integral part of the machine tool or it may be wired to the machine. There are three different means of reading the tape: (1) *electromechanical*, (2) *photoelectric*, and (3) *pneumatic*.

Electromechanical Reader

The output of an electromechanical reader is controlled by mechanical fingers. These fingers are operated through the punched holes in the tape.

Photoelectric Reader

A photoelectric reader utilizes the photoelectric principle of a light beam activating sensitive photoelectric cells. A concentrated light beam passing through the punched holes permits fast reading of the information on the tape.

Photoelectric readers are applied on numerical controls that are designed for continuous-path machines. Since continuous-path machining generally requires a series of cuts to form an angle, radius, or other irregular shape, photoelectric readers are used for such directions.

A machine control unit may be equipped with *memory storage*. The tape reader in this MCU reads and stores one instruction in advance and thus conserves the waiting time that otherwise is required while the instructions are read from the tape. Memory storage is also known as *buffer storage*. As one cut is being taken, the MCU stores the next set of instructions the tape reader is reading. At the completion of one cut, the instructions for the next cut are instantly available.

Pneumatic Reader

A pneumatic reader depends on a flow of a fluid (in this case, air) to flow through the punched holes in a tape. The flow activates electromechanical switches. A pneumatic reader requires precise alignment of the tape reader air jets over the tape columns. This type of reader is slower than either the electromechanical or photoelectric readers.

TAB SEQUENTIAL BASIC FORMAT PROGRAMS

Three basic formats for programming follow. These formats include: tab sequential, word address, and fixed blcok programs.

Tab sequential is a basic NC format used for point-to-point NC applications. It may also be adapted to continuous-path contour programming. The code produced on the tape is *tabbed* by the tape punch typewriter to give separate information about axis positioning, machining, and other functions. The MCU is able to differentiate in the electrical sections between X, Y, and Z axis positioning, tooling functions, and machining processes.

In tab sequential, the information follows a particular sequence. The first information relates to the positioning or operation step. A tab code follows to separate this information from X axis information.

The next tab code separates X positioning from Y positioning. The third tab separates Y positioning information from machining M functions. Additional tab codes are used to separate Z positioning and/or rotational axis positioning (if required).

Tab sequential programming is used in the following four examples. These examples apply to drilling and milling.

Single-Axis, Single Machining Process

Example 1: Point-to-Point Program. Drilling processes are a good example of point-to-point machining. The zero reference point for numerical control machining is usually planned to be off the workpiece. The spindle is usually positioned by manual control over the zero point.

An example of a single-axis, single machining process for drilling a part is shown in Figure 36-32-15. The part, represented by the line drawing (Figure 32-15A), requires the drilling of four holes along the X axis. For convenience, the zero reference point is located 1" from the left work surface. The drilling of the four holes involves five program sequences. Four positioning sequences (Figure 32-15B) are required for drilling. Sequence #5 returns the spindle to the starting position for the next workpiece. The sequences involve +X positioning for the holes and -X positioning to return the drill (spindle) to the starting

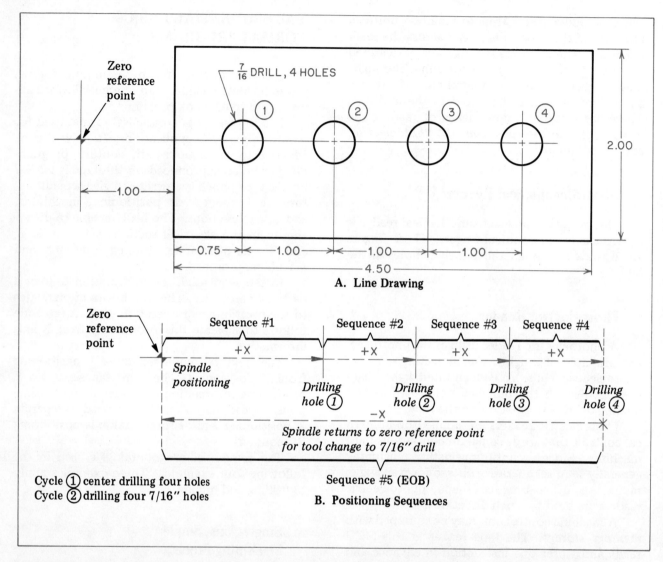

FIGURE 32-15 Example 1: Tab Sequential Numerical Control Program for Drilling a Part

position after the tape run. The total of the +X positioning movements must equal the -X return movement in order to move the spindle back to the point of origin.

A drawing often includes notes, which must be considered in programming. An example of such notes is included in Figure 32-15 for cycle 1 and cycle 2.

A simple numerical control tape program form is completed. The tab sequential numerical control tape program for producing the drilled plate is illustrated in Figure 32-16.

HOW TO PROGRAM A SINGLE-AXIS, SINGLE-MACHINING PROCESS

Console Presets

1. The tool mode switch is set to the automatic position. The cutting tool (in this example, the center drill and the 7/16" diameter drill in turn) is to machine and retract automatically at each position.

SYSTEMS, PRINCIPLES, PROGRAMMING 657

Company: C. G. Whitehurst Machine Tool Works									
Dept. 169			Part Drilled plate				Part # 16		Oper. #1-A
Prepared by Date CTO-82			Remarks Tool mode switch - auto Feed rate - Hi Backlash - takeup #2				Tools Center drill 7/16 diameter drill		
Ck'd by Date TPO-82									
Tape #12206									
Sheet 1 of 1									
Seq. #	Tab or EOB	+ or -	(X) Increment	Tab or EOB	+ or -	(Y) Increment	Tab or EOB	(M) Function	EOB	Instructions
									EOB	
0	RWS								EOB	Change tool; Load; Start
1	TAB		1750						EOB	
2	TAB		1000						EOB	
3	TAB		1000						EOB	
4	TAB		1000						EOB	
5	TAB	−	4750	TAB			TAB	02	EOB	

FIGURE 32-16 Example 1: Tab Sequential Numerical Control Tape Program for Drilling Process

2. The tool positioning rate (feed rate) is set at the high (Hi) rate of speed.
3. Lead screw backlash compensation is provided. (In this example, a #2 compensation provides the degree of precise positioning required.)

Beginning of Program, RWS Code, and Sequencing Statements

The NC program (Figure 32-16) begins with the end-of-block (EOB) code. This code is followed by the RWS code. Before there are any program statements, the sequence statement #0 identifies the rewind stop (RWS) code instruction to the tape reader. The tape is stopped at this point after rewinding and the program is recycled.

1. Sequence statement #1 gives the spindle (table) movement along the X axis from the reference point to the center of the first hole to be center drilled. The 1.75″ movement is indicated as 1750. The first hole is center drilled automatically.
2. Sequence #2 gives the spindle movement (1000) along the X axis from the first hole to the second hole. The second hole is center drilled automatically.
3. Sequence #3 provides positioning information or the spindle movement (1000)

from the second hole to the third hole. The third hole is center drilled automatically.

4. Sequence #4 prescribes positioning information (1000) to move the spindle from the center of the third center-drilled hole to the fourth hole. The fourth hole is center drilled automatically.
5. Sequence #5 gives the -X increment (-4750). This positioning information moves the spindle back along the X axis from the fourth center-drilled hole to the reference (origin) point.

Further, the M02 function instructs the tape reader to rewind the tape. The tape does not call for any drilling at the point of origin.

6. The center drill is replaced with the 7/16" (10.9mm) drill.
7. Sequences #1 through #5 are repeated. The four center-drilled holes are drilled to the required 7/16" (10.9mm) diameter. At the end of this cycle, the 7/16" (10.9mm) drill is replaced with the center drill.
8. The M02 function signifies an end-of-program instruction.

X and Y Axes and Tool Changes

Example 2: Point-to-Point Program. An example of point-to-point positioning along the X and Y axes, with tool changes for drilling eight holes of three different sizes, is illustrated in Figures 32-17A and B. The part drawing (Figure 32-17A) shows positioning sequences. The tab sequential program (Figure 32-17B) for producing the tape accompanies the part drawing.

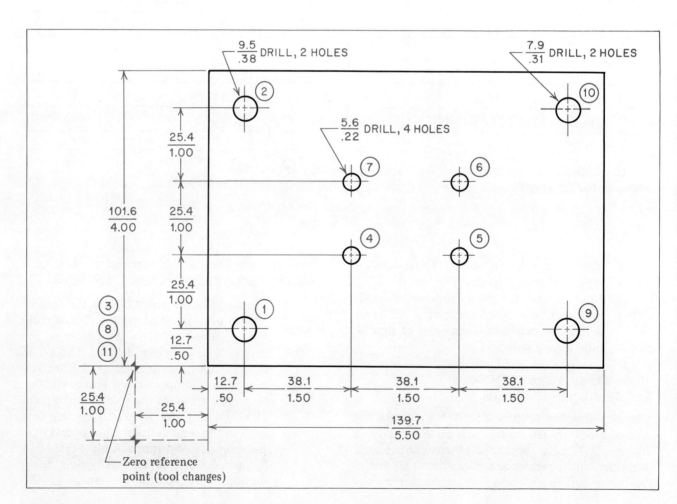

FIGURE 32-17A Example 2: Part Drawing Showing Positioning Sequences for Point-to-Point Positioning along X and Y Axes, with Drilling Tool Changes

SYSTEMS, PRINCIPLES, PROGRAMMING 659

HOW TO PROGRAM FOR TWO AXES AND TOOL CHANGES

Console Presets

1. The tool switch is set at the automatic position.
2. The positioning feed rate switch is set at the high rate.
3. The backlash switch is set at #2.

Beginning of Program, RWS Code, and Sequencing Statements

1. The instructions start with the loading of the part. Sequence position 0 has the RWS for recycling the program.
2. Sequence numbers 1 and 2 give the X axis and/or Y axis increments for positioning and drilling the 9.5mm (0.38) diameter holes.
3. Sequence number 3 provides information about positioning the spindle at the original zero point. The M06 code in the M function column stops the control and lights the tool change lamp. The instructions column indicates that the machine tool operator makes a tool change to the 5.6mm (0.22") stub drill. The -X and -Y increments return the spindle to the starting position.
4. Sequence numbers 4 through 7 position the spindle to drill the four 5.6mm (0.22") diameter holes.

Company: C. G. Whitehurst Machine Tool Works									
Dept.	208			Part	Strip plate		Part # EX 2		Oper. #17
Prepared by Date	CTO-82			Remarks			Tools		
Ck'd by Date	JEF-82			Tool mode switch - auto			5.6mm (0.22") Stub drill		
Tape	#12401			Feed rate - Hi			7.9mm (0.31") Stub drill		
Sheet	1 of 4			Backlash #2			9.5mm (0.38") Stub drill		

Seq. #	Tab or EOB	+ or -	(X) Increment	Tab or EOB	+ or -	(Y) Increment	Tab or EOB	(M) Function	EOB	Instructions
									EOB	Load part
0	RWS								EOB	9.5mm Stub drill
1	TAB		38.1	TAB		38.1			EOB	Start program
2				TAB		76.2			EOB	
3	TAB	-	38.1	TAB	-	114.3	TAB	06	EOB	Tool change
										5.6mm Stub drill
4	TAB		76.2	TAB		63.5			EOB	
5	TAB		38.1						EOB	
6				TAB		25.4			EOB	
7	TAB	-	38.1						EOB	
8	TAB	-	76.2	TAB	-	88.9	TAB	06	EOB	Tool change
										7.9mm Stub drill
9	TAB		152.4	TAB		38.1			EOB	
10				TAB		76.2			EOB	
11		-	152.4	TAB	-	114.3	TAB	02	EOB	

FIGURE 32-17B Example 2: Tab Sequential Numerical Control Tape Program for Point-to-Point Positioning along X and Y Axes, with Drilling Tool Changes

660 NC SYSTEMS AND MACHINE PROCESSING

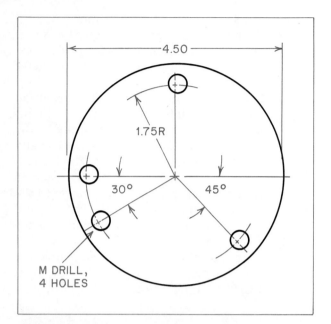

FIGURE 32–18A Example 3: Part to Be Programmed for Drilling Application Using Rotary Table

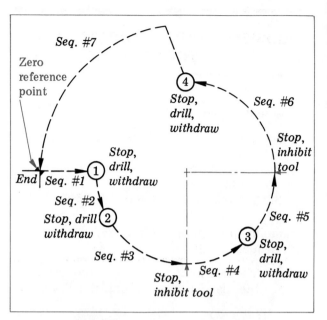

FIGURE 32–18B Example 3: Schematic Sequence Statements for Drilling Application Using Rotary Table

Company: C. G. Whitehurst Machine Tool Works									
Dept. 169		Part Angle positioning plate			Part # 208 NM		Oper. #4	
Prepared by Date CT0-82		Remarks Tool mode switch-auto Feed rate - Hi Backlash #2			Tools M Drill			
Ck'd by Date TP0-82								
Tape #13301								
Sheet 1 of 1								
Seq. #	Tab or EOB	+ or -	(X) Increment	Tab or EOB	+ or -	(Y) Increment	Tab or EOB	(M) Function	EOB	Instructions
									EOB	
0	RWS								EOB	Change tool; Load; Start
1	TAB		1500						EOB	
2	TAB			TAB		3000	TAB	54	EOB	
3	TAB			TAB		6000	TAB	546	EOB	
4	TAB			TAB		4500	TAB	54	EOB	
5	TAB			TAB		4500	TAB	546	EOB	
6	TAB			TAB		9000	TAB	54	EOB	
7	TAB	-	1500	TAB		9000	TAB	0254	EOB	

FIGURE 32–18C Example 3: Tab Sequential Numerical Control Tape Program for Drilling Using Rotary Table

SYSTEMS, PRINCIPLES, PROGRAMMING 661

5. Sequence number 8 (-X and -Y increments) returns the spindle to the zero reference point. Again, the M06 code shows a tool change to a 7.9mm (0.31") stub drill.
6. Sequence numbers 9 and 10 provide information for drilling the two 7.9mm (0.310") diameter holes.
7. Sequence number 11 (-X and -Y increments) returns the spindle to the start of the cycle. The M02 function initiates a tape rewind. A new part is loaded and the 7.9mm (0.31") drill is replaced with the 9.5mm (0.380") diameter stub drill. The new program is started in sequence 1.

Third Axis Function

Example 3: Tab Sequential Program. The third axis function normally controls a spindle component, a Z axis movement of a quill or knee, or an accessory such as a rotary table. Figures 32-18A, B, and C provide an example of a drilling application using a rotary table for the third axis. An angle positioning plate is to be programmed for drilling (Figure 32-18A). Seven sequences are involved in drilling the four holes (Figure 32-18B). Note from the numerical control tape program (Figure 32-18C) that there are seven sequence numbers in the block.

Console presets cover setting the tool switch at automatic, feed rate at high, and backlash switch at #2. The control programming for the three axes in this example starts at a reference point that falls on the horizontal centerline 1.00" from the left machined edge.

HOW TO PROGRAM FOR THREE AXES

1. Sequence #1 moves the spindle (cutting tool) from the reference point along the X axis (centerline) to the center of the first hole. The first hole is drilled automatically.
2. Sequence #2 contains a 54 code in the M function column. The M54 code informs MCU that the numerical information is for the third (Z) axis. Movement of the rotary axis (3000) positions the spindle at the 30° location. The second hole is drilled automatically.
3. Sequence #3 moves the rotary table 60°. The 6000 represents 0.01° of arc for each motor step along the Z increment. Note code 546 in sequence #3. It is a combination of the code 54 (third axis) and code 56 (tool inhibit). The "5" in both codes is entered only once to produce the code 546.
4. Sequence #4 continues the movement of the rotary table 45° (4500) to move the spindle and drill to the 105° position. The third hole is drilled automatically.
5. Sequence #5 includes a tool inhibit code because the fourth hole position is 135° counterclockwise from the third hole. The rotary motion requires two sequences. Sequence #5 advances the tool 45°.
6. Sequence #6 continues the rotary motion another 90° (9000) to the 135° position. The fourth hole is drilled automatically.
7. Sequence #7 performs the 0254 function of rewinding the tape and returning the spindle (drill) to the reference point. The X increment is in the fourth quadrant.

Note: In the system described, one complete revolution about the rotary axis (Z) totals 36,000. The X axis increments add to zero.

Milling Processes Involving Speed/Feed Changes

Example 4: Tab Sequential Program. An example of using tab sequential programming in a milling process is provided in Figures 32-19A, B, and C. The part drawing (Figure 32-19A) shows the position of the zero reference point. Three sequences (Figure 32-19B) are involved in the procedure for milling the elongated slot. Three new features—M55 (override milling feed rate), M52 (quill down), and M53 (quill up)—are included in the tab sequential program (Figure 32-19C).

The common practice in milling is to use a rapid traverse to move the table almost to the starting position. Once the cutting tool is positioned, the appropriate cutting feed rate is set. In the case of NC milling machines, the feed rate may be set manually or controlled by the tape.

662 NC SYSTEMS AND MACHINE PROCESSING

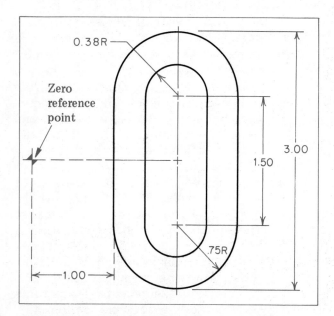

FIGURE 32-19A Example 4: Part Drawing of Tab Sequential Program for Milling Process

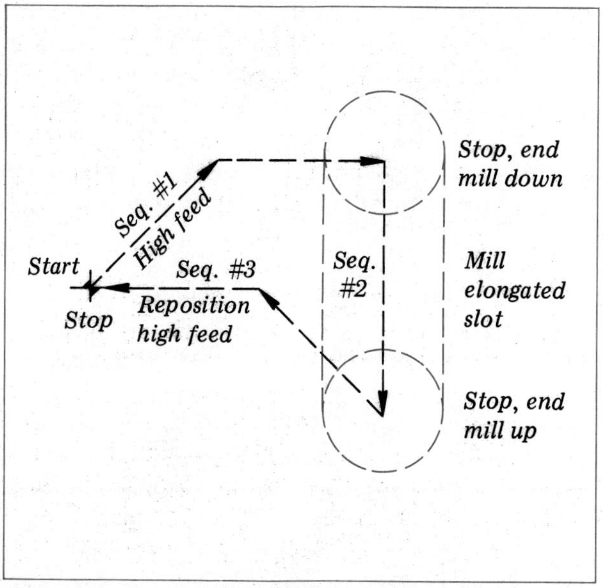

FIGURE 32-19B Example 4: Schematic of Tab Sequence Statements for Milling Process

The feed rate in this example and the tool switch at the off position are preset from the MCU console. The end mill is moved at high speed in sequence #1 from the external reference starting point (0) to position for end milling.

The M55 code causes the (F) rapid movement for initial positioning to override the milling feed rate. Another factor to be considered is the down-feed rate for entry to the end mill and then the continuous feeding to mill the starting hole

Company: C. G. Whitehurst Machine Tool Works										
Dept. 169		Part Elongated cover					Part # EX4		Oper. #6
Prepared by Date CTO-82		Remarks Tool mode switch-off Feed rate - 6 ipm Backlash #1					Tools 0.750" End mill		
Ck'd by Date JEF-82		~					~		
Tape #12206		~					~		
Sheet 1 of 2		~					~		
Seq. #	Tab or EOB	+ or −	(X) Increment	Tab or EOB	+ or −	(Y) Increment	Tab or EOB	(M) Function	EOB	Instructions
									EOB	
0	RWS								EOB	
1	TAB		1750	TAB		750	TAB	55	EOB	
2	TAB			TAB	−	1500	TAB	52	EOB	
3	TAB	−	1750	TAB		750	TAB	02535	EOB	

FIGURE 32-19C Example 4: Tab Sequential Numerical Control Tape Program for Milling Part

at the end of the elongated slot. NC machine tools are designed to permit the change from rapid traverse to a slow quill (and cutting tool) down-feed and finally to a still different table feed rate.

The milling is completed in sequence #2 and the end mill clears the workpiece. Sequence #3 returns the spindle (end mill) to the starting reference point by using rapid traverse.

HOW TO PROGRAM FOR MILLING PROCESSES INVOLVING SPEED/FEED CHANGES

1. Sequence #1 calls for positioning at high feed rate along the X and Y axes to the point where the milling operation begins.
2. Sequence #2 actuates the spindle so that the end mill feeds through the workpiece. The Y axis movement to mill the elongated slot is also identified.
3. Sequence #3 provides information for removing the end mill (upward movement of the quill) from the workpiece. Further directions are given for the rapid traverse (or high speed return) table movement to reposition the spindle at the reference starting point. Instructions are indicated to rewind the tape.

WORD ADDRESS FORMAT PROGRAMS

Word address is another NC format. The word address format requires a single letter code (A to Z) to deliver an address. Thus, control information is differentiated by the single letter address. By contrast, tab sequential format provides control information by a specific sequence that is separated by tab codes.

Sample Coded Functions

In word address format, specific instructions are given by adding a numeric code to the letter address word. Both the letter and the numeric code are used in word address. For example, a feed rate of 8 ipm is identified in a word address program as F8.

Word address format programs use the letter G for preparatory functions; M, for miscellaneous functions; N, for program sequence numbers; and F, for feed rates. Common letter words of X, Y, and Z for axis distances and + or − directions are included. Examples of coded functions in word address format are as follows:

—A G81 preparatory function calls for the cycling of a milling machine spindle (or the quill of a drilling machine) to perform a milling operation;
—A rapid-traverse feed rate may be programmed into the F address by adding the coded feed rate;
—At the end of a program, the M02 instruction cancels the first sequence so that the tool is not actuated;
—In using the third axis, information is programmed into the Z address;
—Information to indicate circular interpolation for a contouring program is provided through I and J data.

Application to Continuous-Path Machining

Word address may also be used for continuous-path machining. The G01 preparatory function information may be used to call up an interpolation cycle. Interpolation cycles on some NC machines require sending pulses to stepping motors. The stepping motor on the X or Y axis "steps" according to a pattern of tape punches passing the tape reader. Radii and angles may be approximated by the cuts in incremental steps. The cuts depend on the rate and relationship of movement of the motors. The motors may be stepped separately or together.

FIXED BLOCK FORMAT PROGRAMS

Fixed block is a third common type of NC format. As the term implies, each tape block in a fixed block format is the same length. All spaces in a fixed block format must be filled with a symbol. The features of a 20-digit fixed block program tape are illustrated in Figure 32-20.

—Tape rows 1, 2, and 3 relate to the sequence number (from 000 to 999);
—Tape row 4 is a one-digit preparatory G function the same as the G function in word address (programming by the fixed block

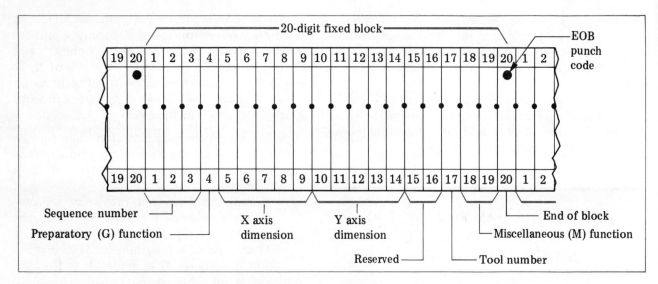

FIGURE 32-20 Features of a 20-Digit Fixed Block Program Tape

method follows each manufacturer's programming handbook of code digits for specific preparatory functions);
— Tape rows 5, 6, 7, 8, and 9 relate to an X axis dimension (decimal point values begin after the first digit (tape row 5) without using a decimal point);
— Tape rows 10, 11, 12, 13, and 14 provide similar information for a Y axis dimension;
— Tape rows 15 and 16 are reserved (the two rows are filled with zeros);
— Tape row 17 is a one-digit tool number (the coded information operates an automatic tool changer or provides a signal indicator for the machine operator to change the tool);
— Tape rows 18 and 19 provide spaces for a two-digit numeric portion of the miscellaneous M function;
— Tape row 20 is the EOB (end-of-block) code for carriage return.

MIRROR IMAGE PROGRAMMING FEATURE

Symmetrical and right-hand and left-hand parts may be efficiently programmed from a single tape by using a *mirror image (axis inversion) feature*. A programmed mirror image produces a reverse duplicate of a part, as illustrated in Figure 32-21. The machine control unit (MCU) is programmed to direct the X and/or Y axis mirror image to electrically reverse the direction of travel. All positive direction (+) movements around a program zero become negative (−). All negative direction movements become positive.

SAFE PRACTICES RELATING TO NUMERICAL CONTROL MACHINE TOOLS

— Check the position of the workpiece to reference point zero to ensure that each feature is dimensionally accurate in relation to the outer surfaces of the workpiece.
— Test the work-holding device to be sure the part is securely fastened.
— Reduce the overhang of each cutting tool or the extension of the spindle. A shortened position provides greater rigidity for machining.
— Study tool position heights, especially when multilevel surfaces are to be machined. There must be adequate clearance between cutter, workpiece, and work-holding device to permit safe slide movement between all processes, including loading and unloading.
— Set cutting speeds and feed rates for each tool process within limits recommended for each process. Although processes are tape programmed, the operator still needs to observe the cutting conditions.
— Read the tape printout. Check for axis positioning accuracy (spindle location) and correctness of each machining process.

SYSTEMS, PRINCIPLES, PROGRAMMING 665

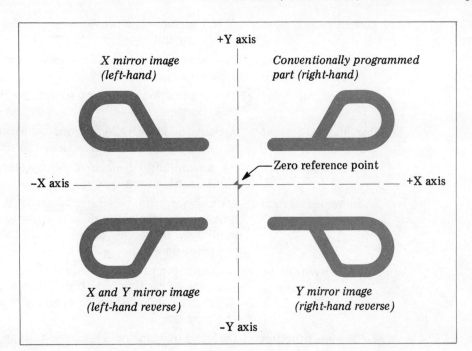

FIGURE 32-21 Mirror Image to Produce Reverse Duplicate of Part

—Make a dry run for safety checks, spindle positioning, and each machining process.
—Check the appropriateness of the cutting fluid, its condition, and the adequacy of flow for each machining process.
—Observe and listen to each machining process to establish that each tool is cutting properly.
—Inspect machined parts regularly to see that dimensional accuracy and required surface finishes are maintained.

—Check in advance on how the MCU and machine tool may be stopped immediately in an emergency.
—Use safety goggles or a protective eye shield and follow personal safety precautions.
—Remove burrs carefully. Use standard safety practices for cutters, work-holding devices, and machine tool processes.

TERMS USED WITH NUMERICAL CONTROL MACHINE TOOLS

Absolute data	Data for measurements that are made from a fixed starting point or zero reference point.
Address	A number, name, or label that provides a method for the NC machine operator or the NC program to identify information or to refer to a position location.
Binary code	A code in which each allowable position permits a choice between two alternatives.
Block	Words grouped together as a unit to provide complete information for a cutting operation. One or more rows of punched holes separated from other words by an end-of-block character.
Block address format	An address code or system of identifying words that specifies the format as well as the meaning of the words in a block.

Closed loop system	A NC system that provides output feedback for comparison to input command.
Continuous-path system (contour control)	The continuous and independent control of two or more instantaneous tool motions.
Control signal	The application of energy to actuate the device that makes corrective changes.
Fixed sequential format	A plan for identifying a word by location in the tape block. Presentation of all words in a particular order in a block.
Hardware	An assembling of mechanical, magnetic, electrical, and electronic components into a system.
Incremental data	Data from which each movement is referenced from the prior movement. (Only the data that represents a change from the immediately preceding data is used. Each move is referenced from the prior one.)
Interpolator	A device for producing smooth curves or straight lines. (The path and rate of cutting tool travel or machine tool slide movement is defined by establishing intermediate points between programmed end points.)
Miscellaneous function	On/off functions of a machine tool as related to work-holding and machining processes.
Point-to-point positioning	A positioning control system in which the controlled motion requires no path control in moving from one end point to the next.
Preparatory function	A command for changing the mode of operation of the control.
Reproducibility	The ability of a numerical control system to maintain input/output precision over a long time span.
Tab sequential format	A method of identifying a word by the number of characters in the block preceding the word. (The first character in each word is a tab character.)
Tool function	A separate tool change command. The automatic or manual selection of a tool by a command.
Zero offset	The ability of a numerical machine tool control to shift the zero point on an axis over a specified range.
Part program	A written program describing the part, together with a description of the sequence in which it is to be machined.

SUMMARY

Numerical control machine tools provide increased productivity while maintaining high-precision accuracy.

Numerical controls deal with tool positioning along two or more axes, tool transfers, machining setups, and sequencing operations; controlling speeds, feeds, and coolant; diagnoses of processes; and other machining controls.

Once a tape is programmed accurately and is dry run and then tested by a complete cycling to machine the first part, successive machining processes are cycled automatically without the possibility of operator error.

Tape instructions are unvarying in controlling successive processes. Thus, system repeatability and reliability are ensured.

Preset tools and the use of tool changers increase the range of machining capability of NC machine tools.

> Conventional machine tools may be retrofitted with numerical control systems.

Machining processes require either a constant movement (cycling) or an intermittent movement.

> Every movement and process is programmed in NC to control impulses of energy. These impulses activate drive mechanisms and systems.

A closed loop NC system feeds a signal back to the MCU. The signal provides an accuracy check of the machine movement.

> An open loop system provides controls of movements without feedback.

Part drawings for NC use the rectangular (cartesian) coordinate system. Each point and feature may be identified along two or three mutually perpendicular planes (axes) in relation to a zero reference point.

> Rotational axes describe circular motion around each normal axis. Rotational motion may be provided for two, three, four, and five axes.

The binary system consists of a two-word vocabulary. Binary numbers and binary notations control the energy pulses to produce every programmed slide, tooling, and machine motion required to produce a part.

> A NC word language contains dimension and nondimension words and functions.

Nondimension words relate to sequence number (N), preparatory function (G), feed rate (F), spindle speed (S), tool selection (T), and miscellaneous function (M).

> Some words have been standardized in NC to identify particular functions —for example, program stop, spindle CCW, and automatic acceleration.

Linear interpolation permits stepping motors to control spindle positioning in a straight-line movement.

> Circular interpolation permits generating a circular arc (99.99° maximum) in one block.

Incremental positioning utilizes + and − directions to position a tool from the location of one process to the next process.

> Absolute or coordinate measurements are made from the same reference point.

A floating zero point may be used as an absolute zero. The floating zero may be located at any point that makes programming easier.

> Manual programming involves the writing of the program, processing of the manuscript by tape preparation, correcting printout information on the tape, and in-feeding to the MCU.

The MCU actuates movement systems, devices, tools, and mechanisms to numerically control the machine tool.

> The tape consists of channels in which holes are punched to control specific machine functions. A complete entity of coded information is contained in a block.

Tape readers are classified according to the form of energy used (electromechanical, photoelectric, or pneumatic).

> Tab sequential format programs produce a tab code that separates one block of information from the next—for example, X positioning from Y positioning, from machining (M) functions, and so on.

The sum of the positioning movements along one axis must equal the return movement along the same axis in order to return the spindle to the original zero reference point.

Tab sequential programs are adapted to machining in two, three, or more axes, including tool, speed, feed rate, and other changes.

The specific sequence of control information in a tab sequential format is separated by tab codes.

A word address format uses a single letter code to differentiate code information.

In a word address format, all activities are programmed by a code letter and a numerical word. For example, a G81 preparatory function controls the cycling of a machine spindle and M02 is a miscellaneous function code for cancelling a sequence so that a tool is not actuated.

Fixed block format programs utilize a constant number of spaces that must be filled in a block.

A mirror image (axis inversion) feature permits machining a reverse duplicate by using a single tape. The MCU may be programmed to reverse the direction of travel to produce a symmetrical or right- or left-hand part.

Personal safety precautions as followed for conventional machine setups, machining, and inspection are required with NC machine tools.

Added safety practices must be observed for emergency stopping of the MCU and NC machine tool. Working clearances are especially important for freedom to position the spindle and each cutting tool without interference.

UNIT 32 REVIEW AND SELF-TEST

1. State four functions, performed by a craftsperson on a conventional machine tool, that may be manually programmed for numerical control.
2. Describe briefly two advantages (excluding repeatability) of NC machine tools over conventional machine tools.
3. Cite three tooling-up and machining functions that may be performed on NC machine tools.
4. Make a simple sketch showing the point values of X and Y coordinates in quadrants I through IV.
5. Differentiate between a fixed zero reference point and a floating zero point.
6. List the main steps required to manually program a NC machine tool to produce a specific part.

7. a. Prepare a simple sketch to show the programming sequence for drilling the three holes in the drill plate shown in the part drawing below. Locate the zero reference point along the centerline, 1.00" from the left end of the workpiece.
b. List the console presets.
c. Prepare the statements for producing the tape according to the tab sequential format.

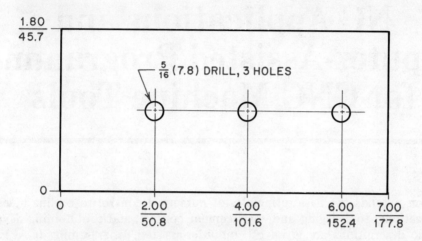

8. a. State the functions performed by the letters G, F, and M in word address format programs.
b. List examples of code numbers used with each letter.
c. Give an application of each word format.
9. Describe mirror image.
10. List two personal safety features that apply particularly to the operation of NC machine tools.

SECTION TWO

NC Applications and Computer-Assisted Programming for CNC Machine Tools

This section extends the basic principles of numerical control to on-line machine tool applications, tool gaging and management control, and tool-holding devices. Then, basic descriptions are given of computer-assisted programming (CAP) and computer numerical control (CNC). Step-by-step procedures are included for planning and preparing a computer-assisted program manuscript. The section concludes wtih descriptions of major components in a sophisticated CNC machining center. The current trend in the machine tool industry is toward CNC machining centers. In the future, these centers may be adapted to complete computer-aided manufacturing (CAM) systems.

UNIT 33

NC Basic Machine Tools and CNC Programming Applications

NC MACHINES AND CONTROLS

Five of the most common categories of machining processes include (1) drilling, (2) turning, (3) milling, (4) grinding, and (5) boring. The machine tools by which these processes are performed may be accurately controlled by numerical information. This information is usually fed to a machine control unit (MCU) by a punched tape or a magnetic tape. The MCU accepts the dimensional, tool control, and other information and processes it through electrical signals. The electrical signals govern the direction, distance, and rate of movements of the prime movers of the NC machine tool. Other auxiliary functions such as coolant flow, spindle direction, and backlash compensation are included in the NC program to produce a part according to design specifications.

Each machine tool is designed with a positioning control system. Drilling machines require movement of a table from one point location to another—that is, *point-to-point positioning*. In milling processes, such as in contour milling, the motion and path of a tool usually needs to be controlled at all times. A *continuous-path control system* is required in many applications for simultaneous motion in two or more axes. A *combination control system* is designed for point-to-point as well as continuous-path machining processes.

Common Control Features

Regardless of the type of control system, there are certain basic control features, as follows:

- Electronic components generate heat and therefore require temperature controls from 50°F to 120°F (10°C to 48°C), humidity control up to 95%, and the use of fans and air conditioning;
- The control panel area must include additional space for the machine-tool builder to mount additional or parallel machine function control switches and other accessories;
- The control system incorporates safeguards to ensure an accurate read-in of the tape;
- The NC system is able to control linear and rotary movements in any combination and simultaneously for all axes of machine motion (the smallest increment of motion, depending on the machine tool and processes, ranges from 0.000020" or 0.0005mm to 0.0001" or 0.0025mm, and the smallest increment of rotary motion ranges from 1.396 seconds of arc, or 0.000001 of a revolution, to 3.6 seconds of arc);
- The overall accuracy of the control system with an increment of motion of 0.0001", operating at a constant temperature, should be within ±0.0002" (0.005mm) (finer accuracies are required for machine tools with smaller increments of motion);
- The feedback system permits the control system to compare the tape command with the actual NC machine positioning or other functions.
- Backlash compensation is provided regardless of direction of motion from the previous position;
- Complete documentation is provided for programming, operating, and maintaining the NC system.

NC Drilling Machines

The three broad classes of drilling machines are (1) *simple* (bench or pedestal), (2) *general*, and (3) *complex* (multi-operation). All machines have a spindle to hold and drive a cutting tool; a stationary or movable table; and a mechanism for feeding a drill, reamer, or other cutting tool into the workpiece.

Simple NC Drilling Machine (Two Axes). The X and Y axes movements of a tape-controlled drilling machine are shown in Figure 33-1. The axes represent the movements of the table. In the case of a simple drilling machine, the positioning of the spindle axis (Z) is controlled mechanically or manually. The X and Y movements may be independent or simultaneous.

The control system provides a *complete signal* so that the X axis process may be cycled mechanically. The NC system may include a sequence number display to guide the operator on the operations being performed. The number following the N code on the visual display may be compared with the program printout.

NC machines may be provided with a control panel display, which indicates the current di-

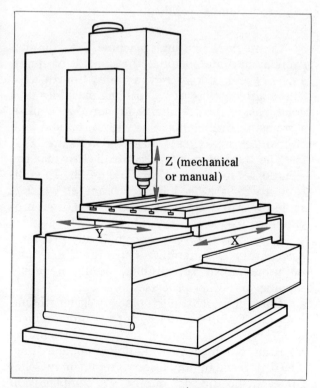

FIGURE 33-1 Tape-Controlled Drilling Machine (Two Axes)

mensions on the tape, the machine slide position, or both. The tape program on some NC machines may be interrupted to insert additional machine operations, to move the tool away from the point of machining for inspection, and to carry on other troubleshooting and checking procedures. Manual data input controls are used for moving the tool and repositioning it at the exact machining point.

General NC Drilling Machine. The classification of a general drilling machine differs from a simple drilling machine in that the design permits the drilling machine to be used for some basic milling processes. Thus, continuous-path machining capability is added to the point-to-point system. Contour milling requires tape-controlled feed rates. The MCU reads the coded number on the tape and outputs it to the machine tool.

The NC system for a general drilling machine may include continuous-path NC units for linear or circular interpolation. Arcs up to 90° may be generated within a single quadrant from tape commands in a single block. The following control features may also be included:

—Spindle speed selection, which stores the coded set of digits and then relays the information to the speed changing mechanism on call;
—Tool select feature, which provides for the automatic selection, changing, and storing of tools for multiple-sequence operations;
—Mirror image, which, through axis symmetry switches, permits the sign reversal of dimensional tape information and thus permits a single tape to be used to produce mirror image parts;
—Auxiliary function display, which, through a function selector switch, permits the operator to select an auxiliary function such as the feed number, preparatory function number, feed rate, spindle speed code, and tool number for display.

Complex NC Drilling Machine. A complex drilling machine includes the control of depth axis (Z) and, if required, a rotary fourth axis. The complex class of NC drilling machines uses combination positioning and contour control systems. A drilling machine with turret and tape control for three axes is illustrated in Figure 33-2.

The third axis is tape controlled so that the spindle is programmed for final depth. In addition, there is a *feed engage point* at which a rapid-traverse feed is changed to a slower programmed feed.

The control system automatically programs preparatory G functions. Thus, the Z axis may be programmed for drilling, boring, reaming, tapping, and other processes.

Since most cutting tools vary in length and may be of a different length than programmed, *tool length compensation* for general machining on some NC machines requires manual setting. The difference in dimensions is set up on switches for each specific tool number. A compensation dial is automatically activated when the tool is selected.

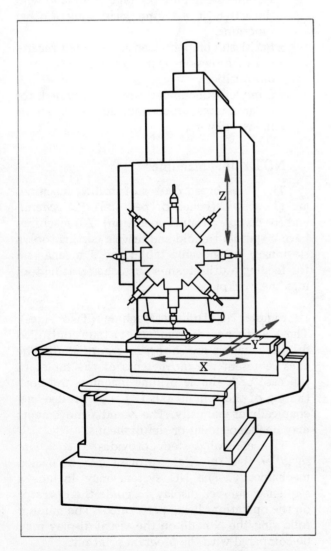

FIGURE 33-2 Tape-Controlled Drilling Machine with Turret (Three Axes)

NC Turning Machines (Point-to-Point Positioning)

The engine lathe and the turret lathe (bar and chucking machine) are the two basic types of numerical control turning machines to be discussed. NC principles and machine control units are adaptable to vertical turret and other types of lathes.

NC Engine Lathe (Two Axes). A two-axis NC positioning system may be used for general turning processes that do not require taper turning, thread cutting, or contour forming. Where contours are required, the NC system must have continuous-path capability. The general NC features, including backlash compensation as described earlier, apply to turning machines. Four additional features increase the control or machine operation capability:

- *Sequence number display*, which provides the NC machine operator with a visual display that may be checked with the program printout (an operation coded with N and a three-digit number is displayed at the time the given sequence of operations is being performed);
- *Dimension display*, which visually indicates either the dimensions on tape, or the actual slide position, or both;
- *Manual data input*, which permits the interruption of a tape program for insertion of additional machine operations, movement of the cutting tool out of position for checking and inspection of the workpiece, and troubleshooting and checking procedures;
- *Tape-controlled spindle speed selector*, which makes it possible to program for multiple-speed operations (the control unit stores the code and digits and relays the commands as required to the speed changing device on the lathe).

NC Engine Lathe (Continuous-Path System). The continuous-path system permits the machining of tapers, threads, radii, and other contour cutting processes. All previously described features for the engine lathe apply to lathes with a continuous-path NC system. In addition, contour cutting requires the feed rates to be tape controlled. The control unit reads the specified rate from the coded number on the tape and processes it. Other features that are included with a continuous-path system are the following:

- *Linear or circular interpolation*, which provides for straight-line or smooth-curve movement of a cutting tool in machining a contour (while free-form curves and circles may be produced by linear interpolation NC systems, a contour surface may be produced more efficiently and with a higher-quality surface finish by using a circular interpolation feature);
- *Internal and external thread cutting*, which is available on numerical contouring control to eliminate manual processes and cumulative errors from thread to thread;
- *Tape-controlled spindle speed selector and auxiliary function display*, a combination of features serving similar functions as described for applications on drilling machines and for drilling processes.

NC Bar and Chucking Machine and Machining Centers. Production turning machines may be tape controlled for two, three, and four axes, depending on the complexity of the machining processes. Some of the common combinations include the following tape controls:

- Simultaneous control of carriage and cross slide along two axes (the turret position selection may also be tape controlled);
- Control of carriage, cross slide, and turret ram (three axes), with turret position selection;
- Simultaneous control of any two of the three axes, with indexing of stations on the turret ram or square tool turret;
- Simultaneous tape control of four axes.

The NC features for an engine lathe with continuous-path control apply to a turret lathe and machining centers. Some typical features include a sequence number display, dimension display, manual data input, tool select, spindle speed control, and auxiliary function display. In addition, a *tool offset* feature permits the cutting tool to be moved a preset distance in the automatic tape mode. Tool offset compensates for changes in tool length due to resharpening a cutter or to a position change of the tool in a different holder. The NC machine operator dials in required offsets so that any one tool may be programmed to pick up any given offset.

NC Milling Machines (Two, Three, and Four Axes)

When classified according to position of the cutter driving spindle, milling machines are of two types: horizontal and vertical. These types may be programmed for tape control of at least two axes of simultaneous motion. Control of three axes and four axes is general. A horizontal milling machine with tape control for three axes is shown in Figure 33-3.

A NC system with continuous-path capabilities is practical for angle and other contour milling jobs that are common in jobbing shops. A continuous-path NC system with three axes of control has controls that relate to M functions of table movement, continuous-path controls, and linear and circular interpolation for generating arcs. Additional features, such as sequence number display, dimension display, manual data input, tool selection, and spindle speed control, are available to improve control capability. An auxiliary function display may be included in conjunction with a function selector switch. The operator is able to select any auxiliary function and display it. Auxiliary functions relate to sequence number, preparatory function number, feed rate or spindle speed code, and tool number.

If a considerable amount of drilling, boring, reaming, and tapping is done, as in the case of universal vertical turret milling machines, a combination of point-to-point and continuous-path systems is desirable. Multiple-operation milling machines are designed with a combination positioning and continuous-path system. This design provides simultaneous control of all axes, accurate positioning of the spindle and cutting tool, built-in cycles for the depth axis, and continuous path for curved or contour machining. All of the previously described basic control requirements and features of the continuous-path control system apply to the combination system. Predetermined fixed cycles of the spindle axis (Z) are required. Drilling, boring, reaming, tapping, and other processes are programmed on tape under the G preparatory function code. The appropriate digits call out a specified cyclic action.

NC Boring Machines

There are two general classifications of boring machines. The first class includes vertical turret lathes and boring mills. The cutting tool, when once positioned for a cut, remains stationary. The workpiece revolves while the cutting tool feeds in to take a cut. In the second classification, the action is reversed. The workpiece is moved into position while the cutting tool revolves and is fed in the direction into the workpiece. Horizontal boring machines and the more widely used jig borer are included in the second class.

Three axes of control are common for horizontal boring machines. The three axes are the longitudinal (X), cross feed or transverse (Y), and the depth axis (Z). Other boring mills have four axes of control. The controls, systems, displays, features, and functions for milling machines are applicable to boring machines. The preparatory G function may be programmed on tape for predetermined fixed cycles of the spindle (Z) axis. The specific cyclic action is called out as required for drilling, boring, reaming, and other operations.

NC Jig Borer. The jig borer is regarded by industry as an extremely precise machine tool. The accuracy of positioning for longitudinal and cross travel is within 0.000030" in any inch or 0.0008mm (0.8μm) in any 30mm. The compound slide is square (along its full travel) to within 0.000060" or 0.0015mm (1.5μm). The alignment of the spindle travel in 5" (127mm) is 0.000090" or 0.0023mm (2.3μm).

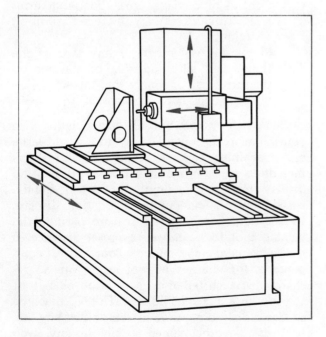

FIGURE 33-3 Tape-Controlled Horizontal Milling Machine (Three Axes)

The NC jig borer shown in Figure 33-4 has an 11" × 18" (280mm × 450mm) table travel. The jig borer is equipped with point-to-point positioning on X and Y axes. The rotary table is controlled by a preset indexer. The table is actuated by the NC system's auxiliary functions. The rotary table may serve as a fully numerically controlled axis. Variable point-to-point, varying angle positioning, and controlled feed rates are provided. Contour positioning is an optional feature. The manufacturer's specifications identify the following features as common to this particular NC jig borer:

— Complete solid-state integrated circuit system;
— Photoelectric tape reader with capacity of 125 characters per second;
— Temperature-controlled operating range from 50°F to 120°F (10°C to 48°C);
— Auto, single, and manual modes;
— Data input of 1" (25.4mm), 8-track perforated tape (EIA standards RS-227);
— Word address, variable block format (EIA RS-244 and RS-274B);
— Automatic backlash compensation;
— Miscellaneous functions M00 through M99;
— Reference, set, and grid zeroing;
— Absolute programming;
— Test mode and test circuits;
— Incremental feed and programmed feed rates;
— 100% manual feed rate override in 10% steps;
— Milling control of continuous, one-axis-at-a-time capability in a straight line parallel to the X and Y axis;
— Spindle cycle control;
— Multiple depth selection.

The jig borer has program data resolution accuracy of 0.00001" (0.001mm). Movements are provided by DC servomotors. Parity checking for accuracy is provided. Position may be read from machine scales and dials. The measuring methods involve lead screw and rotary resolvers. The machine is equipped for mirror image (X-Y plane).

NC Machining Centers

A single numerically controlled machine tool center performs a combination of many operations in one setup. Machining centers are designed to face, end, profile, and contour mill; drill, ream, bore, and tap; and carry on other operations for all axes of the machine in order to produce intricately designed parts. NC machining centers have tape control for three, four, and five axes. It is possible to move the column, spindle head, table traverse, and table rotation and tilt simultaneously under tape control. The NC system may be point-to-point, continuous-path, or a combination system.

The basic control features include simultaneous movements along all axes, programmable feed rates, spindle speeds, and fixed cycle. Typical additional features that increase the control capability or machine operation include: sequence number display, auxiliary function display, manual data input, tool gaging and selection, and tool length compensation. NC machining centers with circular interpolation are capable of generating arcs up to 90° in one quadrant from single-block tape commands. An addition is available on some circular interpolation features for generating 360° of arc from single-block tape commands. Like the controls on milling and other NC machine tools, the spindle depth axis and the feed engage point are programmed. Spe-

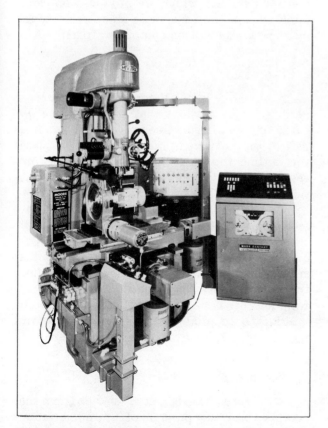

FIGURE 33-4 NC Jig Borer (Equipped with a Rotary Table) and Machine Control Unit

cific cyclic preparatory functions are programmed to control the axis for particular operations and to call out the action as required.

Productivity is also increased by the tool compensation feature. Cutting tools of different length than programmed may be set up on a compensation dial, which is automatically activated as each tool is selected.

NC DATA PROCESSING

NC data processing requires mathematical computations by the programmer and checking of tool travel, feed rates, work-to-cutter clearances, speeds, and so on. Special machine tool functions (tooling requirements, coolant flow, predetermined packaged operation sequences), appropriate miscellaneous or preparatory functions, and codes must be established. Thus, a considerable amount of data processing is completed in advance of the actual machining.

Off-Line Data Processing

Off-line data processing is completed prior to using a control unit in combination with a machine tool to produce the part described through programming. Off-line data processing, when once completed, may be used repeatedly in tooling up for any number of manufacturing processes in which the control data is required.

On-Line Data Processing

On-line data processing relates to translating information from a control tape into input signals that feed directly to a machine tool. The data processing takes place in a time interval that is so minute as to almost coincide with the machine tool's execution of a process. On-line data processing must be repeated each time the machine tool is set up for a new part to be manufactured.

The input to on-line data processing is often a part of programming language describing desired machining. The instructions are transcribed to punched cards or a control tape or are entered into a computer manually. The cards may be transcribed onto magnetic tape on a large-scale computer.

A previously written computer program executes a sequence of instructions for processing the input of part programming. Computer processing results in information that may either be put on a control tape or be fed directly to a numerical control unit at the machine center. The computer program gives the part programmer a printout of the control tape contents and, in many cases, indicates errors.

As the machine control unit reads the control tape, the information is processed on-line. A series of input signals are produced. The signals go directly to the machine control unit to call out control path movement and to execute appropriate miscellaneous and preparatory functions. The combination of the control unit and machine tool has the capability and reliability to manufacture the part repeatedly.

Part Programming

After studying the design features and specifications of a part to be manufactured, a part programmer decides on the part programming language and the processes, tooling, and setups required to machine the part. If there are geometric calculations involved in calculating the tool path and tool offset movement, these calculations may be included in the program for solution by the computer. The part programmer's statements, which are produced on cards, punched tape, or a magnetic tape, are processed by a previously written computer program.

Processor

A *processor* is an NC computer program that performs computations that are *workpiece oriented* or tool offsets. The processor program, as part of the software of numerical control, places the cutting tools on the workpiece. The processor program ignores all control unit information and items that are machine tool oriented.

Additional information required to manufacture a part relates to controls such as spindle speeds, feed rates, spindle direction, coolant requirements, tool selection, and other items that are *machine tool oriented*. The processor assumes that further processing of additional information from another computer program will need to be interlocked to produce the appropriate control tape for the ultimate manufacture of the part.

Postprocessor

A *postprocessor* is a computer program that accepts a processor program of information about the tool located on a part, machining setups, and machine processes. A postprocessor program makes additional computations to ensure compatibility of information among the MCU, NC machine tool, and the part specifications, including tolerances and machine limitations. The product of postprocessing is a control tape or information that may be supplied to produce a control tape.

The greatest advantage of postprocessing is one of economics. Changes to or adaptations of a computer program from one MCU or NC machine to another are readily made. Another advantage is that the postprocessor information may be postprocessed for multiple applications on additional MCU and NC machine tool applications.

There are five major elements in each postprocessor. These elements are: (1) input, (2) motion, (3) auxiliary, (4) output, and (5) control.

Input Element. This element reads the cutter location. It checks other output information of the processor computer program and makes a diagnostic printout of nonprocessable information. Accepted input information is transferred for subsequent postprocessing.

Motion Element. This element of the postprocessor relates to characteristics of the machine tool and geometric functions. Machine dynamics deal with items such as feed rate, cornering velocities, and tolerances.

The geometry portion translates all coordinate information into the two-, three-, four-, or five-axis coordinates, depending on the processes and the machine tool. Further, the geometry portion ensures that when a combination of linear and rotary motions is executed, the path is geometrically accurate and meets the tolerance requirements of the job. Geometry also deals with tool and workpiece clearance and safe operating spaces.

Auxiliary Element. This element is associated with miscellaneous and preparatory functions. When the part programmer needs auxiliary information, the postprocessor searches the computer memory to locate the appropriate control tape code for the desired function. The auxiliary element passes the coded information along to the output element to be incorporated at a specified time.

Certain programmer statements are interpreted to establish conditions within the postprocessor. Examples of these statements include spindle speed and directions, coolant status, tolerances, and feed rates.

Output Element. Information from the motion element is accepted and converted in the output element into codes. The codes output directly to a control tape or they may be produced in a form and format to be readily converted to the input control tape for the MCU.

The output element also accepts miscellaneous and preparatory information from the auxiliary function element. This information is coded and is available for output at a specified time. A printed listing of the tape information, comments, and diagnostic data is generated by the output routine.

Control Element. This element controls the timing for the processing of the information and the appearance of any diagnostic data in proper sequence.

COMPUTER-ASSISTED NUMERICAL CONTROL PROGRAMMING

Thus far, numerical control has been considered in terms of manual programming. All cutter movements and machine setups and machine functions were listed in sequence in the manuscript. All cutter positions were calculated by the programmer and specified on the manuscript. A tape was prepared, a printout was edited, and a control tape was prepared. All of this information was translated in the MCU to command pulses to the NC machine tool. At the end of the program, the machine was reloaded. The processing cycle between the MCU and the NC machine tool was repeated.

Numerical control may also be considered in terms of *computer-assisted programming* (CAP). This method of writing a program is used for continuous-path and other operations requiring complex and time-consuming calculations. CAP eliminates the need to make numerical coordinate data calculations. CAP calculations are made rapidly and are error-free. A number of different processor languages are used for CAP, depending on the complexity of the computations and the combinations of machining processes.

Computer-Assisted Programming (CAP) Language

CAP requires the use of symbols and modified English words, each of which has a precise meaning. The language is simplified to reduce the number of entries in the manuscript and to simplify writing the program. However, different equipment manufacturers use different program-oriented languages. Two of the widely used and powerful languages are APT, for *Automatically Programmed Tools*, and Compact II®. Less powerful languages are used to control basic machine tools such as a lathe, drilling machine, grinding machine, or milling machine or to produce a particular part on a single machine. The events and additional input in computer-assisted programming for producing a part are illustrated in Figure 33-5.

Program Editing

Point-to-point programs may be edited to correct mistakes or to change the part program. When a tape is produced and a printout is provided, the printed letters and numerals, which are coded on the tape by a pattern of punched holes, are easily edited. The numerical data on the printout reports exact cutter positions and machining operations.

Continuous-path programs may be edited directly when the numerical positions of the cutter describe exact cutter positions for machining. When compensation must be made for cutter offsets, nose radius effects, and difficult-to-follow cutter positions, it is necessary to use other methods of tape editing. One method plots the path of the cutting tool on paper. Another editing method displays the cutter path on a cathode ray tube (CRT). When a sample part run is performed, the machine tool is run through the programmed cycle. No part is used in this *dry run*. Errors or changes are noted and the tape is corrected accordingly.

Corrections and changes may be made in the part program while it is stored in computer memory. A corrected tape is then made from the corrected stored part program. The corrected

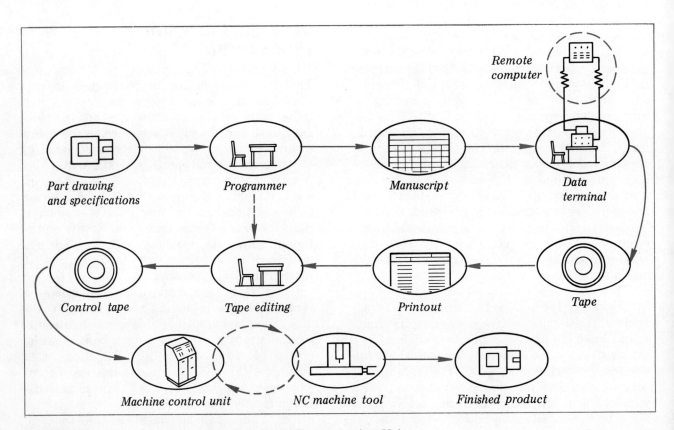

FIGURE 33-5 Events in Computer-Assisted Programming Using Additional Remote Computer Input

tape becomes the permanently stored tape and is used, when required, to reenter the program into computer memory. A tape may also be edited after the operation is started for changes in cutter speed, feed rate, cutter position, or small errors.

A *parity check* permits automatically checking a tape for malfunction errors of the tape punch. Depending on the tape format used, each horizontal row on the tape must have either an odd or an even number of holes. The parity check on the EIA RS-244-A tape format has an odd number of holes; the EIA RS-358 format, an even number of holes. If a row does not have the required odd or even holes, the control system stops to indicate an error.

FUNCTIONS OF THE MACHINE CONTROL UNIT (MCU)

The instructions within the part program are converted in the machine control unit (MCU) into a form of energy that controls the machine tool. Two basic types of machine control units are available: (1) *hard-wired* integrated circuits and (2) *soft-wired* computer numerical control (CNC) units.

Hard-Wired Units

Hard-wired units use *digital logic packages* mounted on plug-in printed circuit boards (PCB) in a fixed and permanent arrangement. PCB connectors receive the circuit boards and are also wired together to connect the electronic components in a permanent, fixed manner. One hard-wired unit controls one type of machine tool. Input signals are derived from the tape reader as the tape is run through each time a part is to be machined. The input signals, in turn, activate the machine control functions.

Soft-Wired Units

Each CNC unit is able to control more than one type of machine tool. Soft-wired CNC units have a control, or *executive*, program adapted to control a particular type of machine tool. The executive program is entered by the builder in the computer memory. The executive program is a computer-based control system capable of executing the commands of the program for a part. The program is occasionally modified by the computer or machine tool builder.

MCU Design Features and Functions

An edited punched tape is run through a tape reader in order to enter the program into the computer and to store it in computer memory. One or more part programs may be stored within the computer storage capacity. One feature of a computer is *random access memory* (RAM). This feature permits any stored program to be called up when needed.

Duplicates of part programs are kept as a permanent record in some storage medium—for example, a punched tape or a *diskette*. A single diskette of 7" (178mm) diameter has the capacity to store the equivalent of 2,000 or 3,000 feet (600 or 900 meters) of punched tape.

When a CNC unit has a *cathode ray tube* (CRT), the part programmer or machine tool operator may visually view positioning and operational information. The CRT is capable of displaying axes positions for all machine slide movements and other machine functions. Messages to the operator may be programmed for display on the CRT screen at a specified time. Other functions displayed on a CRT screen include:

—Part numbers of all programs stored in computer memory;
—Used and available capacity for further storage;
—Compensations for cutter radius, tool lengths, tool offsets, and tool fixture offsets;
—Diagnostic information when the CNC units are able to isolate and identify malfunctions in the numerical control system;
—Part program information when editing.

FUNCTIONS OF A COMPUTER (CPU) IN COMPUTER-ASSISTED PROGRAMMING

The computer used for computer-assisted programming has the computational ability to generate numerical part program data in a useful form by the MCU of a machine. Some computers are located a distance from the MCU. Other smaller computers (often called *minicomputers*) are an integral unit of the MCU.

CNC computers store part programs and process them to generate output signals for the control of a machine tool. Computers use a two-digit binary notation. The two binary digits correspond to two conditions of operation of the electronic components: on or off, charged or discharged, positive or negative charge, and conducting or nonconducting.

The *central processing unit* (CPU) includes all the circuitry controlling the processing and execution of instructions entered into the computer. The circuits relate to basic memory for storage and retrieval and *logic*. The four basic arithmetical processes of addition, subtraction, multiplication, and division are accomplished by adding positive and/or negative numbers. Simple computations are made in billionths of a second.

Input information may be entered into a computer in several forms: punched tape, magnetic tape, diskette, or signals from other computers. Similarly, *output information* may be received in these same forms or as printout sheets or electrical signals that control an NC machine tool operation. Buffer storage is provided on some machines so that the stored advance information is available at the point where it is needed.

ROLE OF THE PROGRAMMER IN COMPUTER-ASSISTED PART PROGRAMMING

The Automatically Programmed Tools (APT) system was one of the first computer processor languages. APT, Compact II®, and many other different processor languages are used depending on the machine tool or the complexity of producing parts requiring several machines to be incorporated in a complex machining center. In each instance, the part programmer must plan and prepare a program with a thorough understanding of the processor language for the function of each word and the precise usage of it in programming the machining of a part. Since APT is one of the most complete processor languages, it will be used later in a shop problem.

Program Planning

The part programmer first views the part to be machined in terms of axes, type of cutting tool, the NC machine tool, and the path of the cutting tool. For purposes of providing an example of computer-assisted programming, the part drawing in Figure 33-6 illustrates a workpiece to be turned on a NC lathe. In this instance, the spindle axis (longitudinal feed) is Z and the transverse (in and out feed) axis is X. Plus and minus coordinates are used. The part drawing uses incremental coordinates for dimensioning.

After viewing the part to be machined, the part programmer prepares a programming layout, as shown in Figure 33-7. The cutting tool has a nose radius of 0.047″ (1.2mm). This layout identifies each process of straight turning

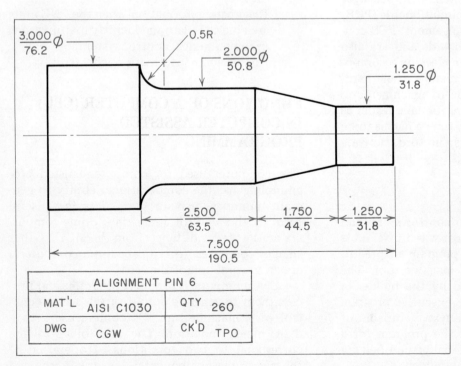

FIGURE 33-6 Part Drawing of Workpieces to Be Turned on a NC Lathe

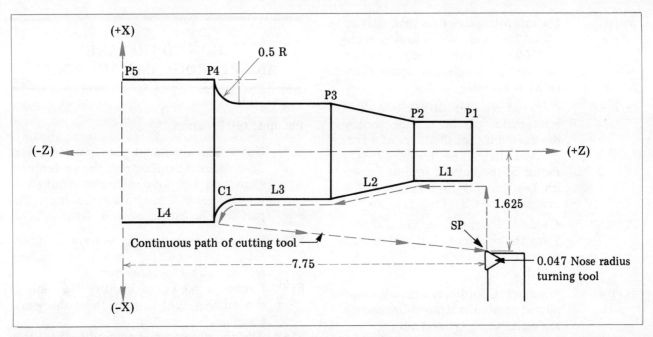

FIGURE 33-7 Part Programming Layout with Turned Surfaces, End Points, and Starting Points

(L1), angular turning (L2), straight turning (L3), and radius turning (C1), as well as the reference surface (L4). The starting and ending points (P1,) P2, P3, and P4) for each cut are indicated. The reference point (P5) and the starting (SP) point of the cutting tool are given. This reference point is 0.250″ from the end of the workpiece and 1.625″ from the centerline.

Manuscript Preparation

After completing a sketch of the programming layout, the part programmer writes the manuscript by preparing a series of statements. An APT program uses four types of statements:

—*Motion* statements to describe cutting tool position,
—*Geometry* statements to describe the features of the part,
—*Postprocessor* statements to identify machine tool and control system data,
—*Auxiliary* statements to provide additional information that is not given in any other statement.

The statements must follow grammatical rules for construction of statements and words, punctuation, and spelling. When the APT system is used, the spelling of an APT word must be exactly as it is spelled in the APT *system dictionary*. This dictionary specifies the only form the computer understands. The computer will indicate an inaccurately spelled word by an undefined symbol.

The following allowable words to be used for programming the turned part shown in Figure 33-6 are selected from and are written according to an APT dictionary. The spelling of each word to be used in programming the part is accompanied with a brief identification of the process, function, or location. An application is then given to show the usage of the word. The program planner first lists postprocessor and auxiliary statements in the manuscript:

PARTNØ — *Part number*. For information. The statement must be the first statement or appear immediately after END.

MACHIN — *Machine identification*. Example: MACHIN/HARDINGE 6. The computer is to make a control tape for a Hardinge turning machine number 6, equipped with a compatible input system.

INTØL — *Inside tolerance*. Example: INTØL/.001. The computer is to stay within 0.001″ on the inside of curves in making straight-line approximations of curves.

ØUTØL — *Outside tolerance.* Example: ØUTØL /.001. The computer is to stay within 0.001" on the outside of curves in making straight-line approximations of curves.

CUTTER — *Cutter.* Example: CUTTER/.94. The nose radius of the cutting tool is designated by the diameter of a theoretical circle. The 0.047" (3/64") radius of the cutting tool as shown on the drawing in Figure 33-6 is designated as CUTTER/.094.

CØØLNT — *Coolant.* Example: CØØLNT/ØN. Turn the coolant on. It will remain on until CØØLNT/ØFF or STOP command.

CLPRNT — *Print out.* Coordinate dimensions of all end points and straight-line moves are called to be printed out.

FEDRAT — *Feed rate.* Example: FEDRAT/4, IPM. Feed rate in all directions, including Z, will be 4" per minute.

SPINDL — *Spindle.* Example: SPINDL/400, RPM. Start spindle at 400 RPM. Spindle stays on until SPINDL/ØFF, STØP, or END.

Postprocessor and auxiliary statements are followed by geometry and then motion statements:

PØINT — *Point.* Example: P2 = PØINT/6.25, 2.20, 1.50. P2 is the point with coordinates X = 6.25, Y = 2.20, and Z = 1.50.

LINE — *Line.* Example: L1 = LINE/P1, P2. Line 1 is the line through points P1 and P2. Example: L3 = LINE/P3, RIGHT, TANTØ, C1. Line 3 is a line, tangent to circle 1 (C1), right of P3, through points P3 and P4.

The program and machining cycle are completed by using closing statements:

CØØLNT — *Coolant.* Example: CØØLNT/ØFF. Turn the coolant off.

FINI — *Part program is completed.* FINI must be the only word in the statement.

HOW TO PREPARE AN APT PROGRAM MANUSCRIPT

Planning the Program

Note: The alignment pin shown in Figure 33-6 is used to illustrate the preliminary steps and the kind of input required to prepare an APT program manuscript. The part is to be turned on a NC lathe.

STEP 1 Study the part drawing in terms of design features, machining processes, tooling, and machine characteristics.

STEP 2 Prepare a sketch to identify the path of the cutting tool in machining the workpiece.

STEP 3 Add the programming layout to the sketch (as shown in Figure 33-7).

STEP 4 Determine the sequence of each APT word to be used in the program.

STEP 5 Check an APT language dictionary for the correct spelling of each word.

Prepare the Program Manuscript

STEP 1 Prepare the computer-assisted program in APT processor language. Start with postprocessor and auxiliary statements—for example,

PARTNØ   ALIGNMENT PIN NØ6
MACHIN/HARDINGE 6
INTØL/.001
ØUTØL/.001
CUTTER/.094
CØØLNT/ØN
CLPRNT
FEDRAT/4, IPM
SPINDL/400, RPM

STEP 2 Add the geometry statements in APT language—for example,

SP = PØINT/7.75, -1.625
P1 = PØINT/7.5, -.613
P2 = PØINT/6.25, -.613
P3 = PØINT/4.5, -1.0
P4 = PØINT/2.0, -1.5
P5 = PØINT/0, -1.5

```
L1 = LINE/P1, P2
L2 = LINE/P2, P3
C1 = CIRCLE/2.5, -1.5, .5
L3 = LINE/P3, RIGHT, TANTØ, C1
L4 = LINE/P4, P5
```

STEP 3 Follow the geometry statements with the motion statements—for example,

```
FRØM/SP
GØ/TØ, L1
GØLFT/L1, TØ, L2
GØLFT/L2, PAST, L3
GØRGT/L3, TANTØ, C1
GØFWD/C1, PAST, L4
GØTØ/SP
```

STEP 4 Conclude the program with two closing statements—for example,

```
CØØLNT/ØFF
FINI
```

BASIC COMPONENTS OF A CNC SYSTEM

The maximum productivity of a machining center depends on the versatility of the computer numerical control system. For example, the Acramatic® CNC system, shown in Figure 33-8, is designed with solid-state circuitry and a minicomputer. This combination of hardware and software systems provides flexibility of application to a complete CNC machining center (Figure 33-9). The CNC unit may be adapted at a later date to a complete computer-aided manufacturing center (CAM). The unit is designed to allow for the addition of advanced features, data input/output devices, monitors, and sensors; tool path modification calculations; and tool management information. The CNC system (Figure 33-8) has four basic components: (1) tape reader, (2) cathode ray tube (CRT) display, (3) keyboard, and (4) manual and NC panel.

Tape Reader

The tape reader in a CNC system console may be of the photoelectric type, as shown in Figure 33-10. This reader is designed for forward and reverse direction operation and has a reading capability of 150 letters per second (lps). The tape reader has an automatic search and rewind capability and take-up spools.

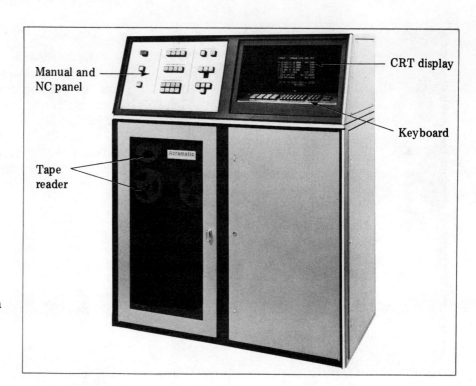

FIGURE 33-8 CNC System for a Machining Center or a Complete Computer-Aided Manufacturing (CAM) Center

684 NC SYSTEMS AND MACHINE PROCESSING

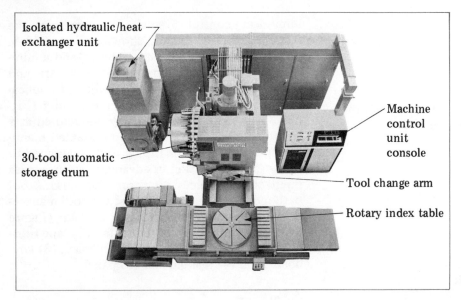

FIGURE 33-9 Complete CNC Machining Center with Rotary Index Table, 30-Tool Automatic Storage, and Tool Change Arm

Cathode Ray Tube (CRT)

The cathode ray tube (CRT) provides the operator with a visual display. The screen permits information to be displayed in an easy-to-read format. The readings are clearly visible from normal operating positions around the machine.

The CRT display provides part program information, error messages, diagnostic instructions, operating data, and machining status messages. Information is displayed on the CRT as the program progresses.

Keyboard for CNC System

The keyboard component, illustrated in Figure 33-11, is *alphanumeric*. The keyboard permits the CNC operator to communicate with the computer control system. The keyboard provides for human element input as required to operate NC machining centers. Inputting, deleting, correcting data, and running diagnostics are initiated through the keyboard.

FIGURE 33-10 Photoelectric Reader with Bi-Directional Operation

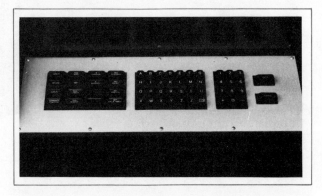

FIGURE 33-11 Keyboard for CNC System

DESIGN FEATURES OF A CNC SYSTEM

Multiple-Axis Contouring

The Acramatic® CNC system (Figure 33-8) uses linear interpolation for three-axis (X, Y, Z) contouring. These axes may be programmed to move simultaneously with a table indexing (B) axis. Additional axis movements are available.

Fixed Machining Cycles

Defined EIA machining cycles—for example, drill, ream, bore, and tap, including dwell—are used to initiate a complete automatic sequence of machining processes within one programmed block of information. Other systems require a separate block for each function in the cycle.

A package of EIA machining cycles saves programming, tape punching, and reader time. A part program may also be stored with less memory and tape length.

Buffer Storage

The machine tool operates from one block of information executed in the control. At the same time, up to 400 characters of data are read and held in advance in buffer storage.

Linear and Circular Interpolation

Any two axes (X-Y, X-Z, Y-Z) may be moved simultaneously. All three axes (X, Y, Z) may be moved in a straight-line path at conventional or rapid feed rates. High-precision angle cuts may be machined. A curve may be machined by cutting a series of short, straight-line segments along an arc.

The CNC system has the capability to machine arcs in a two-axis plane in any circular quadrant. Program input of start point, end point, and centerline of circular arc is used. Instantaneous switching is provided from circular and linear interpolation modes.

Axis Inversion

A keyboard key is used to invert the sign of the X and/or Y dimensional data. The axis inversion feature permits using a single tape to machine symmetrical or right- or left-hand parts.

Programmed Slide Directions and Movement Mode

Directions of linear slide movement along X, Y, and Z axes (planes) may be programmed. The direction is controlled by preceding programmed directions by a + or − sign. Programs may be written in the absolute or incremental dimensional system or in a combination of these systems. Programmed instructions automatically switch to the required code.

The Acramatic® CNC system accepts punched tape programmed to either EIA RS-244 or (ASCII) EIA RS-358 standard. Inch-standard or metric programs for each controlled axis are initiated by a control panel push button. Minimum input units of 0.001″ or 0.001mm (as selected) are automatically displayed on the CRT.

Tape-Controlled Programming

Tape-controlled programming features include:

— Tape-controlled feed rates, programmed in ipm or mm/min;
— Tape-controlled spindle speeds, programmed directly in RPM;
— Rotary table index positions, programmed directly in degrees (noninterpolated rotary axis movement is simultaneous with any X-Y-Z plane movement);
— Preselected program pockets or tool numbers, tape controlled for random tool selection;
— Machine slides, programmed for axis dwell that may be varied from 0.01 to 99.99 seconds in 0.01 second increments (the X, Y, Z slides may be kept motionless in a cut to permit a higher-quality surface finish and greater accuracy for certain machining operations);
— Program interruption, permitting a manual tool change.

Other Standard CNC Features

A few other CNC system features include:

— Slash (/) code, which is an operator selection feature;
— Block delete, which determines whether or not blocks of programmed data, prefixed by the code, will be ignored;

- Two levels of manual control (an operator may use the keyboard to modify the programmed rate by 1% increments from 1% to 999% or, with the *percent potentiometer* in the control, may reduce programmed feed values from 100% to 10%);
- Spindle keyboard override which permits on-site adjustments for material hardness or cutting action through keyboard modification of the programmed spindle speed (spindle speed changes are possible in 1% increments from 1% to 999%);
- Gage height feature, which establishes the limit may be programmed as a six-digit R word);
- Data search, which permits running the tape in forward or reverse direction (special block or a sequence number may be quickly located by data search);
- Manual power axis feed, which permits independent control of each linear axis (three feed ranges may be manually selected from 0.1 ipm to 1.0 ipm, 1.0 to 10.0 ipm, and 10.0 to 100.0 ipm);
- Incremental jog, which permits independent control of each linear axis (the control may be in low, medium, or high feed ranges);
- Target point align, which provides automatic positioning of X, Y, and Z axes (the axes are positioned to respective limit switches);
- Position set, which controls floating zero capacity (correct part-to-part relationship is established without physical movement of the workpiece);
- Inhibit feature, which permits the operator to interrupt machine activities (the operator is provided with on-site capability to stop axis moves, automatic tool changes, spindle rotation, and coolant operation during the workpiece program tryout);
- Push button emergency stop, which initiates stoppage of power to servodrives and feed commands;
- Dry run feature, which allows the program to be cycled through a noncutting tryout at higher than programmed feed rates in order to cycle the numerical control machine tool through all movements and machining processes and check the program before a part is produced.

EXPANDING PRODUCTIVITY AND FLEXIBILITY OF A CNC SYSTEM

The flexibility and productivity of a NC machine tool, machining center, or computer-assisted machining system may be expanded through additional compensation features for cutter diameter, tool length storage, and workpiece irregularities. A cutter diameter compensation (CDC) feature permits the program to be altered mechanically for variations in cutter diameter. Compensation may be made for undersized or oversized cutter diameters. The keyboard permits compensation in increments of 0.0001" up to a maximum of ±1.0000". The CDC feature is operable in linear and circular interpolation modes, but only in the X-Y plane.

The tool length storage/compensation feature permits accurate tool depth adjustment. This feature eliminates the preset tooling stage. The feature also permits storing a six-digit length of tool dimension in core memory for each available tool pocket. Tool length compensation may be made for variables in increments of 0.0001" up to a maximum of 99.9999".

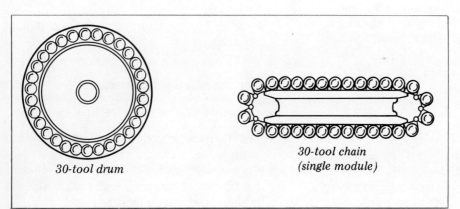

FIGURE 33-12 Automatic Tool Drum and Tool Chain

BASIC MACHINE TOOLS, CNC PROGRAMMING APPLICATIONS 687

Variations in setup or irregularities of workpieces may be compensated for by an assignable tool length trim feature. This feature is available for groups of 16 program selectable tool trims. The trim range is ±0.0001" to ±1.0000".

PRESETTING NUMERICAL CONTROL CUTTING TOOLS

The cutting tools for machining a specific part on a NC machine tool are usually preset for length. The complete set of preadjusted tools are then stored for future use. When needed, the cutting tool set is placed in the NC changer, a tool drum, or a tool chain. Figure 33-12 shows an automatic 30-tool drum and a single-module 30-tool chain.

Tool Length Gage

The length a cutting tool extends from its toolholder may be measured by a *tool length gage*. Figure 33-13 provides one example of a NC *dial indicator gage* for tool length and diameter. The vertical column includes a series of accurately spaced collars. These collars are usually 1" (25.4mm) apart. A micrometer head with 1" (25.4mm) travel is used to cover all measurements in increments of 0.0001" (0.002mm) within the range of the gage.

Each cutting tool is set in its holder. The distance the cutting tool projects beyond the holder is adjusted to meet job requirements.

Electronic Tool Gage With Digital Readout

An *electronic tool gage* provides highly discriminating tool length measurements and adjustments. Figure 33-14 shows an electronic tool gage for length and diameter with digital readout. An advantage over dial indicator types is the ability of the electronic gage to provide instantaneous readout in either customary inch or metric units of measure.

The digital tool length gage may be used for the off-line gaging of tool lengths and diameters. A tool is gaged for length by inserting it into the gage spindle socket. The feeler tip of the tool gage is then brought into contact with the tool point or edge. The enter button is pressed for storage in memory of the CNC system. Figure 33-15 shows an application of the electronic tool

FIGURE 33-13 NC Dial Indicator Gage for Tool Length and Diameter

FIGURE 33-14 Electronic Tool Gage for Length and Diameter with Digital Readout

688 NC SYSTEMS AND MACHINE PROCESSING

FIGURE 33-15 Application of Electronic Tool Gage in Setting an Insert-Tooth Milling Cutter for Diameter

gage in setting an insert-tooth milling cutter for diameter.

When used in conjunction with hard-wired numerical control systems, the electronic tool gage permits tool dimensions to be transferred into the control by means of compensation switches, a keyboard, or punched tape entry of tool data. All tool data, including storage location, tool identification number, tool length, and cutter diameter compensation, are preselected off-line.

SAFE PRACTICES FOR COMPUTER NUMERICAL CONTROL MACHINE TOOLS

—Follow the same safe operating procedures as for conventional work-holding and cutter-holding setups and machining processes.
—Establish whether the MCU console has an emergency work table and/or spindle-positioning stop control. Use the stop control as needed or whenever there is possibility of a personal accident or damage to the cutting tools, machine, table, or other accessories.
—Check the clearance of the work table, vises, clamps, or spindle when manually positioning the work table or spindle. This clearance check is required before positioning data is entered in the MCU console.
—Check the cutter and workpiece clearance before lowering the quill (and cutting tool) on a drill press or vertical mill.
—Reduce the overhang of a quill to ensure maximum rigidity.
—Check the quill movement to be sure a cutter may be raised to clear the workpiece when the quill-up control is actuated by the MCU.
—Adjust the rapid traverse of the quill as it approaches the workpiece to prevent the cutter from running into it.
—Read the tape after it is prepared. Check the printout against the program manuscript.
—Verify the tape input by a dry run on the machine tool.
—Observe each operation as it is performed on the NC machine tool, including checks for accuracy.
—Maintain a duplicate, permanent tape record in a separate storage location.

TERMS USED WITH COMPUTER-ASSISTED NUMERICAL CONTROL MACHINE TOOLS

APT   A computer system for Automatically Programmed Tools. A multi-axis contouring program. A collection of computer programs in which a shortened English language vocabulary enables the part programmer to direct a cutter and its path to produce a required part.

Automatic programming	Use of a digital computer to transform what a part programmer is able to program into a computer programming activity.
Block count readout	NC cabinet display showing the number of blocks that have been read from the tape, counting each block as it is read.
Computer program (numerical control)	Computer instructions for processing a part program that result ultimately in a control tape.
Digital input data	Pulses, digits, or other coding elements that supply information to a machine control.
Digital signal	The presence or absence of information content.
Encoding	Translating to a coded form with limited loss of information.
End-point control	Continuous automatic analysis of the final product. Changing the process as required for quality control.
Floating zero	The movement of the zero reference point on an axis. Establishing the zero point at any point in the travel.
Integrated circuit	An electronic circuit. An inseparable assembly of components contained in a small structure.
Off-line data processing	Computations of a geometric and mathematical nature, tool offsets, feed codes, paths, and others; miscellaneous and preparatory function tasks. Generation of a control tape.
On-line data processing	Translation of the information on a control tape into input signals on a NC machine tool. (On-line data processing is repeated each time the control tape is used to produce a part.)
Processor	A computer program that performs computations that are part oriented, computes tool offsets, and places the cutting tool on the part. A portion of the computer that controls input and output devices. A unit that operates on received, stored, and transmitted data.
Postprocessor	A computer program that translates the results of a basic computer program into a control tape.
Readout	A numerical display showing the actual position of a machine slide or tool.
Readout (command)	A display showing the absolute position. An absolute position derived from a position command. Readout information taken directly from the dimension storage command or as a summation of command departures.

SUMMARY

Electronic components within a NC system require controlled temperatures between 50°F and 120°F (10°C to 48°C) and humidity control up to 95%.

NC system controls for linear motions may be set in increments from 0.000020'' (0.0005mm) to 0.0001'' (0.0025mm).

Rotary motion is controlled within a range from 1.3 to 3.6 seconds of arc.

Complete documentation is provided for programming, operating, and maintaining each particular NC system.

A control panel display visually indicates tape dimensions for controlling movements.

A NC machine tool program may be interrupted to add operations, troubleshoot, and perform checking procedures.

Continuous-path machining capability may be added to a point-to-point system for general drilling machines.

A feed engage point may be programmed. This point permits changing the rapid traverse to the starting point of the process to a slower feed.

Two-axis NC systems may be used for general straight turning processes.

Backlash compensation mechanisms on NC machine tools permit accurate positioning and maintain machine reliability.

A tape-controlled spindle speed selector on a NC engine lathe stores and relays commands to the speed-changing device on the machine.

A continuous-path system permits taper cutting, threading, and contour turning.

Features of NC engine lathes are applicable to general turret lathe machining. The features include but are not limited to: dimension display, sequence number display, manual data input, tool select, tool offset, spindle speed and feed controls, and auxiliary function display.

A combination NC system permits simultaneous control of all axes, positioning of spindle, built-in cycles for depth axis, and continuous-path control.

NC jig borers may be positioned within 0.000030" for longitudinal and traverse travel. The machine may be equipped with a rotary table to serve as a fully numerically controlled axis. Jig borers are generally designed for point-to-point positioning on the X and Y axes.

NC machining centers combine the processes performed on a number of machines into one center. Such a center may have tape control of three, four, or five axes and simultaneous movement of the spindle head, table traverse, table rotation, and table tilt.

NC inspection machines provide a visual dimension display or a printout of variations from a preset size.

On-line data processing requires the translation of information from a control tape that feeds input signals directly to a machine tool.

Off-line data processing permits part programming in advance. The data are then used repeatedly for any number of manufacturing processes.

A processor program is workpiece and tool oriented to place the cutting tool on the workpiece. Other functions relate to programming processes that are machine tool and numerically control oriented and additional information that needs to be interlocked to produce a required part.

Postprocessor programs initiate additional computations. These programs ensure compatibility among the processor program, machine tool, and part specifications.

Computer-assisted numerical control programming (CNC) eliminates time-consuming numerical coordinate data and other complex calculations.

APT is a multi-axis contouring program. It uses an abridged English word vocabulary to direct the path of a cutter, to completely program all machining processes, and to make dimensional accuracy checks.

A parity check permits a review of tape information. Errors in tape preparation and other possible malfunctions are identified.

A machine control unit (MCU) may be hard wired to permanently connect electronic components to control one type of machine tool. Soft-wired CNC units include a control (executive) program entered into computer memory.

> The cathode ray tube (CRT) of a CNC unit provides visual positioning, operating, storage, diagnostic, and other control information.

A control processing unit (CPU) executes output instructions from a computer, as required.

> The part programmer must thoroughly understand the processor language (APT, Compact II®, or other) and its precise application in programming to machine a part.

A particular program language dictionary is used, depending on the system. The correct spelling of words and specific symbols are the only forms the computer accepts.

> A programmed layout drawing may be prepared to identify each process and starting, ending, and reference point.

Postprocessor and auxiliary statements are entered into a CNC manuscript in APT (or other language) format.

> CNC systems are designed with combination hard-ware and soft-ware flexibility. The CNC capability is extended by adding advanced data input/output devices, monitors, sensors, tool path modification, tool management, and other functions.

Advanced CNC systems may be adapted to a complete computer-aided manufacturing (CAM) center.

> CNC standard design features provide for multiple-axis contouring, fixed machining cycles, linear and circular interpolation, buffer storage, axis inversion, programming slide direction, and movement mode.

CNC system tape-controlled programming relates to feed rates, spindle speeds, preselected tool numbers, axis dwell, inhibited functions, and rotary axis movement.

> A few other CNC system features include: block delete, slash (/) code, spindle keyboard override, data search, incremental jog, zero position set, dry run, and emergency stoppage.

CNC system productivity may be extended by the addition of cutter diameter compensation, tool length compensation, and tool length trim features.

> NC cutting tools may be preset with instruments on a tool length gage or with an electronic tool gage with digital readout.

Standard personal, workpiece, cutting tool, and machine tool safety procedures apply to NC and CNC machine tool setups, operations, and inspection.

> Added precautions must be taken to provide sufficient clearance between cutting tools, work-holding devices, and the part to be machined, especially during positioning movements.

Verification of dimensions and work processes requires a review of the tape printout, a dry run, cycling through the production of the first part, and inspection.

UNIT 33 REVIEW AND SELF-TEST

1. Identify three basic control features that apply to all NC machine tools.
2. Indicate three functions that may be displayed on a control panel.
3. Cite one advantage of using a circular interpolation feature to generate curved surfaces in comparison to a linear interpolation feature.
4. Identify six features that are common to NC jig borers.
5. Differentiate between on-line and off-line data processing.
6. State the reason for parity check.
7. Identify two major features for computers used in computer-assisted programming.
8. The contour on follower cam No. 12 as shown on the part drawing below is to be end milled on a NC vertical milling machine. A 1" diameter end mill is to be used.
   a. Make a program planning sketch. Indicate the sequence of events.
   b. Write the program in the APT system. Group the statements according to functions.

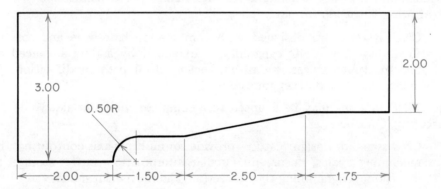

9. State three CNC machine functions that may be tape controlled.
10. Explain the function of a cutter diameter compensation (CDC) feature on a CNC machine tool.
11. State two safety precautions that should be taken with tape information.

PART NINE

Tool Design and Tooling Production

SECTION ONE   Applications of Design Principles
　　　　UNIT 34   Producing Jigs and Fixtures, Dies, Cutting Tools, and Gages
SECTION TWO   Construction and Measurement of Precision Production Tooling
　　　　UNIT 35   Jig Boring, Jig Grinding, and Universal Measuring Machines and Processes

SECTION ONE

Applications of Design Principles

Preceding units have related to the technology and the accompanying processes that underlie the use of hand, cutting, and layout tools; measuring instruments; machine tool work-holding devices, setups, and operation; and NC and CNC applications. This technology and the companion skills are applied by the toolmaker to construct production tooling. The most common categories of such tooling include: jigs and fixtures, punches and dies, special cutting tools, and gages.

This section builds upon the basic information and applications covered in earlier units for general tooling with cutting tools, gages, and jigs and fixtures. Three basic types of jigs and fixtures (open, closed, and indexing) and blanking, piercing, bending, forming, and drawing dies are covered.

Shearing
Notching
Extruding

While work-holding/positioning devices and cutting tools were described earlier according to functions on conventional machine tools, additional equipment and tooling requirements are applied in this section to NC machine tools. NC turret lathes, drilling machines, and milling machine applications provide examples. Selected design geometry for both high-speed steel and carbide cutting tools is related to NC machining processes.

UNIT 34

Producing Jigs and Fixtures, Dies, Cutting Tools, and Gages

POSITIONING AND CLAMPING (WORK-HOLDING) METHODS

The term *positioning* is used to refer to the dimensional relationship between a workpiece and a cutting tool. Consideration is given in positioning to characteristics of the workpiece. The shape (straight, circular, or contour), surface condition (plane, curved, or rough), size, and construction features are considered for both positioning and clamping the workpiece.

Locating devices in jigs are designed with regard to a basic reference plane. The machine table provides a reference plane surface. The table prevents cutting forces from moving the workpiece. However, longitudinal and cross feed forces may cause the workpiece to shift. In general practice, three *locators* are required in jig and fixture work to establish the same location of the workpiece. Figure 34-1 shows a typical work-positioning setup in which the workpiece is positioned on the machine table against locator pins in relation to the plane of the base, the longitudinal movement of the machine table, and the cross feed movement.

Locating from Circular Surfaces

Round surfaces require location from an axis of a circular section. Round workpieces are usually located and clamped by *concentric location*. Piece parts (workpieces) that are drilled or that are to be located according to holes in the workpiece may be located by *conical location*. Machine centers, nesting beveled holders, and locating pins (cones) provide common locating devices. V-blocks provide another common locating device.

Locating from Irregular Surfaces

Weldments, castings, and forgings often have irregular surfaces. In order to prevent deformation (stress forces that alter the shape) when clamping forces are applied, it is necessary to shim or jack the part. *Adjustable rest pins* are used to compensate for the condition of the workpiece surface. Poppets and other adjustable locators provide for surface and size variations. Once positioned, adjustable locators are usually secured by some form of lock nut. *Sight location* is used, particularly for first-operation positioning. Sight features (scribed lines, punch marks, or sight holes) are incorporated in the tooling design for rough location of the part.

TYPES OF LOCATING DEVICES AND METHODS

Locating devices have been standardized to the extent that they are commercially available in a wide variety of shapes and sizes. The devices that follow are *stock items*.

Pin, Button, and Plug Locators

Pins are used in jig, fixture, and die work. Pins are generally used for horizontal location. Larger sizes of pins are often called *plugs*. Shorter pins, generally used for vertical location, are known as *buttons*. Three common applications of button, plug, and pin locators are shown in Figure 34-2A, 34-2B, and 34-2C.

When two holes are used for locating, it is common practice to use two pins of different lengths or to use a *diamond locating pin*. The shape of this pin is a variation of the round pin. Two sides are relieved to provide for any center-to-center dimensional variations.

Rest Plates and Pads

Rest pads and plates are used for work support and positioning. Two common applications of rest plates and pads are illustrated in Figure 34-3. Rest pads and plates are large bearing surfaces on or against which workpiece surfaces may be positioned. A pad may be grooved to permit easy chip removal and proper seating.

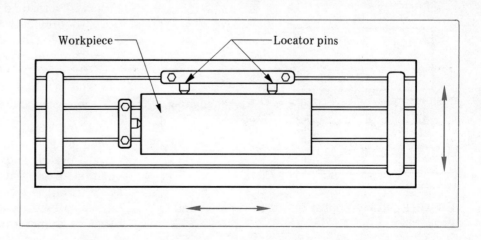

FIGURE 34-1 Workpiece Positioned on Machine Table against Locator Pins as Reference Points

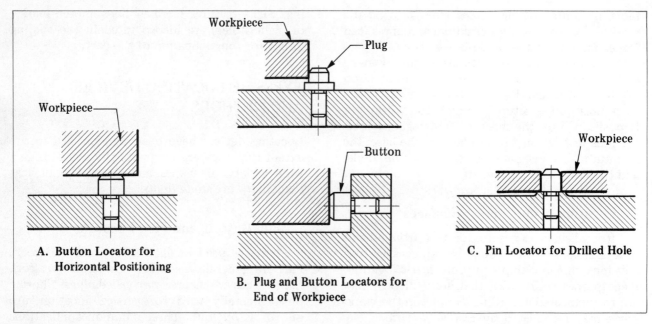

FIGURE 34-2 Applications of Button, Plug, and Pin Locators

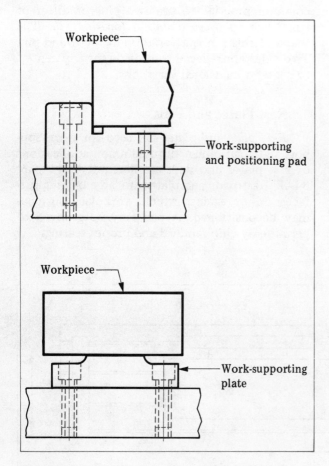

FIGURE 34-3 Applications of Rest Plates and Pads

Locating Nests

Locating nests are areas that receive a piece part (workpiece) and locate it without the need of any supplementary device. A nest accommodates size variations, permits easy removal of the piece part, and provides adequate space for chip control.

PRINCIPLES OF CLAMPING AND COMMON DESIGN FEATURES

Clamps fulfill four requirements:

—To hold the workpiece securely so that it will withstand the cutting forces;
—To provide a quick-acting device for efficient loading and unloading;
—To clamp without distortion or damage to the workpiece;
—To provide positive clamping, which resists vibration and chatter.

Therefore, certain conditions must be observed in clamping. Clamping forces must be directed toward nesting and locating surfaces. Clamps are positioned so that cutting forces are directed away from the clamping setup. Clamping forces must also be directed against areas of the workpiece or a pad that can safely handle the re-

JIGS AND FIXTURES, DIES, CUTTING TOOLS, GAGES **697**

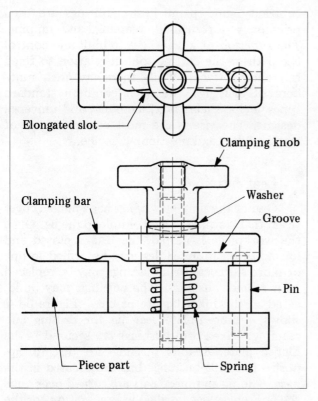

FIGURE 34-4 Main Features of a Strap Clamp

quired holding force. The clamp design permits the loading of the workpiece in only one (the correct) way.

Strap Clamps

Strap clamps work on a lever principle. While there are many designs of strap clamps, depending on the application, the main features are shown in Figure 34-4. The elongated slot and the groove permit the strap clamp to be slid away from the piece part for easy work removal or loading.

The strap clamp is generally tightened by a hand knob, hexagon nut, or knurled handle. More complex tooling requires a cam-locking device, pneumatic tightening, or other nonhand actuating clamping method.

Quick-Acting Swing and Cam Clamps

Swing clamps pivot so that the clamping end may be swung over the workpiece. Once this position is reached, the lever rests at a fixed height to apply the correct clamping force. *Eccentric and spiral cams* are other forms of quick-locking devices. The end of the cam is designed as a spiral or an eccentric. The cam forces the piece part to nest against pins or other work-locating device. Once the piece part is in position, the cam action provides the holding force necessary to machine the part. Figure 34-5 shows a quick-acting swing and cam clamp, which combines the features of the standard strap clamp and the quick action of the cam clamp.

Latch and Wedge Clamps

Latch clamps have the advantage of speed in clamping a piece part in position. One of the simplest forms is a thumbscrew latch. Figure 34-6 illustrates an example of a latch clamp application. A quarter turn of the thumbscrew locks a jig leaf in position.

Wedge clamps depend on a taper angle and levers. The wedge angle provides the holding force. The lever ensures that the wedge is locked securely to maintain a constant holding force against the piece part.

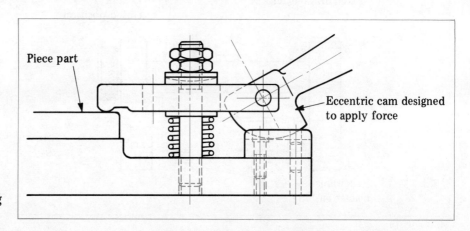

FIGURE 34-5 Quick-Acting Swing and Cam Clamp

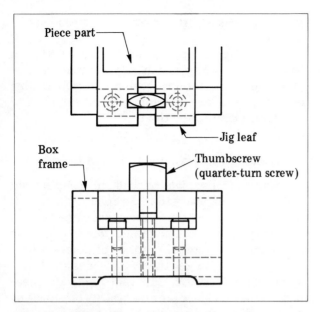

FIGURE 34-6 Example of a Latch Clamp Application

DRILL JIG DESIGN

Holes are incorporated in the design of most manufactured products. Holes may be cast, punched, flame cut, or sawed. Holes may be drilled, reamed, bored, counterbored, or threaded. Holes may be laid out and drilled on drilling, milling, turning, and other machine tools. Holes may be programmed and produced automatically on NC machines.

Drill jigs are still widely used in manufacturing to obtain precise hole location and accuracy. Accuracy relates to dimensional tolerances, concentricity, and parallelism. Drill jigs are used principally for drilling, reaming, and tapping. The function of a drill jig is to limit and control the path of the cutting tool in relation to fixed reference surfaces. While there are great numbers of drill jigs, the most common standard types include leaf, box, indexing, and universal designs. The vise is the most common type of work-holding and positioning device.

Leaf Jigs

One of the simplest *leaf jigs* has a hinged cover (leaf) that is swung to load or unload the jig. Once the workpiece is nested, the leaf is closed and locked in position. Holes may be drilled in one or more surfaces. Drill bushings may be replaced with reamer bushings. The bushings may be located in the hinged plate or base. Figure 34-7 shows a hinged-plate leaf jig for drilling and reaming. Fixed drill bushings are secured in the hinged plate. Reamer bushings are directly opposite the drill bushings but are located in the base. The jig and piece part are turned over once the holes have been drilled in preparation for the reaming process.

Box Jigs

Box jigs, as the name suggests, are box shaped. Bushings may be included on one, two, or more sides. Some box jigs are designed with jig feet on opposite sides of the part. Once a hole is machined, the jig and piece part are turned to the next machining position. When the second hole is machined, the jig and piece part may again

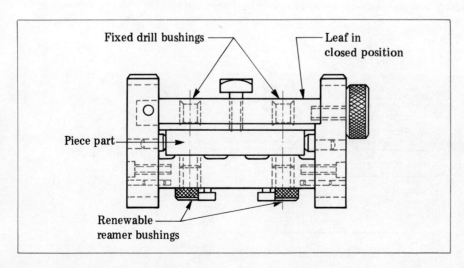

FIGURE 34-7 Leaf Jig for Drilling and Reaming

be rotated to expose the third jig face to the tooling in the machine spindle.

A *channel jig* is an adaptation of a box jig. The jig is designed in a three-sided channel form.

Open-Plate Jigs

Open-plate jigs may or may not have legs. The main construction feature is a plate. The plate is designed with liner bushings that permit the use of replaceable bushings for drilling, reaming, or tapping. Plate jigs are designed for carrying on a single machining process at one loading.

Indexing Jigs

Indexing jigs permit angle setting of a piece part for positioning successive holes in a circular pattern. The indexing may be done by using a standard indexing device. The piece part may be held in a standard chuck mounted on a graduated indexing device. The part is indexed for each location. A drill bushing, secured in an overarm on the knee of a drill fixture for the indexing mechanism, may be adjusted for height and position.

Universal Fixtures

Universal fixtures are commercially available for adaptation to particular hole-forming processes. For example, the fixture illustrated in Figure 34-8 is designed to be a drill jig and features a double post, rack and pinion, and simple clamping and release mechanism. The fixture is open through the guide post bosses. This opening permits parts to project through the fixture and to extend to the right and left of the base. This particular fixture is commercially available with a vertical movement range from 1/2" (12.7mm) on the 3-1/2" (88.9mm) opening height model to 1" (25.4mm) on the 9" (228.6mm) height model.

Vertical movement is precisely aligned—that is, the head and base are accurately positioned. Wear on bushings and cutting tools is eliminated. The downward movement of the lever clamps the piece part. The upward stroke releases the work-holding force and clears the piece part and fixture.

The *shut height* provides for a wide range of workpiece sizes. The adapter position of the head may be changed by turning the lever until the rack is disengaged from the pinion. Once the

FIGURE 34-8 Double-Post, Rack and Pinion Fixture

new adapter height position is established, the rack and pinion are reengaged. Figure 34-9 shows four basic types of fixture heads.

The manual hand crank on some drill jigs and fixtures is replaced with a *rotary actuator*. The fixture shown in Figure 34-10 is pneumatically operated (automated) by a rotary actuator. The construction and design features of the actuator are illustrated by the line drawing. The actuator is used to convert fluid (air) energy that moves a piston rod in a cylinder (linear motion) into rotary motion. The rotary motion substitutes for the manual motion normally required to open and close the drill jig. The rotary actuator is adaptable to advanced automation techniques and NC applications.

Vise Jigs

Short run, low-production drilling is often performed using a cam-actuated vise. Special jig plates are secured to the solid vise jaw. The nesting feature fits between the two jaws.

DRILL JIG DESIGN CONSIDERATIONS

Design considerations are influenced by the required machining specifications. Rigidity, simplicity, clamping forces, chip control, and jig feet and bushings are five prime concerns of the designer, toolmaker, and machine operator.

700 TOOL DESIGN AND TOOLING PRODUCTION

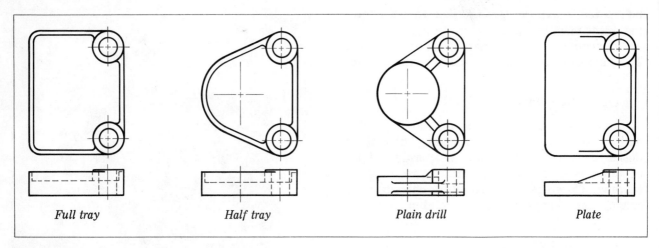

FIGURE 34-9 Four Basic Types of Fixture Heads

Rigidity

Rigidity relates to the ability of the jig to withstand work-holding and cutting forces. The jig must also be designed to support the piece part so that it does not bend during machining.

Simplicity and Clamping Forces

The jig is designed to nest the piece part so that it may be easily loaded and, when machined, removed. Clamping forces are directed toward the nesting feature. The locating points must permit loading in one way only.

Chip Control

Chip control requires, where practical, that a segmental type of chip be produced in preference to long, stringy continuous chips. Adequate space is provided between the piece part and the drill bushing to allow chips to flow freely between the work and the bushing plate. Jigs are provided with corner relief to help prevent chip buildup in corners.

Drill Jig Feet (Legs)

A drill jig seats better on a plane reference surface when the base is cut away to form legs (feet). There are usually four feet. They are ground to form the plane reference surface on the jig. The bases on the different sides on box jigs are also designed with legs.

Press-Fit and Renewable Bushings

In drilling, reaming, and tapping operations in jig work, bushings serve to position the cutting tool and to guide it. A clearance between 0.0005″

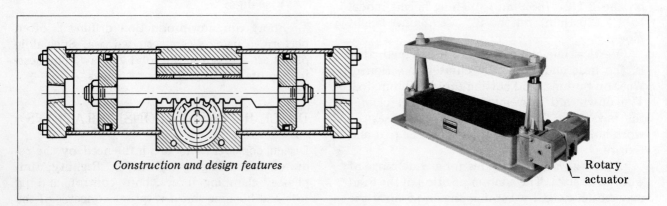

FIGURE 34-10 Fixture Pneumatically Operated (Automated) by a Rotary Actuator

(0.0125mm) and 0.001" (0.025mm) clearance is provided between the cutting tool and the bushing diameter. Larger clearance tends to cause drill margins to chip and to produce machining inaccuracies.

The basic types of press-fit and renewable bushings are shown in Figure 34-11. *Press-fit wearing bushings* (Figure 34-11A) are designed for permanent press-fit applications. The *headless type* is for flush mounting with the jig plate. This type is used for light axial loads. The permanent press-fit *head type* resists the effects of heavy axial loads. The drill plate may be counterbored to permit the head to be flush with the jig plate.

Renewable wearing bushings (Figure 34-11B) are used with *liners* on long production runs and when it is necessary to change bushing size to perform more than one operation in a hole. The outside diameter of the *slip type* is ground to a slip fit with the liner. The bushing is changed without removing the lock screw. The *fixed type* of slip renewable bushing is used for long production, single process (drilling or reaming) runs. The worn bushing is quickly replaced by removing the lock screw. The lock screw recess secures the new bushing in place.

Bushings are hardened to provide a wear-resisting surface. Bushings are ground externally and internally and are lapped to within a concentricity of 0.0003" (0.007mm). ANSI bushing classifications and size standards are used. There are many modifications of bushings—for example, thin-walled bushings for close hole drilling and double-guide bushings for aligning the starting and body pilots of a reamer that requires such guiding.

ADDITIONAL JIG CONSTRUCTION FEATURES

The components of a jig are usually assembled mechanically. The parts are machined before assembly. Worn parts may be replaced. Jigs constructed in this manner are designed with *dowels* for alignment and *fillister and socket-head cap screws* to secure the components. Dowel pin reamers, which are generally ground slightly undersize, are used to ream dowel pin holes. Dowels are press fit into matching parts.

Drill jigs are sometimes designed with welded components. Such jigs are stress-relieved before the components are machined to size.

NC MACHINING TECHNOLOGY IN JIG WORK

Some of the functions performed by drill jigs since the beginning of interchangeable manufacturing are now being done by precision drilling, boring, and other NC machine tools. Precision point-to-point and continuous-path positioning capability makes it possible to precisely locate hole positions. On NC drilling and boring machines, once the processes are programmed and a tape is produced, just a simple positioning fixture may be required to reference all slide movements and machining processes that are controlled by the tape. While jig borers and precision boring and NC machine tools eliminate the need for some drill jigs in production work, drill jigs are still widely used.

Headless type P press-fit bushing for flush mounting

Head type H press-fit bushing

Type S slip renewable bushing

Type S/F slip and fixed renewable bushing

FIGURE 34-11 Basic Types of Press-Fit and Renewable Bushings

A. Permanent Press-Fit Bushings

B. Renewable Bushings Used with Liners

FIXTURE DESIGN

A *fixture* is a work-holding device strapped securely to a machine tool. The fixture holds a piece part in a fixed-position relationship with one or more cutting tools during machining operations. Welding and assembly fixtures are used where parts are positioned in relation to each other in order to be fabricated.

Fixtures are generally classified as follows:

- According to the machine tool and type of operation performed—for example, face milling, straddle milling, and slotting fixtures used with a milling machine;
- According to the manner in which the workpiece is clamped—for example, hand clamping for pneumatic (power) clamping fixtures (automatic fixtures permit loading and unloading according to fixed cycles of machining processes);
- According to the techniques of locating the part—for example, V-block fixtures and center fixtures;
- According to the method of feeding the workpiece to the cutter—for example, indexing fixtures for permitting parts to be rotated (indexed) to the next position during the machining cycle and rotary drum fixtures for securing parts around a drum and permitting separate operations to be performed as the drum is rotated.

Vise Fixtures

Standard machine vises are widely used as fixtures. The jaws may be fitted with inserts designed to accommodate regular or irregular-shaped parts. Workpieces may be held in special jaws for machining surfaces in a horizontal plane. Surfaces may also be machined in second and third planes by swiveling a vise on its base or by positioning a vise and workpiece at a vertical angle when a compound angle vise is used.

Milling Fixtures

Milling fixtures are rigidly held to the work table. A part is secured in a fixture as it is moved past one or more cutting tools, as for gang or straddle milling. Most milling fixtures include a base plate for strapping to the table; nesting areas for supporting and holding the workpiece; locating points; and gaging surfaces. The base plate is usually slotted and keyed to ensure alignment with the table T-slots.

Milling fixtures are designed for permitting rigid clamping while at the same time not obstructing cutter movement. *Feeler (set block) surfaces* are designed as part of the fixture base. Set block surfaces are machined lower than the cutter depth so that a feeler gage may be used to set the cutter. Caution must be used in setting a cutter with a feeler gage. The cutter must be checked in advance for concentricity between the bore and the ground peripheral teeth. Cutter runout produces an inaccurate depth setting. Also, the machine spindle must be stopped in setting the cutter to depth.

Universal and Progressive Fixtures

Families of parts are held in *universal fixtures*. Workpieces in *progressive fixtures* are located in different positions and are moved progressively between machining stations until all processes are completed. The progressive machining stations on a rotary fixture are shown in Figure 34-12.

Magnetic and Vacuum Checks

Magnetic chucks are a practical clamping device for ferrous metal parts. Newer low-voltage magnetic chucks provide increased heat-free holding power and minimized distortion of the workpiece. Magnetic chuck fixtures are designed with

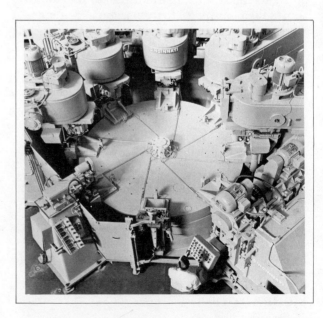

FIGURE 34-12 Progressive Machining Stations on a Rotary Fixture

JIGS AND FIXTURES, DIES, CUTTING TOOLS, GAGES 703

fixed workpiece stops against which cutting forces are directed.

Vacuum chucks, like magnetic chucks, have many advantages over mechanical, hydraulic, and other clamping devices. A vacuum chuck provides a nondistorting and nondamaging force on thin or fragile materials. One prime advantage is that, since less clearance space is required for vacuum chucking because clamping parts are not used, smaller-diameter cutters may be used.

Applications of two principal vacuum chucks are illustrated in Figures 34-13A and B. The rectangular-shaped, top-plate vacuum chuck (Figure 34-13A) is used primarily for conventional and numerical control milling machines, surface grinders, jig borers, and similar machine tools. The rotary vacuum chuck (Figure 34-13B) is adapted to faceplate work. This chuck is made for lathes and other turning machines, rotary abrasive grinding, and similar machining processes.

Vacuum chucks operate with a series of *valve ports* (openings). Each port is controlled with a valve screw to open or seal a port. Sectioning or controlling the vacuum area is thus permitted. The knurled nut on the vacuum fixture is turned to vary the holding power. After machining, the part is released by pushing the release valve. Vacuum chucks are especially adapted to hold ferrous metals and a wide variety of nonferrous and nonmagnetic materials.

Turning Machine Fixtures

Turning machine fixtures are specially designed for holding castings, forgings, and many irregular-shaped parts that cannot be held by conventional chucks. While general principles of fixture making apply, a number of other design and machine operator considerations are necessary:

—Since fast-revolving workpieces produce centrifugal forces, clamps and other work-holding devices must operate under excessive force conditions without loosening;

—Static and/or dynamic fixture balancing must be done before the workpiece is rotated under power (counterweights may be required to prevent vibration);

—Projections and sharp corners must be eliminated because a rotating fixture may be dangerous;

—Cutting forces require that the workpiece be gripped on the largest diameter of the workpiece (thin workpiece sections may need to be supported against cutting tool forces).

The use of chucks with standard jaws and horizontal rocking jaws was discussed earlier in relation to work-holding devices for turret lathe work. Other jaws are designed to rock in a vertical direction. The vertical jaws (on a three-jaw chuck) grip a cylindrical surface in six equally

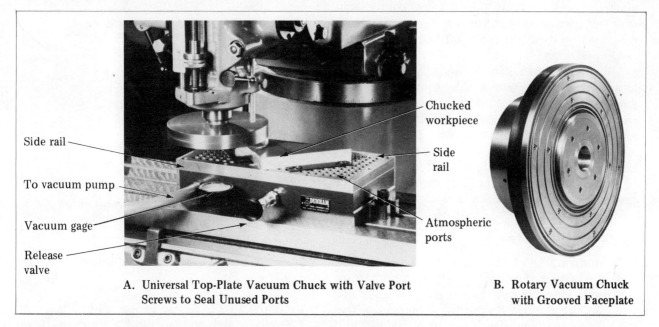

A. Universal Top-Plate Vacuum Chuck with Valve Port Screws to Seal Unused Ports

B. Rotary Vacuum Chuck with Grooved Faceplate

FIGURE 34-13 Applications of Two Principle Vacuum Chucks

spaced sections. Thus, the gripping force is equally distributed. Wrap-around jaws are used to minimize distortion by gripping over a greater area. These jaws are bored to within a thousandth of an inch of the turned diameter of the workpiece in order to evenly distribute the gripping force.

Faceplate Fixtures

Simple *faceplate fixtures* employ locating pins and require direct strapping using the faceplate slots and standard clamps. Other fixtures use a right angle bracket on which the workpiece is secured. Angle-type fixtures require balancing with a counterpoise to compensate for weight imbalance and centrifugal forces.

METAL DIE-CUTTING AND FORMING PROCESSES

The punch press is another machine tool that continues to make significant contributions to the manufacture of interchangeable parts. Parts are cut to internal and external sizes and are formed into simple and complex shapes by die-cutting processes such as blanking, piercing, lancing, bending, forming, cutting off, shaving, trimming, and drawing.

National security, national economy, and standards of everyday living depend on the output of parts, components, and devices that are manufactured by using punches and dies. Ferrous and nonferrous parts are economically mass-produced by dies that are identified by the processes performed.

Single or combined processes may be required to produce or assemble a simple or a complex part. The processes are referred to as *die cutting* and *die forming*. The cutting and forming tools are identified as *punches* and *dies*. The stamping and forming machines on which punches and dies are used are simply called *presses*, *punch presses*, or *power presses*. Categories of such presses and major components are discussed later.

Blanking Processes

Blanking refers to the process of cutting out (usually from a sheet, strip, or roll) a part to a specified shape and size. A punch of the required size and shape forces against the stock and, with the aid of a mating die section, cuts a blank. The blank is moved by successive cuttings through the die section cavity. The relieved sides of the die permit the blanks to fall free and be removed. On the return stroke, the excess material (strip stock) surrounding the punch is removed as the punch moves through a stripper plate. Subsequent pieces are blanked by positioning the stock against a stop pin and repeating the cutting and return stroke cycle. Figure 34-14 illustrates design features of a plain blanking punch and die set.

Piercing Processes

Piercing usually refers to the stamping of one or more holes in a strip of material. The holes are design features of the part that may be blanked simultaneously or at a different time.

Lancing Processes

During *lancing*, a workpiece is cut on two or three sides and formed along one or more sides. A blanking process may follow lancing.

Cutting-Off and Notching Processes

Cutting off refers to separating a blanked workpiece from the scrap strip. Cutting off is generally the final step in progressive die work.

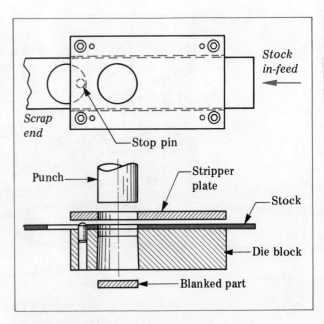

FIGURE 34-14 Design Features of a Plain Blanking Punch and Die Set

Cutting-off processes are straight-line cuts. *Notching* is used to remove material from one or both edges of the strip. Notching serves to remove excess material so that a metal formed part flows as it is being drawn or formed to shape.

Shaving and Trimming Processes

Shaving removes a very thin (small) area of metal from a previously cut edge. Shaving cuts the part to close dimensional tolerances and thus produces a smooth finished edge. *Trimming* is used to remove surplus, irregularly formed material from around the edges of a shaped or drawn part. A smooth accurate surface is produced. Both shaving and trimming are second operations.

POWER PRESSES FOR STAMPING AND FORMING PARTS

Punch presses range from simple bench presses for stamping small, fine parts that are a fractional part of an inch or centimeter to automated systems for large-volume, integrated production units. Such presses are designed with quick-change bolsters; floor-mounted, self-contained materials handling devices; and automated controls for mass-producing complicated panels and forms. While the sizes, tooling, and controls differ, there are a number of common features. The following descriptions of a few basic types of presses provide an overview of the range, capabilities, and characteristics of these machines.

Inclinable Presses

Inclinable presses are designed for metal cutting and forming operations that include blanking, forming, bending, drawing, assembling, and combination progressive die processes. The inclinable press illustrated in Figure 34-15 is available in standard sizes from 22 tons to 250 tons. The smaller sizes are adapted to production requirements for small and medium-sized stamped parts. High-speed gap frame presses, which are adaptations of the inclinable press, are operated at higher speeds up to 1,200 strokes per minute. Continuous automated feeding is employed.

Inclinable presses may be fitted with optional feed devices such as single or double roll, dial, and transfer devices. A combination air-friction clutch and brake are provided for fast safety stops. All models are available with variable-speed drives. The presses up to 110 tons are of cast Meehanite construction. Larger-capacity presses are of welded steel frame construction.

Single-Action Top-Drive Presses

Single-action top-drive presses are adapted to press operations that require dies with greatly varying heights. The construction of the operating surfaces between the two vertical outer frame members permits medium-sized parts to be fed through the press from front to back. Figure 34-16 illustrates a straight-side, single-action, eccentric gear design press that is widely used for blanking, piercing, stamping, forming, and similar operations on medium-sized parts. This press is available in capacities from 200 tons to 2,000 tons for use in aircraft, automotive, appliance, hardware, furniture making, and farm machinery parts production. The press, as illustrated, has a 300-ton capacity, 120″ × 84″ (3048mm × 2134mm) die area, and *rolling bolsters* that permit rapid, accurate die changes. The rolling bolster (with die attached) is moved into the press

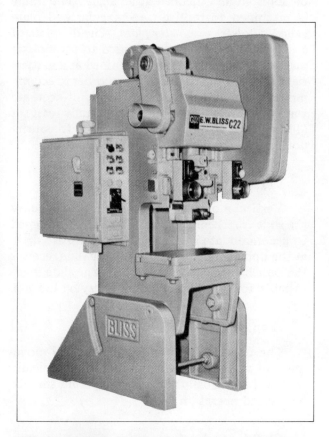

FIGURE 34-15 Inclinable Press for Blanking, Forming, and Assembly Operations

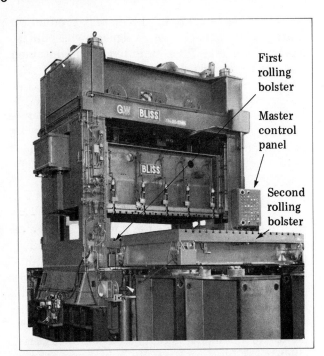

FIGURE 34-16 Single-Action, Eccentric Gear Design Press for Blanking, Piercing, Stamping, Forming, and Similar Operations

and automatically positioned and clamped in place. The bolsters are self-powered, interlocked, and controlled at the master panel.

Hydraulic Presses

Hydraulic presses serve functions similar to the mechanical types. Hydraulic presses are available from small, general-purpose machines to extrusion presses of 35,000 tons capacity. The machines are designed to utilize die slides, feed mechanisms, sliding or rolling bolsters, and many other accessories. Protection is provided against overloading either the dies or the press.

A universal electric control permits speed change and reversal of ram position. A stop button on the cycle control permits instant stopping at any point in the cycle. Die setup is aided by inching with push buttons. The press stops on release of the button. The hydraulic components are protected against wear by a positive filtering and cooling system that provides clean fluid at all times.

Special-Purpose Presses

In addition to standard presses, metal products manufacturing requires *special-purpose presses*. A few examples include: *cold extrusion presses*, for extrusion and piercing operations; *automatic foil presses*, designed for long feed lengths of wide but very thin material; *mechanical forging presses*, built to absorb heavy work load forces imposed by forging operations; and *welding presses*, to speed production of body parts and appliance panels that are to be permanently assembled by welding.

MAJOR COMPONENTS OF THE OPEN-BACK INCLINABLE PRESS

There are five basic components of an inclinable press. These components include the *frame and bed*, *slide (ram)*, *clutch*, *crankshaft*, and *variable-speed drive*.

Frame and Bed

The frame and bed on the inclinable press is mounted on vertical uprights. The frame may be adjusted and inclined at an angle. The frame is machined to receive a rectangular bed plate called a *bolster* or *bolster plate*. The die plate of a die set is mounted and secured to the bolster plate of the bed. Ways are machined at a vertical right angle to the bolster plate in order to accommodate the slide (ram) and to permit accurate vertical travel. The frame is generally cast of Meehanite and is ruggedly constructed to continuously withstand cutting and forming forces.

Vertical Slide (Ram)

Vertical movement on a press is produced by the slide (ram) component. The slide is designed to accommodate a crankshaft or eccentric drive at the upper end. The bottom of the slide receives the punch section of the die set. The slide is adjustable for positioning dies of varying heights.

Crankshaft

The crankshaft produces the vertical movement of the ram. The crankshaft is dynamically balanced and usually nongeared to permit smooth high-speed operation.

Clutch Mechanism

The clutch is generally a combination of a friction clutch and a brake. The clutch is crankshaft mounted and requires only a fraction of an inch travel between full clutch engagement and full brake. Some heavier models have an *unsticker feature*. A press that is stuck on bottom or just past bottom dead center may be freed by turning the press over under power, using this specially designed clutch. A reversing feature is used when the press stops before dead center.

Variable-Speed Drive

The variable-speed drive permits the operator to fine tune the press speed to the particular process and other cutitng and/or forming requirements. Some models are equipped with a mechanical or an eddy-current type of drive. While electrical drives permit speed adjustments from zero to maximum, as the speed is reduced, the horsepower decreases. Mechanical drives produce almost a constant drive but are limited to a particular ratio, such as 2:1 or 3:1.

Standard speed ranges are provided from 45 strokes per minute on presses with 8″ to 10″ (203.2mm to 254mm) lengths of stroke to 750 strokes on smaller presses with 1/2″ (12.7mm) length of stroke. High-speed presses of from 22 tons to 60 tons capacity may be operated at 1,200 strokes per minute. Press speeds are read on some models on a tachometer. Stepless speed adjustments may be made on the eddy-current type of clutch and the variable pitch pulley system from a machine control unit console while the press is running.

Optional Equipment

Roll feeds are available in either single roll feeds or double roll feeds. *Single roll feeds* push or pull the stock through the dies. This design is used where the scrap is cut up by the die. *Double roll feeds* provide for positive control of the strip at both entry and exit ends of dies. Single and double roll feed drives are either rack-and-pinion or lever type. The lever type is used for shorter feed lengths and higher press speed ranges.

Other feed accessories include *special roll finishes* (chrome plated, milled, sand blasted, or coated). Shear-type, press-actuated *scrap cutters* are available. *Slide feeds* and other special arrangements for front-to-back feeding, spray-type automatic *stock oilers*, *power run-in rolls* to start a new strip into the die, *strip-straightening devices*, and motor-driven and unpowered *coil winding equipment* are examples of additional feed accessories.

STROKE, SHUT HEIGHT, AND DIE SPACE

The toolmaker deals with stroke, shut height, and die space in constructing a die and in setting it up on the press for initial tryout. The *stroke* represents the reciprocating motion of the ram. Stroke is the distance between the position of the ram at the top and the bottom ends of a complete cycle. The stroke is the dimensional distance in inches or millimeters.

The *shut height* is measured with the ram at the bottom of the stroke and the stroke adjustment up. The shut height of a die must be less than the shut height of the press. The *die space* denotes the area available on the press for mounting dies.

DIE CLEARANCE AND CUTTING ACTION

The cutting action of punches and dies is similar to the cutting action of single- and multiple-point cutters used on other machine tools. The punch forces the workpiece against the die face. At the point where the *elastic limit* of the material is exceeded, the metal flows due to *plastic deformation*. At the start of the flow, depending on the clearance between the punch and die and other characteristics of the material, a minute radius is turned and a fine burr is produced. As the punched *slug* moves into and passes the land of the die, a burnishing action takes place on the sides. The burr side of the blanked part (slug) is on the upper side toward the punch. Conversely, the burred side of the stock is on the bottom surface next to the die face.

Die Clearance

Die clearance relates to the planned difference in size between the cutting edges of the punch and die. Clearance is most generally expressed in relation to each side. Clearance is determined by the thickness of stock, working properties of the material, and the nature of the blanked edges. Too great a clearance produces a larger roll-over edge and unsafe burred edges. Insufficient cutting clearance produces excessive cutting forces, reduces tool wear life, and may cause the die to fracture.

The size of the die opening establishes the dimensional size of the blank. In the case of a sheared hole, the size of the pierced hole is the size of the punch. The clearance is placed on the punch when the slug is to be blanked to the required size. The clearance is placed on the die when the pierced hole is to be the required punched part.

Angular Work Clearance

Angular work clearance is represented by the number of degrees per side below the narrow cutting land on the die. The angular work clearance permits a stamped part that fits tightly on the die land to be moved through the die. The tight fit is due to springback produced by grain structures being stressed below the elastic limit. Recommended angular clearances for band machining a punch and die from the same piece were given in an earlier unit on band machining.

Stripper Plates and Excess Stock Removal

As a punch cuts through the metal, the slug remains in the die while the strip clings to the punch. The excess stock is removed from the punch by using a *stripper plate*.

The two basic types of stripper plates are shown in Figure 34-17. The first type of stripper plate is fixed and is attached to the die block (Figure 34-17A). The bottom of the stripper plate is designed to provide a track for the stock. After a part is punched, on the return stroke, the stock and punch move a short distance toward the stripper plate. The stripper plate holds the stock as the punch continues its upward movement. Once freed, the strip is then moved to its next position as the stamping cycle continues.

The second type of stripper plate is mounted on the punch assembly and fitted around the punch with springs (Figure 34-17B). Once the blank is punched and the punch leaves the die block, the stripper plate forces the stock off the punch. The thickness of the stripper plate and the size of springs to use depend on the nature and thickness of the stock, the part size, the complexity of the part itself, and the surface finish of the punch.

Cutting Forces and Shapes of Cutting Faces

Cutting force for punch press operations is identified as the total amount of force required to completely blank a workpiece. While the designer is concerned with cutting forces within the capacity of the press, the toolmaker considers changes that may be needed to reduce cutting forces of a punch and die.

Cutting forces may be reduced by altering the heights of punches when several are included in the design. Another common technique is to grind the punch at a slight shear angle to the horizontal plane. Shear reduces the cutting forces.

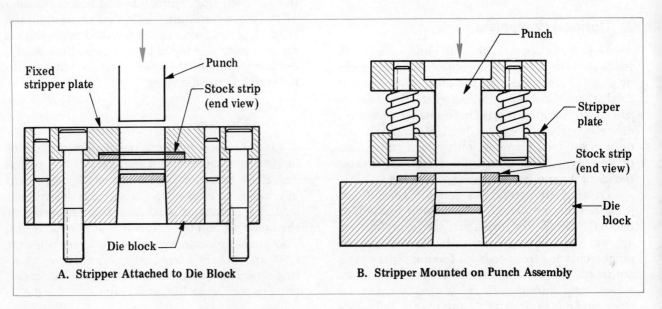

FIGURE 34-17 Basic Types of Stripper Plates

Punches and dies may also be ground with a double-angle shear. For blanking operations, the double-angle shear is on the die plate; for piercing, on the punch. Shear distorts the work material.

DIE SETS AND PUNCH AND DIE MOUNTINGS

Standardized *die sets* are widely used to maintain precise alignment between the punch and die. A standardized two-post die set is shown in Figure 34-18. A die set consists of a base that is flanged for mounting on the bolster plate of the press. The base is called the *die shoe*. *Guide pins* are pressed in the base and serve to keep the punch and die aligned. The guide pins ride in *guide bushings* in the top plate of the die set. The top plate is known as the *punch holder*. The punch and die sections are mounted between the die shoe and punch holder in the *die area*. The shut height of the die set is the vertical distance between the bottom surface of the die shoe and the top face of the punch holder when the die is in a closed position.

Die sets are commercially available with or without a shank on the punch. The guide pins and guide bushings may be centered, mounted at the back end of the die set, or positioned at diagonal corners. Die sets are made with two or four guide pins and bushings. When the shank-type holder is used, the punch is secured to the ram by two tightening nuts in front of the ram.

Standard die sets are available in commercial and precision grades of accuracy. Grade usually relates to the quality of fit between the guide pins and the bushings. The tolerances for the precision grade are finer and provide for extremely accurate alignment between multiple punches and corresponding holes in a die block.

The die shoe and punch holder in standard die sets are made of a high-quality cast iron or cast steel. The cast iron die sets are used where the processes produce a limited amount of shock. Excessive shock would crack the cast iron parts. Cast steel die sets are capable of withstanding greater shock loads. Large and special die sets, which are subjected to sever shock loads, are constructed of rolled steel. Manufacturers' catalogs are available to give die set specifications and ordering information.

CONVENTIONAL, COMPOUND, AND COMBINATION DIES

Inverted Dies

The dies discussed thus far were designed with the die block secured to the die shoe and the punch fastened to the punch holder. The punch and die positions are reversed on *inverted dies*. In the inverted die set shown in Figure

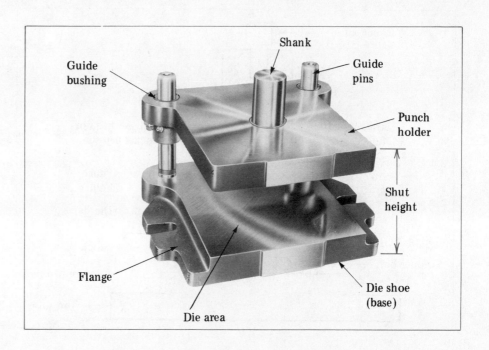

FIGURE 34-18
Standardized Two-Post Die Set

710 TOOL DESIGN AND TOOLING PRODUCTION

34-19, the punch is doweled and fastened to the die shoe. Similarly, the die section is doweled and held securely to the punch holder. Note in this design that on the return stroke a stripper pad with an ejector plug in the die forces the blank to be ejected. The ejector is moved in the die cavity by a knockout rod.

Inverted dies are used to keep the cutting edges clear of chips to minimize grinding. Also, each blank is removed as it is cut instead of being forced through the die.

Progressive Dies

Progressive dies are used when two or more press operations are performed for each stroke. The strip is moved from one station to another. At each station, the part is formed progressively. At the final stage, a completed part is produced. After the strip has once been cycled through each station, each ram stroke produces a finished part.

A simple two-station progressive die is illustrated in Figure 34-20. The strip is fed from the right side. In sequencing operations, piercing is placed first. Pierced holes permit accurate locating for subsequent operations. Bending and forming operations are planned for the last stations. In Figure 34-20, the first operation is performed by a piercing punch. On the next stroke, the workpiece is advanced to the next station where the pilot on the punch centers in the pierced hole. The rectangular blanking punch cuts the part to shape. At the first station, the strip is positioned by the primary die stop. For the second stroke, the strip is advanced to the die-stop pin. At each successive stroke thereafter, the strip is positioned against the die-stop pin. The scrap strip emerges from the left side of the die. The stripper plate is mounted on the die block. A two-post rear position die set is used for the punch and die mounting.

Additional stations involving a greater number of piercing, forming, and blanking operations are common. The final product may be of regular or irregular shape. On some progressive dies, the part is often blanked before the final operation. The blank is then forced back into the strip and carried along to subsequent stations for bending or forming.

Progressive dies take the place of a number of separate dies. A whole sequence of operations may be performed simultaneously on the same strip. One main disadvantage with parts that have small supporting areas is that the final part may become dished as it is forced through the die block.

Compound Dies

Compound dies perform two or more operations at one station during one press stroke. Figure 34-21 shows the setup for piercing and blank-

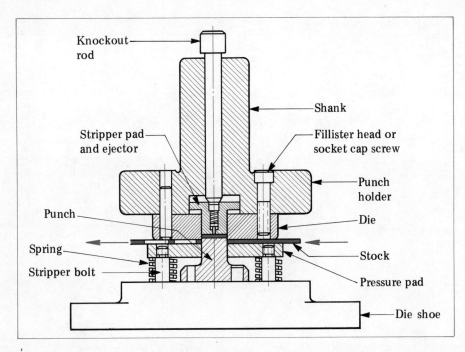

FIGURE 34-19 Example of an Inverted Die Set

JIGS AND FIXTURES, DIES, CUTTING TOOLS, GAGES 711

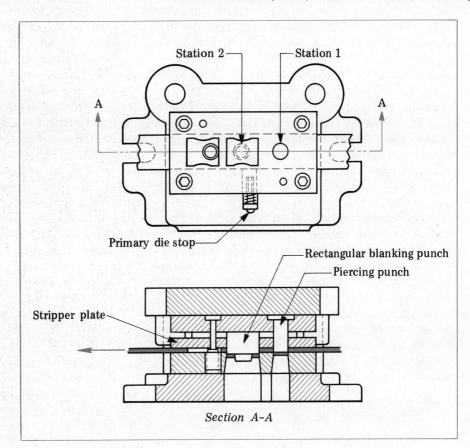

FIGURE 34-20 Example of a Simple Two-Station Progressive Die

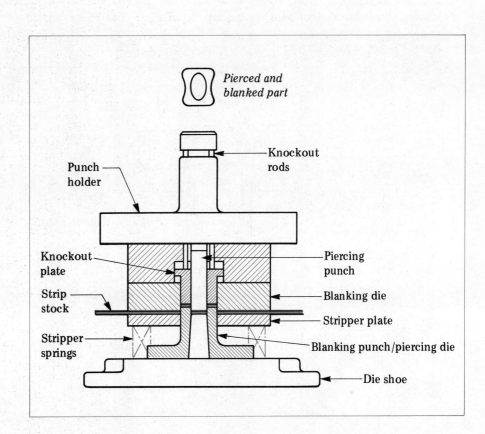

FIGURE 34-21 Example of a Compound Die for Piercing and Blanking a Part at One Station

ing a part at one station of a compound die. Punching and blanking elements are mounted directly opposite each other on the punch plate and die shoe of the die set. One of the simplest compound dies is a cutting die for a washer. The hole must be concentric with the outside diameter. The blanking punch for the outside diameter is also the piercing die for the concentric hole. This part of the cutting tool is mounted on the die shoe. The blanking die for the outside diameter and the piercing punch are mounted on the punch plate.

The stripper plate on a combination blanking punch and piercing die is assembled on the die shoe. A knockout plate is included as part of the punch plate assembly. The knockout plate forces each blanked washer out of the blanking die.

Compound dies are used for stamping processes where parts (with pierced holes) are to be held to close dimensional tolerances and must be flat. Compound dies require less die space so that smaller presses may be used.

Combination Dies

Combination dies are used when a cutting and noncutting operation are combined. For example, the combination die shown in Figure 34-22 may be used to produce a stamped and formed part. The outside rectangular slug is blanked first. As the die sections continue to move together on the downward stroke of the ram, the cup is formed by the forming punch in the die shoe moving the blanked part into the hollowed blanking punch in the punch section.

On the upward stroke, the stripper plug in the blanking punch and forming die (mounted on the punch plate) forces the formed cup out of the die. The blanking die and forming punch are mounted on the die shoe. The blanking punch for the outside rectangular form is designed internally to become the forming die section. This blanking punch/forming die combination is mounted on the punch holder section of the die set.

SHEET METAL BENDING, FORMING, AND DRAWING DIES

Bending refers to the shaping of metal from one plane to another and completely across the part. For example, a part made of flat steel plate is bent at a right angle across its width.

Forming deals with the flow of material along a curved path, usually a closed path. The formed part takes the shape of the punch or die.

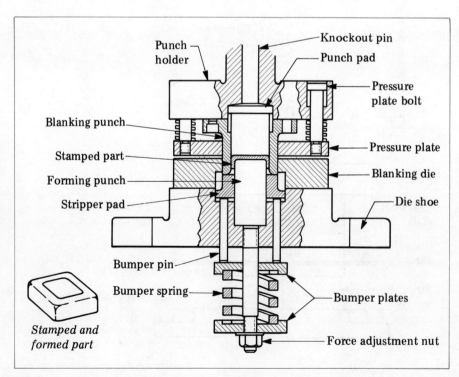

FIGURE 34-22 Example of a Combination Die

Drawing is the process of producing shell forms. These forms may be cylindrical, square, or rectangular in shape. The sides may be square or at an angle.

Bending Dies

Where practical, metal is formed by *bending dies* parallel to the grain direction produced in rolling the stock. The metal is bent without fracturing. In bending, the workpiece is usually held against the die block by a pressure pad. The edge over which the bend is to be formed on the die is rounded to the required internal radius. The forming (leading) edge of the punch is rounded. As the punch moves downward, the metal is caused to flow in the area between the die and punch. This action is sometimes called *wiping* and the die, a *wiping die*.

Large parts, such as panels, are often formed by *press-brake bending dies* and *forming dies*. The long punch and die sections, which conform to the shape to be formed, are mounted on a type of machine tool called a *press brake*. Examples of sheet metal press-brake bends and punch and die forms are illustrated in Figure 34-23.

In most metal bending operations, elastic stresses develop at the bend areas. As a result, the part tends to spring back to produce a smaller bend. This problem is overcome by *overbending*—that is, the part is bent through a greater angle so that it springs back to the required angle. Elastic stress varies according to the composition of the metal, thickness, radius of the bend, part size, and the design of the bend.

Pressure Pads. *Pressure pads* are used in die work to control the position of a part being formed so that sharp, intricate details may be produced. Pressure pads are applied above or under the strip and/or workpiece, depending on the bend and die design. Force is applied against the pressure pad by springs, hydraulics, or fluids (air). While springs are commonly used, the force against the pressure pad is not as constant as the other methods. As the springs are compressed, the force increases so that it becomes difficult to control the stretching and tearing of metal.

Curling Dies

Many parts require a raw blanked edge to be rolled for safety, for appearance, or for strengthening the edge. Curled edges are formed by a group of dies known as *curling dies*. These dies are used on soft, ductile metals that may be rolled. Tempered metals tend to form irregular curves. Usually a forming lubricant is applied to help the metal flow over the highly polished edge curling surfaces of the punch and die. A common design of a curling die is illustrated in Figure 34-24.

Embossing Dies

Embossing dies (also called coining dies) produce surfaces in which a detail is impressed on one side and raised on the other side of a stamped part. Embossing dies are produced on tracer milling machines; three-dimensional contouring machines; and embossing, die-sinking, and other detailing machines. The punch and die are machined with the same details but with allowance for the thickness of the metal. Pressure pads and/or ejector pins are incorporated in the die design to control metal flow during embossing and for ease in removing the stamped part. Embossing may be done in a single-process die, or the operation may be included at one of the stations of a progressive die.

Drawing Dies

Drawing dies form a round cup or shell from a round blank. The blank is centered over a

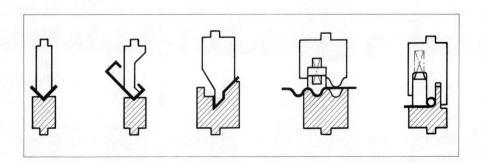

FIGURE 34-23 Examples of Sheet Metal Press-Brake Bends and Forms Using Brake Bending and Forming Dies

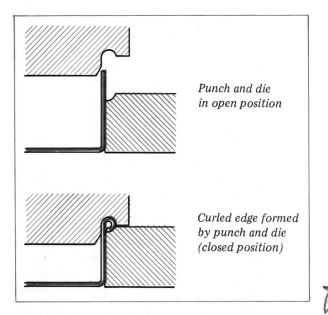

FIGURE 34-24 Common Design of a Curling Die

drawing die of the required outside diameter. The punch, which conforms to the inside shape and dimension, is vented so that no vacuum is created to hold the drawn part to the punch on the return stroke. As the formed cup tends to spring back when it clears the bottom of the drawing die, the cup is held in the die recessed hole area and disengages on the return stroke. A drawing die for forming a shell is illustrated in Figure 34-25. In this case, the pressure pad over the workpiece controls the flow of metal.

Drawing dies are classified as shallow when the drawn shell is less than half the diameter of the part. Deeper draws require the metal to be confined and force to be applied on the blank. These conditions cause the metal to flow without wrinkling or distortion. Pressure-loaded pads, which extend over the nesting area, hold the metal against the die face as the metal is drawn into the die. Exceedingly deep draws require a blanked part to be drawn in a number of stages.

Some products require the shell to be reduced in thickness as it is drawn to shape (lengthened). Such reduction processes call for forcing the shell through a *tight die*. Drawn parts often call for heat treating during drawing stages so that the metal remains ductile.

FACTORS AFFECTING METAL FLOW

Punch Radius

The punch radius is an important factor in terms of pressure required to draw and the possibility of thinning or tearing the metal at the bottom radius of the shell. A small radius requires greater forming forces. Similarly, the larger the radius on the draw ring (die), the freer the metal flows plastically. Too large a radius may cause wrinkles to form. Too small a radius tends to produce uneven thinning of the shell wall. The amount of force exerted by the pressure pad is controlled as the workpiece is drawn to shape. The frictional force must be adequate to permit drawing without wrinkling.

Lubricants

Frictional force is overcome in some applications by using a lubricant. A number of specially prepared lubricants provide chemical protection films. Examples include chlorinated oils, sulfurized oils, soap-fat compounds, and soluble mineral oils. These lubricants are recommended for both stamping and deep drawing press operations. Other lubricants such as chalk and graphite act as physical separators for the workpiece and die sections.

Lubricants increase the wear life of dies, improve the quality of both the process and product, and increase production efficiency. Some lubricants also provide rust protection. Most lubricants leave a residue and parts require vapor or liquid degreasing. Some lubricants produce stains on nonferrous metals. Lubricants are applied to the preformed blank by roller coating, swabbing, spraying, and brushing.

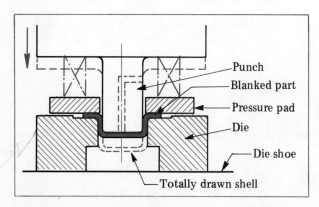

FIGURE 34-25 Example of a Drawing Die for Forming a Shell

Part Material

Consideration must be given in drawing operations to the ductility and yield strength of the material. *Ductility* refers to the ability of the metal to change shape without fracturing. *Yield strength* relates to the property of the metal to the point at which permanent change in shape takes place. These two metal properties must be considered by the designer. Consideration must also be given when selecting the correct material to the following factors:

- Shape of the shell (cylindrical, rectangular, conical);
- Material thickness;
- Amount of work material to be reduced for each draw;
- Grain size and other properties of the material;
- Surface finish;
- Other stamping and forming processes (such as flanging requirements).

Metals handbooks provide manufacturers' recommendations on the classes of commercially available low-carbon hot and cold rolled sheet or strip steels and other nonferrous metals. Mechanical properties such as yield strength, reduction percent in area, and hardness values are also given in trade tables.

WORK-HOLDING DEVICES FOR NC AND CNC MACHINE TOOLS

The principles of design for strap clamps, work supports, and locators for conventional machine tools apply to tooling for NC machine tools. One main difference is that these work-holding devices are brought together as an assembly. The assembled parts can be relocated to accommodate different sizes of workpieces. The strap clamp assembly illustrated in Figure 34-26 shows a NC work-holding setup using conventional parts. The unit is designed for use directly on the table. Normally, such parts are designed for mounting on a fixture base plate. On NC machine tools, the table may replace the base plate.

Special NC Adjustable Vises

Standard vises used on drilling and milling machines may be used to hold workpieces for NC machining. Some vises are fitted with an ad-

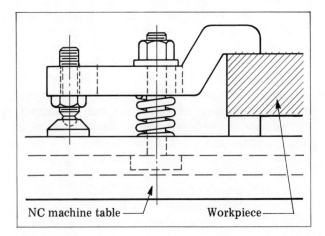

FIGURE 34-26 Strap Clamp Assembly for NC Work-Holding Setup

justable part locator, which is attached to the vise. The adjusting feature accommodates variations in workpiece size.

Vises designed especially for NC machine tools consist of three major parts, as shown in Figure 34-27. The solid jaw and a side aligning plate serve as stationary work locators. The movable jaw is the clamping device. All three parts are keyed to slots and are secured to the machine table.

Grid-Pattern Base Plates

Base plates are built to be table mounted. Base plates contain a grid pattern of holes. One

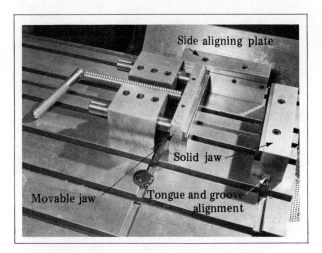

FIGURE 34-27 Adjustable Vise Designed for NC Machine Tools

set of holes receives hardened and ground dowel pins. The dowel pins serve as locators. The dowel pin holes are jig bored to an accuracy of 0.0002″ (0.05mm). Alternate holes are tapped to receive hold-down screws or studs. This type of base plate provides numerous locations for the locating pins. The designer or programmer usually identifies locator positions for workpiece setup. Slide movements are positioned from a starting reference point. Auxiliary fixtures and work-holding devices may be mounted on the grid-pattern base plate, or *grid plate.*

Some NC machine tool builders provide a *grid system* as standard work-holding tooling. The grid system consists of the base plate with a fixed pattern of locating and screw holes. Special angle blocks, or *knees*, of various sizes are included. Knees are designed with the same dowel hole and screw hole pattern as the base plate. Adjustable jacks are provided. They, too, may be screwed into the base plate.

The base plate hole series is marked along X and Y axes. As the part is programmed, the dowel locations are identified on a grid drawing. Information is also given for locations of stops, clamps, and the workpiece. Positioning the machining processes are then programmed in conformance with the location on the workpiece. One advantage of this system is that a record is kept of the setup and it can be precisely duplicated any number of times after disassembly.

The tool setup sheet for the NC machine tool or machining center craftsperson provides information for assembling the tooling components. If a tool management program is followed, information about pregaged and measured cutting tool lengths, diameters, and other features are programmed into the tape. The assembled tool components are arranged in turret stations or in an automatic storage drum. NC machining centers are designed with one or more automatic storage chains. A 48-tool chain storage matrix is pictured in Figure 34-28. Tool storage permits random and shortest-path tool selection. Random select allows tool usage in any operation sequence and as frequently as required.

Turret tooling on turning centers in a number of manufacturing systems are set up so that every tool tip around the turret lies in the same plane. Figure 34-29 shows the tip alignment of cutting tools on the vertical turret for inside operations and the single-point, outside-diameter cutting tools on the crown turret. Each tool clears the workpiece with minimal retraction. The inside-diameter cutting tools are preset by means of a tool setting gage.

FIGURE 34-28 48-Tool Chain Storage Matrix for Random, Shortest-Path Selection and Automatic Switching between Chains

FIGURE 34-29 Tool Tip Alignment for NC Turret Lathe Application

BASIC CUTTING TOOLS FOR NUMERICAL CONTROL MACHINING

Twist Drills

NC drilling operations are performed without the aid of jigs and drill bushings. Therefore, the accuracy of hole location, concentricity, size, and quality depends on the drill point. The sharpness of the drill, rigidity, machining conditions, uneven cutting lip angles, a drill running out-of-true, and other problems of drill tool geometry all affect accuracy.

NC drilling requires a drill point that is self-centering. The spiral-point drill provides a self-centering drill point. The point design is produced by a Spiropoint® drill sharpener in which a generating system grinds the point with an accurate spiral.

Spotting drills and center drills are often used to precede a drilling process for accurately starting standard twist drills. The use of spotting drills and center drills requires a second drilling operation.

A design factor that affects drilling accuracy is shank length and torsional rigidity. Drills used for NC machining should be as short as practical. A collet-type holder is available to grip a twist drill on the flute portion. Twist drills are also engineered with shorter lengths for added strength and greater drilling accuracy.

Holes that are larger than 1" (25.4mm) diameter are often drilled with spade drills. Spade drills are capable of drilling accurate holes from the solid. These drills are of stubby design which adds to their rigidity and ability to withstand drilling forces. The small dead center on the spade drill minimizes end thrust. Chips are broken as they are formed by including chip breakers in the tool point geometry.

Milling Cutters

Cutting speeds, feed rate, and depths of cut are generally higher on NC milling machines and machining centers. Therefore, the milling cutters must be more rigidly constructed and must maintain their cutting edges. The current trend is toward insert-type milling cutters of tungsten carbide. Solid carbide end mills are commercially available for NC machining. Cutters are designed with axial adjustment of the inserts. Finer-pitch insert-type cutters permit faster feeds without increasing the chip load.

Hole Threading Tools (Taps)

NC tapping operations require a tap design that prevents chip accumulation in the flutes and has the greatest possible cross-sectional body area and strength. Gun or spiral-point taps are used on through holes. The spiral point produces a shearing, cutting action. The chips are projected ahead of the tap. When blind holes are tapped with gun or spiral-point taps, care must be taken to provide space below the threads where the chips may accumulate.

Ductile and soft metals are tapped with spiral-fluted taps. The helix of the spiral flute tends to lift chips out of the threaded hole. Bottoming taps are used for blind holes. A torque-limiting adjustment is important when tapping blind holes on NC machines.

In NC tapping operations, the feed rate is manually selected or programmed. If the feed rate of the NC machine does not correspond accurately with the lead of the tap, the feed rate is set at slightly less. The tap floats axially to compensate for any minor lead error. At the end of the tapping process, as the spindle reaches the preset depth, the tap driver spindle reverses and allows the tap to feed out.

CONVENTIONAL AND NC TOOL CHANGE AND SETUP METHODS

Quick-Change Tooling Systems

Quick-change tooling systems for rotating cutting tools permit throwaway inserts to be indexed to a new cutting edge. Holders are designed so that the new cutting edge is in the same relative position with the workpiece and the original reference point.

Quick-change toolholders consist basically of a master toolholder, which becomes an integral part of the spindle design. Adapters are then designed to adapt various cutting and holding tools to the master toolholder. One design consists primarily of a taper spindle nose and quick-locking/releasing screw for the mating spindle and adapter. A second design incorporates an eccentric cam and a grooved area in the adapter. Figure 34-30 shows how a tool is aligned and locked in the adapter by turning the eccentric cam with an Allen or other wrench. The tool is ejected when the eccentric is turned in the opposite direction. Power drawbars provide another mechanical method of securing cutting tools in adapters.

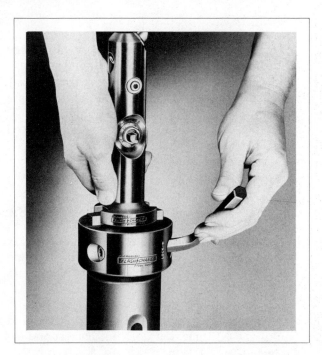

FIGURE 34-30 Locking Tool in Adapter with an Eccentric Cam Movement

FIGURE 34-31 NC Machining Center with Tool Positioning Capability to Drill, Ream, Bore, Counterbore, Tap, and Mill

The quick-change tooling system employed on Moore jig borers consists of a taper-shank adapter with a fast-lead square thread end. The adapter is seated in the spindle by a quick turn. The process is reversed to remove the adapter. Such an adapter is used for lighter cutting operations.

Quick-change tooling systems for stationary cutting tools were discussed in detail in relation to turning machines. Toolholders and tool shanks are adjustable on the machined slides of quick-change tool posts. Different cutting tools may be accurately and quickly positioned and changed. Toolholders may be preset to within 0.001" (0.02mm).

NC Tool Positioning

Conventional machine tools require the manual positioning of tools. NC machine tools permit cycling the tools once they are preset for height and, where applicable, the diametral dimensions are established. The NC machining center shown in Figure 34-31 has a tool positioning capability to automatically drill, ream, bore, counterbore, tap, and mill. Each operation is selected and controlled by tape. Hole depths are automatically controlled. A single tool may be used to cut to several depths during the cycle without manual adjustment.

The turret drilling machine may be used as an example of automatic tool cycling. The cycle covers rapid traverse to position the spindle, power feed control, rapid spindle retraction, automatic indexing, adjustable feed rates, and automatic tap reversal. The combination of inside- and/or outside-diameter tooling and cuts that can be performed on a 12-station NC turret lathe is shown in Figure 34-32. The NC turret tooling at each of the twelve stations is shown in Figure 34-33. On this machine tool, the turret is placed alongside of rather than aligned axially with the spindle axis. The turret moves left and right (at a right angle to the spindle/workpiece) and axially (parallel to the workpiece centerline). Tapers and contours are cut by numerically controlling both axes.

Each station on the turret may be tooled for either internal or external cutting operations. The turret may be indexed in either direction. The two basic toolholders that are required for this 12-station NC turret lathe are illustrated in Figure 34-34. One toolholder is for inside-

JIGS AND FIXTURES, DIES, CUTTING TOOLS, GAGES 719

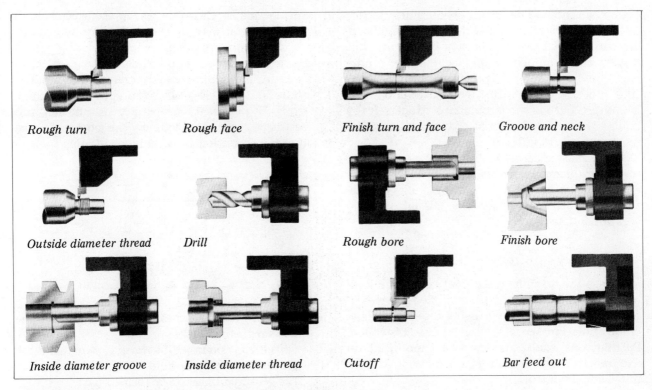

FIGURE 34-32 Combination of Inside- and Outside-Diameter Tooling and Cuts on a 12-Station NC Turret Lathe

FIGURE 34-33 12-Station NC Turret Tooling

FIGURE 34-34 Two Basic Toolholders for 12-Station NC Turret Lathe

diameter tools. The other toolholder is for outside-diameter tools. Solid-shank cutter holders with insert-type cutters may be slid into the holder and clamped. Roughing cutters, whose edge locations are established to within ±0.003" (0.07mm), require no further adjustments. Finish insert cutters are brought to exact size by dialing the required amount of tool offset on the machine control unit console.

As discussed earlier in the unit, the turret tooling setup places the edges of all inside-diameter end working tools in a single plane. Outside-diameter cutters automatically fall into known locations.

Manually Controlled Tool Changes

Once the machine is automatically cycled, all machine functions are tape controlled. However, for trial cutting or for changing inserts, the machine functions may be manually controlled. Pushbuttons, dials, and switches, mounted on the operator's panel on the control console, permit manual control.

TOOL PRESETTING

End Cutting Tools

Presetting means that the end of a cutting tool is set longitudinally in relation to a master reference surface. The diameter of the tool establishes the radial relationship. Tool management systems permit precise, rapid measurement of tool length and feed information for programming as related to the end cutting surface and the diameter. Digital readouts on the measuring machines provide instant dimensional measurement. A commercial universal presetting height gage may also be used to set various tool lengths. A micrometer head attached to an arm that may be positioned at 1" intervals on the height gage permits adjustments to be made within ±0.001" (0.02mm) along the 1" (25.4mm) travel of the micrometer spindle.

Radial Cutting Tools

Boring bars require presetting to required bore diameters prior to use. Precision presetting spindles provide a method of accurately holding the boring bar and adapter so that the cutter teeth edges may be indicated. The preset spindle fixture and inserted cutter are mounted on a surface plate. Gage blocks representing the radius dimension may be set at the required radius.

Precision presetting machines, using a tool-setting microscope and optical reading scales, are used for measurements in increments of 0.0001" (0.002mm) in either horizontal or vertical direction. The optical comparator is another tool setting/measuring machine. The comparator requires a stage that is suitable for holding the tool adapter.

SAFE PRACTICES FOR APPLICATIONS OF PRODUCTION TOOLING

—Inspect every punch and die part against possible hardening cracks.
—Check the tolerance between a punch and die. No undue forces should be applied to produce the required blanked or formed part.
—Make sure the die design and press conditions permit a stamped and/or formed part to be ejected properly.
—Check the shut height of the die set to ensure there is adequate clearance to safely carry on the power press operations.
—Make certain all safety guards and protective devices on the press are operable and in place before power is applied.
—Take standard machine tool safety precautions.
—Balance work-holding fixtures dynamically when the fixtures are used on turning machines. Centrifugal forces must be reduced to minimum for safe operation at machining speed.
—Check to see there are no projections and/or sharp corners on rotating fixtures.
—Secure machining guards before a fixture (and spindle) is rotated by power.
—Use adequate clamping force to prevent a rotating fixture from moving during a machining process.
—Check the valves under the workpiece and adjust the pressure to ensure that the holding forces on a vacuum fixture are adequate to hold the workpiece against cutting forces.
—Check the valve chuck areas. Wipe the chuck face to free chips and foreign matter from around the valve openings.
—Use end and side plates on magnetic fixtures to absorb the cutting forces.

JIGS AND FIXTURES, DIES, CUTTING TOOLS, GAGES 721

—Cover the faces of base plates and other work-holding/positioning parts of a grid system as protection against chips.
—Recheck the tip alignment of cutting tools for turret lathe applications to be sure all cutting edges are in the same plane.
—Use only self-centering drill types when drill points on NC machine tools are to precisely locate and form holes.
—Select taps for NC threading with maximum tooth strength and flutes that force chips ahead of the tap and out of the workpiece.
—Operate a NC machine tool by manual control to cycle all operations before a production run. Check for operating clearances, positioning, surface finish, and dimensional accuracy.
—Check the accuracy of all preset end and radial cutting tools.
—Follow standard hand tool, layout, and machine safety precautions.
—See that all hand and eye protection devices are in place and operating before starting sheet metal blanking and forming operations.
—Handle metal sheets, strips, and rolls; setups; and stamped and scrap metals carefully. The sharp edges can cause serious cuts.

TERMS USED WITH PRODUCTION TOOLING

Positioning	The dimensional relationship between the cutting tool and a specified reference point on the workpiece. The process of moving a table and mounted workpiece to be aligned at a particular reference point.
Piece part	In production work, a workpiece or part whose shape is to be altered by one or more machining processes. (The term is applied synonymously with *part* and *workpiece*.)
Locator (jigs and fixtures)	Pins or other members against which a piece part rests in order to perform another process at a specific distance from the reference surface. Work-positioning members against which successive parts are brought to establish reference surfaces for second operations.
Rest plates	Parts used in jig and fixture design as bearing surfaces for positioning a piece part.
Leaf, box, and open-plate jigs	Three basic types of drill jigs used primarily for hole forming processes such as drilling, reaming, and tapping.
Shut height (jigs and fixtures)	The clamping range of a fixture for holding a piece part. (The shut height is adjustable within the range of the fixture.)
Press-fit bushing	Hardened bushing insert that is pressed into position in a jig. Flush-type (headless) or head-type bushing used as a cutting tool guide for hole forming operations.
Liner	A bushing inserted into a jig or fixture for purposes of permitting renewable wearing bushing to be replaced quickly. A locator for a bushing used for hole forming.
Classification of fixture	Grouping of fixtures according to machine tool applications such as milling, turning, and assembling.
Blanking, piercing, and cutting off (die work)	Three basic sheet metal production stamping processes.
Inclinable presses	A family of power production machines that utilize punches and dies to cut, form, and emboss parts by automatic cold metal stamping processes.

722 TOOL DESIGN AND TOOLING PRODUCTION

Roll feed (power press)	A power feed attachment that controls the automatic feeding of roll stock into a die.
Shut height (press work)	The opening at the bottom of the ram stroke when the machine stroke adjustment is in the up position.
Elastic limit	The point within the metal at which forming or cutting tools cause the metal to flow mechanically.
Stripper (stripper plate)	A plate extending a short distance from the stock. A plate that holds the stock on the return ram stroke so that the punch or die may be freed and the stock positioned to produce the next part.
Progressive and compound dies	Classifications of punch and die tooling in which two or more press operations are performed each stroke.
Pressure pad	A component of a die that forces against a blanked part to control movement (metal flow) during a forming or drawing process.
Embossing (press work)	The process of impressing a design into one side of a stamping while simultaneously raising a duplicate (but altered by the metal thickness) reverse reading design on the opposite side.
Shallow and deep draw	Punch and die tooling for forming a shell by drawing stock through a die between the die cavity and the outside surface of the punch. (Deep draw refers to a succession of drawings. The initial draw is equal to about one-half the diameter of the blank.)
Grid system	NC positioning and holding system. Precision-machined base plate and accessories that use a particular pattern of holes for locating and holding workpieces in horizontal, vertical, and angular planes. Plates and knees with dowel (locator) pins and screw holes alternately and precision spaced. (These holes provide accurate reference points over each grid surface.)
Tip alignment	A system of presetting machine tooling in which the point of each turret tool is set in the same plane.
Spade drill (NC tool point geometry)	Drill of stubby design that provides chip breakers in the cutting edges and lips.
Quick-release adapter	A machine spindle adapter that adapts standard cutting tool shanks for accurately positioning and locking. An adapter designed with a fast-lead thread or cam for quickly locking a cutting tool shank in an adapter.
Manual tool positioning	An operating feature built into NC machine tool programming that permits the craftsperson to manually operate controls on the operator's panel. Operator control during NC trial cutting and insert changing processes.
Presetting spindle	A precision cutter holding device (fixture) that permits cutting tools, particularly precision boring bars, to be accurately preset.

SUMMARY

Locating devices in jigs are related to a basic reference plane.

    Three locators are generally used in jig and fixture work to establish the piece part in the same location each time.

Pins (plugs) are used in jig, fixture, and die work.

    A least two pins are used in combination to fix the location of jig or die elements in the fixture or die set.

Rest pads provide bearing surfaces against which piece parts are positioned.

> Locating nests are areas formed to hold a piece part in a specific location for clamping and machining.

Quick-acting swing and cam clamps are based on lever principles for applying force to securely hold a part in position in a jig or fixture.

> Drill jigs limit the path of a cutting tool to cut concentrically and parallel and to produce dimensionally accurate holes.

Open-plate jigs are primarily for single-hole drilling, reaming, or tapping processes.

> Box jigs permit nesting a piece part within the jig and performing multiple operations with the jig positioned on any one of its sides.

Leaf jigs use a hinged plate to open the jig, secure the workpiece, and position and guide a cutting tool.

> Indexing jigs provide for setting a piece part a specified circular distance for each successive operation.

Universal fixtures are commercially available to house the various components of a drill jig.

> The shut height of a universal fixture may be adjusted to accommodate the height of the jig or fixture components and the piece part.

Segmental chips are preferred in jig operations to prevent interference in loading, unloading, and performing the machining operation.

> Drilling, reaming, tapping, and other hole-forming operations in jig work depend on the use of press-fit and renewable bushings. Drill jig bushings are designed headless for flush, permanent mounting or with heads for changing to different processes at the same location or for replacement when worn.

The inside and outside diameters of bushings are ground concentrically.

> Dowels are used for accurately aligning mating parts. Components are usually secured with fillister and socket-head cap screws.

Fixtures are secured to a machine table and nest a workpiece in a fixed position in relation to one or more cutting tools.

> Fixtures are classified according to machine tools, operations, clamping method, and feeding technique.

Magnetic and vacuum chucks serve as positioning and holding fixtures.

> Turning machine fixtures require balancing to overcome centrifugal forces caused by the position or weight of the fixture.

Sheet metal stamping processes provide for the economical manufacture of interchangeable parts.

> Blanking, piercing, cutting off, shaving, and trimming dies perform metal cutting operations.

Piercing refers to the stamping of holes in a strip or plate.

> Shaving is a second stamping operation. When parts are shaved, a small amount of material is cut to produce a part to close dimensional tolerances.

The inclinable press is a basic production machine tool. Strip and sheet metals may be fed automatically through cutting, forming, and assembly dies.

> Power press ram faces and bolster plates are designed for mounting commercial die sets and for adjusting the machine shut height.

The major components of an open-back inclinable press include a frame that accommodates the bolster plate, vertical slide (ram), clutch mechanism, variable-speed drive, and crankshaft.

Die clearance refers to the dimensional difference between the size of the punch and the die. The die opening represents the size of the blanked part. The size of the punch establishes the diameter of a pierced hole.

Angular work clearance begins below the die land to permit stamped parts to clear the die.

Stripper plates are either fitted around a punch or hold the stock near the die face as the punch is withdrawn on the up ram stroke.

The faces of multiple punch sections are generally ground at different heights or at double angles to reduce the cutting forces.

Commercially available cast iron and cast steel die sets include: a die shoe, guide pins, guide bushings, and the punch holder. Die sets are furnished in two grades, depending on the quality of fit of the guide pins and bushings.

Ejectors are used to remove (eject) stamped parts from punches and dies.

A progressive die replaces a number of separate dies in performing a series of operations simultaneously on the same strip or sheet.

A compound die performs two or more stamping and/or forming operations at one station during one press stroke.

Bending, forming, drawing, and embossing dies are used for mass production of parts that require other than a flat, plane surface.

Provision is made to overbend when elastic stresses cause a formed stamped part to spring back.

Pressure pads control the force on a stamped part so that it may be formed to a required shape without wrinkling or distortion.

Raw, blanked edges may be curled for appearance, safety, and strengthening.

Embossing dies produce a depressed design that is raised and reversed on the second side of a stamped part.

Shallow and deep draw dies form a shell or cup between the inside surface of the die and the outside surface of the punch.

Metal flow is affected by the radius of the punch, characteristics and thickness of the metal, the nature of the forming process, the finish of the die surfaces, speed, and the lubricant used.

NC vises are especially designed for table alignment and positioning of the solid jaw, a right angle reference (locator) surface, and the movable jaw.

Grid patterns on base plates and knees permit precise work positioning and mounting. A grid system provides for aligning workpieces along X, Y, and Z axes.

Some turret lathe systems require tool point positioning in one plane.

NC drills are designed to be self-centering and to drill concentrically, parallel, and true diameters.

Milling cutter design for NC operations requires greater rigidity and the wider use of renewable carbide inserts.

Taps for NC applications are machined with fast chip removal flutes and a more rigid body to withstand greater torque.

Compensation must be made in the feed rate of a NC tapping operation for the lead of a tap.

NC spindle adapters permit the use of many shank designs on tools. A quick-locking/unlocking device is provided in the design of a tool adapter.

Once cutting tools for NC applications are measured for tip length, hole depths are programmed.

Turret lathes are designed for horizontal, angular, and vertical turret positions. Vertical-mounted turrets move at a right angle and parallel to the spindle axis. Tapers and contours are produced by simultaneously controlling movement along X and Y axes.

Machine functions on NC machines are usually manually controlled for trial cutting and for changing inserts.

NC tool management systems may be programmed to compensate for tool length and tool diameter variations.

Precision presetting spindles provide a fixturing device for setting boring bars to required boring diameters.

Presetting machines provide greater flexibility and more precise tool setting measurements than fixture devices. Setting accuracies within 0.0001" (0.002mm) are made using tool-setting microscopes.

Standard hand, machine, tool, and personal safety precautions must be followed when making trial runs on sheet metal blanking and forming production tooling, jigs and fixtures, and cutting tools.

Extra precautions are to be taken to properly handle sheets, rolls, strips, and scrap for power press operations.

ANSI safety codes must be followed for all machine tools, particularly power presses.

UNIT 34 REVIEW AND SELF-TEST

1. a. Identify two forms of locators that are commonly used in jig and fixture work.
   b. State the function served by each locator.
2. State two design features that differentiate a box jig from an indexing jig.
3. a. List two different classifications of jigs or fixtures.
   b. Give an example in each classification.

4. a. Name circled parts 1 through 9 of the combination shell blanking and drawing die shown below.
   b. State the functions of parts 2, 4, 6, 7, and 10.

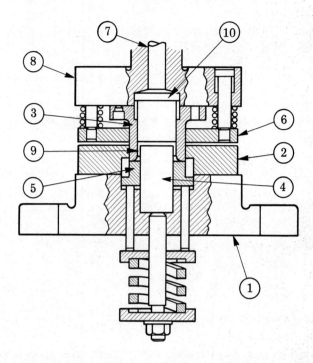

5. Differentiate between (a) blanking and forming dies, (b) progressive and compound dies, and (c) forming and embossing dies.
6. Explain briefly the effect of punch radius on metal flow for drawing processes.
7. Describe briefly one method that is used for tool point setting for a NC turret lathe.
8. Cite two milling cutter design features that are especially important to NC milling setups and operations.
9. Identify three features of a modern NC production machine tool.
10. Describe one tool presetting method for end cutting and radial cutting tools.
11. State two personal safety precautions that the toolmaker takes in testing a punch and die on a power press.

SECTION TWO

Construction and Measurement of Precision Production Tooling

Three of the most widely used machine tools for the laying out, machining, and measuring of production tooling include jig boring, jig grinding, and universal measuring machines. Older models of these high-precision machines are being retrofitted for NC and CNC modes of control, operation, and readout. Toolmaking techniques are examined in this section for the building and measuring of tooling commonly used on conventional and NC machine tools.

UNIT 35

Jig Boring, Jig Grinding, and Universal Measuring Machines and Processes

PRECISION JIG BORING MACHINE TOOLS, PROCESSES, AND ACCESSORIES

The jig borer is especially adapted to machine tool operations requiring extreme dimensional measurement accuracies. Linear and machining accuracies are expressed in terms of millionths of an inch (0.025µm). Repeatability in maintaining dimensional accuracy and complete positioning and machining controls make the jig borer an ideal machine tool for precision machining.

Jig borers are widely used in toolrooms for the construction of jigs and fixtures, punches and dies, special cutting tools, gages and other fine machine work. The basic processes include centering, drilling, reaming, through and step boring, counterboring, and contouring. Holes as small as 0.013" (0.33mm) may be drilled without spot drilling. The spindles are rigidly constructed for power and strength in taking heavy hogging cuts. For example, the power, strength, and rigidity of a hole-hog tool in taking a 1/2" (12.7mm) cut on a jig borer is shown in Figure 35-1.

Since a great variety and number of toolroom applications require constant tool changing, a special shank design permits easy removal and replacement of cutting tools and noncutting locating and positioning devices. The Moore taper-shank, fast-lead square thread tool illustrated in Figure 35-2 ensures accurate, quick-change tooling and permits the removal and

728 TOOL DESIGN AND TOOLING PRODUCTION

FIGURE 35-1 Hole-Hog Tool Taking a 1/2" (12.7mm) Cut on a Jig Borer

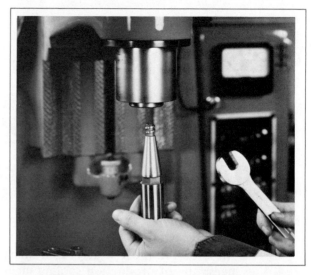

FIGURE 35-2 Taper-Shank, Fast-Lead Square Thread Tool for Accurate, Quick-Change Tooling

replacement of a tool to repeat the hole size on any diameter. Most jig borers have an infinitely variable spindle speed. Speed is read directly on a tachometer.

Jig Boring Tools and Accessories

Precision Boring Chuck. The most common accessory on the jig borer is the precision boring chuck, shown in Figure 35-3. This particular model permits fast, repeatable, accurate tool changes. Hole diameters may be increased 0.001" (0.02mm) per graduation. Vernier settings permit hole diameter increments of 0.0001" (0.002mm) per graduation. This accessory is available with inch or metric graduations. Models of boring heads that were adaptable to milling machine operations were covered in earlier units.

Tool Bits. In addition to straight and offset high-speed steel and carbide-tip tool bits, other bits are available for small-hole boring. The diameters for sets of small-hole bits range from 0.040" to 0.170" (1.0mm to 4.3mm) for high-speed steel and from 1/4" to 7/8" (6mm to 22mm) for carbide-tip bits.

Microbore® Boring Bars. A Microbore® boring bar series is available for jig boring work. The Microbore® boring bar, illustrated in Figure 35-4, is used for roughing and finishing operations on a jig borer where close dimensional limits and accuracy are to be maintained. The micrometer dial permits size changes to be read easily. These boring bars are available in overlapping sizes from 3/4" to 4-5/16" (19mm to 109.5mm) diameter. Complete bars are furnished in high-speed steel and carbide.

Boring Bar Extenders and Adapters. Since a great variety to shank designs for drills, reamers, chucks, and other cutting and holding tools are

FIGURE 35-3 Precision Boring Chuck

used on jig borers, adapters are required to use such tools. For example, adapters are made to hold tools with Morse tapers, standard collets, and National Standard tapers.

Chucks and Collets. Key and keyless chucks are used. Chuck shanks may be straight fitted for insertion in collets or tapered for easy mounting in the quill. Collets are designed for use with spotting drills, conventional drills, reamer drills, and end reamers. The hole tolerance on standard-quality collets is +0.0005″ (0.01mm) and -0.0000″. The hole concentricity of the shank to the collet hole is 0.001″ (0.02mm). The precision quality is finer. The hole tolerance is +0.00005″ (0.001mm); the hole concentricity, 0.0002″ (0.005mm). Collet hole sizes range from 1/8″ to 49/64″ diameter. Collet sets are also available in metric sizes in standard and precision quality.

Drilling Tools. Short-shank drills are used in jig drilling. These drills provide maximum working height under the spindle and great rigidity and drilling accuracy. The short shank has a flat area for securely holding the drill with a setscrew and thus preventing its turning during drilling.

Reamer drills are ground undersize for use with precision end reamers. The combination of reamer drill and end reamer provides a fast, convenient method of locating, drilling, and sizing holes on jig borers. Precision end reamers are furnished in sets in diameter sizes ranging from 1/8″ through 1″ (12mm through 25mm).

Spotting Tools. High-speed drills are used for general spotting. The tip sizes are from 1/16″ to 9/32″ diameter in inch-standard sizes (1.6mm to 4.8mm for metric diameters). A special high-speed steel *spotter* drill, which looks like a single cutting edge countersink, is used to spot very small holes. Center drills are used as standard hole-centering tools.

Tapping Heads. One widely used model of tapping head permits the tapping of holes up to 5/16″ (6.4mm) in mild steel and 1/4″ (6.4mm) in tool steel. The tapping head permits tooling without changing the position of the workpiece. This instant-reversing, speed-reducing tapping attachment is held in a standard collet or drill chuck.

Locating Microscopes. Jig borer setups and other workpiece setting needs require the use of a locating microscope. The microscope is readily adapted to locating edges, irregular shapes, and holes that are too small to be located with an indicator. The application of a locating microscope on a jig borer is shown in Figure 35-5. This

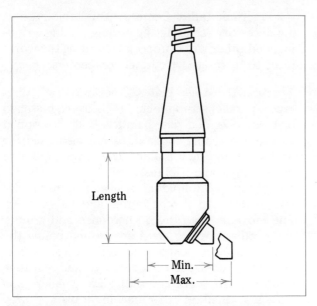

FIGURE 35-4 Microbore® Boring Bar for Roughing and Precision Finishing Operations on a Jig Borer

FIGURE 35-5 Application of a Locating Microscope on a Jig Borer

microscope, with suitable attachments, is interchangeable for application on the jig borer, jig grinder, and universal measuring machine.

Digital Readout and Printer. The digital readout and printer accessory verifies measurement accuracy for the X and Y axes to assure precise hole location. Readout modes are available for inch-standard screw, metric-standard screw, or switchable inch/metric screws. Readout accuracies to 0.00005" (0.0012mm) are provided. The X and Y channel readouts are ± through seven digits.

The digital readout accessory includes positive backlash take-up, manual zero reset, and entering of slide travel distance for reference to readout. Visual signals indicate problems of digital readout overflow and loss of power, when all numbers return to zero and there is no counting until the unit is reset.

The readout printer prints the X or Y axes individually. Each printing is identified by a sequential number to 999 and may be manually reset to zero. The printout is switchable between an inch format and a metric format.

Table Mounting Accessories

Rotary Tables. Standard, precision, and ultraprecise rotary tables permit indexing within tolerances of ±12 seconds, ±6 seconds, and ±2 seconds of arc, respectively. Rotary tables are available for hand-operated mechanical indexing. Rotary tables for NC and CNC machine tools are driven by stepping motors that eliminate mechanical indexing. The addition of preset indexer units to the mechanically operated rotary table provides control of angular location in automated or repetitive operations.

The stepping motor moves through a precise, discrete angle for each electrical pulse it receives. As many as 500 pulses per second may be received. A preset indexer power supply can trigger up to 99,999 pulses and can feed the desired number to the stepping motor to program wide angles. A *tilting rotary table* (Figure 35-6) is available with a tilt axis dial and rotary axis dial capable of direct readings to 1 second of arc. A disengageable worm permits quick alignment during setup and engagement for precise angle positioning.

Center and Micro-Sine Table. A tailstock and adjustable center permit work on vertically mounted rotary tables to be held between centers. Boring, grinding, inspection, and other operations may be performed. The tailstock is adjustable laterally and vertically.

A *micro-sine table* is a jig borer accessory used to position and hold workpieces for accurately machining or measuring angles. A set of adjusting rods permits angular adjustments for fine settings from 0° through 95°. Once the table is set, the gaging rods are removed and the table is secured in position. The table is provided with a fine-pitch screw for up or down adjustments.

Precision Index Center. This table mounting accessory facilitates jig boring, milling, grinding, and other work processes requiring the workpiece to be rotated accurately on centers.

Parallels, Angle Plates, Precision Vise. Precision parallel setup blocks, angle irons, matched box parallels, and riser parallels that are applied to other machine tool setups are also used for jig boring and jig grinding. The same applies to bolts, straps, and heel rests. While the vise is still a common work-holding device, a more precise vise is designed for jig borers and jig grinders. The movable jaw rides on hardened and ground guide rails. Surface areas are ground below the chuck jaw faces to hold thin workpieces. A fine-pitch vise screw permits accurate and sensitive tightening of workpieces that are smaller than parts machined on other machine tools.

Measuring and Locating Accessories. The standard indicator set consists of an indicator, indicator holder, edge finder, and line finder. A

FIGURE 35-6 Tilting Rotary Table for Jig Boring Work

spindle-mounting collet fits into the jig borer spindle. The indicator holder leg is held and secured in the collet concentric with the spindle bore. Indicator holders have a universal ball-joint arrangement and may be designed for fine adjustment. Indicators are of the dial indicator type with readings to within 50 millionths of an inch (0.00005″). Metric dial indicators are also available.

HOW TO ACCURATELY POSITION AND SET UP A WORKPIECE ON A JIG BORER

Aligning a Work Edge by Using an Indicator

STEP 1 Clean the base of the workpiece. Check for burrs and nicks. Do the same with the jig borer table.

STEP 2 Position the workpiece on parallels and as near the center of the table as practical.

STEP 3 Select and mount an indicator holder in a collet in the spindle. Add the indicator to the holder so that the leg almost touches the edge to be indicated.

STEP 4 Bring the indicator leg into contact with the machined edge to produce a reading on the dial face.

STEP 5 Turn the machine spindle slowly in one and then the opposite direction for only a fraction of a turn. Stop the turning when the highest reading is registered.

Note: This step is taken to establish that the axis of the dial indicator is at 90° to the work edge.

STEP 6 Turn the dial indicator scale to the zero position.

STEP 7 Move the table slowly and carefully longitudinally. Note any variation in the dial indicator reading and the direction (±) of movement. The setup for aligning an edge parallel to the longitudinal (X) axis by using a dial indicator is shown in Figure 35-7.

STEP 8 Use a soft-face hammer to tap the workpiece gently away from the dial indicator.

Note: Any abrupt force or jamming against the leg of the indicator may cause damage to the instrument.

STEP 9 Tighten the setup when a zero reading is indicated.

STEP 10 Recheck the alignment to be sure the workpiece has not shifted during tightening.

Aligning a Work Edge by Using Parallel Setup Blocks

STEP 1 Set up the straight edge on the front or back face of the table.

STEP 2 Make the usual checks for burrs, nicks, and cleanliness of surface on each parallel or setup block, table, and workpiece.

STEP 3 Place two setup blocks between a machined edge of the workpiece (either on the front or back side, depending on the nature of the operation to be performed) and the location of the straight edge.

STEP 4 Select appropriate clamps, screws, and heel rests. Secure the workpiece to the table.

STEP 5 Recheck the setup to see that each setup block or the workpiece has not shifted. Then, remove the setup blocks.

Aligning a Work Edge by Using a Parallel

STEP 1 Proceed to set up the workpiece on parallels and to mount a dial holder and indicator in the spindle.

STEP 2 Adjust the dial indicator with the point as close as possible to the center of the machine spindle.

FIGURE 35-7 Aligning an Edge Parallel to Longitudinal (X) Axis by Using a Dial Indicator

STEP 3 Bring the indicator into contact with the edge of the workpiece until there is a reading.

STEP 4 Test to see that the indicator is at a right angle to the edge of the workpiece (Figure 35-8). Set the dial reading at zero.

STEP 5 Raise the dial indicator and spindle so that the point clears the top edge of the workpiece (Figure 35-9).

STEP 6 Rotate the spindle 180°. Secure a parallel against the edge of the workpiece.

STEP 7 Turn the spindle 180°. Lower the spindle carefully. Note any reading change from the zero setting. Adjust the table longitudinally for any variation.

Note: A zero reading in both indicator positions (180° to each other) indicates the axis of the spindle and the edge of the workpiece are aligned.

DESIGN FEATURES OF A LOCATING MICROSCOPE

A locating microscope is used when small or partial holes, slots, or irregular contours serve as reference points. A locating microscope with magnification power of 20X or 50X locates edges, contours, and holes that are too small to be indicated with a dial indicator.

The reticle on the microscope has a number of concentric circles and two pairs of cross lines. These vary between 0.0025" on the 20X magnification to 0.001" on the 50X magnification. On the 20X microscope, six concentric circles range in diameter from 0.005" to 0.030" in increments of 0.005". Another seventeen circles continue from 0.030" to 0.200" in increments of 0.01". Finally, eight additional concentric circles are spaced 0.020" apart from 0.200" to 0.360".

HOW TO USE A LOCATING MICROSCOPE

STEP 1 Mount the locating microscope with an adapter in the machine spindle. Determine the magnification (20X or 50X).

STEP 2 Mount, position, and lightly clamp the workpiece to the table.

STEP 3 Position the hole, edge, or contour within the range of the microscope.

STEP 4 Determine the position of the workpiece in relation to the concentric circles in the reticle.

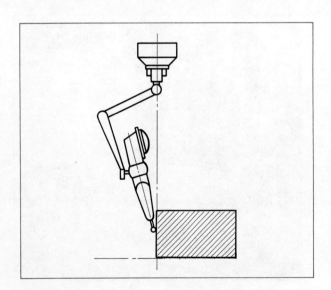

FIGURE 35-8 Dial Indicator (Locator) Set at Right Angle to Edge of Workpiece

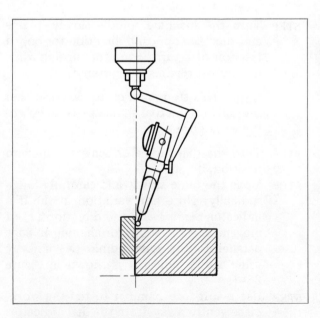

FIGURE 35-9 Dial Indicator Raised to Clear Workpiece and Rotated 180° (Setting Checked against Face of Parallel)

Note: The cross lines are used to zero in on the intersection of scribed lines on the workpiece when identifying the center point.

STEP 5 Move the table longitudinally or the cross feed travel handwheel to position the workpiece concentrically with the concentric circles within the microscope.

HOW TO SET UP AND USE A JIG BORER

Positioning the Workpiece and Drilling

STEP 1 Place the work on parallels. Align the work with respect to X and Y coordinates.

STEP 2 Align the axis of the machine spindle with the starting reference point.

STEP 3 Set the longitudinal and cross feed micrometer collars at zero.

STEP 4 Determine the machining sequence steps and tooling changes in advance.

Note: It is good shop practice to make a sequence sketch showing the positioning requirements for each basic process.

STEP 5 Position the spindle at the X and Y axes for the first hole. Center drill or use a spotting drill.

STEP 6 Position the spindle and center drill at the location of each hole. Center or spot drill each hole location.

STEP 7 Return the spindle to the starting point. Make a tool change.

STEP 8 Position and drill all holes of a particular size. Change drill sizes as may be required.

Note: Standard practices are followed for using a smaller pilot-size drill and successively larger-size drills for hole diameters over approximately 1/4".

STEP 9 Check the work alignment and the setup to be sure the workpiece does not shift if any heavy rough machine drilling takes place.

Boring Concentric, Parallel Holes

STEP 1 Select a Microbore® boring head or a boring chuck and a straight or offset boring tool. Replace the drill chuck with the boring chuck and cutting tool or boring bar setup.

STEP 2 Set the cutting speed (spindle RPM) according to the job requirements for rough boring.

STEP 3 Take a series of roughing cuts. Rough bore all holes to within 0.003" to 0.005" (0.08mm to 0.1mm) of finish size.

STEP 4 Replace the rough boring tool with a finish boring tool for the final cut. Increase the spindle speed. Decrease the tool feed.

Note: The depth of cut must be within the range of the cutting tool and setup. Any deflection of the cutting tool produces a bell-mouth and an irregular-shaped hole.

STEP 5 Use a solid plug, leaf taper gage, or other measuring instrument to check the bore size. The surface finish must meet the specified requirements.

Note: Measurements must be taken at room (a constant 68°F, 29°C) temperature, particularly when the workpiece must be held to extremely precise dimensional tolerances.

Note: Leaf taper gages may be used for measuring trial cuts when boring or grinding holes to size. The leaves are short enough to permit inserting into the hole when the cutter is withdrawn, without changing the work position. Leaf taper gages may also be used to check a through hole for bell-mouth, taper, and out-of-roundness.

PRECISION JIG GRINDING MACHINE TOOLS AND PROCESSES

There are many common elements between a jig borer and a jig grinder. The jig grinder, however, deals primarily with grinding operations where extremely precise dimensional accuracy and surface finish requirements must be met. Grinding is done using a vertical spindle. The jig grinder is used to grind straight and tapered holes and to grind contour forms. Such forms combine straight, angle, round, radii, and tangent surfaces. Since a jig grinder is an abrasive machining machine tool, the process may be used on soft or hardened metals.

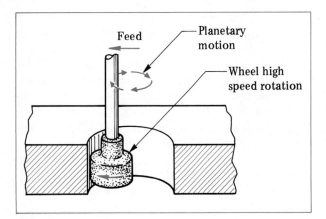

FIGURE 35-10 Out-Feed Grinding

The jig borer permits the slide motion to position a workpiece accurately with the main spindle axis. The advantage of planetary grinding is that radii, holes and cylinders, and tangent and chordal surfaces can be located, ground, measured, and finished to precise size.

In *wipe grinding*, the workpiece is fed past the wheel without any oscillating motion. By contrast, in *chop grinding* (Figure 35-11), the grinding wheel has an oscillating movement. Metal is ground as the fast-revolving wheel oscillates and the work is fed past it. Chop grinding is used in contour grinding for fast stock removal.

Plunge grinding (Figure 35-12) is done with the bottom edge of the grinding wheel and is used for rapid stock removal. The wheel spindle travels in a planetary path with the wheel traveling at a high rate of speed. The wheel is set radially at the required diameter. Feeding is axially into the workpiece.

Taper grinding requires the wheel axis to be inclined at the required taper angle. Straight-sided wheels are usually used in jig grinding tapers.

Shoulder grinding requires grinding with a concaved-end grinding wheel. Increments of down-feed are controlled by a positive stop or precision depth stop on the machine.

Jig grinders are widely used in toolrooms for finishing punches, dies, jigs, fixtures, special cutting tools, and gages to exact size and high-quality surface finish. Jig grinding eliminates many earlier hand fitting processes.

Grinding in a Horizontal Plane

Jig grinders provide versatility in planetary jig grinding in a horizontal plane. They are used for out-feed, wipe, chop, plunge, taper, and shoulder grinding.

In *out-feed grinding* (Figure 35-10), the grinding wheel spindle moves in a planetary path at a slow rate of rotation. The axis of the wheel spindle moves at a high grinding wheel speed within the planetary path. The diameter of a hole is enlarged by continually *out-feeding* the wheel while grinding.

Slot Grinding in a Vertical Plane

Many tooling requirements call for the grinding of angles, slots, and corners that cannot be generated by vertical-spindle grinding wheel action. A *slot grinding attachment* (Figure 35-13) is used in a slot grinding operation to produce corners, radii, slots, and concave sections.

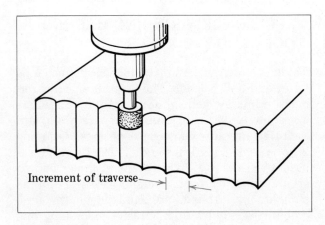

FIGURE 35-11 Chop Grinding

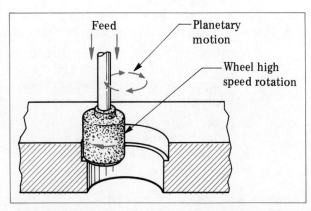

FIGURE 35-12 Plunge Grinding with Bottom Edge of Wheel

JIG BORING, JIG GRINDING, UNIVERSAL MEASURING MACHINES 735

The wheel motion is vertical as contrasted with planetary and horizontal motion used to grind straight, cylindrical, and chordal sections of regular shape. As shown in Figure 35-13, the grinding wheel face is dressed at a double angle. The wheel and attachment are set at the proper angle. A taper-setting feature permits grinding draft on both flanks of the angle at the same time. The inside tapered contour surfaces are ground by vertical oscillating motion of the high-speed revolving grinding wheel.

DESIGN FEATURES OF JIG GRINDERS

Base and Column

The major features on jig grinders, such as the 11″ × 18″ (280mm × 450mm) precision jig grinder illustrated in Figure 35-14, include a base with machined ways. The base supports and provides for the transverse movement of the cross slide. The cross slide is machined for transverse movement and provides the ways on the top surface for longitudinal table motion. The column is designed to provide for the vertical movement of the spindle housing and components. These parts and the main spindle block are made of Meehanite cast iron. The V-ways are double to permit maximum guide surfaces under extremes of travel or weight load. The V-ways are hardened and lapped for permanence of accuracy.

Main Spindle Assembly

The main spindle assembly contains coarse and fine out-feed controls. The spindle may be inclined to grind tapered holes with straight-sided wheels. Segmental grinding features permit grinding contours. The spindle is temperature controlled and contains the air power for the air grinding head. The grinding controls on the main spindle assembly of a jig grinder are shown in Figure 35-15.

Hole depth movement is controlled by an adjustable positive stop. A graduated dial and vernier located on the spindle housing permit direct reference to depth locations. A micrometer stop on the column may also be used to control hole depth.

Jig Grinding Heads

Grinding heads are available for speeds ranging from 9,000 RPM for grinding operations requiring power and rigidity for large, deep holes. The 40,000 RPM heads are designed for high thrust capacity for bottom, face, and shoulder grinding. The 175,000 RPM air-driven heads are applied to grinding ultrasmall holes. These heads are self-cooling. The exhaust air tends to keep the work cool. Dimensional errors caused by thermal expansion are thus minimized.

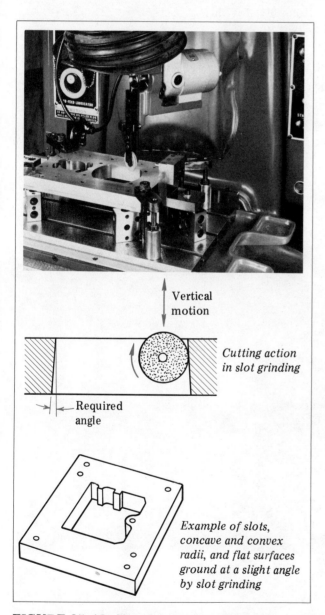

FIGURE 35-13 Slot Grinding Attachment Positioned for Slot Grinding Operation

736 TOOL DESIGN AND TOOLING PRODUCTION

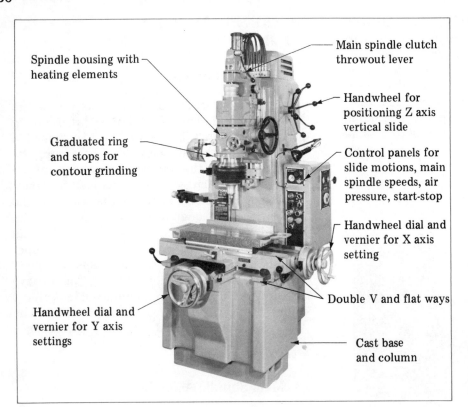

FIGURE 35-14 Major Features of an 11″ × 18″ (280mm × 450mm) Precision Jig Grinder

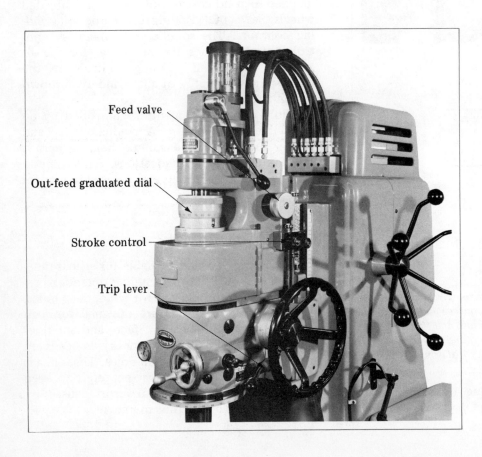

FIGURE 35-15 Grinding Controls on Main Spindle Assembly of a Jig Grinder

DRESSING ATTACHMENTS FOR JIG GRINDING WHEELS

The radius angle, cross slide, spherical socket, angle, and pantograph dressing attachments are each designed for dressing jig grinding wheels to particular shapes. The *radius angle grinding attachment* provides a method of dressing angular and circular forms and cupping the bottom face on vertical-axis wheels. Some of the typical wheel forms include radius forming at the bottom or top edge, formed angles, and radius blending with two tangent angles.

The *cross slide dressing attachment* is used to produce a full 180° concave radius, partial convex radius, two angles, radius tangent to two different angles, and 90° corner radii on the faces of slot grinding wheels.

The *spherical socket dressing attachment* moves a diamond point across the wheel face. The face of a spherical socket wheel is dressed using an angle adapter and angle slide dresser.

The *angle-type dresser* uses a diamond nib for dressing angles on shank-mounted wheels and wheels used for spherical socket grinding.

The *pantograph dressing attachment* utilizes a 10X-size template to dress complex forms. Coordinates are used to dress the front of the wheel. The wheel form is ground to location.

Each dressing attachment, except for the angle-type dresser, is designed for use with a basic universal wheel dresser unit. This unit has a mounting arm, ram assembly, and a sliding ram. The ram supports and positions each dressing attachment. The attachments are mounted in the end face and are quickly adjusted by rack and pinion. Precise micrometer adjustments of 0.001" to 0.0002" (0.025mm to 0.005mm) and finer vernier adjustments of 0.000020" (0.5µm) are made. The repeatability of positioning is within 0.0002" (0.005mm).

FACTORS INFLUENCING WHEEL SELECTION

Wheel selection is based on the factors governing all precision grinding. Hardened high-alloy steels are ground with a wheel having a hard abrasive grain and a weak bond. Soft, low-tensile strength metals require a hard abrasive grain with a strong bond. Open-spaced abrasive grains increase the cutting power of the wheel. The shank on mounted wheels should be as short as practical to assure rigidity. The largest size of grinding wheel should be used. The diameter should be at least 75% of the hole diameter.

WHEEL AND DIAMOND-CHARGED MANDREL SPEEDS

As a general rule, grinding wheels operate efficiently at speeds of 6,000 sfm (1,828 m/min). The spindle speed RPM is found by dividing the sfm by the circumference of the wheel diameter. Therefore, the spindle speed range of jig grinders is from 9,000 to 175,000 RPM. Diamond-charged mandrels, which are used for small-hole grinding, are generally operated at 1,500 sfm (457 m/min).

MATERIAL ALLOWANCE FOR GRINDING

General grinding considerations apply to jig grinding. Also, since hardened materials and complex shapes are ground on the jig grinder, adequate material allowance must be made to compensate for distortion resulting from heat-treating processes. The stock allowance for hole sizes smaller than 1/2" (12.7mm) is from 0.004" to 0.008" (0.1mm to 0.2mm). Hole sizes larger than 1/2" are left from 0.008" to 0.015" (0.2mm to 0.4mm) undersize.

HOW TO DRESS A JIG GRINDER WHEEL

Out-Feed Grinding Processes

STEP 1   Dress the top and bottom face of the wheel by hand with an abrasive stick dresser.

STEP 2   Dress the wheel diameter with a sharp diamond held in a nib holder.

STEP 3   Relieve the cutting face so that about one-fourth of the width remains for cutting.

Plunge Grinding Processes

STEP 1   Repeat steps 1 and 2 as for out-feed grinding.

STEP 2 Use an abrasive stick to dress a concave form in the bottom face. Leave a narrow, flat rim for faster and higher-quality grinding.

HOW TO GRIND HOLES ON A JIG GRINDER

Setting Up and Locating the Workpiece

STEP 1 Mount the workpiece on parallels and position it in the center of the table area.

STEP 2 Place strap clamps on the workpiece over the area of the work-supporting parallels. Apply a slight force on the clamps so that the workpiece is temporarily held in place.

Note: Many jobs require simply seating the workpiece in a precision vise.

STEP 3 Use the straight edge, edge finder, microscope, or indicator method of accurately locating the workpiece.

Note: If the workpiece is distorted due to heat treating, judgment needs to be exercised to locate the workpiece so that all holes and surfaces clean up during grinding.

Note: Heat-treated workpieces are often positioned by indicating two or more holes. The average location of a group of holes is selected.

General Grinding Sequence

The grinding on a workpiece may require more than one process. For instance, a workpiece may have a combination of hole forms such as through straight or tapered cylindrical holes, blind (bottoming) holes, or holes with shoulders. A general grinding sequence is followed.

STEP 1 Rough grind all through holes first.

Note: The temperature of the workpiece must be controlled in order to proceed with the finish grinding of close tolerance holes.

STEP 2 Dress the grinding wheel for finish grinding. Finish grind all holes that can be machined with the same grinding head to avoid changing grinding heads.

STEP 3 Grind all holes that are closely related in importance as a continuous series.

STEP 4 Grind stepped or shoulder surfaces only once to avoid making more than one depth setting.

HOW TO GRIND A TAPERED HOLE

Holes in dies and other toolroom work have a straight section (land or margin) followed by a slight taper. Generally, a layout dye is used for a small distance from the top face. As the grinding progresses, it is possible to see where the taper grinding ends. On small-diameter holes, the taper is often ground before the straight portion. The distance (height) to the top face of the workpiece to which the hole is bored to become the required finish diameter is calculated. The taper is ground to within the calculated height. For general internal taper grinding, the taper angle setting may be established by using the graduations on the taper setting plate. Extremely accurate angle settings require checking the spindle with an indicator.

Using the Taper Setting Plate

STEP 1 Loosen the spindle taper adjustment screw and tighten the other adjusting screw. Continue until the required taper angle is reached on the taper setting plate.

Using a Dial Indicator

STEP 1 Determine the required taper. If given in degrees, convert this value into customary inch or metric linear measurements.

STEP 2 Position a precision angle plate or a master square on the table.

STEP 3 Mount an indicator on the machine spindle.

STEP 4 Bring the indicator point into contact with the square (vertical) surface of the angle plate. Set the reading at zero.

STEP 5 Set the spindle down-feed dial at zero. Move the spindle 1".

STEP 6 Note any variation and the direction in which there is any movement of reading from the original indicator dial reading.

STEP 7 Loosen and tighten the opposite adjusting screws until the difference in indicator dial readings from the starting point to the 1″ height equals the offset calculated to produce the required angular setting.

Note: The adjusting screws must be tightened sufficiently to hold the spindle at the required angle. If too little force is applied, there is the possibility the spindle will grind out-of-round holes. If too much force is used, there may be binding in the vertical movement of the spindle.

Note: The spindle is reset for straight hole grinding by reversing the preceding steps.

HOW TO GRIND SHOULDER AREAS

STEP 1 Select the largest size of grinding wheel that meets the job requirements.

Note: The wheel diameter should be large enough to extend from the outer vertical surface of the shoulder to beyond the periphery of the hole. It is difficult to produce a precisely ground, flat shouldered surface if the wheel diameter extends across the entire shoulder and hole surface.

STEP 2 Dress the bottom face with an abrasive stick to produce a slightly concave face. True and dress the cylindrical face of the wheel.

Caution: Position the machine safety shield before truing, dressing, or grinding.

Caution: Protect the spindle housing, table, and front slides with molded machine bibs. Use machine aprons to further protect machined surfaces.

Note: If possible, the outer bottom edge is dressed to a slight radius that blends the side and bottom faces. This dressing strengthens the wheel edge and tends to prevent scoring.

STEP 3 Position the wheel to start the cut on the face of the workpiece.

STEP 4 Set the depth stop at the required shoulder depth.

STEP 5 Rough grind the outside diameter of the shoulder area and the shoulder itself at the same time.

STEP 6 Dress the wheel for finish grinding.

STEP 7 Finish grind the outside diameter and the shoulder to depth.

STEP 8 Check the dimensional measurements and surface finish.

STEP 9 Disassemble the setup. Return all work-holding, tool-holding, and machine accessories to proper storage areas. Clean each item and check for burrs and nicks. Leave the machine in an operable condition.

UNIVERSAL MEASURING MACHINES AND ACCESSORIES

The fine tolerances between punches and dies; the precise fitting of components in different mechanisms; the precision requirements of fixtures, cutting tools, gages, and gaging systems; and ever higher precision requirements between mating parts of mechanical, pneumatic, and other movements mean that finer measurements are required. Such measurements are in millionths of an inch ($0.025\mu m$) and angular dimensioning, within seconds of arc.

The universal measuring machine is a widely used measuring device. The universal measuring machine illustrated in Figure 35-16 is equipped with a motorized lead screw drive. This universal measuring machine may also be equipped for numerical control. Additional accessories, similar to jig borer and jig grinder accessories, are available for preselect positioning and readout and printer information. The standard model is equipped with manually operated lead screws for table positioning.

The extreme accuracies to which these machines measure demand that all components of a machine are held to equally precise tolerances at 68°F (20°C). The total accumulative positioning accuracy of the X and Y axes is 35 millionths of an inch ($0.9\mu m$). The greatest amount of positioning error in any 1″ (30mm) is 0.000015″ ($0.4\mu m$). The straightness of travel longitudinally is 25 millionths of an inch ($0.6\mu m$). The cross travel accuracy is 15 millionths of an inch. The spindle accuracy in terms of trueness of rotation is 0.000005″ ($0.15\mu m$).

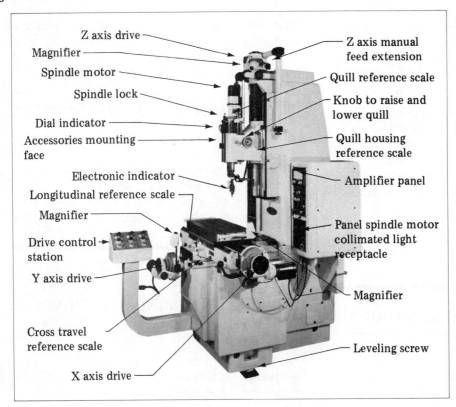

FIGURE 35-16 Major Components of a Universal Measuring Machine with Motorized Lead Screw Drive

Part drawings, using coordinate systems of measurement, are commonly used in jig boring, jig grinding, and machine measurement. NC and CNC universal measuring machines are programmed for two- or three-axis measurements. The Z axis relates to vertical spindle travel.

In addition to accuracy of movement (geometric accuracy), as measured in either a vertical or horizontal plane, squareness of travel is built into the universal measuring machine. There is extreme accuracy in the parallelism of the table to the X and Y axes through the full travel. Squareness of the quill vertical travel and trueness of spindle affect the degree of measurement accuracy. Backlash compensation features are incorporated in the machine design.

Bidirectional Gaging System

The bidirectional type of gaging system permits dimensions to be measured without having to rotate the gaging probe. Also, the probe diameter is not subtracted from an observed measurement. The bidirectional gaging system consists of a series of gage head inputs with probe tips, centering controls, and measuring scale. The gaging system is switchable between inch and metric measurements.

Locating and TV Microscopes

The locating microscope was identified earlier in relation to jig borers and jig grinders. It is used to locate edges, contours, irregular shapes, and holes where it is impractical to use an indicator. Figure 35-17 shows a locating microscope and work setup on a universal measuring machine.

The *TV microscope* is one of the newer advances in micromeasurement technology. The TV microscope employs closed-circuit television for microscopic observation and measurement. The system provides images that have exceptional fidelity and accuracy. Details may be magnified up to 2500X. The TV microscope is particularly adapted for inspection and measurement of actual components of microminiature parts and circuits.

Angular Measurement Instruments

While conventional angle dividing heads and precision rotary tables are used on universal measuring machines, a *small angle divider* permits greater versatility. For measurement and layout purposes, the small angle divider has the capacity to quickly and accurately divide a circle into

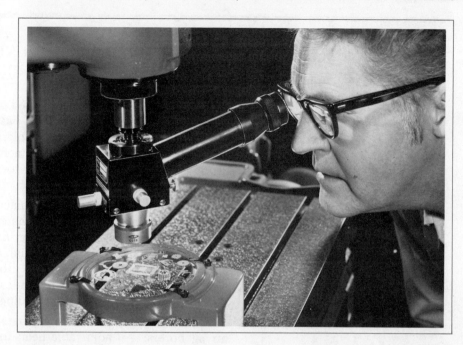

FIGURE 35-17 Locating Microscope and Work Setup on Universal Measuring Machine

12,960,000 parts. Measurements of ±10 seconds of arc are read directly on a vernier dial.

Other precision angular measurement instruments include the index center, rotary table and tailstock, and the micro-sine table. The precision *spin table* is power operated. The table turns at infinitely variable speeds from 5 to 100 RPM. The spin table is suited for inspection of roundness to tolerances within 0.000005″ (0.15μm).

SAFE PRACTICES FOR OPERATING JIG BORERS, JIG GRINDERS, AND MEASURING MACHINES

—Position the leg of a dial indicator carefully against the edge of a workpiece on a jig borer or a jig grinder. Use care in moving the table to obtain an indicator reading.
—Place parallels and setup blocks so that through holes may be drilled, reamed, bored, or tapped without cutting into these positioning tools.
—Limit the amount of force required to strap workpieces securely on parallels on a jig borer, jig grinder, or accessory table. Excessive or uneven force may cause distortion or cracking if the part is hardened.
—Avoid clamping work so tightly in a precision vise as to cause the part to be sprung.
—Check work alignment regularly when comparatively heavy cuts are taken on the jig borer or grinder. This action ensures the workpiece has not shifted.
—Use the shortest possible shank or diamond mandrel length to reduce wheel overhang.
—Dress the side and bottom wheel faces to a small corner radius. This technique provides a stronger grinding wheel edge and tends to prevent the scoring of ground shoulder surfaces.
—Cycle the NC positioning movements and machining processes for programming accuracy before actual part production.
—Check the sharpness of each cutting tool, the accuracy of grinding the cutting edges, and the tool height.
—Check machining speeds and feeds and continuously observe each machining operation. Be ready for immediate shutdown, if necessary.
—Position the jig grinder safety shield for visibility of the grinding process and for eye protection.
—Maintain a constant room and workpiece temperature of 68°F (20°C) for ultraprecise measurements.
—Use machine bibs and aprons to protect housings, tables, slides, graduated collars, and other machine elements from abrasive particles and dust.

TERMS USED WITH JIG BORERS, JIG GRINDERS, AND UNIVERSAL MEASURING MACHINES

Jig borer	A high-precision machine tool used primarily in toolrooms for the production of punches and dies, jigs and fixtures, special cutting tools, gages, and work involving tolerances in millionths of an inch and equivalent metric units. A machine tool adapted to precision drilling, boring, reaming, shoulder forming, indexing, and precision measurement and inspection.
Jig grinder	An extremely precise grinding machine for grinding production tooling components. A grinding machine tool for regular and intricate internal and external form grinding. A precision grinder for generating straight, circular, chordal, and contour surfaces by planetary horizontal and vertical axial movement and high-speed grinding wheel action. A precision grinder for grinding vertical relief on dies and cutting tools by slot grinding in a vertical plane.
Microbore® boring bar	Trade name of a series of boring bars used for precision operations on jig borers. An adjustable head for rough boring and precision finishing operations.
Spotting tools (jig borers)	Short-shank center drills and correctly pointed flat-lip drills. Drills that produce precisely centered holes.
Leaf taper gages	Flat, thin metal gage pieces. Accurately ground gage pieces for measuring diametral sizes, concentricity, parallelism, and bellmouth of bored and/or ground holes.
Wipe grinding, plunge grinding, shoulder grinding	Three precision grinding processes. Abrasive machining processes requiring a combination of planetary motion, the high-speed revolving of a grinding wheel in the planetary path, and vertical movement of the spindle.
Out-feed grinding	Enlarging a hole by continually feeding the grinding wheel outward.
Chop grinding	Grinding with an oscillating spindle as the workpiece is fed past a high-speed revolving grinding wheel.
Slot grinding	Abrasive machining jig grinding operation using a vertical reciprocating motion and slot grinding attachment. Grinding internal and external vertical slots, grooves, radii, and angles, as in the vertical faces of a die.
Diamond-charged mandrel	Simple form of grinding tool designed for collet/chuck mounting. (Body diameter is reduced over an area long enough to grind an internal hole to depth. The mandrel body is charged with diamond particles.) A grinding wheel used principally for finish grinding extremely small hole diameters.
Taper setting plate	An angle-graduated plate for small angle setting of the spindle.
Pantograph dresser	Grinding wheel dresser that forms the grinding face by guiding a diamond nib in a fixed path according to a template. A jig grinder attachment using a linkage system to reproduce a required form from a greatly magnified template.
Micro-sine table	A design of a precise sine table that permits precise angle positioning and serves as an adjustable fixture. (Highly accurate adjusting or measuring rods are used to make angle settings from $0°$ to $95°$.)

Universal measuring machine	A highly precise measuring instrument capable of making dimensional measurements and measuring positions and surfaces in millionths of an inch (0.025μm).
Locating microscope	A spindle-mounted locating accessory. An optical measuring device with which features of a workpiece are positioned in reference to the spindle. (Cross lines and concentric circles are used to locate a specific part feature.)
TV microscope (universal measuring machine)	Micromeasurement device for magnifying part details up to 2500X in closed-circuit TV. An advanced system for visual inspection of microminiature parts, circuits, and other conventional-size parts.

SUMMARY

Jig borers are one of industry's most accurate positioning and fine-machining machine tools.

Precision centering, drilling, reaming, straight and small-angle taper boring, shoulder forming, and tapping operations are commonly performed on the jig borer.

Jig borers are also used for measurement and layout work and all machining processes associated with jig and fixture, punch and die, special tooling, instrument making, and gage work.

Common jig boring tools and accessories include precision boring chucks, Microbore® boring bars, extenders, adapters, chucks, and collets.

Small hole sizes may be tapped using an instant-reversing, speed-reducing tapping attachment.

Hole centers are spotted with center drills, standard drills ground for spotting, and spotter drills.

A locating microscope permits precise location of edges, irregular shapes, and holes that cannot be positioned with an indicator.

Digital readouts verify measurement accuracy along X and Y axes. Readout equipment covers backlash take-up, manual zero reset, and digital readout overflow.

Rotary tables provide for angular indexing within tolerances of ±12 seconds, ±6 seconds, and ±2 seconds, depending on the precision quality.
The tilting rotary table has the capacity to set an angle to 1 second of arc.

The micro-sine table uses gaging rods for accurate work angle setups from 0° to 95°.

The precision index center permits the rotation of a workpiece on centers for boring, milling, grinding, and other processes.

The indicator, line finder, and microscope are widely used on jig borers and jig grinders for aligning work edges.

The locating microscope is used to align reference points and surfaces and for other measurement processes.

Leaf taper gages permit checking hole sizes, concentricity, bell-mouth, and out-of-roundness without changing the tool (spindle) position.

Jig grinders provide the versatility and extreme accuracy required to grind the cutting and forming tools, gages, instruments, and precision parts produced on the jig borer.

Jig grinders are used for out-feed, wipe, chop, plunge, taper, and shoulder grinding.

Angles, slots, corners, and other intricate forms may be slot ground in a vertical plane.

Hole depths are controlled by an adjustable positive stop, graduated dial and vernier, or micrometer stop on the column.

The speeds of jig grinding heads range from 9,000 RPM for heavy grinding operations to 175,000 RPM for grinding ultrasmall holes.

Wheel dressing attachments for jig grinders include radius angle, cross slide, spherical socket, angle, and pantograph attachments.

Extremely small holes are ground with diamond-charged mandrels.

Jig grinding wheels are dressed concave on the flat face for plunge grinding.

The cutting face of a grinding wheel is relieved to about one-fourth of its width.

The jig grinder spindle may be set to grind a tapered hole by either the taper setting plate or the dial indicator method.

Shouldered areas are ground using the outside diameter and concaved-end face of the grinding wheel.

Universal measuring machines are used for ultraprecise measurements on flat, plane surfaces; cylinders; irregular contours; and complex single and compound angle settings.

The accuracy to which the various components of the measuring machine are machined and other construction features ensure that the measurement capacity is in the millionths of an inch range.

Angular measurements and inspection setups often require the use of a small angle divider. With this accessory, direct angle readings are made to ±10 seconds of arc.

Safe operating practices applied to general machine tools and to the use of precision measuring equipment are observed when jig borers are used.

UNIT 35 REVIEW AND SELF-TEST

1. List four classifications of toolroom work that are performed on a jig borer.
2. Identify two important design features of drilling tools that are required for jig boring work.
3. a. Identify the function served by a micro-sine table.
   b. Describe briefly how a micro-sine table is set.
4. a. List the steps for setting up a locating microscope on a jig borer to center hole A in the figure shown below. The hole has been drilled and reamed to size.
   b. Tell how hole B is manually positioned for centering, drilling, and boring.

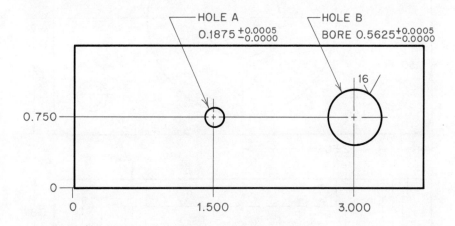

5. State two conditions or factors that differentiate slot grinding from plunge grinding.
6. Identify two methods of manually controlling spindle depth movement on a jig borer or jig grinder.
7. a. Describe the design features of a pantograph dressing attachment for a jig grinder.
   b. Explain how the attachment is operated.
8. List the general sequence of grinding steps that are followed on a jig grinder.

9. Set up a work processing plan to shoulder grind the part shown in the figure below.

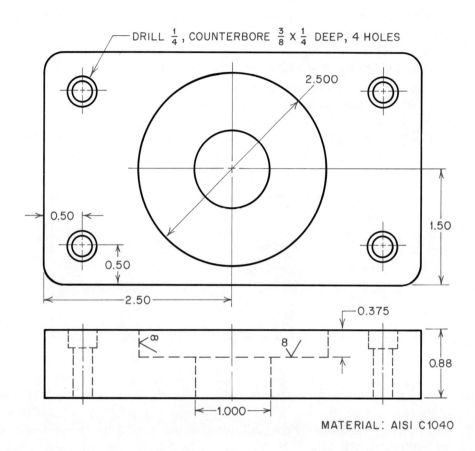

10. Describe two applications of fixture, punch and die, or gage parts that may be measured on a universal measuring machine.
11. Tell what function a spin table serves when used on a universal measuring machine.
12. State two personal safety precautions to observe when using either a jig borer or a jig grinder.

PART TEN

Production of Industrial and Cutting Tool Materials

SECTION ONE   Structure and Classification of Metals
          UNIT 36   Manufacture, Properties, and Classification of Industrial Materials
SECTION TWO   Basic Metallurgy and Heat Treatment of Metals
          UNIT 37   Heat Treating and Hardness Testing

SECTION ONE

Structure and Classification of Materials

Industrial materials include ferrous and nonferrous metals and alloys, cemented carbides, ceramics, and industrial diamonds. This section deals with these materials from the standpoint of their use in the manufacture of metal and nonmetallic parts and in other applications to cutting tools.

The reduction of iron ores to produce pig iron and processes for producing cast irons and cast steels are described first. Consideration is then given to the characteristics, properties, applications, and manufacture of common ferrous and nonferrous metals and alloys, cemented carbides, and ceramic and diamond cutting tools. The next section relates this technology to heat-treating processes and hardness testing. Application is then made of the effects of hardening, tempering, annealing, normalizing, and stress relieving on the microstructure of metals; materials testing equipment and processes are also covered.

Classification systems for iron, steel, aluminum, and alloys are also described. These systems are based on standards of the Society of Automotive Engineers (SAE), the American Iron and Steel Institute (AISI), and the Aluminum Association (AA).

UNIT 36

Manufacture, Properties, and Classification of Industrial Materials

The selection of industrial materials for a workpiece or a cutting tool must take into consideration the physical properties, such as hardenability, machinability, and ductility, for which engineering and design data establish conditions of use. The composition and structure of a pure metal may be changed to provide qualities in a resulting product to meet engineering requirements,

MANUFACTURE, PROPERTIES, AND CLASSIFICATION OF INDUSTRIAL MATERIALS

such as red hardness, tensile strength, wear resistance, and elastic limit, through the addition of alloying elements. Steels, nonferrous metals, and alloys are classified according to their composition and physical properties. This unit describes the manufacture, properties, and classification of industrial materials that are widely used in the machine and metal trades.

MANUFACTURE OF PIG IRON AND CAST IRON

The most commonly used metals in the machine trades are grouped under cast irons, carbon steels, alloy steels, and nonferrous metals and alloys. One of the first steps in making cast iron and steel is to produce pig iron.

Producing Pig Iron

The properties of metals in each ferrous group may be changed by adding various alloying elements. The mechanical and chemical properties need to be combined to produce a metal to serve a specific purpose. The basic metal form is *pig iron*. Pig iron is produced in a *blast furnace* that is charged with an iron ore, coke, and limestone. There are four principal iron ores:

—Hematite,
—Limonite,
—Magnetite,
—Taconite.

Each ore varies in the amount of iron that may be removed. The processing of raw materials in a blast furnace to produce pig iron is illustrated in Figure 36-1.

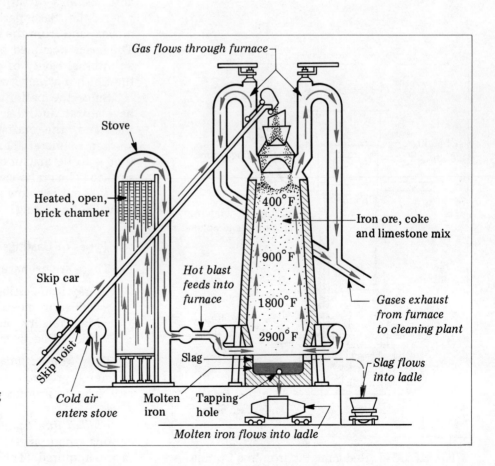

FIGURE 36-1 Processing Raw Materials in a Blast Furnace to Produce Pig Iron

Blast furnaces operate by feeding a hot air blast into the furnace at approximately 1000°F (538°C). The hot air causes the coke to burn vigorously. In the process, carbon monoxide is produced and removes oxygen from the heated iron ore. The reduced, spongy mass of iron moves on down through the charge to collect at the bottom of the furnace.

The limestone in the charge serves as a *flux*. The purpose of a flux is to unite with the silica and sulphur impurities of the iron ore. The product is called a *slag*. The slag also moves down the furnace and collects at the bottom. Since slag is lighter than iron, it floats on top of the iron. The slag is drawn off separately at the slag spout level of the furnace into a ladle truck for disposal. The molten iron is tapped off from the lower tap hole into another pouring ladle. The molten iron (pig iron) is either further processed for steel making or is cast into *pigs*. The pigs are subsequently reprocessed in foundries to manufacture castings. The blast furnaces are operated continuously except for shutdown for overhaul purposes.

Producing Cast Iron

Cast iron is produced by further refining the pig iron in a *cupola furnace*. The charge used is scrap iron, pig iron, and layers of coke. The reducing of pig iron in a cupola furnace to produce cast iron is shown in Figure 36-2. The furnace operates by feeding air from an area above the slag spout up through the burning layers of coke. The molten iron and slag flow through the charge and collect at the bottom of the furnace. The slag is drawn off through the slag spout.

The molten iron is tapped into mixing ladles and then poured into a *sand mold* to produce a cast workpiece. A sand mold is prepared from a pattern (Figure 36-3A) that forms the internal cavity in the sand. The poured metal produces a cast iron casting of a particular form and size. After cooling, the casting (Figure 36-3B) is removed from the mold. The *sprue*, *runner*, *gages*, *riser*, and *baked core* are broken away and removed. The casting is then rough ground at places where the gates and riser were located. Next, the casting is tumbled. Tumbling removes any *flashings* of surplus metal between sections where the flash parts join. Tumbling also removes sand from the inside and outside surfaces. The space occupied by the baked core produces an internal cavity of a desired shape and size in the finish machined workpiece (Figure 36-3C).

Since the patterns used to make the molds have a draft in order to permit withdrawal from the sand, the craftsperson must consider the amount of material to remove from each surface that is to be machined in order to clean up each surface. The craftsperson must note the fact that the first cut is below the hard surface of the outer scale of a casting.

Types of Castings

There are primarily four types of castings:

—Gray iron castings,
—Chilled iron castings,
—Malleable iron castings,
—Alloyed castings.

Gray iron castings are the most widely used in metal products manufacture. This kind of cast iron is produced from pig iron and scrap steel.

Chilled iron castings are produced for applications where an exceptionally hard outer surface is required. The surface of the molten cast iron is cooled as rapidly as possible. The grain

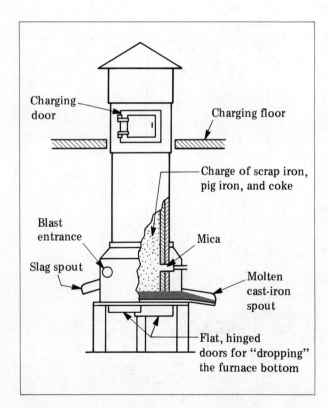

FIGURE 36-2 Reducing Pig Iron in a Cupola Furnace to Produce Cast Iron

MANUFACTURE, PROPERTIES, AND CLASSIFICATION OF INDUSTRIAL MATERIALS 751

structure produced by this quick chilling (cooling) provides an extremely hard surface.

Malleable iron castings are heat treated after they are solidified. Heat treating anneals a casting to improve machinability and to produce a cast iron that can withstand shock. A special grade of pig iron and foundry scrap is used to produce malleable cast iron.

Alloyed castings are extensively used where special properties, in addition to the properties of cast iron, are required. Selected alloys, such as nickel, chromium, and molybdenum, are added to the cast iron.

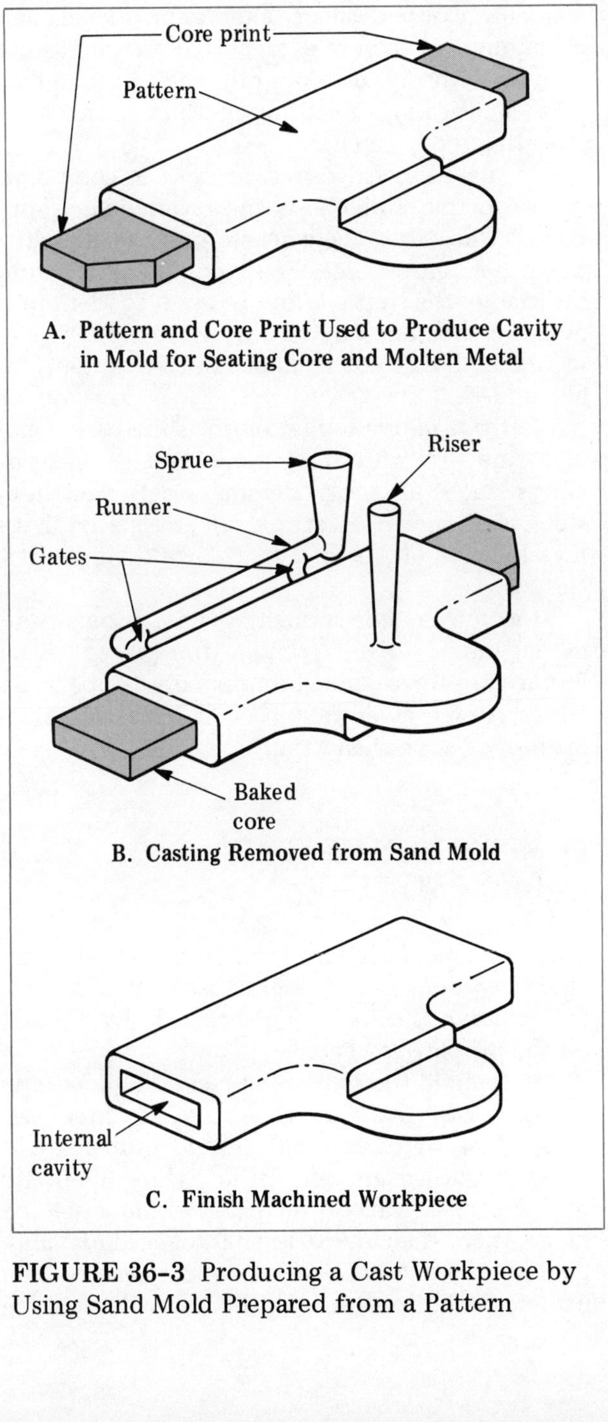

BASIC STEEL MANUFACTURING PROCESSES

Steel-making processes require the impurities within pig iron to be *burned out*. The removal of impurities is done in one of four types of furnaces:

—Open hearth furnace,
—Bessemer converter,
—Electric furnace,
—Oxygen process furnace.

Open Hearth Process

As much as 80% of all steel is produced by the open hearth process, which was introduced in 1908. The dish-shaped *hearth* is charged with limestone and scrap steel. When the scrap is melted, molten pig iron is poured in. The impurities are burned out by sweeping fuel and hot gas over the molten metal. The burned gases are drawn off. The direction of the flames in the furnace is changed at regular intervals. The limestone in the furnace unites with the impurities and rises to the surface of the molten metal.

The process continues for a number of hours. The molten steel is drawn off into a ladle that has the capacity to take the *heat* (the batch of steel to be poured). The slag floating on the metal is drawn off through a spout in the ladle.

Alloying is done by adding fixed amounts of the required alloying metals—for example, tungsten, molybdenum, and manganese—to the molten steel. After mixing, the alloyed steel is poured into *ingot molds*. When the steel within a mold solidifies, the mold is stripped from the solid steel mass, which is called an *ingot*. The ingot is moved into a *soaking pit* so that the complete ingot is slowly brought to a uniform temperature. Next, the heated ingot is rolled into specific sizes and shapes.

Bessemer Converter Process

The Bessemer converter today produces a very small percent of the steel manufactured as compared to the earlier production of over 90% of all steel manufactured. The converter consists of a lined circular shell that is swiveled to receive

A. Pattern and Core Print Used to Produce Cavity in Mold for Seating Core and Molten Metal

B. Casting Removed from Sand Mold

C. Finish Machined Workpiece

FIGURE 36-3 Producing a Cast Workpiece by Using Sand Mold Prepared from a Pattern

a charge of molten iron. The trunnion mountings are designed to provide a blast of air at high velocity. The blast flows through holes in the bottom when the furnace is moved to vertical position. The oxygen in the air burns out the impurities.

After being subjected to the air blast for 10 to 15 minutes, the converter is tilted to pour the metal into a mixing ladle in which alloying elements are added. The alloyed steel is then poured into ingots, soaked, and rolled in much the same manner as the open hearth steel.

Electric Furnace Process

The electric furnace permits precise control over the steel-making process. The producing of fine alloy and tool steels in an electric furnace is shown in Figure 36-4. The furnace operates by lowering three carbon electrodes to strike an arc with the scrap metal loaded in the furnace. The quality of scrap steel, the alloying elements, the amount of oxygen used, and the heat are carefully regulated. The heat generated by the electric arc produces the molten steel.

Chromium, tungsten, nickel, or other alloying materials are added to produce the required alloy steel. After the heat (molten steel) is complete, the furnace is tilted in a pouring position. The molten metal is *teemed* (poured from the ladle) into ingots for subsequent soaking and rolling operations.

Basic Oxygen Process

The oxygen process is one of the newer steel-making processes. A high-pressure stream of pure oxygen is directed over the top of the molten metal. The furnace is charged with about 30% scrap metal. Then, molten pig iron is poured into the furnace. Finally, the required fluxes are added.

The making of steel in a basic oxygen process furnace is illustrated in Figure 36-5. The furnace is covered with a water-cooled hood. The oxygen lance is lowered to within 5 to 8 feet above the surface of the molten metal. The force of the oxygen at a pressure of 140 to 160 pounds per square inch at a flow rate of 5,000 to 6,000 cubic feet per minute creates a churning high-temperature movement. This action continues until the impurities are burned out. Then, the flame drops, the oxygen is shut off, and the lance is removed. The entire process takes almost 50 minutes at a steel production rate of about 300 tons per hour.

The furnace is tapped by tilting it. After alloying materials are added to the mixing ladle, the molten metal is poured into ingots. The same processing is followed to form metal stock.

CHEMICAL COMPOSITION OF METALS

Metals are identified according to a natural element. Copper, silver, aluminum, nickel, and chromium are a few examples. Pure metals are alloyed with one or more other elements to provide qualities in a resulting product to meet engineering requirements.

In metallurgical terms, an *alloy* is composed of two or more elements and has metallic properties. One of the elements must be metallic; the other, either metallic or nonmetallic. Commercially, the term *alloy* is used to denote a metallic substance of two of more metallic elements. One of these elements must be intentionally added.

Ferrous alloys contain iron as the base metal and one or more metallic elements. Ferrous alloys such as molybdenum steel, vanadium steel, and nickel steel are all steels that have metallic elements added to change their properties.

Nonferrous alloys such as brass, bronze, and monel metal do not contain iron (except as an impurity). Brass and bronzes are copper-base alloys. Brass is an alloy of copper and zinc; bronze, of copper and tin.

COMPOSITION OF PLAIN CARBON STEEL

The principal elements in plain carbon steel include: *carbon, manganese, phosphorous, silicon,* and *sulphur*. Each element has a decided effect on the properties of steel.

Carbon is the hardening element and exerts the greatest influence on the physical properties. The effect of carbon on certain properties of steels is shown graphically in Figure 36-6. As the percent of carbon increases within a specific range, there is an increase in hardness, hardenability, wear resistance, and tensile strength. The higher the carbon content, the lower the melting

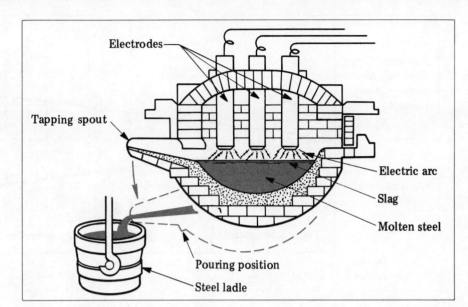

FIGURE 36-4 Producing Fine Alloy and Tool Steels in an Electric Furnace

point. Any additional carbon above the 0.85% point has limited effect on hardness, although wear resistance increases. The graph shows four general ranges of steel with carbon content within the 0 to 1.3% range. Low-carbon steels contain from 0.02% to 0.30% carbon by weight. Medium-carbon steels contain from 0.30% to approximately 0.60% carbon. High-carbon steels fall within the 0.60% to 0.87% carbon content range. The tool steels begin at the 0.87% carbon content point.

A small quantity (0.30% to 0.60%) of manganese is usually added during manufacture to serve as a purifier in removing oxygen from the steel. Oxygen in steel makes it weak and brittle. Manganese also mixes with sulphur, which is an impurity. Manganese is added to steel to increase the tensile strength, toughness, hardenability, and resistance to shock. Manganese is added in amounts from 0.06% to over 15% to produce certain physical properties. A 1.5% to 2.0% addition to high-carbon steel produces a deep-hardening steel. Such a steel must be quenched in oil. The addition of 15% manganese produces hard wear-resistant steels.

The addition of small amounts of phosphorous and sulphur to low-carbon steels increases machinability and decreases shrinkage. Amounts

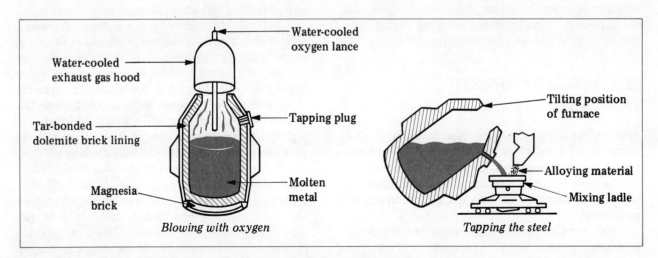

FIGURE 36-5 Making Steel in a Basic Oxygen Process Furnace

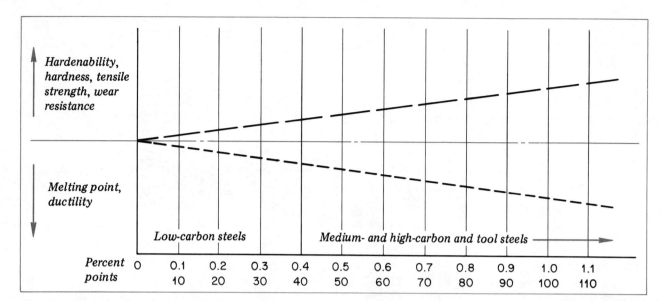

FIGURE 36-6 Effect of Carbon on Certain Properties of Steels

above 0.60% are considered undesirable because the phosphorous causes failures due to shock and vibration.

The presence of from 0.10% to 0.30% silicon makes steel sound as it is cast or hot rolled. Between the 0.60% to 2.0% range, silicon becomes an alloying element. Silicon is usually used in combination with manganese, molybdenum, or chromium. As an alloying element, silicon increases tensile strength, toughness, and hardness penetration.

The addition of from 0.08% to 0.30% sulphur to low-carbon steel increases its machinability. Screw stock for automatic screw machine work is a sulphurized free-cutting steel. Otherwise, sulphur is considered an impurity in steel. As an impurity, sulphur causes cracking during rolling or forging at high temperatures.

MECHANICAL PROPERTIES OF METALS

Mechanical properties are associated with the behavior of a metal as it is acted upon by an external force. The properties of a metal determine the extent, if any, to which it can be hardened, tempered, formed, pulled apart, fractured, or machined.

Hardness is defined as the property of a metal to resist penetration. This property is controlled by heat treating in which a machined or forged tool may be shaped while soft and then hardened.

Hardenability represents the degree to which a metal hardens through completely to its center when heat treated. A low hardenability means the surface layer hardens but the metal is softer toward the center. *Brittleness* is related to hardness. Brittleness refers to the degree to which a metal part breaks or cracks without deformation. *Deformation* is the ability of a material to flex and bend without cracking or breaking.

Ductility refers to the ability of a metal to be bent, twisted, drawn out, or changed in shape without breaking. For example, a deep draw shell requires sheet metal with high ductility to permit drawing without fracturing the metal. The ductility of a metal is usually expressed as a percentage of reduction in area or elongation.

Malleability relates to the ability of a metal to be permanently deformed by rolling, pressing, or hammering. *Toughness* of a metal refers to the property that enables it to withstand heavy impact forces or sudden shock without fracturing.

Fusibility is the ease with which two metals may be joined together when in a liquid state. A high-fusibility metal can be easily welded. In such instances, the term *weldability* is used.

Machinability relates to the ease or difficulty with which a given material may be worked with a cutting tool. After extensive testing of materials under control conditions, machinability ratings have been established. These ratings are published in handbook tables as percentages. AISI 1112 cold-drawn steel is assigned a rating of 100%. Other metals are rated in percentages in comparison to the machinability of AISI 1112 steel. Difficult-to-machine metals are rated below 100%. Easy-to-machine metals are rated above 100%.

Machinability properties need to be considered in terms of the following factors:

—Physical properties of the material (tensile strength, rigidity, hardness, grain structure) and chemical properties;
—Type of chip formation produced during the machining process;
—Cutting characteristics that are influenced by the use or absence of a particular type of cutting fluid;
—Power required in taking a cut.

Each factor influences the degree to which material may be machined to produce a required surface finish and a dimensionally accurate part.

CLASSIFICATION OF STEELS

There are two general categories of steel: plain carbon steels and alloy steels. The plain carbon steels are grouped in three ranges of carbon content as low-carbon steel, medium-carbon steel, and high-carbon steel.

Plain Carbon Steels

Low-Carbon Steel. The range of carbon by weight is from 0.02% to 0.30%. Low-carbon steels may not be hardened except by adding carbon to permit hardening the outer case while the area below remains soft. The range of carbon is from 0.08% to 0.30% in low-carbon steels that are commonly used for manufacturing parts that do not require hardening. Low-carbon steels such as machine steel and cold-rolled steels are used in manufacturing bolts and nuts, sheet steel products, bars, and rods.

Medium-Carbon Steel. The carbon content is from 0.30% to 0.60%. With this amount of carbon, it is possible to harden tools, such as hammers, screw drivers, and wrenches, that may be drop forged or machined.

High-Carbon Steel. The carbon content range for high-carbon and tool steels is from 0.60% to 1.50%. These steels are adapted for edge cutting tools such as punches, dies, taps, and reamers.

Alloy Steels

The addition of alloying elements to steel increases the tensile strength, hardenability, toughness, wear abrasion, red hardness, and corrosion resistance. The most important effects of alloying elements on the properties of steel are indicated in Table 36–1.

SAE AND AISI SYSTEMS OF CLASSIFYING STEELS

Two main systems of classifying steels have been devised. These systems provide for standardization in producing each alloy and in designating each alloy. The systems were developed by the Society of Automotive Engineers (SAE) and the American Iron and Steel Institute (AISI). Standard construction grades of carbon and alloy steels are designated according to the basic chemical composition of each metal.

Four- and Five-Digit Designations

A four-digit series of code numbers is used in both the SAE and the AISI systems. Certain alloys are identified by a five-digit series of code numbers. One of two letters is used in cases where an element has been varied from the normal content of the steel. The prefix X denotes a variation of manganese or sulphur. The prefix T denotes a variation of manganese in the 1300-range steels.

The first digit of the SAE and AISI classification systems indicates the *basic type of steel*, as follows:

1 carbon
2 nickel
3 nickel-chrome
4 molybdenum
5 chromium
6 chromium-vanadium
7 tungsten
8 nickel-chromium-molybdenum
9 silicon-manganese

The second digit classifies the steel or alloy within a particular *series*. Table 36-2 shows the SAE/AISI classifications of standard carbon and alloy steels by series. A steel designated with a 10XX number is in the nonsulphurized plain carbon steel series. An alloy steel numbered 40XX is in the molybdenum steel series that has

0.20% or 0.25% molybdenum content. A zero in the second digit indicates there is no major alloying element.

The third and fourth digits indicate the middle of the carbon content range—that is, the *average percent of carbon*—in the steel. The percent of carbon is indicated by the third and fourth digits as *points*. A point is the same as 0.01% carbon. Thus, a 0.20% carbon content is expressed by the third and fourth digits as 20 points. Fig-

TABLE 36-1 Effects of Alloying Elements on Properties of Steel

Effect on Properties of Steel	Alloying Metallic Element											
	Carbon (C)	Chromium (Cr)	Manganese (Mn)	Molybdenum (Mo)	Nickel (Ni)	Tungsten (T)	Vanadium (V)	Phosphorus (P)	Silicon (Si)	Sulphur (S)	Cobalt (Co)	Lead (Pb)
Increases												
tensile strength	X	X	X	X	X							
hardness	X	X										
wear resistance	X	X	X		X	X						
hardenability	X	X	X	X	X		X					
ductility		X										
elastic limit		X		X								
rust resistance		X			X							
abrasion resistance	X	X										
toughness		X			X	X	X					
shock resistance		X			X	X						
fatigue resistance							X					
Improves machinability										X		X
Decreases												
ductility	X	X										
toughness											X	
Raises critical temperature		X			X						X	
Lowers critical temperature			X	X								
Produces												
red hardness				X		X					X	
fine grain structure			X				X					
oil-hardening properties		X	X	X	X							
air-hardening properties			X	X								
Reduces deformation			X	X								
Deoxidizer agent			X						X			
Desulphurizer agent			X									
Eliminates blow holes								X				
Causes												
hot shortness										X		
cold shortness								X				
Facilitates rolling and forging			X						X			

ure 36-7 provides two examples of the use of SAE and AISI code numbers.

In a few instances, a five-digit numerical code is used. For example, in the SAE system, a designation of 71250 means the steel is a tungsten steel (7 as the first digit) with 12% tungsten (12 as the second and third digits) and 0.50% carbon (50 points as the fourth and fifth digits).

TABLE 36-2 SAE/AISI Classifications of Standard Carbon and Alloy Steels by Series

Series	Type	Composition or Special Treatment*
Classification of Carbon Steels		
10XX	Plain carbon	Nonsulphurized
11XX	Free-machining	Resulphurized
12XX		Rephosphorized and resulphurized
Classification of Alloy Steels		
13XX	Free-machining, manganese	Mn 1.75%
2XXX	Nickel	
23XX	Nickel	Ni 3.50%
25XX		Ni 5.00%
3XXX	Nickel-chromium	
31XX	Nickel-chromium	Ni 1.25% Cr 0.65%
30XXX		Corrosion- and heat-resisting
40XX	Molybdenum	Mo 0.20% or 0.25%
41XX	Chromium-molybdenum	Cr 0.50% or 0.95% Mo 0.12% or 0.20%
43XX	Nickel-chromium	Ni 1.80% Cr 0.50% or 0.80% Mo 0.25%
44XX	Molybdenum	Mo 0.40%
45XX		Mo 0.52%
46XX	Nickel	Ni 1.80% Mo 0.25%
47XX	Nickel-chromium	Ni 1.05% Cr 0.45% Mo 0.20% or 0.35%
48XX	Nickel	Ni 3.50% Mo 0.25%
50XX	Chromium	Cr 0.25%, 0.40%, or 0.50%
50XXX		C 1.00% Cr 0.50%
51XX		Cr 0.80%, 0.90%, 0.95%, or 1.00%
51XXX		C 1.00% Cr 1.05%
52XXX		C 1.03% Cr 1.45%
61XX	Chromium-vanadium	Cr 0.60% to 0.95% V 0.10% min. to 0.15% min.
7XXX and 7XXXX	Tungsten	
8XXX	Triple-alloy	Ni 0.30% to 0.55% Cr 0.40% to 0.50% Mo 0.12% to 0.35%
92XX	Silicon-manganese	Si 2.00% Mn 0.85% Cr 0 or 0.35%
93XX	Triple-alloy	Ni 3.25% Cr 1.20% Mo 0.12%
94XX		Ni 0.45% Cr 0.40% Mo 0.12%
98XX		Ni 1.00% Cr 0.80% Mo 0.25%

*Percents represent the average of the range of a particular metal element. Only the predominant alloying elements (excluding carbon) are identified.

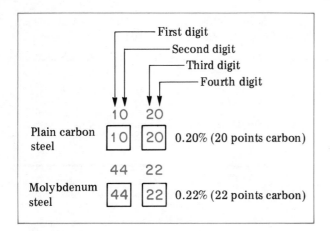

FIGURE 36-7 Examples of SAE and AISI Code Numbers

TABLE 36-3 Identification of Major Groups of Tool and Die Steels

Letter Code		Type of Tool and Die Steels	
W		Water-hardening tool steel	
S		Shock-resisting tool steel	
O			Oil-hardening
A		Cold-worked tool steel	Medium alloy, air-hardening
D			High-carbon, high-chromium
H	1-19	Hot-worked tool steel	Chromium types
	20-39		Tungsten types
	40-59		Molybdenum types
M		High-speed tool steel	Molybdenum types
T			Tungsten types
L		Special-purpose tool steel	Low-alloy types
F			Carbon-tungsten types
P	1-19	Mold Steel	Low-carbon types
	20-39		Other types

The letter H after the SAE or AISI code number specifies that the alloy steel meets specific hardenability standards. Steel manufacturers and handbook tables provide craftspersons with additional information about H designations, composition, properties, and applications of steels designated by SAE and AISI codes.

Coding of Tool and Die Steels

There are eleven major groups of tool and die steels. These groups are identified by a letter. In some cases, subgroups are designated by using a numeral following a letter. Table 36-3 provides the codes for identification of the tool and die steel groups.

Additional Features of AISI System

A prefix letter is sometimes used with the code number to identify the method used to produce the steel. The four basic prefix letters are: B, for Bessemer carbon steel; C, for general open hearth carbon steel; D, for acid open hearth steel; and E, for electric furnace alloy steel. Where there is no letter prefix to the AISI number given in a table, the steel is predominantly open hearth.

UNIFIED NUMBERING SYSTEM (UNS)

While only the SAE and the AISI systems have been discussed here, various trade associations, professional societies, standards organizations, and private industries have developed different numbering and coding systems. The Unified Numbering System (UNS) represents a joint effort by the American Society for Testing and Materials (ASTM), the SAE, and others to correlate the different numbering systems being used for commercial metals and alloys. The important point to remember about a UNS number is that it provides an *identification, not a specification.* The specifications are to be found in manufacturers' literature, trade journals, and handbooks.

Sixteen code letters are used in the Unified Numbering System. Each letter is followed by five digits. The digits run from 00001 through 99999. Each of the sixteen code letters identifies certain metals and alloys. The UNS code letters for ferrous and nonferrous metals and alloys are identified in Table 36-4. The numbers in the UNS groups conform, wherever possible, to the code numbers of the other systems. For example, an AISI or SAE 1020 plain carbon steel has a corresponding UNS number of G10200. An SAE or AISI 4130 alloy steel has a UNS identification of G41300. An M33 tool steel has a UNS code of T11330. Handbook tables provide

TABLE 36-4 Unified Numbering System (UNS) for Ferrous and Nonferrous Metals and Alloys

UNS Series Designation		Metals and Alloys Group
Letter	Range of Numbers	
Ferrous Metals and Alloys		
D		Specified mechanical property steels
F		Cast irons
G		AISI and SAE carbon and alloy steels (except tool steels)
H	00001	AISI H-steels
J	to	Cast steels (except tool steels)
K	99999	Miscellaneous steels and ferrous alloys
S		Heat- and corrosion-resistant (stainless) steels
T		Tool steels
Nonferrous Metals and Alloys		
A		Aluminum and aluminum alloys
C		Copper and copper alloys
E		Rare earth and rare earth-like metals and alloys
L	00001	Low-melting metals and alloys
M	to	Miscellaneous nonferrous metals and alloys
P	99999	Precious metals and alloys
R		Reactive and refractory metals and alloys
Z		Zinc and zinc alloys

the full range of UNS numbers for plain carbon, alloy, and tool steels in the SAE and AISI systems.

ALUMINUM ASSOCIATION (AA) DESIGNATION SYSTEM

A system similar to the SAE, AISI, and UNS classifications for steels and alloys has been developed by the Aluminum Association (AA) for wrought aluminum and aluminum alloys. A four-digit numerical code is used. The first digit identifies the alloy type. For example, a code of 1 indicates an aluminum of 99.00% or greater purity. The numerical code 2 indicates a copper-type alloy; 3, manganese; 4, silicon; 5, magnesium; 6, magnesium and silicon; 7, zinc; and 8, an element other than identified by codes 1 through 7. The number 9 is unassigned at the present time.

The second digit shows the control over one or more impurities. A second digit of zero indicates there is no special impurities control. The third and fourth digits in the 1 series give the amount of aluminum above 99.00%, to the nearest hundredth of a percent. The same digits in the 2 through 8 series are used to identify different alloys in the group. The prefix X is used for experimental alloys. When standardized, the prefix is dropped.

The letters F, O, H, W, and T are used as *temper designations*. These letters follow the four digits (separated by a dash) to designate the temper or degree of hardness, as follows:

F   hardness as fabricated
O   annealed (for wrought alloys only)
H   strain-hardened (for wrought alloys only)
W   solution heat-treated
T   thermally treated

A temper designation may include a numerical code. For example, H designations with one or more digits indicate strain-hardened only (H1), strain-hardened and then annealed (H2), and strain-hardened and then stabilized (H3). A second digit is used to indicate the final degree of strain hardening from 0 to 8—for example, H16. A third digit, when used, gives the variation of a two-digit H temper—for example, H254.

Numerals 2 through 10 have been assigned in the AA system to indicate specific sequences of annealing, heat treating, cold working, or aging, as follows:

- T2 annealed (cast products only)
- T3 solution heat-treated and then cold-worked
- T4 solution heat-treated and naturally aged to a stable condition
- T5 artificially aged only
- T6 product heat-treated and then artificially aged
- T7 product heat-treated and then stabilized
- T8 product heat-treated, cold-worked, and then artificially aged
- T9 product heat-treated, artificially aged, and then cold-worked
- T10 product artificially aged and then cold-worked

VISIBLE IDENTIFICATION OF STEELS

There are two common methods of identifying steels. One method involves the use of the manufacturer's *color code markings*. The other method requires *spark testing*.

Color Code Markings

Some steel producers paint at least one end of steel bars. A color code is supplied by the

TABLE 36-5 Characteristics of Sparks Generated by Spark Grinding (Iron, Steel, Steel Alloys)

Stream	Wrought Iron	Gray Cast Iron	White Cast Iron	Annealed Malleable Iron	Machine Steel (AISI 1020)	Carbon Tool Steel	High Speed Steel (18-4-1)
Volume	Large	Small	Very small	Moderate	Large	Moderately large	Small
Length	Long	Short	Short	Short	Long	Long	Long
Color close to wheel	Straw	Red	Red	Red	White	White	Red
Streaks near end of stream	White	Straw	Straw	Straw	White	White	Straw
Quantity of spurts	Very few	Many	Few	Many	Few	Very many	Extremely few
Nature of spurts	Forked	Fine, repeating	Fine, repeating	Fine, repeating	Forked	Fine, repeating	Forked

steel manufacturer. Stock is cut from the unpainted end in order to preserve the painted identification color code.

Spark Test

A simple identification test is to observe the color, spacing, and quantity of sparks produced by grinding. The visible characteristics of steel, cast iron, and alloy sparks depend largely on carbon content. A high-carbon tool steel, when spark tested, produces a large quantity of fine, repeating spurts of sparks. These sparks are white in color at the beginning and ending of the stream. By contrast, a spark test of a piece of wrought iron stock produces a few spurts of forked sparks. The color varies from straw color close to the wheel to white near the end of the stream.

The characteristics of sparks generated by a spark grinding test are illustrated in Table 36-5.

The spark patterns provide general information about the type of steel, cast iron, or alloy steel.

CHARACTERISTICS OF IRON CASTINGS

Cast iron is an alloy of iron and carbon. The carbon content varies from 1.7% up to 4.5%, with varying amounts of silicon, manganese, and sulphur. The physical properties of cast iron depend on the amount of one of two forms of carbon that is present. *Graphite or free carbon* is one form. *Combined carbon or cementite* (iron carbide) is the second form. The five broad classes of cast iron in general use are as follows:

—Gray cast iron,
—White cast iron,
—Chilled cast iron,

	Austenitic Manganese Steel	Stainless Steel (Type 410)	Tungsten Chromium Die Steel	Stellite	Cemented Tungsten Carbide	Nickel
Stream						
Volume	Moderately large	Moderate	Small	Very small	Extremely small	Very small
Length	Long	Long	Average	Short	Very short	Short
Color close to wheel	White	Straw	Red	Orange	Light orange	Orange
Streaks near end of stream	White	White	Straw; blue white	Orange	Light orange	Orange
Quantity of spurts	Many	Moderate	Many	None	None	None
Nature of spurts	Fine, repeating	Forked	Fine, repeating			

—Alloy cast iron,
—Malleable iron castings.

Ductile iron (nodular cast iron) is a newer type of cast iron used in automotive and industrial applications where factors of high tensile strength and ductility are important.

Gray Cast Iron

Gray iron castings are widely used in machine tool, farm implement, automotive, and other industries. The American National Standard (ANS) specifications G25.1-1964 and the American Society for Testing and Materials (ASTM) standards A48-64 (1971) consider cast iron castings in two groups. The first group of gray iron castings (classes 20, 25, 30, and 35 A, B, and C) have excellent machinability, are comparatively easy to manufacture, and have a low elasticity and a high damping capacity. The second group of gray iron castings (classes 40, 45, 50, and 60 B and C) are more difficult to machine and manufacture, have a lower damping capacity than the first group and have a higher elasticity.

A 100X metallurgical microscope magnification of the graphite structure in gray cast iron is shown in Figure 36-8. The flake form of the graphite is clearly identifiable.

High-strength cast iron castings are produced by the *Meehanite-controlled process*. Some of the more important properties of Meehanite castings were cited earlier in connection with wear-resisting properties on applications such as the cast bed, frame, and column of jig borers, jig grinders, and other machine tools. Meehanite castings may also be produced with heat-resisting, corrosion-resisting, and other combinations of physical properties.

White Cast Iron

White cast iron has a silvery-white fracture. Castings of white cast iron are very brittle, with almost a zero ductility. White cast iron castings have less resistance to impact loading and a comparatively higher compressive strength per square inch than gray cast iron. Nearly all of the carbon in the casting is in the chemically combined carbon (cementite) form. White cast iron is used principally for the production of malleable iron castings and for applications that require a metal with high wear- and abrasive-resistance properties.

A 500X magnification photomicrograph of white cast iron is shown in Figure 36-9 and displays cementite (iron carbide) in the light areas and fine pearlite in the dark areas. Pearlite is comprised of alternating layers of pure iron (ferrite) and cementite.

Chilled Cast Iron

Chilled cast iron is used for products that require wear-resisting surfaces. The surfaces are designated as chilled cast iron. The hard surfaces are produced in molds that have metal chills for the rapid cooling of the outer surface. Rapid cooling causes the formation of cementite and white cast iron.

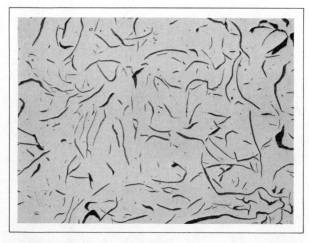

FIGURE 36-8 Flake Graphite Structure in Gray Cast Iron (100X Magnification)

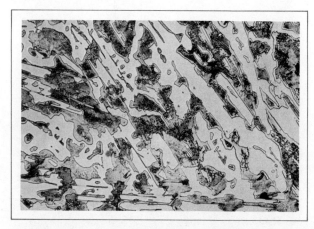

FIGURE 36-9 Cementite (Light Areas) and Fine Pearlite (Dark Areas) in White Cast Iron (500X Magnification Photomicrograph)

Alloy Cast Iron

Alloy cast iron results when sufficient amounts of chromium, nickel, molybdenum, manganese, and copper are added to cast iron castings to change the physical properties. The alloying elements are added to increase strength; to produce higher wear resistance, corrosion resistance, or heat resistance; or to change other physical properties.

Extensive use is made of alloy cast irons for automotive engine, brake, and other systems; for machine tool castings; and for additional applications where high tensile strength and resistance to scaling at high temperatures are required.

Malleable Iron Castings

Grades and properties of malleable iron castings are specified according to ANS G48.1-1969 and ASTM A47-68 specifications. These specifications relate to tensile strength, yield strength, and elongation.

Malleable iron is produced by heat treating. In the process, the cast iron is annealed or *graphitized*. Graphitization produces graphite (temper carbon aggregates). A 100X magnification of temper carbon aggregates in malleable cast iron is shown in Figure 36-10.

Hard, brittle, white cast iron castings are first produced from pig iron and scrap. These castings are then given an annealing heat treatment during which time the temperature is slowly increased over as much as a two-day period up to 1650°F (898.9°C). Then, the temperature is dropped slowly over an equal cooling time.

Malleable iron castings are widely used in industrial applications that require a highly machinable metal, great strength (as compared with other cast irons and nonferrous metals) and ductility, and good resistance to shock.

Cupola Malleable Iron. Malleable iron is also produced by the cupola method. The metal has good fluidity, produces sound castings, and possesses the property of being suited to galvanizing. Therefore, cupola malleable iron is used for making valves, pipe fittings, and similar parts.

Pearlitic Malleable Iron. Pearlitic malleable irons are used where a greater strength or wear resistance is needed than is provided by general malleable iron or in place of steel castings or forgings. There are a number of forms to pearlitic malleable iron. Some forms are engineered to resist deformation. Other forms permit deformation before breaking and are used for crankshafts, camshafts, and differential housings; ordinance equipment; and for other machine parts.

Pearlitic malleable irons contain some combined carbon. The carbon form is produced by stopping the heat treatment of regular malleable iron during production before all of the combined carbon is transformed to graphite form. The carbon may also be changed by reheating regular malleable iron above the transformation range.

Ductile Iron (Nodular Cast Iron)

Ductile iron is a relatively new type of cast iron. It is also known as *spheroidal graphite* because the graphite is present in ball-like form in contrast to the flake-like form in regular gray cast iron. A 100X magnification of spheroidal graphite in nodular cast iron is shown in Figure 36-11. The spheroidal graphite structure produces a casting of high tensile strength and ductility. The spheroidal structure results from the addition of small amounts of magnesium or cerium-bearing alloys and special processing. A few other advantages of nodular cast iron are as follows:

—Toughness is between the toughness of cast iron and steel;
—Shock resistance is comparable to ordinary grades of carbon steel;
—Melting point and fluidity are similar to the properties of high-carbon cast irons;

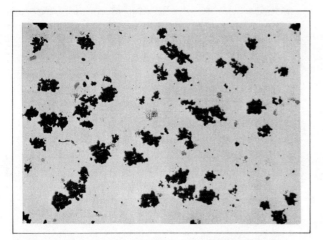

FIGURE 36-10 Temper Carbon (Graphite) Aggregates in Malleable Cast Iron (100X Magnification)

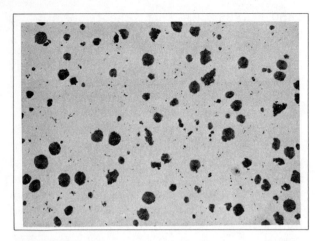

FIGURE 36-11 Spheroidal (Ball-Like) Graphite in Nodular (Ductile) Cast Iron (100X Magnification)

—Pressure tightness under high stress is excellent;
—Castings can be softened by annealing;
—Castings can be hardened by normalizing and air cooling or by oil quenching and drawing;
—Wear-resisting surfaces can be cast in molds containing metal chills;
—Surface hardening by flame or induction methods is feasible;
—Parts may be machined with the same ease as gray iron castings.

Nodular cast iron is used in the automotive industry for cylinder heads, crankshafts, and pistons; in the heavy machinery field; and in the machine tool industry for chuck bodies, forming dies, and other industrial applications.

CHARACTERISTICS OF STEEL CASTINGS

Steel castings are used for parts that must withstand shock and heavy loads and that must be tougher and stronger than cast iron, malleable iron, or wrought iron. Steel castings are usually found in heavy construction equipment, petroleum and mining industries, railroad industries, power production, and other applications where strength is essential.

Steel castings are produced by four common methods: (1) open hearth furnace method, where large tonnage parts, such as turbine crankcases, are required; (2) electric furnace method, for small-lot production of steels of a wide variety of analyses; (3) high-frequency induction furnace method, for small-quantity production of highly selective steels, such as high-alloy steels of special composition; and (4) the converter method, for foundries that have almost continuous pouring.

There are two general classes of steel castings: (1) carbon steel and (2) alloy steel. Carbon steel castings are broken up into three groups: (1) low carbon, (2) medium carbon, and (3) high carbon.

Carbon Steel Castings

Carbon steel castings contain the elements of carbon, manganese, phosphorous, and sulphur. The composition of these elements determines the group (low, medium, or high carbon) in which the cast steel falls and its properties. Handbook tables are available and give the chemical composition and physical properties of heat-resistant and corrosion-resistant steel castings. The percent of each element is given according to type and grade. Minimum physical properties follow in tables for tensile strength, yield point, and elongation.

Low-carbon steel castings have a carbon content below 0.20% with most of the castings produced in the 0.16% to 0.19% range. Medium-carbon steel castings have between 0.20% and 0.50% carbon. High-carbon steel castings have a carbon content above 0.50%.

Alloy Steel Castings

Alloy steel castings are distinguished from carbon steel castings by the addition of sufficient amounts of special alloying elements to obtain or increase specific desirable qualities. There are two groups of alloy steels: (1) low-alloy steels in which the alloy content is less than 8% and (2) high-alloy steels with more than 8% alloying elements.

The addition of alloying elements such as manganese, molybdenum, vanadium, chromium, and nickel in conjunction with heat treatments produces a wide variety of alloy cast steels. Tables of engineering and design data are contained in handbooks. Alloying elements are added to steel to affect one or more of the following physical properties:

—Hardenability,
—Distortion reduction and prevention of cracks by decreasing the rate of cooling during the hardening operation,

—Resistance to hardness reduction when being tempered,
—Material strength that may be increased through heat treatment,
—Abrasion resistance at regular and increased temperatures,
—Corrosion resistance at regular and high temperatures,
—Machinability,
—Toughness.

Alloy steels require heat treatment in order to utilize the potential properties of the alloy. There are three categories of alloy steels: (1) construction, (2) special, and (3) tool.

Construction Alloy Steels. Construction grades of alloy steels are used for structural members and have a low total alloy content in the range from 0.25% to 6.0%. These grades are commercially available in four basic forms. The forms include standard bar shapes (square, round, flat, and hexagon); structural members (channels and angles); flat forms (sheets, plates, and strips); and forging forms (bars and billets).

Special Alloy Steels. Special alloy steels are designed for particular applications, such as equipment parts that become tougher and harder with service. These alloy steels include properties such as heat resistance and corrosion resistance.

Alloy Tool Steels. Alloy tool steels have a total alloy content in the range between 0.25% to over 38%. Within this range, there are hundreds of different alloy tool steels. Characteristics and properties of these steels are found in trade handbooks and manufacturers' tables.

Alloy tool steels are widely used for cutting tools such as drills and milling cutters and stamping tools such as punches and dies. The important quality of alloy tool steels is their higher hardenability. Thus, the hardness of alloy tool steels penetrates (hardens) more deeply than plain carbon tool steels. Alloy tool steels are identified according to whether they may be hardened in water (water-hardening), air (air-hardening), or oil (oil-hardening).

COMMON ALLOYING ELEMENTS AND THEIR EFFECTS

Over 25 elements may be added to the production of alloy steels. Each alloying element may be included singly or in combination. The effect of carbon as one principal element has already been discussed. The effects of seven other major alloying elements on the properties of alloy steels are discussed next.

Manganese

Manganese is an element second in importance to carbon. The presence of manganese in amounts from 0.30% to 1.50% increases the hardenability, strength, toughness, and shock resistance of the alloy steel. Manganese steels are used for machine and other parts that are subjected to severe service.

When a considerable amount of manganese is added (from 10% to 14%), the alloy steel is capable of acquiring a ductility by quenching in water at a high temperature. As alloy steels are cold worked, they become *strain hardened*. The steels become extremely tough and hard by continued use. Strain-hardened steels are applied to crushing, grinding, and heavy-duty equipment.

Nickel

The effect of alloying nickel is to increase the wear-resistance, toughness, strength, and corrosion-resistance properties of steels. Nickel decreases the machinability of steel and has a limited effect in increasing hardenability. Nickel causes work hardening and reduces tool wear life.

Nickel-base alloys machine best when in a cold-drawn or *as-rolled* state. The cutting tool must be fed continuously during a cut. Otherwise, the cutting tool edges become dull and the material work hardens. The cutter and cutting area should be well flooded throughout the machining operation. Hardening and annealing affect machinability.

Nickel steels find wide application in machine parts that are subjected to repeated stress and shock and in structural rails and plates used in construction work. Nickel is introduced in 3% to 3.7% quantities.

Nickel-chrome steels include a small quantity of both nickel and chromium. The product is an alloy that resists distortion and fracture under severe working conditions such as are found when farm implements, automotive drive parts, and earth-moving equipment are used.

Stainless (corrosion resistant) steels require up to 22% nickel alloyed with chromium. The chromium content range is between 10% to 25%.

A minimum of 11% chromium is required to make the steel resistant to atmospheric corrosion. The three main types of stainless steels are: (1) *martensitic*, (2) *ferritic*, and (3) *austenitic*.

The martensitic type has around 1.0% or more carbon content. This stainless steel is hardenable when heated to a high temperature and then quenched in oil or air.

The ferritic type has an exceedingly limited amount of carbon. This type of stainless steel is essentially a soft steel alloy containing 11% or more chromium.

The austenitic stainless steels contain nickel in addition to chromium. While these steels cannot be hardened by quenching, they may work harden until they are almost as hard as hardened martensitic stainless steel. The chrome-nickel stainless steels generally have a white appearance.

The martensitic and ferritic types of stainless steel are magnetic. Austenitic steels range from magnetic when in a work-hardened condition to nonmagnetic when soft or annealed.

Chromium

Chromium (chrome) is essentially a hardening agent. Chromium steels are widely used in the machine industries for bearing races, ball bearings, springs, gears, shafts, and other parts that require resistance to corrosion, toughness, wear, and hardenability. Usually, in the construction grade, the chromium content range is from 0.30% to 1.60%. Chromium is used with steels within the carbon content range of 0.20% to 1.30%.

Chromium steels require proper hardening and tempering to provide significantly increased tensile strength and yield strength as compared to corresponding annealed steel. However, the ductility is reduced. High-speed steels and tool and die steels require the addition of chromium as a principal alloying element.

Molybdenum

Molybdenum steels are used for machine, aircraft, automotive, marine, and other parts where an alloy with toughness and shock-resistance properties is required. Molybdenum adds ductility and toughness and improves the heat treatment, hardenability, and resistance to softening at high temperatures. Molybdenum tool and die steels contain up to 9% molybdenum.

When alloyed with other elements in comparatively small amounts, the alloy steel produced is identified by the major elements—for example, nickel-molybdenum, chromium-molybdenum, and so on.

Vanadium

The addition of from 0.03% to 0.20% vanadium produces a fine grain structure in alloy steels. Within this range, the vanadium increases tensile strength, wear resistance, impact toughness, and yield strength, with no loss in ductility. Vanadium is generally alloyed with chromium to produce chromium-vanadium steels. Parts, such as springs and splined shafts, made of chromium-vanadium steel are heat treated. Vanadium is also mixed in combination with other elements such as cobalt, tungsten, and molybdenum.

High-speed steels are manufactured by adding up to 5% vanadium. Vanadium imparts the property of retarding softening at high working temperatures. This property is significant with high-speed cutting tools because temperatures produced at the cutting edges would otherwise soften the cutting tool. The use of high-speed steels for drilling, milling, turning, and other single- and multiple-point applications is the foundation for all machine and metal products manufacturing.

Cobalt

Cobalt steels are another classification of high-speed steels. The addition of from 5% to 12% cobalt significantly increases the *red hardness* and the wear resistance of the steel while there may be a decrease in impact toughness and/or resistance to shock. Red hardness means the cutting tool retains its hardness up to lower red heat temperatures.

A general application of cobalt is in producing a nonferrous nickel-aluminum-cobalt alloy known as *Alnico*. This alloy is used for high-quality permanent magnets.

Tungsten

Tungsten is noted for its ability to produce a fine, dense grain structure in alloy tool steels. Tungsten also improves the heat treatment and high wear-resistance qualities. These properties make alloy tool steels able to be ground or formed to a keen cutting edge and to retain hardness at high temperatures.

Cemented carbide cutting tools contain tungsten carbide, which is cemented together with cobalt. The advantages of cemented carbides are the high initial hardness, ability of the cutting edge to retain this hardness at red heat (around 1700°F (926°C), increased cutting speeds and efficiency, and prolonged tool wear life.

GROUPS OF NONFERROUS ALLOYS

The base metal is used to identify a group of nonferrous metals. A few important groups of nonferrous metals that are commonly used in jobbing shops are described next. The five groups include copper-, aluminum-, zinc-, magnesium-, and nickel-base alloys.

Copper-Base Alloys

The brasses and bronzes widely used in industry are copper-base alloys. Brasses are predominantly made of copper and zinc. Bronzes are a mixture of copper and tin. Both alloys may be cast, rolled, wrought, or forged. Other alloying elements such as aluminum, manganese, silver, nickel, and antimony are added. High-strength, high-hardness, and age-hardened bronzes are more difficult to machine than free-cutting brasses.

Silicon, manganese, and beryllium are often combined with copper-base alloys. Beryllium in combination with copper forms a hard compound that increases hardness and wear resistance. However, tool wear life is noticeably decreased. Beryllium possesses age-hardening properties.

Aluminum-Base Alloys

Alloying elements and heat-treating processes are added to aluminum to increase its tensile strength. Other properties such as machinability, weldability, and ductility are also developed by adding specific alloying elements. Copper, manganese, chromium, iron, nickel, zinc, and titanium are some of the common alloying elements.

Nickel-Base Alloys

Nickel is noted for its resistance to corrosion from natural elements and many acids. Nickel-base alloys have from 60% to 99% nickel. Nickel is used for parts that are subjected to and must resist corrosive action. In its pure state, nickel is used for plating other metal parts.

One of the common nickel alloys is known as *Monel metal*. This nickel alloy is composed of approximately 65% nickel, 30% copper, and 5% of a few other elements. High-temperature-resistant space age alloys such as Inconel and Waspaloy, depend on nickel as the base metal. Copper, aluminum, iron, manganese, chromium, tungsten, silicon, and titanium are different elements that are alloyed with nickel-base alloys.

Zinc-Base Alloys

Die castings are usually made of zinc-base alloys. The low melting temperature of zinc-base die cast metal alloys of from 725°F to 788°F (385°C to 420°C), the ability to flow easily under pressure into die casting dies, and the quality to which details may be produced make zinc-base alloys exceptionally practical for die casting. Zinc alloys are also rolled into sheets for manufacturing sheet metal products.

Zinc is alloyed with aluminum or small amounts of copper, iron, magnesium, tin, lead, and cadmium to produce the sixteen common zinc-base alloys. Die cast metal alloys fall into six groups: zinc-base; tin-base; lead-base; aluminum-base; magnesium-base; and copper, bronze, or brass alloys. The physical properties of each alloy is affected by the combination and quantity of each element.

Magnesium-Base Alloys

Extreme care must be exercised when machining magnesium parts. Magnesium burns at the low temperature of 600°F (315.6°C). Therefore, cutting temperatures must be kept well below this temperature to prevent chips from igniting.

One great advantage of magnesium is its lightness—almost two-thirds the weight of aluminum. Aluminum, copper, iron, nickel, manganese, tin, zinc, and silicon are metal elements commonly used to produce magnesium-base alloys. The melting temperature range for magnesium-base alloys is from 860°F to 1250°F (460°C to 676°C).

APPLICATIONS OF INDUSTRIAL MATERIALS

Cast Alloy Cutting Tools

Cast alloys are considered as one of the special alloy classifications. The other classification includes *wrought alloys*, which are generally not as hard as the cast alloys and contain up to almost 35% iron. Cast alloys require less than 1% iron.

Because of the extreme hardness of cast alloys, they are used as cutting materials. Cast alloys are known by trade names such as *Stellite*, *Tantung*, and *Rexalloy*. Cast alloys have a cobalt base and are nonferrous (no iron is present, except as an impurity in the raw materials).

One of the wide fields of application of cast alloys is in cutting tools for machining metals. Cast alloys are produced as cutting tool tips to be brazed on tool shanks, as removable tool bits, and as disposable cutting tool inserts. The extreme hardness of cast alloys requires that they be ground to size with aluminum oxide abrasive wheels. The principal elements in cast alloy cutting tools are cobalt (35% to 55%), chromium (25% to 35%), tungsten (10% to 25%), carbon (1.5% to 3%), and nickel (from 0 to 5%). Figure 36-12 shows two examples of tool bits with brazed cast alloy tips.

One big advantage of cast alloy cutting tools is their ability to cut at high cutting speeds and high cutting temperatures. Due to the high red-hardness property, cast alloy cutting tools perform better at temperatures above the temperatures for high-speed cutters. The temperature range is from 1100°F to 1500°F (593°C to 815°C). In terms of cutting speeds, cast alloy cutting tools are capable of machining at speeds above the highest speeds used with high-speed cutters and the lowest practical speeds for carbide cutting tools.

Tool shank support is important. Most cast alloys are brittle and are not capable of withstanding heavy impact forces as are carbon steel and high-speed cutters. There are, however, special grades of cast alloys that have impact-rupture strength comparable to high-speed steel cutters. Cast alloys are also used in noncutting applications as turbine blades, wear surfaces on machines, and for conveyors in heat-treating furnaces.

Cemented Carbide Cutting Tools

Cemented carbide cutting tools have been discussed in relation to their special characteristics. Among these characteristics are the ability to retain hardness at red heats up to approximately 1700°F (930°C), high initial hardness, and the advantage of being able to machine at speeds two to four times faster than may be used for high-speed steel cutters.

Cemented carbide cutting tools are made of two main materials: tungsten carbide and cobalt. Special properties are obtained by the addition of titanium and titanium carbides. The term *cemented* means that during production cobalt is used to cement the carbide grains together.

Cemented carbides are cast to required shapes. Cemented carbides do not require further heat treatment. The cast form is very hard. The cast shapes are used as disposable inserts, chip breakers, cutting tool tips that are brazed on shanks, and solid smaller cutting tools such as end mills, drills, and reamers.

The properties of cemented carbides are affected by cobalt as a principal ingredient. Cobalt affects hardness, wear resistance, and tool life. Increasing the amount of cobalt increases the hardness of the cutting tool. The brittleness is also increased, producing a decreased ability to resist shock and a lower impact toughness. The alloying elements also affect grain structure. A finer grain structure increases the hardness; a coarser structure decreases hardness.

Carbide manufacturers have established a classifying system for cemented carbide machining applications. Of eight general designations, three are used for machining conditions for cast iron and nonferrous metals and five classifications are used for steel and steel alloys. In addition, the carbide industry has six classifications related

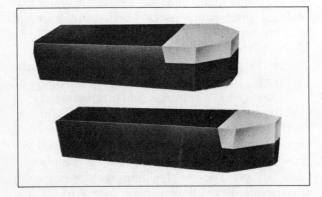

FIGURE 36-12 Tool Bits with Brazed Cast Alloy Tips

TABLE 36-6 Carbide Industry Classification System

Material to Be Machined	Cemented Carbide Classification	Application
Cast iron and nonferrous metals	C-1 C-2 C-3	Medium roughing to finishing cuts Roughing cuts High-impact dies
Steels and steel alloys	C-4 C-5 C-6 C-7 C-8	Light finishing cuts at high speeds Medium cuts at medium speeds Roughing cuts Light finish cuts Heavy roughing cuts and general-purpose machining cuts

to wear and impact applications. Table 36-6 shows the carbide industry classification system by material to be machined and by application. The operator needs to consult each manufacturer's supply catalog to determine the numbering system used for a particular cemented carbide, the properties of the cutting tool, and the performance of a particular grade.

Ceramic Cutting Tools

Ceramic cutting tools are produced from metal oxide powders. These powders are formed into shape by cold pressing and sintering or by hot pressing. Regardless of the method of forming, the pressed blanks are not as strong as the carbide-shaped cutting tools. Ceramic cutting tools are made of aluminum oxide with a metallic binder to improve impact strength. Silicon oxide and magnesium oxide are added.

Figure 36-13 shows a commonly used ceramic disposable insert with a chip breaker secured on a rigid, solid-shank toolholder. The ceramic insert is harder than a cemented carbide cutting tool and has high wear resistance. However, ceramic cutting tools are brittle, have low impact resistance, and shatter easily. The hardness of ceramic cutting tools lies below the hardness of a diamond and above the hardness of a sapphire.

Two great advantages of ceramic cutting tools are: (1) the cutter is not affected by the temperature at the cutting edge and (2) hot metal chips do not tend to fuse to the cutter. Thus, cutting fluids are not required for cutting purposes. However, cutting fluids may be needed to prevent distortion of the workpiece. When cutting fluids are used, a liberal and continuous flow must be provided. Otherwise, any intermittent cooling may cause the cutter to fracture or shatter.

The hardness and wear life properties of ceramic cutting tools make them ideal for machining hard and hardened steels. Cutting speeds up to from two to four times faster than the speeds used for cemented carbides may be used for ceramic cutting tools. The surface finish that may be produced by taking light finishing cuts at high speeds is of a quality that eliminates the need for grinding.

Aluminum oxide/titanium carbide ceramic cutting tools use a newer cutting tool material that combines many features of regular oxide cutting tools with its ability to resist thermal and impact shock. This combination of proper-

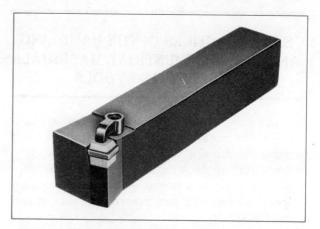

FIGURE 36-13 Ceramic Disposable Insert with Chip Breaker Secured on a Solid-Shank Toolholder

ties permits this oxide family of cutting tools to be used for milling, turning, and other machine tool operations that are not possible with other cutting tools.

Diamond Cutting Tools

Industrial-quality natural diamonds and manufactured diamonds provide the hardest known cutting tool material. Hardness and extreme resistance to heat make the diamond ideal for machining soft, low-strength, and highly abrasive materials. The diamond has low strength and shock resistance. Another limitation is to be found in the geometry of the diamond as a single crystal or as a cluster of single-crystal cutting edges. Since the strength, hardness, and wear resistance is related to the orientation of each diamond crystal, the performance may not be as accurately controlled as desired.

This problem of orientation of the crystal is eliminated by bonding fine diamond crystals to form solid tool shapes. The cutting tools in this diamond family are known as *sintered polycrystalline tools*. They are available in forms for brazing on solid shanks, as tips for bonding to carbide tool inserts, and as solid inserts.

Polycrystalline cutting tool inserts have the advantage over the single-crystal cutting edge in that the diamond crystals are randomly oriented. This feature provides for uniform cutting. Interestingly, the wear-resistance property of diamond cutting tools provides a wear life, based on wear resistance alone, in a ratio of from 10:1 to over 400:1 over carbide cutting tools.

SAFE PRACTICES IN THE HANDLING AND USE OF INDUSTRIAL MATERIALS AND CUTTING TOOLS

—Grind sharp fins and edges on castings and tumble if the molding sand is not removed from outside and inside surfaces.
—Spark test or check the color code or other identification of a material against the required specifications of a part before it is laid out and machined.
—Follow standard bench and machine safety practices when grinding, laying out, or machining castings, bar and sheet stock, and forgings.
—Use a supporting stand and protecting screen to shield long, overhanging lengths of materials when cutting off stock.
—Use a hoist to lift and position heavy and awkwardly designed workpieces on a machine tool.
—Locate, mount, and strap heavy workpieces securely in place before removing the sling or hoist.
—Stand in a safe position away from a heavy workpiece that is being lifted by a mechanical or electrical hoist.
—Select the kind of cutting tool that is appropriate for the material and conditions under which machining is to be done. Consider the part design and whether the cutting action is intermittent with sharp impact or continuous.
—Check the machining impact conditions against the cutting tool manufacturer's recommendations, especially when carbide, ceramic, cast alloy, and other cutting tool materials that may fracture under severe impact are used.
—Avoid quenching a cast alloy cutting tool after dry grinding. Quenching shock may cause the cutter to fracture or shatter.
—Use a liberal, constant flow of coolant when machining with ceramic cutting tools. Intermittent cooling may cause the cutter to fracture or shatter.
—Select a rigid tool shank on which carbide, ceramic, and cast alloy inserts may be correctly seated and securely held.
—Cut under the hard outer scale of cast iron castings on the first cut.
—Have a chloride-base fire extinguisher available for any possible fire when machining magnesium and zirconium parts that may ignite.
—Take machining cuts at the fpm, depths, and feeds recommended by the manufacturers of carbide, ceramic, cast alloy, diamond, and other cutting tool materials.
—Make provisions in advance—that is, before start-up—for the desired chip formation (continuous or discontinuous), chip disposal (particularly when operating at high speeds), and positioning of guards and protective devices.

TERMS USED WITH INDUSTRIAL AND CUTTING TOOL MATERIALS

Pig iron	A ferrous metal produced by reducing an iron ore in a blast furnace charged with a limestone flux and coke. A base metal requiring further refining to produce cast irons, steels, and alloys.
Cupola	A furnace in which a charge of scrap iron, pig iron, and coke are treated. A reducing furnace that produces cast iron.
Alloyed cast iron castings	The addition of an alloying element such as chromium or molybdenum to cast iron. Cast iron containing one or more additional elements to produce required properties.
Open hearth, Bessemer, electric, and basic oxygen processes	Four main types of furnaces and processes for reducing pig iron and scrap iron to steel. Processes for burning off impurities and removing slag from a molten iron mixture. Processes for producing steels with required physical properties and chemical compositions.
Element (steel and alloy making)	A pure metallic or nonmetallic material found in nature or laboratory-produced. A material combined with one or more metallic or nonmetallic materials to produce desired properties in an alloy.
Hardenability, machinability, ductility	Physical properties of industrial manufacturing and cutting tools. Factors to consider in selecting material for a workpiece or a cutting tool. Characteristics of materials for which engineering and design data establish conditions of use.
Alloying	The addition of one or more pure metallic or nonmetallic elements to a base element for purposes of improving properties. Changing the composition and structure of a metal or nonmetal to affect properties such as red hardness, corrosion resistance, toughness, and strength.
Metal classification systems	Identification of steels, nonferrous metals, and alloys according to standards established by SAE, AISI, ASTM, AA, and UNS. Code numbers and letters used in a particular combination to identify the composition and physical properties of metals and alloys.
Aluminum temper designation	Specific letters used to provide information about the strain hardening, cold working, aging, and heat treating of aluminum.
Graphite structure	Flake and spheroidal forms of carbon. A form of carbon that affects grain structure and hardness. (Flake graphite in cast iron improves machinability. Spheroidal graphite adds tensile strength and ductility, as in the case of nodular iron.)
Cementite	An iron carbide. A form of combined carbon that produces hardness, toughness, brittleness, and low ductility in white cast iron.
Alloy steel castings	Steel castings with special metallic alloying elements other than carbon. Casting steels whose properties are altered by the addition of one or more metallic alloying elements.
Martensitic, ferritic, and austenitic stainless steels	Three basic types of stainless steels whose properties are established by the percent of carbon and other alloying elements.

Magnesium, chromium, molybdenum, tungsten, and cobalt	Metal elements added to steel to form alloys that possess specific properties. Metal elements that increase physical properties of steels and alloys such as wear resistance, toughness, red hardness, and hardenability.
Copper-, aluminum-, magnesium-, zinc- and nickel-base alloys	Five important groups of nonferrous metal alloys that are widely used in the machine and metal trades and in industrial manufacturing. (The base metal designates the series of nonferrous alloys produced.)
Cast alloys	Cobalt-base alloys with limited traces of iron as an impurity. Extremely hard disposable cutting tool inserts, brazed tips, and removable tool bits. Alloys used for noncutting applications requiring a material with high red-hardness properties.
Ceramic cutting tools	A cutting material harder than cemented carbides but softer than the diamond. Metallic oxide powders that are pressed to shape and sintered or hot pressed. Cutting tools that may be operated at high cutting speeds from two to four times faster than cemented carbides.
Polycrystalline diamond cutting tools	Bonding of fine diamond crystals to provide a many-sided, randomly oriented crystal structure to control and improve cutting efficiency and wear life.

SUMMARY

The making of pig iron is the first stage in the manufacture of cast iron, steels, and alloys.

> Pig iron and scrap iron are reduced in a cupola to produce gray, white, chilled, malleable, and alloyed iron castings.

Steels are produced by refining pig iron and scrap to remove slag and other impurities and control the carbon content.

> Open hearth, Bessemer converter, electric, and oxygen process furnaces are the principal producers of steels and alloys. Steel ingots are soaked, rolled, and formed into specific sizes and shapes.

Ferrous alloys have iron as the base metal and one or more metallic elements.

> Nonferrous alloys do not contain iron (except as an impurity).

Carbon steels are made of iron, a fixed percent of carbon, and other principal elements, including carbon, manganese, phosphorous, silicon, and sulphur.

> The carbon content in tool steels starts at 0.87%.

Manganese, phosphorous, silicon, and sulphur are added to carbon steel to improve mechanical properties such as machinability, hardenability, resistance to shock, grain structure, ductility, and fusibility.

> Low-, medium-, and high-carbon steels are the three basic groupings of plain carbon steels. The range of carbon content across these three grades is from 0.02% to 0.30% for low-carbon steels to 0.60% to 1.30% for the hardenable high-carbon and tool steels.

The first digit of the SAE and AISI steel classifications identifies the metal element that dominates in the composition.

> The second digit classifies a steel in a particular series. A zero indicates there is no major alloying element.

The third and fourth digits represent the average percent of carbon in the material.

H letters and subnumerals follow the numerical classification to indicate hardenability properties.

The Unified Numbering System (UNS) correlates the major steel and alloy classification systems. Code letters identify metals and alloys. Each letter is followed by a five-digit number from 00001 through 99999. The numbers indicate the quantity of the base metal and other added metal elements in the steel or alloy.

The Aluminum Association (AA) has developed a similar classification to SAE/AISI for wrought aluminum and aluminum alloys. The letters F, O, H, W, and T (separated by a dash) follow the four-digit system. The letters relate to temper and heat-treating processes.

Two simple, common shop methods are used to identify the kind of steel. One method is to use the manufacturer's color marking on the end of drawn or rolled bars. The other method requires spark testing to identify a steel by the color, spacing, and quantity of sparks in a pattern produced by grinding.

Five broad, commercially manufactured, cast iron castings include gray, white, chilled, alloy, and malleable cast irons.

Malleable iron is produced by heat treating white cast iron. The grain structure changes in the process to produce temper carbon aggregates.

Pearlitic forms of malleable iron have great strength and wear resistance and may be engineered to permit deformation without fracture.

Nodular cast irons contain graphite in a spheroidal structure. This structure is produced by the addition of small amounts of magnesium or cerium-bearing alloys. Nodular cast irons possess physical properties between the properties of cast iron and steel.

Steel castings are identified as low-, medium-, or high-carbon and alloy steel.

Alloy steel castings are considered low alloy when the alloying element is less than 8%. High-alloy steel castings have a higher percent.

Alloying elements added to steels and nonferrous metals and heat treatment affect one or more physical properties.

The most important physical properties are: hardenability, distortion reduction, machinability, strength, abrasion resistance, corrosion resistance, ductility, and toughness.

Alloy steels are produced in three general categories: construction, special, and tool steels.

The properties of alloy steels are affected by eight major elements: carbon, manganese, nickel, chromium, molybdenum, vanadium, cobalt, and tungsten.

Stainless steels contain up to 22% nickel alloyed with chromium.

The three types of stainless steels include: martensitic, ferritic, and austenitic.

Widely used nonferrous alloys have as the base metal: copper, aluminum, nickel, zinc, and magnesium.

One or more metallic elements are alloyed with the base metal to produce nonferrous alloys that have properties altered from the base metal.

Zinc-base alloys are widely used in die casting. The low melting temperature, flowability under force, and ability to reproduce fine cast details makes this group ideal for production die casting of intricate parts.

Cast alloys have a cobalt base, are nonferrous, and contain additions of chromium, tungsten, and carbon. These nonferrous metals possess the ability to cut at high speeds and high temperatures of 1100°F to 1500°F (593°C to 815°C).

Special grades of alloys have impact-rupture strength. Cast alloys are used for cutting tools and other applications requiring high resistance to abrasion and heat resistance.

Cemented carbides contain tungsten carbide and cobalt. Cobalt cements the grains together.

Cobalt increases hardness, wear resistance, tool life, and brittleness.

Carbides are classified by manufacturers for machining conditions governing three groups for cast iron and ferrous metals and five groups for steel and steel alloys.

Ceramic cutting tools are harder than cemented carbides. They are produced from metal oxide powders by cold pressing and sintering or hot forming into cutter inserts and other blank forms.

Cutting speeds are from two to four times faster than the speeds used for carbides. Hard and hardened steels may be machined with ceramic cutting tools.

Polycrystalline diamond cutting tools utilize the physical properties of fine diamond crystals. These crystals are bonded in a randomly oriented pattern where the cutting action is not affected by the features of a single crystal.

Safe practices must be observed when lifting heavy castings and cumbersome sheets and when working on long bars of stock.

Added safety precautions must be taken when machining with hard and brittle carbides, cast alloys, ceramic, and diamond cutting tools. The factors of increased speeds and cutting under conditions of high stock removal rates and temperatures require provision for materials removal, adequate guards, and the use of personal protective devices.

UNIT 36 REVIEW AND SELF-TEST

1. Describe briefly how pig iron is produced.
2. State three factors, exclusive of the composition and methods of production, that affect the machinability of metals.
3. Refer to a steel manufacturer's catalog or handbook tables of standard carbon steels.
   a. Give the percent range of the elements for carbon steels (1) 1020 (C1020), (2) 1040 (C1040), and (3) 1095 (C1095).
   b. Rate each of the three carbon steels according to the factors of (1) hardness, (2) hardenability, (3) ductility, and (4) wear resistance. Use A for lowest, B for medium, and C for highest.
4. Identify three different properties of steel that are affected by the addition of (1) chromium, (2) manganese, and (3) tungsten.

5. a. List three different types of tool and die steels.
   b. Indicate the letter (and where applicable the number range) of the selected steels.
   c. Give the (1) hardening temperature range and (2) tempering range for a T4 tungsten high-speed tool steel. Refer to a steel manufacturer's data sheets or a handbook table.

6. a. State two differences between cupola and pearlitic malleable iron.
   b. Give an application of each of the two different malleable irons.
   c. List three physical properties of cast iron castings that are changed by alloys to produce alloy cast irons.

7. a. State two advantages of using alloy tool steels over plain carbon tool steels.
   b. List the quenching baths that are used for hardening different alloy tool steels.

8. Indicate the effects on alloy steels of adding (a) 12% manganese and (b) 20% vanadium.

9. State three safety precautions to take when machining with carbide, ceramic, and cast alloy cutting tools.

SECTION TWO

Basic Metallurgy and Heat Treatment of Metals

This section deals with concepts of basic metallurgy and heat treatment of metals. The structure of steel, its behavior, and the transformation products that result from heating and cooling are described. Phase diagrams and transformation curves are used to explain changes in the microstructure and properties of steels and alloys.

Technical material is provided in relation to the critical points and temperature changes in the transformation of grain structures, the relationship of carbon content to hardenability and depth of hardness, the severity and rate of cooling and time on hardness, and the use of quenching media in interrupted quenching processes. Heat treatment of metals is then covered in terms of the following technology and processes:

—Definitions and applications of heat-treating terms;
—Hardening and tempering of carbon tool steels and high-speed steels;
—Annealing, spheroidizing, and normalizing metals;
—Subzero treatments to stabilize precision tooling and instruments;
—Surface hardening through case-, flame-, and induction-controlled depth hardening;
—Problems in heat treating;
—Hardness testing equipment and measurement systems;
—A sampling of heat-treating processes for selected nonferrous metals.

UNIT 37

Heat Treating and Hardness Testing

Metallurgy is concerned with the technology of producing metals or alloys from raw materials and alloying elements and preparing metals for use. *Heat treatment* is the combination of heating and cooling operations applied to a metal or alloy in the solid state. The purpose of heat treatment is to produce desired conditions or physical properties in the metal or alloy.

HEATING EFFECTS ON CARBON STEELS

Fully annealed carbon steel consists of the element iron and the chemical compound iron carbide. In metallurgical terms, the element iron is known as *ferrite;* the chemical compound *iron carbide* is *cementite.* Cementite is made up of 6.67% carbon and 93.33% iron. There are also traces in carbon steel of impurities such as phosphorous and sulphur.

Depending on the amount of carbon in the steel, the ferrite and cementite are present in certain proportions as a *mechanical mixture.* The mechanical mixture consists of alternate layers or bands of ferrite and cementite. When viewed under a microscope, the mechanical mixture looks like mother-of-pearl. The mixture is known as *pearlite.* With only traces of impurities, pearlite contains about 0.85% carbon and 99.15% iron. Thus, a fully annealed 0.85% carbon steel consists entirely of pearlite. This steel is known as a *eutectoid steel.* A steel having less than 0.85% carbon is a *hypo-eutectoid steel.* This steel has an excess of cementite over the amount that is required to mix with the ferrite to form pearlite. A steel with 0.85% or more carbon content is known as a *hyper-eutectoid steel.* In the fully annealed state, a hyper-eutectoid steel has an excess of cementite so that both cementite and pearlite are present.

Variations in hardness and other mechanical properties occur when the atomic cell structure of steel is changed through heating and cooling within certain *critical points and rates.* There are two types of atomic structures in steel: *alpha* and *gamma.* Steels that are at a temperature below the critical point have an atomic cell structure of the alpha type. Steels heated above the critical point have a gamma atom structure. Alpha iron *mixes* with carbide. Gamma iron *absorbs* the carbon-iron compound into solid solution. The critical points at which changes take place vary according to the composition of the steel. Hardness and strength variations are produced by changing the atoms and cell structure of the iron atoms from alpha to gamma and then back again to alpha.

The lower critical point for steels is considered to be 1333°F (722.8°C). The upper critical point increases to 1670°F (910°C) for the very low-carbon steels. The lower critical point is used for steels with carbon contents within the 0.8% to 1.7% range. Heat treating takes place when the temperature of a steel is raised above the critical point and the temperature is then reversed and controlled for the required degree rate of cooling.

Effects of Temperature and Time

Stresses occur in metals at the critical points when a part, or section of a part that is not of equal mass or shape, is heated unevenly or too fast. Usually a part is *preheated* to around 800°F (427°C). Complicated and intricately shaped parts are preheated again to 1250°F (677°C). Preheating permits a better transition from alpha to gamma iron at the critical point.

Steels are heated from 50°F to 200°F (10°C to 93°C) above the critical point to cause all the constituents to go into solid solution. The added temperature also allows the solution to remain in this state a few additional seconds during the cooling cycle between the time the workpiece is removed from the furnace and quenched. This added temperature and short interval of time permit a complete phase change in the metal.

An *iron-carbon phase diagram* contains general information about the full range of hardening temperatures for carbon steels and is used to show the changes in the microstructure of steel that occur during heat treating. Figure 37-1 is an iron-carbon phase diagram with critical temperatures, heat-treating processes, and grain structures. The diagram shows that when carbon steel in a fully annealed state is heated above the lower critical point—between 1335°F to 1355°F (724°C to 735°C)—the alternate bands of ferrite and pearlite that make up the pearlite begin to merge into one another. The process continues above the critical point until the pearlite completely dissolves to form the iron in an *austenite state.* If an excess of ferrite or cementite is present in the steel, they begin to dissolve into the austenite if the temperature of the steel continues to rise. Ultimately, only austenite is present. The diagram shows that above certain temperatures the excess ferrite or cementite is completely dissolved in the austenite. The temperature at which this transformation occurs is known as the *upper critical point.* This temperature is affected by the carbon content of the steel.

Effects of Slow Cooling on Carbon Steels

The transformation from alpha to gamma iron to the point where the iron is completely austenitic is reversed upon cooling. However, the upper and lower critical points occur at slightly lower temperatures to change from gamma to alpha iron. At room temperature, the structure of the original fully annealed carbon steel part will return to the same proportions of ferrite or cementite and pearlite that were present before heating and cooling. No austenite will be present.

Effects of Rapid Cooling on Carbon Steels

As the rate of cooling from an austenitic state is increased, the temperature at which the austenite changes to pearlite decreases below the slow transformation temperature of 1300°F (704°C). As the cooling rate is increased, the laminations of the pearlite (formed by the austenite as it is being transformed), become finer.

When a carbon steel is quenched at the *critical cooling rate* or faster, a new *martensite structure* is formed. The critical cooling rate is the

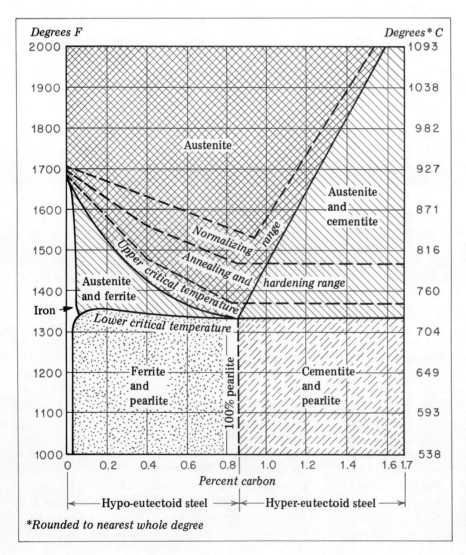

FIGURE 37-1 Iron-Carbon Phase Diagram with Critical Temperatures, Heat-Treating Processes, and Grain Structures

cooling rate at which there is a sudden drop in the transformation temperature. The martensite structure formed has angular, needle-like crystals and is very hard.

APPLICATION OF IRON-CARBON PHASE DIAGRAM INFORMATION

Characteristics of Martensite

Maximum hardness is produced in steel when the grain structure after quenching is transformed from austenite to martensite in a matter of seconds. It is important to quench carbon steels from hardening temperature to below the 1000°F (538°C) temperature in one second or less to prevent austenite from beginning to transform to pearlite: The hardening process avoids the formation of pearlite grains. If the cooling is slower, any pearlite that is formed mixes with the martensite and reduces hardenability throughout the steel. Martensite is the hardest and most brittle form of steel. Figure 37-2 shows the grain structure of martensite (820X magnification). Martensite in pure form is a super-saturated solid solution of carbon in iron and contains no cementite.

Depth of Hardness and Heat-Treating Problems

Hardenability was considered in the previous unit in relation to the uniformity of hardness from the outer surface to the center of a part. Hardenability may be improved by adding alloying elements.

When the rate of cooling during quenching is too fast, a number of problems emerge, such as warpage, internal stresses, and different types of fractures. Common heat-treating problems and their probable causes are given in Table 37-1. Usually no problems are encountered with the hardening of workpieces of uniform cross section up to 1/2" (12.7mm) thickness. These workpieces may be hardened throughout with complete transformation from austenite to martensite. As the thickness increases, additional attention must be paid to the selection of the steel, quenching medium, and cooling time to ensure that there will be uniform hardness or stabilizing of the grain transformation to martensite.

Working Temperatures for Steel Heat-Treating Processes

Information normally contained on an iron-carbon phase diagram is interpreted into a practical guide in Figure 37-3, which shows the working temperatures for carbon steel heat-treating processes. Generalized information is provided according to percent of carbon content, heat-treating processes, and metallurgical properties within different heat ranges. The carbon steel range is from 0.02% to 1.7% carbon. Heat-treating processes range from a subzero cooling temperature of –300°F (–184°C) to forging temperatures ranging from approximately 2500°F (1371°C) for low-carbon steels to 2250°F (1237°C) for 1.7% high-carbon steels to lower temperatures.

Note that the black heat range is from 0°F to 1000°F (538°C); red heat, from 1000°F to 2050°F (1121°C); and white heat, from 2050°F to 2900°F (1593°C). Heat-treating processes for carbon steels are carried on from the subzero range for stabilizing grain structure through normalizing processes starting around 1700°F (927°C) for steels with a minimum of 0.02% carbon. The higher carburizing temperatures are shown graphically above the normalizing range.

FIGURE 37-2 Grain Structure of Martensite (820X Magnification)

TABLE 37-1 Common Heat-Treating Problems and Probable Causes

Heat-Treating Problem	Probable Cause
Circular cracks	—Uneven heating in hardening
Vertical cracks; dark-colored fissures	—Steel burned beyond use
Hard and soft spots	—Uneven or prolonged heating; uneven cooling
Hard and soft spots with tendency to crack	—Tool not moved about continuously in quenching fluid
Soft places	—Tool dropped to bottom of fluid tank
Surface scale; rough decarburized surface	—Heated steel surfaces exposed to effects of an oxidizing agent
Excessive strains (workpiece with cavities and holes)	—Formation of steam or gas bubbles as an insulating film in cavities and pockets
Coarse grain; quench cracking; tool fracture in use	—Overheating of tool
Area around tongs softer than remainder of part	—Tongs not preheated
Quench cracks (straight-line fractures from surface to center)	—Overheating during austenite stage —Incorrect quenching medium and/or nonuniform cooling —Incorrect selection of steel —Time delays in a high-stress state between hardening and tempering
Surface cracks formed by high internal stresses	—Unrelieved stresses produced by high surface temperature generated by prior machining operations such as grinding
Stress cracking	—Failure to temper a part before grinding —Reducing surface hardness through subsequent machining —Forming a hardened crust by machining at a high temperature and immediately quenching area with cutting fluid (coolant)
Undersized machining to clean up decarburized surfaces	—Failure to allow sufficient stock to permit grinding all required surfaces to clean up

Aging and Growth of Steel

A small amount of the austenite is sometimes retained when carbon steels are subjected to severe, extremely rapid cooling. All austenite is not transformed into martensite through quenching. The tendency over a period of time is for the austenite to be transformed into martensite without further heating or cooling. This process, called *aging*, results in the *growth* of the steel with the increase in volume. Additional stresses are also introduced. Therefore, control rates are established for preheating, heating to the lower and upper critical points, cooling, and quenching to provide optimum hardening conditions.

Critical Points in Hardening

The two important stages in hardening steel are heating and quenching. The steel is heated above its transformation point to produce an entirely austenitic structure. The steel is then quenched at a rate faster than the critical rate

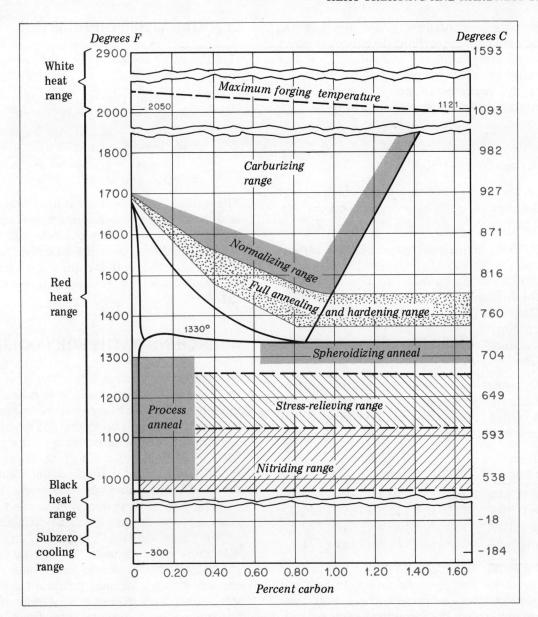

FIGURE 37-3 Working Temperatures (Interpreted from Iron-Carbon Diagram Data) for Carbon Steel Heat-Treating Processes

required to produce a martensitic structure. It should be noted that the critical rate depends on the carbon content, the amount of other alloying elements, and the grain size of the austenite. The hardness of steel depends on the carbon content of the martensitic steel.

Pearlite is transformed into austenite as steel is being heated at a transformation point called the *decalescence point*. At this point, while the steel continues to be heated and the surrounding temperature becomes hotter, the steel continues to absorb heat without any significant change in temperature.

The process is then reversed in cooling when the austenite is transformed into pearlite at the *recalescence point*. The steel gives out heat at this point so that the temperature rises momentarily instead of continuing to drop.

The control points of decalescence and recalescence have a direct relationship to the hardening of steel. These critical points vary for different kinds of steel, requiring different hardening temperatures. For hardening, the temperature must be sufficient to reach the decalescence point to change the pearlite to austenite. Also, the steel must be cooled rapidly enough before it reaches the recalescence point to prevent the transformation from austenite to pearlite.

Hardening Temperatures for Carbon Tool Steels

The hardening temperature to which a steel is heated before quenching depends on the following factors:

— Amount of excess ferrite (0.85% carbon content or more) or cementite (less than 0.85% carbon content) to be dissolved in the austenite,
— Maximum grain size and required physical properties,
— Chemical composition of the steel,
— Geometry of the tool.

Tool steel manufacturers' charts contain specifications. As a rule-of-thumb, based on the carbon content in the steel, the following general guidelines are provided:

Carbon Content	Hardening Temperature Range
0.65 to 0.80%	1450 to 1550°F (788 to 843°C)
0.80 to 0.95%	1410 to 1460°F (766 to 793°C)
0.95 to 1.10%	1390 to 1430°F (754 to 777°C)
over 1.10%	1380 to 1420°F (749 to 771°C)

In general, the highest temperature in the range produces deeper hardness penetration and increases strength. Conversely, the lowest temperature in the range decreases the hardness depth but increases the ability of the steel to resist splitting forces.

It is important to know whether the temperature measurement system in use measures the temperature of the furnace and/or the workpiece. Adequate heating time must therefore be allowed to permit the workpiece to heat up to the decalescence point. A workpiece pyrometer shows a rise until the decalescence (critical) point is reached and a slight fluctuation until the transformation in the structure of the steel is completed. At the completion of the structural changes in the steel, the workpiece temperature begins to rise again.

LIQUID BATHS FOR HEATING

Heating baths are generally used for the following purposes:

— To control the temperature to which a workpiece may be heated,
— To provide uniform heating throughout all sections of a workpiece,
— To protect the finished surfaces against oxidation and scale.

The molten liquid baths that are widely used for steel hardening and tempering operations include sodium chloride, barium chloride, and other metallic salt baths, as well as lead baths. Lead baths are extensively used for quantity heat treating of small tools and for heating below 1500°F (815°C).

QUENCHING BATHS FOR COOLING

Quenching baths serve to remove heat from the steel part being hardened. The rate of cooling must be faster than the critical cooling rate. Hardness partly depends on the rate at which the heat is extracted. The composition of the quenching bath determines the cooling rate. Therefore, different kinds of baths are used, depending on the steel and the required heat treatment.

The two most common quenching baths are fresh, soft water baths; oils of different classes, or oil-water solutions. Brine and caustic soda solutions are also used. High-speed steels are generally cooled in a lead or salt bath. Air cooling serves as a quenching medium for high-speed steel tools that require a slow rate of cooling. Oil quenching is usually used for applications requiring rapid cooling at the highest temperature and slower cooling at temperatures below 750°F (399°C).

Oil Quenching Baths

Oil quenching permits hardening to depth while minimizing distortion and the possibility of cracking of standard steels. Alloy steels are normally oil quenched. Prepared mineral oils have excellent quenching qualities and are chemically stable and cost effective. Vegetable, animal, and fish oils are also used alone or in combination.

One advantage of quenching oils is that they provide fast cooling in the initial stages, followed by slower cooling during the final stages and lower temperatures. This action prevents the steel from cracking. Quenching oils are maintained within a given temperature range of from 90°F to 140°F (32°C to 60°C).

Water Quenching Baths

Carbon steels are hardened by quenching in a bath of fresh, soft water. The bath temperature must be maintained within a 70°F to 100°F (21°C to 38°C) range so that subsequent workpieces are cooled at the same rate. The temperature of the water must be kept constant within the range because the water temperature seriously affects the cooling rate and hardness penetration.

Design features in workpieces, such as holes, cavities, pockets, and other internal corners, result in uneven cooling when quenched in fresh water. Gas bubbles and an insulating vapor film in cavities produce uneven cooling, excessive internal stresses, and greater danger of cracking. The addition of rock salt (8% to 9%) or caustic soda (3% to 5%) to a fresh water quenching bath prevents gas pockets and vapor films from forming. Care must be taken to use a clean, uncontaminated, soft water bath. The quantity must be sufficient to dissipate the heat rapidly and permit the workpiece to be agitated (moved) within the bath. As a general rule, thicker sections of a heated part are immersed first. Movement of the bath or workpiece within the bath permits cooling at a uniform rate and reduces the possibility of a vapor film from forming on certain surfaces.

Molten Salt Quenching Baths

High-speed steels are generally quenched in a molten salt bath in preference to oil quenching. A molten salt bath produces maximum hardness and minimum cooling stresses. Such stresses result in distortion and possible cracking.

Molten salt quenching baths for high-speed steel are maintained at temperatures of 1100°F to 1200°F (593°C to 649°C). After quenching, the hardened part is tempered or drawn in another molten salt bath within a temperature range of 950°F to 1100°F (510°C to 593°C). A general-purpose tempering temperature for high-speed cutting tools is 1050°F (566°C).

QUENCHING BATH CONDITIONS AFFECTING HARDENING

Mention has been made several times about the need to maintain the quenching bath within a specific temperature range. Another condition to consider is the need for movement between the bath and the workpiece so that the temperature throughout the bath remains constant. Under these conditions, cooling proceeds uniformly on all exposed surfaces and completely through the part.

A more desirable practice than agitating the workpiece in the quenching medium is to reverse the process. If the tank permits, the bath is thoroughly agitated while the workpiece is held still. This kind of fluid motion lessens the danger of warping the workpiece during heat treating.

INTERRUPTED QUENCHING PROCESSES

Interrupted quenching processes are used to obtain greater toughness and ductility for a given hardness and to overcome internal stresses that result in quench cracks and distortion. These problems are encountered in general hardening practices.

Austempering, martempering, and *isothermal quenching* are three interrupted methods. The quenching begins at a temperature above the transformation point and proceeds at a rate that is faster than the critical rate. The important fact is that the cooling is interrupted at a temperature above the one at which martensite starts to form.

The steel is maintained at a constant temperature for a fixed time to permit all sections within a part and the external surfaces to reach the same temperature. Transformation of the structure of the steel takes place uniformly for temperature and time throughout the workpiece. Interrupted quenching requires a larger quantity of heat to be absorbed and dissipated without increasing the temperature of the bath.

Austempering

Austempering is a patented heat-treating process in which steels (chiefly with 0.60% or higher carbon content) are quenched in a bath at a constant temperature between 350°F to 800°F (176°C to 427°C) at a higher rate than the critical quenching rate. The quenching action is interrupted when the steel temperature reaches the bath temperature.

The steel is held for a specified time at this temperature. The austenitic structure changes to a *bainite structure*, which resembles a tempered martensite structure normally produced by quenching a steel and drawing its temper at 400°F (204°C) or more. The austempered part will have much greater ductility and toughness. In austempering, the steel is quenched rapidly so that there is no formation of pearlite. The steel is also held at the same transformation temperature to ensure that all austenite is transformed to bainite.

Martempering

Martempering is a heat-treating process that is especially adapted to higher alloyed steels. Martempering produces a high-hardness structure without the problems of internal stresses, which are accompanied in some cases by quench cracks and dimensional changes in the part.

The first rapid quench takes place in martempering at a specific temperature above the transformation point to a temperature above the one where martensite forms. The temperature is held at this point in order to equalize it throughout the part. The workpiece is then removed from the bath and is cooled in air. Martensite begins to form at a uniform rate in a matrix of austenite. The soft austenite tends to absorb some of the internal stresses produced by the formation of martensite.

Isothermal Quenching

Like austempering, isothermal quenching is a process in which steel is rapidly quenched from above the transformation point down to a temperature that is above the one at which martensite forms. The temperature is held constant, usually at 450°F (232°C) or above, until all the austenite is transformed to bainite.

At this stage, the steel is immersed in another bath and the temperature is raised to a higher specified temperature. After being held at this higher temperature for a definite period, the workpiece is cooled in air. In isothermal quenching, tempering takes place immediately after the steel structure is changed to bainite and before the workpiece is air cooled to normal temperature.

HEAT-TREATING EQUIPMENT

Heat-treating processes require gas, oil-fired, or electrical furnaces that are especially regulated for temperature and, in certain cases, atmosphere. Safety devices are built into the equipment. Exhaust ducts, hoods, close-down valves, and other protective units are requirements of each installation.

Caution: Safety precautions are to be observed for furnace lighting, safety shut off valves and switches, exhausting fumes, performance of all heat treatment processes, and the use of personal protective devices.

Heat-Treating Furnace Controls

Furnace temperatures are usually controlled by a *thermocouple* and a *pyrometer*. The thermocouple transmits information from the hot area of the furnace to the pyrometer. The pyrometer converts minute electrical energy generated by the thermocouple into a temperature measurement.

A simple installation of an activated thermocouple and an indicator dial pyrometer for measuring, indicating, and controlling temperatures is illustrated in Figure 37-4. The effect of heat in the furnace on the dissimilar metal parts in the thermocouple is to produce small amounts of electrical energy. This varying amount of energy, depending on changes in temperature, is translated on a calibrated temperature scale of the pyrometer.

The pyrometer controls the furnace temperature by setting the instrument at a required temperature. Solenoid and other types of valves and controls are used to regulate the heat energy supply when the fixed temperature is reached. The furnace temperature is maintained at a constant specified temperature by calling for more heat as the furnace requires.

Pyrometers may have direct temperature sensing and reading controls or a combination of sensing and reading controls and a recording instrument. Control and temperature measurement

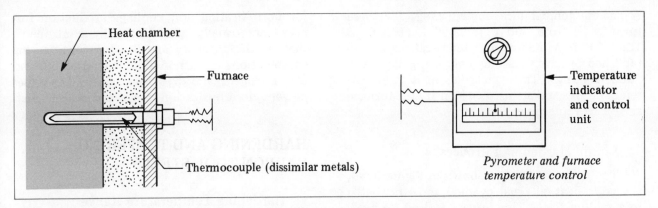

FIGURE 37-4 Thermocouple and Pyrometer for Measuring, Indicating, and Controlling Temperatures

information provides an important record of heating and cooling processes, particularly when continuous and prolonged temperature controls are required.

Basic Heat-Treating Furnaces

General toolroom hardening, tempering, other heat treatment, forging, and heating processes are performed with small gas-fired or electric heat treatment furnaces. Figure 37-5 shows these two basic types of furnaces with temperature-regulating/indicating controls. The *gas-fired furnace* (Figure 37-5A) is available either for manual regulation or it may be secured with temperature-indicating controls. The *electric furnace* (Figure 37-5B) has its temperatures regulated by electronic controls and readouts on the digital unit.

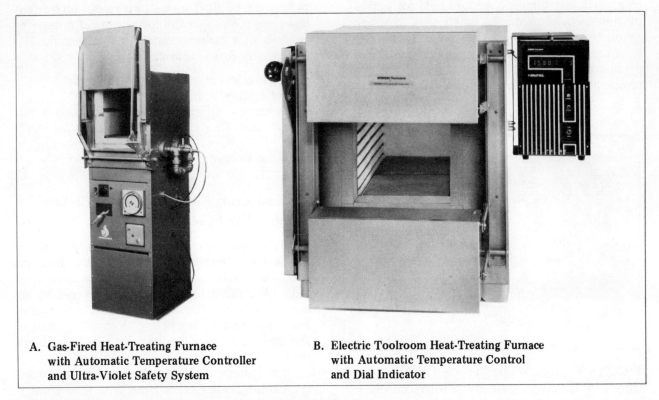

A. Gas-Fired Heat-Treating Furnace with Automatic Temperature Controller and Ultra-Violet Safety System

B. Electric Toolroom Heat-Treating Furnace with Automatic Temperature Control and Dial Indicator

FIGURE 37-5 Basic Types of Heat-Treating Furnaces

A general temperature control range for these furnaces is from 300°F to 2300°F (148°C to 1260°C). In some furnaces, temperatures are established by using marking crayons, pellets, and liquid coatings. These materials melt at known temperatures and are used to identify the furnace and/or workpiece temperature.

Liquid Hardening Furnaces

The *pot-type furnace*, shown in Figure 37-6, is used to heat salt, lead, cyanide, and other baths to a molten state. This furnace is used for heat-treating processes that require a part to be immersed and heated to a specified temperature. In the case of surface hardening, the outer core that is to be carburized is held in the molten bath for a specified period of time. The time is established by the required depth of the casehardening.

Production Heat-Treating Furnaces

Two broad groupings of production heat-treating furnaces provide for batch and continuous processing. *Batch furnaces* provide for the heat treatment of quantity parts and process them as a batch. Batch processing is programmed for mass production. Workpieces are brought up to a particular temperature and held for periods of time depending on the batch heat treatment requirements.

Conveyor furnaces are used for continuous cycling operations. There are different zones of temperature in the furnace to regulate the rate for both heating and cooling. Production furnaces are usually atmospherically controlled—that is, the furnaces are flooded with varying combinations of carbon dioxide, carbon monoxide, nitrogen, and hydrogen gases. The gases prevent decarburization, scale, and surface rust.

HARDENING AND TEMPERING CARBON TOOL STEELS

Hardening Temperature Range

The hardening temperature is the maximum temperature to which a steel is heated before being quenched for purposes of hardening. The hardening temperature is above the lower critical point of a given steel. The iron-carbon diagram, shown earlier in Figure 37-1, provided generalized information about the full range of hardening temperatures for carbon steels. The dotted line for steels having a carbon content up to 0.85% showed the hardening range to start at temperatures above the upper critical points. The starting hardening temperature for steel with a carbon content greater than 0.85% was below the upper critical temperature (point). The temperature to which a steel is raised above the upper critical point (for steels with a carbon content less than 0.85%) or the lower critical point (for steels with a carbon content higher than 0.85%) depends on three factors:

—Chemical composition of the metal,
—Required maximum grain size,
—Amount of excess ferrite (for steels below 0.85% C) or excess cementite (for steels above 0.85% C) that is to be dissolved in the austenite.

Tempering Temperatures for Carbon and High-Speed Steels

Tempering, or *drawing*, reduces the brittleness in hardened steel and removes internal strains produced by the sudden cooling in a quenching bath. A cutting tool or part is tempered by reheating it when the steel is in a fully hardened condition and then cooling. Reheating changes the grain structure to one with a reduced hardness.

The general range of temperatures for tempering carbon tool steel is from 300°F to 1050°F (149°C to 565°C). Tempering tables are available

FIGURE 37-6 Pot-Type Furnace (Without Hood) for Heating Liquid Baths with Console Containing Blower and Safety Systems Attached

TABLE 37-2 Selected Tempering Temperatures, Colors, and Typical Applications of Carbon Steels

Tempering Degrees °F	Tempering Degrees °C	Temper Color	Typical Tool Applications
380	193	Very light yellow	Single-point cutting tools and machine centers requiring maximum hardness
430	221	Light straw	Multiple-point milling cutters, drills, reamers, hollow mills; forming tools
450	232	Pale straw-yellow	Twist drills and screw machine centering tools
470	243	Dark straw	Thread rolling dies, punches, stamping and forming dies, hacksaw blades
490	254	Yellow-brown	Taps over 1/2" (12.7mm); threading dies for tool steels; shearing blades
510	266	Spotted red-brown	Machine taps under 1/4" (6.4mm); general threading dies
530	277	Light purple	Hand and pneumatic punches; scribing tools
550	288	Dark purple	Cold chisels and blunt wedge-shape cutting tools
570	299	Dark blue	Cutting/forming tools requiring minimum hardness
590	310	Pale blue	Torque tools: wrenches, screwdrivers, hammer faces
640	338	Light blue	Generally, noncutting tools requiring a minimum hardness

for recommended hardness requirements for steel parts, hardening temperatures and time, quenching baths, and tempering temperatures. Table 37-2 gives the tempering temperatures for carbon steels that have been hardened at temperatures from 1350°F to 1550°F (732°C to 843°C).

Alloy tool steels are hardened by heating to temperatures between 1500°F to 1900°F (815°C to 1065°C). High-speed tool hardening temperatures are between 2150°F to 2450°F (1193°C to 1343°C). High-speed parts and tools are tempered at 1000°F to 1100°F (538°C to 593°C).

Tempering Colors

A film of oxide forms on the surface of steel that is heated in an oxydizing atmosphere. As the temperature increases above about 400°F (204°C), the surface of the steel starts to change to a light straw color at 425°F (218°C). As different temperatures are reached, the temper colors change, as listed in Table 37-2. At 640°F (338°C), the temper color of steel is a light blue. Above the tempering color range, the heat colors of steel change from a black red to a white heat around 2400°F (1315°C). While some machinists and toolmakers temper tools by color, the more dependable method of tempering is by using a temperature-controlled bath or furnace.

Tempering Baths

Oil Baths. Many tools are tempered in an oil bath that is heated uniformly to a required temperature. Heavy tempering oils may be heated to temperatures between 650°F to 700°F (343°C to 371°C). A partially heated steel part is immersed in the oil. Then, both the steel part and oil are heated to the tempering temperature. The part is next dipped in a tank of caustic soda, followed by quenching in a hot water bath. Oil bath tempering is generally limited to temperatures of 500°F to 600°F (260°C to 316°C).

Salt Baths. Salt baths are recommended for high-speed steel tempering that occurs between 1000°F to 1050°F (538°C to 593°C). Specifications often recommend the use of salt baths above 350°F (173°C) for efficiency and economy. The furnace temperature and part are increased gradually to the tempering range.

Lead Baths. Parts may be tempered by using a lead or lead alloy bath to heat the steel to the required tempering temperature. The workpiece is preheated and immersed in the bath, which has already been brought to the tempering temperature. The part is left in the bath until it reaches the tempering temperature. It is then removed and cooled.

Double Tempering

The tempering operation is often repeated on high-speed steel tools. *Double tempering* ensures that an untempered martensite, which remains in the steel after the first tempering operation, becomes tempered after double tempering. The martensite structure is also relieved of internal strains.

In double tempering, the high-speed steel part is brought to its tempering temperature and held at this temperature for a period of time. The part is then cooled to room temperature. Next, the part is reheated to and held at the original tempering temperature for another period and again cooled to room temperature.

HOW TO HARDEN AND TEMPER CARBON TOOL STEELS

Preliminary Steps to Heat Treatment

STEP 1 Check the blueprint for the classification of the steel part and any other hardening and tempering specifications.

STEP 2 Refer to a table of hardening and tempering data to establish the hardening temperature range, the tempering temperature range, the quenching medium, and any other special heat-treating conditions.

Note: See table in the Appendix for data on heat-treating temperatures for selected grades and kinds of steel.

STEP 3 Check the temperature of the quenching medium

Note: Water-hardening steels are cooled best when the water temperature is around 60°F (16°C). The cooling temperature for oil baths is held between 100°F to 140°F (38°C to 60°C).

STEP 4 Determine the desired hardness and toughness from the specifications or technical table.

STEP 5 Determine the tempering temperature.

STEP 6 Establish the soaking periods at the hardening and the tempering temperatures.

Note: Parts that are 1/4" (6mm) thick normally do not require soaking. Parts from 1/4" to 1" (25.4mm) thick require a soaking period of from three minutes to one hour. Thicker parts require longer periods of time for the internal grain structure transformation to occur.

Hardening Carbon Tool Steels

STEP 1 Set the furnace controls to the required hardening temperature.

STEP 2 Light the furnace if it is gas- or oil-fired. Otherwise, turn the electrical furnace on.

STEP 3 Use a pair of tongs and place the part as near the center of the furnace as practical.

STEP 4 Heat the part uniformly and slowly bring it up to the hardening temperature range.

Note: The fuel and air mixture are adjusted so that the furnace atmosphere is neutral to partly oxidizing.

STEP 5 Allow the part to soak at the hardening temperature as required.

Note: If the part is heated above the hardening temperature range, it is necessary to cool the part in air to a temperature below 1000°F (538°C). The part is then brought back up to the correct hardening temperature. Unless this precaution is taken, a coarse-grain hardened steel is produced with decreased toughness.

STEP 6 Remove the part. Quickly quench it in the appropriate quenching bath.

Note: The part must be moved continuously in an up-and-down motion so that the bath is agitated. Prior to removing the part from the furnace, the tong jaws are heated to prevent soft spots in the area where the part is handled. A slicing motion is used with thin, flat parts to prevent uneven cooling and warping.

HEAT TREATING AND HARDNESS TESTING 789

STEP 7 Check the hardness with hardness testing equipment.

Note: A simple file test may also be used. A corner of a file is slowly moved over the workpiece to see whether it bites. The amount of bite is used to judge the degree of hardness.

Note: The hardened part is very hard and brittle at this stage and is under tremendous internal strains. Therefore, the part is tempered as soon as possible to avoid fracturing.

Tempering Carbon Tool Steels

Using Furnace Equipped with Temperature-Indicating Controls

STEP 1 Bring the furnace to the preset tempering temperature.

STEP 2 Place the workpiece in the center of the furnace.

STEP 3 Heat the workpiece uniformly and slowly bring it to tempering temperature.

Note: Small cutting tools and parts are usually tempered by using an auxiliary heated plate. The tool is held against the plate so that the heat travels from an end or side opposite the cutting edge. Once the correct color is reached, the small cutting tool is quenched to prevent the cutting edge from being tempered to a softer degree than required.

STEP 4 Keep the workpiece at this tempering temperature for the required soaking time.

STEP 5 Remove the part from the furnace. Allow it to cool in air or quench it in an appropriate bath.

STEP 6 Determine the hardness either by a simple file test or more exactly by using a hardness tester.

Using Manually Controlled Furnace

STEP 1 Clean the surface coating or discoloration produced while hardening the part.

Note: Generally, an abrasive cloth is used to remove the scale and produce a bright steel surface.

STEP 2 Use a slow heating flame. Place the part in the furnace so that it is positioned for uniform heating. At the same time, the colors as they start to appear must be seen clearly.

Caution: Since the tong jaws are heated before handling the workpiece, wear safety gloves.

Note: It may be necessary to change the position of the workpiece to be sure it heats uniformly.

STEP 3 Pay special attention to the formation of the temper color just preceding the required temper color.

Note: The temper colors are found in steel suppliers' catalogs and trade handbooks.

STEP 4 Remove the part from the furnace when it reaches the required color.

Note: The part may be cooled either in still air or by quenching.

STEP 5 Test for hardness.

HEAT TREATMENT OF TUNGSTEN HIGH-SPEED STEELS

The hardening temperature for tungsten high-speed steels is from 2200°F to 2500°F (1204°C to 1371°C). Hardening at such a high temperature usually requires one preheating stage. Preheating is particularly important for cutters with thick bodies as compared to the amount of material around cutting teeth. Preheating helps to avoid internal strains. Preheating is done at temperatures below the critical point of the steel, within the 1500°F to 1600°F (816°C to 871°C) range.

At such a temperature, the tool may safely be left in the furnace to heat through uniformly and to bring it up to hardening temperature. Cutters and other cutting tools that are thicker than 1" (25.4mm) often require a second preheating.

The length of time the tool is left at the hardening temperature affects cutting tool efficiency. Therefore, it is important to accurately control the furnace temperature. Such controls make it possible to secure the best possible grain structure and tool surface.

Quenching Tungsten High-Speed Steels

High-speed steel tools are generally quenched in oil. A tool is moved in the bath to prevent a poor, heat-conducting gas film from forming on the tool. An oil quench permits uniform cooling at the required rate.

Salt baths are also used for quenching. Salt baths are particularly adapted for tools and parts that have complex sections or areas where hardening cracks may develop.

High-speed tools are sometimes quenched in a lead bath or by air cooling. Small sections may be cooled in still air; large, heavier sections require a stream of dry compressed air.

Tempering High-Speed Steel Tools

The drawing temperature for high-speed tools is from 900°F to 1200°F (482°C to 649°C). The temperature is higher for single-point cutting tools used on lathe work than for multiple-tooth milling cutters and form tools. Cobalt high-speed steel tools are tempered between 1200°F and 1300°F (649°C to 704°C).

Once the tool or part is heated to the required drawing temperature, it is held at this temperature until heated uniformly throughout its mass. The tool is allowed to cool in dry, still air away from drafts. The tool is not quenched for tempering. Quenching produces internal strains that may later cause the part to fracture.

ANNEALING, NORMALIZING, AND SPHEROIDIZING METALS

Annealing

Annealing is a heat-treating process that requires heating and cooling to:

—Induce softening,
—Remove internal strains and gases,
—Reduce hardening resulting from cold working,
—Produce changes in mechanical properties such as ductility and toughness and magnetic characteristics,
—Form definite grain structures.

The process is also applied to softer metals to permit additional cold working.

The steel is heated to a temperature near the critical range. It is held at this elevated temperature for a period of time and then cooled at a slow rate. Carbon steels are fully annealed by heating *above the upper critical point* for steels with less than 0.85% carbon content (hypo-eutectoid steels). Steels with more than 0.85% carbon content (hyper-eutectoid steels) are heated slightly *above the lower critical point*.

The heated steel is held at this temperature until the part is uniformly heated throughout. It is then slowly cooled to 1000°F (538°C) or below. The result is the formation of pearlite and a layer-like grain structure.

Normalizing

The purpose of *normalizing* is to put the grain structure of a steel part into a uniform, unstressed condition of proper grain size and refinement to be able to receive further heat treatment. The iron-base alloy is heated above the transformation range and then cooled in still air.

Depending on the composition of the steel, normalizing may or may not produce a soft, machinable part. By normalizing low-carbon steel parts, the steel is usually placed in best condition for machining. Also, distortion due to carburizing or hardening is decreased. The practice for producing medium- and high-carbon steels with good machinability is to normalize and anneal.

Spheroidizing

The *spheroidizing of steels* is defined as a heating and cooling process that produces a rounded or globular form of carbide. Steels are spheroidized to increase their resistance to abrasion and to improve machinability, particulary of high-carbon steels that require continuous cutting operations. Low-carbon steels are spheroidized to increase certain properties—for example, strength—before other heat treatment.

Spheroidizing requires heating steel *below the lower critical point*. The part is held at this temperature for a time and then cooled slowly to around 1000°F (538°C) or below. High-carbon steels are spheroidized by heating to a temperature that alternately rises between a temperature inside the critical range and one that is outside the critical range. Tool steels are spheroidized by heating the part slightly above the critical range. The part is held at this temperature for a period of time and then cooled in the furnace.

HOW TO ANNEAL, NORMALIZE, AND SPHEROIDIZE

Annealing Steel

STEP 1 Determine the upper critical temperature of the steel.

STEP 2 Adjust the pyrometer from 25°F to 30°F (-4°C to -1°C) above the upper critical temperature.

STEP 3 Prepare the furnace.

STEP 4 Place the part in the furnace so that it may be heated uniformly.

Caution: Use personal safety protective devices for all furnace and heat treatment processes.

STEP 5 Bring the workpiece to the annealing temperature.

STEP 6 Allow the workpiece to soak in the furnace.

Note: A general practice is to soak the part at the upper critical temperature an hour for each 1" (25.4mm) of thickness.

STEP 7 Shut down the furnace.

Note: The part may be slowly cooled in the furnace. Another method is to remove the part and immediately pack it in ashes or lime for slow cooling.

Normalizing Steel

STEP 1 Follow the same procedures for setting the furnace temperature and placing the workpiece in the furnace for uniform heating as for annealing (steps 1 through 6).

STEP 2 Bring the workpiece to the upper critical temperature.

STEP 3 Soak the workpiece at the normalizing temperature one hour for each 1" (25.4mm) of thickness.

STEP 4 Remove the part from the furnace. Allow the part to slowly cool in still air. Avoid drafts.

Note: Thin workpieces generally are packed in ashes or lime to prevent possible hardening in air and to retard the cooling rate.

Spheroidizing Steel

STEP 1 Determine the lower critical temperature of the part.

STEP 2 Set the pyrometer to about 30°F (-1°C) lower than the lower critical temperature.

STEP 3 Start the furnace. Position the part for heating.

STEP 4 Bring the workpiece up to temperature. Allow it to soak for one or more hours depending on the thickness.

STEP 5 Shut down the furnace. Allow the part to slowly cool at a rate of 50°F to 100°F (10°C to 38°C) per hour. Bring the part temperature down to around 1000°F (538°C).

STEP 6 Remove the part at this temperature from the furnace. Bring the temperature down by cooling the part in still air.

CASEHARDENING PROCESSES

Casehardening relates to the process of increasing the carbon content to produce a thin *outer case* that can be heat treated to harden. Impregnating the outer surface with sufficient amounts of carbon to permit hardening is called *carburizing*. When the carburized part is heat treated, the outer case is hardened, leaving a tough, soft inner core. Casehardening refers to the carburizing and hardening processes. The three general groups of casehardening processes are (1) carburizing, (2) carbonitriding, and (3) nitriding.

Carburization

Iron or steel is carburized when heated to a temperature below the melting point in the presence of solid, liquid, or gaseous carbonaceous materials. These materials liberate carbon when heated. The steel gradually takes on the carbon by penetration, diffusion, or absorption around the outer surface. This action produces a *case or zone* that has a higher carbon content at the outer surface. When a carburized part is heated to hardening temperature and quenched, the outer core acts like a high-carbon steel and becomes hard and tough.

Casehardening produces a steel having surface properties of a hardened high-carbon steel while the steel below the case has the properties of a low-carbon steel. Thus, there are two heattreating processes. One is suitable for the case; the other, for the core. Following an initial heating and slow cooling, a casehardened part is reheated to 1400°F to 1500°F (760°C to 816°C). It is then quenched in oil or water and given a final tempering.

Pack Hardening (Carburizing)

The purpose of *pack hardening* is to protect the delicate edges or finished surfaces of workpieces and to encourage uniform heating and contact with a carbonaceous material. Pack hardening also prevents scale formation and minimizes the danger from warping or cracking.

Pack carburizing requires the steel part to be enclosed in a box. Carbonaceous material such as carbonates, coke, and hardwood charcoal and oil, tar, and other binders are packed around the part in the box. Since the carburizing materials are inflammable, the box is sealed with a refractory cement to permit the gases generated within the box to escape while air is prevented from entering.

Penetration Time. The box is heated to carburizing temperature between 1500°F to 1800°F (816°C to 962°C), usually within the average temperature range of 1650°F to 1700°F (899°C to 927°C). Naturally, the rate of carbon penetration increases at the higher temperature in the range. The approximate *penetration time* for depths of 0.030″ to 0.045″ (0.75mm to 1.14mm) is four hours. It generally takes eight hours to penetrate to 1/16″ (1.6mm) and 24 hours to penetrate 1/8″ (3.2mm).

Carbon monoxide is released from the carbonaceous materials at the carburizing temperatures. The carbon in the mixture is absorbed by the heated surface of the steel part. The carburizing process continues until the desired penetration depth is reached. The box is then removed from the furnace and allowed to cool. When the parts are taken out of the box, they are cleaned by wire brush, tumbled, or sand blasted.

Heat Treating Carburized Parts. Once cooled from the carburizing temperature, the parts are reheated to the hardening temperature of the outer case (approximately 1430°F or 777°C) and quenched. This treatment produces a steel with a hardened case and a tough, soft low-carbon steel core.

A finer surface grain structure is produced by *double quenching*, which requires heating to the hardening temperature of the low-carbon steel core (1650°F or 899°C) and quenching. The coarse grain structure produced is then refined by reheating. The part is brought up to the hardening temperature of the case (about 1430°F or 777°C) and quenched. Double quenching combines the good wearing qualities of a hard case and toughness.

HOW TO CARBURIZE

One commercially prepared carbonaceous substance used in the *pack method* of carburizing is Kasenit®. While nonpoisonous and noninflammable, the fumes must still be exhausted. The part to be carburized is packed in a container with the carbonaceous substance. The box is then covered with a well-vented cover.

A *dip method* is also used where only a shallow case of a few thousandths of an inch (several hundredths of a millimeter) is required. Step-by-step procedures follow for both the pack and dip methods of carburizing.

Pack Method of Carburizing
for Case Depths to 0.015″ (0.4mm)

STEP 1 Cover the part, which is placed in a container, with at least a 1″ (25.4mm) or thicker layer of carbonaceous material. Place a well-vented cover on the container.

Note: About 1-1/2″ (38mm) is left between parts that are packed for carburizing.

STEP 2 Place the container in the furnace so that it may heat uniformly. Heat the container and part to 1650°F (899°C).

STEP 3 Soak the part at this temperature for the length of time recommended by the manufacturer of the carbonaceous material, depending on the required depth of case.

STEP 4 Remove the cooled parts and clean them.

STEP 5 Heat each part to its correct critical temperature.

STEP 6 Quench in oil or water, depending on the steel and part specifications.

Note: The tongs should be dry and preheated.

STEP 7 Temper the part, if required.

Dip Method of Carburizing for Shallow Depths

STEP 1 Heat the workpiece to approximately 1650°F (899°C).

STEP 2 Roll the workpiece in a casehardening compound such as Kasenit® until the material fuses around the outer surfaces.

STEP 3 Reheat the part to 1650°F (899°C). Hold at this temperature for a short time. Quench the part immediately in clean, cool water.

STEP 4 Reheat to temper, if required.

Note: Parts that are casehardened to just a few thousandths of an inch are not tempered.

Greater depths of casehardening may be obtained by bringing the workpiece up to 1650°F (899°C), redipping the part in the compound, and quenching.

Liquid Carburizing

Liquid carburizing has the advantage of not requiring carbonaceous materials to be packed around the workpiece. Liquid baths have a faster and more uniform penetration with minimum distortion. Where sections of a part are to be selectively carburized, the remaining portions may be copper plated. When the whole piece is immersed in the liquid bath, the copper plate inhibits carburization.

Salt bath furnaces are usually designed to be heated by electrodes that are immersed in the bath. The bath is stirred to ensure uniform temperature. The liquid carburizing baths are molten mixtures of cyanides, chlorides, and carbonates. The composition of the mixture depends on the quantity of carbon and/or nitrogen to be absorbed by the steel.

Liquid carburizing temperatures range from 1550°F to 1700°F (843°C to 927°C). In general, it takes about two hours at an average temperature of 1650°F (899°C) to penetrate to a depth of 0.020″ (0.5mm). Deeper case depths require proportionally greater periods of time to penetrate.

After carburizing in the molten salt bath, parts may be quenched directly in water, brine, or oil, depending on the job requirements. The parts are then tempered as required.

Caution: A number of safety precautions are stated later in the unit. They must be strictly observed. Most liquid carburizing, cyaniding, and nitriding salts are hazardous. Cyanides are extremely poisonous internally and in open wounds and scratches. When heated, cyanide fumes are toxic. Salt baths require careful, direct venting to outdoors.

HOW TO CARBURIZE IN A LIQUID BATH

STEP 1 Preheat the parts to be carburized to about 800°F (410°C). Also heat the jaws of the tongs.

STEP 2 Bring the carbonaceous material to carburizing temperature.

Caution: Strictly observe all safety precautions. Use personal safety equipment. Follow regulations concerning working around liquid salt baths and quenching tanks.

STEP 3 Suspend the dry parts in the liquid bath for the period of time required to produce the depth of case.

STEP 4 Remove the carburized part with a dry pair of tongs. Immediately quench in water (or oil, if oil hardening steel).

STEP 5 Temper the part as required.

Gas Carburizing

Methane (natural gas), propane, and butane are three gaseous hydrocarbons (carbon-bearing gases) used for *gas carburizing*. In continuous carburizing furnaces, the parts are heated to carburizing temperature in a horizontal rotary type or vertical pit type of gas carburizer. The carbon-bearing gases are mixed with air and other specially prepared diluent gases. The carburizing gases are fed continuously to the carburizing retort of the furnace. The spent gases are also exhausted.

The parts are soaked in the carburizing chamber. The soaking temperature and time depend on the required depth of case. Gas carburizing temperatures of around 1700°F (927°C) produce an absorption rate for the first 0.020" to 0.030" (0.5mm to 0.75mm) depth of case during a four-hour period.

The horizontal rotary type of gas carburizing furnace, as shown in Figure 37-7, permits direct quenching from the retort section of the furnace. Parts used as roller bearings, bolts, and shafts (up to forgings), which are not damaged by tumbling, may be carburized in this retort furnace.

The vertical pit type of gas carburizer is better adapted for long workpieces, gears, and form parts. While parts remain on a stationary rotor, the gases circulate around them. Single and small-lot quantities of parts are usually gas carburized in a horizontal muffle-type furnace. When carburized parts are allowed to cool first, they are reheated to hardening temperature and then quenched. The preferred method is for direct quenching from the carburizing retort. Otherwise, a surface oxidation builds up.

Cyaniding (Liquid Carbonitriding)

Cyaniding is a casehardening process that is used generally for limited depth case hardening to about 0.020" (0.5mm). The amount of carbon required in the surface case establishes the properties of cyanide salts in the salt bath (cyanide, chlorate, and chloride salts).

Cyaniding requires temperatures above the critical range of 1400°F to 1600°F (760°C to 871°C) for steels. Lower temperatures in the range are used for a limited case depth. As a general rule, case depths of from 0.003" to 0.005" (0.08mm to 0.13mm) require a half-hour soaking. Depths of 0.005" to 0.010" (0.13mm to 0.25mm) take from 60 to 70 minutes. The soaking time at cyaniding temperatures for depths of 0.015" (0.38mm) is two hours.

Parts are quenched in an appropriate water, brine, or oil bath. Tempering where required, is done at temperatures between 250°F to 300°F (120°C to 150°).

Carbonitriding

Carbon and nitrogen are introduced into the surface of steel by a dry (gas) cyaniding process known as *carbonitriding*. Carbonitrided surfaces possess greater hardenability and are harder and more wear resistant than carburized surfaces.

Parts are soaked at between 1350°F to 1650°F (732°C to 899°C) in a gaseous atmosphere composed of carburizing gas and ammonia. The ammonia produces the nitrogen. Carbon and nitrogen are introduced into the heated parts. Case depths of 0.030" (0.74mm) require soaking at temperatures of 1600°F (871°C) for four to five hours. A lower temperature is used when a higher proportion of nitrogen is required or for

FIGURE 37-7 Large Production, Continuous Rotary Furnace for Carburizing, Carbonitriding, and Other Hardness Processes

a shallower case. In such instances, a 0.005" to 0.010" (0.13mm to 0.25mm) case depth may be produced at a temperature around 1450°F (788°C) within 90 minutes. Generally, carbonitrided parts are quenched in oil to prevent distortion and to gain maximum hardness.

Nitriding

Nitriding is a casehardening process of producing surface hardening by absorption of nitrogen, without quenching. Special alloy steels are heated in an atmosphere of ammonia or in contact with a nitrogenous material. An exceptionally hard surface is produced on machined and heat-treated parts. Nitriding permits hardening carbon alloy steels beyond conventional hardness readings.

The alloy steels contain chromium, vanadium, molybdenum, and aluminum as nitride-forming elements. Nitriding requires heating the part below the lower critical temperature in a protected nitrogenous atmosphere or a salt bath.

Gas Nitriding. Ammonia gas is circulated through a gas furnace chamber. The parts are heated in an air-tight drum to a temperature between 900°F to 1150°F (482°C to 621°C). The nitrogen in the ammonia gas, which decomposes at this temperature, combines with the alloying elements in the steel. The hard nitrides that form produce a harder surface than may be obtained from other heat treatment processes.

Gas nitriding is particularly adapted to increase the hardness of parts that are hardened and ground. Many high-speed steel cutting tools are nitrided with shallow case depths of 0.001" to 0.003" (0.025mm to 0.075mm). The core of these tools is not affected by nitriding.

There is limited, if any, distortion due to the fact that gas-nitrided parts require no quenching. The process is slower than other casehardening processes. Depths of 0.020" (0.5mm) require two to three days.

Salt Bath Nitriding. Cutting tools such as taps, drills, reamers, milling cutters, and dies are often nitrided. Nitriding increases the surface hardness, fatigue and wear resistance, tool wear life, durability, and corrosive resistance (except on stainless steels).

The liquid salt bath is brought to temperature between 900°F to 1100°F (482°C to 593°C), depending on the tool requirements. The tool (or part) is suspended in the molten nitriding salt for the required period of soaking time.

Caution: With all heat-treating processes, particularly processes that require carburizing and cyaniding salts, it is necessary to observe all safety requirements. These requirements apply to personal items, materials handling, furnace operation, the exhausting of poisonous fumes, and all processing steps.

SPECIAL SURFACE HARDENING PROCESSES

Flame hardening and induction hardening are considered as two special surface hardening processes. These processes require heating without using a furnace. The purpose in each instance is to harden particular surfaces from a skin surface to depths up to 1/4" (6.4mm).

Flame Hardening

Flame hardening requires heating the surface layer of an iron-base alloy above the transformation temperature range. A high-temperature flame is used. The part is immediately quenched. Flame hardening is especially adapted to large steel forgings, castings, and unusually large parts where it is either not practical to heat treat in regular furnaces or the parts must be finish machined prior to hardening. Figure 37-8 shows an application of flame hardening to gear teeth.

The surface to be hardened is quickly heated with an oxyacetylene torch or air-fuel gas burner.

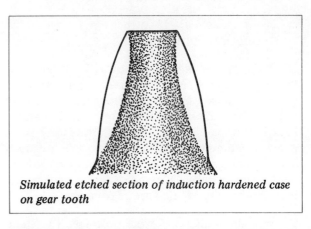

Simulated etched section of induction hardened case on gear tooth

FIGURE 37-8 Application of Flame Hardening to Gear Teeth

The flame is concentrated in the specific area to be hardened. The surface is immediately quenched by quenching jets in back of the torch or burner. A water or compressed-air quenching medium is used, depending on the type of steel. Tempering is recommended close to the hardening process, where practical.

Medium-carbon and many alloy steels may be flame hardened. Plain carbon steels of from 35 to 60 points of carbon are especially adapted to flame hardening without checking or cracking. The higher-carbon steels are capable of providing greater hardness. However, careful quenching controls are required.

Similar methods of flame heating are used for tempering. A special low-temperature flame head follows immediately behind the quench to localize the heating area, which is brought to the required tempering temperature.

Parts may be flame hardened by the *spinning method*, the *stationary spot method*, or by combining the rotation of the workpiece with the longitudinal movement of the flame head.

Flame hardening is particularly suited to the following applications:

- —Large parts and irregular surfaces where it is impractical to surface harden selected areas in a furnace efficiently;
- —Parts requiring harder wear-resistant surfaces where through-hardening processes produce extensive stresses and distortion;
- —Cost-effective applications where carbon steels may be used instead of the more costly high-carbon steels;
- —Machine surfaces (such as bed ways, gears, spline shafts, and cams), forming rolls and dies, and other production parts.

Induction Hardening

Induction hardening is another localized heat-treating process. Electrical heating is required to bring a piece of steel to the required hardening temperature and subsequently quenching it either in a liquid or air bath. Induction heating is especially adapted to parts that require localized and controlled depth of hardening and irregularly contoured surface. The major advantages to using induction hardening are as follows:

- —A short heating cycle produces the same temperature in seconds that normally requires from 30 to 80 times longer to heat in conventional furnaces;
- —There is no tendency to produce oxidation or scaling or decarburization;
- —The depth and localized zones of hardening may be exactly controlled;
- —The automatic heating and quenching cycles permit close controls of the degree of hardness;
- —Warpage or distortion are reduced to a minimum;
- —Carbon steels may be substituted for higher-cost alloy steels;
- —Stress control is possible by localized heating to relieve internal stresses;
- —Welded and brazed design features may be added prior to heat treating particular sections of a workpiece;
- —Long parts may be heat treated more efficiently than by conventional furnace methods;
- —In gear applications of induction hardening, the teeth may be machined and shaved in a soft-annealed or normalized condition (bushings and inserts can be assembled before hardening the gear teeth).

Principle of Induction Heating. The process begins when a metal part is placed inside and close to an applicator coil. The part may be held in a fixed position, turned, or fed through the coil. As a high-frequency electrical current passes through the coil, the surface of the steel is raised to a temperature above the critical temperature in a matter of seconds.

The heated part is then quenched in oil, water, or air. The hardness is localized to the surface. The hardness depth is controlled by the length and intensity of the heating cycle. High-frequency currents are used for localized and surface hardening. Low frequencies are employed for through heating, deep hardening, and large workpieces.

One controlling factor on standard types of steels that may be induction hardened is that the carbon content must permit hardening to the required degree by heating and quenching. Low-carbon steels with a carburized case and plain medium-carbon and high-carbon steels may be induction hardened. Cast irons with a percent of carbon in combined form may also be induction hardened. Induction heat treating of alloy steels is generally limited to shallow hardening types that are not affected by high-stressing and possible cracking induced by the required severe quench.

Induction Surface Hardening. To summarize, smaller and thinner section workpieces and shallow hardening depths require a high heating frequency while internal surfaces may readily be heated by shaping the applicator coil to match the cross section. Bored holes of small diameter are often heated by moving an applicator coil or the work so that the heating zone passes through the opening.

The frequency range for induction heating is from 1kHz to 2MHz per second (1,000 cycles to 2,000,000 cycles per second). The higher frequencies permit shallow hardening to depths from a few thousandths of an inch to 1/16" (1.5mm). The lower frequencies are used for hardening depths from 1/16" to 1/4" (1.5mm to 6.4mm). The lower frequencies (180 to 3,000 cycles per second) are used for through applications.

Quenching After Induction Heating. Heated parts may be quenched by immersing in a liquid bath, by liquid spraying, or by self-quenching. *Self-quenching* is associated with rapid absorption of heat by a large mass of surrounding metal, instead of by using a quenching medium. Self-quenching is generally confined to small, simply designed parts. The exact degree of hardness may be automatically timed for heating and quenching. Induction coils or standard furnaces may also be used to temper parts as required.

SUBZERO TREATMENT OF STEEL

Steels are subjected to a *subzero treatment* in order to stabilize a part and to prevent changes in form or size over a period of time. In conventional hardening, many steels retain some austenite—that is, the sudden cooling from hardening temperatures does not transform all austenite to hard martensite. Under natural aging conditions, the austenite is ultimately transformed to martensite. Subzero treatment reduces the transformation time to a few hours.

There are some risks in subzero treatments as the austenite, which serves to cushion the possibility of cracking due to hardening, is removed. Therefore, subzero treatment requires close control and a series of repeated cycles of chilling and final tempering.

HOW TO STABILIZE GAGES AND OTHER PRECISION PARTS BY SUBZERO COOLING

STEP 1   Cool the hardened and ground gage, sine bar, gage blocks, or other precision part uniformly to $-120°F$ ($-84°C$).

STEP 2   Place the subzero-cooled part in boiling water, oil, or a salt bath.

Note: Precision tools such as thread and plug gages, gage blocks, and other highly accurate parts are immersed for about two hours.

STEP 3   Repeat the cooling and quenching cycle.

Note: Steps 1 and 2 may be repeated two or three times to eventually transform practically all of the austenite to martensite.

STEP 4   Temper according to standard tempering practices for the specific metal in the part.

STEP 5   Proceed to finish grind or lap the subzero-treated part to the required dimensional size and exact form.

Note: Subzero treatment produces a slight increase in size. Therefore, many close-fitting machine parts are subzero treated before a final grinding or other finishing process. Otherwise, the minute increase in the size of close-fitting parts may cause them to seize under drastic changes of temperature.

HARDNESS TESTING

A hardened steel is tested to define its capacity to resist wear and deformation. The hardness of steel, alloys, and other materials may also be used to establish other properties and performance.

Hardness testing is performed using measuring instruments of two basic types. The first set of instruments measures the depth of a penetrator for a given load. A few of the instruments that operate under this principle include: *Rockwell*, *Brinell*, *Vickers*, *Knoop*, and *microhardness testers*. The second set of instruments measures the height of rebound of a given weight hammer of a special shape when dropped from a fixed height. *Scleroscope testers* represent this type of instrument. Each of the two types of instruments is covered in this unit.

Most instrument manufacturers indicate hardness by number scales. The number is related to the size and shape of the penetrator, load, height, and indentation or rebound. A table in the Appendix provides numbers within each major system. These numbers permit conversion of hardness values from one system to another. For example, a Rockwell C (R_C) hardness of 50 is equivalent to 513 on the Vickers scale, 67 on a scleroscope, and BHN 481 on a Brinell scale when a 10mm (0.400") diameter carbide ball and a 3,000 kg (6,600 lb) load are used. A 75.9 reading is recorded on the Rockwell A (R_A) scale using a diamond penetrator 50 kg (132 lb) load. The Appendix Table for Rockwell C (R_C) hardness numbers 68 through 20 is based on extensive tests with carbon and alloy steels primarily in a heat-treated condition. The hardness numbers may be reliably applied to tool steels and alloy steels in annealed, normalized, and tempered condition. Additional Rockwell B and other hardness scale numbers are provided in handbooks for unhardened or soft temper steels, gray and malleable cast irons, and nonferrous alloys.

ROCKWELL HARDNESS TESTING

Tests are made with the Rockwell hardness tester by applying two loads on the part to be tested. The first load is the *minor load;* the second, the *major load.* The Rockwell tester measures the linear depth of penetration, as shown in Figure 37-9. The difference in depth of penetration produced by applying a minor load and then a major load is translated into a hardness number. This number may be read directly on the indicator dial of a portable tester, a manual or motorized bench model (Figure 37-10), or on a digital display unit.

A shallow penetration indicates a high degree of hardness and a high hardness number. A deep penetration indicates a softer degree of hardness and a low hardness number. In general, the harder the material, the greater its ability to resist deformation.

Sizes and Types of Penetrators

Different penetrator points are used on Rockwell testers, depending on the hardness of the material. Diamond- and ball-point penetrators used on Rockwell hardness testers are illustrated in Figure 37-11. Hard materials such as

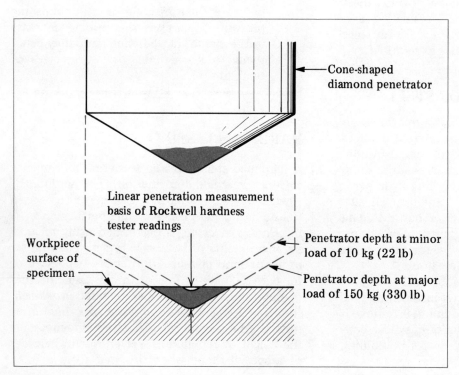

FIGURE 37-9 Linear Penetration Measured by a Rockwell Hardness Tester

hardened steels; white, hard cast irons; and nitrided steels require a cone-shaped diamond. The trademark of this form of penetrator is registered as Brale®. A hardened steel ball-shaped point is used for unhardened steels, cast irons, and nonferrous metals.

C and A 120° Diamond Penetrator Points. The C Brale penetrator is used with a major load of 150 kilograms of force (kgf) or 330 pounds of force (lb/f) for hard materials. The *Rockwell C scale* is used with the C Brale penetrator.

Cemented carbides, shallow casehardened steels, and thin steels require the use of the A Brale penetrator and a 60 kgf (132 lb/f). The hardness is read on the *Rockwell A scale*.

Ball Penetrator Points. The second basic type of penetrator has a ball-shaped point and is made of hardened steel. The two diameters are 1/16" and 1/8" (1.5mm and 3.0mm). A *Rockwell B scale* is used for hardness readings of unhardened steels, cast irons, and nonferrous metals. The major load with the 1/16" (1.5mm) diameter ball is 100 kg (220 lb). The minor load is 10 kg (22 lb). The red dial hardness numbers are used for the R_B scale.

Hardness Scales

Table 37-3 furnishes information about standard, special, and superficial Rockwell hardness scales, hardened steel ball-point and conical-shaped diamond-point penetrators, applications, major kg (lb/f) loads, scale symbols, and dial color. The superficial scales are used on parts where only a light, shallow penetration is permitted. Casehardened parts, nitrided steels, thin metal sheets, and highly finished surfaces are measured for hardness by using a lighter load. Rockwell superficial hardness scales are used to measure such loads. The *Rockwell N scale* requires the use of a special N Brale penetrator; the *Rockwell T scale*, a 1/16" (1.5mm) diameter hardened steel ball penetrator. The 15, 30, and 45 prefixes indicate the major kilogram load in each case. The minor load is 3 kg (6.6 lb).

Features of Rockwell Hardness Tester

The main features of a Rockwell hardness tester with standard anvil forms are identified in Figure 37-12. A small-diameter plane anvil is shown on the instrument. "V," small- and large-diameter, and roller designs are furnished with this instrument. Cylindrical parts are supported on the V-type centering anvil. Tubing is usually mounted on a mandrel to prevent damage to the walls. Long, overhanging parts are supported by an adjustable jack rest so that the surface of the workpiece is in a horizontal plane with the anvil.

The practice under actual testing conditions is to take readings at three different places along the workpiece. The hardness number is the average of the three readings.

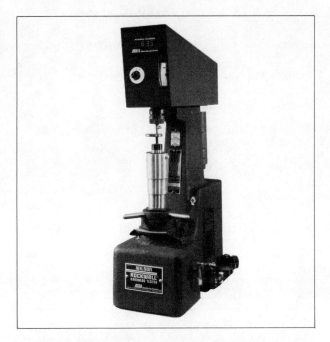

FIGURE 37-10 Motorized Rockwell Hardness Tester for Applications with Standard, Special, and Superficial Hardness Scales

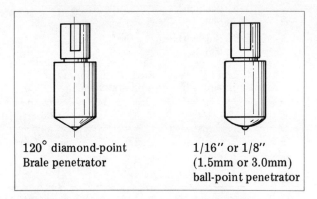

FIGURE 37-11 Diamond- and Ball-Point Penetrators Used on Rockwell Hardness Testers

HOW TO TEST FOR METAL HARDNESS BY USING A ROCKWELL HARDNESS TESTER

STEP 1 Select and mount the appropriate penetrator and anvil in the hardness tester.

Note: A diamond penetrator is used for hardened steel. The 1/16" (1.5mm) diameter hardened steel ball point is generally used for cast irons, soft steels, and nonferrous metals.

Note: The diamond penetrator is used with the R_C scale; the hardened steel ball, with the R_B scale.

STEP 2 Turn the dial on the indicator to position the indicator hand for starting a test.

Note: On some indicators, a dot on the dial face shows the starting position.

STEP 3 Place the workpiece or specimen on the anvil.

TABLE 37-3 Rockwell Hardness Scales, Penetrators, Major Loads, and Reading Dials for Typical Hardness Test Applications

Scale	Type of Penetrator	Major Load (kgf)	Dial Color (Numbers)	Typical Hardness Test Applications
			Standard Scales	
B	1/16" (1.5mm) ball	100	Red	Soft steels, malleable iron, copper and aluminum alloys
C	Brale (diamond)	150	Black	Steel, deep casehardened steel, hard cast iron, pearlitic malleable iron, other materials harder than B-100
			Special Scales	
A	Brale (diamond)	60	Black	Thin steel, shallow casehardened steel, cemented carbides
D	Brale (diamond)	100	Black	Thin steel, medium casehardened steel, pearlitic malleable iron
E	1/8" (3.0mm) ball	100	Red	Cast iron, aluminum and magnesium alloys, bearing metals
F	1/16" (1.5mm) ball	60	Red	Thin soft sheet metals, annealed copper alloys
G	1/16" (1.5mm) ball	150	Red	Malleable iron, phosphor bronze, beryllium copper, other metals within an upper hardness limit of G-92
H	1/8" (3.0mm) ball	60	Red	Aluminum, lead, zinc
K	1/8" (3.0mm) ball	150	Red	Harder grades of aluminum, lead, and zinc

L, M, P, R, S and V scales with 1/4" (6.35mm) or 1/2" (12.70mm) diameter balls and major loads of 60, 100, or 150 kgf are used on very soft and thin materials.

			Superficial Hardness Scales	
15 N	Brale (diamond)	15	(N) Green	Casehardened and nitrided parts, thin metal sheets, highly finished surfaces, metal parts requiring lighter loads of 15, 30, or 45 kgf
30 N		30		
45 N		45		
15 T	1/16" (1.5mm) ball	15	(T) Green	Soft steels, cast iron, nonferrous metals capable of withstanding maximum loads of 15, 30, or 45 kgf
30 T		30		
45 T		45		

HEAT TREATING AND HARDNESS TESTING

Note: Overhanging and long parts must be supported by a jack rest. The work surface must rest solidly on the anvil to prevent damage to the penetrator or an inaccurate reading. A mandrel is used to support tubing walls against damage.

Note: Any scale or oxidation must be removed from the surface to be tested before mounting.

STEP 4 Raise the anvil or lower the penetrator until it just touches the workpiece.

STEP 5 Apply a 10 kg (22 lb) minor load. This load is shown on the indicator dial. Set the hardness reading (indicator) dial at zero.

STEP 6 Apply the appropriate major load.

Note: A Rockwell hardness scale table should be consulted to establish the major load for the type of penetrator and hardness testing requirements.

STEP 7 Reduce the kgf of the major load to the setting of the minor load.

STEP 8 Read the hardness on the appropriate color scale.

Note: The hardness reading on the barrel dial of a portable hand tester is read through a plastic magnifier.

STEP 9 Release the minor load. Move the workpiece or specimen to a second location. Then, repeat steps 2 through 8 to obtain a second hardness reading.

STEP 10 Take a third hardness reading in a third location.

STEP 11 Average the three hardness readings.

Note: The average reading is the hardness reading of the workpiece.

Note: The accuracy of the instrument is checked regularly. Test blocks of given hardness are provided for such checking.

BRINELL HARDNESS TESTING

The Brinell hardness tester produces an impression by pressing a hardened steel ball under a known applied force into the part to be tested. A microscope is then used to establish the diameter of the impression. The ball is 0.394"

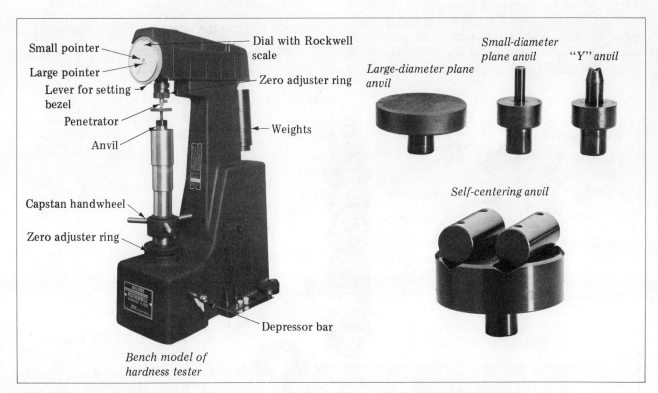

FIGURE 37-12 Main Features of a Rockwell Hardness Tester with Standard Anvil Forms

(10mm) diameter. It may be made of hardened steel or carbide, or it may be identified as a Hultgren ball. The main features of a Brinell hardness tester with a measuring microscope are shown in Figure 37-13.

A Brinell hardness number (Bhn) is identified on a Brinell table according to the diameter of the impression and the specified applied load. The Brinell hardness value equals the applied load in kilograms divided by the square millimeter area (impression). The load for hardened steel parts is generally 3,000 kg (6,600 lb). A standard load of 500 kg (1,100 lb) is used for hardness testing nonferrous metals. Lower numbers on the scale are assigned for softer metals and deeper impressions. Expressed as a formula,

$$Bhn = \frac{\text{applied load in kg (L)}}{\text{diameter of impression}}$$
$$= \frac{L}{\frac{\pi D}{2} \times (D - D^2 - d^2)}$$

Recommended Applications of Brinell Hardness Testing

The range of Brinell hardness testing is between Bhn 150 for low-carbon annealed steels to Bhn 740 for hardened high-carbon steels. A carbide ball is required for upper ranges of hardness to Bhn 630. A Hultgren ball accommodates hardness testing up to Bhn 500. The hardened steel ball-point penetrator is adapted to hardnesses to Bhn 450. The Brinell hardness tester works best on nonferrous metals, soft steels, and hardened steels through the medium-hard tool steel range.

HOW TO TEST FOR METAL HARDNESS BY USING A BRINELL HARDNESS TESTER

STEP 1 Select the load. Adjust the air regulator until the required load is obtained.

STEP 2 Place the part to be tested on the anvil. Raise the setup to within 5/8" (15mm) of the penetrator ball.

STEP 3 Pull out the plunger control to bring the penetrator ball to the work surface to apply the load (Figure 37-14).

STEP 4 Maintain a steady load for at least 15 seconds for iron or steel; 30 seconds, for nonferrous metals.

Note: Generally, a standard 500 kg (1,100 lb) load is applied for nonferrous metals in the Bhn 26-100 hardness range. The standard load for steel and iron is 3,000 kg (6,600 lb).

STEP 5 Release the load and remove the part.

Note: The penetrator and ball retract in readiness for the next test.

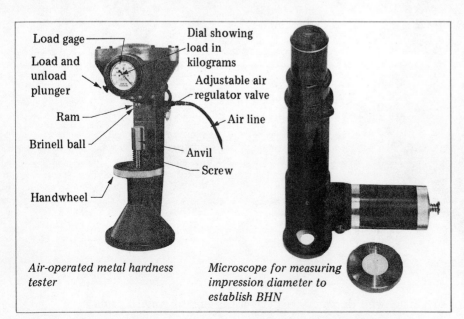

FIGURE 37-13 Main Features of a Brinell Hardness Tester with a Measuring Microscope

HEAT TREATING AND HARDNESS TESTING 803

FIGURE 37-14 Operation of Plunger Control to Apply Selected Load

STEP 6 Measure the diameter of the impression with a Brinell or hardness testing microscope.

STEP 7 Use a table to establish the Brinell hardness number (Bhn). Otherwise, calculate the Bhn by formula according to the diameter of the impression and the load.

VICKERS HARDNESS TESTING

The Vickers hardness test requires a square-based diamond pyramid whose sides are at an angle of 136°. A load of from 5 kg to 120 kg (11 lb to 264 lb) is applied generally for 30 seconds. The diagonal length of the square impression is measured.

The Vickers hardness number equals the applied load divided by the area of the pyramid-shaped impression. The Vickers test is very accurate. It is adapted to large sections as well as thin sheets.

SCLEROSCOPE HARDNESS TESTING

The scleroscope instrument measures hardness in terms of the elasticity of the workpiece. There are two types of scleroscopes. One type has a direct-reading *scale*. The other type has a direct *dial* recording. Figure 37-15 shows the main features of a direct-reading scleroscope. Scleroscopes are used for hardness testing of ferrous and nonferrous metals. Unlike the Brinell and Rockwell hardness testers, no crater is produced.

A diamond hammer, dropped from a fixed height, rebounds. The rebound distance varies in proportion to the hardness of the metal being tested. The harder the metal, the higher the rebound distance. This movement is either read on a scale or a dial.

The scale is calibrated from the average rebound height of a tool steel of maximum hardness that is divided into 100 parts. The range of rebounds is from 95 to 105. The dial of a direct-reading scleroscope is shown in Figure 37-16. Readings are carried beyond 100 to 120. This range covers super-hard metals. Note that the conversion Rockwell C and Brinell hardness values are given on the dial face; a 3,000 kg (6,600 lb) load and a 10mm diameter ball are used.

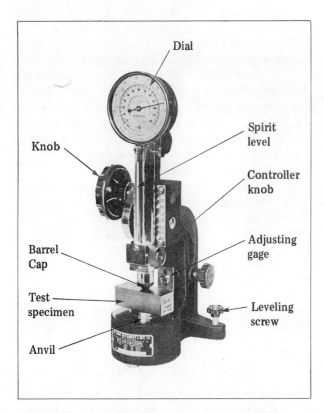

FIGURE 37-15 Main Features of a Direct-Reading Scleroscope

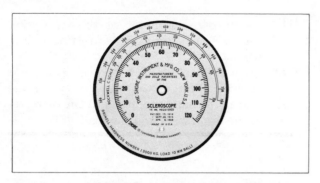

FIGURE 37-16 Direct-Reading Scleroscope Dial with Equivalent Rockwell C and Brinell Hardness Numbers

HOW TO TEST FOR METAL HARDNESS BY USING A SCLEROSCOPE

STEP 1 Level the scleroscope

STEP 2 Turn the knob to bring the barrel cap in firm contact with the part being tested. This contact is maintained throughout the test.

STEP 3 Draw the hammer to the up position.

STEP 4 Release the hammer with the control knob. Note the scale or dial reading to which the hammer rebounds.

STEP 5 Change the position of the workpiece on the anvil. Repeat steps 2 through 4 to obtain a second reading.

STEP 6 Average the rebound dial readings for several tests at different positions along the part. The average number represents the scleroscope hardness reading.

Note: A regular check of the accuracy of the instrument should be made against the reference test bar supplied by the instrument maker.

MICROHARDNESS TESTING

The microhardness tester permits hardness testing of minerals, abrasives, extremely hard metals, very small or thin precision parts, and thinly hardened surfaces. Some of these materials and parts cannot be tested by other methods. Microhardness testers are also used to determine the hardness of grains in the microstructure of the material.

Hardness tests are made with a diamond penetrator. The penetrator is pressed into the specimen with extremely light loads of from 25 to 3,600 grams (0.7 oz. to 7.9 lb). Minute impressions are produced under appropriate low loads.

The microscope feature of the tester permits measuring the fine impression. The depth of penetration is determined by the applied load and the hardness of the material. A number value is assigned to each degree of hardness. The number expresses the relationship between the applied load and the long dimension of the indentation.

Microhardness testers produce a minute indentation that, in most instances, does not damage the work surface. Conversion charts are available from instrument builders for comparative hardness values in other systems.

KNOOP HARDNESS TESTING

The Knoop hardness test is adapted to the same hard materials and thin parts as the Microhardness tester. The load range in grams is also from 25 to 3,600 grams (0.7 oz. to 7.9 lb). The plane surfaces must be free from scratches.

A *Knoop indentor* is used. The indentor is a diamond in an elongated pyramid form. The indentor acts under a fixed load that is applied for a definite period of time to produce an indentation that has long and short diagonals. The Knoop indentor is used in a fully automatic, electronically controlled Tukon tester.

Knoop hardness numbers are used. The numbers are equal to the load in kilograms divided by the area of the indentation in square millimeters. Tables are available that give the indentation number corresponding to the long diagonal and for a given load.

SAFE PRACTICES FOR HEAT TREATING AND HARDNESS TESTING

—Allow the workpiece to heat up. Check the pyrometer and heat measurement system. The

workpiece must be brought up to the critical temperature and heated uniformly.
— Avoid bringing the part to a higher temperature than the critical point or quenching at a more severe rate than is recommended by the steel manufacturer. Overheating produces excessive internal strains when the part is cooled. These strains may cause brittleness, fracturing, or shattering a hardened part.
— Use quenching oils with a flash point sufficiently high to permit safe cooling without the danger of fire.
— Add a rock salt (8% to 9%) or caustic soda (3% to 5%) to a fresh water quenching bath to prevent the formation of gas pockets or a film when quenching an irregularly shaped part or one with holes or cavities.
— Agitate the workpiece in the quenching medium to prevent steam or an insulating vapor film from forming. These conditions produce uneven cooling and the danger of cracking from excessive internal stresses.
— Use safety gloves and face protection when working around a furnace with hot metals and when quenching. Stand to one side when plunging and agitating a workpiece in the quenching bath, particularly when using an oil.
— Start the exhaust fan for all heat-treating operations. Check to be sure the fumes hood is exhausting the gases and excess heat.
— Check the furnace system to see that if the exhaust system fails, there is a shutdown of the main gas/oil supply valve.
— Close the loading door on the jacket enclosure on molten bath furnaces to control and direct the fumes and for protection against spattering.
— Make sure that the workpiece and/or tongs that are to come into contact with molten cyanide are clean and dry. Cyanide, mixtures of nitrate, nitrite, and some other salts spatter and explode violently if any one of them in a molten state comes in contact with water.
— Heat a frozen cyanide bath by electrodes from the top down to ensure that any fumes escape as the bath melts from the top on down through the mass. If electrodes are not used, it is important that a wedge be inserted in the bath before it is frozen. The wide side is on the bottom. The narrow, tapered end extends beyond the top surface. When completely cooled, the wedge is removed. The wedge opening permits the gases to escape readily when the bath is heated.
— Wear asbestos gloves and apron, safety glasses, and a face shield when carrying on heat-treating operations.
— Follow manufacturer's recommendations and safety codes.
— Use an adjustable jack rest for overhanging parts on hardness testing instruments. The area must rest solidly on the hardness tester anvil to obtain an accurate reading and to prevent damage to the penetrator.

TERMS USED IN HEAT TREATING AND HARDNESS TESTING

Critical points of steel	Temperatures at which alpha and gamma structures are formed in steels. Lower and upper temperatures at which transformations of grain structure occur. Upper and lower temperatures affecting hardness and other properties of steels.
Martensite	A grain structure produced by rapidly transforming austenite by quenching a steel that is heated to the hardening temperature to below 1000°F (538°C). The hardest and most brittle form of steel.
Subzero, black, red, and white heats	A range of working temperatures from stabilizing subzero -120°F (-84°C) to forging/welding temperatures around 2900°F (1593°C).
Decalescence point	A point in the heat-treating cycle where pearlite is transformed into austenite. The temperature at which steel continues to absorb heat without any significant change in temperature.

Hardening temperature	The temperature at which a steel part is heated or soaked before quenching. A temperature range that depends on carbon content and composition (including alloying elements).
Austempering, martempering, and isothermal quenching	Interrupted methods of quenching. Interrupting the cooling that begins above the transformation point. Interrupted cooling at a temperature above the one at which martensite begins to form.
Liquid hardening furnaces	Furnaces equipped to heat salt, lead, carbonaceous, and other materials to molten state. Furnaces used for the immersion of parts to be heat treated in a molten liquid bath. Furnaces for bringing parts up to a specified temperature for harding, carburizing, or tempering processes.
Transformation range (ferrous metals)	A temperature zone from which a workpiece is raised or lowered to change the grain structure. (For example, the transformation range of steel is the temperature interval on cooling in which the austenite changes to martensite.)
Tempering	A transformation temperature range below which a hardened or normalized steel part is reheated, followed by cooling at a specified rate.
Annealing	Heating and cooling metals for purposes of softening, producing a definite microstructure, removing internal stresses, or altering mechanical or physical properties.
Normalizing	A heat-treating process of producing a uniform grain structure for a steel part to be able to receive further heat treatment.
Spheroidizing	Heating and cooling a metal to produce a globular form of carbide. A heat-treating process to increase strength, resistance to abrasion, or other properties.
Casehardening	A series of heat-treating processes to produce a hardenable outer case. The introduction into a steel surface of carbon, carbon-nitrogen, or nitrogen by penetration, diffusion, or absorption. Carburizing, carbonitriding, and nitriding processes affecting the surface hardness of metals.
Liquid carburizing	Introducing carbon into a solid iron-base alloy by heating the workpiece above the transformation temperature range while in contact with a liquid carbonaceous material.
Nitriding	Producing an exceptionally hard surface by heating an alloy steel in a nitrogenous atmosphere or salt bath below the lower critical temperature. Absorption of nitrogen into the outer case and producing a very hard surface without quenching.
Flame hardening	A surface hardening process of localized heating of an iron-base alloy above the transformation temperature range, with an immediate quench.
Induction hardening	Use of an electrical coil and current to quickly raise the surface temperature above the critical temperature of the steel and then quenching in an appropriate bath.
Subzero temperature stabilizing	Reducing the temperature of steel to the $-100°F$ to $-120°F$ ($-73°C$ to $-84°C$) range to transform all austenite to martensite. A series of subzero temperatures to normal cooling temperatures and final tempering.
Major and minor loads	Standardized initial and subsequent loads applied on a hardness tester to a diamond or steel ball-shaped penetrator. Two reference

HEAT TREATING AND HARDNESS TESTING 807

	points from which the linear depth of penetration is established to measure hardness.
Hardness testing scales	Numerical values related to depth or area penetration of a hardness tester penetrator for a series of fixed loads, according to material. Bhn, Rockwell, Vickers, Knoop, scleroscope, microhardness and other hardness testing measurement systems. Numerical values that may be converted from one system to the equivalent in another system.
Scleroscope	A hardness testing instrument where a hammer of known weight is dropped from a fixed height and rebounds. An instrument that records the (hardness) rebound of a hammer according to the elasticity of the metal.
Microhardness tester	A precision hardness testing instrument that uses a diamond penetrator under very light loads to obtain a hardness value for extremely hard minerals, metals, and grain microstructures.

SUMMARY

Variations in hardness and other mechanical properties of steel are made possible by changing the cell structure through heating and cooling within certain critical points and rates.

> A hard martensite grain structure is formed by quenching steel that is heated to its hardening temperature at its critical cooling rate. The austenite grains are transformed to martensite.

An excess rate of cooling during quenching produces internal stresses that may result in fractures and warpage.

> Heat-treating processes for steel are carried on from subzero temperatures through the black, red, and white heat ranges. Normalizing processes start at 1700°F (927°C). Higher temperature ranges are required for carburizing.

As steels age, unless the grain structure is stabilized, there is a tendency for the steel to grow as the austenite is transformed into martensite.

> Pearlite is transformed into austenite at the decalescence point. The reverse process takes place at the recalescence point during cooling.

Steel hardening temperatures before quenching are determined by the amount of ferrite or cementite to be dissolved into austenite, grain size, required physical properties, composition, and tool geometry.

> The higher temperatures in the hardening range of steel produce deeper hardness penetration.

Heating baths serve to heat parts uniformly, control the heating temperature, and protect the surface.

> Quenching baths remove heat from a part to be hardened at a rate of cooling that is faster than the critical cooling rate.

Quenching baths of air; fresh, soft water; oils; oil-water solutions; lead; and salt are commonly used with ferrous metals.

> Oil baths permit maximum depth hardening with minimum distortion and cracking. Oils provide fast cooling in the initial stages followed by slower cooling at the lower temperatures.

Fresh water baths must maintain a constant 70°F to 100°F (21°C to 38°C) temperature to uniformly cool a number of workpieces.

808 PRODUCTION OF INDUSTRIAL AND CUTTING TOOL MATERIAL

Salt baths produce maximum hardness and minimum cooling stresses in high-speed steels.

There must be a movement between a quenched part and the bath to permit uniform cooling and to prevent soft spots.

Interrupted quenching permits holding the temperature of steel being hardened for a fixed time to provide for the uniform transformation of the grain structure. This process overcomes internal stresses and produces greater toughness and ductility.

Austempering, martempering, and isothermal quenching are three basic processes of interrupted quenching.

The thermocouple and pyrometer provide controls for furnace and workpiece temperatures.

Gas, oil-fired, and electrical furnaces are generally used for heat-treating and simple forging operations.

Pot-type furnaces are used for liquid baths such as salt, lead, cyanide, chlorides, and nitrides.

The iron-carbon diagram provides basic information about transformation temperatures for changes in grain composition and structure and critical temperatures for hardening, annealing, and normalizing for steels with varying percents of carbon content.

Tempering tables identify temperatures and/or color to which hardened parts are drawn and quenched to relieve internal strains and to improve toughness and other physical properties.

Tempering baths may be oil, salt, lead, or some other medium, depending on the hardened material.

High-speed steels are usually double tempered to change all untempered martensite.

Parts that are to be hardened are slowly and uniformly brought up to and soaked in the hardening temperature range.

There must be continuous motion between the workpiece and the quenching medium during hardening and tempering.

Hardened parts require tempering as close as possible after the hardening process.

High-speed steels are usually preheated at temperatures below the critical point of the steel (1500°F to 1600°F (816°C to 871°C).

Tungsten high-speed steels are generally quenched in oil or a salt bath.

High-speed steel tools are drawn at temperatures from 900°F to 1200°F (482°C to 649°C). Cobalt high-speed steel tools are tempered between 1200°F to 1300°F (649°C to 704°C).

Annealing requires controlled heating and cooling to reduce hardness, improve mechanical properties, and reduce internal stresses. Steels are heated above the upper critical point (when less than 0.85% carbon) or above the lower critical point (for hypereutectoid steels).

Normalizing results in a uniform grain size and refinement and an unstressed structure to receive further heat treatment.

Spheroidizing produces a globular grain structure. The process increases properties such as machinability, strength, and resistance to abrasion.

HEAT TREATING AND HARDNESS TESTING

Spheroidizing requires heating the part to its lower critical temperature. The part is allowed to soak for a time and to be brought down to 1000°F (538°C) before being removed from the furnace and further cooled in still air.

Casehardening produces a thin outer case by increasing the carbon content of iron or steel. Carbon is taken in by penetration, diffusion, or absorption. Carburized parts may be surface hardened while retaining the properties of a low-carbon steel.

Pack carburizing requires heating a part that is surrounded by a carbonaceous material and binders in a container. Penetration time depends on the part material depth of case required and temperature.

Carburized parts are reheated to hardening temperature and quenched. Once hardened, parts are then tempered.

Liquid baths permit faster and more uniform penetration with minimum distortion.

Parts to be carburized are preheated to 800°F (410°C), suspended in a molten liquid bath, brought to hardening temperature, quenched, and tempered.

Gaseous hydrocarbons are used for gas carburizing. Parts are soaked at temperatures around 1700°F (927°C) at the rate of absorption required to penetrate to a specific depth.

Cyaniding produces a hardenable outer case by heating a steel part in a molten bath at critical temperatures of 1400°F to 1600°F (760°C to 871°C) for a stated time and then quenching, followed by tempering.

Carbonitriding is a gas process of introducing carbon and nitrogen into the surface of steel for greater hardenability and wear resistance.

Nitriding permits surface hardening carbon alloy steels beyond the general hardness range. No quenching is required. The steel core is not affected by nitriding.

Flame hardening requires localized heating of the surface layer above the transformation range. The heated area is immediately quenched. Tempering must follow as soon as possible.

Induction hardening is especially adapted to localized and controlled depth hardening. Hardening is closely regulated by automatic short heating and quenching cycles. High-frequency current, depending on depth requirements, is used for heating.

Subzero steel treatments accelerate aging and stabilize the grain structure, hardness, and other physical properties of precision tools, gages, and parts. Parts are cooled to temperatures of -120°F (-84°C), followed by immersion in a boiling water or an oil or salt bath. Standard tempering processes follow hardening.

Hardness testers determine the degree of hardness and establish other properties and performance characteristics associated with hardness.

Conversion tables permit interpretation of hardness values among scales used by different instrument makers.

Numerical values for hardness are based on either the penetration of a ball of known size from a fixed height and load or the rebound of a hammer.

Multiple scales are used on the Rockwell hardness tester, depending on the hardness range of ferrous and nonferrous metals and other characteristics.

The Brinell hardness tester uses a microscope to establish the diameter of an impression. A hardened steel ball, carbide ball, or Hultgren ball are used for different metals and hardness measurements.

Vickers hardness testing is extremely accurate and is adapted to parts that require light loads from 5kg to 120kg (11 lb to 264 lb).

The scleroscope tester measures hardness by the elasticity of the hardened part.

Microhardness testers measure the hardness of grains in minerals, abrasives, and extremely hard steels. Exceptionally fine loads of 25 to 3,600 grams (0.7 oz. to 7.9 lb) are applied. A microscope feature is required to measure the fine surface indentation.

The Knoop hardness tester measures the long and short diagonals produced by an elongated pyramid-form diamond using an extremely fine load.

Standard personal and equipment safe practices are followed for heat treating and hardness testing. Work areas where there are toxic fumes and poisonous bath mixtures, high temperatures, and other safety hazards require that added precautions are taken.

Heat and flame protective gloves and an apron shield, safety goggles, and hood should be used. Exhaust systems and emergency shutdown valves must be operable.

UNIT 37 REVIEW AND SELF-TEST

1. Tell how a study of the iron-carbon diagram is related to everyday applications in the heat treating of metals.
2. a. Give two probable causes of quench cracks in hardened metals.
   b. Identify two causes of stress cracks (surface failure) produced by machining a hardened workpiece.
3. State three advantages of using molten baths for heat treatment processes.
4. a. Describe martempering.
   b. Cite two desirable features associated with this process.
5. a. State the function of a thermocouple and pyrometer unit.
   b. List three different types of toolroom furnaces that are used for hardening and tempering processes.
6. a. Name two types of tempering baths and give the heat-treating temperature range of each bath.
   b. Identify the bath materials that are used in each case.
7. a. State the purpose of nitriding.
   b. List three properties of cutting tools that are improved by nitriding.
8. Give three advantages of flame hardening over conventional furnace hardening.
9. a. State the function of the subzero temperature treatment of steel.
   b. Give two applications of this process.
10. a. Explain the difference between a Rockwell C scale and a Rockwell superficial hardness scale.
    b. Give an application of each scale.
    c. Indicate the type of penetrator for the R_C and superficial scales.

11. a. Harden and temper the SAE 1041 aligning block shown in the illustration. The part is to be drawn to a Bhn 243 hardness. Consult handbook tables to establish the (1) quenching medium, (2) hardening and tempering temperatures, and (3) soaking time.
   b. Test the block after the surfaces are finish ground to check the Bhn 243 hardness. The hardness may be tested on any hardness tester available. Values other than Bhn are to be checked against the required Bhn 243 hardness.
   c. List three personal safety and/or precaution measures to take in carrying out the heat-treating operations.
   Note: Another part may be used or the processes simulated by preparing a step-by-step production plan for the aligning block.

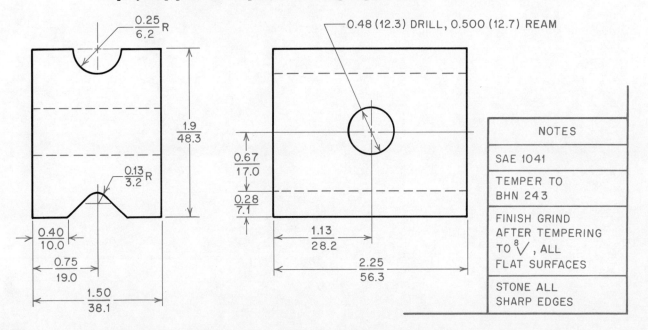

12. a. Make a second hardness test of the aligning block shown in the illustration. (A Rockwell hardness test may be substituted, if this process was not used previously.)
    b. Compare the Bhn scale value in test item 11 with the R_C scale reading.
    c. Check the accuracy of the hardness readings against a handbook table.
    d. Establish the comparable scleroscope and Vickers hardness numbers from the table.

13. List three precautions to take to avoid common problems such as soft spots, uneven hardening, and fracturing.

APPENDIX

Handbook Tables

APPENDIX
Handbook Tables

TABLE A-1 American Standard Surface Texture Values

Dimensional Measurement	Measurement Values*									
	Microinches	μm	Microinches	μm	Microinches	μm	Microinches	μm		
Roughness height (reference scales)	1	0.03	13	0.33	50	1.27	200	5.08		
	2	0.05	16†	0.41	63†	1.60	250†	6.35		
	3	0.08	20	0.51	80	2.03	320	8.13		
	4	0.10	25	0.63	100	2.54	400	10.16		
	5	0.13	32†	0.81	125†	3.18	500†	12.70		
	6	0.15					600	15.24		
	8†	0.20	40	1.02	160	4.06	800	20.32		
							1000†	25.40		
	Inches	mm	Inches	mm	Inches	mm	Inches	mm	Inches	mm
Standard roughness width cutoff	0.003	0.08	0.010	0.25	0.100	2.54	0.300	7.62	1.000	25.4
	Inches	mm	Inches	mm	Inches	mm	Inches	mm		
Waviness height (reference scales)	0.00002	0.0005	0.0001	0.0025	0.001	0.025	0.010	0.25		
	0.00003	0.0008	0.0002	0.0050	0.002	0.050	0.015	0.38		
			0.0003	0.0076	0.003	0.076	0.020	0.51		
	0.00005	0.0013	0.0005	0.0127	0.005	0.127	0.030	0.76		
	0.00008	0.0020	0.0008	0.0203	0.008	0.203				

*SI values are rounded to the number of decimal places for each measurement.
†Values are recommended.

TABLE A–2 Microinch and Micrometer (μm) Ranges of Surface Roughness for Selected Manufacturing Processes

Roughness Height in Microinches and Micrometers (μm)*

Manufacturing Process	4000 (101.60)	3000 (76.20)	2000 (50.80)	1000 (25.40)	500 (12.70)	250 (6.35)	125 (3.18)	63 (1.60)	32 (0.81)	16 (0.41)	8 (0.20)	4 (0.10)	2 (0.05)	1 (0.03)	0.5 (0.01)
Flame cutting			▒	▒	■	■	▒								
Snagging			▒	▒	■	▒	▒								
Sawing			▒	▒	■	■	■	■	▒						
Planing, shaping				▒	▒	■	■	■	■	▒					
Drilling						▒	■	■	■	▒					
Electrical discharge machining						▒	■	■	■	▒					
Milling (chemical)					▒	▒	■	■	■	■	▒				
Milling (rough)				▒	▒	■	■	■	■	▒	▒				
Broaching								■	■	▒					
Reaming								■	■	▒					
Boring, turning (finish)								▒	■	■	■	▒	▒		
Turning (rough)		▒	■	■	■	■	■	■	▒						
Barrel finishing								▒	■	■	■	▒	▒		
Electrolytic grinding									▒	■	■	▒			
Burnishing (roller)									▒	■	■	▒			
Grinding (commercial)								▒	■	■	■	■	▒	▒	
Grinding (finish)									▒	■	■	■	■	▒	▒
Honing									▒	■	■	■	▒		
Polishing										▒	■	■	■	▒	
Lapping										▒	■	■	■	▒	
Superfinishing										▒	■	■	■	▒	
Sand casting			▒	■	■	▒									
Hot rolling			▒	■	■	▒									
Forging				▒	▒	■	■	▒							
Mold casting (permanent)						▒	■	■	▒						
Extruding						▒	▒	■	■	▒					
Cold rolling (drawing)							▒	■	■	▒	▒				
Die casting							▒	■	■	▒					

Code ■ General manufacturing (average) surface finish range
      ▒ Higher or lower range produced by using special processes

*Values rounded to nearest second place μm decimal

TABLE A-3 Conversion Table for Optical Flat Fringe Bands in Inch and SI Metric Standard Units of Measure

Number of Fringe Bands	Equivalent Measurement Value*		
	Microinches	Inches	Millimeters
0.1	1.2	0.0000012	0.000029
0.2	2.3	0.0000023	0.000059
0.3	3.5	0.0000035	0.000088
0.4	4.6	0.0000046	0.000118
0.5	5.8	0.0000058	0.000147
0.6	6.9	0.0000069	0.000176
0.7	8.1	0.0000081	0.000206
0.8	9.3	0.0000093	0.000235
0.9	10.4	0.0000104	0.000264
1.0	11.6	0.0000116	0.000294
2.0	23.1	0.0000231	0.000588
3.0	34.7	0.0000347	0.000881
4.0	46.3	0.0000463	0.001175
5.0	57.8	0.0000578	0.001469
6.0	69.4	0.0000694	0.001763
7.0	81.0	0.0000810	0.002056
8.0	92.5	0.0000925	0.002350
9.0	104.1	0.0001041	0.002644
10.0	115.7	0.0001157	0.002938
11.0	127.2	0.0001272	0.003232
12.0	138.8	0.0001388	0.003525
13.0	150.4	0.0001504	0.003819
14.0	161.9	0.0001619	0.004113
15.0	173.5	0.0001735	0.004407
16.0	185.1	0.0001851	0.004700
17.0	196.6	0.0001966	0.004994
18.0	208.2	0.0002082	0.005288
19.0	219.8	0.0002198	0.005582
20.0	231.3	0.0002313	0.005876

*For a general approximation, one band may be considered to be one microinch or 0.0003mm.

TABLE A-4 Constants for Setting a 5″ Sine Bar (0°1′ to 15°60′)*

Minutes	0°	1°	2°	3°	4°	5°	6°	7°	8°	9°	10°	11°	12°	13°	14°	15°
0	0.00000	0.08725	0.17450	0.26170	0.34880	0.43580	0.52265	0.60935	0.69585	0.78215	0.86825	0.95405	1.0395	1.1247	1.2096	1.2941
1	.00145	.08870	.17595	.26315	.35025	.43725	.52410	.61080	.69730	.78360	.86965	.95545	.0410	.1261	.2110	.2955
2	.00290	.09015	.17740	.26460	.35170	.43870	.52555	.61225	.69875	.78505	.87110	.95690	.0424	.1276	.2124	.2969
3	.00435	.09160	.17885	.26605	.35315	.44015	.52700	.61370	.70020	.78650	.87255	.95835	.0438	.1290	.2138	.2983
4	.00580	.09310	.18030	.26750	.35460	.44155	.52845	.61510	.70165	.78790	.87395	.95975	.0452	.1304	.2152	.2997
5	0.00725	0.09455	0.18175	0.26895	0.35605	0.44300	0.52985	0.61655	0.70305	0.78935	0.87540	0.96120	1.0466	1.1318	1.2166	1.3011
6	.00875	.09600	.18320	.27040	.35750	.44445	.53130	.51800	.70450	.79080	.87685	.96260	.0481	.1332	.2181	.3025
7	.01020	.09745	.18465	.27185	.35895	.44590	.53275	.61945	.70595	.79225	.87825	.96405	.0495	.1346	.2195	.3039
8	.01165	.09890	.18615	.27330	.36040	.44735	.53420	.62090	.70740	.79365	.87970	.96545	.0509	.1361	.2209	.3053
9	.01310	.10035	.18760	.27475	.36185	.44880	.53565	.62235	.70885	.79510	.88115	.96690	.0523	.1375	.2223	.3067
10	0.01455	0.10180	0.18905	0.27620	0.36330	0.45025	0.53710	0.62380	0.71025	0.79655	0.88255	0.96830	1.0538	1.1389	1.2237	1.3081
11	.01600	.10325	.19050	.27765	.36475	.45170	.53855	.62520	.71170	.79795	.88400	.96975	.0552	.1403	.2251	.3095
12	.01745	.10470	.19195	.27910	.36620	.45315	.54000	.62665	.71315	.79940	.88540	.97115	.0566	.1417	.2265	.3109
13	.01890	.10615	.19340	.28055	.36765	.45460	.54145	.62810	.71460	.80085	.88685	.97260	.0580	.1431	.2279	.3123
14	.02035	.10760	.19485	.28200	.36910	.45605	.54290	.62955	.71600	.80230	.88830	.97405	.0594	.1446	.2293	.3137
15	0.02180	0.10905	0.19630	0.28345	0.37055	0.45750	0.54435	0.63100	0.71745	0.80370	0.88970	0.97545	1.0609	1.1460	1.2307	1.3151
16	.02325	.11055	.19775	.28490	.37200	.45895	.54580	.63245	.71890	.80515	.89115	.97690	.0623	.1474	.2322	.3165
17	.02475	.11200	.19920	.28635	.37345	.46040	.54725	.63390	.72035	.80660	.89260	.97830	.0637	.1488	.2336	.3179
18	.02620	.11345	.20065	.28780	.37490	.46185	.54865	.63530	.72180	.80800	.89400	.97975	.0651	.1502	.2350	.3193
19	.02765	.11490	.20210	.28925	.37635	.46330	.55010	.63675	.72320	.80945	.89545	.98115	.0665	.1516	.2364	.3207
20	0.02910	0.11635	0.20355	0.29070	0.37780	0.46475	0.55155	0.63820	0.72465	0.81090	0.89685	0.98260	1.0680	1.1531	1.2378	1.3221
21	.03055	.11780	.20500	.29220	.37925	.46620	.55300	.63965	.72610	.81230	.89830	.98400	.0694	.1545	.2392	.3235
22	.03200	.11925	.20645	.29365	.38070	.46765	.55445	.64110	.72755	.81375	.89975	.98545	.0708	.1559	.2406	.3250
23	.03345	.12070	.20795	.29510	.38215	.46910	.55590	.64255	.72900	.81520	.90115	.98685	.0722	.1573	.2420	.3264
24	.03490	.12215	.20940	.29655	.38360	.47055	.55735	.64400	.73040	.81665	.90260	.98830	.0737	.1587	.2434	.3278
25	0.03635	0.12360	0.21085	0.29800	0.38505	0.47200	0.55880	0.64540	0.73185	0.81805	0.90405	0.98970	1.0751	1.1601	1.2448	1.3292
26	.03780	.12505	.21230	.29945	.38650	.47345	.56025	.64685	.73330	.81950	.90545	.99115	.0765	.1615	.2462	.3306
27	.03925	.12650	.21375	.30090	.38795	.47490	.56170	.64830	.73475	.82095	.90690	.99255	.0779	.1630	.2477	.3320
28	.04070	.12800	.21520	.30235	.38940	.47635	.56315	.64975	.73615	.82235	.90830	.99400	.0793	.1644	.2491	.3334
29	.04220	.12945	.21665	.30380	.39085	.47780	.56455	.65120	.73760	.82380	.90975	.99540	.0808	.1658	.2505	.3348
30	0.04365	0.13090	0.21810	0.30525	0.39230	0.47925	0.56600	0.65265	0.73905	0.82525	0.91120	0.99685	1.0822	1.1672	1.2519	1.3362
31	.04510	.13235	.21955	.30670	.39375	.48070	.56745	.65410	.74050	.82665	.91260	.99825	.0836	.1686	.2533	.3376
32	.04655	.13380	.22100	.30815	.39520	.48210	.56890	.65550	.74190	.82810	.91405	.99970	.0850	.1700	.2547	.3390
33	.04800	.13525	.22245	.30960	.39665	.48355	.57035	.65695	.74335	.82955	.91545	1.0011	.0864	.1714	.2561	.3404
34	.04945	.13670	.22390	.31105	.39810	.48500	.57180	.65840	.74480	.83100	.91690	.0016	.1729	.1729	.2575	.3418
35	0.05090	0.13815	0.22535	0.31250	0.39955	0.48645	0.57325	0.65985	0.74625	0.83240	0.91835	1.0039	1.0893	1.1743	1.2589	1.3432
36	.05235	.13960	.22680	.31395	.40100	.48790	.57470	.66130	.74770	.83385	.91975	.0054	.0907	.1757	.2603	.3446
37	.05380	.14105	.22825	.31540	.40245	.48935	.57615	.66270	.74910	.83530	.92120	.0068	.0921	.1771	.2617	.3460
38	.05525	.14250	.22970	.31685	.40390	.49080	.57760	.66415	.75055	.83670	.92260	.0082	.0935	.1785	.2631	.3474
39	.05670	.14395	.23115	.31830	.40535	.49225	.57900	.66560	.75200	.83815	.92405	.0096	.0949	.1799	.2645	.3488
40	0.05820	0.14540	0.23265	0.31975	0.40680	0.49370	0.58045	0.66705	0.75345	0.83960	0.92545	1.0110	1.0964	1.1813	1.2660	1.3502
41	.05965	.14690	.23410	.32120	.40825	.49515	.58190	.66850	.75485	.84100	.92690	.0125	.0978	.1828	.2674	.3516
42	.06110	.14835	.23555	.32265	.40970	.49660	.58335	.66995	.75630	.84245	.92835	.0139	.0992	.1842	.2688	.3530
43	.06255	.14980	.23700	.32410	.41115	.49805	.58480	.67135	.75775	.84390	.92975	.0153	.1006	.1856	.2702	.3544
44	.06400	.15125	.23845	.32555	.41260	.49950	.58625	.67280	.75920	.84530	.93120	.0168	.1020	.1870	.2716	.3558
45	0.06545	0.15270	0.23990	0.32700	0.41405	0.50095	0.58770	0.67425	0.76060	0.84675	0.93230	1.0182	1.1035	1.1884	1.2730	1.3572
46	.06690	.15415	.24135	.32845	.41550	.50240	.58915	.67570	.76205	.84820	.93405	.0196	.1049	.1898	.2744	.3586
47	.06835	.15560	.24280	.32990	.41695	.50385	.59060	.67715	.76350	.84960	.93550	.0210	.1063	.1912	.2758	.3600
48	.06980	.15705	.24425	.33135	.41840	.50530	.59200	.67860	.76495	.85105	.93690	.0225	.1077	.1926	.2772	.3614
49	.07125	.15850	.24570	.33280	.41985	.50675	.59345	.68000	.76635	.85250	.93835	.0239	.1091	.1941	.2786	.3628

APPENDIX 817

TABLE A-4 Constants for Setting a 5" Sine Bar (0°1' to 15°60') (Continued)

Minutes	0°	1°	2°	3°	4°	5°	6°	7°	8°	9°	10°	11°	12°	13°	14°	15°
50	0.07270	0.15995	0.24715	0.33425	0.42130	0.50820	0.59490	0.68145	0.76780	0.85390	0.93975	1.0253	1.1106	1.1955	1.2800	1.3642
51	.07415	.16140	.24860	.33570	.42275	.50960	.59635	.68290	.76925	.85535	.94120	.0267	.1120	.1969	.2814	.3656
52	.07565	.16285	.25005	.33715	.42420	.51105	.59780	.68435	.77070	.85680	.94260	.0281	.1134	.1983	.2828	.3670
53	.07710	.16430	.25150	.33865	.42565	.51250	.59925	.68580	.77210	.85820	.94405	.0296	.1148	.1997	.2842	.3684
54	.07855	.16580	.25295	.34010	.42710	.51395	.60070	.68720	.77355	.85965	.94550	.0310	.1162	.2011	.2856	.3698
55	.08000	.16725	.25440	.34155	.42855	.51540	.60215	.68865	.77500	.86110	.94690	.0324	.1176	.2025	.2870	1.3712
56	.08145	.16870	.25585	.34300	.43000	.51685	.60355	.69010	.77645	.86250	.94835	.0338	.1191	.2039	.2884	.3726
57	.08290	.17015	.25730	.34445	.43145	.51830	.60500	.69155	.77785	.86395	.94975	.0353	.1205	.2054	.2899	.3740
58	.08435	.17160	.25875	.34590	.43290	.51975	.60645	.69300	.77930	.86540	.95120	.0367	.1219	.2068	.2913	.3754
59	.08580	.17305	.26028	.34735	.43435	.52120	.60790	.69445	.78075	.86680	.95260	.0381	.1233	.2082	.2927	.3768
60	0.08725	0.17450	0.26170	0.34880	0.43580	0.52265	0.60935	0.69585	0.78215	0.86825	0.95405	1.0395	1.1247	1.2096	1.2941	1.3782

5" sine bar applications of constant values to different sizes of sine bars, plates, and chucks

Sine Bar Size (Inches)	Constant for 5" Sine Bar
10	Multiply by 2
20	Multiply by 4
2-1/2	Multiply by 0.5
3	Multiply by 0.6
4	Multiply by 0.8

*Complete tables of sine bar constants from 0°1' to 59°60' are included in engineering and trade handbooks.

TABLE A-5 Suggested Tapping Speeds (fpm) and Cutting Fluids for Commonly Used Taps and Materials

	Carbon Steel Taps				High-Speed Steel Taps				Cutting Fluid
	Hand Tap		Gun Tap		Hand Tap		Gun Tap		
	\multicolumn{8}{c}{Thread System}	
Material	Coarse	Fine	Coarse	Fine	Coarse	Fine	Coarse	Fine	
	\multicolumn{8}{c}{Feet per Minute}	
Stainless steel	20	25	20	25	15	20	20	25	Sulphur-base oil
Drop forgings					50	55	60	65	Soluble oil or sulphur-base oil
Aluminum and die castings	35	40	35	40	70	90	80	100	Kerosene and lard oil
Mild steel	35	40	35	40	70	80	90	100	Sulphur-base oil
Cast iron					80	90	100	110	Dry or soluble oil
Brass	70	80	70	80	160	180	200	220	Soluble or light oil

TABLE A-6 Suggested Starting Speeds and Feeds for High-Speed Steel End Mill Applications

	Materials					
	Heat-Resistant Cobalt-Base Alloys, High Tensile Steels (50-55C)		Heat-Resistant Austenitic Alloys, High Tensile Steels (46-50C)		Heat-Resistant Nickel-Base Alloys, High-Strength Stainless Steels, High-Strength Titanium Alloys	
	Types and Styles of End Mills					
	Premium Cobalt High-Speed Steel 2 or More Flutes					
Diameter of End Mills (in Inches)	Speed 5-10 sfpm RPM	Feed Chip Load per Tooth	Speed 10-15 sfpm RPM	Feed Chip Load per Tooth	Speed 15-20 sfpm RPM	Feed Chip Load per Tooth
1/16	*	*	*	*		
3/32	*	*	*	*	611-815	.0002-.0005
1/8	*	*	*	*	456-611	.0002-.0005
3/16	*	*	204-306	.0002-.0005	306-407	.0002-.0005
1/4	76-153	.0002-.001	153-230	.0002-.001	229-306	.0002-.001
5/16	61-122	.0002-.001	122-183	.0002-.001	183-244	.0002-.001
3/8	51-102	.0002-.001	102-153	.0002-.001	153-203	.0002-.001
7/16	44-88	.0005-.001	88-132	.0005-.001	131-175	.0005-.002
1/2	38-76	.0005-.001	76-115	.0005-.001	115-153	.0005-.002
9/16	34-68	.0005-.002	68-104	.0005-.002	104-136	.0005-.002
5/8	31-61	.0005-.002	61-92	.0005-.002	92-122	.0005-.002
11/16	28-56	.0005-.002	56-84	.0005-.002	84-111	.0005-.002
3/4	26-51	.0005-.002	51-76	.0005-.002	76-102	.001-.004
13/16	24-47	.001-.003	47-71	.001-.003	71-94	.001-.004
7/8	22-44	.001-.003	44-65	.001-.003	65-87	.001-.004
15/16	20-40	.001-.003	40-62	.001-.003	62-81	.001-.004
1	19-38	.001-.003	38-58	.001-.003	58-76	.001-.004
1-1/8	34	.0015-.004	34-51	.0015-.004	51-68	.0015-.005
1-1/4	31	.0015-.004	31-46	.0015-.004	46-61	.0015-.005
1-3/8	28	.0015-.004	28-42	.0015-.004	42-55	.0015-.005
1-1/2	26	.0015-.004	26-38	.0015-.004	38-51	.002 Up
1-5/8	24	.002 Up	35	.002 Up	35-47	.002 Up
1-3/4	22	.002 Up	32	.002 Up	32-43	.002 Up
1-7/8	20	.002 Up	30	.002 Up	30-40	.003 Up
2	19	.002 Up	29	.003 Up	29-38	.003 Up
2-1/8	18	.003 Up	28	.003 Up	36	.003 Up
2-1/4	17	.003 Up	26	.003 Up	34	.003 Up
2-3/8	16	.003 Up	25	.003 Up	32	.003 Up
2-1/2	15	.003 Up	23	.003 Up	30	.003 Up
2-5/8	15	.003 Up	22	.003 Up	29	.003 Up
2-3/4	14	.003 Up	21	.003 Up	28	.003 Up
2-7/8	14	.003 Up	20	.003 Up	27	.003 Up
3	13	.003 Up	19	.003 Up	26	.003 Up

*Use solid carbide end mills in small-diameter applications for materials that are harder than 46 C Rockwell.
Note: The speeds and feeds apply also to cavity and slotting cuts up to the depth equal to the diameter of the end mill. Deeper cuts require these values to be decreased.
Table adapted with permission from The Cleveland Twist Drill Co.

TABLE A-6 *Suggested Starting Speeds and Feeds for High-Speed Steel End Mill Applications (Continued)*

	Materials					
	High-Strength Stainless Steels, High Tensile Steels (40-46C) Medium-Strength Titanium Alloys		Heat-Resistant Ferritic-Base Alloys, Medium-Strength Stainless Steels, Unalloyed Titanium Tool Steels (30-40C)		Machine Steel, Hard Brass and Bronze, Electrolytic Copper, Mild Steel Forgings (20-30C)	
	Types and Styles of End Mills					
	High-Speed Steel End Mills or Premium Cobalt High-Speed Steel Types, 2 or More Flutes		High-Speed Steel End Mills, 2 or More Flutes			
Diameter of End Mills (in Inches)	Speed 20-40 sfpm RPM	Feed Chip Load per Tooth	Speed 40-60 sfpm RPM	Feed Chip Load per Tooth	Speed 60-80 sfpm RPM	Feed Chip Load per Tooth
1/16	1222-2444	.0002-.0005	2444-3667	.0002-.0005	3667-4888	.0002-.0005
3/32	815-1629	.0002-.0005	1629-2750	.0002-.0005	2750-3259	.0002-.0005
1/8	611-1222	.0002-.0005	1222-1833	.0002-.0005	1833-2440	.0002-.001
3/16	407-815	.0002-.0005	815-1222	.0002-.0005	1222-1625	.0002-.001
1/4	306-611	.0002-.001	611-917	.0002-.001	917-1222	.0005-.002
5/16	244-489	.0002-.001	489-733	.0002-.001	733-978	.0005-.002
3/8	203-407	.0005-.002	407-611	.0005-.002	611-815	.001-.003
7/16	175-349	.0005-.002	349-524	.0005-.002	524-698	.001-.003
1/2	153-306	.0005-.003	306-458	.001-.003	458-611	.001-.003
9/16	136-272	.0005-.003	272-412	.001-.003	412-543	.001-.004
5/8	122-244	.001-.004	244-367	.001-.004	367-489	.001-.004
11/16	111-222	.001-.004	222-337	.001-.004	337-444	.001-.004
3/4	102-203	.001-.004	203-306	.001-.004	306-407	.001-.004
13/16	94-189	.001-.004	189-284	.001-.004	284-379	.002-.006
7/8	87-175	.001-.004	175-262	.002-.006	262-349	.002-.006
15/16	81-163	.001-.004	163-246	.002-.006	246-326	.002-.006
1	76-153	.002-.006	153-229	.002-.006	229-306	.002-.006
1-1/8	68-136	.002-.006	136-204	.002-.006	204-272	.002-.006
1-1/4	61-122	.002-.006	122-183	.002-.006	183-244	.003 Up
1-3/8	55-111	.002-.006	111-167	.003 Up	167-222	.003 Up
1-1/2	51-102	.003 Up	102-153	.003 Up	153-204	.003 Up
1-5/8	47-94	.003 Up	94-141	.003 Up	141-188	.003 Up
1-3/4	43-87	.003 Up	87-131	.003 Up	131-175	.003 Up
1-7/8	40-81	.003 Up	81-122	.003 Up	122-163	.003 Up
2	38-76	.003 Up	76-115	.003 Up	115-153	.003 Up
2-1/8	36-72	.003 Up	72-108	.003 Up	108-144	.003 Up
2-1/4	34-68	.003 Up	68-102	.003 Up	102-136	.003 Up
2-3/8	32-64	.003 Up	64-97	.003 Up	97-128	.003 Up
2-1/2	30-61	.003 Up	61-92	.003 Up	92-122	.003 Up
2-5/8	29-58	.003 Up	58-88	.003 Up	88-116	.003 Up
2-3/4	28-56	.003 Up	56-83	.003 Up	83-111	.003 Up
2-7/8	27-53	.003 Up	53-80	.003 Up	80-106	.003 Up
3	26-51	.003 Up	51-76	.003 Up	76-102	.003 Up

Note: The speeds and feeds apply also to cavity and slotting cuts up to the depth equal to the diameter of the end mill. Deeper cuts require these values to be decreased.
Table adapted with permission from The Cleveland Twist Drill Co.

	Materials					
	Cast Iron, Mild Steel, Half-Hard Brass and Bronze		Brass, Bronze, Alloyed Aluminum, Abrasive Plastics		Aluminum, Plastics, Wood	
	Types and Styles of End Mills					
	High-Speed Steel End Mills, 2 or More Flutes Surface Treatment Helpful in C. I. Applications		High-Speed Steel End Mills of High-Helix Type, 1 to 6 Flutes			
Diameter of End Mills (in Inches)	Speed 80-100 sfpm RPM	Feed Chip Load per Tooth	Speed 100-200 sfpm RPM	Feed Chip Load per Tooth	Speed 200-600 sfpm RPM	Feed Chip Load per Tooth
1/16	4888-6111	.0002-.0005	6111-12222	.0002-.0005	12222 Up	.0002-.0005
3/32	3259-4073	.0002-.0005	4073-8146	.0002-.0005	8146 Up	.0002-.0005
1/8	2440-3056	.0002-.001	3056-6112	.0002-.001	6112 Up	.0002-.001
3/16	1625-2037	.0002-.001	2037-4074	.0002-.001	4074-12222	.0002-.001
1/4	1222-1528	.0005-.002	1528-3056	.0005-.002	3056-9168	.0005-.002
5/16	978-1222	.0005-.002	1222-2444	.0005-.002	2444-7332	.0005-.002
3/8	815-1019	.001-.003	1019-2038	.0005-.003	2038-6114	.0005-.002
7/16	698-873	.001-.003	873-1746	.0005-.003	1746-5238	.0005-.002
1/2	611-764	.001-.003	764-1528	.0005-.003	1528-4584	.0005-.002
9/16	543-678	.001-.004	678-1356	.0005-.004	1356-4071	.0005-.003
5/8	489-611	.001-.004	611-1222	.0005-.004	1222-3666	.0005-.003
11/16	444-555	.001-.004	555-1110	.0005-.004	1110-3330	.0005-.003
3/4	407-509	.002-.006	509-1018	.001-.006	1018-3054	.001-.004
13/16	379-469	.002-.006	469-938	.001-.006	938-2814	.001-.004
7/8	349-436	.002-.006	436-872	.001-.006	872-2616	.001-.004
15/16	326-407	.002-.006	407-814	.001-.006	814-2442	.001-.004
1	306-382	.002-.006	382-764	.002 Up	764-2292	.002 Up
1-1/8	272-340	.003 Up	340-680	.002 Up	680-2040	.002 Up
1-1/4	244-306	.003 Up	306-612	.002 Up	612-1836	.002 Up
1-3/8	222-278	.003 Up	278-556	.002 Up	556-1668	.002 Up
1-1/2	204-255	.003 Up	255-510	.003 Up	510-1530	.002 Up
1-5/8	188-235	.003 Up	235-470	.003 Up	470-1410	.002 Up
1-3/4	175-218	.003 Up	218-436	.003 Up	436-1308	.002 Up
1-7/8	163-204	.003 Up	204-408	.003 Up	408-1224	.003 Up
2	153-191	.003 Up	191-382	.003 Up	382-1146	.003 Up
2-1/8	144-179	.003 Up	179-358	.003 Up	358-1074	.003 Up
2-1/4	136-170	.003 Up	170-340	.003 Up	340-1020	.003 Up
2-3/8	128-161	.003 Up	161-322	.003 Up	322-966	.003 Up
2-1/2	122-153	.003 Up	153-306	.003 Up	306-918	.003 Up
2-5/8	116-145	.003 Up	145-290	.003 Up	290-870	.003 Up
2-3/4	111-139	.003 Up	139-278	.003 Up	278-834	.003 Up
2-7/8	106-132	.003 Up	132-264	.003 Up	264-792	.003 Up
3	102-127	.003 Up	127-254	.003 Up	254-762	.003 Up

Note: The speeds and feeds apply also to cavity and slotting cuts up to the depth equal to the diameter of the end mill. Deeper cuts require these values to be decreased.
Table adapted with permission from The Cleveland Twist Drill Co.

TABLE A-7 ISO Metric Screw Thread Tap Drill Sizes in Millimeter and Inch Equivalents with Probable Hole Sizes and Percent of Thread

ISO Metric Tap Size	Recommended Metric Drill				Closest Recommended Inch Drill			
	Drill Size* (mm)	Inch Equivalent	Probable Hole Size (Inches)	Probable Percent of Thread	Drill Size	Inch Equivalent	Probable Hole Size (Inches)	Probable Percent of Thread
M1.6 × 0.35	1.25	0.0492	0.0507	69				
M1.8 × 0.35	1.45	0.0571	0.0586	69				
M2 × 0.4	1.60	0.0630	0.0647	69	#52	0.0635	0.0652	66
M2.2 × 0.45	1.75	0.0689	0.0706	70				
M2.5 × 0.45	2.05	0.0807	0.0826	69	#46	0.0810	0.0829	67
M3 × 0.5	2.50	0.0984	0.1007	68	#40	0.0980	0.1003	70
M3.5 × 0.6	2.90	0.1142	0.1168	68	#33	0.1130	0.1156	72
M4 × 0.7	3.30	0.1299	0.1328	69	#30	0.1285	0.1314	73
M4.5 × 0.75	3.70	0.1457	0.1489	74	#26	0.1470	0.1502	70
M5 × 0.8	4.20	0.1654	0.1686	69	#19	0.1660	0.1692	68
M6 × 1	5.00	0.1968	0.2006	70	#9	0.1960	0.1998	71
M7 × 1	6.00	0.2362	0.2400	70	15/64	0.2344	0.2382	73
M8 × 1.25	6.70	0.2638	0.2679	74	17/64	0.2656	0.2697	71
M8 × 1	7.00	0.2756	0.2797	69	J	0.2770	0.2811	66
M10 × 1.5	8.50	0.3346	0.3390	71	Q	0.3320	0.3364	75
M10 × 1.25	8.70	0.3425	0.3471	73	11/32	0.3438	0.3483	71
M12 × 1.75	10.20	0.4016	0.4063	74	Y	0.4040	0.4087	71
M12 × 1.25	10.80	0.4252	0.4299	67	27/64	0.4219	0.4266	72
M14 × 2	12.00	0.4724	0.4772	72	15/32	0.4688	0.4736	76
M14 × 1.5	12.50	0.4921	0.4969	71				
M16 × 2	14.00	0.5512	0.5561	72	35/64	0.5469	0.5518	76
M16 × 1.5	14.50	0.5709	0.5758	71				
M18 × 2.5	15.50	0.6102	0.6152	73	39/64	0.6094	0.6144	74
M18 × 1.5	16.50	0.6496	0.6546	70				
M20 × 2.5	17.50	0.6890	0.6942	73	11/16	0.6875	0.6925	74
M20 × 1.5	18.50	0.7283	0.7335	70				
M22 × 2.5	19.50	0.7677	0.7729	73	49/64	0.7656	0.7708	75
M22 × 1.5	20.50	0.8071	0.8123	70				
M24 × 3	21.00	0.8268	0.8327	73	53/64	0.8281	0.8340	72
M24 × 2	22.00	0.8661	0.8720	71				
M27 × 3	24.00	0.9449	0.9511	73	15/16	0.9375	0.9435	78
M27 × 2	25.00	0.9843	0.9913	70	63/64	0.9844	0.9914	70
M30 × 3.5	26.50	1.0433						
M30 × 2	28.00	1.1024						
M33 × 3.5	29.50	1.1614						
M33 × 2	31.00	1.2205						
M36 × 4	32.00	1.2598						
M36 × 3	33.00	1.2992						
M39 × 4	35.00	1.3780						
M39 × 3	36.00	1.4173						

Formula for Metric Tap Drill Size:

$$\frac{\text{basic major diameter}}{(\text{mm})} - \frac{\% \text{ thread} \times \text{pitch (mm)}}{76.980} = \frac{\text{drilled hole size}}{(\text{mm})}$$

Formula for Percent of Thread:

$$\frac{76.980}{\text{pitch (mm)}} \times \left(\frac{\text{basic major diameter}}{(\text{mm})} - \frac{\text{drilled hole size}}{(\text{mm})}\right) = \text{percent of thread}$$

*Reaming is recommended to the drill size as given.
Table adapted with permission from TRW: Greenfield Tap & Die Division and Geometric Tool Division

TABLE A-8 Constants for Calculating Screw Thread Elements (Inches) for ISO Metric Screw Threads (0.25mm to 6mm Pitches)

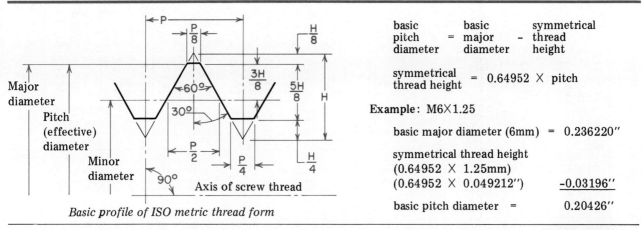

Basic profile of ISO metric thread form

basic pitch diameter = basic major diameter − symmetrical thread height

symmetrical thread height = 0.64952 × pitch

Example: M6×1.25

basic major diameter (6mm) = 0.236220″

symmetrical thread height
(0.64952 × 1.25mm)
(0.64952 × 0.049212″)   −0.03196″

basic pitch diameter = 0.20426″

Pitch (P)		Symmetrical Thread Height $2\left(\frac{3H}{8}\right) = \frac{0.64952P}{\text{(Inches)}}$	Height of Sharp V Thread $H = 0.866025P$ (Inches)	Double Height Internal Thread $2\left(\frac{5H}{8}\right) = \frac{5H}{4}$ $1.08253P$ (Inches)
mm	Inches			
0.25	0.009842	0.006392	0.008523	0.010654
0.3	0.011811	0.007671	0.010229	0.012786
0.35	0.013779	0.008950	0.011933	0.014916
0.4	0.015748	0.010229	0.013638	0.017048
0.45	0.017716	0.011507	0.015342	0.019178
0.5	0.019685	0.012786	0.017048	0.021310
0.6	0.023622	0.015343	0.020457	0.025571
0.7	0.027559	0.017900	0.023867	0.029833
0.75	0.029527	0.019178	0.025571	0.031964
0.8	0.031496	0.020457	0.027276	0.034095
1	0.039370	0.025572	0.034096	0.042619
1.25	0.049212	0.031964	0.042619	0.053273
1.5	0.059055	0.038357	0.051143	0.063929
1.75	0.068897	0.044750	0.059667	0.074583
2	0.078740	0.051143	0.068191	0.085238
2.5	0.098425	0.063929	0.085239	0.106548
3	0.118110	0.076715	0.102287	0.127858
3.5	0.137795	0.089501	0.119335	0.149167
4	0.157480	0.102286	0.136382	0.170477
4.5	0.177165	0.115072	0.153430	0.191786
5	0.196850	0.127858	0.170478	0.213096
5.5	0.216535	0.140644	0.187526	0.234406
6	0.236220	0.153430	0.204574	0.255715

TABLE A-9 Major, Pitch, and Minor Diameters (in mm) of Internal SI Metric Threads for Fine, Medium, and Coarse Fits

Nominal Diameter (mm)	Pitch (mm)	Close (Fine) Fit = 4H5H					Medium Fit = 6H					Loose (Coarse) Fit = 7H			
		Major Diameter Min. (must clear)	Pitch Diameter Min.*	Pitch Diameter Max.†	Minor Diameter Min.	Minor Diameter Max.	Major Diameter Min. (must clear)	Pitch Diameter Min.*	Pitch Diameter Max.†	Minor Diameter Min.	Minor Diameter Max.	Major Diameter Min. (must clear)	Pitch Diameter Min.*	Pitch Diameter Max.†	Minor Diameter Min.	Minor Diameter Max.
3.0	0.5	3.0	2.675	2.738	2.459	2.571	3.0	2.675	2.775	2.459	2.599	2.800	2.675	2.459	2.639
3.5	0.6	3.5	3.110	3.181	2.850	2.975	3.5	3.110	3.222	2.850	3.010	3.250	3.110	2.850	3.050
4.0	0.7	4.0	3.545	3.620	3.242	3.382	4.0	3.545	3.663	3.242	3.422	3.695	3.545	3.242	3.466
4.5	0.75	4.5	4.013	4.088	3.688	3.838	4.5	4.013	4.131	3.688	3.878	4.163	4.013	3.688	3.924
5.0	0.8	5.0	4.480	4.560	4.134	4.294	5.0	4.480	4.605	4.134	4.334	4.640	4.480	4.134	4.384
6.0	1.0	6.0	5.350	5.445	4.917	5.107	6.0	5.350	5.500	4.917	5.153	5.540	5.350	4.917	5.217
7.0	1.0	7.0	6.350	6.445	5.917	6.107	7.0	6.350	6.500	5.917	6.153	6.540	6.350	5.917	6.217
8.0	1.25	8.0	7.188	7.288	6.647	6.859	8.0	7.188	7.348	6.647	6.912	7.388	7.188	6.647	6.982
9.0	1.25	9.0	8.188	8.288	7.647	7.859	9.0	8.188	8.348	7.647	7.912	8.388	8.188	7.647	7.982
10.0	1.5	10.0	9.026	9.138	8.376	8.612	10.0	9.026	9.206	8.376	8.676	9.250	9.026	8.376	8.751
11.0	1.5	11.0	10.026	10.138	9.376	9.612	11.0	10.026	10.206	9.376	9.676	10.250	10.026	9.376	9.751
12.0	1.75	12.0	10.863	10.988	10.106	10.371	12.0	10.863	11.063	10.106	10.441	11.113	10.863	10.106	10.531
14.0	2.0	14.0	12.701	12.833	11.835	12.135	14.0	12.701	12.913	11.835	12.210	12.966	12.701	11.835	12.310
16.0	2.0	16.0	14.701	14.833	13.835	14.135	16.0	14.701	14.913	13.835	14.210	14.966	14.701	13.835	14.310
18.0	2.5	18.0	16.376	16.516	15.294	15.649	18.0	16.376	16.600	15.294	15.744	16.656	16.376	15.294	15.854
20.0	2.5	20.0	18.376	18.516	17.294	17.649	20.0	18.376	18.600	17.294	17.744	18.656	18.376	17.294	17.854
22.0	2.5	22.0	20.376	20.516	19.294	19.649	22.0	20.376	20.600	19.294	19.744	20.656	20.376	19.294	19.854
24.0	3.0	24.0	22.051	22.221	20.752	21.152	24.0	22.051	22.316	20.752	21.252	22.386	22.051	20.752	21.382
27.0	3.0	27.0	25.051	25.221	23.752	24.152	27.0	25.051	25.316	23.752	24.252	25.386	25.051	23.752	24.382
30.0	3.5	30.0	27.727	27.907	26.211	26.661	30.0	27.727	28.007	26.211	26.771	28.082	27.727	26.211	26.921
33.0	3.5	33.0	30.727	30.907	29.211	29.661	33.0	30.727	31.007	29.211	29.771	31.082	30.727	29.211	29.921
36.0	4.0	36.0	33.402	33.592	31.670	32.145	36.0	33.402	33.702	31.670	32.270	33.777	33.402	31.670	32.420
39.0	4.0	39.0	36.402	36.592	34.670	35.145	39.0	36.402	36.702	34.670	35.270	36.777	36.402	34.670	35.420
42.0	4.5	42.0	39.077	39.277	37.129	37.659	42.0	39.077	39.392	37.129	37.799	39.477	39.077	37.129	37.979
45.0	4.5	45.0	42.077	42.277	40.129	40.659	45.0	42.077	42.392	40.129	40.799	42.477	42.077	40.129	40.979
48.0	5.0	48.0	44.752	44.964	42.587	43.147	48.0	44.752	45.087	42.587	43.297	45.177	44.752	42.587	43.487
52.0	5.0	52.0	48.752	48.964	46.587	47.147	52.0	48.752	49.087	46.587	47.297	49.177	48.752	46.587	47.487
56.0	5.5	56.0	52.428	52.652	50.046	50.646	56.0	52.428	52.783	50.046	50.796	52.878	52.428	50.046	50.996
60.0	5.5	60.0	56.428	56.652	54.046	54.646	60.0	56.428	56.783	54.046	54.796	56.878	56.428	54.046	54.996
64.0	6.0	64.0	60.103	60.339	57.505	58.135	64.0	60.103	60.478	57.505	58.305	60.578	60.103	57.505	58.505
68.0	6.0	68.0	64.103	64.339	61.505	62.135	68.0	64.103	64.478	61.505	62.305	64.578	64.103	61.505	62.505

*GO gage pitch diameter
†NO-GO gage pitch diameter

APPENDIX 825

TABLE A-10 Size and Tolerance Limits of Unified Miniature Screw Threads

Size Designation*	Metric Pitch (mm)	Metric-Standard External Threads (mm)							Metric-Standard Internal Threads (mm)							Lead Angle at Basic Pitch Diameter
		Major Diameter		Pitch Diameter		Minor Diameter		Minor Diameter		Pitch Diameter		Major Diameter		
		Max.ᵃ	Min.	Max.ᵃ	Min.	Max.ᵇ	Min.ᶜ	Min.ᵃ	Max.	Min.ᵃ	Max.	Min.ᵈ	Max.ᶜ	
.30 UNM	.080	.300	.284	.248	.234	.204	.183	.217	.254	.248	.262	.306	.327	5° 52'
.35 UNM	.090	.350	.333	.292	.277	.242	.220	.256	.297	.292	.307	.356	.380	5° 37'
.40 UNM	.100	.400	.382	.335	.319	.280	.256	.296	.340	.335	.351	.407	.432	5° 26'
.45 UNM	.100	.450	.432	.385	.369	.330	.306	.346	.390	.385	.401	.457	.482	4° 44'
.50 UNM	.125	.500	.479	.419	.401	.350	.322	.370	.422	.419	.437	.509	.538	5° 26'
.55 UNM	.125	.550	.529	.469	.451	.400	.372	.420	.472	.469	.487	.559	.588	4° 51'
.60 UNM	.150	.600	.576	.503	.483	.420	.388	.444	.504	.503	.523	.611	.644	5° 26'
.70 UNM	.175	.700	.673	.586	.564	.490	.454	.518	.586	.586	.608	.713	.750	5° 26'
.80 UNM	.200	.800	.770	.670	.646	.560	.520	.592	.668	.670	.694	.814	.856	5° 26'
.90 UNM	.225	.900	.867	.754	.728	.630	.586	.666	.750	.754	.780	.916	.962	5° 26'
1.00 UNM	.250	1.000	.964	.838	.810	.700	.652	.740	.832	.838	.866	1.018	1.068	5° 26'
1.10 UNM	.250	1.100	1.064	.938	.910	.800	.752	.840	.932	.938	.966	1.118	1.168	4° 51'
1.20 UNM	.250	1.200	1.164	1.038	1.010	.900	.852	.940	1.032	1.038	1.066	1.218	1.268	4° 23'
1.40 UNM	.300	1.400	1.358	1.205	1.173	1.040	.984	1.088	1.196	1.205	1.237	1.422	1.480	4° 32'

Size Designation*	Threads Per Inch	Inch-Standard External Threads (0.001″)							Inch-Standard Internal Threads (0.001″)							Lead Angle at Basic Pitch Diameter
		Major Diameter		Pitch Diameter		Minor Diameter		Minor Diameter		Pitch Diameter		Major Diameter		
		Max.ᵃ	Min.	Max.ᵃ	Min.	Max.ᵇ	Min.ᶜ	Min.ᵃ	Max.	Min.ᵃ	Max.	Min.ᵈ	Max.ᶜ	
.30 UNM	318	.0118	.0112	.0098	.0092	.0080	.0072	.0085	.0100	.0098	.0104	.0120	.0129	5° 52'
.35 UNM	282	.0138	.0131	.0115	.0109	.0095	.0086	.0101	.0117	.0115	.0121	.0140	.0149	5° 37'
.40 UNM	254	.0157	.0150	.0132	.0126	.0110	.0101	.0117	.0134	.0132	.0138	.0160	.0170	5° 26'
.45 UNM	254	.0177	.0170	.0152	.0145	.0130	.0120	.0136	.0154	.0152	.0158	.0180	.0190	4° 44'
.50 UNM	203	.0197	.0189	.0165	.0158	.0138	.0127	.0146	.0166	.0165	.0172	.0200	.0212	5° 26'
.55 UNM	203	.0217	.0208	.0185	.0177	.0157	.0146	.0165	.0186	.0185	.0192	.0220	.0231	4° 51'
.60 UNM	169	.0236	.0227	.0198	.0190	.0165	.0153	.0175	.0198	.0198	.0206	.0240	.0254	5° 26'
.70 UNM	145	.0276	.0265	.0231	.0222	.0193	.0179	.0204	.0231	.0231	.0240	.0281	.0295	5° 26'
.80 UNM	127	.0315	.0303	.0264	.0254	.0220	.0205	.0233	.0263	.0264	.0273	.0321	.0337	5° 26'
.90 UNM	113	.0354	.0341	.0297	.0287	.0248	.0231	.0262	.0295	.0297	.0307	.0361	.0379	5° 26'
1.00 UNM	102	.0394	.0380	.0330	.0319	.0276	.0257	.0291	.0327	.0330	.0341	.0401	.0420	5° 26'
1.10 UNM	102	.0433	.0419	.0369	.0358	.0315	.0296	.0331	.0367	.0369	.0380	.0440	.0460	4° 51'
1.20 UNM	102	.0472	.0458	.0409	.0397	.0354	.0335	.0370	.0406	.0409	.0420	.0480	.0499	4° 23'
1.40 UNM	85	.0551	.0535	.0474	.0462	.0409	.0387	.0428	.0471	.0474	.0487	.0560	.0583	4° 32'

*Boldface denotes preferred series.
ᵃBasic dimension.
ᵇUse minimum minor diameter of internal thread for mechanical gaging.
ᶜReference only. The thread tool form is relied on for this limit.
ᵈReference only. Use the maximum major diameter of the external thread for gaging.

TABLE A-11 Basic Formulas and Dimensions of General-Purpose Acme Screw Threads (ANSI)

Design Features →	Threads per Inch	Pitch	Height of Thread (Basic)	Total Height of Thread	Thread Thickness (Basic)	Width of Flat — Crest of Internal Thread (Basic)	Width of Flat — Root of Internal Thread
Formulas →	n	P = 1/n	P/2	P/2 + 1/2 allowance*	P/2	0.3707P	0.3707P − (0.259 × allowance)*
Basic dimensions (decimal-inch values rounded to four places)	16	.0625	.0313	.0362	.0313	.0232	.0206
	14	.0714	.0357	.0407	.0357	.0265	.0239
	12	.0833	.0417	.0467	.0417	.0309	.0283
	10	.1000	.0500	.0600	.0500	.0371	.0319
	8	.1250	.0625	.0725	.0625	.0463	.0411
	6	.1667	.0833	.0933	.0833	.0618	.0566
	5	.2000	.1000	.1100	.1000	.0741	.0689
	4	.2500	.1250	.1350	.1250	.0927	.0875
	3	.3333	.1667	.1767	.1667	.1236	.1184
	2-1/2	.4000	.2000	.2100	.2000	.1483	.1431
	2	.5000	.2500	.2600	.2500	.1853	.1802
	1-1/2	.6667	.3333	.3433	.3333	.2471	.2419
	1-1/3	.7500	.3750	.3850	.3750	.2780	.2728
	1	1.0000	.5000	.5100	.5000	.3707	.3655

*The allowance for 10 or more threads per inch is 0.010″; for less than 10 threads per inch, 0.020″.

TABLE A-12 Suggested Starting Points for Cutting Speeds on Turret Lathes for Basic External Cuts

Machinability Class	Nature of Cut	Turning — Overhead and Cross Slide (Carbide)	Turning — Roller Turners (High-Speed Steel)	Turning — Roller Turners (Carbide)	Facing (Carbide)	Forming (High-Speed Steel)	Forming (Carbide)	Cutting Off (High-Speed Steel)	Cutting Off (Carbide)
A	Roughing	490	150	490	490	150	490	120	500
A	Finishing	560			560				
B	Roughing	420	120	420	420	120	420	100	500
B	Finishing	490			490				
C	Roughing	320	90	320	320	90	320	80	400
C	Finishing	420			420				
D	Roughing	245	70	245	245	70	245	60	400
D	Finishing	320			320				
E	Roughing	220	50	220	220	50	220	50	300
E	Finishing	230			230				
F	Roughing	Max.			Max.	250	Max.	Max.	Max.
F	Finishing	Max.			Max.				

TABLE A-13 Suggested Starting Feed Rates on Ram-Type Turret Lathes for External Cuts

Depth of Cut	Turning - Overhead and Cross Slide - Hexagon Turret	Turning - Overhead and Cross Slide - Square Turret	Turning - Roller Turners	Facing	Forming - Steel	Forming - Cast Iron	Cutting Off - High-Speed Steel	Cutting Off - Carbide Insert
				Feed per Revolution of Spindle				
1/16"	0.030"	0.019"	0.030"	0.017"				
1.5mm	0.8mm	0.5mm	0.8mm	0.4mm				
1/8"	0.018"	0.019"	0.030"	0.017"				
3mm	0.5mm	0.5mm	0.8mm	0.4mm				
1/4"	0.012"	0.012"	0.018"	0.010"	0.004"	0.006"		
6mm	0.3mm	0.3mm	0.5mm	0.3mm	0.1mm	0.2mm		
3/8"	0.0075"	0.0075"	0.012"	0.006"			Stock to 1" (25.4mm) diameter 0.0045" (0.1mm)	Machinability Class: A, B, F 0.003" 0.1mm C, D 0.004" 0.1mm E 0.005" 0.1mm
9.5mm	0.2mm	0.2mm	0.3mm	0.2mm				
1/2"	0.0045"	0.0045"	0.0075"	0.0045"				
12.7mm	0.1mm	0.1mm	0.2mm	0.1mm				
5/8"		0.0045"	0.0045"	0.0045"	0.003"			
15.6mm		0.1mm	0.1mm	0.1mm	0.1mm			
3/4"								
19mm								
1"					Over 3/4" (19mm) Cut 1: 0.004" 0.1mm Cut 2: 0.006" 0.2mm	Over 3/4" (19mm) Cut 1: 0.004" 0.1mm Cut 2: 0.0075" 0.2mm	Stock over 1" (25.4mm) diameter 0.0065" (0.2mm)	
25.4mm								

Note: Millimeter values are rounded to nearest 0.1mm.

TABLE A-14 Suggested Starting Points for Cutting Speeds on Turret Lathes for Basic Internal Cuts

Machinability Class	Starting Cutting Speeds in Surface Feet per Minute (sfpm)					Reaming (HSS Reamers)*
	Drilling (HSS Drills)			Boring (Carbide Cutters)		
	Regular Drill	Core Drill				
		Cored Hole	Drilled Hole			
A	120	120	150	Roughing	490	60
				Finishing	560	
B	80	100	120	Roughing	420	50
				Finishing	490	
C	60	80	90	Roughing	350	40
				Finishing	420	
D	50	60	70	Roughing	260	30
				Finishing	320	
E	50	50	50	Roughing	175	25
				Finishing	230	
F	200	200	250	Roughing	Max.	120
				Finishing	Max.	

*Finishing reamers that remove only 0.003" (0.1mm) to 0.015" (0.4mm) total stock

TABLE A-15 Allowances and Tolerances on Reamed Holes for General Classes of Fits

Class of Fit	*Allowances and Tolerances	*Nominal Diameters					
		Up to 1/2"	9/16" through 1"	1-1/16" through 2"	2-1/16" through 3"	3-1/16" through 4"	4-1/16" through 5"
A	High limit (+) Low limit (−) Tolerance	.0002 .0002 .0004	.0005 .0002 .0007	.0007 .0002 .0009	.0010 .0005 .0015	.0010 .0005 .0015	.0010 .0005 .0015
B	High limit (+) Low limit (−) Tolerance	.0005 .0005 .0010	.0007 .0005 .0012	.0010 .0005 .0015	.0012 .0007 .0019	.0015 .0007 .0022	.0017 .0007 .0024
	Allowances and Tolerances for Forced Fits							
F	High limit (+) Low limit (+) Tolerance	.0010 .0005 .0005	.0020 .0015 .0005	.0040 .0030 .0010	.0060 .0045 .0015	.0080 .0060 .0020	.0100 .0080 .0020
	Allowances and Tolerances for Driving Fits							
D	High limit (+) Low limit (+) Tolerance	.0005 .0002 .0003	.0010 .0007 .0003	.0015 .0010 .0005	.0025 .0015 .0010	.0030 .0020 .0010	.0035 .0025 .0010
	Allowances and Tolerances for Push Fits							
P	High limit (−) Low limit (−) Tolerance	.0002 .0007 .0005	.0002 .0007 .0005	.0002 .0007 .0005	.0005 .0010 .0005	.0005 .0010 .0005	.0005 .0010 .0005
	Allowances and Tolerances for Running Fits							
X	High limit (−) Low limit (−) Tolerance	.0010 .0020 .0010	.0012 .0027 .0015	.0017 .0035 .0018	.0020 .0042 .0022	.0025 .0050 .0025	.0030 .0057 .0027
Y (average machine work)	High limit (−) Low limit (−) Tolerance	.0007 .0012 .0005	.0010 .0020 .0010	.0012 .0025 .0013	.0015 .0030 .0015	.0020 .0035 .0015	.0022 .0040 .0018
Z (fine tool work)	High limit (−) Low limit (−) Tolerance	.0005 .0007 .0002	.0007 .0012 .0005	.0007 .0015 .0008	.0010 .0020 .0010	.0010 .0022 .0012	.0012 .0025 .0013

*These inch-standard measurements may be converted to equivalent metric-standard measurements.

TABLE A-16 ANSI Dimensions (in Inches) for Woodruff Keys and Keyseats (ANSI B 17.2, R1978)

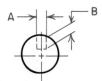 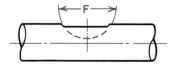

Woodruff Key Number	Nominal Size Key	Keyseat Shaft					Key above Shaft	Keyseat Hub	
		Width* A†		Depth B† +.005 -.000	Diameter F		Height C† +.005 -.005	Width D +.002 -.000	Depth E +.005 -.000
		Min.	Max.		Min.	Max.			
202	1/16 × 1/4	.0615	.0630	.0728	.250	.268	.0312	.0635	.0372
202.5	1/16 × 5/16	.0615	.0630	.1038	.312	.330	.0312	.0635	.0372
302.5	3/32 × 5/16	.0928	.0943	.0882	.312	.330	.0469	.0948	.0529
203	1/16 × 3/8	.0615	.0630	.1358	.375	.393	.0312	.0635	.0372
303	3/32 × 3/8	.0928	.0943	.1202	.375	.393	.0469	.0948	.0529
403	1/8 × 3/8	.1240	.1255	.1045	.375	.393	.0625	.1260	.0685
204	1/16 × 1/2	.0615	.0630	.1668	.500	.518	.0312	.0635	.0372
304	3/32 × 1/2	.0928	.0943	.1511	.500	.518	.0469	.0948	.0529
404	1/8 × 1/2	.1240	.1255	.1355	.500	.518	.0625	.1260	.0685
305	3/32 × 5/8	.0928	.0943	.1981	.625	.643	.0469	.0948	.0529
405	1/8 × 5/8	.1240	.1255	.1825	.625	.643	.0625	.1260	.0685
505	5/32 × 5/8	.1553	.1568	.1669	.625	.643	.0781	.1573	.0841
605	3/16 × 5/8	.1863	.1880	.1513	.625	.643	.0937	.1885	.0997
406	1/8 × 3/4	.1240	.1255	.2455	.750	.768	.0625	.1260	.0685
506	5/32 × 3/4	.1553	.1568	.2299	.750	.768	.0781	.1573	.0841
606	3/16 × 3/4	.1863	.1880	.2143	.750	.768	.0937	.1885	.0997
806	1/4 × 3/4	.2487	.2505	.1830	.750	.768	.1250	.2510	.1310
507	5/32 × 7/8	.1553	.1568	.2919	.875	.895	.0781	.1573	.0841
607	3/16 × 7/8	.1863	.1880	.2763	.875	.895	.0937	.1885	.0997
707	7/32 × 7/8	.2175	.2193	.2607	.875	.895	.1093	.2198	.1153
807	1/4 × 7/8	.2487	.2505	.2450	.875	.895	.1250	.2510	.1310
608	3/16 × 1	.1863	.1880	.3393	1.000	1.020	.0937	.1885	.0997
708	7/32 × 1	.2175	.2193	.3237	1.000	1.020	.1093	.2198	.1153
808	1/4 × 1	.2487	.2505	.3080	1.000	1.020	.1250	.2510	.1310
1008	5/16 × 1	.3111	.3130	.2768	1.000	1.020	.1562	.3135	.1622
1208	3/8 × 1	.3735	.3755	.2455	1.000	1.020	.1875	.3760	.1935
609	3/16 × 1-1/8	.1863	.1880	.3853	1.125	1.145	.0937	.1885	.0997
709	7/32 × 1-1/8	.2175	.2193	.3697	1.125	1.145	.1093	.2198	.1153
809	1/4 × 1-1/8	.2487	.2505	.3540	1.125	1.145	.1250	.2510	.1310
1009	5/16 × 1-1/8	.3111	.3130	.3228	1.125	1.145	.1562	.3135	.1622
610	3/16 × 1-1/4	.1863	.1880	.4483	1.250	1.273	.0937	.1885	.0997
710	7/32 × 1-1/4	.2175	.2193	.4327	1.250	1.273	.1093	.2198	.1153
810	1/4 × 1-1/4	.2487	.2505	.4170	1.250	1.273	.1250	.2510	.1310
1010	5/16 × 1-1/4	.3111	.3130	.3858	1.250	1.273	.1562	.3135	.1622
1210	3/8 × 1-1/4	.3735	.3755	.3545	1.250	1.273	.1875	.3760	.1935

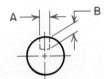 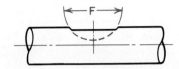

Woodruff		Keyseat Shaft					Key above Shaft	Keyseat Hub	
		Width* A†		Depth B†	Diameter F		Height C†	Width D	Depth E
Key Number	Nominal Size Key	Min.	Max.	+.005 -.000	Min.	Max.	+.005 -.005	+.002 -.000	+.005 -.000
811	1/4 × 1-3/8	.2487	.2505	.4640	1.375	1.398	.1250	.2510	.1310
1011	5/16 × 1-3/8	.3111	.3130	.4328	1.375	1.398	.1562	.3135	.1622
1211	3/8 × 1-3/8	.3735	.3755	.4015	1.375	1.398	.1875	.3760	.1935
812	1/4 × 1-1/2	.2487	.2505	.5110	1.500	1.523	.1250	.2510	.1310
1012	5/16 × 1-1/2	.3111	.3130	.4798	1.500	1.523	.1562	.3135	.1622
1212	3/8 × 1-1/2	.3735	.3755	.4485	1.500	1.523	.1875	.3760	.1935
617-1	3/16 × 2-1/8	.1863	.1880	.3073	2.125	2.160	.0937	.1885	.0997
817-1	1/4 × 2-1/8	.2487	.2505	.2760	2.125	2.160	.1250	.2510	.1310
1017-1	5/16 × 2-1/8	.3111	.3130	.2448	2.125	2.160	.1562	.3135	.1622
1217-1	3/8 × 2-1/8	.3735	.3755	.2135	2.125	2.160	.1875	.3760	.1935
617	3/16 × 2-1/8	.1863	.1880	.4323	2.125	2.160	.0937	.1885	.0997
817	1/4 × 2-1/8	.2487	.2505	.4010	2.125	2.160	.1250	.2510	.1310
1017	5/16 × 2-1/8	.3111	.3130	.3698	2.125	2.160	.1562	.3135	.1622
1217	3/8 × 2-1/8	.3735	.3755	.3385	2.125	2.160	.1875	.3760	.1935
822-1	1/4 × 2-3/4	.2487	.2505	.4640	2.750	2.785	.1250	.2510	.1310
1022-1	5/16 × 2-3/4	.3111	.3130	.4328	2.750	2.785	.1562	.3135	.1622
1222-1	3/8 × 2-3/4	.3735	.3755	.4015	2.750	2.785	.1875	.3760	.1935
1422-1	7/16 × 2-3/4	.4360	.4380	.3703	2.750	2.785	.2187	.4385	.2247
1622-1	1/2 × 2-3/4	.4985	.5005	.3390	2.750	2.785	.2500	.5010	.2560
822	1/4 × 2-3/4	.2487	.2505	.6200	2.750	2.785	.1250	.2510	.1310
1022	5/16 × 2-3/4	.3111	.3130	.5888	2.750	2.785	.1562	.3135	.1622
1222	3/8 × 2-3/4	.3735	.3755	.5575	2.750	2.785	.1875	.3760	.1935
1422	7/16 × 2-3/4	.4360	.4380	.5263	2.750	2.785	.2187	.4385	.2247
1622	1/2 × 2-3/4	.4985	.5005	.4950	2.750	2.785	.2500	.5010	.2560
1228	3/8 × 3-1/2	.3735	.3755	.7455	3.500	3.535	.1875	.3760	.1935
1428	7/16 × 3-1/2	.4360	.4380	.7143	3.500	3.535	.2187	.4385	.2247
1628	1/2 × 3-1/2	.4985	.5005	.6830	3.500	3.535	.2500	.5010	.2560
1828	9/16 × 3-1/2	.5610	.5630	.6518	3.500	3.535	.2812	.5635	.2872
2028	5/8 × 3-1/2	.6235	.6255	.6205	3.500	3.535	.3125	.6260	.3185
2228	11/16 × 3-1/2	.6860	.6880	.5893	3.500	3.535	.3437	.6885	.3497
2428	3/4 × 3-1/2	.7485	.7505	.5580	3.500	3.535	.3750	.7510	.3810

*Values for width A represent the *maximum* keyseat (shaft) width that will assure the key will stick in the keyseat (shaft). The *minimum* keyseat width permits the largest shaft distortion acceptable when assembling a maximum key in a minimum keyseat.
†Dimensions A, B, C, and D are taken at the side intersection.

TABLE A-17 SI Metric Spur Gear Rules and Formulas for Required Features of 20° Full-Depth Involute Tooth Form

Required Feature	Rule	Formula*
Module (M)	Divide the pitch diameter by the number of teeth	$M = \dfrac{D}{N}$
	Divide the outside diameter by the number of teeth plus 2	$M = \dfrac{O}{N+2}$
Number of teeth (N)	Divide the pitch diameter by the module	$N = \dfrac{D}{M}$
Pitch diameter (D)	Multiply the module by the number of teeth	$D = M \times N$
Outside diameter (O)	Add the pitch diameter and twice the module	$O = D + 2M$
	Multiply the number of teeth plus 2 by the module	$O = M \times (N+2)$
Addendum (A)	Multiply the module by 1.000	$A = M \times 1.000$
Clearance (S)	Multiply the module by 0.250	$S = M \times 0.250$
Whole depth (W^2)	Multiply the module by 2.250	$W^2 = M \times 2.250$
Circular pitch (P_c)	Multiply the module by 3.1416	$P_c = M \times 3.1416$
	Divide the pitch diameter by the number of teeth and multiply by 3.1416	$P_c = \dfrac{D}{N} \times 3.1416$
Tooth thickness (T)	Multiply the module by 1.5708	$T = M \times 1.5708$
Center distance (C)	Multiply the module by the number of teeth in both gears and divide by 2	$C = \dfrac{M \times (N^1 + N^2)}{2}$
	Add the pitch diameters of the two gears and divide by 2	$C = \dfrac{D^1 + D^2}{2}$

*Dimensions are in millimeters.

TABLE A-18 ANSI Spur Gear Rules and Formulas for Required Features of 20° and 25° Full-Depth Involute Tooth Forms

Required Feature	Rule	Formula*
Diametral pitch (P_d)	Divide the number of teeth by the pitch diameter	$P_d = \dfrac{N}{D}$
	Add 2 to the number of teeth and divide by the outside diameter	$P_d = \dfrac{N+2}{O}$
	Divide 3.1416 by the circular pitch	$P_d = \dfrac{3.1416}{P_c}$
Number of teeth (N)	Multiply the diametral pitch by the pitch diameter	$N = P_d \times D$
	Multiply the diametral pitch by outside diameter and then subtract 2	$N = (P_d \times O) - 2$
Pitch diameter (D)	Divide the number of teeth by the diametral pitch	$D = \dfrac{N}{P_d}$
	Multiply the addendum by 2 and subtract the product from the outside diameter	$D = O - 2A$
Outside diameter (O)	Add 2 to the number of teeth and divide by the diametral pitch	$O = \dfrac{N+2}{P_d}$
	Add 2 to the number of teeth and divide by the quotient of the number of teeth divided by the pitch diameter	$O = \dfrac{N+2}{N/D}$
Circular pitch (P_c)	Divide 3.1416 by the diametral pitch	$P_c = \dfrac{3.1416}{P_d}$
	Divide the pitch diameter by the product of 0.3183 times the number of teeth	$P_c = \dfrac{D}{0.3183 \times N}$
Addendum (A)	Divide 1.000 by the diametral pitch	$A = \dfrac{1.000}{P_d}$
Working depth (W^1)	Divide 2.000 by the diametral pitch	$W^1 = \dfrac{2.000}{P_d}$
Clearance (S)	Divide 0.250 by the diametral pitch	$S = \dfrac{0.250}{P_d}$
Whole depth of tooth (W^2)	Divide 2.250 by the diametral pitch	$W^2 = \dfrac{2.250}{P_d}$
Thickness of tooth (T)	Divide 1.5708 by the diametral pitch	$T = \dfrac{1.5708}{P_d}$
Center distance (C)	Add the pitch diameters and divide the sum by 2	$C = \dfrac{D^1 + D^2}{2}$
	Divide one-half the sum of the number of teeth in both gears by the diametral pitch	$C = \dfrac{1/2\,(N^1 + N^2)}{P_d}$

*Dimensions are in inches.

TABLE A-19 Selected Standard Shapes, Design Features, and General Toolroom Applications of Grinding Wheels (Based on ANSI B 74.2-1974)

Standard Shapes, Dimensions*, and Grinding Surfaces	Design Features	Principal Applications	
		Nature of Grinding	Grinding Category
Type 1 straight wheel	Cylindrical with concentric bore	Peripheral	—Surface —Cylindrical —Tool and cutter —Offhand —Portable machine
Type 2 cylindrical wheel	Mounted on the diameter, in a chuck, or on a faceplate	Side	Surface (vertical spindle)
Type 5 straight wheel	Recessed on one side	Peripheral	—Cylindrical —Internal —Surface (horizontal spindle)
Type 6 straight-cup wheel	Equal wall thickness from back to grinding face	Side	—Tool and cutter —Portable machine
Type 7 straight wheel	Recessed two sides	Peripheral	—Cylindrical —Surface (horizontal spindle)
Type 11 flaring-cup wheel	Thickness back wall tapered outward to grinding face	Side	—Tool and cutter —Portable machine

Standard Shapes, Dimensions*, and Grinding Surfaces	Design Features	Principal Applications	
		Nature of Grinding	Grinding Category
Type 12 dish wheel	Dressed on the side or "U" face	—Side —"U" face	Tool and cutter
Type 13 saucer wheel	Saucer-shaped equal cross section	—Peripheral —Saw tooth shaping or sharpening	Tool and cutter
Type 21 straight wheel (relieved)	Both sides relieved, leaving a flat area	Peripheral	Cylindrical
Type 22 straight wheel (relieved and recessed)	One side relieved to a flat area and second side recessed	Peripheral	Cylindrical

Types 23, 24, 25, and 26 provide further modifications of straight wheel relieved and recessed combinations

*Key for letter dimensions:

- A Radial width of flat at periphery
- D Outside diameter
- E Thickness at hole
- F Depth of recess on one side
- G Depth of recess on second side
- H Hole (bore) diameter
- J Diameter of outside flat
- K Diameter of inside flat
- N Depth of relief on one side
- O Depth of relief on second side
- P Diameter of recess
- R Peripheral radius
- T Overall thickness
- U Width of edge
- W Rim (wall) thickness at grinding face

TABLE A-20 Standard Types of Diamond Grinding Wheels

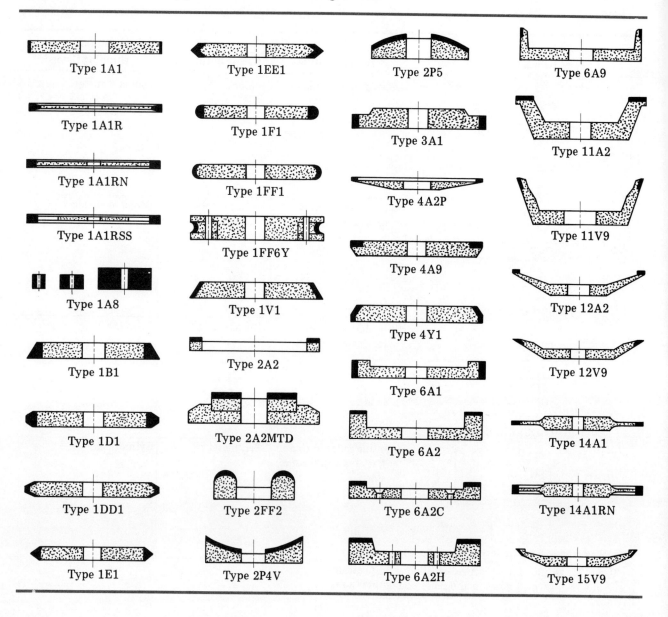

TABLE A-21 Conversion of Surface Speeds (sfpm) to Spindle Speeds (RPM) for Various Diameters of Grinding Wheels (1″ to 72″ or 25.4mm to 1828mm)

Wheel Diameter (mm)	Surface Speed in Feet per Minute (sfpm)										Wheel Diameter (Inches)
	4,000	4,500	5,000	5,500	6,000	6,500	7,000	7,500	8,000	8,500	
	Spindle Speeds in Revolutions per Minute (RPM)										
25.4	15,279	17,189	19,098	21,008	22,918	24,828	26,737	28,647	30,558	32,467	1
50.8	7,639	8,594	9,549	10,504	11,459	12,414	13,368	14,328	15,278	16,238	2
76.2	5,093	5,729	6,366	7,003	7,639	8,276	8,913	9,549	10,186	10,822	3
101.6	3,820	4,297	4,775	5,252	5,729	6,207	6,685	7,162	7,640	8,116	4
127.0	3,056	3,438	3,820	4,202	4,584	4,966	5,348	5,730	6,112	6,494	5
152.4	2,546	2,865	3,183	3,501	3,820	4,138	4,456	4,775	5,092	5,411	6
177.8	2,183	2,455	2,728	3,001	3,274	3,547	3,820	4,092	4,366	4,638	7
203.2	1,910	2,148	2,387	2,626	2,865	3,103	3,342	3,580	3,820	4,058	8
228.6	1,698	1,910	2,122	2,334	2,546	2,758	2,970	3,182	3,396	3,606	9
254.0	1,528	1,719	1,910	2,101	2,292	2,483	2,674	2,865	3,056	3,247	10
304.8	1,273	1,432	1,591	1,751	1,910	2,069	2,228	2,386	2,546	2,705	12
355.6	1,091	1,228	1,364	1,500	1,637	1,773	1,910	2,046	2,182	2,319	14
406.4	955	1,074	1,194	1,313	1,432	1,552	1,672	1,791	1,910	2,029	16
457.2	849	955	1,061	1,167	1,273	1,379	1,485	1,591	1,698	1,803	18
508.0	764	859	955	1,050	1,146	1,241	1,337	1,432	1,528	1,623	20
558.8	694	781	868	955	1,042	1,128	1,215	1,302	1,388	1,476	22
609.6	637	716	796	875	955	1,034	1,115	1,194	1,274	1,353	24
660.4	588	661	734	808	881	955	1,028	1,101	1,176	1,248	26
711.2	546	614	682	750	818	887	955	1,023	1,092	1,159	28
762.0	509	573	637	700	764	828	891	955	1,018	1,082	30
812.8	477	537	597	656	716	776	836	895	954	1,014	32
863.6	449	505	562	618	674	730	786	843	898	955	34
914.4	424	477	530	583	637	690	742	795	848	902	36
965.2	402	452	503	553	603	653	704	754	804	854	38
1016.0	382	430	478	525	573	620	669	716	764	812	40
1066.8	366	409	454	500	545	591	636	682	732	775	42
1117.6	347	390	434	478	521	564	608	651	694	737	44
1168.4	333	375	416	458	500	541	582	624	666	708	46
1219.2	318	358	398	438	478	517	558	597	636	676	48
1524.0	255	287	319	350	387	414	446	478	510	542	60
1828.8	212	239	265	291	318	345	371	398	424	451	72

TABLE A-21 Conversion of Surface Speeds (sfpm) to Spindle Speeds (RPM) for Various Diameters of Grinding Wheels (1" to 72" or 25.4mm to 1828mm) (Continued)

Wheel Diameter (mm)	Surface Speed in Feet per Minute (sfpm)									Wheel Diameter (Inches)
	9,000	9,500	10,000	12,000	12,500	14,200	16,000	16,500	17,000	
	Spindle Speeds in Revolutions per Minute (RPM)									
25.4	34,377	36,287	38.196	45,836	47,745	54,240	61,116	63,025	64,935	1
50.8	17,188	18,143	19,098	22,918	23,875	27,120	30,558	31,510	32,465	2
76.2	11,459	12,096	12,732	15,278	15,915	18,080	20,372	21,010	21,645	3
101.6	8,595	9,072	9,549	11,459	11,940	13,560	15,278	15,755	16,235	4
127.0	6,876	7,258	7,640	9,168	9,550	10,850	12,224	12,605	12,985	5
152.4	5,729	6,048	6,366	7,639	7,960	9,040	10,186	10,505	10,820	6
177.8	4,911	5,183	5,456	6,548	6,820	7,750	8,732	9,005	9,275	7
203.2	4,297	4,535	4,775	5,729	5,970	6,780	7,640	7,880	8,115	8
228.6	3,820	4,032	4,244	5,092	5,305	6,030	6,792	7,000	7,215	9
254.0	3,438	3,629	3,820	4,584	4,775	5,425	6,112	6,300	6,495	10
304.8	2,864	3,023	3,183	3,820	3,980	4,520	5,092	5,250	5,410	12
355.6	2,455	2,592	2,728	3,274	3,410	3,875	4,366	4,500	4,640	14
406.4	2,149	2,268	2,387	2,865	2,985	3,390	3,820	3,940	4,060	16
457.2	1,910	2,016	2,122	2,546	2,655	3,015	3,396	3,500	3,605	18
508.0	1,719	1,814	1,910	2,292	2,390	2,715	3,056	3,150	3,245	20
558.8	1,562	1,649	1,736	2,084	2,170	2,465	2,776	2,865	2,950	22
609.6	1,433	1,512	1,591	1,910	1,990	2,260	2,546	2,625	2,705	24
660.4	1,322	1,395	1,468	1,762	1,840	2.090	2,352	2,425	2,495	26
711.2	1,228	1,296	1,364	1,637	1,705	1,940	2,182	2,250	2,320	28
762.0	1,146	1,210	1,274	1,528	1,595	1,810	2,056	2,100	2,165	30
812.8	1,074	1,134	1,194	1,432	1,495	1,695	1,910	1,970	2,030	32
863.6	1,011	1,067	1,124	1,348	1,405	1,595	1,796	1,855	1,910	34
914.4	954	1,007	1,061	1,273	1,330	1,510	1,698	1,750	1,805	36
965.2	904	955	1,006	1,206	1,260	1,430	1,608	1,660	1,710	38
1016.0	860	908	956	1,146	1,195	1,355	1,528	1,575	1,625	40
1066.8	818	863	908	1,090	1,140	1,295	1,464	1,500	1,545	42
1117.6	780	824	868	1,042	1,085	1,235	1,388	1,432	1,475	44
1168.4	750	791	832	1,000	1,040	1,180	1,332	1,370	1,410	46
1219.2	716	756	796	956	995	1,130	1,272	1,315	1,350	48
1524.0	574	606	638	774	795	905	1,020	1,050	1,080	60
1828.8	477	504	530	637	665	755	849	875	905	72

TABLE A-22 Feed and Depth of Cut Factors for Calculating Cutting Speeds for Turning Ferrous Metals

Feed (Inch)	Feed (mm)	Feed Factor (F_f)	Depth of Cut (Inch)	Depth of Cut (mm)	Depth of Cut Factor (F_d)
0.002	0.05	1.50	0.005	0.13	1.50
0.003	0.08	1.50	0.010	0.25	1.42
0.004	0.10	1.50	0.016	0.41	1.33
0.005	0.13	1.44	0.031	0.79	1.21
0.006	0.15	1.34	0.047	1.19	1.15
0.007	0.18	1.25	0.062	1.57	1.10
0.008	0.20	1.18	0.078	1.98	1.07
0.009	0.23	1.12	0.094	2.39	1.04
0.010	0.25	1.08	0.100	2.54	1.03
0.011	0.28	1.04	0.125	3.18	1.00
0.012	0.30	1.00	0.150	3.81	0.97
0.013	0.33	0.97	0.188	4.78	0.94
0.014	0.36	0.94	0.200	5.08	0.93
0.015	0.38	0.91	0.250	6.35	0.91
0.016	0.41	0.88	0.312	7.92	0.88
0.018	0.45	0.84	0.375	9.53	0.86
0.020	0.50	0.80	0.438	11.13	0.84
0.022	0.55	0.77	0.500	12.70	0.82
0.025	0.64	0.73	0.625	15.88	0.80
0.028	0.70	0.70	0.688	17.48	0.78
0.030	0.75	0.68	0.750	19.05	0.77
0.032	0.81	0.66	0.812	20.62	0.76
0.035	0.89	0.64	0.938	23.83	0.75
0.040	1.02	0.60	1.000	25.40	0.74
0.045	1.14	0.57	1.125	28.58	0.73
0.050	1.27	0.55	1.250	31.75	0.72
0.060	1.52	0.50	1.375	34.93	0.71

Note: Table is applicable to shaper and planer work on ferrous and nonferrous metals.

TABLE A-23 Equivalent Hardness Conversion Numbers for Steel

			Hardness Measurement Systems							
			Brinell			Rockwell		Rockwell Superficial		
			10mm Ball, 3,000 kgf Load			A Scale, 60 kgf Load	D Scale, 100 kgf Load	15N Scale, 15 kgf Load	30N Scale, 30 kgf Load	45N Scale, 45 kgf Load
Rockwell C	Vickers	Shore Scleroscope	Standard Ball	Hultgren Ball	Carbide Ball	Diamond Penetrator		Superficial Diamond Penetrator		
68	940	97				85.6	76.9	93.2	84.4	75.4
67	900	95				85.0	76.1	92.9	83.6	74.2
66	865	92				84.5	75.4	92.5	82.8	73.3
65	832	91			739	83.9	74.5	92.2	81.9	72.0
64	800	88			722	83.4	73.8	91.8	81.1	71.0
63	772	87			705	82.8	73.0	91.4	80.1	69.9
62	746	85			688	82.3	72.2	91.1	79.3	68.8
61	720	83			670	81.8	71.5	90.7	78.4	67.7
60	697	81		613	654	81.2	70.7	90.2	77.5	66.6
59	674	80		599	634	80.7	69.9	89.8	76.6	65.5
58	653	78		587	615	80.1	69.2	89.3	75.7	64.3
57	633	76		575	595	79.6	68.5	88.9	74.8	63.2
56	613	75		561	577	79.0	67.7	88.3	73.9	62.0
55	595	74		546	560	78.5	66.9	87.9	73.0	60.9
54	577	72		534	543	78.0	66.1	87.4	72.0	59.8
53	560	71		519	525	77.4	65.4	86.9	71.2	58.6
52	544	69	500	508	512	76.8	64.6	86.4	70.2	57.4
51	528	68	487	494	496	76.3	63.8	85.9	69.4	56.1
50	513	67	475	481	481	75.9	63.1	85.5	68.5	55.0
49	498	66	464	469	469	75.2	62.1	85.0	67.6	53.8
48	484	64	451	455	455	74.7	61.4	84.5	66.7	52.5
47	471	63	442	443	443	74.1	60.8	83.9	65.8	51.4
46	458	62	432	432	432	73.6	60.0	83.5	64.8	50.3
45	446	60	421	421	421	73.1	59.2	83.0	64.0	49.0

Hardness Measurement Systems

Rockwell C	Vickers	Shore Scleroscope	Brinell 10mm Ball, 3,000 kgf Load			Rockwell		Rockwell Superficial		
						A Scale, 60 kgf Load	D Scale, 100 kgf Load	15N Scale, 15 kgf Load	30N Scale, 30 kgf Load	45N Scale, 45 kgf Load
			Standard Ball	Hultgren Ball	Carbide Ball	Diamond Penetrator		Superficial Diamond Penetrator		
			Equivalent Hardness Numbers for Conversion among the Systems							
44	434	58	409	409	409	72.5	58.5	82.5	63.1	47.8
43	423	57	400	400	400	72.0	57.7	82.0	62.2	46.7
42	412	56	390	390	390	71.5	56.9	81.5	61.3	45.5
41	402	55	381	381	381	70.9	56.2	80.9	60.4	44.3
40	392	54	371	371	371	70.4	55.4	80.4	59.5	43.1
39	382	52	362	362	362	69.9	54.6	69.9	58.6	41.9
38	372	51	353	353	353	69.4	53.8	79.4	57.7	40.8
37	363	50	344	344	344	68.9	53.1	78.8	56.8	39.6
36	354	49	336	336	336	68.4	52.3	78.3	55.9	38.4
35	345	48	327	327	327	67.9	51.5	77.7	55.0	37.2
34	336	47	319	319	319	67.4	50.8	77.2	54.2	36.1
33	327	46	311	311	311	66.8	50.0	76.6	53.3	34.9
32	318	44	301	301	301	66.3	49.2	76.1	52.1	33.7
31	310	43	294	294	294	65.8	48.4	75.6	51.3	32.5
30	302	42	286	286	286	65.3	47.7	75.0	50.4	31.3
29	294	41	279	279	279	64.6	47.0	74.5	49.5	30.1
28	286	41	271	271	271	64.3	46.1	73.9	48.6	28.9
27	279	40	264	264	264	63.8	45.2	73.3	47.7	27.8
26	272	38	258	258	258	63.3	44.6	72.8	46.8	26.7
25	266	38	253	253	253	62.8	43.8	72.2	45.9	25.5
24	260	37	247	247	247	62.4	43.1	71.6	45.0	24.3
23	254	36	243	243	243	62.0	42.1	71.0	44.0	23.1
22	248	35	237	237	237	61.5	41.6	70.5	43.2	22.0
21	243	35	231	231	231	61.0	40.9	69.9	42.3	20.7
20	238	34	226	226	226	60.5	40.1	69.4	41.5	19.6

Note: Hardness numbers correspond to values for carbon and alloy steel (ASTM specifications E 140-71).

TABLE A-24 Recommended Heat Treatment Temperatures for Selected Grades and Kinds of Steel

AISI or SAE Number	Hardening Temperature Range °F	Hardening Temperature Range °C	Annealing Temperature Range °F	Annealing Temperature Range °C	Normalizing Temperature Range °F	Normalizing Temperature Range °C	Quenching Medium
1030	1550/1600	843/871	1525/1575	829/857	1625/1725	885/941	
1040	1500/1550	816/843	1475/1525	802/829	1600/1700	871/927	
1050	1475/1525	802/829	1450/1500	788/816	1550/1650	843/899	
1060	1450/1500	788/816	1425/1475	774/802	1500/1600	816/871	
1070	1450/1500	788/816	1425/1475	744/802	1500/1600	816/871	Water or brine
1080	1400/1450	760/788	1375/1425	746/774	1475/1575	802/857	
1090	1400/1450	760/788	1375/1425	746/774	1475/1575	802/857	
1132	1550/1600	843/871	1525/1575	829/857	1625/1725	885/941	
1140	1500/1550	816/843	1475/1525	802/829	1600/1700	871/927	
1151	1475/1525	802/829	1450/1500	788/816	1550/1650	843/899	
1330	1525/1575	829/857	1500/1550	816/843	1600/1700	871/927	Oil or water
1340	1500/1550	816/843	1475/1525	802/829	1550/1650	843/899	Oil or water
3140	1475/1525	802/829	1475/1525	802/829	1550/1650	843/899	Oil
4028	1550/1600	843/871	1525/1575	829/857	1600/1700	871/927	Oil or water
4042	1500/1550	816/843	1475/1525	802/829	1550/1650	843/899	Oil
4063	1475/1525	802/829	1450/1500	788/816	1550/1650	843/899	Oil
4130	1550/1600	843/871	1525/1575	829/857			Oil or water
4140	1525/1575	829/857	1500/1550	816/843			Oil
4150	1500/1550	816/843	1475/1525	802/829			Oil
4340	1500/1550	816/843	1500/1550	816/843			Oil
5046	1500/1550	816/843	1475/1525	802/829			Oil
5130	1550/1600	843/871	1500/1550	816/843			Oil or water
5145	1525/1575	829/857	1475/1525	802/829			
5160	1500/1550	816/843	1450/1500	788/816	1600/1700	871/927	
50100	1450/1500	788/816	1400/1450	760/788			
51100	1475/1525	802/829	1400/1450	760/788			
52100	1500/1550	816/843	1400/1450	760/788			
6150	1550/1600	843/871	1500/1550	816/843			Oil
8630	1525/1575	829/857	1500/1550	816/843			
8645	1500/1550	816/843	1475/1525	802/829			
8660	1475/1525	802/829	1450/1500	788/816			
8735	1525/1575	829/857	1500/1550	816/843			
8742	1525/1575	829/857	1500/1550	816/843			
9260	1550/1600	843/871	1500/1550	816/843			
9840	1500/1550	816/843	1475/1525	802/829			
9850	1500/1550	816/843	1475/1525	802/829			

APPENDIX 843

TABLE A-25 Melting Points of Metals and Alloys in General Shop Use

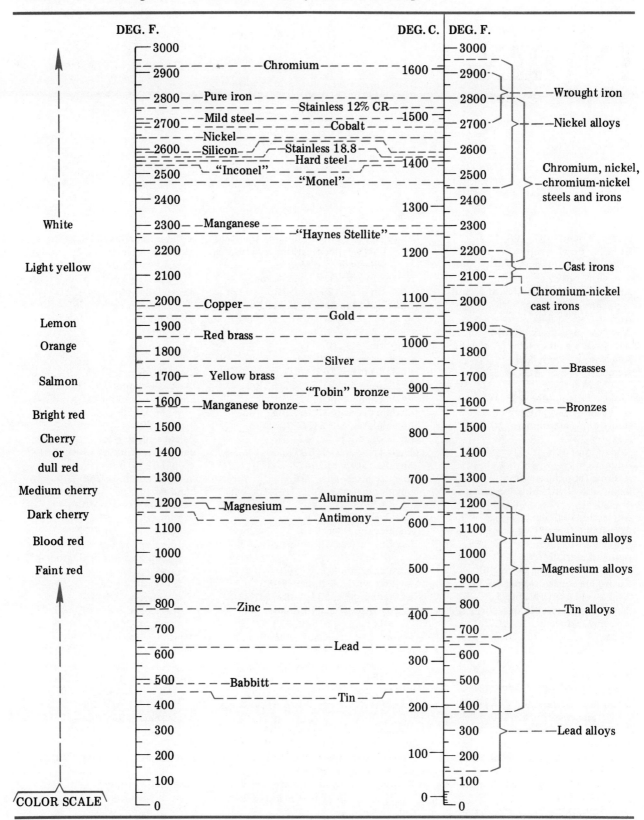

INDEX

Abrasive band, 407-408, 412
Abrasive flows, 459
Abrasive machining, 447-448, 456-457, 462
Abrasives, 445-447, 458-459, 460, 469, 472, 482
Abrasive stick, 548
Absolute measurement, 651, 655
Acceptable quality level (AQL), 99
Acheson, Edward, 446
Acme screw threads, 126-131, 134
"Acorn" die holder, 168, 169
Actuator, rotary, 699, 700
Acute angle attachment, 10
Adapters, 244
Addendum, 333
Addendum circle, 333
Address (computer), 665
Adjustable floating-blade reamers, 185, 194
Adjustable limit snap gage, 87
Adjustable thread ring gages, 87-88
Adjustable toolholder, 167-168, 174
Aging of steel, 780
Air spindle grinding device, 621, 632
Air wedge configuration, 53, 58
AISI steel classifications, 755-758
Alloyed cast iron castings, 751, 763
Alloying elements, 765-767
Alloys, 752, 755, 756-757, 764-768, 771, 772
Alloy steels, 755, 764-765, 771
Alnico, 766
Alpha iron structure, 777
Aluminum Association (AA) designation system, 759-760
Aluminum-base alloys, 767, 772
Aluminum oxide, 446, 460
Aluminum oxide abrasive wheels, 491-497
Aluminum temper designations, 759-760, 771
American National Form screw threads, 89-90

American National Standards Institute (ANSI) specifications for grinding wheels, 474-475
American Standard Stub Acme screw threads, 129-130, 135
American Standard Unified Miniature (UNM) Screw Thread Series, 123-124, 134
Amplifier, 76, 79
Angle dresser, 549, 559
Angle fixtures, 162
Angle gage blocks, 33-35, 38
Angle of incidence, 42
Angle plate accessories, 534
Angle plates, 8, 534, 537, 538
Angle of reflection, 42
Angle-type dresser, 737
Angular face, 586-587
Angular increment, 315
Angular measurement, 10-20, 33-37, 48-49
Angular shaping, 421-422
Angular shoulder, 540
Angular surfaces, grinding of, 533-539, 541
Angular symbol, 108, 109
Angular work clearance, 708
Annealing, 380, 387, 790-791, 806
APT (Automatically Programmed Tools), 678, 680-683, 688
Arbors, 244, 245, 600-601, 631
Arc of contact, 468
Area of contact (grinding), 468-469, 487
Arithmetical average (AA), 108-110
ASA B46.1 Surface Texture Standards, 106
Austempering, 783, 784, 806
Austenite, 778-779, 780-782
Austenitic stainless steel, 766, 771
Autocollimator, 26, 38
Automatic bar and chucking machine, 139-140, 151. *See also* Turret lathes
Automatic programming, 689

Automatic sizing, 448, 462
Auxiliary element, 677
Average outgoing quality limit (AOQL), 99
Axes, 644-645
  programming for, 655-663, 685
Axial advance, 345
Axial pitch, 345
Axial runout, 30, 38

Back-facing cuts, 186, 194
Backlash eliminator, 180, 194
Back rests (cylindrical grinders), 569, 579-580, 589
Bainite, 784
Balance turning tool, 166
Balancing, grinding wheel, 501-503, 510
Ball-end end mill, 253-254, 263
Band filing, 404-407, 412
Band grinding, 407-409, 412
Band machines, 366-411
  components of, 372-376
  filing on, 404-407
  grinding and polishing on, 407-409
  saw band welding on, 376-383
  sawing on, 390-404
  speeds and feeds of, 385
  types of, 366-369
Band machinist, 368, 386
Band polishing, 407-408, 412
Bands, interference, 52, 53-54
Band tension indicator, 374, 387
Bar machines, 141, 151, 673. *See also* Turret lathes
Bar turners, 162-165, 171-173, 174, 178, 179, 202-203
Base circle, 315, 333
Baseline dimensioning, 272-273, 274, 281
Base plates, grind-pattern, 715-716
Basic oxygen steel manufacturing, 752, 753, 771
Basic pitch diameter, 91, 92-94

INDEX 845

Basic size, 88, 91
Batch furnaces, 786
Baths:
    quenching, 782-784, 790, 797
    tempering, 787-788
Bearing area, 108
Beasly disk grinder, 456
Bed-type vertical milling machine, 234-235, 236, 248
Bench centers, 30, 38
Bench comparator, 10, 20
Bending (die work), 712, 713
Bessemer converter steel manufacturing, 751-752, 771
Bevel gears, 327, 339-344, 356-358
Bevel protractors, 10-11
Bidirectional gaging system, 740
Bilateral tolerance, 84
Binary code, 646-647, 665
Binary code decimal (BCD) system, 646-647
Binary notation, 646
Binocular-head microscope, 46
Black granite, 6-10, 19
Black heat, 780, 805
Blanking, 327, 704, 721
Block, tape, 654, 665
Block address format, 654, 665
Block count readout, 689
Bolsters, 705, 706
Bonds (grinding wheels), 473-474, 484, 498, 558
Boring, 268-271, 276-278, 674-675, 727-733
Boring bars, 728-729, 742
Boring cuts, 184-185
Boring head set, 242-243, 269, 280
Boring machines, numerically controlled, 674
Boring toolholders, 169, 170
Box jigs, 698-699, 721
Box parallels, 8, 9, 20
Box tools, 165, 223
Brake-controlled dressing device, 501, 510
Bridge cross slide, 146
Brinell hardness testing, 801-803
Brittleness, 754
Broaching, 122-123, 134, 325-326
Broken-out section, 193
Brush analyzer, 114
Buffing, 461-462
Burr proofing, 3
Bushings, 701
Buttons, 695
Butt welding, 377-380

Calibration, 24, 38
Caliper bars, 5
Cambering, 387
Cam clamps, 697
Cam followers, 314-315, 320
Cam lead, 315, 320
Cam milling, 312-319

Cam motion, 312-313, 320
Cam profile, 315
Cam rise, fall, and swell, 217, 225
Cams, 217-218, 219, 303, 312-319
Carbide cutting tools, 423-425
Carbide-tooth shell mills, 620
Carbonitriding, 794-795
Carbon steel, 752-754, 755, 764, 777-784, 786-789
Carburization, 791-794
Carriage feeds, 147
Carrier wheels, 373-374
Cartridge gaging head, 76, 77, 79
Casehardening, 791-792, 806
Cast alloy cutting tools, 768, 772
Casting methods (screw thread production), 121, 134
Cast iron, 750-751, 761-764
Cathode ray tube (CRT), 679, 684
Cemented carbide cutting tools, 768-769
Cementite, 761, 771, 777
Center cutting, 255, 264
Center drilling, 183
Centering gauge, 600, 611
Centerless grinding, 446, 447, 453-454, 463
Centerline, 107
Center plate, 373, 386
Centralizing Acme threads, 129, 135
Central processing unit (CPU), 679-680
Centrifugal separator, 507
Ceramic cutting tools, 769-770, 772
Certificate of accuracy, 24, 38
Chamfer cutting edge, 627-628, 632
Chamfering tool, 172
Change gears, 305-308
Change gear tables, 307-308, 320
Channel, 5
Channel jigs, 699
Chasing method (screw thread production), 121, 134
Chatter, tool, 206
Chilled iron castings, 750, 762
Chip control, 165
Chip load per tooth, 257
Chop grinding, 734, 742
Chordal addendum, 333
Chordal thickness, 333
Chromium, 766, 772
Chucking machines, 141, 151, 673. See also Turret lathes
Chucking tooling, 204-205, 208
Chuck jaws, 159-161
Chucks, 17, 156-162, 520, 528, 535, 536, 544, 569, 702-703, 729
Chuck V-blocks, 521
Circular interpolation, 650, 685
Circular pitch, 333, 337-339
Circular symbol, 108, 109
Circular thickness, 333
Clamping, 694-697
Clamps, 150, 428, 520, 528, 697

Classes of fits (holes), 88-89, 103
Classes of tolerances (holes), 85, 103
Clearance, 334
Clearance angle, 601-602, 603-607, 611, 615
Clearance fit, 88, 89
Closed loop NC system, 643-644, 666
Coarse-pitch worm gears, 344
Coated flat, 50, 58
Cobalt, 766, 772
Cobalt high-speed steel end mills, 254-255
Cold extrusion presses, 706
Cold rolling, 326
Collapsing taps, 186-187, 195
Collet chuck end mill holder, 256, 264
Collet chucks, 156-158
Collets, 244, 729
Color code markings, 760-761
Column-type pneumatic comparator, 71-72, 80
Combination dies, 712
Combination turner and end former, 180, 194
Combined cuts, 155
Comparator charts, 68, 69
Comparators. See High amplification comparators
Comparator specimens, 111-113
Comparator stands, 77-78
Compound cross slide, 144-146, 152
Compound dies, 710-712, 721
Compound gearing, 307-308, 320
Compound sine plate, 16-17, 20
Computer-assisted programming (CAP), 677-683
Computerized numerically controlled (CNC) vertical milling machine, 235, 236
Computer numerical control (CNC) system, 679, 683-687, 715-720
Computers, See Numerical control systems
Concave face, 587
Concave lenses, 43-44, 59
Concentration number, 498, 509
Concentricity, 29-30, 38
Concentric location, 695
Cone-point diamond cutting tool, 550, 551, 560
Cones, 479
Conical location, 695
Contact, 108
Continuous-path control system, 663, 666, 671, 673, 678
Contour measuring projector, 70
Contour projectors, 62, 67-70, 79
Contour sawing, 390, 393-397, 398
Contour shaping, 422, 438
Control charts, 100-102, 103
Control element, 677
Control limits, 100-102
Controlling point (thread measurement), 91, 103

846 INDEX

Control setting chart (butt welder), 377–379, 387
Convex face, 587
Convex lenses, 43, 58
Conveyor furnaces, 786
Coolant-feeding drill, 183
Coolants, cutting, 503–508
Coordinate axes, 644
Coordinate chart, 56, 59, 274, 281
Coordinate (absolute) measurement, 651, 665
Coordinate system, rectangular, 644
Copper-base alloys, 767, 772
Core drill, 183
Core drilling, 184, 194
Corner radius, 586
Counterbores, 222
Cross feed control, 568–569, 589
Cross-feed handwheels, 524–525
Cross feeds, 522, 530
Cross lines, 48–50
Cross rail (planer), 429, 438
Cross slide dressing attachment, 737
Cross slides, 144–147, 199–200, 224
Cross-sliding hexagon turret boring, 185
Cross-sliding turret, 144–145, 152, 203. *See also* Hexagon turret
Crush dressing, 492, 554–557, 560
Crush roll, 492, 509, 554–557
Crush truing, 587
Cube, 9, 10
Cubic boron nitride (CBN) grinding wheels, 497, 500–501
Cubit stick, 6, 7
Cupola furnace, 750, 771
Curling dies, 713
Cutaway sections, 189
Cutoff cuts, 181
Cutoff sawing, 390, 391
Cutoff shear, 377, 387
Cuts, machine:
 external, 155, 177, 178–182
 internal, 155, 177, 182–188
Cutter buildup, 202, 208
Cutter clearance angle, 601–602, 603–607, 611
Cutter clearance gage method, 606–607
Cutter grinding arbor, 600–601
Cutter slide block, 164–165
Cutter and tool grinder, 455–456, 463, 594–631
 attachments for, 597–600
 design features of, 595–597
 preparation of, 607–609
Cutting coolants, 503–508
Cutting edge, 601–602
Cutting fluids:
 automatic screw machines, 218–220
 end mills, 256
 turret lathes, 150
 vertical band machines, 384–385
Cutting force (die work), 708
Cutting off, 558–559, 560, 704–705, 721

Cutting-off tools, 169–170
Cutting plane line, 189–191
Cutting planes, 189
Cutting speeds and feeds:
 automatic screw machines, 218
 boring tools, 271
 end mills, 257–258
 grinding wheels, 570–571
 planers, 435
 turret lathes, 178, 188–189, 190
 vertical band machines, 385
Cutting-tool holders, 169, 170
Cutting tools:
 industrial materials for, 768–770
 mounting of, 170–172
 numerically controlled, 687–688
 shaping with, 423–425
 types of in NC machining, 717
 vertical milling, 252–256
Cyaniding, 794
Cyclonic separator, 507
Cylindrical gaging head, 76, 77, 79
Cylindrical grinders, 451–453, 463, 565–570, 574
Cylindrical grinding, 564–588
 problems and corrective actions in, 573–575
Cylindrical grinding attachment, 598
Cylindrical grinding wheels, 477–478, 496–497, 576
Cylindrical surfaces, 54–55

Data, operational, 570, 571, 589
Data processing, 676–677
Datum, 36, 38, 56, 273, 281
Datum planes, 36, 38
Decalescence point, 781, 805
Dedendum, 333
Deep draw, 714, 722
Deep-hole drilling, 184
Deep-length collets, 156, 157, 173
Deep-length step chucks, 156, 157
Deformation, 754
Detent, 201, 207
Deviation per six inches of length, 28, 38
Dial indicator "drop" method, 606
Dial indicators, 26–32, 63, 267, 268, 738–739. *See also* Mechanical comparators
Diametral measurement, 280
Diametral pitch, 329, 334, 337–339
Diametral setting, 174
Diamond-charged mandrels, 737, 742
Diamond cluster dressers, 491, 492–493, 509
Diamond cutting tools, 550, 551, 560, 770
Diamond dressers, 492–495, 550–551
Diamond-edge saw bands, 403–404, 412
Diamond grinding wheels, 497–501
Diamond locating pin, 695
Diamond-plated dressing block, 492, 493, 509

Die casting, 121
Die clearance, 707–708
Die-cutting and forming processes, 704–705
Die heads, 169, 181
Die holder, 168, 169
Dies, 709–714
Die sets, 709
Die shoe, 709
Die space, 707
Die and tap cutting method (screw thread production), 121
Digital input data, 689
Digital readout, 246–247, 249, 519, 530, 689, 730
Digital signal, 689
Dimensional grinding, 524–525, 530
Dimensional stability, 3, 19
Dimensional tolerance, 84, 88, 103, 198–199
Dimensioning:
 baseline, 272–273, 274, 281
 ordinate, 36, 38, 273, 274, 280
 tabular, 56, 59, 274–275, 281
Dimension words, 649
Dip carburizing, 793
Direct-reading optical measuring system, 241
Discontinuous surfaces, 521, 530
Dish wheel, 624, 632
Disk-cutting attachment, 397, 412
Disk grinder, 456
Distortion, 522, 530
Dog shaft carriers, 214, 225
Double dial test indicator, 63
Double-housing planers, 426, 438
Double quenching, 792
Double-sampling plan, 99
Double tempering, 788
Dovetail milling, 263, 297–299
Dovetail milling cutter, 297, 300
Dovetail shaping, 422, 438
Dovetail slides, 297, 300
Dowels, 701
Down-feed attachment, power, 239, 249
Down-feed handwheels, 524–525
Down feeds, 522
Drag angle, 494, 509
Dragging, 186, 194
Draw-in collet attachment, 600
Drawing (die work), 713, 714–715
Drawings:
 baseline dimensioning in, 272–273, 274, 281
 bevel gearing, 339–341
 cutting planes and sectioning in, 189–193
 designations for Acme screw threads in, 127–129
 numerical control, 644–645
 ordinate dimensioning in, 36, 38, 273, 274, 280
 piece part specifications in, 198, 199, 207

INDEX 847

reading of, 36-37
spur gearing, 331
symbols in, 108, 109
tabular dimensioning in, 56, 59, 274-275, 281
worm gear, 346
Dressing, wheel, 491-495, 547-556, 584-587, 737
Dressing stick, 491
Dressing traverse rate, 497, 509
Drill chuck, standard, 168
Drill holders, 183
Drilling cuts, 183-184
Drilling machines, numerically controlled, 671-672
Drill jigs, 698-701
Drills, 222, 717, 729
Driver plate method, 132-133
Drive systems (band machines), 372-373
Driving fork, 361
Dry grinding, 506
Ductile iron, 763-764
Ductility, 715, 754, 771
Dwell, 206, 208, 573, 589
Dynamic measurement, 75

Eccentric column, 597, 610
Eccentric relief, 621-622, 631
Edges, grinding of, 525-526, 527
Ejector stop, 156
Elastic limit, 707, 722
Electrical comparators, 62, 75, 80
Electric furnace, 785
Electric furnace steel manufacturing, 752, 753, 771
Electroband machining, 409-410, 413
Electrochemical grinding machine, 457
Electromagnetic chucks, 520
Electronic comparators, 62, 75-78, 80
Electronic tool gage, 687-688
Embossing dies, 713, 722
Encoding, 689
End-of-block (EOB) code, 654
End cutting tools, 720
End facing, 180
End-feed centerless grinding, 454, 463
End former, 180
End mill grinding attachment, 599
End mill holders, 243, 244, 255-256
End mills, 253-262, 311, 620-623
End-point control, 689
Ends, grinding square, 526-528
End standard, 5
End teeth, 255
End-of-work machining tools, 216, 217, 225
Engine lathes, 138, 673
English module, 334
Eutectoid steel, 777
Expandable reamers, 185
Expanding collets, 157
Extended grinding wheel spindle, 600
Extended-nose collet, 156, 173
External cuts, 155, 177, 178-182

External sawing, 390-392
Extruding, 327

Face edge, 604
Face grinding, 582
Faceplate fixtures, 161-162, 174, 704
Face width, 334
Facing cuts, 180, 186
Feed control levers, 233, 234
Feed cycle change gears, 212-214, 225
Feed train, 147
Ferritic stainless steel, 766, 771
Ferrous alloys, 752, 765-767
Field, 45, 59
File bands, 405
Filler plate, 373, 386
Filleted shoulder, 539, 540, 544
Fine-feed cross feed attachment, 524, 529
Fine-pitch worm gears, 344
Finish. *See* Surface texture
Fini-shear, 327
Finish grinding, 542-543
Fit, 88-90, 103
Fitch, Stephen, 139
Fixed block format, 663-664
Fixed-center turret, 144, 152
Fixed-size gages, 84-87, 103
Fixed-table band machine, 368, 386
Fixture plates, 162
Fixtures, 161-162, 174, 699, 702-704
Flame hardening, 795-796, 806
Flanges, 204-205
Flaring-cup wheels, 603, 609
Flash, 379, 387
Flat field optical system, 46
Flat grinding method, 609, 611
Flat plug gages, 86, 87
Flats. *See* Optical flats
Flat surface:
 calibration of, 26-29, 53-54
 generation of, 23-25
 planing of, 431-432
 problems in producing, 521-522
Flaws, 108
Flexation, 410, 413
Floating reamer holder, 185, 194
Floating zero point, 651, 689
Flood coolant system, 504-505, 510
Flour sizes, 472
Flow pneumatic comparator, 71-72, 80
Flux, 750
Fly-tool cutter holder, 242
Focal point, 43, 59
Following size, 124-126, 134
Footstock (cylindrical grinder), 566-568, 589
Form dressing, 492, 509
Formed cutters, 148
Form grinding, 451, 546-559
Forming (die work), 712, 713
Forming cuts, 180-181
Form relieved cutters, 623-625
Free-abrasive grinding machine, 456-457

Frequency-distribution curve, normal, 97-98, 103
Friction sawing, 398-402, 412
Fringe bands, 58. *See also* Interference bands
 conversion table for, 55
Fringe pattern, 50, 58
Full-ball radius diamond dressing tool, 550, 551, 560
Full-section view, 189
Furnaces, heat-treating, 784-786
Fusibility, 754

Gage amplifier, 76
Gage block accessories, 5-6
Gage block combination, computation of, 13-14
Gage block height, computation of, 12-14
Gage blocks, 2-5, 12, 538
 angle, 33-35, 38
Gage maker's tolerances, 85
Gages:
 fixed size, 84-87, 103
 height, 26, 31-32, 46-47
 high amplification comparators, 62-80
Gaging heads, 70, 76-77, 80
Gaging spindle plugs, 73, 74-75
Gamma atom structure, 777
Gas carburizing, 793-794
Gas-fired furnace, 785
Gashed-end teeth, 255, 264
Gashing (worm gear teeth), 359, 361
Gas nitriding, 795
Gear burnishing, 327
Gear cutter sharpening attachment, 598-599
Gear grinding machines, 456
Gear hobbing, 324, 359, 361
Gear milling, 350-360
Gears, 323-347
 bevel, 327, 339-344, 356-358
 helical, 337-339, 340, 355-356
 herringbone, 337
 rack and pinion, 335-337, 353-355
 spur, 331-335, 351-353
 worm, 344-347, 358-360
Gear shapers, 324-325
Gear shaving, 326
Gear tooth generators, 327
Gear tooth measurement, 328-331, 343
Gear tooth vernier caliper, 328, 329
Gear train method, 133
General-purpose Acme thread classes, 127-129, 134
Gooseneck planer toolholder, 433, 438
Grade numbers, 95-96
Grain depth of cut, 468
Grain size (grinding wheels), 472-473, 482-483
Grain structure (grinding wheels), 473
Granite precision measurement products, 6-10, 19

848 INDEX

Graphite, 761, 763, 771
Graphitization, 763
Gray iron castings, 750, 762
Grid system, 716, 722
Grinding, precision:
   centerless, 446, 447, 453-454, 463
   cylindrical, 451-453, 463, 477-478, 496-497, 564-588
   form, 451, 546-559
   plunge, 451, 463
   surface, 448-450, 463
   tool and cutter, 455-456, 463, 594-631
Grinding coolants, 503-508
Grinding fluids, 503-508
Grinding machines, 443-631
   abrasive, 447-448, 456-457, 462
   cutter and tool, 594-631
   cylindrical, 564-588
   early, 446-447
   electrochemical, 457
   gear and disk, 456
   horizontal-spindle surface, 446-450, 513-559
   jig, 733-739, 742
   microfinishing, 457-461, 463
   precision, 448-456, 462
Grinding mandrel, 600, 737, 742
Grinding method (screw thread production), 122, 134
Grinding wheels, 467-509
   accessories for, 519-521
   balancing of, 501-503, 510
   computing speeds of, 480-481, 570-571
   conditions affecting grade of, 467-469
   correction of problems with, 484-486
   dressing, 491-495, 547-556, 584-587, 737
   faces for, 474, 476
   marking system of, 471-474, 558, 560
   peripheral, 469-470, 474-478, 479, 487
   preparation of, 491-501
   selection of, 481-484, 486
   side, 469, 478-479, 487
   truing, 496-497, 500, 584-587
   types, uses, and standards of, 469-471
Grit sizes, 472-473, 482-483
Groove milling, 257
Ground thread taps, 92
Growth of steel, 780

Half-section view, 191, 192
H and L pitch diameter limits, 94, 103
Hampson, John, 24
Hand of helix, 303, 320
Hand-scraping, 24
Handwheels, 233, 234, 524-525
Hardenability, 754, 771, 779

Hardness, 754, 779
Hardness testing, 797-804
Hardness testing scales, 799, 807
Hardware (computer), 666
Hard-wired machine control units, 679
Harmonic motion, 313, 314, 320
Headstock (cylindrical grinder), 566-567, 589
Headstock spindle tooling, 156-162, 173
Heat conductivity, 399, 412
Heat treatment, 777-797
   equipment for, 784-786
   problems and probable causes of, 780
   quenching baths in, 782-784
Height gages, 26, 31-32, 46-47
Heim, L. R., 446
Helical form, 303
Helical gears, 337-339, 340, 355-356
Helical milling, 303-312
Helix angle, 303, 304
Helix lead, 303, 304-305
Herringbone gears, 337
Hexagon turret, 170-171, 172, 179, 181, 194, 200-201. 205
Hexagon turret boring, 184-185, 194
Hexagon turret facing, 180
High amplification comparators, 62-80
   electrical, 62, 75, 80
   electronic, 62, 75-78, 80
   mechanical, 62-64
   mechanical-optical, 62, 64-67
   multiple gaging, 62
   optical, 62, 67-70, 79
   pneumatic, 62, 70-75, 79, 80
High amplification dial indicators. See Dial indicators, Mechanical comparators
High-helix end mill, 254, 264
High-speed cylindrical grinding, 573
High-speed head attachment, 239-240, 249
High-speed sawing, 402-403, 412
High-tool velocity band machines, 369
Hobbing, 324, 359, 361
Hole forming, 267-268, 275-278, 280
Holes:
   classes of fits of, 88-89, 90, 103
   classes of tolerances of, 85, 103
   measurements for, 32-33, 36, 37, 56-57
   specifications for, 91-94
   troubleshooting with drilling of, 222
Hollow ground angle, 611
Hollow mills, 216, 223, 225
Honing, 457-459, 463
Honing stones, 458
Horizontal milling machines, 285-319, 350-360
   cam milling on, 312-319
   dovetail milling on, 297-299
   gear milling on, 350-360

   helical milling on, 303-312
   keyseat milling on, 292-297
   numerically controlled, 674
   sawing on, 286-291
   slotting on, 291-292
Horizontal shaper, 419-420, 438
Horizontal shaping, 421-422
Horizontal-spindle surface grinder, 446-450, 513-559
   design features of, 514-519
   form grinding on, 546-559
   grinding of angular surfaces on, 533-539, 541
   grinding of vertical surfaces and shoulders on, 539-543
   setup of, 523
   wheel and work-holding accessories for, 519-521
Hot rolling, 326
Hydraulic presses, 706
Hydraulic tracing accessory, 369
Hyper-eutectoid steel, 777
Hypo-eutectoid steel, 777
Hypoid gears, 325

Idler crushing unit, 554
Illumination, 44
Incidence, angle of, 42
Inclinable presses, 705, 706-707, 721
Incremental cuts, 315, 320
Incremental measurement, 650-651, 666
Increments, 522, 530
Indexing, 194
Indexing attachment, 600
Indexing features, 162
Indexing jigs, 699
Induction hardening, 796-797, 806
Industrial magnifiers, 47-48
In-feed centerless grinding, 453, 463
In-feeding, 573, 589
Inferometers, optical, 3
Infinitely variable spindle speeds, 233, 248
Input element, 677
Inserts, metal, 8, 9
Insert-type saw band guides, 376
Inspection grade angle gage blocks, 33
Integrated circuit, 679, 689
Interference bands, 52, 53-56, 58
   conversion table for, 55
Interference fit, 88, 89, 103
Internal cuts, 155, 177, 182-188
Internal cylindrical grinding machines, 452-453
Internal grinding, 582-584
Internal grinding spindle, 598
Internal recessing, 186, 194
Internal sawing, 392-395
Interpolation, 12, 650, 685
Interrupted quenching processes, 783-784
Inverted dies, 709-710
Involute curve, 334-335

INDEX 849

Involute gear cutter, 350-351, 361
Iron, 749-751, 761-764. *See also* Metals
Iron-carbon phase diagram, 778, 779-782
ISO metric threads, 94-96
Isothermal quenching, 783, 784, 806

Jacobs, Charles, 446
Jaws, chuck, 159-161
Jaw serrations, 159, 174
Jig borer, 231, 674-675, 727-733, 742
Jig grinders, 733-739, 742
Jigs, 694-695, 698-701

Keyseat milling, 292-297, 300
Keyseats, 292-293
Keyway-cutting end mills, 254, 264
Keyway end milling, 262
Knee tools, 223
Knife-edge bands, 410-411, 413
Knocking off, 195
Knoop hardness testing, 804
Knurling tool head, 167
Knurling tools, 216, 222, 223

Laboratory masters, 33, 34
Lagged down, 206, 208
Laminated blocks, 521
Lancing (die work), 704
Land, 602
Land interference, 632
Lapping, 459-460, 463
Latch clamps, 697
Lathes. *See* Engine lathes, Turret lathes
Law of Refraction, 42-43
Lay, 107
Layout practices (surface plate work), 23-36
Lay symbols, 108, 109
Lead, 49-50, 91
Lead angle, 345
Lead baths (tempering), 788
Leader and follower thread-chasing attachment, 147, 148, 152
Leading-on attachment, 147, 148, 152, 194
Leading side, 124-126, 134
Lead screw attachments, 147-148
Leaf jigs, 698, 721
Leaf taper gage, 733, 742
Lenses, refraction principles applied to, 43-44
Leveling, 150
Lever gaging head, 76-77, 79
Light, monochromatic, 50-51, 58
Light beam lever, 65-66
Light rays, 42-44
Light waves, 42-44, 58
  measurement with, 51-53
Limiting diameters, 123-124, 135
Lineal pitch, 335, 353, 361

Lineal thickness, 335
Linear interpolation, 650, 685
Linear measurement, 2-10, 19, 31-33, 37
Line-graduated tools, 84
Line grinding, 408-409, 413
Liners, 701, 721
Liquid baths, 782
Liquid carbonitriding, 794
Liquid carburizing, 793, 806
Liquid hardening furnaces, 786, 806
Loads, major and minor, 798, 806
Lobe, 315, 320
Locating devices, 695-696
Locating microscopes, 729-730, 732-733, 740, 743
Locating nests, 696
Locators (jigs and fixtures), 695, 721
Lock bolts, 150
Long stop, 156
Loops, programming, 643-644
Lot size, 198, 207
Lubricity, maximum, 219, 225
Lumps, 172

Machinability, 754-755, 771
Machinability classes, 188, 189
Machine control unit (MCU), 679
Machine cuts:
  external, 155, 177, 178-182
  internal, 155, 177, 182-188
Machining centers, numerically controlled, 675-676
Magazine feeding, 211, 225
Magna-sine, 34
Magna-vise clamps, 520, 528
Magnesium-base alloys, 767, 772
Magnetic chucks, 520, 528, 535, 536, 544, 569, 702-703
Magnetic separator, 507
Magnetic sine plate, 17-18
Magnetic sine tables, 535, 538-539, 544
Magnetic work-holding accessories, 534-536
Magnification, 44
Magnification power, 44
Magnifiers, 47-48
Main turrets, 147
Major loads, 798, 806
Malleability, 754
Malleable iron castings, 751, 763
Mandrel, grinding, 600, 737, 742
Manganese, 765
Martempering, 783, 784, 806
Martensite, 779, 780-782, 805
Martensitic stainless steel, 766, 771
Master flats, 7, 50, 58
Mathematics of angular measurement, 11-15
Maudslay, Henry, 23-24
Maximum waviness height, 108
Mean line, 107

Measurement:
  angular, 10-20, 33-37, 48-49, 740-741
  coordinate (absolute), 651, 665
  diametral, 280
  dynamic, 75
  gear tooth, 328-331, 343
  incremental, 650-651, 666
  linear, 2-10, 19, 31-33, 37
  optical, 42-44
  precision, 10-20, 33-38, 41-58, 62-79
Measurement instruments:
  angle plates, 8, 534, 537, 538
  dial test indicators, 26-32, 63, 267, 268, 738-739. *See also* Mechanical comparators
  fixed-size gages, 84-87, 103
  gage blocks, 2-5, 12, 33-35, 38, 538
  gear tooth vernier caliper, 328, 329
  granite accessories, 8-10
  height gages, 26, 31-32, 46-47
  high amplification comparators, 62-79
  microscopes, 41, 44-50, 56-58, 116
  optical flats, 41, 50-58
  optical height gages, 46-47
  parallels, 8, 9, 20
  protractors, 10-11
  sine bars, 12-15, 538
  sine plates, 15, 16-18, 20, 534
  surface finish analyzers, 114-116
  surface plates, 7-8, 12, 23, 24, 26
Mechanical comparators, 62-64
Mechanical-optical comparators, 62, 64-67
Metallurgy, powder, 327
Metals, 748-805
  alloys, 752, 755, 756-757, 764-768, 771, 772
  composition of, 752-754
  cutting tool, 768-769
  hardening of, 780-784, 786-789, 795-797
  hardness testing of, 797-804
  iron, 749-751, 761-764
  mechanical properties of, 754-755
  sheet, 712-715
  steel, 751-805
Metal-slitting saws, 287-288, 300
Metric gage blocks, 4, 5
Microbore® boring bars, 728, 742
Micro-clamps, 5
Microfinishing, 457-461, 463
Microform grinder, 557, 560
Microhardness testing, 804, 807
Microinch, 58, 107
Micro-interferogram, 3
Micrometer table positioning attachment, 600
Microscopes:
  binocular-head, 46

850 INDEX

locating, 729-730, 732-733, 740, 743
shop, 44-46
surface finish, 116
toolmaker's measuring, 46, 47, 48-50
TV, 740
zoom, 46
Micro-sine table, 730, 742
Milling cutters, grinding of, 615-626
Milling fixtures, 702
Milling machines. *See* Horizontal milling machines, Vertical/horizontal milling machines, Vertical milling machines
Milling method (screw thread production), 122, 134
Mineral oils, 219
Minor loads, 798, 806
Mirror image programming feature, 664
Miscellaneous function, 649, 666
Mist coolant system, 505-506
Miter gears, 339
Mode, 233, 247
Modified surface plate, 8, 19
Module systems of gearing, 334
Mold casting, 121
Molten salt quenching baths, 783
Molybdenum, 766, 772
Moly grease, 161
Monel metal, 767
Monochromatic light, 50-51, 58
Monolight, 50-51, 58
Motion element, 677
Multidirectional symbol, 108, 109
Multiple cuts, 155
Multiple-flute end mills, 254, 264
Multiple gaging comparators, 62
Multiple-scale comparators, 79
Multiple-spindle automatic screw machines, 212, 225
Multiple-start threads, 131-133, 134
Multiple turning head, 141, 143, 171, 179
Multipurpose vise, 245, 249

National Acme Thread Form, 127-129, 134
Natural tolerance limit, 98
Natural trigonometric functions, 12
table of, 13
Necking cutter block, 181
Nickel, 765-766
Nickel-base alloys, 767, 772
Nitriding, 795, 806
Nodular cast iron, 763-764
Nominal grade (grinding wheel), 467, 487
Nominal size, 88
Nominal surface, 107
Nondimension words, 649
Nonferrous alloys, 752, 767
Nonpositive cam types, 313-314, 320

Normal circular pitch, 337-339
Normal diametral pitch, 337-339
Normal frequency-distribution curve, 97-98, 103
Normalizing, 790-791, 806
Norton, Charles, 446
Notching, 391, 392, 705
Numerical control (NC) systems, 637-688
advantages of, 639-641
binary code in, 646-647, 665
data processing in, 676-677
disadvantages of, 641
drawings for, 644-645
machines in, 670-676
programming of, 639, 642, 646-665, 676-683
Numerical control word language, 649-650
Numerically controlled machines, 670-676, 715-720
Numerically controlled machining centers, 675-676

Object, 42, 59
Off-line data processing, 676, 689
Offset, 601, 604, 605, 606, 611
Offset boring head, 268-270, 276-278
Offset turning toolholder, 169, 170
Oil baths (tempering), 787
Oil quenching baths, 782-783, 790
One minute of arc, 11, 20
On-line data processing, 676, 689
Open hearth steel manufacturing, 751, 771
Open loop NC system, 644
Open-plate jigs, 699, 721
Open-side planers, 426
Open-side shaper-planers, 426, 438
Operational data, 570, 571, 589
Optical comparators, 62, 67-70, 79
Optical flats, 41, 50-58
Optical height gage, 46-47
Optical inferometers, 3
Optical lever, 65-66, 79
Optical measurement, 42-44
Optical measuring system, 241
Ordinate dimensioning, 36, 38, 273, 274, 281
Ordinate table, 36, 38
Out-feed grinding, 734, 742
Output element, 677
Overbending (die work), 713
Overhead turning from the hexagon turret, 179, 194
Overlapping disk balancing ways, 502, 510
Overshoot, 650
Oxygen steel manufacturing process, 752, 753, 771

Pack hardening, 792-793
Pantograph dresser, 553-554

Pantograph dressing attachment, 737, 742
Parabolic motion, 313, 320
Parallax errors, perpendicular viewing to reduce, 51, 52
Parallel balancing ways, 501, 502
Parallelism, checking for, 26-27, 50, 51
Parallels, 8, 9, 20
Parallel separation planes, 51, 58
Parallel symbol, 108, 109
Partial section, 191-193
Part programming, 666, 676, 680-683
Pearlite, 777-782
Pearlitic malleable iron, 763
Penetrator points, 798-799
Peripheral grinding, 469-470, 474-478, 479, 487
Perma-clamps, 520, 528
Permanent form relief, 623-624, 632
Permanent magnetic sine plate, 17-18
Perma-sine, 17-18, 20
Perpendicular symbol, 108, 109
Piece part drawing specifications, 198, 199, 207
Piercing (die work), 704, 721
Pig iron, 749-750, 771
Pinion, 335, 339
Pins, 695
Pin-type gages, 86
Pitch circle, 332-333
Pitch diameter, 91, 92-94, 333
Pivoted guide plate, 149
Plain cross slide, 144, 152
Planer gage, 430, 438
Planers, 425-437
components of, 427-428
cutting speeds for, 435
cutting tools and holders for, 433-435
designs of, 425-427
drive and control systems of, 428
work-holding devices for, 428-430
Planes, standard, 23-24
Planing, 431-437
Plastic deformation, 121
Plastic zone, 399, 412
Platen, 427, 438
Plug gages, 85-87
Plugs, 434, 479, 695
Plunge grinding, 451, 463, 565, 571, 589, 734, 742
Pneumatic comparators, 62, 70-75, 79, 80
Pocket magnifiers, 48
Point of contact, determination of, 52-53
Point-to-point positioning, 655-656, 658, 666, 671, 678
Polishing, 461
Polycrystalline diamond cutting tools, 770, 772
Porosity (grinding wheels), 473
Positioning, 621, 694

INDEX

Positive cam types, 313-314, 320
Positive-stop collets, 156
Postprocessor, 676-677, 689
Pot-type furnace, 786
Powder metallurgy, 327
Power down-feed attachment, 239, 249
Precision expanding collets, 157-158
Precision grinding, 448-456, 462
    centerless, 446, 447, 453-454, 463
    cylindrical, 451-453, 463, 477-478, 496-497, 564-588
    form, 451, 546-559
    plunge, 451, 463
    surface, 448-450, 463
    tool and cutter, 455-456, 463
Precision measurement, 10-20, 33-38, 41-58, 62-79. See also Measurement instruments
Preparatory function, 649, 666
Press brake, 713
Presses, punch, 705-707
Press-fit bushings, 701, 721
Pressure angle, 333
Pressure pads, 713, 722
Pressure-type pneumatic comparator, 73-75, 80
Probe gaging head, 76-77, 79
Processor, 676, 689
Production heat-treating furnaces, 786
Profile, 107
Profilometer, 115, 116
Programming (computer), 639, 642, 646-665, 676-683
    APT in, 678, 680-683, 688
    CAP, 677-679
    editing in, 678-679
    fixed block format in, 663-664
    NC word language in, 649-650
    part, 666, 676, 680-683
    tab sequential format in, 655-656, 666
    tape preparation and correction in, 653-654
    word address format in, 663
Progressive dies, 710, 722
Progressive fixtures, 702
Progressive plug gage, 86
Protractor eyepiece, 46, 47, 49
Protractors, 10-11
Pulse weight, 650
Punch presses, 705-707
Punch radius (die work), 714
Pusher block, 180
Pyrometer, 784, 785

Quadrants, 645
Quality control, 83-103
    charts for, 100-102
    methods and plans for, 96-100
    principles and measurement practices of, 83-96
Quartz, optical, 50
Quenching baths, 782-784, 790, 797

Quick-acting slide tool, 180, 194
Quick change tooling system, 242, 249
Quill feed controls, 237, 248

Rack, 335
Rack indexing attachment, 353, 361
Rack milling, 353-355
Rack milling attachment, 353, 361
Rack and pinion gears, 335-337
Rack and pinion table drive, 516-517
Radial cutting tools, 720
Radial symbol, 108, 109
Radius and angle dresser, 549-550, 552-553, 560
Radius angle grinding attachment, 737
Radius diamond dressing tools, 550-551, 560
Radius dresser, 547-549, 559
Radius grid chart, 69
Radius grinding attachment, 599-600
Rail and side heads (planer), 429, 438
Ram, 236
Ram-type turret lathe, 141-144, 147-148, 152, 162, 163, 164, 200. See also Turret lathes
Random access memory (RAM), 679
Rapid traverse, 147
Reach-over cross slide, 146, 152
Readout, digital, 246-247, 249, 519, 530, 689, 730
Reamer drills, 729
Reamer holder, self-aligning floating, 168
Reamers, 185, 222, 627-629
Reaming cuts, 185
Rear tool post cutters, 170-171
Recalescence point, 781
Recessing tool head, 167
Reciprocating speed, 459
Rectangular coordinate system, 644
Red heat, 780, 805
Reed comparators, 64-67
Reed comparator scales, 66
Reeds, 65
Reference flats, 50
Reference roll (crush dressing), 554-556, 560
Reflection, angle of, 42
Refraction, 42-44, 58
Reinforced grinding wheel, 558-559, 560
Relieving, 544
Repeatability, 24, 38, 63, 640
Reproducibility, 666
Rest plates, 695, 696, 721
Retainer, 460, 463
Reversible cylindrical plug gages, 86
Rewind stop (RWS) code, 657
Right-angle attachment, 240, 241, 249
Ring gages, 85
Rise, 315, 320
Robert's planer, 425
Rocking jaws, 159, 160, 174

Rockwell hardness testing, 798-801
Roller back box tool, 166, 174
Roller rest taper turner, 148, 149, 152
Roller-type bar turner, 163-164
Roller-type (high-speed) saw band guides, 375-376
Roll feeds, 707, 722
Rolling method (screw thread production), 121, 134
Roll jaw stud, 202, 207
Roll-thread gages, 87
Root circle, 333
Root mean square (RMS) average, 110-111
Rotary actuator, 699, 700
Rotary cross slide milling head, 241
Rotary tables, 246, 730
Rotational axes, 645
Rotation speed, 459
Rough grinding, 542, 544
Roughing end mill, 254
Roughness, 107. See also Surface roughness
Roughness height, 108, 111
Roughness width, 107
Roughness width cutoff, 108, 111
Roundness, 29-30, 38
Running fits, 88, 89
Saddle-type turret lathe, 141, 144-145, 148, 152, 200-201. See also Turret lathes
SAE steel classifications, 755-758
Salt bath nitriding, 795
Salt baths (tempering), 787-788
Sampling plans, 98-100, 103
Saw band guide, 375-376, 387
Saw band welding, 376-383
Saw guide post, 374-375, 387
Sawing:
    horizontal milling machine, 286-291, 300
    vertical band machine, 390-404
Scleroscope hardness testing, 803-804, 807
Screw machines, 211-225
    accessories for, 220
    classifications of, 211-212
    components of, 212-215
    guidelines for tool design and selection, 217
    operation of, 217
    tooling for, 215-216
    troubleshooting for, 220-224
Screw-slotting cutters, 288, 300
Screw threads, 89-96, 120-133
    Acme, 126-131, 134, 135
    cutting of, 120, 124-126, 130-133, 181-182
    interchangeability of, 89-90
    measurement of, 91-92
    multiple-start, 131-133, 134
    production of, 121-123
    specifications of, 92-94

852 INDEX

square, 124-126
  tolerances of, 95-96
  UNM, 123-124, 134
Scriber, 5, 6
Scru-broaching, 122-123, 134
Secondary clearance machining set-up, 311, 320
Secondary turrets, 147
Second machine operations, 220, 225
Sectioning, 189-193, 195
Section lining, 191-193, 195
Self-opening toolholders, 169, 174
Self-quenching, 797
Sequencing of cuts, 177-178, 182, 187-188, 194, 205
Sequential-sampling plan, 99-100
Serial taps, 92
Serrated planer tool bit, 434, 438
Shallow draw, 714, 722
Shank toolholders, 169, 170
Shaper attachment, 278-279
Shaper-planers, 426, 438
Shapers, 418-425, 437
Shaping:
  horizontal, 421-422
  vertical, 420-421
  vertical band machine, 367, 386
  vertical milling machine, 271-272, 278-279
Shaping cuts, 390, 391
Shaping tool set, 242, 243
Shaving (die work), 705
Sheet metal (die work), 712-715
Shell end mill holder, 244
Shell mills, grinding of, 618-620
Shop microscope, 44-46
Short-lead helixes, 308, 320
Shoulder grinding, 539-543, 734, 739, 742
Shoulder runout, 30
Shut height, 699, 707, 721, 722
Side grinding, 469, 478-479, 487
Side-hung cross slide, 146, 152
Side milling, 264
Side turning from square turret, 178-179, 194
Sight location, 695
Silicon carbide, 446, 460
Silicon carbide abrasive wheels, 491-497
SI Metric module, 334
Simple sine plate, 16, 20
Sine angle plate, 534
Sine bar, 12-15, 20, 538
Sine bar attachment, 600
Sine block, 15, 16, 20
Sine chuck, 17, 535, 536
Sine dresser, 549, 560
Sine plates, 15, 16-18, 20, 534
Sine tables, 245-246, 535, 538-539, 544
Sine value, computation of, 12-14
Single cuts, 155
Single-point cutters, 181-182

Single-point thread cutting, 187
Single-sampling plan, 99, 100
Single-spindle automatic screw machine, 212, 225
Sizing, automatic, 448, 462
Slabbing, 367
Slabbing mills, 254
Slag, 750
Slide block, 164-165
Slide rods, 373
Slide tools, 165-169, 170, 172, 180, 184
Slide tool turning head, 167
Slitting, 286
Slitting saws, 287, 614-615
Slot grinding, 734-735, 742
Slotter, vertical, 418
Slotting, 286, 291-292, 300, 391, 395, 412
Snap gage, 5, 6, 86-87
Soft-wired machine control units, 679
Solid jaws, 159, 161
Solid positive stop, 156
Solid reamers, 185
Solid taps, 186
Spacers, 244-245
Spade drilling, 184, 194
Spade drills, 183, 717, 722
Sparking-out pass, 631
Spark out, 544, 589
Sparks, 542, 544
Spark test, 760-761
Speed control relief valve (planer), 428, 438
Speed preselector, 144
Spherical socket dressing attachment, 737
Spherical surfaces, 54-55
Spheroidal graphite, 763
Spheroidizing, 790-791, 806
Spider, 460, 463
Spindle axis, 644
Spindle head, 236-239, 248
Spindles, 212, 231-233, 238, 650-652
Spindle tapers, 238
Spiral bevel gear generators, 327
Spiral-edge saw bands, 403, 412
Spline shaping, 422
Splitting, 395, 412
Spotting tools, 729, 742
Spring back rests, 579-580, 589
Spring-type planer toolholders, 433
Spur gears, 331-335, 351-353
Square-end end mill, 253-254, 264
Squareness, 28-29
Square shoulder, 539
Square thread, 124-126
Square turret, 164, 170, 172, 178-179, 205
Square turret boring, 185
Square turret facing, 180
Stability, dimensional, 3, 19

Staggered-tooth milling cutters, 615-618
Stainless steels, 765-766
Stamping, 327
Standard planes, 23-24
Start drill, 183, 194
Stationary die head, self-opening, 169
Steel, 751-805
  carbon, 752-754, 755, 764, 777-784, 786-789
  castings of, 764-765
  classification of, 755-758
  critical points of, 777-778, 805
  hardening of 780-784, 786-789, 795-797
  hardness testing of, 797-804
  heat treatment of, 777-797
  manufacturing processes of, 751-752
  stainless, 765-766
  subzero treatment of, 797
  tungsten high-speed, 789-790
  visible identification of, 760-761
Steel gage blocks, 2-4
Steep-taper boring, 185
Step blocks, 9, 20
Step chuck closers, 158, 159
Step chucks, 158, 159, 173
Stock and feeding, 221, 226
Stock feed mechanism, 214
Stock stop adjustment, 221
Stop dog assembly, 199-201, 207
Stop pins, 428, 429
Stop rod, 146
Stop roll, 147, 152, 199-200, 207
Stops, 199-201
Stop screws, 200, 207
Straight boring, 184-185
Straight edge, 8-9
Straight extension toolholder, 169, 170
Straight pitch line, 353, 361
Straight turning with a bar turner, 179
Strap clamps, 428, 697
Stripper plate, 708, 722
Stub Acme screw threads, 129, 135
Stub arbors, 245, 631
Stub collet holder, adjustable, 168, 169
Stub-length end mill, 253, 264
Subzero heat, 780, 805
Subzero temperature stabilizing, 797, 806
Successive cuts, 155
Sulphur cutting oils, 219-220
Superfinishing, 460-461, 463
Surface, 107
Surface comparators, 111-113
Surface finish, 107, 114-116, 199
Surface finish analyzers, 114-117
Surface finish microscope, 116
Surface finish viewer, 112-113
Surface flatness, 53-54
Surface grinding, 448-450, 463, 474-477, 496. *See also* Horizontal-spindle surface grinder

Surface grinding attachment, 598
Surface plates, 7-8, 12, 19, 23, 24, 26
Surface plate work, layout and measurement practices of, 23-37
Surface roughness, 107-114
   measurements of, 108-113
   relationship of to common production methods, 113-114
Surface texture, 106-118
   computations and measurements for, 108-114
   measuring instruments for, 114-116
   terms and symbols used with, 106-108, 109
Swarf separators, 506-507, 510
Swasey, Ambrose, 139
Swing clamps, 697
Swivel head, 233, 234
Swivel-point test indicator, 26, 38
Swivel-type magnetic chuck, 535, 544
Swivel vise, 245, 534
Symbols:
   lay, 108, 109
   tolerance, 95-96, 103

Table mounting accessories, 730-731
Table movement increments (rack gear cutting), 353, 361
Table offsetting (bevel gear cutting), 357, 361
Tab sequential format, 655-656
Tabular dimensioning, 56, 59, 274-275, 281
Tange bar, 596, 610
Tap drills, 91-94
Tape, numerical control, 653-654
Tape-controlled programming, 685
Tape punch machine, 654
Taper attachment, 149, 185
Tape readers, 654-655, 683
Tapered end mills, 254
Taper end mill holder, 243, 244
Taper guide bar, 148
Taperlock cylindrical plug gage, 86
Taper plug gages, 86, 87
Taper reamers, 185
Tapers:
   boring of, 185
   grinding of, 580-582, 584, 734
   maintaining accuracy of, 180
   measurement of, 35-36
   turning of, 148-149, 179-180
Taper setting plate, 738, 742
Taper-shank holder, 243
Tap holder, releasing, 168
Tap limits, 94, 95
Tapping, 186-187
Tapping heads, 729
Taps, 92, 186-187, 630, 717
Tarry, 573, 589
Temper designations, 759-760, 771
Tempering, 786-790, 806
Tempering baths, 787-788

Tensioning (vertical band machine), 384, 387
Tenthset boring heads, 243, 249
Test indicators. See Dial indicators, Mechanical comparators
Thermal conductivity, 399, 412
Thermocouple, 784, 785
Thread angles, 68, 124-126, 134
Thread chart, 69
Thread chasing, 121, 132, 134
Thread-chasing attachments, 147-148
Thread-chasing dial, 132
Thread grinding wheels, 122
Thread lead, 49-50, 91
Thread pitch, 49-50
Thread ring gages, 87-88
Threads, screw, 89-96, 120-133
   Acme, 126-131, 134, 135
   cutting of, 120, 124-126, 130-133, 181-182
   interchangeability of, 89-90
   measurement of, 91-92
   multiple-start, 131-133, 134
   production of, 121-123
   specifications of, 92-94
   square, 124-126
   tolerances of, 95-96
   UNM, 123-124, 134
Three-dimensional cutting, 367, 386
Three-plate method, 24, 25, 38
Throat diameter (worm gear), 346
Through-feed centerless grinding, 454, 463
Through-the-wheel coolant system, 505, 510
Tilting work center, 68, 79
Tip alignment, 716, 722
Tolerance, 84, 85, 88, 103, 198-199
Tolerance symbols, 95-96, 103
Tool changers, 641
Tool changing, programming for, 658-661
Tool chatter, 206
Tool circle, 201, 207
Tool and cutter grinder, 455-456, 463, 594-631
   attachments for, 597-600
   design features of, 595-597
   preparation of, 607-609
Tool finisher attachment, 408, 412
Tool function, 649, 666
Tool gage, electronic, 687-688
Tool geometry, 423, 425
Tool heads:
   planers, 420, 438
   turret lathes, 167
Toolholders:
   drills, 183
   end mills, 243, 244, 255-256
   planers, 433-435
   turret lathes, 167-169, 170
   vertical milling machines, 267
Toolholder seat (planer), 433, 439
Tool length gage, 687

Toolmaker's flats, 7, 58
Toolmaker's knee, magnetic, 534-535
Toolmaker's measuring microscope, 46, 47, 48-50
Tool presetting, 720
Toolroom angle gage blocks, 33
Tooth rest, 601, 602-603, 606, 611
Total indicated runout (TIR), 29
Toughness, 754
Tracer milling, 234, 235, 248
Tracking (vertical band machine), 384, 387
Transfer gage, 28
Traverse grinding, 565, 571, 589
Traverse rate, 497, 509
Trigonometric functions, natural, 12
   table of, 13
Trilock plug gage, 86
Trimming (die work), 705
Trip levers, 214, 215
Troubleshooting (automatic screw machines), 220-224, 225
True bore, 205, 208
Truing (grinding wheels), 496-497, 500, 584-587
T-slot cutting, 263
Tungsten, 766-767, 772
Tungsten high-speed steels, heat treatment of, 789-790
Turning centers, 140-141, 152
Turning cuts, 178-180
Turning machine fixtures, 703-704
Turning machines. See Screw machines, Turret lathes
Turret assembly (vertical milling machines), 233, 236-237, 248
Turret clamps, 150
Turret lathes, 138-207
   attachments for, 147-149
   care and maintenance of, 150-151
   components of, 146-147
   external cutting with, 178-182
   headstock spindle tooling for, 155-162
   history of, 138-139
   internal cutting with, 182-188
   modern design changes in, 139-141
   mounting of cutters with, 170-172
   ram-type, 141-144, 147-148, 152, 162, 163, 164, 200
   saddle-type, 141, 144-145, 148, 152, 200-201
   setups for, 198-207
   universal bar equipment on, 162-165
Turrets, 144, 147, 152, 164, 170-171, 172, 178-179, 181, 194, 200-201, 205
Turret slide, 215, 224
Turret tooling, 223, 226
TV microscope, 740, 743
Twist drilling, 184
Twist drills, 183, 717

Undercutting, 186, 540, 544
Unified Form screw threads, 89-90
Unified Miniature (UNM) screw threads, 123-124, 134
Unified Numbering System (UNS), 758-759
Uniform motion, 312-313, 320
Uniform rise, 315
Uniform rise cams, 316-319
Unilateral tolerance, 84
Universal bar equipment, 162-165, 174
Universal clapper box toolholders, 433-434, 439
Universal cross slide, 144, 152
Universal cylindrical grinding machines, 452, 463, 565-569, 576-579, 582-583, 588
Universal fixtures, 699, 702
Universal horizontal shaper, 419-420, 438
Universal measuring machines, 739-741, 743
Universal planers, 427
Universal precision vise, 533-534, 544
Universal right angle, 8, 9, 20, 28
Universal tool and cutter grinder, 455-456
Universal turning systems, 140-141, 142
Unsticker feature, 707
Upsetting, 121

Vacuum chucks, 703
Vanadium, 766
Van Keuren wire diameters, 329-331
V-blocks, 9-10, 29-30, 521, 534, 537-538
Vehicle, 460
Vernier control knob, 373
Vertical band machines. *See* Band machines
Vertical/horizontal milling machines, 233-235

Vertical milling machines, 230-264
  attachments for, 238-241
  cutting tools for, 252-256
  design features of, 231-233
  major components of, 235-238
  numerically controlled, 674
  tooling accessories for, 242-245
  work-holding and work positioning devices for, 245-247
Vertical milling processes:
  boring, 268-271, 276-278
  end milling, 253-262
  hole forming, 267-268, 275-276, 280
  shaping, 271-272, 278-279
Vertical shaper, 418-419, 438
Vertical shaper head, 239, 248
Vertical shaping, 420-421
Vertical slotter, 418
Vertical-spindle lapping machine, 460
Vertical-spindle rotary grinder, 446, 447
Vertical-spindle surface grinder, 450
Vertical-spindle surface grinding wheels, 478-479
Vertical surfaces, grinding of, 539-543
Vibration, machine, 206
Vibratory finishing machine, 456, 463
Vickers hardness testing, 803
Viewing screen of optical comparator, 67
Virtual image, 44, 59
Vise fixtures, 702
Vise jigs, 699
Vises, 245, 249, 267, 292, 293, 533-534, 536-537, 544, 715, 730

Walker, O. S., 446
Warner, Reed, 139
Warping, 522, 530
Water quenching baths, 783
Waviness, 108, 111
Waviness width, 108

Wear allowance, 84-85
Wedge clamps, 697
Weldability, 754
Welding presses, 706
Wheel depth of cut, 468
Wheel forming accessories, 546-548
Wheel head, 516, 529, 568, 589, 597, 604
White cast iron, 762
White heat, 780, 805
White light, 50, 58
Whitworth, Joseph, 23-24
Whole depth, 334
Wipe grinding, 734, 742
Wiping die, 713
Woodruff keyseat, 292, 294-297, 300
Word address format, 663
Words, dimension and nondimension, 649
Work head (cutter and tool grinder), 596-597, 610
Working depth, 334
Working flats, 50, 58
Work rest blade, 453, 463
Work roll (crush dressing), 554-556, 560
Work stresses, 522, 530
Work tables (band machines), 373
Work-wheel interface, 506, 510
Worm gears, 344-347, 358-360
Worm thread, 130, 135
Worm thread gage, 130
Wrap-around jaws, 161
Wringing, 4
Wrought alloys, 768

Yield strength, 715

Zero coordinates, 36, 273, 281
Zero offset, 666
Zinc-base alloys, 767, 772
Zone reinforcing (cutoff wheels), 558
Zoom microscope, 46